S0-BXN-345

A HISTORICAL CATALOGUE OF SCIENTISTS AND SCIENTIFIC BOOKS

GARLAND REFERENCE LIBRARY
OF THE HUMANITIES
(VOL. 495)

A HISTORICAL CATALOGUE OF SCIENTISTS AND SCIENTIFIC BOOKS

From the Earliest Times to the Close of the Nineteenth Century

Robert Mortimer Gascoigne

GARLAND PUBLISHING, INC. • NEW YORK & LONDON
1984

Library of Congress Cataloging in Publication Data

Gascoigne, Robert Mortimer, 1918–
A historical catalogue of scientists and scientific books.

(Garland reference library of the humanities ; vol. 495)
Includes index.
1. Scientists—Biography. 2. Scientific literature—Handbooks, etc. 3. Science—History—Handbooks, manuals, etc.
I. Title. II. Series.
Q141.G38 1984 509.2'2 [B] 84-48013
ISBN 0-8240-8959-6 (alk. paper)

Printed on acid-free, 250-year-life paper
Manufactured in the United States of America

AD OMNIPOTENTIS DEI GLORIAM
ET HOMINUM SANITATEM
(From one of the titles)

CONTENTS

INTRODUCTION

This catalogue is designed to serve both as a general reference work and as a research tool for historians of science. It is, in the first place, a chronological list of 13,300 persons who were of some degree of significance in the development of science up to the close of the nineteenth century and for whom biographical information is available in reference books. The index provides an alphabetical approach, and so the work can be used as a biographical dictionary of first resort, giving basic information with references to places where fuller accounts can be found. Second, it is a catalogue of the most widely available books written by persons in the chronological list, and in this respect it can be regarded as a portrayal of a large sample of the book literature of science from its beginnings to the end of the nineteenth century with an emphasis on the more significant books. (Periodicals and periodical articles are not included except in special cases.)

The list of authors (the term "author" will be used for all the persons in the chronological list even if none of their works is included) was compiled from the reference works cited in Table 1 together with the library survey mentioned below. Besides the reference works indicated there are, of course, many other, more specialized, sources of biographical and bibliographical information which can be used for particular authors in the context of particular fields, periods, or countries, but none of these are cited. The authors included are, in general, those whose work appeared to be scientifically significant by the standards of their own time (in the early periods many astrologers and alchemists of the better sort are included) and who had at least one book or several periodical articles to their credit. A considerable number of the persons listed in Poggendorff's *Handwörterbuch* and some of the other reference works used were excluded because they were judged to be too minor. (The curve for the distribution of scientific—or other—significance probably follows something like an inverse-square relationship with the consequence that the numbers of persons at the lowest levels of significance are very large.) In each entry the author's name, dates, and country are followed by symbols that designate the reference works giving biographical and bibliographical information about him. The number and type of the symbols give a rough indication of his degree of significance.

Though there is an attempt at completeness in regard to the authors included—i.e., completeness down to a fairly low level of significance—there is not, nor can there be, any such attempt with the book titles; here the criterion is availability. The titles were collected from a survey of the relevant holdings of the chief libraries of Australia, using a subject approach. Specialized libraries such as those of natural history museums, botanic gardens, and astronomical observatories were included. (The locations of the books will be given in a separate publication entitled *Sources for the History of Science in Australian Libraries.)*

The resources of the Australian libraries, taken together, are considerable but for holdings of early works they cannot of course be compared with the great libraries of Europe and North America. This deficiency (which does not apply to the ancient and medieval periods, for which modern editions are used) is to a large extent overcome by the availability of much material in microform—notably the Readex Corporation's series "Landmarks of Science" and the books listed in the well-known catalogues of Pollard and Redgrave and of Wing—and as reprint editions and edited texts. (The latter term is used to include twentieth-century critical editions, translations, collections, and selections.) A major proportion of all the reprints and edited texts that have been published for the field is included.* There is, thus,

---

*The field of history of science has shared in the general surge of reprint publishing in recent decades. A count of the number of reprints in the present work reveals that during the late 1940s and early 1950s the average number published per year was about four; in the late 1950s the rate of publication began to increase, and by the late 1960s it was about 100 per year. But after 1970 the rate decreased as rapidly as it had risen, and by the late 1970s it was down to about five per year.

an emphasis on availability: the books included are likely to be in many libraries or to have been reproduced in microform or reprinted. And availability, roughly speaking and with many exceptions, should be an indication of significance.

All kinds of scientific literature, such as textbooks and popularisations, are included and there are special sections for autobiography, biography, and history of science. Unintentionally but unavoidably there is some bias towards books in English, especially translations into English. Many long titles had to be abbreviated but details likely to be of interest to the historian were retained as far as possible. Numerous books were found in the libraries by authors for whom no biographical information is available, not even dates (and there are a considerable number of such books in the "Landmarks of Science" series); these were excluded except for a few special cases. Of the books by known authors the only categories that were systematically excluded—because of their very large numbers—were highly specialized descriptive works published after 1850 in geology, botany, and zoology. Books published after 1899 are not included except for collected works, editions of correspondence, edited texts generally, and some autobiographies.

The subject range is much the same as that of the *Dictionary of Scientific Biography* but somewhat more restricted. Geography (other than physical and astronomical geography) is excluded as well as anthropology and the human sciences generally. In Part 1 (ancient and medieval periods), important figures in the history of medicine are included but otherwise medicine is excluded or, like technology, included only for its relevance to science.

The authors are listed chronologically by birth date. In cases, common in the early periods, where the birth date is unknown, the date is estimated from whatever biographical information is available. Persons born after 1859 are not included. For a few of the authors, birth and death dates are given though no biographies of them were found. In such cases the dates were taken from library catalogues; the existence of such dates indicates that

---

With the edited texts there is no such phenomenon: Here the average number increased with only small fluctuations from about nine per year in the late 1940s to about thirty-five per year in the late 1970s.

biographies exist somewhere, probably as obituary notices in periodicals.

In Part 3 (eighteenth and nineteenth centuries) the entries are classified into sections under the headings of the main sciences, with a chronological arrangement of authors within each section. Many authors appear in more than one section and in such cases cross references are given. In most sections, in addition to the author sequence, there is also an "imprint sequence" for anonymous works, works published by institutions and, in a few special cases, by authors whose birth dates are unknown and cannot be estimated. Here the chronological arrangement is by the books' imprint dates.

The classification of authors in Part 3 is far from perfect but in such matters perfection is not to be had. Each author is classified according to his main field or fields but, especially in the earlier periods, many authors extended their activities into other fields as well. Those who spread their interests widely are put into a special category entitled Various Fields; their number decreases over time and by the nineteenth century there were not many of them. In most cases the classification is based on biographical information but for some sections—Science in General, Biology in General, Autobiography, Biography, and History of Science—which are different in character from the other sections, it is based on the book titles.

After each author's name and dates an indication is given of the country in which he worked, which of course is not always the same as his native country. If he worked (as distinct from studied or travelled) in more than one country for a considerable period the fact is indicated. The abbreviations used are explained in Table 2. Countries are defined territorially in terms of the boundaries existing today except in the case of Germany where they are taken to be those of 1900. This procedure necessitates some deliberate anachronisms, for example, a person who worked in Brussels in the seventeenth century is said to have worked in Belgium though Belgium did not then exist as a political entity; likewise there are some British colonists in North America in the early eighteenth century who are said to have worked in the United States. For most of eastern Europe and some other regions, cities are designated rather than countries because of numerous political changes.

For further details of the arrangement of the catalogue, the notes at the heads of the various sections may be consulted.

The information contained herein lends itself readily to statistical analysis relating especially to the size and composition of the scientific community at different times. Such an analysis will be published elsewhere. Here it may be mentioned that from the middle of the seventeenth century, the number of scientists increased exponentially with a doubling time of fifty years.

Finally, it may be remarked that in a work of this kind there will inevitably be errors, especially as some of the numerous sources used were less than satisfactory. Much effort has been put into trying to reduce the number of errors, but doubtless many remain.

R.M.G.
School of History and Philosophy of Science
University of New South Wales
Sydney

TABLE 1. SYMBOLS USED

*	An asterisk indicates that a title of one or more of the particular author's works is recorded in another section. For example the annotation "Also Astronomy* and Physics" at the end of an entry for an author in the Mathematics section of Part 3 means that the author is also listed in Sections 3.04 (Astronomy) and 3.06 (Physics) and that a title of one or more of his works appears under his name in the Astronomy section.
#	This symbol indicates that an author's collected works or correspondence or both are listed under his name in the section so designated.
BLA	*Biographisches Lexikon der hervorragenden Ärzte aller Zeiten und Völker.* 2nd ed. 5 vols & supplement. Berlin/Vienna, 1929–35; reprinted Munich/Berlin, 1962. Includes many biologists and other kinds of scientists who were medical men by profession. It also includes some scientists (not only biologists) who were not medical men.
BLAF	*Biographisches Lexikon der hervorragenden Ärzte der letzten fünzig Jahre. Hrsg. und bearb. von I. Fischer. Zugleich Fortsetzung des Biographischen Lexikons der hervorragenden Ärzte aller Zeiten und Völker.* 2 vols. Berlin, 1932–33; reprinted Munich/Berlin, 1962. Includes many biologists.
BMNH	British Museum (Natural History). *Catalogue of the Library.* 5 vols. London, 1903–15. One of the largest natural history libraries in the world. It covers botany, zoology, mineralogy, and geology and also extends to some degree into various fields of experimental biology.

DSB	*Dictionary of Scientific Biography.* 14 vols with supplement and index. New York, 1970–80.
DSB Supp.	Vol. XV of the DSB.
G	General biographical dictionaries, universal and national. This symbol on its own is used only for well-known figures who can be found in any appropriate biographical dictionary or encyclopaedia. For most authors one or more of the following symbols are used. These references are nearly all taken from Hyamson, A.M., *A Dictionary of Universal Biography.* 2nd ed. London, 1951. (Biographical dictionaries generally do not include individuals until after their death. Many authors in the last parts of the chronological sequences of Part 3 of the present work are not found in biographical dictionaries because the relevant dictionaries were published before their death.)
GA	*Nouvelle Biographie Universelle* (or *Générale;* title varies). 46 vols. Paris, 1852–66; reprinted in 23 vols, Copenhagen, 1963–69.
GB	*Encyclopaedia Britannica.* 11th ed. 28 vols. Cambridge, 1910–11. This edition is generally better for biographies of pre-1900 figures than later editions.
GC	*Allgemeine Deutsche Biographie.* 56 vols. Leipzig, 1875–1912; reprinted Berlin, 1967–71.
GD	*Dictionary of National Biography.* 63 vols. London, 1885–1900; reprinted in 21 vols, 1908–09 and later.
GD1, GD2, etc.	——— Supplements, 1901 onward.
GE	*Dictionary of American Biography.* 20 vols. New York, 1928–36; reprinted in 10 vols, n.d. [1961?]
GE1, GE2	——— Supplements 1 and 2. 2 vols. 1944 and 1958; reprinted in 1 vol.
GF	See references given in Hyamson, A.M., *A Dictionary of Universal Biography.* 2nd ed. London, 1951. (In addition to the five works listed above, Hyamson also refers to many smaller biographical dictionaries.)

HNC	*Historiae Naturalis Classica.* A series of facsimile reprints of important works in botany and zoology published from 1960 onward by Verlag J. Cramer, Lehre/Weinheim. Each reprint in the series is numbered. Over a hundred have been published.
ICB	*ISIS Cumulative Bibliography. A Bibliography of the History of Science formed from ISIS Critical Bibliographies 1–90, 1913–65.* Vols 1 and 2 (Personalities). London, 1971. The subsequent volumes were not used.
Klebs	Klebs, A.C. *Incunabula Scientifica et Medica.* Hildesheim, 1963 (reprint in book form of a periodical article of 1938). A short-title list of over a thousand incunabula in some three thousand editions. Gives references leading to locations in European and American libraries.
L	See RM/L.
Mort.	Morton, L.T. *Garrison and Morton's Medical Bibliography. An Annotated Check-list of Texts Illustrating the History of Medicine.* 2nd ed. London, 1954; revised 1965. Used only for the biological sciences and for medicine in the ancient and medieval periods.
OKEW	*Ostwalds Klassiker der Exacten Wissenschaften.* A series initiated by the well-known chemist and historian of science, Wilhelm Ostwald, and published from 1889 onward by Verlag W. Engelmann, Leipzig (and now by Akademische Verlagsgeschellschaft, Frankfurt, a.M.). Each volume in the series contains editorial matter and consists of a German translation if the original text is not in German. Each volume is numbered; a new series began in 1965. Over 250 volumes have been published.
P	Poggendorff, J.C. *Biographisch-literarisches Handwörterbuch zur Geschichte der Exacten Wissenschaften, Enthaltend Nachweisungen über Lebensverhältnisse und Leistungen von Mathematikern, Astronomen, Physikern, Chemikern, Mineralogen, Geologen, usw., aller Völker und Zeiten.* Vols I–IV. Leipzig, 1863–1904; reprinted Amsterdam, 1965. Many biologists are also included but not biological literature. Vols I and II (published

	1863) constitute the first alphabetical sequence, Vol. III (in two parts, published 1898) the second, and Vol. IV (in two parts, published 1904) the third. There are supplements at the end of Vols I/II and IV. The subsequent volumes (published 1926 and later) were not used, except for some death dates.
PR	Pollard, A.W., and Redgrave, G.R. *A Short-title Catalogue of Books printed in England . . . 1475–1640.* London, 1926; reprinted 1956. Gives locations in British and a few American libraries. University Microfilms have produced microfilms of many of the books listed.
RM	Title included in Series 1 of the "Landmarks of Science" series produced on microcards by the Readex Microprint Corporation. (Series 2, which comprises mainly periodicals but with some monographs, is not included, being still in progress.)
RM/L	Title included in the "Landmarks of Science" series and also found in the libraries.
Sart.	Sarton, G. *Introduction to the History of Science.* 3 vols in 5. Baltimore, 1927–48 and reprints. An exhaustive coverage from the earliest times to the end of the fourteenth century.
Still.	Stillwell, M.B. *The Awakening Interest in Science during the First Century of Printing, 1450–1550. An Annotated Checklist of First Editions.* New York, 1970. A selection of several hundred first editions with documentation and references leading to locations in European and American libraries.
Wing	Wing, D.G. *Short-title Catalogue of Books printed in England . . . 1641–1700.* 3 vols. New York, 1945–51. Gives locations in British, American, and a few Continental libraries. University Microfilms have produced microfilms of many of the books listed.
X	Title not found in the libraries but included for bibliographical reasons. (This symbol is used only for pre-1900 titles, not for reprints, etc.)

TABLE 2. COUNTRIES AND CITIES

Explanations of the abbreviations used, with some notes on cities whose names have changed.

Agram	Now Zagreb in Yugoslavia
Aus.	Austria
Belg.	Belgium
Breslau	Regarded as part of Germany from the mid-eighteenth century. Now Wroclaw in Poland
Br.	Britain. Regarded as comprising England, Wales, Scotland, and all of Ireland
Brünn	Now Brno in Czechoslovakia
Budweis	Now České Budějovice in Czechoslovakia
Cape Obs.	The astronomical observatory at the Cape of Good Hope
Christiania	Now Oslo
Czernowitz	Or Cernauti. Now Chernovtsy in the U.S.S.R.
Danzig	Now Gdansk in Poland
Den.	Denmark
Dorpat	Or Jurjew. Now Tartu in the U.S.S.R.
Fin.	Finland
Fr.	France
Ger.	Germany
Hermannstadt	Now Sibiu in Romania
Holl.	Holland
Hung.	Hungary
It.	Italy
Jassy	Now Iasi in Romania
Kielze	Now Kielce in Poland
Klausenberg	Now Cluj in Romania
Lemberg	Now Lvov in the U.S.S.R.
Mitau	Now Jelgava in the U.S.S.R.
Nor.	Norway
Olmütz	Now Olomouc in Czechoslovakia
Port.	Portugal

Posen	Now Poznan in Poland
Reval	Now Tallinn in the U.S.S.R.
Russ.	Russia
Swed.	Sweden
Switz.	Switzerland
Tarnopol	Now Ternopol in the U.S.S.R.
Teschen	Now Cieszyn in Poland
Thorn	Now Torun in Poland
Wilno	Now Vilnius in the U.S.S.R.

PART 1

ANCIENT AND MEDIEVAL PERIODS

Modern editions only are included here. Fifteenth and sixteenth-century editions are in Part 2.1.

The chronological arrangement of the authors is generally only approximate because of insufficiency of biographical information.

1.1 ANTIQUITY

1.11 Babylonian

1.111 Author

1. KIDENAS THE CHALDEAN. ca. 300 B.C. ICB. Astronomy.

1.112 Collections of Texts

2. *MATHEMATICAL, meteorological and chronological tablets from the temple library of Nippur*. Philadelphia, 1906. Ed. by H.V. Hilprecht.
3. *MATHEMATISCHE Keilschrifttexte*. 3 vols. Berlin, 1935-37. *(Quellen und Studien zur Geschichte der Mathematik, Astronomie und Physik*. Abt. A3.) Ed. by O. Neugebauer.
4. *TEXTES mathématiques babyloniens*. Leiden, 1938. Ed. by F. Thureau-Dangin.
5. *MATHEMATICAL cuneiform texts*. New Haven, Conn., 1945. Ed. by O. Neugebauer and A. Sachs.
6. *ASTRONOMICAL cuneiform texts*. 3 vols. London, 1955. Ed. by O. Neugebauer.
7. *LATE Babylonian astronomical and related texts*. Providence, R.I,, 1955. Ed. by A. Sachs.

1.12 Egyptian

1.121 Collections of Texts

8. *THE AKHMIM papyrus*. Paris, 1892. Ed. by J. Baillet. Mathematical texts.
9. *THE PETRIE papyri: Hieratic papyri from Kahun and Gurob*. 2 vols. London, 1897. Ed. by F.L. Griffith. Mathematical texts.
10. THE RHIND papyrus.
    1. *Ein mathematisches Handbuch der alten Aegypter: Papyrus Rhind des British Museum*. Leipzig, 1877. Ed. by A. Eisenlohr. Reprinted, Wiesbaden, 1972.
    2. *Facsimile of the Rhind mathematical papyrus*. London, British Museum, 1898.
    3. *The Rhind mathematical papyrus*. Liverpool, 1923. Ed. by T.E. Peet.

4. *The Rhind mathematical papyrus.* 2 vols. Mathematical Association of America, 1927-29. Ed. with mathematical commentary and bibliography by A.B. Chace at al.

11. *MATHEMATISCHER Papyrus des staatlichen Museums der schönen Künste in Moskau.* Berlin, 1930. (*Quellen und Studien zur Geschichte der Mathematik, Astronomie und Physik.* Abt. A1.) Ed. by W.W. Struve.

12. *DEMOTIC mathematical papyri.* Providence, R.I., 1972. Ed. by R.A. Parker.

13. *A VIENNA demotic papyrus on eclipse- and lunar-omina.* Providence, R.I., 1959. Ed. by R.A. Parker.

14. *EGYPTIAN astronomical texts.* 3 vols. Providence, R.I., 1960-69. Ed. by O. Neugebauer and R.A. Parker.

1.13 Greek and Roman

1.131 Authors

15. THALES. 625?-547? B.C. DSB; ICB; Sart.; P II; GA; GB. Natural philosophy.

16. ANAXIMANDER. ca. 610-ca. 545 B.C. DSB; ICB; Sart.; P I; GA; GB. Natural philosophy.

17. ANAXIMENES. fl. ca. 546 B.C. DSB; ICB; Sart.; P I; GA; GB. Natural philosophy.

18. CLEOSTRATUS. late 6th century B.C. ICB; Sart.; GA. Astronomy.

19. PYTHAGORAS. ca. 560-ca. 480 B.C. DSB; ICB; Sart.; P II; GA; GB. Natural philosophy. Mathematics. Theory of music. Astronomy. See also 169/4,5.

20. DEMOCEDES OF CROTONA. 550-460 B.C. ICB; Sart.; BLA & Supp. (DEMOKEDES); GA. Medicine.

21. ALCMAEON OF CROTONA. b. ca. 535 B.C. DSB; ICB; Sart.; BLA & Supp. (ALKMAEON); GA. Medicine. Natural philosophy.

22. HERACLITUS. fl. ca. 500 B.C. DSB; ICB; Sart.; GA; GB. Natural philosophy.

23. ACRON. fl. ca. 480 B.C. Sart.; GA. Medicine. Anatomy.

24. PARMENIDES. ca. 515-after 450 B.C. DSB; ICB; Sart.; P II; GA; GB. Natural philosophy.

25. ANAXAGORAS. ca. 500-ca. 428 B.C. DSB; ICB; Sart.; P I; GA; GB. Natural philosophy.

26. HIPPOCRATES OF CHIOS. fl. ca. 460 B.C. DSB; ICB; Sart.; P I; GA. Mathematics. Astronomy.

27. ARCHELAUS. 5th century B.C. ICB; Sart.; GA; GB. Natural philosophy.

28. DIOGENES OF APOLLONIA. 5th century B.C. ICB; Sart.; P III; BLA; GA; GB. Natural philosophy. Anatomy.

29. HICETAS OF SYRACUSE. 5th century B.C. DSB; Sart.; P II (NICETAS). Astronomy.

30. HIPPASOS OF METAPONTUM. 5th century B.C. ICB. Pythagorean mathematician. Thought to be the discoverer of incommensurability.

31. LEUCIPPUS. 5th century B.C. DSB; ICB; Sart.; P I; GA; GB. Natural philosophy.

32. OCELLUS OF LUCANIA. ca. 5th century B.C. BLA; GA; GB. Pythagorean philosopher. The work *De universi natura* is wrongly attributed to him.

33. OENOPIDES OF CHIOS. 5th century B.C. DSB; Sart.; GA. Astronomy. Mathematics.

34. PHERECYDES OF LEROS. 5th century B.C. ICB; GA; GB. Cosmology.

35. EMPEDOCLES. ca. 492-ca. 432 B.C. DSB; ICB; Sart.; P I; BLA & Supp.; GA; GB. Natural philosophy. See also 15831/2.
    1. *Fragmenta*. Bonn, 1852. Ed. by H. Stein. RM.
    2. *The proem of Empedocles' "Peri physios": Towards a new edition of all the fragments*. Amsterdam, 1975. By N. van der Ben.

36. ZENO OF ELEA. ca. 490-ca. 425 B.C. DSB; ICB; Sart.; GA; GB. Natural philosophy.

37. EURYPHRON OF CNIDOS. mid-5th century B.C. Sart.; BLA & Supp.; GA. Medicine. Anatomy.

38. HIPPON. mid-5th century B.C. ICB; Sart. Natural philosophy.

39. PHILOLAUS OF CROTONA. mid-5th century B.C. DSB; ICB; Sart.; P II; BLA; GA; GB. Natural philosophy. Astronomy.

40. EUCTEMON. fl. 432 B.C. DSB; ICB; Sart. Astronomy.

41. PHAINOS. fl. ca. 430 B.C. P II. Astronomy.

42. THEODORUS OF CYRENE. b. ca. 465 B.C. DSB & Supp.; Sart.; GA. Mathematics.

43. ANTIPHON. late 5th century B.C. DSB; ICB; Sart. Natural philosophy. Mathematics.

44. BRYSON OF HERACLEA. late 5th century B.C. DSB; Sart. Mathematics.

45. METON OF ATHENS. fl. 432-414 B.C. DSB; ICB; Sart.; P II; GA. Astronomy.

46. DEMOCRITUS. ca. 460-ca. 370 B.C. DSB; ICB; Sart.; P I; GA; GB. Natural philosophy. Mathematics.

47. HIPPOCRATES OF COS. 460-ca. 370 B.C. DSB; ICB; Sart.; Mort.; BLA & Supp.; GA; GB. Medicine. Also under Arabic Translations (472)*, Medieval Latin Translations (789)*, and Early Modern Editions (988)*.
    1. *Les oeuvres complètes*. 10 vols. Paris, 1839-61. Ed. by E. Littré. Greek text with French trans. Reprinted, New York, 1961.

Partial Editions

    2. *The genuine works*. 2 vols. London, Sydenham Society, 1849. Trans. with introd. and notes by F. Adams. RM/L. Several reprintings.

3. *Opera quae feruntur omnia.* 2 vols. [No more publ.] Leipzig, 1894-1902. Ed. by H. Kühlewein.
4. [Selections.] 4 vols. London, 1923-31. (Loeb classical library.) Greek text with English trans. by W.H.S. Jones.
5. *The medical works. A new translation.* Oxford, 1950. By J. Chadwick and W.N. Mann.

Individual Works

6. *L'ancienne médecine.* Paris, 1948. French trans. with introd. and comm. by A.J. Festugière.
7. *Du régime.* Paris, 1967. French trans. by R. Joly.
8. *Über die heilige Krankheit.* Berlin, 1968. Greek text with German trans. and comm. by H. Grensemann.

48. PHILISTION OF LOCRI. fl. ca. 400 B.C. ICB; Sart.; BLA. Anatomy. Physiology.

49. HIPPIAS OF ELIAS. fl. 400 B.C. DSB; ICB; Sart.; GA; GB. Mathematics.

50. POLYBUS OF COS. fl. ca. 390 B.C. Sart.; BLA; GA. Medicine.

51. PLATO. 427-348/7 B.C. DSB; ICB; Sart.; P II; G. Philosophy. (Modern editions of his works are not included.) Also under Arabic Translations (478)*, Medieval Latin Translations (790)*, and Early Modern Editions (997).

52. LEODAMUS OF THASOS. fl. ca. 380 B.C. DSB. Mathematics.

53. THEAETETUS. ca. 417-369 B.C. DSB; ICB; Sart. Mathematics.

54. ARCHYTAS OF TARANTUM. fl. ca. 375 B.C. DSB; ICB; Sart.; P I; GA; GB. Mathematics. Acoustics.

55. ECPHANTUS OF SYRACUSE. early 4th century B.C. Sart. Astronomy.

56. LEO. early 4th century B.C. DSB; Sart. (LEON). Mathematics.

57. THYMARIDAS. early (?) 4th century B.C. DSB; Sart. Mathematics.

58. SPEUSIPPUS. ca. 408-339 B.C. DSB; ICB; Sart.; GA; GB. Philosophy. Zoological classification. Also under Early Modern Editions (1009).

59. CHRYSIPPUS OF CNIDOS. fl. ca. 365 B.C. ICB; Sart.; BLA & Supp. Medicine. Anatomy.

60. EUDOXUS OF CNIDOS. ca. 400-ca. 347 B.C. DSB; ICB; Sart.; P I; GA; GB. Astronomy. Mathematics.
    1. *Die Fragmente.* Berlin, 1966. Ed. with German trans. and comm. by F. Lasserre.

61. XENOCRATES OF CHALCEDON. 396/5-314/3 B.C. DSB; ICB; Sart.; GA; GB. Philosophy. Mathematics.

62. MENAECHMUS. fl. ca. 350 B.C. DSB; ICB; Sart. Mathematics.

63. PHILIP OF OPUS. mid-4th century B.C. ICB; Sart. Mathematics. Astronomy.

64. HERACLIDES OF PONTUS. ca. 388-ca. 315 B.C. DSB Supp.; ICB; Sart.; P I; GA; GB. Astronomy.

65. ARISTOTLE. 384-322 B.C. DSB; ICB; Sart.; P I; G. Philosophy and science generally. Also under Arabic Translations (468)*,

Medieval Latin Translations (786)*, and Early Modern Editions (982)*. See also Index.

1. *Opera*. 5 vols. Berlin, 1831-70. The Bekker ed., publ. under the auspices of the Berlin Academy. RM/L.

Other modern editions are not included.

66. ARISTAEUS. fl. ca. 350-330 B.C. DSB; Sart. Mathematics. See also 2304/2.

67. THEOPHRASTUS OF ERESOS. ca. 371-ca. 287 B.C. DSB; ICB; Sart.; P II; GA; GB. Philosophy. Botany. Mineralogy. Also under Early Modern Editions (999)*. See also Index.

1. *Quae supersunt opera et excerpta*. 5 vols. Leipzig, 1818-21. Ed. by J.G. Schneider. Greek text with Latin trans. RM/L.
2. *Opera quae supersunt omnia graeca*. 3 vols. Leipzig, 1854-62. Ed. by F. Wimmer.

2a. ——— Another ed. with addition of Latin trans. 1 vol. Paris, 1866. Reprinted, Frankfurt, 1964.

Individual Works

3. *Metaphysics*. Oxford, 1929. With trans., comm., and introd. by W.R. Ross and F.H. Fobes.
4. *Enquiry into plants [Historia plantarum] and minor works on odours and weather signs*. 2 vols. London, 1916; reprinted 1948-49. (Loeb classical library.) By A. Hort. Greek texts with English trans.
5. *De causis plantarum*. London, 1976. (Loeb classical library.) With an English trans. by B. Einarson and G.K.K. Link.
6. *On stones*. Columbus, Ohio, 1956. Greek text with English trans. and comm. by E.E. Caley and J.F.C. Richards.
7. *De lapidibus*. Oxford, 1965. Ed. with introd., trans., and comm. by D.E. Eichholz.
8. *De igne*. Assen, 1971. Ed. by V. Coutant. Greek text with English trans.
9. *De ventis*. Notre Dame, Ind., 1975. Ed. with introd., trans., and comm. by V. Coutant and V.L. Hichenlaub.

68. CALLIPPUS. b. ca. 370 B.C. DSB; Sart.; GA. Mathematics. Astronomy.

69. ARISTOXENUS. late 4th century B.C. DSB; ICB; Sart.; P I; GA; GB. Theory of music.

70. DINOSTRATUS. late 4th century B.C. DSB; Sart.; P I. Mathematics.

71. DIOCLES (or ANTIGONE) OF CARYSTUS. late 4th century B.C. DSB; ICB; Sart.; BLA & Supp. (DIOKLES); GA; GB (ANTIGONUS, also DIOCLES). Medicine.

72. MENON. late 4th century B.C. ICB; Sart.; BLA. Medicine.

73. THEUDIUS OF MAGNESIA. late 4th century B.C. DSB; Sart. Mathematics.

74. EUDEMUS OF RHODES. fl. 320 B.C. DSB; ICB; Sart.; GA. Philosophy. History of science.

75. DICAEARCHUS OF MESSINA. fl. 310 B.C. DSB; ICB; Sart.; GA; GB. Geodesy. Geography.

76. EPICURUS. 341-270 B.C. DSB; ICB; Sart.; P I; GA; GB. Philosophy. See also 2010/7,9.

1. *Epicurea*. Leipzig, 1887. Ed. by H. Usener. RM.
2. *Epicurus*. Oxford, 1926. Greek text with English trans. by C. Bailey. Reprinted, New York, 1970.

77. PRAXAGORAS OF COS. b. ca. 340 B.C. DSB; ICB; Sart.; BLA; GA. Anatomy. Physiology.
    1. *The fragments of Praxagoras of Cos and his school*. Leiden, 1958. Ed. and trans. by F. Steckerl. Includes the fragments of Phylotimus, Plistonicus, and Xenophon.

78. AUTOLYCUS OF PITANE. fl. ca. 300 B.C. DSB; ICB; Sart.; P I; GA; GB. Astronomy. Geometry. Also under Early Modern Editions (1015)*.
    1. *De sphaera quae movetur et De ortu et occasu libri*. Hamburg, 1877. Ed. by R. Hoche. RM.
    2. *De sphaera quae movetur liber, De ortibus et occasibus libri duo*. Leipzig, 1885. Ed. with Latin trans. by F. Hultsch.
    3. *Rotierende Kugel, und Aufgang und Untergang der Gestirne*. Leipzig, 1931. (OKEW 232)
    4. *On a moving sphere, and On risings and settings*. Beirut,1971. Greek texts and English trans. by F. Bruin and A. Vondjidis.

79. STRATO OF LAMPSACUS. d. ca. 270 B.C. DSB; ICB; Sart.; GA. Natural philosophy.

80. HEROPHILUS. ca. 335-ca. 280 B.C. DSB; ICB; Sart.; BLA; GA. Anatomy. Physiology.

81. EUCLID. fl. ca. 295 B.C. DSB; ICB; Sart.; P I (EUKLIDES); GA (EUCLEIDES); GB. Mathematics. Also under Medieval Latin Translations (787)* and Early Modern Editions (996)*. See also Index.
    1. *Opera omnia*. 8 vols. & supp. Leipzig, 1883-1916. Ed. by J. L. Heiberg and H. Menge.

The *Elements*

2. *The thirteen books of Euclid's Elements*. 3 vols. Cambridge, 1908. Trans. with introd. and comm. by T.L. Heath. 2nd ed., 3 vols., ib., 1926; reprinted, New York, 1956.
3. *Die Elemente*. 5 vols. Leipzig, 1933-37. (OKEW 235, 236, 240, 241, 243)
4. *Elementa. Post I.L. Heiberg*. 2 vols. Leipzig, 1969-70. Ed. by E.S. Stamatis.
5. *The first book of Euclid's Elements, with a commentary based principally upon that of Proclus Diadochus*. Cambridge, 1905. By W.B. Frankland.
6. *Euclid in Greek, Book 1*. Cambridge, 1920. With introd. and notes by T.L. Heath.
7. *The first book of Euclid's Elements*. Leiden,1965. Ed. with glossary by E.J. Dijksterhuis.
8. *Les livres arithmétiques*. Paris, 1961. French trans. of Books 7-9 with introd. and notes by J. Itard.

Other Works

9. *Datorum liber*. Oxford, 1803. Ed. by S. Horsley.
10. *Euclid's book on divisions of figures, with a restoration*. Cambridge, 1915. By R.C. Archibald.
11. *L'optique et La catoptrique*. Paris, 1959. French trans. with introd. and notes by P. ver Eecke.

82. TIMOCHARIS. early 3rd century B.C. Sart. Astronomy.

83. ARATUS OF SOLI. ca. 310-ca. 240 B.C. DSB; ICB; Sart.; P I; GA; GB. Astronomy. Also under Early Modern Editions (987)*. See also Index.
    1. *A literal translation of the astronomy and meteorology of Aratus.* Lewes, 1895. By C.L. Prince. RM.
    2. *Callimachus, Lycophron, Aratus.* London, 1921. (Loeb classical library.) By G.R. Mair. Greek text with English trans.
    3. *Phenomena.* Lexington, Ky., 1975. Trans. by J. Lamb.

84. ARISTARCHUS OF SAMOS. ca. 310-230 B.C. DSB; ICB; Sart.; P I; GA; GB. Mathematics. Astronomy. Also under Early Modern Editions (1002)*.
    1. *Aristarchos über die Grossen und Entfernungen der Sonne und des Mondes.* Freiberg, 1854. German trans. and comm. by A. Nokk. RM.
    2. *Aristarchus ... A history ... together with Aristarchus's treatise on the sizes and distances of the sun and moon.* Oxford, 1913. A new Greek text with trans. and notes by T.L. Heath. Reprinted 1959.

85. ARISTYLLUS. fl. ca. 270 B.C. DSB; Sart.; P I; GA. Astronomy.

86. CTESIBIUS (or KTESIBIOS). fl. 270 B.C. DSB; ICB; Sart.; P I; GA. Mechanical technology.

87. ERASISTRATUS. b. ca. 304 B.C. DSB; ICB; Sart.; BLA & Supp.; GA. Anatomy. Physiology.

88. PERSEUS. fl. 3rd century (?) B.C. DSB; Sart. Mathematics.

89. EUDEMUS OF ALEXANDRIA. mid-3rd century B.C. Sart.; BLA & Supp.; GA. Anatomy.

90. NICOMEDES. fl. ca. 250 (?) B.C. DSB; Sart.; P II (NIKOMEDES). Mathematics.

91. PHILINUS OF COS. fl. ca. 250 B.C. DSB; BLA. Medicine.

92. PHILO OF BYZANTIUM. fl. ca. 250 B.C. DSB. Under PHILON: ICB; Sart.; GA; GB. Mechanics. Also under Arabic Translations (477)*.
    1. *Mechanicae syntaxis libri IV et V.* Berlin, 1893. Ed. by R. Schoene.
    2. *Pneumatica.* Wiesbaden, 1974. English trans. with introd. and notes by F.D. Prager.

93. ARCHIMEDES. 287-212 B.C. DSB; ICB; Sart.; P I; GA; GB. Mathematics. Mechanics. Also under Medieval Latin Translations (785)* and Early Modern Editions (1016)*. See also Index.
    1. *Opera omnia cum commentariis Eutocii.* 3 vols. Leipzig, 1880-81. Ed. by J.L. Heiberg.
    1a. ——— Another ed. Iterum edidit I.L. Heiberg. 3 vols, ib., 1910-15. Reprinted, 3 vols, Stuttgart, 1972.
    2. *The works. Edited in modern notation ...* [With supplement:] *The Method of Archimedes, recently discovered by Heiberg.* 2 vols. Cambridge, 1897-1912. By T.L. Heath. Reprinted in 1 vol., New York, 1953.
    3. *Les oeuvres complètes, suivies des commentaires d'Eutocius d'Ascalon,* 2nd ed. 2 vols. Liège, 1960. French trans. with introd. and notes by P. ver Eecke.

Individual Works

4. *Des théorèmes méchaniques; ou, De la méthode (Ephodiques). Traité nouvellement découvert....* Paris, 1907. French trans. by T. Remach. RM.
5. *Geometrical solutions derived from mechanics. Recently discovered....* Chicago, 1909. Introd. by D.E. Smith.
6. *The Arenarius.* Leiden, 1956. Ed. by E.J. Dijksterhuis.

German translations of several of his works are included in OKEW--vols 201-203, 210, 213.

94. CONON OF SAMOS. fl. 245 B.C. DSB; Sart.; P I (KONON); GA; GB. Astronomy.

95. ERATOSTHENES. ca. 276-ca. 195 B.C. DSB; ICB; Sart.; P I; GA; GB. Geography. Mathematics. See also 15661/1.
1. *Catasterismorum reliquae.* 1878. Ed. by C. Robert. Reprinted, Berlin, 1963.

96. DOSITHEUS. late 3rd century B.C. DSB; GA; GB. Mathematics. Astronomy.

97. APOLLONIUS OF PERGA. 260-200 B.C. DSB; ICB; Sart.; P I; GA; GB. Mathematics. Astronomy. Also under Early Modern Editions (1023)*. See also 2304/1.
1. *Apollonii Pergaei quae graece extant, cum commentariis antiquis.* 2 vols. Leipzig, 1891-93. Ed. with Latin trans. by J.L. Heiberg. RM/L.
2. *Die Bücher De sectione spatii.* 1827. Ed. by W.A. Diesterweg. Reprinted, Wiesbaden, 1969.
3. *Treatise on conic sections.* Cambridge, 1896. Ed. with introd. by T.L. Heath. Reprinted, New York, 1961.
4. *Les coniques.* Bruges, 1924. French trans. with introd. and notes by P. ver Eecke.
5. *Die Kegelschnitte.* Munich/Berlin, 1926. German trans. by A. Czwalina. Reprinted, Darmstadt, 1967.

98. BOLOS OF MENDES (or BOLOS THE DEMOCRITEAN). fl. ca. 200 B.C. DSB; ICB. Natural history.

99. DIONYSODORUS. 3rd-2nd century B.C. DSB; Sart. Mathematics.

100. DIOCLES. fl. ca. 190 B.C. DSB & Supp.; Sart.; P I (DIOKLES). Mathematics. Also under Arabic Translations (469)*.

101. ARRIAN. early 2nd century B.C. Sart. Meteorology.

102. HYPSICLES OF ALEXANDRIA. early 2nd century B.C. DSB; ICB; Sart.; P I; GA. Mathematics. Astronomy.
1. *Die Aufgangszeiten der Gestirne.* Göttingen, 1966. Ed. with German trans. by V. de Falco and M. Krause.

103. SERAPION OF ALEXANDRIA. early 2nd century (?) B.C. Sart.; BLA. Medicine.

104. ZENODORUS. early 2nd century B.C. DSB; ICB; Sart. Mathematics.

105. NICANDER OF COLOPHON. fl. 185-135 (or ca. 275) B.C. ICB; Sart.; Mort.; BLA (NIKANDROS); GA; GB. Medicine. Natural history. Also under Early Modern Editions (1013).

106. GALLUS, Gaius Sulpicius. fl. 170-164 B.C. GA; GB. A Roman astronomer.

107. SELEUCUS THE BABYLONIAN. mid-2nd century B.C. ICB; Sart. Astronomy. A follower of Aristarchus.

108. HIPPARCHUS. fl. 147-127 B.C. DSB Supp.; ICB; Sart.; P I; GA; GB. Astronomy. Also under Early Modern Editions (1030)*. See also 15571/1.
    1. *The geographical fragments*. London, 1960. Ed. with introd. and comm. by D.R. Dicks.

109. ZENO OF SIDON. ca. 150-ca. 70 B.C. DSB; GB. Mathematics. Logic.

110. KRATEVAS (or CRATEVAS). fl. 100 B.C. Sart. (under C); Mort.; BLA; GA (under C). Botany. Materia medica.

111. THEODOSIUS OF BITHYNIA (or OF TRIPOLIS). fl. ca. 100 B.C. DSB; Sart.; P II. Under T. OF TRIPOLIS: ICB; GA; GB. Mathematics. Astronomy. Also under Early Modern Editions (1017)*. See also 1015/1.
    1. *Les sphériques*. Paris, 1927. French trans. with introd. and notes by P. ver Eecke.
    2. *Spherik*. Leipzig, 1931. (OKEW 232) Ed. with German trans. by A. Czwalina.

112. PSEUDO-PETOSIRIS. 2nd-1st century B.C. DSB; ICB. Astrology.

113. POSIDONIUS. ca. 135-ca. 51 B.C. DSB; Sart.; P II. Under POSEIDONIOS: ICB; GA; GB. Philosophy. Earth sciences.

114. ASCLEPIADES OF BITHYNIA. ca. 130-40 B.C. DSB; ICB; Sart.; Mort.; BLA & Supp. (ASKLEPIADES); GA; GB. Medicine.

115. CASSIUS DIONYSIUS. fl. ca. 88 B.C. Sart. Botany. Materia medica.

116. GEMINUS OF RHODES. fl. ca. 70 B.C. DSB; Sart.; P I; GA. Astronomy. Mathematics. Also under Early Modern Editions (1011).
    1. *Elementa astronomiae*. Leipzig, 1898. Ed. by C. Manitius.
    2. *Elementorum astronomiae capita I, III-VI, VIII-XVI*. Leiden, 1957. Ed. with a glossary by E.J. Dijksterhuis.
    3. *Introduction aux phénomènes*. Paris, 1975. Ed. with French trans. by G. Aujac.

117. FIGULUS, Publius Nigidius. ca. 98-44 B.C. ICB; Sart.; GA; GB. Astrology.

118. LUCRETIUS CARUS, Titus. ca. 95-ca. 55 B.C. DSB; ICB; Sart.; GA; GB. Natural philosophy. Also under Early Modern Editions (985)*. See also 15512/1.
    1. *De rerum natura*. There are several modern editions and translations.

119. SOSIGENES. mid-1st century B.C. DSB; Sart.; P II; GA; GB. Astronomy.

120. VITRUVIUS POLLIO, Marcus. d. ca. 25 B.C. DSB Supp.; ICB; Sart.; P II; GA; GB. Architecture. Also under Early Modern Editions (1000)*.
    1. *On architecture*. 2 vols. London, 1970. (Loeb classical library.) Latin text with English trans by F. Granger.

121. ANAXILAUS OF LARISSA. fl. 28 B.C. DSB; ICB; BLA & Supp.; GA; GB. Materia medica. Alchemy. Magic.

122. MUSA, Antonius. fl. 23 B.C. ICB; Sart.; Mort.; BLA & Supp. (ANTONIUS MUSA); GA. Medicine. Also under Early Modern Editions (1019).

123. NICOLAUS OF DAMASCUS. b. ca. 64 B.C. DSB; ICB (NICOLAOS DAMASCENOS); Sart.; GA; GB (DAMASCENUS, N.). Botany. Also under Arabic Translations (474)*.

124. STRABO. 64/3 B.C-ca. A.D. 25. DSB; ICB; Sart.; P II; GA; GB. Physical, astronomical, and descriptive geography. Also under Early Modern Editions (980).
    1. *The geography*. 3 vols. London, 1854-57. Trans. with notes by H.C. Hamilton and W. Falconer. RM/L. Several reprints.
    2. *Geography*. 8 vols. London, 1917-32. (Loeb classical library.) Greek text with English trans. by H.L. Jones.
    3. *Strabonis Geographia*. Bonn, 1968. Ed. by W. Aly.

125. HYGINUS, Caius Julius. fl. 28 B.C.-A.D. 10. Sart.; GA; GB. Astronomy. Also under Early Modern Editions (989)*. See also 1041.
    1. *Higinus astronomus*. Pisa, 1976. Contains the text of *De astronomia*. Ed. by F. Serra.

126. CELSUS, Aulus Cornelius. 25 B.C.-A.D. 50. DSB; ICB; Sart.; Mort.; BLA & Supp.; GA. Medicine. Also under Early Modern Editions (992). (Titles not included.)

127. DOROTHEUS OF SIDON. early 1st century. DSB Supp.; ICB; Sart. (Vol. 3). Astrology.
    1. *Carmen astrologicum*. Leipzig, 1976. English trans. of the Arabic version together with fragments of his writings in Greek and Latin. By D. Pingree.

128. MANILIUS, Marcus. early 1st century. DSB; ICB; Sart.; P II; GB. Astrology. Also under Early Modern Periods (986)*. See also 1038 and 15522/1.
    1. *Astronomicon*. 5 vols. London, 1903-31. Ed. by A.E. Housman. Reprinted, 5 vols in 2, Hildesheim, 1971.
    2. *Astronomica*. London, 1977. (Loeb classical library.) Latin text with English trans. by G.P. Goold.

129. RUFUS OF EPHESUS. early 1st century. DSB; ICB; Sart.; Mort.; BLA; GA. Medicine. Anatomy.
    1. *Oeuvres*. Paris, 1879. Greek text with French trans. by C. Daremberg and E. Ruelle. Reprinted, Amsterdam, 1963.

130. SCRIBONIUS LARGUS. early 1st century. ICB; Sart.; Mort.; BLA; GB (LARGUS, S.). Pharmacy.

131. ALEXANDER OF MYNDOS. fl. ca. 25-50. DSB. Zoology.

132. SENECA, Lucius Annaeus. ca. 2 B.C.-A.D. 65. DSB; ICB; Sart.; P II; GA; GB. Philosophy. Earth sciences. Also under Early Modern Editions (991)*.
    1. *Naturalium quaestionum libri VIII*. Leipzig, 1907. Ed. by A. Gercke. Reprinted, Stuttgart, 1970.
    2. *Physical science in the time of Nero, being a translation of the "Questiones naturales."* London, 1910. By J. Clarke. RM/L.
    3. *Questions naturelles*. Paris, 1930. Ed. by P. Oltramare.

4. *Naturales quaestiones*. London, 1972. (Loeb classical library.) Latin text with English trans. by T.H. Corcoran.

133. CLEOMEDES. 1st century. DSB; ICB; Sart.; P I (KLEOMEDES); GA. Astronomy. Also under Early Modern Editions (1004). See also 1039/1.
   1. *De motu circulari corporum caelestium*. Leipzig, 1891. Greek text with Latin trans. by H. Ziegler.
   2. *Die Kreisbewegung der Gestirne*. Leipzig, 1927. (OKEW 220) German trans. by A. Czwalina.

134. DEMOCRITUS THE ALCHEMIST. 1st (or 3rd?) century. Sart. Alchemy.

135. MARY THE COPT. 1st century. ICB. Alchemy.

136. TEUCRUS OF BABYLON. 1st century (?). ICB. Astrology.

137. MENECRATES. mid-1st century. ICB; Sart.; BLA (MENEKRATES); GA. Pharmacy.

138. THEMISON OF LAODICEA. mid-1st century. ICB; Sart.; BLA. Medicine.

139. AGATHINUS, Claudius. fl. ca. 50. DSB; Sart.; BLA. Medicine.

140. ARETAEUS OF CAPPADOCIA. fl. ca. 50. DSB; ICB; Sart.; Mort.; BLA & Supp.; GA; GB. Medicine. (Titles not included.)

141. ATHENAEUS OF ATTALIA. fl. ca. 50 (?). DSB; Sart.; BLA & Supp.; GA. Medicine.

142. DIOSCORIDES. fl. 50-70. DSB; ICB; Sart.; P I; Mort; BLA & Supp.; GA. Pharmacy. Botany. Also under Early Modern Editions (993)*. See also Index.

143. HERO OF ALEXANDRIA. fl. 62. DSB; P I; GB. Under HERON: Sart.; GA. Applied mathematics. Pneumatics. Also under Early Modern Editions (1032)*.
   1. *Opera quae supersunt omnia*. 5 vols. Leipzig, 1899-1914. Greek text with German trans. by W. Schmidt et al. RM/L.
   2. *The pneumatics*. London, 1851. Trans. for and ed. by B. Woodcroft. RM/L. Reprinted, ib., 1971.
   3. *Metrica*. Leiden., 1964. Ed. by E.M. Bruins.

144. PLINY (THE ELDER). (Full name: GAIUS PLINIUS SECUNDUS.) ca. 23-79. DSB; ICB; Sart.; P II; BLA; GA; GB. Natural history. Also under Early Modern Editions (979)*. See also Index.
   1. *Naturalis historiae libri XXXVII*. 6 vols. Leipzig, 1875-1906. Ed. by L.Janus and C. Mayhoff. Reprinted, Stuttgart, 1967.
   2. *In G. Plinii Secundi "Naturalis historiae" libros indices*. 2 vols. Gotha, 1857-58. By O. Schneider. Reprinted, 2 vols in 1, Hildesheim, 1967.
   3. *Natural history*. 10 vols. London, 1942-63. (Loeb classical library.) Latin text with English trans. by H. Rackham .
   4. *Naturgeschichte*. 1853-55. German trans. by C.F.L. Strack. Reprinted, 3 vols, Darmstadt, 1968.
   5. *Histoire naturelle*. Paris, 1947-. Ed. with French trans. and comm. by J. Beaujeu.

Selections from the *Natural history*

6. *Chrestomathia Pliniana*. Berlin, 1857. Ed. by L. Urlichs.
7. *De medicina libri tres*. Leipzig, 1875. Ed. by V. Rose.
8. *Chapters on chemical subjects*. 2 vols. London, 1929-32. Ed. with trans. and notes by K.C. Bailey.
9. *Naturalis historiae liber secundus*. Aberdeen, 1936. Ed. by D.J. Campbell. Pliny's cosmology.

145. OPPIAN. late 1st century. ICB; GA; GB. Author of a work on fish and fishing. Also under Early Modern Editions (994).

146. FRONTINUS, Sextus Julius. ca. 40-103. ICB; Sart.; P I; GA; GB. Engineering. Surveying. Also under Early Modern Editions (998).

147. BALBUS. fl. ca. 100. DSB; Sart.; GA. Surveying.

148. HYGINUS. fl. ca. 100. Sart.; GB. Surveying.

149. MENELAUS OF ALEXANDRIA. fl. 100. DSB & Supp.;ICB; Sart.; P II; GA. Mathematics. Astronomy. Also under Arabic Translations (473)* and Early Modern Editions (1027).

150. NICOMACHUS OF GERASA. fl. ca. 100. DSB; ICB; Sart.; P II (NIKO-MACHOS); GA; GB. Arithmetic. Theory of music. Also under Arabic Translations (475)* and Early Modern Editions (1024).
    1. *Introductionis arithmeticae libri II*. Leipzig, 1866. Ed. by R. Hoche.
    2. *Introduction to arithmetic*. New York, 1926. Trans. by M.L. D'Ooge.

    See also 169/2.

151. PAMPHILUS THE BOTANIST. (So called because he can be confused with contemporaries of the same name.) fl. 100. Sart.; BLA. Botany.

152. ARCHIGENES. fl. 98-117. DSB; Sart.; BLA & Supp.; GA. Medicine.

153. SORANUS OF EPHESUS. fl. 98-138. DSB; ICB; Sart.; Mort.; BLA; GB. Medicine. Also under Early Modern Editions (1021). (Titles not included.)

154. ANTYLLUS. early 2nd century. ICB; Sart.; Mort.; BLA & Supp. Surgery.

155. MARINUS THE ANATOMIST. fl. ca. 130. Sart.; BLA. Anatomy.

156. THEON OF SMYRNA. fl. ca. 130. DSB; Sart.; P II; GA. Mathematics. Astronomy.
    1. *Exposition des connaissances mathématiques utiles pour la lecture de Platon*. Paris, 1892. Trans. by J. Dupuis. Reprinted, Brussels, 1966.

157. PTOLEMY (or CLAUDIUS PTOLEMAEUS). ca. 100-ca. 170. DSB; ICB; Sart.; P II; GA; GB. Astronomy. Astrology. Geography. Physics. Also under Early Modern Editions (990)*. See also Index.
    1. *Opera quae exstant omnia*. Publication began in 1898 but the work is still incomplete. Individual volumes are cited below.

Almagest

2. *Syntaxis mathematica*. 2 parts. Leipzig, 1898-1903. (*Opera*. Vol. 1.) Ed. by J.L. Heiberg.

3. *Handbuch der Astronomie*. 2 vols. Leipzig, 1912-13. German trans. with notes by K. Manitius. 2nd ed., with introd. and corrections by O. Neugebauer, ib., 1963.
4. *Ptolemy's catalogue of stars. A revision of the "Almagest"*. Washington, D.C., 1915. By C.H.F. Peters and E.B. Knobel.
5. *The Almagest*. Chicago, 1952. (Great books of the western world. 16) Trans. by R.C. Taliaferro.

Geographia

6. *Geographia*. 2 vols. Leipzig, 1843-45. Ed. by C.F.A. Nobbe. Reprinted in 1 vol., Hildesheim, 1966.
7. *Geographia*. 2 vols. Paris, 1883-1901. Ed. by C. Müller. A partial ed.
8. *Geography*. New York, 1932. English trans. by E.L. Stevenson.

Tetrabiblos

9. *Apotelesmatica*. Leipzig, 1957. (*Opera*. Vol. 3, part 1.) Ed. by F. Boll and A. Boer.
10. *Tetrabiblos*. Cambridge, Mass., 1940. (Loeb classical library.) Greek text with English trans. by E.F. Robbins.

Other Works

11. *Opera astronomica minora*. Leipzig, 1907. (*Opera*. Vol. 2.) Ed. by J.L. Heiberg.
12. *De iudicandi facultate et animi principatu*. Leipzig,1961. (*Opera*. Vol. 3, part 2.) Ed. by F. Lammert.
13. *Pseudo-Ptolemaei Fructus sive centiloquium*. Leipzig, 1961. (*Opera*. Vol. 3, part 2.) Ed. by A. Boer.

158. ANONYMUS LONDINENSIS (A Greek author of unknown identity.) fl. ca. 150. Sart.; Mort. Medicine. (Titles not included.)

159. GALEN. 129/130-199/200. DSB; ICB; Sart.; P I; Mort.; BLA & Supp.; GA; GB. Medicine. Also under Arabic Translations (471)*, Medieval Latin Translations (788)*, and Early Modern Editions (984)*. See also Index.

1. *Opera omnia*. 20 vols in 22. Leipzig, 1821-33. (*Medicorum graecorum opera*. Vols 1-20.) Ed. by C.G. Kühn. Greek text with Latin trans. Reprinted, Hildesheim, 1964-65.

Selections

2. *Oeuvres anatomiques, physiologiques et médicales*. 2 vols. Paris, 1854-56. Trans. by C. Daremberg. RM/L.
3. *Scripta minora*. 3 vols. Leipzig, 1884-93. Ed. by J. Marquardt et al. Reprinted, Amsterdam, 1967.
4. *On psychology, psychopathology, and function and diseases of the nervous system*. Basel, 1973. Trans. by R.E. Siegel.

Individual Works

5. *On anatomical procedures*. London, 1956. Trans. with introd. and notes by C. Singer.
6. *On anatomical procedures: The later books*. Cambridge, 1962. Trans. by W.L.H. Duckworth.
7. *De usu partium corporis humani*. 2 vols. Leipzig, 1907-09. Ed. by G. Helmreich. Reprinted, Amsterdam, 1968.
8. *On the usefulness of the parts of the body*. 2 vols. Ithaca, N.Y., 1968. Trans. with introd. and comm. by M.T. May.

9. *On the natural faculties.* London, 1916. (Loeb classical library.) Greek text with English trans. by A.J. Brock.
10. *On the parts of medicine.* Berlin, 1969.
11. *De theriaca ad Pisonem.* Florence, 1959. Latin text with Italian trans. and introd. by E. Coturri.
12. *A translation of Galen's Hygiene (De sanitate tuenda).* Springfield, Ill., 1951. By R.M. Green.
13. *De temperamentis.* Leipzig, 1904. Ed. by G. Helmreich. Reprinted, Stuttgart, 1969.
14. *On the passions and errors of the soul.* Columbus, Ohio, 1963. Trans. by P.W. Hawkins.
15. *Protrepici quae supersunt.* Berlin, 1894; reprinted 1963. Ed. by G. Kaibel.
16. *Einführung in die Logik.* Berlin, 1960. Trans. and comm. by J. Mau.
17. *Institutio logica. English translation.* Baltimore, 1964. With introd. and comm. by J.S. Kieffer.
18. *On the affected parts.* Basel, 1976. Trans. with notes by R.E. Siegel.

160. ALEXANDER OF APHRODISIAS. fl. ca. 200. DSB; ICB; Sart.; P I; BLA & Supp.; GA; GB. Philosophy. Also under Arabic Translations (467)*, Medieval Latin Translations (784)*, and Early Modern Editions (1001)*.
1. ...*A study of the "De mixtione."* Leiden, 1976. By R.B. Todd. Includes text with trans. and comm.

His commentaries on Aristotle are in the collection *Commentaria in Aristotelem graeca* (212) and his minor works are in the *Supplementum Aristotelicum* (213).

161. SEXTUS EMPIRICUS. fl. ca. 200. DSB; ICB; Sart.; BLA; GA; GB. Philosophy. Medicine.

162. ACHILLES TATIUS. fl. ca. 200 (or later). Sart.; P I; GA; GB. Astronomy. See also 1030/1.

163. SERENUS SAMMONICUS, Quintus. d. 212. ICB; Sart.; BLA; GB. Medicine. Natural history. Also under Early Modern Editions (995).

164. AELIANUS, Claudius. ca. 170-235. ICB; Sart. Zoology. Also under Early Modern Editions (1026)*.
1. *On the characteristics of animals.* 3 vols. London, 1958-59. (Loeb classical library.) Latin text with English trans. by A.F. Scholfield.

See also 233/1.

165. CENSORINUS. fl. 238. DSB; ICB; Sart.; GA. Astrology. Encyclopaedism. Also under Early Modern Editions (1006)*.

166. DIOPHANTUS OF ALEXANDRIA. fl. 250. DSB Supp.; ICB; Sart.; P I; GA; GB. Mathematics. Also under Early Modern Editions (1031)*.
1. *Diophantus.... A study in the history of Greek algebra.* Cambridge, 1885. By T.L. Heath. Contains English trans. of his writings.
2. *Le traité des nombres polygones.* Mâcon, 1911. Trans. by G. Massoutié. RM.
3. *Les six livres arithmétiques, et Le livre des nombres polygones.* Bruges, 1926. Trans. with introd. and notes by P. ver Eecke. Reprinted, Paris, 1959.

167. ANATOLIUS OF ALEXANDRIA. fl. ca. 269. DSB; Sart.; GA. Mathematics.

168. SPORUS OF NICAEA. late 3rd century. DSB; Sart. Mathematics.

169. IAMBLICHUS OF CHALCIS. ca. 250-ca. 330. DSB; ICB; Sart.; GA; GB. Philosophy. Also under Early Modern Editions (1008).
    1. *De communi mathematica scientia*. Leipzig, 1891. Ed. by N. Festa. Reprinted, Stuttgart, 1975.
    2. *In Nicomachi "Arithmeticam" introductionem*. Leipzig, 1894. Ed. by H. Pistelli. Reprinted, Stuttgart, 1975.
    3. *Theologumena arithmeticae*. Leipzig, 1922. Ed. by V. de Falco. Reprinted, Stuttgart, 1975.
    4. *De vita Pythagorica*. Leipzig, 1937. Ed. by L. Deubner. Reprinted, Stuttgart, 1975.
    5. *Pythagoras. Legende. Lehre. Lebensgestaltung*. Zurich, 1963. Greek text with German trans. by M. von Albrecht.

170. ZOSIMUS OF PANOPOLIS. fl. ca. 300. DSB; ICB; Sart.; P II. Alchemy.

171. ADAMANTIUS SOPHISTA. early (?) 4th century. Sart.; BLA & Supp. Author of a treatise on winds.

172. PAPPUS OF ALEXANDRIA. fl. 300-350. DSB; ICB; Sart.; P II; GA; GB. Mathematics. Astronomy. Geography. Also under Arabic Translations (476)* and Early Modern Editions (1034). See also 1002/1 and 1023/1.
    1. *Collectionis quae supersunt*. 3 vols. Berlin, 1875-78. Ed. by F. Hultsch. Greek text with Latin trans. and comm.
    2. *La collection mathématique*. 2 vols. Bruges, 1933. Trans. with introd. and notes by P. ver Eecke.

173. FIRMICUS MATERNUS, Julius. fl. 330-354. DSB; ICB; Sart.; P I; GA; GB. Astrology. Also under Early Modern Editions (1007)*. See also 1038.
    1. *Matheseos libri VIII*. 2 vols. Leipzig, 1897-1913. Ed. by W. Kroll and F. Skutsch.
    2. *Ancient astrology, theory and practice*. Park Ridge, N.J., 1975. Trans. by J.R. Brim of item 1.

174. HELIODORUS OF LARISSA. 4th century (or earlier). Sart.; P I; GA. Optics. Also under Early Modern Editions (1035)*.

175. DAMIANUS. 4th century (?). Sart. Optics. Son or pupil of Heliodorus of Larissa.

176. SERENUS OF ANTINOUPOLIS. 4th century (?). DSB; ICB; Sart. Mathematics. Also under Early Modern Editions (1029). See also 1023/1.
    1. *Opuscula*. Leipzig, 1896. Greek text with Latin trans. by J.L. Heiberg.
    2. *Le livre De la section du cylindre, et le livre De la section du cône*. Paris/Bruges, 1929. Trans. with introd. and notes by P. ver Eecke. Reprinted, Paris, 1969.

177. CHALCIDIUS. 4th (or 5th) century. DSB (CALCIDIUS); ICB; Sart.; GA. Author of a commentary on Plato's *Timaeus* which was highly influential in the Middle Ages. Also under Early Modern Editions (1018).

178. ORIBASIUS. ca. 325-ca. 400. DSB; ICB; Sart.; Mort.; BLA & Supp.; GA. Medicine. Also under Early Modern Editions (1020).

179. THEON OF ALEXANDRIA. fl. 364-377. DSB; ICB; Sart.; P II; GA. Mathematics. Astronomy. Also under Early Modern Editions (1014).

180. PAUL OF ALEXANDRIA. fl. 378. DSB; ICB (PAULOS); Sart. Astrology. Also under Early Modern Editions (1033).
   1. *Elementa apotelesmatica*. Leipzig, 1958. Ed. by A. Boer with addition of astronomical interpretations by O. Neugebauer.
   See also 198/1.

181. HEPHAESTION OF THEBES. fl. ca. 381. Sart. Astrology.
   1. *Apotelesmatica*. 2 vols. Leipzig, 1973-74. Ed. by D. Pingree.

182. NEMESIUS. fl. 390-400. DSB; ICB; Sart.; Mort.; BLA; GA; GB. Medicine. Physiology. Also under Early Modern Editions (1028).

183. MACROBIUS, Ambrosius Theodosius. fl. ca. 400-430. DSB; ICB; Sart.; GA; GB. "One of the leading popularizers of science in the Latin West." Also under Early Modern Editions (983)*.

184. MARTIANUS CAPELLA. ca. 365-440. DSB; ICB. Under CAPELLA: Sart.; GA; GB. Author of textbooks of mathematics, science, and other subjects which were much used during the Middle Ages. Also under Early Modern Editions (1012).

185. HYPATIA. ca. 370-415. DSB; ICB; Sart.; P I; GA; GB. Mathematics. "The first woman in history to have lectured and written critical works on the most advanced mathematics of her day." Daughter, pupil, and assistant of Theon of Alexandria.

186. SYNESIUS OF CYRENE. ca. 370-ca. 414. DSB; ICB; Sart.; P II; GA; GB. Astronomy. Physics.

187. OLYMPIODORUS. fl. 412-425. DSB; ICB; Sart.; P II; GA; GB. Alchemy.

188. MARCELLUS THE EMPIRICIST. b. 379. ICB; Sart.; BLA; GA. Medicine. Botany.
   1. *De medicamentis*. Leipzig, 1916. Ed. by M. Niedermann. 2nd ed., with German trans., 2 vols. Berlin, 1968.

189. NONNUS OF PANOPOLIS. ca. 380-450. ICB; GA; GB. Astrology.

190. APULEIUS BARBARUS (or PLATONICUS). 5th century (?) ICB (APULEIUS, Pseudo-); Sart.; BLA & Supp. Supposed author of a herbal; see 1036.

191. DOMNINUS OF LARISSA. 5th century. DSB; Sart.; GF. Mathematics.

192. PROCLUS. 410-485. DSB; ICB; Sart.; P II (PROKLUS); GA. Philosophy. Mathematics. Astronomy. Also under Medieval Latin Translations (791)* and Early Modern Editions (1010)*. See also Index.
   1. *In primum Euclidis "Elementorum" librum commentarii*. Leipzig, 1873. Ed. by G. Friedlein. Greek text with Latin trans. Reprinted, Hildesheim, 1967.
   2. *Les commentaires sur le premier livre des "Eléments" d'Euclide*. Bruges, 1948. Trans. with introd. and notes by P. ver Eecke.
   3. *A commentary on the first book of Euclid's "Elements."* Princeton, N.J., 1970. Trans. with introd. and notes by G.R. Morrow.

4. *Hypotyposis astronomicarum positionum.* Leipzig, 1909. Ed. with German trans. by C. Manitius.
5. *Institutio physica.* Leipzig, 1912. Ed. with German trans. by A. Ritzenfeld.
6. *Commentaire sur le "Timée."* 2 vols. Paris, 1966-67. Trans. with notes by A.J. Festugière.

193. VICTORIUS OF AQUITANIA. fl. 457-465. ICB; Sart. Astronomy. Mathematics.

194. AMMONIUS, Son of Hermes. d. ca. 520. DSB; Sart. Philosophy. Astronomy. Mathematics. His commentaries on Aristotle and Porphyry are in the collection *Commentaria in Aristotelem graeca* (212).

195. JULIANUS OF LAODICEA. fl. ca. 500. Sart. Astronomy/astrology.

196. METRODORUS. fl. ca. 500. Sart. Mathematics.

197. PRISCIANUS OF CAESAREA. fl. 500. ICB; Sart.; GA; GB. Author of a work on weights and measures. Also under Early Modern Editions (981).

198. HELIODORUS, Son of Hermias. fl. 498-509. ICB; Sart. Astronomy.
1. *In Paulum Alexandrinum commentarium.* Leipzig, 1962. Ed. by A. Boer with addition of astronomical interpretations by O. Neugebauer and D. Pingree.

199. TIMOTHY OF GAZA. fl. 491-518. ICB. Zoology.
1. *On animals. Fragments of a Byzantine paraphrase of an animal book of the fifth century A.D.* Leiden, 1949. Trans. with introd. and commentary by F.S. Bodenheimer and A. Rabinowitz.

See also 233/1.

200. ANTHEMIUS OF TRALLES. d. ca. 534. DSB; ICB; Sart.; P I; GA; GB. Architecture. Engineering. Mathematics.

201. BOËTHIUS, Anicius Manlius Severinus. ca. 480-524/5. DSB; ICB; Sart.; P I; GA; GB. Encyclopaedic learning, including logic, the mathematical sciences, and the theory of music. "The broadcaster of much Greek knowledge to many generations who used Latin." Also under Early Modern Editions (1003). See also 1195.
1. *De institutione arithmetica libri II. De institutione musica libri V. Accedit geometria quae fertur Boetii.* Leipzig, 1867. Ed. by G. Friedlein.
2. *Geometrie II: Ein mathematisches Lehrbuch des Mittelalters.* Wiesbaden, 1970. Ed. by M. Folkerts.

202. EUTOCIUS OF ASCALON. b. ca. 480. DSB; ICB; Sart.; P I; GA. Mathematics. Also under Early Modern Editions (1025)*. See also Index.

203. JOHN PHILOPONUS. early 6th century. DSB. Under PHILOPONUS: ICB; Sart.; GA; GB. Philosophy, including cosmology and the principles of dynamics. "The first thinker to undertake a comprehensive and massive attack on the principal tenets of Aristotle's physics and cosmology."
1. *Ausgewählte Schriften.* Munich, 1967. Trans. with introd. and notes by W. Böhm.

His commentaries on Aristotle are in the collection *Commentaria in Aristotelem graeca* (212).

204. ISIDORUS OF MILETUS. fl. 532. DSB; Sart.; GA. Architecture. Mathematics.

205. AËTIUS OF AMIDA. fl. ca. 540. DSB; ICB; Sart.; Mort.; BLA & Supp.; GA; GB. Medicine. Also under Early Modern Editions (1022).

206. DIONYSIUS EXIGUUS. ca. 500-ca. 560. ICB; Sart.; GA; GB. Mathematics. Astronomy.

1.132 Anonymous Works

207. *AETNA*. A Latin poem dealing with geological phenomena. Date uncertain.
   1. *Aetna*. Oxford, 1901. Ed. and trans. by R. Ellis.
   2. *Incerti auctoris Aetna*. Cambridge, 1965. Ed. by F.R.D. Goodyear.

208. *CODEX Constantinopolitanus palatii veteris. No. 1*. 3 vols. Leiden, 1964. A mathematical textbook of uncertain date. Greek text with English trans. by E.M. Bruins.

209. *PHYSIOLOGUS*. A bestiary dating from the second century. (cf. entry 524.)
   1. *Der "Physiologus" nach den Handschriften G und M*. Meisenheim am Glau, 1966. Greek text. Ed. by D. Offermanns.

1.133 Collections of Texts

210. THE PRESOCRATICS.
   1. *Die Fragmente der Vorsokratiker*. 9th ed. 3 vols. Berlin, 1952; reprinted 1960. Ed. by H. Diels and W. Kranz.
   1a. ——— *Ancilla to the presocratic philosophers: A complete translation of the fragments in Diels' "Fragmente der Vorsokratiker."* Cambridge, Mass., 1948. By K. Freeman.

211. THE PYTHAGOREANS.
   1. *Philosophorum Pythagoreorum collectionis specimen*. Chicago, 1941. Ed. by H.A. Brown.
   2. *Pitagorici: Testimonianze e frammenti*. 3 vols. Florence, 1958-64. Ed. by M. Timpanaro-Cardini.
   3. *The Pythagorean texts of the Hellenistic period*. Åbo, 1965. Ed. by H. Thesleff.

212. *COMMENTARIA in Aristotelem graeca*. 23 vols. Berlin, 1882-1902. RM (under Aristotle)/L. Edited by various scholars under the auspices of the Berlin Academy. Texts contained in this collection are noted under their authors in the present catalogue.

213. *SUPPLEMENTUM Aristotelicum*. 3 vols. Berlin, 1885-1903. RM. Edited by various scholars under the auspices of the Berlin Academy. A supplement to the above *Commentaria* and to the Berlin Academy's edition of Aristotle's collected works (65/1). Texts contained in this collection are noted under their authors in the present catalogue.

214. *RERUM naturalium scriptores graeci minores*. Vol. 1. [No more publ.] Leipzig, 1877. Ed. by O. Keller.

215. MATHEMATICIANS.
   1. *Mathematici graeci minores.* Copenhagen, 1927. Ed. by J.L. Heiberg.
   2. *Greek mathematical works: Selections illustrating the history of Greek mathematics.* 2 vols. London, 1939-41. (Loeb classical library.) Ed. by I. Thomas.

216. ASTRONOMERS and Astrologers.
   1. *Catalogus codicum astrologorum graecorum.* 11 vols. Brussels, 1898-1936. Ed. by F. Cumont et al.
   2. *Sphaera: Neue grieschische Texte und Untersuchungen zur Geschichte der Sternbilder.* Leipzig, 1903. RM/L. By F.J. Boll. Reprinted, Hildesheim, 1967.
   3. *Greek horoscopes.* Philadelphia, 1959. By O. Neugebauer and H.B. Van Hoesen.

217. SURVEYORS.
   1. *Gromatici veteres ... (Die Schriften der römischen Feldmesser).* 2 vols. Berlin, 1848-52. Ed. by F. Blume, K. Lachmann, et al. Reprints: Rome, 1960?, and Hildesheim, 1967.
   2. *Corpus agrimensorum romanorum.* Leipzig, 1913. Ed. by C. Thulin. Reprinted, Stuttgart, 1971.

218. ALCHEMISTS.
   1. *Collection des anciens alchimistes grecs.* 3 vols. Paris, 1887-88. Ed. by M. Berthelot. RM/L. Reprinted in 1 vol., London, 1963.
   2. *Catalogue des manuscrits alchimiques grecs.* 8 vols. Brussels, 1924-32. By J. Bidez.
   3. *Archéologie et histoire des sciences. Avec publication nouvelle du papyrus grec chimique de Leyde.* Paris, 1906. By M. Berthelot. RM/L. Reprinted, Amsterdam, 1968.

219. HERMES TRISMEGISTUS. DSB; ICB. A legendary personage. The *Corpus Hermeticum*, the best known of the numerous Greek writings extant under this name, was probably written in the second century A.D. Also under Early Modern Editions (1005)*.
   1. *The theological and philosophical works.* Edinburgh, 1882. Trans. with notes etc. by J.D. Chambers. RM.
   2. *The divine Pymander, and other writings.* Edinburgh, 1882. Trans. by J.D. Chambers. Reprinted, New York, 1972.
   3. *Hermetica.* 4 vols. Oxford, 1924-36. Ed. with English trans. and notes by W. Scott.
   4. *Corpus Hermeticum.* 4 vols. Paris, 1945-54. Ed. by A.D. Nock. French trans. by A.J. Festugière,

220. PHYSICIANS.
   1. *Medicorum graecorum opera quae extant.* 26 vols in 28. Leipzig, 1821-33. Ed. by C.G. Kühn. Reprinted, Hildesheim, 1964-65.
   2. *Corpus medicorum graecorum.* Many vols. Leipzig, 1908-. Ed. under the auspices of the Berlin and other German academies.
   3. *Corpus medicorum latinorum.* Many vols. Leipzig, 1915-. Ed. under the auspices of the Berlin and Leipzig Academies.
   4. *Greek medicine: Being extracts illustrative of medical writers from Hippocrates to Galen.* London, 1929. Ed. and trans. by A.J. Brock. Reprinted, New York, 1972.

221. *CORPUS* *hippiatricorum graecorum.* 2 vols. Leipzig, 1924-27. Ed. by E. Oder and C. Hoppe. Reprinted, Stuttgart, 1971.

1.2 THE MIDDLE AGES

1.21 Byzantine

1.211 Authors

222. ALEXANDER OF TRALLES. ca. 525-608. DSB; ICB; Sart.; Mort.; BLA & Supp.; GA; GB. Medicine. Also under Early Modern Editions (1044). (Titles not included.)

223. THEOPHYLACTUS SIMONCATTA. d. ca. 630. ICB; Sart.; GA.
    1. *Questiones physicae, et Epistolae*. Paris, 1835. Ed. by J.F. Boisonnade. Greek texts with Latin trans.

224. STEPHANUS OF ALEXANDRIA. early 7th century. DSB; ICB; Sart. Mathematics. Astronomy. Alchemy.

225. PAUL OF AEGINA. fl. 640. DSB; ICB (PAULOS); Sart. (PAULOS); Mort.; BLA (PAULUS). Medicine. Surgery. Also under Early Modern Editions (1045). (Titles not included.)

226. SEVERUS SĒBŌKHT. mid-7th century. ICB; Sart. Astronomy. Geography. A Syriac writer.

227. SHIRAKATSÍ, Anania (or ANANIAS OF SHIRAK). ca. 620-ca.680. DSB; ICB (ANANIA SHIRAKAZI). Mathematics. Astronomy. An Armenian writer.

228. EZECHIEL. ca. 673-727. P I; GA. Astronomy. An Armenian writer.

229. HELIODORUS THE ALCHEMIST. early 8th century. ICB; Sart. Alchemy.

230. THEOPHRASTUS CHRISTIANUS. 8th (or 9th) century. ICB. Alchemy.

231. LEO THE MATHEMATICIAN (or LEO THE PHILOSOPHER). ca. 790-after 869. DSB; Sart. (LEON). Mathematics. Astrology/astronomy.

232. HERO OF BYZANTIUM (or HERO THE YOUNGER). fl. ca. 938. ICB; Sart. (HERON); P I; GA (HERON). The mathematical sciences. Also under Early Modern Editions (1049)*.

233. CONSTANTINUS VII, Porphyrogenitus (Emperor of the East). 905-959. Sart.; GA; GB. Patron of the arts and sciences. Also under Early Modern Editions (1046).
    1. *Excerptorum Constantini de natura animalium libri duo*. Berlin, 1885. (*Supplementum Aristotelicum*. [See entry 213.] Vol. 1, part 1.) Ed. by S.P. Lambros. A zoological encyclopaedia based upon Aristotle in the epitome of Aristophanes of Byzantium, and upon Aelian, Timothy of Gaza, and other writers. Traditionally ascribed to the Emperor Constantinus.

234. PSELLUS, Michael (or Constantine). 1018-1078. DSB; ICB; Sart.; P II; BLA; GA; GB. Encyclopaedist in philosophy and science. Also under Early Modern Editions (1043).

235. MICHAEL OF EPHESUS. late 11th century. ICB. His commentaries on Aristotle's zoological works are in the collection *Commentaria in Aristotelem graeca* (212).

236. SETH, Simeon. fl. 1071-1080. Sart.; BLA (SIMEON SETH). Botany.
 1. *Syntagma de alimentorum facultatibus*. Leipzig, 1868. Ed. by B. Langkavel.

237. CAMATEROS, Joannes. mid-12th century. Sart. Astronomy/astrology.

238. NOVGORODETS, Kirik. b. 1110. ICB. A Russian author of a work on time reckoning.

239. NICEPHORUS BLEMMYDES. 1197/8-1272. ICB; Sart.; GA. Alchemy. Geography.

240. MYREPSUS, Nicholas. fl. 1222-1280. ICB; Sart.; GA. Under NICOLAUS M.: Mort.; BLA. Pharmacy. Also under Early Modern Editions (1047).

241. PACHYMERES, George. 1242-ca. 1310. ICB; Sart.; GA; GB. Best known as a historian but he also wrote an influential textbook on the mathematical sciences.

242. PLANUDES, Maximus. ca. 1255-1305. DSB; ICB; Sart.; P III (MAXIMUS P.); GA; GB. A polymath who wrote an important textbook of mathematics.
 1. *Das Rechenbuch*. 1865. Greek text. Ed. by C.I. Gerhardt. Reprinted, Wiesbaden, 1973.

243. PEDIASIMUS, Joannes. late 13th-early 14th century. ICB; Sart. Mathematics.

244. MOSCHOPOULOS, Manuel. fl. 1295-1316. ICB; Sart.; GA; GB. Mathematics.

245. PHILES, Manuel. ca. 1275-ca. 1345. ICB; Sart.; GA; GB.
 1. *De proprietate animalium*. Paris, 1851. Ed. by F. Dübner. Greek text with Latin trans.

246. BARLAAM (DE SEMINARA), Bernard. ca. 1290-ca. 1348. ICB; Sart.; P I; GA. Mathematics. Astronomy. Also under Early Modern Editions (1048).
 1. *Traités sur les éclipses de soleil de 1333 et 1337*. Louvain, 1977. History of the texts, critical eds, French trans. and comm. by J. Mogenet and A. Tihon.

247. GREGORAS, Nicephorus. 1295-1359. ICB; Sart.; GA; GB. Astronomy. Mathematics. Best known as a philosopher and historian.

248. CHRYSOCOCCES, George. fl. 1335-1346. Sart.; GA. Astronomy.

249. RHABDAS, Nicolaus Artabasdus. fl. 1341. Sart. Mathematics.

250. NEOPHYTUS PRODROMENUS. 14th century. Sart. Botany.

251. ISAAC ARGYRUS. ca. 1312-after 1372. Sart.; P I. Mathematics. Astronomy.

252. THEODORUS MELITENIOTES. fl. 1360-1388. Sart. Astronomy.

1.212 Anonymous Works

253. *EIN BYZANTINISCHES Rechenbuch des frühen 14. Jahrhunderts*. Vienna, 1968. Ed. with trans. and comm. by K. Vogel.

1.22 Arabic
(also Syriac and Persian)

1.221 Authors

254. KHĀLID IBN YAZĪD IBN MU'ĀWIYA. d. 704. ICB; Sart. Alchemy.

255. AL-NAUBAKHT. d. ca. 776. Sart. Astronomy/astrology. Engineering.

256. AL-FAZĀRĪ, Ibrāhīm. d. ca. 777. Sart. (IBRĀHĪM AL-FAZĀRĪ). Astronomy/astrology. Father of Muḥammad ibn Ibrāhīm al-Fazārī (258) with whom he is sometimes confused.

257. YA'QŪB IBN ṬĀRIQ. fl. 767-778. d. ca. 796. DSB; Sart. Astronomy.

258. AL-FAZĀRĪ, Muḥammad ibn Ibrāhīm. fl. ca. 760-790. d. ca. 800. DSB; Sart. (MUḤAMMAD IBN IBRĀHĪM AL- FAZĀRĪ); P III; GA (AL-FEZARI). Astronomy.

259. AL-AṢMA'Ī, 'Abd al-Malik ibn Quraib. 739/40-ca. 831. Sart. Anatomy. Zoology.

260. AL-FAḌL IBN NAUBAKHT. d. ca. 815. Sart. Astronomy/astrology.

261. MĀSHĀ'ALLĀH. (Latin form: MESSAHALA.) fl. 762-ca. 815. DSB; ICB; Sart. Astronomy/astrology.
   1. *Chaucer and Messahalla on the astrolabe.* London, 1929; 1932 (revised). By R.T. Gunther. Texts and trans.

262. 'UMAR IBN AL-FARRUKHĀN, al-Ṭabarī. fl. 762-812. DSB; Sart. Astronomy/astrology.

263. JĀBIR IBN HAYYĀN. fl. ca. 800 (?). DSB; ICB; Sart.; BLA & Supp. (DSCHĀBIR). Alchemy. See also 1107.

264. MUḤAMMAD IBN 'UMAR, Abū Bakr. (Latin form: OMAR TIBERIADIS.) fl. ca. 800. Sart. Astronomy/astrology. Also under Early Modern Editions (1071).

265. IBN MĀSAWAIH, Abū Zakarīyyā Yūḥannā. (Latin form: MESÜE THE ELDER.) ca. 776-857. ICB; Sart.; BLA (JAHJA BEN MASEWEIH); GA (MESUEH). Medicine. Also under Early Modern Editions (1058).

266. YAḤYĀ IBN ABĪ MANṢŪR. d. 832. DSB; Sart.; P I (ABUL MANSUR); P III (AL MAUṢILĪ). Astronomy.

267. IBN MAṬĀR, al-Hajjāj ibn Yūsuf. fl. 786-833. ICB; Sart. (AL-HAJJĀJ). The first translator of Euclid's *Elements* into Arabic and one of the first translators of Ptolemy's *Almagest*.

268. ABŪ 'ALĪ AL-KHAIYĀT, Yaḥyā ibn Ghālib. (Latin form: ALBOHALI.) d. ca. 835. Sart. Astronomy/astrology. Also under Early Modern Editions (1080).

269. AḤMAD AL-NAHĀWANDĪ. d. ca. 840. Sart. Astronomy.

270. AL-JĀḤIẒ, Abū 'Uthmān 'Amr ibn Baḥr. ca. 776-868/9. DSB; ICB; Sart.; GA (AMROU BEN BAHR); GB. Zoology.

271. AL-KHWĀRIZMĪ, Abū Ja'far Muḥammad ibn Mūsā. fl. 800-847. DSB; ICB (AL-KHUWĀRIZMĪ); Sart.; P III (AL HOVĀREZMĪ). Mathematics. Astronomy. Geography.

1. *The algebra of Mohammed ben Musa*. London, 1931. Ed. and trans. by F. Rosen.
2. *The astronomical tables of al-Khwārizmī*. Copenhagen, 1962. Trans. with comm. by O. Neugebauer.
3. *Algorismus. Das frühestes Lehrbuch zum Rechnen mit indischen Zahlen*. Aalen, 1963. Ed. by K.Vogel.

See also 795/1.

His works are often catalogued under Muḥammed ibn (or Mohammed ben) Mūsā.

272. JOB OF EDESSA. fl. 817-832. Under AYYŪB AL-RUHĀWĪ AL-ABRASH: ICB; Sart. One of the main translators of medical works from Greek into Arabic. His own writings are in Syriac.
 1. *Encyclopaedia of philosophical and natural sciences as taught in Baghdad about A.D. 817; or, Book of treasures*. Cambridge, 1935. Syriac text and trans. by A. Mingana.

273. 'ALĪ IBN 'ĪSĀ, al-Aṣṭurlābī. early 9th century. ICB; Sart. Astronomy. Author of one of the earliest Arabic treatises on the astrolabe.

274. SAHL IBN BISHR, Abū 'Uthmān. (Latin form: ZAHEL or ZEHEL.) early 9th century. ICB; Sart. Astrology. Also under Early Modern Editions (1068),

275. AL-MA'MŪN, 'Abdallāh. (Abbasid Caliph.) 786-833. ICB; Sart.; P I (ALMAMUN); GA (ALMAMOUN). Patron of science, especially of astronomical projects such as star catalogues. He also established an academy for translation of scientific books.

276. ABŪ MA'SHAR, al-Balkhī, Ja'far ibn Muḥammad. (Latin form: ALBUMASAR.) 787-886. DSB; ICB; Sart.; P I (ALBUMASAR); GA (ALBUMAZAR). Astrology. Also under Medieval Latin Translations (793)* and Early Modern Editions (1062)*.
 1. *De revolutionibus nativitatum*. Leipzig, 1968. Ed. by D. Pingree.

277. AL-JAWHARĪ, al-'Abbās ibn Sa'īd. fl. ca. 830. DSB; Sart. (AL-'ABBĀS IBN SA'ĪD). Mathematics. Astronomy.

278. AL-FARGHĀNĪ, Abu'l-'Abbās Aḥmad ibn Muḥammad ibn Kathīr. (Latin form: ALFRAGANUS.) d. after 861. DSB; ICB; Sart.; P I (ALFRAGAN), P III (AL FERGANI); GA (ALFRAGANUS). Astronomy. Also under Early Modern Editions (1066)*.

279. SANAD IBN 'ALĪ, Abū al-Ṭaiyib. d. after 864. Sart. Astronomy. Mathematics.

280. ḤABASH AL-ḤĀSIB, Aḥmad ibn 'Abdallāh al-Marwazī. d. ca. 870. DSB; ICB (AL-MARWAZĪ); Sart.; P III (AL ḤĀSIB). Trigonometry. Astronomy.

281. AL-MARWARRŪDHĪ, Khālid ibn 'Abd al-Malik. fl. 833. Sart. Astronomy.

282. AL-KINDĪ, Abū Yūsuf Ya'qub ibn Isḥāq al-Ṣabbāḥ. ca. 801-ca. 866. DSB Supp.; ICB; Sart.; P I (ALCHINDI). "The first Arab philosopher." Encyclopaedic learning, including mathematics and science. Also under Early Modern Editions (1072).
 1. *Buch über die Chemie des Parfüms und die Destillationen*. Leipzig, 1948. Arabic text with German trans. by K. Garbers.

2. *The medical formulary*. Madison, Wis., 1966. Arabic text with English trans. by M. Levey.

283. AL-ṬABARĪ, Abu'l-Ḥasan 'Alī ibn Sahl Rabbān. ca. 808-ca. 861. DSB; ICB; Sart. (SAHL AL-ṬABARĪ). Author of the first major medical encyclopaedia in Arabic.

284. ḤUNAYN IBN ISḤĀQ, al-'Ibādī, Abū Zayd. (Latin name: JOHANNITIUS). 808-873. DSB Supp.;ICB; Sart.; P III (ḤONEIN BEN ISḤĀK); BLA & Supp. (HUNAIN). Medicine. The foremost translator of Greek medical works. Also under Early Modern Editions (1057).
1. *The book of the ten treatises on the eye*. Cairo, 1928. Arabic text with English trans. by M. Meyerhof.
2. *Ein Kompendium der aristotelischen Meteorologie in der Fassung des Ḥunain ibn Isḥāq*. Amsterdam, 1975. Arabic text and German trans. by H. Daiber.

285. BANŪ MŪSĀ. (i.e. Sons of Mūsā: three brothers--Muḥammad, Aḥmad, and al-Ḥasan--known under the one name.) 9th century. DSB; ICB; Sart.; P III (ABŪGA'FAR); GA (MUSA IBN SHAKIR). Mathematics. Astronomy.

286. 'UṬĀRID IBN MUḤAMMAD, al-Ḥāsib. 9th century. Sart. Author of the earliest Arabic lapidiary.

287. ABŪ SA'ĪD AL-DARĪR, al-Jurjānī. mid-9th century. ICB (al-JURJĀNĪ); Sart. Astronomy. Mathematics.

288. MOSES BAR KĒPHĀ. ca. 813-903. ICB; Sart. Zoology. A Syriac writer.

289. ABŪ ḤANĪFA, al-Dīnawarī. ca. 820-895. Under AL-DĪNAWARĪ, Abū Ḥanīfa: ICB; Sart. Botany.
1. *The book of plants*. Wiesbaden, 1974. Ed. by B. Lewin.

290. AL-MĀHĀNĪ, Abū 'Abd Allāh Muhammad ibn 'Īsā. fl. ca. 860. d. ca. 880. DSB; Sart.; P III. Mathematics. Astronomy.

291. IBN QUṬAYBA, Abū Muḥammad 'Abdallāh ibn Muslim al-Dīnawarī al-Jabalī. 828-884 (or 889). DSB; ICB; Sart. Encyclopaedist.
1. *The natural history section from a 9th-century "Book of useful knowledge."* Paris, 1949. Trans. by L. Kopf.

292. 'ABBĀS IBN FIRNĀS. d. 887. DSB. Astrology/astronomy. He helped to disseminate oriental science in western Islam.

293. ḤĀMID IBN 'ALĪ, Abū al-Rabī, al-Wāsiṭī. late (?) 9th century. Sart. Astronomy. A maker of astrolabes.

294. AL-TAMĪMĪ, Muḥammad ibn 'Umail. late 9th century. ICB. Alchemy.

295. YŪSUF AL-KHŪRĪ. late 9th century. Sart. Mathematics. A translator from Syriac into Arabic.

296. THĀBIT IBN QURRA, al-Ṣābi' al-Ḥarrānī. 836-901. DSB; ICB; Sart.; P III (AL HARRĀNĪ); BLA (TABIT). Mathematics. Astronomy. Mechanics. Medicine. Also under Medieval Latin Translations (798)* and Early Modern Editions (1073). See also 475/1.

297. QUSṬĀ IBN LŪQĀ, al-Ba'labakkī. fl. 860-900. DSB; ICB; Sart. Medicine. Mathematical sciences. An important translator.

298. ABŪ KĀMIL, Shujā' ibn Aslam. ca. 850-ca. 930. DSB; ICB (SHUJĀ' IBN ASLAM); Sart. Mathematics.

1. *The algebra ... in a commentary by Mordecai Finzi.* Madison, Wis., 1966. Hebrew text, English trans. and comm. with special reference to the Arabic text. By M. Levey.
2. *On the pentagon and decagon.* Tokyo, 1971. Trans. by M. Yadegari and M. Levey.

299. ISḤĀQ IBN ḤUNAYN, Abū Ya'qūb. d. 910. DSB; ICB; Sart.; BLA (ISHAK BEN HONEIN). Medicine. An important translator of philosophical and mathematical works.

300. AL-RAZI, Abū Bakr Muhammad ibn Zakariyā. (Latin form: RHAZES.) ca. 854-925 (or 935). DSB; ICB; Sart.; P II (RHASES); Mort. (RHAZES); BLA & Supp. (ABU BEKR ... EL-RAZI); GA (RHAZES). Medicine. Alchemy. Also under Early Modern Editions (1053).

301. AL-BATTĀNI, Abū 'Abd Allāh Muḥammad. (Latin forms: ALBATENIUS, ALBATEGNIUS, etc.) fl. 858-929. DSB; ICB; Sart.; P I (ALBATEGNUS); GA; GB (ALBATEN). Astronomy. Mathematics. Also under Early Modern Editions (1078)*. See also 1066/1.
1. *Opus astronicum.* 3 vols. Milan, 1899-1907. Arabic text and Latin trans. by C.A. Nallino. RM (under Albategnius). Reprints: 2 vols, Frankfurt, 1969; 3 vols, Hildesheim, 1971.

302. AL-NAYRĪZĪ, Abu'l-'Abbas al-Faḍl ibn Ḥatim. fl. ca. 897. DSB; P III (AL-TEBRĪZĪ). Under AL-NAIRĪZĪ: ICB; Sart. Geometry. Astronomy.

303. IBN WAḤSHIYYA, Abū Bakr Aḥmad. ca. 860-ca. 935. DSB; ICB (IBN AL-WAḤSHĪYA); Sart. Agronomy. Botany. Alchemy. Magic.

304. ABŪ BAKR AL-ḤASAN IBN AL-KHAṢĪB. (Latin form: ALBUBATHER.) fl. 900. Sart. Astrology. Also under Early Modern Editions (1064).

305. IBN AL-ADAMĪ. (Alternative name: MUḤAMMAD IBN AL-ḤUSAIN IBN ḤAMID.) fl. ca. 900. Sart.; P III (AL-ADAMI). Astronomy.

306. AḤMAD IBN YŪSUF. fl. ca. 900-905. d. 912/3 (?) DSB; Sart. Mathematics.

307. IBN AMĀJŪR, Abū al-Qāsim 'Abdallāh, al-Turkī. fl. 885-933. Sart. Astronomy.

308. AL-BALKHĪ, Abū Zaid Ahmad ibn Sahl. d. 934. ICB; Sart. Mathematics. Geography.

309. AL-FĀRĀBĪ, Abū Naṣr Muḥammad. (Latin forms: ALFARABIUS, ABUNAZAR, etc.) ca. 870-950. DSB; ICB; Sart.; P I (ALFARABIUS); GA (ALFARABIUS); GB. Philosophy. Theory of music. Also under Early Modern Editions (1081)*. See also 15480.
1. *Article on vacuum.* Leiden, (date?). Ed. by N.Lugal and A. Sajili.
2. [In Russian. *Selected mathematical works.*] Alma-Ata (U.S.S.R.), 1972.

310. SINĀN IBN THĀBIT IBN QURRA, Abū Sa'id. ca. 880-943. DSB; Sart. Medicine. Astronomy. Mathematics.

311. AL-HAMDĀNĪ, Abū Muḥammad al-Ḥasan. 893?-after 951? DSB; ICB; Sart.; GB. Astronomy. Geography. Chemical technology.
1. *Die beiden Edelmetalle, Gold und Silber.* Uppsala, 1968. Arabic text with German trans. by C. Toll.

312. AL-DIMASHQĪ, Abū 'Uthmān Sa'id ibn Ya'qūb. fl. 908-932. ICB; Sart. (ABŪ 'UTHMĀN). Translator of Greek mathematical works.

313. AL-'IMRĀNĪ, 'Alī ibn Aḥmad. d. 955/6. Sart. Mathematics. Astrology.

314. ISAAC ISRAELI. d.955 (?) DSB; ICB; Sart.; Mort. (ISAAC JUDAEUS); BLA (ISAAC JUDAEUS); GB; GF (ISRAELI, Isaac). Medicine. A Jew who wrote in Arabic. Also under Early Modern Editions (1061).

315. AL-KHĀZIN, Abū Ja'far Muḥammad ibn al-Ḥasan, al-Khurāsānī. d. ca. 965. DSB; Sart. (ABŪ JA'FAR). Astronomy. Mathematics.

316. AL-ṢŪFĪ, Abu'l-Ḥusayn 'Abd al-Raḥmān ibn 'Umar, al-Rāzī. 903-986. DSB; ICB; Sart. ('ABD AL-RAḤMĀN); P I (ALSUFI). Astronomy.
    1. *An Islamic book of constellations.* Oxford, Bodleian Library, 1965.

317. IBRĀHĪM IBN SINĀN IBN THĀBIT IBN QURRA. 908-946. DSB; ICB; Sart. Mathematics. Astronomy.

318. AL-KARABĪSĪ, Aḥmad ibn 'Umar. 10th century (?) ICB. Mathematics.

319. IBN AL-MUTHANNĀ, Aḥmad. 10th century. ICB. Astronomy. Also under Medieval Latin Translations (795)*.
    1. *Commentary on the astronomical tables of al-Khwârizmî.* New Haven, Conn., 1967. Two Hebrew versions ed. and trans. with comm. by B.R. Goldstein. (The Arabic original is lost.)

320. AL-ṬABARĪ, Abu'l-Ḥasan Aḥmad ibn Muḥammad. mid-10th century. DSB; ICB; Sart. (AḤMAD AL-ṬABARĪ); GF. Medicine.

321. AL-QABĪṢĪ, Abū al-Ṣaqr 'Abd al-Azīz. (Latin form: ALCABITIUS.) fl. ca. 950. DSB; Sart.; P I (ALCABITIUS). Astrology. Also under Early Modern Editions (1055)*.

322. IBN HIBINTĀ. fl. ca. 950. DSB. Astrology/astronomy.

323. AL-UQLĪDISĪ, Abu'l-Ḥasan Aḥmad ibn Ibrāhīm. fl. 952. DSB. Arithmetic.
    1. *The arithmetic.* Dordrecht, 1978. Trans with notes by A.S. Saidan.

324. AL-MAJŪSĪ, Abu'l-Ḥasan 'Alī ibn 'Abbās. (Latin form: HALY ABBAS.) ca. 920-994. DSB; Sart. ('ALĪ IBN 'ABBĀS); Mort. (HALY ABBAS); BLA & Supp. (ALI BEN EL-ABBAS). Medicine. Also under Early Modern Editions (1065).

325. RABĪ' IBN ZAID, al-Usquf. fl. 961. Sart. Astrology.

326. IBN AL-A'LAM. d. 985. Sart. Astronomy.

327. MĀSAWAIH AL-MĀRDĪNĪ. (Latin form: MESUË THE YOUNGER.) 925-1015. ICB; Sart.; BLA (MESUË). Medicine. Also under Early Modern Editions (1050).

328. AL-HARĀWĪ, Abū Manṣūr Muwaffaq ibn 'Ali. fl. 961-976. ICB. Under ABŪ MANṢŪR MUWAFFAK: Sart.; Mort.; BLA. Pharmacy.

329. AL-ṢĀGĀNĪ, Abū Ḥāmid Aḥmad ibn Muḥammad, al-Asṭurlābī. d. 990. Sart.; P III (AL-UṢṬURLĀBĪ). Astronomy. A maker of astrolabes.

330. NAẒĪF IBN YUMAN. d. ca. 990. Sart. Mathematics. Translator of Greek mathematical works.

331. AL-TAMĪMĪ, Abū 'Abdallāh Muḥammad ibn Aḥmad ibn Sa'īd. fl. 970-980. Sart. Pharmacy.
    1. *Über die Steine: Das 14. Kapital aus dem "Kitāb al-mursid."* Freiburg, 1976. Arabic text with German trans. and comm. by J. Schönfeld.

332. AL-KHUWĀRIZMĪ, Abū 'Abd Allāh Muḥammad ibn Aḥmad. fl. 975. DSB; ICB; Sart. (MUḤAMMAD IBN AḤMAD AL-KHWĀRIZMĪ). Encyclopaedist.

333. AL-ZAHRĀWĪ, Abu'l-Qāsim Khalaf ibn 'Abbās. (Latin forms: ABULCASIS, ALBUCASIS.) ca. 936-ca. 1013. DSB; ICB; Sart. (ABŪ-L-QĀSIM); Mort. (ALBUCASIS); BLA (ABUL KASIM); GA (ALBUCASIS). Surgery. Medicine. Pharmacy. Also under Early Modern Editions (1051).

334. ABU'L QAFĀ' AL-BŪZJĀNĪ, Muḥammad ibn Muḥammad. 940-997. DSB; ICB; Sart.; P I (ABULWEFÄ), P III (AL BŪZGANI). Mathematics. Astronomy.

335. 'ALĪ IBN 'ĪSĀ, al-Kaḥḥāl. (Latin form: JESU HALY.) ca. 940-1010. ICB; Sart.; Mort.; BLA & Supp. Famous as an oculist; he wrote on the anatomy and physiology of the eye. Also under Early Modern Editions (1070).

336. ABŪ ḤAYYĀN, al-Tauḥīdī. d. ca. 1009. ICB. Zoology.

337. ABŪ AL-FATH, Maḥmūd ibn Muḥammad, al-Iṣfahānī. fl. ca. 982. Sart. Mathematics.

338. IBN JULJUL, Sulaymān ibn Ḥasan. 944-ca. 994. DSB; Sart. History of medicine.

339. IBN YŪNUS, Abu'l-Ḥasan 'Alī ibn 'Abd al-Raḥman, al-Ṣadafī. d. 1009. DSB; ICB; Sart.; P I, III; GA (ALĪ IBN YOUNIS). Astronomy. Mathematics.

340. AL-SIJZĪ, Abū Sa'īd Aḥmad ibn Muḥammad ibn 'Abd, al-Jalīl. ca. 945-ca. 1020. DSB; ICB; Sart.; P III (AL SINGARI). Geometry. Astrology/astronomy.

341. AL-QŪHĪ (or AL-KŪHĪ), Abū Sahl Wayjan ibn Rustam. fl. ca. 970-1000. DSB; ICB (WAIJĀN IBN RUSTAM); Sart. (AL-KŪHĪ); P III (AL KŪHĪ). Mathematics. Astronomy.

342. AL-KHUJANDĪ, Abū Maḥmūd Ḥāmid ibn al-Khiḍr. d. 1000. DSB; ICB; Sart.; P III (AL HOGENDĪ). Mathematics. Astronomy.

343. AL-MAJRĪṬĪ, Abu'l-Qāsim Maslama ibn Aḥmad, al-Faraḍī. d. ca. 1007. DSB; ICB; Sart. (MASLAMA IBN AḤMAD). Astronomy. One of the earliest Hispano-Muslim scientists of importance.

344. AL-KARAJĪ (or AL-KARKHĪ), Abū Bakr ibn Muḥammad. fl. ca. 1000. DSB; ICB; Sart. (AL-KARKHĪ); P III (AL KARHĪ). Mathematics.

345. KUSHYĀR IBN LABBĀN IBN BASHAHRĪ, Abu'l-Ḥasan, al-Jīlī. fl. 971-1029. DSB; ICB; Sart.. Astronomy. Trigonometry. Arithmetic.
    1. *Principles of Hindu reckoning*. Madison, Wis., 1965. Trans. with introd. and notes by M. Levey and M. Petruck.

346. IBN AL-HAYTHAM, Abū 'Alī al-Ḥasan. (Latin form: ALHAZEN.) 965-ca. 1040. DSB; ICB; Sart. (IBN AL-HAITHAM); P I (ALHAZEN), P III (AL HAITAM); GA (ALHAZEL). Optics. Astronomy. Mathematics. Also under Early Modern Editions (1079)*.

347. MANṢŪR IBN 'ALI IBN 'IRAQ, Abū Naṣr. d. ca. 1036. DSB. Under ABŪ NAṢR MANṢUR: ICB; Sart. Mathematics. Astronomy.

348. AL-BĪRŪNĪ (or AL-BĒRŪNĪ), Abū Rayḥān Muḥammad ibn Aḥmad. 973-after 1050. DSB; ICB; Sart.; P I (ALBIRUNIUS); GA (ABU-RIHAN); GB. Astronomy. Mathematics. Geography.
   1. [In Russian. *Selected works*.] Tashkent (U.S.S.R.), 1957-.
   2. *On transits*. Beirut, 1959. Trans. by M. Saffouri and A. Ifran.
   3. *The determination of the coordinates of positions for the correction of distances between cities*. Beirut, 1967. Trans. by Jamil Ali.
   4. *The exhaustive treatise on shadows*. 2 vols. Aleppo, 1976. Trans. and comm. by E.S. Kennedy.

349. IBN AL-ṢAFFĀR. d. 1035. Sart. Mathematics. Astronomy.

350. IBN AL-SAMḤ, Abū al-Qāsim Aṣbagh ibn Muḥammad. 979-1035. ICB; Sart. Mathematics. Astronomy.

351. IBN SĪNĀ, Abū 'Alī al-Ḥusayn ibn 'Abdallāh. (Latin form: AVICENNA.) 980-1037. DSB Supp.; Sart. Under AVICENNA: ICB; P I; Mort.; BLA & Supp.; GA; GB. Philosophy, including natural philosophy. Classification of the sciences. Medicine. Also under Medieval Latin Translations (797)* and Early Modern Editions (1052)*.
   1. *De congelatione et conglutinatione lapidum: Being sections of "Kitāb al-shifā."* Paris, 1927. The Latin and Arabic texts ed. with an English trans. of the latter by E.J. Holmyard and D.C. Mandeville.
   2. *Le livre de science*. 2 vols. Paris, 1955-58. Trans. from the Persian by M. Achena and H. Massé. Contents: Vol. 1, Logic and metaphysics; Vol. 2, Physics and mathematics.
   3. *The life of Ibn Sīnā*. Albany, N.Y., 1974. Critical ed. and annotated trans. by W.E. Gohlman.

352. AL-BAGHDĀDĪ, Abu Manṣūr 'Abd al-Qāhir ibn Ṭāhir. d. 1037. DSB Supp. Under IBN ṬAHIR: ICB; Sart. Arithmetic. Best known as a theologian.

353. IKHWĀN AL-ṢAFA'. (i.e. Brethren of Purity.) DSB Supp.; Sart. An association founded at Basra about 983 whose writings include, as well as philosophy, a summary of the scientific knowledge of the time.

354. IBN AL-ḤUSAIN, Abū Ja'far Muḥammad. early 11th century. Sart.; P III (AL HOSEIN). Mathematics.

355. IBN AL-LAITH, Abū al-Jūd. early 11th century. ICB; Sart. (ABŪ-L-JŪD); P III (AL LEIT). Mathematics.

356. ABU'L HASAN, 'Alī ibn Abī'l-Rijāl. (Latin form: ALBOHAZEN.) fl. 1016-1040. Under IBN ABĪ AL-RIJĀL: ICB; Sart. Under ABEN-RAGEL: P I. Astrology. Also under Early Modern Editions (1060).

357. AL-JAYYĀNĪ, Abū 'Abd Allāh Muḥammad ibn Mu'ādh. ca. 990-after 1079. DSB; ICB. Mathematics. Astronomy.

358. IBN RIDWĀN, Abu'l-Hasan 'Alī ibn 'Alī ibn Ja'afar, al-Miṣrī. 998-1061 (or 1069). DSB. Under 'ALĪ IBN RIDWĀN: ICB; Sart. Medicine. Also under Early Modern Editions (1069).

359. IBN BUṬLĀN, Abu'l-Ḥasan al-Mukhtār. ca. 1000-1068. DSB; ICB; Sart.; BLA (ABUL-HASAN ... IBN BOYLAN). Medicine. Also under Early Modern Editions (1075). (Titles not included.)

360. AL-NASAWĪ, Abu'l-Hasan 'Alī ibn Aḥmad. fl. 1029-1044. DSB; ICB; Sart.; P III (under AL). Arithmetic. Geometry.

361. AL-KĀTĪ, Abū al-Ḥakīm Muḥammad. fl. 1034. ICB; Sart. (AL-KĀTHĪ). Alchemy.

362. IBN WĀFID, Abū al-Muṭarrif 'Abd al-Raḥman. (Latin forms: ABEN-GUEFIT, ALBENGUEFIT, etc.) fl. 1008-1075. DSB; Sart. (IBN AL-WĀFID); BLA. Pharmacy.

363. 'UBAID ALLĀH IBN KHALAF, Abū Marwān, al-Istijī. 11th century. ICB. Author of an astrological treatise which was translated into Spanish in the 13th century by order of King Alfonso el Sabio.

364. IBN ṢA'ID, Abū al-Qāsim, al-Andalusī. 1029-1070. ICB (SA'ĪD IBN AḤMAD); Sart. Astronomy. History, including history of science.

365. AL-ZARQĀLĪ, Abū Isḥāq Ibrāhīm ibn Yaḥyā al-Naqqāsh. (Latin forms: AZARQUIEL, ARZACHEL.) d. 1100. DSB; ICB; Sart.; P I (ARZACHEL), P III (AL ZERKĀLĪ); GA (ARZACHEL). Astronomy. Author of the Toledan Tables. Also under Early Modern Editions (1077).

366. AL-KHAYYĀMĪ (or AL-KHAYYĀM), Ghiyāth al-Dīn Abu'l-Fath 'Umar. (Also known as OMAR KHAYYĀM.) 1048?-1131? DSB; Sart.; P III (AL HAYYĀMĪ). Under OMAR KHAYYĀM: ICB; GB. Mathematics. Astronomy.

367. MUẒAFFAR AL-ASFUZĀRĪ. d. before 1122. Sart. Mathematics. Physics.

368. AL-ṬUGHRĀ'Ī. d. ca. 1121. ICB; Sart. Alchemy.

369. IBN SARĀBĪ. (Latin form: SERAPION THE YOUNGER.) 11th-12th century. ICB; Sart.; BLA (SERAPION). Medical botany. Also under Early Modern Editions (1054).

370. AL-KHARĀQĪ, 'Abd al-Jabbār ibn Muḥammad. d. 1132. ICB; Sart. Mathematics. Astronomy. Geography.

371. IBN AL-TILMĪDH, Amīn al-Dawla Abu'l-Ḥasan Hibat Allāh ibn Sa'id. ca. 1073-1165. DSB; Sart. Medicine. Pharmacy.

372. AL-KHĀZINĪ, Abu'l-Fatḥ 'Abd al-Raḥmān. fl. ca. 1115-ca. 1130. DSB; ICB; Sart. Astronomy. Mechanics. Scientific instruments.

373. ABU'L-BARAKĀT, Hibat Allāh, al-Baghdādī. 1080-after 1164/5. DSB; ICB. Principles of dynamics.

374. IBN ZUHR, Abū Marwān 'Abd al-Malik. (Latin forms: AVENZOAR, ABHOMERON.) ca. 1092-1162. DSB; Sart. Under AVENZOAR: ICB; Mort; BLA; GA; GB. Medicine. Pharmacy. Medical botany. One of the most outstanding physicians of Moorish Spain. Also under Early Modern Editions (1063).

375. IBN BĀJJA, Abū Bakr Muḥammad. (Latin name: AVEMPACE.) d. 1138/9. DSB; ICB (AVEMPACE); Sart.; P III (BEN BAGEH). Philosophy, including natural philosophy and principles of dynamics.

376. AL-BADĪ' AL-AṢṬURLĀBĪ. d. 1139/40. Sart.; ICB (HIBAT ALLĀH IBN AL-ḤUSAIN). Astronomy.

377. IBN AL-SURĀ, Najm al-Dīn Abū al-Futūḥ. d. ca. 1153. ICB. Mathematics.

378. AL-GHĀFIQĪ, Abū Ja'far Aḥmad ibn Muḥammad. d. ca. 1164. ICB; Sart. Botany.

379. JĀBIR IBN AFLAḤ, Abū Muḥammad, al-Ishbīlī. (Latin form: GEBER.) fl. 1145. DSB; Sart. Astronomy. Mathematics. The Latinised name Geber has often led to him being confused with the alchemist Jābir ibn Ḥayyān (see 1107). Also under Early Modern Editions (1076).

380. IBN RUSHD, Abu'l-Walīd Muḥammad. (Latin name: AVERROES.) 1126-1198. DSB; Sart. Under AVERROES; ICB; P I; Mort.; BLA; GA; GB. Astronomy. Medicine. Best known as a philosopher. Also under Medieval Latin Translations (796)* and Early Modern Editions (1056).

381. SAMŪ'ĪL IBN 'ABBĀS, Abū Naṣr. fl. 1163-1174. Sart. Mathematics.

382. MAIMONIDES, Moses. 1135 (or 1138) -1204. DSB; ICB; Sart.; BLA & Supp.; GA; GB; GF. Medicine. Best known as a philosopher and theologian. Also under Hebrew authors (487) and Early Modern Editions (1059).
    1. [*Yad-Hachazakah.*] *Die astronomischen Kapitel in Maimonidis Abhandlung über die Neumondsheiligung.* Berlin, 1882? German trans. and comm. by J. Hildesheimer. RM (under Moses ben Maimon)
    2. *L'explication des noms de drogues: Un glossaire de matière médicale.* Cairo, 1940. First ed. of text. With French trans., comm., and index by M. Meyerhof.
    3. *Treatise on poisons and their antidotes.* Philadelphia, 1966. Arabic text with trans. by S. Muntner.

383. 'ABD AL-MALIK AL-SHĪRĀZĪ, Abū al-Ḥusain. late 12th century. Sart. Mathematics. Astronomy.

384. IBN AL-'AWWĀM, Abū Zakariyyā Yaḥyā ibn Muḥammad. late 12th century. DSB; Sart. Agronomy.

385. AL-SAMAW'AL IBN YAḤYĀ AL-MAGHRIBĪ. d. ca. 1180. DSB; ICB. Mathematics.

386. IBN ARFA' RA'SAHU, Abū al-Ḥasan 'Alī ibn Musa. d. 1196/7. Sart. Alchemy.

387. IBN AL-YĀSMĪNĪ, Abū Muḥammad. d. 1204. Sart. Mathematics.

388. AL-ṬŪSĪ, Sharaf al-Dīn al-Muẓaffar. d. 1213/4. DSB; Sart. (AL-MUẒAFFAR AL-ṬŪSĪ). Mathematics. Astronomy.

389. FAKHR AL-DĪN AL-RĀZĪ. 1149-1210. ICB (AL-RĀZĪ, Fakhr al-Dīn); Sart.; GA; GB. Mathematics. Astronomy. Encyclopaedism.

390. AL-BIṬRŪJĪ, Nūr al-Dīn, Abū Isḥāq, al-Ishbīlī. (Latin form: ALPETRAGIUS.) fl. ca. 1190. DSB Supp.; ICB; Sart.; P I (ALPETRAGIUS). Astronomy. Also under Medieval Latin Translations (794)* and Early Modern Editions (1074).
    1. *On the principles of astronomy.* 2 vols. New Haven, Conn., 1971. Arabic and Hebrew texts with trans. by B.R. Goldstein.

391. AL-JAZARĪ, Badi' al-Zamān Abu'l-'Izz Ismā'īl ibn al-Razzāz. fl. 1181-1206. DSB Supp.; ICB; Sart. Mechanical technology.
    1. *The book of knowledge of ingenious mechanical devices, by Ibn al-Razzāz al-Jazarī.* Dordrecht, 1974. Trans. and annotated by D.R. Hill.

392. KAMĀL AL-DĪN IBN YŪNUS. 1156-1242. ICB (IBN YŪNUS); Sart. Mathematics.

393. AL-SAMARQANDĪ, Najīb al-Dīn Abū Ḥāmid Muḥammad. d. 1222. DSB; Sart. (NAJĪB AL-DĪN). Medicine. Pharmacy.
    1. *The medical formulary.* Philadelphia, 1967. Arabic text with trans. by M. Levey and N. Al-Khaledy.

394. MUḤAMMAD AL-ḤAṢṢĀR, Abū Zakarīyā (or Abū Bakr). 12th or 13th century. Sart. Mathematics.

395. MUḤAMMAD IBN AL-ḤUSAIN. 12th-13th century. Sart. Mathematics.

396. ABU MUḤAMMAD ABDU'L-LATIF. (Latin form: ABDOLLATIF.) 1162-1231. Mort.; BLA. Natural history.

397. ABŪ-AL-'ABBĀS, al-Nabātī. 1165/6 (or 1171/2)-ca. 1240. Sart. Botany.

398. IBN AL-SĀ'ĀTĪ. d. ca. 1230. Sart. Mechanical technology.

399. IBN AL-QIFṬĪ, Abū-al-Ḥasan 'Ali ibn Yūsuf. 1172/3-1248. ICB; Sart. Biographer of Muslim physicians, scientists, and philosophers.

400. AL-JAWBARĪ, 'Abd al-Raḥīm ibn 'Umar. fl. 1216-1220. ICB (AL-JAUBARĪ); Sart. A critic of the frauds and deceptions of alchemists. His writings are valuable for the history of Arabic alchemy and technology.

401. IBN AL-ṢŪRĪ, Manṣūr ibn Abī Faḍl ibn 'Ali Rashīd. 1177/8-ca. 1242. Sart. Botany.

402. QAIṢAR IBN ABĪ-AL-QĀSIM. 1178/9 (or 1168/9)-1251. Sart. Mathematics. Astronomy. Engineering.

403. IBN AL-ṢALĀḤ. 1181/2-1245. Sart. Astronomy.
    1. *Zur Kritik der Koordinatenüberlieferung im Sternkatalog des "Almagest."* Göttingen, 1975. Arabic text with German trans. and introd. by P. Kunitzsch.

404. AL-TĪFĀSHĪ, Shihāb al-Dīn Abu'l-'Abbās Aḥmad ibn Yūsuf. 1184-1253/4. DSB; Sart. Mineralogy.

405. IBN AL-BAYṬĀR, Ḍiyā' al-Dīn Abū Muḥammad 'Abdallāh, al-Mālaqī. ca. 1190-1248. DSB; BLA & Supp.; GA (ABEN-BITAR). Under IBN AL-BAIṬĀR: ICB; Sart. Pharmacy.

406. (Entry cancelled.)

407. ISMĀ'ĪL IBN IBRĀHĪM AL-MĀRIDĪNĪ, Abū-al-Ṭahir. (Often called IBN FALLŪS.) 1194-1239/40 (or 1252). Sart. Mathematics.

408. ABŪ 'ALĪ AL-ḤASAN, al-Marrākushī. d. 1262. ICB; Sart. (AL-ḤASAN). Mathematics. Astronomy.

409. AL-ABHARĪ, al-Mufaḍḍal ibn 'Umar. d. ca. 1263. ICB; Sart. Astronomy.

410. AL-ṬŪSĪ, Muḥammad ibn Muḥammad ibn al-Ḥasan. (Also known as NAṢIR AL-DĪN.) 1201-1274. DSB; ICB; Sart. (NAṢIR AL-DĪN AL-ṬŪSĪ); P II (NASSIR-EDDIN), P III (AL-ṬŪSĪ); GA (NAZIR ED-DIN). Mathematics. Astronomy. Mineralogy. See also 1085* and 1083.

411. AL-QAZWĪNĪ, Zakariyā ibn Muḥammad ibn Maḥmud, Abū Yaḥyā. ca. 1203 -1283. DSB; ICB; Sart. "The Muslim Pliny": author of two big encyclopaedic compilations, one on natural history and the other on geography.

412. IBN BADR, Muḥammad ibn 'Umar. (Latin form: ABENBEDER.) 13th century (?) ICB; Sart. Algebra.

413. AL-'URḌĪ, Mu'aiyad al-Dīn, al-Dimashqī. mid-13th century. ICB; Sart. Astronomy.

414. IBN AL-LUBŪDĪ, Abū Zakarīyā. 1210/11-after 1267. Sart. Mathematics. Astronomy. Medicine.

415. AL-QARĀFĪ, Shihāb al-Dīn Aḥmad ibn Idrīs. d. 1285. ICB; Sart. (Vol. 3). Optics (explanation of the rainbow).

416. BAYLAK AL-QIBJĀQĪ. fl. 1242-1282. DSB; ICB (AL-QIBJĀQĪ); Sart. (BAILAK AL-QABAJĀQĪ). Mineralogy.

417. MUḤYĪ AL-DĪN AL-MAGHRIBĪ. fl. ca. 1260-1265. DSB; ICB (IBN ABĪ AL-SHUKR); Sart. Trigonometry. Astronomy. Astrology.

418. ABŪ-AL-FARAJ IBN AL-'IBRĪ. (Also known as BAR HEBRAEUS.) 1226-1286. ICB (BAR HEBRAEUS); Sart.; GA. Encyclopaedist.
    1. *Le livre de l'ascension de l'esprit sur la forme du ciel et de la terre. Cours d'astronomie rédigé en 1279 par Gregoire Aboulfaraq, dit Bar Hebraeus.* 2 vols. Paris, 1899. Syriac text with French trans. by F. Nau. RM (under Bar Hebraeus).

419. JAMĀL AL-DĪN. fl. 1267. ICB; Sart. Astronomy.

420. IBN AL-NAFĪS, 'Alā al-Dīn Abu'l-Ḥasan 'Alī. d. 1288. DSB; ICB; Sart.; Mort.; BLA & Supp. (IBN EL-NEFIS). Medicine. Physiology.

421. IBN AL-QUFF, Amīn al-Dawlah Abū al-Faraj. 1233-1286. DSB; Sart. Medicine. Surgery. Physiology.
    1. *Anatomie und Chirurgie des Schädels.* Berlin, 1971. By O. Spies.

422. ABŪ-AL-QĀSIM, Muḥammad ibn Aḥmad, al-'Irāqī. late 13th century. ICB; Sart. Alchemy.
    1. *The book of knowledge acquired concerning the cultivation of gold.* Paris, 1923. Arabic text with trans. and introd. by E.J. Holmyard.

423. IBN AL-'AṬṬĀR, Abū al-Munā ibn Abī Naṣr, al-Kūhīn. late 13th century. ICB; Sart. (AL-KŪHĪN AL-'AṬṬĀR). Pharmacy.

424. MUḤAMMAD IBN ABĪ BAKR, al-Fārisī. late 13th century. Sart. Astronomy.

425. QUṬB AL-DĪN, al-Shīrāzī. 1236-1311. DSB; ICB; Sart. Astronomy. Optics. Medicine.

426. AL-SAMARQANDĪ, Shams al-Dīn Muḥammad. fl. 1276. DSB; ICB; Sart. Logic. Mathematics. Astronomy.

427. IBN BĀṢA, Abū 'Alī al-Ḥasan ibn Muḥammad. d. 1316. ICB; Sart. Maker of astronomical instruments.

428. KAMĀL AL-DĪN, Abu'l-Ḥasan Muḥammad, al-Fārisī. d. 1320. DSB; ICB (AL-FĀRISĪ); Sart. Optics. Mathematics.

429. IBN AL-BANNĀ', al-Marrākushī. 1256-1321. DSB; ICB; Sart.; P III (AL MAROKESCHI). Mathematics.

430. ṢALĀḤ AL-DĪN IBN YŪSUF. fl. 1296. Sart.; BLA. Anatomy and physiology of the eye.

431. 'ABDALLĀH IBN 'ALĪ, al-Kāshānī. fl. 1300. Sart. Chemical technology.

432. NIẒĀM AL-A'RAJ. fl. 1305-1312. Sart. Mathematics. Astronomy.

433. AL-JAGHMĪNĪ, Maḥmud ibn Muḥammad ibn 'Umar. d. 1344/5 (?) ICB; Sart. Astronomy/astrology.

434. AL-MIZZĪ, Abū 'Abdallāh Muḥammad ibn Aḥmad. 1291-1349. Sart. Maker of astronomical instruments and author of treatises on their use.

435. AL-JILDAKĪ, 'Izz al-Dīn 'Alī ibn Aidamur ibn 'Ali. fl. 1339-1342. ICB; Sart. "The last important Arabic writer on alchemy."

436. IBN FAḌL ALLĀH, al-'Umarī. 1301-1349. ICB; Sart. Natural history. Geography.

437. IBN AL-SHĀṬIR, 'Alā' al-Dīn Abu'l-Ḥasan 'Alī ibn Ibrāhīm. ca. 1305-ca. 1375. DSB; ICB; Sart. Astronomy.

438. MUḤAMMAD IBN AL-JAZŪLĪ, Shams al-Dīn. fl. 1344. Sart. Author of treatises on the use of the astrolabe.

439. JOSEPH BEN JOSEPH NAḤMIAS. mid-14th century. Sart. Astronomy.

440. IBN AL-DURAIHIM, Tāj al-Dīn 'Alī ibn Muḥammad. 1312-1360. Sart. Zoology.

441. AL-KHALĪLĪ, Shams al-Dīn Abū 'Abdallāh Muḥammad. fl. ca. 1365. DSB Supp.; Sart. Astronomy. Mathematics.

442. AL-UMAWĪ, Abū 'Abdallāh Ya'īsh, al-Andalusī. fl. 1373. DSB. Arithmetic.

443. TAQĪ AL-DĪN IBN 'IZZ AL-DĪN, al-Ḥanbalī. late 14th century. ICB; Sart. Astronomy. Mathematics.

444. AL-MĀRIDĪNĪ, 'Abd Allāh ibn Khalīl ibn Yūsuf. d. 1406. ICB; Sart. Astronomy.

445. AL-DAMĪRĪ, Muḥammad ibn Musa. 1341-1405. DSB; ICB; Sart.; GA (DOMAIRI). Zoology

446. IBN AL-HĀ'IM, Shihāb al-Dīn Abū al-'Abbās Aḥmad. 1352 (or 1355) -1412. Sart. Author of many textbooks of arithmetic and algebra.

447. IBN AL-MAJDĪ. 1358/9 (or 1365)-1447. Sart. Astronomy.

448. 'ABD AL-WĀḤID IBN MUḤAMMAD. fl. 1394. ICB; Sart. Astronomy.

449. IBN ZURAIQ (or ZARĪQ), Muḥammad ibn 'Alī. fl. ca. 1400. Sart. Astronomy.

450. AL-KĀSHĪ (or AL-KĀSHĀNĪ), Ghiyāth al-Dīn Jamshīd Mas'ūd. d. 1429. DSB; ICB; Sart. (JAMSHĪD IBN MAS'ŪD); P III (GIYAT). Astronomy Mathematics.

1. *The planetary equatorium of Jamshīd Ghiyāth al-Dīn al-Kāshī.* Princeton, N.J., 1960. Persian text with trans. and comm. by E.S. Kennedy.

451. QĀḌĪ ZĀDA, Al-Rūmī. (Also known as ṢALĀḤ AL-DĪN.) ca. 1364-ca. 1436. DSB; ICB; Sart. Mathematics. Astronomy.

452. MANṢŪR IBN MUḤAMMAD. fl. 1396-1423. Sart. Medicine. Anatomy.

453. ULUGH BEG. 1394-1449. DSB; ICB; P II; GB. Astronomy. Also under Early Modern Editions (1084)*. See also 1085.

454. IBN MANṢŪR, Muḥammad. 15th century. ICB. Mineralogy.

455. MAḤMŪD SHĀH KHULJĪ. 15th century. Sart. Astronomy. Also under Early Modern Editions (1083)*.

456. AL-QALAṢĀDĪ (or AL-QALṢĀDĪ), Abu'l-Ḥasan 'Alī. 1412-1486. DSB; ICB; Sart.; P III (AL KALṢADĪ). Arithmetic. Algebra. The last important Muslim mathematician of Spain.
 1. *Traduction du "Traité d'arithmétique" d'Aboûl-Haçan Alî Ben Mohammed Alkalçadî.* Rome, 1859. By F. Woepcke.

457. MUḤAMMAD IBN ABĪ AL-FATḤ, al-Miṣri, al-Ṣufī. d. ca. 1494. ICB. Gnomonics.

458. IBN MĀJID, Shihāb al-Dīn Aḥmad ibn Mājid. late 15th century. DSB; ICB. Navigation. Astronomy.

459. AL-IṢFAHĀNĪ, Ghiyāth al-Dīn 'Alī Ḥusainī. fl. 1466-1494. ICB. Natural history.

460. SIBT AL-MĀRIDĪNĪ. d. 1527. ICB. Gnomonics.

461. SĪDĪ 'ALĪ RE'ĪS. 16th century. ICB. Maker of nautical instruments.

462. AL-GHASSĀNĪ, Abū Muḥammad al-Qāsim. 1533-1604. ICB. Pharmacy. Botany.

463. AL-ANṬĀKĪ, Dā'ūd ibn 'Umar. d.1599. ICB. Author of an encyclopaedia of medicine, natural history, and the occult sciences.

464. AL-KHALKHALĪ, Ḥusain ibn al-Ḥusanī. d. 1605. ICB. Astronomy.

465. 'ALĀ AL-DĪN AL-MANṢŪR. fl. 1574-1595. ICB. Astronomy.

466. AL-AMULĪ, Bahā-ed-Dīn Muḥammed ben al-Ḥosein. 1547-1622. P III (under AL). Mathematics.

1.222 Arabic (and Syriac) Translations of Ancient Greek Works

467. ALEXANDER OF APHRODISIAS. (Main entry: 160)
 1. *The refutation of Alexander of Aphrodisias of Galen's treatise on the theory of motion.* Islamabad, 1965? Trans. with introd., notes, and an ed. of the Arabic text by N. Rescher. and M.E. Marmura.

468. ARISTOTLE. (Main entry: 65)
 1. *The Arabic version of Aristotle's "Meteorology."* Beirut, 1967. A critical ed. by C. Petraitis.
 2. *Tract comprising extracts from Aristotle's book of animals.* Cambridge, 1967. Arabic text ed. and trans. by J.N. Mattock.
 3. *"Generation of animals." The Arabic translation.* Leiden, 1971. Ed. by J. Brugman.

469. DIOCLES. (Main entry: 100)
   1. *"On burning mirrors." The Arabic translation of the lost Greek original*. Berlin/New York, 1976. Ed. with trans. and comm. by G.J. Toomer.

470. DOROTHEUS OF SIDON. See main entry--127.

471. GALEN. (Main entry: 159)
   1. *"On medical experience." First edition of the Arabic version*. London, 1944. Ed. by R. Walzer with English trans.

   The Berlin Academy is publishing a series entitled *Corpus medicorum graecorum. Supplementum orientale* comprising editions of the Arabic versions of various works of Galen, with German or English translations. The individual volumes are generally listed under Galen and are published by Akademie-Verlag, Berlin.

472. HIPPOCRATES. (Main entry: 47)

   The Cambridge Middle East Centre is publishing a series entitled *Arabic technical and scientific texts*, several volumes of which are editions of the Arabic versions of various works of Hippocrates, with English translations. They are generally listed under Hippocrates and are published by Heffer, Cambridge.

473. MENELAUS OF ALEXANDRIA. (Main entry:149)
   1. *Sphoericorum libri tres*. Oxford, 1758. Latin trans. made from the Hebrew and Arabic versions by Edmond Halley and publ. after his death by G. Costard.

474. NICOLAUS OF DAMASCUS. (Main entry: 123)
   1. *On the philosophy of Aristotle*. Leiden, 1965. Fragments of the first books trans. from the Syriac with introd. and comm. by H.J.D. Lulofs.

475. NICOMACHUS OF GERASA. (Main entry: 150)
   1. *Tābit b. Qurra's arabische Übersetzung der "Arithmetike eisagoge" des Nikomachus von Gerasa*. Beirut, 1959. Ed. by W. Kutsch.

476. PAPPUS OF ALEXANDRIA. (Main entry: 172)
   1. *The commentary of Pappus on Book X of Euclid's "Elements."* Cambridge, Mass., 1930. Arabic text ed. and trans. by W. Thomson. Reprinted, New York, 1968.

477. PHILO OF BYZANTIUM. (Main entry: 92)
   1. *Le livre des appareils pneumatiques et les machines hydrauliques*. Paris, 1902. Arabic text and French trans. by Baron Carra de Vaux. RM.

478. PLATO. (Main entry: 51)
   1. *Corpus Platonicum Medii Aevi*. [Series 2] *(Plato arabus)*. London, 1951-. Ed. by R. Klibansky. (Series 1 is *Plato latinus*--see 790.)

479. *THE BOOK of medicines*.
   1. *Syrian anatomy, pathology and therapeutics; or, The book of medicines*. 2 vols. London, 1913. Syriac text ed. and trans. by E.A.W. Budge. The work is a Syriac translation or adaption of the original Greek, made perhaps in the 6th century (see Sart. Vol. 1, 309).

1.23 Hebrew

1.231 Authors

480. NEHEMIAH, *Rabbi*. 2nd century. ICB. Author of the first Hebrew book on geometry.

481. ASAPH JUDAEUS. 7th century? ICB; Sart.; BLA Supp. (ASAF). Author of a treatise on medicine, probably the first in Hebrew.

482. DUNASH IBN TAMIM. 10th century. ICB. Author of a treatise on the armillary sphere.

483. ḤISDAI IBN ISAAC IBN SHAPRŪT. 910-975. ICB. Medicine. Translator of medical works into Hebrew.

484. DONNOLO, Shabbathai ben Abraham. 913-after 982. ICB; Sart.; BLA; GF. Medicine.

485. IBN EZRA, Abraham ben Meir. (Arabic name: ABŪ ISḤĀQ IBRĀHIM AL-MĀJID IBN EZRA.) (Latin name: AVENARE.) ca. 1090-ca. 1165. DSB; ICB (ABRAHAM BEN EZRA); GA; GB (ABENEZRA); GF. Mathematics. Astrology/astronomy. Also under Early Modern Editions (1086).
    1. *Le livre des fondements astrologiques, précédé de Le commencement de la sapience des signes*. Paris, 1977. Trans., introd., and notes by J. Halbronn.

486. ABRAHAM BAR ḤIYYA HA-NASI. (Latin names: SAVASORDA or ABRAHAM JUDAEUS.) fl. 1133-1136. DSB; ICB; Sart.; P I (A. BEN CHIJA); GF (A. BAR CHIJA). Mathematics. Astronomy. Also under Early Modern Editions (1087)*.

487. MAIMONIDES, Moses. 1135 (or 1138)-1204. Main entry: 382. He should also be included in this section because, although nearly all his main works were written in Arabic (the rest in Hebrew), they were soon translated into Hebrew and had more influence in their Hebrew, and sometimes Latin, versions than did the Arabic originals.

488. JACOB ANAṬOLI. ca. 1194-ca. 1256? Sart. Translator of Arabic astronomical works into Hebrew.

489. IBN TIBBON, Moses ben Samuel. fl. 1240-1283. DSB; ICB; Sart.; GB. Translator of numerous philosophical, scientific, and medical works from Arabic into Hebrew.

490. GERSHON BEN SOLOMON. late 13th century. ICB; Sart.; GF.
    1. *The gate of heaven*. Jerusalem, 1953. Trans. and ed. by F.S. Bodenheimer. An encyclopaedic work including astronomical and other natural phenomena.

491. IBN TIBBON, Jacob ben Machir. (Latin name: PROPHATIUS JUDAEUS.) ca. 1236-1305. DSB; ICB; Sart.; P III (PROPHATIUS). Astronomy. Translator of scientific works from Arabic into Hebrew.

492. LEVI BEN ABRAHAM BEN ḤAYYIM. 1246-1315. ICB; Sart.; GF. Author of a compendium of various sciences, including mathematics and physics.

493. ISAAC BEN JOSEPH ISRAELI. fl. 1310- 1330. Sart. Astronomy.
    1. *Liber Jesod olam; seu, Fundamentum mundi*. Berlin, 1846-48. Hebrew text with partial German trans. by B. Goldberg.

494. QALONYMUS BEN QALONYMUS. 1286/7-after 1328. Sart.; ICB (KALONYMOS BEN MEIR). Translator of scientific works from Arabic into Hebrew.

495. LEVI BEN GERSON. 1288-1344. DSB; ICB; Sart.; GB; GF (GERSON, Levi ben). Mathematics. Astronomy.
   1. *The astronomical tables*. Hamden, Conn., 1974. Ed. by B.R. Goldstein.

496. BONFILS, Immanuel ben Jacob. fl. 1340-1377. ICB; Sart.; GF. Mathematics. Astronomy.

497. JACOB BONET. fl. 1361. Sart. Astronomy.

498. SOLOMON BEN DAVID DAVIN. late 14th century. Sart. Translator of astronomical and astrological works from Latin into Hebrew.

499. JOSEPH BEN ISAAC IBN WAQAR. fl. 1357-1395. Sart. Astronomy.

500. CRESCAS, Hasdai ben Abraham. ca. 1340-1412. DSB; ICB; Sart.; GB; GF. Philosophy, including criticism of Aristotelian physics. Also under Early Modern Editions (1088).
   1. *Critique of Aristotle*. Cambridge, Mass., 1929. An ed. and trans. of the section of the Hebrew work *Or Adonai*, dealing with natural science. By H.A. Wolfson.

501. JACOB CARSONO. fl. 1375-1378. Sart. Astronomy. Translator of astronomical works from Arabic and Latin into Hebrew.

502. SOLOMON BEN ELIJAH. fl. 1374-1386. Sart. Translator of astronomical works from Greek into Hebrew.

503. EPHRAIM GERONDI. fl. ca. 1390. Sart.; ICB (GERONDI). Mathematics.

504. ISAAC ALḤADIB. fl. 1391-1426. Sart. Astronomy.

505. SOLOMON ABIGDOR. b. 1384 (or 1378). Sart. Translator of astronomical works from Latin into Hebrew.

506. COMTINO, Mordecai ben Eliezer. 1402-1482. ICB; GF. Mathematics. Astronomy.

507. FINZI, Mordecai (or Angelo) ben Abraham. d. 1476. GF. Mathematics. Astronomy. Translator from Arabic into Hebrew. See also 298/1.

508. MIZRAḤI, Elijah ben Abraham. ca. 1455-ca. 1525. GF. Mathematics.

1.24 Latin

(also Western European Vernaculars)

1.241 Authors

509. ISIDORE OF SEVILLE. ca. 560-636. DSB; ICB; Sart.; P III; BLA; GA; GB. Encyclopaedist. Also under Early Modern Editions (1101).
   1. *Etymologiarum sive originum libri*. 2 vols. Oxford, 1911. Ed. by W.M. Lindsay.
   2. *Traité de la nature*. Bordeaux, 1960. Ed. and trans. by J. Fontaine.

510. ALDHELM. ca. 640-709. ICB; Sart.; P I (ADELMUS); GA; GB; GD. Author of *De cyclo paschalis* and mathematical works.

511. BEDE. 672/3-735. DSB; ICB; Sart.; P I (BEDA); GA; GB; GD. Time reckoning and chronology. Cosmology. Natural history. Also under Early Modern Editions (1162)*.
    1. *Opera de temporibus*. Cambridge, Mass., 1941. Ed. by C.W. Jones.

512. FERGIL (or VIRGILIUS) OF SALZBURG. ca. 710-784. ICB; Sart.; P II (VERGILIUS); GD. Cosmography.

513. HRABANUS MAURUS. ca. 776-856. ICB (RABANUS); Sart.; Mort. (RABANUS); BLA & Supp.; GA; GC (RABAN MAUR). Encyclopaedist. Also under Early Modern Editions (1089)*.

514. DUNGAL. fl. 811-827. GA; GD. Astronomy.

515. DICUIL. fl. ca. 816-825. ICB; Sart.; GA; GB; GD. Astronomy. Geography.

516. WALAFRID STRABO. ca. 809-849. ICB; Sart.; BLA; GB; GC. Botany. Also under Early Modern Editions (1178)*.

517. BALD. 9th-10th century. ICB; Sart. Author of a herbal.

518. GERBERT (later Pope Sylvester II). ca. 945-1003. DSB; Sart. Under SYLVESTER II: ICB; P II. Under SILVESTER II: GA; GB; GC. Mathematics.

519. ABBO OF FLEURY. ca. 945-1004. Sart.; ICB (ABBO FLORIACENSIS); P I (ABBON). Mathematics. Astronomy.

520. BERNELINUS. fl. ca. 1000. Sart. Mathematics.

521. BYRHTFERTH (or BRIDFERTH). fl. ca. 1000. ICB; Sart.; GA; GD. Mathematics.
    1. *Manual (A.D. 1011)*. London, 1929. Ed. with introd. and trans. by S.J. Crawford.

522. ERACLIUS. fl. ca. 1000. ICB; Sart. (HERACLIUS); GA. Chemical and other technology.

523. ADELBOLD OF UTRECHT. ca. 960-1027. ICB; Sart.; GA; GC; GF. Mathematics.

524. THEOBALDUS. fl. 1022-1035. ICB. Also under Early Modern Editions (1154).
    1. *Physiologus*. Leiden, 1972. Ed. by P.T. Eden. A bestiary (cf. entry 209).

525. HERMANN THE LAME. 1013-1054. DSB; Sart.; ICB (HERMANNUS CONTRACTUS); GA (H. OF SUABIA); CB; CC (H. OF REICHENAU). Astronomy. Mathematics.

526. FRANCO OF LIEGE. fl. 1047-1083. Sart. Mathmatics. Astronomy.

527. OLIVER OF MALMESBURY. fl. 1066. Sart.; P II; GD. Astrology.

528. WILLIAM OF HIRSAU. d. 1091. ICB (WILHELM); Sart.; GA; GC. Astronomy. Theory of music.

529. LOSINGA, Robert. d. 1095. P II (ROBERT); GD. Astronomy.

530. CONSTANTINE THE AFRICAN. fl. 1065-1085. DSB; ICB; Sart.; Mort.; BLA & Supp. Medicine. Translator--the first major figure in

the transmission of Graeco-Arabic science to the West. Also under Early Modern Editions (1181).

531. MARBODE. 1035-1123. ICB; Sart.; P II; BLA & Supp.; GA. Author of a lapidary. Also under Early Modern Editions (1179)*.

532. GERLAND. fl. 1081-1084. Sart. Author of a computus and a treatise on the abacus.

533. LAMBERT THE CANON (or LAMBERT OF SAINT OMER). d. 1125. Sart.; GA.
   1. *Liber floridus*. Ghent, 1968. Ed. by A. Derolez. An encyclopaedia, compiled ca. 1120.

534. PEDRO ALFONSO. 1062-1110. ICB; Sart. Astronomy. Geography.

535. RALPH (or RAOUL or RADOLF) OF LAON. d. 1131. Sart. Author of a treatise on the abacus.

536. WALCHER OF MALVERN. d. ca. 1135. Sart. Astronomy.

537. HONORIUS OF AUTUN (or HONORIUS AUGUSTODUNENSIS). d. soon after 1150. ICB; Sart.; GA; GC. Cosmography. Also under Early Modern Editions (1100).
   1. *Clavis physicae*. Rome, 1974. Ed. by P. Lucentini.

538. WILLIAM OF CONCHES. ca. 1080-ca. 1154. ICB (GUILLAUME); Sart.; GA; GD. Natural philosophy. Also under Early Modern Editions (1185)*.

539. PHILIP OF THAON. fl. 1119-1125. Sart. Author of a computus, a bestiary, and a lapidary.

540. ALBERT OF SAMARIA. early 12th century. ICB (ADALBERTUS). Pharmacy.

541. NICOLAUS SALERNITANUS. early 12th century. ICB; Sart.; Mort.; BLA. Pharmacy. Also under Early Modern Editions (1095).

542. THEOPHILUS (PRESBYTER). early 12th century. DSB; ICB; Sart.; GA. Chemical and other technology.
   1. *On divers arts*. Chicago, 1963. Trans. with introd. and notes by J.G. Hawthorne and C.S. Smith.

543. ADELARD OF BATH. fl. 1116-1142. DSB; ICB; Sart.; P I; GA; GB; GD. Natural philosophy. Translator of mathematical and scientific works from the Arabic. "One of the pivotal figures in the conversion of Greek and Arabic learning into Latin." Also under Early Modern Editions (1115)*.
   1. *Quaestiones naturales*. Rapallo, 1965. Ed. by S. Balossi et al. Latin text with Italian trans.

544. STEPHEN OF ANTIOCH. fl. 1127. DSB; ICB (STEPHANUS); Sart.; BLA (STEPHANUS). Translator of Arabic medical works into Latin. He had some connection with Salerno.

545. THIERRY OF CHARTRES (or THEODORICUS CARNOTENSIS). d. ca. 1155. DSB; Sart. (THEODORIC). Natural philosophy. An important teacher of mathematics.

546. HUGH OF ST. VICTOR. ca. 1096-1141. DSB; ICB (HUGUES); Sart.; GA; GB. Classification of the sciences. Geometry. Also under Early Modern Editions (1149). See also 15470/1.
   1. *The Didascalion: A medieval guide to the arts*. New York, 1961. Trans. with introd. and notes by J. Taylor.

2. *Opera propaedeutica: Practica geometriae, De grammatica, Epitome Dindimi in philosophiam.* South Bend, Ind., 1966. Ed. by R. Baron.

547. HILDEGARD OF BINGEN. 1098-1179. DSB; ICB; Sart.; BLA & Supp.; GA; GB; GC. Cosmology. Natural history. Medicine. Also under Early Modern Editions (1187).

548. HUGH OF SANTALLA. fl. 1119-1151. Sart. Translator of Arabic works. See also 795/1.

549. JAMES OF VENICE (or IACOBUS VENETICUS GRECUS). fl. 1128-1147. DSB; Sart. Important as a translator of many of Aristotle's works from the Greek.

550. PLATEARIUS, Matthaeus. fl. 1130-1150. ICB; Sart.; Mort.; BLA. Author of the earliest known herbal to appear in France. Also under Early Modern Editions (1171).

551. PLATO OF TIVOLI. fl. 1134-1145. DSB; Sart.; GA. Translator from Arabic and Hebrew.

552. GUNDISSALINUS, Dominicus. fl. 1135-1153. DSB. Under GUNDISALVO: ICB; Sart. Philosophy, including classification of the sciences. Translator of scientific works from the Arabic.

553. JOHN OF SEVILLE. fl. 1135-1153. ICB; Sart. Translator from Arabic.

554. HERMANN THE DALMATIAN. fl. 1138-1143. ICB (H. DE CARINTHIA); Sart. Translator from Arabic. See also 787/2.

555. RAYMOND OF MARSEILLES. fl. ca. 1140. DSB; ICB; Sart. Astronomy.

556. ROBERT OF CHESTER. fl. 1141-1150. ICB; Sart.; GD (CHESTER, R.) Translator of Arabic mathematical works.

557. BERNARD SILVESTRE (or BERNARD OF TOURS). fl. ca. 1145. DSB; ICB (BERNARDUS S.); Sart. Cosmogony.
   1. [*Cosmographia.*] *De mundi universitate libri duo; sive, Megacosmus et microcosmus.* Innsbruck, 1876. Ed. by C.S. Barach amd J. Wrobel. Reprinted, Frankfurt, 1964.
   2. *The Cosmographia.* New York, 1973. Trans. with introd. and notes by W. Wetherbee.

558. ARTEPHIUS (or ARTEFI). fl. ca. 1150. ICB; Sart.; P I. Alchemy.

559. ODDE. 12th century. ICB. Astronomy.

560. GERARD OF CREMONA. ca. 1114-1187. DSB Supp.; ICB; Sart.; P I (GHERARDO); BLA & Supp. "The most prolific translator of scientific and philosophical works from Arabic in the Middle Ages." See also 605.

561. EUGENE THE AMIR. fl. 1154-1160. Sart. Mathematics. Astronomy. Translator from Greek and Arabic into Latin.

562. RICHARD (or NICHOLAS) OF SALERNO. late 12th century. Sart.; BLA (NICOLAUS PHYSICUS). Anatomy.

563. GILES OF CORBEIL. ca. 1140-ca. 1222. ICB (GILLES); Sart.; BLA & Supp. (AEGIDIUS). Medicine. Also under Early Modern Editions (1148).

563A. MARIUS. Known only from the following work, written ca. 1160.

1. *On the elements*. Berkeley, Cal., 1976. Latin text of *De elementis* with trans. and introd. by R.C. Dales.

564. ROGER OF HEREFORD. fl. ca. 1170-ca. 1180. DSB; ICB; Sart.; GD. Astronomy. Astrology.

565. DANIEL OF MORLEY. fl. 1170-1190. ICB; Sart.; GD (MORLEY). Cosmogeny. A student of Arabic science.

566. MAURUS, *Magister*. d. 1214. ICB; Sart.; BLA. Medicine. Anatomy. One of the best known members of the School of Salerno.

567. ALEXANDER NECKAM. 1157-1217. Sart. Under NECKAM: ICB; GA; GB; GD. Encyclopaedist.
   1. *De utensilium nominibus*. Liverpool, 1857. Ed. by T. Wright.
   2. *De naturis rerum*. London, 1863. Ed. by T. Wright. A popular encyclopaedia.

568. ALFRED OF SARESHEL (or ALFREDUS ANGLICUS). 12th-13th century. ICB; Sart.; GD (ALFRED, Anglicus). Author of *De motu cordis* and translator of Arabic works.

569. DAVID OF DINANT. 12th-13th century. ICB. Anatomy. Embryology.

570. URSO SALERNITANUS. d. 1225. ICB; Sart. (URSO OF CALABRIA); BLA & Supp. Medicine. Natural philosophy.
   1. *Libellus de conmixtionibus elementorum*. Stuttgart, 1975. Ed. by W. Sturner.

571. HARPESTRAENG, Henrik. ca. 1164-1244. DSB; ICB (HENRICUS DACUS); Sart.; BLA & Supp. Medicine. Pharmacy.

572. GROSSETESTE, Robert. ca. 1168-1253. DSB; ICB; Sart.; GA; GB; GD. Natural philosophy. Optics.
   1. *Epistolae*. London, 1861. Ed. by H.R. Luard.
   2. *Commentarius in VIII libros "Physicorum" Aristotelis*. Boulder, Colo., 1963. Ed. by R.C. Dales.

573. FIBONACCI, Leonardo (or LEONARD OF PISA). ca. 1170-after 1240. DSB; ICB (LEONARDO DA PISA); Sart.; P I; GA (LEONARD); GB (LEONARDO BONACCI). Mathematics.
   1. *Scritti*. 2 vols. Rome, 1857-62. Ed. by B. Boncompagni. RM. Contents: Vol. 1, *Liber abbaci*; Vol. 2, *La practica geometriae*.

574. MICHAEL SCOT. fl. 1217-1235. DSB; ICB (SCOT, M.); Sart.; GA. Under SCOTT, M.: GB; GD. Astrologer, encyclopaedist, and translator of Arabic works. Also under Early Modern Editions (1128)*. See also 794/1.

575. JORDANUS DE NEMORE. fl. 1220. d. 1237. DSB; ICB; Sart.; P II (NEMORARIUS). Mechanics. Mathematics. Also under Early Modern Editions (1169)*.
   1. *Jordanus de Nemore and the mathematics of astrolabes: De plana spera*. Toronto, 1978. Ed. with introd., trans., and comm. by R.B. Thomson.

576. ALEXANDER OF VILLEDIEU. d. ca. 1240. ICB; Sart. Author of *Carmen de algorismo*, a textbook of arithmetic which had a wide influence.

577. ELIA DA ASSISI. ca. 1182-1253. ICB. Alchemy.

578. WILLIAM OF AUVERGNE (or GUILIELMUS ARVERNUS). ca. 1185-1249. DSB; ICB (GUILLAUME); Sart. Cosmology. Best known as a philosopher and theologian--"the first great scholastic."

579. GERARD OF BRUSSELS. early 13th century. DSB; ICB; Sart. Geometry. Kinematics,

580. GERNARDUS. early 13th century. ICB; Sart. Author of a textbook of arithmetic.

581. WILLIAM THE CLERK. early 13th century. ICB (GUILLAUME LE CLERC); Sart. Author of a bestiary.

582. GARLAND, John. fl. 1202-1252. Sart.; GB; GD. Alchemy.

583. ARNOLD THE SAXON. fl. ca. 1225. Sart. Author of *De finibus rerum naturalium*, an encyclopaedia.

584. WILLIAM THE ENGLISHMAN. fl. 1220-1231. DSB; Sart. Astrology/ astronomy.

585. BARTHOLOMAEUS ANGLICUS (sometimes referred to incorrectly as GLANVILLE, Bartholomew de). fl. 1220-1240. ICB; Sart.; BLA Supp.; GD (GLANVILLE, B. de). Encyclopaedist. Also under Early Modern Editions (1094)*.
   1. *Mediaeval lore ... being classified gleanings from the encyclopaedia of Bartholomaeus Anglicus on the properties of things*. London, 1893; reprinted, 1905. By R. Steele.

586. JOHN OF PALERMO. fl. 1221-1240. DSB; ICB (GIOVANNI DA PALERMO). Translator from Arabic.

587. RICHARD OF WENDOVER. d. 1252. ICB; Sart.; BLA (RICHARDUS ANGLICUS); GD. Medicine. Anatomy.

588. VILLARD DE HONNECOURT. b. ca. 1190. DSB; ICB; Sart. Architecture.

589. VINCENT OF BEAUVAIS (or VINCENTIUS BELLOVACENSIS). ca. 1190-ca. 1264. DSB; ICB; Sart.; BLA; GA; GB. Encyclopaedist. Also under Early Modern Editions (1133)*.

590. FREDERICK II OF HOHENSTAUFEN. (Holy Roman Emperor.) 1194-1250. DSB; ICB; Sart.; GA; GB; GC. Zoology.
   1. *De arte venandi cum avibus*. 2 vols. Graz, 1969. Introd. and comm. in German by C.A. Willemsen.

591. JOHN OF LONDON (first of the name). fl. 1210-1252. Sart. A teacher of astronomy and meteorology at Oxford. "One of the most fervent expounders of the new Aristotle."

592. ALBERTUS MAGNUS (or ALBERT THE GREAT). ca. 1200-1280. DSB; ICB; Sart.; P I; Mort.; BLA & Supp.; GA; GB; GC. The natural sciences generally, and especially the biological sciences, as well as philosophy and theology. Also under Early Modern Editions (1097)*.
   1. *Opera omnia*. 38 vols. Paris, 1890-99. Ed. by A. Borgnet.
   2. *Opera omnia*. 40 vols. Cologne, 1951-. Ed. by B. Geyer.
   3. *Libellus de alchimia*. Berkeley, Cal., 1958. English trans. with introd. and notes by V. Heines. Attribution doubtful.
   4. *Book of minerals*. Oxford, 1967. Trans. by D. Wyckoff.
   See also 781 and 15710/1.

593. BONATTI (or BONATUS), Guido (or GUIDO DE FORLI). ca. 1200-ca. 1297. Sart.; P I. Astrology. Also under Early Modern Editions (1163)*.

594. LATINI, Brunetto. ca. 1200-1295. ICB; Sart.; P I; GA; GB; GF. Also under Early Modern Editions (1113).
   1. *Li livres dou trésor*. Paris, 1863. Ed. by P. Chabaille. RM. A popular encyclopaedia.

595. PARIS, Matthew. ca. 1200-1259. ICB; Sart.; P II (MATTHAEUS, P.); GA; GB; GD. Mathematics. Cartography.

596. SACROBOSCO, Johannes de (or JOHN OF HOLYWOOD). ca. 1200-ca. 1250. DSB; ICB; Sart.; P I (JOANNES DE S.); GB; GD (HOLYWOOD). Author of textbooks on mathematics and astronomy. His *De sphaera* was the standard textbook of astronomy through the Middle Ages and was still in use as late as the 17th century. Also under Early Modern Editions (1103)*. See also Index.
   1. *The "Sphere" of Sacrobosco and its commentators*. Chicago, 1949. Ed. by L. Thorndike.

597. THOMAS OF CANTIMPRÉ (or OF BRABANT). ca. 1200-after 1276. DSB; ICB; Sart.; BLA; GF. Encyclopaedist. Also under Early Modern Editions (1112).
   1. *Liber de natura rerum*. Berlin, 1973. Editio princeps. By H. Boese.

598. WILLIAM OF SALICETO. ca. 1201-1277. ICB (GUGIELMO); Sart.; Mort. (SALICETO); BLA & Supp. (GUILIELMO). Surgery. Anatomy. Also under Early Modern Editions (1104).

599. BORGOGNONI (OF LUCCA), Theodoric. ca. 1205-1298. DSB; ICB (THEODORIC); Sart.; BLA & Supp. Medicine. Surgery. Also under Early Modern Editions (1172).

600. WALTER OF METZ. fl. 1246. Sart. Author of *L'image du monde*, an encyclopaedic treatise. Also under Early Modern Editions (1174).

601. LINUS, *Pater*. 13th century. ICB. A teacher of the mathematical sciences.

602. SIMON OF COLOGNE. 13th century. ICB (SIMEON VON KÖLN); Sart. Alchemy.

603. LEOPOLD OF AUSTRIA. mid-13th century. ICB; Sart. Astronomy. Meteorology. Also under Early Modern Editions (1159).

604. RUFFO, Giordano. mid-13th century. DSB; Sart. (JORDAN RUFFO). Veterinary medicine. Also under Early Modern Editions (1166).

605. GERARD OF SABBIONETTA. fl. 1250. Sart.; P I (GHERARDO); GA. (He is sometimes called Gherardo Cremonese and so can be confused with Gerard of Cremona.) Astronomy/astrology. Also under Early Modern Editions (1099). See also 1103/1.
   1. *Theorica planetarum*. Berkeley, Cal., 1942. Ed. by F.J. Carmody.

606. ULRICH OF STRASBOURG (or ULRICUS DE ARGENTINA, or ULRICH ENGELBERTI). d. ca. 1278. DSB; Sart. Natural philosophy.

607. ROBERT KILWARDBY d. 1279. Sart. Under KILWARDBY: ICB; GA; GB; GD. Natural philosophy.
   1. *De ortu scientiarum*. London, 1976. Ed. by A.G. Judy.

608. MOERBEKE, William of (or GUILLELMUS DE MOERBEKA). fl. 1235-1278.

DSB; Sart. Under WILLIAM: ICB; GC; GF. An important translator of philosophical and scientific works from Greek into Latin. See also 784/1 and 786/3.

609. RISTORO (or RESTORO) D'AREZZO. ca. 1215-after 1282. DSB; ICB; Sart. Author of *Della composizione del mondo*, an encyclopaedic treatise.

610. BACON, Roger. ca. 1219-ca. 1292. DSB; ICB; Sart.; P I; BLA Supp; GA; GB; GD. Natural philosophy. Optics. Astrology/astronomy. Also under Early Modern Editions (1184)*.
   1. *Opera quaedam hactenus inedita*. Vol. 1. [No more publ.?] London, 1859. Ed. by J.S. Brewer.
   2. *Opera hactenus inedita*. 16 vols. Oxford, 1905-40. Ed. by R. Steele et al.
   3. *The Opus majus*. 3 vols. London, 1897-1900. The Latin text. Ed. by J.H. Bridges. Reprinted, Frankfurt, 1964.
   4. *The Opus majus. A translation*. 2 vols. Philadelphia, 1928. By R.B. Burke. Reprinted, New York, 1962.
   5. *Part of the Opus tertium*. Aberdeen, 1912. Ed. by A.G. Little. Reprinted, Farnborough, 1966.
   6. *De retardatione accidentium senectutis*. Oxford, 1928. Ed. by A.G. Little and E. Withington. Reprinted, Farnborough, 1969.

611. ALFONSO EL SABIO. (King of León and Castile.) 1221-1284. DSB; ICB; Sart.; P I. Under ALPHONSO: GA; GB; GC. Patron of science and learning. He sponsored many translations from the Arabic, especially of astronomical works, and was responsible for a new edition of the Toledan Tables, known as the Alfonsine Tables. Also under Early Modern Editions (1146)*. See also Index.
   1. *Lapidario. Codice original*. Madrid, 1881. A facsimile of the 13th-century MS. RM.

612. ALDEROTTI, Taddeo (or THADDAEUS FLORENTINUS). 1223-1295. DSB; ICB; Sart.; BLA; GA. Medicine.

613. AQUINAS, Thomas. ca. 1225-1274. DSB; Sart.; G. Under THOMAS: ICB; P II. Natural philosophy. Famous as a philosopher and theologian. "For a man not usually recognized as a scientist, he made noteworthy contributions to medieval science." -DSB Also under Early Modern Editions (1090)*. See also Index.
   1. *Opera omnia*. Several editions are widely available.
   2. *The division and methods of the sciences: Questions V and VI of his Commentary on the "De Trinitate" of Boethius*. 3rd rev. ed. Toronto, 1963. Trans. with introd. and notes by A. Maurer.
   3. *In octo libros "Physicorum" Aristotelis expositio*. Turin, 1954. Ed. by M. Maggiolo.
   4. *Commentary on Aristotle's "Physics."* New Haven, Conn., 1963. Trans. by R.J. Blackwell. et al.
   5. *In Aristotelis libros "De caelo et mundo," "De generatione et corruptione," "Meteorologicorum" expositio*. Turin, 1952. Ed. by R.M. Spiazzi.
   6. [Selected works.] *An introduction to the philosophy of nature*. 3rd rev. ed. St. Paul, Minn., 1956. Compiled by R.A. Kocourek. Includes *The principles of nature* and *Commentary on Books I and II of Aristotle's "Physics."*

614. JOHN XXI, *Pope* (or PETRUS HISPANUS). 1226?-1277. ICB; Sart.; Mort.; BLA & Supp. (PETRUS H.); GA; GB. Medicine. Zoology. Also under Early Modern Editions (1123).

615. CAMPANUS OF NOVARA (or CAMPANO, Giovanni). d. 1296. DSB & Supp.; Sart. Under CAMPANO, G.: ICB; P I; GA. Mathematics. Astronomy. Also under Early Modern Editions (1182). See also 996/1 and 1195.
   1. *Theorica planetarum*. Madison, Wis., 1971. Ed. with English trans., introd., and comm. by F.S. Benjamin and G.J. Toomer.

616. JUDAH BEN MOSES HA-KOHEN. fl. 1259-1277. Sart. One of the Jewish scholars employed by King Alfonso (611) to translate astronomical works from Arabic into Spanish. He was also one of the main authors of the Alfonsine Tables.

617. ISAAC IBN SID (or ISAAC HA-ḤAZZAN). fl. 1263-1277. Sart. The most important of the Jewish scholars employed as translators by King Alfonso. He was also one of the main authors of the Alfonsine Tables.

618. GERARD OF SILTEO (or SILETO). fl. 1264. DSB. Astronomy.

619. PETER PEREGRINUS. fl. 1269. DSB; ICB (PIERRE DE MARICOURT); Sart. (PETER THE STRANGER); GB (PETER OF MARICOURT). Magnetism. Also under Early Modern Editions (1183).

620. PECHAM, John. ca. 1230/35-1292. DSB; Sart. (JOHN PECKHAM). Under PECKHAM, J.: ICB; P II; GA; GB; GD. Optics. Also under Early Modern Editions (1144).
   1. *Tractatus de perspectiva*. St. Bonaventure, N.Y., 1972. Ed. by D.C. Lindberg.
   2. *John Pecham and the science of optics: Perspectiva communis*. Madison, Wis., 1970. Ed. with introd., trans., and notes by D.C. Lindberg.

621. WITELO. ca. 1230/35-after ca. 1275. DSB; ICB; Sart; P II (VITELLO); BLA; GC. Optics. Natural philosophy. Also under Early Modern Editions (1188). See also 1079 and 1834/2.

622. LULL, Ramon. ca. 1232-1316. DSB; ICB; Sart.; P I; GA; GC (LULLY). Polymath and encyclopaedist. Though he was not an alchemist many alchemical works have been falsely attributed to him. Also under Early Modern Editions (1120)*. See also 1698/2 and 2048/1.
   1. *Opera omnia*. Vols. I-VI and IX (VII and VIII were never publ.) Mainz, 1721-42. Ed. by J. Salzinger. Reprinted, Frankfurt, 1965.
   2. *Opuscula*. Palma, 1744. Reprinted, 3 vols, Hildesheim, 1971-73.
   3. *Quattuor libri principiorum*. Mainz, 1721. Reprinted, Wakefield (U.K.), 1969.

623. ANIANUS, *Magister*. late 13th century. ICB; Sart.; P I. Author of *Computatus manualis*. Also under Early Modern Editions (1155).
   1. *Le comput manuel*. Paris, 1928. Ed. by D.E. Smith.

624. BERNARD OF VERDUN (or BERNARDUS DE VIRDUNO). late 13 th century. DSB; ICB; Sart. Astronomy.

625. PETER OF ST. OMER. late 13th century (?) ICB (PETRUS DE SANCTO AUDEMARO); Sart. Chemical technology.

626. RUFINUS. late 13th century. DSB; ICB. Botany.
    1. *The herbal.* Chicago, 1946. Ed. by L. Thorndike.

627. GILES (or AEGIDIUS) OF LESSINES. ca. 1235-1304 or later. DSB; ICB (GILLES); Sart.; GA; GF (GILLES). Astronomy.

628. ROBERT THE ENGLISHMAN (first of the name). fl. 1272. ICB (ROBERTUS ANGLICUS); Sart.; GD (see under the second of the name, fl. 1326). Astronomy. Also under Early Modern Editions (1131).

629. SAMUEL HA-LEVI. fl. 1276. Sart. Another of the Jewish scholars employed as translators by King Alfonso (cf. 616 and 617).

630. ABRAHAM OF TOLEDO. fl. 1277. Sart. Another of the Jewish scholars employed by King Alfonso.

631. ARNALD OF VILLANOVA. ca. 1240-1311. DSB; ICB; Sart. (ARNOLD); P I; BLA & Supp.; GA; GB (ARNOLDUS). Medicine. Astrology. (A number of alchemical works attributed to him are of doubtful authenticity.) Also under Early Modern Editions (1105)*.

632. BERNARD OF LA TREILLE (or OF TRILIA). ca. 1240-1292. DSB; Sart. Astronomy.

633. SIGER OF BRABANT. ca. 1240-ca. 1282. DSB; ICB; Sart.; GB; GF. Philosophy, including natural philosophy.
    1. *Questions sur la "Physique" d'Aristote.* Louvain, 1941. By P. Delhaye.

634. PETER OF AUVERGNE. d. 1304. ICB; Sart. Natural philosophy. Also under Early Modern Editions (1170).

635. GILES OF ROME (or GILES COLONNA). ca. 1245-1316. DSB; ICB (AEGIDIUS DE COLUMNA); Sart.; GA (COLONNA, Aegidio). Natural philosophy. Embryology. Also under Early Modern Editions (1109).

636. HENRY BATE OF MALINES. 1246-ca. 1310. DSB; ICB (BATE DE MALINES, H.); Sart. Astronomy. Also under Early Modern Editions (1153).
    1. *Speculum divinorum et quorundam naturalium.* Louvain, 1960-. Critical ed. by E. van de Vyer. "The most important of Bate's works--a veritable philosophic and scientific encyclopaedia."

637. JOHN OF LONDON (second of the name). b. ca. 1246. fl. 1267. GD. Mathematics.

638. DIETRICH VON (or THEODERIC OF) FREIBERG. ca. 1250-ca. 1310. DSB; ICB; Sart.; P II (THEODORICH); GC. Optics. Natural philosophy.
    1. *Opera omnia.* Hamburg, 1977-. Ed. by B. Mojsisch.

639. WALTER OF ODINGTON. fl. ca. 1280-ca. 1330. DSB; Sart.; P II (ODINGTON); GD (W. OF EVESHAM). Alchemy. Theory of music. Mathematics.

640. BARTHOLOMEW OF PARMA. fl. 1286-1297. Sart. Astrology.

641. JOHN OF SICILY. fl. 1290. DSB; Sart. Astronomy.

642. LANFRANK OF MILAN. fl. ca. 1290. ICB; GA (LANFRANC). Under LANFRANCHI: Sart.; Mort.; BLA. Surgery. Anatomy. Also under Early Modern Editions (1135).

643. PETER PHILOMENA OF DACIA (or PETRUS DACUS, PETRUS DANUS) fl. 1290-

-1300. DSB; ICB (PETRUS DE DACIA); Sart.; GA (DACE, P. de). Mathematics. Astronomy.

644. WILLIAM OF SAINT-CLOUD. fl. 1292-1296. DSB; Sart. Astronomy.

645. ABANO, Pietro d'. 1257-ca. 1315. DSB; ICB (PIETRO D'ABANO); Sart.; P I; Mort. (PETRUS); BLA & Supp.; GA; GB. Medicine. Astrology. Also under Early Modern Editions (1093)*.

646. BARTOLOMMEO DA VARIGNANA. d. 1318. ICB; Sart.; BLA (VARIGNANA). Anatomy.

647. DINO DEL GARBO. ca. 1260-1327. ICB (DI GARBO); Sart. Under GARBO: BLA & Supp.; GA. Medicine. Also author of *De ponderibus et mensuris*. Also under Early Modern Editions (1157).

648. HENRY OF MONDEVILLE. ca. 1260-ca. 1320. DSB; Sart.; GA (HERMOND-AVILLE). Under MONDEVILLE; ICB; Mort.; BLA. Surgery. Anatomy.
    1. *Die Anatomie*. Berlin, 1889. Ed. by J.L. Pagel.

649. JOHN OF JANDUN. d. ca. 1328. ICB (JEAN); Sart.; GC. Natural philosophy. Also under Early Modern Editions (1111)*.

650. DANTE ALIGHIERI. 1265-1321. The famous poet. ICB; Sart.; G.
    1. *Quaestio de aqua et terra*. The attribution of this work to Dante is not certain. All MSS are now lost so all editions are based on the one printed at Venice in 1508 and ed. by G.B. Moncetti.
    1a. ——— *Edizione principe del 1508 riprodotta in facsimile*. Florence, 1905. Critical ed. with introd.,etc., by G. Boffito. Includes trans. into Italian, French, Spanish, English, and German.
    1b. ——— *Dissertazione critica sull'autenticita*. Modena, 1907. By V. Biagi. Text and comm.
    1c. ——— English trans. by C.L. Shadwell. Oxford, 1909.

651. MAUDITH (or MANDUIT, etc.), John. fl. 1305-1346. Sart.; GD. Astronomy. Trigonometry.

652. JACOB OF FLORENCE. fl. 1307. ICB (JACOPO); Sart. Mathematics.

653. ALARD VON DIEST. fl. ca. 1307-1315. ICB. Astronomy.

654. CECCO D'ASCOLI (or STABILI, Cecco Francesco). 1269-1327. ICB; Sart.; P I; BLA & Supp. (STABILI); GA; GB. Astrology. Also under Early Modern Editions (1106).

655. ANDALO DI NEGRO. ca. 1270-after 1342. ICB; Sart.; P II (NEGRO). Astronomy. Also under Early Modern Editions (1116).

656. SIMON OF GENOA. 1270-1330. Sart.; BLA. Medical botany. Also under Early Modern Editions (1096).

657. TORRIGIANI, Torrigiano dei. ca. 1270-ca. 1350. Sart.; BLA. Medicine. Author of a commentary on Galen which was widely influential for two centuries. Also under Early Modern Editions (1160).

658. BURLEY (or BURLAEUS), Walter. ca. 1275-ca. 1345? DSB; ICB; Sart.; P I (BURLEIGH); GA; GD. Natural philosophy, including the theory of motion. Also under Early Modern Editions (1091)*.
    1. *De sensibus*. Munich, 1966. Ed. by H. Shapiro and F. Scott.
    2. *De formis*. Munich, 1970. Ed. by F.J. Down Scott.

659. MONDINO DE' LUZZI. ca. 1275-1326. DSB; ICB; Sart.; Mort.; BLA & Supp. Anatomy. Also under Early Modern Editions (1124)*.

660. GEOFFROI OF MEAUX. fl. 1310-1348. Sart. Astrology.

661. PIERRE VIDAL. fl. 1311-1318. Sart. Astronomy.

662. DASTIN (or DASTYN or DAUSTIN), John. fl. 1320. ICB; Sart.; GA; GB. Alchemy.

663. FRANCIS OF MARCHIA. fl. 1320-1344. DSB; Sart. Natural philosophy, including the theory of motion.

664. SILVATICO, Matteo. d. 1342. Under MATTHAEUS SYLVATICUS: Sart.; BLA. Medical botany. Also under Early Modern Editions (1114).

665. BERTUCCIO, Nicolò. d. 1347. BLA & Supp.; Sart. (BERTRUCCIO). Medicine. Anatomy. Also under Early Modern Editions (1177).

666. MERLE (or MORLEY), William. d. 1347. Sart.; GD. Meteorology--"Perhaps the oldest systematic record of the weather."

667. HOLKOT, Robert. d. 1349. ICB; P I; GA. Under HOLCOT: Sart.; GD. Astrology.

668. JOHN OF DUMBLETON. d. ca. 1349. DSB; Sart. Under DUMBLETON: ICB; GD. Natural philosophy, including the theory of motion.

669. GUIDO DA VIGEVANO. ca. 1280-after 1345. ICB; Sart.; BLA (VIGEVANO). Medicine. Anatomy.

670. JOHN OF GADDESDEN. ca. 1280-1361. ICB; Sart.; Mort. Under GADDESDEN: BLA & Supp.; GA; GD. Medicine. Also under Early Modern Editions (1164).

671. NICCOLO DA REGGIO. ca. 1280-ca. 1350. ICB; Sart. Translator of Greek medical texts (especially of Galen and Hippocrates)--"the first to reveal Greek medicine without Muslim alterations."

672. PAOLO DELL' ABBACO. 1281-1373. ICB; Sart. (PAOLO DAGOMARI); P I (ABBACO); GA (DAGOMARI). Mathematics.

673. FRANCIS OF MEYRONNES. ca. 1285-ca. 1330. DSB; Sart.; GB. Natural philosophy, including the theory of motion. Also under Early Modern Editions (1122).

674. OCKHAM, William of. ca. 1285-1349. DSB; ICB; Sart.; GA; GB; GC; GD. Philosophy, including natural philosophy. Also under Early Modern Editions (1125)*.

675. PETRUS BONUS (or BONUS LOMBARDUS, or BUONO LOMBARDO OF FERRARA). fl. 1323-1330. DSB; ICB; Sart. (BUONO). Alchemy. Also under Early Modern Editions (1189)*. See also 1199.

676. CHAULIAC, Guy de. ca. 1290-ca. 1368. DSB; ICB; Sart.; Mort. (GUY DE C.); BLA & Supp.; GA (CAULIAC). Medicine. Surgery. Also under Early Modern Editions (1127).
    1. *The Middle English translation of Guy de Chauliac's "Anatomy". With Guy's essay on the history of medicine*, Lund, 1964. Ed. by B. Wallner.

677. DONDI, Jacopo (or Giacomo) de'. ca. 1290-1359. ICB; Sart.; P I; BLA & Supp.; GA. Author of *Planetarium* (astronomical tables), *De fluxu maris*, and *De modo conficiendi salis ex aquis*. Also under Early Modern Editions (1092).

678. GENTILE DA FOLIGNO. ca. 1290-1348. ICB; Sart.; BLA & Supp. Medicine. Also under Early Modern Editions (1108).

679. ROBERT THE ENGLISHMAN (second of the name). fl. 1326. Sart. (ROBERT OF YORK); GD. Meteorology. Astrology. Alchemy.

680. JOHN THE CANON. fl. 1329. ICB; Sart.; GD (CANON, J.). Natural philosophy. Also under Early Modern Editions (1119)*.

681. RICHARD OF WALLINGFORD. ca. 1292-1336. DSB; ICB; Sart.; GD. Mathematics. Astronomy. Horology.
 1. *An edition of his writings*. 3 vols. Oxford, 1976. By J.D. North. With introd., trans., and comm.

682. BRADWARDINE, Thomas. ca. 1295-1349. DSB; ICB; Sart.; P I; GA; GB; GD. Mathematics. Natural philosophy, including the principles of dynamics. Also under Early Modern Editions (1168).
 1. *Tractatus de proportionibus*. Madison, Wis., 1955. Ed. and trans. by H.L. Crosby.

683. BURIDAN, Jean. ca. 1295-ca. 1358. DSB; ICB; Sart.; GA; GB. Natural philosophy, including the principles of dynamics. Also under Early Modern Editions (1137)*.
 1. *Tractatus de consequentia*. Louvain/Paris, 1976. Critical ed. by H. Hubein.

684. YPERMAN, Jan. 1295-1331. ICB; Sart.; Mort.; BLA & Supp.; GC. Surgery.

685. BREDON, Simon. ca. 1300-ca. 1372. DSB; ICB; Sart.; P I. Mathematics. Astronomy. Medicine.

686. JOHN OF LIGNÈRES (or JOHANNES DE LINERIIS). fl. 1320-1350. DSB; Sart. (JEAN DE LINIERES). Astronomy. Mathematics. Also under Early Modern Editions (1150).

687. JOHN OF SAXONY (or JOHN DANK, DANCK, DANCO, etc.) fl. 1323-1361. DSB; Sart. Astronomy. Also under Early Modern Editions (1151). See also 1055.

688. JOHN OF MURS. fl. ca. 1325-1350. DSB; Sart.; P II (MEURS). Under MURIS, J. de: ICB; GA. Mathematics. Astronomy. Theory of music. Also under Early Modern Editions (1173A).

689. JOHN OF GENOA. fl. 1332-1337. Sart. Astronomy.

690. URBANO OF BOLOGNA. fl. 1335. ICB; Sart. Natural philosophy. Also under Early Modern Editions (1165).

691. FIRMIN DE BEAUVAL. fl. 1335-1345. Sart. Meteorology. Astrology. Also under Early Modern Editions (1152).

692. SUNO KAROLI DE SUECIA. fl. 1340-1344. ICB. Lectured on astronomy at the University of Paris.

693. SWINESHEAD, Richard. fl. 1340-1355. DSB; ICB; Sart.; P II (SUISET); GD. Natural philosophy, including the principles of dynamics. Also under Early Modern Editions (1132). See also 1301.

694. JOHN ARDERNE. 1307-1380. Sart.; Mort. (JOHN OF A.); BLA & Supp. (ARDERN, J.). Under ARDERNE, J.: ICB; GA; GD. Surgery.

695. CONRAD (or KONRAD) OF MEGENBERG. ca. 1309-1374. ICB; Sart.;

BLA (KUNRAT); GC. Natural history. Astronomy. The first important scientific writer in German. Also under Early Modern Editions (1117).

1. *Das Buch der Natur. Die erste Naturgeschichte in deutscher Sprache*. Stuttgart, 1861. Ed. by F. Pfeiffer. Reprinted, Hildesheim, 1962 and 1971. (Listed under Konrad.)

696. CANACCI, Rafaele. 14th century. ICB. Algebra.

697. NATTERGAL, Peder. 14th century. ICB. Astronomy.

698. TRESBENS, Bartomeu de. 14th century. ICB. Astrology.

699. HEYTESBURY, William. fl. 1330-1371. DSB; ICB; Sart.; GD. Logic. Kinematics. Also under Early Modern Editions (1140).

700. ESTWOOD, John (or JOHN OF EASTWOOD or OF ESCHENDEN). fl. 1340-1370. Sart. (ASHENDON). Astrology. Also under Early Modern Editions (1158).

701. DOMINICUS DE CLAVASIO (or CLAVAGIO, CLIVAXO, etc., or DOMINICUS PARISIENSIS). fl. 1346-1357. DSB; ICB; Sart. Mathematics. Natural philosophy.

702. JOHN OF RUPESCISSA. fl. 1350. Sart.; ICB (RUPESCISSA); BLA (ROQUETAILLADE, Jean de la). Alchemy. Also under Early Modern Editions (1180)*.

703. THEMO (or THIMON) JUDAEUS. fl. 1349-1361. ICB; Sart. Mechanics. Optics. Astronomy. Also under Early Modern Editions (1139).

1. *L'oeuvre astronomique*. Paris, 1973. Ed. with comm. by H. Hugonnard-Roche.

704. ALBERT OF SAXONY. ca. 1316-1390. DSB; ICB; Sart.; P I. Natural philosophy, including the principles of dynamics. Also under Early Modern Editions (1121)*.

705. CIRIA, Amilcar. ca. 1318-1378. P I. Astronomy.

706. DONDI (DALL' OROLOGIO), Giovanni de'. 1318-1389. DSB; ICB; Sart.; P I; BLA; GA. Horology. Astronomy. Medicine.

1. *The planetarium*. London, 1974. Trans. by G.H. Baillie.

707. PEDRO IV. (King of Aragon.) 1319-1387. ICB; Sart. In 1359 he ordered new astronomical tables to be prepared, as the Alfonsine Tables had become inadequate. The new tables were completed by Dalmatius (715).

708. ORESME, Nicole. ca. 1320-1382. DSB; ICB; Sart.; GA; GB. Mathematics. Natural philosophy. Also under Early Modern Editions (1129).

1. *Der Algorismus proportionum*. Berlin, 1868. Ed. by E. Curtze.
2. *Quaestiones super geometriam Euclidis*. Leiden, 1961. Ed. by H.L.L. Busard.
3. *De proportionibus proportionum, and Ad pauca respicientes*. Madison, Wis., 1966. Ed. with introd. and notes by E. Grant.
4. *Le livre du ciel et du monde*. Madison, Wis., 1968. Ed. with trans. and introd. by A.D. Menut.
5. *Nicole Oresme and the medieval geometry of qualities and motions: A treatise on the uniformity and difformity of intensities known as Tractatus de configurationibus qualitatum et motuum*. Madison Wis., 1968. Ed. with introd., trans., and comm. by M. Clagett.

6. *Nicole Oresme and the kinematics of circular motion: Tractatus de commensurabilitate vel incommensurabilitate motuum celi*. Madison, Wis., 1971. Ed. with introd., trans., and comm. by E. Grant.

709. CHILLINGWORTH, John (first of the name). fl. 1360. GD. Mathematics. Astronomy. Astrology.

710. SEDACER, Guillem. d. 1382/3. Sart. Alchemy.

711. REDE, William. d. 1385. Sart.; GD. Astronomy.

712. JOHN OF CASALI. ca. 1325-1375. Sart.; ICB (CASALI, G. di). Mathematics. Mechanics. Also under Early Modern Editions (1175).

713. HENRY OF HESSE. 1325-1397. DSB; P I (HEINRICH). Under LANGENSTEIN, H.: ICB; Sart.; GA. Astronomy. Optics.

714. FLAMEL, Nicolas. 1330-1418. Sart.; P I; GA; GB. A supposed alchemist. Alchemical writings were falsely ascribed to him in the 17th century--cf. entry 2790.

715. DALMATIUS. fl. 1359-1381. ICB; Sart. (DALMAU CES-PLANES). Astronomer employed by King Pedro IV of Aragon--see entry 707.

716. MARSILIUS OF INGHEN. d. 1396. DSB; ICB; Sart. Natural philosophy, including the theory of motion. Also under Early Modern Editions (1138)*.

717. CYBO OF HYÈRES. late 14th century. ICB; Sart. Naturalistic drawings of plants and animals.

718. JACOPO DA SAN MARTINO. late 14th century. Mechanics. Also under Early Modern Editions (1143).

719. JOHN OF TINEMUE. late 14th century. ICB (JOHANNES DE TINEMUE). Mathematics.

720. WERKWOTH, Thomas. late 14th century. ICB. Astronomy.

721. BRAY, John. fl. 1377. GD. Botany.

722. BERNARD OF TREVISAN. fl. ca. 1378. DSB; Sart. (B. OF TREVES); P I; GA. Alchemy.
   1. *Oeuvre chimique*. Paris, 1976.

723. CHAUCER, Geoffrey. ca. 1343/4-1400. The famous poet. DSB; ICB; Sart.; G. Astronomy. Alchemy. Also under Early Modern Editions (1186).
   1. *A treatise on the astrolabe*. London, 1872. Ed. by W.W. Skeat.
   2. *On the astrolabe*. Oxford, 1931. Ed. by R.T. Gunther.
   3. *The equatorie of the planetis*. Cambridge, 1955. Ed. by D.J. Price.

724. BLASIUS OF PARMA. ca. 1345-1416. DSB; ICB (BIAGIO PELACANI); Sart. (PELACANI). Optics. Statics. Kinematics. Astronomy. Also under Early Modern Editions (1142).

725. DOMBELAY, John. fl. 1384-1386. Sart. Alchemy.

726. NICHOLAS OF LYNN. fl. 1386. ICB; Sart; GD. Astronomy.

727. AILLY, Pierre d'. (or PETRUS DE ALLIACO). 1350-1420. DSB; ICB; Sart.; P I; GA; GB. Astronomy/astrology. Geography. Also

under Early Modern Editions (1145)*.
1. *Ymago mundi*. 3 vols. Paris, 1930. Latin text and French trans. by E. Buron.

728. SOMER, John. fl. 1380-1409. Sart.; GD. Author of a calendar with astronomical tables; many copies are extant.

729. BRYTTE (or BRITTE, BRUTE, etc.), Walter. fl. 1390. DSB; Sart. (BRIT); GD (BRIT). Astronomy.

730. LEONARD OF MAURPERG. fl. 1394. Sart. Alchemy.

731. NICHOLAS OF DINKELSBÜHL. ca. 1360-1433. Sart. Lectured at the University of Vienna on mathematical, astronomical, and physical subjects, 1390-1405.

732. JACOPO DA FORLI (or GIACOMO DELLA TORRE). ca. 1360-1413. ICB; Sart.; BLA (TORRE, G. della). Mechanics. Medicine. Also under Early Modern Editions (1110).

733. ORTOLFF VON BAYRLANDT. fl. ca. 1400. ICB; Sart.; Mort.; BLA. Pharmacy. Also under Early Modern Editions (1130).
1. *Das Arzneibuch*. Stuttgart, 1963. Ed. by J. Follan. The first German pharmacopeia.

734. VENOD (or WENOD), Johann. fl. ca. 1400. ICB. Author of an illustrated text on distillation, ca. 1420.

735. WIMANDUS DE RUFFO CLIPEO. fl. ca. 1400. Sart. Alchemy.

736. JACOBUS ANGELUS DE ULMA. fl. 1402. ICB; Sart. (JACOB ENGELHART). Astronomy. Also under Early Modern Editions (1161).

737. DATI, Gregorio. 1363-1436. GA. Mathematics. Astronomy. Also under Early Modern Editions (1098).

738. FUSORIS, Jean. ca. 1365-1436. DSB Supp.; ICB; Sart. Author of astronomical treatises and tables, and maker of instruments and clocks.

739. PAUL OF VENICE. ca. 1370-1429. DSB; Sart. Natural philosophy, including the principles of dynamics. Best known as a logician. Also under Early Modern Editions (1126)*.

740. ANGELO DA FOSSOMBRONE. ca. 1375-1402. P I (FOSSOMBRUNO). Dynamics. Also under Early Modern Editions (1141).

741. HUGH OF SIENA (or UGO DA SIENA or HUGO SENENSIS). 1376-1439. Sart. Under BENZI, Ugo: ICB; BLA & Supp. Medicine. Also under Early Modern Editions (1134).

742. GHIBERTI, Lorenzo. ca. 1378-1455. The famous sculptor. ICB; GA; GB. Author of a treatise on optics.

743. ANSELMO, Giorgio. d. 1440. P I. Mathematics. Astronomy.

744. CHILLINGWORTH, John (second of the name). d. 1445. GD. Astronomy.

745. PETRUS DE MONTE ALCINO. d. ca. 1448. ICB. Astronomy.

746. JOHN SIMONIS OF SELANDIA. early 15th century. DSB. Astronomy.

747. BELDOMANDI, Prosdocimo de. ca. 1380-1428. ICB; Sart.; P III. Astronomy. Mathematics. Theory of music. Also under Early Modern Editions (1147).

748. TACCOLA, Mariano di Jacomo. 1381-ca. 1455. DSB; ICB (MARIANO TACCOLA, J.); Sart. Mechanical technology.
    1. *De machinis. The engineering treatise of 1449.* 2 vols. Wiesbaden, 1971. Latin text with introd. and comm. by G. Scaglia. (Listed under Jacomo Vanni, Mariano di, called Taccola.)
    2. *Mariano Taccolo and his book "De ingeneis."* Cambridge, Mass., 1972. By F.D. Prager and G. Scaglia.

749. LA FONTAINE, Jean de. b. 1381. Sart.; GA. Alchemy. Also under Early Modern Editions (1176).

750. JOHN OF GMUNDEN. ca. 1382-1442. DSB; ICB (JOHANNES VON GMUNDEN); P I (JOANNES DE GAMUNDIA). Mathematics. Astronomy. Also under Early Modern Editions (1102).

751. SAVONAROLA, (Giovanni) Michele. 1384-1462. ICB; Sart.; P II; BLA. Medicine. Alchemy. Also under Early Modern Editions (1136).

752. VILLENA, Enrique de. 1384-1434. ICB; GB. Astrology.

753. GAETANO DI THIENE. 1387-1465. Mechanics. Also under Early Modern Editions (1118).

754. SIMON DE LENDENARIA. fl. 1418-1434. A logician who dealt with kinematics. Also under Early Modern Editions (1167).

755. LOUFFENBERG, Heinrich. ca. 1391-1460. ICB; Mort.; BLA; GC. Medicine.

756. TOSCANELLI (DAL POZZO), Paolo. 1397-1482. DSB; ICB; P II; GA. Astronomy. Geography.

757. MARROWE, George. fl. 1437. GD. Alchemy.

758. BRZYMALA, Andrzej. d. 1466. ICB. Astronomy.

759. HARTLIEB, Johannes. d. 1468. ICB; GC. Author of a herbal.

760. JOHANNES LAURATIUS DE FUNDIS. fl. 1428-1473. DSB. Astronomy. Astrology.

761. CUSA, Nicholas (or NICOLAUS CUSANUS or NIKOLAUS VON CUSA). ca. 1401-1464. DSB; ICB; P II (NICOLAUS); GA; GB; GC. Mathematics. Best known as a philosopher. Also under Early Modern Editions (1156)*. See also 15589/1.
    1. *Of learned ignorance.* London, 1954. Trans. by G. Heron.
    2. *Philosophisch-theologische Schriften.* Vienna, 1964. Ed. with introd. by L. Gabriel. German trans. and comm. by D. and W. Dupré.
    3. *De coniecturis.* Hamsburg, 1971. Ed. with German trans. by J. Koch and W. Happ.

762. ALMANUS OF AUGSBURG. 15th century. Known only from the following.
    1. *The Almanus manuscript.* London, 1972. Facsimile of a horological MS with trans. and notes by J.H. Leopold.

1.242 Anonymous Works

763. *AGNUS castus: A Middle English herbal.* Uppsala, 1950. Reconstructed from various MSS and ed. by G. Brodin.

764. *ALPHITA: A medico-botanical glossary.* Oxford, 1887. Ed. by J.L. Mowat. Apparently it dates from ca. 1400.

765. *THE BOOK of beasts, being a translation from a Latin bestiary of the twelfth century.* London, 1955. Trans. and ed. by T.H. White.

766. *LE CALENDRIER de Cordoue.* Leiden. 1961. Ed. by R. Dozy. A new ed. with an annotated French trans. by C. Pellat.

767. *DE SIMPLICI medicina: Kräuterbuch-Handschrift aus dem letzen Viertel des 14. Jahrhunderts.* Basel, 1961. Ed. by A. Pfister.

768. *L'HERBIER de Moudon: Un recueil de recettes médicales, de la fin du 14e siècle.* Aarau, 1938. Ed. by P. Aebischer and E. Oliver.

769. *EIN KOMMENTAR zur Physik des Aristoteles, aus der Pariser Artistenfakultät um 1273.* Berlin, 1968. Ed. with an introd. by A. Zimmermann. (Sometimes listed under: Paris. Université. Faculté des Arts.)

770. *LIBRO d'abaco.* Lucca, 1973. Ed. with an introd.(in Italian) by G. Arrighi.

771. *MACER floridus. De viribus herbarum. A Middle English translation.* Uppsala, 1949. Ed. by G. Frisk. The *Macer floridus* probably dates from ca. 1100. It is sometimes attributed to Odo de Meung (fl. ca. 1100). Also under Early Modern Editions (1192).

772. *A MEDIEVAL bestiary.* Boston, 1971. Trans. and introd. by T.J. Elliot.

773. *ROSARIUM philosophorum. Le rosaire des philosophes.* Paris, 1973. French trans. with introd. and notes by E. Perrot. An alchemical work written in the first half of the 14th century and first printed in 1550.

774. *SECRETUM secretorum.* "A minor encyclopaedia of the middle range of medieval learning." Popular in medieval and early modern times and often attributed (wrongly) to Aristotle. (See Sart. Vol. 1, 556.) Also under Early Modern Editions (1190)*.

1.243 Collections of Texts

775. *POPULAR treatises on science written during the Middle Ages, in Anglo-Saxon, Anglo-Norman, and English.* 1841. Ed. by T. Wright. Reprinted, London, 1965.

776. *RARA mathematica. A collection of treatises on the mathematics and subjects connected with them, from ancient unedited manuscripts.* London, 1839; 2nd ed., 1841. By J.O. Halliwell-Phillipps.

777. *LATIN treatises on comets, between 1238 and 1368.* Chicago, 1950. Ed. by L. Thorndike.

778. *THE SCIENCE of mechanics in the Middle Ages.* Madison, Wis., 1959. Ed. by M. Clagett. Contains texts from many authors.

779. *THE MEDIEVAL science of weights.* Madison. Wis., 1952. Ed. by E.A. Moody and M. Clagett. Contains texts from several authors.

780. *TURBA philosophorum; or, Assembly of the sages.* London. 1896.

Trans. by A.E. Waite. A collection of some seventy anonymous tracts on alchemy, dating probably from the 12th century. Reprinted. New York, 1970.

781. *THE MEDIEVAL philosophy and biology of growth ... with translated texts of Albertus Magnus and Thomas Aquinas.* Bloomington, Ind., 1971. By J. Cadden.

782. *ANATOMICAL texts of the earlier Middle Ages.* Washington, 1927. By G.W. Corner.

783. THE SCHOOL of Salerno. fl. 11th-13th centuries. DSB (SALERNITAN ANATOMISTS); Sart. Also under Early Modern Editions (1193)*. See also 544 and 566.
   1. *Collectio salernitana; ossia, Documenti inediti.* 5 vols. Naples, 1852-59. Ed. by G.E.T. Henschel et al.
   2. *Magistri salernitani nondum editi.* Turin, 1901. Ed. by P. Giacosa.

1.244 Translations of Ancient Greek Works

784. ALEXANDER OF APHRODISIAS. (Main entry: 160)
   1. *Commentaire sur les "Météores" d'Aristote. Traduction de Guillaume de Moerbeke.* Louvain, 1968. Critical ed. by A.J. Smet.

785. ARCHIMEDES. (Main entry: 93)
   1. *Archimedes in the Middle Ages.* 3 vols. Madison, Wis., 1964-78. Ed. by M. Clagett.

786. ARISTOTLE. (Main entry: 65)
   1. *Aristotles latinus.* Many vols. Paris, 1951-. Ed. by L. Minio-Paluello et al.
   2. *Les Méthéores.* Uppsala, 1945. First publn of a 13th-century French trans. By R. Edgren.
   3. *"De anima" in the version of William of Moerbeke and the commentary of St. Thomas Aquinas.* London, 1951. With English trans. by K. Foster.
   4. *Problemata varia anatomica.* Lawrence, Kansas, 1968. Ed. from the MS by L.R. Lind.

787. EUCLID. (Main entry: 81)
   1. *Euclidis latine facti fragmenta veronensia.* Milan, 1964. Ed. by M. Geymonat.
   2. *The translation of the "Elements" of Euclid from the Arabic into Latin by Hermann of Carinthia (?).* Leiden 1968. Ed. by H.L.L. Busard. Reprinted, Amsterdam. 1977.
   3. *Ein neuer Text des Euclides latinus: Faksimiledruck der Handschrift Lüneburg.* Hildesheim, 1970. Ed. with an introd. by M. Folkerts.

788. GALEN. (Main entry: 159)
   1. *Burgundio of Pisa's translation of Galen's "Peri kraseon," De complexionibus."* Berlin, 1976. (Galenus latinus. 1) Ed. with an introd., etc., by R.J. Durling.

789. HIPPOCRATES. (Main entry: 47)
   1. *Les amphorismes Ypocras de Martin de Saint-Gille, 1362-65.* Cambridge, Mass., 1954. A medieval French trans., probably from the Latin, of the *Aphorisms* of Hippocrates.

790. PLATO. (Main entry: 51)
 1. *Corpus Platonicum Medii Aevi.* [Series 1] *(Plato latinus).* London, 1940-. Ed. by R. Klibansky. (Series 2 is *Plato arabus*--see 478.)

791. PROCLUS. (Main entry: 192)
 1. *Die mittelalterliche Übersetzung der ... Procli ... "Elementatio physica."* Berlin, 1958. Ed. by H. Boese.

792. *TEXTES latins et vieux français relatifs aux Cyranides.* Liège, 1942. Ed. by L. Delatte. The *Cyranides* are an anonymous Greek compendium of the virtues of stones, plants, and animals.

1.245 Translations of Arabic Works

793. ABŪ MA'SHAR (Albumasar). (Main entry: 276)
 1. *Albumasaris De revolutionibus nativitatum.* Leipzig, 1968. Ed. by D. Pingree.

794. AL-BIṬRŪJĪ (Alpetragius). (Main entry: 390)
 1. *De motibus caelorum. Critical edition of the Latin translation of Michael Scot.* Berkeley, Cal., 1952. Ed. by F.J. Carmody.

795. IBN AL-MUTHANNĀ, Aḥmad. (Main entry: 319)
 1. *El commentario de ibn al-Mutanna a las tablas astronomicas de al-Jwarizmi. Estudio y edición critica del texto latino en la version de Hugo Sanctallensis.* Madrid/Barcelona, 1963. By E. Millás Vendrell.

796. IBN RUSHD (Averroës). (Main entry: 380)
 1. *Corpus commentariorum Averrois in Aristotelem.* Cambridge, Mass., 1949-. Ed. by H.A. Wolfson et al.

797. IBN SĪNĀ (Avicenna). (Main entry: 351)
 1. *De congelatione et conglutinatione lapidum.* See main entry, item 1.

798. THABIT IBN QURRA. (Main entry: 296)
 1. *The astronomical works.* Berkeley, Cal., 1960. Latin texts. Ed. by F.J. Carmody.

1.3 OTHER CIVILIZATIONS

(To the End of the Eighteenth Century)

1.31 Indian

1.311 Authors

799. ĀTREYA. 6th century B.C. (?) ICB; Sart.; BLA & Supp. Medicine.

800. SUŚRUTA. 6th century B.C. (?) ICB; Sart.; Mort.; BLA & Supp. Medicine. (Titles not included.)

801. VṚDDHA GARGA 1st century A.D. ICB; Sart. Astrology/astronomy.

802. CHARAKA. fl. ca. 150. ICB; Sart. (CARAKA); Mort.; BLA & Supp. Medicine. (Titles not included.)

803. YAVANEŚVARA. fl. 150. DSB. Translator of a Greek work on astrology into Sanskrit.

804. GARGA. 2nd-3rd century. ICB. Astrology.

805. SPHUJIDHVAJA. fl. 269. DSB. Astronomy/astrology.
    1. *The Yavanajātaka.* 2 vols. Cambridge, Mass., 1976-78. Ed., trans., and comm. by D. Pingree.

806. PAULIŚA. 4th or 5th century. DSB; Sart. Astronomy.

807. LĀṬADEVA. fl. ca. 505. DSB. Astronomy.

808. ĀRYABHAṬA (first of the name). b. 476. DSB; ICB; Sart.; P III; GA. Mathematics. Astronomy.
    1. *The Āryabhaṭīya, with the commentary Bhaṭadīpikā of Paramādīçvara.* Leiden, 1874. Ed. by H. Kern. Reprinted, Osnabrück, 1973.
    2. *The Āryabhaṭīya, an ancient Indian work on mathematics and astronomy.* Chicago, 1930. Trans. with notes by W.E. Clark.

809. VARĀHAMIHIRA. early 6 th century. DSB; ICB; Sart.; P III (DARĀHA-MIHIRA). Astronomy. Astrology.
    1. *The Pañcasiddhāntikā.* 2 parts. Copenhagen, 1970. Ed. and trans. by O. Neugebauer and D. Pingree.

810. YATIVṚṢABHA. 6th century. DSB. Cosmography. Mathematics.

811. VĀGBHAṬA. early 7th century. ICB; Sart.; BLA. Medicine. (Titles not included.)

812. BHĀSKARA (first of the name). fl. 629. DSB; ICB. Astronomy.
    1. *Mahābhāskarīya. With the Bhāsya of Govindasvāmin and the supercommentry Siddhāntadīpikā of Parameśvara.* Madras, 1957. Critical ed. with introd. and appendices by T.S. Kuppanna Sastri.

813. BRAHMAGUPTA. 598-after 665. DSB; ICB; Sart.; P I; GA. Astronomy. Mathematics.
   1. *The Khaṇḍakhādyaka (an astronomical treatise). With the commentary of Bhaṭṭotpala.* New Delhi, 1970. Ed. and trans. by B. Chatterjee.
   See also 866.

814. HARIDATTA (first of the name). fl. 683. DSB. Astronomy.

815. CANDRAŃANDANA. 8th century. BLA Supp. Medicine.

816. LALLA. 8th century. DSB. Astronomy.

817. VRINDA. 8th or 9th century. ICB; Sart.; BLA (VRNDA). Medicine.

818. MĀDHAVAKARA. 8th or 9th century. Sart. Medicine.

819. KANAKA. fl. ca. 775-820. DSB. Astrology/astronomy.

820. DRDHABALA. 9th century. BLA Supp. Medicine.

821. MAHĀVĪRA. 9th century. DSB; ICB; Sart. Mathematics.
   1. *The Gaṇita-sāra-sañgraha.* Madras, 1912. With English trans. and notes by M. Rañgācārya

822. ŚRĪDHARA. 9th century. DSB; ICB; Sart. Mathematics.
   1. *The Pātīgaṇita With an ancient Sanskrit commentary.* Lucknow, 1959. Ed. with introd., trans., and notes by K.S. Shukla.

823. VAṬEŚVARA. b. 880. DSB. Astronomy.

824. MUÑJĀLA. fl. 932. DSB. Astronomy.

825. VIJAYANANDA (or VIJAYANANDIN). fl. 966. DSB; ICB. Astronomy.

826. ŚRĪPATI. fl. 1039-1056. DSB; ICB. Astronomy. Astrology. Mathematics.

827. DAŚABALA. fl. 1055. DSB. Astronomy.

828. BRAHMADEVA. fl. ca. 1092. DSB. Astronomy.

829. ŚATĀNANDA. fl. 1099. DSB; Sart. Astronomy.

830. HEMACANDRA. 1088-after 1169. ICB; Sart. Botany.

831. ḌALLANA. 12th century (?) Sart.; BLA Supp. Medicine.

832. BHĀSKARA (second of the name). b. 1115. DSB; ICB; Sart.; P III. Astronomy. Mathematics.
   1. *Bija ganita; or, The algebra of the Hindus.* London, 1813. Trans. by E. Strachey.
   2. *The Gunitadhia; or, A treatise on astronomy. With a commentary entitled Mitachshara.* Calcutta, 1842. Ed. by L. Wilkinson.
   3. *Līlāvatī.* Allahabad, 1967. Trans. by Colebrooke with notes by H.C. Banerji.
   4. *Līlāvatī. With Kriyākramakarī of Sankara and Nārāyana. Being an elaborate exposition of the rationale of Hindu mathematics.* Hoshiapur, 1975. Critical ed. with introd. by K.V. Sarma.
   See also 866.

833. ARUṆADATTA. fl. ca. 1200. BLA Supp. Medicine.

834. CAṆGADEVA. fl. 1205. Sart. Mathematics.

835. NĀRĀHARI. fl. 1235-1250. Sart. Pharmacy.

1. *Die indischen Mineralien ... Narahiri's Râĝanighantu Varga XIII.* Leipzig, 1882. Sanskrit text with German trans. and notes by R. von Garbe. Reprinted, Hildesheim, 1974.

836. ŚĀRṄGADHARA. 13th century or earlier. ICB; Sart. Chemical pharmacy.

837. MAHĀDEVA. fl. 1316. DSB. Astronomy.

838. NĀRĀYAṆA. fl. 1356. DSB; ICB; Sart. Mathematics.
1. *Bījagaṇitāvataṃsa.* Part 1. Lucknow, 1970. Critical ed. by K.S. Shukla.

839. MAHENDRA SŪRI. fl. 1370. DSB. Astronomy.

840. MADANAPĀLA. fl. 1375. Sart. Pharmacy.

841. MERUTUṄGA. fl. 1386. Sart. Alchemy.

842. PARAMEŚVARA (or PARAMĀDĪÇVARA). ca. 1380-ca. 1460. DSB; ICB. Astronomy. See also 808/1 and 812/1.

843. NĪLAKAṆṬHA. 1444-ca. 1501. DSB; ICB. Astronomy.

844. MAKARANDA. fl. 1478. DSB. Astronomy.

845. KEŚAVA. fl. 1496. DSB. Astronomy.

846. GANEŚA. b. 1507. DSB. Astronomy.

847. BHĀVAMIŚRA (or MIŚRABHĀVA). 16th century. ICB; BLA Supp. Medicine.

848. MANTREṢVARA. 16th century. ICB. Astrology.

849. DINAKARA. b. ca. 1550. DSB. Astronomy.

850. ACYUTA PIṢĀRATI. ca. 1550-1621. DSB. Astronomy.

851. RĀGHAVĀNANDA ŚARMAN. fl. 1591-1599. DSB. Astronomy

852. RAÑGANĀTHA. fl. 1603. DSB. Astronomy.

853. MATHURĀNĀTHA ŚARMAN. fl. 1609. DSB. Astronomy.

854. ĀNANDARĀYA MAKHI. early 17th century. ICB. Medicine.

855. NĀGEŚA. fl. ca. 1630. DSB. Astronomy.

856. HARIDATTA (second of the name). fl. 1638. DSB. Astronomy.

857. MUNĪŚVARA VIŚVARŪPA. b. 1603. DSB. Astronomy. Mathematics.

858. KAMALĀKARA. b. 1610. DSB. Astronomy.

859. KṚṢṆA. fl. 1653. DSB. Astronomy.

860. JAYASHIṂHA II, *Maharaja.* 1686-1743. DSB; ICB (JAI SINGH II). Patron of science. He established five astronomical observatories, had Arabic works translated into Sanskrit and star catalogues prepared.

861. JAGANNĀTHA. fl. ca. 1720-1740. DSB. Employed by Jayashiṃha II (860) to translate astronomical and mathematical works from Arabic into Sanskrit.
1. *Samrāt siddhānta.* 3 vols. New Delhi, 1967-69. Ed. by A.R.S. Sharma et al. A Sanskrit version of Ptolemy's *Almagest*, trans. from the Arabic.

Authors of Undetermined Date

862. AGNIVEŚA. ICB; BLA Supp. Medicine.

863. KANĀDA, Maharishi. ICB. Physics.

864. ĀRYABHAṬA (second of the name). fl. between 950 and 1100. DSB; ICB. Astronomy.

865. CANDRAṬA. fl. between the 11th and 14th centuries. BLA Supp. Medicine.

1.312 Texts and Collections

866. *ALGEBRA, with arithmetic and mensuration, from the Sanskrit of Brahmagupta and Bhāscara.* London, 1817. Trans. by H.T. Colebrooke.

867. *THE SŪRYASIDDHĀNTA: A textbook of Hindu astronomy. Edited with the Tattvamrita Sanskrit commentary, notes, etc.* Benares, 1946. By S.K. Chaudhary. It dates possibly from the fifth century.

868. *INDIAN science and technology in the eighteenth century. Some contemporary European accounts.* Delhi. 1971. Ed. by Dharampal.

1.32 Chinese

1.321 Authors

869. PIEN CH'IAO. early 5th century B.C. Sart.; BLA. Medicine.

870. SHIH SHEN. 4th century B.C. ICB. Author of a star catalogue, the oldest in the Orient.

871. SHUN YU-I. b. ca. 216 B.C. ICB; Sart. Medicine.

872. CH'UN-YU I. b. 205 B.C. ICB; BLA. Medicine.

873. CHANG T'SANG. fl. 165-142 B.C. ICB; Sart. Mathematics.

874. HUAI NAN TZU. d. 122 B.C. ICB; Sart. Cosmogony. Alchemy.

875. LO HSIA HUNG. fl. 140-104 B.C. Sart. Astronomy.

876. KENG SHOU-CH'ANG. mid-1st century B.C. Sart. Mathematics.

877. LIU HSIN. d. A.D. 22. ICB; Sart. Astronomy.

878. WANG CH'UNG. 27-97. Sart. Natural philosophy.

879. CHIA KHUEI. 30-101. ICB; Sart. Astronomy.

880. CHANG HENG. 78-139. ICB; Sart. Astronomy.

881. WEI PO-YANG. 2nd century. ICB. Alchemy.

882. TS'AI YUNG. 133-192. ICB. Astronomy.

883. LIU HUNG. fl. ca. 156-ca. 189. ICB; Sart. Astronomy.

884. CHANG CHUNG-CHING. fl. ca. 200. Sart.; BLA (CHANG KI). Medicine.

885. HSU YUEH. fl. ca. 200. Sart. Mathematics.

886. LU CHI. fl. ca. 200. Sart. Mathematics. Astronomy.

887. HSU SHIH. early 3rd century. ICB. Alchemy.

888. HUA T'O. ca. 190-265. ICB; Sart.; BLA (HWA Y'UO). Surgery.

889. LIU HUI. fl. ca. 250. DSB; ICB; Sart. Mathematics.

890. WANG FAN. ca. 229-after 264. Sart. Astronomy. Mathematics.

891. HUANG FU-MI. late 3rd century. ICB; Sart.; BLA. Medicine.

892. CHI HAN. fl. ca. 300. Sart. Botany.

893. WANG SHU-HO. fl. ca. 300. Sart.; BLA. Medicine.

894. KO HUNG. ca. 281-361? ICB; Sart.; BLA. Medicine. Alchemy.
   1. *Alchemy, medicine, and religion in the China of A.D. 320: The Nei P'ien of Ko Hung.* Cambridge, Mass., 1966. Trans. by J.R. Ware.

895. CHIANG CHI. fl. ca. 385. Sart. Astronomy.

896. SUN TZU. 4th or 5th century. ICB; Sart.; P III (SUN-TSE). Mathematics.

897. CH'IEN LO-CHIH. fl. ca. 436. Sart. Astronomy.

898. HO CH'ENG-T'IEN. fl. ca. 443. Sart. Astronomy.

899. TSU CH'UNG-CHIH. ca. 429-ca. 500. DSB; Sart. Mathematics.

900. T'AO HUNG-CHING. 451-536. ICB; Sart.; BLA. Medicine. Pharmacy. Alchemy.

901. HSIA-HOU YANG. mid-6th century. ICB; Sart. Arithmetic.

902. CHEN-LUAN. fl. 566. Sart. Mathematics.

903. CHANG CH'IU-CHIEN. late 6th century (?) Sart. Mathematics.

904. SUN SSU-MO. 581-682. ICB; Sart. Medicine.

905. CH'AO YUAN-FANG. fl. 605-609. ICB; Sart.; BLA. Medicine.

906. CH'U-T'AN CHUAN. fl. 618. Sart. Astronomy.

907. WANG HSIAO-T'UNG. fl. ca. 625. Sart. Mathematics.

908. FU JEN-CHUN. fl. 626. Sart. Astronomy.

909. LI SHUN-FENG. mid-7th century. Sart. Mathematics. Astronomy.

910. I CHING. late 7th century. ICB; Sart. Astrology.
   1. *The astrology of I Ching.* London,1976. Trans. by W.K. Chu. Ed. with comm. by W.A. Sherrill.

911. I-HSING. 672-717. ICB; Sart.; P III (YIH-HING). Mathematics. Astronomy.

912. CH'U-T'AN HSI-TA. early 8th century. Sart. Author of an astrological work which was important in transmitting Indian mathematics to China.

913. NANKUNG YUEH. early 8th century. ICB. Astronomy.

914. WANG TAO. fl. 752. ICB; Sart. Medicine.

915. WANG PING. fl. 761. Sart.; BLA. Medicine.

916. WU SHU. 947-1002. Sart. Author of the earliest known Chinese encyclopaedia. It includes astronomy and natural history.

917. CHANG SSU-HSUN. fl. 979. Sart. Astronomy.

918. CHANG PO-TUAN. 983-1082. ICB. Alchemy.

919. TOU P'ING. early 11th century (?) Sart. Author of a distillation book.

920. WANG WEI-TE. fl. 1027. Sart. Medicine.

921. SHEN KUA. 1031-1095. DSB; ICB; Sart. Mathematics. Astronomy. Magnetism. Earth sciences. Medicine.

922. CHOU-TS'UNG. fl. 1065. Sart. Astronomy.

923. P'ANG AN-SHIH. fl. 1090. Sart. Medicine.

924. SU-SUNG. fl. 1092. Sart. Astronomy.

925. CH'IEN-I. fl. 1093. Sart.; BLA. Medicine.

926. CHU I-CHUNG. early 12th century. Sart. Author of a distillation book.

927. TU, WAN. fl. 1126. Known only from the following work.
   1. *Stone catalogue of Cloudy Forest.* Berkeley, Cal., 1961. Comm. and synopsis by E.H. Schafer.

928. FAN CH'ENG-TA. 1126-1193. ICB; Sart. Natural history. Geography.

929. TS'AI CH'EN. 1167-1230. Sart. Mathematics.

930. LI CHIH. 1192-1279. DSB; Sart. (LI YEH). Mathematics.

931. CH'IN CHIU-SHAO. ca. 1202-ca. 1261. DSB; Sart.; P III (TSIN KIN TSCHAN). Mathematics.
   1. *Chinese mathematics in the thirteenth century: The Shu-shu chiu-chang of Ch'in Chiu-shao.* Cambridge, Mass., 1973. By U. Libbrecht.

932. HSU HENG. 1209-1281. Sart. Astronomy.

933. CH'EN CHING-I. fl. 1256. Sart. Botany.

934. KUO SHOU-CHING. 1231-1316. ICB; Sart; GA (CO-CHEOU-KING). Astronomy.

935. YANG HUI. fl. 1261-1275. DSB; Sart. Mathematics.

936. CHU SHIH-CHIEH. fl. 1280-1303. DSB; ICB; Sart. Mathematics.

937. CH'EN KUAN-WU. fl. 1331. Sart.; ICB (SHANG-YANG-TZU). Alchemy.

938. KAO HSIANG-HSIEN. 14th century (?) ICB. Alchemy.

939. CHAO YU-CH'IN. mid-14th century. Sart. Mathematics. Astronomy.

940. LIU CHI. 1311-1375. Sart. Astronomy.

941. TING CHU. fl. 1355. Sart. Arithmetic.

942. YEN KUNG. fl. 1372 Sart. Mathematics.

943. CHU HSIAO. d. 1425. ICB; Sart. Botany.

944. LI SHIH-CHEN. 1518-1593. DSB; ICB; BLA. Pharmacy. Natural history.

945. KUNG T'ING HSIEN. late 16th century. ICB. Medicine.

946. PIN KUE. fl. 1593. P III. Mathematics.

947. HSU KUANG-CH'I (or HSU, Paul). 1562-1633. ICB. The first translator of European scientific works into Chinese. He worked with Matteo Ricci.

948. SUNG YING-HSING. fl. 1615-1644. ICB.
   1. *T'ien-kung k'ai-wu. Chinese technology in the seventeenth century.* University Park, Penn., 1966. A compendium of technology, written in 1637. Trans. by E-tu Zen Sun and Shiou-chuan Sun.

949. WANG HSI-SHAN. 1628-1682. DSB; ICB. Astronomy.

950. MEI WEN TING. 1633-1721. ICB. Mathematics. Astronomy.

1.322 Anonymous Works

951. *CHIU chang suan shu. Neun Bücher arithmetischer Technik. Ein chinesisches Rechenbuch für den praktischen Gebrauch aus der frühen Hanzeit.* Brunswick, 1968. (OKEW, n.F. 4) German trans. and comm. by K. Vogel. The work dates possibly from the first or second centuries B.C.

952. *T'IEN-wen t'u: The Soochow astronomical chart.* Ann Arbor, Mich., 1945. By W.C. Rufus and Hsing-chih Tien. A star map prepared in 1193 and engraved on stone in 1247; still extant.

953. *THE ASTRONOMICAL chapters of the "Chin shu."* Paris, 1966. With amendments, trans., and notes by Ho Peng Yoke. The *Chin shu* is the official history of the Chin dynasty and deals with the period 265-419.

1.33 Japanese

1.331 Authors

954. KWANROKU. fl. 602. Sart. "A Korean bonze who came to Japan in 602 and taught the Japanese how to make a calendar in the Chinese way."

955. WAKE HIROYO. fl. ca. 800. Sart. Medicine.

956. HIROIZUMI MONOBE. 785-860. Sart. Medicine.

957. FUKUYOSHI OMURA. fl. 834-848. Sart. Surgery.

958. NICHIGAKU. 10th century. ICB. Astrology.

959. YASUYORI TAMBA. fl. 982. Sart. Medicine.

960. ABE SEIMEI. d. 1005. Sart. Astronomy.

961. FUJIWARA MICHINORI. d. 1159. Sart. Mathematics.

962. KOREMUNE TOMOTOSHI. fl. 1282. Sart.; BLA. Medical botany.

963. SHIBUKAWA, HARUMI. 1639-1715. DSB. Astronomy.

964. SEKI, TAKAKAZU. 1642?-1708. DSB. "The greatest of the traditional Japanese mathematicians."
   1. *Senshu. (Collected works).* Osaka, 1974. In Japanese with summaries and explanations in English.

965. NAKAMURA, YOZAEMON. late 17th century. ICB. Mathematics.

966. TEKEBE, KATAHIRO. late 17th century. ICB. Mathematics.

967. GENDATS, MATSU-WOKA. fl. ca. 1700. GA. Botany.

968. AJIMA NAONOBU. ca. 1732-1798. DSB; ICB (AJIMA CHOKUYEN and AJIMA MANZO). Mathematics.

969. SUGITA, GEMPAKU. 1733-1817. ICB; BLA. A physician who was one of the first translators of European scientific works into Japanese.
   1. *The dawn of western science in Japan. (Rangaku kotohajime).* Tokyo, 1969. Trans. by Ryozo Matsumoto. Sugita's reminiscences, written in 1815, which illustrate the introduction of European medicine and science into Japan.

970. ASADA GORYU. 1734-1799. DSB; ICB. Astronomy.

971. TAKAHASHI, YOSHITOKI. late 18th century. ICB. Astronomy.

972. INO, TADATAKA. 1745-1818. DSB; ICB. Astronomy. Geodesy. Cartography.

973. AIDA YASUAKI (or AIDA AMMEI). 1747-1817. DSB. Mathematics.

974. MATSUNAGA, SADANOJO. 1751-1795. ICB. Mathematics.

975. SHIZUKI, TADAO. 1760-1806. DSB; ICB. Natural philosophy.

Author of Undetermined Date

976. OHARA, RIMEI. ICB. A practitioner of the traditional Japanese mathematics.

1.34 Central American

1.341 Authors

977. BADIANUS, Juannes. mid-16th century. ICB. An Aztec herbal, 1552. "Produced in the College of Santa Cruz in Mexico City by Martin de La Cruz and Juannes Badianus, native Mexican Indians."

978. TOVAR, Juan de. ca. 1546-ca. 1626.
   1. *The Tovar calendar, an illustrated Mexican manuscript ca. 1585.* New Haven, Conn., 1951. Reprod. with comm., etc., by G. Kubler and C. Gibson.

PART 2

EARLY MODERN PERIOD
(ca. 1450-1700)

2.1 PRINTED EDITIONS OF
ANCIENT AND MEDIEVAL WORKS

Within each section the authors are listed in the order in which their works were first printed: the date after each author's name is the date of the earliest printing of any of his works (generally in Latin translation in the case of Greek and Arabic authors). These dates however are often uncertain and not much reliance can be put on them.

Except for reprints and edited texts, the libraries were not a main bibliographical source for this section, hence the symbol X is not used.

2.11 Greek and Roman

2.111 Authors

979. PLINY. (Main entry: 144) 1469. Klebs 786, 787; Still. 487, 684, 788, 873.
   1. *Naturalis historia*. Venice, 1469. The first printed ed. RM. At least forty-six eds had been publ. by 1550.
   2. *A summarie of the antiquities and wonders of the worlde, out of the sixtene first books of Plinie*. London, 1566. PR 20031 and two later eds.
   3. *The historie of the world*. 2 vols. London, 1601. Trans. by P. Holland. PR 20029 and two later eds. RM.
   3a. ——— *Selections from "The history of the world."* London, 1962. Ed. by S. Turner.
   Later editions and translations are not included.

980. STRABO. (Main entry: 124) 1469. Sart. 1, 228; Klebs 935; Still. 893.

981. PRISCIANUS OF CAESAREA. (Main entry: 197) 1470. Klebs 806; Still. 792, 875.

982. ARISTOTLE. (Main entry: 65) 1472.
   Opera. 1472. Klebs 82, 83; Still. 570, 571.
   1. [*Opera graece.*] 5 vols. Venice, 1495-98. The first printed Greek ed. RM.
   2. *Opera quaecumque impressa hactenus exsisterunt omnia*. Basel, 1531. Greek text. Ed. by Erasmus. RM.
   3. *Opera quae exstant*. 11 vols. Frankfurt, 1584-87. Greek text. Ed. by F. Sylburg. RM.
   Physica. 1472. Klebs 93; Still. 736.
   4. *A. Hyperii compendium "Physices" Aristoteleae*. London, 1583. PR 758.

5. *Commentarium Collegii Conimbricensis Societatis Jesu, in octo libros "Physicorum" Aristotelis ... prima (secunda) pars. Qui nunc primum graeco Aristotelis contextu....* 2 vols. Cologne, 1596. Greek and Latin, with commentary. RM.

De anima. 1472. Klebs 84; Still. 572.

6. *Copulata super libros "De anima" Aristotelis cum textu iuxta doctrinam ... Thome de Aquino*. Cologne, 1492. RM.

Problemata. 1473. Klebs 95; Still. 583.

7. *Problemata*. London, 1583. PR 761.
8. *The problems*. London, 1595. PR 762 and three further eds.
9. *The problems*. London, 1649. Wing A3691 and later eds.

Parva naturalia. 1473. Klebs 92; Still. 580.

De coelo et mundo. 1473. Klebs 87; Still. 25.

De generatione et corruptione. 1474. Klebs 88; Still. 576, 733.

Meteorologica. 1474. Klebs 91; Still. 577, 735.

De animalibus. 1476. Klebs 85; Still. 573.

10. ... *libri de animalibus*. Venice, 1476. Trans. by T. Gaza. The first printed Latin ed. Contains three works as in item 11, below. RM.
11. *De historia animalium. De partibus animalium et earum causis. De generatione animalium*. Venice, 1545. Trans. by T. Gaza. RM.

983. MACROBIUS, Ambrosius Theodosius. (Main entry: 183) 1472. Sart. 1, 385; Klebs 638; Still. 74.

1. *Opera*. London, 1694. Wing M229.

984. GALEN. (Main entry: 159) 1473. Sart. 1, 302; Klebs 432, 433; Still. 374-377, 641-644.

About 700 editions of Galen's various individual works were published between 1473 and 1600, nearly all as Latin translations. The *Opera* were printed in Latin in 1490 and in Greek in 1525.

None of the early modern editions are included in the RM series. PR record editions of several of the individual works, most of them translated into Latin by Linacre, and, with Wing, various selections and epitomes in Latin or English.

985. LUCRETIUS. (Main entry: 118) 1473. Sart. 1, 205; Klebs 623; Still. 672, 773.

1. *De rerum natura*. Verona, 1486. RM. Second printed ed. Later eds are numerous (especially in the late 17th century, cf. Wing).

986. MANILIUS, Marcus. (Main entry: 128) 1473. Sart. 1, 237; Klebs 661; Still. 75.

1. *The sphere*. London, 1675. Wing M432.
2. *The five books. Containing a system of the ancient astronomy and astrology*. London, 1697. Wing M430 and another ed.

987. ARATUS OF SOLI. (Main entry: 83) 1474. Sart. 1, 157; Klebs 77; Still. 23, 730.

1. *Phaenomena*. Oxford, 1672. Wing A3596.

988. HIPPOCRATES OF COS. (Main entry: 47) ca. 1475. Sart. 1, 99; Klebs 517-522; Still. 405-422, 656-662.

A large number of editions of the various individual works of the Hippocratic Corpus were published from ca. 1475 onwards through the 16th century, the great majority as Latin translations. The *Opera* were printed in Latin in 1525 and in Greek in 1526.

1. *Opera quae extant graece et latine*. Venice, 1588. Ed. by G. Mercuriale. RM.

PR and Wing record numerous editions and translations from 1567 onwards.

989. HYGINUS. (Main entry: 125) 1475. Sart. 1, 226; Klebs 527,528; Still. 67.

1. *Poeticon astronomicon*. Venice, 1482. Second printed ed. RM.

990. PTOLEMY. (Main entry: 157) 1475. Sart. 1, 274; Klebs 812-814; Still. 94-97, 212-215.

1. *Omnia quae extant opera, "Geographia" excerpta*. Basel, 1541. RM.
2. *Omnia quae extant opera, praeter "Geographicam."* Basel, 1551. Ed. by E.O. Schreckenfuchs. RM.

Tetrabiblos

3. *Here begynneth the compost of Ptholomeus*. London, [1532?] PR 20480 and later eds.
4. *Libri quattuor compositi Syro fratri. Eiusdem ... Centum dicta*. Nuremberg, 1535. The first printed Greek ed. With Latin trans. Ed. by J. Camerarius. RM.

Other Works

5. *Liber analemmate*. Rome, 1562. On gnomonics. Ed. by F. Commandino. Latin trans. from the Arabic. RM.
6. *Harmonicorum libri tres ... nunc primum graece editus*. Oxford, 1682. Wing P4149. Ed. by John Wallis. With Latin trans.

Editions of the *Geography* are not included.

991. SENECA. (Main entry: 132) 1475. Sart. 1, 248; Klebs 904.

1. *Naturalium quaestionum libri VII. M. Fortunati in eosdem libros annotationes*. Venice, 1522. RM.

992. CELSUS. (Main entry: 126) 1478. Klebs 260; Still. 331, 610.

993. DIOSCORIDES. (Main entry: 142) 1478. Sart. 1, 259; Klebs 342, 343; Still. 354, 618.

1. *Opera quae extant omnia. Ex nova interpretatione J.A. Saraceni*. Frankfort, 1598. Greek text with Latin trans. The best of the early eds. RM.
2. *De medicinali materia*. Strasbourg, 1529. RM.
3. *Kräuterbuch*. Frankfurt, 1610. Reprinted, Munich, 1968.
4. *The Greek herbal, illustrated by a Byzantine A.D. 512. Englished by John Goodyear A.D. 1655*. Oxford, 1934. Ed. and first printed by R.T. Gunther. Reprinted, New York, 1959.

994. OPPIAN. (Main entry: 145) 1478. Klebs 710; Still. 680.

995. SERENUS SAMMONICUS, Quintus. (Main entry: 163) 1480? Sart. 1, 325; Klebs 914.

996. EUCLID. (Main entry: 81) 1482. Sart. 1, 154; Klebs 383; Still. 162-164, 750.

1. *Elementa.* Venice, 1482. The first printed ed. Latin trans. from the Arabic by J. Campanus (i.e. Campanus of Novara). RM.
2. *Elementa.* Basel, 1533. Greek text ed. by S. Grynaeus. The editio princeps. RM.
3. *Elementa. Solo introduttore delle scientie mathematice ... ridotto per ... Nicolo Tartalla* [i.e. Tartaglia]. Venice, 1543. RM.
4. *Elementorum libri XV. Una cum scholiis antiquis.* Pesaro, 1572. Latin trans. Ed. by F. Commandino. RM.

Later editions are very numerous (cf. PR and Wing) and are included only when they were edited or translated by an important mathematician, the entries being made under the names of the mathematicians concerned and indexed under Euclid.

Other Works

5. *Optica et catoptrica. Nunquam antehac graece aedita. Eadem latine reddita per I. Penam* [i.e. J. Pena]. Paris, 1557. RM.
6. *Data ... C. Hardy ... edidit, latine vertit, scholiisque illustravit. Adiectus est Marini philosophi commentarius.* Paris, 1625. RM.

997. PLATO. (Main entry: 51) 1482. Sart. **1**, 113; Klebs 785; Still. 91, 486.

998. FRONTINUS, Sextus Julius. (Main entry: 146) 1483. Sart. **1**, 255; Klebs 422, 423; Still. 842, 843.

999. THEOPHRASTUS. (Main entry: 67) 1483. Sart. **1**, 143; Klebs 958; Still. 533, 534, 701, 702.
   1. *De historia plantarum ... Accesserunt I.C. Scaligeri ... animadversiones, et R. Constantini annotationes.* Amsterdam, 1644. Greek text with Latin trans. RM.

1000. VITRUVIUS. (Main entry: 120) 1483. Klebs 1044; Still. 899.
   1. *De architectura.* Rome, 1483-95. The first printed ed. RM.

1001. ALEXANDER OF APHRODISIAS. (Main entry: 160) 1488. Sart. **1**, 319; Klebs 44,45; Still. 726.
   1. *Quaestiones naturales, de anima, morales.* Venice, 1536. RM.

1002. ARISTARCHUS. (Main entry: 84) 1488. Sart. **1**, 157; Klebs 1012.1, Still. 24.
   1. *De magnitudinibus et distantiis solis et lunae.* Oxford, 1688. Wing A3681. First ed. of the Greek text, together with Commandino's Latin trans. (publ. 1572) and the commentary of Pappus. Ed. by John Wallis.

1003. BOETHIUS. (Main entry: 201) 1488. Klebs 191, 192. Still. 148, 149.

1004. CLEOMEDES. (Main entry: 133) 1488. Sart. **1**, 211; Klebs 280; Still. 43.

1005. HERMES TRISMEGISTUS. (Main entry: 219) 1492. Klebs 510, 511.
   1. *Opuscula cum fragmentis.* London, 1611. PR 13218. Greek and Latin.
   2. *The divine Pymander.* London, 1650. Wing H1565. RM.
   3. *His second book.* London, 1657. Wing, H1567.

1006. CENSORINUS. (Main entry: 165) 1497. Klebs 262; Still. 611.
 1. *De die natali*. Cambridge, 1695. Wing C1664.

1007. FIRMICUS MATERNUS. (Main entry: 173) 1497. Sart. **1**, 354; Klebs 404, 405. Still. 56.
 1. *De nativitatibus*. See entry 1038.

1008. IAMBLICHUS OF CHALCIS. (Main entry: 169) 1497. Klebs 529.

1009. SPEUSIPPUS. (Main entry: 58) 1497. Klebs 529.1.

1010. PROCLUS. (Main entry: 192) 1498. Sart. **1**, 403; Klebs 405.1, 807, 1012.1; Still. 210, 211, 877.
 1. *Sphaera*. Paris, 1534. Trans. by Linacre. RM. First publ. in 1499 in *Scriptores astronomici veteres* (1038).
 2. *The descripcion of the sphere or frame of the worlde*. London, 1550. PR 20399.
 3. *Sphaera*. London, 1620. PR 20398. Greek and Latin.

1011. GEMINUS OF RHODES. (Main entry: 116) 1499. Sart. **1**, 212; Still. 60

1012. MARTIANUS CAPELLA. (Main entry: 184) 1499. Sart. **1**, 407; Klebs 668; Still. 77

1013. NICANDER OF COLOPHON. (Main entry: 105) 1499. Sart. **1**, 158; Klebs 343.1; Still. 460.

1014. THEON OF ALEXANDRIA. (Main entry: 179) 1499. Klebs 405.1; Still. 118, 244.

1015. AUTOLYCUS OF PITANE. (Main entry: 78) 1501. Still. 28, 145.
 1. *De sphaera quae movetur liber, et Theodosii Tripolitae "De habitationibus" liber ... His additae sunt Maurolyci annotationes*. Rome, 1587. RM.

1016. ARCHIMEDES. (Main entry: 93) 1503. Sart. 1, 170; Still. 140-142, 731, 732.
 1. *Opera ... per Nicolaum Tartaleam ...* Venice, 1543. The first printed ed. of the *Opera*. RM. (Some of the individual works were printed in 1503.)
 2. *Opera, quae quidem extant, omnia ... Adiecta quoque sunt Eutocii Ascalonitae ... commentaria*. Basel, 1544. The first printed Greek ed. With Latin trans. RM.
 3. *Opera. Per Is. Barrow*. London, 1675. Wing A3621.

Individual Works

 4. *De iis quae vehuntur in aqua libri duo. A F. Commandino in pristinum nitorem restituti et commentariis illustrati*. Bologna, 1565. RM.
 5. *Arenarius, et Dimensio circuli*. Oxford, 1676. Wing A3622.

1017. THEODOSIUS OF BITHYNIA. (Main entry: 111) 1518. Still. 243.
 1. *Sphaericorum libri tres*. Paris, 1558. Greek text with Latin trans. Ed. by J. Pena.
 1a. ——— Another ed. Oxford, 1707. By J. Hunt. RM.
 2. *Sphaerica*. London, 1675. Wing T857A.

1018. CHALCIDIUS. (Main entry: 177) 1520. Sart. **1**, 352.

1019. MUSA, Antonius. (Main entry: 122) 1528. Sart. **1**, 231; Still. 459.

1020. ORIBASIUS. (Main entry: 178) 1528. Still. 462-465.

1021. SORANUS OF EPHESUS. (Main entry: 153) 1528. Still. 526-528.

1022. AETIUS OF AMIDA. (Main entry: 205) 1534. Sart. 1, 434; Still. 260.

1023. APOLLONIUS OF PERGA. (Main entry: 97) 1537. Sart. 1, 174; Still. 139.
1. *Conicorum libri IV. Una cum Pappi Alexandrini lemmatibus, et commentariis Eutocii Ascalonitae, Sereni Antinonensis philosophi libri duo, nunc primum in lucem editi.* Bologna, 1566. Latin trans. and commentary by F. Commandino. RM.
2. *Conicorum libri V, VI, VII ... Additus in calce Archimedis "Assumptorum" liber, ex codicibus arabicis mss.* Florence, 1661. The first printed ed. of Books V-VII. Latin trans. from the Arabic. Ed. by G.A. Borelli. RM.
3. *Conica: methodo nova illustrata.* London, 1675. Wing A3536. Ed. by Isaac Barrow.
4. *De sectione rationis.* Oxford, 1706. Latin trans. from the Arabic by Edmund Halley. RM.
5. *Conicorum libri octo, et Sereni Antissensis "De sectione cylindri et coni" libri duo.* Oxford, 1710. Ed. by Edmund Halley. RM.

1024. NICOMACHUS OF GERASA. (Main entry: 150) 1538. Still. 199.

1025. EUTOCIUS OF ASCALON. (Main entry: 202) 1544. Still. 165, 751.

1026. AELIANUS, Claudius. (Main entry: 164) 1556? Sart. 1, 326.
1. *De historia animalium.* Lyons, 1562. RM.
2. *His various history.* London, 1665. Wing A679.

1027. MENELAUS OF ALEXANDRIA. (Main entry: 149) 1558. Sart. 1, 254; Still. 196.

1028. NEMESIUS. (Main entry: 182) 1565. Sart. 1, 373.

1029. SERENUS OF ANTINOUPOLIS. (Main entry: 176) 1566. Sart. 1, 353.

1030. HIPPARCHUS. (Main entry: 108) 1567? Sart. 1, 194.
1. *In Arati et Eudoxi "Phaenomena" libri III. Eiusdem liber asterismorum Achillis Statii* [i.e. Achilles Tatius (162)] *in Arati "Phaenomena." Arati vita, et fragmenta aliorum veterum in eius poema.* Florence, 1567. Greek texts. RM.

1031. DIOPHANTUS OF ALEXANDRIA. (Main entry: 166) 1575. Sart. 1, 336.
1. *Arithmeticorum libri sex, et De numeris multianglis liber unus.* Paris, 1621. Greek texts with Latin trans. Ed. by C.G. Bachet de Méziriac. RM.
1a. ——— Another ed. Toulouse, 1670. With notes added by P. de Fermat.

1032. HERO OF ALEXANDRIA. (Main entry: 143) 1575. Sart. 1, 208.
1. [*Pneumatica.*] *Spiritalium liber.* Urbini, 1575. Latin trans. by F. Commandino. RM.
2. [*Automata.*] *De gli automati, overo machine se moventi.* Venice, 1589. Trans. from the Greek by B. Baldi. RM.

1033. PAUL OF ALEXANDRIA. (Main entry: 180) 1586. Sart. 1, 367.

1034. PAPPUS OF ALEXANDRIA. (Main entry: 172) 1588. Sart. 1, 337.

1035. HELIODORUS OF LARISSA. (Main entry: 174)
  1. *Capita opticorum*. Cambridge, 1670. Wing H1375.

2.112 Anonymous Works

1036. *HERBARIUM Apulei*. Sart. **1**, 296; Klebs 505; Still. 401.
  1. *Herbarium Apulei*. [Rome, ca. 1481-83.] Sometimes attributed to Apuleius Barbarus (190).
  1a. ——— *The herbal of pseudo-Apuleius*. Leiden, 1935. Facsimiles of the 9th-century MS and the first printed ed.

1037. *DE UNIVERSI natura*. Cambridge, 1670. Wing O125. Probably written ca. 150 B.C. Wrongly attributed to Ocellus (32).

2.113 Collections of Texts

1038. *SCRIPTORES astronomici veteres*. Venice, 1499. Sart. **2**, 354; Klebs 405; Still. 56. Contains Greek texts or Latin trans. of works by Aratus, Firmicus Maternus, Manilius, and Proclus. Often catalogued under Firmicus Maternus whose *De nativitatibus* is the first item in the collection.

1039. ASTRONOMERS and geographers.
  1. *Procli de sphaera liber. Cleomedis de mundo, sive circularis inspectionis meteorum, libri duo. Arati Solensis phaenomena sive apparentia. Dionysii Afri descriptio orbis habitabilis*. Basel, 1547. Sart. **1**, 404. RM (under Proclus).

1040. *ARISTOTELIS et Theophrasti historiae, cum de natura animalium, tum de plantis et earum causis*. 2 vols. Lyons, 1552. RM (under Aristotle).

1041. *ASTRONOMICA veterum scripta isagogica graeca et latina*. Heidelberg, 1589. Contains works by Proclus, Aratus, Leontius Mechanicus, Avienus, and Hyginus. RM.

2.12 Byzantine

1042. GREGORAS, Nicephorus. (Main entry: 247) 1498. Sart. **3**, 952/3; Klebs 1012.1; Still. 862.

1043. PSELLUS, Michael. (Main entry: 234) 1499. Klebs 811; Still. 490.

1044. ALEXANDER OF TRALLES. (Main entry: 222) 1504. Sart. **1**, 453; Still. 265.

1045. PAUL OF AEGINA. (Main entry: 225) 1510. Sart. **1**, 479; Still. 472-475.

1046. CONSTANTINUS VII. (Main entry: 233) 1530. Sart. **1**, 656; Still. 346.

1047. MYREPSUS, Nicholas. (Main entry: 240) 1541. Sart. **2**, 1094.

1048. BARLAAM, Bernard. (Main entry: 246) 1564. Sart. **3**, 586.

1049. HERO OF BYZANTIUM. (Main entry: 232) 1572. Sart. 1, 632.
   1. *Heronis mechanici Liber de machinis bellicis, necnon Liber de geodasia*. Venice, 1572. Latin trans. by F. Barozzi (i.e. Barocius). RM.

2.13 Arabic

1050. MĀSAWAIH AL-MĀRDĪNĪ (Mesuë the Younger). (Main entry: 327) 1471. Sart. 1, 728; Klebs 680, 681; Still. 553, 554.

1051. AL-ZAHRĀWĪ (Albucasis or Abulcasis). (Main entry: 333) 1471. Sart. 1, 681; Klebs 5; Still. 255, 256, 257.

1052. IBN SĪNĀ (Avicenna). (Main entry: 351) 1472. Sart.1, 711; Klebs 131-136; Still. 287-290, 588-590.
   1. *Opera*. 2 vols. Venice, 1508. Philosophical and scientific works only. Reprinted, Frankfurt, 1961. Also reprinted under the title *Opera philosophica*, Louvain, 1961.

1053. AL-RĀZĪ (Rhazes or Rhasis). (Main entry: 300) 1472. Sart. 1, 609; Klebs 392.1, 825-827; Still. 500-504, 689, 690.

1054. IBN SERĀBĪ (Serapion the Younger). (Main entry: 369) 1473. Sart. 2, 229; Klebs 911, 913; Still. 556.

1055. AL-QABĪṢĪ (Alcabitius). (Main entry: 321) 1473. Sart. 1, 669; Klebs 41; Still. 1.
   1. *Libellus isagogicus ... a Joanne Saxonie editum*. Venice, 1512. An astrological work. The trans. was done in 1144. RM (under Alcabitius).

1056. IBN RUSHD (Averroes). (Main entry: 380) 1474. Sart. 2, 359; Klebs 128-130; Still. 285, 286, 587, 737, 738.

1057. ḤUNAYN IBN ISḤĀQ (Joannitius). (Main entry: 284) ca. 1476. Sart. 1, 612; Klebs 534; Still. 425.

1058. IBN MĀSAWAIH (Mesuë the Elder). (Main entry: 265) 1481. Sart. 1, 574; Still. 552.

1059. MAIMONIDES. (Main entry: 382; cf. 487) 1481. Sart. 2, 376-379; Klebs 642-644; Still. 444, 445, 674.

1060. ABU'L ḤASEN (Albohazan). (Main entry: 356) 1485. Sart. 1, 715; Klebs 35; Still. 6.

1061. ISAAC ISRAELI. (Main entry: 314) 1487. Sart. 1, 639; Klebs 535; Still. 427.

1062. ABŪ MA'SHAR AL-BALKHĪ (Albumasar). (Main entry: 276) 1488. Sart. 1, 568; Klebs 37-39; Still. 7,8,9. The following works are listed under Albumasar.
   1. *Flores astrologiae*. Augsburg, 1488. Reprinted, Leipzig, 1928.
   2. *Introductorium in astronomiam*. Augsburg, 1489. RM.
   3. *De magnis coniunctionibus*. Augsburg, 1489. Ed. by J. Angelus. RM.

1063. IBN ZUHR (Avenzoar). (Main entry: 374) 1490. Sart. 2, 233; Klebs 127; Still. 283, 284.

1064. ABŪ BAKR AL-HASAN IBN AL-KHASĪB (Albubather). (Main entry: 304) 1492. Sart. 1, 603; Klebs 36; Still. 5.

1065. AL-MAJUSI (Haly Abbas). (Main entry: 324) 1492. Sart. 1, 677; Klebs 498; Still. 400, 653.

1066. AL-FARGHĀNĪ (Alfraganus). (Main entry: 278) 1493. Sart. 1, 567; Klebs 51; Still. 13.
   1. *Rudimenta astronomica Alfragani. Item Albategnius* [i.e. al-Battānī] *De motu stellarum ... Omnia cum demonstrationibus geometricis et additionibus Joannis de Regiomonte.* Nuremberg, 1537. RM (under al-Farghānī).

1067. MĀSHĀ'ALLĀH (Messahala). (Main entry: 261) 1493. Sart. 1, 531; Klebs 814.2; Still. 78, 859.
   1. *Messahalae ... astrologi libri tres.* Nuremberg, 1549. Ed. by J. Heller. RM.

1068. SAHL IBN BISHR (Zahel). (Main entry: 274) 1493. Sart. 1, 569; Klebs--see under Zahel.

1069. IBN RIDWĀN. (Main entry: 358) 1496. Sart. 1, 729.

1070. 'ALĪ IBN 'ISĀ (Jesu Haly). (Main entry: 335) 1499. Sart. 1, 731; Klebs 494.2; Still. 426.

1071. MUHAMMAD IBN 'UMAR (Omar Tiberiadis). (Main entry: 264) 1503. Sart. 1, 568.

1072. AL-KINDĪ. (Main entry: 282) 1507. Sart. 1, 559; Still. 15, 266, 727, 728.

1073. THĀBIT IBN QURRA. (Main entry: 296) 1518. Still. 117, 242.

1074. AL-BITRŪJĪ (Alpetragius). (Main entry: 390) 1531. Sart. 2, 400; Still. 12.

1075. IBN BUTLĀN. (Main entry: 359) 1531. Sart. 1, 730.

1076. JĀBIR IBN AFLAH. (Main entry: 379) 1534. Sart. 2, 206; Still. 68, 181.

1077. AL-ZARQĀLĪ (Arzachel). (Main entry: 365) 1534. Sart. 1, 759; Still. 17.

1078. AL-BATTĀNĪ (Albatenius or Albategnius). (Main entry: 301) 1537. Sart. 1. 603; Still. 10, 126.
   1. *Mahometis Albatenii De scientia stellarum.* Bologna, 1645. RM (under Albategnius).
   See also 1066/1.

1079. IBN AL-HAYTHAM (Alhazen). (Main entry: 346) 1542. Sart. 1, 721; Still. 125, 717.
   1. *Opticae thesaurus ... Eiusdem liber De crepusculis et nubium ascensionibus. Item Vitellonis ... libri X* [i.e. Witelo's *Perspectiva*]. Basel, 1572. Ed. by F. Risner. RM (under Alhazan). Reprinted, New York, 1972.

1080. ABŪ 'ALI AL-KHAIYĀT (Albohali). (Main entry: 268) 1546. Sart. 1, 569.

1081. AL-FĀRĀBI (Alfarabius). (Main entry: 309) 1638. Sart. 1, 628.

1. *Opera omnia quae latina lingua conscripta reperiri potuerunt.* Paris, 1638. Reprinted, Frankfurt, 1969.

Seventeenth-Century Editions by John Greaves (2097)

1082. ANONYMOUS PERSA.
   1. *De siglis Arabum et Persarum astronomicis.* London, 1648. Wing G1795. Astronomical tables of the Arabs and Persians.

1083. MAḤMŪD SHĀH KHULJĪ. (Main entry: 455) Sart. 2, 1006.
   1. *Astronomia quaedam ex traditione Shah Cholgii Persae. Una cum hypothesibus planetarum....* London, 1650. Wing C3922, G1796, M261. Persian text with Latin trans. A commentary on a late 14th-century elaboration of the astronomical tables of al-Ṭūsi (Nāṣir al-Dīn).

1084. ULUGH BEG. (Main entry: 453)
   1. *Epochae celebriores astronomis, historicis, chronologicis, Chataiorum, Syro-Graecorum, Arabum, Persarum, Chorasmiorum usitatae, ex traditione Ulug Beigi.* London, 1650. Wing U24.
   2. *Tabulae longitudinis ac latitudinis stellarum figarum ex observatione Ulug-Beighi.* Oxford, 1665. Wing U23. Ed. T. Hayle.

1085. AL-ṬŪSĪ. (Main entry: 410)
   1. *Binae tabulae geographicae, una Nassir Eddini Persae* [i.e. Naṣir al-Dīn, i.e. al-Ṭūsī], *altera Ulug Beigi Tatari* [i.e. Ulugh Beg]. London, 1652. Wing N233.

2.14 Hebrew

1086. IBN EZRA, Abraham (Avenare). (Main entry: 485) 1482. Klebs 3,4; Still. 2,3,4.

1087. ABRAHAM BAR ḤIYYA (Savasorda). (Main entry: 486) late 15th century. Sart. 2, 207.
   1. *Sphaera mundi.* Basel, 1546. Reprinted, Amsterdam, 1968.

1088. CRESCAS, Hasdai. (Main entry: 500) 1555. Sart. 3, 1448.

2.15 Latin

2.151 Author Sequence

1089. HRABANUS MAURUS. (Main entry: 513) 1467. Klebs 524; Still. 491, 687.
   1. *Opus de universo.* Strasbourg, 1467. RM. An encyclopaedia of universal knowledge.

1090. AQUINAS, Thomas. (Main entry: 613) ca. 1470. Sart. 2, 918; Klebs 960-967; Still. 119, 799.
    1. *Expositio super libros "Posteriorum analyticorum" Aristotelis.* Venice, [1477?] RM.

1091. BURLEY, Walter. (Main entry: 658) ca. 1470. Sart. 3, 563; Klebs 232, 233; Still. 743, 744.
    1. *In "Physicam" Aristotelis expositio et quaestiones.* Venice, 1501. Reprinted, Hildesheim, 1972.

1092. DONDI, Jacopo de'. (Main entry: 677) ca. 1470. Sart. 3, 1670; Klebs 349; Still. 355.

1093. ABANO, Pietro d'. (Main entry: 645) 1471. Sart. 3, 445; Klebs 773-776; Still. 478-480, 683.
    1. *Claviculae Salomonis; seu, Philosophia pneumatica.* 1567. Reprinted, Bilfingen, 1971.

1094. BARTHOLOMAEUS ANGLICUS. (Main entry: 585) 1471? Sart. 2, 587; Klebs 149-154; Still. 595.
    1. *De proprietatibus rerum.* Westminster, 1495. PR 1536. First ed. in English. Trans. by J. Trevisa.
    1a. ——— Another ed. London, 1535. PR 1537.
    1b. ——— *Translation by J. Trevisa. A critical text.* 2 vols. Oxford, 1975.
    2. *Batman uppon Bartolome his booke "De proprietatibus rerum", enlarged and amended.* London, 1582. PR 1538. Reprinted with an introd. and a new index, Hildesheim, 1972.
    3. *De genuinis rerum coelestium, terrestrium et inferarum proprietatibus.* Frankfurt, 1601. RM. Reprinted with the title *De rerum proprietatibus*, New York, 1964.

1095. NICOLAUS SALERNITANUS. (Main entry: 541) 1471. Sart. 2, 239; Klebs 703; Still. 461.

1096. SIMON OF GENOA. (Main entry: 656) 1471/72. Sart. 2, 1085; Klebs 920; Still. 523.

1097. ALBERTUS MAGNUS. (Main entry: 592) 1472. Sart. 2, 941-943; Klebs 13-27; Still. 261, 262, 566-568, 722, 807.
    1. *De animalibus.* Rome, 1478. RM.
    2. *De phisico auditu.* Venice, 1494. RM.
    3. *De generatione et corruptione.* Venice, 1495. RM.
    4. *De celo et mundo.* Venice, 1495. RM.
    5. *De mineralibus.* Venice, 1495. RM.

1098. DATI, Gregorio. (Main entry: 737) 1472. Klebs 326.

1099. GERARD OF SABBIONETTA. (Main entry: 605) 1472. Sart. 2, 987; Klebs 874.1; Still. 63.

1100. HONORIUS OF AUTUN. (Main entry: 537) 1472? Sart. 1, 749; Klebs 523.

1101. ISIDORE OF SEVILLE. (Main entry: 509) 1472. Sart. 1, 471; Klebs 536, 537; Still. 180, 665, 850.

1102. JOHN OF GMUNDEN. (Main entry: 750) ca. 1472. Klebs 556; Still. 184.

1103. SACROBOSCO, Johannes de. (Main entry: 596) 1472. Sart. 2, 618; Klebs 874, 875; Still. 69, 70, 185.

1. *Sphera mundi.* [Bound with] *Gerardi Cremonensis "Theorica planetarum."* Venice, 1478. RM.
2. *La sfera del mondo di M.F. Giuntini col testo di M.G. Sacrobosco.* Lyons, 1582. RM. cf. item 1548/1.

1104. WILLIAM OF SALICETO. (Main entry: 598) 1472 or earlier. Sart. 2, 1079; Klebs 484-487; Still. 509-511, 694.

1105. ARNALD OF VILLANOVA. (Main entry: 631) 1473. Sart. 2, 897-899; Klebs 98-108; Still. 27, 272-279, 816, 817.
1. *Opera omnia.* Basel, 1585. RM.

1106. CECCO D'ASCOLI. (Main entry: 654) ca. 1473. Sart. 3, 644; Klebs 259.

1107. GEBER. 1473. Sart. 2, 1043-1045; Klebs 440, 441; Still. 428, 851, 852. An unknown author (or group of authors) of several alchemical works of the 13th and 14th centuries. Until the 20th century Geber was assumed to be identical with the Arabic alchemist Jābir ibn Hayyān (fl. ca. 800). This is no longer thought to be the case: see the DSB article on Jābir ibn Hayyān (see also ICB). However the RM series, and libraries generally, catalogue Geber's works under the Arabic name.
1. *Incipit liber Geber.* [Venice? 1475?] RM.
2. *De alchimia.* Nuremberg, 1541. RM.
2a. ——— *Die Alchemie des Geber.* Berlin, 1922. Trans. and comm. by E. Darmstaeder.
3. *The works of Geber. Englished by R. Russell.* London, 1678. Wing J54 and a later ed. Reprinted with an introd. by E.J. Holmyard, ib., 1928.

See also 15773/6.

1108. GENTILE DA FOLIGNO. (Main entry: 678) 1473. Sart. 3, 850/1; Klebs 444-453; Still. 380-386.

1109. GILES OF ROME. (Main entry: 635) 1473. Sart. 2, 923 et seq.; Klebs 360-367; Still. 719.
1. *Commentaria in octo libros "Physicorum" Aristotelis.* Venice, 1502. Reprinted, Frankfurt, 1968.

1110. JACOPO DA FORLI. (Main entry: 732) 1473. Sart. 3, 1195; Klebs 546-550; Still. 429-431.

1111. JOHN OF JANDUN. (Main entry: 649) 1473. Sart. 3, 539; Klebs 552, 553.
1. *Super octo libros Aristotelis "De physico" auditu subtilissimae quaestiones.* Venice, 1551. Reprinted, Frankfurt, 1969.

1112. THOMAS OF CANTIMPRÉ. (Main entry: 597) 1473. Sart. 2, 594; Klebs 969, 970.

1113. LATINI, Brunetto. (Main entry: 594) 1474. Sart. 2, 927; Klebs 589.

1114. SILVATICO, Matteo. (Main entry: 664) 1474. Klebs 919; Still. 522, 697.

1115. ADELARD OF BATH. (Main entry: 543) 1475. Sart. 2, 168; Klebs 8.
1. *Quaestiones naturales.* Louvain, 1475. RM.

1116. ANDALO DI NEGRO. (Main entry: 655) 1475. Sart. 3, 647; Klebs 63; Still. 808.

1117. CONRAD OF MEGENBERG. (Main entry: 695) 1475. Sart. 3, 819; Klebs 300; Still. 345, 614.

1118. GAETANO DI THIENE. (Main entry: 753) 1475. Klebs 425-429; Still. 754-758.

1119. JOHN THE CANON. (Main entry: 680) 1475. Sart. 3, 739; Klebs 553; Still. 768.
   1. *Quaestiones super Aristotelis "Physica."* St. Albans, 1481. PR 14621.

1120. LULL, Ramon. (Main entry: 622) 1475. Sart. 2, 901 et seq.; Klebs 627-630; Still. 856, 857.
   1. *Ars generalis ultima.* 1645. Reprinted, Frankfurt, 1970.
   2. *Ars brevis.* 1669. Reprinted, Frankfurt, 1970.

1121. ALBERT OF SAXONY. (Main entry: 704) 1476. Sart. 3, 1431; Klebs 29-31; Still. 11, 723-725.
   1. *Sophismata.* Paris, 1490. RM.
   2. *Quaestiones in libros "De celo et mundo."* Venice, 1492. RM.

1122. FRANCIS OF MEYRONNES. (Main entry: 673) 1476. Sart. 3, 530; Klebs 66.2; Still. 58.

1123. JOHN XXI, *Pope* (Petrus Hispanus). (Main entry: 614) 1476. Sart. 2, 891; Klebs 748, 749.

1124. MONDINO DE' LUZZI. (Main entry: 659) ca. 1476. Sart. 3, 844; Klebs 688; Still. 454, 679.
   1. *Anothomia.* Pavia. 1478. RM.

1125. OCKHAM. William of. (Main entry: 674) 1476. Sart. 3, 554-556; Klebs 706; Still. 782.
   1. *Philosophia naturalis.* Rome, 1637. Reprinted, Vaduz (Liechtenstein), 1970

1126. PAUL OF VENICE. (Main entry: 739) 1476. Klebs 732-735; Still. 681, 784.
   1. *Summa philosophiae naturalis.* Venice, 1503. Reprinted, Hildesheim, 1974. (Listed under Venetus, Paulus.)

1127. CHAULIAC, Guy de. (Main entry: 676) 1477. Sart. 3, 1693; Klebs 491-497; Still. 398, 399, 650, 651.

1128. MICHAEL SCOT. (Main entry: 574) 1477. Sart. 2, 581; Klebs 899-900; Still. 453, 678.
   1 *Liber phisionomie.* Venice, 1485. RM (under Scott, M.)

1129. ORESME, Nicole. (Main entry: 708) 1477. Sart. 3, 1496; Klebs 29.9, 714; Still. 201, 783.

1130. ORTOLFF VON BAYRLANDT. (Main entry: 733) 1477. Sart. 3, 1206; Klebs 715, 716; Still. 466.

1131. ROBERT THE ENGLISHMAN (first of the name). (Main entry: 628) 1477. Sart. 2, 993; Klebs 850; Still. 884.

1132. SWINESHEAD, Richard. (Main entry: 693) ca. 1477. Sart. 3, 738; Klebs 943; Still. 794.

1133. VINCENT OF BEAUVAIS. (Main entry: 589) 1477. Sart. 2, 931; Klebs 1036, 1037; Still. 550, 712. His works are generally listed under Vincentius *Bellovacensis*.
   1. *The myrrour of the worlde*. Westminster, 1481. PR 24762 and a later ed. Trans. and publ. by W. Caxton.
   2. *The myrrour and dyscrypcyon of the worlde*. London, [1529?] PR 24764.
   3. *Speculum maius*. 4 vols. Douai, 1624. Reprinted, Graz, 1964.

1134. HUGH OF SIENA (Ugo Benzi). (Main entry: 741) 1478. Sart. 3, 1196; Klebs 997-1003; Still. 300-309.

1135. LANFRANK OF MILAN. (Main entry: 642) 1479. Sart. 2, 1081; Klebs 585-587; Still. 437, 438.

1136. SAVONAROLA, (Giovanni) Michele. (Main entry: 751) 1479. Sart. 3, 1197/8; Klebs 882-886; Still. 512-516.

1137. BURIDAN, Jean. (Main entry: 683) ca. 1480? Sart. 3, 545; Still. 742.
   1. *Quaestiones super octo "Physicorum" libros Aristotelis*. Paris, 1509. RM. Reprinted, Frankfurt, 1964.
   2. *In "Metaphysicam" Aristotelis quaestiones*. Paris, 1518. Reprinted, Frankfurt, 1964.

1138. MARSILIUS OF INGHEN. (Main entry: 716) ca. 1480. Sart. 3, 1436/7; Klebs 667; Still. 777, 778.
   1. *Quaestiones subtilissimi super octo libros "Physicorum" secundum nominalium viam*. Lyons, 1518. Reprinted, Frankfurt, 1964.

1139. THEMO. (Main entry: 703) 1480. Sart. 3, 1540; Klebs 959; Still. 798.

1140. HEYTESBURY, William. (Main entry: 699) 1481. Klebs 512-515; Still. 762-765.

1141. ANGELO DA FOSSOMBRONE. (Main entry: 740) ca. 1482. Klebs 70; Still. 729.

1142. BLASIUS OF PARMA. (Main entry: 724) 1482. Sart. 3, 1566; Klebs 739; Still. 785.

1143. JACOPO DA SAN MARINO. (Main entry: 718) 1482. Still. 767.

1144. PECHAM, John. (Main entry: 620) 1482? Sart. 2, 1029; Klebs 738; Still. 205.

1145 AILLY, Pierre d'. (Main entry: 727) 1483. Sart. 3, 1148; Klebs 766-770; Still. 88,89.
   1. *Tractatus de imagine mundi* [and other works]. Louvain, 1483. RM.
   2. *De eis quae in tribus regionibus aeris fiunt*. Leipzig, 1495. RM.

1146. ALFONSO EL SABIO. (Main entry: 611) 1483. Sart. 2, 840; Klebs 50; Still. 14.
   1. *Celestium motuum tabulae, nec non stellarum fixarum longitudines ac latudines Alfontii tempore ad motus veritatem ... reductae*. Venice, 1483. RM.

1147. BELDOMANDI, Prosdocimo de. (Main entry: 747) 1483. Klebs 167; Still. 146.

1148. GILES OF CORBEIL. (Main entry: 563) 1483. Sart. 2, 441; Klebs 464-466; Still. 391, 648.

1149. HUGH OF ST. VICTOR. (Main entry: 546) 1483. Klebs 207.1.

1150. JOHN OF LIGNÈRES. (Main entry: 686) 1483. Sart. 3, 651; Klebs 167.1; Still. 182.

1151. JOHN OF SAXONY. (Main entry: 687) 1483. Sart. 3, 677; Klebs 41.3, 50.1; Still. 48.

1152. FIRMIN DE BEAUVAL. (Main entry: 691) 1485. Sart. 3, 658; Klebs 406; Still. 753.

1153. HENRY BATE OF MALINES. (Main entry: 636) 1485. Sart. 2, 995; Klebs 4.1; Still. 823.

1154. THEOBALDUS. (Main entry: 524) 1487. Klebs 956, 957.

1155. ANIANUS. (Main entry: 623) 1488. Sart. 2, 992; Klebs 71; Still. 134.

1156. CUSA, Nicholas. (Main entry: 761) 1488. Klebs 700; Still. 81, 198, 779.
   1. *Opera*. Strasbourg, 1488. Reprinted, 2 vols, Berlin, 1967.
   2. *Opera*. Paris, 1514. Reprinted, 3 vols, Frankfurt, 1962.

1157. DINO DEL GARBO. (Main entry: 647) 1489. Sart. 3, 837/8; Klebs 336; Still. 351-353.

1158. ESTWOOD, John. (Main entry: 700) 1489. Sart. 3, 673; Klebs 381; Still. 52, 360, 749.

1159. LEOPOLD OF AUSTRIA. (Main entry: 603) 1489. Sart. 2, 996; Klebs 601; Still. 71, 771.

1160. TORRIGIANI, Torrigiano dei. (Main entry: 657) 1489. Sart. 3, 839; Klebs 983; Still. 538.

1161. JACOBUS ANGELUS DE ULMA. (Main entry: 736) 1490. Sart. 3, 1145; Klebs 69.

1162. BEDE. (Main entry: 511) 1491. Sart. 1, 511; Klebs 164.
   1. *De natura rerum et temporum ratione libri duo*. Basel, 1529. RM.
   2. *Axiomata philosophica*. London, 1592. PR 1777.
   3. *Opera*. London, 1693. Wing B1658.

1163. BONATTI (or BONATUS), Guido. (Main entry: 593) 1491. Sart. 2, 989; Klebs 195; Still. 31.
   1. *Decem tractatus astronomiae*. Augsburg, 1491. Ed. by J. Angelus. RM.

1164. JOHN OF GADDESDEN. (Main entry: 670) 1492. Sart. 3, 882; Klebs 424.

1165. URBANO OF BOLOGNA. (Main entry: 690) 1492. Sart. 3, 519; Klebs 1006; Still. 802.

1166. RUFFO, Giordano. (Main entry: 604) 1493. Sart. 2, 1076; Klebs 868; Still. 508.

1167. SIMON DE LENDENARIA. (Main entry: 754) 1494. Klebs 514.2; Still. 793.

1168. BRADWARDINE, Thomas. (Main entry: 682) 1495. Sart. 3, 671; Klebs 208, 209; Still. 152, 153, 740.

1169. JORDANUS DE NEMORE. (Main entry: 575) 1496. Sart. 2, 615; Klebs 563; Still. 187, 769.
1. *De ponderibus propositiones XIII et earumdem demonstrationes*. Nuremberg, 1533. Ed. by Peter Apian. RM.
2. *Opusculum de ponderositate*. Venice, 1565. Ed. by N. Tartaglia. RM.

1170. PETER OF AUVERGNE. (Main entry: 634) 1497. Klebs 771.

1171. PLATEARIUS, Matthaeus. (Main entry: 550) 1497. Sart. 2, 241; Klebs 911.2.

1172. BORGOGNONI, Theodoric. (Main entry: 599) 1498. Sart. 2, 655; Klebs 494.

1173. GROSSETESTE, Robert. (Main entry: 572) 1500. Sart. 2, 585; Klebs 851; Still. 176, 761.

1173A. JOHN OF MURS. (Main entry: 688) 1500? Sart. 3, 655/6; Klebs 554; Still. 183.

1174. WALTER OF METZ. (Main entry: 600) 1501. Sart. 2, 591.

1175. JOHN OF CASALI. (Main entry: 712) 1505. Sart. 3, 739; Still. 760.

1176. LA FONTAINE, Jean de. (Main entry: 749) 1505? Sart. 3, 1135; Still. 841.

1177. BERTUCCIO, Nicolo. (Main entry: 665) 1509. Sart. 3, 847; Still. 313, 602.

1178. WALAFRID STRABO. (Main entry: 516) 1510. Sart. 1, 570; Still. 698.
1. *Hortulus*. Vienna, 1510. Describes plants grown in the garden of the abbey of Reichenau.
1a. ——— *Des Walafrid von Reichenau "Hortulus."* Munich, 1926. Ed. by K. Sudhoff et al. (Listed under Strabo,W.)

1179. MARBODE. (Main entry: 531) 1511. Sart. 1, 765.
1. *De lapidibus pretiosis*. Freiburg i.B., 1531. RM.

1180. JOHN OF RUPESCISSA. (Main entry: 702) 1514. Sart. 3, 1573; Still. 433, 853.
1. [*De consideratione quintae essentiae rerum omnium.*] *La vertu et propriété de la quinte essence de toutes choses*. 1549. Reprinted, Milan, 1971.

1181. CONSTANTINE THE AFRICAN. (Main entry: 530) 1517. Sart. 1, 769.

1182. CAMPANUS OF NOVARA. (Main entry: 615) 1518. Sart. 2, 986; Still. 39.

1183. PETER PEREGRINUS. (Main entry: 619) 1520 or earlier. Sart. 2, 1031; Still. 787, 871.

1184. BACON, Roger. (Main entry: 610) 1530? (or 1541) "In spite of his many writings, no incunabula are known and early sixteenth-century imprints are rare." -Still. Sart. 2, 963; Still. 591, 820.

1. *This boke doth treate all of the best waters artyfycialles.* London. [1530?] PR 1180. Attribution incorrect?
2. *De retardandis senectutis accidentibus.* Oxford, 1590. PR 1181.

2a. ——— *The cure of old age.* London, 1683. Wing B372.

3. *The mirror of alchemy.* London, 1597. PR 1182. Attribution doubtful.
4. *Specula mathematica.* Frankfurt, 1614. Ed. by J. Combach. RM.
5. *Perspectiva.* Frankfurt, 1614. Ed. by J. Combach. RM.
6. *The famous historie of Fryer Bacon.* London, 1627. PR 1183 and another ed.
7. *Frier Bacon his discovery of the miracles of art.* London, 1659. Wing B373.
8. *Opus majus.* London, 1733. Ed. by S. Jebb. The 1st ed. RM.

1185. WILLIAM OF CONCHES. (Main entry: 538) 1531. Sart. 2, 198.

1. *Dialogus de substantiis physicis.* Strasbourg, 1567. Reprinted, Frankfurt, 1967. (Listed under Guillaume de C.)

1186. CHAUCER, Geoffrey. (Main entry: 723) 1532 (His non-scientific works were printed much earlier). Sart. 3, 1424; Still. 832.

1187. HILDEGARD OF BINGEN. (Main entry: 547) 1533. Sart. 2, 387; Still. 404, 655.

1188. WITELO. (Main entry: 621) 1535. Sart. 2, 1028; Still. 254.

1189. PETRUS BONUS. (Main entry: 675) 1546. Sart. 3, 752.

1. *Pretiosa margarita novella.* See entry 1199.

2.152 Imprint Sequence

1190. *SECRETUM secretorum.* (Main entry: 774) ca. 1472. Sart. 1, 556/7; Klebs 96; Still. 584.

1. *The secrete of secretes of Arystotle.* London, 1528. PR 770. The attribution to Aristotle is incorrect. Reprinted, New York, 1970.
2. *Secretum secretorum. Nine English versions.* 2 vols. Oxford, 1977. Ed. by M.A. Manzalaoui.

1191. ALBERTUS MAGNUS. Supposititious works.

1. *Secreta mulierum et virorum, cum commento.* Cologne, ca. 1475. Klebs 26; Still. 262. There are many later eds with varying titles. PR 273 record a London ed. of 1485.

1a. ——— *Daraus man alle Heimligkeit dess weiblichen Geschlechts erkennen kan....* Frankfurt, 1581. Reprinted, Stuttgart, 1966.

2. *Liber aggregationis; seu, Liber secretorum de virtutibus herbarum, lapidum, et animalium quorundam.* Ferrara, 1477. Klebs 18; Still. 261. Many later eds. PR 258 record an undated London ed.

2a. ——— *The boke of secretes.* London, n.d. (mid-16th century). PR 260 and eight later eds up to 1627.

2b. ——— ——— Reprint of 1560 ed. New York, 1969.

2c. ——— *The book of secrets of the virtues of herbs, stones, and certain beasts; also, A book of the marvels of the world.* Oxford, 1973. Ed. by M.R. Best.

1192. *MACER floridus.* (Main entry: 771) 1477.
   1. *Macer floridus de virtutibus herbarum carmen.* Naples, 1477. Sart. **1**, 765 (under Odo of Meung, to whom the work is sometimes attributed); Klebs 636, 637; Still. 442.
   1a. ——— Another ed. Geneva, 1495. Reprinted, ib., 1970.
   1b. ——— *Macer floridus. De materia medica.* Frankfurt, 1540. Ed. by J. Cornarius. RM.

1193. SCHOOL of Salerno. (Main entry: 783) 1480. Klebs 829-831; Still. 498, 499.
   1. *Regimen sanitatis salerni.* PR (21596-21610) record several eds of three English trans of 1528, 1607, and 1613; cf. Wing V384. The 1607 trans. has been reprinted: New York, 1920.

1194. *GEOMETRIA deutsch.* Nuremberg, 1489. Sart. **3**, 642; Klebs 454. The original possibly derives from the 12th or 13th century.

1195. *TETRAGONISMUS; id est, Circuli quadratura. Per Campanum, Archimedem ... atque Boetium. Ed. Luca Gaurico.* Venice, 1503. Sart. **2**, 986. RM (under Campano Novarese [i.e. Campanus of Novara]).

1196. *ANATOMIA porci.* Lyons, 1523. Sart. **1**, 770; **2**, 237; Still. 569. A 12th-century account of the dissection of a pig. Sometimes attributed to Nicholas Copho.

1197. ARISTOTLE. Suppositious works.
   1. [*De astronomia.*] *De cursione lune. Here begynneth the course of the moone.* London, [1530?] PR 768 and another ed.
   2. *Aristotles master-piece.* London, 1684. Wing A3689 and two later eds.
   3. *Aristotle's legacy.* London, 1699. Wing A3688A.

1198. *ALGORITHMUS demonstratus.* Nuremberg, 1534. Sart. **2**, 616. Derives from the 13th century.

1199. *PRETIOSA margarita novella.*
   1. *Pretiosa margarita novella de thesauro ac pretiosissimo philosophorum lapide. Ed. J. Lacinius.* Venice, 1546. Sart. **3**, 752. A condensation by Giano Lacinio of an alchemical work originally written by Petrus Bonus in 1330.
   1a. ——— *The new pearl of great price.* London, 1894; reprinted 1963. Trans. by A.E. Waite.

2.2 AUTHORS AND BOOKS OF THE PERIOD
ca. 1450–1700

2.21 Author Sequence

Except for special cases mathematical practitioners (i.e. exponents of practical mathematics) are not included after the early sixteenth century, nor are mathematical textbooks (which are very numerous) unless the author also published original work or the textbook had some special significance. Instrument makers and horologists are not included except in special cases.

1200. DECEMBRIO, Pietro Candido. 1399–1477. It. ICB; Klebs; GA. Author of *Tierbuch*, 1460; *Genitura hominis*, 1474.

1201. PUFF (VON SCHRICK), Michael. 1400–1473. Aus. Klebs (SCHRICK); Still.; BLA; GC (SCHRICK). Author of a distillation book, 1477; many later editions.

1202. ASHTON, Thomas de. b. 1403. Br. GD. Alchemy.

1203. ALBERTI, Leone Battista. 1404–1472. It. The famous Renaissance figure. DSB; ICB; Klebs; Still.; P I; G. He wrote on various aspects of the physical sciences and natural history. His best known work is the following.
    1. *De re aedificatoria*. Florence, 1485. Written ca. 1452 and first printed in 1485.
    1a. ——— *L'architectura*. 2 vols. Milan, 1966. Latin text with Italian trans. by G. Orlandi.

1204. NORTON, Thomas. fl. 1436–1477. Br. ICB; GD. Alchemy.
    1. *The ordinall of alchimy*. The original English version, written in 1477, was first printed in 1652 in Ashmole's *Theatrum chemicum Britannicum* (2220/1). A Latin trans. was published at Frankfort in 1618.
    1a. ——— *Facsimile reproduced from "Theatrum chemicum Britannicum" with Ashmole's annotations*. London, 1928. Introd. by E.J. Holmyard.
    1b. ——— *The ordinal of alchemy*. London, 1975. Ed. by J. Reidy.

1205. SALADINO DA ASCOLI. fl. 1441–1463. It. ICB; P II; BLA. Pharmacy.

1206. BUONINCONTRO (or BONINCONTRO), Lorenzo. 1410–ca. 1502. It. Klebs (BONINCONTRI); P I; GA. Astronomy.

1207. FRANCESCA. Piero della. ca. 1415–1492. It. The Renaissance artist. DSB; ICB; G. Mathematics.
    1. *Trattato d'abaco*. Pisa, 1970. Ed. by G. Arrighi. (Listed under Franceschi.)

1208. RIPLEY, George. 1415-1490. Br. ICB; P II; GD. Alchemy. (PR list a work written in 1471 but not printed until 1591.) See also 2831.

1209. BATECUMBE (or BADECUMBE), William. d. 1487? Br. GA; GD. Astronomy.

1210. BIANCHINI (or BLANCHINUS), Giovanni. fl. 1458. It. ICB; Klebs; Still.; P I. Astronomy.

1211. GEORGE OF HUNGARY. ca. 1422-1502. Klebs (GEORGIUS).
 1. *Arithmeticae summa*. 1499. X. Reprinted, Nieuwkoop, 1965. The first arithmetic printed in the Netherlands.

1212. PEURBACH (or PURBACH), Georg von. 1423-1461. Aus. DSB Supp.; Klebs; Still.; P II; GA; GC. Mathematics. Astronomy.
 1. *Theoricae novae planetarum*. Nuremberg, 1474. RM.
 2. *Epitoma in "Almagestum" Ptolemaei*. See 1226/3.
 3. *Libellus de quadrato geometrico*. See 1226/5.
 See also 1433/1 and 2010/8.

1213. MARLIANI, Giovanni. ca. 1425-1483. It. DSB; ICB; Klebs; Still.; P II. Mechanics. Study of heat.

1214. PONTANO, Giovanni Gioviano. 1427-1503. It. ICB; P II; GA; GB. Astronomy.

1215. LEONICENO, Niccolò. 1428-1524. It. DSB; ICB; Klebs; Still.; Mort.; BLA & Supp. One of the chief pioneers in the recovery and editing of Greek medical texts.

1216. VERSOR, Johannes. d. ca. 1485. Ger. ICB; Klebs; GC. Natural philosophy.

1217. CARRARA, Giovanni Michele Alberto. d. 1490. It. ICB; Klebs; BLA; GA. Author of *De constitutione mundi* and medical works.

1218. ANGULO, Luis de. d. before 1493. Fr. GF. Astrology.

1219. CATO, Angelus. fl. 1450-1495. It. Klebs (CATONE); Still. Astronomy.

1220. PFLAUM, Jakob. fl. 1450-1500. Ger. Klebs; Still. Astronomy.

1221. KETHAM, Joannes de. fl. 1460. d. ca. 1490. Ger. ICB; Klebs; Still.; Mort.; BLA. Anatomy.
 1. *Fasiculus medicinae*. Venice, 1491. X. Reprinted in facsimile with English trans. by C. Singer. Milan, 1924. The first anatomical book to be illustrated.

1222. MANFREDI, Girolamo (or Hieronymo). ca. 1430-1493. It. ICB; Klebs; Still.; BLA. Anatomy.

1223. WALTHER, Bernhard. 1430-1504. Ger. P II; GB; GC. Astronomy. Assistant to Regiomontanus.

1224. ARNALDUS DE BRUXELLA. ca. 1432-ca. 1492. ICB. Alchemy.

1225. FICINO, Marsiglio. 1433-1499. It. The Platonic philosopher. ICB; Klebs; Still.; P I; BLA Supp.; GA; GB. Astrology, etc.

1226. REGIOMONTANUS (or MÜLLER), Johannes. 1436-1476. Ger./Aus./It./Hung. DSB; ICB; Klebs; Still.; P II; GB; GC (MÜLLER). Astronomy. Mathematics.

1. *Calendarium novum*. Nuremberg, 1474. X. Another ed., Venice, 1476. RM.
1a. ——— German ed. Nuremberg, 1474. X. Reprinted, Leipzig, 1937.
2. *Tabulae directionum profectionumque*. Augsburg, 1490. X. Another ed., Venice, 1504. RM. Another ed., Wittenberg, 1584.
3. *I. Regiomontani et G. Purbachii Epitoma in "Almagestum" Ptolemaei*. Venice, 1496. Begun by Peurbach and completed after his death by Regiomontanus ca. 1462, but not printed until 1496.
4. *De triangulis omni modis*. Nuremberg, 1533. First ed. RM.
4a. ——— *Regiomontanus on triangles*. Madison, Wis., 1967. Ed. with introd. and notes by B. Hughes. Includes a facsimile of the Latin ed. of 1533.
5. *Scripta ... de torqueto, astrolabio, armillari, regula magna Ptolemaica, baculoque astronomico, et observationibus cometarum ...* [To which is added] *Libellus m. G. Purbachii quadrato geometrico*. Nuremberg, 1544. RM.
6. *Opera collectanea. Faksimiledrucke von neun Schriften*. Osnabrück, 1972. Ed. by F. Schmeidler.

1227. MARTINI, Francesco di Giorgio (or FRANCESCO DI SIENA). 1439-1501. It. DSB; ICB. Architecture. Mechanical technology.
1. *La praticha di geomoetria*. Florence, 1970. Ed. by G. Arrighi.

1228. ADAM, Jehan. late 15th century. ICB. Author of an arithmetic, 1475.

1229. WAGNER, Ulrich. d. ca. 1490. Ger. ICB; Klebs; Still. Author of *Rechenbuch*, 1482, the first German arithmetic to be printed.

1230. ARTALDO, Giovanni. d. 1493. ICB; Klebs. Author of *De motu gravium et levium*, 1500.

1231. LEWIS OF CAERLEON. d. 1494? Br. ICB; GD (under CHARLTON, Lewis, with whom he is sometimes confused). Astronomy. Mathematics.

1232. TOLHOPF (or TOLOPHUS), Johann. d. 1503. Ger. ICB. Astronomy.

1233. BRICOT, Thomas. fl. 1475. Fr. Klebs; Still. Natural philosophy. Principles of dynamics.

1234. ECK, Paulus. fl. 1479. Klebs; Still. Astrology.

1235. ZERBI. Gabriele de. ca. 1440-1505. It. ICB; Klebs; Still.; BLA (ZERBIS); GA. Anatomy.

1236. WODKA, Mikolaj. ca. 1442-1494. Poland. ICB. Astronomy.

1237. BENIVIENI, Antonio. 1443-1502. It. DSB; ICB; Still.; BLA & Supp. Anatomy.

1238. FERRARI (or GALATEO), Antonio. 1444-1517. It. ICB (GALATEO); P I; GA. Earth sciences.

1239. BRUDZEWO, Albertus de. 1445-1497. Klebs (ALBERTUS DE B.); Still. Astronomy.

1240. DATI, Giuliano. 1445-1524. It. Klebs; GA. Astronomy.

1241. PACIOLI (or PATIULUS). Luca. ca. 1445-1517. It. DSB; ICB; Klebs; Still.; P II; GA (PACCIOLI). Mathematics.
 1. *Divina proportione*. Venice, 1509. RM.

1242. PAUL OF MIDDELBURG. 1445-1534. Holl./It. P II; GA; GC. Under PAULUS: ICB; Klebs; BLA. Astrology.

1243. HUNDT (or HUND or CANIS), Magnus. 1449-1519. Ger. DSB; Klebs; Still.; BLA; GA. Anatomy.

1244. AUSLASSER, Vitus. late 15th century. Ger. ICB. Author of a herbal, 1479.

1245. SANCT CLIMENT, Francesch. late 15th century. Spain. ICB; Klebs (FRANCESCH); Still. Author of the first arithmetic printed in Spain, 1482.

1246. GRANOLLACHS, Bernat de. late 15th century. Klebs; Still. Author of *Lunarium, 1485-1550*. 1485; many later eds.

1247. GEORGES DE BRUXELLES. late 15th century. Fr. Klebs; Still. Natural philosophy. Principles of dynamics.

1248. TRISMOSIN, Solomon. fl. 1473-1520. Ger. ICB; GC. Alchemy.
 1. *Splendor solis. Alchemical treatises*. London, 1920. With introd. and notes by J.K.
 2. *La toison d'or; ou, La fleur des trésors*. Paris, 1975. Text of the French ed. of 1612. Ed. by B. Musson.

1249. FABER VON BUDWEIS, Wenzel. fl. 1475-1495. Ger. Klebs; Still. Astronomy/astrology.

1250. WILLEM GILLISZOON OF WISSEKERKE. fl. 1476-1492. Holl. ICB (GILLIZOON); Klebs; Still.; P I (AEGIDIUS, Wilhelm). Astronomy.
 1. *Liber desideratus super celestium motuum indagatione sine calculo*. 1494. X. Reprinted, Nieuwkoop, 1965. (Dutch classics on history of science. 12) Generally listed under Aegidius, G.

1251. LEONARDI, Camillo. fl. 1480. It. Klebs; Still.; BMNH (LEONARDUS). Mineralogy. Astronomy.
 1. *Speculum lapidum*. Venice, 1502. RM.
 1a. ——— *The mirror of stones. In which the nature ... of more than 200 different jewels, precious and rare stones are distinctly described*. London, 1750. RM/L.

1252. VERNIAS (DA CHIETI), Nicoletto. fl. 1480-1490. It. Klebs (VERNIA); Still. Natural philosophy.

1253. SANTRITTER, Johannes. fl. 1480-1492. Ger./It. Klebs; GC. Author of *Canones in tabulas Alphonsi*, 1492.

1254. CHUQUET, Nicolas. fl. 1484-1493. Fr. DSB; ICB. Algebra.

1255. BORGHI, Pietro. fl. 1484-1494. It. ICB; Klebs; Still. P I (BORGO). Arithmetic.
 1. *Arithmetica*. Venice, 1484. In Italian. X. Reprinted, Munich, 1964.

1256. CUBE, Johan von. fl. 1484-1503. Ger. ICB; BMNH; BLA & Supp. (CUBA); GC. Author of a herbal, 1485.

1257. JOHANNES VON GLOGAU. d. 1507. Cracow. Klebs; P I (JOANNES DE G.), P III (GLOGOVIENSIS, J.). Mathematics. Astronomy. Astrology.

1258. DANTE, Pietro Vincenzo. d. 1512. It. P I. Astronomy.

1259. BENEDETTI, Alessandro. ca. 1450-1512. It. DSB; ICB; Klebs; Still.; BLA & Supp.; GA. Anatomy.

1260. BRUNSCHWIG (or BRUNSWYCK, etc.), Hieronymus. ca. 1450-ca. 1512. Strasbourg. DSB; ICB; Klebs; Still.; Mort.; BLA & Supp. (BRAUNSCHWEIG). Anatomy. Pharmacy.
   1. *Liber de arte distillandi de simplicibus.* Strasbourg, 1500. X.
   1a. ——— *Grosses Buch der Destillation.* Ib., 1512. X. Reprinted, Leipzig, 1972.
   1b. ——— *The vertuose boke of distyllatyon.* London, 1527. PR 13435. Reprinted, New York, 1971.
   2. *Apoteck für den gemainen Man.* Strasbourg, 1507. X.
   2a. ——— *A most excellent homish apothecarye.* London. 1561. PR 13433. X. Reprinted, Amsterdam, 1968.

1261. POLLICH (VON MELLERSTADT), Martin. ca. 1450-1513. Ger. ICB; Klebs; Still.; BLA; GA. Astrology.

1262. SIMON DE PHARES. ca. 1450-after 1499. Fr. DSB; ICB; GA (PHARES). Astrology.

1263. VITTORI (DA FAENZA), Leonello dei. ca. 1450-1520. It. ICB; BLA (VETTORI). Anatomy.

1264. ZACUTO (or ZACUT), Abraham bar Samuel bar Abraham. ca. 1450-ca. 1522. Spain. DSB; ICB; Klebs; Still.; GA. Astronomy. Astrology

1265. LEONARDO DA VINCI. 1452-1519. It. DSB; ICB; P I; Mort.; BLA Supp. G. Anatomy. Mechanics. Mathematics. Geology. Mechanical technology.

Anatomy

   1. *Il trattato della anatomia.* 3 vols. Rome, 1962.
   2. *The anatomical, physiological, and embryological drawings.* New York, 1952. With trans., notes, etc., by C.D. O'Malley and J.B. Saunders.
   3. *Dessins anatomiques: Anatomie artistique, descriptive, et functionelle.* Paris, 1961. Ed. by P. Huard.
   4. *Anatomical drawings from the Queen's collection at Windsor Castle.* Los Angeles, 1976.

Notebooks

   5. *The notebooks.* London, 1906. Trans. and ed. by E. MacCurdy. Reprinted 1938 and 1956.
   6. *Selections from the notebooks.* London, 1952. Ed. by I.A. Richter.
   7. *The Madrid codices. National Library, Madrid.* 5 vols. New York, 1974.

1266. SAVONAROLA, Girolamo. 1452-1498. It. The preacher and reformer. ICB; Klebs; Still.; P II; G. Astrology.

1267. STÖFFLER, Johann. 1452-1531. Ger. ICB; Klebs; Still.; P II; GA; GC. Astronomy.

1. *Elucidatio fabricae ususque astrolabii*. Oppenheim, 1513.
1a. ——— *Traité de la composition et fabrique de l'astrolabe et de son usage*. Paris, 1560. RM.
2. *Tabulae astronomicae*. Tübingen, 1514. RM.
3, *In Procli Diadochi ... "Sphaeram mundi" ... commentarius*. Tübingen, 1534. RM.

1268. TORNI, Bernardo. 1452-1497. It. ICB; Klebs; Still. Dynamics.

1269. ANGELUS (or ENGEL), Johannes. 1453?-1512. Aus. DSB; P I. Under ENGEL: ICB; Klebs; Still. Astronomy/astrology. Wing (M3039) records an English trans., made in 1655, of an astrological work (which is incorrectly attributed to Johann Müller, i.e. Regiomontanus). See also 1062/3 and 1163/1.

1270. BEROALDO, Filippo. 1453-1505. It. Klebs; BLA; GA.
1. *Annotationes centum [in Plinium]*. Brescia, 1496.

1271. AUGURELLI, Giovanni Aurelio. 1454-1537. It. P III; GA. Alchemy.

1272. BARBARO, Ermolao. 1454-ca. 1493. It. ICB; Klebs; Still.; P I; BMNH (BARBARUS. H.); BLA; GA. Mathematics. Astrology. Author of commentaries on the writings of Pliny, Dioscorides and other ancients.

1273. NOVARA, Domenico Maria. 1454-1504. It. DSB; ICB (MARIA DE NOVARA); Klebs; P II. Astronomy.

1274. LE FÈVRE (D'ÉTAPLES) (or FABER STAPULENSIS), Jacques. ca. 1455-1536. Fr. Klebs; Still.; P I; GA; GB. Arithmetic. Astronomy.
1. *The prefatory epistles and related texts*. New York, 1972. Ed. by E.F. Rice.

1275. LICHTENBERGER, Johann. 1458-1510. Ger. ICB; Klebs; Still.; P I; GA; GB; GC. Astrology.

1276. MAXIMILIAN I. (Holy Roman Emperor.) 1459-1519. Aus. ICB; GA; GB; GC.
1. *Das Tiroler Fischereibuch*. 2 vols. Graz, 1967. Ed. by W. Hohenleiter. Written in 1504, apparently by Maximilian himself. He was a patron of the sciences and the arts.

1277. CALANDRI, Filippo. late 15th century. It. Klebs; Still. Author of an arithmetic, 1491.

1278. CODRONCHI, Masino. late 15th century. It. Klebs (MESSINUS); Still. Author of *De motu locali*, 1494.

1279. TARTARET, Pierre. fl. 1490. Fr. Klebs; Still. Natural philosophy. Principles of dynamics.

1280. AUGUSTONE, Giovanni Basilo. fl. 1490-1500. Klebs; Still. Astrology.

1281. SCHREIBER (or GRAMMATEUS), Heinrich. fl. 1490-1510. Ger. Still.; P III. Mathematics. Astronomy. Author of the first German textbook of algebra.

1282. REISCH, Gregor. d. 1525. Ger. ICB; Klebs; Still.; P II.
1. *Margarita philosophica*. Strasbourg, 1503. X. A popular encyclopaedia which included mathematics, astronomy, and optics. Many later eds. Reprinted, Düsseldorf, 1973.

1283. BERENGARIO DA CARPI, Giacomo. ca. 1460-1530. It. DSB; ICB; Still.; Mort.; BLA & Supp. Anatomy.
   1. *Isagogae breves ... in anatomia humani corporis*. Bologna, 1522. X.
   1a. ——— *A description of the body of man*. London, 1664. Wing B1959.
   1b. ——— *A short introduction to anatomy*. Chicago, 1959. Trans. with introd. and notes by L.R. Lind and P.G. Roofe.

1284. MANSSON, Peder. ca. 1460-1534. ICB. Chemical technology.

1285. TORRELLA, Girolamo. ca. 1460-1496. Klebs; Still. Astrology.

1286. KÖBEL (or KOBEL or KOBELIN), Jacob. ca. 1462-1533. Ger. DSB; Still.; P I; GC. Mathematics.

1287. MANARDO, Giovanni. 1462-1536. It. DSB. Under MANARDUS. J.: ICB; BMNH. Under MANARDI, G.: Still; BLA. Botany.

1288. POMPONAZZI, Pietro. 1462-1525. It. DSB; ICB; Still.; P II; GA; GB. Natural philosophy.
   1. *De naturalium effectuum causis*. Basel, 1567. X. Reprinted, Hildesheim, 1970.

1289. TRITHEMIUS (or TRITHEIM), Johannes. 1462-1516. Ger. ICB; GA; GB; GC. Alchemy.

1290. WIDMAN (or WEIDEMAN or WIDEMAN), Johannes. ca. 1462-after 1498. Ger. DSB; Still.; GC. Mathematics.

1291. ACHILLINI, Alessandro. 1463-1512. It. DSB; ICB; Klebs; Still.; P I; BLA & Supp.; GA; GB. Anatomy.

1292. VIRDUNG, Johann. 1463-ca. 1540. Ger. ICB; GC. Astronomy.

1293. FERRO (or FERREO), Scipione. 1465-1526. It. DSB; P I; GA. Mathematics.

1294. AUGUSTINUS OLOMUCENSIS (or A. MORAVUS). ca. 1466-1513. Olmütz. Klebs; P I. Astronomy/astrology.

1295. ROSENBACH, Johann. 1467-1537. Ger. ICB. Astrology.

1296. RÜLEIN, Ulrich (or ULRICH RÜLEIN VON CALW). ca. 1467-1523. Ger. DSB; ICB; BLA. Geology. See also 2768/1.

1297. WERNER, Johann. 1468-1522. Ger. DSB; ICB; Still.; P II; GC. Astronomy. Mathematics. Earth sciences. See also 1320/3.

1298. MAIOR, John (or MAIR, Jean). 1469-1550. Br./Fr. DSB; ICB (MAJOR). Under MAIR: GA; GB; GD. Natural philosophy.

1299. THOMAS, Alvarus. early 16th century. ICB. Mathematics. Mechanics.

1300. BONET DE LATTES, Gabriel. fl. 1493-1534. Fr. Klebs; BLA; GA. Astrology/astronomy.

1301. POLITUS, Bassanus. fl. 1500. Still. Natural philosophy. Author of a commentary on the work of Swineshead.

1302. BOLOGNINI, Angelo. fl. 1508-1536. It. BLA & Supp. Anatomy.

1303. THOMAZ, Alvaro (or ALVARO TOMAS). fl. 1509-1513. Fr. DSB; Still. (THOME). Mathematics. Dynamics.

1304. PANTEO, Giovanni Agostino. fl. 1520. It. Klebs; Still. Alchemy.

1305. STABIUS, Johann. d. 1522. Ger./Aus. Klebs; P II; GC. Mathematics. Astrology.

1306. SUPERCHIO, Valerio. d. 1540. ICB; Klebs. Astronomy.

1307. ABIOSI, Giovanni Battista. ca. 1470-1523. Klebs; Still. Alchemy.

1308. BOUVELLES, Charles. ca. 1470-ca. 1553. Fr. DSB. Under BOUELLES: Still.; P I. Under BOVILLUS: ICB; GA. Mathematics.

1309. CIRUELO, Pedro Sánchez. 1470-1554. Spain. DSB; ICB; Klebs; Still.; P I; GA (CIRVELO). Mathematics.

1310. DULLAERT, Jean. ca. 1470-1513. Fr. DSB; Still. Natural philosophy. Dynamics.

1311. HEYNFOGEL (or HEINFOGEL), Conrad. ca. 1470-ca. 1530. Ger. P I. Astronomy.

1312. NIFO, Agostino. ca. 1470-1538. It. DSB; ICB; Klebs; Still.; GA; GB. Natural philosophy.

1313. STIBORIUS (or STÖBERL), Andreas. ca. 1470-1515. Ger./Aus. P II; GC. Astronomy.

1314. VERGILIUS, Polydorus. ca. 1470-1555. It./Br. ICB; Klebs; Still.; GB; GD. Author of *De inventoribus rerum*, 1499. "A popular medley of information, frequently reprinted. Relates to inventors and covers a wide variety of topics." -Still.

1315. DÜRER, Albrecht. 1471-1528. Ger. The famous artist. DSB; ICB; Still.; P I; G. Mathematics.
   1. *Underweysung der Messung*. Nuremberg, 1525. X. Facsimile reprint with English trans., entitled *Geometry and perspective*. Zurich, 1966.
   2. *Vier Bücher von menschlicher Proportion*. Nuremberg, 1528. X. Reprinted, Zurich, 1969, and London, 1970.

1316. RUSCELLI, Girolamo (known as ALEXIS OF PIEDMONT). 1471-ca. 1565. It. DSB Supp.; ICB (ALESSIO DA PIEMONTE); GA. Chemical technology.

1317. CHAMPIER, Symphorien. 1472-1539. Fr. ICB; Klebs; Still.; BMNH; BLA & Supp.; GA. Botany.

1318. CLICHTOVE, Josse. 1472-1543. Fr. ICB; Still.; P I; GA; GB. Optics. Arithmetic.

1319. GRISOGONO, Federico. 1472-1538. Dalmatia. DSB. Astrology.

1320. COPERNICUS, Nicolas. 1473-1543. Thorn, etc. DSB; ICB; Still.; P I & Supp.; G. Astronomy. See also Index.
   1. *De lateribus et angulis triangulorum, tum planorum rectilineorum, tum sphaericorum*. Wittenberg, 1542. Forms part of the first book of *De revolutionibus*. Ed. by G.J. Rheticus. RM.
   2. *De revolutionibus orbium coelestium*. Nuremberg, 1543. RM. There are several reprints and translations into English, French, and German.

3. *Three Copernican treatises: The commentariolus of Copernicus; The letter against Werner; The narratio prima of Rheticus.* New York, 1939. 2nd ed., 1959. 3rd ed., 1971. Trans. with introd. and notes by E. Rosen.
4. *Gesamtausgabe.* 2 vols. Munich, 1944-49. Ed. by F. Kubach.
5. *Complete works.* London, 1972-. Ed. by P. Czartoryski.

1321. BONAFIDE, Francesco. 1474-1558. It. BLA (BUONAFEDE); GA. Founder of the botanic garden at Padua, the first of its kind.

1322. CURTIUS, Matthaeus. ca. 1474-1544. It. ICB; BLA (CORTI). Anatomy. See also 1449/1.

1323. PEYLICK, Johann. 1474-1522. Ger. ICB; Klebs; Still.; BLA. Physiology. Anatomy.

1324. RUEL, Jean. 1474-1537. Fr. DSB; ICB; Still. (RUELLE); BMNH (RUELLIUS); BLA (RUELLE); GA. Botany.

1325. TUNSTALL, Cuthbert. 1474-1559. Br. DSB; ICB; Still.; P II (TONSTALL); GA; GB; GD. Mathematics. (Title in PR.)

1326. GAURICO, Luca. 1476-1558. It. Still.; P I; GA. Astronomy. Astrology.
1. *Tractatus astrologicus.* Venice, 1552. RM. See also 1195.

1327. DELFINO (or DELPHINUS), Federigo. 1477-1547. It. P I; BLA; GA. Astronomy.

1328. MAGGI, Bartolommeo. 1477 (or 1516) -1552. It. DSB; BLA; GA. Anatomy.

1329. SCHÖNER, Johannes. 1477-1547. Ger. DSB; Still. Under SCHONER: P II; GC. Astronomy. Geography.

1330. DUBOIS, Jacques (or SYLVIUS, Jacobus). 1478-1555. Fr. DSB; ICB (SYLVIUS); Still.; BLA & Supp.; GA. Anatomy, human and comparative.
1. *De mensibus mulierum et hominis generatione.* Basel, 1557.

1331. FRACASTORO, Girolamo. ca. 1478-1553. It. DSB; ICB; Still.; P I; BLA & Supp.; GA; GB. Natural philosophy. Astronomy, etc. Best known as a physician.
1. *Homocentria.* [To which is added] *De causis criticorum dierum.* Venice, 1538. RM.
2. *Opera omnia.* Ib., 1555. RM. 2nd ed., 1574. Another ed., Lyons, 1591.

1332. OVIEDO Y VALDES, Gonzalo Fernandez. 1478-1557. Spain/Central America. ICB; BLA; GA; GB. Under FERNANDEZ DE OVIEDO: Mort.; Still. Author of the first natural history of Central America, 1526.

1333. CALCAGNINI, Celio. 1479-1541. It. Still.; P I; GA. In a work of 1544 (possibly written before 1525) he presents "the thesis that the earth, having been given an impetus by nature, continues to revolve upon its own axis but inclines from side to side." -Still.

1334. LACHER, Ambrosius. early 16th century. Ger. P I. Author of *Tabulae resolutae de motibus planetarum,* 1511.

1335. WEIDITZ, Hans. fl. 1500-1536. Ger. ICB; GF. Botanical illustration.

1336. PERLACH, Andreas. fl. 1500-1550. Aus. Still. Astronomy.

1337. CORONEL, Luis Nuñez. d. 1531. Spain/Fr. DSB; Still. Natural philosophy. Dynamics.

1338. TZWYVEL. Dietrich. d. 1544. Ger. GC. Mathematics.

1339. VITALI, Ludovico. d. 1554. It. P II. Astronomy.

1340. BIRINGUCCIO, Vannocio. 1480-ca. 1539. It. DSB; ICB; Still.; P I; GA. Metallurgy.
1. *De la pirotechnia*. Venice, 1540. RM.
1a. ——— *The Pirotechnia*. New York, 1943; reprinted 1959. Trans. with introd. and notes by C.S. Smith and M.T. Gnudi.

1341. ENCLEN DE CUSA, Johannes. b. 1480. ICB. Author of *Algorismus proiectilium*, 1502.

1342. MONTAÑA, Bernardino. ca. 1480-ca. 1551. Spain. ICB; BLA. Author of the first anatomy in Spanish.

1343. NAUSEA, Fridericus. ca. 1480-1550. Aus. BMNH; GA; GC. Author of *De locustis*, 1544.

1344. ORTEGA, Juan de. ca. 1480-ca. 1568. Spain. DSB; ICB; Still. (JUAN DE O.). Mathematics.

1345. ZIEGLER, Jacob. 1480-1549. Ger. ICB; P II; GA; GC. Astronomy.

1346. TORRE, Marcantonio della. 1481-1511. It. DSB; BLA. Anatomy.

1347. TANNSTETTER (VON THANNAU), Georg. 1482-1535. Ger. P II; GC. Mathematics. Astronomy/astrology.

1348. CONTARINI, Gasparo. 1483-1542. It. P I; GA. Author of *De elementis et eorum mixtionibus*, 1548.

1349. GIOVIO, Paolo. 1483-1552. It. ICB; BMNH; GA; GB. Author of *De piscibus romanis*, 1524.

1350. LANGER, Johann. 1484-1548. Ger. P I. Astronomy.

1351. SCALIGER (or BORDONIUS), Julius Caesar. 1484-1558. It./Fr. The classical scholar. DSB; ICB; P II; BMNH; BLA; GA; GB. Natural philosophy. Botany. See also 999/1.
1. *In libros "De plantis" Aristoteli inscriptos commentarii*. Geneva, 1566. RM. The 1st ed. was at Paris in 1556.
2. *Commentarii et animadversiones in sex libros "De causis plantarum" Theophrasti*. Geneva, 1566. RM.

1352. BILDIUS, Beatus, *Rhenanus*. 1485-1547. BMNH. Author of a commentary on Pliny's *Natural history*.

1353. LANGE, Johann. 1485-1565. Ger. BMNH (LANGIUS); BLA & Supp.; GA; GC. Botany.

1354. MASSA, Niccolo. 1485-1569. It. DSB; ICB; Still.; Mort.; BLA; GA. Anatomy.

1355. AGRIPPA (VON NETTESHEIM), (Heinrich) Cornelius. 1486-1535. Several countries. DSB; ICB; P I; BLA & Supp.; GA; GB; GC. Magic. Alchemy. Best known as a Neoplatonic philosopher.
1. *De incertitudine et vanitate scientiarum atque artium declamatio*. Antwerp, 1530. X.

1a. ——— *Of the vanitie and uncertaintie of artes and sciences*. Northridge, Cal., 1974. Ed. by C.M. Dunn.
2. *Opera*. 2 vols. Lyons, 1531. RM. Reprinted, Hildesheim, 1970.
3. *De occulta philosophia*. Cologne, 1533. X. Reprinted, Graz, 1968.
3a. ——— *Three books of occult philosophy or magic*. Chicago, 1898. X. Reprinted, London, 1971.
See also 15483/1.

1356. CORDUS, Euricius. 1486-1535. Ger. DSB; Still.; BMNH; BLA; GA; GC. Botany.

1357. DUBRAVIUS, Janus. 1486-1553. Olmütz. BMNH; GA. Ichthyology.
1. *De piscinis et pisciū*. [Zurich?], 1559. RM.

1358. MICHIEL, Marcantonio. ca. 1486-1552. It. ICB. Astrology.

1359. KRATZER, Nicholas. 1487-ca. 1550. Ger./Br. ICB; GC; GD. Mathematics. Astronomy.

1360. LAX, Gaspar. 1487-1560. Spain/Fr. DSB; P I. Mathematics.

1361. OELLINGER, Georg. 1487-1557. ICB. Botany.
1. *Herbarium des Georg Oellinger, anno 1553 zu Nürnberg: 51 Original-Aufnahmen*. Salzburg, 1949. Selections from the drawings in a MS in the library of Erlangen University. By H. Retzlaff.

1362. STIFEL (or STYFEL, etc.), Michael. ca. 1487-1567. Ger. DSB; ICB; Still.; P II; GA (STIEFEL); GC. Mathematics.

1363. GLAREANUS (or LORITUS), Heinrich. 1488-1563. Switz. ICB; Still.; P I; GA; GC (LORITI). Mathematics.

1364. BRUNFELS, Otto. ca. 1489-1534. Ger. DSB; Still.; BMNH; BLA & Supp.; GA. Botany.
1. *Herbarum vivae cicones*. 2 vols. Strasbourg, 1530. RM. Another ed., 2 vols in 1, ib., 1532.
2. *Contrafayt Kreuterbuch*. 2 vols. Strasbourg, 1532. X. Reprinted, Wurzburg, 1965.

1365. HARTMANN, Georg. 1489-1564. Ger. DSB; P I; GC. Instrument making. Mathematics.

1366. RUDOLFF, Christoff. early 16th century. Aus. DSB; Still.; P III; GC. Mathematics. Author of the first comprehensive work in German on algebra.

1367. ULSTADT, Philipp. early 16th century. Switz. DSB; ICB; Still.; BLA; GC (ULSTED).
1. *Coelum philosophorum* Fribourg, 1525. X. Another ed., Lyons, 1572. Another ed., Strasbourg, 1630. A popular distillation book which went through more than twenty eds.

1368. GHALIGAI (or GALIGAI), Francesco. fl. ca. 1515. It. ICB; Still.; P I. Mathematics.

1369. LA ROCHE, Estienne de. fl. ca. 1520. Fr. DSB; Still. (ROCHE). Author of the first French arithmetic to be printed.

1370. ABATI, Baldo Angelo. fl. ca. 1530. It. BMNH (ABBATIUS); BLA. Zoology.

1371. ATROCIANUS, Johann. d. ca. 1540. Ger. BMNH; GA. Botany.

1372. VOEGELIN, Johannes. fl. 1520. d. 1549. Aus. Still.; GC. Mathematics. Astronomy.

1373. RÖSSLIN (or RÖSLIN or RHODION), Eucharius (the younger). d. 1553/4. Ger. ICB; BMNH; BLA.
 1. *On minerals and mineral products: Chapters on minerals from his "Kreutterbuch"* [1533]. Berlin, 1978. Critical text, English trans., and comm. by J.S. Belkin and E.R. Caley.

1374. CELAYA, Juan de. ca. 1490-1558. Spain/Fr. Natural philosophy. Logic. Dynamics.

1375. FRIES, Laurenz. ca. 1490-1531. Ger. ICB; Still.; BLA & Supp. (FRISIUS); GC. Anatomy.

1376. GHINI, Luca. ca. 1490-1556. It. DSB; ICB; BLA; GA. Botany.

1377. GILLES (or GILLIUS), Pierre. 1490-1555. Fr. GA. Zoology.

1378. MELLO, Francisco de. 1490-1536. Port. DSB; P II. Mathematics.

1379. PIGHIUS, Albert. ca. 1490-1542. Holl./It. P II; GA; GC. Astronomy.

1380. BUTEO, Johannes. ca. 1492-ca. 1570. Fr. DSB; P I & Supp. (BOTEO); GA. Mathematics.

1381. DORSTENIUS, Theodoricus (or DORSTEN, Dietrich). 1492-1552. Ger. BMNH; BLA & Supp. Botany.
 1. *Botanicon*. Frankfurt, 1540.

1382. RIES (or RIS, RISZ, etc.), Adam. 1492-1559. Ger. DSB. Under RIESE: ICB; Still.; P II. Mathematics.

1383. WOTTON, Edward. 1492-1555. Br. DSB; BMNH; BLA; GD. Zoology.
 1. *De differentiis animalium*. Paris, 1552. RM.

1384. MONARDES, Nicolas Bautista. ca. 1493-1588. Spain. DSB; ICB; BMNH; Mort.; BLA; GA. Natural history.
 1. *The three bookes written in the Spanishe tonge*. London, 1577. PR 18005. X. Trans. of the three parts of Monardes' natural history of America which were publ. at Seville in 1569-74.
 1a. ——— Another issue, entitled *Joyfull newes out of the newe founde worlde*. London, 1577. PR 18005A and two later eds. RM. Reprinted, ib., 1925.

1385. PARACELSUS (or HOHENHEIM, Theophrastus Philippus Aureolus Bombastus von). ca. 1493-1541. Switz./Ger./Aus. DSB; ICB; Still.; P II; Mort.; BLA & Supp.; GA; GB; GC (HOHENHEIM). Natural philosophy. Cosmology. Iatrochemistry. Magic. See also Index.

Individual Works and Small Collections

 1. *Volumen medicinae paramirum....* Baltimore, 1949. Trans. from the original German by K.F. Leidecker.
 2. *Liber de nymphis, sylphis ... et de caeteris spiritibus*. Bern, 1960. Ed. by R. Blaser.
 3. *Das Buch der Erkenntnis*. Berlin, 1964. Ed. from the MS by K. Goldammer.
 4. *The archidoxes of magic*. London, 1975. Introd. by S. Skinner.

Many 17th-century English translations are listed by PR and Wing.

Collected Works and Selections

5. *Bücher und Schriften*. 10 vols. Basel, 1589-90. Ed. by J. Huser. The first authoritative collected ed. X. Reprinted, 10 vols in 5, Hildesheim, 1971. Another ed., 2 vols, Strasbourg, 1616.
6. *Opera omnia medico-chemico-chirurgica ... Editio novissima*. 3 vols. Geneva, 1658. RM/L.
7. *Sämtliche Werke*. 14 vols. Munich, 1922-33. Ed. by K. Sudhoff. The definitive ed.
8. *The hermetic and alchemical writings*. 2 vols. London, 1894. Trans. and ed. by A.E. Waite. Reprinted, Berkeley, Cal., 1976.
9. *Four treatises*. Baltimore, 1941. Trans. and ed. by H.E. Sigerist et al.
10. *Die Geheimnisse: Ein Lesebuch aus seinen Schriften*. Leipzig, 1941. Ed. by W.E. Peuckert.
11. *Selected writings*. New York, 1951. Trans. by N. Guterman.
12. *Werke*. 5 vols. Basel, 1965-68. Ed. by W.E. Peuckert. Contains the most important works.

1386. AGRICOLA, Georgius (or BAUER, Georg). 1494-1555. Ger. DSB; ICB; Still.; P I; BMNH; BLA & Supp.; GA; GB; GC. Metallurgy. Mining.

1. *Bermannus; sive, De re metallica dialogus*. Basel, 1530. RM.
2. *De mensuris et ponderibus*. Ib., 1533. RM.
3. *Opuscula. (De ortu et causis subterraneorum. De natura eorum quae effluunt ex terra. De natura fossilium. De veteribus et novis metallis. Bermannus. Interpretatio Germanica vocorum rei metallicae.)* Ib., 1546. RM/L.

3a. ——— *De natura fossilium. (Textbook of mineralogy.)* New York, 1955. English trans. by M.C. Bandy.

4. *De animantibus subterraneis*. Basel, 1549. RM.
5. *De re metallica libri XII*. Ib., 1556. RM. Reprinted, Brussels, 1967.

5a. ——— *De re metallica. Translated....* London, 1912. English trans. with introd. and notes by H.C. Hoover. Reprinted, New York, 1950.

6. *Ausgewählte Werke*. 12 vols. Berlin, 1955-. Modern German trans. of most of his works. Ed. by H. Prescher.

1387. BRONKHORST (or NEOMAGUS or NOVIMAGUS), Jan. 1494-1570. Holl./Ger. P I; GA (BRONCHORST). Mathematics.

1388. FINE (or FINÉ), Oronce. 1494-1555. Fr. DSB Supp.; ICB; Still.; P I (FINAEUS); GA (FINAEUS). Astronomy. Mathematics.

1. *Protomathesis: Opus varium*. Paris, 1532. RM.
2. *Opere*. Venice, 1587. RM. Italian trans.--presumably made from the separate works as there appears to be no collected ed. in the original Latin.

See also 2753.

1389. MAUROLICO, Francesco. 1494-1575. It. DSB; ICB; Still.; P II (MAUROLYKUS); GA. Mathematics. Astronomy. Optics.

1. *Photismi de lumine et umbra*. Naples, 1611. X. There may have been an ed. at Venice in 1575.

1a. ——— *The Photismi de lumine: A chapter in late medieval optics*. New York, 1940. Trans. by H. Crew.
2. *Prologi sive sermones quidam de divisione artium, de quantitate, de proportione*. Messina, 1968. Ed. by G. Bellifemine.
See also 1015/1.

1390. SCHEUBEL (or SCHEYBL), Johann. 1494-1570. Ger. ICB; Still.; P II. Mathematics.

1391. SOTO. Domingo de. 1494-1560. Spain. DSB. Natural philosophy. Dynamics.

1392. APIAN (or BIENEWITZ or BENNEWITZ), Peter. 1495-1552. Ger. DSB; ICB; Still.; P I, III; GA; GC. Astronomy. Geography.
1. *Cosmographicus liber*. Landshut, 1524. RM.
1a. ——— *La cosmographie*. Paris, 1551. Trans. by Gemma Frisius. RM.
2. *Astronomicon Caesareum*. Ingolstadt, 1540. X. Reprinted, Leipzig, 1967.
See also 1169/1.

1393. VICARY, Thomas. ca. 1495-ca. 1561. Br. ICB; BLA; GD. Anatomy.
1. *Anatomie of the bodie of man*. 1548. No copies of the 1st ed. are extant. PR record an ed. of 1577; this was reprinted (with a life of Vicary) at London in 1888 and again in 1930. PR also record seven further eds up to 1633. Though Vicary's name appears on the title-page of the 1577 ed. the work is actually a transcript of a 14th-century MS (see GD).

1394. AGRICOLA, Johann. 1496-1570. Ger. BMNH; BLA; GC. Author of a herbal, 1539.

1395. BIONDO, Michel-Angelo. 1497-1565. It. Still.; BLA & Supp.; GA (BLONDUS). Astrology.

1396. FERNEL, Jean François. 1497?-1558. Fr. DSB; ICB; Still.; P I; Mort.; BLA & Supp.; GA; GB. Physiology.
1. *Monalosphaerium*. Paris, 1526. RM. A mathematical and astronomical work.
2. *Universa medicina*. Ib., 1567. Another ed., Leiden, 1645.
Wing (F807 onwards) records English trans of two of his medical works.

1397. MELANCHTHON, Philipp. 1497-1560. Ger. The Protestant reformer. ICB; Still.; P II; G. Astronomy.

1398. PORZIO (or PORTIUS or PORTA), Simone. 1497-1554. It. ICB; P II; BLA; GA; GB. Author of works on the principles of natural bodies, the question of the efficacy of alchemy, the colours of eyes, etc.

1399. BOCK, Jerome (or TRAGUS, Hieronymus). 1498-1554. Ger. DSB; ICB; Still.; BMNH; BLA & Supp.; GA (TRAGUS). Botany.
1. *Kreuterbuch*. Strasbourg, 1539. X.
1a. ——— Another ed., ib., 1551. X. Reprinted, Munich, 1964.

1400. OSIANDER, Andreas. 1498-1552. Ger. DSB; ICB; P II; GA; GB; GC. Astronomical and mathematical publishing.
1. *Schriften und Briefe, 1522 bis März 1525*. Gütersloh, 1975. Ed. by G. Müller.

1401. LAGUNA (or LACUNA), Andrés. 1499-1560. Spain. ICB; BMNH; Mort.; BLA; GA. Anatomy.

1402. RINGELBERGH, Joachim Sterck van. ca. 1499-1536. Belg./Ger./Fr. Still.; P II; GA; GF. Astronomy.
   1. *Institutiones astronomicae.* Cologne. 1528. RM.

1403. POBLACION, Juan Martinez. fl. 1530. Spain/Fr. P II. Author of *Compendium de usu astrolabii*, 1527.

1404. ROJAS SARMIENTO, Juan de. fl. 1530. ICB; Still. Mathematics.

1405. ROSETTI, Gioanventura. fl. 1530-1548. It. Still. (ROSETO).
   1. *Plictho de larte de tentori.* Venice, 1548. X. The first book on dyeing.
   1a. ——— *The Plictho: Instructions in the art of the dyers.* Cambridge, Mass., 1969. Trans. by S.M. Edelstein and H.C. Borghetty.

1406. NEGKER, Jobst Jost de. d. ca. 1545. Belg. ICB; GF. Anatomy.

1407. RISTORI, Giuliano. d. 1556. It. P II. Astronomy.

1408. DAL POZZO, Francesco. d. 1564. ICB. Anatomy.

1409. AFFAITATI (or AFFAYDATUS), Fortunio. ca. 1500-1550. It. Still.; P I. Author of *Physicae et astronomicae considerationes*, 1549 (includes a discussion of magnetism).

1410. BESSON, Jacques. ca. 1500-1575. Fr. ICB; P I & Supp. Instrument maker.
   1. *Théatre des instrumens mathématiques et mécaniques.* Geneva, 1594.

1411. BRASAVOLA, Antonio Musa. 1500-1555. It. ICB; Still.; P I; BMNH; Mort.; BLA & Supp. (BRASSAVOLA); GA. Pharmacy.

1412. CAMERARIUS, Joachim (first of the name). 1500-1574. Ger. ICB; P I; BMNH; GA; GB; GC. Arithmetic. Astronomy. Translator of Greek scientific works into Latin.
   1. *Opuscula.* Basel, 1536. RM.

1413. CARCANO, Francesco (called SFORZINO). 1500-1580. It. BMNH. Ornithology.

1414. CORNARIUS (or HAGENBUT), Janus. 1500-1558. Ger. ICB; BMNH; BLA & Supp. (HAGENBUT). Natural history. See also 1192/1b.

1415. DRYANDER (or EICHMANN), Johann. 1500-1560. Ger. Still.; P I; Mort.; BLA (EICHMANN); GA (EICHMANN); GC. Astronomy. Anatomy.

1416. DUVAL, Robert. ca. 1500-1567. Fr. P I & Supp.; GA. Chemistry.

1417. FELICIANO (DA LAZISIO), Francesco. ca. 1500-1563. It. Still.; GA. Author of a widely used textbook of arithmetic and geometry.

1418. MARANTA, Bartolomeo. ca. 1500-1571. It. ICB; BMNH; BLA; GA. Botany.

1419. ORTA, Garcia da. ca. 1500-ca. 1568. Port./India. DSB; ICB; BMNH; BLA; GA. Botany.
   1. *Coloquios dos simples e drogas e cousas medicinais da India.* Goa, 1563. X.

1a. ——— *Aromatum et simplicium aliquot medicamentorum apud Indos nascentium historia*. Antwerp, 1567. Trans. by C. de L'Ecluse. RM. Reprinted, Nieuwkoop, 1963. (Dutch classics on history of science. 6)
1b. ——— *Colloquies on the simples and drugs of India*. London, 1913. Trans. by C. Markham. RM.

1420. PICTORIUS, Georg. ca. 1500-1569. Ger. ICB; BMNH; BLA. Natural history.

1421. ROBINS, John. ca. 1500-1558. Br. P II; GD. Astronomy. Astrology.

1422. RUEFF, Jacob. 1500-1558. Switz. ICB; Mort.; BLA; GC (RUF). Embryology.
1. *De conceptu et generatione hominis*. Zurich, 1554. X. Another ed., Frankfurt, 1580.
1a. ——— Another ed., Frankfurt, 1587. X. Reprinted, New York, 197-?

1423. TARTAGLIA (or TARTALEA), Niccolò. 1500-1557. It. DSB; ICB; Still.; P II; GA; GB. Mathematics. Mechanics.
1. *La nuova scientia*. Venice, 1537. RM. Another ed., ib., 1558. Another ed., ib., 1583.
2. *Questi ed invenzioni diverse*. Venice, 1546. X. Another ed. ib., [1588?]. Reprint of 1546 ed., Brescia, 1959.
2a. ——— *Three bookes of colloquies concerning the arte of shooting, etc*. London, 1588. PR 23689. X.
See also Index.

1424. CARDANO, Girolamo. 1501-1576. It. DSB; ICB; Still.; P I; BLA & Supp.; GA; GB. Mathematics. Mechanics. Natural philosophy. See also 15502/1 and 15608/1.
1. *Practica arithmetice et mensurandi singularis*. Milan, 1539. RM/L.
2. *Artis magnae; sive, De regulis algebraicis*. Nuremberg, 1545. X.
2a. ——— *The great art; or, The rules of algebra*. Cambridge, Mass., 1968. Trans. and ed. by T.R. Witmer.
3. *De subtilitate*. Nuremberg, 1550. X. Another ed., Basel, 1554. An encyclopaedia of natural science.
4. *De rerum varietate*. Basel, 1557. A supplement to item 3. RM/L.
5. *De propria vita*. Paris, 1643. X.
5a. ——— *The book of my life*. London, 1931. Trans. by J. Stoner. Reprinted, New York, 1962.
6. *Opera omnia*. 10 vols. Lyons, 1663. Ed. by C. Sponi. RM. Reprinted, New York, 1967.

1425. CARPI, Girolamo da. 1501-1556. It. ICB; GA; GB. Anatomy.

1426. CATENA, Pietro. ca. 1501-1577. It. P I. Astronomy.

1427. FUCHS, Leonhard. 1501-1566. Ger. DSB Supp.; ICB; Still.; BMNH; BLA & Supp.; GA; GB; GC. Botany.
1. *De historia stirpium*. Basel, 1542. RM/L. Another ed., ib., 1545. Another ed., ib., 1549.
1a. ——— *New Kreuterbuch*. Ib., 1543. X. Reprinted, Munich, 1964.

1b. ——— *L'histoire des plantes.* Lyons, 1550. RM.
See also 1482/1.

1428. MATTIOLI, Pietro Andrea Gregorio. 1501-1577. It. DSB; ICB; Still.; BMNH; BLA & Supp.; GA. Botany.
1. *Commentarii in libros sex Pedacii Dioscorides Anazarbei "De medica materia."* Venice, 1554. RM.
1a. ——— French trans. by A. Du Pinet, Lyons, 1572. RM.
2. *Apologia adversus Amathum Lusitanum cum censura in eiusdem enarrationes.* Venice, 1558. RM.
3. *Compendium de plantis omnibus.* Venice, 1571.
3a. ——— Another ed., entitled *De plantis epitome utilissima.* Frankfurt, 1586. RM/L.
4. *Opera quae extant omnia.* Frankfurt, 1598. RM.

1429. MILICH, Jacob. 1501-1559. Ger. P II. Astronomy.

1430. WILLICH (or WILKE or WILD), Jodocus. 1501-1552. Ger. P II; BLA; GC. Arithmetic. Anatomy.

1431. EDWARDES, David. 1502-ca. 1542. Br. DSB; ICB; GD (EDGUARD). Anatomy.
1. *In anatomicen introductio.* London, 1532. PR 7483. X.
1a. ——— *Introduction to anatomy, 1532.* London, 1961. Facsimile reprint with English trans. and introd. by C.D. O'Malley and K.F. Russell.

1432. EGENOLPHUS, Christianus. 1502-1555. Ger. BMNH; GC (EGENOLF). Author of a herbal, 1535-36.

1433. NUÑEZ (SALACIENSE), Pedro. 1502-1578. Port. DSB; ICB (NUNES); Still. (NUNES); P II; BLA; GA; GB. Mathematics. Cosmography.
1. *De arte atque ratione navigandi.* Coimbra, 1573. RM (under NUNES). The volume also contains his commentary on Peurbach's *Theoricae novae planetarum* and several other small works, including *De crepusculis.*

1434. BASSANTIN (or BASSANTOUN), James. ca. 1504-1556 (or 1568). Br./Fr. P I & Supp.; GA; GD. Astronomy. Mathematics.

1435. MATHESIUS, Johann. 1504-1565. Ger. GC. Under MATTHESIUS: P II; BMNH. Mining. Metallurgy.
1. *Sarepta, oder, Bergpostill, sampt der Joachimsthalischen kurtzen Chroniken.* Nuremberg, 1562. RM.

1436. ARETIUS, Benedict. 1505-1574. Switz. P I & Supp.; BMNH; GA; GC. Botany.

1437. ESTIENNE (or STEPHANUS), Charles. ca. 1505-1564. Fr. A member of the Parisian family of printers and publishers. DSB; ICB; Still.; BMNH; Mort.; BLA & Supp.; GA; GB. Anatomy. Natural history.
1. *De re hortensi.* Paris, 1535. RM.
2. *Sylva.* Paris, 1538. RM.
3. *De dissectione partium corporis humani.* Paris, 1545. RM/L.
4. *Praedium rusticum.* Paris, 1554. X.
4a. ——— Trans. and revised by J. Liebault as *L'agriculture et maison rustique.* Paris, 1564. X. Another ed., Lyons, 1689. RM.

1438. EUSTACHI, Bartolomeo. ca. 1505-1574. It. DSB; ICB; BMNH; Mort.;

BLA & Supp.; GA. Anatomy.
1. *Tabulae anatomicae*. Rome, 1714. RM. Ed. by G.M. Lancisi. Engraved illustrations that were lost following his death and not found until after 1700. Another ed., Amsterdam, 1722. Another ed., Leiden, 1744. Reprint of 1714 ed., Modena, 1968.
See also 13968/2.

1439. GASSER, Achilles Pirmenius. 1505-1577. Ger. ICB; BLA & Supp.; GA; GC. Anatomy.

1440. GUINTER (or ANDERNACH or WINTHER), Joannes. ca. 1505-1574. Ger./Fr. DSB; ICB (GUENTHER); Still. (GÜNTHER); BLA & Supp. (GUENTHER). Anatomy.

1441. LEMNIUS, Levinus. 1505-1568. Holl. ICB (LEMMENS); P I; BMNH; BLA (LEMMENS); GA. Author of a work on astrology as well as the following.
1. *Occulta naturae miracula*. Antwerp, 1559, and later eds. X.
1a. ——— *The secret miracles of nature*. London, 1658. Wing L1044. X.

1442. RAIMONDI, Annibale. b. 1505. It. P III. Author of a treatise on the tides and of a proposal to reform the calendar.

1443. BUCHANAN. George. 1506-1582. Br. ICB; GA; GB; GD. Astronomy.

1444. DELFINO (or DELPHINUS), Giovanni Antonio. ca. 1506-after 1560. It. P I. Astronomy.

1445. LONGOLIUS, Gilbert. 1507-1543. Holl. ICB; BLA. Under LONGUEIL: GA; GC. Zoology.
1. *Dialogus de avibus*. Cologne, 1544. Ed. by William Turner.

1446. RONDELET, Guillaume. 1507-1566. Fr. DSB; ICB; BMNH; BLA; GA. Ichthyology.
1. *Libri de piscibus marinis*. 2 vols in 1. Lyons, 1554-55.
See also 1641/2.

1447. GEMMA (-FRISIUS), Reiner. 1508-1555. Belg. DSB; ICB; Still.; P I; BLA & Supp.; GA; GC. Mathematics. Astronomy. Cosmography.
1. *De principiis astronomiae et cosmographiae*. Antwerp, 1530. X. Another ed., Cologne, 1578. RM.
2. *De radio astronomico et geometrico*. Antwerp, 1545. RM.
See also 1392/1a.

1448. GUIDI, Guido (or VIDIUS, Vidus). 1508-1569. It./Fr. DSB; ICB (VIDIUS); Still.; Mort.; BLA & Supp.; GA (VIDIUS). Anatomy.

1449. HESELER, Baldasar. 1508-1567. ICB.
1. *Andreas Vesalius' first public anatomy at Bologna, 1540: An eyewitness report by Baldasar Heseler ... together with his notes on Matthaeus Curtius' lectures on "Anatomia Mundini."* Uppsala, 1959. Ed. with introd., trans., and notes by R. Eriksson.

1450. PICCOLOMINI, Alessandro. 1508-1578. It. ICB; Still.; P II; GA. Astronomy. Natural philosophy. Dynamics.
1. *De la sfera del mundo*. [With] *De le stelle fisse*. Venice, 1540. RM. Another ed., ib., 1559.
1a. ——— *La sphère du monde*. Paris, 1550.

1b. ——— *De sphaera*. Basel, 1568. Latin trans.
2. *L'instrumento de la filosofia*. Rome, 1551. RM.
3. *Della grandezza della terra et dell'acque*. Venice, 1558. RM. A Latin trans. is included in item 1b.

1451. TURNER, William. 1508-1568. Br. DSB; ICB; Still.; BMNH; BLA; GA; GB; GD. Botany. Zoology.
1. *Libellus de re herbaria*. London, 1538. PR 24358. X.
1a. ——— Reprinted with a biographical introd. by B.D. Jackson. London, 1877. X. This item, together with item 3a, were reprinted by the Ray Society. London, 1965.
2. *Avium praecipuarum quarum apud Plinium et Aristotelem mentio est ... historia*. Cologne, 1544.
3. *The names of herbes in Greke, Latin, Englishe, Duche and Frenche*. London., 1548/9. PR 24359. X.
3a. ——— Reprint ed. by J. Britten. London, 1882. X. See note to item 1a.
4. *A new herball*. London, 1551. PR 24365 and later eds.
PR record some other titles.

1452. COMMANDINO, Federico. 1509-1575. It. DSB; P I; GA. Important translator and editor of the classics of Greek mathematics (see Index). Apart from commentaries on them, his main work is the following.
1. *Liber de centro gravitatis solidorum* Bologna, 1565. RM/L.

1453. TAISNIER, Jean. 1509-ca. 1565. Belg./It. ICB; P II; GA. Astronomy/astrology. Magnetism.
1. *De natura magnetis et ejus effectibus*. Cologne, 1562. X. Reprinted, London, 1966.

1454. TELESIO, Bernardino. 1509-1588. It. DSB; ICB; P II; GA; GB. Natural philosophy.
1. *De rerum natura iuxta propria principia*. Rome, 1565. X. Reprint of the 3rd ed. (Naples, 1586), Hildesheim, 1971.
2. *Varii de naturalibus rebus libelli*. Venice, 1590. X. Reprinted, Hildesheim, 1971.

1455. VINET, Elie. 1509-1587. Fr. P II; GA. Astronomy. Mathematics.

1456. RECCHI, Nardo Antonio. mid-16th century. It./Spain. BMNH; BLA; GA. Botany.

1457. VASSÉ, Loys. fl. 1535-1560. Fr. ICB. Anatomy.

1458. FIGULUS, Carl. fl. 1540. Ger. GA. Zoology. Botany.

1459. RYFF, Walter Hermann. fl. 1541-1551. Ger. ICB; BMNH; BLA. Anatomy. Pharmacy.
1. *Das ist des Menschen wahrhafftige Beschreybung; oder, Anatomi*. Strasbourg, 1541.

1460. GUEVARA, Alonso Rodriguez de. fl. 1545-1559. Spain. ICB; BLA. Anatomy.

1461. FAUSTO, Sebastiano da Longiano. fl. 1550. It. GA.
1. *Meteorologia*. Venice, 1542. In Italian. RM.

1462. MADAUER, Bartholomäus. d. 1579. Ger. P II. Astronomy.

1463. QUERCU, Leodegarius a (or DUCHESNE, Léger). d.1588. Fr. BMNH; Natural history.

1464. CAIUS, John. 1510-1573. Br. DSB; ICB; Still; BMNH; BLA (KAYE); GA; GB; GD. Natural history.
1. *De canibus Britannicis. De rariorum animalium et stirpium historia. De libris propriis.* 3 parts. London, 1570. PR 4346. X.
1a. ——— *Of Englishe dogges.* London, 1576. PR 4347. X. Reprinted, Amsterdam, 197-?
2. *The works.* Cambridge, 1912. RM. Ed. by E.S. Roberts. Includes a biography by J. Venn.

1465. COLOMBO, (Matteo) Realdo. ca. 1510-1559. It. DSB; ICB; BMNH (COLUMBUS); BLA & Supp.; GA (COLUMBUS). Anatomy. Physiology.
1. *De re anatomica.* Venice, 1559. RM.

1466. DU PINET, Antoine. ca. 1510-ca. 1565. Fr. BMNH; GA. Botany.
1. *Historia plantarum.* Lyons, 1561. RM. Another ed., ib., 1567. See also 1428/1a.

1467. FUSCH, Remacle. ca. 1510-1587. Belg. ICB. Under FUCHS: BMNH; BLA; GA. Botany.

1468. GEMINUS (or LAMBRIT or LAMBERT), Thomas. ca. 1510-1562. Br. DSB; ICB; Still.; GD (GEMINI). Anatomy.
1. *Compendia totius anatomie delineatio.* London. 1545. PR 11714. X.
1a. ——— Another ed. (including an English trans.), ib., 1553. PR 11716. X. Reprinted, ib., 1959. With an introd. by C.D. O'Malley.

1469. GOSSELIN (or JOSSELIN), Jean. ca. 1510-1604. Fr. P I; GA. Astronomy/astrology.

1470. INGRASSIA, Giovanni Filippo. ca. 1510-1580. It. DSB; ICB; Mort.; BLA & Supp.; GA. Anatomy.

1471. MICHIEL, Pietro Antonio. 1510-1576. ICB. Author of a herbal.

1472. MIZAULD (or MIZALDUS), Antoine. ca. 1510-1578. Fr. ICB (MISALDUS); P II & Supp.; BMNH; GA. Astrology. Herbal pharmacy. A very prolific author.

1473. PALISSY, Bernard. ca. 1510-ca. 1590. Fr. DSB; ICB; P II; GA; GB. Natural history. Chemical technology. Earth sciences.
1. *Discours admirables de la nature des eaux et fontaines.* Paris, 1580. RM.
1a. ——— *The admirable discourses.* Urbana, Ill., 1957. Trans. by A. La Rocque.
2. *Oeuvres.* Paris, 1777. RM. Ed. by B. Faujas de Saint-Fond and N. Gobet.
3. *Oeuvres complètes.* Paris, 1843. Ed. by P.A. Cap. X. Reprinted, ib., 1961.

1474. PARÉ, Ambroise. 1510?-1590. Fr. The famous surgeon. DSB; ICB; Still.; Mort.; BLA & Supp.; GA; GB.
1. *Les oeuvres.* Paris, 1575. Many later eds. X.
1a. ——— *Opera.* Paris, 1582. Latin trans. X.
1b. ——— ——— *The workes. Translated out of the Latin and compared with the French.* London, 1649. Wing P349. Trans. by T. Johnson. X. Reprinted, Pound Ridge, N.Y., 1968.

2. *Oeuvres complètes*. 3 vols. Paris, 1840-41. Ed. by J.F. Malgaigne. RM.
3. *Selections from the works*. London, 1924. By D.W. Skinner.
4. *Des monstres et prodiges. Edition critique*. Geneva, 1971. Ed. by J. Céard.

1475. POSTEL, Guillaume. 1510-1581. Fr./It. ICB (POSTELLUS); P II; GA. Astronomy. Cosmography. Alchemy.
1. *Son interprétation du candélabre de Moyse*. Nieuwkoop, 1966. Contains facsimiles of the Hebrew original and the Latin trans. (both Venice, 1548). Ed. by F. Secret. An alchemical work.
2. *De universitate*. Paris, 1552. An astronomical and cosmographical work. X. Another ed., Leiden, 1635. RM.

1476. RECORDE, Robert. ca. 1510-1558. Br. DSB; ICB; Still.; P II; GA; GB; GD. Mathematics.
1. *The castle of knowledge, a treatise on astronomy and the sphere*. London, 1556. PR 20796. RM. Another ed., ib., 1596. Other titles in PR and Wing.

1477. ZACHAIRE, Dénis. 1510-1556. Fr. ICB; P II (ZECAIRE). Alchemy.

1478. AMATUS LUSITANUS (or RODRIGUES, João). 1511-1568. Port./It. DSB (LUSITANUS, A.); ICB; Still. (AMATUS RODRIGUEZ); BMNH; BLA & Supp.; GA. Botany. Anatomy. Best known as a physician. See also 1428/2.

1479. REINHOLD, Erasmus. 1511-1553. Ger. DSB; P II; GA; GC. Astronomy.

1480. SCHEGK (or SCHEGKIUS or SCHEGGIUS), Jakob. 1511-1587. Ger. DSB; BLA. Natural philosophy.

1481. SCHRECKENFUCHS, Erasmus Oswald. 1511-1579. Ger. P II. Astronomy.
1. *Commentarii in "Sphaeram" Ioannis de Sacro Bosco*. Basel, 1569. RM. See also 990/2.

1482. SERVETUS, Michael. 1511?-1553. Fr. DSB; ICB; Still.; Mort.; BLA & Supp.; GA; GB. Physiology.
1. *In Leonardum Fuchsium apologia*. Lyons, 1536. X. Reprinted, 1909.
2. *A translation of his geographical, medical, and astrological writings*. Philadelphia, 1953. By C.D. O'Malley. With introd. and notes.

1483. AMICO, Giovanni Battista. ca. 1512-1538. It. Still.; P I. Astronomy.

1484. ANGUILLARA, Luigi. ca. 1512-1570. It. DSB; ICB; BMNH; BLA; GA. Botany.

1485. BORRO (or BORRI or BORRIUS), Girolamo. 1512-1592. It. DSB Supp. Natural philosophy. Dynamics.
1. *Dialogo del flusso e reflusso del mare*. Lucca, 1561. RM.

1486. CONSTANTINUS, Robertus. 1512-1605. Fr. BMNH; GA (CONSTANTINE). Botany. See also 999/1.

1487. VALERIUS (or WOUTERS), Cornelius. 1512-1578. Belg. P II. Under WOUTERS: GC; GF. Astronomy.

1488. DALÉCHAMPS, Jacques. 1513-1588. Fr. DSB; BMNH (ALECHAMPS); BLA & Supp.; GA. Botany.
   1. *Historia generalis plantarum.* 2 vols. Lyons, 1586-87.

1489. FERRIER (or FERRARIUS or FERRERIUS), Auger. 1513-1588. Fr. P I; BLA; GA. Astronomy. (PR and Wing record an English trans. of one of his works.)

1490. MEURER, Wolfgang. 1513-1585. Ger. P II. Meteorology.

1491. SCHRÖTER, Johann. 1513-1593. Aus./Ger. P II; BLA. Astronomy. Astrology.

1492. CORTUSO, Giacomo Antonio. 1513-1603. It. ICB; GA (CORTUSI, Jacobo A.). Botany.

1493. FLOCK, Erasmus. 1514-1568. Ger. P I. Astronomy.

1494. PLACOTOMUS (or BRETSCHNEIDER), Johann. 1514-1576/7. Danzig. P II; BLA; GC. Pharmacy. Distillation.

1495. RHETICUS (or JOACHIM), Georg. 1514-1574. Ger./Cracow. DSB; ICB; P II (RHAETICUS); GA (JOACHIM); GB; GC. Mathematics. Astronomy.
   1. *De libris "Revolutionum ... " Nicolai Copernici ... narratio prima.* Danzig, 1540. X. Reprinted, Osnabrück, 1965. For English trans. see 1320/3. See also 1320/1.

1496. SALVIANI, Ippolito. 1514-1572. It. DSB; ICB; BMNH; BLA; GA. Ichthyology.

1497. VESALIUS, Andreas. 1514-1564. Belg./It. DSB; ICB; Still.; Mort.; BLA & Supp.; GA; GC. Anatomy.
   1. *De humani corporis fabrica.* Basel, 1543. RM/L. Reprinted, Brussels, 1964, and Budapest, 1968. 2nd ed., Basel, 1555.
   2. *Epitome.* Basel, 1543. X.
   2a. ——— *The Epitome. Translated from the Latin.* New York, 1949. By L.R. Lind. Reprinted, Cambridge, Mass., 1969.
   3. *Epistola rationem modumque propinandi radicis Chynae decocti ... pertractans.* Basel, 1546. RM.
   4. *Opera omnia anatomica et chirurgica.* 2 vols. Leiden, 1725. Ed. by H. Boerhaave and B.S. Albini. RM/L.

   Selections and Correspondence

   5. *The bloodletting letter of 1539. An annotated translation.* London, [1944?]. By J.B. Saunders and C.D. O'Malley.
   6. *Illustrations from the works of A. Vesalius.* Cleveland, 1950. By J.B. Saunders and C.D. O'Malley. With notes and trans.
   7. *Vesalius on the human brain.* London, 1952. By C. Singer. With notes and trans.
   8. *Préface d'André Vésale à ses livres sur l'anatomie, suivie d'une lettre à Jean Oporinus, son imprimeur.* Brussels, 1961.
   See also 1449/1 and 15531/1.

1498. BRACESCO (or BRACESCHI), Giovanni. mid-16th century. It. P I. Alchemy.

1499. AGRIPPA, Camillo. fl. 1538-1598. It. ICB; P I; GA.

1. *Dialogo sopra la generatione de venti, baleni, tuoni, fiumi, laghi, valli e montagne.* Rome, 1584. RM.

1500. ASKAM, Anthony. fl. 1553. Br. Under ASCHAM: ICB; GA; GD. Astrology. Botany. (Titles in PR.)

1501. CANANO, Giovan Battista. 1515-1579. It. DSB; ICB; Still.; Mort.; BLA & Supp.; GA (CANANI). Anatomy.
1. *Musculorum humani corporis picturata dissectio.* Ferrara, 1541. X. Reprinted with Italian and English trans. and notes by G. Muratori, Florence, 1962.

1502. CORDUS, Valerius. 1515-1544. Ger. DSB; ICB; Still.; P I; BMNH; Mort.; BLA & Supp.; GA; GC. Botany. Pharmacy.

1503. RAMUS, Petrus (or LA RAMÉE, Pierre de). 1515-1572. Fr. DSB; ICB; P II. Under RAMÉE: GA; GB. Logic and methodology. Mathematics. See also Index.
1. *Geometriae libri XXVII.* Basel, 1569. X.
1a. ——— *The elements of geometrie.* London, 1590. PR 15250. Trans. by T. Hood. X.
1b. ——— *Via regia ad geometriam. The way to geometry.* London, 1636. PR 15251. Trans. and "much enlarged" by W. Bedwell.
2. *Arithmeticae libri II et geometriae XXVII.* Frankfurt, 1599. X. Another ed., ib., 1627.

Non-mathematical Works

3. *Dialecticae libri duo.* Paris, 1556. X.
3a. ——— *The logike.* Northridge, Cal., 1969. English trans. made in 1574. Ed. by C.M. Dunn.
4. *Scholarum physicarum libri octo.* Paris, 1565. X. Another ed., Frankfurt, 1583; reprinted, ib., 1967.
5. *Scholae in liberales artes.* Basel, 1569. X. Reprinted, Hildesheim, 1970.
6. *Collectaneae praefationes, epistolae, orationes.* Paris, 1577. X. Another ed., Marburg, 1599; reprinted, Hildesheim, 1969.
6a. ——— Reprint of 1577 ed., entitled *Oeuvres diverses.* Geneva, 1971. Text in Latin.

1504. TABERNAEMONTANUS, Jakob Teodor. ca. 1515-1590. Ger. ICB; BLA; GA. Under THEODORUS, Jacobus: BMNH; GC. Botany.

1505. TOXITES (or SCHÜTZ), Michael. ca. 1515-1581. Ger. ICB; BMNH; BLA. Alchemy. Editor and exponent of the works of Paracelsus.

1506. BEYER, Hartmann. 1516-1577. Ger. P I. Astronomy.
1. *Quaestiones novae in libellum "De sphaera" Joannis de Sacro Bosco, collectae ab Ariele Bicardo* [pseud.]. Paris, 1552. Another ed., Frankfurt, 1552.

1507. DODOENS, Rembert. 1516-1585. Belg./Holl. DSB; ICB; P III (DODONAEUS); BMNH; BLA & Supp.; GA. Botany.
1. *Cosmographica in astronomiam et geographiam isagoge.* Antwerp, 1548. X. Reprinted, Nieuwkoop, 1963. (Dutch classics in history of science. 7)
2. *Cruydeboeck.* Antwerp, 1554. X. Reprinted, Leuven, 1968.
2a. ——— *A niewe herball.* Antwerp, 1578. PR 6984. English

trans. by H. Lyte. PR record three later eds published in London.

2b. ——— ——— *Rams little Dodeon. A briefe epitome of the "New herball."* London, 1606. PR 6988. Ed. by W. Ram.

3. *Florum coronariorum odoratarumque nonnullarum herbarum historia.* Antwerp, 1569.
4. *Stirpium historiae ... libri XXX.* Antwerp, 1583.

1508. ENGELHARDT, Valentin. 1516-after 1562. Ger. ICB; P I. Astronomy.

1509. FABRICIUS (or GOLDSCHMIED), Georg. 1516-1571. Ger. P I; BMNH. Metallurgy.

1510. GESNER, Konrad. 1516-1565. Switz. DSB; ICB; Still.; P I; BMNH; BLA & Supp.; GA; GB. Botany. Zoology. Geology.

1. *Historia plantarum.* Basel, 1541. X. Reprinted, Zurich, 1973.
2. *Historia animalium.* 5 vols in 4. Zurich, 1551-87. RM.

2a. ——— *Icones animalium ... quae in "Historia animalium" ... describuntur.* Ib., 1553. X. 2nd ed., 1560.

2b. ——— *Animantium marinorum ordo primus.* Ib., 1553-60.

2c. ——— *Nomenclator aquatilium animantium.* Ib., 1560.

2d. ——— *Vogelbuch.* Ib., 1557. X. Reprinted, ib., 1969. Another ed., ib., 1581.

2e. ——— *Thierbuch.* Ib., 1563. X. Reprinted, Wurzburg, 1965. Another ed., Zurich, 1583.

2f. ——— *Fischbuch.* Ib., 1563. X. Another ed., ib., 1575.

2g. ——— *Deutsche Namen der Fische und Wassertiere.* Ib., 1556. X. Reprinted, Aalen, 1974.

3. *De raris et admirandis herbis quae ... lunariae nominantur.* Zurich, 1555. RM.
4. *Vingt lettres à Jean Bauhin fils, 1563-1565.* Saint-Etienne, 1976. French trans. by A. Sabot.

See also 1666/1 and 1847/1.

1511. BELON, Pierre. 1517-1564. Fr. DSB; ICB; BMNH; BLA & Supp.; GA; GB. Zoology.

1. *L'histoire naturelle des estranges poissons marins.* Paris, 1551. RM.
2. *De aquatilibus,* Ib., 1553. RM.
3. *Les observations de plusieurs singularitez et choses memorables, trouvées en ... pays estranges.* Ib., 1553. RM.
4. *De arboribus.* Ib., 1553. RM.
5. *L'histoire de la nature des oyseaux.* Ib., 1555. RM.
6. *Les remonstrances sur la défault du labour et culture des plantes et de la cognoissance d'icelles.* Ib., 1558. X.

6a. ——— *De neglecta stirpium cultura atque earum cognitatione.* Antwerp, 1589. Latin trans. by C. de L'Ecluse. RM.

1512. FIORAVANTI, Leonardo. 1517-1588. It. ICB; P I; BLA & Supp.; GA. Alchemy.

1. *Dello specchio di scienza universale.* Venice, 1564,. X.

1a. ——— *Miroir universel des arts et sciences.* Paris, 1584. X. 2nd ed., 1586. RM.

1513. HERNANDEZ, Francisco. 1517-1587. Spain/Mexico. DSB; ICB; BMNH; BLA; GA. Natural history.

1514. PELETIER, Jacques. 1517-1582. Fr. DSB; ICB; P II; GA. Mathematics.
 1. *L'arithmétique*. Poitiers, 1549. X. Reprinted, Geneva, 1969.

1515. ZARLINO, Gioseffo (or Giuseppe). 1517 (or 1519) -1590. It. P II; GA; GB; GF.
 1. *De vera anni forma*. Venice, 1580. RM. Re the Gregorian calendar.

1516. HELLER, Joachim. 1518-1590. Ger. P I. Astronomy. See also 1067/1.

1517. KENTMANN, Johann. 1518-1574. Ger. ICB; P I; BMNH; BLA. Natural history.

1518. BOTALLO, Leonardo. ca. 1519-1587/8. It./Fr. DSB; ICB; BLA & Supp.; GA. Anatomy.

1519. BUCKLEY (or BUCLAEUS), William. 1519-1571. Br. P I; GD. Mathematics. (Title in PR.)

1520. CESALPINO, Andrea. 1519-1603. It. DSB Supp.; ICB; P I (CAESALPIN); BMNH (CAESALPINUS); Mort.; BLA & Supp.; GA (CESALPINI); GB (CAESALPINO). Botany. Physiology.
 1. *Quaestionum peripateticarum libri V*. Venice, 1571. X.
 1a. ——— 2nd ed. of Italian trans., ib., 1593. X.
 1b. ——— ——— *Questions péripatéticiennes*. Paris, 1929. French trans. by M. Dorolle from Italian.
 2. *De plantis*. Florence, 1583. "The first true textbook of botany."
 3. *De metallicis*. Rome, 1596. RM.

1521. COLLADO, Luis. mid-16th century. Spain. ICB; BLA. Anatomy.

1522. GIMENO, Pedro. mid-16th century. Spain. ICB; BLA. Anatomy.

1523. LOBERA DE AVILA, Luis. fl. 1551. Spain. ICB; Mort. Anatomy.

1524. PEREZ DE VARGAS, Bernardo. fl. 1560-1568. Spain. DSB. Metallurgy.
 1. *De re metalica*. Madrid, 1569. RM.
 1a. ——— *Traité singulier de métallique*. 2 vols. Paris, 1743. RM.

1525. FORCADEL, Pierre. d. ca. 1576. Fr. ICB; P I; GA. Mathematics.

1526. LILIO, Luigi (or LILIUS, Aloysius). d. 1576. It. P I; GA. Astronomy.

1527. ENTZELT, Christoph. d. 1583. P I; BMNH (ENCELIUS).
 1. *De re metallica*. Frankfurt, 1551. RM.

1528. ACRONIUS (or ATROCIANUS), Johann. 1520-1564. Switz. P I; BLA; GC. Astronomy.

1529. AYRER, Melchior. 1520-1579. Ger. P I; BLA. Mathematics. Chemistry.

1530. DIGGES, Leonard. ca. 1520-1559? Br. DSB; ICB; P I; GA; GD. Mathematics.

1. *A prognostication of right good effect.* London, 1555. PR 6860. X. Reprinted, Oxford, 1926. PR record ten further eds up to 1605.
2. *A geometrical practise, named Pantometria.* London, 1571. PR 6858. X. Another ed.,1591. The work was completed by his son, Thomas Digges.

2a. ——— *The theodelitus and topographical instrument of L. Digges, described by his son, T. Digges, in 1571. Reprinted from the first book of "Pantometria."* Oxford, 1927. Ed. by R.W.T. Gunther.

PR record some other titles.

1531. FIELD, John. ca. 1520-1587. Br. GD. Astronomy. He is said to have been the first Copernican in England. (Titles of two of his ephemerides are in PR under FEILD.)

1532. GALILEI, Vincenzio. ca. 1520-1591. It. The father of the great physicist. DSB; ICB; P I; GA; GF. Theory of music. Acoustics.

1533. GOHORY, Jacques. (Pseudonym: SUAVIS, Leo) 1520-1576. Fr. DSB; ICB; BLA; GA. Botany. Alchemy. "Important as an early disseminator of Paracelsan ideas in France." -DSB.

1534. PICCOLOMINI, Francesco. ca. 1520-1604. It. P II; GA. Natural philosophy.

1535. REYNA, Francisco de la. b. ca. 1520. Spain. DSB; ICB. Physiology.

1536. SUCHTEN (or ZUCHTA), Alexander. ca. 1520-1590? Poland/Ger. DSB; BLA. Iatrochemistry. (Wing records an English trans. of one of his works.)

1537. VALVERDE (or AMUSCO), Juan de. ca. 1520-ca. 1588. Spain/Italy. DSB; ICB; BLA & Supp. (AMUSCO). Anatomy.

1538. WIELAND (or GUILANDINUS), Melchior. ca. 1520-1589. It. DSB. Under GUILANDINUS: BMNH; BLA; GA. Botany.

1539. CALZOLARI, Francesco. 1521-ca. 1600. It. BMNH; GA (CALCEOLARI). Botany.

1540. COSTER, Martin Janszoon. 1521-1592. Holl. ICB; GF. Anatomy.

1541. ALDROVANDI, Ulysse. 1522-1605. It. DSB; ICB; P I & Supp.; BMNH; Mort.; BLA & Supp.; GA; GB. Botany. Zoology. Embryology.

1. *Ornithologia.* 3 vols. Bologna, 1600-03. RM. Another ed., 3 vols, Frankfurt, 1610-35.

1a. ——— *On chickens: The Ornithology (1600), Vol. II, Book XIV.* Norman, Okla., 1963. Trans. with introd. and notes by L.R. Lind.

2. *De animalibus insectis.* Bologna, 1602. RM/L. Another ed., Frankfurt, 1623.
3. *De reliquis animalibus exanguibus, utpote de mollibus crustaceis, testaceis et zoophytis.* Bologna, 1606. RM. Another ed., Frankfurt, 1623.
4. *Quadrupedum omnium bisulcorum historia.* Bologna, 1613. X. Another ed., ib., 1621. RM. Another ed., Frankfurt, 1647.
5. *De piscibus ... et de cetis.* Bologna, 1613. RM. Another ed., Frankfurt, 1647.

6. *De quadrupedibus solidipedibus*. Bologna, 1616. X. Another ed., Frankfurt, 1623.
7. *De quadrupedibus digitatis viviparis ... et ... oviparis*. Bologna, 1637. RM.
8. *Monstrorum historia*. Bologna, 1642.
9. *Museum metallicum*. Bologna, 1648. RM/L.
10. *Dendrologiae naturalis, scilicet arborum historiae, libri II*. Bologna, 1668. RM.

See also 1847/2.

1542. BEUTHER, Michael. 1522-1587. Ger. P I; GA; GC. Astronomy.

1543. FALCO, Jaime Juan. 1522-1594. Spain. P III. Mathematics.

1544. FERRARI, Ludovico. 1522-1565. It. DSB; ICB; P I; GA. Algebra.

1545. WITTEKIND, Hermann. 1522-1603. Ger. P II. Astronomy.

1546. ERASTUS (or LIEBER), Thomas. 1523-1583. Ger./Switz. The theologian. DSB; ICB; P I; BLA & Supp.; GA (LIEBER); GB; GC. In the history of science he is remembered "chiefly as an inexorable and abusive critic of astrology, natural magic, and particularly of Paracelsus and iatrochemistry." -DSB.

1547. FALLOPPIO, Gabriele. 1523?-1562. It. DSB; ICB; P I; Mort.; BLA & Supp.; GA; GB. Anatomy.
1. *Epistolario*. Ferrara, 1970. Ed. by P. Di Pietro.

1548. GIUNTINI, Francesco. 1523-1590. Fr. ICB; P I (JUNCTINUS). Astrology/astronomy.
1. *La sfera del mondo*. Lyons, 1582. RM. See also 1103/2.

1549. VIGENÈRE, Blaise de. 1523-1599. Fr. ICB; P II; GA. Mathematics. Astronomy. Chemistry.

1550. WIGAND, Johannes. 1523-1587. Ger. ICB; BMNH; GC. Natural history.

1551. BACCI (or BACCIO or BACCIUS), Andrea. 1524-1600. It. ICB; P I; BMNH; BLA. Mineralogy. Earth sciences.
1. *L'alicorno discorso*. Florence, 1573. RM.
2. *Le XII pietre pretiose*. Rome, 1587. RM.
2a. ——— *De gemmis et lapidibus pretiosis*. Frankfurt, 1603. RM.

1552. LEOWITZ, Cyprianus von. 1524-1574. Ger. ICB; P I (LEOVITIUS); GA; GC. Astronomy. (PR record London eds of two Latin works on astrology.)

1553. LÜBER, Thomas. b. 1524. Ger. P III. Astrology. Alchemy.

1554. OCCO, Adolf. 1524-1606. Ger. ICB; Mort.; BLA (OCCON); GA. Pharmacy.

1555. ROTA, Gian Francesco. ca. 1524-1558. It. BLA. Anatomy.

1556. SOLENANDER, Reiner. 1524-1601. Ger. BLA & Supp.
1. *De caloris fontium medicatorum causa, eorumque temperatione*. Lyons, 1558. RM.

1557. DUHAMEL, Pasquier (or Pascal). d. 1565. Fr. P I. Astronomy. Mathematics.

1558. STEINMETZ, Moritz. d. 1584. Ger. P II. Astronomy.

1559. ACOSTA, Cristóbal. ca. 1525-ca. 1594. Port./India. DSB; ICB; BMNH; BLA. Natural history.
   1. *Tractado de las drogas y medicinas de las Indias Orientales.* Burgos, 1578. RM (under COSTA).

1560. BLOMEFIELD, Miles. 1525-1574. Br. ICB; GD. Alchemy.

1561. HAGECIUS, Thaddeus. 1525-1600. Prague. ICB; P I (HAGEK). Astronomy.
   1. *Dialexis de novae et prius incognitae stellae inusitatae magnitudinis et splendissimi luminis apparitione.* Frankfurt, 1574. X. Reprinted, Prague, 1967.

1562. KYBER, David. 1525-1553. BMNH. Botany.

1563. PEUCER, Kaspar. 1525-1602. Ger. ICB; P II; BLA; GA; GC. Astronomy.

1564. PICCOLOMINI, Arcangelo. 1525-1586. It. DSB; BLA. Anatomy. Physiology.

1565. SCHÖNFELD, Victorin. 1525-1591. Ger. P II; BLA. Astrology.

1566. TRENCHANT, Jean. b. ca. 1525. Fr. ICB. Mathematics.

1567. BOMBELLI, Rafael. 1526-1572. It. DSB; ICB; P I; GA. Algebra.
   1. *L'algebra.* Milan, 1966. The first complete ed.: it comprises Books I-III from the original 1572 ed. and Books IV and V from a MS source. Ed. by U. Forti.

1568. CHARNOCK, Thomas. 1526-1581. Br. ICB; GD. Alchemy.

1569. HEYDEN (or HEYDIUS), Christian. 1526-1576. Ger. P I. Astronomy.

1570. L'ÉCLUSE, Charles de. 1526-1609. Belg. DSB; BMNH; BLA & Supp. (DE L'ECLUSE); GA. Under CLUSIUS: ICB; GC. Botany.
   1. *Rariorum plantarum historia.* Antwerp, 1601.
   2. *Exoticorum libri decem.* 3 vols. Leiden, 1605. A collection of his works.
   See also 1419/1a and 1511/6a.

1571. RANTZAU (or RANZAU or RANZOW), Heinrich von. 1526-1599. Ger. P II; GA; GC. Astrology.

1572. DEE, John. 1527-1608. Br. DSB; ICB; P I; GA; GD. Mathematics. Astrology. Alchemy. (Titles in PR.)
   1. *The mathematicall praeface.* [Dee's preface to Henry Billingsley's trans. of Euclid, 1570; see entry 1621/1.] Reprinted, New York, 1975.
   2. *The private diary, 1577-1601, and the catalogue of his library of manuscripts.* London. 1842. Ed. by J.O. Halliwell. X. Reprinted, New York, 1968.
   3. *Diary for the years 1595-1601.* London, 1880. Ed. by J.E. Bailey.

1573. LAVATER, Ludwig. 1527-1586. Switz. ICB; P I; GA; GC. Astronomy.

1574. QUACKELBEEN, Willem. 1527-1561. Belg. ICB; GF. Botany.

1575. STADIUS, Joannes. 1527-1579. Belg./Fr. P II; GF (STADE). Astronomy.

1. *Ephemerides novae et exactae ab anno 1554 ad annum 1570.* Cologne, 1556. RM.

1576. BODENSTEIN, Adam von. 1528-1577. Switz. DSB (ADAM OF B.); P I; BLA & Supp.; GA; GC. Alchemy.

1577. LONICERUS (or LONITZER), Adam. 1528-1586. Ger. DSB; ICB; P I; BMNH; BLA; GA; GC. Botany.

1578. WECKER, Johann (or Hans) Jacob. 1528-1586. Switz. BLA; GC.
 1. *De secretis.* Basel, 1560. X. 3rd ed., ib., 1592. RM. At least eight eds appeared.

1579. DURANTE, Castore. 1529-1590. It. ICB; BMNH; BLA & Supp.; GA. Botany.

1580. FABRICIUS, Paul. 1529 (or 1519)-1588. Aus. P I. Astronomy.

1581. NEANDER, Michael. 1529-1581. Ger. DSB; P II; BLA; GA; GC. Mathematics.

1582. PATRIZI, Francesco. 1529-1597. It. DSB; ICB; P II (PATRICIO); GA; GB. Mathematics. Natural philosophy.
 1. *Lettere ed opuscoli inediti.* Florence, 1975. Critical ed. by D.A. Barbagli.

1583. RETHE (or RHETHE), Georg. d. 1586. Ger. ICB. Astronomy.

1584. PENNY, Thomas. d. 1589. Br. BMNH; GA; GD. Botany. Entomology.

1585. NABOD (or NAIBOD), Valentin. d. 1593. Ger./It. P II. Astronomy.

1586. ARANZIO, Giulio Cesare. ca. 1530-1589. It. DSB; Mort.; BLA & Supp.; GA (ARANTIUS). Anatomy. Embryology.

1587. BENEDETTI, Giovanni Battista. 1530-1590. It. DSB; ICB; P I; GA. Mathematics. Statics. Dynamics. Acoustics.
 1. *De gnomonum umbrarumque solarium usu.* Turin, 1574. RM.
 2. *De temporum emendatione opinio.* Turin, 1578. RM.
 3. *Consideratione d'intorno al discorso della grandezza della terra et dell'acqua del ... Sig. Antonio Berga.* Turin, 1579. RM. cf. item 2784/1.
 4. *Diversarum speculationum mathematicarum et physicarum liber.* Turin, 1585. RM.
 See also 2830.

1588. BRUCAEUS, Heinrich. 1530-1593. Ger./It. ICB; P I; BLA & Supp.; GA; GC. Astronomy.

1589. DASYPODIUS, Cunradus. ca. 1530-1600. Strasbourg. DSB; P I & Supp.; GA; GC. Mathematics. Astronomy.
 1. *Brevis doctrina de cometis.* Strasbourg, 1578. RM.
 2. *Heron mechanicus; seu, De mechanicis artibus atque disciplinis.* Strasbourg, 1580. RM.
 3. *Institutionum mathematicarum ... erotemata.* Strasbourg, 1593. RM.
 4. *Protheoria mathematica.* Strasbourg, 1593. RM.

1590. DULCO (or DUCLOS or CLAVEUS), Gaston. ca. 1530-ca. 1600. Fr. GA. Alchemy.

1591. ERCKER (or ERCKNER or ERCKEL), Lazarus. ca. 1530-1594. Ger./Prague. DSB; ICB; BMNH. Chemistry. Metallurgy.

1. *Beschreibung allerfürnemisten mineralischen Erzt und Berckwercksarten*. Prague, 1574. X. Another ed., Frankfurt, 1598. "May be regarded as the first manual of analytical and metallurgical chemistry."

1a. ——— *Treatise on ores and assaying*. Chicago, 1951. Trans. by A.G. Sisco and C.S. Smith.

2. *Aula subterranea ... oder Gründliche Beschreibung derjenigen Sachen so in der Tieffe der Erden wachsen*. Frankfurt, 1672. X. Another ed., ib., 1684. RM.

3. *Das kleine Probierbuch von 1556. Vom Rammelsberge und dessen Bergwerk ein kurzer Bericht von 1565. Das Münzbuch von 1563*. Bochum, 1968. A collection, ed. by P.R. Beierlein.

1592. GARCAEUS (or GARTZE), Johann. 1530-1574. Ger. P I; GC. Astronomy.

1593. GRAMINEUS (or GRAS), Theodor. ca. 1530-1592. Ger. P I. Astronomy.

1594. LATOSZ, Johann. ca. 1530-ca. 1600. Cracow. P III. Astronomy.

1595. MERCURIALE, Girolamo (or Hieronymus). 1530-1606. It. ICB; BMNH; BLA & Supp.; GA. Anatomy. See also 988/1.

1596. NICOT, Jean. ca. 1530-1604. Fr. DSB; ICB; GA. Botany.

1597. RUINI, Carlo. ca. 1530-1598. It. DSB; ICB; Mort.; BLA. Author of a notable work on the anatomy of the horse which gives him a place among the founders of comparative anatomy.

1598. SIMI (or SIMIUS), Niccolo. ca. 1530-1564. It. P II. Astronomy.

1599. BUTRIGARIUS, Hercules. 1531-1612. It. ICB. Under BOTTRIGARI, Ercole: P I; GA. Theory of music. Astronomy.

1600. HERBEST, Benedict. 1531-1593. Cracow. ICB; P I. Mathematics.

1601. MOLETI, Giuseppe. 1531-1588. It. ICB; P II; GA. Astronomy. His *Tabulae Gregorianae* were compiled for the reform of the calendar.

1602. RAMELLI, Agostino. 1531-1600. It./Fr. ICB; P II; GA. Mechanical technology.

1. *Le diverse et artificiose machine*. Paris, 1588. In both Italian and French. RM.

1603. THURNEYSSER (or THURNYSER), Leonard. 1531-1596. Ger. DSB; ICB; P II; BMNH; BLA; GA; GC. Alchemy. Botany.

1604. CURAEUS, Joachim. 1532-1573. Ger. P I; BLA & Supp.; GC.

1. *Libellus physicus, continens doctrina de natura et differentiis colorum, sonorum, odorum, saporum et qualitatum tangibilium*. Wittenberg, 1572. RM.

1605. RULAND, Martin (first of the name). 1532-1602. Ger. P II; BLA; GA. Alchemy.

1606. WILHELM IV. (Landgrave of Hesse.) 1532-1592. Ger. DSB; P II; BMNH. Under WILLIAM: ICB; GA; GB; GC. Astronomy. Botany.

1607. XYLANDER (or HOLTZMANN), Wilhelm. 1532-1576. Ger. P II; GB; GC. Mathematics. Translator of many Greek mathematical works into Latin and of Euclid into German.

1608. DARIOT, Claude. 1533-1594. Fr. BLA.
  1. *Ad astrorum iudicia facilis introductio.* Lyons, 1557. X.
  1a. ——— *A breefe introduction to the astrologicall iudgement of the starres.* London, [1583?]. PR 6275. X. Reprinted, Hildesheim, 1971.

1609. DUDITH (or DUDITIUS), Andreas. 1533-1589. Hung./Ger. DSB; ICB; P I; BLA; GA. Astronomy/astrology. Mathematics.

1610. FABRICI (or FABRIZIO), Girolamo (or FABRICIUS AB AQUAPENDENTE, Hieronymus). ca. 1533-1619. It. DSB; ICB; P I (FABRIZIO AB A.); Mort. (FABRIZZI); BLA & Supp. (FABRICIUS); GA (FABRIZIO). Anatomy. Physiology. Embryology.
  1. *De venarum ostiolis.* Padua, 1603. X. Facsimile ed. with introd., trans., and notes by K.J. Franklin. Springfield, Ill., 1933.
  2. *Opera anatomica.* Padua, 1625.
  3. *Opera omnia anatomica et physiologica.* Leipzig, 1687. RM. Another ed., Leiden, 1738.
  4. *The embryological treatises: The formation of the egg and of the chick (De formatione ovi et pulli); The formed fetus (De formato foetu).* Ithaca, N.Y., 1942; reprinted, 1967. Facsimile ed. with introd., trans., and comm. by H.B. Adelmann.

1611. HEMMINGA, Sixtus ab. 1533-ca. 1586. Holl. P I. Author of *Astrologiae ratione et experientia confutae liber,* 1583.

1612. HILDERICUS (or HILLRICHS) VON VAREL, Edo. 1533-1599. Ger. P I. Astronomy.

1613. PENOT, Bernard George. ca. 1533-ca. 1620. Switz. P II & Supp. Alchemy. Iatrochemistry.

1614. ZABARELLA, Jacopo (or Giacomo). 1533-1589. It. DSB; P II; GA. Natural philosophy. Methodology.

1615. ZWINGER, Theodor (first of the name). 1533-1588. Switz. ICB; BMNH; BLA; GC. Physiology. Anatomy.

1616. CAMERARIUS, Joachim (second of the name). 1534-1598. Ger. BMNH; BLA; GA; GC. Botany.
  1. *Hortus medicus et philosophicus.* Frankfurt, 1588. RM.
  2. *Symbolorum et emblematum ex herbis et animalibus centuriae quattuor.* Nuremberg, 1605.

1617. COITER, Volcher. 1534-1576. Ger./It. DSB; ICB; Mort.; BLA & Supp.; GA; GC. Human and comparative anatomy. Embryology. Physiology.

1618. DORN (or DORNAEUS), Gerard. fl. 1566-1584. Switz./Ger. DSB; P I; BLA; GA. Alchemy/chemistry.

1619. MERLIÈRES (or DEMERLIERIUS), Jean de. d. 1580. Fr. P II. Mathematics.

1620. MUÑOZ (or MUNYOS), Geronimo. d. 1584. Spain. P II. Astrology.

1621. BILLINGSLEY, Henry. d. 1606. Br. P I; GA; GD.
  1. *The elements of geometrie of ... Euclide of Megara. Faithfully (now first) translated into the Englishe toung.* London, 1570. PR 10560. RM. With a preface by John Dee: see 1572/1.

1622. LOCHNER, Zacharias. d. 1608. Ger. P I. Metallurgy. Mathematics.

1623. GEMMA (-FRISIUS), Cornelius. 1535-1577. Belg. P I; BLA & Supp.; GA; GC. Astronomy.

1624. NEODOMUS, Nicolaus. 1535-1578. Ger. P II. Astronomy.

1625. PENA, Pierre. 1535-1605. Fr. BMNH. Botany. (PR record a London ed. of one of his works.)

1626. PEREIRA (or PERERIUS), Benedictus. 1535-1610. It. DSB; P II (PERERA); GA. Author of a widely used textbook of Aristotelian natural philosophy, dating from 1562, and of a work of 1591 which was a trenchant attack on alchemy, natural magic, and astrology.

1627. PORTA, Giambattista della. 1535-1615. It. DSB; ICB; P II; BMNH; GA; GB (DELLA PORTA). Natural magic. Optics. Founder of one of the earliest scientific academies.
1. *Magiae naturalis libri XX*. Naples, 1589. X. Another ed., Hanover, 1619.
1a. ——— *Magia naturalis; oder, Haus-, Kunst-, und Wunder-Buch*. Nuremberg, 1600.
1b. ——— *Natural magick*. London, 1658. Wing P2982. Reprinted, New York, 1957.
2. *De humana physiognomonica*. 1586. X. Another ed., Hanover, 1593.
2a. ——— *Della fisonomia dell'huomo*. Naples, 1610. Reprinted, Milan, 1971.
3. *Phytognomonica*. Naples, 1588. X. Another ed., Frankfort, 1591. Another ed., Rouen, 1650.
4. *De refractione*. Naples, 1593.

1628. RAUWOLF, Leonhard. 1535-1596. Ger. DSB; ICB; BMNH; GA; GC. Botany.
1. *Aigentliche Beschreibung der Raisz ... inn die Morgenländer*. Lauingen, 1582. X. Reprinted, Heidenheim, 1969, and Graz, 1971.
See also 11380/2.

1629. CARCANO LEONE, Giovanni Battista. 1536-1606. It. Mort.; BLA & Supp. Anatomy.

1630. DANTI (or DANTE or DANTES), Egnatio (Pellegrino Rainaldi). 1536-1586. It. DSB; ICB; P I(DANTE); GA (DANTE,I.). Astronomy. Cartography.
1. *Trattato dell'uso et della fabbrica dell'astrolabio*. Florence, 1569. RM.

1631. MONANTHEUIL (or MONANTHOLIUS), Henry de. ca. 1536-1606. Fr. P II; GA. Mathematics.

1632. PLATTER, Felix. 1536-1614. Switz. DSB; ICB; BMNH; BLA (PLATER); GA (PLATER). Anatomy. Botany.
1. *Beloved son Felix: The journal of Felix Platter, a medical student in Montpellier in the sixteenth century*. London, 1961. Trans. by S. Jennett. cf. the entry for his brother, Thomas--1856.
2. *Tagebuch (Lebensbeschreibung)*. Basel, 1976. Ed. by V. Lötscher.

1633. BAROCIUS, Franciscus. 1537-1604. It. DSB. Under BAROZZI: ICB; P I. Mathematics. Astronomy.

1. *Cosmographia*. Venice, 1585. RM (under BAROZZI).
2. *Admirandum illud geometricum problema tredecim modis demonstratum, quod docet duas linas in eodem plano designare, quae nunquam invicem coincidant, etiam si in infinitum protrahantur*. Venice, 1586. RM (under BAROZZI).

See also 1049/1.

1634. BOROUGH, William. 1537-1598. Br. GD.

1. *A discours of the variation of the cumpas*. London,1581. PR 3389 and three later eds. X.

1635. CLAVIUS, Christoph. 1537-1612. It. DSB; ICB; P I; GA; GC. Mathematics. Astronomy. The Gregorian calendar, introduced in 1582, was based on his calculations.

1. *In "Sphaeram" Joannis de Sacro Bosco commentarius*. Rome, 1570. RM. Another ed., 1607.
2. *Epitome arithmeticae practicae*. Rome, 1583. X.

2a. ——— *Aritmetica prattica*. Venice, 1652. RM.

3. *Horologium nova descriptio*. Rome, 1599. RM.
4. *Algebra*. Rome, 1608. X. Another ed., Geneva, 1609. RM.

1636. HORST, Jacob. 1537-1600. Aus./Ger. BMNH; BLA. Botany.

1. *Epistolae philosophicae et medicinales*. Leipzig, 1596.

1637. PRAETORIUS (or RICHTER), Johann. 1537-1616. Ger. P II; GA; GC. Mathematics. Astronomy.

1638. WITTICH, Johannes. 1537-1596. Ger. ICB; BLA; GC.

1. *Bericht von den bezoardischen Steinen*. Leipzig, 1589. RM.

1639. FULKE, William. 1538-1589. Br. BMNH; GA; GB; GD. Meteorology. (Titles in PR.)

1640. GALLUCCI, Giovanni Paolo. 1538-1621? It. P I; GA. Astronomy. A member of one of the earliest scientific academies, founded at Venice in 1593.

1. *Theatrum mundi et temporis*. Venice, 1588. RM.
2. *Speculum Uranicum*. Venice, 1593. RM.

1641. L'OBEL, Mathias de. 1538-1616. Fr./Belg./Br. DSB; ICB (LOBELIUS); BMNH; BLA; GA (LOBEL). Botany.

1. *Plantarum seu stirpium historia*. 2 vols. Antwerp, 1576. RM.
2. *In G. Rondelleti "Pharmaceuticam officinam" animadversiones*. London, 1605. PR 16650. RM.

Other titles in PR and Wing.

1642. RIOLAN Jean (first of the name). 1538-1605. Fr. BLA; GA. Author of *De primis principiis rerum naturalium*, 1571. Best known as a physician (see the DSB entry for his son of the same name).

1643. ACOSTA, José de. ca. 1539-1600. Spain/Peru. DSB; ICB; P I; BMNH; GA; GB. Natural history.

1. *Historia natural y moral de las Indias*. Seville, 1590. RM. Reprinted, Valencia, 1977.

1a. ——— *The naturall and morall historie of the East and West Indies*. London, 1604. PR 94. X. Reprinted, ib., 1880.

1644. AMMAN, Jost. 1539-1591. Switz. ICB; GA; GB; GC. Anatomical illustration.

1645. FLEISCHER, Johannes. 1539-1593. Ger. DSB; P I. Optics.

1646. GABELKHOVER (or GABELSCHOVER, etc.), Oswald. 1539-1616. Ger. P I; BLA (GAEBELKHOUER). Author of *De gemmis et lapidibus pretiosis*, 1603, and *Disputatio de generatione auri et ejus temperamento*, 1603.

1647. GRYNAEUS, Simon. 1539-1582. Switz./Ger. P I. Astronomy.

1648. RUBEUS, Hieronymus. ca. 1539-1607. It. P II; GA (ROSSI, G.). Author of a distillation book, 1582.

1649. BASSO, Sebastian. late 16th century. DSB; P I. One of the chief revivers of the atomic philosophy. Author of *Philosophia naturalis adversus Aristotelem*, 1574.

1650. NORMAN, Robert. fl. 1560-1596. Br. DSB; P II; GD. Magnetism. (Title in PR.)

1651. HEATH, Thomas. fl. 1567-1583. Br. GD. Astronomy. (Title in PR under HETH.)

1652. BREVENTANO. Stefano. d. 1577. It. P I; GA. Author of works on comets, earthquakes, and winds and their origin.

1653. RISNER, Friedrich. d. ca. 1580. Fr. DSB; ICB; P II. Mathematics. Optics.
   1. *Risneri optica cum annotationibus W. Snelli*. Ghent, 1918. Ed. by J.A. Vollgraff. Reprinted, Leiden, 1966.
   See also 1079.

1654. BUSCH, Georg. d. ca. 1590. Ger. P I. Astronomy.

1655. MAPLET, John. d. 1592. Br. ICB; GD.
   1. *A greene forest; or, A naturall historie*. London, 1567. PR 17296. X. Reprinted, ib., 1930. Deals with stones, plants, and animals.
   2. *The diall of destiny*. London, 1581. PR 17295. An astrological work. X.

1656. COSTEO, Giovanni. d. 1603. It. BMNH (COSTAEUS, Joannes); BLA; GA. Botany.
   1. *De universali stirpium natura*. Turin, 1578. RM.

1657. ALBERTI, Salomon. 1540-1600. Ger. DSB; ICB (ALBERT); P III; BMNH (ALBERTUS); BLA; GA; GC. Anatomy. Botany.

1658. BAKER, George. 1540-1600. Br. BLA & Supp.; GD. Pharmacy.
   1. *The composition or making of the ... oil called oleum magistrale, and the third book of Galen*. London, 1574. PR 1209. X. Reprinted, Amsterdam, 1969.

1659. BANISTER, John (first of the name). 1540-1610. Br. ICB; BLA & Supp.; GD. Anatomy.
   1. *The historie of man, sucked from the sappe of the most approved anathomistes*. London, 1578. PR 1359. X. Reprinted, Amsterdam, 1969.
   2. *Works*. 3 vols. London, 1633. PR 1357. X.

1660. BAUDERON, Brice. ca. 1540-1625. Fr. BLA & Supp.; GA.
   1. *Pharmacopea*. 2 parts. London, 1639. PR 1592. X.

1661. CEULEN, Ludolph van. 1540-1610. Holl. DSB; ICB; P I; GA (KEULAN). Mathematics.

1662. RICCI, Ostilio. 1540-1603. It. DSB; ICB. Mathematics.

1663. SCALIGER, Joseph Justus. 1540-1609. Fr./Holl. ICB; P II; BMNH; GA; GC. Astronomy. Best known as a historian and pioneer in chronology. See also 1749/4.

1664. SCULTETUS (or SCHULTZ), Bartholomäus. 1540-1614. Ger. P II; GC. Astronomy.

1665. VIÈTE, François. 1540-1603. Fr. DSB; ICB. Under VIETA: P II; GA; GB. Mathematics.
 1. *Canon mathematicus.* Paris, 1579. X.
 1a. ——— Another ed., entitled *Opera mathematica.* London, 1589. PR 24718. X.
 2. *Algèbre de Viète d'une méthode nouvelle.* Paris, 1636. Ed., with many alterations, by James Hume. RM.
 3. *Opera mathematica.* Leiden, 1646. Ed. by F. van Schooten. RM. Reprinted, Hildesheim, 1970.

1666. BAUHIN, Jean (or Johann). 1541-1613. Switz. DSB; ICB; BMNH; BLA & Supp.; GA; GC. Botany.
 1. *De plantis a divis sanctisve nomen habentibus ... Additae sunt Conradi Gesneri epistolae.* Basel, 1591. RM.
 2. *Historia plantarum universalis.* 3 vols. Yverdon, 1650-51. With J.H. Cherler.
 See also 1826 and 1510/4.

1667. GANZ, David ben Solomon. 1541-1613. Ger. ICB; GF (GANS). Astronomy.

1668. MERCATI, Michele. 1541-1593. It. DSB; ICB; P III; BMNH; BLA; GA. Botany. Mineralogy. "One of the founders of palaeontology."
 1. *Metallotheca. Opus posthumum.* Rome, 1717. Published after discovery of the MS more than a century after it was written. RM.

1669. ALLEN (or ALLAN or ALLEYN), Thomas. 1542-1632. Br. P I; GA; GB; GD. Mathematics. Astrology.

1670. SEVERINUS, Petrus (or SØRENSON, Peder). 1542 (or 1540)-1602. Den. DSB; BLA & Supp. (SOERENSEN). Iatrochemistry. Embryology--"The most eloquent exponent of epigenesis between Aristotle and Harvey." -DSB.

1671. THAL, Johannes. ca. 1542-1583. Ger. ICB; BMNH (THALIUS); BLA; GC. Botany.

1672. FREIGE, Johann Thomas. 1543-1583. Switz. P I. Under FREIGIUS: GA; GC. Mathematics. Disciple and biographer of Ramus.

1673. VAROLIO, Constanzo. 1543-1575. It. DSB; ICB; BLA. Anatomy.

1674. BARLOW, William. 1544-1625. Br. P I, IV; GD.
 1. *The navigators supply.* London, 1597. PR 1445. RM.
 2. *Magneticall advertisements ... concerning the nature and properties of the load-stone.* London, 1616. PR 1442/4. RM. Reprinted, Amsterdam, 1968.
 3. *A briefe discovery of the idle animadversions of Marke*

Ridley upon a treatise entituled "Magneticall advertisements." London, 1618, PR 1443/4. RM. cf. item 1776/2.

1675. DUCHESNE (or QUERCETANUS), Joseph. 1544 (or 1521)-1609. Fr./Switz. DSB; ICB; P II (QUERCETANUS); BLA & Supp. (CHESNE). Iatrochemistry.
 1. *Ad Iacobi Auberti Vindonis "De ortu et causis metallorum" brevis responsio.* Lyons, 1575. RM. Aubert's book is item 1680/1.
 PR record English translations of some of his works, including the above.

1676. GILBERT, William. 1544-1603. Br. DSB; ICB; P I; BLA & Supp.; GA; GB; GD. Magnetism. Cosmology. Earth sciences.
 1. *De magnete.* London, 1600. PR 11883. RM/L. Reprinted, Brussels, 1967.
 1a. ——— *On the loadstone.* New York, 1893; reprinted, 1958. Trans. by P.F. Mottelay.
 1b. ——— *On the magnet.* London, 1900; reprinted, New York, 1958. Trans. by S.P. Thompson.
 2. *De mundo nostro sublunari philosophia nova.* Amsterdam, 1651. RM. Reprinted, ib., 1965.

1677. MENIUS (or MEINE), Matthias. 1544-1601. Ger. P II. Astronomy.

1678. URSTISIUS, Christian. 1544-1588. Switz. P II; GC (WURSTICIUS). Astronomy. Mathematics.
 1. *Elementa arithmeticae.* Basel, 1579. X.
 1a. ——— *The elements of arithmeticke.* London, 1596. PR 24540. Trans. by T. Hood. X.

1679. LUTH, Oluf. d. 1580. Swed. ICB. Astronomy. Author of an adaptation in Swedish of Sacrobosco's *Sphaera.*

1680. AUBERT, Jacques. d. 1586. Fr. P I; BLA. Author of a commentary on Aristotle's *Physics* and of works opposing alchemy.
 1. *De metallorum ortu et causis, contra chemistas.* Lyons, 1575. RM. cf. item 1675/1.

1681. GOSSELIN, Guillaume. d. ca. 1590. Fr. P I; GA. Mathematics.

1682. ROTHMANN, Christoph. d. ca. 1600. Ger. DSB; P II; GC. Astronomy.

1683. BRESSIEU (or BRESSIUS), Maurice. d. 1617. Fr. P I & Supp. Astronomy.

1684. KHUNRATH (or KUNRATH, KUHNRATH, CUNRADIUS, CUNRATHUS), Conrad. fl. ca. 1595. Den. DSB. Iatrochemistry.

1685. FRANKE, Johannes. 1545-1617. Ger. ICB; BLA & Supp. Botany. Chemistry.

1686. GERARD, John. 1545-1612. Br. DSB; ICB; BMNH; GA; GB;GD. Botany. See also 15543/1.
 1. *Catalogus arborum, fruticum ac plantarum ... in horto J. Geradi ... nascentium.* London, 1596. PR 11748. RM.
 2. *The herball.* London, 1597. PR 11750. RM/L.
 2a. ——— Another ed., ib., 1633. PR 11751. Much enlarged by Thomas Johnson.
 2b. ——— Another ed., ib., 1636. PR 11752. Reprinted, ib., 1927 and 1964.

1687. MONTE, Guidobaldo del. 1545-1607. It. DSB; ICB (DEL MONTE); P II; GA (GUIDO UBALDO). Mechanics. Mathematics. Astronomy. "A prominent figure in the renaissance of the mathematical sciences." -DSB.
   1. *Mechanicorum liber*. Pesaro, 1577. RM.
   2. *Planisphaeriorum universalium theorica*. Pesaro, 1579. RM.

1688. TAGLIACOZZI, Gaspare. 1545-1599. It. ICB; BLA & Supp.; GA; GB. Anatomy.

1689. AGUILON (or AGUILLON), François d'. 1546-1617. Belg. DSB; ICB; P I; GA; GB. Optics.

1690. AVELLAR, André do. 1546-ca. 1621. Port. P I. Astronomy.

1691. BRAHE, Tycho. 1546-1601. Den. DSB; ICB; P I; GA; GB. Astronomy. See also Index.
   1. *De nova stella ... 1572, mense Novembri, primum conspecta*. Copenhagen, 1573. RM. Reprinted, Brussels, 1969.
   2. *De mundi aetherei recentioribus phaenomenis*. Uraniborg, 1588. RM.
   3. *Epistolarum astronomicarum liber*. Uraniborg, 1596. RM.
   4. *Astronomiae instauratae mechanica*. Wandsbeck, 1598. X. Reprinted, Stockholm, 1901, and Brussels, 1969.
   4a. ——— *Tycho Brahe's description of his instruments and scientific work*. Copenhagen, 1946. Trans. and ed. by H. Raeder.
   5. *Astronomiae instauratae progymnasmata*. Prague, 1602. X. Reprinted, Brussels, 1969.
   6. *Meteorologiske dagbog*. Copenhagen, 1876.
   7. *Opera omnia*. 15 vols. Copenhagen, 1913-29. Ed. by J.L.E. Dreyer. RM/L. Reprinted, 15 vols, Amsterdam, 1972.

   PR and Wing record English translations of some extracts from his works.

1692. CHAMBER, John. 1546-1604. Br. ICB; GD. Astronomy. (Title in PR.)

1693. DIGGES, Thomas. 1546?-1595. Br. DSB; ICB; P I; GA; GD. Mathematics. (Titles in PR.) See also 1530/2.

1694. SNELL, Rudolph. 1546-1613. Holl. P II; GA; GC. Mathematics.

1695. GUIBERT, Nicolas. ca. 1547-ca. 1620. Fr./It./Ger. DSB; P I; BLA; GA. A vehement opponent of alchemy.

1696. LAUREMBERG, Wilhelm. 1547-1612. Ger. P I; BMNH; BLA. Mathematics. Botany.

1697. STELLIOLA (or STIGLIOLA), Niccolo Antonio. 1547-1623. It. P II; GA. Author of *Il telescopio*, 1627. A member of the Accademia de' Lincei.

1698. BRUNO, Giordano. 1548-1600. It./Fr./Br./Ger. DSB; ICB; P I; GA; GB. Natural philosophy. Cosmology. Magic. (Titles in PR.)
   1. *De progressu et lampade venatoria logicorum*. n.p., 1587. RM.
   2. *De specierum scrutinio et lampade combinatorio Raymundi Lulli*. Prague, 1588. RM.
   3. *De monade numero et figura* ... [With] *De innumerabilibus*,

immenso et infigurabili, seu de universo et mundis. Frankfurt, 1591. RM.
4. *Praelectiones geometriae e Ars deformationum.* Rome, 1964. Unpubl. texts ed. by G. Aquilecchia.
Other individual works are not included.
5. *Opere.* Leipzig, 1830. Ed. by A. Wagner. RM/L.
6. *Opera latine.* 3 vols. Naples, 1879-91. Ed. by F. Fiorentino et al. X. Reprinted, 3 vols in 8, Stuttgart, 1962.
7. *Gesammelte Werke.* 6 vols. Leipzig, 1904-09. Ed. by L. Kuhlenbeck.

1699. NORTON, Samuel. 1548-ca. 1604. Br. GD. Alchemy.

1700. STEVIN, Simon. 1548-1620. Holl. DSB; ICB; P II (STEVINUS); GA; GB; GC. Mathematics. Mechanics. Astronomy. Theory of music.
1. *De thiende.* Leiden, 1585. X. Reprinted, Nieuwkoop, 1965. (Dutch classics on history of science. 15)
1a. ——— *Disme: The art of tenths, or decimall arithmetike.* London, 1608. PR 23264. X.
1b. ——— German trans. Frankfurt, 1965. (OKEW, n.F. 1)
2. *De weeghdaet.* Leiden, 1586. RM.
3. *De beghinselen der weeghconst.* Leiden, 1586. RM.
4. *De beghinselen des waterwichts.* Leiden, 1586. RM.
5. *De havenvinding.* Leiden, 1599. X.
5a. ——— *The haven finding art.* London, 1599. PR 23265. Trans. by E. Wright. X. Reprinted, Amsterdam, 1968.
6. *Les oeuvres mathématiques.* Leiden, 1634. Ed. by A. Girard. RM/L.
7. *The principal works.* 5 vols. Amsterdam, 1955-65. Ed. by E.J. Dijksterhuis.

1701. DYBVAD (or DIBUADIUS), Georg Christoph. 1649-1612. Den. P I. Mathematics. Astronomy. Meteorology.

1702. SAVILE, Henry. 1549-1622. Br. ICB; P II; GA; GB; GD. Mathematics. (Title in PR.)

1703. BELLI, Honorius. late 16th century. It. BMNH (BELLUS); GA. Botany.

1704. VARO, Michel. d. 1586. Switz. ICB (VARRO); P II. Mechanics.

1705. AUBERY, Claude. d. 1596. Fr. P I; BLA & Supp. Author of *De terrae motu*, 1585.

1706. MAGIRUS, Johann (first of the name). d. 1596. Ger. P II; BLA (MAGIRI). Natural philosophy. (PR record a Cambridge ed. of one of his works.)

1707. ADRIAANS (or DRUNAEUS), Gerardus. d. 1601. Belg. P I. Astronomy.

1708. BUONAMICI, Francesco. d. 1603. It. DSB. Natural philosophy. Dynamics.

1709. ZEISIUS, Matthaeus. d. 1607. Ger. ICB. Astronomy.

1710. KRABBE, Johann. fl. 1585-1604. Ger. P I. Astronomy.

1711. HARTGYLL, George. fl. 1594. Br. GD (HARTGILL). Astronomy. (Titles in PR and Wing.)

1712. ANTHONY, Francis. 1550-1623. Br. ICB; P I; BLA & Supp.; GA; GD. Iatrochemistry. (Titles in PR.) See also 2831.

1713. BEGUIN, Jean. ca. 1550-ca. 1620. Fr. DSB; ICB; P I. Iatrochemistry.
 1. *Tyrocinium chymicum.* n.p., 1610. RM. Another ed., Venice, 1643. RM.
 1a. ——— *Les élémens de chymie.* Paris, 1615. X. Another ed., ib., 1624. RM.
 1b. ——— *Tyrocinium chymicum; or, Chymical essays.* London, 1669. Wing B1703. Trans. by R. Russell.

1714. BOODT, Anselmus Boëtius de. ca. 1550-1632. Belg./Prague. DSB; ICB; P I, III; BMNH; BLA & Supp.; GA. Mineralogy. Botany.
 1. *Gemmarum et lapidum historia.* Hanau, 1609. RM.

1715. HABICOT, Nicolas. ca. 1550-1624. Fr. BLA; GA. Anatomy.
 1. *La semaine; ou, Practique anatomique.* Paris, 1610. X. Another ed., ib., 1631.

1716. IMPERATO, Ferrante. 1550-1625. It. BMNH; GA. Natural history.
 1. *Dell'historia naturale.* Naples, 1599. RM. Deals mainly with minerals, also plants and animals.

1717. LINOCIER, Geoffroy. ca. 1550-after 1620. Fr. BMNH; GA. Botany.

1718. MÄSTLIN, Michael. 1550-1631. Ger. DSB; ICB; P II (MOESTLIN); GA; GC. Astronomy.
 1. *Observatio et demonstratio cometae aetheri qui anno 1577 et 1578 ... apparuit.* Tübingen, 1578. RM.
 2. *Epitome astronomiae.* Heidelberg, 1582. X. Another ed., Tübingen, 1597. RM.
 3. *Ausführlicher und gründtlicher Bericht von der ... Jarrechnung.* Heidelberg, 1583. RM.

1719. MONTALTO, (Filotheo) Eliahu (or Elias). 1550-1616. Fr. ICB; P II; BLA Supp.; GF. Physiological optics.

1720. NAPIER, John. 1550-1617. Br. DSB; ICB; P II; GA; GB; GD. Mathematics. See also Index.
 1. *Mirifici logarithmorum canonis descriptio.* Edinburgh, 1614. PR 18349. RM.
 1a. ——— *A description of the admirable table of logarithmes.* London, 1616. PR 18351. Trans. by E. Wright. X. Reprinted, U.S.A., 1969.
 2. *Rabdologiae, seu numerationis per virgulas, libri duo.* Edinburgh, 1617. PR 18357. RM. Reprinted, Osnabrück, 1966.
 3. *Mirifici logarithmorum canonis constructio.* Edinburgh, 1619. X.
 3a. ——— *The construction of the wonderful canon of logarithms.* Edinburgh, 1889. Trans. with notes and a bibliography of Napier by W.R. Macdonald. Reprinted, London, 1966.

1721. ROBIN, Jean. 1550-1629. Fr. BMNH; GA. Botany.

1722. STADIUS, Georg. ca. 1550-1593. Aus. GC. Astronomy.

1723. URSUS, Nicolas Reimarus. ca. 1550-1599. Prague. ICB; P II (REIMARUS URSUS); GC. Mathematics. Astronomy.

1724. SKOMAGER, Hans Rasmussen. 1551-1614. Den. P II. Astronomy. Also author of *De calore*, 1613.

1725. VERANZIO, Fausto. 1551-1617. It. ICB; P II. Mechanical technology.
1. *Machinae novae.* Venice, [1615?]. X. Reprinted?

1726. BÜRGI, Joost. 1552-1632. Ger./Prague. DSB; ICB; P I; GA (BYRGE); GC (BURGI). Mathematics. Astronomy.

1727. CASSERI, Giulio. 1552-1616. It. DSB; BMNH; GA. Under CASSERIO: ICB; Mort.; BLA & Supp. Human and comparative anatomy.
1. *De vocis auditusque organis historia anatomica.* Ferrara, 1600-01. RM.
2. *Tabulae anatomicae.* Venice, 1627. X. Reprinted, New York, 197-?

1728. CATALDI, Pietro Antonio. 1552-1626. It. DSB; ICB; P I; GA. Mathematics.
1. *Trattato del modo brevissimo di trovare la radice quadra delli numeri.* Bologna, 1613. RM.

1729. OWEN, George. 1552-1613. Br. DSB; ICB; GD. Geology.

1730. PLATT, Hugh. 1552-1608. Br. BMNH; GD (PLAT). Botany. (Titles in PR.)

1731. RICCI, Matteo. 1552-1610. China. The famous Jesuit missionary. DSB; ICB; P II; GA; GB. In China "he disseminated Western science [chiefly mathematics, astronomy, and geography] by lecturing, publishing books and maps, and making instruments." See also entry 947.

1732. SARPI, Paolo. 1552-1623. It. The theologian. DSB; ICB; P II; GA; GB. Natural philosophy. He was closely associated with the scientific movement.

1733. SCHOLZ, Lorenz. 1552-1599. Ger. BMNH; BLA (SCHOLTZ VON ROSENAU); GC. Botany.

1734. SWEERTS, Emanuel. 1552-after 1612. Belg. BMNH; GF. Botany.

1735. VALERIO (or VALERI), Luca. 1552-1618. It. DSB; ICB; P II. Mathematics. Mechanics. Member of the Accademia dei Lincei.

1736. VILLALPANDO, Juan Bautista. 1552-1608. Spain/It. DSB; GA. Mathematics. Mechanics. Architecture.

1737. ALPINI, Prospero. 1553-1616. It. DSB; BMNH; BLA & Supp.; GA; GB. Botany.
1. *De balsamo dialogus.* Venice, 1591. X.
1a. ——— *Histoire du baulme.* Lyons, 1619. RM.
2. *De plantis Aegypti.* Venice, 1592. RM.
3. *Historiae Aegypti naturalis.* 2 vols. Leiden, 1735. RM/L. Vol. II is an ed. of *De plantis Aegypti* by J. Vesling who added to it two of his own works, *Paraenses ad rem herbariam* and *Vindicae opobalsami.*

1738. BALDI (D'URBIN), Bernardino. 1553-1617. It. DSB; ICB; P I; GA. Mechanics.
1. *In mechanica Aristotelis problemata exercitationes. Adiecta succinta narratione de autoris vita et scriptis.* Mainz, 1621. RM.

2. *Cronica de' matematici; overo, Epitome dell'istoria delle vite loro.* Urbino, 1707. Completed in 1589 but not publ. until 1707 and then only in part. RM.

See also 1032/2.

1739. MOFFETT (or MUFFET, etc.), Thomas. 1553-1604. Br. DSB; ICB (MOUFET); BMNH; BLA (MOUFET); GA; GD. Entomology.
1. *De jure et praestantia chymicorum medicamentorum.* Frankfurt, 1584. RM.

Other titles (entomological) in PR.

1740. CHRISTMANN, Jacob. 1554-1613. Ger. DSB; ICB; P I; GA; GC. Mathematics. Astronomy.

1741. CORTESI, Giovanni Battista. 1554-1634. It. BLA & Supp.; GA. Anatomy.

1742. DE GROOT, Jan Cornets (or John Hugo) (or GROTIUS, Janus). 1554-1640. Holl. DSB; GF (GROOT, Johan Hugo de). Mechanics.

1743. WALDUNG (or BALDUNG), Wolfgang. 1554-1621. Ger. P II; BMNH; BLA; GC. Natural philosophy. Meteorology.

1744. HOOD, Thomas. fl. 1577-1596. Br. ICB; GD. Mathematics. (Titles in PR.)

1745. LANGHAM, William. fl. 1579-1633. Br. BMNH. Author of a herbal, 1579. (Title in PR.)

1746. LEEUWEN (or LEONINUS), Albert van. d. 1614. Holl. P I. Astronomy.

1747. BONAVENTURA, Federigo. 1555-1602. It. DSB; P III; GA. Meteorology.

1748. KELLEY (or KELLY or TALBOT), Edward. 1555-1595. Br./Prague. P I; GA; GD. Alchemy.
1. *The alchemical writings.* London, 1893. Trans. from the Hamburg ed. of 1676 and ed. with a biographical preface by A.E. Waite. Reprinted, New York, 1970.

1749. MAGINI, Giovanni Antonio. 1555-1617. It. DSB; ICB; P II; GA. Mathematics. Astronomy/astrology. Geography.
1. *Novae coelestium orbium theoricae congruentes cum observationibus N. Copernici.* Venice, 1589. RM.
2. *De astrologica ratione.* Venice, 1607. RM.
3. *Breve instruttione sopra l'apparenze et mirabili effetti dello specchio concavo sferico.* Bologna, 1611. RM.
4. *Confutatio diatribae Ios. Scaligeri de aequinoctiorum praecessione.* Rome, 1617. RM.

1750. WITTICH, Paul. 1555?-1587. Breslau. DSB; ICB; GC. Mathematics.

1751. CALVISIUS (or KALWITZ), Sethus. 1556-1615. Ger. P I; GA; GB; GC. Astronomy. Best known as a musician.
1. *Elenchus calendarii Gregoriani.* Frankfurt, 1612. RM.

1752. ECKSTROM, Heinrich. 1557-1622. Ger. ICB; P III. Astronomy.

1753. HERLICIUS (or HERLICH), David. 1557-1636. Ger. ICB; P I; BLA & Supp. (HERLITZ); GA; GC. Astrology/astronomy. Meteorology.

1754. TANK (or TANCKE), Joachim. 1557-1609. Ger. P II; BLA (TANCK). Alchemy/chemistry.

1755. DU LAURENS. André. 1558-1609. Fr. GA. Under LAURENS: DSB; BLA. Under LAURENTIUS: ICB; BMNH. Anatomy.
    1. *Opera anatomica*. Lyons, 1593.
    2. *Historia anatomica humani corporis*. Paris, 1600. Another ed., Frankfurt, 1600. A widely used textbook.
    3. *Opera omnia anatomica et medica*. Frankfurt, 1627.

1756. ORIGANUS (or TOST), David. 1558-1628. Ger. P II. Astronomy. Astrology. (PR record a London ed. of his ephemerides.)

1757. PAPIUS (or PAPE), Johann. 1558-1622. Ger. P II. Iatrochemistry.

1758. ZALUŽANSKÝ ZE ZALUŽAN, Adam. ca. 1558-1613. Prague. DSB; BMNH; BLA. Botany.

1759. HAPELIUS, Nicolaus Niger. 1559-1622. ICB. Alchemy/chemistry.

1760. PLINIUS, Basilius. d. 1604. Riga. P II. Author of *De colorum natura*, 1599; *De magnete*, 1603; *De ventis*, 1603.

1761. DURET, Claude. d. 1611. Fr. BMNH; GA. Botany.

1762. SCHENCK VON GRAFENBERG, Johann Georg. d. 1620. Ger. BMNH; BLA; GC. Botany.

1763. FIGULUS, Benedictus. fl. 1587-1607. Alchemy.
    1. *Pandora magnalium naturalium ... de benedicto lapidis philosophorum mysterio*. Strasbourg, 1608. X.
    1a. ——— *A golden and blessed casket of nature's marvels*. London, 1893. Trans. by A.E. Waite. X. Reprinted, ib., 1963.

1764. CARGILL, James. fl. 1605. Br. GD. Botany.

1765. BAUHIN, Gaspard (or Caspar). 1560-1624. Switz. DSB; ICB; BMNH; Mort.; BLA; GA; GB. Anatomy. Botany.
    1. *De lapidis Bezaar*. Basel, 1613. RM.
    2. [*Prodromos*] *theatri botanici*. Frankfurt, 1620. RM.
    3. *Vivae imagines partium corporis humani*. Basel, 1620. RM.
    4. [*Pinax*] *theatri botanici; sive, Index in Theophrasti, Dioscoridis, Plinii ... opera*. Basel, 1623. RM. 2nd ed. ib., 1671.

1766. BUTLER, Charles. 1560-1647. Br. BMNH; GA; GD.
    1. *The feminine monarchie; or, A treatise concerning bees*. Oxford, 1609. PR 4192 and two later eds. X. Reprinted, Amsterdam, 1969.

1767. CROLL, Oswald. ca. 1560-1609. Ger./Prague. DSB (CROLLIUS); ICB; P I; BLA & Supp.; GA; GC. Iatrochemistry.
    1. *Basilica chymica*. Frankfurt, 1609. RM.
    1a. ——— Another ed., Leipzig, 1634. Ed. by J. Hartmann.
    1b. ——— ——— *Bazilica chymica ... or, Royal and practical chymistry*. London, 1670. Wing C7022. X.
    1c. ——— ——— *Philosophy reformed and improved*. London, 1657. Wing C7023. X. Trans. of preface.

1768. DOMINIS, Marco Antonio de. 1560-1626. It./Br. DSB; ICB; P I; GA; GB; GD. Optics. Also a work on the tides.
    1. *De radiis visus et lucis in vitris perspectivis et iride*. Venice, 1611. RM.

1769. FABRICIUS (HILDANUS), Wilhelm. 1560-1634. Ger. BLA & Supp.;

GA; GC. Anatomy.
1. *Von der Fürtrefflichkeit und Nutz der Anatomy.* Aarau, 1936. Ed. from the MS by F. de Quervain.
2. *Opera qui extant omnia.* Frankfurt, 1646. Ed. by G. Scheffer. Another ed., ib., 1682.

1770. HARRIOT (or HARIOT), Thomas. ca. 1560-1621. Br. DSB; ICB; P I; GA; GB; GD. Mathematics. Astronomy. Mechanics. Optics.
1. *A briefe and true report of the new found land of Virginia.* London, 1588. PR 12785. Includes natural history. X.
2. *Artis analyticae praxis, ad aequationes algebraicas nova, expedita et generali methodo resolvendas.* London, 1631. PR 12784. RM.
See also 1862.

1771. HOGELANDE, Theobald van. ca. 1560-ca. 1608. Holl. GF. Alchemy.

1772. KHUNRATH, Heinrich. ca. 1560?-1605. Ger. DSB; ICB; P I; GA (KUNRATH). Alchemy.

1773. LAVATER, Heinrich. 1560-1623. Switz. P I; BLA. Natural philosophy. Attributed to him are many theses, presumably of his students, on such topics as earthquakes, the elements, metals, meteorology, comets, etc.

1774. LIBAVIUS (or LIBAU, etc.), Andreas. ca. 1560-1616. Ger. DSB; ICB (LIEBAU); P I; BMNH; BLA; GA; GC. Chemistry.
1. *Alchemia.* Frankfurt, 1597. RM.
1a. ——— *Die Alchemie: Ein Lehrbuch der Chemie.* Weinheim, 1964. The first German trans. By F. Rex.

1775. RENEAULME, Paul de. ca. 1560-1624. Fr. BMNH (RENEALMUS); BLA; GA. Botany.

1776. RIDLEY, Mark. 1560-1624. Br. P IV; GD.
1. *A short treatise of magneticall bodies and motions.* London, 1613. PR 21045. RM.
2. *Magneticall animadversions upon certaine magneticall advertisements from W. Barlow.* London, 1617. PR 21044. X. cf. item 1674/3.

1777. BACON, Francis. 1561-1626. Br. DSB; ICB; P I; GA; GB; GD. "His place in the history of science rests chiefly upon his natural philosophy, his philosophy of scientific method, his projects for the practical organization of science, and the influence of all these upon the science of the later seventeenth century." -DSB. (Titles in PR and Wing.)
1. *The works.* 14 vols. London, 1857-74. Ed. by J. Spedding et al. Reprinted, New York, 1968.

1778. BESLER, Basilius. 1561-1629. Ger. BMNH; BLA & Supp.; GA; GC. Natural history.
1. *Hortus Eyestettensis.* 1713. X. Reprinted, Munich, 1964.

1779. BRIGGS, Henry. 1561-1631. Br. DSB; ICB; P I; GA; GB; GD. Mathematics.
1. *Logarithmorum chilias prima.* London, 1617. PR 3741. RM.
1a. ——— *The first chiliad of logarithmes.* [n.p., 1630?] PR 3742. X.
2. *Arithmetica logarithmica.* London, 1624. PR 3739. X.
2a. ——— *Logarithmicall arithmeticke.* Ib., 1631. PR 3740. X.

3. *Trigonometria Britannica.* Gouda, 1633. Completed after Brigg's death by H. Gellibrand. RM.
3a. ——— *Trigonometria Britannica; or, The doctrine of triangles.* London, 1658. Wing G479. X.
See also 3392/1.

1780. BRY, Johann Theodor de. 1561-1623. BMNH. Botany.

1781. CHALONER, Thomas. 1561-1615 (or 1559-1603). Br. P III (CHALLONER); GA; GD.
1. *A shorte discourse of the most rare vertue of nitre.* London, 1584. PR 4940. X.

1782. FINK, Thomas. 1561-1656. Den. DSB; P I; BLA & Supp. (FINCKE); GA (FINCK); GC (FINCK). Mathmetics. Astrology/astronomy.

1783. GRIENBERGER, Christoph. 1561-1636. Aus./It. ICB; P I. Mathematics. Astronomy.

1784. HÉROARD, Jean. 1561-1627. Fr. BLA; GA. Comparative anatomy.

1785. LANSBERGE, Philip van. 1561-1632. Holl. DSB; P I; GA; GC; GF. Geometry. Astronomy.
1. *Commentationes in motuum terrae diurnum et annum.* Middelburg, 1630. RM.
2. *Tabulae motuum coelestium perpetuae.* Middelburg, 1632.
3. *Opera omnia.* Middelburg, 1663. RM.
See also 1994/1.

1786. PITISCUS, Bartholomeo. 1561-1613. Ger. DSB; ICB; P II; GA; GC. Mathematics. (PR record English translations of some of his works.)

1787. ROOMEN, Adriaan van. 1561-1615. Belg./Ger. DSB; ICB; P I (ADRIANUS ROMANUS); BLA (ROMANUS); GA (ROMAIN). Mathematics.

1788. SANTORIO (or SANCTORIUS), Santorio. 1561-1636. It. DSB; ICB; P II; Mort.; BLA; GA. Physiology.
1. *De statica medicina.* Venice, 1614. X. Another ed., with commentary by M. Lister, London, 1701.
1a. ——— *Medicina statica; or, Rules of health.* London, 1676. Wing S571. X.
1b. ——— *Medicina statica. Being the aphorisms of Sanctorius.* London, 1712. 4th ed., ib., 1728.

1789. TARDE, Jean. 1561-1636. Fr. DSB; P II, IV. Astronomy.

1790. WRIGHT, Edward. 1561-1615. Br. DSB; ICB; P II; GA; GD. Mathematics. Cartography. (Titles in PR.) See also 1700/5a and 1720/1a.

1791. SEVERIN (or LONGOMONTANUS), Christian. 1562-1647. Den. Under LONGOMONTANUS: P I; GA; GB. Astronomy. A disciple of Tycho Brahe.

1792. MALLEOLUS (or HÄMMERLEIN), Isaac. b. ca. 1563. Strasbourg. P II. Mathematics. Astronomy.

1793. SCHWENCKFELD, Caspar. 1563-1609. Ger. P II (SCHWENKFELD); BMNH; BLA. Pharmacy. Natural history.

1794. BELLEVAL, Pierre Richer de. ca. 1564-1632. Fr. DSB; BMNH; BLA; GA. Botany.

1795. FABRICIUS, David. 1564-1617. Ger. P I; GA; GC. Astronomy.
A diligent correspondent of Kepler's.

1796. FRANZ (or FRANTZ), Wolfgang. 1564-1628. BMNH (FRANZIUS).
Zoology.
1. *Historia animalium sacra.* Wittenberg, 1612. X. Another ed., Frankfurt, 1671. RM.
1a. ——— *The history of brutes.* London, 1670. Wing F2094. X.

1797. GALILEI, Galileo. 1564-1642. It. DSB; ICB; P I; G. Mechanics.
Astronomy. See also Index.
1. *De motu.* Unpublished treatise, ca. 1590.
1a. ——— *On motion, and On mechanics.* Madison, Wis., 1960. Trans. of *De motu* and *Le meccaniche* (item 2) with introd. and notes by I.E. Drabkin and S. Drake.
2. *Le meccaniche.* Unpublished treatise, ca. 1600.
2a. ——— *Les méchaniques.* Paris, 1634. Trans. by M. Mersenne. X. Critical ed. by B. Rochot, ib., 1966.
2b. ——— English trans.: see item 1a.
3. *Le operazioni del compasso geometrico et militare.* Padua, 1606. RM.
3a. ——— *Operations of the geometric and military compass.* Washington, D.C., 1978. Trans. with an introd. by S. Drake.
4. *Difesa ... contro alle calunnie ... di B. Capra ... sopra la nuova stella del 1604.* Venice, 1607. RM.
5. *Sidereus nuncius.* Venice, 1610. RM. Another ed., London, 1653. Wing G167. Original ed. reprinted, Brussels, 1967.
5a. ——— *The sidereal messenger, and a part of the preface to Kepler's "Dioptrics" containing the original account of Galileo's astronomical discoveries.* London, 1880. Trans. with introd. and notes by E.S. Carlos. Reprinted, ib., 1960.
5b. ——— Another English trans.: see item 20.
6. *Discorso ... intorno alle cose che stanno in sù l'acqua, ò che in quella si muovono.* Florence, 1612. RM.
6a. ——— *Discourse on bodies in water.* Urbana, Ill., 1960. The Salusbury trans. (see entry 2808) with introd. and notes by S. Drake.
7. *Istoria e dimonstrazioni intorno alle macchie solari ... tre lettere.* 2 parts. Rome, 1613. RM.
7a. ——— English trans.:see item 20.
8. *Discorso sulle comete.* Florence, 1619. X.
8a. ——— English trans.: see entry 2829.
9. *Il saggiatore.* Rome, 1623. RM. Reprinted, Milan, 1965.
9a. ——— English trans.: see item 20.
10. *Dialogo ... sopra i due massimi sistemi del mondo.* Florence, 1632. RM/L. Several 20th-century reprints. Wing records two London eds, both in 1663 (G165 and G168).
10a. ——— *Dialogue on the great world systems.* Chicago, 1953. The Salusbury trans. (see entry 2808) with introd. and notes by G. de Santillana.
10b. ——— *Dialogue concerning the two chief world systems.* Berkeley, Cal., 1953.; 2nd ed., 1967. Trans. by S. Drake with a foreword by A. Einstein.
11. *Lettera a Madama Cristina di Lorena.* Strasbourg, 1636. Written in 1615 but not publ. until after Galileo's condemnation. X.

11a. ——— English trans.: see item 20.

12. *Nov-antiqua sanctissimorum Patrum et probatorum theologorum doctrina. De Sacrae Scripturae testimonio....* Strasbourg, 1636. Written before Galileo's trial but not publ. until after it. The Latin trans. (by E. Deodatus) is printed alongside the Italian original. RM.

13. *Discorsi e dimonstrazioni matematiche intorno a due nuove scienze.* Leiden, 1638. RM. Reprinted, Turin, 1958, and Brussels, 1966.

13a. ——— *Les nouvelles pensées de Galilée.* Paris, 1639. Trans. by M. Mersenne. X. Critical ed. by P. Costabel and M.P. Lerner. 2 vols. Paris, 1973.

13b. ——— *Mathematical discourses and demonstrations touching two new sciences.* London, 1665. Wing G166. X. Evidently a separate publn of one of Salusbury's translations; see entry 2808.

13c. ——— German trans. 3 vols. Leipzig, 1890-91. (OKEW)

13d. ——— *Dialogues concerning two new sciences.* New York, 1914. Trans. by H. Crew and A. de Salvio. Reprinted, Evanston, Ill., 1939, and New York, 1952 and 1963.

13e. ——— *Two new sciences.* Madison, Wis., 1974. Trans. by S. Drake. Includes *Centers of gravity* and *Force of percussion.*

Letters and Notebooks

14. *Due insigni autografi di G. Galilei e E. Torricelli.* Florence, 1908.

15. *Galileo's letter about the libration of the moon.* New York, 1965. A trans. and critical study by S. Prete.

16. *A letter from Galileo ... to Peiresc on a magnetic clock.* Norwalk, Conn., 1967. By B. Dibner and S. Drake.

17. *Galileo's early notebooks: The physical questions.* Notre Dame, Ind., 1977. Trans. from the Latin with comm. by W.A. Wallace.

Collected Works

18. *Opere. Edizione nazionale.* 21 vols. Florence, 1890-1909. RM/L. Reprinted 1929-39 and 1964-66.

Selections

19. *Opere.* Milan, 1953. Ed. by F. Flora.

20. *Discoveries and opinions of Galileo.* Garden City, N.Y., 1957. Trans. with introd. and notes by S. Drake. Contents: *The starry messenger; Letters on sunspots; Letter to Grand Duchess Christina;* and excerpts from *The assayer.*

21. *Scritti letterari.* Florence, 1970. Ed. by A. Chiari.

22. *Opere.* 2 vols. Naples, 1970. Ed. by A. Pacchi.

23. *Galileo against the philosophers.* Los Angeles, 1976. Trans. with introd. and notes by S. Drake.

1798. GAULTIER (DE LA VALETTE), Joseph. ca. 1564-1647. Fr. ICB; P I. Astronomy.

1799. MULERIUS (or MULIERIUS), Nicolaus. 1564-1630. Holl. P II; BLA; GA (MULIERS); GF. Astronomy.

1. *Institutionum astronomicarum libri duo.* Groningen, 1616. RM.

1800. PAAW (or PAVIUS), Petrus. 1564-1617. Holl. BMNH. Under PAUW: BLA; GA; GC. Anatomy. Botany.

1801. TORPORLEY, Nathaniel. 1564-1632. Br. P II; GD. Astronomy. (Title in PR.)

1802. GRAU (or GRAVIUS), Johann. fl. 1599-1605. Ger. P I; BLA. Many dissertations, presumably of his students, are attributed to him, e.g. *De meteoris*, *De metallis*, *De fossilibus*, *De elementis*, etc.

1803. CHIARAMONTI (or CLARAMONTIUS), Scipione. 1565-1652. It. P I; GA. Astronomy. About 1630 he was one of Galileo's main opponents.

1804. GOTHUS, Laurentius Paulinus. 1565-1646. Swed. ICB. Astronomy.

1805. LOEW, Johann Georg. 1565-1610. Switz. P I. Many dissertations, presumably of his students, are attributed to him, e.g. *De principiis corporis naturalis*, *De respiratione*, *De mundo*, *De meteoris*, *De terrae motu*, *De putredine*, etc.

1806. AIGUILLON, François d'. 1566-1617. Belg. GF. Optics.

1807. BIANCANI, Giuseppe (or BLANCANUS, Joseph). 1566-1624. It. ICB; P I; GA. Author of *Clarorum mathematicorum chronologia*, 1615.

1808. FROBENIUS, Georg Ludwig. 1566-1645. Ger. P I; GC (FROBEN). Astronomy. An associate of Tycho Brahe.

1809. GHETALDI (or GHETTALDI), Marino. 1566-1626. It. DSB; ICB; P I. Mathematics.
 1. *Opera omnia.* Zagreb, 1968.

1810. SENDIVOGIUS, Michael. 1566-1636. Cracow/Prague. DSB; ICB; P II. Alchemy.
 1. *Novum lumen chymicum.* Cologne, 1614. X.
 1a. ——— *A new light of alchymie.* London, 1650. Wing S2506 and F2174. Trans. by John French who added "a chymicall dictionary explaining hard places and words met withall in the writings of Paracelsus and other obscure authors." RM/L.

1811. BALLO (or BALLI), Giuseppe. 1567-1640. It. ICB; P I; GA. Mechanics.

1812. BERGIER, Nicolas. 1567-1623. Fr. P I; GA. Astronomy.

1813. COLONNA, Fabio. 1567-1650. It. P I; BMNH; BLA (COLUMNA); GA. Natural history.
 1. [*Phytobasayos*]; *sive*, *Plantarum aliquot historia.* Naples, 1592. RM.
 2. *Minus cognitarum stirpium aliquot.* Rome, 1606. RM.

1814. FEYENS, Thomas. 1567-1631. Belg. ICB; P I (FIENUS); BLA & Supp.; GA (FYENS); GF. Astronomy.

1815. PARKINSON, John. 1567-1650. Br. BMNH; GD. Botany.
 1. *Theatrum botanicum. The theater of plants; or, An herball of large extent.* London, 1640. PR 19302.
 Other titles in PR and Wing.

1816. CAMPANELLA, Tommaso. 1568-1639. It. DSB Supp.; ICB; P I; BMNH; GA; GB. Natural philosophy.
1. *Apologia pro Galileo*. Frankfurt, 1622. RM.
2. *Lettere*. Bari, 1927. Ed. by V. Stampanato.
3. *Opera latina*. Turin, 1975. Ed. by L.Firpo.

1817. HARTMANN, Johannes. 1568-1631. Ger. DSB; P I; BLA & Supp.; GA. Iatrochemistry.
1. *Praxis chymiatrica*. Leipzig, 1633.
See also 1767/1a.

1818. MAIER, Michael. ca. 1568-1662. Ger. DSB; ICB; P II (MAYER); GA. Alchemy.
1. *Symbola aureae mensae duodecim nationum*. Frankfurt, 1617. X. Reprinted, Graz, 1972.
2. *Viatorum; hoc est, De montibus planetarum septem; seu, Metallorum tractatus*. Oppenheim, 1618. RM.
3. *Atlanta fugiens; hoc est, Emblemata nova de secretis naturae chymica*. Oppenheim, 1618. X. Reprinted, Kassel, 1964.
3a. ——— Another ed., entitled *Secretis natura secretorum scrutinum chymicum*. Frankfurt, 1687. RM.
3b. ——— *Atlante fugitive*. Paris, 1969. French trans. by E. Perrot.

1819. BARBA, Alvaro Alonso. 1569-ca. 1640. Quito. DSB; ICB; P I; BMNH; GA. Metallurgy.
1. *El arte de los metales*. Madrid, 1640. RM.
1a. ——— *The first book of the art of metals ... The second book....* London, 1670. Wing B679, 681. X.
1b. ——— *A collection of scarce and valuable treatises upon metals, mines and minerals*. London, 1738. RM. 2nd ed., 1740.
1c. ——— *El arte de los metales, translated....* London, 1923. English trans. by R.E. Douglas.

1820. JORDEN, Edward. 1569-1632. Br. P I; BLA; GA; GD. Chemistry. (Title in PR and Wing.)

1821. RULAND, Martin (second of the name). 1569-1611. Ger./Prague. DSB; BLA. Iatrochemistry. Alchemy.
1. *Lexicon alchemiae; sive, Dictionarium alchemisticum cum obscuriorum verborum et rerum hermeticarum ... explicationem continens*. Frankfurt, 1612. X. Reprinted, Hildesheim, 1964.

1822. CHESNECOPHERUS, Niels. d. 1634. Swed. P I. Mathematics.

1823. DESPAGNET, Jean. fl. 1616-1633. Fr. DSB; GA (ESPAGNET). Alchemy.

1824. ARGOLI, Andrea. 1570 (or 1568)-1657. It. DSB; ICB; P I; GA. Astrology/astronomy.
1. *Ephemeridum iuxta Tychonis hypotheses et coelo deductas observationes. Ab anno 1631. Ad 1655*. Padua, 1638. RM.
2. *Ephemerides annorum L iuxta Tychonis hypotheses et accurate e coelo deductas observationes ab anno 1630 ad annum 1680*. Venice, 1638. RM.
3. *Ptolemaeus parvus in genethliacis junctus Arabibus*. Lyons, 1652. RM. Another ed., ib., 1659. RM.

1825. BRUYN, Niklaas de. 1570-1641. Holl. BMNH. Ichthyology.

1826. CHERLER, Johann Heinrich. 1570-1610. BMNH. Co-author with Jean Bauhin (entry 1666) of *Historia plantarum universalis.*

1827. CYRIAQUE DE MANGIN, Clement. ca. 1570-1642. Fr. P I; GA (CYRIACUS). Mathematics.

1828. FABER, Johann. ca. 1570-ca. 1640. Ger. ICB; BMNH; BLA & Supp.; GA. Anatomy. Zoology. Botany.

1829. HOUTMAN, Frederick de. 1570-1627. Holl. ICB; GA; GF. Author of a catalogue of southern stars.

1830. MINDERER, Raymond. ca. 1570-1621. Fr. P II & Supp.; BLA. Iatrochemistry.

1831. PASSE (or PAS or PASSEUS), Crispin de (the younger). b. 1570. Holl. BMNH.
1. *Hortus floridus.* Arnhem, 1614. In Latin. X.
1a. ——— *A garden of flowers. Translated from Netherlandish.* 2 parts. Utrecht, 1615. PR 19459. X.
1b. ——— ——— *Hortus floridus.* 2 parts. London, 1928-29. Reprint of the English trans. of 1615 together with a trans. from the Latin. By S. Savage.

1832. BLAEU, Willem Janszoon (or CAESIUS, Guilelmus). 1571-1638. Holl. DSB; ICB; P I; GA. Astronomy (he worked with Tycho Brahe in 1595-96). Best known as a cartographer.
1. *Institutio astronomica.* Amsterdam, 1634. X. Another ed., Oxford, 1663. Wing B3108. X.
1a. ——— *Institution astronomique de l'usage des globes et sphères célestes et terrestres. Comprise en deux parties, l'une suivant l'hypothèse de Ptolemée ... l'autre selon l'intention de N. Copernicus.* Amsterdam, 1642. RM.
1b. ——— *A tutor to astronomy.* London, 1654. Wing B3109. X.
Maps and geographical works not included.

1833. BRANCA, Giovanni. 1571-1640. It. P I & Supp.; GA.
1. *Le machine.* Rome, 1629. RM.

1834. KEPLER, Johannes. 1571-1630. Ger./Aus./Prague. DSB; ICB; P I (KEPPLER); G. Astronomy. Optics. See also Index.
1. *Mysterium cosmographicum.* Tübingen, 1596. RM.
1a. ——— *Das Weltgeheimnis.* Augsburg, 1923; reprinted, Munich, 1936. Trans. with an introd. by M. Caspar.
2. *Ad Vitellionem paralipomena, quibus astronomiae pars optica traditur.* Frankfurt, 1604. RM. Reprinted, Brussels, 1968. Referred to by Kepler and others as *Astronomiae pars optica.*
2a. ——— German trans. Leipzig, 1922. (OKEW 198)
3. *De stella nova.* Prague, 1606. RM.
4. *Astronomia nova.* Prague, 1609. RM. Another ed., Leipzig, 1609.; reprinted, Brussels, 1968.
4a. ——— *Neue Astronomie.* Munich, 1929. Trans. with an introd. by M. Caspar.
5. *Dissertatio cum "Nuncio sidereo" nuper ad mortales misso a Galilaeo Galilaeo,* Prague, 1610. RM. Reprinted, Munich, 1964.

5a. ——— *Unterredung mit dem "Sternenboten"....* Prague, 1610. X. Reprinted, Grafeling, 1964.
5b. ——— *Conversation with Galileo's "Sidereal messenger."* New York, 1965. Trans. with introd. and notes by E. Rosen.
6. *Dioptrice.* Augsburg, 1611. RM. Reprinted, Cambridge, 1962.
6a. ——— German trans. Leipzig, 1904. (OKEW 144)
7. *Strena; seu, De nive sexangula.* Frankfurt, 1611. RM.
7a. ——— *The six-cornered snowflake.* Oxford, 1966. Ed. and trans. by C. Hardie.
8. *Nova stereometria doliorum.* Linz, 1615. X.
8a. ——— German trans. Leipzig, 1908. (OKEW 165)
9. *Epitome astronomiae Copernicanae.* 3 parts. Linz, 1618-21. RM.
10. *De cometis.* Augsburg, 1619. RM.
11. *Harmonices mundi.* Linz, 1619. RM. Reprinted, Brussels, 1968.
11a. ——— *Weltharmonik.* Munich, 1939; reprinted 1967. Trans. with an introd. by M. Caspar.
12. *Chilias logarithmorum.* Marburg, 1624. RM.
13. *Tabulae Rudolphinae.* 2 vols in 1. Ulm, 1627. RM/L.
13a. ——— Another ed. with substantial alterations by J.B. Morin, Paris, 1650. X.
13b. ——— ——— *Tabulae Rudolphinae; or, The Rudolphine tables.* London, 1675. Wing K332. RM.
14. *Somnium; seu, Opus posthumum de astronomia lunari.* Frankfurt, 1634. X. Reprinted, Osnabrück, 1969.
14a. ——— *Kepler's dream.* Berkeley, Cal., 1965. Trans. with notes by P.F. Kirkwood.
14b. ——— *Somnium: The dream.* Madison, Wis., 1967. Trans. with comm. by E. Rosen.
15. *Opera omnia.* 8 vols. Frankfurt/Erlangen, 1858-71. Ed. by C. Frish. RM/L. Reprinted, Hildesheim, 1971.
16. *Gesammelte Werke.* Munich, 1937-. Still in progress; will consist of more than 20 vols. Ed. by M. Caspar et al.
17. *Keplers Hochzeitgedicht für Johannes Huldenreich (1590).* Munich, 1976. Ed. by F. Seck.

1835. LAGALLA, Giulio Cesare. 1571 (or 1576)-1624. It. P I; GA. Astronomy.

1836. METIUS, Adriaen. 1571-1635. Holl. DSB; P II; GA. Mathematics. Astronomy.

1837. BAYER, Johann. 1572-1625. Ger. DSB; ICB; P I & Supp.; GA; GC. Astronomy.

1838. DEANE, Edmond. b. ca. 1572. Br. P I. Alchemy/chemistry. (Title in PR.)

1839. DREBBEL, Cornelius. 1572-1633. Holl./Br. The inventor. DSB; ICB; P I; GA; GC; GD. Mechanical, optical, and chemical technology.

1840. GLORIOSO (or GLORIOSI), Giovanni Camillo. 1572-1643. It. P I. Mathematics. Also works on comets, the scale of nature, etc.

1841. HOFMANN, Caspar. 1572-1648. Ger. ICB; BLA. Physiology. He debated with Harvey about the circulation of the blood.

1842. JUNGERMANN, Ludwig. 1572-1653. Ger. ICB; BMNH; GA; GC. Botany.

1843. KECKERMANN, Bartholomew. ca. 1572-1609. Ger. DSB; ICB; P I; GA; GC. Reform of university teaching of mathematics and natural philosophy.

1844. LYDIAT, Thomas. 1572-1646. Br. P I; GA; GD. Astronomy. (Titles in PR.)

1845. MALCOTIUS, Odo. 1572-1615. Belg./It. P II; GF (MAELCOTE). Astronomy.

1846. SENNERT, Daniel. 1572-1637. Ger. DSB; P II; BLA; GA; GC. Chemistry.
- 1. *Epitome naturalis scientiae*. Wittenberg, 1618. X. Another ed., Oxford, 1632. PR 22231. X. Another ed., Oxford, 1653. Wing S2532. RM. Another ed., Oxford, 1664. Wing S2533.
- 2. *De chymicorum cum Aristotelicis et Galenicis consensu ac dissensu*. Wittenberg, 1619. X.
- 2a. ——— *Thirteen books of natural philosophy*. London, 1659. Wing S2544 and two later eds. X. Trans. by N. Culpeper and A. Cole.

1847. TOPSELL, Edward. 1572-1625? Br. ICB; BMNH; GD. Zoology.
- 1. *The historie of foure-footed beastes and serpents*. London, 1607. PR 24123. Largely derived from Gesner.
- 1a. ——— Another ed. of part 2: *The historie of serpents*. Ib., 1608. PR 24124. Another ed., ib., 1658.
- 1b. ——— Another ed. *Whereunto is now added "The theater of insects"... by T. Muffet*. Ib., 1658. Wing G624 (under Gesner). RM.
- 2. *The fowles of heaven; or, History of birdes*. Austin, Texas, 1972. Ed. from the MS by T.P. Harrison and F.D. Hoeniger. An abridged trans. of Aldrovandi's *Ornithology*.

1848. TRADESCANT, John (first of the name). ca. 1572-1638. DSB; GD. Botany.

1849. MAYR, Simon. 1573-1624. Ger. DSB; ICB (MARIUS); P II (MARIUS); GA (MAYER); GC (MAYER). Astronomy.
- 1. *Mundus Jovialis ... hoc est, Quatuor Jovialium planetarum, cum theoria, tum tabulae, propriis observationibus....* Nuremberg, 1614. RM.

1850. SCHEINER, Christoph. 1573-1650. Ger. DSB; ICB; P II; Mort.; BLA; GA; GC. Astronomy. Physiological optics.
- 1. *Sol ellipticus; hoc est, Novum et perpetuum solis contrahi soliti phaenomenon*. Augsburg, 1615. RM.
- 2. *Refractiones coelestes; sive, Solis elliptici phaenomenon illustratum*. Ingolstadt, 1617. RM.
- 3. *Oculus; hoc est, Fundamentum opticum*. Innsbruck, 1619. RM. Another ed., London, 1652. Wing S858. X.
- 4. *Rosa ursina; sive, Sol ex admirando facularum et macularum suarum phaenomeno varius*. Brescia, 1630. RM/L.

1851. TURQUET DE MAYERNE. Theodore. 1573-1655. Fr./Br. DSB; P II. Under MAYERNE: ICB; BLA & Supp.; GA; GD. Iatrochemistry. (Title in Wing.)

1852. AENITIUS, Theophilus. 1574-1631. Ger. P I & Supp. Mineralogy.

1853. FLUDD, Robert. 1574-1637. Br. DSB; ICB; P I; BLA; GA; GB; GD. Hermetic philosophy.
1. *Utriusque cosmi maioris scilicet et minoris metaphysica, physica atque technica historia*. Oppenheim, 1617. X.
1a. ——— *Etude du macrocosme*. Paris, 1907. Trans. with notes by P. Piobb.
2. *Pulsus; seu, Nova et arcana pulsuum historia*. Frankfurt, 1631. The first book to support Harvey's theory.
3. *A philosophical key*. Cambridge, 1970. Ed. from the MS by A.G. Debus.
Other titles in PR and Wing.

1854. GUDMUNDSSON, Jón. 1574-1658. ICB. Natural history of Iceland.

1855. HIRZGARTER, Matthias. 1574-1653. Switz. P I. Astronomy.

1856. PLATTER, Thomas (the younger). 1574-1628. Switz. ICB; BMNH; BLA. Botany.
1. *Journal of a younger brother. The life of Thomas Platter as a medical student in Montpellier at the close of the sixteenth century*. London, 1963. Trans. from the German with an introd. by S. Jennett. cf. the entry for his brother, Felix--1632.
2. *Beschreibung der Reisen durch Frankreich, Spanien, England und die Niederlande, 1595-1600*. 2 vols. Basel, 1968. Ed. by R. Keiser.

1857. PUTEANUS (or PUTTEN, van der), Erycius. 1574-1646. Belg. ICB; P II; GC; GF. Astronomy.

1858. NAGEL, Paul. d. 1621. Ger. P II. Astronomy/astrology.

1859. CASTELLI, Pietro. ca. 1575-1657. It. P I; BMNH; BLA & Supp.; GA. Botany. Zoology. Chemistry.
1. *Incendio del monte Vesuvio*. Rome, 1632. RM.

1860. OUGHTRED, William. 1575-1660. Br. DSB; ICB; P II; GA; GB; GD. Author of numerous textbooks of mathematics (see PR and Wing). His most important book is his *Clavis mathematicae*.
1. *Arithmeticae in numeris et speciebus institutio ... quasi clavis mathematicae est*. London, 1631. PR 18898. X.
1a. ——— 2nd ed. *Clavis mathematicae*. Ib., 1648. Wing O573 and four later eds up to 1698. X.
1b. ——— ——— *The key of the mathematics*. Ib., 1647. Wing O582. Trans. by R. Wood. X.
1c. ——— ——— *The key of the mathematics*. Ib., 1694. Wing O583. Trans. by E. Halley. X.
2. *Opuscula mathematica hactenus inedita*. Oxford, 1677. Wing O586. RM.
See also 2002/1a.

1861. WINSTON, Thomas. 1575-1655. Br. BLA; GD. Anatomy. (Title in Wing.)

1862. AYLESBURY, Thomas. 1576-1657. Br. GA; GD. A patron of mathematicians, including Harriot.

1863. CAUS (or CAUX, CAULS, etc.), Salomon de. 1576-1626. Fr./Br./Ger. ICB; P I & Supp.; GA; GD (DE CAUS). Mechanical technology. (Title in PR.)

1864. CROOKE, Helkiah. 1576-1635. Br. BLA; GD. Anatomy. (Title in PR and Wing.)

1865. NERI, Antonio. 1576-ca. 1614. It. DSB; ICB; P II; GA. Chemical technology.
1. *L'arte vetraria.* Florence, 1612. X.
1a. ——— *The art of glass ... With some observations on the author.* London, 1662. Wing N438. Trans. by C. Merrett. X.
1b. ——— ——— *De arte vitraria libri septem et in eosdem C. Merretti ... observationes et notae.* Amsterdam, 1668.
1c. ——— ——— German trans.: see 2391/2. See also 8651.

1866. SALA, Angelo. 1576-1637. Holl./Ger. DSB; ICB; P II; BLA & Supp.; GA; GC. Chemistry.
1. *Opiologia; ou, Traicté concernant ... l'opium.* The Hague, 1614. X.
1a. ——— *Opiologia; or, A treatise concerning ... opium.* London, 1618. PR 21594. X.
2. *Opera medico-chymica.* Frankfurt, 1647. X. Another ed., ib., 1682.

1867. TERRENTIUS, Johann. ca. 1576-1630. Jesuit missionary in China. ICB (SCHRECK, G.); P II. Author of *Epistolium ex Regno Sinarum ad mathematicos Europaeos*, 1630, and several astronomical works in Chinese.

1868. BRONZERIO, Giovanni Hieronimo. It. 1577-1630. BLA. Zoology.

1869. CLUYT (or CLUTIUS), Auger. 1577-after 1636. Holl. ICB; BMNH; GA. Botany. Zoology.

1870. DYBVAD (or DIBUADIUS), Christoph. 1577-1622. Den./Swed. P I; BLA & Supp. Mathematics.

1871. GULDIN, Paul. 1577-1643. Aus. DSB; ICB; P I; GA (GULDINUS, H.). Mathematics. Mechanics.

1872. HEEK, Jan van (or ECKIUS, Johannes). 1577-ca. 1620. ICB. Astronomy.

1873. LICETI. Fortunio. 1577-1657. It. ICB; P I; BLA; GA. Astronomy. Earth sciences.
1. *De natura primo-movente.* Padua, 1634. RM.
2. *De regulari motu minimaque parallaxi cometarum caelestium.* Udine, 1640. RM.
3. *De luminis natura et efficientia.* Udine, 1640. RM.
4. *De lucidis in sublimi.* Padua, 1641. RM.

1874. RHODIUS (or RHODE), Ambrosius. 1577-1633. Ger. P II. Optics. Astronomy.

1875. SANTINI, Antonio. ca. 1577-1662. It. ICB; P III. Geometry.

1876. STELLUTI, Francesco. 1577-1652. It. DSB; ICB; P II; BMNH; GA. Microscopy. Natural history.
1. *Trattato del legno fossile minerale nouvamente scoperto.* Rome, 1637. RM.

1877. VOSSIUS (or VOSS). Gerhard Johann. 1577-1649. Holl./Br. P II; GA; GB; GC; GD.

1. *De universa matheseos natura et constitutione liber, qui subjungitur Chronologia mathematicorum.* Amsterdam, 1650.

1878. BEAUSOLEIL, Jean du Chatelet de. ca. 1578-ca. 1645. Fr. P I; GA. Alchemy/chemistry. Mineralogy.

1879. CASTELLI, Benedetto. 1578-1643. It. DSB; ICB; P I; GA. Hydrodynamics, Optics.
1. *Riposta alle opposizioni del S. Lodovico delle Colombe e del S. Vincenzio di Grazia contro al trattato del S. Galileo Galilei delle cose che stanno su l'acqua, o che in quella si muovono.* Florence, 1615. RM.
2. *Della misura dell'acque correnti.* Rome, 1628. X.
2a. ——— *On the mensuration of running waters.* London, 1661. Wing C1222. X. Evidently a separate publication of one of Salusbury's translations; see 2808.

1880. HARVEY, William. 1578-1657. Br. DSB; ICB; Mort.; BLA & Supp.; G. Physiology. Anatomy. Embryology. See also Index.
1. *Exercitatio anatomica de motu cordis et sanguinis in animalibus.* Frankfurt, 1628. RM.
1a. ——— Reprint with English trans. and notes by C.D. Leake, London, 1928 and re-issues.
1b. ——— *The movement of the heart and blood in animals.* Oxford, 1957. Trans. by K.J. Franklin.
1c. ——— *An anatomical disputation concerning the movement of the heart and blood in living creatures.* Oxford, 1976. Trans. with introd. and notes by G. Whitteridge.
2. *Exercitatio anatomica de circulatione sanguinis.* Cambridge, 1649. Wing H1087/8. X.
3. *The anatomical exercises concerning the motion of the heart and blood.* London, 1653. Wing H1083 (cf. H1092). X. Another ed., ib., 1673; Wing H1084. Trans. of items 1 and 2.
3a. ——— *The anatomical exercises: De motu cordis, 1628; De circulatione sanguinis, 1649. The first English text of 1653, now newly edited.* London, 1928. By G. Keynes.
4. *Exercitationes anatomicae de motu cordis et sanguinis circulatione.* London, 1660. Wing H1089. Reprint of items 1 and 2. X.
5. *Exercitationes de generatione animalium.* London, 1651. Wing H1091. X. Other eds, Amsterdam, 1651 and 1662.
5a. ——— *Anatomical exercitations concerning the generation of living creatures.* London, 1653. Wing H1085.
6. *De motu locali animalium.* Cambridge, 1959. Unpublished lecture notes, ed. with an English trans. by G. Whitteridge.
7. *Prelectiones anatomiae universalis.* Unpublished lecture notes.
7a. ——— *Lectures on the whole of anatomy.* Berkeley, Cal., 1961. Trans. with notes by C.D. O'Malley et al.
7b. ——— *Anatomical lectures.* Edinburgh, 1964. Ed. with trans. and notes by G. Whitteridge.
8. *Some recently discovered letters of William Harvey, with other miscellanea.* Philadelphia, 1912. Ed. by C.P. Fisher.

Collected Works and Selections

9. *Opera omnia. A Collegio Medicorum Londinensi edita.* 2 vols. London, 1766. RM/L.
9a. ——— *Works.* London, Sydenham Society, 1847. Trans. by R. Willis, with a biography. Reprinted, New York, 1965.
10. *The circulation of the blood and other writings.* London, 1907; reprinted, 1923 and 1952. Extracts from the Willis trans. of the *Works.*
11. *The circulation of the blood. Two anatomical essays by William Harvey, together with nine letters written by him.* Oxford, 1958. Ed. and trans. with notes by K.J. Franklin.

1881. HORST, Gregor. 1578-1636. Ger. P I; BMNH; BLA; GA (HORSTIUS). Pharmacy.

1882. SPIEGEL, Adriaan van den. 1578-1625. Belg./It. DSB; ICB; BMNH (SPIGELIUS); BLA (SPIEGHEL); GA (SPIGELIUS). Botany. Anatomy Embryology.
1. *Opera quae extant omnia.* 2 vols in 3. Amsterdam, 1645. Ed. by J.A. van der Linden.
Wing (V64/5) records an English trans. of one of his anatomical works (or of an extract).

1883. WEICKHARD, Arnold. ca. 1578-1645. Ger. P II; BLA. Pharmacy.

1884. DEE, Arthur. 1579-1651. Br. P I; BLA & Supp.; GA; GD. Alchemy.
1. *Fasciculus chemicus, abstrusae hermeticae scientiae ingressum....* Paris, 1631. X.
1a. ——— *Fasciculus chemicus; or, Chymical collections.* London, 1650. Wing D810. X.

1885. HELMONT, Johannes Baptista van. 1579-1644. Belg. DSB; ICB; P I; BLA & Supp.; GA; GB; GC. Chemistry. Paracelsan philosophy and cosmology.
1. *Opuscula medica inaudita.* Cologne, 1644. X.
1a. ——— 2nd ed., Amsterdam, 1648. X. Reprinted, Brussels, 1966.
1b. ——— Another ed., Frankfurt, 1707. RM.
2. *Ortus medicinae.* Amsterdam, 1648. The collected works. Ed. by his son, F.M. van Helmont. Reprinted, Brussels, 1966.
2a. ——— Another ed., Frankfurt, 1707. RM.
2b. ——— *Oriatrike; or, Physick refined.* London. 1662. Wing H1400. Trans. by J. Chandler.
2c. ——— ——— Re-issue with new title: *Van Helmont's Workes.* Ib., 1664. Wing H1397. X.
2d. ——— *Aufgang der Artzney-Kunst.* Sultzbach, 1683. X. Reprinted, Munich, 1971.
Wing also records English translations of various individual works.

1886. ROBIN, Vespasian. 1579-1662. Fr. GA (ROBIN, Jean). Botany.

1887. BLOCHWITZ, Martin. early 17th century. BLA.
1. *Anatomia sambuci.* Leipzig, 1631. X. Another ed., London, 1650. Wing B3198. X.
1a. ——— *Anatomia sambuci; or, The anatomie of the elder.* London, 1655. Wing B3199 and two later eds. X.

1888. POTIER (or POTERIUS), Michel. early 17th century. Fr./Ger. P II. Alchemy.

1889. BERTEREAU. Martine de, *Madame*. fl. 1601-1640. Fr. GA. Mineralogy.
   1. *La restitution de Pluton ... Des mines et minières de France*. Paris, 1640. RM.

1890. SIZI, Francesco. d. 1618. It./Fr. ICB. Astronomy. Author of *Dianoia astronomica, ottica, fisica*, 1611, an attempt to refute Galileo's telescopic discoveries.

1891. CASTAIGNE (or CASTAGNE), Gabriel de. d. ca. 1630. Fr. P III. Alchemy.

1892. FORSIUS (or HELSINGFORSIUS), Siegfried Aronus (or Aronsen). d. 1637. Swed. ICB; P I. Astronomy. Mineralogy.

1893. ROFFENI, Giovanni Antonio. d. 1643. It. P II. Astronomy. A supporter of Galileo in the controversy about his telescopic discoveries.

1894. BÉRIGARD, Claude Guillermet de. ca. 1580-1663/4. Fr./It. DSB; P I; BLA (BEAUREGARD); GA. Natural philosophy. One of the revivers of atomism.

1895. CAPRA, Baldassar. ca. 1580-1626. It. DSB; P I; BLA; GA. Astronomy. See also 1797/4.

1896. CRÜGER, Peter. 1580-1639. Danzig. P I. Astronomy. Mathematics.

1897. FAULHABER, Johann. 1580-1635. Ger. DSB; P I; GA; GC. Mathematics.

1898. FOSCARINI, Paolo Antonio. ca. 1580-1616. It. P I. He supported the Copernican system in a work published in 1615.

1899. HENRION, Denis (or Didier). ca. 1580-ca. 1632. Fr. DSB; P I; GA. Mathematics. Important for his educational role.

1900. NARDI, Giovanni. ca. 1580-ca. 1655. It. ICB, GA.
   1. *De igne subterraneo physica propulsio*. Florence, 1641. RM.

1901. ODONTIUS (or ZAHN), Johann Kaspar. 1580-1626. Ger. P II. Astronomy. In his early years he was an assistant of Kepler and later corresponded with him.

1902. PEIRESC, Nicolas Claude Fabri de. 1580-1637. Fr. DSB; ICB; P II; GA. A virtuso patron of science and scientists, notably Gassendi. He carried on an extensive correspondence. See also 1797/16 and 2010/2, 2a.

1903. PRIMROSE, James. 1580-1659. Br. BLA (PRIMEROSE); GA; GD.
   1. *Exercitationes et animadversiones in librum "De motu cordis et circulatione sanguinis": Adversus Guilielmum Harveum*. London, 1630. PR 20385. RM.

1904. READ, Alexander. 1580 (or 1586)-1641. Br. BLA & Supp.; GD (REID). Anatomy. (Titles in PR and Wing.)

1905. RIOLAN, Jean (second of the name). 1580-1657. Fr. DSB; ICB; BLA; GA. Anatomy. (Wing records some London editions and translations of his works.)

1906. SEVERINO, Marco Aurelio. 1580-1656. It. DSB; ICB; P II; BMNH; Mort.; BLA; GA. Human and comparative anatomy.

1907. SNEL, Willebrord. 1580-1626. Holl. DSB; ICB (SNELLIUS). Under SNELL: P II; GA; GB. Optics. Mathematics. Astronomy.
1. *De re nummeria*. Leiden, 1613. RM.
2. *Tiphys batavius; sive, Histiodromice de navium cursibus et re navali*. Leiden, 1624. RM.
See also 1653/1.

1908. WENDELIN (or VENDELINUS), Gottfried. 1580-1667. Belg. DSB; ICB; P II; GF. Astronomy.

1909. ASELLI, Gaspare. 1581-1625. It. DSB; ICB; Mort.; BLA & Supp. Under ASELLIO, C.: GA; GB. Anatomy.
1. *De lactibus; sive, Lacteis venis quarto vasorum mesaraicorum genere novo invento*. Milan, 1627. X. Reprinted, Leipzig, 1968, and Milan, 1972.

1910. BACHET DE MÉZIRIAC, Claude Gaspar. 1581-1638. Fr. DSB; ICB; P I & Supp.; GA. Mathematics. See also 1031/1.

1911. CESI (or CESIO or CAESIUS), Bernardo. 1581-1630. It. P I; BMNH (CAESIUS).
1. *Mineralogia*. Lyons, 1636. RM.

1912. CHESNECOPHERUS, Johann. 1581-1635. Swed. P I; BLA & Supp.; GA. About fifty dissertations, presumably of his students, are attributed to him, of which half are on scientific or medical topics, e.g. *De coelo*, *De natura*, *Isagoge meterologica*, *De tempore*, *De mundo*, *De stellis in specie*, *De eclipsi solis et lunae*, *De elementorum qualitatibus*, *De metallis*, etc.

1913. FÉRONCE, Elzéar. 1581-after 1654. Fr. ICB. Astronomy.

1914. GUNTER, Edmund. 1581-1626. Br. DSB; ICB; P I; GA; GB; GD. Mathematics.
1. *Canon triangulorum; sive, Tabulae sinuum et tangentium artificialium*. London, 1620. PR 12516. RM.
1a. ——— *A canon of triangles; or, Tables of artificiall sines and tangents*. London, 1620. PR 12518. X.
2. *De sectore et radio. The description and use of the crosse-staffe*. 2 parts. London, 1623. PR 12520 and three further eds to 1636; cf. Wing G2243. X.
3. *The description and use of His Majesties dials in the White-Hall garden*. London, 1624. PR 12524. X.
4. *Works*. London, 1624. X. 2nd ed., enl. and ed. by S. Foster, ib., 1636. X. Wing (G2239 et seq.) records four further eds up to 1680.

1915. MALAPERT (or MAUPERTUIS), Charles. 1581-1630. Belg. ICB; P II; GA; GF. Mathematics. Astronomy. Meteorology.

1916. NOËL, Etienne (or NATALIS, Stephanus). 1581-1659. Fr. DSB; P I (NATALIS). Natural philosophy. A friend of Descartes and an opponent of Pascal.

1917. RECUPITO, Giulio Cesare. 1581-1647. It. P II. Author of works on the eruption of Vesuvius in 1631 and on earthquakes.

1918. SANTORELLI, Antonio. ca. 1581-1653. It. P II; BLA. Author of works on the eruption of Vesuvius in 1631.

1919. SCHOOTEN, Frans van (first of the name). 1581-1646. Holl. P II; GF. Mathematics. He was a teacher of Huyghens.

1920. ANDERSON, Alexander. 1582-1619. Br./Fr. P I; GA; GB; GD. Mathematics.

1921. BAINBRIDGE, John. 1582-1643. Br. P I; GA; GB; GD. Astronomy. (Title in PR.)

1922. BALIANI, Giovanni Battista. 1582-1666. It. DSB; ICB; P I & Supp. Mechanics. A correspondent of Galileo and Mersenne.
1. *De motu naturali gravium solidorum.* Genoa, 1638. RM.

1923. BERNEGGER, Matthias. 1582-1640. Strasbourg. P I; GA; GC. Mathematics.
1. *Annotationi sopra'l trattato dell'instrumento delle proportione del Sig. Galileo Galilei.* Bologna, 1655. RM.

1924. BETTINI, Mario. 1582-1657. It. P I; GA. Mathematics.

1925. GRASSI (or GRASSO), Oratio. 1582-1654. It. ICB; P I; GA. Astronomy. Also a work on the rainbow. See also 2829.

1926. HOFMANN, Lorenz. ca. 1582-1630. Ger. P I. Iatrochemistry.

1927. REY, Jean. ca. 1582/3-1645 or later. Fr. DSB; ICB; P II; GA. Chemistry.
1. *Essays.... Sur la recherche de la cause pour laquelle l'estain et le plomb augmentent de poids quand on les calcine.* Bazas, 1630. X. Reprinted with introd. and notes by D. McKie, London, 1951.
1a. ——— English trans. Edinburgh, 1904 and re-issues. (Alembic Club reprints. 11)
1b. ——— German trans. Leipzig, 1909. (OKEW. 172)

1928. VALOIS, Jacques. 1582-1654. ICB. Astronomy.

1929. WERENBERG, Jacob. 1582-1623. Ger. P II. Author of *Disputationes meteorologicae*, 1605, and *Theoremata de loco et vacuo*, 1606.

1930. BURSER, Joachim. 1583-1639 (or 1593-1649). Ger./Den. ICB; P I; BLA; GA. Botany. Also author of *De fontium origine*, 1639, and *Introductio ad scientiam naturalem*, 1659.

1931. MORIN, Jean Baptiste. 1583-1656. Fr. DSB; ICB; P II; GA. Astronomy/astrology. See also 1994/1.

1932. MUNTING, Henrik. 1583-1658. Holl. BLA; GA; GF. Botany.

1933. PAULI, Adrian. ca. 1583-1622. Danzig. P II. The following works , attributed to him (all from the period 1613-1620), are probably his students' theses; *De motu*, *De coelo*, *De succini natura*, *De calore*, *De natura lucis*, *De metallis*, *De cometis*, *De origine fontium*, *De mari ejusque natura*.

1934. PETAU, Dénis. 1583-1652. Fr. P II. Under PETAVIUS: GA; GB. Astronomy. Best known as a chronologist.

1935. REMMELIN, Johann. 1583-1632. Ger. ICB; BLA (RÜMELIN). Anatomy.

1936. BAERLE, Gaspard van. 1584-1648. Holl. ICB (BARLAEUS, C.); P III; GA; GC; GF. Terrestrial magnetism.

1937. FERRARI, Giovanni Battista. 1584-1655. It. BMNH; GA. Botany.

1938. FINELLI, Filippo. ca. 1584-after 1649. ICB. Alchemy. Astrology. Natural history.

1939. GUIDUCCI, Mario. 1584-1646. It. ICB; P I. Astronomy. A disciple of Galileo and a member of the Accademia dei Lincei. See also 2829.

1940. SAINT VINCENT, Gregorius. 1584-1667. Belg. DSB; ICB; P I (GREGORIUS); GA; GC (GREGORY); GF. Mathematics.

1941. VERNIER, Pierre. 1584-1638. Fr. DSB; P II; GA; GB. Inventor of the vernier scale, the original applications of which were in astronomy.

1942. KRESLIN, Georg. fl. 1618-1623. Ger. P I. Astronomy.

1943. HABRECHT, Isaak. d. 1633. Strasbourg. P I. Astronomy.

1944. ZOBOLI, Alfonso. d. 1640. It. P II. Astronomy/astrology.

1945. BARTHOLIN, Caspar. 1585-1629. Den. DSB; P I; BLA & Supp.; GA; GB. Anatomy.
1. *Exercitatio de stellis.* Wittenberg, 1606. An astrological work. X. 3rd ed., ib., 1609. Title varies. RM.
2. *Anatomicae institutiones corporis humani.* Wittenberg, 1611. X. Another ed., Oxford, 1633. PR 1535. X.
2a. ——— Another ed. (the first to be revised by his son, Thomas Bartholin) entitled *Institutiones anatomicae.* Leiden, 1641. X.
2b. ——— Another ed. (the second revised by Thomas Bartholin). Leiden, 1645. X.
2c. ——— ——— *Institutions anatomiques.* Paris, 1647. RM.
2d. ——— For a later version see 2211/1.
3. *Enchiridion physicum ex priscis ac recentioribus philosophis.* Strasbourg, 1625.

1946. BROZEK, Jan. 1585-1652. Cracow. DSB; ICB (BROSCIUS). Mathematics.
1. *Aristoteles et Euclides defensus contra Petrum Ramum.* Cracow, 1638. X.
1a. ——— Another ed., to which is added *Duae disceptationes de numeris perfectis.* Danzig, 1652. RM.

1947. CESI, Federico. 1585-1630. It. DSB; ICB; P I; BMNH; GA. Botany. Founder of the Accademia dei Lincei.
1. *Apiarium.* Rome, 1625. RM.

1948. DORISY, Jean. ca. 1585-1657 (or 1652). Fr. P I & Supp.; GA. Author of *Quaestiones curiosae de ventorum origine, de accessu et recessu maris,* 1646.

1949. LAUREMBERG, Peter. 1585-1639. Ger. ICB; P I; BMNH; BLA. Under LAURENBERG: GA; GC. Astronomy. Botany. Anatomy.
1. *Apparatus plantarius primus.* Frankfurt, 1632. RM.

1950. MÜLLER, Philipp. 1585-1659. Ger. P II. Astronomy. Chemistry.

1951. MYDORGE, Claude. 1585-1647. Fr. DSB; ICB; P II; GA. Mathemics. Optics. A close friend of Descartes.

1952. SCHWENTER, Daniel. 1585-1636. Ger. ICB; P II; GC. Mathematics.

1953. TASSE, Johann Adolph. 1585-1654. Ger. P II; GC (TASSIUS). Mathematics. Astronomy. Geography.

1954. BARTOLETTI (or BERTOLETTI), Fabrizio. 1586-1630. It. ICB; P II; BLA & Supp. Anatomy. Chemistry.

1955. CABEO, Niccolo. 1586-1650. It. DSB; P I. Magnetism.
   1. *Philosophia magnetica.* Ferrara, 1629. RM.

1956. CYSAT, Johann Baptist. ca. 1586-1657. Ger. DSB; P I. Astronomy.

1957. LA BROSSE, Guy de. ca. 1586-1641. Fr. DSB; ICB; BMNH; BLA (BROSSE); GA (BROSSE). Botany. Chemistry. Founder of the Jardin des Plantes in Paris in 1626.

1958. MAGNI, Valeriano. 1586-1661. Several cities in central Europe. DSB; ICB; GA; GC (MAGNUS). Investigation of the vacuum. (His *Demonstratio ocularis* of 1647 was the first printed account of the barometric experiment.)

1959. RENAUDOT, Théophraste. 1586-1653. Fr. ICB; BLA & Supp.; GA; GB. "Intelligencer" and founder of a bureau d'adresse in Paris in 1630; see under his son Eusèbe Renaudot (2123). Best known as the founder of France's first newspaper.

1960. TURNER, Peter. 1586-1652. Br. DSB; GD. Mathematics.

1961. ZUCCHI, Niccolo. 1586-1670. It. DSB; P II. Optics. Astronomy.

1962. AROMATARI, Giuseppe degli. 1587-1660. It. DSB; BMNH; BLA; GA. Plant embryology.

1963. FABRICIUS, Johann. 1587-1615? Ger./Holl. P I. The discoverer of sun spots.
   1. *De maculis in sole observatis.* Wittenberg, 1611. RM.

1964. FROIDMONT, Liebert. 1587-1653. Belg. GA; GF. Under FROMONDUS: ICB; P I. Astronomy. Meteorology.
   1. *Meteorologicorum libri sex.* Antwerp, 1627. RM. PR record an Oxford ed. of 1639 and Wing two London eds of 1655 and 1670. See also 1994/1.

1965. JUNGIUS, Joachim. 1587-1657. Ger. DSB; ICB; P I & Supp. (JUNG); BMNH (JUNG); BLA (JUNG); GA (JUNGE); GC. The mathematical sciences. Methodology. Chemistry. Botany.

1966. REINESIUS, Thomas. 1587-1667. Ger. P II; BLA; GA; GC. Iatrochemistry.

1967. URSINUS (or BEHR), Benjamin. 1587-1633. Ger./Prague/Aus. P II; GC. Mathematics.

1968. VONDEL, Joost van den. 1587-1679. Holl./Ger. ICB; GA; GB; GC. Mineralogy.

1969. ALSTED, Johann Heinrich. 1588-1638. Ger. DSB; ICB; P I; GA; GB; GC. Natural philosophy. Best known as a theologian and educationalist.

1970. AMBROSINI, Bartolomeo. 1588-1657. It. BMNH; BLA; GA; GB. Zoology. Botany.

1971. BEECKMAN, Isaac. 1588-1637. Holl. DSB; ICB; P I; GF. Mechanics. Various aspects of physics. Meteorology.

1. *Journal tenu par Isaac Beeckman de 1604 à 1634.* 4 vols. The Hague, 1939-53. Ed. with an introd. and notes by C. de Waard.

1972. BRAMER, Benjamin. 1588-1652. Ger. DSB; P I. Mathematics.

1973. GRANDAMI, Jacques. 1588-1672. Fr. P I; GA. Astronomy.
1. *Nova demonstratio immobilitatis terrae petita ex virtute magnetica.* La Flèche, 1645. RM.
2. *Tractatus de eclipsibus solis et lunae.* Paris, 1668.

1974. HOBBES, Thomas. 1588-1679. Br. The philosopher. DSB; ICB; P I; G. Geometry. Optics.
1. *Dialogus physicus de natura aeris* ... [With] *De duplicatione cubi.* London, 1661. Wing H2229. RM.
2. *De principiis et ratiocinatione geometrarum.* London, 1666. Wing H2270. An anonymous work which Wing attributes to William Hobbes. RM.
3. *Thomas White's "De mundo" examined.* London, 1976. Trans. from the Latin by H.W. Jones.

Other titles in Wing. See also 2355/1a.

1975. LOSS, Peter. 1588-1639. Danzig. P I. The eight works attributed to him (from the period 1627-36) are apparently his students' theses. They deal chiefly with topics in natural philosophy and also with mineralogy.

1976. MERSENNE, Marin. 1588-1648. Fr. DSB; ICB; P II; GA; GB. Natural philosophy. Acoustics. Optics. Mechanics. "Intelligencer" par excellence. "Architect of the European scientific community." See also Index.
1. *La verité des sciences, contre les sceptiques ou Pyrrhoniens.* Paris, 1626. X. Reprinted, Stuttgart, 1969.
2. *Questions inouyes; ou, Récréation des sçavans.* Paris, 1634. X. Reprinted, Stuttgart, 1972.
3. *Questions harmoniques.* Paris, 1634. X. Reprinted, Stuttgart, 1972.
4. *Harmonie universelle, contenant la théorie et le pratique de la musique.* 2 parts. Paris, 1636-37. X. Reprinted, 3 vols, Paris, 1965.

4a. ——— *Harmonicorum libri XII.* Paris, 1636. X.

4b. ——— ——— Another ed., ib., 1648. X. Reprinted, Geneva, 1972.

4c. ——— *An edited translation of the fourth treatise of the "Harmonie universelle".* Thesis, University of Rochester, N.Y., 1972. By R.F. Williams.

5. *Cogitata physico-mathematica.* Paris, 1644. RM.
6. *Universae geometriae mixtaeque mathematicae synopsis.* Paris, 1644. RM.
7. *Novarum observationum physico-mathematicarum tomus III.* Paris, 1647. X. Another ed., ib., 1667. RM.
8. *L'optique et la catoptrique.* Paris, 1651. RM.
9. *Correspondance.* Many vols. Paris, 1932-. Ed. by Mme P. Tannery, C. de Waard, et al.

1977. PASCAL, Etienne. 1588-1651. Fr. The father of Blaise Pascal. DSB; P II. mathematics.

1978. WORM, Ole. 1588-1654. Den. DSB; ICB; P II; BMNH; BLA; GA.

Natural history.
1. *Museum Wormianum.* Leiden, 1655. The catalogue of his museum.

1979. CASRÉE (or CASRAEUS), Pierre. 1589-1664. Fr. P I. Mechanics.

1980. FABRICIUS, Georg Andreas. 1589-1645. Ger. P I. Author of *Theatridium physicum*, *Speculum astronomicum*, and *De origine montium*.

1981. RICHARD, Claude. 1589-1664. Fr. P II; GA. Mathematics.

1982. DELAMAIN, Richard (the elder). fl. 1610-1645. Br. DSB; GD. Mathematics. (Titles in PR.)

1983. BONACCORSI, Bartolommeo. fl. 1618-1656. It. ICB; BLA (BUONACORSI). Physiology.

1984. CELLARIUS, Conrad. fl. 1619-1636. Ger. P I. Author of *Positiones meteorologicae*, 1627, and *Institutiones physicae*, 1632.

1985. CLAVE, Estienne de. fl. 1624-1646. Fr. BLA; GA. Chemistry.

1986. SCHMUCK (or SCHMUCKER), Martin. d. 1640. Ger. P II; BLA. Chemistry.

1987. POTIER (or POTERIUS), Pierre. d. after 1640. Fr./It. P II. Chemistry.

1988. VIGNA, Domenico. d. 1647. It. BMNH. Botany.

1989. BARANZANO, Giovanni Antonio. 1590-1622. It./Fr. DSB; P I (B., Redemptus); GA. Astronomy.

1990. BURGERSDIJK (or BURGERSDICIUS), Frank. 1590-1635. Holl. DSB; GA; GB; GC. Natural philosphy.
1. *Idea philosophiae naturalis.* Leiden, 1622. X. Another ed., ib., 1652.

1991. DURET (or DURRET), Noël. 1590-ca. 1650. Fr. P I; GA. Astronomy.

1992. FRANCK, Johan. 1590-1661. Swed. P III (FRANK); BMNH (FRANCKENIUS). Botany.
1. *Speculum botanicum.* 1659. X. Reprinted, Stockholm, 1973.
2. *Botanologia. Nunc primum edita, praefatione historica.* Uppsala, 1877. By R.F. Fristedt.

1993. GONZALEZ, Francisco. ca. 1590-1661. Spain/It. P I. Author of *Philosophia de physico audito*, 1645.

1994. LANSBERGE. Jacob. 1590-1657. Holl. Son of Phillipe Lansberge (entry 1785). GA; GF.
1. *Apologia pro "Commentationibus" P. Lansbergii in motum terrae diurnum et annum: Adversus L. Fromondum ... et J.B. Morinum.* Middelberg, 1633. RM.

1995. MAGNENUS, Johann Chrysostom. ca. 1590-1679? Fr./It. DSB; BLA (MAGNEN); GA (MAGNEN). Natural philosophy. One of the revivers of atomism.
1. *Democritus reviviscens; sive, De atomis.* Pavia, 1646. RM.

1996. NORWOOD, Richard. 1590-1665. Br. DSB; ICB; P II; GA; GD. Mathematics. (Titles of his textbooks are in PR and Wing.)

1997. ROSS (or ROSSE), Alexander. 1590-1654. Br. ICB; P II; GD. Astronomy (anti-Copernican). (Titles in PR and Wing.)

1998. STRAUSS, Johann. 1590-1630. Ger. P II. Astronomy.

1999. CAVENDISH, Charles. 1591-1654. Br. ICB; GD1. Mathematics.

2000. DELMEDIGO, Joseph Solomon. 1591-1655. It. ICB; GB; GF. Mathematics. Astronomy.

2001. DESARGUES, Girard. 1591-1661. Fr. DSB; ICB; P I; GA. Geometry.
1. *Brouillon project d'une atteinte aux événements des rencontres d'un cone avec un plan*. Paris, 1639. X.
1a. ——— German trans. Leipzig, 1922. (OKEW. 197)
2. *Manière universelle pour poser l'essieu etc. aux cadrans au soleil*. Paris, 1643. X.
2a. ——— *Universal way of dyaling*. London, 1659. Wing D1127. X.
3. *Oeuvres*. 2 vols. Paris, 1864. Ed. with comm. and biography by N. Poudra. RM/L.
4. *L'oeuvre mathématique*. Paris, 1951. Ed. with comm. and biography by R. Taton.

2002. LEURECHON, Jean. ca. 1591-1670. Fr. DSB; P I; GA. Mathematics. Astronomy.
1. *La récréation mathématique*. Pont-à-Mousson, 1624. X. 5th ed., Paris, 1660.
1a. ——— *Mathematicall recreations*. London, 1633. PR 15530. (Later eds in Wing.) Trans. by W. Oughtred. X.

2003. SAXONIUS (or SAXE or SACHSE), Peter, 1591-1625. Ger. P II. Astronomy.

2004. SCHALL (VON BELL), Johann Adam. 1591-1666. Jesuit missionary in China. ICB; P II; GA; GB; GC. Author of many books on astronomy and mathematics in Chinese.

2005. ARRIGHETTI, Andrea. 1592-1671. It. P I. Fluid mechanics. A disciple of Galileo.

2006. BONITUS, Jacobus. 1592-1631. Holl. ICB; BMNH; BLA & Supp.; GF. Natural history.

2007. BRENDEL, Zacharias. 1592-1638. Ger. P I; BLA.
1. *Chimia in artis formam reducta*. Jena, 1641. X. Another ed., Leiden, 1671. RM.

2008. FINKE (or FINCK), Jacob. 1592-1663. Den. P I. Author of *De thermoscope*, 1655, and other works dealing with heat and cold.

2009. FRANCKENIUS, Gottfried. 1592-1654. Olmütz. P I. Optics.

2010. GASSENDI, Pierre. 1592-1655. Fr. DSB; ICB; P I; GA; GB. Natural philosophy. Astronomy. Mechanics. See also Index.
1. *Exercitationes paradoxicae adversus Aristoteleos*. Grenoble, 1624. X. Another ed., Amsterdam, 1649.
1a. ——— *Dissertations en forme de paradoxes contre les Aristotéliciens*. Paris, 1959. Critical ed. with French trans. and notes by B. Rochot.
2. *De Nicolai Claudii Fabricii de Peiresc ... vita*. Paris, 1641. X.
2a. ——— *The mirrour of true nobility and gentility, being*

the life of ... N.C. Fabricius, lord of Peiresk. London, 1657. Wing G295. X.

3. *De apparente magnitudine solis humilis et sublimis epistolae quatuor.* Paris, 1642. RM.
4. *De motu impresso a motore translato epistolae duae.* Paris, 1642. RM.
5. *Disquisitio metaphysica; seu, Dubitationes et instantiae adversus Renati Cartesii metaphysicam et responsa.* Amsterdam, 1644. X.

5a. ——— *Recherches métaphysiques.* Paris, 1962. Critical ed. with trans. and notes by B. Rochot.

6. *Institutio astronomica juxta hypotheseis tam veterum quam Copernici et Tychonis.* Paris, 1647. RM. Wing (G291 et seq.) records two London eds and two Cambridge eds.
7. *De vita et moribus Epicuri.* Lyons, 1647. X. Reprinted, Amsterdam, 1968.
8. *Tychonis Brahi ... vita. Acessit N. Copernici, G. Peurbachii et J. Regiomontani ... vitae.* Paris, 1654. RM.
9. *Philosophiae Epicuri syntagma.* The Hague, 1659. Another ed., London, 1660. Wing G296. X.
10. *Lettres familières à François Luillier pendant l'hiver 1632-33.* Paris, 1944. Ed. by B. Rochot with introd., notes, and index.
11. *Opera omnia.* 6 vols in 3. Lyons, 1658. RM/L. Reprinted, Stuttgart, 1964.
12. *Selected works.* New York, 1972. Ed. and trans. by C.B. Brush.

2011. SCHICKARD, Wilhelm. 1592-1635. Ger. DSB; ICB; P II; GA; GC. Astronomy. A friend and correspondent of Kepler.

2012. BACKHOUSE, William. 1593-1662. Br. ICB; GD. Alchemy.

2013. BAGWELL, William. b. 1593. Br. GA; GD. Astronomy. (Title in Wing.)

2014. DAVISON (or DAVISSON), William (or Guillaume). 1593-1669. Fr. DSB; GD. Under DAVIDSON: ICB; P I; BLA. Iatrochemistry.

1. *Philosophia pyrotechnica ... seu, Curriculus chymiatricus.* Paris, 1635. RM.

2015. FRANCKENBERG, Abraham von. 1593-1652. Ger. P I; GA; GC. Alchemy.

2016. GENERINI, Francesco. ca. 1593-1663. It. ICB; P I. Astronomy.

2017. LAET, Jan de. 1593-1649. Belg./Holl. ICB; P I; BMNH; BLA; GF (DE LAET). Natural history.

1. *Novus orbis; seu, Descriptiones Indiae Occidentalis.* Leiden, 1633. RM.
2. *Correspondence of John Morris with Johannes De Laet (1634-1649).* Assen, 1970. Ed. by J.A.F. Bekkers.

2018. MARCHETTI, Pietro de. 1593 (or 1589)-1673. It. BLA; GA. Anatomy.

2019. RHO (or RHAUDENSIS), Jacob (or Giacomo). 1593-1638. Jesuit missionary in China. ICB; P II. Author of books on astronomy and mathematics in Chinese.

2020. TULP, Nicolaas. 1593-1674. Holl. DSB; ICB; BMNH; BLA; GA. Anatomy.

2021. WHITE, Thomas. 1593-1676. Br. DSB; GA; GD. Natural philosophy. (Titles in Wing.) Best known as a theologian. See also 1974/3.

2022. BACK, Jacob van de. 1594-1658. Holl. ICB; BMNH; BLA & Supp.; GF. Physiology.

2023. BEVERWIJK, Johan van. 1594-1647. Holl. ICB; BLA & Supp.; GA; GF. Anatomy.

2024. FOSTER, Samuel. d. 1652. Br. P I; GA; GD. Mathematics. (Titles in PR and Wing.) See also 1914/4.

2025. WELPER, Eberhard. fl. ca. 1630. Strasbourg. P II. Astronomy.

2026. BEAUGRAND, Jean. ca. 1595-1640. Fr. DSB. Mathematics.

2027. GIRARD (or GERARDUS), Albert. 1595-1632. Holl. DSB; ICB; P I; GA. Mathematics. See also 1700/6.

2028. LEOTAUD, Vincent. 1595-1672. Fr. P I; GA. Mathematics. Magnetism.
1. *Magnetologia*. Lyons, 1668. RM.

2029. LINUS, Franciscus. 1595-1675. Br./Belg. ICB (LINE); P I; GD (LINE). Mathematics. Pneumatics. Optics. (Titles in Wing.) See also 2355/1a.

2030. MARCI (VON KRONLAND), Johannes Marcus. 1595-1667. Prague. DSB; ICB; P II; BLA; GA (KRONLAND); GC. Mechanics. Optics.
1. *Thaumantias liber de arcu coelesti deque colorum apparentium natura, ortu et causis*. Prague, 1648. X. Reprinted, ib., 1968. Re the colours of the spectrum.

2031. NIEREMBERG, Juan Eusebio. 1595-1658. Spain. BMNH; GA; GB.
1. *Historia naturae*. Antwerp, 1635.

2032. DESCARTES. René du Perron. 1596-1650. Fr./Holl. DSB; ICB; P I; G. Natural philosophy. Methodology. Mathematics. Mechanics. Optics. Physiology. See also Index.
1. *Discours de la méthode ... plus la Dioptrique, les Météores, et la Géometrie, qui sont des essais de cete méthode*. Leiden, 1637. RM. Many reprints and translations which, however, do not often include the appended three essays.
1a. ——— *Discourse on method, Optics, Geometry, and Meteorology*. Indianapolis, 1965. Trans. with an introd. by P.J. Olscamp.
1b. ——— *La géometrie*. Paris, 1927.
1c. ——— *Geometry*. New York, 1954. Trans. by D.E. Smith.
2. *Principia philosophiae*. Amsterdam, 1644. RM. Another ed., ib., 1650.
3. *Compendium musicae*. Utrecht, 1650. X. Reprinted, New York, 1968.
4. *Traité de l'homme*. First published in 1662 in Latin trans. (item 4a). The original French text was published in 1664 (item 4b).
4a. ——— *De homine*. Leiden, 1662. RM. Trans. by F. Schuyl. Another ed., ib., 1664.
4b. ——— *Traité de l'homme et de la formation du foetus*. Paris, 1664. Ed. by C. Clerselier. X.
4c. ——— ——— 2nd ed., to which is added *Le monde, ou*

Traité de la lumière. Paris, 1677. RM. Latin ed., Amsterdam, 1677.

4d. ——— ——— *Treatise on man*. Cambridge, Mass., 1972. French text with trans. and comm. by T.S. Hall.

5. *Traité de la mécanique* ... [With] *L'abregé musique ... mis en françois*. Paris, 1668. RM.
6. *Opuscula posthuma, physica et mathematica*. Amsterdam, 1701. RM.

Correspondence and Collected Works

7. *Epistolae*. Amsterdam, 1668. RM/L.
8. *Correspondence of Descartes and Constantyn Huyghens, 1635-47*. Oxford, 1926. Ed. from MSS by L. Roth.
9. *Correspondance*. Many vols. Paris, 1936-. Ed. by C. Adam and G. Milhaud.
10. *Oeuvres*. 12 vols. Paris, 1897-1913. Ed. by C. Adam and P. Tannery. RM/L. Reprinted, ib., 1956-57.
11. *The philosophical works*. 2 vols. Cambridge, 1911-12. Trans. by E.S. Haldane and G.R.T. Ross. Incomplete. Reprinted, 2 vols in 1, New York, 1955.

2033. EICHSTADT, Lorenz. 1596-1660. Ger. P I; BLA. Astronomy.

2034. GLISSON, Francis. 1596-1677. Br. DSB; ICB; Mort.; BLA & Supp.; GA; GD. Anatomy. Physiology. (Titles in Wing.)

2035. LA CHAMBRE, Marin Cureau de. 1596-1669. Fr. ICB; P I (CUREAU); BLA & Supp. (CUREAU); GA. Optics. Various other fields. Member of the Académie des Sciences.

1. *Nouvelles pensées sur les causes de la lumière, du désbordement du Nil, et de l'amour d'inclination*. Paris, 1634. RM.
2. *Traité de la connoissance des animaux*. Paris, 1647. RM.

2a. ——— *A discourse of the knowledge of beasts*. London, 1657. Wing L131. X.

3. *La lumière*. Paris, 1657. RM.
4. *Discours sur les causes du désbordement du Nil*. Paris, 1665. RM.

2036. PETIT, Pierre. ca. 1596-1677. Fr. DSB; ICB; P II; GA. Astronomy. Magnetism. Heat and cold.

2037. GELLIBRAND, Henry. 1597-1636. Br. DSB; P I; GA; GD. Geomagnetism. (Titles in PR and Wing.) See also 1779/3.

2038. LA FAILLE, (Jean) Charles. 1597-1652. Belg./Spain. DSB. Under FAILLE: ICB; P I; GF. Statics.

2039. MAGIOTTI, Raffaelo. 1597-1656. It. DSB; P II. Fluid mechanics.

2040. ODIERNA, Giovanni Battista. 1597-1660. It. DSB; ICB; P I (HODIERNA); GA (HODIERNA). Astronomy. Optics. Meteorology. Comparative anatomy.

2041. RHEITA, Anton Maria Schyrlaeus de. 1597-1660. Belg. DSB; P II; GA. Astronomy.

2042. TREW (or TREU), Abdias. 1597-1669. Ger./Switz. ICB; P II; GC. Astronomy/astrology (anti-Copernican).

2043. VOORST. Adolf van. 1597-1663. Holl. BMNH (VORSTIUS); BLA (VORSTIUS); GA. Botany.

2044. BILLICH, Anton Günther. 1598-1640. Ger. P I; BLA; GA. Iatrochemistry.

2045. CAVALIERI, (Francesco) Bonaventura. 1598?-1647. It. DSB; ICB; P I; GA. Mathematics.
   1. *Lo specchio ustorio; overo, Trattato delle settioni coniche.* Bologna, 1632. RM.
   2. *Geometria indivisibilibus continuorum nova quadam ratione promota.* Bologna, 1635. X. 2nd ed. 1653. RM.
   3. *Trattato della ruota planetaria perpetua.* Bologna, 1646. RM.

2046. DANESI, Luca. 1598-1672. It. P I. Author of *Trattato di meccaniche cavato di Galileo*, 1649.

2047. DENYS, Nicolas. 1598-1688. Fr./Canada. ICB; GF. Ornithology.

2048. GERHARD, Johann. 1598-1657. Ger. BLA. Alchemy.
   1. *Commentatio ... in apertorium Raimundi Lulli ... de lapide philosophorum.* Tübingen. 1641. RM.

2049. HARDY, Claude. ca. 1598-1678. Fr. DSB; P I; GA. Mathematics. See also 996/6.

2050. KRÜGER, Oswald. ca. 1598-1665. Wilno. P I. Mathematics. Optics. Also *De vacuo*, 1648.

2051. MELLAN, Claude. 1598-1688. Fr. ICB. Made a map of the moon, 1634/5.

2052. REGIUS (or VAN ROY), Heinrich. 1598-1679. Holl. ICB; P II; BLA (ROY); GA (ROY). Physiology.

2053. RICCOLI, Giambattista. 1598-1671. It. DSB; ICB; P II; GA. Astronomy. Geography. "One of the most ardent opponents of the Copernican system."
   1. *Almagestum novum, astronomicam veterem novamque complectens.* Bologna, 1651. RM.
   2. *Astronomiae reformatae.* 2 vols. Bologna, 1665. RM.
   3. *Argomento fisico-mattematico ... contro il moto diurno della terra.* Bologna, 1668. RM.

2054. VESLING, Johann. 1598-1649. It. DSB; BMNH; BLA; GA; GC. Botany. Human and comparative anatomy. Embryology. (Wing records an English translation of a work on human anatomy.) See also 1737/3.

2055. BOBART, Jacob (first of the name). 1599-1680. Br. ICB; BMNH; GA; GD. Botany. (Titles in Wing, who conflates the works of father and son.)

2056. ROLFINCK, Guerner (or Werner). 1599-1673. Ger. DSB; P II; BMNH; BLA; GA. Anatomy. Botany. Chemistry.

2057. ZEISOLD, Johannes. 1599-1677. Ger. P II. Scholastic natural philosophy. One of his works was *De absurditate et varietate novae physices.*

2058. STOLCIUS, Daniel. fl. 1620-1630. Prague/Ger./Br. ICB. Alchemy.
   1. *Hortulus hermeticus.* Frankfort, 1627.

2059. HUME, James. fl. 1630-1639. Br./Fr. ICB; GD. Mathematics. See also 1665/2.

2060. CARDOSO, Fernando. fl. 1631-ca. 1640. Spain. P I; BLA. Author of a work on Mount Vesuvius, the eruption of 1631 and its causes, and the origin of earthquakes and tempests.

2061. DAUCOURT, Bonaventure. fl. 1633. Fr. P III; GA. Geology.

2062. PLATTES, Gabriel. fl. 1638-1640. Br. ICB; BMNH; GD. Chemistry.
   1. *A discovery of infinite treasure*. London, 1639. PR 19998/9. RM.
   1a. ——— Another ed., entitled *A discovery of subterraneall treasure, viz. of all manner of mines and minerals ... The art of melting, refining and assaying of them ... with a perfect way to make colours*. Ib., 1639. PR 20000. Wing records three further eds up to 1684. X.

2063. LOCATELLI, Lodovico (or Luigi). d. 1637. It. P I; BLA; GA. Iatrochemistry.

2064. HÉRIGONE, Pierre. d. ca. 1643. Fr. DSB; P I. Mathematics.
   1. *Cursus mathematicus*. 6 vols. Paris, 1634-42.

2065. FANTUZZI (or FANTUTIUS), Giovanni. d. 1646. It. P I; GA. Scholastic natural philosophy.

2066. ZOBEL, Friedrich. d. 1647. Ger. P II. Iatrochemistry.

2067. BIRINGUCCI, Carlo. d. 1648. It. P I. Astrology.

2068. RENIERI (or REINERIUS or REINIERI), Vincenzo. d. 1648. It. P II. Astronomy. A disciple of Galileo.

2069. LE TENNEUR, Jacques Alexandre. d. after 1652. Fr. DSB. Mathematics. Mechanics.

2070. GUIRAND, Claude. d. 1657. Fr. P I. Mechanics. Optics. Acoustics.

2071. LIONNE, Artus de. d. 1663. Fr. ICB; P III. Mathematics.

2072. AGGIUNTI (or ADJUNCTUS), Niccolo. 1600-1635. It. ICB; P I. Reputed to be the discoverer of capillarity.

2073. BARTSCH, Jacob. 1600-1633. Strasbourg. P I. Astronomy. Kepler's son-in-law and one-time assistant.
   1. *Usus astronomicus planisphaerii stellati*. Strasbourg, 1624. RM.

2074. BERTI, Gasparo. ca. 1600-1643. It. DSB. Devised an experiment which led to Torricelli's famous barometric experiment.

2075. BOND, Henry. ca. 1600-1678. Br. P I. Geomagnetism. (Titles in PR and Wing.)

2076. CARCAVI, Pierre de. ca. 1600-1684. Fr. DSB; ICB; P I; GA. "Intelligencer." Carried on an extensive correspondence. Friend of Huygens, Fermat, Pascal, etc. Member of the Académie des Sciences.

2077. CURTZ (or CURTIUS), Albert. ca. 1600-1671. Ger. P I. Astronomy.

2078. HARTLIB, Samuel. ca. 1600-1662. Ger./Br. DSB; ICB; GA; GB; GC; GD. "Intelligencer" and stimulator of science in England in the 1640s and 1650s. (Titles in Wing.)

2079. JOHNSON, Thomas. ca. 1600-1644. Br. DSB; ICB; BMNH; GD. Botany. (Titles in PR.) See also 1474/1b and 1686/2a.

2080. LALOUVÈRE, Antoine de. 1600-1664. Fr. DSB; ICB; P I (LOUBÈRE); GA (LA LOUBÈRE). Mathematics. Mechanics.

2081. LANGREN, Michael Florent van. ca. 1600-1675. Belg. DSB; P I; GF. Astronomy (chiefly selenography).

2082. MONTMOR, Henri Louis Habert de. ca. 1600-1679. Fr. DSB. Patron of scientists and philosophers. The weekly meetings at his mansion (after the death of Mersenne) evolved into the Académie Montmor.

2083. SCHRÖDER, Johann. 1600-1664. Ger. P II; BMNH. Iatrochemistry. Zoology.
1. *Pharmacopoea medico-chymica*. Ulm, 1641. X. Another ed., 1721.
1a. ——— *The compleat chymical dispensatory*. London, 1669. Wing S898.

2084. VLACQ (or VLACK or VLACCUS), Adriaan. 1600-1666/7. Holl. DSB; P II; GA; GC. Mathematics.
1. *Trigonometria artificialis; sive, Magnus canon triangulorum logarithmicus*. Gouda, 1633.
2. *Tabulae sinuum, tangentium et secantium, et logarithmi sin. tan. et numerorum*. Gouda, 1636. X. Another ed., Frankfurt, 1790.

2085. AUBIGNÉ DE LA FOSSE (or ALBINEUS), Nathan. 1601-after 1669. Switz. P I. Chemistry.
1. *Bibliotheca chemica*. Geneva, 1653. RM.

2086. DEBEAUNE, Florimond. 1601-1652. Fr. DSB. Under BEAUNE: ICB; P I; GA. Mathematics.
1. *La doctrine de l'angle solide*. [With] *Inventaire de sa bibliothèque*. Paris, 1975. (Listed under BEAUNE, F. de.) Ed. from the MSS by P. Costabel with introd. and comm.

2087. FERMAT, Pierre de. 1601-1665. Fr. DSB; ICB; P I, III; GA; GB. Mathematics. See also Index.
1. *Varia opera mathematica*. Toulouse, 1679. RM. Also includes some letters. Reprinted, Brussels, 1969.
2. *Oeuvres*. 4 vols. Paris, 1891-1912. Ed. by P. Tannery and C. Henry.
German translations of three of his works are in OKEW.

2088. GREIFF, Friedrich. 1601-1668. Ger. P I; BLA; GA. Iatrochemistry.

2089. MAIGNAN (or MAGNANUS), Emanuel. 1601?-1676. Fr./It. DSB; P II; GA. Optics and various other branches of experimental physics.

2090. MONTALBANI, Ovidio. 1601-1671. It. P II; BMNH; BLA; GA. Astronomy/astrology. Meteorology. Botany.

2091. PATIN, Guy. 1601-1672. Fr. ICB; BLA & Supp.; GA. His letters, which cover the period 1645 to 1672, range over many topics

but are chiefly concerned with medicine and related sciences.
1. *Lettres choisies*. Paris, 1683. X. Many later eds.
1a. ——— Another ed., Cologne, 1691. Includes 300 additional letters.

2092. RIVINUS (or BACHMANN), Andreas. 1601-1656. Ger. P I; BLA; GA. Physiology. Botany.

2093. BILLY, Jacques de. 1602-1679. Fr. DSB; P I; GA. Mathematics. Astronomy.

2094. BOSSE, Abraham. 1602-1676. Fr. DSB; ICB; P I; GA. Geometry.

2095. CIERMANS, Jean. 1602-1648. Belg. P I; GC; GF. Mathematics.

2096. FONTANA, Francesco. 1602-1656. It. ICB; P I; GA. Astronomy. A renowned telescope maker.
1. *Novae celestium terrestriumque rerum observationes*. Naples, 1646.

2097. GREAVES (or GRAVE or GRAVIUS), John. 1602-1652. Br. ICB; P I; BMNH; GA; GB; GD. Astronomy. Best known for his editions of the works of Ulugh Beg and other medieval Muslim astronomers (see entry 1082 et seq.). (Titles in Wing.)

2098. GUERICKE, Otto von. 1602-1686. Ger. DSB; ICB; P I; GA; GB; GC. Pneumatics--inventor of the air pump and discoverer of the elasticity of air.
1. *Experimenta nova (ut vocantur) Magdeburgica de vacuo spatio*. Amsterdam, 1672. RM. Reprinted, Aalen, 1962.
1a. ——— German trans. Leipzig, 1894. (OKEW. 59)
1b. ——— *Neue (sogenannte) Magdeburger Versuche über den leeren Raum*. Düsseldorf, 1968. Ed. and trans. by H. Schimank et al. Also contains letters and other source material.

2099. KIRCHER, Athanasius. 1602-1680. Ger./It. DSB; ICB; P I; BMNH; Mort.; BLA & Supp.; GA; GB; GC. "Some forty-four books and more than 2000 extant letters and manuscripts attest to the extraordinary variety of his interests and to his intellectual endowments. His studies covered practically all fields both in the humanities and the sciences." -DSB.
1. *Magnes; sive, De arte magnetica*. Rome, 1641. RM.
2. *Ars magna lucis et umbrae*. Rome, 1646. RM.
3. *Musurgia universalis*. 2 vols. Rome, 1650. X. Reprinted, 2 vols in 1, Hildesheim, 1970.
4. *Itinerarium exstaticum quo mundi opificium....* Rome, 1656. X. Another ed., Würzburg, 1660. RM.
5. *Mundus subterraneus*. 2 vols. Amsterdam, 1665. RM/L. 3rd ed., ib., 1678.
6. *Magneticum naturae regnum*. Amsterdam, 1667. RM.
7. *Phonurgia nova*. Kempten, 1673. RM. Reprinted, New York, 1966.
8. *Physiologia Kircheriana experimentalis*. Amsterdam, 1680. RM.
See also 2151/3 and 2474/3,3a.

2100. LILLY, William. 1602-1681. Br. ICB; P I; GA; GB; GD. Astrology. (Titles in Wing.)
1. *Mr. W. Lilly's history of his life and times ... Written*

by himself. London, 1715. Ed. with notes by E. Ashmole. X.

1a. ——— *The last of the astrologers. The autobiography of William Lilly*. Menston, 1973. With introd. and notes by K.M. Briggs.

2101\. MORET, Theodor. 1602-1667. Prague. ICB (MORETUS); P II; GF (MORETUS). Various aspects of physics.

2102\. NOFERI, Cosimo. ca. 1602-1660. It. P II. Mathematics. A disciple of Galileo.

2103\. ROBERVAL, Gilles Personne de. 1602-1675. Fr. DSB; ICB; P II; GA; GB. Mathematics. Mechanics. Member of the Académie des Sciences.
1. *Huygens et Roberval: Documents nouveaux*. Leiden, 1879. Ed. by C. Henry.

2104\. SPERLING, Otto. 1602-1681. Den. ICB; BLA; GA; GC. Botany.

2105\. ALPINUS, Alpinus. 1603-1637. BMNH. Botany.

2106\. DIGBY, Kenelm. 1603-1665. Br. DSB; ICB; P I; BMNH; GA; GB; GD. Natural philosophy. Botany. Alchemy.
1. *Two treatises. In the one of which the nature of bodies ... is looked into....* Paris, 1644. Wing D1448. RM. Reprinted, Stuttgart, 1970. Wing records four later London eds.
2. *A discourse concerning the vegetation of plants*. London, 1661. Wing D1432.
3. *A choice collection of rare chymical secrets and experiments in philosophy*. London, 1682. RM.

3a. ——— Another ed., entitled *Chymical secrets*. London, 1685. Wing D1422. X.

4. *Private memoirs, written by himself*. 2 parts. London, 1827-28. Ed. by H.H. Nicolas.

Other titles in Wing.

2107\. GOLDMAYR, Andreas. 1603-1664. Ger. P I; GA. Astronomy. Astrology.

2108\. JONSTON, John (or Johann or Jan). 1603-1675. Br./Holl./Ger. DSB; ICB; P I; BMNH; BLA; GA; GD (JOHNSTONE). Natural history.
1. *Dendrographias; sive, Historiae naturalis de arboribus et fructibus*. Frankfurt, 1622. RM.
2. *Thaumatographia naturalis*. Amsterdam, 1632. RM.
3. *Historiae naturalis de quadrupedibus libri*. Frankfurt, 1650-53.

3a. ——— *A description of the nature of four-footed beasts*. Amsterdam, 1678. Wing J1015A. X.

4. *Naturae constantia*. Amsterdam, 1652. X.

4a. ——— *A history of the constancy of nature*. London, 1657. Wing J1016. X.

5. *A history of the wonderful things of nature*. London, 1657. Wing J1017. X. A trans. of four of his works: *De piscibus et cetis*, 1649; *De avibus*, 1650; *De serpentibus et draconibus*, 1653; and item 3, above.

2109\. LINEMANN, Albert. 1603-1653. Ger. P I. Astronomy.

2110\. MALVASIA, Cornelio. 1603-1664. It. P II. Astronomy/astrology.

2111. MYNSICHT, Hadrian (or Adrian). 1603-1638. Holl. BLA; GA; GC. Iatrochemistry. (Wing records an English trans. of one of his works.)

2112. PAULLI, Simon. 1603-1680. Den. DSB; BMNH; BLA; GA. Botany. Human and comparative anatomy.

2113. BOATE, Gerard. 1604-1650. Holl./Br. P III (BOOT); BMNH; GA; GD. Natural philosophy. Natural history. (Titles in Wing.)

2114. COURCIER, Pierre. 1604-1692. Fr. P I. Mathematics. Astronomy.
1. *Astronomia practica; sive, Motuum caelestium praxes per astrolabia....* Nancy, 1653. RM.

2115. DECKER, Ezechiel de. ca. 1604-ca. 1646. Holl. ICB. Arithmetic.
1. *Tweede deel van de nieuwe tel-konst.* Gouda, 1627. X. Reprinted, Nieuwkoop, 1964. (Dutch classics in history of science. 10)

2116. ENT, George. 1604-1689. Br. DSB; P I; BLA & Supp.; GA; GD. Physiology. Comparative anatomy. (Titles in Wing.)

2117. GLAUBER, Johann Rudolph. 1604-1670. Ger./Holl. DSB; ICB; P I; BLA & Supp.; GA; GB; GC. Chemistry.
1. *Furni novi philosophici.* 5 parts. Amsterdam, 1646-49. X.
1a. ——— *A description of new philosophical furnaces; or, A new art of distilling.* London, 1651. Trans. by John French. RM.
2. *Opera chymica.* Frankfurt, 1658. RM.
3. *Proposal for printing all the works of J.R. Glauber.* London, 1687. Wing G847. X.
4. *The works.* London, 1689. Wing G845. Trans. by C. Packe.

2118. KOHLHANS, Johann Christoph. 1604-1677. Ger. P I; GC. Optics.

2119. MICHELINI, Famiano. 1604-1665. It. DSB; P II. Hydraulic engineering.

2120. PAGAN, Blaise François de. 1604-1665. Fr. P II; GA. Astronomy.

2121. WALAEUS, Joannes (or WALE, Jan de). 1604-1649. Holl. ICB; BMNH; BLA (WALE); GF. Physiology.

2122. KELLER, Michael. fl. 1636. Ger. P I. Theory of music.

2123. RENAUDOT, Eusèbe. d. 1679. Fr. BLA; GA. "Intelligencer."
1. *Première (-quatriesme) centurie des questions traitées ez conférences du Bureau d'adresse.* 4 vols. Paris, 1634-41. Ed. by Théophraste Renaudot (entry 1959) and Eusèbe Renaudot. X.
1a. ——— Another ed. *Recueil général des questions traictées ....* 4 vols. Paris, 1655-56. Ed. by E. Renaudot. X.
1b. ——— ——— *A general collection of discourses of the virtuosi of France upon questions of all sorts of philosophy and other natural knowledge.* London, 1664. Wing R1034.
1c. ——— *Cinquiesme et dernier tome du Recueil général des questions traittées....* Paris, 1655. Ed. by E. Renaudot. X.
1d. ——— ——— *Another collection of philosophical conferences.* London, 1665. Wing R1033A. X.

2124\. AMBROSINUS, Hyacinthus. 1605-1671. It. BMNH; GA (AMBROSINI, Giacinto). Botany.

2125\. BAUSCH, Johann Lorenz. 1605-1665. Ger. P I; BMNH; BLA & Supp.; GA; GC. Mineralogy. In 1652 he founded a society from which the Academia Caesarea Leopoldina Naturae Curiosorum originated in 1672.

2126\. BOULLIAU, Ismael. 1605-1694. Fr. DSB; P I; GA. Astronomy. Mathematics. "One of the last reputable scholars to maintain confidence in astrology." -DSB.
1. *Astronomia philolaica. Opus novum, in quo motus planetarum per novam ac veram hypothesim demonstrantur.* Paris, 1645. RM.

2127\. BROWNE, Thomas. 1605-1682. Br. The famous physician and author. DSB; ICB; BMNH; BLA & Supp.; GA; GB; GD. Science in general. Natural history.
1. *Pseudoxia epidemica; or, Enquiries into very many received tenets and commonly presumed truths.* London, 1646. Wing B5159. Reprinted, Menston (U.K.), 1972. Wing records six later eds as well as other titles.

2128\. FRENICLE DE BESSY, Bernard. ca. 1605-1675. Fr. DSB; ICB; P I; GA. Mathematics. Member of the Académie des Sciences.

2129\. FROMM, Georg. 1605-1651. Den. P I. Astronomy.

2130\. HAAK, Theodore. 1605-1690. Br. DSB; ICB; P I; GC; GD. "Intelligencer" and stimulator of the scientific movement in England in the 1640s and later.

2131\. HORTENSIUS (or ORTENSIUS or VAN DER HOVE), Martinus. 1605-1639. Holl. DSB; P I; GF. Astronomy.

2132\. PLATER, Felix. 1605-1671. Switz. P II; BLA. Many dissertations, presumably of his students, are attributed to him (from the period 1634-1646), e.g. *De visu, De meteoris in genere et speciatim de ignitis, De stellis in genere, De motu, De maris aestu, De meteoris aquaeis, De iride,* etc.

2133\. SCHLEGEL, Paul Marquart. 1605-1653. Ger. ICB; BLA & Supp. Physiology.

2134\. BARRELIER, Jacques. 1606-1673. Fr. BMNH; GA. Botany.

2135\. CARAMUEL (Y LOBKOWITZ), Juan. 1606-1682. Spain and other countries. DSB; ICB; P I; GA; GC. Mathematics.

2136\. CONRING, Hermann. 1606-1681. Ger. P I; BMNH; BLA & Supp.; GA; GC. Chemistry. Physiology.

2137\. CORNUT, Jacques Philippe. 1606-1651. Fr. BMNH; GA. Botany.
1. *Canadensium plantarum ... historia.* Paris, 1635. X. Reprinted, New York, 1966. The first book on North American flora.

2138\. DONATI, Antonio. 1606-1659. It. BMNH; BLA. Natural history.

2139\. MICHAELIS, Johann. 1606-1667. Ger. P II; BLA. Iatrochemistry.

2140\. WINTHROP, John (first of the name). 1606-1676. U.S.A. DSB; ICB; P II; GB; GD; GE. Science in general. "Although Winthrop contributed little to the history of scientific thought, he was the first scientific investigator of note in British America." -DSB.

2141. BESLER, Michael Rupertus. 1607-1661. Ger. BMNH; BLA; GA; GC. Natural history.

2142. ERICI, Johann. 1607-1686. Dorpat. P I; GA. Many dissertations, presumably of his students, are attributed to him (from the period 1638-1650), e.g. *De theoria solis*, *De astronomiae natura*, *De coelo*, *De mundo*, *De meteoris*, *De natura elementorum in genere*, *De diebus canicularibus*, etc.

2143. FABRI (or FABRIUS), Honoré. 1607-1688. Fr./It. DSB; ICB; P I; BMNH; GA. Mathematics. Mechanics. Astronomy. Physiology. Natural history.
1. *Tractatus physicus de motu locali*. Lyons, 1646. RM.

2144. LOESELIUS, Johannes. 1607-1655. Ger. BMNH; BLA. Botany.

2145. NOTTNAGEL, Christoph. 1607-1666. Ger. P II. Astronomy. His works include *De originibus astronomiae*, 1650.

2146. TURRE, Georgio di. 1607-1688. It. BMNH; BLA. Botany.

2147. BARTOLI, Daniello. 1608-1685. It. DSB; P I; BLA; GA; GB. Several aspects of physics.
1. *Del suono, de' tremori armonici, e dell'udito*. Rome, 1679. RM.
2. *Del ghiaccio e della coagulatione*. Rome, 1681. RM.

2148. BATE, George. 1608-1668. Br. ICB; P I; GD. Pharmacy. (Titles in Wing.)

2149. BORELLI, (Giovanni) Alfonso. 1608-1679. It. DSB; ICB; P I; Mort.; BLA & Supp. (BORRELLI); GA; GB. Mathematics. Astronomy. Mechanics. Physiology. Vulcanology. Member of the Accademia del Cimento.
1. *Euclides restitutus; sive, Prisca geometriae elementa, brevius et facilius contexta....* Pisa, 1658. RM.
2. *Theoricae Mediceorum planetarum ex causis physicis deductae*. Florence, 1666. RM.
3. *De vi percussionis*. Bologna, 1667. RM. Another ed., Leiden, 1686.
4. *Historia et meteorologia incendii Aetnaei anno 1669*. Regio, 1670. RM/L.
5. *De motu animalium*. 2 parts. Rome, 1680-81. RM. Another ed., 2 vols, The Hague, 1743.
5a. ——— German trans. Leipzig, 1927. (OKEW. 221)
See also 1023/2.

2150. MORAY, Robert. 1608?-1673. Br. DSB; ICB; P II; GA (MURRAY); GD (MURRAY). A prominent virtuoso who was the first president of the Royal Society of London and whose influential connections ensured its survival during its difficult early years.

2151. SCHOTT, Gaspar (or Kaspar). 1608-1666. Ger./It. DSB; P II; GA; GC. "His contribution was essentially that of an editor who prepared the researches of others for the press without adding much of consequence. Still, he did much to popularize the achievements of contemporary physicists." -DSB.

1. *Mechanica hydraulico-pneumatica.* Würzburg, 1657. RM/L.
2. *Magia universalis naturae et artis ... pars I. optica; II. acoustica; III. mathematica; IV. physica.* 4 vols. Würzburg, 1657-58. RM/L. Another ed., Bamberg, 1677.
3. *Pantometrum Kircherianum; hoc est, Instrumentum geometricum novum ... a A. Kirchero.* Würzburg, 1660. RM.
4. *Cursus mathematicus.* Würzburg, 1661. RM.
5. *Mathesis Caesarea.* Würzburg, 1662. RM.
6. *Physica curiosa; sive, Mirabilia naturae et artis.* 2 vols. Würzburg, 1662. RM/L. 2nd ed., 1667.
7. *Anatomia physico-hydrostatica fontium ac fluminum.* Würzburg, 1663. RM.
8. *Technica curiosa; sive, Mirabilia artis.* Nuremberg, 1664. RM.
9. *Organum mathematicum.* Würzburg, 1668. RM.

2152. TORRICELLI, Evangelista. 1608-1647. It. DSB; ICB; P II; GA; GB. Mathematics. Mechanics. Originator of the famous barometric experiment. See also 2074.
1. *Opera geometrica.* Florence, 1644. RM.
2. *Opere.* 4 vols in 5. Faenza, 1919. Ed. by G. Lauria.
3. *Opere scelte.* Turin, 1975. Ed. by L. Belloni.
4. *Lezioni accademiche.* Florence, 1715. Ed. by T. Bonaventuri.
5. *Due insigni autografi di G. Galilei e E. Torricelli.* Florence, 1908.

2153. TRADESCANT, John (second of the name). 1608-1662. Br. DSB; ICB; BMNH; GD. Botany.
1. *Musaeum Tradescantium; or, A collection of rarities.* London, 1656. Wing T2005. X. Reprinted, Oxford, 1925.

2154. DIEMERBROECK, Ijsbrand van. 1609-1674. Holl. BLA & Supp.; GA; GF. Anatomy.
1. *Anatome corporis humani.* Utrecht, 1672. Another ed., Lyons, 1679. RM.

1a. ——— *The anatomy of human bodies.* London, 1689. Wing D1415. X.

2. *Opera omnia.* Utrecht, 1685.

2155. HALE, Matthew. 1609-1676. Br. ICB; P I; GA; GB; GD. Some aspects of physics.
1. *An essay touching the gravitation or non-gravitation of fluid bodies.* London, 1673. Wing H244. X.
2. *Difficiles nugae; or, Observations touching the Torricellian experiment.* London, 1674. RM.
3. *Observations touching the principles of natural motions, and especially touching rarefaction and condensation.* London, 1677. Wing H252. RM.

2156. ARTIGA (or ARTIEDA), Francisco Antonio. fl. 1630-1692. Spain. P III. Astronomy.

2157. JOSSELYN, John. fl. 1633-1675. Br./U.S.A. ICB; BMNH; GA; GD; GE. Natural history. (Titles in Wing.)

2158. HORVÁTH, Andreas. fl. 1635-1656. Hung. ICB. Entomology.

2159. PRIEZAC, Salomon de, *Sieur* de Saugues. fl. 1639-1666.
1. *L'histoire des éléphants.* Paris, 1650. RM.

2160. BLAGRAVE, Joseph. 1610-1682 (or 1616-1679). Br. P I; GA; GD. Astrology/astronomy. (Titles in PR.)

2161. BOURDELOT, Pierre Michon. 1610-1685. Fr. DSB. Founded the Académie Bourdelot which in the 1640s, and again in the period 1664-1684, played an important part in the dissemination of scientific ideas and information in Paris.

2162. CHABRÉ, Dominique. 1610-ca. 1667. Switz. BMNH (CHABRAEUS); BLA; GA. Botany.

2163. CHRISTIANI, David. 1610-1688. Ger. P I. Astronomy. Geography.

2164. CRABTREE, William. 1610-1644. Br. DSB; ICB; P I; GD. Astronomy.

2165. CUNITZ, Maria (married name: VON LÖVEN). ca. 1610-1664. Ger. P I. Under CUNITIA: GA; GB. Astronomy.

2166. DIVINI, Eustachio. 1610-1685. It. DSB; ICB; P I & Supp.; GA. Astronomy. "Among the first to develop technology for the production of scientifically designed optical instruments."
1. *Lettera ... intorno alle macchie nuovamente scoperte ... nel pianta di Giouve con suoi cannocchiali.* Rome, 1666. RM.

2167. ESCHOLT, Mikkel Pedersön. ca. 1610-1669. Nor. DSB. Geology. (Wing records an English trans. of his *Geologia norvegica.*)

2168. JOHNSON, William. ca. 1610-1665. Br. DSB. Chemistry. (Titles in Wing.)

2169. LAGUS, Daniel. ca. 1610-1678. Danzig. P I & Supp. Astronomy.

2170. LE FEBVRE, Nicaise (or Nicolas). 1610-1669. Fr./Br. DSB; ICB (LE FÈVRE); P I; GD (LE FEVRE). Chemistry.
1. *Traicté de la chimie.* 2 vols. Paris, 1660. X.
1a. ——— *A compleat body of chymistry.* London, 1664. Wing L925.
Listed under LE FÉVRE by Wing and some libraries.

2171. MARKGRAF, George. 1610-1644. Ger./Brazil. DSB; ICB (MARCGRAVE); P II (MARGGRAF); BMNH (MARCGRAVIUS); GA (MARGGRAG); GC (MARCGRAF). Natural history of Brazil.

2172. REMBRANDSZ, Dirk. 1610-1682. Holl. P II. Astronomy.

2173. ROBERT, Nicolas. 1610-1684. Fr. BMNH; GA; GF. Natural history illustration.

2174. SENGUERD, Arnold. 1610-1667. Holl. P II; BLA. Natural philosophy.

2175. STAMPIOEN, Jan Janszoon. 1610-1689. Holl. DSB; ICB; GF. Mathematics.

2176. WEBSTER, John. 1610-1682. Br. DSB; BMNH; GD. Chemistry.
1. *Metallographia; or, An history of metals ... ores and minerals.* London, 1671. Wing W1231. RM/L.

2177. WILLSFORD, Thomas. b. 1610. Br.
1. *Arithmetick.* London, 1656. Wing W2874. X.
2. *Nature's secrets; or, The ... history of the generation of meteors.* London, 1658. Wing W2875.

2178. WIRSUNG, Johann Georg. ca. 1610-1643. Ger./It. ICB; BLA; GC. Anatomy.

2179. HEVELIUS, Johannes. 1611-1687. Danzig. DSB; ICB; P I; GA; GB; GC. Astronomy. Instrument making.
1. *Selenographia; sive, Lunae descriptio.* Danzig, 1647. Reprinted, New York, 1968.
2. *Descriptio cometae anno 1665 exorti.* Danzig, 1666. RM.
3. *Cometographia, totam naturam cometarum ... exhibens.* Danzig, 1668. RM.
4. *Machina coelestis. I. Organographia; sive, Instrumentum astronomicorum omnium delineatio et descriptio. II. Rerum uranicarum observationes.* Danzig, 1673-79. X. Reprinted, Leipzig, 1969, and Osnabrück, 1969.

See also 2448/3 and 2676.

2180. PELL, John. 1611-1685. Br./Holl. DSB; ICB; P II; GA; GB; GD. Mathematics. (Title in Wing.)

2181. PERRAULT, Pierre. 1611-1680. Fr. DSB; ICB. The chief founder of experimental hydrology.
1. *De l'origine des fontaines.* Paris, 1674. RM.
1a. ——— *On the origin of springs.* New York, 1967. Trans. by A. La Roque.

2182. PISO, Willem (or LE POIS, Guillaume). ca. 1611-1678. Holl./Brazil. DSB; ICB; BMNH; BLA; GA (PISON). Natural history of Brazil.

2183. TRABER, Zacharias. 1611-1679. Aus. P II. Optics.
1. *Nervus opticus; sive, Tractatus theoricus ... opticam, catoptricam, dioptricam, distributus.* Vienna, 1675.

2184. ARNAULD, Antoine. 1612-1694. Fr. The Jansenist theologian. DSB; ICB; GA; GB. Mathematics.

2185. BALDNER, Leonhard. 1612-1694. Strasbourg. ICB. Zoology.

2186. DEUSING, Anton. 1612-1666. Holl. P I; BLA & Supp.; GA; GC; GF; Astronomy. Pneumatics. Iatrochemistry.

2187. GASCOIGNE, William. ca. 1612-1644. Br. DSB; P I; GD. Optics. Astronomy--improvement of the telescope and invention of the micrometer.

2188. LEICHNER, Eckard. 1612-1690. Ger. BLA; GA. Physiology. Natural history.

2189. TACQUET, Andreas. 1612-1660. Belg. DSB; ICB; P II; GA; GC; GF. Mathematics.
1. *Cylindricorum et annularium.* Antwerp, 1651. RM.
2. *Elementa [Euclidea] geometriae ... Quibus accedunt selecta ex Archimede "Theoremata."* Antwerp, 1654. X. Many eds and trans. "Although little more than a paraphrase of parts of Euclid and Archimedes, the book was distinguished by its clarity and order." -DSB.
2a. ——— *The elements of Euclid.* [place? ca. 1700] Trans. by William Whiston. X. 10th ed., Dublin, 1772.
3. *Arithmeticae theoria et praxis.* Louvain, 1656. X. *Editio secunda correctior*, Antwerp, 1665. RM.
4. *Opera mathematica.* Antwerp, 1668. RM/L.

2190. HIGHMORE, Nathaniel. 1613-1685. Br. DSB; ICB; BMNH; Mort.; BLA; GA; GD. Anatomy. "Generation." (Title in Wing.)

2191. MEURSIUS, Joannes. 1613-1654. BMNH. Botany.

2192. NICERON, Jean François. 1613-1646. Fr. DSB; ICB; P II; GA. Geometrical optics.

2193. PERRAULT, Claude. 1613-1688. Fr. DSB; ICB; P II; BMNH; BLA; GA. Zoology. Plant and animal physiology. Mechanical technology. Best known as an architect. Member of the Académie des Sciences.
1. *Mémoires pour servir à l'histoire naturelle des animaux.* See entry 2817.
2. *Recueil de plusieurs machines de nouvelles inventions.* Paris, 1700. RM.

2194. PETTUS, John. 1613-1690. Br. GD. Metallurgy. (Titles in Wing.)

2195. CORNELIO, Tommaso. 1614-1684. It. P I; BLA. Some aspects of physics. Physiology.

2196. HEEREBORD, Adriaan. ca. 1614-after 1660. Holl. P I.
1. *Philosophia naturalis.* Leiden, 1665. X. Another ed., Oxford, 1665; Wing H1362 and five later eds. X.

2197. HELMONT, Franciscus Mercurius van. 1614-1699. Belg. The son of the famous chemist. ICB; BLA & Supp.; GA; GF. Chemistry. (Wing records some English translations of his works.) See also 1885/2.

2198. MERRETT, Christopher. 1614-1695. Br. DSB; ICB; P II; BMNH; BLA; GA; GD. Natural history. Glass making. See also Index.
1. *The art of glass.* See 1865/1a.
2. *Pinax rerum naturalium Britannicarum, continens vegetabilia, animalia, et fossilia in hoc insula inchoatus.* London, 1666. Wing M1839 and another ed. X.

2199. SCHNEIDER, Conrad Viktor. 1614-1680. Ger. Mort.; BLA. Anatomy.

2200. SYLVIUS (or DUBOIS), Franciscus de le Boë. 1614-1672. Holl. DSB; ICB; P II; BLA & Supp. (BOË); GA. Anatomy. Iatrochemistry.
1. *Disputationum medicarum pars prima.* Amsterdam, 1663.
2. *Praxeos medicae idea nova.* 4 vols. Leiden, 1671-74. X.
2a. ——— *A new idea of the practice of physic.* London, 1675. Wing S6338. X.

2201. WHARTON, Thomas. 1614-1673. Br. DSB; Mort.; BLA; GD. Anatomy. (Title in Wing.)

2202. WILKINS, John. 1614-1672. Br. DSB; ICB; P II; GA; GB; GD. "The chief promoter of the new science in England--not only by virtue of his writings, but also owing to his personal encouragement of individuals and his success in the shaping of scientific organization before and after the official formation of the Royal Society." -DSB.
1. *The discovery of a world in the moone.* London, 1638. PR 25640. X. Reprinted, New York, 1973.
1a. ——— *A discourse concerning a new world. In two bookes.*

(The first book: the third impression. The second book: now first published). 2 parts. London, 1640, PR 25641. X. Wing (W2186/7) records two eds in 1684. X.

2. *Mathematicall magick.* London, 1648. Wing 2198/9 and two later eds.
3. *The mathematical and philosophical works.* London, 1708. Contains a biography and bibliography. 2nd ed., 2 vols, 1802; reprinted, ib., 1970.

Other titles in Wing.

2203. ODERICUS, Thomas. d. 1657. It. P II. Astrology.

2204. FOLIUS, Caecilius. 1615-1650. It. ICB; Mort. (FOLLI); BLA & Supp. (FOLLI). Anatomy. Physiology.

2205. GLASER, Christophe. ca. 1615-1672? Fr. DSB; ICB; P I; GA; GB. Chemistry.
1. *Traité de la chimie.* Paris, 1663. X. Another ed., Lyons, 1676. RM.
1a. ——— *The compleat chymist; or, A new treatise of chymistry.* London, 1677. Wing G843. X.

2206. MAGIRUS, Johann (second of the name). 1615-1697. Ger. P II. Astrology.

2207. RENALDINI (or RINALDINI), Carlo. 1615-1698. It. P II. Physics, especially thermometry and the study of heat and cold. Member of the Accademia del Cimento.

2208. SCHOOTEN, Frans van (second of the name). ca. 1615-1660. Holl. DSB; ICB; P II; GA; GC. Mathematics. See also 1665/3.

2209. SORBIÈRE, Samuel. 1615-1670. Fr. ICB; GA. Secretary of the Académie Montmor.
1. *Relation d'un voyage en Angleterre.* Paris, 1664. X.
1a. ——— *A journey to England.* London, 1700. Wing S4697. X.
1b. ——— *A voyage to England, containing many things relating to the state of learning ... Also observations on the same voyage by Dr. T. Sprat.* 2 parts. London, 1709-08. X. Sprat's *Observations* were first publ. in 1665: see 2457/1.
2. *A journey to London in the year 1698 ... Written originally in French by Monsieur Sorbière, and newly translated into English.* London, 1698. Wing S4698. X. Actually written by William King, not by Sorbière.

2210. ZANONI, Giacomo (or Jacopo). 1615-1682. It. BMNH; GA. Botany.
1. *Istoria botanica.* Bologna, 1675. RM.
2. *Rariorum stirpium historia. Ex parte olim edita, nunc ... ampliata.* Bologna, 1742. Ed. by C. Montius. RM.

2211. BARTHOLIN, Thomas. 1616-1680. Den. DSB; ICB; P I; BMNH; Mort.; BLA & Supp.; GA; GB. Anatomy. Physiology.
1. *Anatomia. Ex Caspari Bartholini parentis "Institutionibus."* Leiden, 1651. (cf. item 1945/2) RM. Other eds: The Hague, 1655; Leiden, 1684 and 1686.
1a. ——— *Anatomy.* London, 1663. Wing B976 and another ed. X.
2. *De lacteis thoracis in homine brutisque.* Copenhagen, 1652. X. Another ed., London, 1652. Wing B978. X.

2a. ——— *The anatomical history concerning the lacteal veins of the thorax.* London, 1653. Wing B975. X.
3. *Historiarum anatomicarum rariorum centuria.* 3 vols. Copenhagen, 1654-61.
4. *Opuscula nova anatomica de lacteis thoracis et lymphaticis vasis, uno volumine comprehensa.* Copenhagen, 1670. RM.
See also 1945/2a,b.

2212. CULPEPER, Nicholas. 1616-1654. Br. ICB; BMNH; BLA & Supp.; GA; GD. Herbal pharmacy. Astrology.
1. *A physicall directory; or, A translation of the "London Dispensatory" made by the Colledge of Physicians ... with many hundred additions.* London, 1649. Wing C7540 and another ed.
1a. ——— *Pharmacopeia Londinensis; or, The "London Dispensatory" further adorned.* London, 1653. Wing C7525 and twelve further eds up to 1695.
2. *The English physician; or, An astrologo-physical discourse of the vulgar herbs of this nation.* London, 1652. Wing C7500 and fourteen later eds up to 1695. Subsequent eds up to 1850 with various changes of title, the last title being *The complete herbal.*
3. [Reprints. 3 vols in 1.] *The complete herbal ... The English physician enlarged* [the 1653 ed.], *and Key to physic.* Birmingham, 1953.
See also 1846/2a.

2213. FRENCH, John. 1616-1657. Br. P I; GA; GD. Chemistry. (Titles in Wing.) See also 1810/1a and 2117/1a.

2214. GRAINDORGE (or GRANDORGAEUS), André. 1616-1676. Fr. ICB; P I; BLA; GA. Embryology. Also author of *Dissertatio de natura ignis, lucis et colorum,* 1664. In 1664 he founded a (short-lived) Académie de Physique in Caen.

2215. HORST, Johann Daniel. 1616-1685. Ger. P I; BLA. Chemistry. Anatomy.

2216. KERSEY, John. 1616-ca. 1690. Br. GA; GD. Author of a highly regarded treatise on algebra. (Title in Wing.)

2217. LANGLI, Willem. 1616-1668. Holl. BLA & Supp. Embryology.

2218. VOLCKAMER, Johann Georg (first of the name). 1616-1693. Ger. P II; BMNH; BLA (VOLKAMER); GC. Botany.

2219. WALLIS, John. 1616-1703. Br. DSB; ICB; P II; GA; GB; GD. Mathematics. See also Index.
1. *A discourse of gravity and gravitation.* London, 1675. Wing W574. RM.
2. *A defence of the Royal Society and the "Philosophical transactions" ... in answer to the cavils of Dr. William Holder.* London, 1678. Wing W573.
3. *A treatise of angular sections.* Oxford, 1684. Wing W614.
4. *A treatise of algebra.* Oxford, 1685. Wing W613. RM/L.
5. *Opera mathematica.* 3 vols. Oxford, 1693-99. Wing W596/7. RM/L. Reprinted, Hildesheim, 1972.
Many other titles in Wing.

2220. ASHMOLE, Elias. 1617-1692. Br. DSB; ICB; P I; GA; GB; GD. Astrology. Alchemy.

1. *Theatrum chemicum Britannicum.* London, 1652. Wing A3987. In English. Contains annotated extracts from the writings of English alchemists. X. Reprinted, New York, 1967, and Hildesheim, 1968.
2. *Catalogue of library.* London, 1694. Wing A3981. X.
3. *The diary and will.* Oxford, 1927. Ed. from the MSS by R.T. Gunther.
4. *His autobiographical and historical notes, his correspondence, and other contemporary sources relating to his life and work.* 5 vols. Oxford, 1966. Ed. with a biographical introd. by C.H. Josten.

See also Index.

2221. CASATI, Paolo. 1617-1707. It. P I; GA. Mechanics. Optics. Also works on heat, etc.
1. *Vacuum proscriptum.* Geneva, 1649. RM. On the Torricellian experiment and the non-existence of a vacuum.
2. *Mechanicorum libri octo.* Lyons, 1684. RM/L.
3. *Hydrostaticae dissertationes.* Parma, 1695. RM.
4. *Opticae dissertationes.* Parma, 1705. RM.

2222. CUDWORTH, Ralph. 1617-1688. Br. DSB; ICB; GA; GB; GD. Philosophy. "His attempt to combine Neoplatonism and the mechanical philosophy is important in understanding the background to Newton and the early Newtonians." -DSB.
1. *The true intellectual system of the universe. The first part.* [No more publ.] London, 1678. Wing C7471. Another ed., 4 vols, ib., 1820; ed. by T. Birch.

2223. GOEDAERT, Johannes. 1617-1668. Holl. DSB; BMNH; GA. Entomology.
1. *Metamorphosis et historia naturalis insectorum.* 3 parts. Middelburg, 1662-70.

1a. ——— *J. Godartius. Of insects.* York, 1682. Wing G1003. Trans. with notes by M. Lister. X.

1b. ——— *J. Goedartius. De insectis.* London, 1685. Wing G1002. Ed. with notes by M. Lister. RM/L.

2224. MATTHAEUS A S. JOSEPH (or FOGLIA, Petrus). ca. 1617-1691. BMNH. Botany.

2225. MEY, Joannes de. 1617-1678. BMNH. Zoology.

2226. MOORE, Jonas. 1617-1679. Br. ICB; P II; GA; GD. Patron of Flamsteed and instigator of the establishment of the Greenwich observatory. Author of several works on various aspects of practical mathematics (see Wing). His chief work is his *New system.*
1. *A new system of the mathematicks.* 2 vols. London, 1681. Wing M2579.
2. *Works.* London, 1681. Wing M2562A. X.
3. *Bibliotheca matematica.* London, 1684. Wing M2567. A catalogue of his extensive library. X.

2227. WARD, Seth. 1617-1689. Br. DSB; ICB; P II; GA; GB; GD. Astronomy. (Titles in Wing.)

2228. WHARTON, George. 1617-1681. Br. DSB; GD. Astrology/astronomy. (Titles in Wing.)

2229. ANTHELME, Voituret. ca. 1618-1683. Fr. DSB; ICB. Astronomy.

2230. BLONDEL, (Nicolas) François. 1618-1686. Fr. DSB; ICB; P I; GA. Mechanics. Engineering. Architecture. Member of the Académie des Sciences.
1. *Histoire du calendrier romain*. Paris, 1682. X. Another ed., The Hague, 1684.
2. *L'art de jetter les bombes*. Paris, 1683. X. Another ed., The Hague, 1685. RM.

2231. CHARAS, Moïse. 1618-1698. Fr. ICB; P I; BMNH; BLA & Supp.; GA. Chemistry. Zoology. Member of the Académie des Sciences.
1. *Nouvelles expériences sur la vipère, où l'on verra une description exacte de toutes ses parties, la source de son vénin, etc.* Paris, 1669. X.
1a. ——— *New experiments upon vipers*. London, 1670. Wing C2037 and two later eds. X.
2. *Pharmacopée royale*. Paris, 1676. X.
2a. ——— *The royal pharmacopoea, galenical and chemical*. London, 1678. Wing C2040.
See also 2817/1c.

2232. COLLINS, Samuel. 1618-1710. Br. BMNH; BLA; GD. Human and comparative anatomy. (Title in Wing.)

2233. GRIMALDI, Francesco Maria. 1618-1663. It. DSB; ICB; P I; GA; Astronomy. Optics.
1. *Physico-mathesis de lumine, coloribus, et iride*. Bologna, 1665. X. Reprinted, ib., 1963, and London, 1966.

2234. HOLWARDA (or PHOCYLIDES), Johann Fokkens. 1618-1651. Holl. P I. Astronomy.
1. *Panselenos; id est, Dissertatio astronomica*. Franeker, 1649. RM.

2235. HORROCKS, Jeremiah. 1618-1641. Br. DSB; ICB; P I (HORROX); GA; GB; GD. Astronomy. (Titles in Wing.) See also 15580/1.

2236. LEONARDO DA CAPUA. ca. 1618-1695. It. P I. Chemistry. He founded a scientific academy in Naples.

2237. MOUTON, Gabriel. 1618-1694. Fr. DSB; P II. Mathematics. Astronomy.

2238. MYLON, Claude. ca. 1618-ca. 1660. Fr. DSB. Secretary of the Académie Parisienne (a continuation of Mersenne's group) and an "intelligencer" during the 1650s.

2239. OLDENBURG, Henry. ca. 1618-1677. Br. DSB; ICB; P II; GA; GD. Indefatigable "intelligencer" and secretary of the Royal Society from 1662 until his death. In 1665 he established the *Philosophical transactions*, the first purely scientific journal.
1. *Correspondence*. Several vols. Madison, Wis., 1965-. Ed. and trans. by A.R. Hall and M.B. Hall.
See also 2367/1 and 2486/3a.

2240. VOSSIUS (or VOSS), Isaac. 1618-1689. Holl./Br. ICB; P II; BMNH; GA; GC; GD.; GF. A versatile savant; his main scientific works are *De lucis natura et proprietate*, 1662, *De Nili et aliorum fluminum origine*, 1666, and the following.

1. *De motu marium et ventorum*. The Hague, 1663. RM.
1a. ——— *A treatise concerning the motion of the seas and winds*. London, 1677. Wing V706. RM.

2241. ZWELFER, Johann. 1618-1668. Aus. DSB; P II; BLA. Pharmacy. Chemistry.

2242. DATI, Carlo. 1619-1679. It. ICB; P I; GA. Author of *Lettera ... sulla vera storia della cicloide e della famosissima esperienza dell'argento vivo*, and *Sull'invenzione degli occhiali*.

2243. LANGE, Christian. 1619-1662. Ger. P I; BLA. Chemistry.

2244. MARCHE, Caspar. 1619-1677. Ger. P II. Astrology/astronomy. Member of the Leopoldina Academy.

2245. MERCATOR (or KAUFFMAN), Nicolaus. ca. 1619-1687. Den./Br. DSB; ICB; P II & Supp.; GA. Mathematics. Astronomy. (Titles in Wing.)

2246. RICCI, Michelangelo. 1619-1682. It. DSB; ICB; P II. Mathematics.

2247. RUPERT, *Prince*. 1619-1682. Br., etc. ICB; P II (RUPRECHT); GA; GB; GD. A virtuoso and Fellow of the Royal Society.

2248. SCHENK, Johann Theodor. 1619-1671. Ger. BMNH; BLA. Botany. Zoology.

2249. SCHUYL, Florentinus. 1619-1669. Holl. BMNH; BLA. Botany. See also 2032/4a.

2250. WING, Vincent. 1619-1668. Br. DSB; P II; GD. Astronomy.
1. *Urania practica; or, Practical astronomy*. London, 1649. Wing W2994. With W. Leybourne.
2. *Astronomia Britannica*. London, 1669. Wing W2986. RM.
Other titles in Wing.

2251. TACHENIUS, Otto. fl. 1640-1670. Ger./It. DSB; P II; BLA; GC. Iatrochemistry.
1. *Hippocrates chimicus*. Venice, 1666. X.
1a. ——— English trans. *Hippocrates chymicus*. London, 1677. Wing T89. X.
2. *Antiquissimae Hippocraticae medicinae clavis*. Brunswick, 1668. X.
2a. ——— English trans. *Clavis*. London, 1677. Wing T97. X.

2252. ARVIDI, Andreas. fl. 1644-1651. Dorpat. P III. Meteorology.

2253. MOLINS (or MULLINS, etc.), William. fl. 1648-1680. Br. GD (MOLINES, James). Anatomy. (Titles in Wing.)

2254. BEUTEL, Tobias. fl. 1651-1683. Ger. P I. Astronomy.

2255. NICOLS, Thomas. fl. 1659. Br. BMNH; GD.
1. *A lapidary; or, The history of precious stones*. Cambridge, 1652. Wing N1145. RM.

2256. TITI, Placido. d. 1668. It. P II. Astronomy/astrology.

2257. LOMBARD, Carl. d. 1669. Ger. P I. Many dissertations, presumably of his students, are attributed to him (from the period 1657-66), e.g. *De mundo*, *De materiae proprietatibus*,

De vacuo, *De divisione motus*, *De tempore*, *De qualitatibus occultis*, *De continuo*, etc.

2258. JONCQUET, Dionysius. d. 1671. Fr. BMNH. Botany.

2259. BUOT, Jacques. d. ca. 1675. Fr. DSB; P I. Astronomy. Member of the Académie des Sciences.

2260. CASTIGLIONE, Giovanni Onorio. d. 1679. It. BLA. Pharmacy.
1. *Prospectus pharmaceuticus*. Milan, 1668. RM.

2261. FORNEL, Jöns. d. 1679. Swed. P I. Astronomy. Optics.

2262. MARIOTTE, Edme. d. 1684. Fr. DSB; ICB; P II & Supp.; GA; GB. Mechanics (incl. fluid mechanics). Optics (incl. physiological optics). Pneumatics. Heat and cold. Meteorology. Plant physiology. Methodology. Member of the Académie des Sciences.
1. *Traité de la percussion ... dans lequel les principales règles du mouvement ... sont démonstrées*. Paris, 1673. RM.
2. *Traités du mouvement des eaux et des autres corps fluides*. Paris, 1686. RM.
2a. ——— *The motion of water and other fluids*. London, 1718. Trans. by J.T. Desaguliers. RM.
3. *Oeuvres*. 2 vols. Leiden, 1717. RM/L.

2263. BETTS, John. d. 1695. Br. BLA & Supp.; GA; GD. Physiology. (Title in Wing.)

2264. BATHURST, Ralph. 1620-1704. Br. ICB; P I; BLA; GA; GD. Physiology.

2265. BIE, Alexander de. ca. 1620-1690. Holl. GF. Astronomy.

2266. BONET, Théophile. 1620-1689. Switz. ICB; BLA; GA. Anatomy. See also 2279/1a.

2267. BOREL, Pierre. ca. 1620-1671. Fr. DSB; ICB; P I; BLA; GA. Natural history. Chemistry. Microscopy. His works include *De vero telescopii inventore*, 1655.
1. *Bibliotheca chimica; seu, Catalogus librorum philosophicorum hermeticorum*. Paris, 1654. X. Reprinted, Hildesheim, 1969.
2. *Discours nouveau prouvant la pluralité des mondes*. Geneva, 1657. RM.

2268. BOYM, Michael. 1620-1659. BMNH. Author of *Flora Sinensis*, 1656.

2269. BROUNCKER, William. 1620-1684. Br. DSB; ICB; P I; GA; GD. Mathematics. President of the Royal Society.

2270. BRUYN, Jan van. 1620-1675. Holl. P I; GA; GF. Author of *De corporum gravitate et levitate* and *Epistola ad J. Vossium de lucis causa et origine*.

2271. CHARLETON, Walter. 1620-1707. Br. DSB; ICB; BMNH; BLA (CHARLTON); GA; GD. Natural philosophy. Physiology. Zoology.
1. *Physiologia Epicuro-Gassendo-Charltoniana; or, A fabrick of science natural, upon the hypothesis of atoms*. London, 1654. Wing C3691. Reprinted, New York, 1966.
2. *Natural history of nutrition, life, and voluntary motion*. London, 1659. Wing C3684.

3. *Onomasticon zoicon.* London, 1668. Wing C3688. RM.
3a. ——— 2nd ed. *Exercitationes de differentiis et nominibus animalium.* Oxford, 1677. Wing C3672. X.
Other titles in Wing.

2272. CRAANEN, Theodorus. 1620-1690. Holl./Ger. BLA & Supp.; GA. Physiology.

2273. EVELYN, John. 1620-1706. Br. The diarist. DSB; ICB; BMNH; GA; GB; GD. Silviculture and horticulture.
1. *Sylva; or, A discourse of forest trees.* London, 1664. Wing E3516. RM. 2nd ed., 1670.
2. *A philosophical discourse of earth, relating to the culture and improvement of it for vegetation.* London, 1676. Wing E3507. RM.

2274. EYSSON, Henrique. 1620-1690. Holl. BLA; GA; GF. Anatomy.

2275. HOW, William. 1620-1656. Br. BMNH; GD (HOWE). Botany.
1. *Phytologia Britannica.* London, 1650. Wing H2956. RM.

2276. KIRSTEN, Michael. 1620-1678. Ger. P I; BLA & Supp.; GA. Chemistry.

2277. MORISON, Robert. 1620-1683. Br. DSB; ICB; BMNH; GA; GD. Botany. (Titles in Wing.)

2278. PICARD, Jean. 1620-1682. Fr. DSB; ICB; P II; GA. Astronomy. Geodesy. Member of the Académie des Sciences.
1. *Mesure de la terre.* Paris, 1671. X.
1a. ——— Another ed. *Degré du méridien entre Paris et Amiens, déterminé par la mesure de M. Picard, et par les observations de Mrs de Maupertuis, Clairaut, Camus, Le Monnier.* Paris, 1740. RM (under title).

2279. ROHAULT, Jacques. 1620-1675. Fr. DSB; ICB; P II; GA. "The leading advocate and teacher of Descartes's natural philosophy among the first generation of French Cartesians." -DSB.
1. *Traité de physique.* Paris, 1671. X. 6th ed., 1683. RM.
1a. ——— *Tractatus physicus.* Geneva, 1674. Trans. by T. Bonet. X. Another ed., with notes by A. Le Grand, London, 1682. Wing R1871.
1b. ——— *Physica.* London, 1697. Wing R1870. Latin trans. with Newtonian annotations by Samuel Clarke. X. 2nd ed., 1701. X. 3rd ed., 1710. 4th ed., 1718.
1c. ——— *System of natural philosophy. Illustrated with Dr. S. Clarke's notes, taken mostly out of Sir Isaac Newton's philosophy.* 2 vols. London, 1723. Trans. by John Clarke. Reprinted, New York, 1969.
2. *Oeuvres posthumes.* Paris, 1682. RM.

2280. STEPHENS, Philip. ca. 1620-1679. Br. BMNH. Botany. (Title in Wing.)

2281. THÉVENOT, Melchisédech. 1620-1692. Fr. DSB; ICB; P II; GA. An "intelligencer" and patron of science who was at the centre of the scientific community in Paris from about 1655 until his death. Member of the Académie des Sciences from 1685. See also 2470/5.

2282. WEPFER, Johann Jakob. 1620-1695. Switz. DSB; ICB; P II; BMNH; BLA. Anatomy. Physiology. Botany,

2283. BOURDELIN, Claude. ca. 1621-1699. Fr. DSB; ICB; P I; BLA. Chemistry. Member of the Académie des Sciences.

2284. CELSIUS, Magnus Nicolaus. 1621-1679. Swed. P I. Astronomy.

2285. DECHALES (or DE CHALLES, etc.), Claude François Milliet. 1621-1678. Fr. DSB; P I (DESCHALES); GA (CHÂLES). Mathematics. Mechanics.
1. *Traitté du mouvement local, et du ressort.* Lyons, 1682. RM.

2286. HORNE (or HORNIUS), Jan van. 1621-1670. Holl. DSB; BLA; GA (HOORNE); GF. Anatomy.

2287. PITTON, Jean Scholastique. 1621-1689. Fr. BMNH; GA. Natural history.

2288. STANSEL, Valentin. 1621-after 1694. Brazil. P II. Astronomy.

2289. WILLIS, Thomas. 1621-1675. Br. DSB; ICB; P II; Mort.; BLA & Supp.; GA; GB; GD. Anatomy.
1. *Cerebri anatome. Cui accessit Nervorum descriptio et usus.* London, 1664. Wing W2823/4. RM.
1a. ——— *The anatomy of the brain and nerves.* London, 1681. X. Reprinted, 2 vols, Leicester, 1965.
2. *De anima brutorum.* Oxford, 1672. Wing W2825/9.
2a. ——— *Two discourses concerning the soul of brutes.* London, 1683. Wing W2856. X.
3. *Opera omnia.* 2 vols. Geneva, 1680. X. Another ed., Amsterdam, 1682. Ed. by G. Blasius.
Other titles in Wing.

2290. AUZOUT, Adrien. 1622-1691. Fr. DSB; P I; GA. Astronomy. Member of the Académie des Sciences.

2291. BAYLE, François. 1622-1700 (or 1709). Fr. P I & Supp.; BMNH; BLA; GA. Physiology.

2292. BÜTHNER, Friedrich. 1622-1701. Danzig. P I. Astronomy. Astrology.

2293. EIRENAEUS PHILALETHES. b. ca. 1622. GD. Pseudonym of an unidentified alchemist, possibly George Starkey (entry 2370); see the articles on Starkey in DSB and GD. See also 2831.

2294. HOFFMANN, Moritz. 1622-1698. Ger. ICB; BMNH; BLA (HOFMANN); GA. Human and comparative anatomy. Botany.

2295. MENTZEL (or MENZEL), Christian. 1622-1701. Ger. ICB; P II; BMNH; GA; GC. Optics. Botany. Comparative anatomy. Member of the Leopoldina Academy.

2296. NEWTON, John. 1622-1678. Br. P II; GA; GD. Author of a textbook of astronomy and numerous textbooks of mathematics. (Titles in Wing.)

2297. PANCKOW (or PANCOVIUS), Thomas. 1622-1665. BMNH. Author of a herbal, 1654.

2298. PECQUET, Jean. 1622-1674. Fr. DSB; ICB; P II & Supp.; Mort.; BLA; GA. Anatomy. Member of the Académie des Sciences.
1. *Experimenta nova anatomica.* Paris, 1651. X.
1a. ——— *New anatomical experiments.* London, 1653. Wing P1045. X.

2299. ROOKE, Lawrence. 1622-1662. Br. DSB; ICB; P II; GA; GD. Astronomy.

2300. SLUSE (or SLUZE or SLUSIUS), René François de. 1622-1685. Belg. DSB; ICB; P II; GC; GF. Mathematics.

2301. STREETE, Thomas. 1622-1689. Br. DSB. Astronomy. (Titles in Wing.)

2302. VARENIUS, Bernhardus. 1622-1650. Holl. DSB; ICB (VAREN); P II; GA; GB; GC. Physical geography.
   1. *Geographia generalis*. Amsterdam, 1650. X. Another ed., ib., 1664. Another ed., Cambridge, 1672 and 1681. Wing V106/7.
   1a. ——— *Cosmography and geography*. London, 1652. Wing V101. X. Another ed., 1682. Wing V102. Another ed., 1693. Wing V104. RM.

2303. VAUGHAN, Thomas. 1622-1666. Br. ICB; BLA; GB; GD. Alchemy. He used the pseudonym EUGENIUS PHILALETHES.
   1. *The works*. London, 1919. Ed. with introd. and notes by A.E. Waite. X. Reprinted, New Hyde Park, N.Y., 1968.

2304. VIVIANI, Vincenzo. 1622-1703. It. DSB; ICB; P II; GA. Mathematics. A disciple of Galileo and a member of the Accademia del Cimento.
   1. *De maximis et minimis geometrica divinatio in quintum librum "Conicorum" Apollonii Pergaei, adhuc desideratum*. Florence, 1659. RM.
   2. *De locis solidis secunda divinatio geometrica in quinque libros ... amissos Aristaei*. Florence, 1701. RM.
   3. *Vita di Galileo*. [With] *Il processo di Galileo*. Milan, 1954. Ed. by F. Flora.
   See also 3389/1.

2305. ANGELI, Stefano degli. 1623-1697. It. DSB; P I. Mathematics. Mechanics.
   1. *De infinitis parabolis, de infinitisque solidis ex variis rotationibus ipsarum, partiumque earundem genitis*. Venice, 1659. RM.
   2. *Accessionis ad stereometriam et mechanicam*. Venice, 1662. RM.
   3. *Della gravita dell'aria e fluidi*. Padua, 1671. RM.

2306. CHILDREY, Joshua. 1623-1670. Br. DSB; P I; BMNH; GA; GD. Natural history. Meteorology. Astrology. (Titles in Wing.)

2307. DU HAMEL, Jean Baptiste. 1623-1706. Fr. DSB; P I; BMNH; GA; GB. Astronomy. Secretary of the Académie des Sciences from its foundation in 1666 until 1697.
   1. *Elementa astronomica*. Paris, 1643. X. Another ed., London, 1654. Wing D2497. X. Another ed., Cambridge, 1665. Wing D2498. X.
   2. *Astronomia physica*. Paris, 1660. RM.
   3. *De meteoris et fossilibus*. Paris, 1660. RM.
   4. *De consensu veteris et novae philosophiae ubi ... de principiis rerum excutiuntur et de principiis chymicis*. Paris, 1663. X. Another ed., Oxford, 1669. Wing D2496. X.
   5. *Regiae Scientiarum Academiae historia*. Paris, 1698. RM.

2308. ELSHOLTZ, Johann Sigismund. 1623-1688. Ger. ICB; P I; BMNH; BLA & Supp.; GA; GC. Chemistry. Botany.
1. *Destillatoria curiosa; das ist, Curiose ... Destillir-Kunst.* Berlin, 1674. X.
1a. ——— *The curious distillatory.* London, 1677. Wing E638. RM.
2. *De phosphoris observationes.* Berlin, 1676. X. Another ed., ib., 1681.

2309. ESCHINARDI, Francesco. 1623-after 1699. It. P I; GA. Optics. Mechanics.

2310. LUBIENITZKI (or LUBIENSKI or LUBIENIECKI), Stanislaw. 1623-1675. Poland. ICB; P I; GA.
1. *Theatrum cometicum.* Amsterdam, 1668. RM.

2311. MEGERLIN, Peter. 1623-1686. Switz. P II. Astronomy.

2312. PASCAL, Blaise. 1623-1662. Fr. DSB; ICB; P II; G. Mathematics. Hydrostatics. The barometric experiment. See also Index.
1. *Lettres de A. Dettonville* [pseud.] *contenant quelques-unes de ses inventions de géometrie.* Paris, 1659. X. Reprinted, London, 1966.
2. *Traité de l'équilibre des liqueurs, et de la pesanteur de la masse de l'air.* Paris, 1663. RM. Reprinted, ib., 1956. 2nd ed., 1664.
3. *Oeuvres.* 5 vols. The Hague, 1779. Ed. by C. Bossut. Another ed., Paris, 1819. RM.
4. *Oeuvres.* 14 vols. Paris, 1904-14. Ed. by L. Brunschvicq and P. Boutroux. Reprinted, Vaduz, 1965.
5. *Oeuvres complètes.* Paris, 1954. 2nd ed., 1962. Ed. by J. Chevalier. Includes some scientific works which are not in item 4.
6. *Oeuvres complètes.* Paris, 1964-. Ed. by J. Mesnard.
7. *The physical treatises.* New York, 1937. Trans. by I.H.B. Spiers; introd. and notes by F. Barry. Reprinted, 1973.

2313. POWER, Henry. 1623-1668. Br. DSB; ICB; P II; GD. His only published work is the following.
1. *Experimental philosophy.* London, 1664. Wing P3099. RM. Deals with microscopy, atmospheric pressure, and magnetism. Reprinted with addition of Power's notes, corrections, and emendations; ed. with an introd. by M.B. Hall. New York, 1966.

2314. VERBIEST, Ferdinand. 1623-1688. Jesuit missionary in China. ICB; P II; GA; GC; GF. Author of many books in Chinese on astronomy, mathematics, and physics.
1. *Correspondance.* Brussels, 1938. Ed. by H. Josson and L. Willaert.

2315. BILS, Louis de. 1624-1671. Holl. ICB; BLA & Supp.; GA. Anatomy. (Wing records an English trans. of one of his works.)

2316. DICKINSON, Edmund. 1624-1707. Br. ICB; BLA; GA; GD. Alchemy.

2317. GUARINI, Camillo Guarino. 1624-1683. It. P I; GA; GB. Astronomy.

2318. LANGE (or LANGIUS), Wilhelm. 1624-1682. Den. P I; GA. Astronomy.

2319. NEWCASTLE, *Duchess of* (or CAVENDISH, Margaret). 1624?-1674. Br. Under CAVENDISH: GA; GD. Authoress of *Philosophical and physical opinions*, 1655, *Observations upon experimental philosphy*, 1666, etc. (Titles in Wing.)

2320. VELSCHIUS, Georgius Hieronymus. 1624-1676. Ger. BMNH; BLA (WELSCH); GC (WELSCH). Zoology.

2321. MÜLLER, Johann d. 1671. Ger. P II. Astronomy.

2322. HECKER, Johann. d. 1675. Danzig. P I. Astronomy.

2323. MOELLENBROCCIUS, Valentinus Andreas. d. 1675. Ger. BMNH; BLA. Zoology.

2324. STAHL, Peter. d.1675. Br. ICB. Chemistry.

2325. MARCHANT, Nicolas. d. 1678. Fr. DSB; BMNH. Botany. Member of the Académie des Sciences.

2326. PLACENTINUS, Johann. d. 1683. Ger. P II. Astronomy. Also author of *De oceani fluxu et refluxu*, 1654, and *De calore et motu membranorum corporis humani*, 1659.

2327. BAKER, Thomas. ca. 1625-1690. Br. P I; GA; GD. Mathematics. (Titles in Wing.)

2328. BARTHOLIN, Erasmus. 1625-1698. Den. DSB; ICB; P I; BLA & Supp.; GA; GB. Mathematics. Astronomy. Optics.
1. *De figura nivis*. Copenhagen, 1661. RM.
2. *De cometis anni 1664 et 1665*. Copenhagen, 1665. RM.
3. *Experimenta crystalli islandici disdiaclastici, quibus mira et insolita refractio detegitur*. Copenhagen, 1669. RM.
3a. ——— German trans. Leipzig, 1922. (OKEW. 205)
4. *De naturae mirabilibus quaestiones academicae*. Copenhagen, 1674. RM.

2329. BLASIUS, Gerardus. ca. 1625-1692. Holl. BMNH; Mort. (BLAES); BLA & Supp. (BLAES). Human and comparative anatomy.
1. *Anatome animalium terrestrium variorum*. Amsterdam, 1681. RM.
See also 2289/3.

2330. BUONO, Paolo del. 1625-1659. It./Aus. DSB; P I. Some aspects of experimental physics. He was a correspondent of the Accademia del Cimento and kept it informed of scientific developments in central Europe.

2331. CASSINI, Gian Domenico (or Jean Dominique). 1625-1712. It./Fr. DSB; ICB; P I; GA; GB. Astronomy. Geodesy. Member of the Académie des Sciences.
1. *Découverte de deux nouvelles planètes autour de Saturne*. Paris, 1678. RM.

2332. COLLINS, John. 1625-1683. Br. DSB; ICB; P I; GA; GD. A mathematical practitioner (his books are listed in Wing) whose importance in the history of science is as an "intelligencer" in the field of the mathematical sciences. Between 1662 and 1677 he carried on an extensive correspondence with many of the most outstanding scientists of the time. See also 2478/4.

2333. FABRICIUS, Wolfgang Ambrosinus. 1625-1653. Ger. BMNH; BLA; GA. Author of *De signaturis plantarum*, 1653.

2334. HELVETICUS (or SCHWEITZER), Johann Friedrich. 1625 (or 1630)-1709. Holl. P I; BLA & Supp.; GA; GF. Alchemy.

2335. LUDWIG (or LUDOVICI), Daniel. 1625-1680. Ger. P I; BLA. Chemistry.

2336. MENGOLI, Pietro. 1625-1686. It. DSB; ICB; P II; GA. Mathematics.

2337. MORLAND, Samuel. 1625-1695. Br. DSB; ICB; P II; GA; GD. Invention of scientific instruments. Mechanical technology. (Titles in Wing.)

2338. REISEL, Salomon. 1625-1702. Ger. P II. Thermometry and other aspects of experimental physics. Botany. Zoology. Member of the Leopoldina Academy.

2339. WEIGEL, Erhard. 1625-1699. Ger. P II; GA; GC. Astronomy.

2340. WITT, Jan de. 1625-1672. Holl. DSB; ICB; P II. Mathematics.

2341. BORRICHIUS (or BORCH), Olaus. 1626-1690. Den. DSB; ICB; P I; BMNH; BLA & Supp. (BORCH); GA. Chemistry (and history of chemistry). Botany.

2342. COLES, William. 1626-1662. Br. ICB; BMNH. Under COLE: GA; GD. Botany. (Titles in Wing.)

2343. DANFORTH, Samuel. 1626-1674. U.S.A. ICB. Astronomy. (Title in Wing.)

2344. FABER, Johann Matthaeus. 1626-1702. Ger. ICB; P I; BMNH; BLA. Botany. Member of the Leopoldina Academy.

2345. LEYBOURN, William. 1626-1700? Br. P I; GA; GD. Author of numerous textbooks on various aspects of practical mathematics, many of which went through several editions (see Wing). His main work is his *Cursus mathematicus* which incorporates the substance of his earlier textbooks.
1. *Cursus mathematicus. Mathematical sciences in nine books.* London, 1690. Wing L1911. RM/L.
2. *Mathematical institutions.* London, 1704. RM.
See also 2250/1.

2346. LICHTNER, Johann Christoph. 1626-1687. Ger. P I. Optics.

2347. MARGGRAV (or MARCGRAF), Christian. 1626-1687. Holl. P II; BLA. Chemistry.

2348. MUNTING, Abraham. 1626-1683. Holl. BMNH; BLA; GA; GF. Botany.

2349. POSNER, Kaspar. 1626-1700. Ger. P II. Chemistry.

2350. REDI, Francesco. 1626-1697/8. It. DSB; ICB; P II; BMNH; Mort.; BLA & Supp.; GA. Zoology. "Generation." Member of the Accademia del Cimento.
1. *Osservazioni intorno alle vipere.* Florence, 1664. RM.
2. *Esperienze intorno alla generazione degl' insetti.* Florence, 1668. RM.
2a. ——— *Experimenta circa generationem insectorum.* Amsterdam, 1671. RM/L.

2b. ——— *Experiments on the generation of insects.* Chicago, 1909. Trans. by M. Bigelow from the 5th impression (1688) of the Italian ed.
3. *Esperienze intorno a diverse cose naturali.* Florence, 1671. RM.
4. *Osservazioni intorno agli animali viventi che si trovano negli animali viventi.* Florence, 1684. RM.
5. *A letter of Francesco Redi concerning some objections made upon his "Observations about vipers."* London, 1673. Wing R663. X.
6. *Lettere.* Naples, 1748. RM.
7. *Opusculorum....* 2 vols. Amsterdam, 1685-86. X. Another ed., Leiden, 1729. RM.
8. *Opere.* 5 vols. Florence, 1724-27. Another ed., 7 vols, Naples, 1740-41. RM.
9. *Scritti di botanica, zoologica e medicina.* Milan, 1975. Ed. by P. Polito.

2351. SALLO, Denis de. 1626-1669. Fr. DSB; P II; GA; GB. In 1665 he established the first scholarly periodical, the *Journal des Sçavans* (which still continues). In its early years it contained many scientific articles.

2352. SHAKERLEY, Jeremy. 1626-ca. 1655. Br. DSB; GD. Astronomy. (Titles in Wing.)

2353. BALL, William. 1627-1690. Br. ICB; GD. Astronomy.

2354. BORRI (or BORRO or BURRHUS), Giuseppe Francesco. 1627-1695. Several countries. ICB; P I & Supp.; BLA & Supp. (BORRO); GA. Alchemy.

2355. BOYLE, Robert. 1627-1691. Br. DSB; ICB; P I; GA; GB; GD. Natural philosophy. Physics. Chemistry.
1. *New experiments physico-mechanicall touching the spring of the air.* Oxford, 1660. Wing B3998. RM.
1a. ——— 2nd ed. *Whereunto is added a defence of the authors explications of the experiments, against the objections of F. Linus and T. Hobbes.* Oxford, 1662. Wing B3999. RM/L. 3rd ed., London, 1682. Wing B4000.
1b. ——— *A continuation of new experiments physico-mechanicall touching the spring and weight of the air. The first part.* Oxford, 1669. Wing B3934. RM/L.
1c. ——— ——— *The second part.* London, 1682. Wing B3935. RM/L.
2. *Certain physiological essays.* London, 1661. Wing B3929. RM/L. 2nd ed., 1669. Wing B3930. RM/L.
3. *The sceptical chymist.* London, 1661. Wing B4021. RM/L. 2nd ed., Oxford, 1680. Wing B4022. There are several 20th-century reprints of both eds.
4. *Some considerations touching the usefulnesse of experimental naturall philosophy.* Oxford, 1663. Wing B4029. RM/L.
4a. ——— *Some considerations ... The second tome.* Oxford, 1671. Wing B4031.
5. *Experiments and considerations touching colours.* London, 1664. Wing B3967/8. RM. Reprinted, New York, 1964.
6. *Occasional reflections upon several subjects.* London, 1665. Wing B4005.

7. *New experiments and observations touching cold.* London, 1665. Wing B3996/7. RM.
8. *Hydrostatical paradoxes.* Oxford, 1666. Wing B3985. RM.
9. *The origine of formes and qualities.* Oxford, 1666. Wing B4014/5. RM/L.
10. *Tracts.* Oxford, 1670. Wing B4056/8. X.
11. *Tracts.* London, 1671. Wing B4059. RM.
12. *An essay about the origine and virtues of gems.* London, 1672. Wing B3947. RM. Reprinted, New York, 1972.
13. *Tracts.* London. 1672. Wing B4060/1. RM/L.
14. *Essays of the strange subtilty, great efficacy,* [and] *determinate nature of effluviums.* London, 1673. Wing B3951/2. RM/L.
15. *Tracts.* London, 1674. Wing B4053. X.
16. *Tracts.* London, 1674. Wing B4054/5. RM/L.
17. *Experiments, notes, etc., about the mechanical origine or production of divers particular qualities.* Oxford, 1675. Wing B3976. RM. Another ed., ib., 1676. Wing B3977. X. The work consists of eleven tracts.

17a. ——— Reprint of tracts (j) and (k): *Two tracts on electricity and magnetism, reprinted from the rare editions of 1675 and 1676.* London, 1898. Another reprint of the same: *Electricity and magnetism, 1675-6.* Oxford, 1927. Reprint of tract (k): *Mechanical origine or production of electricity.* New York, 1945.

18. *The aerial noctiluca.* London, 1680. Wing B3925. RM. Re phosphorus.
19. *New experiments and observations made upon the icy noctiluca.* London, 1681/2. Wing B3995. RM.
20. *Experiments and considerations about the porosity of bodies.* London, 1684. Wing B3966. RM.
21. *A disquisition about the final causes of natural things.* London, 1688. Wing B3946. RM.
22. *Experimenta et observationes physicae. Wherein are briefly treated of several subjects relating to natural philosophy.* London, 1691. Wing B3959. RM/L.
23. *The general history of air, designed and begun.* London, 1692. Wing B3981. RM.

Collected Works

24. *The philosophical works. Abridged, methodized, and disposed under general heads.* 3 vols. London, 1725. Ed. by Peter Shaw.
25. *The works.* 5 vols. London, 1744. Ed. by Thomas Birch, with a biography. RM/L.

25a. ——— Another ed., 6 vols, ib., 1772. Reprinted, Hildesheim, 1965-66.

Further titles in Wing (minor works, separates, Latin translations, etc.). See also 2707 and 15403/1.

2356. FRANCISI, Erasmus. 1627-1694. Ger. ICB; BMNH; GA. Natural history.

2357. GADBURY, John. 1627-1704. Br. ICB; GA; GD. Astrology. (Titles in Wing.)

2358. LYSERUS, Michael. 1627-1660. Den. BMNH; BLA & Supp. Anatomy.

2359. MORTON, Charles. 1627-1698. Br./U.S.A. ICB; GD; GE. Author of a work which can be regarded as marking the beginning of the scientific tradition in British America, namely *Compendium physicae*, a much-copied manuscript textbook of natural philosophy dating from ca. 1686.

2360. MOXON, Joseph. 1627-1700. Br. ICB; GD. Author of several textbooks on various aspects of practical mathematics (see Wing), the most notable of which is the following.
1. *A tutor to astronomy and geography ... the use of both the globes, coelestial and terrestrial.* London, 1659. Wing M3021 and six later eds up to 1699. Reprinted, New York, 1968.

2361. RAY, John. 1627-1705. Br. DSB; ICB; P II; BMNH; GA; GB; GD. Botany. Zoology. See also 2460 and 2834.
1. *Catalogus plantarum circa Cantabrigiam nascentium.* Cambridge, 1660. Wing R383. RM.
1a. ——— English trans. by A.H. Ewan and C.T. Prime. Hitchin (U.K.), 1975.
2. *Observations ... made in a journey ... With a catalogue of plants not native of England ... Whereunto is added a brief account of Francis Willughby his voyage through a good part of Spain.* London, 1673. Wing R399. RM/L.
3. *Methodus plantarum nova.* London, 1682. Wing R396. Reprinted, Weinheim, 1962 (HNC. 26)
3a. ——— Revised ed. *Methodus plantarum emenda.* London, 1703.
4. *Historia plantarum.* 3 vols. London, 1686-1704. Wing R394/5. RM/L.
5. *Synopsis methodica stirpium Britannicarum.* London, 1690. Wing R406. X.
5a. ——— 2nd ed., ib., 1696. Wing R407.
5b. ——— 3rd ed., ib., 1724. X. Reprinted, ib., 1973.
6. *The wisdom of God manifested in the works of Creation.* London, 1691. Wing R410. X. Later eds up to 1827.
7. *Miscellaneous discourses concerning the dissolution and changes of the world.* London, 1692. Wing R397. RM/L. Reprinted, Hildesheim, 1968.
7a. ——— 2nd ed., revised. *Three physico-theological discourses.* London, 1693. Wing R409. Several later eds.
8. *Synopsis methodica animalium quadrupedum et serpentini generis.* London, 1693. Wing R405. RM.
9. *Historia insectorum.* London, 1710. RM.
10. *Synopsis methodica avium et piscium.* London, 1713. RM.
Other titles in Wing.

Correspondence, etc.

11. *Philosophical letters ... To which are added those of Francis Willughby.* London, 1718. Ed. by W. Derham. RM/L.
12. *Select remains.* London, 1760. Ed. by W. Derham, RM/L.
13. *The correspondence.* London, 1848. (Ray Society publn. 14) Ed. by E. Lankester. RM/L.
14. *Further correspondence.* London, 1928. (Ray Society publn. 114) Ed. by R.W.T. Gunther.

2362. RUMPF, Georg Eberhard. 1627-1702. Ger./East Indies. ICB (RUMPH); P II; BMNH; GC; GF (RUMPHIUS). Botany. Zoology. See also 15439/1.
1. *Het Amboinsche kruid-boek ... Herbarium Amboinense.* 6 vols and index. Amsterdam, 1741-55. In Latin. Reprinted, 7 vols, New York, 1965.

2363. SACHS (VON LEWENHAIMB), Philipp Jacob. 1627-1672. Breslau. P II; BMNH; BLA. Botany. Zoology. Mineralogy. Member of the Leopoldina Academy.

2364. BROWNE, William. ca. 1628-1678. Br. BMNH; GA; GD. Botany.

2365. CONCIUS, Andreas. 1628-1682. Ger. P I. Various fields.

2366. HUDDE, Jan. 1628-1704. Holl. DSB; ICB; P I; GA; GF. Mathematics.

2367. MALPIGHI, Marcello. 1628-1694. It. DSB; ICB; P II; BMNH; Mort.; BLA & Supp.; GA; GB. Human and comparative anatomy (incl. microscopic anatomy). Plant anatomy. "The founder of histology." From 1667 he was a correspondent of the Royal Society and thereafter all his works were published in London under its auspices.
1. *Dissertatio epistolica de bombyce.* London, 1669. Wing M349. Ed. by H. Oldenburg. RM.
2. *Dissertatio epistolica de formatione pulli in ovo.* London, 1673. Wing M350. RM.
3. *Anatome plantarum.* 2 vols. London, 1675-79. Wing M345. RM. Reprinted, Brussels, 1968.
3a. ——— German trans. Leipzig, 1901. (OKEW. 120)
4. *Opera omnia.* 2 vols. London, 1686-87. Wing M342/4. RM/L.
5. *Opera posthuma.* London, 1697. Wing M352. Includes his autobiography. RM/L.
6. *Opere scelte.* Turin, 1967. Ed. by L. Belloni.
7. *The correspondence.* 5 vols. Ithaca, N.Y., 1975. Ed. by H.D. Adelmann.
Other titles in Wing.

2368. MILLINGTON, Thomas. 1628-1704. Br. DSB; ICB; GD. The reputed discoverer of sexuality in plants.

2369. SLADE, Mathew. 1628-1689. Br./Holl. BLA; GD. Embryology. He used the pseudonym Theodore Aldes.

2370. STARKEY, George. 1628-1665. Br. DSB; ICB; GD. Iatrochemistry. Alchemy. Many of the titles attributed to him in Wing were published under the pseudonym Eirenaeus Philalethes (see entry 2293) and it is not certain whether Starkey was their author; see the articles on Starkey in DSB and GD.

2371. BARBETTE, Paul. 1629-1699. Holl. BLA & Supp.; GA. Anatomy. (Wing lists an English trans. of one of his works.)

2372. BAYFIELD, Robert. 1629-1690. Br. BLA & Supp.; GA; GD. Anatomy. (Title in Wing.)

2373. COMMELIN. Jan. 1629-1692. Holl. BMNH; GA; GF. Botany.

2374. GLASER, Johann Heinrich. 1629-1679. Switz. DSB; BMNH; BLA & Supp. Anatomy. Physiology. Botany.

2375. HUYGENS (or HUYGHENS), Christian. 1629-1695. Holl./Fr. DSB; ICB; P I; G. Mathematics. Mechanics. Astronomy. Optics. Member of the Académie des Sciences. See also Index.
1. *De circuli magnitudine inventa.* [With] *Problematum quorundam illustrium constructiones.* Leiden, 1654. RM.
2. *Systema Saturnium.* The Hague, 1659. RM.
3. *Horologium oscillatorium.* Paris, 1673. RM. Reprinted, Brussels, 1966.
3a. ——— German trans. Leipzig, 1913. (OKEW. 192)
4. *Traité de la lumière.* Leiden, 1690. RM. Reprinted, Paris, 1920; London, 1966; Brussels, 1967.
4a. ——— German trans. Leipzig, 1890. (OKEW. 20)
4b. ——— *Treatise on light.* London, 1912. Trans. by S.P. Thompson. RM/L. Reprinted, New York, 1962.
5. [*Cosmotheoros*]; *sive, De terris coelestibus earumque ornatu conjecturae.* The Hague, 1698. RM.
5a. ——— *The celestial worlds discover'd.* London, 1698. Wing H3859. RM. Reprinted, London, 1968.
5b. ——— *Nouveau traité de la pluralité des mondes.* Paris, 1702. RM.
6. *Opuscula postuma.* Leiden, 1703. RM.
7. German trans. of two unpublished works: *De motu corporum ex percussione* and *De vi centrifuga.* Leipzig, 1903. (OKEW. 138)

Correspondence and Collected Works

8. *Huygens et Roberval: Documents nouveaux.* Leiden, 1879. Ed. by C. Henry.
9. *Leibnizens und Huygens Briefwechsel mit Papin.* Berlin, 1881. Ed. by E. Gerland. X. Reprinted, Wiesbaden, 1966.
10. *Opera mechanica, geometrica, astronomica et miscellanea.* Leiden, 1715. Ed. by W.J. 'sGravesande. RM.
11. *Oeuvres complètes, publiées par la Société Hollandaise des Sciences.* 22 vols. The Hague, 1888-1950. RM/L.

2376. KING, Edmund. 1629-1709. Br. BLA & Supp.; GD. Comparative anatomy.

2377. PUGET, Louis de. 1629-1709. BMNH. Microscopy.

2378. SCILLA, Agostino. 1629-1700. It. DSB; P II; BMNH. Geology.
1. *La vana speculazione disingannata dal senso.* Naples, 1670. X. "One of the classics of geology."
1a. ——— Latin trans. entitled *De corporibus marinis quae defossa reperiuntur.* Rome, 1747. RM.

2379. SEGER, Georg. 1629-1678. Ger. BLA. Comparative anatomy.

2380. TOWNELEY, Richard. 1629-1707. Br. DSB; ICB; P II (TOWNLEY). Pneumatics. Meteorology. Astronomy.

2381. MACKAILE, Matthew. fl. 1657-1696. Br. GD. Chemistry. Also author of *Terrae prodromus theoricus* (a criticism of Burnet's *Theory of the earth*). (Titles in Wing.)

2382. COLSON, Lancelot. fl. 1660-1680. Br. GA (COELSON); GD. Alchemy. Astrology. (Title in Wing.)

2383. BELLUCCIUS, Thomas. d. 1671. It. BMNH. Botany.

2384. GAYANT, Louis. d. 1673. Fr. DSB; BLA. Human and comparative anatomy. Member of the Académie des Sciences.

2385. SUETONIO, Agostino. d. 1685. It. P II. Astronomy.

2386. MORETTI, Gaetano. d. 1697. It. P III. Astronomy.

2387. BARROW, Isaac. 1630-1677. Br. DSB; ICB; P I; GA; GB; GD. Geometry. Optics. See also Index.
1. *Euclidis "Elementorum" libri XV breviter demonstrati.* 2 vols. Cambridge, 1655-57. Wing E3392A. RM. Wing records three later eds up to 1685. Another ed., London, 1711.
2. *Lectiones geometricae.* London, 1670. Wing B940. RM.
2a. ——— *Geometrical lectures.* London, 1735. Trans. by Edmund Stone. Revised by Isaac Newton. RM.
2b. ——— *The geometrical lectures.* Chicago, 1916. Trans. with notes by J.M. Child.
3. *Lectiones opticae et geometricae.* London, 1674. Wing B945. RM.
4. *Lectio ... in qua theoremata Archimedis de sphaera et cylindro per methodum indivisibilium investigata....* London, 1678. RM.
5. *Lectiones mathematicae,* London, 1684. Wing B943/4. X.
5a. ——— *The usefulness of mathematical learning.* London, 1734. Trans. by J. Kirkby. RM/L. Reprinted, London, 1970.
6. *Works.* 4 vols. London, 1683-87. Wing B925 and two later eds. X. Another ed., 3 vols, ib., 1716.
7. *The mathematical works.* Cambridge, 1860. Ed. by W. Whewell. RM/L. Reprinted, Hildesheim, 1973.
Other titles in Wing.

2388. DAVISI, Urbano. ca. 1630-ca. 1700. It. P I & Supp. Astronomy.

2389. GOTTIGNIEZ, Gilles François de. 1630-1689. Belg./It. ICB; P I; GA; GF. Astronomy. Mathematics.

2390. HUET, Pierre Daniel. 1630-1721. Fr. ICB; GA; GB. One of the most learned men of his time. His numerous and varied interests included mathematics and science, especially astronomy and anatomy.

2391. KUNCKEL (VON LÖWENSTERN), Johann. 1630 (or 1638)-1702. Ger./Swed. DSB; ICB; P I; GA; GB (KUNKEL); GC. Chemistry. Member of the Leopoldina Academy.
1. *Nützliche Observationes; oder, Anmerckungen von den fixen und flüchtigen Salzen....* Hamburg, 1676. X.
1a. ——— *Utiles observationes; sive, Animadversiones de salibus....* London, 1678. Wing L2818B. X.
2. *Ars vitraria experimentalis; oder, Vollkommene Glasmacher-Kunst.* Frankfurt/Leipzig, 1679. X. Includes a German trans. of Neri's work, *L'arte vetraria* (1865/1) and of Merrett's notes thereon (1865/1a) with Kunckel's remarks on Merrett. See also 8651.
2a. ——— 2nd ed., ib., 1689. X. Reprinted, Hildesheim, 1971.
3. *Collegium physico-chymicum experimentale; oder, Laboratorium chymicum.* Hamburg, 1716. Reprinted, Hildesheim, 1975.

2392. LOVELL, Robert. 1630?-1690. Br. P I (LOVEL); BMNH; GD. Natural history.
1. [*Pambotanologia*]; *sive, Enchiridion botanicum; or, A compleat herball.* Oxford, 1659. Wing L3243. X.

2. [*Panzooryktologia*]; *sive, Panzoologicomineralogia; or, A compleat history of animals and minerals.* Oxford, 1661. Wing L3246. RM/L.
3. [*Panoryktologia*]; *sive, Pammineralogicon; or, An universal history of minerals.* Oxford, 1661. Wing L3245. RM.

2393. MALLET, Alain Manesson. ca. 1630-ca. 1706. Fr. P II; GA. Author of *Description de l'univers, contenant les différens systèmes du monde*, 5 vols, Paris, 1683.

2394. NIEUHOF, Johan. 1630-1672. Holl. BMNH; GA.
1. *Legatio Batavica ad ... Sinae Imperatorem. Historiarum narratione....* Amsterdam, 1668. RM. Contains a description of the natural history and other features of China.
1a. ——— *An embassy from the* [Dutch] *East-India Company ... to the ... Emperour of China.* London, 1669. Wing N1152/3. RM.

2395. RICHER, Jean. 1630-1696. Fr. DSB; ICB; P II; GA. Astronomy. Member of the Académie des Sciences.

2396. RUDBECK, Olof (first of the name). 1630-1702. Swed. DSB; ICB; BMNH; Mort.; BLA & Supp.; GA. Anatomy. Botany.
1. *Nova exercitatio anatomica.* Uppsala, 1653. X. Reprinted with an English trans. by A.E. Nielsen. Uppsala, 1930.

2397. SHARROCK, Robert. 1630-1684. Br. DSB; ICB; BMNH; GD. Botany. (Titles in Wing.)

2398. SINCLAIR (or SINCLARE), George. ca. 1630-1696. Br. P II; GD. Some aspects of physics. Also textbooks of mathematics and astronomy.
1. *The hydrostaticks.* Edinburgh, 1672. Wing S3854.
1a. ——— Another ed., entitled *Natural philosophy improven by new experiments.* Edinburgh, 1683. Wing S3855. X.
Other titles in Wing.

2399. SPOLE, Anders. 1630-1699. Swed. ICB; P II (SPOLIUS). Astronomy.

2400. TOLL, Jacob. ca. 1630-1696. Holl./Ger. P II. Under TOLLIUS: GA; GC; GF. Alchemy/chemistry.

2401. COCKER, Edward. 1631-1675. Br. ICB; GB; GD. A teacher of arithmetic. (Titles in Wing.) His main textbook, *Arithmetick*, first published posthumously in 1678, was highly regarded and went through more than a hundred editions.

2402. LANA (or LANA TERZI), Francesco de. 1631-1687. It. ICB; P I; GA; GF. Various aspects of experimental physics. In 1686 he founded an academy in Brescia.
1. *Prodromo; overo, Saggio di alcune invenzioni nuovo premesso all'arte maestra.* Brescia, 1670. RM.
2. *Magisterium naturae et artis. Opus physicomathematicum.* 3 vols. Brescia, 1684-92. RM.

2403. LOWER, Richard. 1631-1691. Br. DSB; ICB; Mort.; BLA & Supp.; GA; GD. Physiology.
1. *Tractatus de corde.* London, 1669. Wing L3310. RM. Wing

records a 2nd ed. in 1670 and a 4th in 1680. X. 7th ed., Leiden, 1740.
Other titles in Wing.

2404. NEEDHAM, Walter. ca. 1631-ca. 1691. Br. BLA; GD. Human and comparative anatomy. (Title in Wing.)

2405. STERBEECK, Frans van. 1631-1693 (or 1683). Belg. BMNH; GA; GF. Botany.

2406. WILSON, George. 1631-1711. Br. ICB.
1. *A compleat course of chymistry*. London, 1691. X. Another ed., ib., 1699. Wing W2892. RM.

2407. BALDUIN, Christoph Adolph. 1632-1682. Ger. P I; GA; GC. Chemistry.

2408. CROUCH, Nathaniel (pseudonym: BURTON, R.). ca. 1632-ca. 1725. Br. BMNH; GD (BURTON). Author of *The general history of earthquakes ... By R.B.* [i.e. R. Burton], 1694. (Titles in Wing.)

2409. GALLOIS, Jean. 1632-1707. Fr. DSB; P I; GA. Editor of the *Journal des Sçavans* from 1666 to 1674. (It had been founded in 1665 by De Sallo; see 2351.) Member of the Académie des Sciences and for a short time its secretary.

2410. HARTMANN, Sigismund Ferdinand. 1632-1681. Prague. P I. Mathematics. Catoptrics.

2411. HOBOKEN, Nicolas. 1632-1678. Holl. P I; BLA. Human and comparative anatomy.

2412. JONGHE, Ignaz de. 1632-1692. Belg. P I. Mathematics.

2413. LEEUWENHOEK, Antoni van. 1632-1723. Holl. DSB; ICB; P I; BMNH; Mort.; BLA & Supp. Under LEUWENHOEK: GA; GB; GF. Microscopy. "Generation" Physiology.
1. *Arcana naturae detecta*. Delft, 1695. RM. Reprinted, Brussels, 1966.
2. *Continuatio "Arcanorum naturae detectorum."* Delft, 1697. RM. Reprinted, Brussels, 1966.
3. *Alle de brieven. The collected letters*. Several vols. Amsterdam, 1939-. In Dutch and English.
4. *Opera omnia*. 4 vols. Leiden, 1715-22. RM/L. Reprinted, Hildesheim, 1971.
5. *Select works*. 2 vols. London, 1798-1807. Trans. from Dutch and Latin by S. Hoole.
6. *Antony van Leeuwenhoek and his "little animals."* New York, 1958. Numerous texts ed. and trans. by C. Dobell.
7. *The discovery of unicellular life*. Waltham, Mass., 1954. Excerpts from his letters to the Royal Society, 1674 and 1676. Foreword by A.J. Kluyver.
8. *On the circulation of the blood*. Nieuwkoop, 1962. Facsimile of the Latin text of his 65th letter to the Royal Society (1688) with an English trans. (Dutch classics in history of science. 2)

See also 2606/2.

2414. MAJUS (or MAY), Heinrich. 1632-1696. Ger. P II. Natural philosophy (*Physicae veteris et novae ... principia* ... 1688).

Several disputations from the period 1673-81, presumably of his students, are attributed to him; they deal with such topics as lightning, thunder, wind, comets, gravity, etc.

2415. MAPPUS, Marcus. 1632-1701. Strasbourg. BMNH; BLA (MAPP); GA. Botany.

2416. RÉGIS, (Pierre) Sylvain. 1632-1707. Fr. ICB; P II; GA. Cartesian natural philosophy. Member of the Académie des Sciences.
1. *Cours entier de philosophie; ou, Système général selon les principes de M. Descartes.* 3 vols. Amsterdam, 1691. X. Reprinted, New York, 1970.

2417. SEIGNETTE, Elie. 1632-1698. Fr. ICB. Chemistry.

2418. SPINOZA, Benedict de. 1632-1677. Holl. The famous philosopher. ICB; G.
1. *Stelkonstige reeckening van den regenboog.* 1687. X.
1a. ——— *Algebraic calculation of the rainbow.* Nieuwkoop, 1963. (Dutch classics on history of science. 5)

2419. STUBBE, Henry. 1632-1676. Br. ICB; BLA (STUBBES); GA; GD (STUBBS). Redoubtable opponent of the Royal Society around 1670. (Titles in Wing.)

2420. WREN, Christopher. 1632-1723. Br. The celebrated architect. DSB; ICB; P II; G. Mathematics. Mechanics.

2421. BOCCONE, Paolo (or Silvio). 1633-1704. It. ICB; P I (BOCCONI); BMNH; GA. Natural history, chiefly botany.
1. *Icones et descriptiones rariorum plantarum Siciliae, Melitae, Galliae, et Italiae.* Oxford, 1674. Wing B3385. X.
2. *Recherches et observations naturelles.* Amsterdam, 1674. RM. Chiefly mineralogy.
3. *Museo di fisica e di esperienze variato.* Venice, 1697. RM.
4. *Museo di piante rare della Sicilia, Malta, Corsica, Italia, Piemonte, e Germania.* Venice, 1697.

2422. CIAMPINI, Giovanni Guistino. 1633-1698. It. ICB; P I; GA. Astronomy. In 1677 he founded the Accademia Fisico-Matematica Romana.

2423. CLAUDER, Gabriel. 1633-1691. Ger. P I; BLA; GA; GC. Zoology. Member of the Leopoldina Academy.

2424. COLEY, Henry. 1633-1707. Br. GA; GD. Astrology/astronomy. (Titles in Wing.)

2425. CROONE, William. 1633-1684. Br. DSB; ICB; P I; Mort.; BLA & Supp.; GA; GD. Physiology. Embryology. (Title in Wing.)

2426. DRELINCURTIUS (or DRELINCOURT), Charles. 1633-1697. Fr. BLA; GA; GB. Anatomy. Embryology.

2427. GIORDANI (or GIORDANO), Vitale. 1633-1711 (or 1691). It. P I; GA. Mechanics.

2428. HERBINUS, Johann. 1633-1676. Poland/Ger. P I; GA; GC.
1. *Dissertatio de admirandis mundi cataractis supra et subterraneis.* Amsterdam, 1678. RM.

2429. HEURAET, Hendrik van. 1633-1660? Holl. DSB. Mathematics.

2430. MARCHETTI, Alessandro. 1633-1714. It. ICB; P II; GA. Mathematics. Mechanics.
   1. *Exercitationes mechanicae.* Pisa, 1669. RM.

2431. MONTANARI, Geminiano. 1633-1687. It. DSB; ICB; P II; GA. Astronomy. Meteorology. Various aspects of experimental physics.
   1. *Pensieri fisico-matematici sopra alcune esperienze fatti in Bologna nell'Accademia filosfica ... intorno diversi effetti de' liquidi.* Bologna, 1667.

2432. AMMAN, Paul. 1634-1691. Ger. BMNH; BLA & Supp.; GA; GB; GC. Botany.
   1. *Character plantarum naturalis à fine ultimo videlicet fructificatione desumptus.* Leipzig. 1676. X. Another ed., Frankfurt/Leipzig, 1685.

2433. DODART, Denis. 1634-1707. Fr. DSB; P I; BMNH; BLA; GA. Botany. Physiology. Member of the Académie des Sciences. See also 2817/1c.

2434. FOGEL, Martin. 1634-1675. Ger. ICB; P I; BMNH; BLA. Botany.

2435. HAUNOLD, Johann Sigismund. 1634-1711. Breslau. P I. Natural history. A wealthy patron of science and possessor of a large natural history collection.

2436. MAJOR, Johann Daniel. 1634-1693. Ger. ICB; P II; BMNH; BLA; GA. Comparative anatomy. Botany. Mineralogy. Member of the Leopoldina Academy.

2437. TILING, Matthias. 1634-1685. Ger. P II (TILLING); BMNH; BLA. Chemistry. Botany. Anatomy.

2438. LAMY, Guillaume. fl. 1668-1682. Fr. DSB. Natural philosophy. Anatomy. Physiology.

2439. DENYS, Guillaume. d. ca. 1680. Fr. P I.
   1. *L'art de naviger perfectionné par la cognoissance de la variation de l'aimant; ou, Traicté de la variation de l'aiguille aimentée.* Dieppe, 1666. RM.

2440. COMIERS, Claude. d. 1693. Fr. P I & Supp; GA. Various fields.

2441. TERZAGO, Paolo Maria. d. 1695. It. ICB (TERZAGHI); P II & Supp.; BLA. Natural history. Member of the Nobile Collegio dei Fisici in Milan.

2442. LE GRAND, Antoine. d. 1699. Belg./Br. GA; GD; GF. Cartesian natural philosophy.
   1. *Historia naturae, variis experimentis et ratiociniis elucidata.* London, 1673. Wing L951. X. Another ed., Nuremberg, 1678. RM. Another ed., London, 1680.
   2. *Curiosus rerum abditarum naturaeque arcanorum perscrutator.* Nuremberg, 1681. RM.
   3. *An entire body of philosophy according to the principles of the famous Renate des Cartes, in three books ... I. The institution ... II. The history of nature ... III. A dissertation on the want of sense and knowledge in brute animals.* London, 1694. Wing L950. Trans. by R. Blome from three of Le Grand's works. RM.
   See also 2279/1a.

2443. BECHER, Johann Joachim. 1635-1682. Ger./Aus./Holl./Br. DSB; ICB; P I; BMNH; BLA; GA; GB; GC. Chemistry.
1. *Parnassus medicinalis illustratus; oder. Ein neues ... Thier-, Kräuter- und Berg-Buch.* Ulm, 1663.
2. *Institutiones chimicae prodromae.* Frankfurt, 1664. RM.
3. *Actorum laboratorii chymici Monacensis; seu, Physicae subterraneae libri duo.* Frankfurt, 1671.
3a. ——— [First supplement.] *Experimentum chymicum novum.* Ib., 1671. X.
3b. ——— *Supplementum secundum in "Physicam subterraneam."* Ib., 1675.
3c. ——— *Chymisches Laboratorium.* Ib., 1680. A German trans. by Becher himself of 3, 3a, and 3b, and also of 2. X. Reprinted, Wurzburg, 1969.
4. *Chymischer Glücks-Hafen.* Frankfurt, 1682. A collection of many chemical processes. X. Reprinted, Hildesheim, 1974.
Other titles in Wing. See also 6982/1.

2444. BURNET, Thomas. ca. 1635-1715. Br. DSB; ICB; P I; GA; GB; GD. Cosmogeny. Geology. See also Index.
1. *Telluris theoria sacra ... Libri duo priores.* 2 vols. London, 1681. Wing B5948. RM.
1a. ——— *The theory of the earth. The first two books.* Ib., 1684. Wing B5950. RM/L. Another ed., 1690. Wing B5951. X. 2nd ed., 1691. Wing B5952. 3rd ed., 1697. Wing B5953.
1b. ——— *Libri duo posteriores.* 2 vols in 1. Ib., 1689. Wing B5949.
1c. ——— ——— *The two last books.* Ib., Wing B5954.
1d. ——— *The sacred theory of the earth.* Ib., 1816. Another reprint, ib., 1965.
2. *A review of "The theory of the earth" and of its proofs.* London, 1690. Wing B5945.
3. *An answer to the late exceptions made by Mr. Erasmus Warren against "The theory of the earth."* London, 1690. Wing B5942.
4. *A short consideration of Mr. Erasmus Warren's defence of his exceptions against "The theory of the earth."* Ib., 1691. Wing B5947.
Other titles in Wing.

2445. CAMPANI, Giuseppe. 1635-1715. It. DSB; ICB; P I; GA. Astronomy. Maker of optical instruments.

2446. COLE, William. 1635-1716. Br. ICB; BLA & Supp.; GA; GD. Anatomy. Physiology. (Titles in Wing.)

2447. HILL, Abraham. 1635-1721. Br. ICB; GD. Treasurer of the Royal Society, 1663-65 and 1679-1700.

2448. HOOKE, Robert. 1635-1702. Br. DSB; ICB; P I; Mort.; GA; GB; GD. Mechanics. Microscopy. Optics. Physiology. Geology. Scientific instruments. Curator of experiments for the Royal Society from 1662 onwards, and its secretary, 1677-82.
1. *Micrographia; or, Some physiological descriptions of minute bodies made by magnifying glasses.* London, 1665. Wing H2620. RM/L. Several 20th-century reprints.

1a. ——— *Micrographia restaurata, or the copperplates ... reprinted and fully explained.* London, 1745.
2. *An attempt to prove the motion of the earth.* London, 1674. Wing H2613. RM.
3. *Animadversions on the first part of "Machina coelestis" of ... J. Hevelius ... with an explication of some instruments made by R. Hooke.* London, 1674. Wing H2611. RM/L.
4. *A description of helioscopes and some other instruments made by R. Hooke.* London, 1676. Wing H2614. RM/L.
5. *Lampas; or, Descriptions of some mechanical improvements of lamps and waterpoises. Together with some other physical and mechanical discoveries.* London, 1677. Wing H2616. RM/L.
6. *Lectures de potentia restutiva or of spring, explaining the power of springing bodies.* London, 1678. Wing H2619. RM/L.
7. *Lectures and collections: Cometa, Microscopium.* London, 1678. Wing H2618. RM/L.
8. *The posthumous works, containing his Cutlerian lectures and other discourses.* London, 1705. Ed. by R. Waller. RM. Reprinted: New York, 1969; Hildesheim, 1970; London, 1971.
9. *Philosophical experiments and observations of ... Robert Hooke ... and other eminent virtuso's in his time.* London, 1726. Ed. by W. Derham. RM. Reprinted, ib., 1967.
10. *The diary, 1672-1680.* London, 1935. Ed. by H.W. Robinson and W. Adams. Reprinted, ib., 1968.

Other titles in Wing. See also 2834 and 9073/1,1a.

2449. KIRCHMAIER, Georg Caspar. 1635-1700. Ger. P I; BMNH; BLA; GA. His numerous works covered many fields but chiefly chemistry. Member of the Leopoldina Academy.
1. *De phosphoris et natura lucis necnon de igne.* Wittenberg, 1680.

2450. LACHMUND, Friedrich. ca. 1635-1676. Ger. P I; BMNH. Mineralogy. Zoology.
1. *Oryctographia Hildesheimensis; sive, Admirandorum fossilium ... descriptio.* Hildesheim, 1669. RM.

2451. MONTALBANI, Marc Antonio. 1635-1695. It. P II. Mineralogy.

2452. MORIN (DE SAINT-VICTOR), Louis. 1635-1715. Fr. GA. Botany. Member of the Académie des Sciences.

2453. MUSITANO, Carlo. 1635-1714. It. ICB; P II; BLA; GA. Chemistry.

2454. REYHER (or REIHER), Samuel. 1635-1714. Ger. P II; GA; GC. Numerous works in various fields.

2455. RHEEDE TOT DRAAKESTEIN, Heinrich Adrian van. 1635-1691. Holl. BMNH; GF (REEDE). Botany.
1. *Hortus Indicus Malabaricus.* 6 vols. Amsterdam, 1678-1703.
1a. ——— *Horti Malabarici pars prima ... Nunc primum classium, generum, et specierum characteres Linnaeanas et indice Linnaeano adauxit J. Hill.* London, 1774.

2456. SOUTHWELL, Robert. 1635-1702. Br. ICB; P II; GD. Various fields.

2457. SPRAT, Thomas. 1635-1713. Br. DSB; ICB; P II; GA; GB; GD. Apologist for the Royal Society and "the new experimental philosophy."
1. *Observations on Monsieur Sorbier's "Voyage into England."* London, 1665. Wing S5035 and three further eds. X. cf. item 2209/1.
2. *The history of the Royal Society of London.* London, 1667. Wing S5032. RM/L. Critical ed. by J.I. Cope and H.W. Jones, St. Louis, 1959.
2a. ——— 2nd ed., ib., 1702. 3rd ed., 1722. 4th ed., 1734.

2458. STURM, Johann Christoph. 1635-1703. Ger. P II; GA; GC. Mathematics. Astronomy.

2459. VIEUSSENS, Raymond. ca. 1635-1715. Fr. DSB; ICB; P II; Mort.; BLA & Supp.; GA. Anatomy.

2460. WILLUGHBY, Francis. 1635-1672. Br. DSB; BMNH; GB; GD. Zoology. In consequence of his early death all the books under his name were published, and partly written, by his collaborator, John Ray.
1. *Ornithologiae libri tres.* London, 1676. Wing W2879. RM/L.
1a. ——— *Ornithology.* Ib., 1678. Wing W2880. RM.
2. *De historia piscium.* Oxford, 1686. Wing W2877. RM/L.
See also 2361/2,11.

2461. GLANVILL, Joseph. 1636-1680. Br. DSB; ICB; P I; GA; GB; GD. Apologist for the Royal Society and "the new experimental philosophy."
1. *The vanity of dogmatizing.* London, 1661. Wing G834. X. Critical ed. by S. Medcalf, Brighton (U.K.), 1970.
2. *Scepsis scientifica; or, Confest ignorance the way to science.* London, 1665. Wing G827. X. Reprinted, ib., 1885.
3. *Plus ultra; or, The progress and advancement of knowledge since the days of Aristotle.* London, 1668. Wing G820. X. Reprinted, Gainesville, Florida, 1958.
4. *Philosophia pia; or, A discourse of the religous temper and tendencies of the experimental philosophy which is profest by the Royal Society.* London, 1671. Wing G817. X. Reprinted, Hildesheim, 1970.
5. *Essays on several important subjects in philosophy and religion.* London, 1676. Wing G809. X. Reprinted, New York, 1970, and Stuttgart, 1970.
Other titles in Wing. See also 2679/1d.

2462. GOTTWALDT, Christopher. 1636-1700. Ger. BMNH; BLA; GA. Zoology.

2463. PARDIES, Ignace Gaston. 1636-1673. Fr. DSB; P II; GA. Mechanics. Optics.
1. *Élémens de géometrie.* Paris, 1671. X.
1a. ——— *Elementa geometriae.* Oxford, 1694. Wing P347. X.
1b. ——— *Short ... elements of geometry.* London, 1701. X. Trans. by J. Harris. 3rd ed., 1705.
2. *Discours de la connoissance des bestes.* Paris, 1672. X. Reprinted, New York, 1972.
3. *Oeuvres.* Lyons, 1696. X. Another ed., ib., 1725. RM.

2464. SIMPSON, William. 1636-1680. Br. ICB. Natural philosophy. Chemistry. (Titles in Wing.)

2465. BREYNIUS, Jacobus B. 1637-1697. Ger. BMNH; GA; GC. Botany.

2466. CESTONI, Giacinto. 1637-1718. It. DSB; ICB; BLA & Supp.; GA. Zoology. "Generation."

2467. MAGALOTTI, Lorenzo. 1637-1712. It. DSB; ICB; P II; GA. Secretary of the Accademia del Cimento.
1. *Lettere scientifiche ed erudite*. Florence, 1721. RM.
See also 2812/1.

2468. NEILE, William. 1637-1670. Br. P II (NEIL); GD. Mathematics.

2469. RENTSCH (or RENSCHEL), Johann Wolfgang. 1637-1690. Ger. P II. Many disputations, presumably of his students, are attributed to him, including several on astronomical topics in the period 1661-76.

2470. SWAMMERDAM, Jan. 1637-1680. Holl. DSB; ICB; BMNH; Mort.; BLA; GA; GB. Entomology. Anatomy. Physiology. Embryology.
1. *Historia insectorum generalis; ofte, Algemeene verhandeling van de bloedeloose dierkens*. Utrecht, 1669. RM.
1a. ——— *Histoire générale des insectes*. Utrecht, 1682. RM.
1b. ——— *Historia insectorum generalis*. Leiden, 1685. Latin trans. of the Dutch original.
2. *Miraculum naturae; sive, Uteri muliebris fabrica*. Leiden, 1672. X. Another ed., London, 1680. Wing S6234. X.
3. *Ephemeri vita; of, Afbeeldingh....* Amsterdam, 1675. X.
3a. ——— *Ephemeri vita; or, The natural history and anatomy of the ephemeron*. London, 1681. Wing S6233. Trans. with a preface by E. Tyson. X.
4. *Bybel der natuure; of, Historie der insecten ... Biblia naturae; sive, Historia insectorum*. 2 vols. Leiden, 1737-38. Parallel text in Dutch and Latin. A much more extensive work than item 1. Ed. from the MS, with a biographical preface, by H. Boerhaave. RM/L.
4a. ——— Reprint of the Latin text. 2 vols. Leiden, 1837.
4b. ——— *The book of nature; or, The history of insects*. London, 1758. Trans. by T. Flloyd. Revised, with notes, by J. Hill. RM/L.
5. *The letters of Jan Swammerdam to Melchisedec Thevenot*. Amsterdam, 1975. Ed. with an English trans. and biographical sketch by G.A. Lindeboom.

2471. BERNARD, Edward. 1638-1696. Br. P I; GA; GD. Astronomy. Also works on the mathematics and astronomy of antiquity. (Title in Wing.)

2472. BIDLOO, Lambert. 1638 (or 1633)-1724. Holl. BMNH; GF. Botany.

2473. BOURDON, Aimé. 1638-1706. Fr. BLA & Supp.; GA. Anatomy.

2474. BUONANNI, Filippo. 1638-1725. It. DSB; ICB (BONANNI); P I; BMNH; Mort. (BONANNI); BLA; GA. Zoology. Microscopy.
1. *Ricreazione dell'occhio e della mente nell'osservazione delle chiocciole*. Rome, 1681. RM.
1a. ——— 2nd ed. *Recreatio mentis et oculi in observatione animalium testaceorum*. Ib., 1684. RM/L.

2. *Observationes circa viventia quae in rebus non viventibus reperiuntur. Cum micrographia curiosa....* Rome, 1691. RM.
3. *Museum Collegii Romani Kircherianum descriptum.* Rome, 1709. X.

3a. ——— Another ed. *Rerum naturalium historia ... in Museo Kircheriano.* Ib., 1788.

2475. EIMMART, Georg Christoph. 1638-1705. Ger. P I; GA; GC. Astronomy.

2476. FAGON, Guy Crescent. 1638-1718. Fr. ICB; BLA & Supp.; GA. Botany. Superintendent of the Jardin des Plantes in Paris.

2477. FONTANA, Carlo. 1638 (or 1634)-1714. It. P I; GA.
1. *Utilissimo trattato dell'acque correnti.* Rome, 1696. RM.

2478. GREGORY, James. 1638-1675. Br. DSB; ICB; P I; GA; GB; GD. Mathematics. Optics. Astronomy.
1. *Optica promota.* London, 1663. Wing G1912. RM.
2. *Vera circuli et hyperbolae quadratura.* Padua, 1667. RM.
3. *Exercitationes geometricae.* London, 1668. Wing G1909. RM.
4. *James Gregory tercentenary memorial volume, containing his correspondence with John Collins and his hitherto unpublished mathematical manuscripts.* London, 1939. Ed. by H.W. Turnbull.

2479. KNAUT, Christophor. 1638-1694. Ger. BMNH. Botany.

2480. MAGNOL, Pierre. 1638-1715. Fr. DSB; BMNH; GA. Botany. Member of the Académie des Sciences.

2481. MALEBRANCHE, Nicolas. 1638-1715. Fr. The philosopher. DSB; ICB; P II; GA; GB. Cartesian natural philosophy. Mathematics. Mechanics. Member of the Académie des Sciences.
1. *Oeuvres complètes.* 21 vols. Paris, 1958-70.

2482. MARTINEZ, Crisostomo. 1638-1694. Spain. DSB. Anatomy (incl. microscopic anatomy).

2483. MEIBOM, Heinrich. 1638-1700. Ger. P II; Mort.; BLA. Anatomy.

2484. RUYSCH, Frederik. 1638-1731. Holl. DSB; ICB; BMNH; Mort.; BLA; GA (RUISCH). Anatomy. Botany. Zoology.
1. *Dilucidatio valvularum in vasis lymphaticis et lacteis.* The Hague, 1665. X. Reprinted, Nieuwkoop, 1964. (Dutch classics on history of science. 11)
2. *Observationum anatomico-chirurgicarum centuria. Accedit Catalogus rariorum quae in Museo Ruyschiano asservantur.* Amsterdam, 1691.
3. *Opera anatomico-medico-chirurgica.* 3 vols. Amsterdam, 1725-37.

2485. SHERLEY, Thomas. 1638-1678. Br. ICB; BLA (SHIRLEY); GD.
1. *A philosophical essay declaring the probable causes whence stones are produced in the greater world.* London, 1672. Wing S3523. X.

2486. STENSEN, Niels (or STENO, Nicolaus). 1638-1686. Den./It. DSB; Mort.; BLA. Under STENO: ICB; P II; BMNH; GA; GB; GC. Anatomy. Geology. Member of the Accademia del Cimento.
1. *Elementorum myologiae specimen; seu, Musculi descriptio geometrica.* Florence, 1667. RM.

2. *Discours sur l'anatomie du cerveau*. Paris, 1669. X.
2a. ——— *Lecture on the anatomy of the brain*. Copenhagen, 1965. Trans. with an introd. by G. Scherz.
3. *De solido intra solidum naturaliter contento dissertationis prodromus*. Florence, 1669. RM.
3a. ——— *The prodromus to a dissertation concerning solids naturally contained within solids*. London, 1671. Wing S5409. Trans. by H. Oldenburg. RM.
3b. ——— *The prodromus ... concerning a solid body enclosed by process of nature within a solid*. New York, 1916 ; reprinted, 1968. Trans. with introd. and notes by J.G. Winter.
3c. ——— German trans. Leipzig, 1923. (OKEW. 209)
4. *The earliest geological treatise, 1667. Translated from* [the appendix to] *"Canis Carchariae dissectum caput."* London, 1958. Latin text and trans. with introd. and notes by A. Garboe.
5. *Disputatio physica de thermis*. Montecatini, 1966. Ed. by G. Scherz.
6. *Epistolae, et epistolae ad eum datae*. 2 vols. Copenhagen, 1952. Ed. by G. Scherz.
7. *Opera philosophica*. 2 vols. Copenhagen, 1910. Ed. by V. Maar. RM/L.
8. *Geological papers*. Odense, 1969. Latin text with English trans. by A.J. Pollock. Ed. by G. Scherz.

2487. TOZZI, Luca. 1638-1717. It. ICB; P II; BLA; GA. Iatrochemistry.

2488. WALTHER, Michael. 1638-1692. Ger. P II; GC. Astronomy.

2489. BARLOW (or BOOTH), Edward. 1639-1719. Br. GD. Meteorology. Also in Part 3: Meteorology*.

2490. BROMELIUS (or BROMELL), Olaus. 1639-1705. Swed. BMNH. Botany.

2491. KIRCH, Gottfried. 1639-1710. Ger. DSB; P I; GA; GC. Astronomy.

2492. LISTER, Martin. 1639-1712. Br. DSB; ICB; P I; BMNH; BLA; GA; GB; GD. Zoology. Geology.
1. *Historiae animalium Angliae tres tractatus*. London, 1678. Wing L2523. RM/L.
2. *Historiae sive synopsis methodicae conchyliorum....* 6 parts in 2 vols. London, 1685-92. Wing L2523A/4. X. For a 19th-century index to this work see 12871/2.
3. *A journey to Paris in the year 1698*. London, 1699. Wing L2525/7. X. Reprinted with notes, and a biography and bibliography, by R.P. Stearns, Urbana, Ill., 1967.
Other titles in Wing. See also 1788/1 and 2223/1a,1b.

2493. PORZIO, Lucantonio. 1639-1723. It. ICB; P II; BLA; GA. Human and comparative anatomy. Physiology.

2494. SUTHERLAND, James. 1639?-1719. Br. BMNH. Botany. (Title in Wing.)

2495. ZORN, Bartholomaeus. 1639-1717. Ger. BMNH; BLA. Botany.

2496. GALLET, Jean Charles. late 17th century. Fr. P I. Astronomy.

2497. MAETS, Karl Ludwig van. fl. 1668-1700. Holl. P II. Chemistry.

2498. GRÜNDEL (or GRIENDEL or GRENDEL) VON ACH, Johann Franz. fl. 1670-1680. Ger. P I. Microscopy.

2499. MATTE LA FAVEUR, Sébastien. fl. 1670-1684. Fr. ICB. Iatrochemistry.

2500. CHÉRUBIN (D'ORLÉANS), ... *Père*. fl. 1671-1681. Fr. ICB; P I. Optics. Mechanics.
1. *Effets de la force de la contiguité des corps*. Paris, 1679. RM.

2501. LORENZINI, Stefano. fl. 1678. It. BMNH; BLA. Comparative anatomy.

2502. LOHMEIER, Philipp. d. 1680. Ger. P I. Meteorology.

2503. BROECKHUISEN, Benjamin van. d. ca. 1686. Holl./Br. BLA (BROEKHUYZEN). Physiology.
1. *Rationes philosophico-medicae*. The Hague, 1687.

2504. BONFIOLI, Orazio Maria. d. 1702. It. P I.
1. *De immobilitate terrae*. Bologna, 1667. RM.

2505. MEZZAVACCA, Flaminio. d. 1704. It. P II. Astronomy. Also *De terraemotu*, 1672.

2506. HONOLD, Jacob. d. 1727. Ger. P I. Astronomy.

2507. BOHN, Johannes. 1640-1718. Ger. DSB; P I (BOHNE); Mort.; BLA & Supp.; GA; GC. Physiology.

2508. DALENCÉ, Joachim. ca. 1640-1707? Fr./Belg. DSB; ICB (ALENCÉ). Astronomy. Magnetism. Physical instruments.
1. *Traité de l'aiman*. 2 parts. Amsterdam, 1687.
2. *Traittez des barometres, thermometres et notiometres ou hygrometres*. Amsterdam, 1688. RM.

His works are commonly catalogued under ALENCÉ.

2509. HAGENDORN, Ehrenfried. 1640-1692. Ger. P I; BMNH; BLA. Chemistry. Zoology. Member of the Leopoldina Academy.

2510. HANNEMAN, Johann Ludwig. ca. 1640-1724. Ger. P I; BMNH; BLA. Alchemy/chemistry. Member of the Leopoldina Academy.

2511. KERCKRING, (Thomas) Theodor. 1640-1693. Holl./Ger. ICB; P I; Mort.; BLA & Supp.; GA; GC. Anatomy. Embryology.

2512. KYLLING, Peder. 1640-1696. Den. BMNH. Botany.

2513. LA HIRE, Philippe de. 1640-1718. Fr. DSB; ICB; P I; GA. Mathematics. Astronomy. Geodesy. Member of the Académie des Sciences. "For nearly half a century, one of the principal animators of scientific life in France." -DSB. Also in Part 3: Mathematics, Astronomy*, Physics, and Geodesy.
1. *Nouveaux élémens des sections coniques, les lieux géométriques, la construction ou effection des équations*. Paris, 1679. RM.
2. *Traité de gnomonique*. Paris, 1682. X.

2a. ——— *Gnomoniques; or, The art of drawing sun-dials*. London, 1685. Wing L181A. X.

3. *Sectiones conicae*. Paris, 1685. RM.
4. *Mémoires de mathématique et de physique*. Paris, 1694. RM.

2514. LAMY, Bernard. 1640-1715. Fr. DSB; P I; GA. Mathematics. Mechanics.

2515. MOHR, Georg. 1640-1697. Den. DSB; ICB. Mathematics.
1. *Euclides Danicus.* Amsterdam, 1672. X.
1a. ——— German trans. by J. Pál, Copenhagen, 1928.

2516. OZANAM, Jacques. 1640-1718. Fr. DSB; ICB; P II; GA. A successful populariser of mathematics. Member of the Académie des Sciences.
1. *Cours de mathématiques.* 5 vols. Paris, 1693. X.
1a. ——— *Cursus mathematicus; or, A compleat course of the mathematics.* London, 1712.
2. *Récreations mathématiques et physiques.* 2 vols. Paris, 1694. X.
2a. ——— Another ed., 4 vols, ib., 1778. Ed. by Montucla.
2b. ——— ——— *Recreations in science and natural philosophy.* 4 vols. London, 1803. Trans. by C. Hutton. X. Other eds, 1851 and 1854.

2517. PLOT, Robert. 1640-1696. Br. DSB; ICB; P II; BMNH; GA; GB; GD. Natural history. Secretary of the Royal Society, 1682-84 and 1692.
1. *The natural history of Oxford-shire.* Oxford, 1676. Wing P2585. Another ed., ib., 1677. Wing P2586. RM/L. 2nd ed., ib., 1705.
2. *De origine fontium.* Oxford, 1684. Wing P2582. X. Another ed., ib., 1685. Wing P2583. RM.
3. *The natural history of Stafford-shire.* Oxford, 1686. Wing P2588. Reprinted, Manchester, 1973.
Other titles in Wing.

2518. TESTI, Ludovico. ca. 1640-1707. It. P II; BLA. Chemistry.

2519. TILLANDZ, Elias. 1640-1693. Fin. BMNH; BLA. Botany.

2520. BARNER, Jacob. 1641-1686. Ger./It. P I; BLA; GA. Chemistry.

2521. BOBART, Jacob (second of the name). 1641-1719. Br. ICB; BMNH; GA; GD. Botany. (Title in Wing, who conflates the works of father and son.)

2522. GRAAF, Reinier de. 1641-1673. Holl. DSB; ICB; BMNH; Mort.; BLA & Supp.; GA. Anatomy. Physiology.
1. *De succi pancreatici natura et usu.* Leiden, 1664. X. Another ed., ib., 1671.
1a. ——— *De succo pancreatico; or, A physical and anatomical treatise of the nature and office of the pancreatick juice.* London, 1676. Wing G1463. Trans. by C. Packe. X.
2. *De virorum organis generationi inservientibus.* Leiden, 1668.
3. *De mulierum organis generationi inservientibus.* Leiden, 1672. Reprinted, Nieuwkoop, 1965. (Dutch classics on history of science. 13)
4. *Opera omnia.* Leiden, 1677. RM/L. Another ed., Amsterdam, 1705.
5. *On the human reproductive organs.* Oxford, 1972. An annotated trans. of items 2 and 3. By H.D. Jocelyn and B.P. Setchell.

2523. GREW, Nehemiah. 1641-1712. Br. DSB; ICB; P I; BMNH; BLA; GA; GB; GD. Plant anatomy.

1. *The anatomy of vegetables begun.* London, 1672. Wing G1946. RM.
2. *An idea of a phytological history propounded, together with a continuation of the "Anatomy of vegetables."* London, 1673. Wing G1951. RM/L.
3. *The comparative anatomy of trunks.* London, 1675. Wing G1947.
4. *Experiments in consort of the luctation arising from the affusion of several menstruums upon all sorts of bodies.* London, 1678. Wing G1950. X. Reprinted, Cambridge, 1962.
5. *Musaeum Regalis Societatis; or, A catalogue of the natural and artificial rarities belonging to the Royal Society ... Whereunto is subjoyned the comparative anatomy of stomachs and guts.* London, 1681. Wing G1952 and four later eds.
6. *The anatomy of plants. With an idea of a philosophical history of plants. And several other lectures read before the Royal Society.* London, 1682. Wing G1945. RM/L. Reprinted, New York, 1965.
7. *Cosmologia sacra; or, A discourse of the universe as it is the creature and kingdom of God.* London, 1701. RM.

Other titles in Wing.

2524. GRIM (or GRIMM), Hermann Niklas. 1641-1711. Several countries. P I; BLA & Supp.; GA. Chemistry.

2525. HIÄRNE, Urban. 1641-1724. Swed. DSB. Under HJÄRNE: ICB; P I; BLA; GA. Chemistry. Mineralogy

2526. KIRCHMAIER, Sebastian. 1641-1700. Ger. P I; GA. Various fields.

2527. LANGENMANTEL, Hieronymus Ambrosius. 1641-1718. Ger. ICB; P I. Various fields, chiefly astronony. Member of the Leopoldina Academy.

2528. MAUROKORDATOS, Alexandros. 1641-1709 (or 1636-1711). Constantinople. Under MAUROCORDATO: ICB; BLA.

1. *Pneumaticum instrumentum circulandi sanguinis; sive, De motu et usu pulmonum.* Bologna, 1664. X. Reprinted, Florence, 1965.

2529. MAYOW, John. 1641-1679. Br. DSB; ICB; P II; Mort.; BLA; GA; GB; GD. Physiology. Chemistry.

1. *Tractatus quinque medico-physici. Quorum primus agit de sal-nitro et spiritu nitro-aero....* Oxford, 1674. Wing M1537. RM.

1a. ——— *Chemical experiments and opinions extracted from a work published in the last century.* Oxford, 1790. Ed. and trans. by T. Beddoes. RM.

1b. ——— German trans. Leipzig, 1901. (OKEW. 125)

1c. ——— English trans. Edinburgh, 1907 and re-issues. (Alembic Club reprints. 17)

2. *Opera omnia medico-physica, tractatibus quinque comprehensa.* The Hague, 1681. RM.

Other titles in Wing.

2530. SIBBALD, Robert. 1641-1722. Br. BMNH; BLA & Supp.; GB; GD. Natural history of Scotland. (Titles in Wing.)

2531. WAGNER, Johann Jacob. 1641-1695. Switz. P II; BMNH; BLA; GA. Natural history, especially botany.

2532. ZAHN, Johann. 1641-1707. Ger. P II. Author of *Oculus artificialis teledioptricus*, 1685, a history of optical instruments.

2533. BRIGGS, William. 1642-1704. Br. P I; BLA; GA; GD. Anatomy and physiology of the eye. (Titles in Wing.)

2534. BROWNE, John. 1642-1702. Br. ICB; BMNH; BLA & Supp.; GA; GD. Anatomy.
1. *A compleat treatise of the muscles, as they appear in the humane body.* London, 1681. Wing B5126. RM. Latin trans., ib., 1684. Wing B5127. X.
1a. ——— Another ed. *Myographia nova; or, A graphical description of all the muscles in the humane body.* London, 1697. Wing B5128. X. Reprinted, New York, 197-?

2535. CHOUET, Jean Robert. 1642-1731. Switz. P I; GA. Various fields.

2536. CRAUSE (or KRAUSE), Rudolph Wilhelm. 1642-1718. Ger. P I; BMNH (CRAUSIUS); BLA (under both C and K). Chemistry. Botany.

2537. CYPRIAN, Johann. 1642-1723. Ger. P I; BMNH. Zoology.

2538. GADROIS (or GADROYS), Claude. ca. 1642-1678. Fr. P I.
1. *Discours sur les influences des astres, selon les principes de M. Descartes.* Paris, 1671. RM.
2. *Le système du monde, selon les trois hypothèses.* Paris, 1675. RM.

2539. NEWTON, Isaac. 1642-1727. Br. DSB; ICB; P II; G. Mathematics. Mechanics. Astronomy. Optics. Natural philosophy. Also in Part 3: Mathematics*, Astronomy*, and Physics*#. See also Index.
1. *Philosophiae naturalis principia mathematica.* London, 1687. Wing N1048/9. RM/L. Reprinted, ib., 1954.
1a. ——— 2nd ed. Cambridge, 1713. Ed. by R. Cotes. RM/L.
1b. ——— 3rd ed. London, 1726. Ed. by H. Pemberton. RM/L. Reprinted, Glasgow, 1871.
1c. ——— ——— *The third edition with variant readings.* 2 vols. Cambridge, 1972. Ed. by A. Koyré and I.B. Cohen.
1d. ——— ——— "The Continental edition", 3 vols in 4, Geneva, 1739-42. Ed. by T. Le Sueur and F. Jacquier. Contains commentaries by D. Bernoulli, L. Euler, C. Maclaurin, et al. RM/L. 2nd ed., 3 vols in 2, ib., 1760. Another ed., Glasgow, 1833.

De Mundi Systemate

2. *De mundi systemate liber.* London, 1728. The early version of what became Book III of the *Principia*. Published after Newton's death. RM.
2a. ——— *A treatise of the system of the world.* London, 1728. RM/L. Reprinted with an introd. by I.B. Cohen, ib., 1969.

Excerpts

3. *Excerpta quaedam e Newtoni "Principiis philosophia naturalis", cum notis variorum.* Cambridge, 1765. RM.

Translations

4. *The mathematical principles of natural philosophy.* 2 vols. London, 1729. Trans. by Andrew Motte. RM/L. Reprinted with an introd. by I.B. Cohen, ib., 1968.

4a. ——— Another ed., 3 vols, ib., 1803. Revised by W. Davis with addition of the *System of the world.* RM/L. Another ed., 3 vols in 1, 1819.

4b. ——— Revised and supplied with an historical and explanatory appendix. Berkeley, Cal., 1934 and reissues. By F. Cajori.

5. *Mathematical principles of natural philosophy.* Vol 1. [No more publ.] London, 1777. Trans. with a commentary by R. Thorp. X. Reprinted with an introd. by I. B. Cohen, ib., 1969.

6. *Principes mathématiques de la philosophie naturelle.* 2 vols. Paris, 1759. Trans. by the Marquise du Châtelet, with a commentary by A.C. Clairaut. X. Reprinted, ib., 1966.

2540. PLUKENET, Leonard. 1642-1706. Br. BMNH; GA; GD. Botany. (Titles in Wing.)

2541. TREIBER, Johann Friedrich. 1642-1719. Ger. P II. Astronomy.

2542. BELLINI, Lorenzo. 1643-1704. It. DSB; ICB; Mort.; BLA & Supp.; GA; GB. Physiology. Anatomy.

2543. DÖRFFEL, Georg Samuel. 1643-1688. Ger. DSB; ICB; P I (DÖRFEL); GA (DOERFEL). Astronomy.

2544. GABRIELI, Piero Maria. 1643-1705. It. ICB; BLA. Mathematics. Founded the Physiocratic Academy of Siena in 1691.

2545. PAULLINI, Christian Franz. 1643-1712. Ger. ICB; BMNH; BLA & Supp.; GA; GC. Botany.

2546. SCHULTZ, Gottfried. 1643-1698. Breslau. P II. Various fields. Member of the Leopoldina Academy.

2547. BROWNE, Edward. 1644-1708. Br. P I; GD. Comparative anatomy. (Titles in Wing.)

2548. COLONNA, Francesco Maria Pompeji. 1644 (or 1649)-1726. It./Fr. P I; GA. Alchemy/chemistry.

2549. ETTMÜLLER, Michael. 1644-1683. Ger. ICB; P I; BMNH; BLA & Supp.; GA; GB; GC. Chemistry. Member of the Leopoldina Academy.

1. *Chimia rationalis ac experimentalis curiosa.* Leiden 1684. X.

1a. ——— *Nouvelle chymie raisonnée.* Lyons, 1693.

2. *Opera omnia.* Lyons, 1685. X. Another ed., London, 1688. Wing E3385. X.

3. *Ettmullerus abridg'd.* London, 1699. Wing E3385A. X.

2550. FOUCHER, Simon. 1644-1696. Fr. ICB; P I; GA; GB. Hygrometry. Best known as a philosopher.

2551. FRANCK VON FRANCKENAU, Georg. 1644-1704. Ger./Den. P I; BMNH; BLA & Supp.; GA; GC. Botany. Comparative anatomy. Member of the Leopoldina Academy.

2552. HANKE, Johannes. 1644-1713. Olmütz. P I. Astronomy.

2553. MENCKE, Otto. 1644-1707. Ger. ICB; P II; GA; GC. In 1682 he founded the *Acta eruditorun*, the first German scholarly periodical, and continued editing it until his death. In its early years it contained many mathematical and scientific articles.

2554. MERCKLEIN, Georg Abraham. 1644-1702. Ger. BMNH; BLA (MERCKLIN); GA. Zoology.

2555. MONFORTE, Antonio di. 1644-1717. It. P II. Mathematics. Astronomy.

2556. PECHLIN, Johann Nicolaus. 1644-1706. Den. P II; BMNH; BLA; GA; GC. Anatomy.

2557. RÖMER, Ole Christensen (or ROEMER, Olaus). 1644-1710. Den./Fr. DSB; ICB; P II; GA; GB. Astronomy. Determination of the velocity of light. Member of the Académie des Sciences.

2558. SALMON, William. 1644-1713. Br. BMNH; GD. Pharmacy. Botany. (Titles in Wing.)

2559. SNAPE, Andrew. b. 1644. Br. Zoology. (Title in Wing.)

2560. THOMAS, Antoine. 1644-1709. Jesuit missionary in China. ICB; P II; GF. Disseminated European mathematics in China.

2561. VOGT, Gottfried. 1644-1682. Ger. P II; BMNH (VOIGT); GA (VOIGT). Zoology.

2562. WALDSCHMIDT. Johann Jacob. 1644-1689. Ger. ICB; P II; BMNH; BLA; GC. Many dissertations, presumably of his students, are attributed to him. They date from the period 1682-88 and deal with a variety of topics in natural philosophy, the physical sciences, and the earth sciences. He was best known as a physician and was a member of the Leopoldina Academy.

2563. ZIMMERMANN, Johann Jacob. 1644-1693. Ger. P II; GC. Astronomy.

2564. DU CLOS, Samuel Cottereau. d. 1715. Fr. P I; BLA; GA. Chemistry. Member of the Académie des Sciences.
1. *Observations sur les eaux minérales de ... France*. Paris, 1675. X.
1a. ——— *Observations on the mineral waters of France*. London, 1684. Wing D2432 and two later eds. X.

2565. ROBINSON, Thomas (of Ousby). d. 1719. Br. BMNH; GD. Geology. Also in Part 3: Geology*.
1. *The anatomy of the earth*. London, 1694. Wing R1718. RM.
2. *New observations on the natural history ... being a philosophical discourse grounded upon the Mosaick system of the Creation and the Flood*. London, 1696. Wing R1719. X.

2566. PACKE, Christopher. fl. 1670-1711. Br. GD. Chemistry. (Titles in Wing.) See also 2117/4 and 2522/1a.

2567. BEUGHEM, Cornelius à. fl. 1678-1710. Ger. ICB; GA. Bibliography.

1. *Bibliographia mathematica*. Amsterdam, 1688. X. Reprinted?
2. *Bibliographia medica et physica*. Amsterdam, 1691. X.

2568. BRAND, Johann Georg. 1645-1703. Ger. P I. Many dissertations from the period 1676-90 on mathematical and astronomical subjects, presumably of his students, are attributed to him.

2569. DETHARDING, Georg. 1645-1712. Ger. P I; BLA. Chemistry.

2570. DIONIS, Pierre. 1645-1718. Fr. ICB; BLA & Supp.; GA. Anatomy.
1. *L'anatomie de l'homme*. Paris, 1690. X.
1a. ——— *The anatomy of humane bodies*. London, 1703. X. 2nd ed., ib., 1716.

2571. FONTANA, Gaëtano. 1645-1719. It. P I; GA. Astronomy.

2572. KELLNER, David. b. ca. 1645. Ger. DSB; P I; BLA. Chemistry.

2573. LEMERY, Nicolas. 1645-1715. Fr. DSB; ICB; P I; BMNH; BLA; GA. Chemistry. Member of the Académie des Sciences.
1 *Cours de chymie*. Paris, 1675. X. 8th ed.,ib., 1693.
1a. ——— *A course of chymistry*. London, 1677. Wing L1038. Trans. by Walter Harris. RM/L.
1b. ——— ——— 2nd ed., ib., 1686. Wing L1039. Trans. by W. Harris from the 5th French ed. X.
1c. ——— ——— 3rd ed., ib., 1698. Wing L1040. Trans. by James Keill from the 8th French ed.
1d. ——— ——— 4th ed., ib., 1720. Trans. by James Keill from the 11th French ed. X.
2. *Pharmacopée universelle*. Paris, 1698. X.
2a. ——— *Pharmacopoeia Lemeriana contracta. Lemery's universal pharmacopoeia abridg'd*. London, 1700. Wing L1042. X.
3. *Traité universelle des drogues simples, mises en ordre alphabétique*. Paris, 1698. X.
3a. ——— Another ed. *Dictionnaire universel des drogues simples*. Paris, 1759.

2574. MÉRY, Jean. 1645-1722. Fr. DSB; ICB; P II; Mort.; BLA; GA. Human and comparative anatomy. Physiology. Member of the Académie des Sciences.

2575. MURALT, Johannes von. 1645-1733. Switz. DSB; ICB; P II; BMNH; BLA; GA; GC. Human and comparative anatomy. Botany.

2576. PFAUTZ, Christoph. 1645-1711. Ger. P II. Astronomy.

2577. SIGÜENZA Y GÓNGORA, Carlos de. 1645-1700. Mexico. DSB; ICB; P II. Astronomy.

2578. WEDEL, Georg Wolfgang. 1645-1721. Ger. DSB; ICB; P II; BMNH (WEDELIUS); BLA. Iatrochemistry. Botany. Zoology.

2579. AUGUSTINUS, Thomas a St. Josepho. 1646-1717. Aus. P I. Mathematics. A correspondent of Leibniz.

2580. FLAMSTEED, John. 1646-1719. Br. DSB; ICB; P I; GA; GB; GD. Astronomy. Also in Part 3: Astronomy*. See also Index.
1. *The doctrine of the sphere, grounded on ... the Copernican system of the world*. London, 1680. Wing F1137. RM/L.
2. *The Gresham lectures*. London, 1975. Ed. from the MS by E. G. Forbes.
Other titles in Wing.

2581. HEIDE (or HEYDE), Anton van der. 1646-ca. 1693. Holl. BMNH; GF. Comparative anatomy.

2582. HERMANN, Paul. 1646 (or 1640)-1695. Holl. BMNH (HERMANNUS); BLA & Supp.; GA. Botany.
1. *Florae ... sive, Enumeratio stirpium horti Lugduno-Batavi.* Leiden, 1690.
2. *Paradisus Batavus, continens plus centum plantas.* Leiden, 1698. Ed. by W. Sherard. X.
2a. ——— Another ed. *Paradisus Batavus; seu, Descriptio rariorum plantarum.* Leiden, 1705.

2583. LEIBNIZ, Gottfried Wilhelm. 1646-1716. Ger. The famous philosopher. DSB; ICB; P I; G. Mathematics. Mechanics. See also Index.
1. *Hypothesis physica nova ... seu, Theoria motus abstracti et concreti.* Mainz, 1671. X. Another ed., London, 1671. Wing L962. X.

Posthumous Works

2. *Protogaea; sive, De prima facie telluris et antiquissimae historiae vestigiis in ipsis naturae monumentis.* Göttingen, 1749. RM/L.
2a. ——— *Protogée; ou, De la formation et des révolutions du globe.* Paris, 1859. Trans. with introd. and notes by B. de Saint-Germain.
3. *Nachgelassene Schriften physikalischen, mechanischen und technischen Inhalts.* Leipzig, 1906.
4. *Reise Journal, 1687-88.* Hildesheim, 1966. Facsimile of the MS.
5. *Marginalia in Newtoni "Principia mathematica", 1687.* Paris, 1973. First publication. By E.A. Fellman.
6. *Ein Dialog zur Einführung in die Arithmetik und Algebra.* Stuttgart, 1976. Ed. from the MS with a German trans. and commentary by E. Knobloch.

Correspondence

7. *A collection of papers which passed between ... M. Leibnitz and Dr.* [Samuel] *Clarke in 1715 and 1716 relating to the principles of natural philosophy and religion.* London, 1717. RM/L.
8. *The Leibniz-Clarke correspondence.* Manchester, 1956. Ed. with introd. and notes by H.G. Alexander.
9. *Correspondance Leibniz-Clarke.* Paris, 1957. Ed. by G. Le Roy.
10. *Virorum celeberr. G.W. Leibnitii et Joh. Bernoulli commercium philosophicum et mathematicum.* 2 vols. Geneva, 1745.
11. *Nouvelles lettres et opuscules inédits.* Paris, 1857. Introd. by L.A. Foucher de Careil. RM. Reprinted, Hildesheim, 1971.
12. *Leibnizens und Huygens Briefwechsel mit Papin.* Berlin, 1881. Ed. by E. Gerland. X. Reprinted, Wiesbaden, 1966.
13. *Der Briefwechsel in der K. öffentlichen Bibliothek zu Hannover.* Hannover, 1895. Ed. by E. Bodemann. X. Reprinted, Hildesheim, 1966.
14. *Der Briefwechsel mit Mathematikern.* Berlin, 1899. Ed. by C.I. Gerhardt. Reprinted, Hildesheim, 1962.

15. *Zwei Briefe: Über das binäre Zahlensystem und die chinesische Philosophie.* Stuttgart, 1968. Ed. with German trans. and commentary by R. Loosen and F. Vonessen.

Collected Works and Selections

16. *Sämtliche Schriften und Briefe.* Many vols. Berlin, 1923-. Still in progress. Ed. by the Berlin Academy. Partial reprint, Hildesheim, 1970-.
17. *Mathematische Schriften.* 7 vols. Berlin/Halle, 1849-63. Ed. by C.I. Gerhardt. X. Reprinted, Hildesheim, 1962 and 1971.

17a. ——— *The early mathematical manuscripts.* Chicago, 1920. Trans. from the Latin texts published by Gerhardt, with critical and historical notes, by J.M. Child.

18. *Über die Analysis des Unendlichen.* Leipzig, 1908. (OKEW. 162) Trans. from the Latin.
19. *Leibniz et la dynamique: Les textes de 1692.* Paris, 1960. Ed. by P. Costabel.

19a. ——— *Leibniz and dynamics: The texts of 1692.* Paris, 1973.

2584. PLUMIER, Charles. 1646-1704. Fr. DSB; BMNH; GA. Botany. Also in Part 3: Botany*.

2585. SCHROECK, Lucas. 1646-1730. Ger. P II; BLA; GC. Zoology. Member, and in 1693 President, of the Leopoldina Academy.

2586. SENGUERD, Wolferd. 1646-1724. Holl. P II; BLA. Pneumatics (he was well known for his air-pump).

2587. BONTEKOE, Cornelius. 1647-1685. Holl./Ger. ICB; P I; BLA & Supp.; GA; GF. Chemistry.

2588. CEVA, Giovanni. ca. 1647-1734. It. DSB; ICB; P I. Mathematics. Mechanics.

2589. GIBSON, Thomas. 1647-1722. Br. GD. Anatomy. (Titles in Wing.)

2590. HARRIS, Walter. 1647-1732 (or 1725). Br. BLA & Supp.; GA; GD. Pharmacy. (Titles in Wing.) See also 2573/1a,1b.

2591. HAUTEFEUILLE, Jean de. 1647-1724. Fr. P I; GA. An outstanding horologist, instrument-maker, and inventor.

2592. MERIAN, Maria Sibylla. 1647-1717. Holl. ICB (MARIAN); BMNH; GA; GC; GF. Botany. Zoology. Also in Part 3: Botany and Zoology*.

1. *Neues Blumen-Buch.* Nuremberg, 1680. X. Reprinted, Berlin, 1968.
2. *Leningrader Aquerelle.* 2 vols. Lucerne, 1974. Facsimiles of watercolours which are now in the Academy of Sciences, Leningrad. Ed. with commentary by E. Ullman.

2593. PAPIN, Denis. 1647-ca. 1712. Fr./Br./It./Ger. DSB; ICB; P II; GA; GB; GC; GD. Pneumatics. Technology (one of the inventors of the steam engine).

1. *A new digester or engine for softening bones.* London, 1681. Wing P309. RM. Reprinted, London, 1966.
2. *A continuation of the "New digester of bones" ... together with some improvements and new uses of the air-pump.* London, 1687. Wing P308. RM.

3. *Fasiculus dissertationum de novis quibusdam machinis, atque aliis argumentis philosophicis.* Marburg, 1695. RM.
4. *Leibnizens und Huygens Briefwechsel mit Papin.* Berlin, 1881. Ed. by E. Gerland. Includes a biography of Papin and related documents. X. Reprinted, Wiesbaden, 1966.

2594. RHYNE, Willem ten. 1647-1700. Holl./East Indies. DSB (TEN RHYNE); ICB; BMNH (RHIJNE); BLA; GA; GF. Botany.

2595. TRIONFETTI, Lelio. 1647-1722. It. GA. Botany.

2596. CEVA, Tomasso. 1648-1737. It. DSB; ICB; P I; GA. Mathematics.

2597. CHIRAC, Pierre. 1648-1732. Fr. BLA & Supp.; GA. A physician who was in charge of the Jardin du Roi. Member of the Académie des Sciences.

2598. CRAMER, Caspar. 1648-1682. Ger. P I. Chemistry.

2599. DUVERNEY, Joseph Guichard. 1648-1730. Fr. DSB; ICB; Mort.; BLA & Supp.; GA. Anatomy. Member of the Académie des Sciences.
1. *Traité de l'organe de l'ouie.* Paris, 1683. X.
1a. ——— *A treatise of the organ of hearing.* London, 1737. Trans. by L. Marshall. X. Reprinted, New York, 1973.

2600. HARTMANN, Philipp Jacob. 1648-1707. Ger. P I; BMNH; BLA; GA. Natural history. Comparative anatomy. History of anatomy. Member of the Leopoldina Academy.
1. *Succini prussici physica et civilis historia.* Frankfurt, 1677. RM. Re amber and its electrical properties.

2601. HOTTON, Petrus. 1648-1709. Holl. BMNH; BLA. Botany. Zoology.

2602. JÜNGKEN (or JUNGKEN), Johann Helfrich. 1648-1726. Ger. P I; BLA. Pharmacy/chemistry. Member of the Leopoldina Academy.

2603. PRESTET, Jean. 1648-1691. Fr. ICB; P II & Supp. (PRESTEL). Mathematics.

2604. VERHEYEN, Philippe. 1648-1710. Belg. ICB; Mort.; BLA; GA; GF. Anatomy.

2605. BAILLET, Adrien. 1649-1706. Fr. GA; GB.
1. *La vie de monsieur Des-Cartes.* 2 vols. Paris, 1691. X. Reprinted, Geneva, 1970, and Hildesheim, 1972.

2606. BIDLOO, Govard (or Godefried). 1649-1713. Holl. DSB Supp.; ICB; BMNH; Mort.; BLA & Supp.; GA. Anatomy.
1. *Anatomia humani corporis.* Amsterdam, 1685. X. Reprinted, Paris, 1972. Another ed., Amsterdam, 1687.
2. *Letter to Antony van Leeuwenhoek about the animals which are sometimes found in the liver of sheep and other beasts.* Nieuwkoop, 1972. (Dutch classics on history of science. 18) Facsimile of the Dutch original of 1698 with English trans., introd., and notes by J. Jansen.

2607. BLOUNT, Thomas Pope. 1649-1697. Br. GA; GB; GD.
1. *A natural history.* London, 1693. Wing B3351. RM.

2608. DUNCAN, Daniel. 1649-1735. Fr. etc. ICB; P I; BLA; GA; GD. Iatrochemistry.
1. *La chymie naturelle; ou, L'explication chymique et méchanique de la nourriture de l'animal.* Paris, 1682. RM.

2609. KAHLER, Johann. 1649-1729. Ger. P I. Astronomy. Earth sciences.

2610. PAPA, Giuseppe del. 1649-1735. It. ICB; P II; BLA. Natural philosophy.

2611. SCHELHAMMER, Günther Christoph. 1649-1716. Ger. P II; BMNH; BLA; GA; GC. Chemistry. Botany. Comparative anatomy. Also author of *De motu mercurii in tubo torricelliano*, 1699. Member of the Leopoldina Academy.

2612. ANGO, Pierre. late 17th century. Fr. P I. Optics.

2613. STERRE, Dionysius van der. d. 1691. Holl. BLA. Embryology.

2614. SÉDILEAU, ... d. 1693. Fr. P II. Astronomy. Meteorology. Member of the Académie des Sciences.

2615. ARNOLD, Christoph. 1650-1695. Ger. P I. Astronomy.

2616. BANISTER, John (second of the name). 1650-1692. Br./U.S.A. DSB; ICB; GA; GD; GE. Botany. Zoology.
   1. *John Banister and his natural history of Virginia, 1678-1692*. Urbana, Ill., 1970. A selection from his works. Ed. by J. and N. Ewan.

2617. BILLBERG, Johann. 1650-1717. Swed. P I. Astronomy.

2618. BLANKAART, Stephen. 1650-1702. Holl. ICB; P I; BMNH; BLA & Supp.; GA (BLANKAERTS). Natural history. Anatomy.
   1. *Anatomia reformata*. Leiden, 1688.

2619. FARDELLA, Michel Angelo. 1650-1718. It. P I; GA; GF. Mathematics.

2620. GOUYE, Thomas. 1650-1725. Fr. P I; GA. Astronomy. Natural history. Member of the Académie des Sciences.

2621. HAM, Jan. b. 1650. Holl. BLA. Microscopy.

2622. JACOBAEUS, Holger (or Oligerus). 1650-1701. Den. ICB; BMNH; BLA & Supp. (JACOBSEN). Natural history. Human and comparative anatomy.

2623. LE MORT, Jacob. 1650-1718. Holl. GA. Under MORT: P II; BLA; GF. Chemistry.

2624. MALÉZIEU, Nicolas de. 1650-1727. Fr. P II; GA. Astronomy. Microscopy.

2625. MARCHANT, Jean. 1650-1738. Fr. DSB. Botany. Member of the Académie des Sciences.

2626. NUCK, Antoni. 1650-1692. Holl. ICB; Mort.; BLA; GA; GC. Anatomy.

2627. TYSON, Edward. 1650/51-1708. Br. DSB; ICB; BMNH; Mort.; BLA; GD. Comparative anatomy.
   1. *Orang-outang sive homo sylvestris; or, The anatomie of a pygmie compared with that of a monkey, an ape, and a man*. London, 1699. Wing T3598. RM. Reprinted, London, 1966.
   Other titles in Wing. See also 2470/3a.

2628. VIGANI, John Francis. ca. 1650?-1713. Br. DSB; ICB; BMNH; GD. Chemistry. (Title in Wing.)

2629. WALLER, Richard. ca. 1650-1715. Br. ICB. Natural history. Secretary of the Royal Society, 1687-1715. See also Index.

2630. DOLAEUS, Johann. 1651-1707. Ger. P I; BLA. Various fields. Member of the Leopoldina Academy.

2631. KAEMPFER, Engelbert. 1651-1716. Ger. DSB; ICB; P I; BMNH; BLA & Supp.; GA; GB; GC. Botany. Also in Part 3: Botany*.

2632. MENTZER, Balthasar. 1651-1727. Ger. P II. Astronomy.

2633. SHARP, Abraham. 1651 (or 1653)-1742. Br. ICB; P II; GA; GD. Astronomy. Flamsteed's assistant at Greenwich. See also 3392/1 and 15602/1.

2634. TSCHIRNHAUS, Ehrenfried Walther. 1651-1708. Ger. DSB; ICB; P II; GA; GC. Mathematics. Also in Part 3: Science in General* and Mathematics.
   1. *Medicina mentis et corporis.* 3 vols in 2. Amsterdam, 1686-87. X. A philosophical work which also contains some mathematics. Reprinted, Hildesheim, 1964.

2635. WURZELBAU (or WURTZELBAU), Johann Philipp von. 1651-1725. Ger. P II; GA; GC. Astronomy.

2636. BION, Nicolas. ca. 1652-1733. Fr. DSB; P I; GA. Outstanding maker of globes, mathematical instruments, etc.
   1. *L'usage des globes célestes et terrestres et des sphères suivant les differens systèmes du monde.* Paris, 1699. X. 5th ed., ib., 1728.
   2. *Traité de la construction et des principaux usages des instruments de mathématique.* Paris, 1709. RM. Another ed., The Hague, 1723.
   2a. ——— *The construction and principal uses of mathematical instruments.* 2nd ed. London, 1758. RM.

2637. HOMBERG, Wilhelm (or Guillaume). 1652-1715. Fr. DSB; ICB; P I; BLA; GA; GB. Chemistry. Member of the Académie des Sciences.

2638. LE CLERC, Daniel. 1652-1728. Fr. ICB; BMNH; BLA; GA. Anatomy. Zoology. Best known as a historian of medicine.
   1. *Bibliotheca anatomica.* Geneva, 1685. A bibliography.

2639. LE FÈVRE, Jean. 1652?-1706. Fr. DSB; P I (LE FÈBVRE); GA. Astronomy. Member of the Académie des Sciences.

2640. PITCAIRN, Archibald. 1652-1713. Br. DSB; ICB; P II; BLA; GA; GB; GD. Physiology. (Titles in Wing.)

2641. RIVINUS (or BACHMANN), Augustus Quirinus. 1652-1723. Ger. DSB (BACHMANN); P II; BMNH; BLA; GA; GC. Botany. Anatomy. Physiology.

2642. ROLLE, Michel. 1652-1719. Fr. DSB; ICB; P II; GA. Mathematics. Member of the Académie des Sciences.

2643. ALBINUS, Bernhard. 1653-1721. Ger./Holl. DSB Supp.; P I; BMNH; BLA; GA; GC. Anatomy. Botany.

2644. BRUNNER (or BRUNN), Johann Conrad von. 1653-1727. Ger. ICB; P I; Mort.; BLA & Supp.; GA; GC. Anatomy. Member of the Leopoldina Academy.

2645. HOFFMANN, Johann Moritz. 1653-1727. Ger. ICB; P I; BLA (HOFMANN); GA. Chemistry. President of the Leopoldina Academy.

2646. MALLEMANS (DE MESSANGES), Claude. 1653-1723. Fr. ICB; P II; GA. Astronomy.

2647. PEYER, Johann Conrad. 1653-1712. Switz. DSB; ICB; BMNH; Mort.; BLA; GA; GC. Physiology. Human and comparative anatomy.

2648. RIDLEY, Humphrey (*incorrectly* Henry). 1653-1708. Br. BLA; GD. Anatomy. (Title in Wing.)

2649. BERNOULLI, Jakob (or Jacques) (first of the name). 1654-1705. Switz. DSB; ICB; P I; GA; GB; GC. Mathematics. Mechanics.
1. *Dissertatio de gravitate aetheris.* Amsterdam, 1683. RM.
2. *Ars conjectandi, opus posthumum. Accedit Tractatus de seriebus infinitis, et Epistola gallice scripta de ludo pilae reticularis.* Basel, 1713. RM/L. Reprinted, Brussels, 1968.
2a. ——— German trans. 2 vols. Leipzig, 1899. (OKEW)
3. *Über unendliche Reihen (1689-1704).* Leipzig, 1909. (OKEW) Trans. from the Latin.
4. *Opera.* Geneva, 1744. Ed. by G. Cramer. RM.
5. *Werke.* Several vols. Basel, 1969-. Ed. by the Naturforschende Gesellschaft in Basel.

2650. HELLWIG, Johann Otto von. 1654-1698. Ger. P I; BLA; GA (HELWIG). Alchemy.

2651. LITTRÉ, Alexis. 1654-1725. Fr. Mort.; BLA. Anatomy. Member of the Académie des Sciences.

2652. MULLENS (or MOULIN, MOLINS, etc.), Allan. 1654-1690. Br. GD (MOLINES). Human and comparative anatomy. (Title in Wing.)

2653. NIEUWENTIJT, Bernard. 1654-1718. Holl. DSB; P II; BLA; GA. Mathematics. Also in Part 3: Science in General* and Mathematics.

2654. VARIGNON, Pierre. 1654-1722. Fr. DSB; ICB; P II; GA. Mathematics. Mechanics. Also in Part 3: Mathematics* and Mechanics*. Member of the Académie des Sciences.
1. *Projet d'une nouvelle méchanique.* Paris, 1687. RM.
2. *Nouvelles conjectures sur la pesanteur.* Paris, 1690. RM.

2655. BEDDEVOLE, Dominique. d. ca. 1692. Switz.? BLA & Supp. Anatomy.
1. *Essais d'anatomie.* Leiden, 1686. X.
1a. ——— *Essayes of anatomy.* Edinburgh, 1691. Wing B1663 and another ed. X.

2656. CLÜVER, Detlef. Ger./Br. d. 1708. P I; GA; GC. Mathematics. Geology. (Title in Wing.)

2657. HERBERSTEIN, Ferdinand Ernst Karl von. d. 1720. Prague. P I; GA. Mathematics. Mechanics.

2658. BEAUMONT, John. d. 1731. Br. GD. Geology.
1. *Considerations on a book entituled "The theory of the earth", published some years since by ... Dr. Burnet.* London, 1693. Wing B1620. RM.
Other titles in Wing.

2659. BARTHOLIN, Caspar Thomesen. 1655-1738 (or 1705). Den. ICB; P I; BMNH; BLA & Supp.; GA. Anatomy.
1. *De inauribus veterum*. Amsterdam, 1676. RM.
2. *Specimen philosophiae naturalis*. 4 parts. 1690-92. X. Another ed., Oxford, 1698. Wing B974. X. Another ed., Oxford, 1703.

2660. GENGA, Bernardino. 1655-1734. It. ICB; Mort.; BLA; GA. Anatomy.
1. *Anatomia chirurgica*. Bologna, 1687. RM.
2. *Anatomia per uso et intelligenza del disegno*. Rome, 1691. X.
2a. ——— *Anatomy ... with regard to the uses therof in designing*. London, 1723. X. Reprinted, New York, 197-?

2661. GUGLIELMINI, Domenico. 1655-1710. It. ICB; P I; BLA; GA. Fluid mechanics. Various other fields.
1. *Aquarum fluentium mensura nova methodo inquisita*. Bologna, 1690. RM.
2. *Della natura de' fiumi, trattato fisico-matematico*. Bologna, 1697. RM.
2a. ——— *Nuova edizione con le annotazioni di E. Manfredi*. Bologna, 1739. RM.
3. *Opera omnia mathematica, hydraulica, medica, et physica: Accessit vita autoris a J.B. Morgagni*. Geneva, 1719. RM.

2662. HAVERS, Clopton. ca. 1655-1702. Br. DSB; Mort.; BLA; GA; GD. Osteology. (Title in Wing.)

2663. LE COMTE, Louis (Daniel). 1655-1728. Jesuit missionary in China where he cultivated astronomy. GA.
1. *Nouveaux mémoires sur l'état présent de la Chine*. Paris, 1696. RM.

2664. WURFBAIN, Johann Paul. 1655-1711. Ger. P II; BMNH; BLA. Zoology. Botany. Member of the Leopoldina Academy.

2665. ZAMBECCARI, Giuseppe. 1655-1728. It. DSB; ICB; BLA. Physiology.

2666. EISENSCHMIDT, Johann Caspar. 1656-1712. Strasbourg. P I; GA; GC. Geodesy (he initiated the dispute about the shape of the earth.)

2667. HALLEY, Edmund. 1656?-1743. Br. DSB; ICB; P I; GA; GB; GD. Astronomy. Geomagnetism. (Titles in Wing.) Also in Part 3: Various Fields*, Astronomy*#, and Geomagnetism. See also Index.

2668. HARDER, Johann Jacob. 1656-1711. Switz. BMNH; BLA; GA; GC. Human and comparative anatomy.

2669. HARTSOEKER, Nicolaus. 1656-1725. Holl./Fr./Ger. DSB; P I; BLA; GA. Optics. Embryology. Also in Part 3: Physics*.
1. *Essay de dioptrique*. Paris, 1694. RM.
2. *Principes de physique*. Paris, 1696. RM.

2670. MOLYNEUX, William. 1656-1698. Br. DSB; ICB; P II; GA; GD. Optics. In 1683 he founded the Dublin Philosophical Society.
1. *Sciothericum telescopium; or, A new contrivance of adapting a telescope to a horizontal dial*. Dublin, 1686. Wing M2406. X.

2. *Dioptrica nova. A treatise of dioptricks.* London, 1692. Wing M2405. RM.

2671. TOURNEFORT, Joseph Pitton de. 1656-1709. Fr. DSB; ICB; P II; BMNH; GA; GB. Botany. Also in Part 3: Botany*. Member of the Académie des Sciences.
1. *Schola botanica; sive, Catalogus plantarum quas ab aliquot annis in Horto Regio Parisiensi studiosis indigitavit.* Amsterdam, 1689. RM.
2. *Élémens de botanique; ou, Méthode pour connoître les plantes.* 3 vols. Paris, 1694. X.
2a. ——— *Institutiones rei herbariae.* 3 vols. Paris, 1700. RM. 3rd ed., ib., 1719.
2b. ——— ——— *The compleat herbal ... With a short account of the life and writings of the author.* 2 vols. London, 1719-30. RM/L.
2c. ——— *Abrégé des "Élémens de botanique."* Avignon, 1739. RM.
3. *Histoire des plantes qui naissent aux environs de Paris.* Paris, 1698. RM.

2672. TOZZI, Bruno. 1656-1743. It. DSB. Botany.

2673. ALMELOVEEN, Theodor Jansson van. 1657-1712. Holl. BMNH; BLA & Supp.; GA. Botany.

2674. CHAZELLES, Jean Mathieu de. 1657-1710. Fr. P I; GA; GB. Astronomy.

2675. CLAYTON, John (first of the name). 1657-1725. Br./U.S.A. ICB. Various fields,
1. *Scientific writings and other related papers.* Charlottesville, Virginia, 1965. Ed. with a short biography by E. Berkeley.

2676. COLB, Christoph. 1657-1689. Ger. P I. Astronomy. Assistant of Hevelius.

2677. CUPANI, Franciscus. 1657-1711. It. BMNH; GA. Natural history.
1. *Hortus catholicus.* Naples, 1696-97.

2678. DERHAM, William. 1657-1735. Br. DSB; ICB; P I; BMNH; GA; GB; GD. Various fields. (Title in Wing.) Best known for his writings on natural theology; see entry 2837 in Part 3. See also Index.

2679. FONTENELLE, Bernard Le Bovier de. 1657-1757. Fr. DSB; ICB; P I; GA; GB. Popularisation of science. From 1697 onwards he was secretary of the Académie des Sciences, a position he filled with great distinction. Also in Part 3: Science in General*#, Mathematics*, Mechanics*, Biography*, and History of Science*. See also 15442/4.
1. *Entretiens sur la pluralité des mondes.* Paris, 1686. RM. Another ed., Amsterdam, 1719.
1a. ——— Reprinted, Paris, 1945 and 1970. Critical eds by R. Shackleton, Oxford, 1955, and A. Calame, Paris, 1966.
1b. ——— *A discourse of the plurality of worlds.* Dublin, 1687. Wing F1411. X.

1c. ——— *A discovery of new worlds.* London, 1688. Wing F1412. Trans. by Mrs. A. Behn. X. Another ed., 1700. Wing F1418. X.

1d. ——— *A plurality of worlds.* London, 1688. Wing F1416. Trans. by J. Glanvill. X. Reprinted, ib., 1929.

1e. ——— ——— 2nd ed., 1695. Wing F1417. X. Another ed., 1702. 4th ed., 1719. Title varies.

1f. ——— *Conversations on the plurality of worlds.* London, 1715. Trans. by W. Gardiner. X.

1g. ——— ——— 2nd ed. *To which is added Mr. Secretary Addison's Oration ... in defence of the new philosophy.* Ib., 1728. X. (i.e. Joseph Addison, the well known writer.) 3rd ed., 1737. Another ed., 1760. RM/L.

2680. LENTILIUS, Rosinus. 1657 (or 1651)-1733. Ger. P I; BLA. Various fields. Member of the Leopoldina Academy.

2681. ROBINSON, Tancred. ca. 1657-1748. Br. BMNH; GD. Natural history, chiefly botany.

2682. SCHRADER, Friedrich. 1657-1704. Ger. P II; BMNH; BLA. Microscopy. Zoology.

2683. STISSER, Johann Andreas. 1657-1700. Ger. P II; BMNH; BLA. Chemistry. Botany.

2684. TITA, Antonius. ca. 1657-1729. It. BMNH. Botany.

2685. VALENTINI, Michael Bernhard. 1657-1729. Ger. P II; BMNH; Mort; BLA; GC. Comparative anatomy. Natural history. Member of the Leopoldina Academy. See also 2731/1.

2686. GARIDEL, Pierre Joseph. 1658-1737. Fr. BMNH; GA. Botany.

2687. KÖNIG, Emanuel. 1658-1731. Switz. DSB; P I; BMNH; BLA; GA; GC. Natural history. Member of the Leopoldina Academy.

2688. LIMMER, Conrad Philipp. 1658-1730. Ger. P I & Supp. Over sixty disputations from the period 1686-1709, presumably of his students, are attributed to him. They deal with various topics in natural philosophy, physics, and astronomy.

2689. MARSIGLI, Luigi Ferdinando. 1658-1730. It. DSB (MARSILI); ICB; P II; BMNH; GA; GB. Geology. Physical geography. Oceanography. Also in Part 3: Geology and Oceanography*.

2690. POMET, Pierre. 1658-1699. Fr. ICB; BMNH; GA. Pharmacy. Botany.

2691. RAU (or RAW), Johannes Jacobus. 1658 (or 1668)-1719 (or 1721) Ger./Holl. BLA. Anatomy.

2692. TRIONFETTI, Giovanni Battista. 1658-1708. It. BMNH. Botany.

2693. ZWINGER, Theodor (second of the name). 1658-1724. Switz. ICB; P II; BMNH; BLA; GC. Botany.

1. *Theatrum botanicum; das ist, Neu vollkommenes Kräuter-Buch.* Basel, 1696. X. Another ed., ib., 1744. RM.

2694. BERGER, Johann Gottfried. 1659-1736. Ger. DSB; ICB; P I; BMNH; BLA & Supp. Physiology.

2695. DALE, Samuel. 1659?-1739. Br. BMNH; BLA & Supp.; GA; GD. Botany. Pharmacology. (Title in Wing.)

2696. GREGORY, David. 1659-1708. Br. DSB; ICB; P I; GA; GB; GD. Mathematics. Astronomy. Optics. Also in Part 3: Mathematics*, Astronomy*, and Physics.
1. *Exercitatio geometrica de dimensione figurarum.* Edinburgh, 1684. Wing G1884. X.
2. *Catoptricae et dioptricae sphericae elementa.* Oxford, 1695. Wing G1883. X.
2a. ——— *Elements of catoptrics and dioptrics.* London, 1715. RM.
3. *David Gregory, Isaac Newton, and their circle: Extracts from David Gregory's "Memoranda", 1677-1708.* Oxford, 1937. Ed. by W.G. Hiscock.

2697. SHERARD, William. 1659-1728. Br. DSB; BMNH; GA; GD. Botany. See also 2582/2.

2698. MEISTER, Georg. late 17th century. BMNH. Botany.
1. *Der orientalisch-indianische Kunst- und Lust-Gärtner.* 1692. Re the botany of India and the Far East. X. Reprinted, Weimar, 1973.

2699. CASE, John. fl. 1680-1700. Br. GA; GD. Embryology. (Titles in Wing.)

2700. POTHENOT, Laurent. d. 1732. Fr. P II. Mathematics. Member of the Académie des Sciences.

2701. BUISSIÈRE (or BUSSIÈRE), Paul. d. 1739. Fr./Br. GA; GD. Anatomy.

2702. BLAESING, David. 1660-1719. Ger. P I. Astronomy.

2703. BOULTON, Richard. 1660-1724. Br. BLA & Supp.; GD. Physiology. (Titles in Wing.)

2704. CRAIG, John. ca. 1660-1731. Br. DSB; P I; GA; GD. Mathematics. (Titles in Wing.)

2705. GAGLIARDI, Domenico. 1660-ca. 1725. It. DSB; ICB; BLA & Supp.; GA. Anatomy

2706. GAUTERON, Antoine. 1660-1737. Fr. P I; BLA; GA. Various fields.

2707. HANCKWITZ, Ambrose Godfrey. 1660-1741. Br. ICB; GD (GODFREY). Chemistry. Employed by Robert Boyle as an "operator" in his laboratory.

2708. HOFFMANN, Friedrich. 1660-1742. Ger. DSB; ICB; P I; BMNH; Mort.; BLA; GA; GB; GC. Chemistry. Physiology. Also in Part 3: Chemistry*# and Physiology.

2709. LAGNY, Thomas Fantet de. 1660-1734. Fr. DSB; P I; GA. Mathematics.

2710. LHWYD, Edward. 1660-1709. Br. DSB; ICB; P I & Supp.; BMNH (LHUYD); GA (LLWYD); GB (LLWYD); GD (LHUYD). Palaeontology. Botany.
1. *Lithophylacii Britannici ichnographia.* London, 1699. Wing L1945/6. Another ed., Oxford, 1760. RM.

2711. PERNAUER, Johann F.A. von. 1660-1731. ICB. Zoology.

2712. RUDBECK, Olof (second of the name). 1660-1740. Swed. BMNH; BLA; GA. Botany. Zoology.

2713. SEIGNETTE, Pierre. 1660-1719. Fr. P II & Supp.; BLA. Chemistry.

2714. STAHL, Georg Ernst. 1660-1734. Ger. DSB; ICB; P II; Mort.; BLA & Supp.; GA; GB; GC. Chemistry. Physiology. Also in Part 3: Chemistry* and Physiology*. See also 15505/1.

2715. L'HOSPITAL, Guillaume François Antoine de. 1661-1704. Fr. DSB; P I (under H); ICB (L'HÔPITAL); GA. Mathematics. Member of the Académie des Sciences.
1. *Analyse des infiniment petits, pour l'intelligence des lignes courbes.* Paris, 1696. RM. 2nd ed., 1715. Another ed., Avignon, 1768.
2. *Traité analytique des sections coniques et de leur usage pour la résolution des équations.* Paris, 1707. RM/L.
2a. ——— *An analytick treatise of conick sections.* London, 1723. Trans by E. Stone.
See also 2738/1.

2716. ODHELSTJERNA, Eric. 1661-1704. Swed. P II,III; BLA (ODHELIUS). Metallurgical chemustry.

2717. POUPART, François. 1661-1709. Fr. ICB; Mort.; BLA; GA. Anatomy. Zoology. Member of the Académie des Sciences.

2718. LEIGH, Charles. 1662-1701? Br. P I; BMNH; BLA; GA; GD. Natural history. (Titles in Wing.)

2719. VOLCKAMER, Johann Georg (second of the name). 1662-1744. Ger. BMNH; BLA (VOLKAMER). Botany.

2720. AMONTONS, Guillaume. 1663-1705. Fr. DSB; P I; GA; GB. Thermometry. Development of scientific instruments. Mechanics.
1. *Remarques et expériences phisiques sur la construction d'une nouvelle clepsidre, sur les baromètres, thermomètres et higromètres.* Paris, 1695. RM.

2721. HELLWIG, Christoph von (pseudonym: KRÄUTERMANN, Valentinus). 1663-1721. Ger. ICB; P I; BMNH (KRAEUTERMANN); BLA; GA (HELWIG). Pharmacy. Chemistry. Botany. "The Thuringian Paracelsus."

2722. LANZONI, Giuseppe. 1663-1730. It. BMNH; BLA; GA. Zoology. Botany.

2723. MÖREN, Johann Theodor. 1663-1702. Ger. P II. Meteorology. Astronomy. Zoology. Member of the Leopoldina Academy.

2724. NICOLAS, Pierre. 1663-1708. Fr. P II; GA. Mathematics.

2725. PETIVER, James. 1663-1718. Br. ICB; BMNH; GA; GD. Natural history. (Titles in Wing.)

2726. SAHME, Christian. 1663-1732. Ger. P II. Astronomy.

2727. BURLET, Claude. 1664 (not 1692)-1731. Fr. BLA. Botany. Member of the Académie des Sciences.

2728. FATIO DE DUILLIER, Nicolas. 1664-1753. Switz./Br. ICB; P I;

GA; GD (FACCIO). Mathematics. Astronomy. Initiator of the famous dispute between Newton and Leibniz.

2729. FALUGI, Virgilio. d. 1707. BMNH. Botany.

2730. RAPHSON, Joseph. d. 1715/6. Br. P II. Mathematics. (Title in Wing.)

2731. CAMERARIUS, Rudolf Jacob. 1665-1721. Ger. DSB Supp.; P I; BMNH; BLA & Supp.; GA; GB; GC. Botany.
1. *Epistola ad Mich. Bernardum Valentini de sexu plantarum.* Tübingen, 1694. RM.
1a. ——— *Über das Geschlecht der Pflanzen.* Leipzig, 1899. (OKEW. 105)

2732. WOODWARD, John. 1665-1728. Br. DSB; P II; BMNH; BLA; GB; GD. Geology. Botany. Also in Part 3: Geology* and Botany.
1. *An essay toward a natural history of the earth.* London, 1695. Wing W3510. RM/L.
2. *Brief instructions for making observations in all parts of the world.* London, 1696. Wing W3509. X. Reprinted, London, 1973.
See also 2737/1.

2733. BARCHUSEN, Johann Conrad. 1666-1723. Holl. DSB; P I; BLA (BARKHAUSEN); GA; GC. Chemistry/alchemy.
1. *Pyrosophia.* Leiden, 1698.
1a. ——— Another ed., entitled *Elementa chemiae.* Ib., 1718. X.
2. *La traité symbolique de la pierre philosophale.* See item 2819/1a.

2734. CONNOR (or O'CONNOR), Bernard. 1666-1698. Br. ICB; BLA & Supp.; GA; GB; GD. Anatomy. Chemistry. (Titles in Wing.)

2735. COWPER, William. 1666-1709. Br. ICB; BMNH; Mort.; BLA & Supp.; GA; GD. Anatomy. (Titles in Wing.)

2736. WOTTON, William. 1666-1726. Br. ICB; BMNH; GA; GB; GD. Geology. Best known as author of the following.
1. *Reflections upon ancient and modern learning.* London, 1694. Wing W3658. X. Reprinted, Hildesheim, 1968.
1a. ——— 2nd ed., 1697. Wing W3659. X. 3rd ed., 1705.

2737. ARBUTHNOT, John. 1667-1735. Br. DSB; ICB; P I; BMNH; BLA & Supp.; GA; GB; GD. Mathematics, especially statistics. Various other fields. Also in Part 3: Various Fields, Mathematics, and Physiology*.
1. *An examination of Dr. Woodward's account of the deluge.* London, 1697. Wing A3601. RM.
Other titles in Wing.

2738. BERNOULLI, Johann (or Jean) (first of the name). 1667-1748. Switz. DSB; ICB; P I; BLA & Supp.; GA; GB; GC. Mathematics. Mechanics. Also in Part 3: Mathematics and Mechanics*.
1. *Die erste Integralrechnung.* Leipzig, 1914. (OKEW. 194) Trans. (from the Latin) of an extract from his correspondence with the Marquis de l'Hospital, 1691-92.
2. *Die Differentialrechnung.* Leipzig, 1924. (OKEW. 211) Trans. of a Latin MS in the library of the University of Basel.

3. *Commercium philosophicum et mathematicum G. Leibnitii et Joh. Bernoullii.* 2 vols. Geneva, 1745. Ed. by Bousquet.
4. *Der Briefwechsel.* Basel, 1955. Ed. by the Naturforschende Gesellschaft in Basel.
5. *Opera omnia.* 4 vols. Geneva, 1742. Ed. by G. Cramer. RM. Reprinted, Hildesheim, 1968.

2739. SACCHERI, (Giovanni) Girolamo. 1667-1733. It. DSB; ICB; P II. Mathematics. Also in Part 3: Mathematics*.

2740. WHISTON, William. 1667-1752. Br. DSB; P II; BMNH; GA; GB; GD. Geology (other fields after 1700). Also in Part 3: Science in General*, Astronomy*, Mechanics*, Physics*, and Autobiography*. See also Index.
1. *A new theory of the earth.* London, 1696. Wing W1696. RM/L. 5th ed., 1737.
2. *A vindication of the new theory of the earth.* London, 1698. Wing W1698. X.
3. *A second defence of the new theory of the earth from the exceptions of Mr. John Keill.* London, 1700. Wing W1697. X. cf. 2744/1.

2741. BAGLIVI, Georgio. 1668-1707. It. DSB; ICB; Mort.; BLA & Supp.; GA. Anatomy. Physiology.

2742. TAUVRY, Daniel. 1669-1701. Fr. BLA; GA. Anatomy. Member of the Académie des Sciences.

2743. COLBATCH, John. 1670-1729. Br. ICB; BLA & Supp.; GA; GD. Iatrochemistry. (Titles in Wing.)

2744. KEILL, John. 1671-1721. Br. DSB; P I; GA; GD. Geology (other fields after 1700). Also in Part 3: Mathematics*, Astronomy*, and Mechanics*. See also Index.
1. *An examination of Dr. Burnet's theory of the earth. Together with some remarks on Mr. Whiston's new theory of the earth.* Oxford, 1698. Wing K132. RM.
2. *An examination of the "Reflections"* [by Burnet] *on the theory of the earth. Together with a defence of the remarks on Mr. Whiston's new theory.* London, 1699. Wing K133. X.

2745. CRAMER, Johann Daniel. 1672-1715. Ger. P I. Pneumatics.

2.22 Imprint Sequence

Except for reprints and edited texts, the libraries were not a main bibliographical source for this section, hence the symbol X is not used.

2746. ALMANACS. Still. 16 records a German item of 1461 and gives references to 15th-century almanacs generally.

2747. *DE COMETIS tractatus*. Beromünster, 1472. Klebs 972.1; Still. 44.

2748. JUDICIUM. Under this heading Klebs records three astrological works published between 1472 and 1496.

2749. ALGORITHMUS. Under this heading Klebs and Still. record several works published between ca. 1475 and 1500.

2750. CALENDARIUM. Under this heading Klebs and Still. record several perpetual calendars published between 1475 and 1490.

2751. *LAPIDAIRE*. Lyons, 1477. Klebs 588.

2752. *ARTE dell'abbaco*. Treviso, 1478. Klebs 115; Still. 144. The first printed arithmetic of known date.

2753. COELESTINUS, Claudius. late 15th century. ICB.
   1. *De his quae mundo mirabiliter eveniunt*. Paris, 1542. Written in 1478. Ed. by Oronce Finé. RM.

2754. PROGNOSTICATIONS. Under this heading Klebs records numerous publications issued between 1478 and 1500. See also Still.

2755. *KALENDER deutsch*. Augsburg, 1481. Klebs 569. Many later eds.

2756. PYTHAGORAS. Supposititious works.
   1. *Ludus*. Padua, 1482. Klebs 820.
   2. *Aurea verba*. *Symbola*. Venice, 1497. Klebs 529.1.

2757. *BAMBERGER Rechenbuch*. Bamberg, 1483. Reprinted, Munich, 1966.

2758. COMPUTUS. Under this heading Klebs and Still. record several items published between 1483 and 1500.

2759. *HERBARIUS*. Mainz, 1484. Klebs 506.1; Still. 402. In Latin, with German synonyms.

2760. *GART der Gesundheit*. Mainz, 1485. Klebs 507; Still. 379. A herbal. Sometimes called *Herbarius zu Deutsch*.

2761. HERBAL.
   1. *Arbolayre*. Besançon, ca. 1487. Klebs 508; Still. 269.
   1a. ——— Another ed., entitled *Le grant herbier en françois*. Paris, 1498. Klebs 508.
   1b. ——— *The grete herball*. Southwark, 1526. PR 13176 (and three later eds up to 1561). RM.

2762. *HORTUS sanitatis*.
   1. *Hortus sanitatis*. Mainz, 1491. Klebs 509; Still. 424, 663. RM (under *Ortus sanitatus*). Deals with birds, fish, and stones as well as plants. Profusely illustrated.
   1a. ——— *The noble lyfe and natures of man, of bestes, serpentys, fowles, and fisshes*. Antwerp, [ca. 1521].

1b. ——— ——— Reprint entitled *An early English version of the "Hortus sanitatis."* London, 1954. Ed. by N. Hudson.

2763. *LUNAIRE en francoys, 1492-1523.* Lyons, 1491. Klebs 632.

2764. ALMANAC.
1. *Kalendrier des bergers.* Paris, 1493.
1a. ——— *The kalendayr of the shyppars.* Paris, 1503. PR 22407. The first printed almanac in English. PR record two further English translations--in 1506 and 1508, both published at London--and fourteen eds of them up to 1625.
1b. ——— ——— *The kalender of shepherdes.* London, 1892. The ed. of 1503 in facsimile with a reprint of the trans. of 1506. Ed. with introd. and glossary by H.O. Skinner.

2765. *ASTROLABII canones.* Venice, 1497/8. Klebs 119. Still. 819.

2766. *LIBELLUS de natura animalium. A fifteenth-century bestiary, reproduced in facsimile.* London, 1958. Introd. by J.I. Davis.

2767. ALMANACKS and Kalendars. Prognostications. Under these headings PR record a large number of English items from ca. 1500 to 1640.

2768. *BERGBÜCHLEIN.*
1. *Bergbüchlein.* Augsburg, 1505. Still. 601, 824. Also cited as *Bergwerkbüchlein*; the title varies in later eds. A prospector's manual sometimes attributed to Ulrich Rühlein von Kalbe (1296).
1a. ——— *Bergwerk- und Probierbüchlein. Translated into English.* New York, 1949. By A.G. Sisco and C.S. Smith. cf. entry 2771.

2769. BLASIUS. (An unidentified author.)
1. *Liber arithmetice practice.* Paris, 1513. Reprinted, New York, 1961.

2770. *HERBOLARIO volgare.* Venice, 1522. Still. 403..

2771. *PROBIERBÜCHLEIN.* Magdeburg, 1524. Still. 876. The earliest dated ed.; there are several other eds, some undated. A manual of assaying and metallurgy. For an English trans. see entry 2768/1a.

2772. BANCKES' herbal.
1. *Here begynneth a new mater, the whiche sheweth and treateth vertues and proprytes of herbes.* London: R. Banckes, 1525. Still. 292; ICB. PR 4720 and nine later eds up to 1560. (PR attribute the work to Walter Cary.) "The first book devoted solely to herbs to be printed in England."
1a. ——— *Facsimile of R. Banckes's herbal of 1525.* New York, 1941. Ed. by S.V. Larkey and T. Pyles.

2773. *THE GRETE herball.* 1526. See entry 2761/1b.

2774. GODFRIDUS. (An unidentified author.)
1. *The knowledge of thynges unknowen apperteynynge to astronomye.* London, [1530?]. PR 11931 and three later eds. Wing (G928 onwards) records ten more up to 1697.

2775. CENEAU, Robert.
 1. *De vera mensurarum ponderumque ratione opus.* Paris, 1535. RM.

2776. *HERBARUM imagines vivae. Der Kreuter lebliche Contrafaytung.* Frankfurt, 1535. Reprinted, New York, 197-?

2777. *AN INTRODUCTION for to lerne to recken with the pen or with the counters.* London, 1539. Still. 179. PR 14118 and four later eds with varying titles. The first arithmetic printed in English.

2778. GABRIELE, Giacomo.
 1. *Dialogo ... nel quale de la sphera, et de gli orti et occasi de le stelle, minutamente si ragiona.* Venice, 1545. RM.

2779. MEMMO, Giovanni Maria. d. 1553.
 1. *Tre libri della sostanza et forma del mondo.* Venice, 1545. RM.

2780. *HERBARUM, arborum, fruticum ... imagines ... una cum eorumdem nomenclaturis....* Frankfurt, 1552. In German as well as Latin. RM.

2781. MONTULMO, Antonio di.
 1. *A ryghte excellente treatise of astronomie.* London, 1554. PR 18054. Trans. by F. van Brunswicke.

2782. BOAISTUAU, Pierre. d. 1566.
 1. *Histoires prodigieuses.* Paris, 1560.
 1a. ——— *Certaine secrete wonders of nature.* London, 1569. PR 10787. Trans. by E. Fenton.

2783. *PHARMACOPOEIA seu medicamentarium pro Reipublica Augustana.* Augsburg, 1564. Reprinted 1927.

2784. BERGA, Antonio.
 1. *Discorso della grandezza dell'acqua e della terra. Contra l'opinione del S. Alessandro Piccolomini.* Turin, 1579. RM. cf. item 1587/3.

2785. *DISCOURS des causes et effects admirables des tremblemens de terre. Par V.A.D.L.C.* Paris, 1580. RM (under C., V.A.D.L.)

2786. SFONDRATI. Pandolfo.
 1. *Causa aestus maris.* Ferraria, 1590. RM.

2787. VALENTINE, Basil. DSB; ICB; P I (BASILIUS V.); BLA & Supp. (BASILIUS V.); GA (BASIL V.). A fictitious author, supposedly of the early fifteenth century. The alchemical and iatrochemical works extant under this name were probably written about 1600 by Johann Thölde.
 1. *Zwölff Schlüssel.* 1599.
 1a. ——— *Les douze clefs de la philosophie.* Paris, 1956; re-issued 1968.
 2. *Von dem natürlichen und übernatürlichen Dingen.* Leipzig, 1603.
 2a. ——— *Of natural and supernatural things.* London, 1670. Wing B1019 and another ed.
 3. *Triumph Wagen Antimonii.* Leipzig, 1604. RM (under BASILIUS V.).

3a. ——— *The triumphant chariot of antimony.* London, 1660. Wing B1021 (and another trans. of 1678). RM (under BASILIUS V.).
3b. ——— *The triumphal chariot of antimony.* London, 1893. Reprinted, ib., 1962.
4. *Letztes Testament.* Jena, 1626.
4a. ——— Wing (B1015 onwards) records several English translations in the period 1657-71.
5. *Chymische Schriften.* 2 vols. Hamburg, 1700. RM.

2788. TAVERNER, John.
1. *Certaine experiments concerning fish and fruite.* London, 1600. PR 23708. Reprinted, New York, 1968.

2789. NAUTONIER, Guillaume.
1. *Mecometrie de leymant; cest a dire, La maniere de mesurer les longitudes par le moyen de l'eymant.* Venes, 1603-04.
1a. ——— *The mecographie of ye loadstone.* Toulouse, 1603. PR 18415.

2790. FLAMEL, Nicolas. (Main entry: 714) Supposititious works.
1. *Le livre des figures hiéroglyphiques.* Paris, 1612. Reprinted with two other alchemical works supposedly by Flamel, Paris, 1970. Introd. by E. Canseliet.

2791. GRAZIA, Vincenzio di.
1. *Considerazioni sopra'l "Discorso" di Galileo Galilei intorno alle cose che stanno in su l'acqua e che in quella si muovono.* Florence, 1613. RM. cf. item 1879/1.

2792. NOLLE, Heinrich.
1. *Theoria philosophiae hermeticae.* Hanau, 1617. RM.

2793. ROYAL College of Physicians. London.
1. *Pharmacopoea Londinensis.* London, 1618. Reprinted 1944. Wing (R2111 onwards) records twelve eds up to 1696. See also entry 2821.

2794. SCHONEVELDT, Stephan von. BMNH.
1. *Ichthyologia et nomenclaturae animalium marinorum, fluviatilium, lacustrium.* Hamburg, 1624. RM.

2795. GUIDIUS, Joannes (the elder).
1. *De mineralibus tractatus.* Venice, 1625. RM.

2796. *MUSAEUM hermeticum.*
1. *Musaeum hermeticum.* Frankfurt, 1625.
1a. ——— *Musaeum hermeticum reformatum et amplificatum.* Frankfurt, 1678.
1b. ——— ——— *The hermetic museum, restored and enlarged.* 2 vols. London, 1893. Reprinted, Graz, 1970.

2797. *ANIMALIUM quadrupedum, avium, delineationes. A book of beasts, birds, etc.* London, [1628?]. PR 653 and another ed.

2798. *A STOREHOUSE of physicall and philosophicall secrets, teaching to distill all manner of oyles from gummes ... hearbs, and mineralls, etc.* 2 parts. London, 1633. PR 23293.

2799. *PHARMACOPOEA Amstelredamensis.* Amsterdam, 1636. Reprinted, Nieuwkoop, 1961. (Dutch classics in history of science. 1)

2800. *CATALOGUS plantarum horti academici Lugduno Batavi*. 1636. Reprinted, Zug (Switz.), 1966. Re the botanic garden of Leiden University.

2801. REMNANT, Richard.
 1. *A discourse or historie of bees*. London, 1637. PR 20879.

2802. *THE PRINCIPLES of astronomy*. London, 1640. PR 20396.

2803. ALMANACS. Under this heading Wing lists a large number of English items dating from 1641 to 1700.

2804. RESTA, Franciscus.
 1. *Meteorologia de igneis, aereis, aqueisque corporibus*. Rome, 1644. RM.

2805. PHAEDRO, George.
 1. *Physicall and chymicall works*. London, 1654. Wing P1955.
 2. *The art of chymistry*. 3rd ed. London, 1674. Wing P1954.

2806. GABRIEL DE DOULLENS, ...
 1. *Tabulae Ambianenses; seu, Theoriae planetarum, tam in forma Tychonica quam Copernicana, per unicam cujusque eclipsim ex proprio centro descriptam, plano-geometrico delineatio*. Paris, 1658. RM.

2807. *PHARMACOPOEIA Belgica; or, The Dutch dispensatory*. London, 1659. Wing P1971.

2808. SALUSBURY, Thomas. ICB. "The Thomas Salusbury who did so much to present Galileo in English dress has never been identified. He certainly died before the second volume of his translations was published, perhaps of the plague in 1665." S. Drake.
 1. *Mathematical collections and translations*. Vol. 1. London, 1661. Wing S517. RM.
 1a. ——— Vol.2. Ib., 1665. Very rare: most copies were destroyed in the great fire of London.
 1b. ——— Vols 1 and 2. In facsimile, with an analytical and bio-bibliographical introduction. 2 vols, London, 1967. By S. Drake.
 Contains English translations of Galileo's chief works, the hydrological treatises of Benedetto Castelli, and several works of other authors, including Kepler, Descartes, and Tartaglia.

2809. BREMBATO, Ottavio.
 1. *La mineralogia*. Bergamo, 1663. RM.

2810. LEVERA, Francesco.
 1. *Prodromus universae astronomiae*. Rome, 1663. RM.
 2. *De inerrantium stellarum viribus*. Rome, 1664. RM.

2811. ROYAL Society of London.
 1. *List of the Fellows*. London, 1663. Wing L2423 and seventeen further eds up to 1694.
 2. *A list of the Royal Society*. London, 1667. Wing L2499 amd five further eds up to 1699.

2812. ACCADEMIA del Cimento. Florence.
 1. *Saggi di naturali esperienze*. Florence, 1666. RM. Reprinted (under the name of Lorenzo Magalotti, the secretary of the Academy), Milan, 1976.

1a. ——— *Essayes of natural experiments.* London, 1684. Trans. by Richard Waller. Wing A161. RM. Reprinted, New York, 1964.
1b. ——— *Tentamina experimentorum naturalium.* 2 vols. Leiden, 1731. RM. Trans. by P. van Musschenbroek who added some writings of his own; see 5818/11.

2813. *OBSERVATIONES anatomici collegii privati Amstelodamensis. Pars prior (1667) et altera (1673).* Facsimile, Nieuwkoop, 1975. (Dutch classics in history of science. 19)

2814. THIBAUT, Pierre (*called* le Lorrain). ICB.
1. *Cours de chymie.* Paris, 1667.
1a. ——— *The art of chymistry.* London, 1668. Wing T891 and another ed.

2815. *PROPOSITIONS for the carrying on a philosophical correspondence.* London, 1670. Wing P3786.

2816. THRUSTON, Malachi.
1. *De respirationis usu primario diatriba.* London, 1670. Wing T1132. Another ed. was published in Leiden in 1671.

2817. ACADÉMIE des Sciences. Paris.
1. *Mémoires pour servir à l'histoire naturelle des animaux.* Paris, 1671. Sometimes attributed to Claude Perrault, a leading figure in the Academy at the time.
1a. ——— *Memoirs for a natural history of animals ... To which is added an account of the measure of a degree of a great circle of the earth.* London, 1688. wing M1667. Trans. by R. Waller.
1b. ——— ——— Another ed. Ib., 1702. Published by the Royal Society. RM.
1c. ——— *Der Herren Perrault, Charras und Dodarts Abhandlungen zur Naturgeschichte der Thiere und Pflanzen.* 3 vols. Leipzig, 1757-58.

2818. APPLEFORD (or APLEFORD), Robert.
1. *Mechanica rerum explicatio non ducit ad atheismum.* Cambridge, 1676. Wing A3581.

2819. *MUTUS liber.*
1. *Mutus liber, in quo tamen tota philosophia hermetica figuris hieroglyphicis depingitur.* La Rochelle, 1677.
1a. ——— *Trésor hermétique, comprenant* [1] *Le livre d'images sans paroles (Mutus liber) où toutes les opérations de la philosophie hermétique sont décrites ... et* [2] *La traité symbolique de la pierre philosophale en 78 figures, par J.C. Barchusen.* Lyons, 1947.
1b. ——— *L'alchimie et son Livre muet (Mutus liber).* Paris, 1967. Introd. and comm. by E. Canseliet.

2820. *A GENERAL index or alphabetical table to all the "Philosophical transactions."* London, 1678. Wing G500.

2821. ROYAL College of Physicians. London.
1. *Pharmacopoeia.* London, 1678. Wing R2106 and four further eds up to 1699. See also 2793.

2822. BERTRAND, ... (A physician of Marseilles.)
1. *Réflexions nouvelles sur l'acide et sur l'alcalie.* Lyons, 1683. RM.

2823. *ASTRONOMY'S advancement of news for the curious: Being a treatise of telescopes and an account of the marvellous astronomical discoveries of late years made ... Done out of French.* London, 1684. Wing A4084. Trans. by Joseph Walker.

2824. *COLLECTANEA chymica: A collection of ten several treatises in chymistry.* London, 1684. Wing C5103. cf. entry 2831.

2825. WARREN, Erasmus. BMNH.
1. *Geologia; or, A discourse concerning the earth before the deluge. Wherein the form and properties ascribed to it, in a book* [by Thomas Burnet] *intituled "The theory of the earth" are excepted against.* London, 1690. Wing W966. RM. cf. items 2444/3,4

2826. *A CATALOGUE of books of two eminent mathematicians ... 21 May, 1691.* London, [1691]. Wing C1287.

2827. *EPISTOLA ad Regiam Societatem Londinensem.* London, 1693. Wing E3169.

2828. PARKER, Gustavus.
1. *An account of a portable barometer, with reasons and rules for the use of it.* London, 1699. Wing P391.
2. *A new account of the alterations of the wind and weather by the discoveries of a portable barometer.* London, 1700. Wing P392.

2.23 Collections of Texts, etc.

2829. *THE CONTROVERSY on the comets of 1618: G. Galilei, H. Grassi, M. Guiducci, J.Kepler.* Philadelphia, 1960. Trans. by S. Drake and C.D. O'Malley.

2830. *MECHANICS in sixteenth-century Italy: Selections from Tartaglia, Benedetti, Guido Ubaldo, and Galileo.* Madison, Wis., 1969. Trans. with notes by S. Drake and I.E. Drabkin.

2831. *COLLECTANEA chemica: Being certain select treatises on alchemy and hermetic medicine.* London, 1893. Includes treatises by Eirenaeus Philalethes (pseud.), Francis Anthony, George Starkey, George Ripley, and an anonymous unknown. Reprinted, London, 1963. cf. entry 2824.

2832. CORRESPONDENCE.
1. *A collection of letters illustrative of the progress of science in England from the reign of Queen Elizabeth to that of Charles the Second.* London, 1841. Ed. by J.O. Halliwell-Phillips. Reprinted, London, 1965.
2. *Correspondence of scientific men of the seventeenth century, including letters of Barrow, Flamsteed, Wallis, and Newton.* 2 vols in 3. Oxford, 1841-42. Ed. by S.P. Rigaud. Reprinted, 2 vols, Hildesheim, 1965.

2833. *RARA arithmetica. A catalogue of the arithmetics printed before*

MDCI. Boston, 1908. By D.E. Smith. *Addenda*. Ib., 1939.

2834. *SALES catalogues of libraries of eminent persons. Vol. 11: Scientists*. London, 1975. Ed. by A.N.L. Munby. Facsimiles of the catalogues of E. Ashmole, R. Hooke, J. Ray, and E. Halley.

PART 3

EIGHTEENTH AND NINETEENTH CENTURIES

3.01 SCIENCE IN GENERAL

Including natural philosophy and philosophy of science.

The term 'physics' (or its equivalent in other languages) occurs in a number of the eighteenth-century titles and it should be remembered that the term then covered all the natural sciences.

2835. TSCHIRNHAUS, Ehrenfried Walter von. 1651-1708. Ger. DSB; ICB; P II; GA; GC. Also Mathematics and Part 2*.
    1. *Gründliche Anleitung zu nützlichen Wissenschaften, absonderlich zu der Mathesis und Physica.* Frankfurt, 1708. X. Reprint of 4th ed. (1729), Stuttgart, 1967.

2836. NIEUWENTIJT, Bernard. 1654-1718. Holl. DSB; P II; BLA; GA. Also Mathematics and Part 2.
    1. *Het regt gebruik der wereldbeschouwingen.* Amsterdam, 1715. X.
    1a. ——— *The religious philosopher; or, The right use of contemplating the work of the Creator ... throughout which all the late discoveries in anatomy, philosophy and astronomy ... are most copiously handled.* 3 vols. London, 1718. Trans. by J. Chamberlayne. X. 3rd ed., 1724.

2837. DERHAM, William. 1657-1735. Br. DSB; ICB; P I; BMNH; GA; GB; GD. Also in Part 2. See also Index.
    1. *Physico-theology; or, A demonstration of the being and attributes of God, from his works of creation.* London, 1713. RM. 2nd ed., 1714. 3rd ed., 1714. Another ed., 1716; reprinted, Hildesheim, 1971. Another ed., 1720. 7th ed., 1727. 11th ed., 1749. Another ed., 1798.
    2. *Astro-theology; or, A demonstration of the being and attributes of God, from a survey of the heavens.* London, 1715. RM/L. 4th ed., 1721. 6th ed., 1741. 7th ed., Glasgow, 1755.

2838. FONTENELLE, Bernard Le Bovier de. 1657-1757. Fr. DSB; ICB; P I; GA; GB. Also Mathematics*, Mechanics*, Biography*, History of Science*, and Part 2*.
    1. *Oeuvres diverses.* 3 vols. Amsterdam, 1701. X. Other eds, ib., 1716 and 1743.
    2. *Oeuvres.* 5 vols. Paris, 1825. Ed. with a biography by J.B.J. Champagnac. RM.
    3. *The achievement of ... Fontenelle.* New York, 1970. Selections. Trans. with an introd. by L.M. Marsak.

2839. POLIGNAC, Melchior de. 1661-1741. Fr. GA.
    1. *Anti-Lucretius; sive, De deo et natura.* Paris, 1747. X. Another ed., Venice, 1749.

2840. MATHER, Cotton. 1663-1728. U.S.A. ICB; GA; GB; GD (under M., Increase); GE.
1. *The Christian philosopher: A collection of the best discoveries in nature, with religious improvements.* London, 1721. X. Reprinted, Gainesville, Florida, 1968.

2841. HARRIS, John. ca. 1666-1719. Br. DSB; ICB; P I; GA; GB; GD. Also Mathematics*, Astronomy*, and Natural History*.
1. *Lexicon technicum; or, An universal English dictionary of arts and sciences.* 2 vols. London, 1704-10. RM/L. Reprinted, New York, 1966. 2nd ed., 1708-10. 4th ed., 1725.

2842. WHISTON, William. 1667-1752. Br. DSB; P II; BMNH; GA; GB; GD. Also Astronomy*, Mechanics*, Physics*, Autobiography*, and Part 2*. See also Index.
1. *Astronomical principles of religion, natural and reveal'd.* London, 1717. RM.

2843. SOUCIET, Etienne. 1671-1744. Fr. P II.
1. *Observations mathématiques, astronomiques, géographiques, chronologiques et physiques, tirées des anciens livres chinois.* 3 vols. Paris, 1729-32. RM.

2844. CLARKE, Samuel. 1675-1729. Br. DSB; ICB; P I; GA; GB; GD. Also Mechanics. See also Index. For his correspondence with Leibniz see 2583/7-9.

2845. GREENE, Robert. 1678?-1730. Br. GD. Also Mathematics*.
1. *The principles of natural philosophy. In which is shewn the insufficiency of the present systems to give us any just account of that science.* Cambridge, 1712. RM.
2. *The principles of the philosophy of the expansive and contractive forces; or, An enquiry into the principles of the modern philosophy.* Cambridge, 1727. RM.

2846. WOLFF, Christian. 1679-1754. Ger. The philosopher. DSB; ICB; P II (WOLF); BMNH; G. Also Mathematics*.
1. *Cosmologia generalis.* Frankfurt/Leipzig, 1731. X. Another ed., 1737. Another ed., by J. Ecole, reprinted, Hildesheim, 1964.
2. *Briefe aus den Jahren 1719-1753. Ein Beitrag zur Geschichte der K. Akademie der Wissenschaft zu St. Petersburg.* St. Petersburg, 1860. X. Reprinted, Hildesheim, 1971.

2847. QUINCY, John. d. 1722. Br. ICB; BLA; GA; GD. Also Chemistry*.
1. *Lexicon physico-medicum; or, A new physical dictionary explaining the difficult terms used in the several branches of the profession, and in such parts of philosophy as are introductory thereunto.* London, 1719. RM. 2nd ed., 1722. 9th ed., 1775.

2848. CHAMBERS, Ephraim. ca. 1680-1740. Br. P I; GA; GB; GD.
1. *Cyclopaedia; or, An universal dictionary of arts and sciences.* London, 1728. RM.
See also 2998/1 and 6986/2a.

2849. REGNAULT, Noël. 1683-1762. Fr. P II; GA.
1. *Les entretiens physiques d'Ariste et d'Eudoxe; ou, Physique nouvelle en dialogues, qui renferme précisément ce qui*

s'est découvert de plus curieux et de plus utile dans la nature. 3 vols. Paris, 1729. X. 7th ed., 5 vols, 1745-50. RM.

1a. ——— *Philosophical conversations; or, A new system of physics by way of dialogue.* 3 vols. London, 1731. Trans. with notes by T Dale. RM.

2. *L'origine ancienne de la physique nouvelle, où l'on voit dans les entretiens par lettres ce que la physique nouvelle a de commun avec l'ancienne.* Paris, 1734. RM.

2850. BERKELEY, George. 1685-1753. Br. The philosopher. DSB; ICB; G. Philosophy of science.

2851. SWEDENBORG, Emanuel. 1688-1772. Swed. DSB; ICB; P II; BMNH (SVEDENBORG); GA; GB. Also Various Fields*.

1. *Opera philosophica et mineralogica.* 3 parts. Dresden, 1734. RM.

1a. ——— Part 1. *Principia rerum naturalium.*

1b. ——— ——— *The principia; or, The first principles of natural things.* 2 vols. London, 1845-46. Trans. by A. Clissold. Reprinted, Bryn Athyn, Pennsylvania, 1976.

1c. ——— ——— *The principia; or, The first principles of natural things. To which are added The minor principia and Summary of the principia.* 2 vols in 1. London, Swedenborg Society, 1912. RM/L.

2. *Miscellaneous observations connected with the physical sciences.* London, 1847. Trans. from the Latin by C.E. Strutt.

2852. PIVATI, Giovanni Francesco. 1689-1764. It. P II. Encyclopedist.

2853. BOUGEANT, Guillaume Hyacinthe. 1690-1743. Fr. P I. GA. Also Zoology*.

1. *Observations curieuses sur toutes les parties de la physique.* 3 vols. Paris, 1719-26-30. (The work is anonymous and Vols 2 and 3 are sometimes attributed to Nicolas Grozelier.) RM (Vol. 1)/L.

2853A BÖRNER, Nicolaus. 1693-ca. 1770. Ger. P I; BLA & Supp.; GA. Natural philosophy.

2854. CARPOV, Jacob. 1699-1768. Ger. P I. Natural philosophy.

2855. GOTTSCHED, Johann Christoph. 1700-1766. Ger. ICB; P I; GA; GB; GC.

1. *Erste Gründe der gesamten Weltweisheit. Darinn alle philosophische Wissenschaften in ihrer natürlichen Verknüpfung abgehandelt werden.* Leipzig, 1733-34. X. Reprinted, Frankfurt, 1965.

2856. WESLEY, John. 1703-1791. Br. The famous evangelist. ICB; G.

1. *A survey of the wisdom of God in the Creation; or, A compendium of natural philosophy.* 2 vols. Bristol, 1763. X. 4th ed., 5 vols, London, 1784.

1a. ——— Another ed.: *A compendium of natural philosophy, being a survey....* 3 vols. London, 1840.

2857. EULER, Leonhard. 1707-1783. Ger./Russ. DSB; ICB. P I; GA; GB;

GC. Also Mathematics*#, Astronomy*, Mechanics*, Physics*, and Cartography*. See also Index.

1. *Lettres concernant le jugement de l'Académie.* Berlin, 1752. X. Another ed., Paris, 1753. X. Reprinted, 1889. Re the König controversy in the Berlin Academy.
2. *Lettres à une princesse d'Allemagne sur divers sujets de physique et de philosophie.* 3 vols. St. Petersburg, 1768-72. RM. Another ed., Geneva, 1775. Another ed., Paris, 1812.

2a. ——— *Briefe über verschiedene Gegenstände aus der Naturlehre.* Leipzig, 1792-94. Trans. by F. Kreis. RM.

2b. ——— *Letters to a German princess on different subjects in physics and philosophy.* 2 vols. London, 1795. Trans. by H. Hunter. 2nd ed., 1802; reprinted, New York, 1966.

2858. LA METTRIE, Julien Offray de. 1709-1751. Fr. DSB; ICB; Mort.; BLA: GA; GB; GC. Also Physiology.

1. *L'homme machine.* Leiden, 1748. RM. Reprinted, Paris, 1921 and 1966. Critical ed. by A. Vartanian, Princeton, 1960.

1a. ——— *Man a machine.* Chicago 1927. Trans. with notes by G.C. Bussey.

2. *L'homme-plante.* Potsdam, 1748. RM. Critical ed. by F. Rougier, New York, 1936.
3. *Oeuvres philosophiques.* 2 vols. Amsterdam, 1753. Reprinted, Hildesheim, 1970.

2859. BERRYAT, Jean. d. 1754. Fr. P I; BLA. Initiator and first editor of *Collection académique* (see 2999).

2860. DELLA TORRE, Giovanni Maria. 1713-1782. It. DSB. Under TORRE: P II; BMNH; GA. Encyclopedist.

2861. DIDEROT, Denis. 1713-1784. Fr. The famous philosophe. DSB; ICB; P I; G.

The *Encyclopédie* and numerous other works are not included.

1. *Mémoires sur différens sujets de mathématique.* Paris, 1748. RM. Deals with musical acoustics, mathematical instruments, etc.

Posthumous Works (Modern Editions)

2. *Supplément au voyage de Bougainville.* Geneva/Lille, 1955. Ed. by H. Dieckmann.
3. *Le rêve de d'Alembert. Texte intégral (d'après la copie inédite de Léningrad).* Paris, 1962. Ed. by J. Varloot.
4. *Éléments de physiologie. Édition critique.* Paris, 1964. Ed. by J. Mayer.

Correspondence and Collected Works

5. *Correspondance.* 12 vols. Paris, 1955-65. Ed. by G. Roth.
6. *Oeuvres complètes.* 20 vols. Paris, 1875-77. Ed. by J. Assézat and M. Tourneux.

Selections

7. *Early philosophical works.* Chicago, 1916. Trans. and ed. by M. Jourdain.
8. *Diderot, interpreter of nature: Selected writings.* London,

1937. Trans. and ed. by J. Stewart and J. Kemp. X. Reprinted, New York, 1963.
9. *Oeuvres.* Paris, 1951. Ed. by A. Billy.
10. *Selected philosophical writings.* Cambridge, 1953. Ed. by J. Lough.
11. *Oeuvres philosophiques.* Paris, 1956. Ed. by P. Vernière.

See also 2866/1a,b.

2862. CONDILLAC, Etienne Bonnot de. 1714-1780. Fr. The philosopher. DSB; ICB; G.
1. *Traité des animaux, où après avoir fait des observations critiques sur le sentiment de Descartes et sur celui de M. de Buffon, on entreprend d'expliquer leurs principales facultés.* Amsterdam, 1755. RM.
2. *La langue des calculs.* Paris, [1798?]. Posthumous.

2863. PLOQUET, Gottfried. 1716 (or 1706)-1790. Ger P II; GA; GC. Natural philosophy.

2864. AMIOT (or AMYOT), Joseph Marie. 1718-1794. Jesuit missionary in China. P I; GA; GB; GF. Descriptions of Chinese science.

2865. POWNALL, Thomas. 1722-1805. Br. P II; GA; GB; GD; GE.
1. *Intellectual physicks: An essay concerning the nature of being and progression of existence.* Bath, 1795. RM.

2866. HOLBACH, Paul Henri Thiry, *Baron* d'. 1723-1789. Fr. DSB; ICB; P I; GA; GB; GC. See also Index.
1. *Système de la nature; ou, Des lois du monde physique et du monde moral. Par M. Mirabaud* [pseud.]. 2 vols. London [really Amsterdam], 1770.

1a. ——— Another ed. *Avec des notes et des corrections par Diderot.* Paris, 1821. X. Reprinted, Hildesheim, 1966.

1b. ——— ——— *The system of nature; or, Laws of the moral and physical world ... with notes by Diderot.* Boston, 1889. Trans. by H.D. Robinson. X. Reprinted, Ann Arbor, 1959.

2867. KANT, Immanuel. 1724-1804. Ger. The famous philosopher. DSB; ICB; P I; G. Also Astronomy*.
1. *Kant's inaugural dissertation of 1770.* [*De mundi sensibilis*]. New York, 1894. Trans. with an introd. by W.J. Eckoff. X. (Deals with the concepts of space and time.) Reprinted, Indianapolis, 1970.
2. *Die metaphysischen Anfangsgründe der Naturwissenschaft.* 1786. X.

2a. ——— *Metaphysical foundations of natural science.* Indianapolis, 1970. Trans. with an introd. by J. Ellington.

3. *Der Organismus.* [*Kritik der Urtheilskraft.* Part 2]. Stuttgart, 1923. Ed. with an introd. by V. Weizsäcker.

2868. KIPPIS, Andrew. 1725-1795. Br. GA; GB; GD.
1. *Observations on the late contests in the Royal Society.* London, 1784.

See also 3106/1.

2869. HUTTON, James. 1726-1797. Br. DSB; ICB; P I; BMNH; GA; GB; GD. Also Physics* and Geology*.

1. *An investigation of the principles of knowledge and of the progress of reason from sense to science and philosophy.* 3 vols. Edinburgh, 1794. RM/L.

2870. JONES, William (second of the name). 1726-1800. Br. BMNH; GA; GB; GD.
1. *An essay on the first principles of natural philosophy. Wherein the use of natural means, or second causes, in the oeconomy of the material world is demonstrated.* Oxford, 1762. RM.
2. *Physiological disquisitions; or, Discourses on the natural philosophy of the elements.* London, 1781. RM/L.

2871. CROKER, Temple Henry. 1730?-1790? Br. GD. Also Physics*.
1. *The complete dictionary of arts and sciences.* London, 1764-66. With Thomas Williams, Samuel Clarke, et al. RM.

2872. GOLDSMITH, Oliver. 1730 (or 1728)-1774. Br. The well known writer. ICB; BMNH; G. Also Natural History*.
1. *A survey of experimental philosophy, considered in its present state of improvement.* 2 vols. London, 1776. RM.

2873. HORNE, George. 1730-1792. Br. GB; GD.
1. *A fair, candid, and impartial state of the case between Sir Isaac Newton and Mr Hutchinson.* Oxford, 1753. (Re the theologian, John Hutchinson, 1674-1737.) RM. Another ed., [Dublin ?], 1756.

2874. MÜLLER, Johann Traugott. 1730-1794. Ger. P II. Bibliography.

2875. SCOTT, James. 1733-1814. Br. GD.
1. *A general dictionary of arts and sciences.* London, 1765-66. RM.

2876. ROZIER, François. 1734-1793. Fr. ICB; P II; BMNH; GA. Editor from 1771 to 1793 of the important journal *Observations sur la physique et l'histoire naturelle.*

2877. BERNARDIN DE SAINT-PIERRE, Jacques Henri. 1737-1814. Fr. ICB; BMNH. Under SAINT-PIERRE: GA, GB.
1. *Etudes sur la nature.* 3 vols. Paris, 1784. X.
1a. ——— *St. Pierre's studies of nature.* 5 vols. London, 1796. Trans. by Henry Hunter. Another ed., Philadelphia, 184-?

2878. BUGGE, Thomas. 1740-1815. Den. P I; GA. Also Astronomy.
1. *Reise til Paris i aarne 1798 og 1799.* Copenhagen, 1799-1800. X.
1a. ——— *Science in France in the revolutionary era.* Cambridge, Mass., 1969. Ed. with introd. and comm. by M.P. Crosland.

2879. HOLLAND, Georg Jonathan von. 1742-1784. Ger. P I; GA.
1. *Réflexions philosophiques sur le "Système de la nature."* Neufchâtel, 1772. (Presumably refers to the work by Holbach, item 2866/1.) X. Another ed., London, 1773.

2880. SENEBIER, Jean. 1742-1809. Switz. DSB; ICB; P II; BMNH; BLA; GA; GB. Also Chemistry* and Plant Physiology*. See also Index.
1. *L'art d'observer.* 2 vols. Geneva, 1775. X.
1a. ——— 2nd ed., entitled *Essai sur l'art d'observer et de faire des expériences.* 3 vols. Ib., 1802. RM/L.

2881. CONDORCET, Marie Jean Antoine Nicolas Caritat, *Marquis* de. 1743-1794. Fr. DSB; ICB; P I,III; GA; GB. Also Mathematics* and Biography*. See also Index.
   1. *Oeuvres*. 12 vols. Paris, 1847-49. Ed. by Mme Condorcet-O'Connor and F. Arago. Does not contain his mathematical writings. X. Reprinted, Stuttgart, 1968.
   2. *Mathématique et société*. Paris, 1974. Selected texts. Ed. with comm. by R. Rashed.
   3. *Selected writings*. Indianapolis, 1976. Ed. with an introd. by K.M. Baker.

2882. LAMÉTHERIE, Jean Claude de. 1743-1817. Fr. DSB; P I; BMNH; BLA (MÉTHERIE); GA. Also Chemistry*, Geology*, and Biology in General*. See also Index.
   1. *Principes de la philosophie naturelle dans lesquelles on cherche à déterminer les degrés de certitude ou de probabilité des connoissances humaines*. Geneva, 1787. RM.

2883. PALEY, William. 1743-1805. Br. DSB; GA; GB; GD.
   1. *Natural theology; or, Evidences of the existence and attributes of the Deity, collected from the appearances of nature*. London, 1802. X. Reprinted, Farnborough, 1970. Other eds: London, 1817; Edinburgh, 1822.
   See also 2904/2 and 15570/1.

2884. REES, Abraham. 1743-1825. Br. P II; BMNH; GA; GD. Encyclopedist.

2885. LAMARCK, Jean Baptiste Pierre Antoine de Monet, *Chevalier* de. 1744-1828. Fr. DSB; ICB; P I; BMNH; GA; GB. Other entries: see 3204.
   1. *Système analytique des connaissances positives de l'homme, restreintes à celles qui proviennent directement ou indirectement de l'observation*. Paris, 1820. X. Another ed., 1830. RM.

2886. REUSS, Jeremais David. 1750-1837. Ger. P II; BLA; GC.
   1. *Repertorium commentationum a societatibus litterariis editarum*. 16 vols. Göttingen, 1801-21. A bibliography, classified by subject and with author indexes, of the papers in the periodical publications of learned societies and academies from their commencement in the seventeenth century until 1800. X. Reprinted, New York, 1961.

2887. GEHLER, Johann Samuel Traugott. 1751-1795. Ger. P I; GA; GC.
   1. *Physikalisches Wörterbuch*. 4 vols. Leipzig, 1787-91. X.
   1a. ——— Another ed. *Neu bearbeitet von Brandes, Gmelin, Horner, Muncke, Pfaff*. 11 vols in 22 & 1 vol. of plates. Leipzig, 1825-45. RM/L.

2888. NICHOLSON, William. 1753-1815. Br. DSB; ICB; P II; BLA; GA; GB; GD. Also Physics* and Chemistry*.
   1. *The British encyclopaedia; or, Dictionary of the arts and sciences*. 6 vols. London, 1809.

2889. TURNER, Richard. 1753-1788. Br. GD.
   1. *An easy introduction to the arts and sciences*. London, 1783. X. Another ed., 1804.

2890. GREGORY, George. 1754-1808. Br. P I; GA; GD.

1. *The economy of nature explained and illustrated on the principles of modern philosophy.* 3 vols. London, 1796. RM/L. 2nd ed., with additions, 3 vols, 1798.
2. *A dictionary of arts and sciences.* 2 vols. London, 1806-07.
3. *Lectures on experimental philosophy, astronomy and chemistry.* 2 vols. London, 1808.

2891. STASZIC, Stanislaw Wawrzyniec. 1755-1826. Poland. DSB Supp.; ICB. Also Geology.
1. [In Polish. *On science, its significance and organisation: Selected letters.*] Cracow, 1952. Ed. by B. Suchodolski.

2892. TILLOCH, Alexander. 1759-1825. Br. DSB; P II; GD. Founder and editor of the *Philosophical magazine.*

2893. GIRTANNER, Christoph. 1760-1800. Ger. DSB; ICB; P I; BLA & Supp.; GA; GC. Also Chemistry*.
1. *Über das Kantische Prinzip für die Naturgeschichte. Ein Versuch diese Wissenschaft philosophisch zu behandeln.* Göttingen, 1796. X. Reprinted, Brussels, 1968.

2894. BARROW, John. 1764-1848. Br. P I; BMNH; GA; GB; GD.
1. *Sketches of the Royal Society and Royal Society Club.* London, 1849. Reprinted, London, 1971.

2895. GOOD, John Mason. 1764-1827. Br. P I; BLA & Supp.; GA; GB; GD.
1. *The book of nature.* 3 vols. London, 1826. X. 2nd ed., 1828.

2896. ERSCH, Johann Samuel. 1766-1828. Ger. "The founder of German bibliography." BMNH; GA; GB; GC.
1. *Allgemeine Encyclopädie der Wissenschaften und Künste, in alphabetischer Folge von genannten Schriftstellern.* Leipzig, 1818-89. (Section 1: 99 vols in 51, 1818-82. Section 2: 43 vols in 23, 1827-88. Section 3: 25 vols in 13, 1827-50.) With J.G. Gruber.

2897. CUVIER, Georges. 1769-1832. Fr. DSB; ICB; P I; BMNH; GA; GB. Also Geology*, Zoology*, Biography*, and History of Science*. See also Index.
1. *Compte rendu des travaux de la Classe des sciences mathématiques et physiques de l'Institut national,* [1803-04]. Paris, n.d. By Cuvier (*Partie physique*) and Delambre (*Partie mathématique*).
2. *Rapport historique sur les progrès des sciences naturelles depuis 1789, et sur leur état actuel. Presenté ... par la Classe des sciences physiques et mathématiques de l'Institut.* Paris, 1810. Ed. by Cuvier. RM. Reprinted, Brussels, 1969.
3. *Histoire des progrès des sciences naturelles depuis 1789 jusqu'à ce jour* [i.e. to 1831]. Paris, 1826-36. *(Oeuvres complètes de Buffon. Complément.)* Another ed., 2 vols, Brussels, 1837-38.
4. *Histoire des sciences naturelles depuis leur origine....* See 15666/2.

2898. AMPÈRE, André Marie. 1775-1836. Fr. DSB; ICB; P I; GA; GB. Also Mathematics*, Physics*#, and Personal Writings. See also Index.
1. *Essai sur la philosophie des sciences; ou, Exposition anal-*

ytique d'une classification naturelle de toutes les connaissances humaines. Paris, 1834. X. Reprinted, Brussels, 1966. Another ed., 1838-43. RM. Another ed., 1843-52; reprinted, Frankfurt, 1972.

2899. KIDD, John. 1775-1851. Br. DSB; P I; BMNH; GB; GD. Also Various Fields and Mineralogy*.
1. *On the adaptation of external nature to the physical condition of man.* London, 1833. (Bridgewater treatises. 2) X. Other eds: 1836 and 1838. 6th ed., 1852.

2900. SCHELLING, Friedrich Wilhelm Joseph von. 1775-1854. Ger. The philosopher. DSB; ICB; G.
1. *Erster Entwurf eines Systems der Naturphilosophie.* Jena, 1799.
2. *Einleitung zu seinem "Entwurf eines Systems der Naturphilosophie"; oder, Über den Begriff der speculativen Physik.* Jena, 1799.
See also 2960/1.

2901. OERSTED, Hans Christian. 1777-1851. Den. DSB; ICB; P II; BLA; GA. Also Physics*#.
1. *Forste indledning til den almindelige naturlaere.* Copenhagen, 1811. RM.
2. *Naturlaerens, mechaniske deel.* Copenhagen, 1844. RM.
2a. ——— *Der mechanische Theil der Naturlehre.* Brunswick, 1851. RM.
3. *Aanden i naturen.* Copenhagen, 1850. RM.
3a. ——— *Der Geist in der Natur.* Munich, 1850-51. RM.
3b. ——— ——— *The soul in nature.* London, 1852. Trans. from the German by L. and J.B. Horner. RM/L. Reprinted, London, 1966.

2902. CRABB, George. 1778-1851. Br. GD.
1. *Universal technological dictionary; or, Familiar explanaations of the terms used in all arts and sciences.* 2 vols. London, 1823.
2. *A technical dictionary; or, A dictionary explaining the terms used in all arts and sciences.* London, 1851.

2903. PRECHTL, Johann Joseph von. 1778-1854. Aus. P II. Also Physics and Chemistry*.
1. *Technologische Encyklopädie.* 20 vols. Stuttgart, 1830-55.
1a. ——— *Supplemente.* 5 vols & atlas. Ib., 1857-69. Ed. by K. Karmarsch. RM.

2904. BROUGHAM, Henry. 1779-1868. Br. The lawyer and politician. P I,III; GA; GB; GD. Also Mechanics*, Physics*, Autobiography*, and Biography*.
1. *A discourse of the objects, advantages and pleasures of science.* London, 1827.
2. *Dissertations on subjects of science connected with natural theology. Being the concluding volumes of the new edition of Paley's work.* 2 vols. London, 1839.
3. *Brougham and his early friends: Letters to James Lock, 1798-1809.* 3 vols. London, 1908. Ed by R.H.M.B. Atkinson.
4. *Works.* Edinburgh, 1872.

2905. CHALMERS, Thomas. 1780-1847. Br. P I; GA; GB; GD. Also Astronomy.

1. *A series of discourses on the Christian revelation, viewed in connection with the modern astronomy.* Glasgow, 1817. X. 6th ed., Edinburgh, 1817. Another ed., 1861.
2. *On the power, wisdom and goodness of God as manifested in the adaptation of external nature to the moral and intellectual constitution of man.* 2 vols. London, 1833. (Bridgewater treatises. 1) X. Other eds: 1835 and 1838,

2906. SCHUBERT, Gotthilf Heinrich von. 1780-1860. Ger. ICB; P II; BMNH; BLA; GC. Other entries: see 3285.
1. *Ansichten von der Nachtseite der Naturwissenschaft.* Dresden, 1808. X. Reprinted, Darmstadt, 1967.
2. [Trans.] *The mirror of nature; or, Manual of science and philosophy.* London, 1854.

2907. SOMERVILLE, Mary Fairfax Greig. 1780-1872. Br. DSB; ICB; P II, III; GB; GD. Also Astronomy*, Geology*, and Autobiography*.
1. *On the connexion of the physical sciences.* London, 1834. RM/L. 3rd ed., 1836 and 1837. 6th ed., 1846. 8th ed., 1849. 9th ed., 1858. 10th ed., corr. and rev. by A.B. Buckley, 1877.
2. *On molecular and microscopic science.* 2 vols. London, 1869. RM/L.

2908. BREWSTER, David. 1781-1868. Br. DSB; P I,III; GA; GB; GD. Also Physics* and Biography*. See also Index.
1. *The Edinburgh encyclopaedia.* 18 vols. Edinburgh, 1830. Ed. by Brewster.
2. *More worlds than one: The creed of the philosopher and the hope of the Christian.* London, 1854. RM/L. Other eds: 1865, 1867, 1876.

2909. TAYLOR, Richard. 1781-1858. Br. P II; BMNH; GD. Editor of scientific journals. See also 3009.

2910. GRANVILLE, Augustus Bozzi. 1783-1872. Br. ICB; P I; BLA; GA; GD. Also Chemistry.
1. *Science without a head; or, The Royal Society dissected.* London, 1830. X. Reprinted, Farnborough, 1969.
2. *The Royal Society in the XIXth century.* London, 1836. RM/L.

2911. NOLAN, Frederick. 1784-1864. Br. GD.
1. *The analogy of revelation and science established.* Oxford, 1833. RM.

2912. MOLL, Gerard. 1785-1838. Holl. DSB; P II. Also Physics.
1. *On the alleged decline of science in England. By a foreigner.* London, 1831. With a foreword by M. Faraday. X. Reprint: see 2918/1.

2913. PARIS, John Ayrton. 1785-1856. Br. P II; BLA; GA; GD. Also Chemistry* and Biography*.
1. *Philosophy in sport made science in earnest: Being an attempt to illustrate the first principles of natural philosophy.* London, 1827. X. 8th ed., 1857.

2914. SEDGWICH, Adam (first of the name). 1785-1873. Br. DSB; P II, III; BMNH; GB; GD. Also Geology*. See also 15519/1.
1. *A discourse on the studies of the University.* Cambridge, 1833. Reprinted, Leicester, 1969. 5th ed., with additions. 1850.

2915. SOUTH, James. 1785-1867. Br. DSB; P II,III; GD. Also Astronomy.
1. *Charges against the President and Councils of the Royal Society.* London, 1830.

2916. BRANDE, William Thomas. 1788-1866. Br. DSB; ICB; P I,III; BMNH; GB; GD. Also Chemistry*, Geology*, and History of Chemistry*.
1. *An introductory discourse delivered in ... the London Institution ... 1819.* London, 1819.
2. *A dictionary of science, literature and art.* London, 1842. RM/L. 2nd ed., 1852. 3rd ed., 1853. Other eds: 3 vols, 1865-67 and 1875.

2917. JOURDAN, Antoine Jacques Louis. 1788-1848. Fr. P III; BMNH; BLA. See also Index.
1. *Dictionnaire raisonné, étymologique, synonymique et polyglotte des termes usités dans les sciences naturelles.* 2 vols. Paris, 1834. Covers the physical and biological sciences. RM/L.

2918. BABBAGE, Charles. 1792-1871. Br. DSB; ICB; P I,III; GA; GB; GD. Also Mathematics*, Geology*, and Autobiography*.
1. *Reflections on the decline of science in England, and on some of its causes.* London, 1830. RM/L. Reprinted, Farnborough, 1969, and Shannon (Ireland), 1972. Another reprint, New York, 1970, to which is added a reprint of item 2912/1.
2. *The ninth Bridgewater treatise: A fragment.* London, 1837. RM/L. 2nd ed., 1838; reprinted, London, 1967.
3. *How to invent machinery.* Manchester, 1899. Ed. by W.H. Atherton. RM.

2919. HERSCHEL, John Frederick William. 1792-1871. Br. DSB; ICB; P I,III; GA; GB; GD. Also Astronomy*# and Physics.
1. *A preliminary discourse on the study of natural philosophy.* London, 1830. Reprinted, New York, 1966. Other eds: 1831, 1832, 1835, 1840, 1851, 1855.
2. *A manual of scientific enquiry. Prepared for the use of Her Majesty's Navy, and adapted for travellers in general.* London, the Admiralty, 1849. Ed. by Herschel. RM/L. 2nd ed., 1851. 3rd ed., 1859. 4th ed., 1861. 5th ed., 1886. cf. item 11848/15.
2a. ——— Reprint entitled *Admiralty manual of scientific enquiry.* Folkestone, 1974.
3. *Essays from the Edinburgh and Quarterly Reviews, with addresses and other pieces.* London, 1857. RM/L.
4. *Familiar lectures on scientific subjects.* London, 1866. Another ed., 1867. Another ed.: *Popular lectures...* , 1873. Another ed.,1876.

2920. VANUXEM, Lardner. 1792-1848. U.S.A. DSB; P II; BMNH; GE. Also Geology*.
1. *An essay on the ultimate principles of chemistry, natural philosophy and physiology.* Part 1 [No more publ.] Philadelphia, 1827. RM.

2921. LARDNER, Dionysius. 1793-1859. Br. ICB; P I & Supp.; BMNH; GA; GB; GD. Other entries: see 3303.

1. *A discourse on the advantages of natural philosophy and astronomy as a part of a general and professional education.* London, 1828. RM.
2. *The museum of science and art.* 12 vols in 6. London, 1854-56. Ed. by Lardner. X. Another ed., 1873.
3. *Common things explained ... From "The museum of science and art."* London, 1856.

2922. McMURTRIE, Henry. 1793-1865. U.S.A. BLA.
1. *Lexicon scientiarum. A dictionary of terms used in the various branches of anatomy, astronomy ... zoology, etc.* 2nd ed. Philadelphia, 1849. Covers the physical and biological sciences. RM.

2923. BABINET, Jacques. 1794-1872. Fr. DSB; P I,III; GA. Also Physics and Meteorology.
1. *Etudes et lectures sur les sciences d'observation et leurs applications pratiques.* Paris, 1855-68. RM.

2924. WHEWELL, William. 1794-1866. Br. DSB; ICB; P II,III; BMNH; GB; GD. Other entries: see 3304.
1. *The philosophy of the inductive sciences, founded on their history.* 2 vols. London, 1840. 2nd ed., 1847; reprinted, New York, 1967.
1a. ——— 3rd ed., 1st part. *History of scientific ideas. Being the first part of "The philosophy of the inductive sciences." The third edition.* 2 vols. London, 1858. Reprinted, ib., 1966.
1b. ——— 3rd ed., 2nd part. *Novum organon renovatum. Being the second part of "The philosophy of the inductive sciences." The third edition, with large additions.* London, 1858. Reprinted, ib., 1966.
1c. ——— 3rd ed., 3rd part. *On the philosophy of discovery, chapters historical and critical. Including the completion of the third edition of "The philosophy of the inductive sciences."* London, 1860. Reprinted, ib., 1966.
2. *Indications of the Creator. Extracts, bearing upon theology, from the history and philosophy of the inductive sciences.* London, 1845. 2nd ed., 1846.
3. *William Whewell's theory of scientific method.* Pittsburgh, 1968. Selected texts. Ed. with an introd. by R.E. Butts.

2925. PARTINGTON, Charles Frederick. d. 1857? Br. BMNH; GD. Also Physics* and Natural History*.
1. *The British cyclopaedia of the arts and sciences.* 2 vols. London, 1835.

2926. GORIANINOV, Pavel Feodorovich. 1796-1865. Russ. ICB; BMNH (GHORYANINOV), BLA. Also Botany. A *Naturphilosoph*; he clashed with K.E. von Baer.

2927. POWELL, Baden. 1796-1860. Br. DSB; P II,III; GD. Also Physics* and History of Science*.
1. *Essays on the spirit of the inductive philosophy, the unity of worlds, and the philosophy of creation.* London, 1855. Reprinted, Farnborough, 1969.

2928. COMTE, (Isidore) Auguste (Marie François Xavier). 1798-1857.

Fr. The philosopher. DSB; ICB; P III; G. Also Mathematics*. See also Index.
1. *Ecrits de jeunesse, 1816-28. Suivis du mémoire sur la cosmogénie de Laplace.* Paris 1970.
2. *Philosophie des sciences. Textes choisis.* Paris, 1974. Ed. by J. Laubier.

2929. BOASE, Henry Samuel. 1799-1883. Br. BMNH; GB; GD. Also Geology*.
1. *The philosophy of nature: A systematic treatise on the causes and laws of natural phenomena.* London, 1860.

2930. YOUNG, John Radford. 1799-1885. Br. P III; GD. Also Mathematics*.
1. *Modern scepticism viewed in relation to modern science, more especially in reference to the doctrines of Colenso, Huxley, Lyell and Darwin, respecting the Noachian deluge, the antiquity of man, and the origin of species.* London, 1865. RM/L.

2931. COURNOT, Antoine Augustin. 1801-1877. Fr. DSB; ICB; P III; GA; GB. Also Mathematics*.
1. *Traité de l'enchainement des idées fondamentales dans les sciences et dans l'histoire.* 2 vols. Paris, 1861. X. Reprinted, Rome, 1968.

2932. WISEMAN, Nicholas Patrick Stephen, *Cardinal.* 1802-1865. Br. GA; GB; GD.
1. *Twelve lectures on the connexion between science and revealed religion.* 2 vols. London, 1836. (The science is chiefly geology.) X. 5th ed., 1853.

2933. HOBLYN, Richard Dennis. 1803-1886. Br. GD.
1. *A dictionary of scientific terms.* London, 1849.

2934. LIEBIG, Justus von. 1803-1873. Ger. DSB; ICB; P I,III; GA; GB; GC. Also Chemistry*. See also Index.
1. *Induction und Deduction.* Munich, 1865.

2935. NICHOL, John Pringle. 1804-1859. Br. P II & Supp.; GD. Also Astronomy*.
1. *A cyclopaedia of the physical sciences.* London, 1857. 2nd ed., 1860. 3rd ed., 1868.

2936. SCHLEIDEN, Matthias Jacob. 1804-1881. Ger. DSB; ICB; BMNH; Mort.; BLA & Supp.; GB; GC. Also Botany* and Cell Theory. See also Index.
1. *Über den Materialismus der neueren deutschen Naturwissenschaft, sein Wesen und Geschichte.* Leipzig, 1863. RM.

2937. GEOFFROY SAINT-HILAIRE, Isidore. 1805-1861. Fr. DSB; BMNH; BLA; GA; GB. Also Zoology* and Biography*.
1. *Notions synthétiques, historiques et physiologiques de philosophie naturelle.* Paris, 1838.

2938. TWINING, Thomas. 1806-1895. Br. BMNH; GD.
1. *Science for the people.* London, 1870.

2939. HUNT, Robert. 1807-1887. Br. P III; GA; GB; GD. Also Chemistry*.
1. *The poetry of science; or, Studies of the physical phenom-*

ena of nature. London, 1848. RM. 2nd ed., 1849. 3rd ed., 1854.
2. *Panthea: The spirit of nature.* London, 1849.

2940. BREWER, Ebenezer Cobham. 1810-1897. Br. GD1.
1. *A guide to the scientific knowledge of things familiar.* London, [1847?]. X. 6th ed., [185-?].

2941. MAILLY, (Nicolas) Edouard. 1810-1891. Belg. P III; BMNH. Also Biography* and History of Science*.
1. *Essai sur les institutions scientifiques de la Grande-Bretagne et de l'Irlande.* 6 parts in 1 vol. Brussels, 1861-67.
2. *L'Espagne scientifique.* Brussels, 1868.

2942. APELT, Ernst Friedrich. 1812-1859. Ger. DSB; P I & Supp.; GC. Also History of Science. Philosophy of Science.

2943. LAING, Samuel. 1812-1897. Br. BMNH; GB; GD1.
1. *Modern science and modern thought.* London, 1885. (Deals with relations between science and religion.) X. 6th ed., 1889; re-issued 1890, 1893, and 1898.
2. *Problems of the future, and other essays.* London, 1890. Another ed., 1892.

2944. FRAUENSTÄDT, (Christian Martin) Julius. 1813-1879. Ger. GC.
1. *Die Naturwissenschaft in ihrem Einfluss auf Poesie, Religion, Moral und Philosophie.* Leipzig, 1855. RM.

2945. MARTIN, Thomas Henri. 1813-1884. Fr. P II,III. Also Biography* and History of Science*.
1. *Les sciences et la philosophie.* Paris, 1869.

2946. SCHOEDLER, Friedrich Karl Ludwig. 1813-1884. Ger. P II,III,IV; Also Chemistry.
1. *Das Buch der Natur, die Lehren der Physik, Chemie, Mineralogie, Geologie, Physiologie, Botanik und Zoologie umfassend.* Brunswick, 1846. 22 eds were publ. up to 1884. X.
1a. ——— *The book of nature: An elementary introduction to the sciences....* 2 parts. London, 1851. Ed. from the 5th German ed. by A. Medlock.
1b. ——— ——— Another ed. *The treasury of science, natural and physical.* London, 1872.
1c. ——— *An introductory manual of the physical sciences, comprising physics, astronomy and chemistry.* London, 1855. Another ed., 1858.
1d. ——— *An introductory manual of the natural sciences, comprising mineralogy, geology, botany, physiology and zoology.* London, 1855.

2947. HUME, Abraham. 1814-1884. Br. GD.
1. *The learned societies and printing clubs of the United Kingdom.* London, 1847. X.
1a. ——— Another ed., with a supplement, ib., 1853. Reprinted, Detroit, 1966.

2948. DU BOIS-REYMOND, Emil Heinrich. 1818-1896. Ger. DSB; ICB; P I,III,IV (under B); BMNH; Mort.; BLA & Supp. (under B); GB; GC. Also Physiology*. See also Index.
1. *Uber die Grenzen des Naturerkennens. (Vortrag).* Leipzig, 1872. X. Another ed., 1884.

2. *Darwin versus Galiani.* Berlin, 1876.
3. *Culturgeschichte und Naturwissenschaft. (Vortrag).* Leipzig, 1878.
4. *Die sieben Welträthsel.* Leipzig, 1880. X. Another ed., 1884.
5. *Über die Übung. (Rede).* Berlin, 1881.
6. *Goethe und sein Ende. (Rede).* Leipzig, 1883.
7. *Reden.* 2 vols. Leipzig, 1886-87. RM/L. 2nd ed., 1912; ed. with a biography by J. Rosenthal.
8. *Vorträge über Philosophie und Gesellschaft.* Berlin, 1974. Ed. with an introd. by S. Wollgast.

2949. STRANGE, Alexander. 1818-1876. Br. P III; GD. Advocate of governmental support of science.

2950. ALBERT, *Prince.* 1819-1861. Br. ICB; GA; GB; GD. Patronage of science.

2951. KINGSLEY, Charles. 1819-1875. Br. The writer. ICB; BMNH; GA; GB; GD. Also Geology* and Zoology*.
1. *Madame How and Lady Why; or, First lessons in earth lore for children.* London, 1870. X. Another ed., 1890.
2. *Scientific lectures and essays.* (*Works.* Vol. 19) London, 1880. Another ed., 1890.

2952. DAWSON, (John) William. 1820-1899. Canada. DSB; P III,IV; BMNH. Also Geology* and Evolution*.
1. *Facts and fancies in modern science: Studies of the relations of science to prevalent speculations and religious belief.* Philadelphia, 1882. RM.

2953. ENGELS, Friedrich. 1820-1895. Ger./Br. The collaborator of Karl Marx. DSB Supp.; ICB.
1. *Herrn Eugen Dühring's Umwälzung der Wissenschaft.* 1878. X.
1a. ——— *Herr Eugen Dühring's revolution in science. Anti-Dühring.* London, 1935 and 1943. Trans. by T. Burns. Another ed., Moscow, 1954.
2. *Dialektik und Natur.* Moscow, 1925. First publication. X.
2a. ——— *Dialectics of nature.* London, 1940 and 1946. Trans. and ed. by C. Dutt. Other eds: Moscow, 1954; New York, 1960.

2954. SPENCER, Herbert. 1820-1903. Br. The philosopher. DSB; ICB; BMNH; GB; GD2. Also Biology in General*, Evolution*, and Autobiography*.
1. *The classification of the sciences.* London, 1864.
2. *Essays: Scientific, political and speculative.* 3 vols. London, 1868-78. Another ed., 1883-88.

2955. TYNDALL, John. 1820-1893. Br. DSB; ICB; P II,III,IV; GB; GD. Also Physics*, Geology*, Microbiology*, and Biography*. See also Index.
1. *On the scientific use of the imagination. A discourse.* London, 1870. RM.
2. *Essays on the use and limit of the imagination in science.* London, 1870. RM/L.
3. *Fragments of science for unscientific people: A series of detached essays, lectures and reviews.* London, 1871. X. 2nd ed., 1871. 4th ed., 1872. 5th ed., 1876. 6th ed.,

2 vols, 1879. 8th ed., 2 vols, 1892; reprinted, Farnborough, 1970.
4. *Address delivered before the British Association assembled at Belfast. With additions.* London, 1874. 7th ed., rev. and with an appended article on scientific materialism, New York, 1875.
5. *New fragments.* London, 1892. RM/L. Reprinted, Farnborough, 1968.

2956. HELMHOLTZ, Hermann Ludwig Ferdinand von. 1821-1894. Ger. DSB; ICB; P I,III,IV; Mort.; BLA & Supp.; GB; GC. Also Mechanics*, Physics*, and Physiology*. See also Index.
1. *Populäre wissenschaftliche Vorträge.* 3 vols. Brunswick, 1865-76. RM/L.
1a. ——— 3rd ed. *Vorträge und Reden.* 2 vols. Ib., 1884.
1b. ——— *Popular lectures on scientific subjects.* [First series] London, 1873. Trans. by E. Atkinson. RM/L. Second series, 1881. Other eds, 2 vols (1st and 2nd series): 1884-89 and 1893.
2. *Das Denken in der Medicin.* Berlin, 1878.
2a. ——— *On thought in medicine. An address.* Baltimore, 1938.
3. *Wissenschaftliche Abhandlungen.* 3 vols. Leipzig, 1882-95. RM/L.

Selections

4. *Abhandlungen zu Philosophie und Naturwissenschaften.* 1896. X. Reprinted, Darmstadt, 1966.
5. *Schriften zur Erkenntnistheorie.* Berlin, 1921. Ed. with notes by P. Hertz and M. Schlick. X.
5a. ——— *Epistemological writings.* Dordrecht, 1977. Trans. by M.F. Lowe.
6. *Popular scientific lectures.* New York, 1962. Selected from item 1b. Ed. with an introd. by M. Kline.
7. *Das Denken in der Naturwissenschaft.* Darmstadt, 1968. Essays selected from item 1a.
8. *Philosophische Vorträge und Aufsätze.* Berlin, 1971. Ed. with introd. and notes by H. Hörz and S. Wollgast.
9. *Selected writings.* Middletown, Conn., 1971. Ed. by R. Kahl.

2957. PRIVAT-DESCHANEL, Augustin. 1821-1883. Fr. P III. Also Physics*.
1. *Dictionnaire générale des sciences théoriques et appliquées.* 2 vols. Paris, 1865-69. With A. Focillon. Supplements, 1882.

2958. VIRCHOW, Rudolf Ludwig Karl. 1821-1902. Ger. DSB; ICB; BMNH; Mort.; BLA & Supp.; GB. Also Cytology* and Biography*.
1. *Die Freiheit der Wissenschaft im modernen Staat. Rede....* Berlin, 1877. X.
1a. ——— *The freedom of science in the modern state. A discourse....* London, 1878.
See also 2974/1.

2959. RENAN, (Joseph) Ernest. 1823-1892. Fr. The writer. ICB; GA; GB.
1. *L'avenir de la science: Pensées de 1848.* Paris, 1890. Written in 1848 but not published until 1890. Reprinted, Paris, 1934.

1a. ——— *The future of science: Ideas of 1848.* London, 1891;
2. *Les sciences de la nature et les sciences historiques: Lettre a Marcellin Berthelot.* [With] *L'avenir de la science, chap. II et XVI.* Princeton, 1944. Ed. by I.O. Wade.

2960. STALLO, John Bernhard. 1823-1900. U.S.A. DSB; P IV; GE. Also Physics*.
1. *General principles of the philosophy of nature, with an outline of some of its recent developments among the Germans; embracing the philosophical systems of Schelling and Hegel, and Oken's system of nature.* Boston, 1848. RM.

2961. BÜCHNER, (Friedrich Karl Christian) Ludwig. 1824-1899. Ger. DSB; P III,IV; BLA (B., Louis); GC. Also Physiology*.
1. *Kraft und Stoff. Empirisch-naturphilosophische Studien.* Frankfurt, 1855. X. Another ed., New York, 1871. 20th ed., Leipzig, 1902. X.
1a. ——— Another ed. Leipzig, [191-?]. Ed. with introd. and notes by W. Bölsche.
1b. ——— *Force and matter. Empirio-philosophical studies.* London, 1864. Ed. by J. Collingwood. 2nd ed. (from the 10th German ed.), 1870. 3rd ed., 1881. 4th ed., 1884.
2. *Aus Natur und Wissenschaft. Studien, Kritiken und Abhandlungen.* Leipzig, 1862. X. 3rd ed., 1874.
3. [Trans.] *Last words on materialism and kindred subjects.* London, 1901. With a biography by his brother, Alex. Büchner. Trans. by J. McCabe.

2962. MARCOU, Jules Belknap. 1824-1898. Fr./Switz./U.S.A. DSB; P III,IV; BMNH; GB; GE. Also Geology* and Biography*.
1. *De la science en France.* Paris, 1869.

2963. HUXLEY, Thomas Henry. 1825-1895. Br. DSB; ICB; BMNH; Mort.; BLA; GB; GD1. Other entries: see 3340.
Many editions of his various works were published after 1900; these are not included. All the works listed below were published at London.
1. *Lay sermons, addresses and reviews.* 1870. RM/L. 2nd ed., 1871. 3rd ed., 1871. 4th ed., 1872. Other eds: 1887 and 1895.
2. *Science primers. Edited by Professors Huxley, Roscoe and B. Stewart.* 1872. A series of elementary textbooks.
2a. ——— *Science primers. Introductory. By Professor Huxley.*
3. *Critiques and addresses.* 1873. Other eds: 1883 and 1890.
4. *American addresses. With a lecture on the study of biology.* 1877. Another ed., 1886.
5. *Science and culture, and other essays.* 1881. RM/L. Another ed., 1882.
6. *Essays upon some controverted questions.* 1892.
7. *Possibilities and impossibilities. Reprinted from "The agnostic annual", with additions.* 1896.
8. *Collected essays.* 9 vols. 1893-94 and re-issues.
Contents: Vol. 1. *Method and results.* Vol. 2. *Darwiniana.* Vol. 3. *Science and education.* Vol. 4. *Science and Hebrew tradition.* Vol. 5. *Science and Christian tradition.* Vol. 6. *Hume, with helps to the study of Berkeley.* Vol. 7.

Man's place in nature. Vol. 8. *Discourses, biological and geological*.. Vol. 9. *Evolution and ethics*. RM/L. Reprinted, Hildesheim, 1969-70.

Selections

9. *Lectures and essays*. 1902.
10. *Twelve lectures and essays*. 1908. Introd. by E. Clodd.
11. *Aphorisms and reflections from the works of T.H. Huxley*. 1908. Selected by Henrietta A. Huxley.
12. *T.H. Huxley on education: A selection*. Cambridge, 1971. Introd. and notes by C. Bibbey.

2964. STORMONTH, James. 1825?-1882. Br. BMNH.
1. *A manual of scientific terms*. Edinburgh, 1879. X. 2nd ed., 1892.

2965. GORE, George. 1826-1908. Br. DSB; P III,IV; GD2. Also Chemistry*.
1. *The art of scientific discovery*. London, 1878.
2. *The scientific basis of national progress, including that of morality*. London, 1882. X. Reprinted, ib., 1970.

2966. HUTTON, Richard Holt. 1826-1897. Br. GB; GD1.
1. *Aspects of religious and scientific thought*. London, 1899. Reprinted, Farnborough, 1971.

2967. BERTHELOT, (Pierre Eugène) Marcellin. 1827-1907. Fr. DSB; ICB; P I,III,IV; GB. Also Chemistry* and History of Chemistry*.
1. *Science et philosophie*. Paris, 1886. RM/L.
See also 2959/2.

2968. COOKE, Josiah Parsons. 1827-1894. U.S.A. DSB; P III,IV; GE. Also Chemistry*.
1. *Scientific culture, and other essays*. 1882.
2. *The credentials of science and the warrant of faith*. New York, 1888. X. 3rd ed., London, 1893.

2969. MIVART, St. George Jackson. 1827-1900. Br. DSB; ICB; BMNH; GB; GD1. Also Zoology* and Evolution*.
1. *Nature and thought*. London, 1882.
2. *Introduction to the elements of science*. London, 1894.
3. *The groundwork of science*. London, 1898.

2970. MURPHY, Joseph John. 1827-1894. Br. P III. Also Geology.
1. *Habit and intelligence ... matter and force. A series of scientific essays*. London, 1869.

2971. LAUGEL, (Antoine) Auguste. 1830-1914. BMNH.
1. *Les problèmes de la nature*. Paris, 1864.

2972. DU BOIS-REYMOND, Paul David Gustav. 1831-1889. Ger. DSB; ICB; P III,IV (under B); GC. Also Mathematics*.
1. *Über die Grundlagen der Erkenntnis in den exacten Wissenschaften*. Tübingen, 1890. Reprinted, Darmstadt, 1966.

2973. BECKER, Bernard Henry. b. 1833.
1. *Scientific London*. London, 1874. Reprinted, ib., 1968.

2974. HAECKEL, Ernst Heinrich Philipp August. 1834-1919. Ger. DSB; ICB; BMNH; Mort.; BLAF; GB. Also Zoology*, Evolution*#, Embryology*, and Personal Writings*.

1. *Freie Wissenschaft und freie Lehre. Eine Entgegnung auf R. Virchow's Münchener Rede über die Freiheit der Wissenschaft im modernen Staat.* Stuttgart, 1878. X. For Virchow's address see 2958/1.

1a. ——— *Freedom in science and teaching.* London, 1879. With a preface by T.H. Huxley. Another ed., 1892.

2. *Der Monismus als Band zwischen Religion und Wissenschaft: Glaubensbekenntniss eines Naturforschers.* Bonn, 1892.

2a. ——— *Monism, as connecting religion and science: The confession of faith of a man of science.* London, 1894. Trans. by J. Gilchrist.

3. *Die Welträthsel: Gemeinverständliche Studien über monistische Philosophie.* Bonn, 1899. RM. Another ed., 1900.

3a. ——— *The riddle of the universe at the close of the nineteenth century.* London, 1900. Trans. by J. McCabe. RM/L. Reprinted, Grosse Pointe, Mich., 1968.

4. *Die Lebenswunder: Gemeinverständliche Studien über biologische Philosophie.* Stuttgart, 1904. RM.

4a. ——— *The wonders of life: A popular study of biological philosophy.* London, 1904. Trans. by J. McCabe.

2975. LUBBOCK, John, *1st Baron* AVEBURY. 1834-1913. Br. DSB; ICB; BMNH; GB; GD3. Other entries: see 3351.

1. *Fifty years of science. Being the address ... to the British Association ... 1881.* London, 1882. 4th ed., 1890. 5th ed., 1890.
2. *The beauties of nature and the wonders of the world we live in.* London, 1892. 3rd ed., 1893.
3. *The use of life.* London, 1894. X. Another ed., 1902.

2976. JEVONS, William Stanley. 1835-1882. Br. DSB; ICB; P III; GB; GD.

1. *The principles of science: A treatise on logic and scientific method.* 2 vols. London, 1874. 2nd ed., 1877; reprinted, New York, 1958. Other eds., 1883 and 1887.
2. *Letters and journal.* London, 1886. Ed. by his wife.

2977. PROCTOR, Richard Antony. 1837-1888. Br. DSB; P III,IV; GB; GD. Also Astronomy*.

1. *Light science for leisure hours: A series of familiar essays on scientific subjects.* London, 1871.
2. *The borderland of science.* London, 1873.
3. *Science byways: A series of familiar dissertations on life in other worlds ... to which is appended an essay entitled "Money for science."* London, 1875. Another ed., 1882.
4. *Wages and wants of science-workers.* London, 1876. X. Reprinted, ib., 1970.
5. *Pleasant ways in science.* London, 1878.
6. *Rough ways made smooth: A series of familiar essays on scientific subjects.* London, 1880.
7. *Familiar science studies.* London, 1882.

2978. GRAHAM, William. 1839-1911. Br. GD2.

1. *The creed of science, religious, moral and social.* London, 1881. 2nd ed., 1884; reprinted, Farnborough, 1971.

2979. PEIRCE, Charles (Santiago) Sanders. 1839-1914. U.S.A. DSB;

ICB; P III,IV; GE. Also Mathematics* and Geodesy.
1. *Collected papers*. 8 vols. Cambridge, Mass., 1931-35, 1958. Ed. by C. Hartshorne et al. Reprinted, 1960.
2. *Essays in the philosophy of science*. New York, 1957. Ed. with an introd. by V. Tomas.
3. *The essential writings*. New York, 1972. Ed. by E.C. More.
4. *Complete published works*. Microfiche edition. Greenwich, Conn., 1977. Ed. by K.L. Ketner.

2980. BROWN, Robert (of Campster). 1842-1895. Br. P III; BMNH; GD1. Also Geology*, Botany*, and Zoology.
1. *Science for all*. 5 vols. London, 1877-82? Ed. by Brown.

2981. RODWELL, George Farrer. 1843-after 1883. Br. P III. Also History of Chemistry*.
1. *A dictionary of science*. London, 1871. RM/L.

2982. BOUTROUX, Emile. 1845-1921. Fr. ICB; GF.
1. *De l'idée de loi naturelle dans la science et la philosophie contemporaines*. Paris, 1895. X. Reprinted, ib., 1949.

2983. CLIFFORD, William Kingdon. 1845-1879. Br. DSB; ICB; P III; GB; GD. Also Mathematics*# and Mechanics*.
1. *The common sense of the exact sciences*. London, 1885. (Left incomplete at his death, the work was added to and revised by R.C. Rowe, and finally completed by Karl Pearson.) RM. 2nd ed., 1885. 3rd ed., 1892. Reprinted, New York, 1946.
2. *Letters and essays*. London, 1879. Ed. by L. Stephen. RM/L.

2984. RICE, William North. 1845-1928. U.S.A. BMNH; GE. Also Geology.
1. *Twenty-five years of scientific progress, and other essays*. New York, 1894.

2985. LANKESTER, Edwin Ray. 1847-1929. Br. DSB; BMNH; GD4. Also Zoology*, Evolution*, and Microscopy. See also Index.
1. *The advancement of science: Occasional essays and addresses*. London, 1890.

2986. ARMSTRONG, Henry Edward. 1848-1937. Br. DSB; ICB; P IV; GD5. Also Chemistry*.
1. *H.E. Armstrong and the teaching of science, 1880-1930*. Cambridge, 1973. Ed. with an introd. by W.H. Brock.
2. *H.E. Armstrong and science education. Selections*. London, 1973. Ed. with an introd. by G. Van Praagh.

2987. LASSWITZ, (Carl Theodor Victor) Kurd. 1848-1910. Ger. ICB; P III,IV. Also History of Science*.
1. *Atomistik und Kriticismus: Ein Beitrag zur erkenntiss-theoretischen Grundlegung der Physik*. Brunswick. 1878. RM.
See also 3311/1a.

2988. ROMANES, George John. 1848-1894. Br. DSB; ICB; BMNH; Mort.; GD. Also Evolution*, Physiology, and Biography*. See also 15607/1.
1. *A candid examination of theism. By Physicus*. London, 1878.
2. *Mind and motion (Rede Lecture, 1885), and Monism*. London, 1895.
3. *Thoughts on religion*. London, 1895. Ed. by Charles Gore.
4. *Essays*. London, 1897. Ed. by C.L. Morgan. Reprinted, Westmead, 1970.

2989\. MÜLLER, Johannes. 1850-1919.
1. *Wissenschaftlichen Vereine und Gesellschaften Deutschlands im neunzehnten Jahrhundert: Bibliographie ihrer Veröffentlichungen*. 2 vols in 3. Berlin, 1883/87-1917. Reprinted, Hildesheim, 1965.

2990\. CARUS, Paul. 1852-1919. U.S.A. DSB Supp.; ICB; GE. Philosophy of science.

2991\. DENIKER, Joseph. 1852-1918. Fr. BMNH. Also Natural History.
1. *Bibliographie des travaux scientifiques (mathématiques, physique et sciences naturelles) publiés par les sociétés savantes de France de 1700 à 1888*. Paris, 1895. Vol. 2 was published in 1922 by R. Descharmes. The work is arranged by place. It was never completed and does not include Paris.

2992\. HOFFMAN, Frank Sargent. 1852-1928.
1. *The sphere of science: A study of the nature and method of scientific investigation*. London, 1898.

2993\. HANNEQUIN, Arthur. 1856-1905.
1. *Essai critique sur l'hypothèse des atomes dans la science contemporaine*. Paris, 1895. 2nd ed., 1899.

2994\. VOLKMANN, Paul Oskar Eduard. 1856-1938. Ger. DSB; P III,IV. Also Physics and History of Physics. Philosophy of science.

2995\. PEARSON, Karl. 1857-1936. Br. DSB; ICB; P IV; GD5. Also Mathematics*, Evolution*, and Biography*.
1. *The grammar of science*. London, 1892. RM/L. 2nd ed., 1900.

See also 2983/1.

2996\. MILHAUD, Gaston Samuel. 1858-1918. Fr. DSB; ICB; P IV. Philosophy of science.
1. *Le rationnel*. Paris, 1898.

Imprint Sequence

2997\. *PHILOSOPHICAL transactions*--Abridgments.
1. *The "Philosophical transactions" and "Collections" to the end of the year 1700, abridg'd and dispos'd under general heads*. Vols 1-3. London, 1705. By John Lowthorp. Several editions were issued.

1a. ——— From 1700 to 1720. Vols 4 and 5. 1721. By Henry Jones.

1b. ——— From 1719 to 1733. Vols 6 and 7. 1734. By John Eames.

1c. ——— From 1732 to 1744. Vols 8 and 9. 1747. By John Martyn.

1d. ——— From 1743 to 1750. Vol. 10. 1756. By John Martyn.

1e. ——— *A general index of all the matters contained in the seven volumes of the "Philosophical transactions" abridged from the beginning to the year 1733 by Mr Lowthorp, Mr Jones, Mr Eames ... and J. Martyn. London, 1735.*

2. *Memoirs of the Royal Society: Being a new abridgment of*

the "Philosophical transactions" ... from 1665 to 1735, inclusive. 10 vols. London, 1738-41. By Baddam.

2a. ——— 2nd ed., covering the period 1665-1740, 10 vols, 1745.

2998. ACADÉMIE des Sciences, *Histoire et mémoires*--Abridgments and extracts.

1. *The philosophical history and memoirs of the Royal Academy of Sciences at Paris; or, An abridgment of all the papers relating to natural philosophy.* 5 vols. London, 1742. Trans. and abridged by J. Martyn and E. Chambers.
2. *Curious remarks and observations in physics, anatomy, chirurgery, chemistry, botany, and medicine. Extracted from the "History and memoirs" of the Royal Academy of Sciences at Paris.* 2 vols. London, 1753-54. By P. Templeman.

2999. *COLLECTION académique.* Many vols. Dijon, later Paris, 1754-87. A massive collection of articles in the non-mathematical sciences, selected from the chief periodicals of the time. The initiator and first editor was Jean Berryat (2859). Later editors included Buffon and Daubenton.

3000. *ENCYCLOPÉDIE méthodique.* 166 vols of text and 40 vols of plates. Paris, 1782-1832. Ed. by Panchoucke and later by Agasse. It is divided by subjects into some forty sections; only the following sections, dealing with the pure sciences, are included in the present catalogue. Entries for each of these sections may be found at the places indicated.

(a) Botanique. 12628.
(b) Chimie. 7150/3 and 7262/3.
(c) Géographie physique. 9026/2.
(d) Histoire naturelle. 13849.
(e) Mathématiques. 3467/2.
(f) Physique. 6026/1.
(g) Système anatomique. 13849/3.

3001. *AN HISTORY of the instances of exclusion from the Royal Society which were not suffered to be argued in the course of the late debates. With strictures of the formation of the Council, and other instances of the despotism of Sir Joseph Banks ... By some members in the minority.* London, 1784.

3002. *PHILOSOPHICAL transactions*--Index.

1. *General index to the "Philosophical transactions" from the first to the end of the seventieth volume.* London, 1787. By P.H. Maty. Subject and author index. X.

3003. IMISON, John. d. 1788. Br. P I; GD.

1. *Elements of science and art: Being a familiar introduction to natural philosophy and chemistry, together with their applications.* London, 1803. X.

1a. ——— Another ed., enlarged and adapted by J. Webster, 1808.

3004. *PHILOSOPHICAL transactions*--Abridgment.

1. *The "Philosophical transactions" ... from their commencement in 1665 to 1800, abridged. With notes and biographic illustrations.* 18 vols. London, 1803-09. By C. Hutton, G. Shaw, and R. Pearson.

3005. *ENCYCLOPAEDIA metropolitana; or, Universal dictionary of knowledge*. 25 vols of text, 3 vols of plates, and an index vol. London, 1817-45. Arranged by subjects. The natural sciences are dealt with in Vols I-VIII, the articles being written by some of the leading British scientists of the time.

3006. ROYAL Society of London.
1. *Catalogue of the library*. London, 1825.
For later catalogues see 3010 and 3025.

3007. *BRIDGEWATER treatises on the power, wisdom and goodness of God as manifested in the Creation*. Eight treatises published in the period 1833-36 under a bequest from the eighth Earl of Bridgewater. The authors and item numbers are as follows.
(1) T. Chalmers. 2905/2.
(2) J. Kidd. 2899/1.
(3) W. Whewell. 4895/1.
(4) C. Bell. 14085/9.
(5) P. M. Roget. 14751/1.
(6) W. Buckland. 9294/3.
(7) W. Kirby. 12788/3.
(8) W. Prout. 7470/1.
See also C. Babbage. 2918/2.

3008. ROYAL Society of London.
1. *Report on the adjudication of the Copley, Rumford and Royal medals, and appointment of the Bakerian, Croonian and Fairchild Lectures. Compiled from the original documents in the archives of the Royal Society*. London, 1834. By James Hudson.

3009. *SCIENTIFIC memoirs, selected from the transactions of foreign academies of science and learned societies, and from foreign journals*. 5 vols. London, 1837-52. Ed. by Richard Taylor. X. (cf. entries 6973 and 11267.) Reprinted, New York, 1966.

3010. ROYAL Society of London.
1. *Catalogue of the scientific books in the library*. London, 1839. Compiled by A. Panizzi. RM/L. For other catalogues see 3006 and 3025.
2. *Catalogue of the miscellaneous manuscripts and of the manuscript letters in the possession of the Royal Society*. 2 parts in 1 vol. London, 1840.

3011. ROYAL Society Club.
1. *Sketch of the rise and progress of the Royal Society Club*. London, 1860. Ed. by W.H. Smyth.

3012. SCHWEIZERISCHE Naturforschende Gesellschaft.
1. *Verzeichniss der Bibliothek*. Bern, 1864. Supplement, 1882.

3013. *SCIENCE lectures for the people*. Series I-XI. Manchester, 1866-81. Several re-issues; various collations. For contents see BMNH. The lectures were given by many prominent British scientists. One of the chief organisers was H.E. Roscoe, professor of chemistry at Manchester.

3014. SMITHSONIAN Institution. Washington.
1. *Catalogue of publications of societies and of periodical works belonging to the Smithsonian Institution*. Washington, 1866.

3015. ACADÉMIE Impériale des Sciences de St.-Pétersbourg.
1. *Catalogue des livres publiés en langues etrangères par l'Académie*. St. Petersburg, 1867.

3016. ROYAL Society of London.

1. *Catalogue of scientific papers*. [1800-1900]. 19 vols and subject index of 3 vols in 4. London, 1867-1925. Comprises the following series, each containing an alphabetical sequence of authors' names, the references to periodicals being listed under the names.
First series: Vols 1-6. Covers the period 1800-63. Published 1867-72.
Second series: Vols 7-8. Covers the period 1864-73. Published 1877-79.
Third series: Vols 9-11. Covers the period 1874-83. Published 1891-96.
Supplementary vol.: Vol. 12. Covers the period 1800-83 for periodicals not hitherto searched. Published 1902.
Fourth series: Vols 13-19. Covers the period 1884-1900. Published 1914-25.
The subject index (for the whole period, 1800-1900) was never completed. It consists of three vols (in four) instead of the seventeen envisaged, and covers only mathematics, mechanics, and physics.

3017. ECOLE Pratique des Hautes Etudes. Paris.

1. *Rapports des directeurs de laboratoires et de conférences, 1868-77*. Paris, 1879.

3018. ROYAL Society of London.

1. *List* [of Fellows] *of the Royal Society*. 1869-70. 1875-76. 1878-95. (After 1895 the list was incorporated in the Society's *Yearbook*.) Lists were issued at irregular intervals from the beginning of the Society (cf. entry 2811).

3019. GREAT Britain. Parliament.

1. *British parliamentary papers. Education, scientific and technical. First and second reports from the Royal Commission on Scientific Instruction and the Advancement of Science, 1871-1872*. X. Reprinted, Shannon (Ireland), 1970.

3020. ACADÉMIE Royale de Belgique. Brussels.

1. *Centième anniversaire de sa fondation, 1772-1872*. 2 vols. Brussels, 1872.
2. *Notices biographiques et bibliographiques concernant les membres*. Brussels, 1874. 3rd ed., 1886. 4th ed., 1896.

3021. MANCHESTER Literary and Philosophical Society.

1. *Catalogue of books in the library*. Manchester, 1875.

3022. SOUTH Kensington Museum. London. (Founded 1857. In 1899 it was divided into the Science Museum and the Victoria and Albert Museum of Art.)

1. *Handbook to the special loan collection of scientific apparatus, 1876*. London, 1876.
2. *Conferences held in connection with the special loan collection of scientific apparatus. 1876*.

2a. ——— *Physics and mechanics*. London, 1876.
2b. ——— *Physics, etc*. London, 1876.
2c. ——— *Chemistry, biology, physical geography, geology, mineralogy and meteorology*. 2 vols. London, 1877.

3. *Free evening lectures.* London, 1878.
4. *Science lectures at South Kensington.* 2 vols. Ib., 1878-79.
5. *Catalogue of the science library.* Ib., 1891. Supp., 1895.

3023. *ENCYKLOPÄDIE der Naturwissenschaften.* 38 vols in 42. Breslau, 1879-1902. Ed. by G. Jäger, A. Kenngott, et al.

3024. R. FRIEDLÄNDER und Sohn. (Publishers.) Berlin.
1. *Bibliotheca historico-naturalis et mathematica.* Berlin, 1880.

3025. ROYAL Society of London.
1. *Catalogue of the scientific books in the library.*
1a. ——— *General catalogue.* London, 1881.
1b. ——— *Transactions, journals, etc.* Ib., 1881.
For earlier catalogues see 3006 and 3010.

3026. *THE INTERNATIONAL scientists' directory.* Boston, 1883. By S.E. Cassino (the publisher). Another ed., 1896.

3027. ROYAL Society of London.
1. *The record of the Royal Society.* London, 1897. The second and later editions were published after 1900.)

3028. ROYAL Societies Club.
1. *Foundation and objects. Rules and by-laws. List of members.* London, 1897.

Twentieth-Century Editions of
Archives, Correspondence, etc.

3029. AKADEMIE der Wissenschaften. Berlin.
1. *Die Registres der Berliner Akademie der Wissenschaften, 1746-1766: Dokumente für das Wirken Leonhard Eulers in Berlin.* Berlin, 1957. Ed. with an introd. by E. Winter.

3030. BAYERISCHE Akademie der Wissenschaften. Munich.
1. *Electoralis Academiae Scientiarum Boicae primordia: Briefe aus der Gründungszeit der Bayerischen Akademie der Wissenschaften.* Munich, 1959. Ed. by M. Spindler.

3031. ROYAL Institution of Great Britain. London.
1. *The archives of the Royal Institution of Great Britain in facsimile. Minutes of the managers' meetings, 1799-1900.* 16 vols. Ilkley, Yorkshire, 1971-. Ed. by F. Greenaway.
2. *The Royal Institution Library of Science, being the Friday Evening Discourses in physical sciences held at the Royal Institution, 1851-1939.*
2a. ——— *Physical sciences* [i.e. physics and chemistry]. 10 vols. Barking, Essex, 1970. Ed. by W.L. Bragg and L. Porter.
2b. ——— *Astronomy.* 2 vols. Ib., 1970. Ed. by B. Lovell.
2c. ——— *Earth sciences.* 3 vols. Ib., 1971. Ed. by S.K. Runcorn.

3032. AKADEMIE der Wissenschaften. Vienna.
1. *Dokumentation zur Österreichischen Akademie der Wissenschaften, 1847-1972.* Vienna, 1972.

3.02 VARIOUS FIELDS

This section includes: (a) Major authors who published in several different fields; their collected works and correspondence, if any, are generally listed here, with references to the entries for them in other sections. (b) Minor authors who published in several different fields; most of these are not entered in any other section. (c) Titles of books whose contents are of a mixed or miscellaneous character, or are marginal to this catalogue.

In contrast to the other sections, the number of persons in this section decreases over time.

3033. HALLEY, Edmond. 1656?-1743. Br. DSB; ICB; P I; GA; GB; GD. Also Astronomy*#, Geomagnetism, and Part 2. See also Index.
   1. *Miscellanea curiosa: Being a collection of some of the principal phaenomena in nature, accounted for by the greatest philosophers of this age. Together with several discourses read before the Royal Society.* 3 vols. London, 1705-07. Edited and partly written by Halley. RM.
   2. *Two voyages made in 1698, 1699 and 1700.* London, 1773.

3034. VALLISNIERI, Antonio. 1661-1730. It. DSB; ICB; P II; BMNH; Mort.; BLA; GA. Also Geology* and Embryology.
   1. *Raccolta di varie osservazioni spettanti all'istoria medica e naturale.* Venice, 1728. RM.

3035. AVERANI, Giuseppe. 1662-1738. It. P I; GA.

3036. CROUSAZ, Jean Pierre de. 1663-1750. Switz. DSB; ICB; P I & Supp.; GA; GB. Also Mathematics*.

3037. ROBERG, Lars (or Laurentius). 1664-1742. Swed. ICB; P II; BMNH; BLA.

3038. COHAUSEN, Johann Heinrich. 1665-1750. Ger. ICB; P I; BLA & Supp.; GA; GC.

3039. ARBUTHNOT, John. 1667-1735. Br. DSB; ICB; BMNH; BLA & Supp.; GA; GB; GD. Also Mathematics, Physiology*, and Part 2*.

3040. CAMPAILLA, Tommaso. 1668-1740. It. ICB; P III; GA.

3041. BALTHASAR, Theodor. fl. 1710. Ger. P I; BLA.

3042. BRUCE, Jacob Daniel. 1670-1735. Russ. ICB; GA. "Russia's first Newtonian."

3043. CYRILLUS, Nicolaus. 1671-1735. It. P I; BLA (CIRILLO); GA (CIRILLO).

3044. MÜLLER, Johann Heinrich. 1671-1731. Ger. P II; GA; GC.

3045. BECKER, Peter. 1672-1753. Ger. P I.

3046. CAMERARIUS, Elias. 1672-1734. Ger. P I; BMNH; BLA & Supp.; GA; GC.

3047. GRANDIN, Charles. ca. 1672-1741. Fr. P III.

3048. CREILING, Johann Conrad. 1673-1752. Ger. ICB; P I; GC.

3049. ETTMÜLLER, Michael Ernst. 1673-1732. Ger. P I; BLA (E., Ernst Michael).

3050. DÖBELIUS (later VON DÖBELN), Johann Jacob. 1674-1743. Swed. P III; BLA.

3051. LOGAN, James. 1674-1751. U.S.A. DSB; ICB; P I & Supp.; BMNH; GD; GE. Also Botany*.

3052. ANDRÉ, Yves Maria. 1675-1764. Fr. P I; GF.

3053. BAIER, Johann Wilhelm. 1675-1729. Ger. P I; BLA & Supp.

3054. LEHMANN, Johann Christian. 1675-1739. Ger. P I.

3055. WOLFART, Peter. 1675-1726. Ger. P II; BMNH; BLA.

3056. JUCH, Hermann Paul. 1676-1756 (or 1736). Ger. P I; BLA.

3057. CONTI, Antonio Schinella. 1677-1749. It. ICB; P I; GA.

3058. HEUCHER, Johann Heinrich von. 1677-1747. Ger. ICB; P I; BMNH; BLA.

3059. MAZZINI, Giovanni Battista. 1677-1743. It. P II; BLA (MAZINI).

3060. ALGÖWER, David. 1678-1737. Ger. P I.

3061. BORGONDIO, Orazio. 1679-1741. It. P I; GA (BURGUNDIO).

3062. LIEBKNECHT, Johann Georg. 1679-1749. Ger. P I; BMNH; GA.

3063. VERDRIES, Johann Melchior. 1679-1735. Ger. P II; BMNH; BLA; GC.

3064. LÖSCHER, Martin Gotthelf. d. 1735. Ger. P III.

3065. ASTRUC, Jean. 1684-1766. Fr. DSB; ICB; P I; BMNH; BLA & Supp.; GA; GB; GF. Also Natural History*.

3066. GRAM, Hans. 1685-1748. Den. P I; GA (G., Jens).

3067. TEICHMEYER, Hermann Friedrich. 1685-1744. Ger. P II; BMNH; BLA; GA; GC.

3068. FÜRSTENAU, Johann Hermann. 1688-1756. Ger. P I; BLA; GA.

3069. SWEDENBORG, Emanuel. 1688-1772. Swed. DSB; ICB; P II; BMNH (SVEDENBORG); GA; GB. Also Science in General*.

3070. KULMUS, Johann Adam. 1689-1745. Danzig. P I; BMNH; BLA.

3071. BOUILLET, Jean. 1690-1777. Fr. P I; BLA; GA.

3072. DESLANDES, André François Boureau. 1690-1757. Fr. P I; GA.

3073. FOLKES, Martin. 1690-1754. Br. DSB; ICB; P I; GA; GB; GD.

3074. KÜHN, Heinrich. 1690-1769. Danzig. P I. Also Geology*.

3075. HAUSEN, Christian August. 1693-1743. Ger. P I.

3076. KELSCH, Michael. 1693-1742. Ger. P I.

3077. MARQUARDT, Konrad Gottlieb. 1694-1749. Ger. P II.

3078. THOMAS, Corbinianus. 1694-1767. Aus. P II.

3079. LOZERAN DU FECH, Louis Antoine. d. 1755. Fr. P I.

3080. ALBRECHT, Johann Sebastian. 1695-1774. Ger. P I; BLA & Supp.; GA.

3081. FISCHER, Daniel. 1695-1746. Hung. P I; BMNH; BLA; GA.

3082. HANOV, Michael Christoph. 1695-1773. Danzig. P I.

3083. RAST, Georg Heinrich. 1695-1726. Ger. P II.

3084. WENZ, Ludwig. 1695-1772. Switz. P II.

3085. CILANO (DE MATERNUS), Georg Christian. 1696-1773. Ger. P I; BLA (MATERN DE CILANO); GA.

3086. QUELLMALZ, Samuel Theodor. 1696-1758. Ger. P II.

3087. THÜMMIG, Ludwig Philipp. 1697-1728. Ger. P II; GA; GC.

3088. MASSUET, Pierre. 1698-1776. Holl. P II; BLA; GA. See also 5818/4.

3089. MAUPERTUIS, Pierre Louis Moreau de. 1698-1759. Fr./Ger. DSB; ICB; P II; GA; GB; GC. Also Mathematics, Astronomy*, Mechanics, Geodesy*, and Embryology*. See also 2278/1a.
1. *Lettre sur le progrès des sciences.* n.p., 1752. RM.
2. *Lettres.* 2nd ed. Berlin, 1753.
3. *Maupertuis et ses correspondents. Lettres inédites.* Montreuil, 1896. Ed. by A. Le Sueur. X. Reprinted, Geneva, 1971.
4. *Oeuvres.* Paris, 1752. X.
4a. ——— Another ed., corr. and enl., 4 vols, Lyons, 1768. RM. Reprinted, Hildesheim, 1965.
5. *Maupertuis, le savant et le philosophe: Présentation et extraits.* Paris, 1964. Ed. by E. Callot.

3090. SARRABAT, Nicolas (pseudonym: DE LA BAÏSSE). 1698-1737. Fr. P II; BMNH (LA BAÏSSE).

3091. FAGGOT, Jacob. 1699-1777. Swed. P I; GA.

3092. HARMENS, Gustav. 1699-1774. Swed. P I & Supp.; BLA.

3093. BRÖNDLUND, Lorenz Jens. 1700-1750. Den. P I.

3094. MORTIMER, Cromwell. ca. 1700-1752. Br. ICB; P II; GD. Secretary of the Royal Society, 1730-52.

3095. BÜCHNER, Andreas Elias. 1701-1769. Ger. P I; BMNH; BLA & Supp.

3096. FERRARI (or FORTUNATUS A BRIXIA), Geronimo. 1701-1754. It. P I.
1. *Philosophia sensuum mechanica methodice tractata.* Brescia, 1735. X. Another ed., 4 vols, 1745-48.

3097. BÉRAUD, Laurent. 1703-1777. Fr. ICB (BÉRAUT); P I; BLA; GA. See also 5983/2.

3098. CALANDRINI, Giovanni Ludovico. 1703-1758. Switz. P I; GA.

3099. MARTIN, Benjamin. 1704?-1782. Br. DSB; ICB; P II; BMNH; GA; GD. Also Mathematics*, Astronomy*, Physics*, and Biography*.

1. *The philosophical grammar: Being a view of the present state of experimental physiology, or natural philosophy.* London, 1735. X. 2nd ed., with additions, 1738.
2. *Bibliotheca technologica; or, A philological library of literary arts and sciences.* London, 1737. RM.
3. *Philosophia Britannica; or, A new and comprehensive system of the Newtonian philosophy, astronomy and geography.* 2 vols. Reading, 1747. RM. 2nd ed., 1759.
4. *A plain and familiar introduction to the Newtonian philosophy.* London, 1751. RM. Another ed., 1754.
5. *The young gentleman and lady's philosophy, in a continued survey of the works of nature and art.* 2 vols. London, 1755. X. 2nd ed., 1772.

3100. TESKE, Johann Gottfried. 1704-1772. Ger. P II.

3101. GENSSANE, ... de. d. 1780. Fr. P I.

3102. PERLICZY, Johann Daniel. 1705-1778. Hung. P III; BLA (PERLITZI).

3103. SPLEISS, Thomas. 1705-1775. Switz. P II,III; GC.

3104. ASCLEPI, Giuseppe Maria. 1706-1776. It. P I. Also Astronomy*.

3105. HILL, John. 1707?-1775. DSB; ICB; P I; BMNH; BLA; GA; GB; GD. Also Astronomy*, Mineralogy*, Natural History*, Botany*, and Zoology*. See also Index.
1. *A review of the works of the Royal Society of London.* London, 1751. RM/L.

3106. PRINGLE, John. 1707-1782. Br. DSB; ICB; P II; BLA; GA; GB; GD.
1. *Six discourses delivered* ... [to the Royal Society] *on the occasion of six annual assignments of Sir Godfrey Copley's medal.* London, 1783. Ed. with a biography by A. Kippis.

3107. STOCK, Johann Christian. 1707-1759. Ger. P II; BLA.

3108. STRÖMER, Märten. 1707-1770. Swed. ICB; P II.

3109. ARENA, Filippo. b. 1708. It. P I; BMNH.

3110. COTHENIUS, Christian Andreas. 1708-1789. Ger. ICB; P I; BMNH; BLA & Supp.

3111. DENSO, Johann Daniel. 1708-1795. Ger. P I; BMNH; GC. See also 8666/1a.

3112. MITCHELL, John. d. 1768. Br./U.S.A. ICB; P II; GD; GE.

3113. ELVIUS, Peter. 1710-1749. Swed. P I; GA.

3114. BOSCOVICH, Ruggiero Giuseppe. 1711-1787. It./Fr. DSB (BOŠKOVIĆ); ICB; P I; GA; GB. Also Mathematics, Astronomy*, Mechanics*, Physics*, and Geodesy*. See also 5532/1 and 15569/1.
1. *Giornale di un viaggio da Constantinopoli in Polonia.* Bassano, 1784. RM.
2. *Opera pertinentia ad opticam et astronomiam , maxima ex parte nova et omnia hususque inedita.* 5 vols. Ib., 1785.

3115. JACQUIER, François. 1711-1788. Fr./It. P I; GA. See also 2539/1d.

3116. LOMONOSOV, Mikhail Vasilievich. 1711-1765. Russ. DSB; ICB; P I; GA; GB. Also Physics and Chemistry*.
   1. [In Russian. *Collected works.*] 10 vols & supp. Moscow, Academy of Sciences, 1950-59.
   2. [In Russian. *Selected philosophical works.*] Moscow, 1950. Ed. by G.S. Vasetskogo.
   3. *Ausgewählte Schriften.* 2 vols. Berlin, 1961. Trans. from the Russian by H. Hösel and E. John.
   4. *Lomonosov: Sa vie, son oeuvre.* Paris, 1967. Selected texts, trans. from the Russian, with introd. and notes, by L. Langevin.

3117. BÄCK, Abraham. 1713-1795. Swed. ICB; P I; BMNH; BLA & Supp.; GA.

3118. KNUTZEN, Martin. 1713-1751. Ger. P I; GA; GC.

3119. BAMMACARI, Niccolo. d. ca. 1778. It. P I & Supp.

3120. GISSLER, Nils. 1715-1771. Swed. P I; BLA & Supp.

3121. LANGER, Georg. 1716-1778. Prague. P I.

3122. MAYER, Andreas. 1716-1782. Ger. P II.

3123. SCHERFFER, Karl. 1716-1783. Aus. P II.

3124. STEPLING, Joseph. 1716-1778. Prague. DSB; ICB; P II; GC.

3125. SCHMID, Nicolaus Ehrenreich Anton. 1717-1785. Ger. P II.

3126. FORBIN, Gaspard François Anne de. 1718-ca. 1780. Fr. P I.

3127. GRÄFENHAHN, Wolfgang Ludwig. 1718-1767. Ger. P I; GA.

3128. MILLER, Gerhard Andreas. 1718-1762. Ger. P II; BLA.

3129. BARBIERI, Ludovico. 1719-after 1756. It. P I.
   1. *Storia del mare, e confutazione della favola, dove scopronsi insigni errori di vari scrittori e specialmente del Signor di Buffon.* Vinegia, 1782. RM.

3130. GADOLIN, Jacob. 1719-1802. Fin. P I.

3131. DELIUS, Heinrich Friedrich von. 1720-1791. Ger. P I; BMNH; BLA & Supp.; GA; GC.

3132. SULZER, Johann Georg. 1720-1779. Ger. P II; GC.

3133. MATSKO, Johann Matthias. 1721-1796. Ger. P II; GA.

3134. SAGNER, Caspar. 1721-1781. Prague. P II.

3135. SILBERSCHLAG, Johann Esaias. 1721-1791. Ger. P II; GC.

3136. BUCK, Friedrich Johann. 1722-1786. Ger. P I.

3137. GOUSSIER, Louis Jacques. 1722-1799. Fr. P I; GA.

3138. KENNEDY, Ildephons. 1722-1804. Ger. P I; GA.

3139. MYLIUS, Christlob. 1722-1754. Ger. P II; BMNH; GC.

3140. SUCKOW (or SUCCOV), Lorenz Johann Daniel. 1722-1801. Ger. P II; GC.

3141. BERGMANN, Joseph (first of the name). 1723-after 1786. Prague. P I.

3142. LÖHE, Johann Conrad. 1723-1768. Ger. P I.

3143. POLANSKY, Nepomuk. 1723-1776. Olmütz. P II.

3144. ARNOLD, Johann Christian. 1724-1765. Ger. P I.

3145. DELARBRE, Antoine. 1724-1811. Fr. P I & Supp; BMNH; GA.

3146. GRANT, Bernhard. 1724-1796. Ger. P I & Supp.

3147. HELFENZRIEDER, Johann Evangelist. 1724-1803. Ger. P I.

3148. LOUS, Christian Karl. 1724-1804. Den. P I.

3149. MEISTER, Albrecht Ludwig Friedrich. 1724-1788. Ger. P II.

3150. CARL, Anton Joseph. 1725-1800. Ger. P I & Supp.; BLA; GA.

3151. KIRCHHOF, Nicolaus Anton Johann. 1725-1800. Ger. P I; GC.

3152. ACKERMANN, Johann Friedrich. 1726-1804. Ger. P I; BLA.

3153. LIMBOURG, Jan Philip de. 1726-1811. Belg. P I; GA.

3154. MORAND, Jean François Clement. 1726-1784. Fr. ICB; P II; BLA; GA.

3155. ROSSIGNOL, Jean Louis. 1726-1817. Fr./It. P II.

3156. ROUX, Augustin. 1726-1776. Fr. P II; BLA; GA.

3157. HAMBERGER, Adolph Friedrich. 1727-1750. Ger. P I; BLA.

3158. JUNG, Johann. 1727-1793. Ger. P I.

3159. ANNONE (or ANNONI), Johann Jacob d'. 1728-1804. Switz. P I.

3160. FRISI, Paolo. 1728-1784. It. DSB; P I; GA; GB. Also Astronomy*, Mechanics*, and Physics. See also Index.
1. *Dissertationum varium....* Lucca, 1759. RM.
2. *Opusculi filosofici.* Milan, 1781. RM.
3. *Opera.* 3 vols. Milan, 1782-85. RM.

3161. HORREBOW, Peter. 1728-1812. Den. P I.

3162. HUNTER, John. 1728-1793. Br. DSB; ICB; P I; BMNH; Mort.; BLA & Supp.; GA; GB; GD. Also Geology*, Zoology*, and Anatomy. See also Index.
1. *Works.* 4 vols & atlas. London, 1835-37. Ed. with notes by J.F. Palmer. Includes a biography.
2. *Essays and observations on natural history, anatomy, physiology, psychology and geology. Being his posthumous papers on those subjects.* 2 vols. London, 1861. RM/L.
3. *Letters ... from John Hunter to Edward Jenner.* London, Royal College of Surgeons, 1976.

3163. MARIVETZ, Etienne Claude (or Clément), *Baron* de. 1728-1794. Fr. ICB; P II; GA.
1. *Physique du monde.* 5 vols & supp. Paris, 1780-87. Includes cosmogony, astronomy, optics, and chemistry. RM.

3164. STATTLER, Benedict. 1728-1797. Ger. P II; GC.

3165. WALDIN, Johann Gottlieb. 1728-1795. Ger. P II; BMNH.

3166. AGRICOLA, Joseph. 1729-1777. Ger. P I & Supp.

3167. SPALLANZANI, Lazzaro. 1729-1799. It. DSB; ICB; P II; BMNH;

Mort.; BLA; GA; GB. Also Natural History*, Embryology*, Physiology*#, and Microbiology. See also 10837/7.

3168. BEAUSOBRE, Louis de. 1730-1783. Fr./Ger. GA; GC.
1. *Dissertations philosophiques, dont la première roule sur la nature du feu, et la seconde sur les différentes parties de la philosophie et des mathématiques.* Paris, 1753. RM/L.

3169. MATANI, Antonio Maria. 1730-1779. It. P II; BMNH; BLA; GA.

3170. GOEZE, Johann August Ephraim. 1731-1793. Ger. BMNH; BLA & Supp. Also Zoology*.
1. *Nützliches Allerley aus der Natur und dem gemeinen Leben.* 6 vols. Leipzig, 1785-88. RM.

3171. WALKER, Adam. 1731-1821. Br. P II; GD.
1. *Analysis of a course of lectures on natural and experimental philosophy.* 2nd ed. [Manchester, 1771?] 6th ed., [London, 1790?] Another ed., [1795?]
2. *A system of familiar philosophy in twelve lectures.* London, 1799. Deals with topics from chemistry, mechanics, physics, and astronomy. RM/L. Another ed., 2 vols, 1802.

3172. DICQUEMARE, Jacques François. 1733-1789. Fr. P I; GA.

3173. PASUMOT, François. 1733-1804. Fr. P II & Supp.; GA.

3174. PRIESTLEY, Joseph. 1733-1804. Br. DSB; ICB; P II; GA; GB; GD; GE. Also Physics*, Chemistry*, and Autobiography*.
1. *Disquisitions relating to matter and spirit.* London, 1777. RM.
2. *Discourses on various subjects.* Birmingham, 1787. RM.
3. *Scientific correspondence.* New York, 1892. Ed. with notes by H.C. Bolton. Reprinted, ib., 1969.
4. *Selections from his writings.* University Park, Penn., 1962. Ed. by I.V. Brown.
5. *Writings on philosophy, science, and politics.* New York, 1965. Ed. by J.A. Passmore.

3175. WIEDEBURG, Johann Ernst Basilius. 1733-1789. Ger. P II; GC.

3176. SCHINZ, Salomon. 1734-1784. Switz. P II; BMNH; BLA.

3177. BERGMAN, Torbern Olof. 1735-1784. Swed. DSB; ICB; P I & Supp. BMNH; GA; GB. Also Chemistry*# and Mineralogy*.

3178. BJERKANDER, Claes. 1735-1795. Swed. P I.

3179. KING, Edward. ca. 1735-1807. Br. P I; BMNH; GA; GD.

3180. MANN, Theodore Augustin. 1735-1809. Br./Belg. ICB; P II; GA; GD.

3181. BERGMANN, Joseph (second of the name). 1736-1803. Ger. P I; BMNH.

3182. FORDYCE, George. 1736-1802. Br. ICB; P I; BMNH; BLA & Supp.; GA; GD. Also Physiology*.

3183. SCHEIBEL, Johann Ephraim. 1736-1809. Ger. P II.

3184. HUBE, (Johann) Michael. 1737-1807. Warsaw. P I; GA.
1. *Vollständiger und fasslicher Unterricht in der Naturlehre, mit allen neuen Entdeckungen.* 4 vols. Leipzig, 1793-94.

3185. ALZATE Y RAMÍREZ, José. 1738-1799. Mexico. DSB; P I; BMNH; GA.

3186. STEIGLEHNER, Cölestin (or Georg Christoph). 1738-1819. Ger. P II; GC.

3187. ABILDGAARD, Peter Christian. 1740-1801. Den. P I; BMNH; BLA & Supp.

3188. ANDRES, Juan. 1740-1817. It. ICB; P I; GA; GB.

3189. CANOVAI, Stanislao. 1740-1811. It. P I & Supp.; GA.
   1. *Elementi di fisica matematica*. Florence, 1788. With G. del Ricco. RM.

3190. AMORETTI, Carlo. 1741-1816. It. P I; BMNH; GA.

3191. BARCA, Alessandro. 1741-1814. It. P I.

3192. GÜSSMANN, Franz. 1741-1806. Aus. P I.

3193. KNOCH, August Wilhelm. 1742-1818. Ger. P I; BMNH.

3194. REISER, Sebastian Jacob Wilhelm. b. 1742. Ger. P II.

3195. PATTERSON, Robert Maskell. 1743-1824. U.S.A. ICB; P II; GE.

3196. PLANER, Johann Jacob. 1743-1789. Ger. P II; BMNH.

3197. ZALLINGER ZUM THURN, Franz Seraphim. 1743-1828. Aus. P II.

3198. ZIMMERMANN, Eberhard August Wilhelm von. 1743-1815. Ger. ICB; P II; BMNH; GC.

3199. DALBERG, Carl Theodor Anton von. 1744-1817. Ger. ICB; P I; GA; GB; GC.

3200. FALCONER, William. 1744-1824. Br. BMNH; BLA; GA; GD. Also Chemistry*.

3201. GERSDORF, Adolph Traugott von. 1744-1807. Ger. P I; GA.

3202. GRUBER, Tobias. 1744-1806. Aus./Prague. P I; BMNH.

3203. HEURLIN, Samuel. 1744-1835. Swed. P I; GA.

3204. LAMARCK, Jean Baptiste Pierre Antoine de Monet, *Chevalier* de. 1744-1829. Fr. DSB; ICB; P I; BMNH; GA; GB. Also Science in General*, Chemistry*, Geology*, Natural History*, Botany*, Zoology*, and Evolution*. See also Index.
   1. *The Lamarck manuscripts at Harvard*. Cambridge, Mass., 1933. Ed. by W.M. Wheeler and T. Barbour.
   2. *Inédits de Lamarck*. Paris, 1972. Ed. by M. Vachon et al.

3205. DEMESTE, Jean. 1745-1783. Belg. P I; GF.

3206. FÉRUSSAC, Jean Baptiste Louis d'Audebard de. 1745-1815. Fr. P I; BMNH; GA.

3207. BIRKHOLZ, Adam Michael. 1746-1818. Ger. P I; BMNH; BLA.

3208. COMPARETTI, Andrea. 1746-1801. It. P I; Mort.; BLA & Supp.; GA.

3209. WESTFELD, Christian Friedrich Gotthard Henning. 1746-1823. Ger. P II; GC.

3210. BURROW, Reuben. 1747-1792. Br. ICB; P I; GA; GD.

3211. HALES, William. 1747-1831. Br. P I; GD.

3212. HOPF, Philipp Heinrich. 1747-1804. Ger. P I.

3213. THOUVENEL, Pierre. 1747-1815. Fr. P II; BLA; GA.
 1. *Mémoire physique et médicinal, montrant des rapports évidens entre les phénomènes de la baguette divinatoire, du magnétisme et de l'électricité.* Paris, 1781. RM/L.

3214. RAPPOLD, Wilhelm Gottlieb. 1748-1808. Ger. P II.

3215. ROUSSEL, Henri François Anne de. 1748-1812. Fr. P III; BMNH.

3216. GOETHE, Johann Wolfgang von. 1749-1832. Ger. The famous writer. DSB; ICB; P I; BMNH; Mort.; G. Also Physics*, Geology, Botany*, and Zoology. See also Index.
 1. *Die Schriften zur Naturwissenschaft.* Many vols. Weimar, 1947-. Ed. under the auspices of the Deutsche Akademie der Naturforscher zu Halle.

3217. ADAMS, George (second of the name). 1750-1795. Br. P I; BMNH; BLA; GA; GD. Also Mathematics*, Astronomy*, Physics*, Microscopy*, and Physiology*.

3218. DENTAND, Pierre Gédéon. 1750-1780. Switz. P I; GA.

3219. GIROD-CHANTRANS, Justin. 1750-1841. Fr. P I; BMNH; GA.

3220. MARTÍ FRANQUÉS (or MARTÍ D'ARDENYA), Antonio de. 1750-1832. Spain. DSB; ICB; P II (MARTÍ Y TRANQUÉS).

3221. MEIDINGER, Carl von. 1750-1820. Aus. P II; BMNH. Also Zoology*.

3222. BERNOULLI, Daniel (second of the name). 1751-1834. Switz. P I; BLA.

3223. LEFÈVRE-GINEAU, Louis. 1751-1829. Fr. P I; GA.

3224. MONGEZ, Jean André. 1751-1788. Fr. P II; BMNH.

3225. PFINGSTEN, Johann Hermann. 1751-1798. Ger. P II; BMNH; BLA.

3226. PICKEL, Johann Georg. 1751-1838. Ger. P II; BLA.

3227. VOIGT, Johann Heinrich. 1751-1823. Ger. P II; BMNH; GC.

3228. DAETZL (or DAEZEL), Georg Anton. 1752-1847. Ger. P I.

3229. DALLA DECIMA, Angelo. 1752-1825. It. P I; BMNH (DECIMA); BLA.

3230. FABBRONI, Giovanni Valentino Mattia. 1752-1822. It. DSB; ICB; P I; BLA (FABRONI); GA (FABRONI).

3231. GAELLE, Meingosus. b. 1752. Aus. P I.

3232. LAMANON, Robert de Paul de. 1752-1787. Fr. P I; BMNH.

3233. MAYER, Joseph. 1752-1814. Prague/Aus. P II.

3234. PICTET, Marc Auguste. 1752-1825. Switz. DSB; P II; BMNH; GA. Also Physics.

3235. GOSSE, Henri Albert. 1753-1816. Switz. P I; BLA & Supp.; GA.

3236. FOSSOMBRONI, Vittorio. 1754-1844. It. P I; GB.

3237. LANGGUTH, Christian August. 1754-1814. Ger. P I; BMNH; BLA.

3238. PILATRE DE ROZIER, Jean François. 1754-1785. Fr. DSB; ICB; P II; GA.

3239. KÖSTLIN, Karl Heinrich. 1755-1783. Ger. P I; BLA.

3240. PRÉVOST, Isaac Bénédict. 1755-1819. Fr. DSB; ICB; P II; BMNH; GA. Also Microbiology.

3241. BÖTCHER, Nicolaus. 1756-1821. Den. P I.

3242. MATHIEU, Charles Léopold. b. 1756. Fr. P II,III.

3243. WELLS, William Charles. 1757-1817. Br./U.S.A. DSB; ICB; P II; BMNH; BLA; GD; GE. Also Meteorology* and Physiology*.
1. *Two essays, one upon single vision with two eyes, the other on dew; A letter to ... Lord Kenyon; and An account of a female of the white race of mankind, part of whose skin resembles that of a negro ... With a memoir of his* [i.e. Wells'] *life, written by himself.* London, 1818.

3244. BOSSI, Luigi. 1758-1835. It. P I.

3245. HEINRICH, Placidus. 1758-1825. Ger. ICB; P I.

3246. MEURER, Heinrich. b. 1758. Ger. P II.

3247. CANALI, Luigi. 1759-1841. It. P I.

3248. BEDDOES, Thomas. 1760-1808. Br. DSB; ICB; P I; BLA & Supp.; GA; GB; GD.
1. *Observations on the nature of demonstrative evidence.* London, 1793.
2. *Contributions to physical and medical knowledge.* Bristol, 1799.
See also Index.

3249. KRAMP, Crétien (or Christian). 1760-1826. Ger./Fr. DSB; ICB; P I; BMNH; BLA; GA.

3250. BISCHOF, Carl August Leberecht. 1762-1814. Ger. P I.

3251. REYNIER, Jean Louis Antoine. 1762-1824. Switz. P II; GA.

3252. RHODE, Johann Gottlieb. 1762-1827. Ger. P II; BMNH; GC. Also Geology.

3253. RICHE, Claude Antoine Gaspard. 1762-1797. Fr. P II; BMNH; BLA; GA.

3254. SILVESTRE, Augustin François de. 1762-1851. Fr. ICB; BMNH; GA.

3255. BRUGMANS, Sebald Justin. 1763-1819. Holl. ICB; P I; BMNH; BLA & Supp.; GA.

3256. HUTH, Johann Sigismund Gottfried. 1763-1818. Ger./Russ. P I.

3257. VIETH, Gerhard Ulrich Anton. 1763-1836. Ger. P II; GC.

3258. NIEUWLAND, Pieter. 1764-1794. Holl. P II; GA.

3259. BAADER, Franz Xaver von. 1765-1841. Ger. P I; BLA; GA; GB; GC.
1. *Sämmtliche Werke.* Leipzig, 1856-60. RM.

3260. KIELMEYER, Karl Friedrich von. 1765-1844. Ger. DSB; ICB; P I; BMNH; BLA; GA; GC. Also Chemistry, Zoology, and Physiology*.

3261. AZAIS, Pierre Hyacinthe. 1766-1845. Fr. P I & Supp.; GA; GB.

3262. HAUFF, Johann Karl Friedrich. 1766-1846. Ger./Belg. P I.

3263. RASCHIG, Christoph Eusebius. 1766-1827. Ger. P II; BLA.

3264. WREDE, Karl Friedrich. 1766-1826. Ger. P II.

3265. LINK, Heinrich Friedrich. 1767-1851. Ger. DSB; P I; BMNH; BLA; GA; GC. Also Botany*.

3266. METZ, Andreas. b. 1767. Ger. P II.

3267. KRIES, Friedrich Christian. 1768-1849. Ger. P I; GC.
1. *Von den Ursachen der Erdheben und von den magnetischen Erscheinungen: Zwey Preiss-Schriften.* Leipzig, 1827. RM.

3268. HUMBOLDT, Alexander von. 1769-1859. Ger./Fr. DSB; ICB; P I, III; BMNH; G. Also Chemistry*, Geology*, Other Earth Sciences*, Natural History*, Botany*, and Physiology*. See also Index.
1. *Ansichten der Natur, mit wissenschaftlichen Erläuterungen.* Tübingen, 1808. X. Reprinted, Stuttgart, 1969. 2nd ed., 2 vols, Stuttgart, 1826. 3rd ed., 1849; re-issued, 1860.
1a. ——— *Aspects of nature, in different lands and different climates. With scientific elucidations.* 2 vols. London, 1849. Trans. by Mrs Sabine. RM/L. Another ed., 1850; reprinted, New York, 1870. Another ed., 1894.
1b. ——— *Views of nature ... with scientific illustrations.* London, 1850. Trans. by E.C. Otté and H.G. Bohn. RM/L.
2. *Kosmos: Entwurf einer physischen Weltbeschreibung.* 5 vols & atlas. Stuttgart, 1845-62. RM/L. Other eds, 4 vols, 1870 and 1877.
2a. ——— *Cosmos: A general survey of the physical phenomena of the universe.* 2 vols. London, 1845-48. Trans. by A. Prichard.
2b. ——— *Cosmos: Sketch of a physical description of the universe.* 4 vols. London, 1846-58. Trans. by Mrs Sabine.
2c. ——— *Cosmos: A sketch of a physical description of the universe.* 5 vols. London, 1849-58. Trans. by E.C. Otté. RM/L. Another ed., 1871-72.

Correspondence

3. *Briefe an Varnhagen von Ense ... 1827-1858.* Leipzig, 1860. X. 2nd ed., 1860. RM. 3rd ed., 1860.
3a. ——— *Letters written ... 1827-1858 to Varnhagen von Ense.* London, 1860. RM/L.
4. *Correspondance inédite scientifique et littéraire ... suivie de la biographie des principaux correspondents de Humboldt.* Paris, 1864. (Part of the *Oeuvres*--see item 10.)
5. *Briefe an C.C.J. Freiherr von Bunsen.* Leipzig, 1969. RM.
6. *Briefe zwischen A. von Humboldt und Gauss.* Leipzig, 1877. Ed. by K. Bruhns. RM.
6a. ——— *Briefwechsel zwischen A. von Humboldt und Gauss.* Berlin, 1977. Newly ed. by K.R. Biermann.
7. *Lettres américaines (1798-1807).* Paris, 1905. Ed. by E.T. Hamy. RM.
8. [In Russian. *Correspondence with scientists and government officials in Russia.*] Moscow, Academy of Sciences, 1962. Ed. by D.I. Sherbakov.
9. *Die Jugendbriefe, 1787-1799.* Berlin, 1973. Ed. by I. Jahn and F.G. Lange.

Collected Works

10. *Oeuvres*. 11 vols. Paris, 1864. X.
11. *Gesammelte Werke*. 12 vols. Stuttgart, 1889. RM/L. Both this and item 10 are far from complete.
12. *Gespräche*. Berlin, 1959. Ed. under the auspices of the Humboldt Commission of the Berlin Academy.

3269. ALTEN, Johann Wilhelm von. b. 1770. Ger. P I; BMNH.

3270. WILDT, Johann Christian Daniel. 1770-1844. Ger. P II.

3271. GÜNTHER, Johann Jacob. 1771-1852. Ger. P I; BLA.

3272. SCHULTES, Joseph August. 1773-1831. Aus./Ger. P II; BMNH; BLA. See also 11418/16d.

3273. GRUITHUISEN, Franz von Paula. 1774-1852. Ger. P I; BLA; GA; GC.

3274. HORNER, Johann Kaspar. 1774-1834. Ger./Switz. P I; GC.

3275. SCHRÖDER, Johann Friedrich Ludwig. 1774-1845. Holl. P II; GF.

3276. CANTOR, Johann Chrysostomus. 1775-1815. Ger. P I.

3277. KIDD, John. 1775-1851. Br. DSB; P I; BMNH; GB; GD. Also Science in General* and Mineralogy*.

3278. AHLSTEDT, Johann Frederick. 1776-1823. Fin. P I.

3279. MUDIE, Robert. 1777-1842. Br. BMNH; GA; GD. Also Earth Sciences*.

3280. PANSNER, (Johann Heinrich) Lorenz von. 1777-1851. Russ. P II; BMNH.

3281. DRAPIEZ, Pierre Auguste Joseph. 1778-1856. Fr./Belg. P III; BMNH; GF (DRAPIER).

3282. MILANO, Michele. 1778-1843. It. P II.

3283. BERGER, Jean François. 1779-1833. Switz. P I.

3284. FLINT, Timothy. 1780-1840. U.S.A. P I; GB; GE.
    1. *Lectures upon natural history, geology, chemistry*, [etc.] Boston, 1833. RM.

3285. SCHUBERT, Gotthilf Heinrich von. 1780-1860. Ger. ICB; P II; BMNH; BLA; GC. A very prolific author. Also Science in General*, Geology*, and Natural History*. See also 6160/3 and 12705/1.

3286. WUCHERER, Gustav Friedrich. 1780-1843. Ger. P II; GC.

3287. BERNOULLI, Christoph. 1782-1863. Switz. P I,III; BMNH; GA; GC.

3288. SCHULZE-MONTANUS, Karl August. 1782-1823. Ger. P II. Also Chemistry*.

3289. PAOLI, Domenico. 1783-1849. It. P II,III.

3290. PALMSTEDT, Carl. 1785-1870. Swed. P II,III.

3291. LAGERHJELM, Per. 1787-1856. Swed. P I,III; BMNH.

3292. PLEISCHL, Adolph Martin. 1787-1867. Prague/Aus. P II,III.

3293. BREDA, Jacques Gisbert Samuel van. 1788-1867. Holl. P I,III; BMNH; GF.

3294. CAP (or GRATACAP), Paul Antoine. 1788-1877. Fr. P III; BMNH; GA. Also Natural History*. See also 1473/3.

3295. RENDU, Louis. 1789-1858. Fr. P II; BMNH; GA.

3296. THILO, Ludwig. 1789-1831. Ger./Switz. P II.

3297. GREEN, Jacob. 1790-1841. U.S.A. DSB; ICB; P III; BMNH; GE. Also Botany*.

3298. HOMBRES-FIRMAS, Louis Augustin d'. 1790-1857. Fr. P I; BMNH; GA.

3299. OLMSTED, Denison. 1791-1859. U.S.A. P II; GA; GB; GE.

3300. PARROT, Johann (Jacob) Friedrich (Wilhelm) von. 1791-1841. Russ. ICB; P II; BMNH; BLA; GA; GC.

3301. CARPI, Pietro. 1792-1861. It. P III.

3302. EGEN, P.N.C.E. 1793-1849. Ger. P I; GC.

3303. LARDNER, Dionysius. 1793-1859. Br. ICB; P I & Supp.; BMNH; GA; GB; GD. Also Science in General*, Mathematics*, Astronomy*, Mechanics*, Physics*, Geology*, and Physiology*. See also Index.

3304. WHEWELL, William. 1794-1866. Br. DSB; ICB; P II,III; BMNH; GB; GD. Also Science in General*, Astronomy*, Mechanics*, Mineralogy, and History of Science*. See also Index.

3305. DAUBENY, Charles Giles Bridle. 1795-1867. Br. DSB; ICB; P I, III; BMNH; GB; GD. Also Chemistry*, Geology*, Climatology*, Botany, and History of Botany*.
   1. *Miscellanies: Being a collection of memoirs and essays on scientific and literary subjects.* 2 vols. Oxford, 1867. RM.

3306. QUETELET, (Lambert) Adolphe (Jacques). 1796-1874. Belg. DSB; ICB; P II,III; GA; GB. Also Mathematics*, Physics*, Earth Sciences*, and History of Science*.
   1. *Sur l'homme et le développement de ses facultés; ou, Essai de physique social.* 2 vols. Paris, 1835. X.
   1a. ——— *A treatise on man and the development of his faculties.* Edinburgh, 1842. Trans. by R. Knox.
   2. *Lettres ... sur la théorie des probabilités appliquée aux sciences morales et politiques.* Brussels, 1846.
   2a. ——— *Letters ... on the theory of probabilities, as applied to the moral and political sciences.* London, 1849. Trans. by O.G. Downes. RM/L.
   3. *Du système social et des lois qui le régissent.* Paris, 1848. RM.
   4. *Physique social; ou, Essai sur le développement des facultés de l'homme.* 2 vols. Brussels, 1869. RM/L.
   5. *Inventaire de la correspondance.* Brussels, 1966. Ed. by L.W. de Donder.

3307. ALMROTH, Nils Wilhelm. 1797-1852. Swed. P I; BMNH.

3308. MORTON, Samuel George. 1799-1851. U.S.A. DSB; P II; BMNH; Mort.; BLA; GA; GE. See also 9364/1.

3309. SCHRÖN, Heinrich Ludwig Friedrich. 1799-1875. Ger. P II,III; BMNH; GC.

3310. BAILLY (DE MERLIEUX), Charles François. 1800-1862. Fr. P I; BNMH.

3311. FECHNER, Gustav Theodor. 1801-1887. Ger. The psychologist. DSB; ICB; P I,III,IV; BMNH; Mort.; BLA & Supp.; GB; GC. Also Physics*.
1. *Nanna; oder, Über das Seelenleben der Pflanzen*. Leipzig, 1848. X. Reprinted with an introd. by K. Lasswitz, Hamburg, 1899.
2. *Professor Schleiden und der Mond*. 2 parts. Leipzig, 1856. RM. Re this and item 1 see DSB, Vol. XII, 175.

3312. GUÉRANGER, Edouard Auguste François. b. 1801. P III; BMNH.

3313. BORENIUS, Henrik Gustaf. b. 1802. Fin. P I,III.

3314. BRAYLEY, Edward William. 1802-1870. Br. P III; BMNH; GD. See also 7297/1.

3315. PFEIL UND KLEIN ELLGUTH, Friedrich Ludwig. 1803-1894. Ger. P IV.

3316. BACHE, Alexander Dallas. 1806-1867. U.S.A. DSB; ICB; P I,III; GB; GE. Also Earth Sciences.

3317. BLANCHET, Rodolphe. b. 1807. Switz. P I; BMNH.

3318. BARNARD, Frederick Augustus Porter. 1809-1889. U.S.A. P III, IV; GB; GE.

3319. HAEDENKAMP, Hermann. 1809-1860. Ger. P I & Supp.

3320. KARSTEN, Hermann. 1809-1877. Ger. P I,III; BMNH; GC.

3321. AIMÉ, Georges. 1810-1846. Fr./Algeria. P I; GF.

3322. GIBBES, Lewis R. 1810-1894. U.S.A. P III; BMNH.

3323. SCHUMACHER, Christian Andreas. 1810-1854. Den. P II,III.

3324. SENFF, Karl Eduard. 1810-1849. Dorpat. P II.

3325. BRAVAIS, Auguste. 1811-1863. Fr. DSB; ICB; P I,III; BMNH; GA. Also Crystallography*.

3326. CARPENTER, William Benjamin. 1813-1885. Br. DSB; ICB; P III; BMNH; BLA & Supp.; GB; GD. Also Botany*, Zoology*, Physiology*, and Microscopy*. See also Index.
1. *Mechanical philosophy, horology and astronomy*. New edition. London, 1857.
2. *Nature and man: Essays scientific and philosophical*. London, 1888. Reprinted, Farnborough, 1970.

3327. MOBERG, Adolf. 1813-1895. Fin. P II,III,IV; BMNH.

3328. SILJESTRÖM, Per Adam. 1815-1892. Swed. P II,III,IV; BMNH.

3329. SCHNEIDER, Jacob. 1818-1898. Ger. P II,III; GC.

3330. VAUGHAN, Daniel. ca. 1818-1879. U.S.A. GE.

3331. BARRAL, Jean Auguste. 1819-1884. Fr. P I,III,IV; BMNH; GA. See also 4850/4.

3332. FIGUIER, Louis Guillaume. 1819-1894. Fr. P I,III; BMNH; GA. Also Geology*, Botany*, Zoology*, Biography*, and History of Science*.

1. *Les merveilles de la science.* 4 vols. Paris, 1868-70.

3333. CHASE, Pliny Earle. 1820-1886. U.S.A. P III,IV; GE.

3334. ULE, Otto Eduard Vincenz. 1820-1876. Ger. P II,III; BMNH; GC.
   1. *Das Weltall: Beschreibung und Geschichte des Kosmos im Entwicklungskampfe der Natur.* 3 vols. Halle, 1850. X. 2nd ed., 1853.
   2. *Warum und weil: Fragen und Antworten aus der Naturlehre.* Berlin, 1868. X. Another ed., 1882.

3335. DUPONCHEL, Adolphe. 1821-after 1900. Fr. P III,IV.

3336. HAUGHTON, Samuel. 1821-1897. Br. P III,IV; BMNH; Mort.: GB; GD1. Also Astronomy*, Mechanics*, Geology*, Climatology*, Natural History*, and Physiology*. See also 6362/1.

3337. HOUGH, Franklin Benjamin. 1822-1885. U.S.A. P III; GE.

3338. TOSCANI, Cesare. 1822-1889. It. P III,IV.

3339. SERPIERI, Alessandro. 1823-after 1883. It. P III.

3340. HUXLEY, Thomas Henry. 1825-1895. Br. DSB; ICB; BMNH; Mort.; BLA; GB; GD1. Also Science in General*, Geology*, Biology in General*, Natural History*, Zoology*#, Evolution*, Physiology*, and Autobiography*. See also Index.

3341. LORENZ (VON LIBURNAU), Josef Roman. 1825-1911. Aus. P III,IV; BMNH.

3342. DEICKE, Hermann Gustav. 1826-after 1896. Ger. P III,IV; BMNH.

3343. HENNESSY, Henry G. 1826-1901. Br. P I,III,IV; GD2.

3344. DELLINGSHAUSEN, Nikolai. 1827-1896. Russ. P III,IV.

3345. DUFOUR, Charles. 1827-1902. Switz. P I,III,IV.

3346. REIS, Paul. 1828-1895. Ger. P III,IV.

3347. ZENKER, Wilhelm. 1829-1899. Ger. P IV; BMNH (ZENKER, G.F.G.).

3348. BACALOGLO, Emmanuel. 1830-after 1864. Bucharest. P III.

3349. BURCKHARDT, Karl Friedrich. 1830-1913. Switz. P I,III,IV; BMNH.

3350. EKMAN, Frederik Laurenz. 1830-1890. Swed. P III,IV; BMNH.

3351. LUBBOCK, John, *1st Baron* AVEBURY. 1834-1913. Br. DSB; ICB; BMNH; GB; GD3. Also Science in General*, Geology*, Botany*, and Zoology*.

3352. DRAPER, John Christopher. 1835-1886. U.S.A. P III.

3353. HINRICHS, Gustavus Detlef. 1836-1923. Den./U.S.A. P III,IV & Supp.; BMNH; BLA. Also Chemistry*.

3354. DRONKE, Adolf. 1837-1898. Ger. P III,IV.

3355. FALB, Rudolf. 1838-1903. Aus. P III,IV & Supp.

3356. FORSSMAN, Lars Arvid. 1842-1890. Swed. P III,IV.

3357. BRAHAM, Philip. 1843-after 1881. Br. P III.

3358. DELAUNEY, Félix Julien. 1843-after 1913. Fr. P IV.

3359. FORSTER, Aimé Julius Théophile. 1843-after 1883. Switz. P III.

3360. SCHIÖTZ, Oscar Emil. 1846-after 1921. Nor. P III,IV; BMNH.

3361. EHRLICH, Paul. 1854-1915. Ger. DSB; ICB; Mort.; BLAF. Haematology. Immunology. Several other biomedical fields.
   1. *Gesammelte Arbeiten zur Immunitätsforschung*. Berlin, 1904. Ed. by Ehrlich himself. RM.
   2. *Collected papers*. 3 vols. London, 1956-60. Includes a bibliography. Ed. by F. Himmelweit.

3362. DEKHUIJZEN, Marinus Cornelis. 1859-1924. Holl. BLAF.

3.03 MATHEMATICS

This section is confined as far as possible to pure mathematics; applied mathematics is included in Mechanics, Physics, Astronomy, and occasionally some other sections. However the section includes many mathematicians who, though predominantly concerned with pure mathematics, also published in one or more fields of applied mathematics.

Textbooks of mathematics are very numerous in all periods. They are included only if the author also published original work, or if the textbook had some special significance. Thus authors who published textbooks only are not included, except in special cases.

3363. LA HIRE, Philippe de. 1640-1718. Fr. DSB; ICB; P I; GA. Also Astronomy*, Physics, Geodesy, and Part 2*.

3364. NEWTON, Isaac. 1642-1727. Br. DSB; ICB; P II; G. Also Astronomy*, Physics*#, and Part 2*(the *Principia*). See also Index.
  1. *Arithmetica universalis*. Cambridge, 1707. Ed. by W. Whiston. X. 2nd ed., rev., London, 1722.
  1a. ——— *Universal arithmetick*. London, 1720. Trans. by Ralphson, rev. by Cunn. RM. 2nd ed., 1728. Another ed., 2 parts, 1769.
  1b. ——— *Arithmétique universelle*. Paris, 1802.
  2. *Analysis per quantitatum series, fluxiones, et differentias*. London, 1711. Ed. by W. Jones. RM/L.
  3. *The method of fluxions and infinite series*. London, 1736. "Translated from the author's Latin original not yet made publick." By J. Colson. RM/L.
  3a. ——— *La méthode des fluxions et des suites infinies*. Paris, 1740. Trans. from Colson's English version by Buffon. X. Reprinted, Paris, 1966.
  4. *Tractatus de quadratura curvarum*. Contained in item 2.
  4a. ——— *Sir Isaac Newton's two treatises: Of the quadrature of curves, and Analysis by equations of an infinite number of terms*. London, 1745. Trans. by J. Stewart. RM.
  4b. ——— German trans. Leipzig, 1908. (OKEW. 164)
  5. *The mathematical works*. 2 vols. New York, 1964-67. Ed. by D.T. Whiteside.
  6. *The mathematical papers*. Cambridge, 1967-. Ed. by D.T. Whiteside.

3365. CHRISTOFORO, Giacinto de. b. 1650. It. P I.

3366. TSCHIRNHAUS, Ehrenfried Walter von. 1651-1707. Ger. DSB; ICB; P II; GA; GC. Also Science in General* and Part 2*.

3367. LORENZINI, Lorenzo. 1652-1721. It. P I; GA.

3368. NIEUWENTIJT, Bernard. 1654-1718. Holl. DSB; P II; BLA; GA. Also Science in General* and Part 2.

3369. VARIGNON, Pierre. 1654-1722. Fr. DSB; ICB; P II; GA. Also Mechanics* and Part2*.
1. *Eclairissemens sur l'analyse des infiniment petits*. Paris, 1725. RM.
2. *Elémens de mathématique*. Paris, 1731. RM.
See also 3370/1.

3370. REYNEAU, Charles René. 1656-1728. Fr. DSB; ICB; P II; GA.
1. *Analyse démontrée*. 2 vols. Paris, 1708. X. 2nd ed., 1736-38; with comments by Varignon.

3371. FONTENELLE, Bernard Le Bovier de. 1657-1757. Fr. DSB; ICB; P I; GA; GB. Also Science in General*#, Mechanics*, Biography*, History of Science*, and Part 2*.
1. *Eléments de la géométrie de l'infini*. Paris, 1727. RM/L.

3372. FISCHER, Anton. 1657-1741. Ger. P I.

3373. LEFEBRE, Tanneguy. 1658-1717. Fr./Switz./Br. P III; GA (LEFEBVRE).

3374. GREGORY, David. 1659-1708. Br. DSB; ICB; P I; GA; GB; GD. Also Astronomy*, Physics, and Part 2*.
1. *Euclidis quae supersunt omnia*. Oxford, 1703.
1a. ——— *Euclid's elements of geometry*. London, 1752. Trans. by E. Stone.

3375. SAURIN, Joseph. 1659-1737. Fr. DSB; P II; GA. Also Mechanics.

3376. DORIA, Paolo Matteo. 1662 (or 1666)-1746. It. P I; GA.
1. *Dialoghi ... si esamina l'algebra ed i nuovi metodi de' moderni*. Amsterdam, 1718. RM.

3377. CROUSAZ, Jean Pierre de. 1663-1750. Switz. DSB; ICB; P I & Supp.; GA; GB. Also Various Fields.
1. *Commentaire sur l'analyse des infiniment petits*. Paris, 1721. RM.

3378. HARRIS, John. ca. 1666-1719. Br. DSB; ICB; P I; GA; GB; GD. Also Science in General*, Astronomy*, and Natural History*.
1. *A new short treatise on algebra*. London, 1702. X. 2nd ed., 1705.

3379. PARENT, Antoine. 1666-1716. Fr. DSB; P II. Also Mechanics.

3380. ARBUTHNOT, John. 1667-1735. Br. DSB; ICB; P I; BMNH; BLA & Supp.; GA; GB; GD. Also Various Fields, Physiology* and Part 2*.

3381. BERNOULLI, Johann (or Jean) (first of the name). 1667-1748. Switz. DSB; ICB; P I; BLA & Supp.; GA; GB; GC. Also Mechanics* and Part 2*#.

3382. MOIVRE, Abraham de. 1667-1754. Br. DSB; ICB (under DE); P II; GA; GB; GD.
1. *The doctrine of chances*. London, 1718. RM/L. 2nd ed., 1738; reprinted, London, 1967. 3rd ed., 1756; reprinted, New York, 1967.
2. *Annuities upon lives*. London, 1725. 2nd ed., 1731. 4th ed., 1752.

3. *Miscellanea analytica de seriebus et quadraturis.* London, 1730. RM/L.

3383. SACCHERI, (Giovanni) Girolamo. 1667-1733. It. DSB; ICB; P II. Also Part 2.
1. *Euclides ab omni naevo vindicatus.* Milan, 1733. X.
1a. ——— *Euclides vindicatus.* Chicago, 1920. Ed. with an English trans. by G.B. Halstead.

3384. ARCHINTO, Carlo. 1669-1732. It. P I.

3385. MAGNITSKY, Leonty Filippovich. 1669-1739. Russ. DSB; ICB. Author of the first Russian textbook of mathematics.

3386. RABUEL, Claude. 1669-1728. Fr. P II; GA.

3387. LE POIVRE, Jacques François. d. 1710. Belg. DSB; ICB.

3388. CHEYNE, George. 1671-1743. Br. DSB; ICB; P I; BLA & Supp.; GA; GD. See also 5476/1.

3389. GRANDI, Guido. 1671-1742. It. DSB; ICB; P I; GA. Also Mechanics*.
1. *Geometrica demonstratio Vivianeorum problematum.* Florence, 1699.
2. *Geometrica demonstratio theorematum Hugenianorum circa logisticam seu logarithmicam lineam.* Florence, 1701.

3390. KEILL, John. 1671-1721. Br. DSB; P I; GA; GD. Also Astronomy*, Mechanics*, and Part 2*. See also Index.
1. *Euclid's "Elements of geometry", from the Latin translation of Commandino.* London, 1723. X. 11th ed., 1772.

3391. MARCHETTI, Angelo. 1674-1753. It. ICB; P II.

3392. SHERWIN, Henry. fl. 1705-1717. Br.
1. *Mathematical tables.* London, 1706. Includes tables by Briggs, Wallis, Halley, and Abraham Sharp. 5th ed., rev. by S. Clark, 1770.

3393. GUISNÉE, N. d. 1718. Fr. P I; GA.

3394. DITTON, Humphrey. 1675-1715. Br. P I; GA; GB; GD. Also Mechanics*.
1. *An institution of fluxions.* London, 1706. RM.
See also 4462/2.

3395. JONES, William (first of the name). 1675-1749. Br. DSB; P I; GA; GD. See also 3364/2.

3396. KLAUSING, Heinrich. 1675-1745. Ger. P I.

3397. HERTTENSTEIN, Johann Heinrich. 1676-1741. Strasbourg. P I.

3398. RICCATI, Jacopo Francesco. 1676-1754. It. DSB; P II; GA; GB. Also Mechanics.

3399. LONGOLIUS, Johann Daniel. 1677-1740. Ger. P I.

3400. GREENE, Robert. 1678?-1730. Br. GD. Also Science in General*.
1. *Geometricae solidorum.* Cambridge, 1712. RM.

3401. HERMANN, Jakob. 1678-1733. Switz./It./Ger./Russ. DSB; P I; GA; GC. Also Mechanics*.

3402. MONTMORT (or MONMORT), Pierre Rémond de. 1678-1719. Fr. DSB;

ICB; P II; GA.
1. *Essay d'analyse sur les jeux de hazard*. Paris, 1708. RM.

3403. WOLFF, Christian. 1679-1754. Ger. The philosopher. DSB; ICB; P II (WOLF); BMNH; G. Also Science in General*.
1. *Mathematisches Lexicon*. Leipzig, 1716. X. Reprinted, Hildesheim, 1965.
2. *Anfangsgründe aller mathematischen Wissenschaften*. 5 vols. 1750. (*Gesammelte Werke*. I) Reprinted, Hildesheim, 1973.
3. *Meletemata mathematico-philosophica*. Halle, 1755. (*Gesammelte Werke*. II. Band 35) Reprinted, Hildesheim, 1974.

3404. COLSON, John. 1680-1760. Br. P I; GD. See also Index.

3405. RICHER DU BOUCHET, Claude. 1680-1756. Fr. P II.

3406. MANFREDI, Gabriello. 1681-1761. It. ICB; P II; GA.

3407. COTES, Roger. 1682-1716. Br. DSB; ICB; P I; GA; GB; GD. Also Astronomy and Physics*. See also Index.
1. *Harmonia mensurarum*. Cambridge, 1722. RM/L.

3408. FAGNANO (DEI TOSCHI), Giulio Carlo. 1682-1766. It. DSB; ICB; P I,III; GA (FAGNANI).

3409. SAUNDERSON, Nicholas. 1682-1739. Br. P II; GB; GD.
1. *The elements of algebra*. 2 vols. Cambridge, 1740.
1a. ——— *Select parts of Saunderson's "Elements of algebra" for the uses of students*. London, 1756. X. 2nd ed., 1761. 4th ed., 1776. 5th ed., 1792.
2. *The method of fluxions*. London, 1756. RM/L.

3410. HANSCH, Michael Gottlieb. 1683-after 1752. Ger./Aus. P I.

3411. NICOLE, François. 1683-1758. Fr. P II; GA.

3412. KOES (or KOSIUS), Friedrich. 1684-1766. Ger. P I; GA.

3413. NAUDÉ, Philipp. 1684-1745. Ger. P II; GA.

3414. KASCHUBE (or CASCHUBIUS), Johann Wenceslaus. d. before 1727. Ger. P I.

3415. TAYLOR, Brook. 1685-1731. Br. DSB; ICB; P II; GA; GB; GD.
1. *Methodus incrementorum directa et inversa*. London, 1715.
2. *New principles of linear perspective*. London, 1715. X. 3rd ed., 1749. Another ed., 1811.

3416. BERNOULLI, Nikolaus (first of the name). 1687-1759. Switz. DSB; ICB; P I; GA; GB; GC.

3417. SIMSON, Robert. 1687-1768. Br. DSB; ICB; P II; GB; GD. See also 15418/1.
1. *Sectiones conicae*. Edinburgh, 1735. 2nd ed., 1750.
1a. ——— *The elements of the conic sections*. Glasgow, 1817.
2. *Euclidis Elementorum libri priores sex*. Glasgow, 1756.
2a. ——— *The Elements of Euclid*. Glasgow, 1756. X. 15th ed., 1811.
3. *Opera quaedam reliqua*. Glasgow, 1776.

3418. STRUYCK, Nicolaas. 1687-1769. Holl. P II (STRUIJCK). Also Astronomy*.
1. *Les oeuvres qui se rapportent au calcul des chances*.

Amsterdam, 1912. "Tirées des *Oeuvres complètes* et traduites du hollandais par J.A. Vollgraff." RM.

3419. BRAGELOGNE, Christophe Bernard de. 1688-1744. Fr. P I; GA.

3420. 'sGRAVESANDE, Willem Jacob. 1688-1742. Holl. DSB; P I; GA (under S). Also Mechanics*#.
1. *Matheseos universalis elementa.* Leiden, 1727. RM.
1a. ——— *The elements of universal mathematics; or, Algebra.* London, 1728.
See also 2375/10.

3421. GOLDBACH, Christian. 1690-1764. Russ. DSB; ICB; P I; GC.
See also 3447/10.

3422. STIRLING, James. 1692-1770. Br. DSB; ICB; P II; GB; GD.
1. *Lineae tertii ordinis Neutonianae.* Oxford, 1717.
2. *Methodus differentialis.* London, 1730.

3423. BERNOULLI, Nikolaus (second of the name). 1695-1726. Russ. DSB; P I; GA; GB; GC.

3424. RIVARD, Dominique François. 1697-1778. Fr. P II; GA. Also Astronomy*.

3425. MACLAURIN, Colin. 1698-1746. Br. DSB; ICB; P II; GA; GB; GD. Also Biography*.
1. *Geometria organica; sive, Descriptio linearum curvarum universalis.* London, 1720.
2. *A treatise of fluxions.* 2 vols. Edinburgh, 1742. RM/L. Another ed., London, 1801.
3. *A treatise of algebra.* London, 1748. RM. Another ed., 1779. 5th ed., 1788.
See also 2539/1d.

3426. MAUPERTUIS, Pierre Louis Moreau de. 1698-1759. Fr./Ger. DSB; ICB; P II; GA; GB; GC. Other entries: see 3089.

3427. MULLER, John. 1699-1784. Br. P II (MÜLLER, John). GD
1. *A mathematical treatise. Containing a system of conic sections, with the doctrine of fluxions and fluents.* London, 1736. RM.

3428. CURY, ... de. d. 1763. Fr. ICB.

3429. STONE, Edmund. d. 1768. Br. P II; GA; GD. See also Index.
1. *A new mathematical dictionary.* London, 1726.

3430. CLAIRAULT, Jean Baptiste. d. ca. 1770. Fr. ICB; P I.

3431. BERNOULLI, Daniel. 1700-1782. Switz./Russ. DSB; ICB; P I; BLA; GA; GB; GC. Also Mechanics* and Physiology.
1. *Specimen theoriae novae de mensura sortis.* Farnborough (U.K.), 1967. A periodical article of 1738 trans. into German and English.
See also 2539/1d and 3447/14.

3432. BRAIKENRIDGE (or BRAKENRIDGE), William. ca. 1700-1762. Br. DSB.

3433. BRESCIA, Fortunato da. 1701-1754. It. P I.

3434. EMERSON, William. 1701-1782. Br. P I; GA; GB; GD. Also

Astronomy*, Mechanics*, and Physics*.
1. *The doctrine of fluxions.* London, 1748. X. 3rd ed., 1768. RM.
2. *The method of increments.* London, 1763. RM.
3. *A treatise of algebra.* London, 1764.

3435. FROBESIUS, Johann Nicolaus. 1701-1756. Ger. P I; GA (FROBES).

3436. KRAFFT, Georg Wolfgang. 1701-1754. Ger./Russ. P I; GA (KRAFT); GC (KRAFT). Also Astronomy and Physics.

3437. MARTINO, Niccolo. 1701-1769. It. ICB; P II.

3438. BAYES, Thomas. 1702-1761. Br. DSB; ICB.
1. *An essay towards solving a problem in the doctrine of chances.* [Periodical article, 1763] German trans. Leipzig, 1908. (OKEW. 169)
2. *Facsimiles of two papers by Bayes. (i) An essay towards solving a problem in the doctrine of chances. (ii) A letter on asymptotic series.* New York, 1963. Ed. by W.E. Deming.

3439. DEPARCIEUX, Antoine. 1703-1768. Fr. DSB; ICB; P I; GA.
1. *Essai sur les probabilités de la durée de la vie humaine.* Paris, 1746. RM.
1a. ——— *Addition a "L'essai sur les probabilités...."* Paris, 1760. RM.

3440. LE SEUR (or LE SUEUR), Thomas. 1703-1770. It. P I. See also 2539/1d.

3441. CASTILLON, Johann (or Giovan Francesco Mauro Melchior). 1704-1791. Switz./Holl./Ger. DSB; P I; GA; GC. See also 5769/8.

3442. CRAMER, Gabriel. 1704-1752. Switz. DSB; ICB; P I; GA. Also History of Mathematics. See also Index.
1. *Introduction à l'analyse des lignes courbes algébriques.* 4 vols. Geneva, 1750. RM/L.

3443. FONTAINE (DES BERTINS), Alexis. 1704-1771. Fr. DSB; P I; GA.

3444. GEBHARDI, Brandanus. 1704-1784. Ger. P I.

3445. MARTIN, Benjamin. 1704?-1782. Br. DSB; ICB; P II; BMNH; GA; GD. Other entries: see 3099.
1. *A new and comprehensive system of mathematical institutions.* 2 vols. London, 1759-64. RM/L.

3446. SEGNER, János András (or Johann Andreas). 1704-1777. Hung./Ger. DSB; ICB; P II; BLA. Also Mechanics and Physics.

3447. EULER, Leonhard. 1707-1783. Ger./Russ. DSB; ICB; P I; GA; GB; GC. Also Science in General*, Astronomy*, Mechanics*, Physics*, and Cartography*. See also Index.
1. *Methodus inveniendi lineas curvas maximi minimive proprietate gaudentes.* Lausanne, 1744. RM/L.
2. *Introductio in analysin infinitorum.* 2 vols. Lausanne, 1748. RM. Reprinted, Brussels, 1967. Another ed., Lyons, 1797.
2a. ——— *Introduction à l'analyse infinitésimale.* 2 vols. Paris, 1796-97. RM/L.
3. *Institutiones calculi differentialis.* St. Petersburg, 1755. Another ed., Zurich, 1787. RM.

4. *Institutiones calculi integralis.* 3 vols. St. Petersburg, 1768-70. RM. Another ed., 4 vols, 1802-05. 3rd ed., 4 vols, 1824-25.
5. *Vollständige Anleitung zur Algebra.* 2 vols. St. Petersburg, 1770. RM.

5a. ——— *Elémens d'algèbre.* 2 vols. St. Petersburg, 1798. RM/L. Another ed., Paris, 1807.

5b. ——— *Elements of algebra.* 2 vols. London, 1797. 5th ed., 1840.

6. *Zwei Abhandlungen über sphärische Trigonometrie.* Leipzig, 1896. (OKEW. 73) Trans. from the French and Latin.
7. *Drei Abhandlungen über die Auflösung der Gleichungen.* Leipzig, 1928. (OKEW. 226) Trans.

Correspondence

8. *Correspondance mathématique et physique de quelques célèbres géomètres du XVIIIème siècle. Précedée d'une notice sur les travaux de L. Euler.* 2 vols. St. Petersburg, 1843. By P.H. Fuss. The correspondence is chiefly with Euler. X. Reprinted, New York, 1968.
9. *Die Berliner und die Petersburger Akademie der Wissenschaften im Briefwechsel L. Eulers.* 3 vols. Berlin, 1959-76. Ed. by A.P. Juskevic and E. Winter.
10. *L. Euler und C. Goldbach: Briefwechsel, 1729-1764.* Berlin, 1965. Ed. by A.P. Juskevic and E. Winter.
11. *The Euler-Mayer correspondence (1751-1755).* London, 1971. Ed. and trans. by E.G. Forbes. Euler's correspondent is the astronomer, Tobias Meyer (1723-1762).

Collected Works

12. *Opuscula varii argumenti.* 3 vols in 2. Berlin, 1746-51. A collection of articles in mathematics, mechanics, astronomy, and physics. RM/L.
13. *Opuscula analytica.* St. Petersburg, 1783. RM.
14. *L'arithmétique raisonnée et démontrée.* Berlin, 1792. Posthumous works trans. by D. Bernoulli with additions by Lagrange. RM.
15. *Commentationes arithmeticae collectae.* 2 vols. St. Petersburg, 1849. Ed. by P.H. Fuss and N.Fuss. Reprinted, Nendeln, 1969.
16. *Opera postuma mathematica et physica anno 1844 detecta.* St. Petersburg, 1862. Ed. by P.H. Fuss and N. Fuss. RM. Reprinted, Nendeln, 1969.
17. *Opera omnia.* Many vols. Leipzig, etc., 1911-. Ed. under the auspices of the Swiss Society for the Natural Sciences. Still in progress.
18. *Manuscripta Euleriana archivi Academiae Scientiarum U.R.S.S.* Several vols. Moscow, 1962-.

3448. RICCATI, Vincenzo. 1707-1775. It. DSB; P II; GA. Also Mechanics.

1. *Opusculorum ad res physicas et mathematicas pertinentium.* 2 vols in 1. Bologna, 1757-62.

3449. ROBINS, Benjamin. 1707-1751. Br. DSB; ICB; P II; GA; GB; GD. Also Mechanics.

3450. BLAKE, Francis. 1708-1780. Br. ICB; P I; GD.

3451. PÖZINGER, Georg Wilhelm. 1709-1753. Ger. P II.

3452. LA CHAPELLE, ... de, *Abbé*. ca. 1710-ca. 1792. Fr. ICB; P I; GA.
   1. *Institutions de géometrie*. 2 vols. Paris, 1746. X. 4th ed., 1765.

3453. SIMPSON, Thomas. 1710-1761. Br. DSB; P II; GA; GB; GD.
   1. *Essays on ... subjects in speculative and mix'd mathematicks*. London, 1740. RM.
   2. *The nature and laws of chance*. London, 1740.
   3. *Mathematical dissertations on a variety of physical and analytical subjects*. London, 1743. RM/L.
   4. *A treatise of algebra*. London, 1745. 7th ed., 1800.
   5. *Elements of geometry*. London, 1747. X. 2nd ed., 1760.
   6. *Trigonometry, plane and spherical*. London, 1748. 2nd ed., 1765.
   7. *The doctrine and application of fluxions*. 2 vols. London, 1750. 2nd ed., 1776. Another ed., 1823.
   8. *Miscellaneous tracts on ... subjects in mechanics, physical astronomy, and speculative mathematics*. London, 1757. RM.

3454. BARTENSTEIN, Lorenz Adam. 1711-1796. Ger. P I & Supp.

3455. BOSCOVICH, Ruggiero Giuseppe. 1711-1787. It./Fr. DSB (BOŠKOVIĆ); ICB; P I; GA; GB. Other entries: see 3114.

3456. BRENDEL, Johann Gottfried. 1712-1758. Ger. P I; BLA & Supp. Also Chemistry.

3457. GUA DE MALVES, Jean Paul de. ca. 1712-1786. Fr. DSB; P I; GA.

3458. KOENIG, (Johann) Samuel. 1712-1757. Holl. DSB; ICB; P I; GA; GC. Also Mechanics. See also 2857/1.

3459. ROBERTSON, John. 1712-1776. Br. P II; GD.

3460. CLAIRAUT, Alexis Claude. 1713-1765. Fr. DSB; ICB; P I; GA; GB. Also Astronomy*, Mechanics, Physics, and Geodesy*. See also Index.
   1. *Recherches sur les courbes à double courbure*. Paris, 1731. RM.
   2. *Elémens de géométrie*. Paris, 1741. RM.
   3. *Elémens d'algèbre*. Paris, 1746. RM. 4th ed., 1768.

3461. MELDERCREUTZ, Jonas. 1713-1785. Swed. P II.

3462. BORZ, Georg Heinrich. 1714-1799. Ger. P I.

3463. DARJES, Joachim Georg. 1714-1791. Ger. P I; GA.

3464. FAGNANO DEI TOSCHI, Giovanni Francesco. 1715-1797. It. DSB; ICB.

3465. CLAIRAUT, ... , *le cadet*. 1716-1732. Fr. Younger brother of A.C. Clairaut. ICB; P I.

3466. LE GUAY DE PRÉMONTVAL, André Pierre. 1716-1764. Fr./Ger. P I; GA (PRÉMONTVAL).
   1. *Discours sur l'utilité des mathématiques*. Paris, 1742.
   2. *Discours sur la nature des quantités que les mathématiques ont pour object*. Paris, 1742.

3. *Discours sur diverses notions préliminaire à l'étude des mathématiques.* Paris, 1743.
4. *Discours sur la nature du nombre.* Paris, 1743.

3467. ALEMBERT, Jean le Rond d'. 1717-1783. Fr. DSB; ICB; P I; GA; GB; GF. Also Astronomy* and Mechanics*#. See also Index.
1. *Opuscules mathématiques; ou, Mémoires sur différents sujets de géometrie, de méchanique, d'optique, d'astronomie, etc.* 8 vols. Paris, 1761-80. RM. Reprinted, New York, 1967.
2. *Encyclopédie méthodique.* [Main entry: 3000] *Mathématiques* ([including] *Hydrostatiques, Optiques, Perspective, Astronomie).* 3 vols & supp. Paris, 1784-92. RM/L.

3468. BAERMANN, Georg Friedrich. 1717-1769. Ger. P I; GA.

3469. STEWART, Matthew. 1717-1785. Br. DSB; ICB; P II; GA; GD. Also Astronomy*.

3470. AGNESI, Maria Gaetana. 1718-1799. It. DSB; ICB; P I; GA; GB. "The first woman in the Western world who can accurately be called a mathematician." -DSB.
1. *Istituzioni analytiche.* 1748. X.
1a. ——— *Analytical institutions.* London, 1801. Trans. by J. Colson.

3471. CHÉZY, Antoine de. 1718-1798. Fr. P I.

3472. GLEIXNER, Franz. 1718-1783. Ger./Prague. P I.

3473. KAESTNER, Abraham Gottlob. 1719-1800. Ger. DSB; ICB; P I; GA; GC. Also Mechanics, Physics, and History of Mathematics*.

3474. LANDEN, John. 1719-1790. Br. DSB; ICB; P I; GA; GB; GD.
1. *The residual analysis: A new branch of the algebraic art.* London, 1764.

3475. KRAFT, Jens. 1720-1765. Den. DSB; P I; GA. Also Mechanics.

3476. PALMQUIST, Frederik. 1720-1771. Swed. P II.

3477. CASALI-BENTIVOGLIA-PALEOTTI, Gregorio Filippo Maria. 1721-1802. It. P I.

3478. JACOBS, Johann. 1721-1800. Ger. P I.

3479. TORELLI, Giuseppe. 1721-1781. It. P II; GA.

3480. HENTSCH, Johann Jacob. 1723-1764. Ger. P I.

3481. CARAVELLI, Vito. 1724-1800. It. ICB; P I.
1. *Opuscoli matematici.* Naples, 1789. RM.

3482. CLEMM, Heinrich Wilhelm. 1725-1775. Ger. P I; GA; GC.

3483. WOLFRAM, Isaac. ca. 1725-after 1780. ICB.

3484. MALTON, Thomas. 1726-1801. Br. ICB; P II; GA; GD.

3485. NICOLAI, Giambattista. 1726-1793. It. P II.

3486. LAMBERT, Johann Heinrich. 1728-1777. Switz./Ger. DSB; ICB; P I; GA; GB; GC. Also Astronomy*, Physics*, and Cartography*.
1. *Die freye Perspektive.* Zurich, 1759. "A masterpiece in descriptive geometry, containing a wealth of geometrical discoveries."

1a. ——— *La perspective, affranchie de l'embarras du plan géométral.* Zurich, 1759. X. Reprinted, Paris, 1977.
2. *Opera mathematica.* 2 vols. Zurich, 1946-48. First publication. Ed. by A. Speiser.

3487. MALLET, Frederik. 1728-1797. Swed. ICB; P II. Also Astronomy.

3488. TESSANEK, Johann. 1728-1788. Prague. P II.

3489. BOUGAINVILLE, Louis Antoine de. 1729-1811. Fr. The explorer. DSB; ICB; P I; GA; GB. Also Natural History*.
1. *Traité du calcul intégral.* 2 vols. Paris, 1754-56.

3490. LORGNA, Antonio Maria. 1730-1796. It. ICB; P I; GA. Also Mechanics.

3491. BERTRAND, Louis. 1731-1812. Switz. ICB; P I; GA.

3492. MALFATTI, Gian Francesco. 1731-1807. It. DSB; ICB; P II,III.

3493. MASERES, Francis. 1731-1824. Br./Canada. DSB; ICB; P II; GA; GD.
1. *Scriptores logarithmici; or, A collection of several curious tracts on the nature and construction of logarithms.* 6 vols. London, 1791-1807.
2. *The doctrine of permutations and combinations.* London, 1795. RM/L.

3494. MAUDUIT, Antoine René. 1731-1815. Fr. P II; GA.

3495. MEYEN, Johann Jacob. 1731-1797. Ger. P II.

3496. KARSTEN, Wenceslaus Johann Gustav. 1732-1787. Ger. P I; GA; GC. See also 5501/3.

3497. BEHN, Friedrich Daniel. 1734-1804. Ger. P I.

3498. CANTERZANI, Sebastiano. 1734-1819. It. P I; GA.

3499. DIONIS DU SÉJOUR, Achille Pierre. 1734-1794. Fr. DSB; P I; GA. Also Astronomy*.

3500. LEJONMARCK (or LEYONMARCK), Gustav Adolph. 1734-1815. Swed. P I.

3501. OBEREIT, Ludwig. 1734-1803. Ger. P II.

3502. CALDANI, Petronio Maria. ca. 1735-1808. It. P I; GA.

3503. FONTANA, Gregorio. 1735-1803. It. P I; GA. Also Mechanics.
1. *Disquisitiones physico-mathematicae.* Pavia, 1780. RM.

3504. LANDERBECK, Nils. 1735-1810. Swed. P I.

3505. VANDERMONDE, Alexandre Théophile (or Alexis, or Abnit, or Charles Auguste). 1735-1796. Fr. DSB; ICB; P II (V., Charles Auguste); P III; GA.

3506. BRING, Erland Samuel. 1736-1798. Swed. DSB.

3507. LAGRANGE, Joseph Louis. 1736-1813. It./Ger./Fr. DSB; ICB; P I; GA; GB. Also Astronomy and Mechanics*. See also Index.
1. *Additions à l'algèbra d'Euler.* Lyons, 1774. X.
1a. ——— German trans. Leipzig, 1898. (OKEW. 103; also 146)
2. *Théorie des fonctions analytiques.* Paris, 1797. RM/L. 2nd ed., 1813; reprinted, ib., 1847 and 1913.

3. *De la résolution des équations numériques de tous les degrés.* Paris, 1798. RM/L.
3a. ——— 2nd ed. *Traité de la résolution....* 1808. 3rd ed., 1826.
4. *Leçons sur le calcul des fonctions.* Lectures given at the Ecole Normale in 1795 and published in the *Journal de l'Ecole polytechnique* in 1804. Published in book form, Paris, 1806; another ed., 1808.
5. *Les leçons élémentaires sur les mathématiques.* Lectures given at the Ecole Normale in 1795 and published in a much enlarged version in the *Journal de l'Ecole polytechnique* in 1812.
5a. ——— *Lectures on elementary mathematics.* Chicago, 1898. Trans. by T.J. McCormack. 2nd ed., 1901.
6. *Oeuvres.* 14 vols. Paris, 1867-92. Ed. by J.A. Serret. RM/L. Reprinted, Hildesheim, 1973.

3508. PFLEIDERER, Christoph Friedrich von. 1736-1821. Ger./Warsaw. P II.

3509. TETENS, Johann Nicolaus. 1736-1807. Ger./Den. P II; GC.

3510. WARING, Edward. ca. 1736-1798. Br. DSB; ICB; P II; GD.
1. *Miscellanea analytica de aequationibus algebraicis et curvarum proprietatibus.* Cambridge, 1762.
2. *Meditationes algebraicae.* Cambridge, 1770. X. 3rd ed., 1782.
3. *Meditationes analyticae.* Cambridge, 1776. 2nd ed., 1785.

3511. HUTTON, Charles. 1737-1823. Br. DSB; ICB; P I; GA; GB; GD. See also Index.
1. *A treatise on mensuration.* Newcastle-upon-Tyne, 1770.
2. *Mathematical tables.* London, 1794. Another ed., 1811.
3. *A mathematical and philosophical dictionary.* 2 vols. London, 1795-96. RM/L. Another ed., 1815. Reprinted, Hildesheim, 1973.
4. *A course of mathematics.* 2 vols. London, 1798-1801. X. 3rd ed., 1800-01. Other eds: 1810-13, 1827, 1837, 1857.
5. *Tracts on mathematical and philosophical subjects.* 3 vols. London, 1812. RM/L.

3512. TEMPELHOF, Georg Friedrich von. 1737-1807. Ger. P II; GC.

3513. VALPERGA DI CALUSO, Tommaso. 1737-1815. It. P II; GA.

3514. BEZOUT, Etienne. 1739-1783. Fr. DSB; ICB; P I; GA.

3515. COUSIN, Jacques Antoine Joseph. 1739-1800. Fr. P I; GA. Also Astronomy*.
1. *Traité de calcul différentiel et de calcul intégral.* Paris, 1796.

3516. KLÜGEL, Georg Simon. 1739-1812. Ger. DSB; P I; GA. Also Astronomy.

3517. LYONS, Israel. 1739-1775. Br. P I; GA; GD. Also Astronomy and Botany.
1. *A treatise of fluxions.* London, 1758.

3518. SCHULTZ, Johann. 1739-1805. Ger. P II; GC.

3519. PLANTIN, Zacharias Zachrisson. d. 1799. Swed. P II.

3520. ANITSCHKOW, Dimitri Sergievich. ca. 1740-1788. Russ. P I & Supp.

3521. CHERNAC, Ladislaus. 1740-1816. Hung. ICB.

3522. FELKEL, Anton. b. 1740. Aus. P I.

3523. GIANELLA, Carlo. 1740-1810. It. P I; GA.

3524. KESAER, Franz Xaver von. 1740-1804. Aus. P I.

3525. LEXELL, Anders Johan. 1740-1784. Russ. DSB; ICB; P I. Also Astronomy*.

3526. LUINI (or LUINO), Francesco. 1740-1792. It. P I.

3527. HINDENBURG, Carl Friedrich. 1741-1808. Ger. DSB; P I.

3528. WILSON, John. 1741-1793. Br. DSB; GD.

3529. CONDORCET, Marie Jean Antoine Nicolas Caritat, *Marquis* de. 1743-1794. Fr. DSB; ICB; P I,III; GA; GB. Also Science in General*# and Biography*. See also Index.
    1. *Essai sur l'application de l'analyse à la probabilité des décisions rendues à la pluralité des voix*. Paris, 1785. RM. Reprinted, New York, 1972.
    2. *Discours sur les sciences mathématiques, prononcé ... 1786*. Paris, 1812.

3530. PESSUTI, Giovacchino. 1743-1814. It. P II; GA.

3531. SCHWAB, Johann Christoph. 1743-1821. Ger. P II.

3532. CUNHA, José Anastácio da. 1744-1787. Port. DSB Supp.; P I; GA.

3533. FERRONI, Pietro. 1744-1825. It. P I.

3534. ROPPELT, Johann Baptist Georg. 1744-1814. Ger. ICB; P II.

3535. VIVORIO, Agostino. 1744-1822. It. P II.

3536. GOLOVIN, Mikhail. d. 1790. Russ. P I; GA.

3537. WESSEL, Caspar. 1745-1818. Den. DSB; ICB; P II.

3538. MONGE, Gaspard. 1746-1818. Fr. DSB; ICB; P II; GA; GB. Also Mechanics, Physics*, and Chemistry. See also Index.
    1. *Géométrie descriptive*. Paris, 1799. RM. 4th ed., 1820. 5th ed., 1827. 7th ed., 1847; reprinted, ib., 1922.
    1a. ——— German trans. Leipzig, 1900. (OKEW. 117)
    2. *Application de l'algèbre à la géométrie*. Paris, 1805. With Hachette. X. Another part (by Hachette), ib., 1813. RM. 5th ed., 1850.
    3. *Application de l'analyse à la géométrie*. Paris, 1807. RM.

3539. NIEUPORT, Charles François le Prudhomme d'Hailly, *Vicomte* de. 1746-1827. Belg. P II; GA; GF.

3540. PFEIFFER, Johann Georg. b. 1746. Ger. P II.

3541. COSSALI, Pietro. 1748-1815. It. ICB; P I; GA.

3542. PLAYFAIR, John. 1748-1819. Br. DSB; ICB; P II; BMNH; GA; GB; GD. Also Physics*# and Geology*.

3543. TINSEAU D'AMONDANS, Charles de. 1748-1822. Fr. DSB.

3544. DELAMBRE, Jean Baptiste Joseph. 1749-1822. Fr. DSB; ICB; P I; GA; GB. Also Astronomy*, Geodesy*, and History of Astronomy*.
1. *Rapport historique sur les progrès des sciences mathématiques depuis 1789, et sur leur état actuel.* Paris, 1810. RM. Reprinted, Amsterdam, 1966.

3545. LAPLACE, Pierre Simon de. 1749-1827. Fr. DSB Supp.; ICB; P I; GA. Also Astronomy*#, Physics, and History of Astronomy*. See also Index.
1. *Théorie analytique des probabilités.* Paris, 1812. Reprinted, Brussels, 1967. 2nd ed., 1814. 3rd ed., 1820.
2. *Essai philosophique sur les probabilités.* Paris, 1814. RM. Reprinted, Brussels, 1967. Another ed., 1840.
2a. ——— *A philosophical essay on probabilities.* New York, 1902. Trans. from the 6th French ed. by F.W. Truscott and F.L. Emory. RM. Reprinted, ib., 1951.
2b. ——— German trans. Leipzig, 1932. (OKEW. 233)

3546. MICHELSEN, Johann Andreas Christian. 1749-1797. Ger. ICB; P II.

3547. TREMBLEY, Jean. 1749-1811. Switz./Ger. P II. Also Astronomy.

3548. VINCE, Samuel. 1749-1821. Br. P II; GD. Also Astronomy*, Mechanics*, and Physics*.
1. *A treatise on fluxions.* 5th ed. Cambridge, 1818.
2. *The elements of the conic sections.* Cambridge, 1781. X. 2nd ed., 1800. 4th ed., 1810.
3. *A treatise on plane and spherical trigonometry.* Cambridge, 1800. X. 2nd ed., 1805. 3rd ed., 1810.

3549. BOURNONS, Rombaut. d. 1788. Belg. P I.

3550. HELLINS, John. d. 1827. Br. GD.

3551. ADAMS, George (second of the name). 1750-1795. Br. P I; BMNH; BLA; GA; GD. Other entries: see 3217.
1. *Geometrical and graphical essays, containing a description of the mathematical instruments....* London, 1791. RM. Another ed., 2 vols, 1813.
1a. ——— *Geometrische und graphische Versuche.* 2 vols. Leipzig, 1795.

3552. BONNYCASTLE, John. 1750?-1821. Br. P I; GA; GD. Also Astronomy*.
1. *An introduction to algebra.* London, 1782. X. 7th ed., 1805.
2. *An introduction to mensuration and practical geometry.* London, 1782. X. 10th ed., 1807.
3. *A treatise on algebra.* 2 vols. London, 1813.

3553. GLENIE, James. 1750-1817. Br. P I (GLENNIE); GD.

3554. L'HUILLIER (or LHUILIER), Simon Antoine Jean. 1750-1840. Switz. DSB; P I; GA.

3555. MASCHERONI, Lorenzo. 1750-1800. It. DSB; ICB; P II; GA; GB.

3556. MORGAN, William. 1750-1833. Br. P II; GD.
1. *The doctrine of annuities and assurances.* London, 1779.

3557. NORDMARK, Zacharias. 1751-1828. Swed. P II.

3558. ROBERTSON, Abraham. 1751-1826. Br. GD. Also Astronomy.

3559. BÜRJA. Abel. 1752-1816. Ger. P I. Also Mechanics*.

3560. LEGENDRE, Adrien Marie. 1752-1833. Fr. DSB; ICB; P I; GA; GB. Also Astronomy* and Geodesy*.
1. *Eléments de géométrie*. Paris, 1794. RM. Many later eds.
1a. ——— New ed. with additions and alterations by A. Blanchet, ib., 1845. X. Later eds: 1855, 1869, 1880, 1881.
1b. ——— *The elements of geometry and trigonometry*. Edinburgh, 1824. Trans. and ed. by D. Brewster.
2. *Essai sur la théorie des nombres*. Paris, 1798. RM. 3rd ed., 2 vols, 1830; reprinted, ib., 1955.
3. *Exercices de calcul intégral*. 3 vols. Paris, 1811-17.
4. *Traité des fonctions elliptiques et des intégrales eulériennes*. 3 vols. Paris, 1825-28. RM.

3561. CARNOT, Lazare Nicolas Marguerite. 1753-1823. Fr. DSB; ICB; P I; GA; GB. Also Mechanics*. See also 15552.
1. *Réflexions sur la métaphysique du calcul infinitésimal*. Paris, 1797. X. 2nd ed., 1813. Reprinted, ib., 1970.
2. *Oeuvres mathématiques*. Basel, 1797. RM.
3. *La géometrie de position*. Paris, 1803.

3562. FERGOLA, Niccolo. 1753-1834. It. P I; GA.

3563. MEUSNIER (DE LA PLACE), Jean Baptiste Marie Charles. 1754-1793. Fr. DSB; ICB; P II; GA. Also Chemistry.

3564. SCHÜBLER, Christian Ludwig. 1754-1820. Ger. P II.

3565. PEZZI, Francesco. d.1813. It. P II.

3566. FUSS, Nicolaus. 1755-1826. Russ. DSB; ICB; P I. Also Astronomy. See also 3447/15,16.

3567. PARSEVAL DES CHÊNES, Marc Antoine. 1755-1836. Fr. DSB.

3568. STRUVE, Jacob. 1755-1841. Ger. P II; GC.

3569. VEGA, Georg von. 1756-1802. Aus. ICB; P II; GC.

3570. HULTÉN, Andreas. 1757-1831. Swed./Ger. P I & Supp.

3571. TEGMAN, Pehr. 1757-1810. Swed. ICB; P II.

3572. ARBOGAST, Louis François Antoine. 1759-1803. Fr. DSB; ICB; P I; GA.
1. *Du calcul des dérivations*. Strasbourg, 1800. RM.

3573. BERGSTEN, Nils Johan. 1759-1837. Swed. P I,III.

3574. GARÇÃO-STOCKLER, Francesco de Borja. 1759-1829. Port. P I; GA.

3575. MANSFIELD, Jared. 1759-1830. U.S.A. GE. His *Essays, mathematical and physical*, of 1801, is considered to be "the first book of original mathematical researches by a native American."

3576. PAOLI, Pietro. 1759-1839. It. ICB; P II & Supp.

3577. SCHÖNBERG, Kurt Friedrich von. b. 1759. Ger. P II.

3578. SIPOS, Paul. 1759-1816. Hung. ICB.

3579. BRISMANN, Carl. 1760-1800. Ger. P I.

3580. KAUSLER, Christian (or Christoph) Friedrich. 1760-1825. Ger. P I.

3581. LOTTERI, Angelo Luigi. 1760-1840. It. P I; GA.

3582. PEYRARD, François. 1760-1822. Fr. P II; GA.

3583. SCHAFFGOTSCH, Franz Ernst. 1760-1809. Prague. P II.

3584. WOOD, James. 1760-1839. Br. P II; GD. Also Mechanics* and Physics*.
 1. *The elements of algebra*. 2nd ed. Cambridge, 1798. Another ed., 1841. 14th ed., 1852.

3585. BENDAVID, Lazarus. 1762-1832. Ger. P I; GA; GC.

3586. ENCONTRE, Daniel. 1762-1818. Fr. ICB; P I; GA.

3587. BÉRARD, Joseph Balthazar. b. 1763. Fr. P I & Supp.; GA.

3588. BRINKLEY, John. 1763-1835. Br. DSB; P I; GA; GD. Also Astronomy*.

3589. HIPP, Karl Friedrich. 1763-1838. Ger. P I.

3590. LOCKHART, James. 1763-1852. Br./Holl. P I.

3591. TRALLES. Johann Georg. 1763-1822. Ger./Switz. P II; GC. Also Astronomy and Physics.

3592. CACCIANINO, Antonio. 1764-1838. It. P I.

3593. LAVERNÈDE, Joseph Esprit Thomas de. 1764-1848. Fr. P I.

3594. DUBOURGUET, J.B.E. fl. 1794-1822. Fr. P I.
 1. *Traités élémentaires de calcul différentiel et de calcul intégral*. 2 vols. Paris, 1810-11.

3595. COLLALTO, Antonio. 1765-1820. It. P I.

3596. IVORY, James. 1765-1842. Br. DSB; P I; GB; GD. Also Astronomy, Mechanics, Physics, and Geodesy.

3597. LACROIX, Sylvestre François. 1765-1843. Fr. DSB; ICB; P I; GA. Also Geography (mathematical and physical).
 1. *Traité élémentaire d'arithmétique*. Paris, 1797. X. 17th ed., 1826. 20th ed., 1848.
 1a. ——— *Elementary treatise on the mathematical principles of arithmetic*. London, 1823.
 2. *Traité du calcul différentiel et du calcul intégral*. 2 vols. Paris, 1797-98. X. 2nd ed., 3 vols, 1810-19.
 3. *Traité élémentaire de trigonométrie*. Paris, 1798. X. 8th ed., 1827. 9th ed., 1837.
 4. *Elémens d'algèbre*. Paris, 1799. X. 2nd ed., 1800. 14th ed., 1825. 19th ed., 1849.
 4a. ——— *Complément des "Elémens d'algèbre."* Paris, 1799. X. 2nd ed., 1801. 5th ed., 1825. 6th ed., 1835.
 5. *Elémens de géométrie*. Paris, 1799. 15th ed., 1837. 18th ed., 1863.
 5a. ——— *Complément des "Elémens de géométrie."* Paris, 1796. X. 5th ed., 1822. 6th ed., 1829. 7th ed., 1840.
 6. *Traité élémentaire de calcul différentiel et de calcul intégral*. Paris, 1802. X. 4th ed., 1828. 5th ed., 1837.
 6a. ——— *An elementary treatise on the differential and*

integral calculus. Cambridge, 1816. Trans. by C. Babbage, J.F.W. Herschel, and G. Peacock.
7. *Essais sur l'enseignement en général et sur celui des mathématiques en particulier*. Paris, 1805. X. 2nd ed., 1816. 3rd ed., 1828.
8. *Traité élémentaire du calcul des probabilités*. Paris, 1816. X. 2nd ed., 1822. 3rd ed., 1833.

3598. MÜLLER, Johann Wolfgang. b. 1765. Ger. P II. Bibliography of mathematics.

3599. NICHOLSON, Peter. 1765-1844. Br. GD.
1. *A popular course of pure and mixed mathematics*. 2 vols. London, 1823.

3600. OSIPOVSKII, Timofei Fedorovich. 1765-1832. Russ. ICB.

3601. PFAFF, Johann Friedrich. 1765-1825. Ger. DSB; ICB; P II; GA; GB.

3602. RUFFINI, Paolo. 1765-1822. It. DSB; ICB; P II; GA.
1. *Opere matematiche*. 3 vols. Palermo, 1953-54. Ed. by E. Bortolotti.

3603. DEGEN, Carl Ferdinand. 1766-1825. Den. P I; GA.

3604. GARNIER, Jean Guillaume. 1766 (or 1760)-1840. Fr./Belg. P I; GA.
1. *Cours d'analyse algébrique*. Paris, 1803. X. 2nd ed., 1814.
2. *Elémens d'algèbre*. Paris, 1803. X. 3rd ed., 1811.
3. *Leçons de calcul différentiel*. 3rd ed. Paris, 1811.
4. *Leçons de calcul intégral*. 3rd ed. Paris, 1812.
5. *Géométrie analytique*. 2nd ed. Paris, 1813.

3605. GURIEF (or GOURIEF), Simon. ca. 1766-1813. Russ. P I.

3606. SERVOIS, François Joseph. 1767-1847. Fr. DSB.

3607. ABBATI, Pietro (*Conte* MARESCOTTI). 1768-1842. It. P III.

3608. ARGAND, Jean Robert. 1768-1822. Fr. DSB; P III.
1. *Essai sur une manière de représenter les quantités imaginaires dans les constructions géométriques*. Paris, 1806. X. 2nd ed., 1874; reprinted, ib., 1971.

3609. FOURIER, Jean Baptiste Joseph. 1768-1830. Fr. DSB; ICB; P I; GA; GB. Also Physics*#.
1. *Analyse des équations déterminés*. Paris, 1831. RM/L.
1a. ——— German trans. Leipzig, 1902. (OKEW. 127)
See also 3750/4.

3610. FRANÇAIS, François (Joseph). 1768-1810. Fr. DSB.

3611. FRANCHINI. Pietro. 1768-1837. It. P I.

3612. GRÜSON (or GRUSON), Johann Philipp. 1768-1857. Ger. P I.
1. *Pinakothek; oder, Sammlung ... Rechnungstafeln*. Berlin, 1798.
2. *Systematischer Leitfaden der reinen Mathematik*. 1822.

3613. WALLACE, William. 1768-1843. Br. DSB; P II; GB; GD.

3614. BARTELS, Johann Martin Christian. 1769-1836. Ger./Russ. P I, III.

3615. HACHETTE, Jean Nicolas Pierre. 1769-1834. Fr. DSB; P I; GA; GB. Also Mechanics and Physics. See also 3538/2.

3616. PRASSE, Moritz von. 1769-1814. Ger. P II.

3617. RAYMOND, Georges Marie. 1769-1839. Fr. P II; GA.

3618. CORANCEZ, Louis Alexandre Olivier de. 1770-1832. Fr. P I; GA.

3619. BUZENGEIGER, Karl Heribert Ignatius. 1771-1835. Ger. ICB; P I.

3620. GERGONNE, Joseph Diaz. 1771-1859. Fr. DSB; ICB; P I,III.

3621. POSELGER, Friedrich Theodor. 1771-1838. Ger. P II. Also Mechanics*.

3622. SCHÖN, Johann. 1771-1839. Ger. P II; GC. Also Meteorology.

3623. SVANBERG, Jöns. 1771-1851. Swed. P II. Also Geodesy*.

3624. FRANCOEUR, Louis Benjamin. 1773-1849. Fr. P I; BMNH; GA. Also Astronomy* and Mechanics*.
1. *Cours complet des mathématiques pures.* 2 vols. Paris, 1809. X. 2nd ed., 1819. 3rd ed., 1828. 4th ed., 1837.
1a. ——— *A complete course of pure mathematics.* 2 vols. Cambridge, 1829-30.

3625. FRIES, Jakob Friedrich. 1773-1843. Ger. DSB; P I; GA; GB; GC. Also Physics.

3626. ROTHE, Heinrich August. 1773-1842. Ger. ICB; P II.

3627. WOODHOUSE, Robert. 1773-1827. Br. DSB; ICB; P II; GA; GD. Also Astronomy*.
1. *The principles of analytical calculation.* Cambridge, 1803.
2. *A treatise on plane and spherical trigonometry.* London, 1809. Another ed., Cambridge, 1822.
3. *A treatise of isoperimetrical problems, and the calculus of variations.* Cambridge, 1810.

3628. BIOT, Jean Baptiste. 1774-1862. Fr. DSB; ICB; P I,III; GA; GB. Also Astronomy*, Physics*, and History of Astronomy*.
1. *Traité analytique des courbes et des surfaces du second degré.* Paris, 1802. X. 6th ed., 1823.

3629. LANCRET, Michel Ange. 1774-1807. Fr. DSB.

3630. MOLLWEIDE, Karl Brandon. 1774-1825. Ger. DSB; P II; GA; GC. Also Astronomy and Physics.

3631. ADRAIN, Robert. 1775-1843. U.S.A. DSB; ICB; GD; GE.

3632. AMPÈRE, André Marie. 1775-1836. Fr. DSB; ICB; P I; GA; GB. Also Science in General*, Physics*#, and Personal Writings*. See also Index.
1. *Considerations sur la théorie mathématique du jeu.* Lyons, 1802. RM.

3633. BOLYI, Farkas (or Wolfgang). 1775-1856. Hung. DSB; ICB; P III.
1. *Wolfgang und Johann Bolyai: Geometrische Abhandlungen.* 2 vols. Leipzig, 1913. Contains German trans. of the theory of parallels of 1804, and of other writings. Ed. by P. Stäckel. Reprinted, 2 vols in 1, New York, 1972.
2. *Briefwechsel zwischen C.F. Gauss und W. Bolyai.* Leipzig,

1899. Ed. by F. Schmidt and P Stäckel. RM/L. Reprinted, New York, 1972.

See also 3763/1.

3634. BOUCHARLAT, Joseph Louis. ca. 1775-1848. Fr. P I.
 1. *Théorie des courbes et des surfaces du second ordre*. Paris, 1810.
 2. *Elémens de calcul différentiel et de calcul intégral*. Paris, 1813. X. 2nd ed., 1820.

3635. DEALTRY, William. 1775-1847. Br. P I; GD.
 1. *The principles of fluxions*. Cambridge, 1810. 2nd ed., 1816.

3636. FRANÇAIS, Jacques Frédéric. 1775-1833. Fr. DSB.

3637. HOËNÉ-WROŃSKI, Józef Maria. ca. 1775-1853. Fr. DSB Supp.; ICB (WROŃSKI); P I; GA (WROŃSKI).

3638. MAURICE, Jean Frédéric Théodore. 1775-1851. Switz./Fr. P II.

3639. SOMERSET, Edward Adolphus Seymour, *11th Duke of*. 1775-1855. Br. GD (SEYMOUR).
 1. *Alternate circles and their connection with the ellipse*. London, 1850.

3640. WAGNER, Johann Jacob. 1775-1841. Ger. GC.
 1. *Mathematische Philosophie*. Ulm, 1851. X. Reprinted, Wiesbaden, 1969. He also wrote some other works on the philosophy of mathematics.

3641. BARLOW, Peter. 1776-1862. Br. DSB; ICB; P I,III; GA; GB; GD. Also Physics*.
 1. *An elementary investigation of the theory of numbers*. London, 1811. RM/L.
 2. *A new mathematical and philosophical dictionary*. London, 1814. RM/L.

3642. GERMAIN, Sophie. 1776-1831. Fr. DSB; ICB; P I,III; GA.

3643. LEONELLI, (Giuseppe) Zecchini. 1776-1847. It./Fr. ICB; P I; GA. Also Physics*.

3644. BRISSON, Barnabé. 1777-1828. Fr. DSB; P III; GA.

3645. GAUSS, Karl Friedrich. 1777-1855. Ger. DSB; ICB; P I,III; GA; GB; GC. Also Astronomy*, Mechanics*, Physics*, and Earth Sciences*. See also Index.
 1. *Demonstratio nova theorematis, omnem functionem algebraicam rationalem integram unius variabilis in factores reales primi vel secundi gradus resolvi posse*. Helmstadt, 1799. RM.
 2. *Disquisitiones arithmeticae*. Leipzig, 1801. RM.
 2a. ——— *Recherches arithmétiques*. Paris, 1807. RM. Reprinted, ib., 1953.
 2b. ——— English trans. New Haven, 1966. By A.A. Clarke.
 3. *Beiträge zur Theorie der algebraischen Gleichungen*. Göttingen, 1849. RM.

Editions of Periodical Articles

 4. *Disquisitiones generales circa superficies curvas*. 1827.
 4a. ——— *Recherches sur la théorie générale des surfaces courbes*. Grenoble, 1855. X. 2nd ed., Paris? 1870; reprinted, Paris, 1967.

4b. ——— *Allgemeine Flächentheorie*. Leipzig, 1889. (OKEW)
4c. ——— *General investigation of curved surfaces*. Princeton, 1902. RM. Reprinted, Hewlett, N.Y., 1965.
5. Articles on the theory of least squares:-
5a. ——— *Méthode des moindres carrés*. Paris, 1855.
5b. ——— *Abhandlungen zur Methode der kleinsten Quadrate*. Berlin, 1887. X. Reprinted, Würzburg, 1964.
5c. ——— *Gauss's work (1803-1826) on the theory of least squares*. Princeton, 1957. Trans. by H. F. Trotter.
6. *Die vier Gauss'schen Beweise für die Zerlegung ganzer algebraischer Functionen in reelle Factoren ersten oder zweiten Grades (1799-1849)*. Leipzig, 1890. (OKEW)
7. *Sechs Beweise des Fundamentaltheorems über quadratische Reste*. Leipzig, 1901. (OKEW)
8. *Nachlass zur Theorie des arithmetisch-geometrischen Mittels und der Modulfunktion*. Leipzig, 1927. (OKEW)

Correspondence, Collected Works, etc.

9. *Briefwechsel zwischen Gauss und Bessel*. Leipzig, 1880. Ed. by G. von Auwers. RM.
10. *Briefwechsel zwischen C.F. Gauss und W. Bolyai*. Leipzig, 1899. Ed. by F. Schmidt and P. Stäckel. RM/L. Reprinted, New York, 1972.
11. *Briefwechsel zwischen C.F. Gauss und C.L. Gerling*. Berlin, 1927. Ed. by C. Schäfer.
12. *C.L. Gerling und C.F. Gauss. Sechzig bisher unveröffentlichte Briefe*. Göttingen, 1964. Ed. by T. Gerardy.
13. *Nachträge zum Briefwechsel zwischen C.F. Gauss und H.C. Schumacher*. Göttingen, 1969. Ed. by T. Gerardy.
14. *Werke*. 12 vols. Göttingen, 1863-1933. Reprinted, Hildesheim, 1973-.
15. *Mathematisches Tagebuch, 1796-1814*. Leipzig, 1976. (OKEW. 256) German trans. by E. Schumann. Historical introd. by K.R. Biermann.

3646. HOFFMANN, Johann Joseph Ignatz von. 1777-1866. Ger. P I,III.

3647. MAGISTRINI, Giambattista. 1777-1849. It. P II. Also Mechanics.

3648. POINSOT, Louis. 1777-1859. Fr. DSB; P II; GA; GB. Also Mechanics*.

3649. REYNAUD, Antoine André Louis. 1777-1844. Fr. P II; GA.

3650. WERNEBURG, Johann Friedrich Christian. 1777-1851. Ger. P II; GC.

3651. GOMPERTZ, Benjamin. 1779-1865. Br. DSB; P III; GD. Also Astronomy.

3652. MURHARD, Friedrich Wilhelm August. 1779-1853. Ger. P II; GA. Also Mechanics* and Physics*.
1. *Litteratur der mathematischen Wissenschaften*. 2 vols. Leipzig, 1797-98. Vol. 1: Arithmetic and geometry. Vol. 2: Geometry and analysis.
1a. ——— *Literatur der mechanischen und optischen Wissenschaften*. 3 vols. Leipzig, 1803-05. Constitutes Vols 3-5 of item 1.

3653. NÜRNBERGER, Joseph Christian Emil. 1779-1848. Ger. P II; GC. Also Astronomy.

3654. SOULIMA, Pavel Akimovich. 1779-1812. Russ. ICB.

3655. BUDAN DE BOISLAURENT, Ferdinand François Désiré. fl. 1803-1853. Fr. DSB.
1. *Nouvelle méthode pour résolution des équations numériques d'un degré quelconque*. Paris, 1807. X. 2nd ed., 1822.
See also 3750/4.

3656. CHRISTMANN, Wilhelm Ludwig. 1780-1835. Ger. P I.

3657. CRELLE, August Leopold. 1780-1855. Ger. DSB; ICB; P I; GA; GC.

3658. LEROY, Charles François Antoine. ca. 1780-1854. Fr. P I,III.
1. *Analyse appliquée à la géométrie des trois dimensions*. Paris, 1829. X. 3rd ed., 1843. 4th ed., 1854.
2. *Traité de géométrie descriptive*. 2 vols. Paris, 1842. Another ed., 1855.
3. *Traité de stéréotomie*. 2 vols. Paris, 1844. X. Another ed., 1866.

3659. SCHWEIKART, Ferdinand Karl. 1780-1859. Ger. DSB; P II.

3660. ATKINSON, Henry. 1781-1829. Br. GD.

3661. BOLZANO, Bernard. 1781-1848. Prague. DSB; ICB; P I; GA; GB; GC.
1. *Beyträge zu einer begründeteren Darstellung der Mathematik*. Prague, 1810. X. Reprinted, Paderborn, 1926, and Darmstadt, 1974.
2. *Rein analytischer Beweis des Lehrsatzes, dass zwischen je zwey Werthen, die ein entgegengesetztes Resultat gewähren, wenigstens eine reelle Wurzel der Gleichung liege*. Prague, 1817. X. Reprinted, Leipzig, 1905. (OKEW. 153)
3. *Paradoxien des Unendlichen*. Leipzig, 1851. X.
3a. ——— *Paradoxes of the infinite*. London, 1950. Trans. with a historical introd. by D.A. Steele.
4. *Theorie der reellen Zahlen im Bolzanos handschriftlichen Nachlasse*. Prague, 1962. First publication of the MS. Ed. with a commentary by K. Rychlik.
5. *Gesamtausgabe*. Stuttgart, 197-? In progress.

3662. BRET, Jean Jacques. 1781-1819. Fr. DSB; P III.

3663. KRAUSE, Karl Christian Friedrich. 1781-1832. Ger. ICB; P I; GA; GB; GC.

3664. PLANA, Giovanni (or Jean). 1781-1864. It. DSB; ICB; P II,III. Also Astronomy, Physics*, and History of Mechanics*.

3665. POISSON, Siméon Denis. 1781-1840. Fr. DSB Supp.; ICB; P II; GA; GB. Also Mechanics* and Physics*.
1. *Recherches sur la probabilité des jugements*. Paris, 1837.

3666. DIESTERWEG, Wilhelm Adolph. 1782-1836. Ger. P I. See also 97/2.

3667. TERQUEM, Olry (first of the name). 1782-1862. Fr. P II,III; GA.

3668. BRIANCHON, Charles Julien. 1783-1864. Fr. DSB; P I,III.

3669. MAZZOLA, Angelo. b. 1783. It. P II.

3670. NOËL, Jean Nicolas. 1783-1867. Belg. P II; GF.

3671. POLLOCK, (John) Frederick. 1783-1870. Br. P III; GB; GD.

3672. QUERRET, Jean Joseph. 1783-1839. Fr. P II.

3673. BESSEL, Friedrich Wilhelm. 1784-1846. Ger. DSB; ICB; P I; GA; GB; GC. Also Astronomy*# and Geodesy*. See also Index.

3674. DUPIN, (Pierre) Charles (François). 1784-1873. Fr. DSB; P I, III; GA; GB. Also Biography*.

3675. AHRENS, Johann Thomas. 1786-1841. Ger. P I.

3676. BAZAINE, Pierre Dominique. 1786 (or 1783)-1838. Russ. P I; GA.

3677. BINET, Jacques (Philippe Marie). 1786-1856. Fr. ICB; P I; GA. Also Astronomy.

3678. BLAND, Miles. 1786-1867. Br. P III; GD. Also Mechanics*.
   1. *Geometrical problems, deducible from the first six books of Euclid, arranged and solved.* Cambridge, 1819. X. 3rd ed., 1827.

3679. HORNER, William George. 1786-1837. Br. DSB; GD.

3680. THOMSON, James (first of the name). 1786-1849. Br. P III; GD.

3681. PAUCKER, Magnus Georg von. 1787-1855. Mitau. P II; GC. Also Astronomy.

3682. PONCELET, Jean Victor. 1788-1867. Fr. DSB; ICB; P II,III; GA; GB. Also Mechanics*.
   1. *Traité des propriétés projectives des figures.* Paris, 1822. RM. 2nd ed., 2 vols, 1865-66.
   2. *Applications d'analyse et de géométrie.* 2 vols. Paris, 1862-64.

3683. BORDONI, Antonio Maria. 1789-1860. It. P I,III. Also Mechanics.

3684. CAUCHY, Augustin Louis. 1789-1857. Fr. DSB; ICB; P I,III; GA; GB. Also Astronomy and Mechanics. See also 15509/1.
   1. *Cours d'analyse.* Paris, 1821.
   2. *Mémoire sur les intégrales définies, prises entre les limites imaginaires.* Paris, 1825. RM.
   2a. ——— German trans. Leipzig, 1900. (OKEW. 112)
   3. *Exercices de mathématiques.* 5 vols. Paris, 1826-30.
   4. *Nouveaux exercices de mathématiques.* Prague, 1835.
   5. *Exercices d'analyse et de physique mathématique.* 4 vols. Paris, 1840-47.
   6. *Oeuvres complètes.* First series, Paris, 1882-1911. Second series, ib., 1905-58.

3685. SARRUS, P.F. fl. 1824-1852. Fr. P II.

3686. MAGNUS, Ludwig Immanuel. 1790-1861. Ger. ICB; P II,III; GC.

3687. MÖBIUS, August Ferdinand. 1790-1868. Ger. DSB; P II,III; GA; GB; GC. Also Astronomy*.
   1. *Gesammelte Werke.* 4 vols. Leipzig, 1885-87. Ed. by R. Baltzer et al. X. Reprinted, Wiesbaden, 1967.

3688. STRONG, Theodore. 1790-1869. U.S.A. P III; GE.

3689. COLLINS, Eduard Albert Christoph Ludwig. 1791-1840. Russ. P I.

3690. FÖRSTEMANN, Wilhelm August. 1791-1836. Danzig. P I.

3691. LE BESGUE, Victor Amédée. 1791-1875. Fr. P III.

3692. PEACOCK, George. 1791-1858. Br. DSB; ICB; P II; GB; GD.
Also Biography*.
1. *A collection of examples of the applications of the differential and integral calculus.* Cambridge, 1820.
2. *A treatise on algebra.* Cambridge, 1830. 2nd ed., rev., 2 vols, 1842-45; reprinted, New York, 1940.
See also 3597/6a and 6144/3.

3693. BABBAGE, Charles. 1792-1871. Br. DSB; ICB; P I,III; GA; GB; GD. Also Science in General*, Geology*, and Autobiography*.
1. *Examples of the solutions of functional equations.* London, 1820.
2. *Tables of the logarithms of the natural numbers from 1 to 108,000.* London, 1827. X. Another ed., 1845.
3. *Analysis of the statistics of the clearing house during the year 1839.* London, 1856.
4. *Babbage's calculating machines, being a collection of papers relating to them.* London, 1889. Ed. by H.P. Babbage.
5. *Charles Babbage and his calculating engines: Selected writings by Babbage and others.* New York, 1961. Ed. with an introd. by P. and E. Morrison.
See also 3597/6a.

3694. DIRKSEN, Enno Heeren. 1792-1850. Ger. P I & Supp.

3695. LOBACHEVSKY, Nikolai Ivanovich. 1792-1856. Russ. DSB; ICB; P I,IV (LOBATSCHEVSKIJ); GA; GB.
1. [Trans. of two periodical articles in Russian, 1835 and 1836.] *Imaginäre Geometrie und Anwendung der imaginäre Geometrie auf einige Integrale.* Leipzig, 1904. Trans. and ed. by H. Liebermann.
2. *Geometrische Untersuchungen zur Theorie der Parallellinien.* Berlin, 1840. X. Reprinted, ib., 1887.
2a. ——— *Etudes géometriques sur la théorie des parallèles ... suivi d'un extrait de la correspondance de Gauss et Schumacher.* Paris, 1866. Trans. by J. Houel. X. Another ed., 1900.
2b. ——— *Geometrical researches on the theory of parallels.* Austin, Texas, 1891. Trans. by G.B. Halsted. Another ed., Chicago, 1914; reprinted, ib., 1942.
3. *Pangéométrie; ou, Précis de géométrie fondée sur une théorie générale et rigoureuse des parallèles.* Kazan, 1855. X. Another ed., Paris, 1905.
3a. ——— German trans. Leipzig, 1902. (OKEW. 130)
4. *Zwei geometrische Abhandlungen.* 2 vols in 1. Leipzig, 1898-99. Trans. from the Russian with commentary and a biography by F. Engel. Reprinted, New York, 1972.
5. [In Russian. *Complete works.*] 5 vols. Moscow, 1946-51.

3696. MASETTI, Giambattista. 1792-1827. It. P II.

3697. MONTFERRIER, Alexandre André Victor Sarrazin de. 1792-1863. Fr. P II,III; GA.

1. *Dictionnaire des sciences mathématiques, pures et appliquées.* 3 vols. Paris, 1834-40. RM.

3698. OHM, Martin. 1792-1872. Ger. P II,III.
1. *Versuch eines vollkommen consequenten Systems der Mathematik.* 9 vols. Nuremberg, 1822-52.
2. *Übung in der Anwendung der Integral-Rechnung.* Nuremberg, 1856.

3699. TUCCI, Francesco Paolo. 1792-1872? It. P II,III.

3700. CHASLES, Michel. 1793-1880. Fr. DSB; ICB; P I,III; GA. Also History of Mathematics*.
1. *Traité de géométrie supérieure.* Paris, 1852. 2nd ed., 1880.
2. *Traité des sections coniques.* Paris, 1865.

3701. GREEN, George. 1793-1841. Br. DSB Supp.; ICB; P I; GD. Also Physics*.
1. *The mathematical papers.* London, 1871. Ed. by N.M. Ferrers. RM/L. Reprinted, New York, 1970.

3702. HILL, Carl Johan Danielsson. 1793-1875. Swed. P I,III.

3703. KULIK, Jacob Philipp. 1793-1863. Aus./Prague. ICB; P I,III.

3704. LARDNER, Dionysius. 1793-1859. Br. ICB; P I & Supp.; BMNH; GA; GB; GD. Other entries: see 3303.
1. *A system of algebraic geometry.* Vol. 1 [No more publ.] London, 1823.
2. *An elementary treatise on the differential and integral calculus.* London, 1825.
3. *The first six books of the "Elements" of Euclid, with a commentary.* London, 1828. X. 11th ed., 1855.

3705. OLIVIER, Théodore. 1793-1853. Fr. P II. Also Mechanics.
1. *Cours de géométrie descriptive.* 2 vols. Paris, 1843-44.

3706. DANDELIN, Germinal Pierre. 1794-1847. Belg. DSB; ICB; P I; GA. Also Mechanics.

3707. POUDRA, Noël Germinal. b. 1794. Fr. P II,III. See also 2001/3.

3708. RIECKE, Friedrich Joseph Pythagoras. 1794-1876. Ger. P II,III.
1. *Mathematische Unterhaltungen.* 3 vols. Stuttgart, 1867-73. X. Reprinted, 3 vols in 1, Wiesbaden, 1973.

3709. SCHULTÉN, Nathanael Gerhard af. 1794-1860. Fin. P II,III.

3710. TAURINUS, Franz Adolph. 1794-1874. Ger. DSB.

3711. DAVIES, Thomas Stephens. 1795-1851. Br. P I; GD.

3712. FRULLANI, Giuliano. 1795-1834. It. ICB; P I.

3713. GUIBERT, Adolphe Pierre Marie. 1795-1864. Fr. P III.

3714. HOLMBOE, Berndt Michael. 1795-1850. Nor. DSB; P I; GA. See also 3762/4.

3715. JACOBI, Karl Friedrich Andreas. 1795-1855. Ger. P I & Supp.

3716. LAMÉ, Gabriel. 1795-1870. Fr. DSB; ICB; P I,III; GA. Also Mechanics* and Physics*.
1. *Examen des différentes méthodes employées pour resoudre les problèmes de géométrie.* Paris, 1818.

2. *Leçons sur les fonctions inverses de transcendantes et les surfaces isothermes.* Paris, 1857.
3. *Leçons sur les coordonnées curvilignes et leurs diverses applications.* Paris, 1859.

3717. ROGG, Ignaz. 1795-1886. Ger. P II,III.
1. *Handbuch der mathematischen Literatur vom Anfänge der Buchdruckerkunst bis ... 1830.* Tübingen, 1830. X. Reprinted, (place? date?). For a continuation of this work see 3808/1.

3718. STEIN, Johann Peter Wilhelm. 1795-1831. Ger. P III.

3719. BIENAYMÉ, Irénée Jules. 1796-1878. Fr. DSB Supp.; P III.

3720. ETTINGSHAUSEN, Andreas von. 1796-1878. Aus. P I,III. Also Physics*.

3721. MOSSBRUGGER, Leopold. 1796-1864. Switz. P II; GC.

3722. QUETELET, (Lambert) Adolphe (Jacques). 1796-1874. Belg. DSB; ICB; P II,III; GA; GB. Other entries: see 3306.
1. *Instructions populaires sur le calcul des probabilités.* Brussels, 1828. X.
1a. ——— *Popular instructions on the calculation of probabilities.* London, 1839. RM.

3723. STEINER, Jacob. 1796-1863. Ger. DSB; ICB; P II,III; GB; GC.
1. *Einige geometrische Betrachtungen.* [Periodical article, 1826] Leipzig, 1901. (OKEW. 123)
2. *Systematische Entwicklung der Abhängigkeit geometrischer Gestalten von einander.* Berlin, 1832. X. Reprinted, 2 vols, Leipzig, 1896. (OKEW. 82,83)
3. *Die geometrischen Constructionen, ausgeführt mittelst der geraden Linie und eines festen Kreises.* Berlin, 1833. X. Reprinted, Leipzig, 1895. (OKEW. 60)
3a. ——— *Geometrical constructions with a ruler, given a fixed circle with its center.* New York, 1950. Trans. by M.E. Stark.
4. *Vorlesungen über synthetische Geometrie.* 2 vols. Leipzig, 1867. X. 2nd ed., 1875. Ed. by C.F. Geiser and H. Schrötter.
5. *Gesammelte Werke.* 2 vols. Berlin, 1881-82. Ed. by K. Weierstrass. Reprinted, Bronx, N.Y., 1971.

3724. BURG, Adam von. 1797-1882. Aus. P I,III,IV. Also Mechanics.

3725. DUHAMEL, Jean Marie Constant. 1797-1872. Fr. DSB; P I,III. Also Mechanics* and Physics.
1. *Cours d'analyse.* 2 vols. Paris, 1840-41. X. 2nd ed., 1847.
2. *Eléments de calcul infinitésimal.* 2 vols. Paris, 1856. 3rd ed., 1874-76. 4th ed., 1886-87.
3. *Des méthodes dans les sciences de raisonnement.* 5 vols. Paris, 1865-73. X. 2nd ed., 1879-86. 3rd ed., 1885-96.

3726. GRUNERT, Johann August. 1797-1872. Ger. P I,III; GA; GC.

3727. LOBATTO, Rehuel. 1797-1866. Holl. P I,III.

3728. MÜLLER, Johann Heinrich Traugott. 1797-1862. Ger. P II,III; GC.

3729. OETTINGER, Ludwig. 1797-1869. Ger. P II,III.

3730. SAINT-VENANT, Adhémar Jean Claude Barré de. 1797-1886. Fr. DSB; P III,IV. Also Mechanics. See also 15565/1.

3731. VINCENT, Alexandre Joseph Hidulphe. 1797-1868. Fr. P II,III.

3732. BOBILLIER, Etienne. 1798-1840. Fr. DSB; P IV. Also Mechanics.

3733. COMTE, (Isidore) Auguste (Marie François Xavier). 1798-1857. Fr. The philosopher. DSB; ICB; P III; G. Also Science in General*. See also Index.
 1. *Philosophy of mathematics.* New York, 1851. Trans. by W.M. Gillespie.

3734. GHIJBEN, Jacob Badon. 1798-1870. Holl. P III.

3735. GUDERMANN, Christoph. 1798-1852. Ger. DSB; P I.
 1. *Theorie der Potential- oder cyklish-hyperbolischen Functionen.* Berlin, 1833.

3736. MATZKA, Wilhelm. 1798-1891. Aus./Prague. P II,III,IV.

3737. NEUMANN, Franz Ernst. 1798-1895. Ger. DSB; ICB; P II,III,IV; BMNH; GA; GB; GC. Also Physics* and Crystallography.
 1. *Beiträge zur Theorie der Kugelfunctionen.* Leipzig, 1878.
 2. *Vorlesungen über die Theorie des Potentials und der Kugelfunctionen.* Leipzig, 1887.

3738. RUTHERFORD, William. 1798-1871. Br. P III; GD.

3739. SCHERK, Heinrich Ferdinand. 1798-1885. Ger. P II,III,IV; GC.

3740. STAUDT, Karl Georg Christian von. 1798-1867. Ger. DSB; ICB; P II,III,IV; GC.
 1. *Die Geometrie der Lage.* Nuremberg, 1847.

3741. UMPFENBACH, Hermann. 1798-1862. Ger. P II; GC.

3742. GERONO, Camille Christophe. 1799-1892. Fr. P III,IV & Supp.

3743. GRÄFFE. Karl Heinrich. 1799-1873. Switz. DSB; ICB; P I,III; GC.

3744. KNAR, Josef. 1799-1864. Aus. P III.

3745. LANDRY, Fortuné. b. 1799. Fr. P III.

3746. MÜLLER, Anton. 1799-1860. Ger./Switz. P II,III.

3747. SCHMIDTEN, Henrik Gerner. 1799-1831. Den. P II.

3748. SCHOLZ, Ernst Julius. 1799-1841. Ger. P II.

3749. SPEHR, Friedrich Wilhelm. 1799-1833. Ger. P II.

3750. YOUNG, John Radford. 1799-1885. Br. P III; GD. Also Science in General*.
 1. *The elements of plane and spherical trigonometry.* London, 1833.
 2. *The theory and solution of algebraical equations of the higher orders.* 2nd ed. London, 1843.
 3. *The analysis and solution of cubic and biquadratic equations.* London, 1842.
 4. *On the theory and solution of algebraical equations, with the recent researches of Budan, Fourier and Sturm.* London, 1895.

3751. FEUERBACH, Karl Wilhelm. 1800-1834. Ger. DSB; ICB; P I; GA; GC.

3752. HOLDITCH, Hamnet. 1800-1867. Br. P I,III.

3753. MAINARDI, Gaspare. 1800-1879. It. P III.

3754. AIRY, George Biddell. 1801-1892. Br. DSB; ICB; P I,III,IV; GB; GD1. Also Astronomy*, Physics*, and Autobiography*.
1. *On the algebraical and numerical theory of errors of observations and the combination of observations.* Cambridge, 1861. RM/L. 2nd ed., London, 1875. 3rd ed., London, 1879.
2. *An elementary treatise on partial differential equations.* London, 1866. Another ed., 1873

3755. CLAUSEN, Thomas. 1801-1885. Ger./Dorpat. DSB; P I,III. Also Astronomy.

3756. COURNOT, Antoine Augustin. 1801-1877. Fr. DSB; ICB; P III; GA; GB. Also Science in General*.
1. *Traité élémentaire de la théorie des fonctions et du calcul infinitésimal.* 2 vols. Paris, 1841. X. 2nd ed., 1857.
2. *Exposition de la théorie des chances et des probabilités.* Paris, 1843.
3. *De l'origine et des limites de la correspondance entre l'algèbre et la géométrie.* Paris, 1847.

3757. OLIVEIRA, Candido Baptista de. 1801-1865. Brazil. P III.

3758. OSTROGRADSKY, Mikhail Vasilievich. 1801-1862. Russ. DSB; ICB; P II & Supp. Also Mechanics.

3759. PLÜCKER, Julius. 1801-1868. Ger. DSB; ICB; P II,III; GB; GC. Also Physics.
1. *System der analytischen Geometrie.* Berlin, 1835.
2. *Die Theorie der algebraischen Curven.* Bonn, 1839.
3. *System der Geometrie des Raumes in neuer analytischer Behandlungsweise.* Düsseldorf, 1846. X. 2nd ed., 1852.
4. *Neue Geometrie des Raumes.* Leipzig, 1868.
5. *Gesammelte wissenschaftliche Abhandlungen.* 2 vols. Leipzig, 1895-96. Vol. 1: Mathematics. Vol. 2: Physics. X. Reprinted, New York, 1972.

3760. RAABE, Joseph Ludwig. 1801-1859. Switz. P II.
1. *Die Differenzial- und Integralrechnung.* 3 vols. Zurich, 1839-47.

3761. TIMMERMANS, Jean Alexis. 1801-1864. Belg. P II,III; GF.

3762. ABEL, Niels Henrik. 1802-1829. Nor. DSB; ICB; P I; GA; GB.
1. *Mémoire sur les équations algébriques.* Christiania, 1824. X. Reprinted, Oslo, 1957.
2. *Untersuchungen über die Reihe....* Christiania, 1826. X. Reprinted, Leipzig, 1895. (OKEW. 71)
3. *Abhandlung über eine besondere Klasse algebraisch auflösbarer Gleichungen.* [Periodical article, 1829] Leipzig, 1900. (OKEW. 111)
4. *Oeuvres complètes.* Christiania, 1839. Ed. by B. Holmboe. RM/L.
5. *Oeuvres complètes. Nouvelle édition.* 2 vols. Christiania,

1881. Ed. by L. Sylow and S. Lie. RM/L. Reprinted, New York, 1965.
See also 15508/1.

3763. BOLYAI, János (or Johann). 1802-1860. Hung. DSB; ICB.
1. *Appendix, scientiam spatii absolute veram exhibens.* First published as an appendix to a work of 1832 by his father, Farkas Bolyai. Reprinted, Budapest, 1902, and Leipzig, 1903.
1a. ——— *The science of absolute space.* Austin, Texas, 1891. Trans. by G.B. Halsted. X. 4th ed., 1896. Reprinted, New York, 1955.
2. *Geometrische Abhandlungen.* See 3633/1.
See also 4084/1.

3764. BRASSEUR, Jean Baptiste. 1802-1868. Belg. P I,III.

3765. CONTI, Carlo. 1802-1849. It. P I.

3766. DROBISCH, Moritz Wilhelm. 1802-1896. Ger. P I,III,IV; GA; GC.

3767. MEYER, Anton. 1802-1857. Belg. P II; GF.

3768. MOTH, Franz Xaver. 1802-1879. Aus. P II,III.

3769. VERDAM, Gideon Jan. 1802-1866. Holl. P II,III.

3770. ANGER, Carl Theodor. 1803-1858. Danzig. P I,III. Also Astronomy.

3771. BELLAVITIS, Giusto. 1803-1880. It. DSB; ICB (BELAVITIS); P I,III.

3772. GRONAU, Johann Friedrich Wilhelm. 1803-1887. Danzig. P I,III, IV.

3773. HYMERS, John. 1803-1887. Br. P III; GD.
1. *A treatise on analytical geometry of three dimensions.* Cambridge, 1830. X. 2nd ed., 1836. 3rd ed., 1848.
2. *A treatise on the integral calculus.* Cambridge, 1831. X. 3rd ed., 1844.
3. *A treatise on the theory of algebraical equations.* Cambridge, 1837. X. 2nd ed., 1840.
4. *A treatise on conic sections and the application of algebra to geometry.* Cambridge, 1837. X. 2nd ed., 1840. 3rd ed., 1845.
5. *A treatise on differential equations and on the calculus of finite differences.* Cambridge, 1839.
6. *A treatise on plane and spherical trigonometry.* 3rd ed. Cambridge, 1847.

3774. LIBRI (-CARRUCCI DELLA SOMMAIA), Guglielmo. 1803-1869. It./Fr. ICB; P I,III; GA. Also History of Mathematics*.

3775. STRASSNITZKI, Leopold Karl Schulz von. 1803-1852. Aus. P II; GC.

3776. STURM, (Jacques) Charles François. 1803-1855. Fr. DSB; ICB; P II; GA; GB. Also Physics.
1. *Mémoire sur la résolution des équations numériques.* [Periodical article, 1835] German trans. Leipzig, 1905. (OKEW. 143)

2. *Cours d'analyse*. 2 vols. Paris, 1857-59. X. 6th ed., 1880. Another ed., 1884.
See also 3750/4.

3777. BUNYAKOVSKY, Viktor Yakovlevich. 1804-1889. Russ. DSB Supp.; ICB (BUNIAKOVSKII); P I,III,IV (BOUNIAKOWSKY).

3778. GREBE, Ernst Wilhelm. 1804-1874. Ger. P I,III.

3779. JACOBI, Carl Gustav Jacob. 1804-1851. Ger. DSB; ICB; P I,III; GA; GB; GC. Also Mechanics*.
1. *Fundamenta nova theoriae functionum ellipticarum*. Königsberg, 1829.
2. *Canon arithmeticus; sive, Tabulae....* Königsberg, 1839. X. 2nd ed., recomputed by W. Patz and ed. by H. Brandt, Berlin, 1956.
3. *Gesammelte Werke*. 7 vols & supp. Berlin, 1881-91. Ed. by C.W. Borchardt et al. RM/L. Reprinted, New York, 1969.
Five of his periodical articles are reprinted in OKEW.

3780. JERRARD, George Birch. 1804-1863. Br. DSB; GD.

3781. VERHULST, Pierre François. 1804-1849. Belg. DSB; ICB; P II.
1. *Traité élémentaire des fonctions elliptiques*. Brussels, 1841.

3782. CASINELLI, Luigi. d. 1846. It. P I.

3783. BRASSINNE. Philippe Emile. b. 1805. Fr. P III.

3784. DIRICHLET, Gustav Peter Lejeune. 1805-1859. Ger. DSB; ICB; P I,III; GA (LEJEUNE-D.); GC. Also Mechanics*. His books are often listed under LEJEUNE-DIRICHLET.
1. *Vorlesungen über Zahlentheorie*. Brunswick, 1863. Ed. by R. Dedekind. X. 3rd ed., 1879. 4th ed., 1894.
2. *Werke*. 2 vols. Berlin, 1889-97. Ed. by L. Kronecker. Reprinted, 2 vols in 1, Bronx, N.Y., 1969.
Several of his periodical articles are reprinted in OKEW.

3785. HAMILTON, William Rowan. 1805-1865. Br. DSB; ICB; P III; GB; GC. Also Mechanics and Physics. See also 15596/1.
1. *Lectures on quaternions*. Dublin, 1853. RM/L.
2. *The elements of quaternions*. London, 1866. 2nd ed., 2 vols, 1899-1901.
3. *The mathematical papers*. Cambridge, 1931-. Still in progress. Ed. for the Royal Irish Academy.

3786. JÜRGENSEN, Christian. 1805-1861. Den. P I,III.

3787. LENTHÉRIC, Jacques. b. 1805. Fr. P III.

3788. LIONNET, François Joseph Eugène. 1805-1884. Fr. P III,IV.

3789. NEWMAN, Francis William. 1805-1897. Br. ICB; P III; GA; GB; GD1.

3790. POTTS, Robert. 1805-1885. Br. GD.

3791. REISS, Michel. 1805-1869. Ger. P III.

3792. SCHELLBACH, Karl Heinrich. 1805-1892. Ger. P II,III,IV.; GC.
1. *Die Lehre von den elliptischen Integralen und den Theta-Functionen*. Berlin, 1864.

3793. TRANSON, Abel Etienne Louis. 1805-1876. Fr. P II,III.

3794. VALLÈS. François. b. ca. 1805. Fr. P III.

3795. BRUUN, Heinrich Wilhelm. 1806-1854. Russ. P I.

3796. DE MORGAN, Augustus. 1806-1871. Br. DSB; ICB; P I (under D); P III (under M); GA; GB; GD.. Also Astronomy*, Biography*, and History of Mathematics*. See also Index.
1. *On the study and difficulties of mathematics*. London, 1830-32. X. 2nd reprint ed., Chicago, 1902. 4th reprint ed., La Salle, Ill., 1943.
2. *The elements of arithmetic*. London, 1830. 4th ed., 1840. 5th ed., 1854.
3. *The elements of algebra*. London, 1835. X. 2nd ed., 1837.
4. *The connexion of number and magnitude: An attempt to explain the fifth book of Euclid*. London, 1836.
5. *An explanation of the gnomic projection of the sphere*. London, 1836.
6. *The elements of trigonometry and trigonometrical analysis*. London, 1837. Another ed., 1849.
7. *An essay on probabilities and on their application to life contingencies*. London, 1838. RM/L.
8. *The differential and integral calculus*. London, 1842.
9. *Trigonometry and double algebra*. London, 1849.
10. *A budget of paradoxes*. London, 1872. RM/L. 2nd ed., by D.E. Smith, 2 vols, Chicago, 1915; reprinted 1954.

3797. DRUCKENMÜLLER, Nicolaus. b. 1806. Ger. P I.

3798. GRAVES, John Thomas. 1806-1870. Br. P III; GD.

3799. KIRKMAN, Thomas Penyngton. 1806-1895. Br. DSB; P III,IV.

3800. LAMARLE, Anatole Henri Ernest. 1806-1875. Belg. P I,III.

3801. MINDING, Ernst Ferdinand Adolf. 1806-1885. Ger./Dorpat. DSB; P II,III,IV.

3802. MURPHY, Robert. 1806-1843. Br. P II & Supp.; GA; GB; GD.
1. *A treatise on the theory of algebraical equations*. London, 1839.

3803. RAMUS, Christian. 1806-1856. Den. P II.

3804. HEINEN, Franz. 1807-1870. Ger. P I,II.

3805. MAYR, Aloys. 1807-1890. Ger. P II,III,IV.

3806. PETZVAL, Joseph. 1807-1891. Aus. P III. Also Physics.
1. *Integration der linearen Differentialgleichungen mit constanten und veränderlichen Coefficienten*. Vienna, 1853-59.

3807. SEYDEWITZ, Franz. 1807-1852. Ger. P II.

3808. SOHNCKE, Ludwig Adolph. 1807-1853. Ger. P II.
1. *Bibliotheca mathematica. Verzeichniss der Bücher über die gesammten Zweige der Mathematik*. Leipzig, 1854. RM. For the predecessor of this bibliography see 3717/1.

3809. STERN, Moritz Abraham. 1807-1894. Ger. P II,III,IV; GC.

3810. BJÖRLING, Emmanuel Gabriel. 1808-1872. Swed. P I,III,IV.

3811. BRETSCHNEIDER, Carl Anton. 1808-1878. Ger. P I,III. Also History of Mathematics*.

3812. COTTERILL, Thomas. 1808-1881. Br. P IV.

3813. DUPRÉ, Athanase Louis Victoire. 1808-1869. Fr. DSB; P III. Also Physics*.

3814. MINICH, Serafino Raffaele. 1808-1883. It. P III.

3815. MONTUCCI, Henri Jean. 1808-1877. Fr. P III.

3816. RICHELOT, Friedrich Julius. 1808-1875. Ger. P II,III.

3817. TORTOLINI, Barnaba. 1808-1874. It. P II,III.

3818. GRASSMANN, Hermann Günther. 1809-1877. Ger. DSB Supp.; ICB; P I,III,IV; GC.
  1. *Die Ausdehnungslehre von 1844*. Berlin, 1862. X. 2nd ed., Leipzig, 1878.
  2. *Gesammelte mathematische und physikalische Werke*. 3 vols in 6. Leipzig, 1894-1911.

3819. LIOUVILLE, Joseph. 1809-1882. Fr. DSB; ICB; P I,III; GA.

3820. PEIRCE, Benjamin. 1809-1880. U.S.A. DSB; ICB; P II,III; GB; GE. Also Astronomy.

3821. BOOTH, James. 1810 (or 1806)-1878. Br. P III; GD.
  1. *A treatise on some geometrical methods*. 2 vols. London, 1873.

3822. KUMMER, Ernst Eduard. 1810-1893. Ger. DSB; ICB; P I,III,IV; GC.

3823. LUCHTERHANDT, August Rudolph. 1810-after 1847. Ger. P I.

3824. ZURRIA, Giuseppe. 1810-1896. It. P III,IV.

3825. ADAMS, Carl. 1811-1849. Switz. P I.

3826. AZZARELLI, Mattia. 1811-1897. It. P IV & Supp.

3827. DENZLER, Wilhelm. 1811-1894. Switz. P I,III.

3828. GALOIS, Evariste. 1811-1832. Fr. DSB; ICB; P I; GB. Also Autobiography*.
  1. *Oeuvres mathématiques*. Paris, 1897. Ed. by J. Picard. RM. Reprinted, Paris, 1951.
  2. *Ecrits et mémoires mathématiques*. Paris, 1962. Critical ed. of his MSS and publications by R. Bourgne and J. Azra.

3829. HART, Andrew Searle. 1811-1890. Br. P III,IV; GD.

3830. HESSE, (Ludwig) Otto. 1811-1874. Ger. DSB; P I,III; GC.
  1. *Vorlesungen über analytische Geometrie des Raumes*. Leipzig, 1861. X. 3rd ed., 1876.
  2. *Vorlesungen aus der analytischen Geometrie der geraden Linie, des Punktes und des Kreises in der Ebene*. Leipzig, 1865. X. 2nd ed., 1873.

3831. MATTHES, Carel Joannes. 1811-1882. Holl. P III; GF.

3832. MICHAELIS, Johann Peter. 1811-1867. Luxembourg. P III.

3833. PRATT, Orson. 1811-1881. U.S.A. GB; GE1.

3834\. STRAUCH, Georg Wilhelm. 1811-1868. Switz. P II,III.
1\. *Die Theorie und Anwendung des sogenannten Variations-calculus.* 2 vols. Zurich, 1849. X. 2nd ed., 1854.

3835\. TRUDI, Nicolo. 1811-1884. It. ICB; P IV.

3836\. AGARDH, John Mortimer. 1812-1862. Swed. P I & Supp. Also Astronomy.

3837\. GÖPEL, Adolph. 1812-1847. Ger. DSB; P I.
1\. *Theoriae transcendentium Abelianarum primi ordinis adumbratio levis.* [Periodical article, 1847] German trans. Leipzig, 1895. (OKEW. 67)

3838\. GRAVES, Charles. 1812-1899. Br. GD1.

3839\. GUILMIN, Charles Marie Adrien. 1812-1884. Fr. P III.

3840\. KERZ, Ferdinand. 1812-1892. Ger. P III,IV. Also Astronomy.

3841\. REUSCHLE, Karl Gustav. 1812-1875. Ger. P II,III.

3842\. SCHÖNEMANN, Theodor. 1812-1868. Ger. P II,III.

3843\. SHANKS, William. 1812-1882. Br. DSB; P III,IV.

3844\. CHIÒ, Felice. 1813-1871. It. P III.

3845\. DIPPE, Martin Christian. 1813-1891. Ger. P I,III.

3846\. GREGORY, Duncan Farquharson. 1813-1844. Br. DSB; P I; GD.
1\. *Examples of the processes of the differential and integral calculus.* Cambridge, 1841. 2nd ed., 1846.
2\. *A treatise on the application of analysis to solid geometry.* Cambridge, 1845. X. 2nd ed., 1852.
3\. *Mathematical writings.* Cambridge, 1865. Ed. by W. Walton.

3847\. LAURENT, Pierre Alphonse. 1813-1854. Fr. DSB; P I. Also Physics.

3848\. MOLINS, Louis François Henri Xavier. 1813-after 1883. Fr. P III.

3849\. STEGMANN, Friedrich Ludwig. 1813-1891. Ger. P II,III.

3850\. TURQUAN, Louis Victor. 1813-after 1881. Fr. P III.

3851\. WALTON, William. 1813-1901. Br. P III,IV.
1\. *A treatise on the differential calculus.* Cambridge, 1845.
See also 3846/3 and 3885/1.

3852\. AOUST, Louis. 1814-1885. Fr. P III,IV.

3853\. BRETON (DE CHAMP), Paul Emile. 1814-1885. Fr. P III.

3854\. CATALAN, Eugène Charles. 1814-1894. Fr./Belg. ICB; P I,III,IV.
1\. *Théorèmes et problèmes de géométrie élémentaire.* 2nd ed. Paris, 1852. X. 5th ed., 1872.

3855\. DONKIN, William Fishburn. 1814-1869. Br. P I,III; GD.

3856\. MAILLARD DE LA GOURNERIE, Jules Antoine René. 1814-1883. Fr. P II,III,IV.
1\. *Traité de géométrie descriptive.* 2 vols. Paris, 1862-64. X. 2nd/3rd ed., 1880-91.

3857\. MALMSTEN, Carl Johan. 1814-1886. Swed. P II,III,IV.

3858\. MÖLLINGER, Otto. 1814-1886. Switz. P II,III. Also Astronomy.

3859. RAWSON, Robert. 1814-1906. Br. P III,IV.

3860. SCHLÄFLI, Ludwig. 1814-1895. Switz. DSB; ICB; P II,III,IV; GC.
1. *Tractatus de functionibus sphaericis.* Bern, 1881.
2. *Gesammelte mathematische Abhandlungen.* 3 vols. Basel, 1950-56.

3861. SYLVESTER, James Joseph. 1814-1897. Br. DSB; ICB; P II,III,IV; GB; GD; GE.
1. *The collected mathematical papers.* 4 vols. Cambridge, 1904-12. Ed. by H.F. Baker. RM/L. Reprinted, New York, 1973.

3862. VIEILLE, Jules Marie Louis. 1814-1880. Fr. P III.
1. *Cours complémentaire d'analyse et de mécanique rationelle.* Paris, 1851. X. Another ed., 1856.

3863. WANTZELL, Pierre Laurent. 1814-1848. Fr. P II.

3864. WIEGAND, August. 1814-1871. Ger. P II,III.

3865. BOOLE, George. 1815-1864. Br. DSB; ICB; P I,III; GB; GD.
1. *The mathematical analysis of logic, being an essay towards a calculus of deductive reasoning.* Cambridge, 1847. Reprinted, Oxford, 1948.
2. *An investigation of the laws of thought on which are founded the mathematical theories of logic and probabilities.* London, 1854. RM/L. Reprinted, New York, 1951.
3. *A treatise on differential equations.* Cambridge, 1859. X. Supplementary vol., compiled from Boole's notes by I. Todhunter, 1865. 2nd ed., 1865. 3rd ed., 1872. 4th ed., 1877. Reprinted, New York, 1959.
4. *A treatise on the calculus of finite differences.* Cambridge, 1860. RM/L. 2nd ed., 1872. 3rd ed., 1880. Reprinted, New York, n.d.

3866. BOYMAN, Johann Robert. 1815-1878. Ger. P I,III.

3867. MEYER, Karl Otto. 1815-1893. Ger. P II,III.

3868. PADULA, Fortunato. 1815-1881. It. ICB; P III.

3869. POPOFF, Alexander Fedorovich. 1815-1879. Russ. P II,III. Also Physics.

3870. SOMOV, Osip Ivanovich. 1815-1876. Russ. DSB; ICB; P II (SOMOFF, J.); P III (SSOMOFF, O.I.). Also Mechanics*.

3871. WEIERSTRASS, Karl Theodor Wilhelm. 1815-1897. Ger. DSB; ICB; P II,III,IV; GC.
1. *Formeln und Lehrsätze zum Gebrauche der elliptischen Functionen.* Göttingen, 1883. X. 2nd ed., Berlin, 1893; reprinted, Wurzburg, 1962.
2. [In Russian. *Letters to Sofie Kovalevskaia, 1871-1891.*] Moscow, 1973. The letters are in German, with Russian translation.
3. *Mathematische Werke.* 7 vols. Berlin, 1894-1927. Reprinted, Hildesheim, 1968.

See also 3723/5.

3872. WHITEHOUSE, Edward Orange Wildman. ca. 1815-after 1875. Br. P III.

3873. FASBENDER, Eduard. 1816-1892. Ger. P III.

3874. FRENET, Jean Frédéric. 1816-1900. Fr. DSB; P III.

3875. HOPPE, Ernst Reinhold Eduard. 1816-1900. Ger. P I,III,IV.

3876. LINDMAN, Christian Frederick. 1816-1901. Swed. P I,III,IV.

3877. OLTRAMARE, Gabriel. 1816-1906. Switz. P II,III,IV.

3878. ROSENHAIN, (Johann) Georg. 1816-1887. Aus./Ger. DSB; P II, III,IV.
  1. *Sur les fonctions de deux variables à quatre périodes....* [Periodical article, 1851] German trans. Leipzig, 1895. (OKEW. 65)

3879. STEEN, Adolph. 1816-1886. Den. P II,III,IV.

3880. TARDY, Placido. 1816-1914. It. P III,IV.

3881. WITTSTEIN, Theodor Ludwig. 1816-1894. Ger. P II,III.

3882. ARNDT, Peter Friedrich. 1817-1866. Ger. P I,III.

3883. BORCHARDT, Carl Wilhelm. 1817-1880. Ger. DSB; P I,III,IV; GC.
  1. *Gesammelte Werke.* Berlin, 1888. Ed. by G. Hettner.
  See also 3779/3.

3884. BRIOT, Charles Auguste Albert. 1817-1882. Fr. DSB; P III,IV. Also Physics*.
  1. *Théorie des fonctions elliptiques.* Paris, 1875. With J.C. Bouquet.
  2. *Théorie des fonctions Abéliennes.* Paris, 1879. RM/L.

3885. ELLIS, Robert Leslie. 1817-1859. Br. P I & Supp.; GD.
  1. *The mathematical and other writings.* Cambridge, 1863. Ed. by W. Walton, with a biography. RM/L.

3886. FROST, Percival. 1817-1898. Br. P IV; GD1. Also Mechanics*.
  1. *A treatise on solid geometry.* Cambridge, 1863. With J. Wolstenholme. 2nd ed., London, 1875. 3rd ed., 1886.

3887. GENOCCHI, Angelo. 1817-1889. It. ICB; P III,IV & Supp.

3888. HELMLING, Peter. 1817-1901. Dorpat. P I,III,IV.

3889. HOUSEL, Charles Pierre. 1817-after 1869. Fr. P III.

3890. JELLETT, John Hewitt. 1817-1888. Br. P III,IV; GD. Also Physics* and Chemistry*.
  1. *An elementary treatise on the calculus of variations.* Dublin, 1850.

3891. PROUHET, (Pierre Marie) Eugène. 1817-1867. Fr. P III.

3892. ROBERTS, Michael. 1817-1882. Br. P IV; GD.
  1. *A tract on the addition of elliptic and hyper-elliptic integrals.* Dublin, 1871.

3893. ROBERTS, William (first of the name). 1817-1883. Br. P IV.

3894. RUBINI, Raffaele. 1817-1890. It. P III,IV.

3895. SCHAAR, Mathieu. 1817-1867. Belg. P II,III; GF.

3896. WEDDLE, Thomas. 1817-1853. Br. P II.

3897. BALTZER, Richard, 1818-1887. Ger. P III,IV; GC.
 1. *Die Elemente der Mathematik.* 2 vols. Leipzig, 1960. X. 5th ed., 1878.
 See also 3687/1.

3898. DESBOVES, Honoré Adolphe. 1818-after 1887. Fr. P III,IV.

3899. DIENGER, Joseph. 1818-1894. Ger. P I,III,IV; GC.

3900. GRANLUND, Jacob Niclas. 1818-1859. Swed. P I & Supp.

3901. HILL, Thomas. 1818-1891. U.S.A. P III; GE. Also Natural History.

3902. JOACHIMSTHAL, Ferdinand. 1818-1861. Ger. DSB; P I,III.

3903. REALIS, Savino. 1818-1886. It. P III,IV.

3904. ARONHOLD, Siegfried Heinrich. 1819-1884. Ger. DSB; P I,III, IV; GC.

3905. BONNET, Pierre Ossian. 1819-1892. Fr. DSB; P I,III,IV.

3906. BOUQUET, Jean Claude. 1819-1885. Fr. DSB; P III,IV. See also 3884/1.

3907. COCKLE, James. 1819-1895. Australia/Br. P IV; GB; GD1.

3908. DAVID, Jean Marie. 1819-1890. Fr. P III,IV Supp.

3909. JANNI, Vincenzo. 1819-1891. It. P IV.

3910. KORALEK (or CORALEQUE), Philipp. 1819-after 1854. Fr. P III.

3911. MARIE, (Charles François) Maximilien. 1819-1891. Fr. P III, IV. Also History of Science*.
 1. *Théorie des fonctions de variables imaginaires.* 3 vols. Paris, 1874-76.

3912. PLARR, Gustav. 1819-1892. Ger. P III,IV.

3913. SALMON, George. 1819-1904. Br. DSB; P III,IV; GB; GD2.
 1. *A treatise on conic sections.* Dublin, 1848. X. 2nd ed., 1850. 3rd ed., London, 1855. 4th ed., 1863. 5th ed., 1869. 6th ed., 1879.
 2. *A treatise on the higher plane curves.* Dublin, 1852. 2nd ed., 1873. 3rd ed., 1879.
 3. *Lessons introductory to the modern higher algebra.* Dublin, 1859. X. 2nd ed., 1866. 3rd ed., 1876. 4th ed., 1885.
 4. *A treatise on the analytic geometry of three dimensions.* Dublin, 1862. X. 2nd ed., 1865. 4th ed., 1882.

3914. SERRET, (Joseph) Alfred. 1819-1885. Fr. DSB; P II,III,IV.
 1. *Cours d'algèbre supérieure.* Paris, 1849. X. 4th ed., 2 vols, 1877-79. 5th ed., 2 vols, 1885.
 2. *Cours de calcul différentiel et intégral.* 2 vols. Paris, 1867-68. X. 2nd ed., 1879-80.

3915. SIEBECK, Friedrich Herrmann. 1819-after 1879. Ger. P II,III.

3916. YOUNG, George Paxton. 1819-1889. Canada. P IV & Supp.; GF.

3917. BAUER, (Conrad) Gustav. 1820-1906. Ger. P III,IV.

3918. BAUER, Carl Wilhelm von. 1820-1894. Ger. P III,IV.

3919. CASEY, John. 1820-1891. Br. P III,IV & Supp.; GD1.

1. *On cubic transformations.* Dublin, 1880.
2. *A sequel to the first six books of the "Elements" of Euclid, containing an easy introduction to modern geometry.* Dublin, 1881. 3rd ed., 1884.
3. *A treatise on the analytical geometry of the line, circle, and conic sections.* Dublin, 1885.
4. *A treatise on spherical trigonometry.* Dublin, 1889.

3920. HARGREAVE, Charles James. 1820-1866. Br. P III; GD.

3921. HEILERMANN, Johann Bernhard Hermann. 1820-1899. Ger. P I,III, IV.

3922. JONQUIÈRES, Ernest (Jean Philippe Fouque) de. 1820-1901. Fr. DSB; ICB; P III,IV.

3923. LEMONNIER, Hippolyte Guillaume. 1820-1882. Fr. P III.

3924. PUISEUX, Victor Alexandre. 1820-1883. Fr. DSB; ICB; P II,III; GA. Also Astronomy.

3925. SCHEFFLER, (August Christian Wilhelm) Hermann. 1820-1903. Ger. P II,III,IV. Also Mechanics and Physics.

3926. TODHUNTER, Isaac. 1820-1884. Br. DSB; P III; GB; GD. Also Mechanics*, Biography*, and History of Mathematics*.
1. *A treatise on the differential calculus.* Cambridge, 1852. 3rd ed., 1860. 4th ed., 1864.
2. *A treatise on the integral calculus.* 1857. X. 2nd ed., 1862. 4th ed., 1874. Another ed., 1878.
3. *Algebra.* 1858. X. 3rd ed., 1862. Other eds, 1883 and 1887.
4. *A treatise on plane co-ordinate geometry.* 1858. X. 5th ed., 1874. 7th ed., 1881.
5. *Examples of analytical geometry of three dimensions.* 1858. X. Another ed., 1878.
6. *Spherical trigonometry.* Cambridge, 1859. 2nd ed., 1863. Another ed., 1878.
7. *Plane trigonometry.* 1859. X. 6th ed., 1876.
8. *An elementary treatise on the theory of equations.* 1861. 3rd ed., 1875.
9. *The "Elements" of Euclid.* 1862. X. Other eds: 1877, 1883.
10. *Researches in the calculus of variations.* London, 1871.
11. *An elementary treatise on Laplace's functions, Lamé's functions, and Bessel's functions.* London, 1875.

See also 3865/3.

3927. ULLHERR, Johann Konrad. 1820-1887. Ger. P III; GC.

3928. BÖKLEN, Georg Heinrich Otto. 1821-1900. Ger. P III,IV.

3929. CAYLEY, Arthur. 1821-1895. Br. DSB; ICB; P I,III,IV; GB; GD1.
1. *An elementary treatise on elliptic functions.* London, 1876. 2nd ed., 1895.
2. *The collected mathematical papers.* 14 vols. Cambridge, 1889-98. Ed. by Cayley himself until his death, and thereafter by A.R. Forsyth. RM/L. Reprinted, New York, 1963.

3930. CHEBYSHEV, Pafnuty Lvovich. 1821-1894. Russ. DSB; ICB; P II, III,IV (TSCHEBYTSCHEW); GB.
1. *Oeuvres de P.L. Tchebychef.* 2 vols. St. Petersburg, 1899-

1907. Ed. by A. Markoff and N. Sonin under the auspices of the St. Petersburg Academy. Reprinted, New York, 1962?

3931. DURÈGE, Jacob Heinrich Karl. 1821-1893. Switz./Prague/Aus. P III,IV.
1. *Theorie der elliptischen Functionen*. 1861. X. 2nd ed., Leipzig, 1868.
2. *Elemente der Theorie der Functionen complexer Variabeln*. 1864. X. 2nd ed., Leipzig, 1873.
3. *Die ebenen Curven dritter Ordnung*. Leipzig, 1871.

3932. HEINE, (Heinrich) Eduard. 1821-1881. Ger. DSB; P I,III.
1. *Handbuch der Kugelfunctionen, Theorie und Anwendungen*. Berlin, 1861. X. 2nd ed., 2 vols, 1878-81; reprinted, 2 vols in 1, Wurzburg, 1961.

3933. LIGOWSKI, Wilhelm Johann Otto. 1821-1893. Ger. P I,III,IV.

3934. MEECH, Levi Witter. 1821-after 1881. U.S.A. P II,III.

3935. MENTION, J. ca. 1821-after 1865. Fr./Russ. P IV.

3936. PARMENTIER, Joseph Charles Théodore. 1821-after 1883. Fr. P III.

3937. RAMACHANDRA, Y. 1821-1880. India.
1. *A treatise on problems of maxima and minima, solved by algebra ... Reprinted by order of the ... East-India Company for circulation in Europe and in India*. London, 1859. Ed. by A. De Morgan.

3938. SEIDEL, Philipp Ludwig von. 1821-1896. Ger. DSB; P II,III,IV; GC. Also Astronomy.

3939. SWELLENGREBEL, Jan Gerard Hendrik. 1821-1854. Holl. P II; GC.

3940. TOWNSEND, Richard. 1821-1884. Br. P III; GD.
1. *Chapters on the modern geometry of the point, line and circle*. 2 vols. Dublin, 1863-65.

3941. WINCKLER, Anton. 1821-1892. Ger./Aus. P II,III,IV.

3942. BAEHR, George Frederik Willem. 1822-1898. Holl. P III,IV; GF. Also Mechanics.

3943. BERGER, Charles Hippolyte. 1822-1869. Fr. P III.

3944. BERTRAND, Joseph Louis François. 1822-1900. Fr. DSB; ICB; P I,III,IV; GA. Also Physics*, Biography*, and History of Science*.
1. *Traité élémentaire d'algèbre*. 2 vols. Paris, 1850. X. 6th ed., 1869-70. 11th ed., 1878-79.
2. *Traité de calcul différentiel et de calcul intégral*. 2 vols. Paris, 1864-70.
3. *Calcul des probabilités*. Paris, 1889.
See also 5545/1.

3945. BIERENS DE HAAN, David. 1822-1895. Holl. ICB; P III,IV; GF (HAAN). Also History of Science*.
1. *Nouvelles tables d'intégrales définies*. 1867. X. Reprinted with an English trans. of the introduction by J.F. Ritt. New York, 1938.

3946. BOURGET, Justin. 1822-1887. Fr. P III,IV. Also Mechanics.

3947. HERMITE, Charles. 1822-1901. Fr. DSB; ICB; P I,III,IV; GA.
1. *Cours d'analyse*. Paris, 1873.
2. *Sur la fonction exponentielle*. 1874.
3. *Sur quelques applications des fonctions élliptiques*. 1885.
4. *Correspondance d'Hermite et de Stieltjes*. 2 vols. Paris, 1905. Ed. by B. Baillaud and H. Bourget.
5. *Oeuvres*. 4 vols. Paris, 1905-17. Ed. by E. Picard.
See also 4063/1.

3948. BETTI, Enrico. 1823-1892. It. DSB; P III,IV. Also Mechanics.
1. *Opere matematiche*. 2 vols. Milan, 1903-15. Ed. by the Accademia dei Lincei.

3949. DAVIDOV, August Yulevich. 1823-1885. Russ. DSB; P I,IV (DAW-IDOFF). Also Mechanics.

3950. EISENSTEIN, Ferdinand Gotthold Max. 1823-1852. Ger. DSB; ICB; P I; GC.
1. *Mathematische Abhandlungen ... Mit einem Vorrede von C.F. Gauss*. Berlin, 1847. X. Reprinted, Hildesheim, 1967.

3951. HELLWIG, Carl Franz. 1823-1898. Ger. P IV.

3952. HOÜEL, (Guillaume) Jules. 1823-1886. Fr. DSB; ICB; P III,IV.
1. *Cours de calcul infinitésimal*. 4 vols. Paris, 1878-81.
2. *Essai critique sur les principes fondamentaux de la géométrie élémentaire*. Paris, 1883.
See also 3695/2a.

3953. KRONECKER, Leopold. 1823-1891. Ger. DSB; ICB; P I,III,IV; GC.
1. *Vorlesungen über Mathematik*. 2 vols in 3. Leipzig, 1894-1903. Ed. under the auspices of the Berlin Academy.
2. *Werke*. 5 vols. Leipzig, 1895-1931. Ed. by K. Hensel. Reprinted, New York, 1968.
See also 3784/2.

3954. LE COINTE, Ignace Louis Alfred. 1823-after 1880. Fr. P III.

3955. ROGNER, Johann. 1823-1886. Aus. P III.

3956. RUFFINI, Ferdinando Paolo. 1823-1908. It. P IV.

3957. RUSSELL, William Henry Leighton. 1823-after 1888. Br. P III,IV.

3958. SCHLÖMILCH, Oscar Xaver. 1823-1901. Ger. P II,III,IV.
1. *Compendium der höheren Analysis*. Brunswick, 1864. X. 3rd ed., 2 vols, 1874-79. 5th ed., 2 vols, 1879-81.
2. *Übungsbuch zum Studium der höheren Analysis*. 2 vols. Leipzig, 1870. X. 2nd ed., 1873-74.

3959. ZEIPEL, Ewald Victor Ehrenhold von. 1823-1893. Swed. P II,III.

3960. ALLMAN, George Johnston. 1824-1904. Br. P III,IV; GD2. Also History of Mathematics*.

3961. BRIOSCHI, Francesco. 1824-1897. It. DSB; P III,IV. Also Mechanics.
1. *Teoria dei determinanti*. Pavia, 1854. X.
1a. ——— *Théorie des déterminants*. Paris, 1856.
2. *Opere matematiche*. 5 vols. Milan, 1901-09.

3962. CODAZZI, Delfino. 1824-1873. It. DSB; P III.

3963. COMBESCURE, Jean Joseph Antoine Edouard. 1824-after 1891. Fr. P III,IV.

3964. MEYER, Martin Hermann. 1824-1856. Ger. P II. Also Physics.

3965. ŽMURKO (or SCHMURKO). Wawrzyniec (or Lorenz). 1824-1889. Lemberg. P III,IV.

3966. CALDARERA, Francesco. 1825-after 1906. It. P IV & Supp.

3967. DESPEYROUS, Charles. ca. 1825-after 1883. Fr. P III.

3968. DOSTOR, Georges J. ca. 1825-after 1883. Fr. P III.

3969. FAÀ DI BRUNO, Francesco. 1825-1888. It. ICB; P III,IV & Supp.
1. *Théorie des formes binaires*. Turin, 1876.
1a. ——— *Einleitung in die Theorie der binären Formen*. Leipzig, 1881.

3970. FAURE, Henri Auguste. 1825-after 1889. Fr. P III,IV.

3971. ROGER, Emile. 1825-after 1880. Fr. P III.

3972. SPOTTISWOODE, William. 1825-1883. Br. P II,III,IV; GB; GD. Also Physics*.

3973. WALKER, John James. 1825-1900. Br. P III,IV.

3974. WASCHTSCHENKO-SACHARTSCHENKO. Michail Jegorovich. 1825-after 1883. Russ. P III.

3975. BATTAGLINI, Giuseppe. 1826-1894. It. P III,IV.

3976. CROFTON, Morgan William. 1826-1915. Br. P IV & Supp.

3977. JEFFERY, Henry Martyn. 1826-1891. Br. P III,IV.

3978. MEISSEL, Daniel Friedrich Ernst. 1826-1890. Ger. ICB; P II, III,IV.

3979. PAINVIN, Louis Félix. 1826-1875? Fr. P III.

3980. PEPIN, Jean François Théophile. 1826-1904. Fr. P III.

3981. POLIGNAC, Alphonse Armand Charles Marie de. 1826-1863. Fr. P III.

3982. RIEMANN, (Georg Friedrich) Bernhard. 1826-1866. Ger. DSB; ICB; P II,III,IV; GB; GC. Also Physics*.
1. *Partielle Differentialgleichungen und deren Anwendung auf physikalische Frage. Vorlesungen*. Brunswick, 1869. Ed. by K. Hattendorff. X. 2nd ed., 1876. 3rd ed., 1882.
2. *Gesammelte mathematische Werke und wissenschaftlicher Nachlass*. Leipzig, 1876. Ed. by R. Dedekind and H. Weber. RM/L.
2a. ——— 2nd ed. with *Nachträge*. 2 vols, ib., 1892-1902. Reprinted, New York, 1953.
2b. ——— *Oeuvres mathématiques*. Paris, 1898. Trans. by L. Laugel. Reprinted, Paris, 1968.
See also 4185/2.

3983. SCHEIBNER, Wilhelm. 1826-1908. Ger. P II,III,IV.

3984. SCHELL, Wilhelm Joseph Friedrich Nikolaus. 1826-1904. Ger. P II,III,IV. Also Mechanics*.

3985. SMITH, Henry John Stanley (or Stephen). 1826-1883. Br. DSB; P III,IV; GB; GD.
    1. *The collected mathematical papers.* 2 vols. Oxford, 1894. Ed. by J.W.L. Glaisher.

3986. SPITZER, Simon. 1826-1887. Aus. P II,III,IV.
    1. *Vorlesungen über lineare Differential-Gleichungen.* Vienna, 1878.

3987. STAMMER, Wilhelm. 1826-after 1904. Ger. P II,III,IV.

3988. TYCHSEN, Nicolai Georg Camillo. 1826-1888. Den. P III.

3989. WERNER, Ferdinand Oscar. 1826-after 1858. Ger./Russ. P II.

3990. WIENER, (Ludwig) Christian. 1826-1896. Ger. DSB; ICB; P II, III,IV; GC. Also Physics.
    1. *Über Vielecke und Vielfläche.* Leipzig, 1864.

3991. WOEPCKE, Franz. 1826-1864. Ger. DSB; P II,III; GA; GC. Also History of Mathematics.

3992. BEEZ, Emil Ludwig Richard. 1827-after 1901. Ger. P III,IV.

3993. BISCHOFF, Johann Nicolaus. 1827-1893. Ger. P III; GC.

3994. LINDELÖF, Lorentz Leonard. 1827-1908. Fin. P III,IV. Also Astronomy.

3995. MOUTARD, Théodore Florentin. 1827-1901. Fr. DSB; P IV.

3996. ROBERTS, Samuel. 1827-1913. Br. P III,IV.

3997. SERRET, Paul Joseph. 1827-1898. Fr. P III,IV.
    1. *La géométrie de direction.* Paris, 1869.

3998. WILLIAMSON, Benjamin. 1827-1916. Br. P III,IV; GF. Also Mechanics*.
    1. *An elementary treatise on the differential calculus.* London, 1872. X. 7th ed., 1889.
    2. *An elementary treatise on the integral calculus.* London, 1874. X. 2nd ed., 1877.

3999. BESANT, William Henry. 1828-1917. Br. P III. Also Mechanics*.
    1. *Notes on roulettes and glissettes.* London, 1870. X. 2nd ed., Cambridge, 1890.

4000. BRUNO, Giuseppe. 1828-1893. It. P III,IV.

4001. DAUG, Hermann Theodor. 1828-1888. Swed. P III,IV.

4002. FREYCINET, Charles Louis de Saulses de. 1828-1923. Fr. ICB; P III; GB. Also Mechanics*.
    1. *Essais sur la philosophie des sciences: Analyse, mécanique.* Paris, 1896. X. 2nd ed., 1900.

4003. HARLEY, Robert. 1828-1910. Br. P III,IV; GD2.

4004. KÜPPER, Karl Josef. 1828-1900. Ger./Prague. P IV & Supp.

4005. LUYNES, Victor Hippolyte de. 1828-after 1877. Fr. P III.

4006. PETERSON, Karl Mikhailovich. 1828-1881. Russ. DSB; P IV.

4007. ALLÉGRET, Alexandre. 1829-after 1881. Fr. P III.

4008. CHRISTOFFEL, Elwin Bruno. 1829-1900. Ger. DSB; P III,IV.

1. *Gesammelte mathematische Abhandlungen*. Leipzig, 1910. Ed. by L. Maurer et el. RM/L.

4009. CLASEN, Bernard Isidore. 1829-1902. Luxembourg. ICB.

4010. ENDE, Heinrich am. 1829-1899. Ger. P III,IV.

4011. FERRERS, Norman Macleod. 1829-1903. Br. P III,IV & Supp.; GD2.
1. *An elementary treatise on trilinear co-ordinates*. 2nd ed. London, 1866. 3rd ed., 1876.
2. *An elementary treatise on spherical harmonics*. London, 1877.
See also 3701/1.

4012. HAYWARD, Robert Baldwin. 1829-1903. Br. P III; GD2.

4013. PICART, Alphonse. 1829-1884. Fr. P III.

4014. SCHROETER, Heinrich Eduard. 1829-1892. Ger. DSB; P II,III,IV; GC.

4015. WOLSTENHOLME, Joseph. 1829-1891. Br. P III,IV; GD. See also 3886/1.

4016. CREMONA, (Antonio) Luigi (Gaudenzio Giuseppe). 1830-1903. It. DSB; ICB; P III,IV & Supp.; GB.
1. *Grundzüge einer allgemeiner Theorie der Oberflächen in synthetischer Behandlung*. Berlin, 1870. Trans. by M. Curtze of two periodical articles of 1863 and 1866.
2. *Le figure reciproche nella statica grafica*. Milan, 1872. X.
2a. ——— *Graphical statics*. Oxford, 1890.
3. *Elementi di geometria proiettiva*. Turin, 1873. X.
3a. ——— *Elements of projective geometry*. Oxford, 1885.
4. *Elementi di calcolo grafico*. Turin, 1874.
5. *Opere matematiche*. 3 vols. Milan, 1914-17. Ed. under the auspices of the Accademia dei Lincei.

4017. ENNEPER, Alfred. 1830-1885. Ger. P III,IV.
1. *Elliptische Functionen: Theorie und Geschichte*. Halle, 1876.

4018. GRETSCHEL, Heinrich Friedrich. 1830-1892. Ger. P III.

4019. HIRST, Thomas Archer. 1830-1892. Br. P I,III,IV; GD1. See also 6475/2a.

4020. MATTHIESSEN, Heinrich Friedrich Ludwig. 1830-1906. Ger. P II, III,IV. Also Physics and Physiology.

4021. PESCHKA, Gustav Adolf von. 1830-after 1883. Aus. P III.

4022. DEDEKIND, (Julius Wilhelm) Richard. 1831-1916. Ger. DSB; ICB; P I,III,IV.
1. *Über die Theorie der ganzen algebraischen Zahlen*. Brunswick, 1964. Reprint of a periodical article of 1877.
2. *Was sind und was sollen die Zahlen?* Brunswick, 1888. X. 2nd ed., 1893.
2a. ——— *Essays on the theory of numbers*. 1901.
3. *Gesammelte mathematische Werke*. 3 vols. Brunswick, 1930-32. Ed. by R. Fricke, E. Noether, and O. Ore. Reprinted, New York, 1969.
See also 3784/1, 3982/2, and 4196/3.

4023. DEWULF, Eugène Edouard Désiré. 1831-after 1886. Fr. P III,IV.

4024. DU BOIS-REYMOND, Paul David Gustav. 1831-1889. Ger. DSB; ICB;

P III,IV (under B); GC. Also Science in General*.
1. *Zwei Abhandlungen über unendliche und trigonometrische Reihen.* [1871, 1874] Leipzig, 1912. (OKEW. 185)
2. *Abhandlung über die Darstellung der Funktionen durch trigonometrische Reihen.* [1876] Leipzig, 1912. (OKEW. 186)
3. *Die allgemeine Functionentheorie.* Tübingen, 1882. Reprinted, Darmstadt, 1968.

4025. McLAREN, John Lord. 1831-1910. Br. P IV; GD2.

4026. MANNHEIM, (Victor Mayer) Amedée. 1831-1906. Fr. DSB; ICB; P III,IV.
1. *Cours de géométrie descriptive.* Paris, 1880.
2. *Principes et développements de géométrie cinématique.* Paris, 1894.

4027. MARTUS, Hermann Carl Eberhard. 1831-after 1903. Ger. P III,IV.

4028. MOTTA-PEGADO, Luiz Porfirio da. 1831-after 1880. Port. P III.

4029. NIEMTSCHIK (or NĚMCZIK), Rudolph. 1831-1876. Aus. P III.

4030. SABININ, Georg. 1831-1909. Russ. P III,IV.

4031. SKŘIVAN, Gustav. 1831-1866. Aus. P III.

4032. TAIT, Peter Guthrie. 1831-1901. Br. DSB; ICB; P III,IV; GB; GD2. Also Mechanics* and Physics*#. See also Index.
1. *An elementary treatise on quaternions.* Oxford, 1867. 2nd ed., Cambridge, 1873. 3rd ed., 1890.

4033. BOUR, Edmond. 1832-1866. Fr. DSB; P III. Also Astronomy and Mechanics.

4034. BUCHWALDT, Frantz Ingstrup. 1832-after 1901. Den. P III,IV.

4035. CASSANI, Pietro. 1832-1905. It. P III; IV.

4036. DILLNER, Göran. 1832-1906. Swed. P III,IV.

4037. DODGSON, Charles Lutwidge (pseudonym: CARROLL, Lewis). 1832-1898. Br. DSB; ICB (CARROLL); GB; GD1.
1. *A treatise on determinants.* London, 1867.
2. *Euclid and his modern rivals.* 2 vols. London, 1879-85. 2nd ed., 1885.
3. *Curiosa mathematica. Part I: A new theory of parallels.* London, 1890. 4th ed., 1895.
3a. ——— *Part II: Pillow-problems.* Ib., 1893. 4th ed., 1895.

4038. ESCHER, Paul Friedrich Joseph. 1832-after 1865. Switz. P III.

4039. FIEDLER, (Otto) Wilhelm. 1832-1912. Ger./Prague/Switz. P III, IV.
1. *Die darstellende Geometrie.* Leipzig, 1871. X. 2nd ed., 1875. 3rd ed., 3 vols, 1885-88.

4040. GILBERT, (Louis) Philippe. 1832-1892. Belg. P III,IV. Also Mechanics.

4041. IMSCHENETSKY, Wassily Grigorievich. 1832-1892. Russ. P III,IV.

4042. KINKELIN, Hermann. 1832-1913. Switz. P I,III,IV.

4043. LIPSCHITZ, Rudolf Otto Sigismund. 1832-1903. Ger. DSB; P I, III,IV.

4044. NEUMANN, Carl Gottfried. 1832-1925. Ger. DSB; P II,III,IV. Also Physics*.
1. *Vorlesungen über Riemann's Theorie der Abel'schen Integrale.* Leipzig, 1865.
2. *Untersuchungen über das logarithmische und Newton'sche Potential.* Leipzig, 1877.

4045. POLIGNAC, Camille Armand Jules Marie de. 1832-1913. Fr. P III, IV.

4046. ROUCHÉ. Eugène. 1832-1910. Fr. P III,IV.
1. *Traité de géométrie.* Paris, 1865. X. 5th ed., 1883. 6th ed., 1891.
See also 4063/1.

4047. SYLOW, (Peter) Ludvig (Mejdell). 1832-1918. Nor. DSB; P III, IV. See also 3762/5.

4048. TUCKER, Robert. 1832-1905. Br. P III,IV. Also Astronomy. See also 4197/1.

4049. VELTMANN, Wilhelm. 1832-1902. Ger. P III,IV.

4050. ZEHFUSS, Johann Georg. 1832-after 1872. Ger. P II,III.

4051. BECKER, Johann Carl. 1833-1887. Switz./Ger. P III,IV.

4052. BERG, Franciscus Johannes van den. 1833-1892. Holl. P III,IV (under both B and V); GF.

4053. CLEBSCH, (Rudolph Friedrich) Alfred. 1833-1872. Ger. DSB; P III; GC. Also Mechanics*.
1. *Theorie der Abelschen Functionen.* Leipzig, 1866.
2. *Theorie der binären algebraischen Formen.* Ib., 1872. RM.
3. *Vorlesungen über Geometrie.* 2 vols. Ib., 1876-91. Ed. by F. Lindemann.

4054. FUCHS, Immanuel Lazarus. 1833-1902. Ger. DSB; P III,IV.

4055. MONRO, Cecil James. 1833-1882. Br. P III.

4056. MOST, Robert. 1833-after 1878. Ger. P III.

4057. RICE, John Minot. 1833-1901. U.S.A. P III.

4058. SCHERING, Ernst Christian Julius. 1833-1897. Ger. P II,III,IV.

4059. UNFERDINGER, Franz Xaver. 1833-1890. Aus. P III.

4060. VÖLLER, Andreas. 1833-1859. Ger. P II; GC.

4061. DE FOREST, Erastus Lyman. 1834-1888. U.S.A. P III; GE.

4062. HATTENDORFF, Karl Friedrich Wilhelm. 1834-1882. Ger. P III. See also 3982/1.

4063. LAGUERRE, Edmond Nicolas 1834-1886. Fr. DSB; P III,IV.
1. *Oeuvres.* 2 vols. Paris, 1898-1903. Ed. by C. Hermite, H. Poincaré, and E. Rouché. Reprinted, New York, 1972.

4064. MEYER, Gustav Ferdinand. 1834-after 1902. Ger. P III,IV.
1. *Vorlesungen über die theorie der bestimmten Integrale zwischen reellen Grenzen.* Leipzig, 1871.

4065. STOLL, Franz Xaver. 1834-1902. Ger. P IV.

4066. VENN, John. 1834-1923. Br. DSB; GD4.

1. *The logic of chance*. London, 1866. 2nd ed., 1876. 3rd ed., 1888.
See also 1464/2.

4067. BELTRAMI, Eugenio. 1835-1899. It. DSB; P III,IV. Also Mechanics*.
1. *Opere matematiche*. 4 vols. Milan, 1902-20.

4068. CASORATI, Felice. 1835-1890. It. P III,IV.

4069. GRUBE, Franz. 1835-1893. Ger. P III. See also 5635/1.

4070. MARTIN. Artemas. 1835-1918. U.S.A. P IV; GE

4071. MATHIEU, Emile Léonard. 1835-1890. Fr. DSB; P III,IV. Also Mechanics* and Physics*.

4072. MEHLER, Ferdinand Gustav. 1835-1895. Ger. P III,IV.

4073. MÉRAY, Hugues Charles Robert. 1835-1911. Fr. DSB; P III,IV.

4074. SAALSCHÜTZ, Louis. 1835-1913. Ger. P III,IV & Supp.

4075. WORPITZKY, Julius Daniel Theodor. 1835-1895. Ger. P III,IV.

4076. DUPUIS, Nathan Fellowes. 1836-after 1910. Canada. P IV.

4077. GRELLE, Friedrich Heinrich. 1836-1878. Ger. P III.

4078. STUDNIČKA, Franz Josef. 1836-1903. Prague. P III,IV.

4079. WEINGARTEN, (Johannes Leonard Gottfried) Julius. 1836-1910. Ger. DSB; P III,IV.

4080. ZINGER, Vassily Jakovlevich. 1836-1907. Russ. P III,IV.

4081. BACHMANN, Paul Gustav Heinrich. 1837-1920. Ger. DSB; ICB; P III,IV.
1. *Die Lehre von der Kreistheilung und ihre Beziehungen zur Zahlentheorie*. Leipzig, 1872. Reprinted, Wiesbaden, 1968.
2. *Vorlesungen über die Natur der Irrationalzahlen*. Ib., 1892.
3. *Zahlentheorie*. 5 vols. Ib., 1892-1923.
4. *Die Arithmetik der quadratischen Formen*. Ib., 1898.

4082. BUGAEV, Nicolai Vasilievich. 1837-1903. Russ. DSB Supp.; ICB; P III,IV (BOUGAJEFF).

4083. CARR, George Shoobridge. 1837-after 1912.
1. *A synopsis of elementary results in pure mathematics ...* [and] *an index to the papers on pure mathematics ... in the principal journals and transactions ... of the present century*. London, 1886.

4084. FRISCHAUF, Johannes. 1837-1924. Aus. P III,IV. Also Astronomy.
1. *Absolute Geometrie nach Johann Bolyai*. Leipzig, 1872.
2. *Elemente der absoluten Geometrie*. Ib., 1876.

4085. GORDAN, Paul Albert. 1837-1912. Ger. DSB; P III,IV.
1. *Vorlesungen über Invariantentheorie*. 2 vols. Leipzig, 1885-87.

4086. GRIFFITHS, John. 1837-after 1903. Br. P IV.

4087. KÖNIGSBERGER, Leo. 1837-1921. Aus./Ger. DSB; P III,IV. Also Biography*.

1. *Vorlesungen über die Theorie der elliptischen Functionen.* Leipzig, 1874.

4088. KORKIN, Alexander Nicolaievich. 1837-1908. Russ. P III,IV.

4089. LETNIKOW, Alexei Vassilievich. 1837-1888. Russ. P III,IV.

4090. McCOLL, Hugh. 1837-1909. Br./Fr. DSB; P III,IV.

4091. MILINOWSKI, Alfons. 1837-1888. Ger. P III,IV.

4092. MOSHAMMER, Karl. 1837-after 1901. Aus. P III,IV.

4093. PIUMA, Carlo Maria. 1837-1912. It. P III,IV.

4094. TILLY, Joseph Marie de. 1837-1906. Belg. DSB; ICB; P III,IV.
1. *Recherches sur les élémens de la géométrie.* Brussels, 1860.
2. *Essai sur les principes fondamentaux de la géométrie et de la mécanique.* Bordeaux, 1878.

4095. ZAJASZKOWSKI. Władysław. 1837-1898. Cracow etc. P III,IV.

4096. ABBOTT, Edwin Abbott. 1838-1926. Br. ICB; GB; GD4.

4097. GRÜNWALD, Anton Karl. 1838-1920. Prague. P III,IV. Also Physics.

4098. GULDBERG, Axel Sophus. 1838-1913. Nor. P III,IV.

4099. HAMBURGER, Meyer. 1838-1903. Ger. P IV & Supp.

4100. HUNYADY, Eugen von. 1838-1889. Hung. P III.

4101. IGEL, Benzion. 1838-1898. Aus. P III,IV.

4102. JORDAN, Camille. 1838-1921. Fr. DSB; ICB; P III,IV.
1. *Traité des substitutions et des équations algébraiques.* Paris, 1870. Reprinted, ib., 1957.
2. *Cours d'analyse.* 3 vols. Paris, 1882-87. 2nd ed., 1893-96.
3. *Oeuvres.* 4 vols. Paris, 1961-64. Ed. by G. Julia and J. Dieudonné.

See also 4182/2.

4103. NÄGELSBACH, Hans Eduard von. 1838-after 1879. Ger. P III.

4104. REYE, (Carl) Theodor. 1838-1919. Switz./Ger. DSB; ICB; P III, IV. Also History of Mathematics*.
1. *Geometrie der Lage.* 2 vols. Leipzig, 1866-68. X. 2nd ed., 2 vols in 1, 1882.

1a. ——— *Lectures on the geometry of position.* New York, 1898.

2. *Synthetische Geometrie der Kugeln und linearen Kugelsysteme.* Leipzig, 1879.

4105. ROUQUET, Pierre Victor. 1838-after 1903. Fr. P IV.

4106. SCHIAPPA MONTEIRO, Alfredo. 1838-after 1882. Port. P III.

4107. STAUDIGL, Rudolf. 1838-1891. Aus. P III.

4108. THIELE, Thorvald Nicolai. 1838-1910. Den. DSB; P III,IV. Also Astronomy.

4109. ALDIS, William Steadman. 1839-1928. Br. P III,IV. Also Physics*.

4110. BARBIER, (Joseph) Emile. 1839-1889. Fr. DSB; P III,IV.

4111. BJÖRLING, (Carl Fabian) Emanuel. 1839-1920. Swed. P III,IV.

4112. BURNSIDE, William Snow. 1839-ca. 1921. Br. P IV.
1. *The theory of equations.* Dublin, 1881. With A.W. Panton. 2nd ed., 1886. Another ed., 1892.

4113. DE LARNE, Daniel. 1839-after 1883. Russ. P III.

4114. ESCARY, Jean. 1839-1900. Fr. P III,IV.

4115. HANKEL, Hermann. 1839-1873. Ger. DSB; P III; GC. Also History of Mathematics*.
1. *Theorie der complexen Zahlensysteme.* Leipzig, 1867.
2. *Untersuchungen über die unendlich oft oszillirenden und unstetigen Funktionen.* Tübingen, 1870. X. Reprinted, Leipzig, 1905. (OKEW. 153)
3. *Elemente der projectivischen Geometrie.* Leipzig, 1875.

4116. MAYER, Christian Gustav Adolph. 1839-1908. Ger. DSB; P III,IV.

4117. PEIRCE, Charles (Santiago) Sanders, 1839-1914. U.S.A. DSB; ICB; P III,IV; GE. Also Science in General*# and Geodesy.
1. *The new elements of mathematics.* 3 vols in 4. The Hague, 1974-76. Ed. by C. Eisele.

4118. PETERSEN, Julius Peter Christian. 1839-1910. Den. DSB; P III, IV.

4119. ROCH, Gustav. 1839-1866. Ger. P III.

4120. ZEUTHEN, Hieronymus George. 1839-1920. Den. DSB; ICB; P III, IV & Supp. Also History of Mathematics*.

4121. AMSTEIN, (Heinrich) Hermann. 1840-after 1901. Switz. P III,IV.

4122. ANDRÉ, (Antoine) Désiré. 1840-after 1904. Fr. P III,IV.

4123. AUGUST, Friedrich Wilhelm Oscar. 1840-1900. Ger. P III,IV.

4124. HOCHHEIM, Karl Adolphe. 1840-1898. Ger. P III,IV.

4125. LAMPE, Karl Otto Emil. 1840-1918. Ger. P III,IV.

4126. LEMOINE, Emile Michel Hyacinthe. 1840-1912. Fr. DSB; ICB; P III,IV.

4127. MacCLINTOCK, Emory. 1840-1916. U.S.A. P III,IV; GE.

4128. MERTENS, Franz Carl Joseph. 1840-1927. Cracow/Aus. P III,IV.

4129. NEUBERG, Joseph Jean Baptiste. 1840-1926. Belg. DSB; ICB; P III,IV.

4130. PURSER, Frederick. 1840-1910. Br. P III.

4131. SCHAPIRA, Hermann Hirsch. 1840-1898. Ger. P III,IV; GC.

4132. TAYLOR, Charles. 1840-1908. Br. P III,IV; GD2.
1. *An introduction to the ancient and modern geometry of conics.* Cambridge, 1881.

4133. THOMAE, Karl Johannes. 1840-1921. Ger. P III,IV.

4134. WHITWORTH, William Allen. 1840-1905. Br. GD2.
1. *Trilinear coordinates and other methods of modern analytical geometry of two dimensions. An elementary treatise.* London, 1866.

2. *Choice and chance: An elementary treatise on permutations, combinations and probability*. Cambridge, 1867. X. 3rd ed., 1878. 4th ed., 1886.

4135. FAIS, Antonio. 1841-1925. It. ICB; P III.

4136. FALK, Matthias. 1841-after 1913. Swed. P III,IV.

4137. FORMENTI, Carlo. 1841-1918. It. P III,IV.

4138. GEER, Peter van. 1841-after 1900. Holl. P III,IV.

4139. JOHNSON, William Woolsey. 1841-1927. U.S.A. P III,IV; GE.

4140. LAISANT, Charles Ange. 1841-1920. Fr. P III,IV; GB.

4141. LAURENT, (Matthieu Paul) Hermann. 1841-1908. Fr. DSB; P III, IV. Also Mechanics*.
1. *Traité du calcul des probabilités*. Paris, 1873.
2. *Traité d'analyse*. 7 vols. Paris, 1885-91.

4142. LEGOUX, Edme Alphonse. 1841-1909. Fr. P IV.

4143. LEVÄNEN, Sakari. 1841-1898. Fin. P III,IV. Also Meteorology.

4144. POCHHAMMER, Leo August. 1841-1920. Ger. P III,IV.

4145. PRYM, Friedrich Emil. 1841-1915. Ger. P III,IV.

4146. SCHRÖDER, (Friedrich Wilhelm Karl) Ernst. 1841-1902. Ger. DSB; P III,IV.
1. *Lehrbuch der Arithmetik und Algebra*. Leipzig, 1873.

4147. STURM, (Friedrich Otto) Rudolf. 1841-1919. Ger. DSB; P III,IV.

4148. THOMÉ, Ludwig Wilhelm. 1841-1910. Ger. P III,IV.

4149. AMIGUES, Edouard Pierre Marie. 1842-1900. Fr. P III,IV & Supp.

4150. BONSDORFF, Ernst Jacob Waldemar. 1842-after 1902. Fin. P III,IV.

4151. BRILL, Alexander Wilhelm von. 1842-1935. Ger. DSB; P III,IV.

4152. CUNNINGHAM, Allan Joseph Champneys. 1842-1928. Br. P IV & Supp.

4153. DARBOUX, (Jean) Gaston. 1842-1917. Fr. DSB; ICB; P III,IV. Also Biography*. See also Index.
1. *Leçons sur la théorie générale des surfaces et les applications géométriques du calcul infinitésimal*. 4 vols. Paris, 1887-96.
2. *Leçons sur les systèmes orthogonaux et les co-ordonnés curvilignes*. Paris, 1898.

4154. LIE, (Marius) Sophus. 1842-1899. Nor./Ger. DSB; ICB; P III, IV; GB; GC.
1. *Die Theorie der Transformationsgruppen*. 3 vols. Leipzig, 1888-93. Reprinted, New York, 1970.
2. *Vorlesungen über Differentialgleichungen mit bekannten infinitesimalen Transformationen*. Leipzig, 1891. Ed. by G. Scheffers. Reprinted, New York, 1967.
3. *Vorlesungen über continuierliche Gruppen*. Leipzig, 1893. Ed. by G. Scheffers. X. Reprinted, New York, 1971.
4. *Über die Grundlagen der Geometrie*. *Sonderausgabe*. Darmstadt, 1967. Reprint of a periodical article.
5. *Gesammelte Abhandlungen*. 6 vols in 11. Leipzig, 1922-37.

Ed. by F. Engel and P. Heegaard. Reprinted,, ib., 1960-.
See also 3762/5.

4155. LONGCHAMPS, Gaston Albert Gohierre de. 1842-1906. Fr. P III,IV.

4156. LUCAS, François Eduard Anatole. 1842-1891. Fr. DSB; ICB; P III.

4157. ROSANES, Jakob. 1842-1922. Ger. DSB; P III,IV.

4158. SALVERT, Marie Adolphe François de. 1842-1918. Fr. P IV.

4159. SOKHOTSKY. Yulian-Karl Vasilievich. 1842-1927. Russ. DSB; P III (SOKOTSKI).

4160. STOLZ, Otto. 1842-1905. Aus. DSB; P III,IV.

4161. TAYLOR, Henry Martyn. 1842-1927. Br. P III,IV; GD4.

4162. WEBER, Heinrich Martin. 1842-1913. Ger. DSB; ICB; P III,IV.
1. *Lehrbuch der Algebra*. 2 vols. Brunswick, 1895-96. X.
2nd ed., 1898-99.
See also 3982/2.

4163. ASCOLI, Giulio. 1843-1896. It. P III,IV.

4164. BRISSE, Charles Michel. 1843-1898. Fr. P III,IV.

4165. D'OVIDIO, Enrico. 1843-1933. It. ICB; P III,IV.

4166. GEISER, Karl Friedrich. 1843-1934. Switz. DSB; P III,IV.

4167. GRAINDORGE, (Louis Arnold) Joseph. 1843-1896. Belg. P III,IV.
1. *Mémoire sur l'intégration des équations aux dérivées partielles des deux premiers ordres*. Brussels, 1872.

4168. HESS, (Adolph) Edmund. 1843-1903. Ger. P III,IV & Supp.

4169. HOZA, Franz. 1843-after 1896. Prague. P III,IV.

4170. KOUTNY, Emil. 1843-1880. Aus. P III.

4171. PÁNEK, Augustin. 1843-1908. Prague. P IV.

4172. PASCH, Moritz. 1843-1930. Ger. DSB; ICB; P III,IV.
1. *Vorlesungen über neuere Geometrie*. Leipzig, 1882.

4173. SCHLEGEL, Stanislaus Ferdinand Victor. 1843-1905. Ger. P III, IV.

4174. SCHWARZ, (Karl Hermann) Amandus. 1843-1921. Switz./Ger. DSB; P III,IV.
1. *Gesammelte mathematische Abhandlungen*. 2 vols. Berlin, 1890.

4175. STAHL, Hermann Bernard Ludwig. 1843-1908. Ger. P III,IV.
1. *Theorie der Abel'schen Functionen*. Leipzig, 1896.

4176. TESAŘ, Josef. 1843-after 1904. Brünn. P III,IV.

4177. ARMENANTE, Angelo. 1844-1878. It. P III.

4178. ASCHIERI, Ferdinando. 1844-1907. It. P III,IV.

4179. BERGER, Alexander Fredrik. 1844-1901. Swed. P III,IV.

4180. ECKHARDT, Friedrich Emil. 1844-1875. Ger. P IV.

4181. GLASHAN, John Stuart Cadenhead. 1844-after 1901. Canada. P IV.

4182. HALPHEN, Georges Henri. 1844-1889. Fr. DSB; ICB; P III,IV.

1. *Traité des fonctions elliptiques et de leurs applications.* 3 vols. Paris, 1886-91.
2. *Oeuvres.* 4 vols. Paris, 1916-24. Ed. by C. Jordan et al.

4183. HOLZMÜLLER, (Ferdinand) Gustav. 1844-1914. Ger. P III,IV.
1. *Einführung in die Theorie der isogonalen Verwandschaften und der conformen Abbildungen.* Leipzig, 1882.

4184. LUEROTH, Jakob. 1844-1910. Ger. DSB; P III,IV.

4185. MANSION, Paul. 1844-1919. Belg. DSB; ICB; P III,IV. Also History of Science.
1. *Théorie des équations aux dérivées partielles du premier ordre.* Paris, 1875.
2. *Principes fondamentaux de la géométrie non-euclidienne de Riemann.* Paris, 1895.

4186. MUIR, Thomas. 1844-1934. Br./South Africa. ICB; P III,IV.
1. *A treatise on the theory of determinants.* London, 1882.
2. *The theory of determinants in the historical order of development.* 4 vols. London, 1890. Reprinted, 4 vols in 2, New York, 1960.

4187. NOETHER, Max. 1844-1921. Ger. DSB; P III,IV.

4188. SIMON, Max. 1844-1918. Ger. P III,IV.
1. *Die Elemente der Arithmetik als Vorbereitung auf die Funktionentheorie.* Strasbourg, 1884.
2. *Die Elemente der Geometrie. Rücksicht auf die absolute Geometrie.* Strasbourg, 1890.
3. *Zu den Grundlagen der nicht-euklidischen Geometrie.* Strasbourg, 1891.

4189. TICHOMANDRITZKY, Matvey Alexandrovich. 1844-after 1913. Russ. P IV.

4190. TOGNOLI, Oreste. 1844-after 1904. It. P IV.

4191. WANGERIN, (Friedrich Heinrich) Albert. 1844-1933. Ger. DSB; P III,IV.

4192. BÄCKLUND, Albert Victor. 1845-1922. Swed. P III,IV.

4193. BESSO, Davide. 1845?-1906. It. P IV.

4194. BIEHLER, Charles. 1845-after 1902. Fr. P III,IV.

4195. BROCARD, Pierre René Jean Baptiste Henri. 1845-1922. Fr./Algeria. DSB; P III,IV. Also Meteorology.

4196. CANTOR, Georg Ferdinand Ludwig. 1845-1918. Ger. DSB; ICB; P III,IV.
1. *Grundlegen einer allgemeinen Mannigfaltigkeitslehre. Ein mathematisch-philosophischer Versuch in der Lehre des Unendlichen.* Leipzig, 1883.
2. *Gesammelte Abhandlungen mathematischen und philosophischen Inhalts.* Berlin, 1930. Ed. by E. Zermelo. Reprinted, Hildesheim, 1962.
3. *Briefwechsel Cantor-Dedekind.* Paris, 1937. Ed. by E. Noether and J. Cavaillès.

4197. CLIFFORD, William Kingdon. 1845-1879. Br. DSB; ICB; P III; GB; GD. Also Science in General* and Mechanics*.

1. *Mathematical papers*. London, 1882. Ed. by R. Tucker. RM/L. Reprinted, New York, 1968.

4198. DIDON, François. 1845-1872. Fr. P III.

4199. DINI, Ulisse. 1845-1918. It. DSB; P III,IV.

4200. EDGEWORTH, Francis Ysidro. 1845-1926. Br. P IV; GD4.

4201. FOURET, Georges François Jean Baptiste. 1845-after 1902. Fr. P III,IV.

4202. JEŘÁBEK, Wenzel. 1845-after 1914. Brünn. P III,IV.

4203. JUNG, Giuseppe. 1845-1926. It. P III,IV.

4204. PELZ, Carl. 1845-1908. Aus./Prague. P III,IV.

4205. PICQUET, Louis Didier Henry. 1845-1925. Fr. P III,IV.

4206. RADICKE, Eduard Albert. 1845-after 1880. Ger. P III.

4207. RIBAUCOUR, Albert. 1845-1893. Fr. DSB; P III,IV.

4208. STAHL, Wilhelm. 1845-1894. Ger. P III,IV.

4209. VERSLUIJS, Jan. 1845-after 1882. Holl. P III.

4210. VOSS, Aurel Edmund. 1845-after 1922. Ger. P III,IV.

4211. BERTINI, Eugenio. 1846-1933. It. DSB; P III,IV.

4212. COLLET, Jean. 1846-after 1905. Fr. P III.

4213. GALL, August von. 1846-1899. Ger. P III,IV.

4214. GARCIA DE GALDEANO Y YANGUAR, Zoel. 1846-1924. Spain. P III,IV.

4215. GUNDELFINGER, Sigmund. 1846-1910. Ger. P III,IV.
    1. *Vorlesungen aus der analytischen Geometrie der Kegelschnitte*. Leipzig, 1895. Ed. by F. Dingeldey.

4216. HEGER, Richard. 1846-1919. Ger. P IV.

4217. HERMES, Johann Gustav. 1846-1912. Ger. P III,IV.

4218. KIEPERT, (Friedrich Wilhelm August) Ludwig. 1846-after 1921. Ger. P IV.

4219. LEBON, (Désiré) Ernest. 1846-1922. Fr. P IV. Also History of Astronomy*.

4220. MITTAG-LEFFLER, Magnus Gustaf. 1846-1927. Swed. DSB; P III,IV.

4221. NIEWENGLOWSKI, Boleslas Alexandre. 1846-after 1928. Fr. P III, IV.

4222. PORETSKY, Platon Sergeevich. 1846-1907. Russ. DSB. Also Astronomy.

4223. SCHOUTE, Pieter Hendrik. 1846-1923. Holl. DSB; P III,IV.

4224. SCHWERING, Karl Maria Johann Gerhard. 1846-1925. Ger. P III,IV.

4225. AFFOLTER, Gabriel Ferdinand. 1847-1926. Switz. P III,IV.

4226. ARZELÀ, Cesare. 1847-1912. It. P III,IV.

4227. ASTOR, Auguste Marie. 1847-after 1901. Fr. P IV.

4228. ELLIOT, Zépherin Victor. 1847-after 1881. Fr. P III.

4229. FARKAS, Julius. 1847-1930. Hung. P III,IV.

4230. FLOQUET, (Achille Marie) Gaston. 1847-1920. Fr. P III,IV.

4231. GARBIERI, Giovanni. 1847-after 1902. It. P III,IV.

4232. GREENHILL, (Alfred) George. 1847-1927. Br. ICB; P III,IV & Supp. Also Mechanics*.
   1. *The differential and integral calculus.* London, 1886. 2nd ed., 1891.
   2. *The applications of elliptic functions.* London, 1892. Reprinted, New York, 1959.

4233. KILLING, Wilhelm Karl Joseph. 1847-1923. Ger. P III,IV.
   1. *Die nicht-euklidischen Raumformen in analytischer Behandlung.* Leipzig, 1885.
   2. *Einführung in die Grundlagen der Geometrie.* 2 vols. Paderborn, 1893-98.

4234. LADD (-FRANKLIN), Christine. 1847-1930. U.S.A. P III; Mort. Also Physiology.

4235. MALET, John Christian. 1847-1901. Br. P III,IV.

4236. PACI, Paolo. 1847-1904. It. P III,IV.

4237. POSSÉ, Constantin Alexandrovich. 1847-after 1919. Russ. P III, IV.

4238. REUSCHLE, Carl. 1847-1909. Ger. P IV.

4239. RINK, Hendrik Jan. 1847-1883. Holl. P III,IV.

4240. SALTEL, Louis Marie. 1847-after 1883. Fr. P III.

4241. SÖDERBLOM, Anders Leonard Axel. 1847-1923. Swed. P IV.

4242. STRINGHAM, (Washington) Irving. 1847-1909. U.S.A. P III,IV; GE. Also History of Mathematics*.

4243. ZOLOTAREV, Egor Ivanovich. 1847-1878. Russ. DSB; ICB.

4244. ANDREJEW (or ANDRÉEFF), Konstantin Alekseievich. 1848-ca. 1923. Russ. P III,IV.

4245. CARDINAAL, Jacob. 1848-1922. Holl. P IV.

4246. DEMARTRES, Gustave Léon. 1848-1919. Fr. P IV.

4247. DIEKMANN, (Franz) Josef (Konrad). 1848-1905. Ger. P III,IV.

4248. FONTENÉ, Georges. 1848-1923. Fr. P IV.

4249. FREGE, (Friedrich Ludwig) Gottlob. 1848-1925. Ger. DSB; ICB; P IV Supp.
   1. *Begriffsschrift, eine der arithmetischen nachgebildete Formelsprache des reinen Denkens.* Halle, 1879. X.
   1a. ——— Reprint entitled *Begriffsschrift und andere Aufsätze.* Hildesheim, 1964.
   2. *Der Grundlagen der Arithmetik.* Breslau, 1884. X. Reprinted, ib., 1934, and Hildesheim, 1961.
   2a. ——— *The foundations of arithmetic.* Oxford, 1950. Trans. by J.L. Austin. 2nd ed., rev., 1953.
   3. *Grundgesetze der Arithmetik.* 2 vols. Jena, 1893-1903. X. Reprinted, 2 vols in 1, Hildesheim, 1962.

3a. ——— *The basic laws of arithmetic*. Berkeley, 1964. Partial trans. by M. Furth.
4. *Kleine Schriften*. Hildesheim, 1967. Ed. by I. Angelelli.
5. *Nachgelassene Schriften und wissenschaftlicher Briefwechsel*. 2 vols. Hamburg, 1969-76. Ed. by H. Hermes et al.
6. *Translations from the philosophical writings*. 2nd ed. Oxford, 1960. By P. Geach and M. Black.

4250. GENESE, Robert William. 1848-1928. Br. P III,IV.

4251. GLAISHER, James Whitbread Lee. 1848-1928. Br. DSB; ICB; P III, IV; GD4. See also 3985/1 and 5105/4a.

4252. GUENTHER, (Adam Wilhelm) Siegmund. 1848-1923. Ger. DSB; P III, IV & Supp. Also Earth Sciences* and History of Mathematics*.

4253. HART, Harry. 1848-after 1883. Br. P III.

4254. JABLONSKI, Edouard Joseph Marie. 1848-1923. Fr. P IV.

4255. MOLLAME, Vincenzo. 1848-1912. It. P III,IV.

4256. NETTO, Eugen. 1848-1919. Ger. DSB; P III,IV.

4257. PELLET, Auguste Eliacin Claude. 1848-1935. Fr. P III,IV.

4258. SCHUBERT, Hermann Cäsar Hannibal. 1848-1911. Ger. DSB; P III, IV.

4259. TANNERY, Jules. 1848-1910. Fr. DSB; ICB; P III,IV.
1. *Introduction à la théorie des fonctions d'une variable*. Paris, 1886.
2. *Eléments de la théorie des fonctions elliptiques*. 4 vols. Paris, 1893-1902. With J. Molk.
3. *Introduction a l'étude de la théorie des nombres et de l'algèbre supérieure*. Paris, 1894. Ed. by E. Borel and J. Drach from Tannery's lectures.

4260. WEYR, Emil. 1848-1894. Prague/Aus. P III,IV; GC.

4261. ZAHRADNIK, Dragutin (or Karl). 1848-1916. Agram. P III,IV.

4262. ARCAIS, Francesco d'. 1849-1927. It. P IV.

4263. CHARVE, Léon. 1849-after 1881. Fr. P III.

4264. DICKSON, James Douglas Hamilton. 1849-1931. Br. P III,IV.

4265. ESCHERICH, Gustav von. 1849-1935. Aus. P III,IV.

4266. FROBENIUS, Ferdinand Georg. 1849-1917. Ger./Switz. DSB; ICB; P III,IV.
1. *Gesammelte Abhandlungen*. 3 vols. Berlin, 1968. Ed. by J.P. Serre.

4267. GEGENBAUER, Leopold. 1849-1903. Aus. P III,IV & Supp.

4268. HOLST, Elling Bolt. 1849-after 1886. Nor. P III,IV & Supp.

4269. KAPTEIJN, Willem. 1849-1927. Holl. P III,IV.

4270. KEMPE, Alfred Bray. 1849-1922. Br. P III,IV.

4271. KLEIN, (Christian) Felix. 1849-1925. Ger. DSB; ICB; P III,IV.
1. *Vorlesungen uber das Ikosaeder und die Auflösung der Gleichungen vom fünften Grade*. Leipzig, 1884.

2. *Vorlesungen über die Theorie des elliptischen Modulfunctionen.* 2 vols. Leipzig, 1890-1901. Ed. by R. Fricke. Reprinted, Stuttgart, 1966.
3. *Nicht-euklidische Geometrie.* 2 vols. Göttingen, 1893.
4. *Einleitung in die höhere Geometrie.* 2 vols. Göttingen, 1893.
5. *Über die hypergeometrische Function.* Göttingen, 1894.
6. *Über lineare Differentialgleichungen der zweiten Ordnung.* Göttingen, 1894.
7. *Riemann'sche Flächen.* 2 vols. Göttingen, 1894.
8. *The Evanston colloquium: Lectures delivered ... 1893, before members of the congress of mathematics held ... in Chicago.* New York, 1894.
9. *Vorträge über ausgewählte Frage der Elementargeometrie.* Leipzig, 1895. Ed. by F. Tägert.

9a. ——— *Famous problems of elementary geometry.* Boston, 1897. Trans. by W.W. Beman and D.E. Smith. Reprinted, New York, 1950.

10. *Über die Theorie des Kreisels.* 4 vols. Leipzig, 1897-1910. With A. Sommerfeld. X. Reprinted, Stuttgart, 1965.

Later monographs not included.

11. *Gesammelte mathematische Abhandlungen.* 3 vols. Berlin, 1921-23. Ed. by R. Fricke and A. Ostrowski. Reprinted, Berlin, 1973.

4272. KOENIG, Julius. 1849-1914. Hung. DSB; P III,IV.

4273. MANGEOT, (François Constant) Stéphane. 1849-after 1921. Fr. P IV & Supp.

4274. SCOTT, Robert Forsyth. 1849-after 1904. Br. P IV.
1. *A treatise on the theory of determinants and their applications.* Cambridge, 1880.

4275. SONIN, Nikolai Yakovlevich. 1849-1915. Russ. DSB; P III,IV (SSONIN). See also 3930/1.

4276. TONELLI, Alberto. 1849-after 1905. It. P IV.

4277. TORELLI, Gabriele. 1849-after 1922. It. P III,IV.

4278. LIOUVILLE, R. fl. 1879-1903. Fr. P IV.

4279. ANGLIN, Arthur Henry. 1850-1934. Br. P IV.

4280. BALL, Walter William Rouse. 1850-1925. Br. ICB (ROUSE BALL); P IV. Also History of Mathematics*.

4281. CROCCHI, Leopoldo. 1850-after 1885. It. P III,IV.

4282. DRASCH, Heinrich. 1850-after 1902. Aus. P IV.

4283. GRAM, Jórgen Pedersen. 1850-1916. Den. P III,IV.

4284. HAMMOND, James. 1850-1930. Br. P IV.

4285. KOVALEVSKY, Sonya. 1850-1891. Swed. DSB; ICB; P III,IV (KOWALEWSKY, Sophie); GB. Also Autobiography*. See also 3871/2.

4286. MAXIMOWITSCH. Wladimir. 1850-1889. Russ. P III,IV.

4287. NANSON, Edward John. 1850-1936. Australia. P IV & Supp.

4288. PRINGSHEIM, Alfred. 1850-1941. Ger. DSB; P III,IV.

4289. STORY, William Edward. 1850-1930. U.S.A. P III,IV; GE.

4290. VALENTINER, Herman. 1850-1913. Den. P IV. See also 6535/2.

4291. CHRYSTAL, George. 1851-1911. Br. DSB; P III,IV; GD2. Also Physics.
   1. *Algebra. An elementary textbook.* 2 vols. Edinburgh, 1886-89. "Profoundly influenced mathematical education."

4292. CRONE, Hans Christian Rasmus. 1851-after 1900. Den. P III,IV.

4293. CZUBER, Emanuel. 1851-1925. Prague/Aus. P III,IV.

4294. DICKSTEIN, Samuel. 1851-1939. Warsaw. DSB; P IV. Also History of Mathematics.

4295. ELLIOTT, Edwin Bailey. 1851-1937. Br. ICB; P III,IV; GD5.
   1. *An introduction to the algebra of quantics.* Oxford, 1895.

4296. GREINER, Max. 1851-after 1903. Ger. P III,IV.

4297. HARNACK, (Carl Gustav) Axel. 1851-1888. Ger. P III,IV; GC.

4298. KRAUSE, Johann Martin. 1851-1920. Ger. P IV.

4299. MACFARLANE, Alexander. 1851-1913. Br./U.S.A. P III,IV. Also Physics.
   1. *The principles of elliptic and hyperbolic analysis.* Boston, 1894.

4300. NASIMOV, Petr Sergeievich. 1851-1901. Russ. P IV.

4301. NEOVIUS, Edvard Rudolf. 1851-1917. Fin. P III,IV.

4302. PETOT, Charles Albert. 1851-1927. Fr. P IV.

4303. PUCHTA, Anton. 1851-1903. Prague/Czernowitz. P III,IV.

4304. RODENBERG, Karl Friedrich. 1851-1933. Ger. P III,IV.

4305. SCHOTTKY, Friedrich Hermann. 1851-1935. Ger./Switz. DSB; P III,IV.

4306. TANNER, Henry William Lloyd. 1851-1915. Br. P III,IV.

4307. TEIXEIRA, Francisco Gomes. 1851-1933. Port. ICB; P III,IV.

4308. WEILER, Adolf. 1851-1916. Switz. P III,IV.

4309. WEILL, Matthieu. 1851-after 1907. Fr. P IV.

4310. ZÜGE, Eduard Heinrich. 1851-1902. Ger. P IV.

4311. ALBEGGIANI, Micheli Luigi. 1852-after 1901. It. P IV.

4312. BIGLER, Ulrich. 1852-after 1898. Switz. P IV.

4313. BURNSIDE, William. 1852-1927. Br. DSB; P IV; GD4.
   1. *Theory of groups of finite order.* Cambridge, 1897.

4314. GRAF (VON WILDBERG), Johann Heinrich. 1852-1918. Switz. P III, IV.

4315. LE PAIGE, Constantin. 1852-1929. Belg. DSB; ICB; P III,IV.

4316. LINDEMANN, (Carl Louis) Ferdinand. 1852-1939. Ger. DSB; P III, IV. See also 4053/3.

4317. MAISANO, Giovanni. 1852-1929. It. P IV.

4318. MININ, Alexander. 1852-after 1883. Russ. P III.

4319. PITTARELLI, Giulio. 1852-1934. It. P IV & Supp.

4320. ROGEL, Franz. 1852-after 1920. Aus./Ger. P IV.

4321. RULF, Wilhelm Friedrich Johann. 1852-after 1899. Prague/Aus. P IV.

4322. SIMONY, Oskar. 1852-1915. Aus. P III,IV.

4323. WEYR, Eduard. 1852-1903. Prague. P III,IV.

4324. CASPARY, Ferdinand. 1853-1901. Ger. P IV & Supp.

4325. FRANKLIN, Fabian. 1853-1939. U.S.A. ICB; P III,IV.

4326. HALSTED, George Bruce. 1853-1922. U.S.A. DSB; P III,IV; GE. See also Index.

4327. HOČEVAR, Franz. 1853-1919. Aus. P III,IV.

4328. JAMET, Emile Victor. 1853-after 1914. Fr. P IV.

4329. LÉVY, Lucien. 1853-1912. Fr. P IV.

4330. MASCHKE, Heinrich. 1853-1908. Ger./U.S.A. P IV; GE.

4331. NEKRASSOFF, Pavel Alekseievich. 1853-1924. Russ. P IV.

4332. PINCHERLE, Salvatore. 1853-1936. It. DSB; ICB; P III,IV.

4333. RICCI-CURBASTRO, Gregorio. 1853-1925. It. DSB; ICB; P III,IV.

4334. RIQUIER, Charles Edmond Alfred. 1853-1929. Fr. P IV.

4335. SAUVAGE, Louis Charles. 1853-after 1909. Fr. P IV.

4336. SCHOENFLIES, Arthur Moritz. 1853-1928. Ger. DSB; P III,IV; BMNH. Also Crystallography.

4337. VANEČEK, Josef Sylvester. 1853(or1848)-1922. Budweis. P III,IV.

4338. AMANZIO, Domenico. 1854-1908. It. P III,IV.

4339. BEYEL, Christian. 1854-after 1923. Switz. P III,IV.

4340. BORDIGA, Giovanni Alfredo. 1854-1929. It. P IV.

4341. GIERSTER, Joseph. 1854-1893. Ger. P III,IV.

4342. HETTNER, (Hermann) Georg. 1854-1914. Ger. P III,IV. See also 3883/1.

4343. KLUG, Leopold. 1854-after 1931. Hung. P III,IV.

4344. LACOUR, (Victor Louis) Emile. 1854-after 1902. Fr. P IV. See also 4355/2.

4345. MacMAHON, Percy Alexander. 1854-1929. Br. DSB; ICB; P IV; GD4.

4346. MANGOLD, Hans Carl Friedrich von. 1854-1925. Ger. P III,IV.

4347. MELLIN, Robert Hjalmar. 1854-1933. Fin. P III,IV.

4348. PAOLIS, Riccardo de. 1854-1892. It. P III(under P), IV(under DE).

4349. PEIRCE, Benjamin Osgood. 1854-1914. U.S.A. DSB; P III,IV; GE. Also Physics.
    1. *Elements of the theory of the Newtonian potential functions.* Boston, 1888.

4350. PEROTT, Joseph de. 1854-after 1903. U.S.A. P IV.

4351. POINCARÉ, (Jules) Henri. 1854-1912. Fr. DSB; ICB; P III,IV.
 Also Astronomy*, Mechanics*, and Physics*.
 1. *Calcul des probabilités: Leçons....* Paris, 1896. Ed. by A. Quiquet.
 2. *Théorie du potentiel Newtonien: Leçons....* Paris, 1899. RM/L.
 Later monographs not included.
 3. *Oeuvres.* 11 vols. Paris, 1916-54. Ed. by J. Darboux.
 See also 4063/1.

4352. PTASZYCKI, Ivan Lvovich. 1854-1912. Russ. P III,IV.

4353. VERONESE, Giuseppe. 1854-1917. It. DSB; P III,IV.

4354. ROBERTS, Ralph A. fl. 1884-1905. Br./U.S.A. P IV.

4355. APPELL, Paul (Emile). 1855-1930. Fr. DSB; ICB; P III,IV.
 Also Mechanics*.
 1. *Théorie des fonctions algébriques et de leurs intégrales.* Paris, 1897. With E.J.B. Goursat.
 2. *Principes de la théorie des fonctions elliptiques et applications.* Paris, 1897. With E. Lacour.

4356. BARBARIN, Paul Jean Joseph. 1855-1931. Fr. ICB; P IV.

4357. BOBEK, Karl Joseph. 1855-1899. Prague. P III,IV.

4358. CAPELLI, Alfredo. 1855-1910. It. P III,IV.

4359. CAPORALI, Ettore. 1855-1886. It. P III,IV & Supp.

4360. CRAIG, Thomas. 1855-1900. U.S.A. P III,IV; GE. Also Mechanics*.

4361. GIUDICE, Francesco. 1855-after 1931. It. P IV.

4362. GRAEFE, (Heinrich Franz Konrad Karl) Friedrich. 1855-1919. Ger. P III,IV.

4363. GUCCIA, Giovanni Battista. 1855-1914. It. DSB; P IV.

4364. HEYMANN, Carl Woldemar. 1855-1910. Ger. P IV.

4365. HOFMANN, Fritz. ca. 1855-after 1890. Ger. P IV Supp.

4366. JUEL, Sophus Christian. 1855-1935. Den. DSB; ICB; P III,IV.

4367. KNOBLAUCH, Johannes. 1855-1915. Ger. P IV.

4368. MACHOVEC, Franz. 1855-1892. Prague. P IV.

4369. PANNELLI, Marino. 1855-1934. It. P IV & Supp.

4370. RAFFY, Louis. 1855-1910. Fr. P IV.

4371. ROHN, Karl Friedrich Wilhelm. 1855-1920. Ger. DSB; P IV.

4372. SCHIFFNER, Franz. 1855-after 1907. Aus. P IV.

4373. SELIWANOFF, Dmitri Fedorovich. 1855-1932. Russ. P IV.

4374. VÁLYI, Gyula (or Julius). 1855-1913. Klausenburg. P IV.

4375. WILTHEISS, Ernst Eduard. 1855-1900. Ger. P IV.

4376. BIANCHI, Luigi. 1856-1928. It. DSB; ICB; P IV.

4377. BRUNEL, Georges Edouard Auguste. 1856-1900. Fr. P IV.

4378. DYCK, Walther Franz Anton von. 1856-1934. Ger. DSB; P III,IV.

4379. FABRY (Charles) Eugène. 1856-after 1923. Fr. P IV.

4380. GRÜNFELD, Emanuel. 1856-after 1903. Aus. P IV.

4381. HILL, Micaiah John Muller. 1856-1929. Br. P III,IV.

4382. HOBSON, Ernest William. 1856-1933. Br. DSB; ICB; P IV; GD5.
1. *A treatise on plane trigonometry.* Cambridge, 1891. 2nd ed., 1897.

4383. McMAHON, James. 1856-1922. U.S.A. P IV.

4384. MARKOV, Andrei Andreevich. 1856-1922. Russ. DSB; ICB; P III,IV.
1. [In Russian. *Differential calculus.*] 2 vols. St. Petersburg, 1889-91. X.
1a. ——— *Differenzenrechnung.* Leipzig, 1896.
2. [In Russian. *Selected works: Theory of numbers; Theory of probability.*] Leningrad, Academy of Sciences, 1951.
See also 3930/1.

4385. MEYER, (Friedrich) Wilhelm Franz. 1856-1934. Ger. DSB; P IV.

4386. PICARD, (Charles) Emile. 1856-1941. Fr. DSB; ICB; P III;IV.
1. *Traité d'analyse.* 3 vols. Paris, 1891-96.
2. *Théorie des fonctions algébriques de deux variables indépendantes.* 2 vols. Paris, 1897-1906.
See also 3947/5.

4387. RUDIO, Ferdinand. 1856-1929. Switz. DSB; P IV. Also History of Mathematics.

4388. RUNGE, Carl David Tolmé. 1856-1927. Ger. DSB; ICB; P IV. Also Physics.

4389. SCHUR, Freidrich Heinrich. 1856-after 1925. Ger. P III,IV.

4390. STIELTJES, Thomas Jan. 1856-1894. Holl./Fr. DSB; ICB; P III,IV.
1. *Oeuvres complètes.* 2 vols. Groningen, 1914-18.
See also 3947/4.

4391. STODÓLKIEWICZ, Aloisius Johann Wladislaus. 1856-after 1922. Warsaw. P IV.

4392. BAUR, Ludwig Heinrich Gustav. 1857-after 1899. Ger. P IV.

4393. BOLZA, Oskar. 1857-1942. U.S.A. DSB; P IV.

4394. BRODÉN, Torsten. 1857-after 1924. Swed. P IV.

4395. DOBRINER, Hermann. 1857-1902. Ger. P IV & Supp.

4396. KANTOR, Seligmann. 1857-after 1904. Prague. P III,IV.

4397. LILIENTHAL, Franz Reinhold von. 1857-ca. 1936. Ger./Chile. P IV.

4398. LYAPUNOV, Aleksandr Mikhailovich. 1857-1918. Russ. DSB; ICB (LIAPUNOV); P III,IV (LJÄPUNOW). Also Mechanics.

4399. MEHMKE, Rudolf. 1857-after 1932. Ger. P III,IV.

4400. MÜLLER, Heinrich Robert Reinhold. 1857-after 1930. Ger. P IV.

4401. PEARSON, Karl. 1857-1936. Br. DSB; ICB; P IV; GD5. Also

Science in General*, Evolution*, and Biography*.
1. *Early statistical papers*. Cambridge, 1948; reprinted 1956.

4402. PIRONDINI, Geminiano. 1857-after 1909. It. P IV.

4403. STAUDE, (Ernst) Otto. 1857-1928. Ger. ICB; P III,IV. Also Mechanics.

4404. STEPHANOS, Kyparissos. 1857-1917. Greece. P III,IV.

4405. AMESEDER, Adolf. 1858-1891. Aus. P III.

4406. BIERMANN, (August Leo) Otto. 1858-1909. Brünn. P IV.
1. *Theorie der analytischen Functionen*. Leipzig, 1887.

4407. BRAMBILLA, Alberto. 1858-after 1899. It. P IV.

4408. BRILL, John. 1858-after 1924. Br. P IV.

4409. DE VRIES, Jan. 1858-after 1922. Holl. P IV.

4410. FINE, Henry Burchard. 1858-1928. U.S.A. DSB; P IV; GE.
1. *The number-system of algebra treated theoretically and historically*. Boston, 1890.

4411. FORSYTH, Andrew Russell. 1858-1942. Br. DSB; ICB; P III,IV; GD6.
1. *A treatise on differential equations*. London, 1885. 2nd ed., 1888.
2. *The theory of differential equations*. 4 vols in 6. Cambridge, 1890-1906.
3. *The theory of functions of a complex variable*. Cambridge, 1893.
See also 3929/2.

4412. GERBALDI, Francesco. 1858-after 1932. It. P IV.

4413. GOURSAT, Edouard Jean Baptiste. 1858-1936. Fr. DSB; P III,IV.
1. *Leçons sur l'integration des équations aux dérivées partielles du premier ordre*. Paris, 1891. Ed. by C. Bourlet.
2. *Leçons sur l'integration des équations aux dérivées partielles du second ordre à deux variables indépendantes*. 2 vols. Paris, 1896-98.
See also 4355/1.

4414. HAENTZSCHEL, Emil Rudolf. 1858-after 1926. Ger. P IV.
1. *Studien über die Reduction der Potentialgleichung auf gewöhnliche Differentialgleichungen*. Berlin, 1893.

4415. JOHNSON, William Ernest. 1858-1931. Br. DSB; GD5.
1. *A treatise on trigonometry*. London, 1889.

4416. KOENIGS, Gabriel Xavier Paul. 1858-1931. Fr. DSB; P IV. Also Mechanics*.
1. *Leçons sur l'agrégation classique de mathématiques*. Paris, 1892.

4417. KRAZER, (Karl) Adolf (Joseph). 1858-1926. Ger. P III,IV.

4418. PEANO, Giuseppe. 1858-1932. It. DSB; ICB; P IV.
1. *Il calcolo geometrico; preceduto dalla Operazioni della logica deduttiva*. Turin 1888.
2. *I principi di geometria*. Turin, 1889.
3. *Opere scelte*. 3 vols. Rome, 1957-59. Ed. by U. Cassina.

4. *Selected works.* London/Toronto, 1973. Trans. and ed. with a biographical sketch and bibliography by H.C. Kennedy.

4419. SCOTT, Charlotte Angas. 1858-after 1906. U.S.A. P IV.
 1. *An introductory account of certain modern ideas and methods in plane analytical geometry.* London, 1894.

4420. AMODEO, Federico. 1859-after 1926. It. P IV.

4421. AUTONNE, Léon. 1859-1916. Fr. P III,IV.

4422. BIOCHE, Charles. 1859-after 1931. Fr. P IV.

4423. BUKREJEFF, Boris Jakovlevich. 1859-after 1901. Russ. P IV.

4424. CESÀRO, Ernesto. 1859-1906. It. DSB; P IV.

4425. DEL PEZZO, Pasquale. 1859-after 1913. It. P IV & Supp.

4426. DEL RE, Alfonso. 1859-after 1918. It. P IV.

4427. DINGELDEY, Friedrich. 1859-1930. Ger. P IV. See also 4215/1.

4428. HEUN, Karl. 1859-1929. Ger. P IV.

4429. HÖLDER, Otto Ludwig. 1859-1937. Ger. DSB; P IV.

4430. HUMBERT, (Marie) Georges. 1859-1921. Fr. DSB; ICB; P IV.

4431. HURWITZ, Adolf. 1859-1919. Ger./Switz. DSB; ICB; P III,IV.
 1. *Mathematische Werke.* 2 vols. Basel, 1932-33.

4432. JENSEN, Johan Ludvig William Valdemar. 1859-1925. Den. DSB.

4433. KOHN, Gustav. 1859-1921. Aus. P IV.

4434. MARTINETTI, Vittorio. 1859-1936. It. P IV.

4435. MARTINS DA SILVA, Joaquim Antonio. 1859-1885. Port. P III,IV.

4436. MAURER, Ludwig. 1859-1927. Ger. P IV. See also 4008/1.

4437. MIMORI, Mamoru. 1859-1932. Japan. ICB.

4438. PICK, Georg Alexander. 1859-after 1929. Prague. P III,IV.

4439. SHATUNOVSKY, Samuil Osipovich. 1859-1929. Russ. DSB.

4440. STROH, Georg Emil. 1859-after 1904. Ger. P IV.

4441. VIVANTI, Giulio. 1859-after 1922. It. P IV.

Imprint Sequence

4442. RALPHSON, Joseph.
 1. *A mathematical dictionary.* London, 1702.
 See also 3364/1a.

4443. UNIVERSITY of Cambridge.
 1. *Introduction to the differential calculus on algebraical principles.* Cambridge, 1825.
 2. *A syllabus of the differential and integral calculus.* Cambridge, 1826.

4444. *ENCYCLOPAEDIA of pure mathematics, forming part of the "Encyclopaedia metropolitana."* London, 1847. cf. entry 3005.

4445. ERLECKE, Albert.
    1. *Bibliotheca mathematica: Systematisches Verzeichniss der bis 1870 in Deutschland auf den Gebieten der Arithmetik, Algebra, Analysis, Geometrie ... erschienenen Werke, Schriften und Abhandlungen ... Erster Band, die encyclopaedisch-mathematische Literatur umfassend.* [No more publ.] Halle, 1873. X. Reprinted, Wiesbaden, 1971.

4446. *ENCYKLOPÄDIE der mathematischen Wissenschaften.* 6 vols in 23. Leipzig, 1898-1904. Ed. under the auspices of the Göttingen, Leipzig, Munich, and Vienna Academies.

3.04 ASTRONOMY

Including celestial mechanics and astrophysics.

4447. LA HIRE, Philippe de. 1640-1718. Fr. DSB; ICB; P I; GA. Also Mathematics, Physics, Geodesy, and Part 2*.
    1. *Tabulae astronomicae.* Paris, 1702. RM.

4448. NEWTON, Isaac. 1642-1727. Br. DSB; ICB; P II; G. Also Mathematics*, Physics*#, and Part 2*(the *Principia*). See also Index.
    1. *A new and most accurate theory of the moon's motion.* London, 1702. A rare pamphlet. X.
    1a. ——— Reprint entitled *Theory of the moon's motion.* Folkestone (U.K.), 1975. Introd. by I.B. Cohen.

4449. FLAMSTEED, John. 1646-1719. Br. DSB; ICB; P I; GA; GB; GD. Also Part 2*. See also Index.
    1. *Historiae coelestis libri duo.* London, 1712. RM.
    2. *Atlas coelestis.* London, 1729. Another ed., 1781.
    3. *A letter concerning earthquakes, written in the year 1693.* London, 1750. RM.
    4. *An account of the Revd John Flamsteed, compiled from his own manuscripts ... To which is added his "British catalogue of stars", corrected and enlarged.* London, 1835. Ed. by F. Bailey. RM/L. Reprinted, London, 1966.

4450. VIGNOLES (or VIGNOLLES), Alphonse des. 1649-1744. Fr./Switz./Ger. P II; GA.

4451. NOËL, François. 1651-1729. Jesuit missionary in China. P II & Supp.; GA; GF.

4452. VILLEMOT, Philippe. 1651-1713. Fr. P II.
    1. *Nouveau système, ou nouvelle explication du mouvement des planètes.* Lyons, 1707. RM.

4453. HALLEY, Edmond. 1656?-1743. Br. DSB; ICB; P I; GA; GB; GD. Also Various Fields*, Geomagnetism, and Part 2.
    1. *Tabulae astronomicae.* London, 1749. Ed. by J. Bevis.
    2. *Correspondence and papers.* Oxford, 1932. Ed. by E.F. MacPike. Another ed., London, 1937.

4454. DESPLACES, Philippe. 1659-1736. Fr. P I.

4455. GREGORY, David. 1659-1708. Br. DSB; ICB; P I; GA; GB; GD. Also Mathematics*, Physics, and Part 2*.
    1. *Astronomiae physicae et geometricae elementa.* Oxford, 1702. RM.
    1a. ——— *The elements of astronomy, physical and geometrical.* 2 vols. London, 1715.
    1b. ——— ——— 2nd ed., 1726. X. Reprinted, New York 1972.

4456. FEUILLÉE, Louis Econches. 1660-1732. Fr. DSB; P I; BMNH; GA (FEUILLET). Also Botany*.

4457. ZUMBACH VON KOESFELD, Lothar. 1661-1727. Ger./Holl. P II; BLA (ZUMBAG); GC.

4458. BIANCHINI (or BLANCHINUS), Francesco. 1662-1729. It. P I; GA; GB.
1. *Astronomicae ac geographicae observationes selectae.* Verona, 1737. Ed. by E. Manfredi. RM.

4459. LAVAL, Antoine François. 1664-1728. Fr. P I.

4460. MARALDI, Giacomo Filippo. 1665-1729. Fr. DSB; P II; GA. Also Geodesy.

4461. HARRIS, John. ca. 1666-1719. Br. DSB; ICB; P I; GA; GB; GD. Also Science in General*, Mathematics*, and Natural History*.
1. *The description and use of the celestial and terrestrial globes.* London, 1703.
2. *Astronomical dialogue between a gentleman and a lady.* London, 1719. RM/L.

4462. WHISTON, William. 1667-1752. Br. DSB; P II; BMNH; GA; GB; GD. Also Science in General*, Mechanics*, Physics*, Autobiography*, and Part 2*. See also Index.
1. *Praelectiones astronomicae.* Cambridge, 1707. RM.
1a. ——— *Astronomical lectures.* London, 1715. RM. Another ed., 1728; reprinted, New York, 1972.
2. *A new method for discovering the longitude.* London, 1714. With H. Ditton.
3. *An account of a surprizing meteor.* London, 1716.

4463. POUND, James. 1669-1724. Br. P II; GD.

4464. SEMLER, Christoph. 1669-1740. Ger. P II; GC.

4465. CLAPIÈS, Jean de. 1670-1740. Fr. P I; GA.

4466. JUNIUS, Ulrich. 1670-1726. Ger. P I.

4467. KIRCH, Maria Margarethe Winkelmann. 1670-1720. Ger. DSB; P I; GA; GC.

4468. PLANTADE, François de. 1670-1741. Fr. P II; GA.

4469. DOPPELMAYR, Johann Gabriel. 1671?-1750. Ger. DSB; P I; GA; GC. Also Physics* and History of Mathematics*.

4470. KEILL, John. 1671-1721. Br. DSB; P I; GA; GD. Also Mathematics*, Mechanics*, and Part 2*. See also Index.
1. *Introductio ad veram astronomiam.* Oxford, 1718. RM. 2nd ed., London, 1721.
1a. ——— *An introduction to the true astronomy.* London, 1721. RM. Other eds: 1730, 1760, 1778 (6th), 1793 (7th).

4471. LOUVILLE, Jacques Eugène d'Allonville, *Chevalier* de. 1671-1732. Fr. P I; GA.

4472. GAMACHES, Etienne Simon de. 1672-1756. Fr. P I; GA.
1. *Astronomie physique.* Paris, 1740. RM.

4473. HODGSON, James. 1672-1755. Br. P I; GA; GD.

4474. PAPKE, Jeremias. 1672-1736. Ger. P II.

4475. BEYER, Johann. 1673-1751. Ger. P I.

4476. GRAHAM, George. ca. 1674-1751. Br. DSB; ICB; P I; GA; GD.

4477. MANFREDI, Eustachio. 1674-1739. It. DSB; ICB; P II; GA.
1. *De gnomone meridiano Bononiensi ad Divi Petronii.* Bologna, 1736. RM.
2. *Istitutzioni astronomiche.* Bologna, 1749. RM.
See also 2661/2a and 4458/1.

4478. MARINONI, Giovanni Giacomo di. 1676-1755. It. GA.

4479. QUENSEL, Conrad. 1676-1732. Swed. P II.

4480. CASSINI, Jacques. 1677-1756. Fr. DSB; PI; GA; GB. Also Geodesy.
1. *Tables astronomiques.* Paris, 1740. RM.
2. *Elémens d'astronomie.* Paris, 1740. RM.

4481. LA HIRE, Gabriel Philippe de. 1677-1719. Fr. DSB; P I. Also Physics.

4482. MAURICE, Antoine. 1677-1756. Switz. P II; GA.

4483. HAYES, Charles. 1678-1760. Br. P I; GA; GD.

4484. SLAVISECK (or SLAWICZEK), Karl. ca. 1678-1735. Jesuit missionary in China. P II.

4485. HORREBOW, Peder Nielsen. 1679-1764. Den. DSB; P I; GA.
1. *Basis astronomiae; sive, Astronomiae pars mechanica.* Copenhagen, 1735. RM.
2. *Operum mathematico-physicorum.* 3 vols. Ib., 1740-41. RM.

4486. KÖGLER (or KEGLER), Ignatius. 1680-1746. Jesuit missionary in China. P I; GA (KOEGLER).

4487. LONG, Roger. 1680-1770. Br. P I; GD.
1. *Treatise on astronomy.* 3 vols. Cambridge, 1742-84. RM/L.

4488. COTES, Roger. 1682-1716. Br. DSB; ICB; P I; GA; GB; GD. Also Mathematics* and Physics*. See also Index.

4489. HADLEY, John (first of the name). 1682-1744. Br. DSB; P I; GA; GD.

4490. GHISLIERI, Antonio. ca. 1685-1734. It. P I.

4491. BAXTER, Andrew. 1686?-1750. Br. GA; GB; GD.
1. *Matho; or, The cosmotheoria puerilis. A dialogue.* 2 vols. London, 1740. An elementary textbook of astronomy. Another ed., 1745.

4492. BRIGA, Melchior della. 1686-1749. It. P I.

4493. RAMUS, Jochum Friderik. 1686-1769. Den. P II.

4494. SIEGESBECK, Johann Georg. 1686-1755. Ger./Russ. P II & Supp.; BMNH; BLA; GC. Also Botany.

4495. STRUYCK. Nicolaas. 1687-1769. Holl. P II (STRUIJCK). Also Mathematics*.
1. *Inleiding tot de algemeene geographie.* Amsterdam, 1740. Astronomical geography.

2. *Vervolg van de beschryung der staartsterren.* Ib., 1753.

4496. WASSENIUS, Birger. 1687-1771. Swed. P II.

4497. WIEDEBURG, Johann Bernhard. 1687-1766. Ger. P II; GC.

4498. DELISLE, Joseph Nicolas. 1688-1768. Fr./Russ. DSB; ICB; P I; GA; GB.

4499. PLUCHE, Noël Antoine. 1688-1761. Fr. DSB; ICB; BMNH; GA. A successful populariser. Also Natural History*.
1. *Histoire du ciel.* 2 vols. Paris, 1739. RM. Another ed., 1748. Another ed., Amsterdam, 1769.
1a. ——— *The history of the heavens.* London, 1740. RM.
2. *Concorde de la géographie des différens âges.* Paris, 1765. X. Another ed., 1772.

4500. ROST, Johann Leonhard. 1688-1727. Ger. ICB; P II; GC.

4501. GAUBIL, Antoine. 1689-1759. Jesuit missionary in China. P I; GA.
1. *Correspondance de Pékin, 1722-1759.* Geneva, 1970. Ed. by R. Simon. Re Chinese astronomy.

4502. MOLYNEUX, Samuel. 1689-1728. Br. DSB; P II; GA; GD.

4503. MAYER (or MEYER). Friedrich Christian. fl. 1726-1730. Russ. P II.

4504. GRAMMATICO, Nicaise. d. 1736. It./Ger. P I; GA.

4505. MACHIN, John. d. 1751. Br. P II; GA; GD.

4506. LANGHANSEN, Christoph. 1691-1770. Ger. P I.

4507. WEIDLER, Johann Friedrich. 1691-1755. Ger. P II; GA; GC. Also History of Astronomy*.
1. *De via curva Mercurii sub sole in rectam convertenda exemplo transitus Mercurii per solem ... anno 1743.* Wittenberg, 1748. RM.
2. *Institutiones astronomiae.* Wittenberg, 1754. RM.

4508. PEZENAS, Esprit. 1692-1776. Fr. DSB; P II; GA.

4509. BRADLEY, James. 1693-1762. Br. DSB; ICB; P I; GA; GB; GD.
1. *Miscellaneous works and correspondence.* Oxford, 1832. Ed. by S.P. Rigaud. RM/L.
1a. ——— *Supplement.* Ib., 1833. Ed. by Rigaud.
1b. ——— Reprint of both items, New York, 1972.
See also 4841/1, 4941/1, and 5216/1.

4510. KIRCH, Christfried. 1694-1740. Ger. DSB; P I; GA.

4511. OUTHIER, Réginald. 1694-1774. Fr. DSB; P II; GA. Also Earth Sciences.

4512. VANDELLI, Francesco. 1694-1771. It. P II.

4513. LEADBETTER, Charles. fl. 1728-1769. Br. GD.
1. *Astronomy; or, The true system of the planets demonstrated.* London, 1727.
2. *Astronomy of the satellites of the Earth, Jupiter, and Saturn.* London, 1729. RM.
3. *Uranoscopia; or, The contemplation of the heavens.* London, 1735. RM.

4. *Mechanik dialling.* London, 1737. X. Another ed., 1773.
5. *New tables of the motions of the planets, the fixed stars and the first satellite of Jupiter.* 2nd ed. Ib., 1742.

4514. BEVIS (or BEVANS), John. 1695 (or 1693)-1771. Br. P I; GA; GD. See also 4453/1.

4515. GSCHWANDTNER, Andreas. 1696-1762. Aus. P I.

4516. HIORTER, Olof Peter. 1696-1750. Swed. ICB; P I (HJORTER).

4517. HONOLD, Matthäus. 1696-1726. Ger. P I.

4518. KIRCH, Christine. ca. 1696-1782. Ger. DSB; P I.

4519. MACCLESFIELD, George PARKER, *2nd Earl of.* 1697-1764. Br. P II; GD (PARKER).

4520. MAIRE, Christopher. 1697-1767. Br./Belg./It. P II; GA; GD.

4521. RIVARD, Dominique François. 1697-1778. Fr. P II; GA. Also Mathematics.
1. *Traité de la sphère.* Paris, 1741. X. 7th ed., 1816.

4522. BOUGUER, Pierre. 1698-1758. Fr. DSB; ICB; P I; GA; GB. Also Mechanics*, Physics*, and Geodesy*.
1. *Entretiens sur la cause de l'inclinaison des orbites des planètes.* Paris, 1748.

4523. KLIMM, Johann Albrecht. 1698-1778. Ger. P I.

4524. MAUPERTUIS, Pierre Louis Moreau de. 1698-1759. Fr./Ger. DSB; ICB; P II; GA; GB; GC. Other entries: see 3089.
1. *Discours sur les différentes figures des astres ... Avec une exposition abrégée des systèmes de M. Descartes et M. Newton.* Paris, 1732. RM. 2nd ed., 1742.
2. *Discours sur la parallaxe de la lune.* Paris, 1741. RM.
3. *Lettre sur la comète.* Paris, 1742. RM.
3a. ——— *A letter upon comets.* London, 1769. RM.
4. *Astronomie nautique; ou, Elémens d'astronomie.* Paris, 1743. RM.

4525. DELISLE DE LA CROYÈRE, Louis. d. 1741. Fr./Russ. P I.

4526. HERTEL, Wolfgang Christian. d. 1742. Ger. P I.

4527. GOIFFON, Joseph. d. 1751. Fr. P I.
1. *Harmonie des deux sphères, céleste et terrestre; ou, L'art de connoître la situation ... par le soleil et par les étoiles.* Paris, 1739. RM.

4528. BECKER, Johann Hermann. 1700-1759. Ger. P I.

4529. BLISS, Nathaniel. 1700-1764. Br. P I; GD.

4530. CELSIUS, Anders. 1701-1744. Swed. DSB; ICB; P I; GA; GB. Also Geodesy*.

4531. EMERSON, William. 1701-1782. Br. P I; GA; GB; GD. Also Mathematics*, Mechanics*, and Physics*.
1. *A system of astronomy.* London, 1769. RM.

4532. GERSTEN, Christian Ludwig. 1701-1762. Ger. P I; GA.

4533. KRAFFT, Georg Wolfgang. 1701-1754. Ger./Russ. P I; GA (KRAFT); GC (KRAFT). Also Mathematics and Physics.

4534. ADELBULNER, Michael. 1702-1779. Ger. P I. Under ADELBUENER: GA; GC.

4535. HARRIS, Joseph. 1702-1764. Br. P I; GD.
1. *The description and use of the globes and the orrery.* London, 1731. X. 5th ed., 1740. 9th ed., 1763.

4536. CLAP, Thomas. 1703-1767. U.S.A. ICB; GE.
1. *Conjectures upon the nature and motion of meteors.* Norwich, Conn., 1781. RM.

4537. HALLERSTEIN (or HALLER VON HALLESTEIN), Augustin. 1703-1774. Jesuit missionary in China. P I; GA.

4538. GODIN, Louis. 1704-1760. Fr. DSB; P I; GA.

4539. MARTIN, Benjamin. 1704?-1782. Br. DSB; ICB; P II; BMNH; GA; GD. Other entries: see 3099.
1. *The theory of comets.* London, 1757. RM.

4540. PERELLI, Tommaso. 1704-1783. It. ICB; P II.

4541. ASCLEPI, Guiseppe Maria. 1706-1776. It. P I. Also Various Fields.
1. *De objectivi micrometri usu in planetarum diametris metiendis. Exercitatio optico-astronomica.* Rome, 1765. RM.

4542. ENGLERT, Johann Wilhelm. 1706-1777. Ger. P III; GA.

4543. ZUCCONI, Ludovico. ca. 1706-1783. It. P II.

4544. EULER, Leonhard. 1707-1783. Ger./Russ. DSB; ICB; P I; GA; GB; GC. Also Science in General*, Mathematics*#, Mechanics*, Physics*, and Cartography*. See also Index.
1. *Theoria motuum planetarum et cometarum.* Berlin, 1744. RM.
2. *Theoria motus lunae, exhibens omnes eius inaequalitates.* St. Petersburg, 1753. RM.
3. *Recherches sur les inégalités de Jupiter et Saturne.* Paris, 1769. RM.

4545. FOUCHY, Jean Paul Grandjean de. 1707-1788. Fr. DSB. Under GRANDJEAN DE FOUCHY: ICB; P I; GA. Also Biography.

4546. HILL, John. 1707?-1775. Br. DSB; ICB; P I; BMNH; BLA; GA; GB; GD. Other entries: see 3105.
1. *Urania; or, A compleat view of the heavens ... in form of a dictionary.* London, 1754. RM.

4547. MATTEUCCI, Petronio. ca. 1708-1800. It. P II; GA.

4548. BIRD, John. 1709-1776. Br. DSB; ICB; P I; GD.

4549. HEINSIUS, Gottfried. 1709-1769. Ger./Russ. P I. Also Mechanics.

4550. HEYN, Johann. 1709-1746. Ger. P I.

4551. KLINKENBERG, Dirk. 1709-1799. Holl. P I.

4552. MARALDI, Giovanni Domenico. 1709-1788. Fr. DSB; P II; GA. Also Geodesy. See also 4568/1.

4553. TRESENREUTER, Christoph Friedrich. 1709-1746. Switz. P II.

4554. ZANOTTI, Eustachio. 1709-1782. It. DSB; P II; GA.

4555. MURDOCH, Patrick. d. 1774. Br. ICB; P II; GD. See also 15401/1.

4556. FERGUSON, James (first of the name). 1710-1776. Br. DSB; P I; GA; GB; GD. Also Physics* and Autobiography*.
  1. *A dissertation upon the phaenomena of the harvest moon. Also, the description and use of a new four-wheel'd orrery, and an essay upon the moon's turning round her own axis.* London, 1747. RM.
  2. *Astronomy, explained upon Sir Isaac Newton's principles.* London, 1756. RM. 2nd ed., 1757. 5th ed., 1772. 12th ed., 1809. Another ed., with additions by D. Brewster, 2 vols, Edinburgh, 1821.
  3. *The young gentleman's and lady's astronomy, familiarly explained.* London, 1768. X.
  3a. ——— 2nd ed. *An easy introduction to astronomy for young gentlemen and ladies.* Ib., 1769. X. 5th ed., 1790.

4557. SHORT, James. 1710-1768. Br. DSB; ICB; P II; GD.

4558. BOSCOVICH, Ruggiero Giuseppe. 1711-1787. It./Fr. DSB (BOŠKOVIĆ); ICB; P I; GA; GB. Other entries: see 3114.
  1. *De lunae atmosphaera.* Rome, 1753. RM.
  2. *De solis ac lunae defectibus.* London, 1760. Another ed., Venice, 1761. RM. Another ed., London, 1830.
  2a. ——— *Les éclipses.* Paris, 1779.

4559. DUNTHORNE, Richard. 1711-1775. Br. P I; GA; GD.

4560. GARIPUY, François Philippe Antoine de. 1711-1782. Fr. P I.

4561. LAGRANGE, Louis. 1711-1783. Fr./It. P I.

4562. PINGRÉ, Alexandre Gui. 1711-1796. Fr. DSB; ICB; P II; GA.
  1. *Cométographie; ou, Traité historique et théorique des comètes.* 2 vols. Paris, 1783-84.

4563. WRIGHT, Thomas (of Durham). 1711-1786. Br. DSB; ICB; GD.
  1. *The use of the globes; or, The general doctrine of the sphere.* London, 1740.
  2. *Clavis coelestis: Being the explanation of a diagram entituled "A synopsis of the universe"; or, The visible world epitomized.* London, 1742. Reprinted, ib., 1967.
  3. *An original theory or new hypothesis of the universe.* London, 1750. RM/L.
  3a. ——— Reprint, London, 1971. Includes the first publication of *A theory of the universe*, written in 1734. Ed. by M.A. Hoskin.
  4. *Second or singular thoughts upon the theory of the universe.* London, 1968. Ed. from the MS by M.A. Hoskin.

4564. RÖNNBERG, Bernhard Heinrich. 1712-1760. Ger. P II.

4565. CLAIRAUT, Alexis Claude. 1713-1765. Fr. DSB; ICB; P I; GA; GB. Also Mathematics*, Mechanics, Physics, and Geodesy*. See also Index.
  1. *Théorie de la lune.* St. Petersburg, 1752. RM. 2nd ed., Paris, 1765.
  2. *Théorie du mouvement des comètes.* Paris, 1760. RM.

4566. COWPER, Spencer. 1713-1774. Br. P I; GA; GB.

4567. KIES, Johann. 1713-1781. Ger. P I.

4568. LACAILLE, Nicolas Louis de. 1713-1762. Fr. DSB; ICB; P I; GA (CAILLE); GB (CAILLE). Also Geodesy.
1. *Coelum australe stelliferum.* Paris, 1763. Ed. by Maraldi. X.
1a. ——— *A catalogue of 9766 stars in the southern hemisphere.* London, 1847. Ed. by F. Baily.
See also 5830/1 and 10502/1.

4569. AUDIFFREDI, Giovanni Battista. 1714-1794. It. P I; GA.
1. *Transitus Veneris ante solem, observati Romae.* Rome, 1762. RM.
2. *De solis parallaxi.* Rome, 1766. RM.

4570. CASSINI DE THURY, César François. 1714-1784. Fr. DSB; ICB; P I; GA; GB. Also Geodesy*.

4571. KYLIAN (or KILIAN), Jacob. 1714-1774. Danzig. P I.

4572. WILSON, Alexander (first of the name). 1714-1786. Br. DSB; P II; GD.

4573. WINTHROP, John (second of the name). 1714-1779. U.S.A. DSB; ICB; P II; GE.
1. *Two lectures on comets.* Boston, 1759. RM.

4574. BRANCAS-VILLENEUVE, André François de. d. 1758. Fr. P I; GA. Also Mechanics*.

4575. BOUIN, Jean Théodor. 1715-1795. Fr. P I.

4576. LE MONNIER, Pierre Charles. 1715-1799. Fr. DSB; ICB; P I; GA; GB. Also Geomagnetism*.
1. *Histoire céleste; ou, Recueil de toutes les observations astronomiques faites par ordre du Roi.* Paris, 1741. RM.
2. *La théorie des comètes.* Paris, 1743. RM.

4577. SEMLER, Christian Gottlieb. 1715-1782. Ger. P II.
1. *Astrognosia nova; oder, Ausführlich Beschreibung des ganzen Fixstern und Planeten Himmels.* Halle, 1742. RM.
2. *Vollständig Beschreibung des Sterns der Weisen.* Halle, 1743. RM.

4578. XIMINEZ, Leonardo. 1716-1786. It. ICB; P II; GA. Also Mechanics.

4579. ALEMBERT, Jean le Rond d'. 1717-1783. Fr. DSB; ICB; P I; GA; GB; GF. Also Mathematics* and Mechanics*#. See also Index.
1. *Recherches sur la précession des equinoxes, et sur la nutation de l'axe de la terre.* Paris, 1749. RM/L. Reprinted, Brussels, 1967.
2. *Recherches sur différens points importans du système du monde.* 3 vols. Paris, 1754-56. RM. Reprinted, Brussels, 1966.

4580. BENCKEN, Christoph Georg. 1717-1787. Riga. P I.

4581. LUDLAM, William. 1717-1788. Br. P I; GD.

4582. STEWART, Matthew. 1717-1785. Br. DSB; ICB; P II; GA; GD. Also mathematics.
1. *A solution of Kepler's problem.* Edinburgh, 1756. RM.

2. *Tracts, physical and mathematical, containing an explication of several important points in physical astronomy.* Edinburgh, 1761. RM.
3. *The distance of the sun from the earth, determined by the theory of gravity.* Edinburgh, 1763. RM/L.

4583. WARGENTIN, Pehr Wilhelm. 1717-1783. Swed. DSB; ICB; P II.

4584. WEISS, Franz. 1717-1785. Hung. P II.

4585. COURTANVAUX, François César le Tellier, *Marquis* de. 1718-1781. Fr. P I; GA.
1. *Journal du voyage sur la frégate "l'Aurore" pour essayer par ordre de l'Académie plusieurs instrumens relatifs à la longitude.* Paris, 1768. RM.

4586. DARQUIER (DE PELLEPOIX), Antoine (or Augustin). 1718-1802. Fr. P I; GA.
1. *Observations astronomiques.* Avignon, 1777. RM.

4587. DURAEUS, Samuel. 1718-1789. Swed. P I.

4588. HORREBOW, Christian. 1718-1776. Den. DSB; P I; GA.

4589. LOYS DE CHÉSEAUX, Jean Philippe de. 1718-1751. Switz. P I.

4590. MAYER, Christian. 1719-1783. Ger. DSB; P II; GA; GC.

4591. GABRY, Pieter. d. 1770. Holl. P I.

4592. ADAMS, George (first of the name). 1720-1773. Br. P I; BMNH; GA; GD. Also Microscopy*.
1. *A treatise describing and explaining the construction and use of globes.* London, 1766.

4593. BROCKARD, Aloys. 1720-1797. Ger. P III.

4594. HARDY, Samuel. ca. 1720-1793. Br. P I.

4595. HELL (or HÖLL), Maximilian. 1720-1792. Aus. DSB; ICB; P I; GA; GC.

4596. HESSE, Wilhelm Gottlieb. 1720-1784. Ger. P I.

4597. SCHENMARK, Nils. 1720-1788. Swed. P II.

4598. FIXLMILLNER, Placidus. 1721-1791. Aus. P I. Under FIXMILLNER: GA; GC.

4599. SHEPHERD, Antony. 1721-1795. Br. P II; GD.
1. *Tables for correcting the apparent distance of the moon and a star from the effects of refraction and parallax.* Cambridge, 1772.

4600. BARKER, Thomas. 1722-1809. Br. P I; GD. Also Meteorology.

4601. LOWITZ, Georg Moritz. 1722-1774. Ger./Russ. P I; GA; GC.

4602. RATTE, Etienne Hyacinthe de. 1722-1805. Fr. P II; GA.

4603. SAINT-JACQUES (DE SILVABELLE), Guillaume de. 1722-1801. Fr. P II; GA.

4604. WALMESELEY, Charles. 1722-1797. Br./Fr. P II; GD.

4605. WIEDEBURG, Basilius Christian Bernhard. 1722-1758. Ger. P II; GC.

4606. LEPAUTE, Nicole Reine, *Madame*. 1723-1788. Fr. ICB; P I; GA.

4607. MAYER, (Johann) Tobias. 1723-1762. Ger. DSB; ICB; P II; GA (MAYER, T.); GB; GC.
1. *Theoria lunae*. London, 1767.
2. *Tabulae motuum solis et lunae novae et correctae. Quibus accedit Methodus longitudinum promota*. London, 1770.
3. *Lehrbuch über die physische Astronomie*. Göttingen, 1805. RM.
4. *The Euler-Mayer correspondence (1751-1755)*. London, 1971. Ed. by E.G. Forbes.
5. *Opera inedita*. Göttingen, 1775. Ed. by G.C. Lichtenberg. X.
5a. ——— *Opera inedita. The first translation of the Lichtenberg edition of 1775*. London, 1971. Ed. by E.G. Forbes.
6. *The unpublished writings*. 3 vols. Göttingen, 1972. Ed. by E.G. Forbes.

4608. PALITZSCH, Johann Georg. 1723-1788. Ger. ICB; P II.

4609. SCANELLI, Cesare. 1723-after 1782. It. P II.

4610. EICHHORN, Johann Aegidius. 1724-1787. Ger. P I.

4611. FERNER, Bengt. 1724-1802. Swed. P I; GA.

4612. JEAURAT, Edme Sébastien. 1724-1803. Fr. DSB; P I; GA.

4613. KANT, Immanuel. 1724-1804. Ger. The famous philosopher. DSB; ICB; P I; G. Also Science in General*.
1. *Allgemeine Naturgeschichte und Theorie des Himmels*. Königsberg, 1755. RM. Reprinted, Leipzig, 1890. (OKEW. 12) Another reprint, Munich, 1971.
1a. ——— *Universal natural history and theory of the heavens*. Ann Arbor, Mich., 1969. Trans. by W. Hastie.
2. *Kant's cosmogony, as in his essay on the retardation of the rotation of the earth and his "Natural history and theory of the heavens."* Glasgow, 1900. Ed. and trans. by W. Hastie. RM/L. Reprinted, New York, 1968 and 1970.

4614. MICHELL, John. 1724?-1793. Br. DSB; ICB; P II; GB; GD. Also Physics* and Geology*.
1. *Proposal of a method for measuring degrees of longitude upon parallels of the equator*. London, 1767. RM.
2. *An inquiry into the probable parallax and magnitude of the fixed stars from the quantity of light which they afford*. London, 1768. RM.

4615. PLANMAN, Anders. 1724-1803. Fin. P II.

4616. RÖHL, Lampert Heinrich. 1724-1790. Ger. P II.

4617. WOLF, Nathanael Matthaeus von. 1724-1784. Danzig. P II; BLA.

4618. D'ARCY (or D'ARCI), Patrick. 1725-1779. Br./Fr. DSB; P I (ARCY); GD. Also Mechanics.

4619. KURGANOV, Nikolai Gavrilovich. ca. 1725-1796. Russ. ICB.

4620. LE GENTIL DE LA GALAISIÈRE, Guillaume Joseph Hyacinthe Jean Baptiste. 1725-1792. Fr. DSB; P I; GA.

4621. MONTIGNOT, Henri. ca. 1725-ca. 1795. Fr. P II.

4622. SCHÖN, Adam Ehregott. 1725-1805. Ger. P II.

4623. BURNEY, Charles. 1726-1814. Br. P I; GA; GB; GD.

4624. GRISCHOW, Augustin Nathanael. 1726-1760. Ger./Russ. P I; GA.

4625. MELANDERHJELM, Daniel. 1726-1810. Swed. ICB; P II; GA.

4626. AMMAN, Caesarius. 1727-1792. Ger. P I.
  1. *Quadrans astronomicus novus descriptus.* Augsburg, 1770. RM.

4627. ANDRÉ, Noël (or CHRYSOLOGUE, *Père*). 1728-1808. Fr. Under CHRYSOLOGUE, Noël André: P I; GA. Also Geology*.

4628. CHAPPE D'AUTEROCHE, Jean Baptiste. 1728-1769. Fr. DSB; ICB; P I; GA.
  1. *Voyage en Sibérie.* Paris, 1768.
  1a. ——— *A journey into Siberia.* London, 1770.
  2. *Voyage fait ... pour éprouver les montres marines inventées par M. Le Roy.* Paris, 1770. Ed. by Cassini. X. For English trans. see item 3a.
  3. *Voyage en Californie pour l'observation du passage de Vénus sur le disque du soleil.* Paris, 1772. Ed. by Cassini. X.
  3a. ——— *A voyage to California to observe the transit of Venus ... Also, A voyage ... to make experiments on Mr. Le Roy's time keepers.* London, 1778. RM. Reprinted, Richmond (U.K.), 1973.

4629. FRISI, Paolo. 1728-1784. It. DSB; P I; GA; GB. Also Various Fields*#, Mechanics*, and Physics.
  1. *De motu diurno terrae.* Pisa, 1756. RM.

4630. LAMBERT, Johann Heinrich. 1728-1777. Switz./Ger. DSB; ICB; P I; GA; GB; GC. Also Mathematics*, Physics*, and Cartography*.
  1. *Cosmologische Briefe über die Einrichtung des Weltbaues.* Augsburg, 1761. RM.
  1a. ——— *Cosmological letters on the arrangement of the world-edifice.* New York, 1976. Trans. by S.L. Jaki.
  1b. ——— *Lettres cosmologiques sur l'organisation de l'univers.* Paris, 1977.
  2. *Insigniores orbitae cometarum proprietates.* Augsburg, 1761. RM.
  2a. ——— German trans. Leipzig, 1902. (OKEW. 133)

4631. MALLET, Frederik. 1728-1797. Swed. ICB; P II. Also Mathematics.

4632. MASON. Charles. 1728-1786. Br. DSB; ICB; P II; GA; GD. Also Geodesy.

4633. POCZOBUT, Martin Odlanicky. 1728-1810. Wilno. P II; GA.

4634. TITIUS (or TIETZ), Johann Daniel. 1729-1796. Ger. DSB; P II; BMNH; GC. Also Physics and Natural History.

4635. DUNN, Samuel. d. 1794. Br. P I; GA; GD.

4636. AUBERT, Alexander. 1730-1805. Br. ICB; P I; GD.

4637. BJÖRNSEN, Stephen. 1730-1798. Den. P I.

4638. BOCHART DE SARON, Jean Baptiste Gaspard. 1730-1794. Fr. DSB; P I; GA.

4639. DOLLOND, Peter. 1730-1820. Br. P I; GA; GD.
1. *Some account of the discovery, made by the late John Dollond, which led to the grand improvement of refracting telescopes.* London, 1789. RM.

4640. MESSIER, Charles. 1730-1817. Fr. DSB; ICB; P II; GA.

4641. MIRUS, Christian Erdmann. 1730-1803. Ger. P II.

4642. PILGRIM, Anton. 1730-1793. Aus. P II.

4643. KORDENBUSCH VON BUSCHENAU, Georg Friedrich. 1731-1802. Ger. P I; BLA.

4644. OLIVER, Andrew. 1731-1799. U.S.A. GE.
1. *An essay on comets.* Salem, New England, 1772. RM.

4645. WOLLASTON, Francis. 1731-1815. Br. DSB; P II; GD.
1. *Fasciculus astronomicus. Containing observations of the northern circumpolar region.* London, 1800.

4646. FESTER, Diderich Christian. 1732-1811. Nor./Den. P I.
1. *Mathematiske og physiske betänkninger over ... cometer.* Copenhagen, 1759. RM.

4647. HELMUTH, Johann Heinrich. 1732-ca. 1815. Ger. P I.
1. *Anleitung zur Kenntniss des grossen Weltbaues.* Brunswick, 1791. RM.

4648. LALANDE, Joseph Jérome Lefrançais de. 1732-1807. Fr. DSB; ICB; P I; GA; GB. Also History of Astronomy*.
1. *Traité d'astronomie.* 3 vols. Paris, 1764. RM. 2nd ed., 1771. 3rd ed., 1792; reprinted, New York, 1965.
2. *Réflexions sur les comètes qui peuvent approcher de la terre.* Paris, 1773. RM.
3. *Abrégé d'astronomie.* Paris, 1774. RM.
4. *Abrégé de navigation, historique, théorique et pratique.* Paris, 1793. RM/L.
5. *Bibliographie astronomique.* Paris, 1803. RM/L. Reprinted, Amsterdam, 1970.

See also 4808/2.

4649. MASKELYNE, Nevil. 1732-1811. Br. DSB; P II; GA; GB; GD.
1. *Instructions relative to the observation of the ensuing transit of the planet Venus.* London, 1768.

4650. RITTENHOUSE, David. 1732-1796. U.S.A. DSB; ICB; P II; GA; GB; GE. Also Physics.

4651. DIXON, Jeremiah. 1733-1779. Br. DSB; ICB; P I.

4652. HENNERT, Johann Friedrich. 1733-1813. Holl. P I,III. Also Mechanics.
1. *Dissertations physiques et mathématiques sur la figure de la terre, les comètes, l'attraction....* Utrecht, 1778. RM.
2. *Dissertationes de uniformitate motus diurni terrae.* St. Petersburg, 1783. With P. Frisi. RM.

4653. HORNSBY, Thomas. 1733-1810. Br. DSB; P I; GD.
1. *The observations ... made with the transit instrument and quadrant at the Radcliffe Observatory, Oxford, in the years 1774 to 1798.* London, 1932. Ed. by H. Knox-Shaw.

4654. HORSLEY, Samuel. 1733-1806. Br. P I; GA; GB; GD. See also 81/9 and 5769/9.

4655. DIONIS DU SÉJOUR, Achille Pierre. 1734-1794. Fr. DSB; P I; GA. Also Mathematics.
1. *Essai sur les comètes en général*. Paris, 1775. RM.
2. *Essai sur les phénomènes relatifs aux disparitions périodiques de l'anneau de Saturne*. Paris, 1776. RM.
3. *Traité analytique des mouvemens apparens des corps célestes*. 2 vols. Paris, 1786-89. RM.

4656. DUVAUCEL, Charles. 1734-1820. Fr. P I; GA.

4657. EULER, Johann Albrecht. 1734-1800. Ger./Russ. ICB; P I; GA. Also Physics*.

4658. GOUDIN, Matthieu Bernard. 1734-1817. Fr. P I; GA.

4659. MATTUSCHKA, Heinrich Gottfried von. 1734-1779. Ger. P II; BMNH; GA; GC. Also Botany.

4660. MITTERPACHER (VON MITTERBURG), Ludwig. 1734-1814. Aus./Hung. P II; BMNH. Also Natural History.
1. *Anfangsgründe der physikalischen Astronomie*. Vienna, 1781. RM.

4661. MONTEIRO DA ROCHA, José. 1734-1819. Port. P II,III; GA.
1. *Mémoires sur l'astronomie pratique*. Paris, 1808.

4662. RUMOVSKY, Stepan Yakovlevich. 1734-1812. Russ. DSB; P II.

4663. WALES, William. 1734?-1798. Br. P II; GD.
1. *The original astronomical observations made in the course of a voyage ... in H.M. ships the "Resolution" and "Adventure" in the years 1772-1775*. London, 1777. With W. Bayly.
2. *Astronomical observations made in the voyages ... for making discoveries in the southern hemisphere*. London, 1788.
See also 11065/2.

4664. GRUBER (or GRUEBER), Leonhard. d. 1810/11. Ger./Aus. P I.

4665. GILLING, Christian Gottlieb. 1735-1789. Ger. P I.

4666. METZGER, Johann. 1735-1780. Ger. P II.

4667. MOUGIN, Pierre Antoine. 1735-1816. Fr. P II; GA.

4668. RECCARD, Gotthilf Christian. 1735-1798. Ger. P II; GC.

4669. BAILLY, Jean Sylvain. 1736-1793. Br. DSB; ICB; P I; GA; GB. Also Biography* and History of Astronomy*.
1. *Essai sur la théorie des satellites de Jupiter*. Paris, 1766. RM.
See also 9068/1a.

4670. BRÜHL, Hans Moritz von. 1736-1809. Ger. P I.

4671. FUNK, Christlieb Benedict. 1736-1786. Ger. P I; BLA. Also Physics.

4672. LAGRANGE, Joseph Louis. 1736-1813. It./Ger./Fr. DSB; ICB; P I; GA; GB. Also Mathematics*# and Mechanics*. See also Index.

4673. BAUDOUIN DE GUEMADEUC, Armand Henri. b. 1737. Fr. P I.

4674. BAYLY, William. 1737-1810. Br. P I; GA; GD. See also 4663/1.

4675. HAMBERGER, Adolph Albrecht. 1737-1785. Ger. P I; BLA. Also Mechanics*.
1. *Dissertatio physica et mathematica qua causae motus planetarum explicantur.* Jena, 1769. A student's thesis. RM.

4676. RAIN, Ignatz. b. 1737. Aus. P III.

4677. HERSCHEL, (Frederick) William. 1738-1822. Br. DSB; ICB; P I; GA; GB; GC; GD. See also 15436/3.
1. *An account of the discovery of two satellites revolving round the Georgian planet.* London, 1787. RM.
2. *The scientific papers.* 2 vols. London, 1912. Ed. with a biographical introd. by J.L.E. Dreyer.

4678. WILLARD, Joseph. 1738-1804. U.S.A. P II; GE.

4679. COUSIN, Jacques Antoine Joseph. 1739-1800. Fr. P I; GA. Also Mathematics*.
1. *Introduction à l'étude de l'astronomie physique.* Paris, 1787.

4680. KLÜGEL, Georg Simon. 1739-1812. Ger. DSB; P I; GA. Also Mathematics.

4681. LYONS, Israel. 1739-1775. Br. P I; GA; GD. Also Mathematics* and Botany.

4682. PICTET, Jean Louis. 1739-1781. Switz. P II.

4683. PLATEN ZU HALLERMUND, Ernst Franz von. 1739-1818. Ger. P II.

4684. PROSPERIN, Erik. 1739-1803. Swed. P II.

4685. BUGGE, Thomas. 1740-1815. Den. P I; GA. Also Science in General*.

4686. LEXELL, Anders Johan. 1740-1784. Russ. DSB; ICB; P I. Also Mathematics.
1. *Disquisitio de investiganda vera quantitate parallaxeos solis, ex transitu Veneris ante discum solis anno 1769.* St. Petersburg, 1772. RM.

4687. MALLET-FAVRE, Jacques André. 1740-1790. Switz. P II; GA.

4688. NOUET, Nicolas Antoine. 1740-1811. Fr. P II; GA.

4689. PIGOTT, Nathaniel. ca. 1740-1804. Br./Fr./Belg. DSB; ICB; P II; GD.

4690. RÖSLER, Gottlieb Friedrich. 1740-1790. Ger. P II.
1. *De cometis.* Tübingen, [1759?]. His doctoral thesis. RM.

4691. SLOP VON CADENBERG, Joseph Anton. 1740-1808. It. P II.

4692. CAROUGE, Bertrand Augustin. 1741-1798. Fr. P I; GA.

4693. CHIMINELLO, Vincenzo. 1741-1815. It. P I. Also Meteorology.

4694. HAHN, Friedrich von. 1741-1805. Ger. P I.

4695. MARQUEZ, Pedro José. 1741-after 1814. It. P II.

4696. ROCHON, Alexis Marie de. 1741-1817. Fr. P II; GA. Also Physics.

4697. USSHER, Henry. 1741-1790. Br. P II; GD.

4698. POITEVIN, Jacques. 1742-1807. Fr. P II; GA.

4699. CAGNOLI, Antonio. 1743-1816. It. P I (C., Andrea) & Supp.; GA.
1. *Trigonometria piana e sferica.* Paris, 1786. X.
1a. ——— *Traité de trigonométrie rectiligne et sphérique ... avec des applications à la plupart des problèmes de l'astronomie.* Paris, 1786. X. 2nd ed., 1808.
2. *Notizie astronomiche adattate all'uso comune.* 2 vols. Modena, 1799. X. 2nd ed., Milan, 1822.

4700. KRAFFT, Wolfgang Ludwig. 1743-1814. Russ. P I; GC (KRAFT). Also Geomagnetism.

4701. LINDQUIST, Johann Henrik. 1743-1798. Fin. P I.

4702. REGGIO, Francesco. 1743-1804. It. P II; GA.

4703. WILLIAMS, Samuel. 1743-1817. U.S.A. P II.

4704. BERNOULLI, Johann (third of the name). 1744-1807. Ger. DSB; ICB; P I; GA; GB; GC. Also Biography.

4705. MÉCHAIN, Pierre François André. 1744-1804. Fr. DSB; ICB; P II; GA. Also Geodesy.

4706. NICANDER, Henric. 1744-1815. Swed. P II.

4707. VILLAS-BOAS, Custodio Gomes de. 1744-1808. Port. P III.

4708. BEITLER (or BEUTLER), Wilhelm Gottlieb Friedrich von. 1745-1811. Mitau. P I.

4709. KÖHLER, Johann Gottfried. 1745-1801. Ger. P I.

4710. SCHROETER, Johann Hieronymus. 1745-1816. Ger. DSB; P II; GB; GC.
1. *Beiträge zu den neuesten astronomischen Entdeckungen.* Berlin, 1788. RM.
2. *Selenotopographische Fragmente, zur genauern Kenntniss der Mondfläche.* 2 vols. Göttingen, 1791-1802. RM/L.
3. *Aphroditographische Fragmente, zur genauern Kenntnis des Planaten Venus.* Helmstedt, 1796. RM.
4. *Beobachtungen des grossen Cometen von 1807.* Göttingen, 1811. RM.

4711. TRIESNECKER, Franz von Paula. 1745-1817. Aus. P II; GC.

4712. PIAZZI, Giuseppe. 1746-1826. It. DSB; P II; GA; GB.
1. *Praecipuarum stellarum inerrantium positiones mediae.* Palermo, 1814.

4713. BODE, Johann Ehlert. 1747-1826. Ger. DSB; ICB; P I; GA; GB; GC.
1. *Kurzgefasste Erläuterung der Sternkunde.* Berlin, 1778. RM.
2. *Von dem neu entdeckten Planaten.* Berlin, 1784. RM.
3. *Kurzer Entwurf der astronomischen Wissenschaften.* Berlin, 1794. RM.
4. *Allgemeine Betrachtungen über das Weltgebäude.* Berlin, 1801. X. 2nd ed., Vienna, 1805. RM.

4714. HAMILTON, James Archibald. 1747-1815. Br. P I; GD.

4715. VIDAL, Jacques. 1747-1819. Fr. P II.

4716. CASSINI, Jean Dominique. 1748-1845. Fr. DSB; ICB; P I; GA; GB. Also Geodesy*, Autobiography*, and History of Astronomy*. See also Index.

4717. DERFFLINGER, Thaddeus. 1748-1824. Aus. P I.

4718. THULIS, Jacques Joseph Claude. 1748-1810. Fr. P II.

4719. AASHEIM, Arnold Nicolaus. 1749-1800. Den. P I.

4720. CALANDRELLI, Giuseppe. 1749-1827. It. DSB; P I; GA.

4721. CESARIS, (Giovanni) Angelo. 1749-1832. It. P I; GA.

4722. DELAMBRE, Jean Baptiste Joseph. 1749-1822. Fr. DSB; ICB; P I; GA; GB. Also Mathematics*, Geodesy*, and History of Astronomy*.
1. *Abrégé d'astronomie; ou, Leçons élémentaires d'astronomie théorique et pratique.* Paris, 1813.
2. *Astronomie théorique et pratique.* 3 vols. Paris, 1814. RM/L.
3. *Tables écliptiques des satellites de Jupiter.* Paris, 1817.

4723. FISCHER, Johann Nepomuk. 1749-1805. Ger. P I.

4724. LAPLACE, Pierre Simon. 1749-1827. Fr. DSB Supp.; ICB; P I; GA. Also Mathematics*, Physics, and History of Astronomy*. See also Index.
1. *Théorie du mouvement et de la figure elliptique des planètes.* Paris, 1784. RM.
2. *Théorie des attractions des sphéroïdes et de la figure des planètes.* Paris, 1785. RM.
3. *Exposition du système du monde.* 2 vols. Paris, 1796. RM. 3rd ed., 1808. 5th ed., 1824. 6th ed., 1835.
3a. ——— *The system of the world.* 2 vols. London, 1809. Trans. by J. Pond. RM/L.
3b. ——— *The system of the world.* 2 vols. Dublin, 1830. Trans. with notes by H.H. Harte.
4. *Traité de mécanique céleste.* 5 vols in 4. Paris, 1799-1825. RM/L. Reprinted, Brussels, 1967. 2nd ed., 1829-39.
4a. ——— *A treatise upon analytical mechanics, being the first book of the "Mécanique céleste."* Nottingham, 1814. Trans. with notes by J. Toplis. RM.
4b. ——— *Elementary illustrations of the celestial mechanics of Laplace.* London, 1821. By T. Young. RM/L.
4c. ——— *A treatise of celestial mechanics.* 2 vols. Dublin, 1822-27. Trans. by H.H. Harte.
4d. ——— *Mécanique céleste. Translated with a commentary.* 4 vols. Boston, 1829-39. By N. Bowditch. RM/L. Reprinted, Bronx, N.Y., 1966.
5. *Oeuvres.* 7 vols in 4. Paris, 1843-47.
6. *Oeuvres complètes.* 14 vols. Paris, 1878-1912. RM/L.

4725. SCHULZE, Johann Karl. 1749-1790. Ger. P II.

4726. TREMBLEY, Jean. 1749-1811. Switz./Ger. P II. Also Mathematics.

4727. VINCE, Samuel. 1749-1821. Br. P II; GD. Also Mathematics*, Mechanics*, and Physics*.
1. *A treatise of practical astronomy.* Cambridge, 1790.

2. *The principles of astronomy*. Cambridge, 1794. Another ed., 1799.
3. *A complete system of astronomy*. 3 vols. Cambridge, 1797-1808. 2nd ed., 1814.
4. *The elements of astronomy*. 2nd ed. Cambridge, 1801.

4728. ADAMS, George (second of the name). 1750-1795. Br. P I; BMNH; BLA; GA; GD. Other entries: see 3217.
    1. *Astronomical and geographical essays*. London, 1789. RM.

4729. BONNYCASTLE, John. 1750?-1821. Br. P I; GA; GD. Also Mathematics*.
    1. *An introduction to astronomy*. London, 1786. 2nd ed., 1787. 3rd ed., 1796.

4730. HERSCHEL, Caroline Lucretia. 1750-1848. Br. DSB; ICB; P I; GA; GB; GC; GD. See also 15590/1.

4731. AGELET, Joseph Lepaute d'. 1751-1788. Fr. P I.

4732. ROBERTSON, Abraham. 1751-1826. Br. GD. Also Mathematics.

4733. BARRY, Roger. 1752-1813. Fr./Ger. P I.

4734. ENGLEFIELD, Henry Charles. 1752-1822. Br. P I; GD.
    1. *On the determination of the orbits of comets*. London, 1793.

4735. KOCH, Julius August. 1752-1817. Danzig. P I.

4736. LEGENDRE, Adrien Marie. 1752-1833. Fr. DSB; ICB; P I; GA; GB. Also Mathematics* and Geodesy*.
    1. *Nouvelles méthodes pour la détermination des orbites des comètes*. Paris, 1805. RM. Supplement, 1806. RM. Another supplement, 1820. RM.

4737. ORIANI, Barnabe. 1752-1832. It. ICB; P II; GA.

4738. BEIGEL, Georg Wilhelm Siegmund. 1753-1837. Ger. P I.

4739. MINTO, Walter. 1753-1796. U.S.A. ICB; P II; GE.

4740. PASQUICH, Johann. 1753-1829. Hung. ICB; P II. Also Mechanics.

4741. PIGOTT, Edward. 1753-1825. Br. DSB; GD.

4742. ELLICOTT, Andrew. 1754-1820. U.S.A. P I & Supp.; GE.

4743. ROELOFS, Arjen. 1754-1828. Holl. ICB.

4744. ZACH, Franz Xaver von. 1754-1832. Aus./Ger. DSB; ICB; P II; GA; GB; GC. Also Geodesy*.
    1. *Tabulae speciales aberrationis et nutationis*. 2 vols. Gotha, 1806-07.

4745. BERTRAND, ... *Abbé*. ca. 1755-1792. Fr. P I; GA.

4746. CASSELLA (or CASELLA or CASSELLI), Giuseppe. 1755-1808. It. P I; GA.

4747. FLAUGERGUES, Honoré. 1755-1835. Fr. P I; GA. Also Physics.

4748. FOKKER, Jan Pieter. 1755-1831. Holl. P I & Supp.; GF.

4749. FUSS, Nicolaus. 1755-1826. Russ. DSB; ICB; P I. Also Mathematics. See also 3447/15,16.

4750. GROOMBRIDGE, Stephen. 1755-1832. Br. P I; GD.

4751. SNIADECKI, Jan Baptista. 1756-1830. Cracow/Wilno. ICB; P II.

4752. DAVID, Martin Alois. 1757-1836. Prague. ICB; P I.

4753. OLBERS, (Heinrich) Wilhelm (Matthias). 1758-1840. Ger. DSB; ICB; P II; BLA; GA; GB; GC.
1. *Abhandlung über die leichteste und bequemste Methode, die Bahn eines Cometen aus einigen Beobachtungen zu berechnen.* Weimar, 1797. RM.
2. *Briefwechsel zwischen W. Olbers und F.W. Bessel.* 2 vols. Leipzig, 1852. Ed. by A. Erman. RM.
3. *Wilhelm Olbers, sein Leben und seine Werke.* 2 vols in 3 & supp. Berlin, 1894-1902. Contains his collected works and correspondence with Gauss. Ed. by C. Schilling.

4754. PACASSI, Johann Baptiste von. 1758-1818. Aus. P II.

4755. SCHUBERT, Friedrich Theodor (or Theophilus). 1758-1825. Russ. ICB (SHUBERT, Feodor Ivanovich); P II; GC.
1. *Traité d'astronomie théorique.* 3 vols. St. Petersburg, 1791. X. Another ed., 1822.

4756. ROHDE, Johann Philipp von. 1759-1834. Ger. P II; GC.

4757. SPÄTH, Johann Leonhard. 1759-1842. Ger. ICB; P II. Also Physics.
1. *Photometrische Untersuchung über die Deutlichkeit mit welcher wir entfernte Gegenstande vermittelst dioptrischer Fernrohre beobachten können.* Leipzig, 1789. RM.

4758. VOGEL, Christian Leberecht. 1759-1816. Ger. P II; GC.

4759. BLAIR, Robert. d. 1828. Br. P I; GD.

4760. ENDE, Ferdinand Adolph von. 1760-1817. Ger. P I; GC.
1. *Über Massen und Steine die aus dem Monde auf die Erde gefallen sind.* Brunswick, 1804. RM.

4761. RÜDIGER, Christian Friedrich. 1760-1809. Ger. P II.

4762. UNGESCHICK, Peter. 1760-1790. Ger. P II; GC.

4763. WURM, Johann Friedrich. 1760-1833. Ger. P II; GC.

4764. LAX, William. 1761-1836. Br. P I; GA; GD.

4765. PONS, Jean Louis. 1761-1831. Fr./It. DSB; ICB; P II; GA; GB.

4766. SCARPELLINI, Feliciano. 1762-1840. It. P II.

4767. SEYFFER, Karl Felix von. 1762-1822. Ger. P II; GC.

4768. BRINKLEY, John. 1763-1835. Br. DSB; P I; GA; GD. Also Mathematics.
1. *Elements of plane astronomy.* Dublin. 1808. X. Another ed., 1819. 6th ed., rev. by T. Luby, 1845.
1a. ——— Another ed., entitled *Astronomy.* Dublin, 1871. Rev. by J.W. Stubbs and F. Brünnow. 2nd ed., London, 1874.

4769. CAMERER, Johann Wilhelm von. 1763-1847. Ger. P I.

4770. GOLDBACH, Christian Friedrich. 1763-1811. Ger./Russ. P I.

4771. HENRY, Maurice. 1763-1825. Fr./Ger./Russ. P I,III.

4772. TRALLES, Johann Georg. 1763-1822. Ger./Switz. P II; GC. Also Mathematics and Physics.

4773. GOODRICKE, John. 1764-1786. Br. DSB; ICB; P I (GOODRIKE); GD.

4774. JUNGNITZ, Longinus Anton. 1764-1831. Ger. P I.

4775. MARSCHALL VON BIEBERSTEIN, Karl Wilhelm. 1764-1817. Ger. P II; GC.

4776. BRYAN, Margaret. fl. 1797-1815. Br. GD. Also Physics*.
  1. *A compendious system of astronomy*. London, 1797. RM/L. 2nd ed., 1799. 3rd ed., 1805.

4777. BERTIROSSI-BUSATA, Francesco. 1765-1824. It. P I.

4778. BOHNENBERGER, Johann Gottlieb Friedrich von. 1765-1831. Ger. P I; GC.

4779. HARDING, Carl Ludwig. 1765-1834. Ger. DSB; P I; GA.

4780. IVORY, James. 1765-1842. Br. DSB; P I; GB; GD. Also Mathematics, Mechanics, Physics, and Geodesy.

4781. PERNY DE VILLENEUVE, J. b. 1765. Fr. P II.

4782. BÜRG, Johann Tobias. 1766-1834. Aus. DSB; P I.

4783. IZARN, Joseph. 1766-after 1834. Fr. P I. Also Physics* and Chemistry*.
  1. *Des pierres tombées du ciel*. Paris, 1805. RM.

4784. LALANDE, Michel Jean Jérome le François de. 1766-1839. Fr. P I; GA.

4785. WALKER, William. ca. 1766-1816. Br. P II; GD.

4786. BOUVARD, Alexis. 1767-1843. Fr. DSB; P I; GA (BOUVART).

4787. CICCOLINI, Lodovigo. 1767-1854. It. P I & Supp.

4788. PASTORFF, Johann Wilhelm. 1767-1838. Ger. ICB; P II.

4789. PEARSON, William. 1767-1847. Br. P II; GD.
  1. *An introduction to practical astronomy*. 2 vols. London, 1824-29.

4790. POND, John. 1767-1836. Br. P II; GA; GB; GD. See also 4724/3a.

4791. DAMOISEAU, (Marie Charles) Théodore de. 1768-1846. Fr. P I; GA.
  1. *Tables de la lune*. Paris, 1824.

4792. HUBER, Daniel. 1768-1829. Switz. P I.

4793. FERRER, José Joaquin de. d. 1818. Cuba/U.S.A. P I.

4794. BREDMAN, Johann. 1770-1859. Swed. P I,III.

4795. RIENKS, Sieds Johannes. 1770-ca. 1844. Holl. ICB.

4796. STARK, Augustin. 1771-1839. Ger. ICB; P II; GC.

4797. BEEK-CALKOEN, Jan Frederik van. 1772-1811. Holl. P I,III; GA (CALKOEN).

4798. CATUREGLI, Pietro. ca. 1772-1833. It. P I.

4799. DICK, Thomas. 1772 (or 1774)-1857. Br. P III; GB; GD.
  1. *The sidereal heavens*. London, 1849. X. Another ed., 1866.

4800. FRITSCH, Johann Heinrich. 1772-1829. Ger. P I.

4801. BOWDITCH, Nathaniel. 1773-1838. U.S.A. DSB; ICB; P I; GB; GE. See also 4724/4d.

4802. BRISBANE, Thomas Macdougall. 1773-1860. Br./Australia. DSB; P I & Supp.; GB; GD.

4803. BURCKHARDT, Johann Karl. 1773-1825. Ger./Fr. P I; GA; GC.

4804. EPPS, James. 1773-1839. Br. P I.

4805. FRANCOEUR, Louis Benjamin. 1773-1849. Fr. P I; BMNH; GA. Also Mathematics* and Mechanics*.
1. *Uranographie; ou, Traité élémentaire de l'astronomie.* Paris, 1812. X. 2nd ed., 1818. 5th ed., 1837.

4806. UTENHOVE VAN HEEMSTEDE, Jacob Maurits Carel van. 1773-1836. Holl. P II; GF.

4807. WOODHOUSE, Robert. 1773-1827. Br. DSB; ICB; P II; GA; GD. Also Mathematics*.
1. *A treatise on astronomy.* Cambridge, 1812. X. Another ed., 2 vols in 3, 1818-23.

4808. BAILY, Francis. 1774-1844. Br. DSB; ICB; P I; GA; GB; GD. Also Biography*.
1. *Astronomical tables and formulae.* London, 1827. RM/L.
2. *A catalogue of those stars in the "Histoire céleste française" of J. de Lalande for which tables of reduction to the epoch 1800 have been published by Professor Schumacher ....* London, 1847.
See also 4564/1a.

4809. BIOT, Jean Baptiste. 1774-1862. Fr. DSB; ICB; P I,III; GA; GB. Also Mathematics*, Physics*, and History of Astronomy*.
1. *Traité élémentaire d'astronomie physique.* 2 vols. Paris, 1802. X. 2nd ed., 3 vols, 1810-11. 3rd ed., 5 vols in 2, 1841-57.

4810. GREGORY, Olinthus Gilbert. 1774-1841. Br. DSB; P I; GA; GB; GD. Also Mechanics* and Physics.
1. *Lessons astronomical and philosophical.* London, 1793. X. 5th ed., 1815.
2. *A treatise on astronomy.* London, 1802.

4811. HEILIGENSTEIN, Conrad von. 1774-1849. Ger. P I.

4812. LYNN, Thomas. 1774-1847. Br. GD.

4813. MOLLWEIDE, Karl Brandon. 1774-1825. Ger. DSB; P II; GA; GC. Also Mathematics and Physics.

4814. PFAFF, Johann Wilhelm Andreas. 1774-1835. Ger. P II.

4815. RICCHEBACH, Giacomo. 1776-1841. It. P II.

4816. SOLDNER, Johann Georg von. 1776-1833. Ger. DSB; ICB; P II; GC. Also Geodesy*.

4817. BENZENBERG, Johann Friedrich. 1777-1846. Ger. DSB; ICB; P I; GA; GC. Also Physics and Geodesy.

4818. BITTNER, Adam. 1777-1844. Prague. P I & Supp.

4819. BRANDES, Heinrich Wilhelm. 1777-1834. Ger. DSB; P I; GC. Also Physics.

4820. GAUSS, Karl Friedrich. 1777-1855. Ger. DSB; ICB; P I,III; GA; GB; GC. Also Mathematics*#, Mechanics*, Physics*, and Earth Sciences*. See also Index.
1. *Theoria motus corporum coelestium in sectionibus conicis solem ambientium.* Hamburg, 1809. RM/L. Reprinted, Brussels, 1968.
1a. ——— *Theory of the motion of the heavenly bodies moving about the sun in conic sections.* Boston, 1857. Trans. by C.H. Davis. RM/L. Reprinted, New York, 1963.
2. *Determinatio attractionis quam ... exerceret planeta....* [Periodical article, 1816] German trans. Leipzig, 1927. (OKEW. 225)

4821. MÜNCHOW, Karl Dieterich von. 1778-1836. Ger. P II.

4822. FARRAR, John. 1779-1853. U.S.A. DSB; P III; GE. Also Physics*.
1. *An elementary treatise on astronomy.* Cambridge, Mass., 1827. RM.

4823. GOMPERTZ, Benjamin. 1779-1865. Br. DSB; P III; GD. Also Mathematics.

4824. INGHIRAMI, Giovanni. 1779-1851. It. P I; GB.

4825. LINDENAU, Bernhard August von. 1779-1854. Ger. DSB; ICB; P I; GA; GC.

4826. NÜRNBERGER, Joseph Christian Emil. 1779-1848. Ger. P II; GC. Also Mathematics.

4827. CACCIATORE, Niccolo. 1780-1841. It. P I; GA.

4828. CHALMERS, Thomas. 1780-1847. Br. P I; GA; GB; GD. Also Science in General*.

4829. HALLASCHKA, Franz Ignaz Cassian. 1780-1847. Brünn/Prague. P I. Also Physics.

4830. SCHUMACHER, Heinrich Christian. 1780-1850. Den. DSB; ICB; P II; GB; GC. Also Geodesy. See also 3645/13, 3695/2a, and 4808/2.

4831. SOMERVILLE, Mary Fairfax Greig. 1780-1872. Br. DSB; ICB; P II, III; GB; GD. Also Science in General*, Geology*, and Autobiography.*.
1. *The mechanism of the heavens.* London, 1831.
2. *A preliminary dissertation on the mechanism of the heavens.* London, 1832.

4832. BRAG, Jonas. 1781-1857. Swed. P III.

4833. LITTROW, Joseph Johann von. 1781-1840. Russ./Aus. ICB; P I; GA; GC.
1. *Theoretische und practische Astronomie.* 3 vols. Vienna, 1821-27.

4834. PLANA, Giovanni (or Jean). 1781-1864. It. DSB; ICB; P II,III. Also Mathematics, Physics*, and History of Mechanics*. See also 4837/1.

4835. WISNIEWSKY, Vincent. 1781-1855. Russ. P II.

4836. BIELA, Wilhelm von. 1782-1856. Aus. DSB; ICB; P I; GC.

4837. CARLINI, Francesco. 1783-1862. It. P I,III.
1. *Observations sur l'écrit de M. Laplace ... intitulé "Sur le perfectionnement de la théorie et des tables lunaires."* Genoa, 1820. With G.A.A. Plana. RM.

4838. KMETH, Daniel. 1783-1825. Hung. ICB; P I; GA.

4839. MATHIEU, Claude Louis. 1783-1875. Fr. P II,III; GA.

4840. SCHERER, Adrian Johann Philipp von. 1783-1835. Switz. P II.

4841. BESSEL, Friedrich Wilhelm. 1784-1846. Ger. DSB; ICB; P I; GA; GC. Also Mathematics and Geodesy*. See also Index.
1. *Fundamenta astronomiae pro anno 1755 deducta ex observationibus ... James Bradley ... per annos 1750-1762 institutis.* Königsberg, 1818. RM.
2. *Astronomische Untersuchungen.* 2 vols. Königsberg, 1841-42. RM/L.
3. *Briefwechsel zwischen Gauss und Bessel.* Leipzig, 1880. RM.
4. *Abhandlungen.* 3 vols. Leipzig, 1875-76. Ed. by R. Engelmann.

4842. CERQUERO, José Sanchez. ca. 1784-1850. Spain. P I.

4843. CRONSTRAND, Simon Anders. 1784-1850. Swed. P I. Also Geodesy.

4844. HANSTEEN, Christopher. 1784-1873. Nor. DSB; P I,III; GA; GB. Also Physics* and Geomagnetism*.

4845. TITTEL, Paul. 1784-1831. Hung. P II,III; GC.

4846. ZUCCARI, Federico. 1784-1817. It. P II.

4847. BISHOP, George. 1785-1861. P III; GD.
1. *Astronomical observations, 1839-51.* London, 1852.

4848. SIMONOFF, Ivan Mikailovich. 1785-1855. Russ. P II.

4849. SOUTH, James. 1785-1867. Br. DSB; P II,III; GD. Also Science in General*.

4850. ARAGO, (Dominique) François (Jean). 1786-1853. Fr. DSB; ICB; P I; GA; GB. Also Physics, Earth Sciences*, and Biography*.
1. *Des comètes.* [Periodical article, 1832]
1a. ——— *Tract on comets.* Boston, 1832. RM.
2. *Leçons d'astronomie.* Paris, 1834. X. 4th ed., 1845.
2a. ——— *Popular lectures on astronomy.* London, 1841.
2b. ——— ——— Reprint of extract: *Popular treatise on comets.* London, 1861.
3. *Astronomie populaire.* 4 vols. Paris, 1854-57. X.
3a. ——— *Popular astronomy.* 2 vols. London, 1855-58.
4. *Oeuvres complètes.* 17 vols. Paris, 1854-62. Ed. by J.A. Barral. RM/L.

4851. BINET, Jacques (Philippe Marie). 1786-1856. Fr. ICB; P I; GA. Also Mathematics.

4852. KUNOWSKY, Georg Karl Friedrich. 1786-1846. Ger. P I.

4853. PAUCKER, Magnus Georg von. 1787-1855. Mitau. P II; GC. Also Mathematics.

4854. SANTINI, Giovanni. 1787-1877. It. P II,III; GB.

4855. VALZ, (Jean Elie) Benjamin. 1787-1867. Fr. P II,III.

4856. GERLING, Christian Ludwig. 1788-1864. Ger. ICB; P I,III; GC.
1. *Briefwechsel zwischen C.F. Gauss und C.L. Gerling.* Berlin, 1927. Ed. by C. Schäfer. X.
2. *C.L. Gerling an C.F. Gauss: Sechzig bisher unveröffentliche Briefe.* Göttingen, 1964. Ed. by T. Gerardy.

4857. RIDDLE, Edward. 1788-1854. Br. P II; GD.

4858. RÜMKER, Karl Ludwig Christian. 1788-1862. Ger./Australia. P II & Supp.; GB; GC.
1. *A preliminary catalogue of fixed stars intended for a prospectus of a catalogue of the stars of the southern hemisphere.* Hamburg, 1832.

4859. SMYTH, William Henry. 1788-1865. Br. P II,III; GD. See also 3011.
1. *A cycle of celestial objects.* 2 vols. London, 1844. RM/L. 2nd ed., by G.F. Chambers, Oxford, 1881.

4860. BOGUSLAVSKY, Palm Heinrich Ludwig von. 1789-1851. Ger. DSB; P I; GA.

4861. BOND, William Cranch. 1789-1859. U.S.A. DSB; P III; GE.

4862. CAUCHY, Augustin Louis. 1789-1857. Fr. DSB; ICB; P I,III; GA; GB. Also Mathematics*# and Mechanics. See also 15509/1.

4863. FALLOWS, Fearon. 1789-1831. Cape Obs. P I; GD.

4864. SCHWABE, Samuel Heinrich. 1789-1875. Ger. DSB; ICB; P II,III; GB; GC.

4865. RICHARDSON, William (second of the name). fl. 1822-1835. Br./ Australia.
1. *A catalogue of 7385 stars, chiefly in the southern hemisphere.* London, 1835.

4866. DENT, Edward John. 1790-1853. Br. P I; GD. Also Physics*.

4867. MÖBIUS, August Ferdinand. 1790-1868. Ger. DSB; P II,III; GA; GB; GC. Also Mathematics*#.
1. *Die wahre und die scheinbare Bahn des Halley'schen Kometen bei seiner Wiederkunft im Jahre 1835.* Leipzig, 1834. RM.
2. *Die Elemente der Mechanik des Himmels.* Leipzig, 1843. RM.

4868. PEREVOSHCHIKOV, Dimitry Vassilievich. 1790-1880. Russ. P II, III (PEREWOSCHTSCHIKOW).

4869. STRATFORD, William Samuel. 1790-1853. Br. P II; GD.

4870. BIANCHI, Giuseppe. 1791-1866. It. P III.

4871. ENCKE, Johann Franz. 1791-1865. Ger. DSB; ICB; P I,III; GA; GB; GC.
1. *Über die Bestimmung einer elliptischen Bahn aus drei vollständigen Beobachtungen.* [Periodical article, 1850] Leipzig, 1903. (OKEW. 141)
2. *Gesammelte mathematische und astronomische Abhandlungen.* Berlin, 1888. RM.

4872. LARGETEAU, Charles Louis. 1791-1857. Fr. P I; GA.

4873. CALANDRELLI, Ignazio. 1792-1866. It. DSB; P III.

4874. HERSCHEL, John Frederick William. 1792-1871. Br. DSB; ICB; P I,III; GA; GB; GD. Also Science in General* and Physics.
1. *A treatise on astronomy*. London, 1833. Other eds: 1834, 1835,1841.
2. *Results of astronomical observations made ... at the Cape of Good Hope, being the completion of a telescopic survey of the whole surface of the visible heavens, commenced in 1825*. London, 1847. RM/L.
3. *Outlines of astronomy*. London, 1849. 2nd ed., 1849. RM/L. 3rd ed., 1850. 4th ed., 1851. 6th ed., 1859. Another ed., 1875.
4. *Herschel at the Cape: Diaries and correspondence, 1834-1838*. Austin, Texas, 1969. Ed. by D.S. Evans et al.
5. *Scientific papers*. 2 vols. London, 1912. Ed. by J.L.E. Dryer.
See also 3597/6a.

4875. KOLLER, Marian. 1792-1866. Aus. P I,III. Also Earth Sciences.

4876. ROBINSON, (John) Thomas Romney. 1792-1882. Br. P III; GB; GD.

4877. SCHWERD, Friedrich Magnus. 1792-1871. Ger. P II,III; GC. Also Physics*.

4878. STAMPFER, Simon. 1792-1864. Aus. P II,III,IV; GC. Also Geodesy.

4879. WACHTER, Friedrich Ludwig. 1792-1817. Ger. ICB; P II,III.

4880. GAUTIER, (Jean) Alfred. 1793-1881. Switz. P I,III.

4880A HENCKE, Karl Ludwig. 1793-1866. Ger. P I,III.

4881. LARDNER, Dionysius. 1793-1859. Br. ICB; P I & Supp.; BMNH; GA; GB; GD. Other entries: see 3303.
1. *Handbook of astronomy*. 2 vols. London, 1855. X. 4th ed., 1875.
2. *Popular astronomy*. London, 1856.

4882. MITCHELL, William. 1793 (or 1791)-1869. U.S.A. P III; GE.

4883. NICOLAI, Friedrich Bernhard Gottfried. 1793-1846. Ger. DSB; P II.

4884. STRUVE, Friedrich Georg Wilhelm. 1793-1864. Russ. DSB; ICB; P II,III; GA; GB; GC. Also Geodesy.
1. *Description de l'observatoire astronomique central de Poulkova*. St. Petersburg, 1845.
2. *Etudes d'astronomie stellaire*. St. Petersburg, 1847.

4885. WALBECK, Henric Johan. 1793-1822. Fin. P II.

4886. WARTMANN, Louis François. 1793-1864. Switz. P II,III.

4887. BERTELLI, Francesco. 1794-1844. It. P I.

4888. CHEVALLIER, Temple. 1794-1873. Br. DSB; P III; GD.

4889. MACLEAR, Thomas. 1794-1879. Br./Cape Obs. DSB; ICB; P III; GD. Also Geodesy*.

4890. MÄDLER, Johann Heinrich von. 1794-1874. Ger./Dorpat. DSB; P II,III; GA; GC. Also History of Astronomy*. See also 4905/1.

4891. NELL DE BRÉAUTÉ, Eléonore Suzanne. 1794-1855. Fr. P II.

4892. POSSELT, Johannes Friedrich. 1794-1823. Ger. P II.

4893. SHEEPSHANKS, Richard. 1794-1855. Br. P II; GD.

4894. WESTPHAL, Johann Heinrich Christoph. 1794-1831. Ger. P II; GC.

4895. WHEWELL, William. 1794-1866. Br. DSB; ICB; P II,III; BMNH; GB; GD. Other entries: see 3304.
1. *Astronomy and general physics considered with reference to natural theology.* London, 1833. (Bridgewater treatises. 3) RM. 3rd ed., 1834. 6th ed., 1837. 7th ed., 1852. 8th ed., 1862.
2. *Of the plurality of worlds.* London, 1853. RM. 2nd ed., 1854. 3rd ed., 1854. 4th ed., 1855. Another ed., 1859.

4896. DUNLOP, James. 1795-1848. Australia. P I,III; GD.
1. *A catalogue of nebulae and clusters of stars in the southern hemisphere.* London, 1828.

4897. FRISIANI, Paolo. 1795-1880. It. P III.

4898. HANSEN, Peter Andreas. 1795-1874. Ger. DSB; ICB; P I,III; GA; GC.
1. *Untersuchungen über die gegenseitigen Störungen des Jupiters und Saturns.* Berlin, 1831.
2. *Fundamenta nova investigationis orbitae verae quam luna perlustrat.* Gotha, 1838.
3. *Theorie der Pendelbewegung.* Danzig, 1853.
4. *Tables de la lune.* London, 1857.
5. *Über die Bestimmung der Bahn eines Himmelskörpers aus drei Beobachtungen.* [Periodical article, 1863] Leipzig, 1903. (OKEW. 141)

4899. PONTÉCOULANT, (Philippe) Gustave (Doulcet) de. 1795-1874. Fr. P II,III.
1. *Théorie analytique du système du monde.* 4 vols. Paris, 1829-46. RM/L. 2nd ed., 1856.

4900. SLAWINSKY, Peter von. 1795-1881. Wilno. P III.

4901. FELDT, Laurentius. b. 1796. Ger. P I.

4902. GALLOWAY, Thomas. 1796-1851. Br. P I; GB; GD.

4903. LOHRMANN, Wilhelm Gotthelf. 1796-1840. Ger. P I.

4904. OBODOVSKII, Aleksander Grigorievich. 1796-1852. Russ. ICB.

4905. BEER, Wilhelm. 1797-1850. Ger. DSB; P I; GA; GC.
1. *Mappa selenographica totam lunae hemisphaeram visibilem complectans.* Berlin, 1834. With J.H. Mädler.

4906. FERGUSON, James (second of the name). 1797-1868. U.S.A. P III.

4907. SAVARY, Félix. 1797-1841. Fr. P II.

4908. CAPOCCI, Ernesto. 1798-1864. It. P III.

4909. COOPER, Edward Joshua. 1798-1863. Br. P III; GD.
1. *Cometic orbits.* Dublin, 1852.

4910. HENDERSON, Thomas. 1798-1844. Br. DSB; P I; GA; GD.

4911. WEISSE, Maximilian von. 1798-1863. Cracow. P II,III; GC.

4912. WROTTESLEY, John. 1798-1867. Br. P II,III; GD.

4913. ABBOTT, Francis. 1799-1883. Australia. P III.

4914. ARGELANDER, Friedrich Wilhelm August. 1799-1875. Fin./Ger. DSB; P I,III; GB; GC.
1. *Untersuchungen über die Bahn des grossen Cometen vom Jahre 1811*. Königsberg, 1822. RM.
2. *Anzeige von einer auf der Königlichen Universitäts-Sternwarte zu Bonn unternommenen Durchmusterung des nördlichen Himmels als Grundlage neuer Himmelscharten*. Bonn, 1856. RM.

4915. DAWES, William Rutter. 1799-1868. Br. DSB; P III; GD.

4916. LASSELL, William. 1799-1880. Br. DSB; ICB; P III; GD.

4917. CALDECOTT, John. 1800-1849. Br./India. P I; GD. Also Earth Sciences.

4918. GAMBART, Jean Félix Adolphe. 1800-1836. Fr. P I; GA.

4919. GRUBB, Thomas. 1800-1878. Br. DSB; P III; GD.

4920. LEHMANN, Jacob Wilhelm Heinrich. 1800-1863. Ger. P I,III.

4921. PARSONS, William, *3rd Earl of* ROSSE. 1800-1867. Br. DSB; ICB; P II,III (ROSSE); GB; GD.
1. *The scientific papers*. Bradford, 1926. Ed. by C. Parsons.

4922. ROSENBERGER, Otto August. 1800-1890. Ger. DSB; P II,III.

4923. AIRY, George Biddell. 1801-1892. Br. DSB; ICB; P I,III,IV; GB; GD1. Also Mathematics*, Physics*, and Autobiography*.
1. *Mathematical tracts on physical astronomy, the figure of the earth, precession and nutation, and the calculus of variations*. Cambridge, 1826. RM/L. 2nd ed., 1831. RM/L. 3rd ed., 1842. 4th ed., 1858.
2. *Gravitation: An elementary explanation of the principal perturbations in the solar system*. London, 1834. RM/L.
3. *Six lectures on astronomy*. London, 1849. 3rd ed., 1856. 6th ed., 1868. 7th ed., 1868. 9th ed., 1877.
4. *Lecture on the pendulum experiments at Harton Pit*. London, 1854. RM.
5. *Account of the observations of the transit of Venus, 1874*. London, 1881. RM/L.
6. *Numerical lunar theory*. London, 1886. RM/L.

4924. CAPELLI, Giovanni. 1801-1877. It. P III.

4925. CLAUSEN, Thomas. 1801-1885. Ger./Dorpat. DSB; P I,III. Also Mathematics.

4926. KNORRE, Karl Friedrich. 1801-1883. Russ. P I,III.

4927. RITTER, Elie. 1801-1862. Switz. P II & Supp.

4928. STEINHEIL, Karl August. 1801-1870. Ger. DSB; ICB; P II,III; GC. Also Physics.

4929. TWINING, Alexander Catlin. 1801-1884. U.S.A. P III; GE.

4930. BOURIS, Georg Constantin. 1802-1860. Greece. P III.

4931. COULVIER-GRAVIER, Rémi Armand. 1802-1868. Fr. P III.

1. *Recherches sur les météores et sur les lois qui les régissent.* Paris, 1859. RM.

4932. GOLDSCHMIDT, Hermann. 1802-1866. Ger./Fr. P I,III; GA; GB; GC.

4933. OLUFSEN, Christian Friis Rottboll. 1802-1855. Den. DSB; P II.

4934. ANGER, Carl Theodor. 1803-1858. Danzig. P I,III. Also Mathematics.

4935. CHALLIS, James. 1803-1882. Br. DSB; P I,III; GD. Also Physics*.
1. *Lectures on practical astronomy and astronomical instruments.* Cambridge, 1879.

4936. DOPPLER, (Johann) Christian. 1803-1853. Aus. DSB; ICB; P I; GA; GC. Also Physics*.

4937. LUBBOCK, John William. 1803-1865. Br. DSB; P I,III; GA; GD.

4938. WOLFERS, Jacob Philipp. 1803-1878. Ger. P II,III; GC.
1. *Tabulae reductionum astronomicarum annis 1860 usque ad 1880 respondentes.* Berlin, 1858.

4939. BIRT, William Radcliff. 1804-1881. Br. DSB; P III. Also Meteorology.

4940. BREMIKER, Carl. 1804-1877. Ger. DSB; P I,III. Also Geodesy.

4941. BUSCH, August Ludwig. 1804-1855. Ger. DSB; P I.
1. *Reduction of the observations made by J. Bradley ... to determine the quantities of aberration and nutation.* Oxford, 1838.

4942. CLARK, Alvan. 1804-1887. U.S.A. DSB; P III,IV; GE.

4943. HODGSON, Richard. 1804-1872. Br. P III.

4944. JAHN, Gustav Adolph. 1804-1857. Ger. P I; GC.

4945. NICHOL, John Pringle. 1804-1859. Br. P II & Supp.; GD. Also Science in General*.
1. *The architecture of the heavens.* London, 1830. X. 9th ed., 1851.
2. *The phenomena and order of the solar system.* Edinburgh, 1838. X. 3rd ed., 1847.
3. *Thoughts on some important points relative to the system of the world.* Edinburgh, 1846.

4946. PETERSEN, Adolph Cornelius. 1804-1854. Ger. P II.

4947. SELANDER, Nils Haquin. 1804-1870. Swed. P II,III.

4948. TAYLOR, Thomas Glanville. 1804-1848. Br./India. P II; GD.

4949. NOBILE, Antonio. fl. 1834-1859. It. P IV.

4950. JOHNSON, Manuel John. 1805-1859. Br. DSB; P III; GD.

4951. LAMONT, Johann von. 1805-1879. Ger./Br. DSB; ICB; P I,III; GB; GC; GD. Also Earth Sciences*.

4952. SANG, Edward. 1805-1890. Br. P III,IV. Also Mechanics.

4953. VICO, Francesco de. 1805-1848. It. P II.

4954. WALKER, Sears Cook. 1805-1853. U.S.A. P II; GB; GE.

4955. ALEXANDER, Stephen. 1806-1883. U.S.A. P III; GE.

4956. COLLA, Antonio. ca. 1806-1857. It. P I.

4957. DE MORGAN, Augustus. 1806-1871. Br. DSB; ICB; P I (under D); P III (under M); GA; GB; GD. Also Mathematics*, Biography*, and History of Mathematics*. See also Index.
1. *The book of almanacs ... by which the almanac may be found for every year ... up to A.D. 2000. With means of finding the day of any new or full moon from B.C. 2000 to A.D. 2000.* London, 1851. 2nd ed., 1871.

4958. HARTNUP, John. 1806-1885. Br. P III.

4959. HEIS, Eduard. 1806-1877. Ger. ICB; P I,III.

4960. MAURY, Matthew Fontaine. 1806-1873. U.S.A. DSB; ICB; P II, III; GA; GB; GE. Also Earth Sciences*.

4961. PEDERSEN, Peder. 1806-1861. Den. P II,III.

4962. PETERS, Christian August Friedrich. 1806-1880. Ger. DSB; P II,III; GC.

4963. WEBB, Thomas William. 1806-1885. Br. ICB; P III; GD.
1. *Celestial objects for common telescopes.* London, 1859. X. 2nd ed., 1868.

4964. BÖHM, Joseph Georg. 1807-1868. Aus./Prague. P I,III.

4965. COOKE, Thomas. 1807-1868. Br. P III; GD.

4966. GOLDSCHMIDT, Carl Wolfgang Benjamin. 1807-1851. Ger. P I. Also Geomagnetism.

4967. HABICHT, Karl Wilhelm Eberhard. b. 1807. Ger. P I.

4968. KAISER (or KEYSER). Frederik. 1808-1872. Holl. DSB; P I,III.

4969. MAIN, Robert. 1808-1878. Br. P III; GD.
1. *Rudimentary astronomy.* London, 1852. 2nd ed., 1869. 3rd ed., 1882.
2. *Practical and spherical astronomy.* Cambridge, 1863.

4970. NASMYTH, James. 1808-1890. Br. DSB; P III; GB; GD.
1. *The moon considered as a planet, a world and a satellite.* London, 1874. With James Carpenter. 2nd ed., 1874. Another ed., 1885.

4971. PRITCHARD, Charles. 1808-1893. Br. DSB; P III,IV; GB; GD.

4972. RESLHUBER, (Augustin) Wolfgang. 1808-1875. Aus. P II,III; GC. Also Earth Sciences.

4973. DREW, John. 1809-1857. Br. P III; GD. Also Meteorology*.

4974. MAUVAIS, Félix Victor. 1809-1854. Fr. P II; GA.

4975. MITCHELL, Ormsby Macknight. 1809-1862. U.S.A. ICB; P III; GB; GE.
1. *The planetary and stellar worlds. A popular exposition of the great discoveries and theories of modern astronomy.* New York, 1848. X.
1a. ——— Another ed. *The orbs of heaven; or, The planetary and stellar worlds.* London, 1851. Later eds: 1852, 1859, 1860, 1862.

4976. PEIRCE, Benjamin. 1809-1880. U.S.A. DSB; ICB; P II,III; GB; GE. Also Mathematics.

4977. ABBADIE, Antoine d'. 1810-1897. Fr. ICB; P III; GB.

4978. NORTON, William Augustus. 1810-1883. U.S.A. P III.
 1. *An elementary treatise on astronomy.* New York, 1839. X.
 1a. ——— 4th ed. *A treatise on astronomy, spherical and practical.* New York, 1867.

4979. PETIT, Frédéric. 1810-1865. Fr. P II,III.

4980. SABLER, Georg Thomas. 1810-1865. Russ./Wilno. P II,III.

4981. SAWITSCH, Alexei Nikolaievich. 1810-1883. Russ./Ger. P II, III; GC.

4982. GILLISS, James Melville. 1811-1865. U.S.A. P I,III; GE. Also Earth Sciences*.

4983. HERRICK, Edward Claudius. 1811-1862. U.S.A. P III; BMNH; GE. Also Zoology.

4984. LE VERRIER, Urbain Jean Joseph. 1811-1877. Fr. DSB; ICB; P I,III; GA; GB. Also Meteorology.
 1. *Développements sur plusieurs points de la théorie des perturbations des planètes.* Paris, 1841. RM.
 2. *Théorie du mouvement de Mercure.* Paris, 1843. RM.
 3. *Recherches sur les mouvements de la planète Herschel.* Paris, 1846.
 4. *Examen de la discussion ... au sujet de la découverte de l'attraction universelle.* Paris, 1869.

4985. LITTROW, Karl Ludwig von. 1811-1877. Aus. P I,III; GC.

4986. LOOMIS, Elias. 1811-1889. U.S.A. DSB; P I,III,IV; GE. Also Earth Sciences.
 1. *Recent progress of astronomy, especially in the United States.* New York, 1850. X. Another ed., 1856.
 2. *An introduction to practical astronomy.* New York, 1855. Another ed., 1861.
 3. *A treatise on astronomy.* New York, 1865. X. Other eds: 1876, 1883, 1889.

4987. SOUSA PINTO, Rodrigo Ribeiro da. 1811-1893. Port. P III,IV.

4988. AGARDH, John Mortimer. 1812-1862. Swed. P I & Supp. Also Mathematics.

4989. DEMBOWSKI, Ercole. 1812-1881. It. DSB; P III.

4990. FLEMMING, Friedrich Wilhelm. 1812-1840. Danzig. P III.

4991. GALLE, Johann Gottfried. 1812-1910. Ger. DSB; ICB; P I,III,IV.
 1. *Verzeichniss der Elemente der bisher berechneten Cometenbahnen.* Leipzig, 1894. RM.

4992. KERZ, Ferdinand. 1812-1892. Ger. P III,IV. Also Mathematics.

4993. LAUGIER, (Paul Auguste) Ernst. 1812-1872. Fr. P I,III; GA.

4994. JACOB, William Stephen. 1813-1862. India. P III; GD.

4995. LUNDAHL, G. 1813-1844. Fin. P I.

4996. PETERS, Christian Heinrich Friedrich. 1813-1890. It./U.S.A. DSB; P II,III,IV; GE.

4997. SCHUBERT, Ernst. 1813-1873. Ger./U.S.A. P II,III.

4998. WACKERBARTH, Athanasius Frans Diderik. 1813-1884. Swed./Br. P III.

4999. YVON-VILLARCEAU, Antoine Joseph François. 1813-1883. Fr. P II,III.

5000. ÅNGSTRÖM, Anders Jonas. 1814-1874. Swed. DSB; P I,III; GB. Also Physics*.

5001. CACCIATORE, Gaetano. 1814-1889. It. P III.

5002. FAYE, Hervé. 1814-1902. Fr. DSB; P I,III,IV. Also Earth Sciences.

5003. GRANT, Robert. 1814-1892. Br. P III; GB; GD1. Also History of Astronomy*.

5004. GÜNTHER, Friedrich Wilhelm. 1814-1869. Ger. P I,III.

5005. KIRKWOOD, Daniel. 1814-1895. U.S.A. DSB; P III,IV; GE.
   1. *Meteoric astronomy: A treatise on shooting-stars, fireballs and aerolites.* Philadelphia, 1867.
   2. *Comets and meteors, their phenomena ... and the theory of their origin.* Philadelphia, 1873.

5006. LYMAN, Chester Smith. 1814-1890. U.S.A. DSB; ICB; P III; GE.

5007. MÖLLINGER, Otto. 1814-1886. Switz. P II,III. Also Mathematics.

5008. GROSSO, Remigio del. d. 1876. It. P III.

5009. BAXENDELL, Joseph. 1815-1887. Br. P III,IV; GD1. Also Meteorology.

5010. DE LA RUE, Warren. 1815-1889. Br. DSB; P III,IV; GB; GD. Also Physics and Chemistry.
   1. *Researches on solar physics.* 3 parts. London, 1865-68.

5011. DRACH, Salomon Moses. 1815-1879. Br. P III.

5012. GRAHAM, Andrew. 1815-1908. Br. P I,III.

5013. LIAGRE, Jean Baptiste Joseph. 1815-1891. Belg. P I,III.

5014. PLANTAMOUR, Emile. 1815-1882. Switz. P II,III. Also Earth Sciences.

5015. BIRMINGHAM, John. ca. 1816-1884. Br. DSB; P III,IV; GD.
   1. *The red stars: Observations and catalogue.* Dublin, 1890.

5016. DELAUNAY, Charles Eugène. 1816-1872. Fr. DSB; P I,III. Also Mechanics*.
   1. *Cours élémentaire d'astronomie.* Paris, 1853. X. 2nd ed., 1855. 4th ed., 1864.

5017. LUTHER, Eduard. 1816-1887. Ger. P I,III.

5018. RIDDLE, John. 1816-1862. Br. P III.

5019. RUTHERFURD, Lewis Morris. 1816-1892. U.S.A. DSB; P III; GE.

5020. SCHWEIZER, (Kaspar) Gottfried. 1816-1873. Ger./Russ. P II, III; GC.

5021. WOLF, (Johann) Rudolf. 1816-1893. Switz. DSB; ICB; P II,III, IV; GC. Also History of Astronomy*.

5022. YARNALL, Mordecai. 1816-1879. U.S.A. P III.

5023. MANN, William. 1817-1873. Cape Obs. P III; GD.

5024. MILLER, William Allen. 1817-1870. Br. DSB; ICB; P II,III; GA; GD. Also Chemistry*.

5025. PENROSE, Francis Cranmer. 1817-1903. Br. P III,IV; GD2.
1. *A method of predicting, by graphical construction, occultations of stars, etc.* London, 1869.

5026. POWALKY, Karl Rudolph. 1817-1881. Ger. DSB; P II,III.

5027. SCHAUB, Franz von. 1817-1871. Aus./Trieste. P II,III; GC.

5028. DESBOVES, Honoré Adolphe. b. 1818. Fr. P III,IV.

5029. FEARNLEY, Carl Frederic. 1818-1890. Nor. P III,IV.

5030. MITCHELL, Maria. 1818-1889. U.S.A. DSB; ICB; P III; GB; GE.

5031. SCHIDLOWSKI, Andrei. 1818-1894. Russ. P III.

5032. SECCHI, (Pietro) Angelo. 1818-1878. It. DSB; ICB; P II,III, IV; GB.
1. *Le soleil.* 4 vols. Paris, 1875-77. (Apparently a trans. from the Italian.)
2. *Le stelle.* Milan, 1877. X.
2a. ——— *Les étoiles: Essai d'astronomie sidérale.* 2 vols. Paris, 1879.

5033. WEYER, Georg Daniel Eduard. 1818-1896. Ger. P II,III,IV.

5034. ADAMS, John Couch. 1819-1892. Br. DSB; ICB; P I,III,IV; GB; GD1.
1. *An explanation of the observed irregularities in the motion of Uranus, on the hypothesis of disturbances caused by a more distant planet....* London, 1846. RM/L.
2. *The scientific papers.* 2 vols. Cambridge, 1896-1900. Ed. by W.G. Adams. RM/L.
3. *Lectures on the lunar theory.* Cambridge, 1900. Reprinted from item 2. Ed. by R.A. Sampson.

5035. ÅSTRAND, Johan Julius. 1819-1900. Nor. P III,IV.

5036. BRORSEN, Theodor J.C.A. 1819-1895. Bohemia. P I,III.

5037. GASPARIS, Annibale de. 1819-1892. It. ICB; P I,III,IV.

5038. KEY, Henry Cooper. 1819-1879. Br. P III.

5039. LINDHAGEN, Daniel Georg. 1819-1906. Russ./Swed. P I,III,IV.

5040. POWELL, Eyre Burton. 1819-after 1875. India. P III,IV Supp.

5041. SMYTH, (Charles) Piazzi. 1819-1900. Br. DSB; P II,III,IV; GB; GD1. Also Meteorology.
1. *Teneriffe: An astronomer's experiment.* London, 1858. RM/L.
2. *Madeira spectroscopic, being a revision of 21 places in the red half of the solar visible spectrum.* Edinburgh, 1882. RM/L.

5042. STRUVE, Otto Wilhelm. 1819-1905. Russ. DSB; P II,III,IV. Also Geodesy.
1. *Librorum in bibliotheca speculae Pulcovensis contentorum catalogus systematicus.* St. Petersburg, 1860. Supp., 1880.

5043. SONNTAG, August. d. 1861. Ger. P III.

5044. CHAUVENET, William. 1820-1870. U.S.A. ICB; P I,III; GE.
1. *A manual of spherical and practical astronomy*. 2 vols. Philadelphia, 1863. Another ed., 1864. 4th ed., London, 1868. 5th ed., Philadelphia, 1874. Other eds: 1887, 1891.

5045. CORNELIUS, Carl Sebastian. 1820-1896. Ger. P I,III,IV; GC. Also Physics*.
1. *Über die Entstehung der Welt*. Halle, 1870. RM.

5046. DÖLLEN, Johann Heinrich Wilhelm. 1820-1897. P I,III,IV.

5047. HOUZEAU (DE LA HAYE), Jean Charles. 1820-1888. Belg. ICB; P I,III,IV.
1. *Bibliographie générale de l'astronomie jusqu'en 1880*. 2 vols in 3. Brussels, 1882-89. With A. Lancaster. RM/L.
1a. ——— Another ed., with introd. etc. in English, London, 1964. By D.W. Dewhirst.

5048. PUISEUX, Victor Alexandre. 1820-1883. Fr. DSB; ICB; P II,III; GA. Also Mathematics.

5049. ROCHE, Edouard Albert. 1820-1883. Fr. DSB; P II,III. Also Earth Sciences.

5050. SWIFT, Lewis. 1820-1913. U.S.A. ICB; P III; GE.

5051. BRÜNNOW, Franz Friedrich Ernst. 1821-1891. Ger./U.S.A./Br. P I,III; GB.
1. *Lehrbuch der sphärischen Astronomie*. Berlin, 1851.
1a. ——— *Spherical astronomy*. London, 1865.
See also 4768/1a.

5052. BURR, Thomas William. 1821-1874. Br. P III.

5053. DUNKIN, Edwin. 1821-1898. Br. P III. Also Biography*.

5054. FRIESACH, Carl. 1821-1891. Aus. P III,IV.

5055. HAUGHTON, Samuel. 1821-1897. Br. P III,IV; BMNH; Mort.; GB; GD1. Other entries: see 3336.
1. *Manual of astronomy*. London, 1855. X. 2nd ed., 1857.

5056. KOVALSKY, Marian Albertovich. 1821-1884. Russ. DSB; ICB; P I, III (KOWALSKI).

5057. MOUCHEZ, Ernest Amédée Barthélémy. 1821-1852. Fr. DSB; P III, IV; GB.

5058. PRAZMOWSKY, Adam. 1821-after 1881. Warsaw/Fr. P II,III.

5059. PRINCE, Charles Leeson, 1821-1899. Br. P III. See also 83/1.

5060. SAPORETTI, Antonio. 1821-1900. It. P III,IV.

5061. SEIDEL, Philipp Ludwig von. 1821-1896. Ger. DSB; P II,III,IV; GC. Also Mathematics.

5062. TEMPEL, Ernst Wilhelm Leberecht. 1821-1889. Fr./It. P II,III, IV.

5063. WICHMANN, Moritz Ludwig Georg. 1821-1859. Ger. P II,III; GC.

5064. ZECH, Julius (August Christoph). 1821-1864. Ger. P II,III; GC.

5065. ARREST, Heinrich Louis d'. 1822-1875. Ger./Den. DSB; P I,III.

5066. DUBOIS, Edmond Paulin. 1822-1891. Fr. P III,IV.

5067. GAUTIER, Etienne Alfrède Emile. 1822-1891. Switz. P I,III,IV.

5068. LUTHER, (Karl Theodor) Robert. 1822-1900. Ger. DSB; P I,III, IV.

5069. PIHL, Olaf Andreas Löwald. 1822-1895. Nor. P III,IV.

5070. SCHWARZ, (Peter Karl) Ludwig. 1822-1894. Dorpat. P III,IV.

5071. SMALLEY, G.R. 1822-1870. Australia. P III.

5072. SPÖRER, Gustav Friedrich Wilhelm. 1822-1895. Ger. DSB; P II, III,IV.

5073. TRETTENERO, Virgilio. ca. 1822-1863. It. P III.

5074. CHACORNAC, Jean. 1823-1873. Fr. P I,III.

5075. GORTON, Sąndford. ca. 1823-1879. Br. P III.

5076. GOUJON, Jean Jacques Emile. 1823-1856. Fr. P I; GA.

5077. HIND, John Russell. 1823-1895. Br. DSB; P I,III,IV; GA; GD.
1. *An introduction to astronomy*. London, 1852. X. 2nd ed., 1863. 3rd ed., 1871.
2. *The comets: A descriptive treatise*. London, 1852.
3. *An astronomical vocabulary*. London, 1852.
4. *The comet of 1856*. London, 1857.

5078. HUBBARD, Joseph Stillman. 1823-1863. U.S.A. P I,III; GE.

5079. LESPIAULT, Frédéric Gaston. 1823-1904. Fr. P III,IV.

5080. PRITCHETT, Carr Waller. 1823-1910. U.S.A. P III.

5081. SCHULTZ, (Per Magnus) Herman. 1823-1890. Swed. P II,III,IV.

5082. FEDORENKO, Ivan Ivanovich. 1824-1888. Russ. P III,IV.

5083. GOULD, Benjamin Apthorp. 1824-1896. U.S.A./Argentina. DSB; ICB; P I,III,IV; GB; GE.

5084. HORNSTEIN, Karl. 1824-1882. Aus./Prague. P I,III,IV.

5085. HUGGINS, William. 1824-1910. Br. DSB; P III,IV; GB; GD2.
1. *Publications of Sir William Huggins's observatory*. 2 vols. London, 1899-1909. Ed. by Sir William and Lady Huggins. The two vols are sometimes catalogued individually.
1a. ——— Vol. I. *Atlas of representative stellar spectra*. 1899. Contains a history and description of the observatory.
1b. ——— Vol. II. *The scientific papers of Sir William Huggins*. 1909. RM.

See also 6438/1a.

5086. JANSSEN, Pierre Jules César. 1824-1907. Fr. DSB; ICB; P III, IV; GB.

5087. NELL, Adam Maximilian. 1824-1901. Ger. P II,III,IV.

5088. OELTZEN, Wilhelm Albrecht. 1824-after 1874. Aus./Fr. P II,III.

5089. RESPIGHI, Lorenzo. 1824-1889. It. DSB; P III,IV.

5090. STRASSER, Gabriel. 1824-1882. Aus. P II,III.

5091. TISSOT, Nicolas Auguste. 1824-after 1894. Fr. P III,IV.

5092. BOND, George Phillips. 1825-1865. U.S.A. DSB; P III; GE.

5093. DAVIDSON, George. 1825-1911. U.S.A. ICB; P III,IV; GE.

5094. DORNA, Alessandro. 1825-1886. It. P III,IV.

5095. MOESTA, Karl Wilhelm (or Carlos Guillelmo). 1825-1884. Ger./Chile. P II,III.

5096. QUETELET, Ernest Adolphe François. 1825-1878. Belg. ICB; P III; GF. Also Earth Sciences.

5097. ROSA, Paolo. 1825-1874. It. P III.

5098. SCHMIDT, (Johann Friedrich) Julius. 1825-1884. Ger./Greece. DSB; ICB; P II,III,IV; BMNH; GC. Also Geology.
1. *Karte der Gebirge des Mondes*. 2 vols. Berlin, 1878.

5099. SCOTT, William. 1825-after 1881. Australia. P III.

5100. SIMON, Charles Marie Etienne Théophile. 1825-1880. Fr. P III.

5101. BREEN, James. 1826-1866. Br. P III; GD.

5102. BROTHERS, Alfred. 1826-after 1881. Br. P III.

5103. CARRINGTON, Richard Christopher. 1826-1875. Br. DSB; ICB; P III; GB; GD.
1. *Observations of the spots on the sun*. London, 1863. RM/L.

5104. DONATI, Giovan Battista. 1826-1873. It. DSB; ICB; P III; GB.

5105. GUILLEMIN, Amédée Victor. 1826-1893. Fr. P III. Also Physics*.
1. *Causeries astronomiques*. Paris, 1861. X. 4th ed., 1865.
2. *Le ciel*. Paris, 1864.
2a. ——— *The heavens: An illustrated handbook of popular astronomy*. London, 1866. Ed. by J.N. Lockyer. 2nd ed., 1867. 7th ed., 1878.
3.*Le soleil*. Paris, 1869. X.
3a. ——— *The sun*. London, 1870. Trans. by T.L. Phipson.
4.*Les comètes*. Paris, 1875.
4a. ——— *The world of comets*. London, 1877. Trans. and ed. by J. Glaisher.

5106. GUSSEW, Matvei Matveievich. 1826-1866. Russ. P III.

5107. LIAIS, Emmanuel. 1826-1900. Fr./Brazil. P III,IV. Also Earth Sciences*.

5108. STONEY, George Johnstone. 1826-1911. Br. DSB; P IV; GD2. Also Physics.

5109. TODD, Charles. 1826-1910. Br./Australia. P III,IV; GD.

5110. WINLOCK, Joseph. 1826-1875. U.S.A. DSB; P II,III; GE.

5111. BOGUSLAWSKI, Georg Heinrich von. 1827-1884. Ger. P I,III. Also Earth Sciences*.

5112. ELLERY, Robert Lewis John. 1827-1908. Australia. P III,IV & Supp.; GD2.

5113. ERCK, Wentworth. 1827-1890. Br. P IV.

5114. FLETCHER, Isaak. 1827-1879. Br. P III.

5115. KLINKERFUES, (Ernest Friedrich) Wilhelm. 1827-1884. Ger. P I, III,IV; GC.
    1. *Theoretische Astronomie*. Brunswick, 1872.

5116. LINDELÖF, Lorentz Leonard. 1827-1908. Fin. P III,IV. Also Mathematics.

5117. OUDEMANS, Jean Abraham Chrétien. 1827-1906. Holl./Java. P II, III,IV; GF. Also Geodesy.

5118. SCHJELLERUP, Hans Carl Frederik Christian. 1827-1887. Den. DSB; P III.

5119. SCHUMACHER, Richard. 1827-1902. Ger./Chile. P IV.

5120. TROUVELOT, Etienne Léopold. 1827-1895. Fr./U.S.A. DSB; P III, IV. Also Zoology.

5121. WEILER, Johann August. 1827-1911. Ger. P II,III,IV.

5122. WOLF, Charles Joseph Etienne. 1827-1918. Fr. DSB; P III,IV. Also History of Science.
    1. *Astronomie et géodésie: Cours professé à la Sorbonne*. Paris, 1891.

5123. WOLFF, Julius Theodor. 1827-1896. Ger. P III,IV.

5124. ELLIS, William. 1828-1916. Br. DSB; P III,IV. Also Earth Sciences.

5125. MARTH, Albert. 1828-1897. Br. P II,III,IV.

5126. NOBLE, William. 1828-after 1903. Br. P III,IV.

5127. RESAL, Henri Amé. 1828-1896. Fr. P II,III,IV. Also Mechanics*.
    1. *Traité élémentaire de mécanique céleste*. Paris, 1875. 2nd ed., 1884.

5128. SCHÖNFELD, Eduard. 1828-1891. Ger. DSB; P II,III; GB.

5129. SOUILLART, Cyrille. 1828-1898. Fr. P III,IV.

5130. WAGNER, August. 1828-1886. Russ. P III,IV.

5131. CAPRON, John Rand. 1829-after 1881. Br. P III.
    1. *Photographs of metallic etc. spectra*. London/New York, 1877.
    2. *Aurorae, their characters and spectra*. London, 1879.

5132. HALL, Asaph. 1829-1907. U.S.A. DSB; P III,IV; GE.

5133. HARTWIG, Ernst Wilhelm. 1829-after 1902. Ger. P I,III,IV.

5134. POGSON, Norman Robert. 1829-1891. Br./India. ICB; P II,III, IV; GD; GF.

5135. ROBERTS, Isaac. 1829-1904. Br. DSB; P IV; GD2.

5136. ŠAFAŘIK (or SCHAFARIK), Adalbert. 1829-1902. Aus./Prague. P III,IV.

5137. STANLEY, William Ford Robinson. 1829-1909. Br. GD2. Also Mechanics*.
    1. *Notes on the nebular theory*. London, 1895.

5138. TENNANT, James Francis. 1829-1915. Br./India. P III,IV.

5139. THOLLON, Louis. 1829-1887. Fr. DSB; P III,IV.

5140. TROWBRIDGE, David. 1829-after 1880. U.S.A. P III.

5141. VOGEL, Eduard. 1829-1856. Ger./Br. P II; GB; GC.

5142. BRUHNS, Karl Christian. 1830-1881. Ger. DSB; P I,III; GC. Also Biography* and History of Astronomy*.

5143. FERGOLA, Emmanuele. 1830-1915. It. P III,IV.

5144. HIRSCH, Adolph. 1830-1901. Switz. P III,IV. Also Earth Sciences.

5145. KARLINSKI, Michal Franciszek. 1830-1906. Cracow. P I,III. Also Meteorology.

5146. KAYSER, Ernst. 1830-1907. Ger. P III,IV.

5147. LESSER, Otto Leberecht. 1830-1887. Ger. P I,III,IV.

5148. MÖLLER, Diderik Magnus Axel. 1830-1896. Swed. DSB; P III,IV.

5149. NEWTON, Hubert Anson. 1830-1896. U.S.A. DSB; P III,IV; GE.

5150. ZENGER, Karel Václav. 1830-1908. Prague. ICB; P IV. Also Meteorology.

5151. BRAUN, Karl. 1831-1907. Ger./It./Hung. P III,IV.

5152. BREDIKHIN, Feodor Aleksandrovich. 1831-1904. Russ. DSB; ICB; P III,IV (BREDICHIN).

5153. BRETT, John. 1831-1902. Br. P III; GD2.

5154. LOEWY, Benjamin. 1831-after 1877. Br. P III.

5155. MASTERMAN, Stillman. 1831-1863. U.S.A. P III.

5156. SIDLER, Georg Joseph. 1831-1907. Switz. P II,III,IV.

5157. STONE, Edward James. 1831-1897. Br. P III,IV; GB; GD.

5158. WHITE, Edward John. 1831-after 1881. Australia. P III.

5159. BOUR, Edmond. 1832-1866. Fr. DSB; P III. Also Mathematics and Mechanics.

5160. CLARK, Alvan Graham. 1832-1897. U.S.A. DSB; P III,IV; GE.

5161. FÖRSTER, Wihelm Julius. 1832-1921. Ger. P I,III.

5162. KRUEGER, (Karl Nicolaus) Adalbert. 1832-1896. Ger./Fin. P I, III,IV.

5163. ROGERS, William Augustus. 1832-1898. U.S.A. P III,IV; GE.

5164. RÜMKER, Georg Friedrich Wilhelm. 1832-1900. Ger. P II,III,IV.

5165. STEINHEIL, Hugo Adolph. 1832-1893. Ger. P III,IV.

5166. STOCKWELL, John Nelson. 1832-1920. U.S.A. P III,IV; GE.

5167. TUCKER, Robert. 1832-1905. Br. P III,IV. Also Mathematics.

5168. FOLIE, François Jacques Philippe. 1833-1905. Belg. ICB; P III,IV.

5169. KOWALCZYK, Johann. 1833-1911. Warsaw. P III,IV.

5170. LOEWY, Maurice. 1833-1907. Aus./Fr. P III,IV; GF.

5171. PERRY, Stephen Joseph. 1833-1889. Br. P III,IV; GD.

5172. GAILLOT, Aimable Jean Baptiste. 1834-1921. Fr. DSB; P IV.

5173. HOEK, Martinus. 1834-1873. Holl. DSB; P I,III.

5174. LANGLEY, Samuel Pierpoint. 1834-1906. U.S.A. DSB; ICB; P III, IV; GB; GE. Also Physics* and Biography*.
 1. *The new astronomy.* Boston, 1888. X. Another ed., 1896.

5175. PAPE, Karl Ferdinand. 1834-1862. Ger. P II,III.

5176. TEBBUTT, John. 1834-1917. Australia. P III,IV; GF.
 1. *Astronomical memoirs, 1853 to 1907.* Sydney, 1908.

5177. TIETJEN, Friedrich. 1834-1895. Ger. P III,IV.

5178. YOUNG, Charles Augustus. 1834-1908. U.S.A. DSB; P III,IV; GE.
 1. *The sun.* New York, 1881. 2nd ed., 1883. 3rd ed., 1888.
 2. *A text-book of general astronomy.* Boston, 1888. X. Another ed., 1899.
 3. *Elements of astronomy.* Boston, 1890. X. Another ed., 1896.

5179. ZÖLLNER, (Johann Karl) Friedrich. 1834-1882. Ger. DSB; P II, III,IV; GB; GC.

5180. KNOTT, George. 1835-1894. Br. P III.

5181. LYNN, William Thynne. 1835-1911. Br. P III.
 1. *Celestial motions.* London, 1884. X. 8th ed., 1894.
 2. *Remarkable comets: A brief survey of the most interesting facts in the history of cometary astronomy.* London, 1893. X. 2nd ed., 1894.
 3. *Remarkable eclipses.* London, 1896.

5182. NEWCOMB, Simon. 1835-1909. U.S.A. DSB; ICB; P III,IV; GB; GE. Also Autobiography*.
 1. *Popular astronomy.* London, 1878. 2nd ed., 1883.
 2. *Researches on the motion of the moon.* Washington, 1878.
 3. *On the recurrence of solar eclipses.* Washington, 1879.
 4. *The elements of the four inner planets and the fundamental constants of astronomy.* 1895.

5183. RADAU, (Jean Charles) Rodolphe. 1835-1911. Fr. P III,IV.

5184. SCHIAPARELLI, Giovanni Virginio. 1835-1910. It. DSB; ICB; P III,IV; GB. Also History of Astronomy*.
 1. *Osservazioni astronomiche e fisiche sull'asse di rotazione e sulla topografia del pianeta Marte.* 6 vols. Rome, 1896-99.
 2. *Opere.* 11 vols. Milan, 1929-43. Ed. by the Specola di Brera. Reprinted, New York, 1968.

5185. WINNECKE, (Friedrich) August (Theodor). 1835-1897. Ger./Russ. P II,III,IV.
 1. *Über die Sonne.* St. Petersburg, 1861.

5186. EASTMAN, John Robie. 1836-1913. U.S.A. P III,IV; GE.

5187. GLEDHILL, Joseph. 1836-1906. Br. P III,IV. See also 5246/1.

5188. HERSCHEL, Alexander Stewart. 1836-1907. Br. P III; GD2.

5189. HOUGH, George Washington. 1836-1909. U.S.A. DSB; P III,IV; GE.

5190. KAM, Nicolaas Matteus. 1836-1896. Holl. P IV.

5191. LOCKYER, Joseph Norman. 1836-1920. Br. DSB; ICB; P III,IV; GB; GD3. Also Physics* and History of Astronomy*.
1. *Elementary lessons in astronomy*. London. 1868. X. Another ed., 1874. 6th ed., 1876. Other eds: 1879, 1883, 1892.
2. *Contributions to solar physics*. London, 1874.
3. *A primer on astronomy*. London, 1884.
4. *The chemistry of the sun*. London, 1887. RM/L.
5. *The meteoritic hypothesis: A statement of the results of a spectroscopic inquiry into the origin of cosmical systems*. London, 1890. RM/L.
6. *The sun's place in nature*. London, 1897.
7. *Inorganic evolution as studied by spectrum analysis*. London, 1900.

See also 5105/2a.

5192. ROMBERG, Hermann. 1836-1898. Ger./Russ. P III,IV.

5193. RUSSELL, Henry Chamberlaine. 1836-1907. Australia. DSB Supp.; P III,IV; GD2. Also Meteorology.

5194. SAFFORD, Truman Henry. 1836-1901. U.S.A. P III,IV; GE.

5195. WILSON, James Maurice. 1836-1931. Br. P III; GD5. See also 5246/1.

5196. CHANDRIKOW, Mitrofan Fedorovich. 1837-after 1902. Russ. P IV.

5197. COPELAND, Ralph. 1837-1905. Br. P III,IV; GD2.

5198. DRAPER, Henry. 1837-1882. U.S.A. DSB; P III,IV; GE.

5199. FRISBY, Edgar. 1837-after 1902. U.S.A. P III,IV.

5200. FRISCHAUF, Johannes. 1837-1924. Aus. P III,IV. Also Mathematics*.

5201. GRUEY, Louis Jules. 1837-1902. Fr. P III,IV.

5202. HARKNESS, William. 1837-1903. U.S.A. DSB; P III,IV & Supp.; GE.
1. *The solar parallax and its related constants*. 1891.

5203. HERSCHEL, John. 1837-after 1883. India. P III.

5204. KORTAZZI, Ivan Jegorovich. 1837-1903. Russ. P III,IV.

5205. LINSSER, Carl. 1837-1869. Russ. P III.

5206. MacCLEAN, Frank. 1837-1904. Br. P IV; GD2.

5207. MURMANN, August. 1837-1872. Prague. P III.

5208. NYRÉN, Magnus. 1837-1921. Russ. P III,IV.

5209. PROCTOR, Richard Antony. 1837-1888. Br. DSB; P III,IV; GB; GD. Also Science in General*.
The following works were all published at London.
1. *Saturn and his system*. 1865.
2. *The stars in twelve maps on the gnomonic projection*. 1866.
3. *Constellation seasons*. 1867.
4. *Sun views of the earth*. 1867.
5. *Half-hours with the telescope*. 1868.
6. *Star atlas*. 1870. X. 2nd ed., 1870. 3rd ed., entitled *New star atlas*, 1873. Another ed., 1882.

7. *Other worlds than ours.* 1870. Another ed., 1872.
8. *The sun.* 1871.
9. *Orbs around us.* 1872.
10. *Essays on astronomy.* 1872.
11. *The moon.* 1873. 2nd ed., 1878.
12. *Transits of Venus.* 1874.
13. *The universe and the coming transits.* 1874.
14. *Our place among infinities.* 1875. Another ed., 1880.
15. *The universe of stars.* 1876? X. 2nd ed., 1878.
16. *A treatise on the cycloid and all forms of cycloidal curves and on their use in astronomy.* 1878.
17. *Flowers of the sky.* 1879.
18. *The expanse of heaven.* 1880.
19. *Myths and marvels of astronomy.* 1880.
20. *The poetry of astronomy.* 1881.
21. *Easy star lessons.* 1881.
22. *Studies of Venus transits.* 1882.
23. *Mysteries of time and space.* 1883.
24. *The universe of suns.* 1884.
25. *Southern skies.* 1889.
26. *Old and new astronomy.* 1892.

5210. REMEIS, Karl. 1837-1882. Ger. P III.

5211. SEARLE, Arthur. 1837-1920. U.S.A. P III,IV; GE.

5212. SIDGREAVES, Walter. 1837-1919. Br. P IV.

5213. STEPHAN. Edouard Jean Marie. 1837-1923. Fr. DSB; P III,IV.

5214. WEISS, Edmund. 1837-1917. Aus. DSB; P II,III,IV & Supp.
1. *Bilder-Atlas der Sternenwelt.* Esslingen, 1888.

5215. ADOLPH, Gottfried Wilhelm Carl. 1838-1890. Ger. P III.

5216. AUWERS, Arthur (Julius Georg Friedrich) von. 1838-1915. Ger. DSB; P III,IV.
1. *Neue Reduktion der Bradley'schen Beobachtungen aus den Jahren 1750 bis 1762.* 3 vols. St. Petersburg, 1882-1903.
2. *Fundamental-Catalogue für Zonenbeobachtungen am Südhimmel und südlicher Polar-Catalog für die Epoche 1900.* Berlin, 1897.

See also 3645/9.

5217. BAKHUIJZEN VAN DE SANDE, Henricus Gerardus. 1838-1923. Holl. P III (under B); P IV (under VAN DE SANDE).

5218. BURNHAM, Sherburne Wesley. 1838-1921. U.S.A. DSB; P III,IV; GE.

5219. CHEYNE, Charles Hartwell Horne. 1838-1877. Br. P III.
1. *An elementary treatise on the planetary theory.* London, 1862. X. 3rd ed., 1883.
2. *The earth's motion of rotation.* London, 1867.

5220. ELGER, Thomas Gwyn. 1838-1897. ICB.
1. *The moon: A full description and map.* London, 1895.

5221. FREEMAN, Alexander. 1838-after 1883. Br. P III. See also 6121/1a.

5222. HILL, George William. 1838-1914. U.S.A. DSB; ICB; P III,IV; GE.

1. *The collected mathematical works*. 4 vols. Washington, 1905-07. Reprinted, New York, 1965. Mathematical astronomy.

5223. JESSE, Otto. 1838-1901. Ger. P III,IV.

5224. NOBILE, Arminio. 1838-1897. It. P III,IV.

5225. TACCHINI, Pietro. 1838-1905. It. DSB; P III,IV. Also Earth Sciences.

5226. THIELE, Thorvald Nicolai. 1838-1910. Den. DSB; P III,IV. Also Mathematics.

5227. WATSON, James Craig. 1838-1880. U.S.A. P III; GE.
 1. *Theoretical astronomy relating to the motions of the heavenly bodies*. Philadelphia, 1869. Other eds: 1877 and 1881.

5228. DUNÉR, Nils Christofer. 1839-1914. Swed. DSB; P III,IV.

5229. FUSS, Victor Friedrich. 1839-after 1901. Russ. P III,IV.

5230. NIESSL VON MAYENDORF, Gustav. 1839-1919. Brünn. P III,IV.

5231. RAYET, Georges Antoine Pons. 1839-1906. Fr. DSB; ICB; P III,IV.

5232. SEARLE, George. 1839-after 1897. U.S.A. P IV.

5233. WIERZBICKI, Daniel Joseph. 1839-1901. Cracow. P III,IV.

5234. BALL, Robert Stawell. 1840-1913. Br. P III,IV; GD3. Also Mechanics*, Autobiography*, and Biography*.
 1. *Astronomy*. London, 1877.
 2. *Elements of astronomy*. Ib., 1880. 2nd ed., 1896.
 3. *The story of the heavens*. Ib., 1885. 2nd ed., 1886.
 4. *Time and tide*. Ib., 1889.
 5. *The cause of an ice age*. Ib., 1891. RM/L. 2nd ed., 1892.
 6. *An atlas of astronomy*. Ib., 1892.
 7. *The story of the sun*. Ib., 1893. 2nd ed., 1897.
 8. *In the high heavens*. Ib., 1893. Another ed., 1895.
 9. *In starry realms*. Ib., 1893.

5235. BRASHEAR, John Alfred. 1840-1920. U.S.A. DSB; ICB; GE.

5236. CARPENTER, James. 1840-1899. Br. P IV. See also 4970/1.

5237. DE WORMS, Henry, *Baron* PIRBRIGHT. 1840-1903. Br. GD2.
 1. *The earth and its mechanisms, being an account of the various proofs of the rotation of the earth*. London, 1862.

5238. HATT, Philippe. 1840-1918. Fr. P IV.

5239. KNORRE, Victor Carl. 1840-1919. Ger. P III,IV.

5240. PARSONS, Laurence, *4th Earl of* ROSSE. 1840-1908. Br. P III, IV (ROSSE); GD2.

5241. SCHMIDT, Carl August von. 1840-1929. Ger. DSB; P III,IV. Also Geophysics.

5242. TALMAGE, Charles George. 1840-1886. Br. P III.

5243. WOSTOKOW, Ivan Anatolievich. 1840-after 1883. Russ. P III.

5244. CHAMBERS, George Frederick. 1841-after 1881. Br. P III. Also Meteorology*.

1. *Handbook of descriptive and practical astronomy.* Oxford, 1867. 3rd ed., 1877. 4th ed., 3 vols, 1889-90.
2. *Cycle of celestial objects.* Oxford, 1881.
3. *Pictorial astronomy.* 1891.
4. *The story of the solar system.* London, 1895.
5. *The story of the stars.* New York, 1895.

See also 4859/1.

5245. COMMON, Andrew Ainslie. 1841-1903. Br. DSB; P IV & Supp.; GD2.

5246. CROSSLEY, Edward. 1841-1904. Br. P III,IV.
1. *Handbook of double stars.* 1879. With J. Gledhill and J.M. Wilson.

5247. ENGELMANN, Friedrich Wilhelm Rudolph. 1841-1888. Ger. P III, IV. See also 4841/4.

5248. GYLDÉN, (Johan August) Hugo. 1841-1896. Swed. P III,IV.
1. *Traité analytique des orbites absolues des huit planètes principales.* 2 vols. 1893-1908.

5249. KNOBEL, Edward Ball. 1841-1930. Br. ICB; P III.

5250. LEVEAU, Gustave. 1841-1911. Fr. P III,IV.

5251. OPPOLZER, Theodor von. 1841-1886. Aus. DSB; ICB; P III,IV; GC. Also Geodesy.
1. *Lehrbuch zur Bahnbestimmung der Cometen und Planeten.* 2 vols. Leipzig, 1870-80. X. 2nd ed., 1882.
1a. ——— *Traité de la détermination des orbites des comètes et des planètes.* 2 vols. Paris. 1886. Trans. from the 2nd German ed. by E. Pasquier.
2. *Canon der Finsternisse.* Vienna, 1887.
2a. ——— *Canon of eclipses.* New York, 1962. Trans. by O. Gingerich and D.H. Menzel.

5252. VOGEL, Hermann Carl. 1841-1907. Ger. DSB; P III,IV.

5253. ANDRÉ, Charles (Louis François). 1842-1912. Fr. DSB; P III, IV. Also Meteorology and History of Astronomy*.

5254. ASTEN, Friedrich Emil von. 1842-1878. Ger./Russ. P III.

5255. BERG, Friedrich Wilhelm Julius von. 1842-after 1902. Wilno/ Warsaw. P III,IV.

5256. BICKERTON, Alexander William. 1842-1929. Br./New Zealand. DSB.

5257. BORRELLY, Alphonse Louis Nicolas. 1842-1926. Fr. P III,IV.

5258. CELORIA, Giovanni. 1842-1920. It. P III,IV.

5259. CLERKE, Agnes Mary. 1842-1907. Br. GB; GD2. Also History of Astronomy*.
1. *The system of the stars.* London, 1890.
2. *The "Concise Knowledge" astronomy.* London, 1898.

5260. FLAMMARION, (Nicholas) Camille. 1842-1925. Fr. DSB; ICB; P III,IV. See also 15813/1.
1. *La pluralité des mondes habités.* Paris, 1862. X. 16th ed., 1871.
2. *Les mondes imaginaires et les mondes réels, voyage astronomique pittoresque.* Paris, 1865. X. 9th ed., 1870.

3. *Histoire du ciel*. Paris, 1867. X. Another ed., 1872.
4. *Contemplations scientifiques*. Paris, 1869.
5. *L'atmosphère*. Paris, 1871. X.
5a. ——— *The atmosphere*. London, 1873.
6. *Astronomie populaire*. Paris, 1880. One of the most successful of all scientific popularisations. Many editions and translations.
7. *Les étoiles et les curiosités du ciel*. Paris, 1882.
8. *La planète Mars et ses conditions d'habitabilité*. 2 vols. Paris, 1892-1902.

5261. GAUTIER, Paul Ferdinand. 1842-1909. Fr. DSB.

5262. KONKOLY THEGE, Miklós von. 1842-1916. Hung. DSB; P III,IV (THEGE *Edler von* KONKOLY, Nicolaus).

5263. LAMEY, Moyeul Charles. 1842-after 1902. Fr. P IV.

5264. LINDEMANN, Eduard. 1842-1897. Russ. P III,IV.

5265. WINKLER, Karl Wilhelm. 1842-1910. Ger. P IV.

5266. ABNEY, William de Wiveleslie. 1843-1920. Br. DSB; P III,IV; GD3. Also Physics*.

5267. ANTON, Ferdinand. 1843-1900. Trieste. P IV & Supp.

5268. BECKER, Ernest Emil Hugo. 1843-1912. Ger. P III,IV.

5269. BEHRMANN, C. 1843-after 1874. Ger. P III.

5270. DAVIS, Arthur Sladen. 1843-after 1881. Br. P III.

5271. DOOLITTLE, Charles Leander. 1843-1919. U.S.A. P IV & Supp.; GE.
1. *A treatise on practical astronomy as applied to geodesy and navigation*. New York, 1885. X. 2nd ed., 1888.

5272. GILL, David. 1843-1914. Br./Cape Obs. DSB; ICB; P III,IV; GD3.

5273. LORENZONI, Giuseppe. 1843-1914. It. DSB; P III,IV. Also Geodesy.

5274. PECHÜLE, Carl Frederik. 1843-1914. Den. P III,IV.

5275. THRÄN, Carl Anton. 1843-after 1902. Ger. P IV.

5276. ARCIMIS ET WERLE, Augusto. 1844-1910. Spain. P III,IV.

5277. BROOKS, William Robert. 1844-1921. U.S.A. DSB; ICB; GE.

5278. FIEVEZ, Charles Jean Baptiste. 1844-1890. Belg. P III,IV.

5279. GEELMUYDEN, Hans. 1844-1920. Nor. P III,IV.

5380. GRUBB, Howard. 1844-1931. Br. DSB; ICB; P III,IV.

5281. KLEIN, Hermann Joseph. 1844-1914. Ger. DSB; P III,IV. Also Meteorology.
1. *Sternatlas*. Leipzig, 1886. X.
1a. ——— *Star atlas*. 1888. 2nd ed., 1893.

5282. NIESTEN, Jean Louis Nicolas. 1844-1920. Belg. DSB; P IV.

5283. PETERS, Carl Friedrich Wilhelm. 1844-1894. Ger. DSB; P III,IV.

5284. RICCÒ, Annibale. 1844-1919. It. DSB; P III,IV. Also Geophysics.

5285. CHRISTIE, William Henry Mahony. 1845-1922. Br. DSB; ICB; P III, IV; GD4.

5286. DARWIN, George Howard. 1845-1912. Br. DSB; ICB; P III,IV; GD3. Also Geophysics.
   1. *The tides and kindred phenomena in the solar system.* London, 1898. RM/L.
   2. *Scientific papers.* 5 vols. Cambridge, 1907-16. RM/L.

5287. FABRITIUS, W. 1845-1895. Russ. P III,IV.

5288. FINCK, Julius (or Fényi). 1845-1927. Hung. P IV.

5289. GORE, John Ellard. 1845-1910. Br. BMNH; GD2.
   1. *Southern stellar objects.* London, 1877.
   2. *Planetary and stellar studies.* Ib., 1888.
   3. *The scenery of the heavens.* Ib., 1890. 2nd ed., 1892.
   4. *Star groups.* Ib., 1891.
   5. *An astronomical glossary.* Ib., 1893.
   6. *The worlds of space.* Ib., 1894.

5290. HALL, Maxwell. 1845-after 1902. Jamaica. P III,IV.

5291. JOHNSON, Samuel Jenkins. 1845-1905. Br. P IV.
   1. *Eclipses past and future.* Oxford, 1874. Another ed., London, 1896.
   2. *Eclipses and transits in future years.* 1889.

5292. LAIS, Giuseppe. 1845-1921. It. P IV. Also Meteorology.

5293. LOHSE, (William) Oswald. 1845-1915. Ger. DSB; P III,IV.

5294. PERROTIN, (Henri) Joseph (Anastase). 1845-1904. Fr. DSB; P III,IV & Supp.

5295. PLUMMER, John Isaac. 1845-after 1902. Br./China. P III,IV.
   1. *Introduction to astronomy.* London, 1876.

5296. RANYARD, Arthur Cowper. 1845-1894. Br. DSB; P III,IV; GD.

5297. SKINNER, Aaron Nichols. 1845-1918. U.S.A. GE.

5298. TISSERAND, François Félix. 1845-1896. Fr. DSB; ICB; P III,IV.
   1. *Traité de mécanique céleste.* 4 vols. Paris, 1889-96.
   2. *Leçons sur la détermination des orbites.* Paris, 1899. Ed. by J. Perchot.

5299. TRÉPIED, Charles. 1845-1907. Fr./Algeria. ICB; P IV.

5300. VALENTINER, (Karl) Wilhelm (Friedrich Johannes). 1845-after 1909. Ger. P III,IV.

5301. WENDELL, Oliver Clinton. 1845-1912. U.S.A. P IV.

5302. ABETTI, Antonio. 1846-1928. It. DSB; P III,IV.

5303. BACKLUND, Jöns Oskar. 1846-1916. Russ. DSB; P III,IV.

5304. BOSS, Lewis. 1846-1912. U.S.A. DSB; P III; GE.

5305. BURTON, Charles Edward. 1846-1882. Br. P III; GD.

5306. CHANDLER, Seth Carlo. 1846-1913. U.S.A. DSB; GE.

5307. HOLDEN, Edward Singleton. 1846-1914. U.S.A. DSB; P III,IV; GE.
   1. *Mountain observatories in America and Europe.* Washington, 1896.

5308. HOLETSCHEK, Johann. 1846-1923. Aus. P III,IV.

5309. PICKERING, Edward Charles. 1846-1919. U.S.A. DSB; P III,IV; GB; GE. Also Physics*.

5310. PORETSKY, Platon Sergeevich. 1846-1907. Russ. DSB. Also Mathematics.

5311. SCHUR, Adolph Christian Wilhelm. 1846-1901. Ger. P III,IV.

5312. TERBY, François Joseph Charles. 1846-1911. Belg. P III,IV; GF.

5313. BLOCK, Eugen. 1847-after 1883. Russ. P III.

5314. DI LEGGE, Alfonso. 1847-after 1902. It. P IV.

5315. FRANZ, Julius Heinrich Georg. 1847-1913. Ger. P III,IV.

5316. HAGEN, John George. 1847-1930. Ger./U.S.A. ICB; P III,IV.
1. *Atlas stellarum variabilium.* Series 1-8. Berlin, 1899-1934.

5317. LINDSAY, James Ludovic. 1847-1913. Br. P III; GB; GD3.

5318. OPPENHEIM, Heinrich. 1847-1896. Ger. P IV.

5319. SCHULHOF, Leopold. 1847-1921. Aus./Fr. P III,IV.

5320. STONE, Ormond. 1847-1933. U.S.A. P III,IV; GE1.

5321. WHITNEY, Mary Watson. 1847-1921. U.S.A. ICB; P IV; GE.

5322. BAILLAUD, (Eduard) Benjamin. 1848-1934. Fr. P III,IV. See also 3947/4.

5323. BRUNS, Ernst Heinrich. 1848-1919. Ger. P III,IV.

5324. CRULS, Louis (or Luiz). 1848-1908. Brazil. P III,IV.

5325. DENNING, William Frederick. 1848-1931. Br. P III.

5326. GLASENAPP, Sergei Pavlovich von. 1848-1937. Russ. ICB (GLAZ-ENAP); P III,IV.

5327. HASSELBERG, (Clas) Bernhard. 1848-1922. Russ./Swed. P III,IV. Also Physics.

5328. HENRY, Paul Pierre. 1848-1905. Fr. DSB; P III,IV.

5329. MILLOSEVICH, Elia Filippo Francesco Giuseppe Maria. 1848-1919. It. P III,IV.

5330. PALISA, Johann. 1848-1925. Aus. DSB; P III,IV.

5331. SANDE, Ernst Frederik Bakhuyzen van de. 1848-1918. Holl. P III (under S); P IV (under VAN DE S.).

5332. SEABROKE, George Mitchell. 1848-after 1883. Br. P III.

5333. WEINEK, Ladislaus. 1848-1913. Ger./Prague. P III,IV.

5334. ZONA, Temistocle. 1848-1910. It. P IV.

5335. COGGIA, Jérôme Eugène. 1849-after 1900. Fr. P III,IV.

5336. DUBJAGO, Dmitry Ivanovich. 1849-1918. Russ. P III,IV.

5337. FINLAY, William Henry. 1849-1924. Cape Obs. P III,IV.

5338. FORBES, George. 1849-1936. Br. ICB; P IV. Also Physics*.
1. *The transit of Venus.* London, 1874.

5339. HENRY, Prosper Mathieu. 1849-1903. Fr. DSB; P III,IV.

5340. LANCASTER, Albert Benoît Marie. 1849-1908. Belg. ICB; P III.
Also Meteorology.
1. *Liste générale des observatoires et des astronomes, des sociétés et des revues astronomiques.* Brussels, 1886. X. 3rd ed., 1890.
See also 5047/1.

5341. NEVILL (or NEISON), Edmund Neville. 1849-1940. Br. ICB; P III, IV (NEISON); GF.
The following works are catalogued under NEISON.
1. *The moon, and the condition and configurations of its surfaces.* London, 1876.
2. *Astronomy.* 1886.

5342. PLUMMER, William Edward. 1849-1928. Br. P III,IV.

5343. SEELIGER, Hugo von. 1849-1924. Ger. DSB; P III,IV.

5344. SEYDLER, August Johann Friedrich. 1849-1891. Prague. P III, IV. Also Physics.

5345. TSERASKY (or CERASKY), Vitold Karlovich. 1849-1925. Russ. DSB.

5346. THACKERAY, William Grasett. fl. 1876-1917. Br. P IV.

5347. DOWNING, Arthur Matthew Weld. 1850-1917. Br. DSB; P III,IV.

5348. GINZEL, Friedrich Karl. 1850-1926. Ger. P III,IV.
1. *Neue Untersuchungen über die Bahn des Olbers'schen Cometen und seine Wiederkehr.* 1881.

5349. LAMP, Ernst August. 1850-1901. Ger. P III,IV.

5350. SCHRAM, Robert Gustav. 1850-1923. Aus. P IV.

5351. VERSCHAFFEL, Aloys. 1850-1933. Belg./Fr. ICB.

5352. BIGOURDAN, (Camille) Guillaume. 1851-1932. Fr. DSB; ICB; P III,IV. Also History of Astronomy.

5353. BOSSERT, Joseph François. 1851-1906. Fr. P III,IV.

5354. GRUBER, Ludwig. 1851-1888. Hung. P III,IV.

5355. HARTWIG, (Carl) Ernst (Albrecht). 1851-1923. Ger. DSB; P III, IV.

5356. HILFIKER, Jakob. 1851-1913. Switz. P III,IV.

5357. KAPTEYN, Jacobus Cornelius. 1851-1922. Holl. DSB; ICB; P IV.

5358. LAGRANGE, Charles Henri. 1851-1932. Belg. P III,IV. Also Geomagnetism.

5359. MAUNDER, Edward Walter. 1851-1928. Br. DSB; P III,IV.

5360. MÜLLER, (Carl Hermann) Gustav. 1851-1925. Ger. DSB; P IV.

5361. PAUL, Henry Martyn. 1851-1931. U.S.A./Japan. P III; GE.

5362. REES, John Krom. 1851-1907. U.S.A. P IV; GE.

5363. WILSON, William Edward. 1851-1908. Br. P IV; GD2.

5364. CALLANDREAU, (Pierre Jean) Octave. 1852-1904. Fr. DSB; P III, IV & Supp.

5365. DOBERCK, August William. 1852-after 1931. Br./China. P III,IV.

5366. DREYER, John Louis Emil. 1852-1926. Br. DSB; ICB; P III,IV; GD4. Also Biography*.

5367. KLOTZ, Otto Julius. 1852-1923. Canada. P IV; GF.

5368. KÖHL, Torvald Heinrich Johan. 1852-after 1903. Den. P III,IV.

5369. LEWITZKY, Grigorii Vassilievich. 1852-1917. Russ. P IV.

5370. PORTER, Jermain Gildersleeve. 1852-1933. U.S.A. P IV; GE.

5371. VENTURI, Adolfo Raffaele Vincenzo. 1852-1915. It. P IV.

5372. VERY, Frank W. 1852-after 1921. U.S.A. P IV.

5373. BALL, Leo Anton Carl de. 1853-1916. Belg./Aus. P III,IV.

5374. BOEDDICKER, Otto. 1853-after 1900. Ger./Br. P III,IV.
   1. *The Milky Way*. 1892.

5375. DESLANDRES, Henri Alexandre. 1853-1948. Fr. DSB; ICB; P IV. Also Physics.

5376. FLINT, Albert Stowell. 1853-1923. U.S.A. GE.

5377. MEYER, Max Wilhelm. 1853-1910. Switz./Ger. P III,IV.

5378. MÜLLER, Adolf. 1853-after 1938. It. P IV.

5379. PETER, Bruno Edmund August. 1853-1911. Ger. P III,IV.

5380. SCHAEBERLE, John Martin. 1853-1924. U.S.A. DSB; ICB; P IV; GE.

5381. SOKOLOFF, Aleksei Petrovich (first of the name). 1853-1910. Russ. P IV.

5382. UPTON, Winslow. 1853-1914. U.S.A. GE.

5383. AMBRONN, (Friedrich Anton) Leopold. 1854-1930. Ger. P IV.

5384. BAILEY, Solon Irving. 1854-1931. U.S.A. DSB; GE1.

5385. BELOPOLSKY, Aristarkh Apollonovich. 1854-1934. Russ. DSB; P IV.

5386. DONNER, Anders Severin. 1854-after 1934. Fin. P IV.

5387. GAUTIER, (Adolph) Raoul. 1854-1931. Switz. P III,IV.

5388. GEDEONOV, Dmitri Danilovich. 1854-1908. Russ. ICB.

5389. GRUSS, Gustav. 1854-1922. Prague. P III,IV.

5390. KREUTZ, Heinrich. 1854-1907. Ger. P III,IV.

5391. LEHMANN FILHÉS, Rudolf. 1854-1914. Ger. P IV.

5392. LINSTEDT, Anders. 1854-after 1903. Dorpat/Swed. P III,IV.

5393. POINCARÉ, (Jules) Henri. 1854-1912. Fr. DSB; ICB; P III,IV. Also Mathematics*#, Mechanics, and Physics*.
   1. *Les méthodes nouvelles de la mécanique céleste*. 3 vols. Paris, 1892-99. RM/L.

5394. STRUVE, (Karl) Hermann. 1854-1920. Russ./Ger. DSB; P III,IV.

5395. WOLFER, Heinrich Alfred. 1854-after 1922. Switz. P IV.

5396. ZELBR, Karl. 1854-1900. Aus./Brünn. P III,IV.

5397. BRENNER, Leo. 1855-after 1905. Trieste. P IV.

5398. COMSTOCK, George Cary. 1855-1934. U.S.A. DSB; P IV; GE1.

5399. DEICHMÜLLER, Friedrich Heinrich Carl. 1855-1903. Ger. P III, IV.

5400. ELKIN, William Lewis. 1855-1933. U.S.A. DSB; P III; GE1.

5401. HEPPERGER, Josef von. 1855-1928. Aus. P IV.

5402. LOWELL, Percival. 1855-1916. U.S.A. DSB; ICB; GE.
  1. *Mars*. Boston, 1895. RM/L.

5403. PUISEAUX, Pierre Henri. 1855-1928. Fr. ICB; P III,IV.

5404. SCHWAB, Franz. 1855-after 1907. Aus. P IV.

5405. SEYBOTH, Jacob Julius. 1855-after 1903. Russ. P IV.

5406. TODD, David Peck. 1855-after 1920. U.S.A. P III,IV.
  1. *A new astronomy*. New York, 1897.

5407. ANGELITTI, (Ascenzo) Filippo. 1856-1931. It. P III,IV.

5408. ENGSTRÖM, Folke August. 1856-1926. Swed. P IV.

5409. GONNESSIAT, François. 1856-1934. Fr. P IV.
  1. *Recherches sur l'équation personelle dans les observations astronomiques de passage*. Paris, 1892.

5410. KEMPF, Paul Friedrich Ferdinand. 1856-1920. Ger. P IV.

5411. KNOPF, Otto Heinrich Julius. 1856-after 1931. Ger. P IV.

5412. KÜSTER, (Karl) Friedrich. 1856-1936. Ger. ICB; P III,IV.

5413. NACCARI, Giuseppe. 1856-after 1923. It. P IV.

5414. SADLER, Herbert. 1856-after 1885. Br. P III,IV.

5415. STECHERT, Carl Friedrich Gottlieb Peter Heinrich. 1856-after 1909. Ger. P IV.

5416. VITKOVSKII, Vassilii Vasil'evich. 1856-1924. Russ. ICB.

5417. WILSING, Johannes. 1856-1943. Ger. DSB; P IV.

5418. BARNARD, Edward Emerson. 1857-1923. U.S.A. DSB; ICB; P IV Supp.; GE.

5419. FLEMING, Williamina Paton. 1857-1911. U.S.A. DSB; P IV & Supp.; GE.

5420. GOTHARD, Eugen von. 1857-1909. Hung. P IV.

5421. HARZER, Paul Hermann. 1857-1932. Ger. P III,IV.

5422. KEELER, James Edward. 1857-1900. U.S.A. DSB; P IV; GE.

5423. NEWALL, Hugh Frank. 1857-1944. Br. DSB; ICB; P IV; GD6.

5424. OPPENHEIM, Samuel. 1857-1928. Aus./Prague. DSB; P III,IV.

5425. PRITCHETT, Henry Smith. 1857-after 1930. U.S.A. P III,IV.

5426. RIGGE, William Francis. 1857-1927. U.S.A. GE.

5427. SCHROETER, Jens Frederik Wilhelm. 1857-after 1921. Nor. P IV.

5428. ESPIN, Thomas Henry E.C. 1858-after 1923. Br. P IV.

5429. GALLE, Andreas Wilhelm Gottfried. 1858-after 1931. Ger. P IV.

5430. HACKENBERG, Josef. 1858-after 1909. Aus. P IV.

5431. HERZ, Norbert. 1858-1927. Aus. P IV.

5432. HOWE, Herbert Alonzo. 1858-1926. U.S.A. P IV; GE.

5433. KOBOLD, Hermann Albert. 1858-after 1930. Ger. P III,IV.

5434. LEAVENWORTH, Francis Preserved. 1858-1928. U.S.A. P IV; GE.

5435. OERTEL, Karl Johannes. 1858-after 1927. Ger. P IV.

5436. PICKERING, William Henry. 1858-1938. U.S.A. DSB; ICB; P III, IV.

5437. SCHEINER, Julius. 1858-1913. Ger. DSB; P IV.
1. *Die Spectralanalyse der Gestirne*. Leipzig, 1890. RM.
1a. ——— *A treatise on astronomical spectroscopy*. Boston, 1894. Trans., rev., and enl. by E.B. Frost. RM/L.

5438. STRUVE, (Gustav Wilhelm) Ludwig. 1858-1920. Russ. DSB; P III, IV. Also Geodesy.

5439. BOCCARDI, Giovanni. 1859-1936. It. ICB.

5440. PLASSMANN, (Eduard Clemens Franz) Joseph. 1859-1940. Ger. ICB; P IV & Supp.

5441. RAMBAUT, Arthur Alcock. 1859-1923. Br. P IV.

5442. SPITALER, Rudolf. 1859-after 1925. Aus./Prague. P IV.

5443. VOGEL, Robert Philippovich Simon. 1859-1920. Russ. P IV.

5444. WISLICENUS, Walter Friedrich. 1859-1905. Ger. P IV.

Imprint Sequence

5445. ROYAL Society of London.
1. *Correspondence concerning the great Melbourne telescope ... 1852-1870*. London, 1871.

5446. ACADÉMIE des Sciences. Paris. Commission du passage de Vénus.
1. *Passage de Vénus du 6 décembre, 1882: Rapports préliminaires*. Paris, 1883.
2. *Recueil de mémoires, rapports et documents rélatifs à l'observation du passage de Vénus sur le soleil*. Ib., 1882-5.

5447. UNITED States. Bureau of Navigation.
1. *Catalogue of "National almanac" library*. 1883.

5448. INTERNATIONAL Conference at Washington for the Fixing of a Prime Meridian and a Universal Day.
1. [Publication.] 1884.

5449. ROYAL Astronomical Society. London.
1. *Catalogue of the library*. 1886.

5450. ASTROPHYSIKALISCHES Observatorium. Potsdam.
1. *Die königlichen Observeratorien für Astrophysik, Meteorologie ... bei Potsdam*. 1890.

5451. CONFÉRENCE Internationale des Étoiles Fondamentales. Paris, 1896.
1. *Procès-verbaux*. Paris, 1896.

3.05 MECHANICS

Including the Cartesian and Newtonian natural philosophies in their relation to mechanics.

Works on celestial mechanics are listed under Astronomy.

5452. ALEXANDRE, Jacques. 1653-1734. Fr. P I; GA; GF.
1. *Traité du flux et reflux de la mer.* Paris, 1726. RM.

5453. VARIGNON, Pierre. 1654-1722. Fr. DSB; ICB; P II; GA. Also Mathematics* and Part 2*.
1. *Nouvelle mécanique ou statique, dont le projet fut donné en 1687.* 2 vols. Paris, 1725. RM/L.
2. *Traité du mouvement et de la mesure des eaux coulantes et jaillissantes. Avec un traité préliminaire du mouvement en général.* Paris, 1725. RM.

5454. FONTENELLE, Bernard Le Bouvier de. 1657-1757. Fr. DSB; ICB; P I; GA; GB. Also Science in General*#, Mathematics*, Biography*, History of Science*, and Part 2*.
1. *Théorie des tourbillons cartésiens, avec des réflexions sur l'attraction.* Paris, 1752.

5455. SAURIN, Joseph. 1659-1737. Fr. DSB; P II; GA. Also Mathematics.

5456. POLHEM, Christopher. 1661-1751. Swed. ICB; P II; GA.

5457. PARENT, Antoine. 1666-1716. Fr. DSB; P II. Also Mathematics.

5458. BERNOULLI, Johann (or Jean) (first of the name). 1667-1748. Switz. DSB; ICB; P I; BLA & Supp.; GA; GB; GC. Also Mathematics and Part 2*#.
1. *Essay d'une nouvelle théorie de la manoeuvre des vaissaux.* Basle, 1714. RM.
2. *Nouvelles pensées sur le systême de M. Descartes et la manière d'en déduire les orbites et les aphélies des planètes.* Paris, 1730.

5459. WHISTON, William. 1667-1752. Br. DSB; P II; BMNH; GA; GB; GD. Also Science in General*, Astronomy*, Physics*, Autobiography*, and Part 2*. See also Index.
1. *Sir Isaac Newton's mathematick philosophy more easily demonstrated.* London, 1716. Reprinted, with an introd. by I.B. Cohen, New York, 1972.

5460. PASCOLI, Alessandro. 1669-1757. It. ICB; P II; BLA.
1. *Del moto che nei mobili si rifonde per impulso esteriore: Trattato fisicomatematico.* Rome, 1723. RM.

5461. GOBERT, Thomas. fl. 1702. Fr. P I.

1. *Nouveau sistème sur la construction et les mouvemens du monde*. Paris, 1703. RM.

5462. GRANDI, Guido. 1671-1742. It. DSB; ICB; P I; GA. Also Mathematics*.
1. *Instituzioni meccaniche*. Florence, 1739. RM.

5463. KEILL, John. 1671-1721. Br. DSB; P I; GA; GD. Also Mathemics*, Astronomy*, and Part 2*. See also Index.
1. *Introductio ad veram physicam*. Oxford, 1702. RM/L. Another ed., Cambridge, 1741.
1a. ——— *An introduction to natural philosophy*. London, 1720. RM/L. 4th ed., trans. from the last ed. of the Latin, 1745.

5464. SAULMON, ... d. 1725. Fr. P II.

5465. CLARKE, Samuel. 1675-1729. Br. DSB; ICB; P I; GA; GB; GD. Also Science in General*. See also Index.

5466. DITTON, Humphrey. 1675-1715. Br. P I; GA; GB; GD. Also Mathematics*.
1. *The general laws of nature and motion, with their application to mechanicks ... Being a part of the great Mr. Newton's "Principles."* London, 1705. RM.
2. *The new law of fluids; or, A discourse concerning the ascent of liquors, in exact geometrical figures, between contiguous surfaces*. London, 1714. RM.
See also 4462/2.

5467. RICCATI, Jacopo Francesco. 1676-1754. It. DSB; P II; GA; GB. Also Mathematics.

5468. PRIVAT DE MOLIÈRES, Joseph. 1677-1742. Fr. DSB; P II (MOLIÈRES); GA (MOLIÈRES).

5469. HERMANN, Jakob. 1678-1733. Switz./It./Ger./Russ. DSB; P I; GA; GC. Also Mathematics.
1. *Phoronomia; sive, De viribus et motibus corporum solidorum et fluidorum*. Amsterdam, 1716. RM/L.

5470. MAIRAN, Jean Jacques d'Ortus de. 1678-1771. Fr. DSB; ICB; P II; GA. Also Physics*, Meteorology*, and Biography*.
1. *Dissertation sur l'estimation et la mesure des forces motrices des corps*. Paris, 1741.

5471. ABAUZIT, Firmin. 1679-1767. Fr./Switz. P I; GA; GB.

5472. LUDOLFF, Hieronymus (first of the name). 1679-1728. Ger. PI; BLA.

5473. MAZIÈRES, Jean Simon. ca. 1679-1761. Fr. P II.
1. *Traité des petits tourbillons de la matière subtile*. Paris, 1627. RM.

5474. ZENDRINI, Bernardino. 1679-1747. P II; GA.

5475. CORRADI D'AUSTRIA, Domenico. 1680-1756. It. P I.

5476. ROBINSON, Bryan. 1680-1754. Br. GD. Also Physiology*.
1. *A letter to Dr Cheyne, containing an account of the motion of water through orifices and pipes*. Dublin, 1735.
2. *A dissertation on the aether of Sir Isaac Newton*. Dublin, 1743. cf. item 5769/13.

5477. CLARKE (or CLARK), John. 1682-1757. Br. GD.
1. *A demonstration of some of the principal sections of Sir Isaac Newton's "Principles of natural philosophy."* London, 1730. RM/L. Reprinted, with an introd. by I.B. Cohen, New York, 1972.
See also 2279/1c.

5478. SWITZER, Stephen. 1682?-1745. Br. GD.
1. *An introduction to a general system of hydrostaticks and hydraulicks, philosophical and practical.* 2 vols. London, 1729.

5479. BOMBARDI, Michael. 1683-1729. Aus. P I.

5480. POLENI, Giovanni. 1683-1761. It. DSB; ICB; P II; GA.

5481. FALCK, Joseph. d. 1737. Ger. P I.

5482. CASTEL, Louis Bertrand. 1688-1757. Fr. DSB; P I; GA; GB. Also Physics*.
1. *Traité de physique sur la pesanteur universelle des corps.* 2 vols. Paris, 1724. RM.
2. *Le vrai systême de physique générale de M. Isaac Newton exposé et analysé en parallèle avec celui de Descartes.* Paris, 1743. RM.

5483. COLDEN, Cadwallader. 1688-1776. Br./U.S.A. DSB; ICB; P I; BLA & Supp.; GA; GB; GD; GE. Also Botany.

5484. 'sGRAVESANDE, Willem Jacob. 1688-1742. Holl. DSB; P I; GA (under S). Also Mathematics*.
1. *Physices elementa mathematica, experimentis confirmata; sive, Introductio ad philosophiam Newtonianiam.* 2 vols. Leiden, 1720-21. X. 3rd ed., 1742.
1a. ——— *Mathematical elements of natural philosophy, confirmed by experiments.* 2 vols. London, 1720-21. Trans. by J.T. Desaguliers. RM/L. 6th ed., 1747.
1b. ——— *Mathematical elements of natural philosophy, prov'd by experiments.* London, 1720. Trans. by John Keill. RM.
1c. ——— *Elémens de physique démontrez mathématiquement et confirmez par des expériences.* Leiden 1746. Trans. by E. de Joncourt. RM.
2. *Philosophiae Newtonianae institutiones.* Leiden, 1723. RM/L.
2a. ——— *An explanation of the Newtonian philosophy, in lectures....* London, 1735. Trans. by E. Stone. 2nd ed., 1741. RM.
3. *Oeuvres philosophiques et mathématiques.* Amsterdam, 1774. Ed. with a biography and bibliography by J.N.S. Allamand. RM.
See also 2375/10.

5485. EAMES, John. d. 1744. Br. P I; GD. See also 2997/1b,1c.

5486. ZANOTTI, Francesco Maria. 1692-1777. It. P II; GA.
1. *Della forza de' corpi che chiamano viva.* Bologna, 1752. RM.

5487. BILFINGER, Georg Bernhard. 1693-1750. Ger. P I; GA; GB; GC.
1. *De causa gravitatis physica generali disquisitio experimentalis.* Paris, 1728.

5488. PEMBERTON, Henry. 1694-1771. Br. DSB; ICB. P II; BLA; GA; GD. Also Chemistry* and Physiology.
1. *A view of Sir Isaac Newton's philosophy*. London, 1728. RM/L. Reprinted with an introd. by I.B. Cohen, New York, 1972.
See also 2539/1b.

5489. PITOT, Henri. 1695-1771. Fr. DSB; ICB; P II; GA.

5490. DEIDIER, ... *Abbé*. 1696-1746. Fr. P I & Supp.; GA.
1. *La mécanique générale, contenant la statique, l'airométrie, l'hydrostatique et l'hydraulique*. Paris, 1741. RM.

5491. BÉLIDOR, Bernard Forest de. 1697/8-1761. Fr. DSB; P I; GA.

5492. BOUGUER, Pierre. 1698-1758. Fr. DSB; ICB; P I; GA; GB. Also Astronomy*, Physics*, and Geodesy*.
1. *De la manoeuvre des vaisseaux; ou, Traité de mécanique et de dynamique*. Paris, 1757. RM.

5493. MAUPERTUIS, Pierre Louis Moreau de. 1698-1759. Fr./Ger. DSB; ICB; P II; GA; GB; GC. Other entries: see 3089.

5494. CAMUS, Charles Etienne Louis. 1699-1768. Fr. DSB; P I; GA; GB.
1. *Traité des forces mouvantes*. Paris, 1722. RM.
See also 2278/1a.

5495. BERNOULLI, Daniel. 1700-1782. Switz./Russ. DSB; ICB; P I; BLA; GA; GB; GC. Also Mathematics* and Physiology.
1. *Hydrodynamica; sive, De viribus et motibus fluidorum commentarii*. Strasbourg, 1738. RM/L.
1a. ——— *Der Hydrodynamik*. 2 vols. Munich, 1965. Trans. by K. Flierl.
1b. ——— *Hydrodynamics*. New York, 1968. Trans. by T. Carmody and H. Kobus.

5496. POLACK (or POLAC), Johann Friedrich. 1700-1771. Ger. P II.

5497. EMERSON, William. 1701-1782. Br. P I; GA; GB; GD. Also Mathematics*, Astronomy*, and Physics*.
1. *The principles of mechanics, explaining and demonstrating the general laws of motion*. London, 1754. X. 2nd ed., 1758. RM/L. Another ed., 1825.
2. *A short comment on Sir I. Newton's "Principia", containing notes upon some difficult places of that excellent book*. London, 1770. RM/L.
3. *Tracts. Containing: I, Mechanics, or the doctrine of motion; II, The projection of the sphere; III, The laws of centripetal and centrifugal force*. London, 1793. RM.

5498. LECCHI, Giovanni Antonio. 1702-1776. It./Aus. P I; GA.
1. *Memorie idrostatico-storiche*. 2 vols. Modena, 1773. RM.
2. *Trattato de' canali navigabili*. Milan, 1776. RM.

5499. FROMOND, Giovanni Claudio. 1703-1765. It. P I; BLA; GA.
1. *Della fluidita de' corpi*. Leghorn, 1754. RM.

5500. SEGNER, János András (or Johann Andreas). 1704-1777. Hung./Ger. DSB; ICB; P II; BLA. Also Mathematics* and Physics.

5501. EULER, Leonhard. 1707-1783. Ger./Russ. DSB; ICB; P I; GA; GB. Also Science in General*, Mathematics*#, Astronomy*, Physics*,

and Cartography*. See also Index.
1. *Mechanica; sive, Motus scientia analytice exposita.* 2 vols. St. Petersburg, 1736. RM/L.
2. *Vollständigere Theorie der Maschinen die durch Reaktion des Wassers in Bewegung versetzt werden.* [Periodical article, 1754] Leipzig, 1911. (OKEW. 182)
3. *Theoria motus corporum solidorum seu rigidorum.* Rostock, 1765. With a preface by W.J.G. Karsten. RM/L.
3a. ——— *Theorie der Bewegung fester oder starrer Körper.* 1853.
4. *Théorie complette de la construction et de la manoeuvre des vaisseaux.* St. Petersburg, 1773. Another ed., Paris, 1776. RM.
4a. ——— *A compleat theory of the construction and properties of vessels.* London, 1776. Trans. by H. Watson. RM.

5502. RICCATI, Vincenzo. 1707-1775. It. DSB; P II; GA; GA. Also Mathematics*.

5503. ROBINS, Benjamin. 1707-1751. Br. DSB; ICB; P II; GA; GB; GD. Also Mathematics*.

5504. HEINSIUS, Gottfried. 1709-1769. Ger./Russ. P I. Also Astronomy.

5505. RICCATI, Giordano. 1709-1790. It. P II; GA.

5506. BERNOULLI, Johann (or Jean) (second of the name). 1710-1790. Switz. DSB; ICB; P I; GA; GB; GC.

5507. MICHELOTTI, Francesco Domenico. 1710-1777. It. P II.

5508. STÜBNER, Friedrich Wilhelm. 1710-1736. Ger. P II; GC.

5509. BOSÇOVICH, Ruggiero Giuseppe. 1711-1787. It./Fr. DSB (BOŠKOVIĆ); ICB; P I; GA; GB. Other entries: see 3114.
1. *Dissertatio de maris aestu.* Rome, 1747. RM.
2. *De centro gravitatis dissertatio.* Rome, 1751. RM.
3. *Philosophiae naturalis theoria.* Vienna, 1758. RM. Another ed., rev., Venice, 1763. RM.
3a. ——— *A theory of natural philosophy.* Chicago, 1922. Trans. from the 1763 ed. by J.M. Child. Reprinted, Cambridge, Mass., 1966.

5510. HEE, Christian. 1712-1781. Den. P I.

5511. KOENIG, (Johann) Samuel. 1712-1757. Holl. DSB; ICB; P I; GA; GB. Also Mathematics. See also 2857/1.

5512. MATHON DE LA COUR, Jacques. 1712-1770 (or 1777). Fr. P II; GA.
1. *Nouveaux éléments de dynamique et de mécanique.* Lyons, 1763. RM.

5513. RUTHERFORD, Thomas. 1712-1771. Br. P II. Also Physics*.
1. *A system of natural philosophy, being a course of lectures in mechanics, hydrostatics, and astronomy.* 2 vols. Cambridge, 1748.

5514. CLAIRAUT, Alexis Claude. 1713-1765. Fr. DSB; ICB; P I; GA; GB. Also Mathematics*, Astronomy*, Physics, and Geodesy*. See also Index.

5515. KRAZ (or KRATZ), Georg. 1713-1766. Ger. P I.

5516. BRANCAS-VILLENEUVE, André François de. d. 1758. Fr. P I; GA. Also Astronomy.
1. *Explication du flux et reflux*. Paris, 1749. RM.

5517. COURTIVRON, Gaspard le Compasseur de Créquy-Montford, *Marquis* de. 1715-1785. Fr. DSB; P I; GA. Also Physics*.

5518. XIMENEZ, Leonardo. 1716-1786. It. ICB; P II; GA. Also Astronomy.

5519. ALEMBERT, Jean le Rond d'. 1717-1783. Fr. DSB; ICB; P I; GA; GB; GF. Also Mathematics* and Astronomy*. See also Index.
1. *Traité de dynamique*. Paris, 1743. RM. Reprinted, Brussels, 1967, and New York, 1968.
1a. ——— German trans. Leipzig, 1899. (OKEW. 106)
2. *Traité de l'équilibre et du mouvement des fluides*. Paris, 1744. RM. Reprinted, Brussels, 1966.
3. *Essai d'une nouvelle théorie de la résistance des fluides*. Paris, 1752. RM. Reprinted, Brussels, 1966.
4. *Oeuvres*. 5 vols. Paris, 1821-22. Ed. by Bastien. RM.
5. *Oeuvres et correspondances inédites*. Paris, 1887. Ed. with introd., notes etc., by C. Henry. X. Reprinted, Geneva, 1967.

5520. GERDIL, Giacinto Sigismondo. 1718-1802. It. P I; GA.
1. *Dissertations sur l'incompatibilité de l'attraction, et de ses différentes loix, avec les phénomènes; et sur les tuyaux capillaires*. Paris, 1754. RM.

5521. KAESTNER, Abraham Gotthelf. 1719-1800. Ger. DSB; ICB; P I; GA; GC. Also Mathematics, Physics, and History of Mathematics*.

5522. SIGORGNE, Pierre. 1719-1809. Fr. DSB; P II; GA.
1. *Institutions newtoniennes; ou, Introduction à la philosophie de M. Newton*. Paris, 1747. RM/L. 2nd ed., 1769.

5523. KRAFT, Jens. 1720-1765. Den. DSB; P I; GA. Also Mathematics.

5524. BOLL, Anton. b. 1721. Prague. P I.

5525. CARACCIOLI, Giovanni. 1721-1798. It. P I.

5526. FELICE, Fortunato Bartolommeo. 1723-1789. Switz. P I; GA.

5527. BONATI, Theodore Maxime. 1724-1820. It. GA.

5528. LE SAGE, George Louis (second of the name). 1724-1803. Switz. DSB; ICB; P I; GA. Also Autobiography*.

5529. SMEATON, John. 1724-1792. Br. The engineer. DSB; ICB; P II; GB; GD.
1. *Experimental enquiry concerning the natural powers of wind and water to turn mills and other machines ... and an experimental examination of the quantity and proportion of mechanic power necessary to be employed in giving different degrees of velocity to heavy bodies from a state of rest. Also new fundamental experiments upon the collision of bodies*. London, 1794. X. 3rd ed., 1813.
2. *Reports made on various occasions in the course of his employment as a civil engineer*. London, 1812. RM.

3. *The miscellaneous papers, comprising his communications to the Royal Society.* London, 1814.
4. *Diary of his journey to the Low Countries.* Leamington Spa, 1938. Ed. from the original MS by A. Titley.

5530. D'ARCY (or D'ARCI), Patrick. 1725-1779. Br./Fr. DSB; P I (ARCY); GD. Also Astronomy.

5531. FRISI, Paolo. 1728-1784. It. DSB; P I; GA; GB. Also Various Fields*#, Astronomy*, and Physics.
1. *Del modo di regolare i fiumi e i torrenti.* Lucca, 1762. RM.
1a. ——— *Traité des rivières et des torrens.* Paris, 1774. Trans. by Deseney.
2. *De gravitate universali corporum.* Milan, 1768. RM.
3. *Instituzioni di meccanica, d'idrostatica, d'idrometria e dell'architettura statica e idraulica.* Milan, 1777. RM.

5532. DIESBACH, Johannes. 1729-1792. Prague/Aus. P I & Supp.
1. *Institutiones philosophicae de corporum communibus attributis, ad mentem cl. Rogerii Boschovichii ... ad unicam virium legem redactae.* Prague, 1767. RM. cf. 5509/3.

5533. BOSSUT, Charles. 1730-1814. Fr. DSB; P I; GA. Also History of Mathematics*.
1. *Recherches sur la construction la plus avantageuse des digues.* Paris, 1754. With Viallet. RM.
2. *Nouvelles expériences sur la résistance des fluides.* Paris, 1777. With d'Alembert and Condorcet. RM/L.
3. *Traité théorique et expérimental d'hydrodynamique.* 2 vols. Paris, 1786-87. RM. Another ed., 1796.
4. *Mémoires de mathématiques.* Paris, 1812. RM.

5534. LORGNA, Antonio Maria. 1730-1796. It. ICB; P I; GA. Also Mathematics.

5535. SALADINI, Girolamo. 1731-1813. It. ICB; P II.

5536. BECKER, Heinrich Valentin. 1732-1796. Ger. P I.

5537. BORDA, Jean Charles. 1733-1799. Fr. DSB; ICB; P I; GA; GB. Also Geophysics*.

5538. HENNERT, Johann Friedrich. 1733-1813. Holl. P I,III. Also Astronomy*.

5539. STRATICO, Simone. 1733-1824. It. P II.

5540. BÜCHER, Christian Bernhard. 1734-1759. Ger. P I.

5541. DU BUAT, Pierre Louis Georges. 1734-1809. Fr. DSB.
1. *Principes d'hydraulique.* Paris, 1779. X.
1a. ——— 3rd ed. *Principes d'hydraulique et de pyrodynamique.* 3 vols. Paris, 1816.

5542. FONTANA, Gregorio. 1735-1803. It. P I; GA. Also Mathematics*.

5543. ZALLINGER ZUM THURN, Jacob Anton. 1735-1813. Aus./Ger. P II; GC.

5544. COULOMB, Charles Augustin. 1736-1806. Fr. DSB; ICB; P I; GA; GB. Also Physics*#.
1. *Essai sur une application des règles "de maximis et minimis" à quelques problèmes de statique.* [Periodical

article, 1773].

1a. ——— *Coulomb's memoir on statics*. Cambridge, 1972. Includes a facsimile with an English trans. By J. Heyman.

2. *Théorie des machines simples*. Paris, 1809. X. 2nd ed., with addition of his memoirs, 1821. RM/L.

5545. LAGRANGE, Joseph Louis. 1736-1813. It./Ger./Fr. DSB; ICB; P I; GA; GB. Also Mathematics*# and Astronomy. See also Index.

1. *Mécanique analitique*. Paris, 1788. RM/L. 2nd ed., 2 vols, 1811-15. 3rd ed., with notes by J. Bertrand, 2 vols, 1853-55. 4th ed., by G. Darboux (includes Bertrand's notes), 2 vols, 1888-89; reprinted, 1965.

5546. HAMBERGER, Adolph Albrecht. 1737-1785. Ger. P I; BLA. Also Astronomy*.

1. *Kurzer Entwurf einer Naturlehre, worinnen alles aus dem einzigen Begriffe dass Kraft nicht anders als Druck sey, erwiesen ist*. Jena, 1780. RM.

5547. LUDEÑA (or LUDENNA), Antonio. 1740-1820. It. P I.

5548. CAMETTI, Ottaviano. d. 1789. It. P I.

1. *Mechanica; seu, Brevis tractatus de motu et aequilibrio*. Pisa, 1768. RM.

5549. ATWOOD, George. 1745-1807. Br. DSB; P I; GA; GB; GD. Also Physics*.

1. *A treatise on the rectilinear motion and rotation of bodies*. Cambridge, 1784. RM/L.

5550. PARKINSON, Thomas. 1745-1830. Br. P II; GA; GD.

1. *A system of mechanics*. Cambridge, 1785. RM.

5551. MONGE, Gaspard. 1746-1818. Fr. DSB; ICB; P II; GA; GB. Also Mathematics*, Physics*, and Chemistry. See also Index.

5552. BREGUET, Abraham Louis. 1747-1823. Switz./Fr. ICB; P I; GA.

5553. COCCOLI, Domenico. 1747-1812. It. P I.

5554. FERRARI, Luigi Maria Bartolommeo. 1747-1820. It. P I.

5555. FABRE, Jean Antoine. 1749-1837. Fr. P I.

1. *Essai sur la théorie des torrens et des rivières*. Paris, 1797. RM.

5556. VINCE, Samuel. 1749-1821. Br. P II; GD. Also Mathematics*, Astronomy*, and Physics*.

1. *The principles of hydrostatics*. 2nd ed. Cambridge, 1800. 3rd ed., 1806. 5th ed., 1820.

5557. DANOW, Gottlob. 1750-1794. Ger. P I.

1. *Beytrag zur Statik*. Berlin, 1780. RM.

5558. BÜRJA, Abel. 1752-1816. Ger. P I. Also Mathematics.

1. *Grundlehren der Hydrostatik*. Berlin, 1790. RM.

2. *Grundlehren der Dynamik*. Berlin, 1791. RM.

5559. AVANZINI, Giuseppe. 1753-1827. It. P I.

5560. CARNOT, Lazare Nicolas Marguerite. 1753-1823. Fr. DSB; ICB;

P I; GA; GB. Also Mathematics*. See also 15552.
1. *Principes fondamentaux de l'équilibre et du mouvement.* Paris, 1803. RM/L.

5561. PASQUICH, Johann. 1753-1829. Hung. ICB; P II. Also Astronomy.

5562. MICHEL, Claude Louis Samson. 1754-1814. Belg. P II; GA.
1. *Essai sur les attractions moléculaires.* Paris/Douay, 1803. RM.

5563. DELANGES, Paolo. d. 1810. It. P I (DESLANGES).
1. *Esperienze ed osservazioni intorno alla pressione delle terre ed alla resistenza de' muri.* Verona, 1779. RM.
2. *Esperienze intorno alla resistenza della sfregamento del legno e de' metalli.* Verona, 1782. RM.
3. *Meccanica pratica in cui si dimostra la maniera di determinare l'equilibrio delle macchine, computando le resistenze degli sfregamenti.* Verona, 1783. RM.

5564. PRONY, (Gaspard Clair François Marie) Riche de. 1755-1839. Fr. DSB; P II; GA; GB.
1. *Mécanique philosophique; ou, Analyse raisonnée des diverses parties de la science de l'équilibre et du mouvement.* Paris, 1800. RM/L.
2. *Recherches physico-mathématiques sur la théorie des eaux courantes.* Paris, 1804. RM.
3. *Leçons de mécanique analytique.* 2 vols. Paris, 1810-15. RM/L.

5565. BUSSE, Friedrich Gottlieb von. 1756-1835. Ger. P I; GC.

5566. GERSTNER, Franz Joseph von. 1756-1832. Prague. P I; GC.
1. *Handbuch der Mechanik.* 3 vols. Prague, 1831. X. 2nd ed., 1832-34.

5567. LANGSDORF, Karl Christian von. 1757-1834. Ger. P I.
1. *Mechanische und hydrodynamische Untersuchungen.* Altenburg, 1782. RM.

5568. WOLTMAN, Reinhard. 1757-1837. Ger. DSB; P II; GC.

5569. METTERNICH, Matthias. 1758-1825. Ger. P II.

5570. MOLLET, Joseph. 1758-1829. Fr. P II; GA.
1. *Hydraulique physique.* Lyons, 1809.

5571. BERNOULLI, Jakob (or Jacques) (second of the name). 1759-1789. Russ. DSB; P I; GA; GB; GC.

5572. WOOD, James. 1760-1838. Br. P II; GD. Also Mathematics* and Physics*.
1. *The principles of mechanics.* Cambridge, 1796. X. 3rd ed., 1803. Another ed., 1824.
2. *Elements of mechanics.* New edition. Cambridge, 1841.

5573. ZENDRINI, Antonio. 1763-1849. It. P II.

5574. EYTELWEIN, Johann Albert C. 1764-1848. Ger. DSB; P I; GA; GC.

5575. GUGLIELMINI, Giovanni Battista. d. 1817. It. P I.

5576. GIRARD, Pierre Simon. 1765-1836. Fr. DSB; ICB; P I; BLA; GA.
1. *Traité analytique de la résistance des solides.* Paris, 1798. RM.

5577. IVORY, James. 1765-1842. Br. DSB; P I; GB; GD. Also Mathematics, Astronomy, Physics, and Geodesy.

5578. BRUNACCI, Vincenzo. 1768-1818. It. P I; GA.

5579. VENTUROLI, Giuseppe. 1768-1846. It. P II.
1. [Trans.] *Elements of the theory of mechanics.* Cambridge, 1822.
2. [Trans.] *Elements of practical mechanics. To which is added a treatise on the principle of virtual velocities.* Cambridge, 1823. Trans. by D. Cresswell.

5580. AUBUISSON DE VOISINS, Jean François d'. 1769-1841. Fr. DSB; P I; BMNH. Also Geology*.
1. *Traité du mouvement de l'eau dans les tuyaux de conduite.* Paris, 1827. X. 2nd ed., 1836. RM.
2. *Traité d'hydraulique.* Paris, 1834. RM. 2nd ed., 1840.

5581. HACHETTE, Jean Nicolas Pierre. 1769-1834. Fr. DSB; P I; GA; GB. Also Mathematics and Physics.

5582. BONFADINI, Jacopo. 1771-1835. It. P I.

5583. POSELGER, Friedrich Theodor. 1771-1838. Ger. P II. Also Mathematics.
1. *Allgemeine Grundsätze von Gleichgewicht und Bewegung.* Berlin, 1824. RM.

5584. CAYLEY, George. 1773-1857. Br. ICB; GF. Founder of the science of aerodynamics.
1. *Aeronautical and miscellaneous note-book (ca. 1799-1826).* Cambridge, 1933.

5585. FRANCOEUR, Louis Benjamin. 1773-1848. Fr. P I; BMNH. Also Mathematics* and Astronomy*.
1. *Traité de mécanique.* Paris, 1800. X. 5th ed., 1825.

5586. GREGORY, Olinthus Gilbert. 1774-1841. Br. DSB; P I; GA; GB; GD. Also Astronomy* and Physics.
1. *A treatise of mechanics.* 3 vols. London, 1806. X. 2nd ed., 1807. 4th ed., 1826.

5587. ALLIX, Jacques Alexandre François. 1776-1836. Fr. P I; GA.
1. *Théorie de l'univers, ou de la cause primitive du mouvement et de ses principaux effets.* 2nd ed., Paris, 1818. RM.

5588. GAUSS, Karl Friedrich. 1777-1855. Ger. DSB; ICB; P I,III; GA; GB; GC. Also Mathematics*#, Astronomy*, Physics*, and Earth Sciences*. See also Index.
1. *Principia generalia theoriae figurae fluidorum in statu aequilibrii.* [Periodical article, 1830]
1a. ——— *Allgemeine Grundlagen einer Theorie der Gestalt von Flüssigkeiten im Zustand des Gleichgewichts.* Leipzig, 1903. (OKEW. 135)
2. *Allgemeine Lehrsätze in Beziehung auf die im verkehrten Verhältnisse des Quadrats der Entfernung wirkenden Anziehungs- und Abstossungs-Kräfte.* [Periodical article, 1840] Leipzig, 1889. (OKEW. 2)

5589. HECHT, Daniel Friedrich. 1777-1833. Ger. DSB; P I.

5590. KATER, Henry. 1777-1835. Br. DSB; P I; GA; GB; GD. Also Geodesy.
1. *A treatise on mechanics.* London, 1830. With D. Lardner. RM/L. Another ed., 1837.

5591. MAGISTRINI, Giambattista. 1777-1849. It. P II. Also Mathematics.

5592. POINSOT, Louis. 1777-1859. Fr. DSB; P II; GA; GB. Also Mathematics.
1. *Eléments de statique.* Paris, 1803. X. 4th ed., 1824. 9th ed., 1848. 11th ed., 1873. 12th ed., 1877.
2. *Théorie des cones circulaires roulants.* Paris, 1853.
3. *La théorie générale de l'équilibre et du mouvement des systèmes.* Paris, 1975. Critical ed. by P. Bailhache.

5593. BROUGHAM, Henry. 1779-1868. Br. The lawyer and politician. P I,III; GA; GB; GD. Also Science in General*#, Physics*, Autobiography*, and Biography*.
1. *Analytical view of Sir Isaac Newton's "Principia."* London, 1855. With E.J. Routh. Reprinted, New York, 1972.

5594. MURHARD, Friedrich Wilhelm August. 1779-1853. Ger. P II; GA. Also Mathematics* and Physics*.
1. *Literatur der mechanischen und optischen Wissenschaften.* 3 vols. Leipzig, 1803-05. See 3652/1a.

5595. SCHWEINS, Franz Ferdinand. 1780-1856. Ger. P II.

5596. BIDONE, Giorgio. 1781-1839. It. P I; GA.

5597. POISSON, Siméon Denis. 1781-1840. Fr. DSB Supp.; ICB; P II; GA; GB. Also Mathematics* and Physics*.
1. *Traité de mécanique.* 2 vols. Paris, 1811. RM/L. 2nd ed., 1833.
1a. ——— *A treatise on mechanics.* 2 vols. London, 1842. Trans. with notes by H.H. Harte. RM/L.
2. *Recherches sur le mouvement des projectiles dans l'air.* Paris, 1839. RM/L.

5598. CISA DI GRESY, Tommaso Asinari. d. 1846. It. P I.

5599. NAVIER, Claude Louis Marie Henri. 1785-1836. Fr. DSB; P II; GA.

5600. BLAND, Miles. 1786-1867. Br. P III; GD. Also Mathematics*.
1. *Elements of hydrostatics.* Cambridge, 1824.

5601. PONCELET, Jean Victor. 1788-1867. Fr. DSB; ICB; P II,III; GA; GB. Also Mathematics*.
1. *Mémoire sur les roues hydrauliques.* Metz, 1826. X. Nouv. éd., rev., corrigée et aug., ib., 1827. RM.

5602. BORDONI, Antonio Maria. 1789-1860. It. P I,III. Also Mathematics.

5603. CAUCHY, Augustin Louis. 1789-1857. Fr. DSB; ICB; P I,III; GA; GB. Also Mathematics*# and Astronomy. See also 15509/1.

5604. HODGKINSON, Eaton. 1789-1861. Br. DSB; P I,III; GB; GD.

5605. BÉLANGER, Jean Baptiste Charles Joseph. 1790-1874. Fr. P I, III.

5606. PIOLA, Gabrio. 1791-1850. It. P II.
1. *Sull'applicazione de principi della meccanica analitica del Lagrange ai principali problemi*. Milan, 1825. RM.

5607. CORIOLIS, Gaspard Gustave de. 1792-1843. Fr. DSB; ICB; P I; GA.
1. *Du calcul de l'effet des machines; ou, Considérations sur l'emploi des moteurs et sur leur évaluation*. Paris, 1829. RM.
2. *Traité de la mécanique des corps solides et du calcul de l'effet des machines*. Paris, 1829. X. 2nd ed., 1844. RM.
3. *Théorie mathématique des effets du jeu de billard*. Paris, 1835. RM.

5608. DELPRAT, Isaac Paul. 1793-1880. Holl. P I,III; GF.

5609. LARDNER, Dionysius. 1793-1859. Br. ICB; P I & Supp.; BMNH; GA; GB; GD. Other entries: see 3303.
1. *Elements of the theory of central forces*. Dublin, 1820.
2. *A treatise on hydrostatics and pneumatics*. London, 1831.

5610. OLIVIER, Théodore. 1793-1853. Fr. P II. Also Mathematics*.

5611. DANDELIN, Germinal Pierre. 1794-1847. Belg. DSB; ICB; P I; GA. Also Mathematics.

5612. WHEWELL, William. 1794-1866. Br. DSB; ICB; P II,III; BMNH; GB; GD. Other entries: see 3304.
1. *An elementary treatise on mechanics*. Cambridge, 1819. 2nd ed., 1824. Another ed., 1828. 5th ed., 1836. 7th ed., 1847.
2. *A treatise on dynamics*. Ib., 1823. RM/L. Another ed., 2 vols, 1832-34.
3. *An introduction to dynamics*. Ib., 1832.
4. *The first principles of mechanics*. Ib., 1832.
5. *Analytical statics*. Ib., 1833.
6. *Mechanical Euclid*. Ib., 1837. X. Another ed., 1838. 5th ed., 1849.
7. *Newton's "Principia", Book 1, Sections I,II,III*. Ib., 1846. Ed. by Whewell. In the original Latin.

5613. JACKSON, Thomas. d. 1837. Br. P I.
1. *Elements of theoretical mechanics*. Edinburgh, 1827. RM.

5614. BLAKE, Eli Whitney (senior). 1795-1886. U.S.A. P III,IV; GE.

5615. LAMÉ, Gabriel. 1795-1870. Fr. DSB; ICB; P I,III; GA. Also Mathematics* and Physics*.
1. *Leçons sur la théorie mathématique de l'élasticité des corps solides*. Paris, 1852. 2nd ed., 1866.

5616. MORIN, Arthur Jules. 1795-1880. Fr. ICB; P II,III; GA.
1. *Notices fondamentales de mécanique*. 2nd ed., Paris, 1855. 3rd ed., 1860.
2. *Notices géométriques sur les mouvements et leur transformations; ou, Eléments de cinématique*. Paris, 1857. X. 3rd ed., 1861.
3. *Hydraulique*. Paris, 1858. X. 2nd ed., 1858.

5617. BRASHMAN, Nikolai Dmitrievich. 1796-1866. Russ. DSB; P I,III (BRASCHMANN).

5618. GARTHE, Caspar. 1796-1876. Ger. P I,III.
 1. *Foucault's Versuch als direkter Beweis der Achsendrehung der Erde angestellt im Dom zu Köln ... und Beschreibung eines neuen Apparats dazu.* Cologne, 1852. X. Reprinted, Wiesbaden, 1969.

5619. PAGANI, Gaspard Michel Marie. 1796-1855. Belg. P II,III; GF.

5620. BURG, Adam von. 1797-1882. Aus. P I,III,IV. Also Mathematics.

5621. DUHAMEL, Jean Marie Constant. 1797-1872. Fr. DSB; P I,III. Also Mathematics* and Physics.
 1. *Cours de mécanique.* 2 vols. Paris, 1845-46. 2nd ed., 1853-54.

5622. HAGEN, Gotthilf Heinrich Ludwig. 1797-1884. Ger. ICB; P I,III.

5623. SAINT-VENANT, Adhémar Jean Claude Barré de. 1797-1886. Fr. DSB; P III,IV. Also Mathematics. See also 5693/1a and 15565/1.

5624. BOBILLIER, Etienne. 1798-1840. Fr. DSB; P IV. Also Mathematics.

5625. GASCHEAU, Gabriel. 1798-1883. Fr. P III; IV Supp.

5626. POTTER, Richard. 1799-1886. Br. P II,III; GD. Also Physics*.
 1. *An elementary treatise on mechanics.* 3rd ed. London, 1855.
 2. *An elementary treatise on hydrostatics.* Cambridge, 1859.

5627. WILLIS, Robert. 1800-1875. Br. DSB; P II,III; GD.

5628. COMBES, Charles Pierre Mathieu. 1801-1872. Fr. DSB; P I,III. Also Physics*.

5629. MOSELEY, Henry. 1801-1872. Br. P III; GD.
 1. *A treatise on hydrostatics and hydrodynamics.* Cambridge, 1830.
 2. *Illustrations of mechanics.* London, 1839.
 3. *A treatise on mechanics applied to the arts.* 2nd ed. London, 1839.

5630. OSTROGRADSKY, Mikhail Vasilievich. 1801-1862. Russ. DSB; ICB; P II & Supp. Also Mathematics.

5631. CHELINI, Domenico. 1802-1878. It. P III.

5632. JACOBI, Carl Gustav Jacob. 1804-1851. Ger. DSB; ICB; P I,III; GA; GB; GC. Also Mathematics*#.
 1. *Vorlesungen über Dynamik.* Berlin, 1866. Ed. by Clebsch.

5633. MOIGNO, François Napoléon Marie. 1804-1884. Fr. P II,III; GA. Also Physics*.
 1. *Leçons de mécanique analytique ... Statique.* Paris, 1868. Reprinted, New York, 1966.

5634. STEICHEN, Michel. 1804-1891. Belg. P II,III.

5635. DIRICHLET, Gustav Peter Lejeune. 1805-1859. Ger. DSB; ICB; P I,III; GA (LEJEUNE-D.); GC. Also Mathematics*. (His books are often listed under LEJEUNE-DIRICHLET.)
 1. *Vorlesungen über die umgekehrten Verhältniss des Quadrats der Entfernung wirkenden Kräfte.* Leipzig, 1876. Ed. by F. Grube. 2nd ed., 1887.

5636. HAMILTON, William Rowan. 1805-1865. Br. DSB; ICB; P III; GB; GD. Also Mathematics*# and Physics. See also 15596/1.

5637. SANG, Edward. 1805-1890. Br. P III,IV. Also Astronomy.

5638. WEISBACH, Julius Ludwig. 1806-1871. Ger. DSB; P II,III; GC.

5639. MENABREA, Luigi Federico. 1809-1896. It. DSB; P II,III,IV; GB.

5640. PRATT, John Henry. 1809-1871. Br./India. DSB; P III; GD.
1. *The mathematical principles of mechanical philosophy, and their application to the theory of universal gravitation.* Cambridge, 1836. RM/L. 2nd ed., 1845.
2. *A treatise on attractions, Laplace's functions, and the figure of the earth.* Cambridge, 1860. X. 3rd ed., 1865.

5641. REDTENBACHER, Ferdinand Jakob. 1809-1863. Switz./Ger. DSB; P II,III; GC.

5642. FROUDE, William. 1810-1879. Br. DSB; ICB; P III; GD.

5643. DIEU, Théodore. 1811-1877. Fr. P III.

5644. SOKOLOV, Ivanov Dmitrievich. b. 1812. Russ. ICB.

5645. DECHER, Georg. 1813-after 1868. Ger. P III.

5646. RICHELMY, Prospero. 1813-1884. It. P III.

5647. TURAZZA, Domenico. 1813-1892. It. P III,IV.

5648. DA SILVA, Daniel Augusto. 1814-1878. Port. ICB.

5649. BARNARD, John Gross. 1815-1882. U.S.A. P III; GE.

5650. DESPEYROUS, Théodore. b. 1815. Fr. P III,IV Supp.
1. *Cours de mécanique.* 2 vols. Paris, 1884-86. Ed. by G. Darboux.

5651. SOMOV, Osip Ivanovich. 1815-1876. Russ. DSB; ICB; P II (SOMOFF, J.); P III (SSOMOFF, O.I.). Also Mathematics.
1. [In Russian. *Rational mechanics.*] 2 vols. St. Petersburg, 1872-74. X.
1a. ——— *Theoretische Mechanik.* Leipzig, 1878. Trans. by A. Ziwet.

5652. DELAUNAY, Charles Eugène. 1816-1872. Fr. DSB; P I,III. Also Astronomy*.
1. *Cours élémentaire de mécanique.* 5th ed. Paris, 1862.
2. *Traité de mécanique rationelle.* Paris, 1856. RM/L.

5653. FROST, Percival. 1817-1898. Br. P IV; GD1. Also Mathematics*.
1. *Newton's "Principia", Sections I,II,III, with notes and illustrations, and a collection of problems principally intended as examples of Newton's methods.* Cambridge, 1854. X. 2nd ed., 1863. 3rd ed., 1880. 4th ed., 1883. 5th ed., 1900.

5654. CELLÉRIER, Charles. 1818-1890. Switz. P III,IV.

5655. GIRAULT, Charles François. b. 1818. Fr. P III,IV Supp.

5656. PRICE, Bartholomew. 1818-1898. Br. P III; GB; GD1.

5657. BASHFORTH, Francis. 1819-1912. Br. P III; GD3.
1. *Tables of remaining velocity, time of flight, and energy*

of various projectiles calculated from the results of experiments made with the Bashforth chronograph. London, 1871.
2. *A mathematical treatise on the motion of projectiles.* London, 1873.
3. *An attempt to test the theories of capillary action.* Cambridge, 1883.
4. *A revised account of the experiments made with the Bashforth chronograph ... with the application of the results to the calculation of trajectories.* Cambridge, 1890.

5658. FOUCAULT, (Jean Bernard) Léon. 1819-1868. Fr. DSB; ICB; P I, III; GA; GB.
1. *Sur divers signes sensibles du mouvement diurne de la terre.* [Periodical articles, 1851-52.] Osnabrück, 1972. Ed. by H. Kangro. See also 5618/1.

5659. STOKES, George Gabriel. 1819-1903. Br. DSB; ICB; P II,III,IV; GB; GD2. Also Physics*#.

5660. SCHEFFLER, (August Christian Wilhelm) Hermann. 1820-1903. Ger. P II,III,IV. Also Mathematics and Physics.

5661. TODHUNTER, Isaac. 1820-1884. Br. DSB; P III; GB; GD. Also Mathematics*, Biography*, and History of Mathematics*.
1. *A treatise on analytical statics.* London, 1853. 3rd ed., 1866. 4th ed., 1874. 5th ed., 1887.

5662. CULMANN, Karl. 1821-1881. Switz. DSB; P III; GC.

5663. HAUGHTON, Samuel. 1821-1897. Br. P III,IV; BMNH; Mort.; GB; GD1. Other entries: see 3336.
1. *A manual of mechanics.* London, 185-? X. Another ed., 1866.

5664. HELMHOLTZ, Hermann Ludwig Ferdinand von. 1821-1894. Ger. DSB; ICB; P I,III,IV; Mort.; BLA & Supp.; GB; GC. Also Science in General*#, Physics*, and Physiology*. See also Index.
1. *Zwei hydrodynamische Abhandlungen: I, Über Wirbelbewegungen.* [1858] *II, Über discontinuirliche Flüssigkeitsbewegungen.* [1868] Leipzig, 1896. (OKEW. 79)
2. *Vorlesungen über die Dynamik discreter Massenpunkte.* Leipzig, 1898. Ed. by O. Krigar-Menzel.
3. *Dynamik continuirlich verbreiteter Massen.* Leipzig, 1902. Ed. by O. Krigar-Menzel.

5665. BAEHR, George Frederik Willem. 1822-1898. Holl. P III,IV; GF. Also Mathematics.

5666. BOURGET, Justin. 1822-1887. Fr. P III,IV. Also Mathematics.

5667. BRESSE, Jacques Antoine Charles. 1822-1883. Fr. ICB; P I,III.

5668. BETTI, Enrico. 1823-1892. It. DSB; P III,IV. Also Mathematics*.

5669. DAVIDOV, August Yulevich. 1823-1885. Russ. DSB; P I,IV (DAWIDOFF). Also Mathematics.

5670. BRIOSCHI, Francesco. 1824-1897. It. DSB; P III,IV. Also Mathematics*.

5671. KIRCHHOFF, Gustav Robert. 1824-1887. Ger. DSB; ICB; P I,III,

IV; GB; GC. Also Physics*# and Chemistry*.
1. *Vorlesungen über mathematische Physik: Mechanik.* Leipzig, 1876. 3rd ed., 1883.

5672. BJERKNES, Carl Anton. 1825-1903. Nor. DSB; ICB; P III,IV. Also Biography*.
1. *Hydrodynamische Fernkräfte. Fünf Abhandlungen.* [1868-80] Leipzig, 1915. (OKEW. 195) Trans. by A. Korn.

5673. GRASHOF, Franz. 1826-1893. Ger. DSB; ICB; P III. Also Physics.

5674. RITTER, (Georg) August (Dietrich). 1826-1908. Ger. P III,IV. Also Physics.
1. *Lehrbuch der analytischen Mechanik.* 2nd ed. Leipzig, 1883.

5675. SCHELL, Wilhelm Joseph Friedrich Nikolaus. 1826-1904. Ger. P II,III,IV. Also Mathematics.
1. *Theorie der Bewegung und der Kräfte.* Leipzig, 1870. 2nd ed., 2 vols, 1879.

5676. CURTIS, Arthur Hill. 1827-after 1885. Br. P III,IV.

5677. MERRIFIELD, Charles Watkins. 1827-1884. Br. P III; GD.

5678. RAZZABONI, Cesare. 1827-1893. It. P III,IV.

5679. WATSON, Henry William. 1827-1903. Br. P III,IV; GD2. Also Physics*.
1. *A treatise on the application of generalised co-ordinates to the kinetics of a material system.* Oxford, 1879. With S.H. Burbury.

5680. WILLIAMSON, Benjamin. 1827-1916. Br. P III,IV; GF. Also Mathematics*.
1. *An elementary treatise on dynamics.* London, 1885.
2. *Introduction to the mathematical theory of the stress and strain of elastic solids.* London, 1894.

5681. BESANT, William Henry. 1828-1917. Br. P III. Also Mathematics*.
1. *A treatise on hydromechanics.* Cambridge, 1859. 2nd ed., 1867. 3rd ed., 1877. 4th ed., 1883.
2. *Elementary Hydrostatics.* London, 1859. X. 17th ed., 1895. Another ed., 1898.
3. *A treatise on dynamics.* London, 1885.

5682. FREYCINET, Charles Louis de Saulses de. 1828-1923. Fr. ICB; P III; GB. Also Mathematics*.
1. *Traité de mécanique rationelle.* Paris, 1858. RM.

5683. RESAL, Henri Amé. 1828-1896. Fr. P II,III,IV. Also Astronomy*.
1. *Traité de mécanique générale.* 4 vols. Paris, 1873-76.

5684. STANLEY, William Ford Robinson. 1829-1909. Br. GD2. Also Astronomy*.
1. *Experimental researches into properties and motions of fluids with theoretical deductions therefrom.* London, 1881.

5685. COLLIGNON, Romain Charles Edouard. 1831-after 1906. Fr. P III, IV.
1. *Traité de mécanique.* 4 vols. Paris, 1874-80. 2nd ed., 1881.

5686. DURRANDE, Antoine Henri. 1831-after 1880. Fr. P III.

5687. ROUTH, Edward John. 1831-1907. Br. DSB; P III,IV; GB; GD2.
1. *A treatise on the dynamics of a system of rigid bodies.* 2 vols. London, 1860. X.
1a. ——— *The elementary part.* 2nd ed., 1868. 3rd ed., 1877. 4th ed., 1882. 5th ed., 1884. 6th ed., 1897.
1b. ——— *The advanced part.* 4th ed., 1884. 5th ed., 1892.
2. *A treatise on the stability of a given state of motion.* London, 1877.
3. *A treatise on analytical statics.* 2 vols. Cambridge, 1891-92. 2nd ed., 1896-1902.
4. *A treatise on the dynamics of a particle.* Cambridge, 1898.
See also 5593/1.

5688. SAINT-LOUP, Louis. 1831-after 1890. Fr. P III,IV.

5689. TAIT, Peter Guthrie. 1831-1901. Br. DSB; ICB; P III,IV; GB; GD2. Also Mathematics* and Physics*#. See also Index.
1. *A treatise on the dynamics of a particle.* Cambridge, 1856. With W.J. Steele. X. 2nd ed., 1865. 4th ed., London, 1878. 5th ed., 1882. Another ed., 1895.
2. *Newton's laws of motion.* London, 1899. RM/L.

5690. BOUR, Edmond. 1832-1866. Fr. DSB; P III. Also Mathematics and Astronomy.

5691. GILBERT, (Louis) Philippe. 1832-1892. Belg. P III,IV. Also Mathematics.

5692. WOOD, De Volson. 1832-after 1883. U.S.A. P III.

5693. CLEBSCH, Rudolph Friedrich Alfred. 1833-1872. Ger. DSB; P III; GC. Also Mathematics*.
1. *Theorie der Elasticität fester Körper.* Leipzig, 1862. RM/L.
1a. ——— *Théorie de l'élasticité des corps solides.* Paris, 1883. Trans. with extensive notes by Barré de Saint-Venant. Reprinted, New York, 1966.
See also 5632/1.

5694. DÜHRING, Eugen Karl. 1833-1921. Ger. The philosopher. P III, IV; GB. Also Physics*.
1. *Kritische Geschichte der allgemeinen Principien der Mechanik.* Berlin, 1873. RM. 3rd ed., Leipzig, 1877; reprinted, Wiesbaden, 1970.
See also 2953/1,1a.

5695. HATON DE LA GOUPILLIÈRE, Julien Napoléon. 1833-1927. Fr. ICB; P III,IV.

5696. MOGNI, Antonio. 1834-after 1882. It. P III.

5697. BELTRAMI, Eugenio. 1835-1899. It. DSB; P III,IV. Also Mathematics*.
1. *Richerche sulla cinematica dei fluidi.* Bologna, 1875.

5698. MATHIEU, Emile Léonard. 1835-1890. Fr. DSB; P III,IV. Also Mathematics and Physics*.
1. *Dynamique analytique.* Paris, 1878.
2. *Théorie de l'élasticité des corps solides.* 2 vols. Paris, 1890.

5699. COTTERILL, James Henry. 1836-1922. Br. P IV & Supp. Also Physics*.

5700. BARDELLI, Giuseppe. 1837-1908. It. P III,IV.

5701. DOLBEAR, Amos Emerson. 1837-1910. U.S.A. P III,IV.
1. *Matter, ether and motion*. London, 1892. X. Another ed., 1899.
2. *Machinery of the universe: Mechanical conception of physical phenomena*. London, 1897.

5702. LÉVY, Maurice. 1838-1910. Fr. DSB; P III,IV. Also Physics.

5703. MACH, Ernst. 1838-1916. Aus./Prague. DSB; ICB; P III,IV. Also Physics*, Physiology, and History of Physics*.
1. *Die Mechanik in ihrer Entwickelung historisch-kritisch dargestellt*. Leipzig, 1883. X.
1a. ——— *The science of mechanics: A critical and historical exposition of its principles*. Chicago, 1893. Trans. from the 2nd German ed. by T.J. McCormack.

5704. GIBBS, (Josiah) Willard. 1839-1903. U.S.A: DSB; ICB; P III, IV & Supp.; GB; GE. Also Physics*#.
1. *The early work of Willard Gibbs in applied mechanics*. New York, 1947. Ed. by L.P. Wheeler et al.

5705. SAINT-GERMAIN, Albert Léon de. 1839-1914. Fr. P III,IV.

5706. SIACCI, Francesco. 1839-1907. It. ICB; P III,IV.

5707. BALL, Robert Stawell. 1840-1913. Br. P III,IV; GD3. Also Astronomy*, Autobiography*, and Biography*.
1. *Experimental mechanics*. London, 1871. Another ed., 1888.
2. *The theory of screws: A study in the dynamics of a rigid body*. Dublin, 1876. RM/L.
3. *Mechanics*. London, 1879. Another ed., 1883.
4. *Dynamics and modern geometry*. Dublin, 1887.

5708. BURMESTER, Ludwig Ernst Hans. 1840-1927. Ger. P III,IV.
1. *Lehrbuch der Kinematik*. Vol. 1 [No more publ.?] Leipzig, 1888.

5709. GIESEN, Anton. 1840-after 1882. Posen. P III. See also 6501/3.

5710. FINGER, Josef. 1841-1925. Aus. P III,IV.

5711. LAURENT, (Matthieu Paul) Hermann. 1841-1908. Fr. DSB; P III, IV. Also Mathematics*.
1. *Traité de mécanique rationelle*. Paris, 1870. X. 2nd ed., 2 vols, 1878. 3rd ed., 2 vols, 1889.

5712. SLUDSKY, Fedor Alekseievich. 1841-1897. Russ. P III,IV. Also Geophysics and Geodesy.

5713. TARLETON, Francis Alexander. 1841-1920. Br. P IV.
1. *An introduction to the mathematical theory of attraction*. London, 1899.

5714. BOBYLEFF, Dmitry Konstantinovich. 1842-after 1900. Russ. P III,IV.

5715. BOUSSINESQ, Joseph Valentin. 1842-1929. Fr. DSB; ICB; P III, IV. Also Physics.

1. *Essai théorique sur l'équilibre des massifs pulvérulents comparé à celui de massifs solides, et sur la poussée des terres sans cohésion.* Brussels, 1876.
2. *L'application des potentiels à l'étude de l'équilibre et du mouvement des solides élastiques.* Lille, 1885.
3. *Théorie de l'écoulement tourbillonnant et tumulteux des liquides.* Paris, 1897.

5716. BUDDE, Emil Arnold. 1842-1921. Ger. P III,IV. Also Physics.
1. *Allegemeine Mechanik der Punkte und starren Systeme.* 2 vols. Berlin, 1890-91.

5717. REYNOLDS, Osborne. 1842-1912. Br. DSB; ICB; P III,IV; GD3. Also Physics*.
1. *Papers on mechanical and physical subjects.* 3 vols. Cambridge, 1900-03.

5718. BOLTZMANN, Ludwig. 1844-1906. Aus./Ger. DSB; ICB; P III,IV. Also Physics*#.
1. *Vorlesungen über die Principe der Mechanik.* 3 vols. Leipzig, 1897-1920.
2. *Uber die Prinzipien der Mechanik.* Leipzig, 1903. RM.

5719. CLIFFORD, William Kingdon. 1845-1879. Br. DSB; ICB; P III; GB; GD. Also Science in General* and Mathematics*#.
1. *Elements of dynamics: An introduction to the study of motion and rest in solid and fluid bodies.* 2 vols. London, 1878-87.

5720. MINCHIN, George. 1845-1914. Br. P III. Also Physics.
1. *A treatise on statics.* London, 1877. 2nd ed., Oxford, 1880. 4th ed., Oxford, 1890.
2. *Uniplanar kinematics of solids and fluids.* Oxford, 1882.
3. *Hydrostatics and elementary hydrokinetics.* Oxford, 1892.

5721. PADOVA, Ernesto. 1845-1896. It. P IV.

5722. LIGIN (or LIGUINE), Valerian Nicolaievich. 1846-1900. Russ. P III,IV.

5723. GREENHILL, (Alfred) George. 1847-1927. Br. ICB; P III,IV & Supp. Also Mathematics*.
1. *A treatise on hydrostatics.* London, 1894.

5724. KENNEDY, Alexander Blackie William. 1847-1928. Br. DSB; ICB; P III,IV; GD4.

5725. LÉAUTÉ, Henry Charles Victor Jacob. 1847-1916. Fr. P III,IV.

5726. OEKINGHAUS, Emil Wilhelm. 1847-ca. 1918. Ger. P IV.

5727. ZHUKOVSKY, Nikolai Egorovich. 1847-1921. Russ. DSB; ICB; P III,IV (SCHUKOWSKI).

5728. KORTEWEG, Diederik Johannes. 1848-1941. Holl. DSB; P III,IV.

5729. MERRIMAN, Mansfield. 1848-1925. U.S.A. P III,IV. Also Geodesy*.
1. *A treatise on hydraulics.* New York, 1889. 4th ed., 1892.

5730. RÉTHY, Moritz. 1848 (or 1846)-1925. Hung. P III,IV.

5731. LAMB, Horace. 1849-1934. Br. DSB; ICB; P III,IV; GD5. Also Physics and Geophysics.

1. *A treatise on the mathematical theory of the motion of fluids.* Cambridge, 1879. 2nd ed., entitled *Hydrodynamics*, 1895.

5732. SPARRE, Magnus Louis Marie de. 1849-after 1919. Fr. P IV.

5733. CERRUTI, Valentino. 1850-1909. It. P III,IV.

5734. HENNEBERG, (Ernst) Lebrecht. 1850-1933. Ger. P III,IV.

5735. HICKS, William Mitchinson. 1850-1934. Br. ICB; P III,IV; GF.

5736. PENNACHIETTI, Giovanni. 1850-1916. It. P IV.

5737. PERRY, John. 1850-1920. Br. P III,IV; GF. Also Physics.
   1. *Spinning tops.* London, 1890.

5738. HUGONIOT, Pierre Henri. 1851-1887. Fr. DSB.

5739. LEWIS, Thomas Crompton. 1851-1929. Br. P III,IV & Supp.

5740. MISCHER, Rudolf. 1851-after 1884. Ger. P III.

5741. GWYTHER, Reginald Felix. 1852-1930. Br. P IV.

5742. PADELLETTI, Dino. 1852-1892. It. P III,IV.

5743. RAUSENBERGER, (Siegmund) Otto. 1852-after 1912. Ger. P III,IV.
   1. *Lehrbuch der analytischen Mechanik.* Leipzig, 1888.

5744. SOMOV, Pavel Ossipovich. 1852-after 1904. Russ./Warsaw. P IV (SSOMOFF).

5745. WORTHINGTON, Arthur Mason. 1852-1916. Br. P III,IV. Also Physics*.
   1. *The dynamics of rotation: An elementary introduction to rigid dynamics.* 1892.
   2. *The splash of a drop. A discourse....* London, 1895.

5746. JAERISCH, Paul Georg Maximilian. 1853-1931. Ger. P III,IV.

5747. BASSET, Alfred Barnard. 1854-1930. Br. P IV. Also Physics*.
   1. *A treatise on hydrodynamics.* 2 vols. Cambridge, 1881. Another ed., 1888.
   2. *An elementary treatise on hydrodynamics and sound.* Cambridge, 1890.

5748. HELE-SHAW, Henry Selby. 1854-1941. Br. ICB (HALE-S.); GD6.

5749. LECORNU, Léon François Alfred. 1854-1940. Fr. DSB; P IV.

5750. POINCARÉ, (Jules) Henri. 1854-1912. Fr. DSB; ICB; P III,IV. Also Mathematics*#, Astronomy*, and Physics*.
   1. *Leçons sur la théorie de l'élasticité.* Paris, 1892.
   2. *Théorie des tourbillons; leçons....* Paris, 1893.

5751. APPELL, Paul (Emile). 1855-1930. Fr. DSB; ICB; P III,IV. Also Mathematics*.
   1. *Leçons sur l'attraction et la fonction potentielle.* Paris, 1892.
   2. *Traité de mécanique rationelle.* 2 vols. Paris, 1893-96.
   3. *Les mouvements de roulement en dynamique.* Paris, 1899.

5752. CRAIG, Thomas. 1855-1900. U.S.A. P III,IV; GE. Also Mathematics.
   1. *The motion of a solid in a fluid, and the vibrations of*

liquid spheroids. New York, 1879.
2. *Elements of the mathematical theory of fluid motion: Wave and vortex motion.* New York, 1879.

5753. DELAUNAY, Nicolas. 1856-1931. Russ. P IV.

5754. MAGGI, Gian Antonio. 1856-after 1931. It. P III,IV. Also Physics.

5755. MORERA, Giacinto. 1856-1909. It. P III,IV.

5756. ANDRADE, Jules Frédéric Charles. 1857-1933. Fr. P IV.

5757. HERTZ, Heinrich Rudolph. 1857-1894. Ger. DSB; ICB; P III,IV; GB; GC. Also Physics*# and Autobiography*.
1. *Die Prinzipien der Mechanik, in neuem Zusammenhänge.* Leipzig, 1894. (Vol. III of his *Gesammelte Werke:* see 6938/2.)
1a. ——— *The principles of mechanics, presented in a new form.* London, 1899. Trans. by D.E. Jones and J.T. Walley. Reprinted, New York, 1956.

5758. KÖTTER, Fritz Wilhelm Ferdinand. 1857-1912. Ger. P IV.

5759. LYAPUNOV, Aleksandr Mikhailovich. 1857-1918. Russ. DSB; ICB (LIAPUNOV); P III,IV (LJÄPUNOW). Also Mathematics.

5760. STAUDE, (Ernst) Otto. 1857-1928. Ger. ICB; P III,IV. Also Mathematics.

5761. SUSLOW, Gavril Konstantinovich. 1857-1935. Russ. P IV.

5762. HESS, Wilhelm Philipp. 1858-after 1930. Ger. P IV.

5763. KOENIGS, Gabriel Xavier Paul. 1858-1931. Fr. DSB; P IV. Also Mathematics*.
1. *Leçons de cinématique.* Paris, 1895.

5764. MESHCHERSKY, Ivan Vsevolodovich. 1859-1935. Russ. DSB; ICB; P IV (MESTSCHERSKY).

Imprint Sequence

5765. RACCOLTA.
1. *Raccolta d'autori che trattano del moto dell'acque.* 3 vols. Florence, 1723. X.
1a. ——— 2nd ed., rev. and enl., 9 vols, ib., 1765-74. RM.
2. *Nuova raccolta d'autori che trattano del moto dell'acque.* 7 vols. Parma, 1766-68. RM.

5766. MOTTE, Andrew. d. 1734. Br. ICB; GD (under M., Benjamin).
1. *A treatise on the mechanical powers, wherein the laws of motion and the properties of those powers are explained.* London, 1727. RM/L.
See also 2539/4.

5767. WRIGHT, John Martin Frederick.
1. *"Principia"* [of Newton] *with notes, examples and deductions. Containing all that is read at the University of Cambridge* [i.e. Sections 1-3]. Cambridge, 1830.
2. *A commentary on Newton's "Principia."* 2 vols. London, 1833. Reprinted, New York, 1972.

3.06 PHYSICS

5768. LA HIRE, Philippe de. 1640-1718. Fr. DSB; ICB; P I; GA. Also Mathematics, Astronomy*, Geodesy, and Part 2*.

5769. NEWTON, Isaac. 1642-1727. Br. DSB; ICB; P II; G. Also Mathematics*, Astronomy*, and Part 2*(the *Principia*). See also Index.

1. *Opticks; or, A treatise of the reflections, refractions, inflections and colours of light.* London, 1704. RM/L. Reprinted, Brussels, 1966.

1a. ——— *Optice; sive, De reflexionibus, refractionibus, inflexionibus et coloribus lucis.* London, 1706. Trans. by S. Clarke and J. Moore. RM/L.

1b. ——— German trans. 2 vols. Leipzig, 1898. (OKEW)

1c. ——— 2nd ed., London, 1718. RM.

1d. ——— ——— Latin trans. Ib., 1719. RM.

1e. ——— 3rd ed. Ib., 1721. RM/L.

1f. ——— 4th ed. Ib., 1730. RM/L. Reprinted, ib., 1931, and New York, 1952.

1g. ——— ——— *Optique. Traduction nouvelle.* 2 vols. Paris, 1787. Trans. by J.P. Marat.

Optical Lectures

2. *Optical lectures read in the publick schools of the University of Cambridge, anno 1669 ... never before printed. Translated ... out of the original Latin.* London, 1728. RM.

3. *Lectiones opticae, annis 1669, 1670, et 1671 in scholis publicis habitae, et nunc primum ex MSS in lucem editae.* London, 1729. RM/L.

4. *The unpublished first version of Isaac Newton's Cambridge lectures on optics, 1670-1672.* Cambridge, 1973. A facsimile of the autograph with an introd. by D.T. Whiteside.

Correspondence

5. *Correspondence of Sir I. Newton and Professor Cotes, including letters of other eminent men.* London, 1850. Ed. with notes by J. Edleston. Reprinted, ib., 1969.

6. *Correspondence of Sir I. Newton and Professor Cotes.* n.p. [1852]. Extracts from the *Journal des savants.* Text in French. Ed. by J.B. Biot.

7. *The correspondence.* Cambridge, 1959-. Ed. under the auspices of the Royal Society by H.W. Turnbull et al.

Collected Works and Selections

8. *Opuscula mathematica, philosophica et philologica.* Lausanne, 1744. Ed. by J. Castillon. RM.

9. *Opera quae extant omnia.* 5 vols. London, 1779-85. Ed. by S. Horsley. (Far from complete.) RM/L. Reprinted, Stuttgart, 1964.
10. *A catalogue of the Portsmouth collection of books and papers written by or belonging to Sir Isaac Newton.* Cambridge, 1888. RM/L.
11. *Papers and letters on natural philosophy, and related documents.* Cambridge, Mass., 1958. Ed. by I.B. Cohen et al. 2nd ed., 1978.
12. *Unpublished scientific papers. A selection from the Portsmouth collection.* Cambridge, 1962. Ed. by A.R. Hall and M.B. Hall.
13. *Sir Isaac Newton's account of the ether. With some additions.* Dublin, 1745. By Bryan Robinson. (cf. 5476/2) RM.
14. *Newton's philosophy of nature. Selections from his writings.* New York, 1953 and 1960. Ed. by H.S. Thayer.

5770. JOBLOT, Louis. 1645-1723. Fr. DSB; ICB. Also Microscopy*.

5771. VATER, Christian. 1651-1732. Ger. P II; BLA; GC. Also Chemistry.

5772. SAUVEUR, Joseph. 1653-1716. Fr. DSB; ICB; P II; GA.

5773. HARTSOEKER, Niklaas. 1656-1725. Holl./Fr./Ger. DSB; P I; BLA; GA. Also Part 2*.
1. *Conjectures physiques.* Amsterdam, 1706. RM.
1a. ——— *Suite des "Conjectures physiques."* 2 vols, ib., 1708.
1b. ——— *Eclaircissements sur les "Conjectures physiques."* Ib., 1710. RM.
2. *Recueil de plusieurs pièces de physique où l'on fait principalement voir l'invalidité du système de Mr. Newton.* Utrecht, 1722. RM.

5774. GOBART, Laurent. ca. 1658-1750. Belg. P I.

5775. JALLABERT, Etienne. 1658-1724. Switz. P I; GA.

5776. GREGORY, David. 1659-1708. Br. DSB; ICB; P I; GA; GB; GD. Also Mathematics*, Astronomy*, and Part 2*,

5777. CARRÉ, Louis. 1663-1711. Fr. P I; GA.

5778. LUTHER, Karl Friedrich. 1663-1744. Ger. P I.

5779. GRAY, Stephen. 1666-1736. Br. DSB; ICB; P I; GA; GD.

5780. HAUKSBEE, Francis (first of the name). ca. 1666-1713. Br. DSB; ICB; P I (HAWKSBEE); GA; GD.
1. *Physico-mechanical experiments on various subjects.* London, 1709. RM. 2nd ed., enl., 1719. RM/L. Reprinted, New York, 1970.
1a. ——— *Expériences physico-mécaniques sur différens sujets.* 2 vols. Paris, 1754. Trans. by Brémond with an introd. by Desmarest. RM/L.

5781. LEUTMANN, Johann Georg. 1667-1736. Ger./Russ. P I.

5782. WHISTON, William. 1667-1752. Br. DSB; P II; BMNH; GA; GB; GD. Also Science in General*, Astronomy*, Mechanics*, Autobiography*, and Part 2*. See also Index.
1. *A course of mechanical, optical, hydrostatical and pneumat-*

ical experiments. To be performed by F. Hauksbee [the younger] *and the explanatory lectures read by W. Whiston.* London, n.d. [1730?]. RM.

5783. GOTTSCHED, Johann. 1668-1704. Ger. P I; BLA.

5784. DUPREZ D'AULNAY, Louis. ca. 1670-1758. Fr. P I.

5785. DOPPELMAYR, Johann Gabriel. 1671?-1750. Ger. DSB; P I; GA; GC. Also Astronomy and History of Mathematics*.
1. *Neu-entdeckte Phaenomena ... welche bey fast allen Cörpern zukommenden electrischen Krafft und dem dabey ... erscheinenden Liecht....* Nuremberg, 1744. RM.

5786. POLINIÈRE (or POLYNIER), Pierre. 1671-1734. Fr. DSB; P II; GA.
1. *Expériences de physique.* Paris, 1709. RM.

5787. SANDEN, Heinrich von. 1672-1728. Ger. P II; BLA.

5788. LEUPOLD, Jacob. 1674-1727. Ger. ICB; P I; BMNH; GA. Also Geology.

5789. LE SAGE, George Louis (first of the name). 1676-1759. Switz. P I; GA.

5790. LA HIRE, Gabriel Philippe de. 1677-1719. Fr. DSB; P I. Also Astronomy.

5791. MAIRAN, Jean Jacques d'Ortus de. 1678-1771. Fr. DSB; ICB; P II; GA. Also Mechanics*, Meteorology*, and Biography*.
1. *Dissertation sur la glace; ou, Explication physique de la formation de la glace, et de ses divers phénomènes.* Paris, 1749. RM.

5792. STANCARI, Vittorio Francesco. 1678-1709. It. P II; GA.

5793. TAGLINI, Carlo. 1679-1747. It. P II & Supp.
1. *Lettere scientifiche sopra vari dilettevoli argomenti di fisica.* Florence, 1747. RM.

5794. VREAM, William. fl. 1710-1727. Br.
1. *A description of the air-pump, according to the late Mr. Hawksbee's best and last improvements.* London, 1717.

5795. GAUGER, Nicolas. ca. 1680-1730. Fr. P I.

5796. HELSHAM, Richard. ca. 1680-1738. Br. P I; GA; GD.
1. *A course of lectures in natural philosophy.* London, 1739. RM/L. 2nd ed., 1743. 3rd ed., 1755.

5797. EHRENBERGER, Bonifaz Heinrich. 1681-1759. Ger. P I; GA.

5798. COTES, Roger. 1682-1716. Br. DSB; ICB; P I; GA; GB; GD. Also Mathematics* and Astronomy. See also Index.
1. *Hydrostatical and pneumatical lectures.* London, 1738. Ed. with notes by Robert Smith, his successor at Cambridge. RM/L. 2nd ed., Cambridge, 1747.

5799. MUYS, Wijer Willem. 1682-1744. Holl. P II; BLA; GA.

5800. DESAGULIERS, John Theophilus. 1683-1744. Br. DSB; ICB; P I; GA; GD. See also Index.
1. *A system of experimental philosophy.* London, 1719. RM.
2. *A course of experimental philosophy.* 2 vols. London, 1734-44. RM/L. 2nd ed., 1745. 3rd ed., 1763.

3. *A dissertation concerning electricity*. London, 1742. RM.

5801. RAMEAU, Jean Philippe. 1683-1764. Fr. The composer. ICB; P II; GA; GB. Acoustics and theory of music.

5802. RÉAUMUR, René Antoine Ferchault de. 1683-1757. Fr. DSB; ICB; P II,III; BMNH; Mort.; GA; GB. Also Chemistry*, Zoology*, and Embryology*.

5803. HASE (or HASIUS), Johann Matthias. 1684-1742. Ger. P I; GA (HAAS).

5804. JURIN, James. 1684-1750. Br. ICB; P I; BLA; GA; GD. Also Physiology.

5805. BARTH, Johann Matthaeus. fl. 1716-1751. Ger. P I.

5806. KRIEGER, Wilhelm. 1685-1769. Aus. P III.

5807. FAHRENHEIT, Gabriel Daniel. 1686-1736. Holl. DSB; ICB; P I; GA; GB; GC.

5808. BUNSEN, Jeremias. 1688-1752. Ger. P I.

5809. CASTEL, Louis Bertrand. 1688-1757. Fr. DSB; P I; GA; GB. Also Mechanics*.
1. *L'optique des couleurs*. Paris, 1740. RM.

5810. FREKE, John. 1688-1756. Br. BLA; GD. Also Chemistry*.
1. *An essay to shew the cause of electricity and why some things are non-electricable*. London, 1746. RM. 2nd ed., 1746.

5811. HAUKSBEE, Francis (second of the name). 1688-1763. Br. DSB; ICB; GD. See also 5782/1.

5812. SMITH, Robert. 1689-1768. Br. DSB; P II; GA; GB; GD.
1. *A compleat system of optics*. 2 parts. Cambridge, 1738. RM/L.
1a. ——— *Cours complet d'optique*. 2 vols. Avignon, 1767.
2. *Harmonics; or, The philosophy of musical sounds*. Cambridge, 1749. X. 2nd ed., 1759.
See also 5798/1.

5813. CONRADI, Johann Michael. d. 1742. Ger. P I.

5814. GERING, Jacob. ca. 1690-1756. Ger. P I.

5815. RICHTER, Georg Friedrich. 1691-1742. Ger. P II.

5816. TRIEWALD (or TRIEVALD), Mårten. 1691-1747. Swed. ICB; P II.
1. *Kort beskrifning om en eld- och luft-machin vid Dannemora grufwar*. Stockholm, 1734. RM.

5817. LOVETT, Richard. 1692-1780. Br. GD.

5818. MUSSCHENBROEK, Petrus van. 1692-1761. Holl. DSB; ICB; P II; BLA; GA; GB.

Editions of his Lecture Course

1. *Epitome elementorum physico-mathematicorum*. Leiden, 1726. X.
2. *Elementa physicae*. Leiden, 1734. X.
3. *Beginsels der natuurkunde*. Leiden, 1736. X. 2nd ed., 1739. RM.
4. [Trans.] *Essai de physique, avec une description de nouv-*

elles sortes de machines pneumatiques et un recueil d'expériences. 2 vols. Leiden, 1736-39. Trans. from the Dutch by P. Massuet. RM.
5. [Trans.] *The elements of natural philosophy.* London, 1744. Trans. from the Latin by J. Colson. RM. Reprinted, 2 vols, Ann Arbor, 1972.
6. *Institutiones physicae.* Leiden, 1748. RM.
7. *Compendium physicae experimentalis.* Leiden, 1762. RM.
8. *Introductio ad philosophiam naturalem.* 2 vols. Leiden, 1762. RM.
9. [Trans.] *Cours de physique expérimentale et mathématique.* 3 vols. Paris, 1769. Trans. by Sigaud de la Fond. RM/L.

Other Works

10. *Physicae experimentalis et geometricae ... dissertationes.* Leiden, 1729.
11. *Tentamina experimentorum naturalium captorum in Academia del Cimento ... Ex italico in latinum sermonem conversa. Quibus commentarios, nova experimenta et orationem de methodo instituendi experimenta physica addidit.* Leiden, 1731. (cf. 2812/1b) RM.

5819. TARTINI, Giuseppe. 1692-1770. It. P II; GA; GB.

5820. BALBI, Paolo Battista. 1693-1772. It. P I; BLA & Supp.

5821. VOLTAIRE, François Marie Arouet de. 1694-1778. Fr. The famous philosophe. DSB; ICB; P II; G.
1. *Elémens de la philosophie de Neuton.* Amsterdam, 1738. Deals mostly with optics. RM/L. Reprinted, Warsaw, 1956.
1a. ——— *The elements of Sir Isaac Newton's philosophy.* 1738. X. Reprinted, London, 1967.
See also 15636/2 and 15901/1.

5822. RIZZETTI, Giovanni. d. 1751. It. P II.

5823. WAGNER, Johann Georg. d. 1756. Ger. P II.
1. *Erforschung der Ursachen von den electrischen Wirkungen.* Liegnitz, 1747. RM.

5824. ALEFELD, Johann Ludwig. 1695-1759/60. Ger. P I.

5825. STAEHELI, Benedict. 1695-1750. Switz. P II; GA (STÄHELIN). Also Botany.

5826. ENGELHARD, Nicolaus. 1696-1765. Ger./Holl. P I.

5827. HOLLMANN, Samuel Christian. 1696-1787. Ger. ICB; P I.

5828. HAMBERGER, Georg Erhard. 1697-1755. Ger. ICB; P I; BMNH; Mort.; BLA; GA; GC. Also Physiology.
1. *Elementa physices, methodo mathematica.* Jena, 1727. X. 3rd ed., 1741. RM.

5829. VIVENS, François de. 1697-1780. Fr. P II.

5830. BOUGUER, Pierre. 1698-1758. Fr. DSB; ICB; P I; GA; GB. Also Astronomy*, Mechanics*, and Geodesy*.
1. *Traité d'optique sur la gradation de la lumière.* Paris, 1760. Publ. posthumously by de Lacaille. RM.
1a. ——— *Optical treatise on the gradation of light.* Toronto, 1961. Trans. with introd. by W.E.K. Middleton.

5831. DIVIŠ, Prokop. 1698-1765. Moravia. DSB; ICB; P I (DIVISCH).

5832. DUFAY, Charles François de Cisternay. 1698-1739. Fr. DSB; ICB; P I; GA.

5833. KLINGENSTIERNA, Samuel. 1698-1765. Swed. DSB; ICB; P I; GA.

5834. MILES, Henry. 1698-1763. Br. ICB; P II; GD.

5835. KLEIST, Ewald Georg von. ca. 1700-1748. Ger. DSB; ICB; P I.

5836. NOLLET, Jean Antoine. 1700-1770. Fr. DSB; P II; GA; GB.
1. *Leçons de physique expérimentale*. 6 vols. Paris, 1743-64. Another ed., Amsterdam, 1754-65. RM.
2. *Essai sur l'électricité des corps*. Paris, 1746. RM. 2nd ed., 1750. Another ed., 1765.
3. *Recherches sur les causes particulières des phénomènes électriques*. Paris, 1749. RM. 3rd ed., 1753.
4. *Lettres sur l'électricité dans lesquelles on examine les dernières découvertes*. 2 vols. Paris, 1753-60. RM. Another ed., 1764.
5. *L'art des expériences; ou, Avis aux amateurs de la physique sur la choix, la construction et l'usage des instruments*. 3 vols. Paris, 1770. X. 2nd ed., 1770. RM.

5837. EMERSON, William. 1701-1782. Br. P I; GA; GB; GD. Also Mathematics*, Astronomy*, and Mechanics*.
1. *The elements of optics*. London, 1768. RM.

5838. KRAFFT, Georg Wolfgang. 1701-1754. Ger./Russ. P I; GA (KRAFT); GC (KRAFT). Also Mathematics and Astronomy.

5839. ROWNING, John. 1701?-1771. Br. DSB; P II; GD.
1. *A compendious system of natural philosophy*. 4 parts. Cambridge/London, 1735-43. X. 8th ed., London, 1779.

5840. WHELER (or WHEELER), Granville. 1701-1770. Br. P II; GD.

5841. BERTHIER, Joseph Etienne. 1702-1783. Fr. P I (BERTIER); P III.

5842. KERY, Franz Borgia. 1702-1768. Aus. P I; P III (KERI).

5843. MARTINE, George. 1702-1741. Br. ICB; BMNH; BLA; GA; GD. Also Physiology*.
1. *Essays and observations on the construction and graduation of thermometers, and on the heating and cooling of bodies*. 2nd ed. Edinburgh, 1772. RM.

5844. SCHILLING, Johann Jacob. b. 1702. Ger. P II.

5845. WEITBRECHT, Josias. 1702-1747. Russ. P II; BLA; GC. Also Anatomy*.

5846. HALL, Chester Moor. 1703-1771. Br. ICB; P I; GD.

5847. SORGE, Georg Andreas. 1703-1778. Ger. P II; GC.

5848. WINKLER, Johann Heinrich. 1703-1770. Ger. P II; GC (WINCKLER). Also Meteorology*.
1. *Gedanken von den Eigenschaften, Wirkungen und Ursachen der Electricität, nebst einer Beschreibung zwo neuer electrischen Maschinen*. Leipzig, 1744. RM.

5849. BELGRADO (or BELLOGRADUS), Giacopo. 1704-1789. It. P I & Supp.; GA.

1. *I fenomeni elettrici*. Parma, 1749. RM.

5850. MARTIN, Benjamin. 1704?-1782. Br. DSB; ICB; P II; BMNH; GA; GD. Other entries: see 3099.
 1. *A new and compendious system of optics*. London, 1740. RM.
 2. *A course of lectures in natural and experimental philosophy*. London, 1743.
 3. *New elements of optics*. London, 1759.

5851. SEGNER, János András. 1704-1777. Hung./Ger. DSB; ICB; P II; BLA. Also Mathematics* and Mechanics.

5852. HOADLEY, Benjamin. 1705-1757. Br. P I & Supp.; GA; GD.
 1. *Observations on a series of electrical experiments*. London, 1756.

5853. MORIN, Jean. 1705-1764. Fr. P II; GA.
 1. *Abrégé de mécanisme universel; ou, Discours et questions physiques*. Chartres, 1735. X. Another ed., Paris, 1740. RM.
 2. *Nouvelle dissertation sur l'électricité des corps*. Paris, 1748. RM.

5854. TRESSAN, Louis Elisabeth de la Vergne de Brouissin, *Comte* de. 1705-1783. Fr. P II; GA.
 1. *Essai sur le fluide électrique, considéré comme agent universel*. Paris, 1786. RM.

5855. CHÂTELET, Gabrielle Emilie le Tonnelier de Breteuil, *Marquise* du. 1706-1749. Fr. DSB; ICB; P I (CHASTELET); GA (DUCHATELET).
 1. *Institutions de physique*. Paris, 1740. RM. Another ed., Amsterdam, 1742.
 See also 2539/6.

5856. DOLLOND, John. 1706-1761. Br. DSB; ICB; P I; GA; GB; GD.
 See also 4639/1.

5857. FRANKLIN, Benjamin. 1706-1790. U.S.A. The famous statesman. DSB; ICB; P I; G. Also Earth Sciences and Autobiography*. See also Index.
 1. *Experiments and observations on electricity, made at Philadelphia*. London, 1751. X.
 1a. ——— *New experiments and observations....* London, 1754. X. 3rd ed., 1760. RM/L.
 1b. ——— *Experiments and observations ... To which are added letters and papers on philosophical subjects*. London, 1769. Ed. by P. Collinson.
 1c. ——— *Benjamin Franklin's experiments: A new edition of Franklin's "Experiments and observations on electricity."* Cambridge, Mass., 1941. Ed. with an introd. by I.B. Cohen.
 2. *The ingenious Dr Franklin: Selected scientific letters*. Philadelphia, 1931 and reprints. Ed. by N.G. Goodman.

5858. EULER, Leonhard. 1707-1783. Ger./Russ. DSB; ICB; P I; GA; GB; GC. Also Science in General*, Mathematics*#, Astronomy*, Mechanics*, and Cartography*. See also Index.
 1. *Tentamen novae theoriae musicae*. St. Petersburg, 1739. RM. Reprinted, New York, 1968.
 2. *Dioptricae*. 3 vols. St. Petersburg, 1769-71. RM/L.

5859. LUDOLFF, Christian Friedrich. 1707-1763. Ger. P I.

5860. SYMMER, Robert. ca. 1707-1763. Br. DSB; P II.

5861. THOMIN, Marc. ca. 1707-1752. Fr. P II.

5862. VERATTI, Giuseppe. 1707-1793. It. P II.

5863. FAULHABER, Christoph Erhard. 1708-1781. Ger. P I; GA.

5864. GRALATH, Daniel. 1708-1767. Danzig. P I (the dates given by P are incorrect); GC.

5865. ARRIGHETTI, Niccolo. 1709-1767. It. P I & Supp.

5866. GESSNER, Johannes. 1709-1790. Switz. DSB; BLA & Supp.; GC. Under GESNER: P I & Supp.; BMNH; GA. Also Geology* and Botany. See also 14659/9.

5867. LANTHÉNÉE, Le Ratz de. d. 1770? Belg. P I.

5868. ROMAS, ... de. d. 1776. Fr. P II.

5869. BOSE, Georg Matthias. 1710-1761. Ger. DSB; P I; BLA; GA; GC.
   1. *Tentamina electrica*. Wittenberg, 1744. RM.

5870. FERGUSON, James (first of the name). 1710-1776. Br. DSB; P I; GA; GB; GD. Also Astronomy* and Autobiography*.
   1. *Analysis of a course of lectures on mechanics, hydrostatics, pneumatics and astronomy*. London, 1761. RM.
   2. *Lectures on select subjects in mechanics, pneumatics, hydrostatics and optics, with the use of the globes....* London, 1764. X. 2nd ed., 1770. 4th ed., 1772. 5th ed., 1776. Another ed. (by D. Brewster), 1805.
   3. *An introduction to electricity*. London, 1770. RM/L. Another ed., 1775.

5871. BASSI, Laura Maria Catterina. 1711-1778. It. P I; GA; GB.

5872. BOSCOVICH, Ruggiero Giuseppe. 1711-1787. It./Fr. DSB (BOŠKOVIĆ); ICB; P I; GA; GB. Other entries: see 3114.
   1. *Dissertationes quinque ad dioptricam pertinentes*. Vienna, 1767. RM.
   2. *Memorie sulli cannocchiali diottrici*. Milan, 1771. RM.

5873. DUTOUR, Etienne François. 1711-1784. Fr. P I; GA.

5874. FLASCHNER, Johannes. 1711-1761. Prague. P I.
   1. *De elemento aeris: Tractatus physico-experimentalis in quo natura, proprietates et effectus ejusdem elementi rationum et experimentorum serie demonstrantur*. Prague, 1748. RM.

5875. FRÄNKLIN, Georg. b. 1711. Aus. P I.

5876. KINNERSLEY, Ebenezer. 1711-1778. U.S.A. DSB; ICB; P I; GE.

5877. LOMONOSOV, Mikhail Vasilievich. 1711-1765. Russ. DSB; ICB; P I; GA; GB. Also Various Fields*# and Chemistry*.

5878. RICHMANN, Georg Wilhelm. 1711-1753. Russ. DSB; ICB; P II; GC.

5879. RIKHMAN, Georg Vil'gel'm. 1711-1753. Russ. ICB. (cf. 5878)

5880. SCARELLA, Joanne Baptista (or Giambattista). 1711-1779. It. P II.

1. *Physicae generalis methodo mathematica tractatae.* 3 vols. Brescia, 1754-57. RM.
2. *De magnete.* Brescia, 1759. RM.

5881. STEINER, Johann Ludwig. 1711-1779. Switz. ICB; P II.

5882. ALGAROTTI, Francesco. 1712-1764. It./Fr. ICB; P I; GA; GB; GF.
1. *Il Newtonianismo per le dame; ovvero, Dialoghi sopra la luce e i colori.* Naples, 1737. RM/L. Reprinted with the title *Dialoghi sopra l'ottica neutoniana*, Turin, 1977.
1a. ——— *Le Newtonianisme pour les dames.* 2 vols. Paris, 1738. RM.
1b. ——— *Sir Isaac Newton's philosophy explain'd for the use of ladies. In six dialogues on light and colours.* London, 1739. RM.
1c. ——— *Sir Isaac Newton's theory of light.* 2 vols. London, 1742. RM.

5883. BRAUN, Josias Adam. 1712-1768. Russ. P I.

5884. GORDON, Andreas. 1712-1751. Br./Ger. P I; GA; GC; GD.

5885. JALLABERT, Jean (or Louis). 1712-1768. Switz. P I; BLA (J., Louis); GA.
1. *Expériences sur l'électricité, avec quelques conjectures sur la cause de ses effets.* Geneva, 1748. RM/L.

5886. MAZÉAS, Guillaume. 1712-1776. Fr. P II. Also Chemistry.

5887. RUTHERFORD, Thomas. 1712-1771. Br. P II. Also Mechanics*.
1. *A system of natural philosophy, being a course of lectures in mechanics, optics, hydrostatics and astronomy.* 2 vols. Cambridge, 1748. RM.

5888. BRANDER, Georg Friedrich. 1713-1783. Ger. ICB; P I; GA; GC.

5889. CLAIRAUT, Alexis Claude. 1713-1765. Fr. DSB; ICB; P I; GA; GB. Also Mathematics*, Astronomy*, Mechanics, and Geodesy*. See also Index.

5890. DALHAM (A S. THERESIA), Florian. 1713-1795. Aus. P I.
1. *Institutiones physicae.* Vienna, 1753. RM.

5891. KNIGHT, Gowin. 1713-1772. Br. P I,III; GD.
1. *An attempt to demonstrate that all the phoenomena in nature may be explained by two simple active principles, attraction and repulsion. Wherein the attractions of cohesion, gravity and magnetism are shown to be one and the same ...* London, 1748. RM.

5892. BEGUELIN, Nicolas de. 1714-1789. Ger. P I; GA.

5893. CHAULNES, Michel Ferdinand d'Albert d'Ailly, *Duc* de. 1714-1769. Fr. P I; GA.
1. *Description d'un microscope et de différents micromètres destinés à mesurer des parties circulaires ou droits avec la plus grande précision.* Paris, 1768.

5894. HAUBOLD, Georg Gottlieb. 1714-1772. Ger. P I.

5895. COURTIVRON, Gaspard le Compasseur de Créquy-Montford, *Marquis* de. 1715-1785. Fr. DSB; P I; GA. Also Mechanics.
1. *Traité d'optique, où l'on donne la théorie de la lumière dans le systeme newtonien.* Paris, 1752. RM.

5896. FRANTZ, Ignatz. 1715-1776. Prague. P III.

5897. JACQUET (or JAQUET) DE MALZET, Louis Sebastian. 1715-1800. Aus. P I.

5898. KRÜGER, Johann Gottlob. 1715-1759. Ger. P I; BLA; GA. Also Geology.

5899. WATSON, William. 1715-1787. Br. DSB; ICB; P II; BLA; GA; GD. Also Botany.
1. *Experiments and observations tending to illustrate the nature and properties of electricity.* London, 1745. X. Another ed., 1746.
1a. ——— *A sequel to the "Experiments and observations...."* London, 1746. RM.

5900. BECCARIA, Giambatista. 1716-1781. It. DSB; ICB; P I; GA; GB.
1. *Dell'elettricismo artificale e naturale.* Turin, 1753. RM.
2. *Electricismo atmosferico.* [2nd ed.] Bologna, 1758. RM.
3. *Experimenta atque observationes quibus electricitas vindex late constituitur atque explicatur.* Turin, 1769. RM.
4. *Elettricismo artificiale.* Turin, 1772. RM/L.
4a. ——— *A treatise upon artificial electricity.* London, 1776. RM.
5. *Della elettricità terrestre atmosferica a cielo sereno.* Turin, 1775. RM.
5a. ——— English trans. appended to item 4a.
See also 10438/1.

5901. BENVENUTI, Carlo. 1716-1789. It. P I; GA.

5902. BISCHOFF, Johann. 1716-1779. Ger. P I.

5903. MANGOLD, Joseph. 1716-1787. Ger. P II.

5904. ROUSSIER, Pierre Joseph. 1716-ca. 1790. Fr. P II; GA.

5905. SECONDAT (DE MONTESQUIEU), Jean Baptiste de. 1716-1796. Fr. P II; GA.
1. *Mémoire sur l'électricité.* Paris, 1746. RM.
1a. ——— *Suite du "Mémoire sur l'électricité."* Paris, 1748. RM.
2. *Observations de physique ... sur l'influence de la pesanteur de l'air dans la chaleur des liqueurs bouillantes et dans leur congelation.* Paris, 1750. RM.

5906. SELVA, Lorenzo. ca. 1716-1800. It. P II.

5907. BIANCONI, Giovanni Lodovico. 1717-1781. It./Ger. P I; BLA; GA; GC.

5907A CARDELL, Franz Paula. 1717-1768. Prague/It. P I.

5908. GAUTIER D'AGOTY, Jacques Fabian. 1717-1785. Fr. P I (GAUTHIER); BMNH; Mort.; BLA & Supp.; GA (GAUTHIER). Also Botany and Anatomy.
1. *Chroa-génésie; ou, Génération des couleurs, contre le systême de Newton.* Paris, 1749.

5909. LE MONNIER, Louis Guillaume. 1717-1799. Fr. DSB; ICB; P I; BMNH; BLA; GA. Also Botany.

5910. CANTON, John. 1718-1772. Br. DSB; ICB; P I; GA; GB; GD. See also 5934/1a.

5911. LACHEMAYR, Karl. 1718-1783. Aus. P I.

5912. PENROSE, Francis. 1718-1798. Br. GD. Also Physiology*.
 1. *A treatise on electricity ... To which is added a short account, how the electrical effluvia act upon the animal frame.* Oxford , 1752.

5913. BIANCHINI, Giovanni Fortunato. 1719-1779. It. P I; BLA; GA.
 1. *Osservazioni intorno all'uso dell'elettricita celeste.* Venice, 1754. RM.

5914. CARDELL, Karl. 1719-1757. Prague. P I.

5915. DEGNER, Johann Michael. 1719-1780. Ger. P I.

5916. HEINZE, Johann Georg. 1719-1801. Ger. P I.

5917. KAESTNER, Abraham Gotthelf. 1719-1800. Ger. DSB; ICB; P I; GA; GC. Also Mathematics, Mechanics, and History of Mathematics*.

5918. MANGIN, ... *Abbé* de. fl. 1749-1772. Fr. P II.

5919. LE ROY, Jean Baptiste. 1720-1800. Fr. DSB; ICB; P I; GA.

5920. REDFERN, Sigismund Ehrenreich von. 1720-1789. Ger. P II.

5921. ZEIHER, Johann Ernst. 1720-1784. Ger./Russ. P II.

5922. MELCHIOR, Johann Albrecht. 1721-1783. Ger. P II.

5923. WILSON, Benjamin. 1721-1788. Br. DSB; ICB; P II; GD.
 1. *An essay towards an explication of the phaenomena of electricity, deduced from the aether of Sir Isaac Newton.* London, 1746. RM.
 2. *A treatise on electricity.* London, 1750. RM/L. 2nd ed., 1752.
 3. *Observations upon lightning and the method of securing buildings from its effects.* London, 1773. RM.
 4. *A series of experiments relating to phosphori and the prismatic colours they are found to exhibit in the dark.* London, 1775. RM.

5924. COMINALE, Celestino. 1722-1785. It. P I.
 1. *Anti-Newtonianismi.* 4 vols. Naples, 1754-56. Contents: Part 1. *In qua Newtoni de coloribus systema ex propriis principiis geometrice evertitur....* Part 2. *In qua rejectis methodo et philosophandi regulis Newtonianis....* RM.

5925. MAGELLAN, Jean Hyacinthe. 1722-1790. Port./Br. DSB; ICB; P II (MAGELHAENS); GA (MAGALHAENS); GD. Also Chemistry.

5926. PAULIAN, Aimé Henri. 1722-1801. Fr. P II; GA.
 1. *Dictionnaire de physique portatif, dans lequel on expose les découvertes les plus intéressantes de Newton et les notions géométriques nécessaires....* Avignon, 1758. X. 2nd ed., 1760.
 2. *Dictionnaire de physique.* 3 vols. Avignon, 1761. RM/L. Apparently a different work from item 1. It ran to at least eight editions.
 3. *L'électricité soumise à un nouvel examen.* Avignon, 1768. RM.

5927. BARBARIGO, Girolamo. 1723-1782. It. P I.
1. *Principi di fisica generale*. Padua, 1780. RM.

5928. BRISSON, Mathurin Jacques. 1723-1806. Fr. DSB; P I; BMNH; GA; GB. Also Zoology*.
1. *Dictionnaire raisonné de physique*. 2 vols & atlas. Paris, 1781. RM/L.
2. *Observations sur les nouvelles découvertes aerostatiques, et sur la probabilité de pouvoir diriger les ballons*. Paris, 1784. RM.
3. *Pesanteur spécifique des corps*. Paris, 1787. RM.
4. *Traité élémentaire ou principes de physique*. 3 vols. Paris, 1789.

5929. KRATZENSTEIN, Christian Gottlieb. 1723-1795. Den./Ger./Russ. ICB; P I; BLA; GA; GC.

5930. MAKO (VON KEREK GEDE), Pal. 1723-1793. Hung./Aus. P II; GA; GC.
1. *Compendiaria physicae institutio*. 2 vols. Vienna, 1762-65. X. 2nd ed., 1766. RM.

5931. AEPINUS, Franz Ulrich Theodosius. 1724-1802. Ger./Russ. DSB; P I; BLA; GA; GB.
1. *Sermo academicus de simultudine vis electricae atque magneticae*. St. Petersburg, 1758. RM.
2. *Tentamen theoriae electricitatis et magnetismi*. Ib., 1759.
See also 6012/1.

5932. BINA, Andrea. b. 1724. It. P I.

5933. HJORTBERG, Gustaf Frederik. 1724-1776. Swed. P I.

5934. MICHELL, John. 1724?-1793. Br. DSB; ICB; P II; GB; GD. Also Astronomy* and Geology*.
1. *A treatise of artificial magnets*. Cambridge, 1750. RM.
1a. ——— *Traités sur les aimans artificiels*. Paris, 1752. Trans. by Rivoire of Michell's treatise together with a similar work by J. Canton.

5935. COCHET, Jean Baptiste. d. 1771. Fr. P I.
1. *La physique expérimentale et raisonnée*. Paris, 1756. RM.

5936. HERBERT, Joseph von. 1725-1794. Aus. P I.
1. *Theoriae phaenomenorum electricorum*. Vienna, 1772. X. Another ed., 1778. RM.

5937. STEGMANN, Johann Gottlieb. 1725-1795. Ger. P II; GC.

5938. ANDERSON, John. 1726-1796. Br. P I; GA; GB; GD.

5939. HUTTON, James. 1726-1797. Br. DSB; ICB; P I; BMNH; GA; GB; GD. Also Science in General* and Geology*.
1. *Dissertations on different subjects in natural philosophy*. Edinburgh, 1792. RM/L.
2. *A dissertation upon the philosophy of light, heat and fire*. Edinburgh, 1794. RM.

5940. LE ROY, Charles. 1726-1779. Fr. DSB; ICB; P I; BLA; GA. Also Chemistry.

5941. MELVILL, Thomas. 1726-1753. Br. DSB; GD.

5942. NAIRNE, Edward. 1726-1806. Br. DSB; P II; GD.

5943. RADICS (or RADITS, RADICH, etc.), Anton. 1726-1773. Hung. P II,III.

5944. BONNE, Rigobert. 1727-1795. Fr. P I.

5945. DELUC, Jean André (first of the name). 1727-1817. Switz./Br./Ger. DSB; P I; BMNH; GA; GB; GD. Also Geology* and Meteorology*.
   1. *Traité élémentaire sur le fluide électrogalvanique*. Paris, 1804. RM.

5946. DEMARCO, Saverio. 1727-after 1785. It. P I.

5947. EBERHARD, Johann Peter. 1727-1779. Ger. P I; BLA & Supp.; GA; GC.

5948. BLACK, Joseph. 1728-1799. Br. DSB; ICB; P I; BLA & Supp.; GA; GB; GD. Also Chemistry*.

5949. FRISI, Paolo. 1728-1784. It. DSB; P I; GA; GB. Also Various Fields*#, Astronomy*, and Mechanics*.

5950. LAMBERT, Johann Heinrich. 1728-1777. Switz./Ger. DSB; ICB; P I; GA; GB; GC. Also Mathematics*, Astronomy*, and Cartography*.
   1. *Les propriétés remarquables de la route de la lumière par les airs et en général par plusieurs milieux réfringens*. The Hague, 1758. RM. Reprinted, Paris, 1977.
   2. *Photometria; sive, De mensura et gradibus luminis, colorum et umbrae*. Augsburg, 1760. RM.
   2a. ——— German trans. 3 vols. Leipzig, 1892. (OKEW)
   3. *Pyrometrie; oder, Vom Maase des Feuers und der Wärme*. Berlin, 1779. RM.

5951. CORTI, Bonaventura. 1729-1813. It. DSB; ICB; P I; BMNH. Also Botany.

5952. DELAVAL, Edward Hussey. 1729-1814. Br. P I; GD.
   1. *An experimental inquiry into the cause of the changes of colours in opake and coloured bodies*. London, 1777. X.
   1a. ——— *Recherches expérimentales sur la cause des changemens de couleurs dans les corps opaques et naturellement colorés*. Paris, 1778. Trans. by Quatremère-Dijonval.

5953. HAMILTON, Hugh. 1729-1805. Br. P I; GA; GD.
   1. *Philosophical essays on the following subjects. 1. On the principles of mechanics. 2. On the ascent of vapours, the formation of clouds, rain and dew ... 3. Observations and conjectures on the nature of aurora borealis and the tails of comets*. Dublin, 1766. RM.
   2. *Four introductory lectures in natural philosophy*. Dublin, 1774.

5954. REIMARUS, Johann Albert Heinrich. 1729-1814. Ger. P II; BLA; GA; GC.

5955. SOCIN, Abel. 1729-1808. Switz./Ger. P II.

5956. TITIUS (or TIETZ), Johann Daniel. 1729-1796. Ger. DSB; P II; BMNH; GC. Also Astronomy and Natural History.

5957. LA BORDE, Jean Baptiste de. d. 1777. Fr. P I (BORDE).

5958. HARTMANN, Johann Friedrich. d. 1800. Ger. P I.

5959. CROKER, Temple Henry. 1730?-1790? Br. GD. Also Science in General*.
1. *Experimental magnetism*. London, 1761. RM.

5960. DÜRNBACHER (or DIRNBACHER), Johann Nepomuk. b. 1730. Olmütz. P I.

5961. INGENHOUSZ, Jan. 1730-1799. Holl./Br. DSB; ICB; P I; BMNH; BLA; GA; GD. Also Botany*.
1. *Nouvelles expériences et observations sur divers objets de physique*. 2 vols. Paris, 1785-89. RM/L.

5962. MEISSNER, Ferdinand. 1730-1784. Ger. P II.

5963. NECKER, Louis. 1730-1804. Switz./Fr. P II; GA.

5964. SIGAUD DE LAFOND, Joseph Aignan. 1730-1810. Fr. DSB; P II (S., Jean René); BLA; GA. Also Chemistry*.
1. *Leçons de physique expérimentale*. 2 vols. Paris, 1767. RM.
2. *Traité de l'électricité*. Paris, 1771. RM/L.
3. *Elémens de physique théorique et expérimentale*. 4 vols. Paris, 1777. RM.
4. *Dictionnaire de physique*. 5 vols. Paris, 1781-82. RM.
5. *Précis historique et expérimental des phénomènes électriques*. Paris, 1781. RM/L.
See also 5818/9.

5965. BAUER, Fulgentius. 1731-1765. Aus. P I & Supp.

5966. BIWALD, Gottlieb Leopold. 1731-1805. Aus. P I.

5967. BUTSCHANY, Matthias. 1731-1796. Ger. P I.

5968. CAVENDISH, Henry. 1731-1810. Br. DSB; ICB; P I; GA; GB; GD. Also Chemistry*. See also 15486/1.
1. *The electrical researches, written between 1771 and 1781*. Cambridge, 1879. Ed. from the published papers and unpublished MSS by J.C. Maxwell. RM/L. Reprinted, London, 1967.
2. *The scientific papers*. 2 vols. Cambridge, 1921. Contents: Vol. 1. *The electrical researches*. Ed. by J.C. Maxwell, rev. by J. Larmor. Vol. 2. *Chemical and dynamical*. Ed. from the published papers and unpublished MSS by E. Thorpe.

5969. BECK, Dominicus. 1732-1791. Aus. P I; GA; GC.

5970. BOHNENBERGER, Gottlieb Christian. 1732-1807. Ger. P I.

5971. BRUGMANS, Antonius. 1732-1789. Holl. P I; GF.
1. *Tentamina philosophica de materia magnetica, ejusque actione in ferrum et magnetem*. Franeker, 1765. RM.
1a. ——— *Philosophische Versuche über die magnetische Materie, und deren Wirkung in Eisen und Magnet*. Leipzig, 1784. RM.
2. *Magnetismus; seu, De affinitatibus magneticis observationes academicae*. Leiden, 1778. RM.
2a. ——— *Beobachtungen über Verwandtschaften des Magnets*. Leipzig, 1781. RM.

5972. EGELL, Ambrosius. 1732-1801. Ger. P I.

5973. GROSS, Johann Friedrich. 1732-1795. Ger. P I.

5974. HAESELER, Johann Friedrich. 1732-1797. Ger. P I.

5975. HORVATH, Johann Baptist. 1732-1799. Hung. P I.

5976. PAUCHTON, Alexis Jean Pierre. 1732-1798. Fr. P II.

5977. RITTENHOUSE, David. 1732-1796. U.S.A. DSB; ICB; P II; GA; GB; GE. Also Astronomy.

5978. WIESE, Christian Ludwig Gustav von. 1732-1800. Ger. P II.

5979. WILCKE (or WILKE), Johan Carl. 1732-1796. Swed. DSB; ICB; P II.

5980. EPP, Franz Xaver von. 1733-1789. Ger. P I.
    1. *Abhandlung von dem Magnetismus der naturlichen Electricität.* Munich, 1778. RM.

5981. PRIESTLEY, Joseph. 1733-1804. Br. DSB; ICB; P II; GA; GB; GD; GE. Also Various Fields*#, Chemistry*, and Autobiography*.
    1. *The history and present state of electricity, with original experiments.* London, 1767. RM/L. 2nd ed., 1769. 3rd ed., 1775; reprinted, New York, 1966.
    1a. ——— *Histoire de l'électricité.* Paris, 1771.
    2. *A familiar introduction to the study of electricity.* London, 1768. X. 3rd ed., 1777.
    3. *The history and present state of discoveries relating to vision, light and colours.* London, 1772. RM/L.

5982. CIGNA, Giovanni Francesco. 1734-1790. It. P I; BLA & Supp.; GA.

5983. EULER, Johann Albrecht. 1734-1800. Ger./Russ. ICB; P I; GA. Also Astronomy.
    1. *Disquisitio de causa physica electricitatis.* St. Petersburg, 1755. RM.
    2. *Dissertationes selectae: J.A. Euleri, P. Frisii et L. Beraud.* St. Petersburg, 1757.

5984. LYON, John. 1734-1817. Br. P I; GD.
    1. *Experiments and observations made with a view to point out the errors of the present theory of electricity.* London, 1780. RM.
    2. *Remarks on the leading proofs offered in favour of the Franklinian system of electricity.* London, 1791. RM.

5985. SCHURER, Jacob Ludwig. 1734-1792. Strasbourg, P II; BLA.

5986. EANDI, Giuseppe Antonio Francesco Geronimo. 1735-1799. It. P I; GA.

5987. RAMSDEN, Jesse. 1735-1800. Br. DSB; P II; GA; GB; GD.

5988. REUSCH, Karl Daniel. 1735-1806. Ger. P II. Also Chemistry.

5989. COULOMB, Charles Augustin. 1736-1806. Fr. DSB; ICB; P I; GA; GB. Also Mechanics*.
    1. [Trans. of periodical articles, 1785-86] *Vier Abhandlungen über die Elektricität und den Magnetismus.* Leipzig, 1890. (OKEW. 13)
    2. *Collection de mémoires relatifs à la physique.* 5 vols. Paris, 1884-91. RM/L.

5990. FUNK, Christlieb Benedict. 1736-1786. Ger. P I; BLA. Also Astronomy.

5991. GABLER, Matthias. 1736-1805. Ger. P I.

5992. PISTOJ, Candido. ca. 1736-ca. 1780. It. P II.

5993. DUFIEU, Jean Férapie. 1737-1769. Fr. P I & Supp.; BLA & Supp.; GA (DUFLEU).
1. *Manuel physique ... d'expliquer les phénomènes de la nature.* Paris, 1758. RM.

5994. BURDACH, Daniel Christian. 1739-1777. Ger. P I; BLA.

5995. ROBISON, John. 1739-1805. Br. DSB; P II; GA; GD.
1. *The articles "Steam" and "Steam engines" written for the Encyclopaedia Britannica* [3rd ed., 1797] . Edinburgh, 1818. Ed. by D. Brewster with notes by J. Watt.
2. *A system of mechanical philosophy.* 4 vols. Edinburgh, 1822. Ed. by D. Brewster. RM/L.
See also 7117/4.

5996. SAN MARTINO, Giambattista da. 1739-1800. It. P II.

5997. SUE, Pierre. 1739-1816. Fr. BLA; GA.
1. *Histoire du galvanisme, et analyse des différens ouvrages publiés sur cette découverte depuis son origine jusqu'à ce jour.* 2 vols. Paris, 1802.

5998. HALLER, D. Antoine. late 18th century. Luxembourg. ICB.

5999. HENLEY (or HENLY), William. d. ca. 1779. Br. P I; GD.

6000. ARALDI, Michele. 1740-1813. It. P I; BLA. Also Physiology.

6001. GARDINI, Francesco Giuseppe. 1740-1816. It. P I; BLA.
1. *De influxu electricitatis atmosphaericae in vegetantia.* Turin, 1784. RM.

6002. SAUSSURE, Horace Bénédict de. 1740-1799. Switz. DSB; ICB; P II; BMNH; GA; GB. Also Geology*.
1. *Essai sur l'hygrométrie.* Neuchâtel, 1783. RM.
1a. ——— German trans. 2 vols. Leipzig, 1900. (OKEW)

6003. BERTHOLON (DE SAINT-LAZARE), Pierre. 1741-1800. Fr. DSB; P I & Supp.; BLA & Supp.; GA.
1. *De l'électricité du corps humain dans l'état de santé et de maladie.* Paris, 1780. RM. 2nd ed., 2 vols, 1786.
2. *De l'électricité des végétaux.* Paris, 1783. RM.
3. *De l'électricité des metéores: Ouvrage dans lequel on traite de l'électricité naturelle en général, et des méteores en particulier.* 2 vols. Paris, 1787. RM.

6004. BOECKMANN, Johann Lorenz. 1741-1802. Ger. P I.

6005. ENFIELD, William. 1741-1797. Br. P I & Supp.; GD.
1. *Institutes of natural philosophy, theoretical and experimental.* London, 1785. RM/L. Another ed, 1809.

6006. RACAGNI, Giuseppe Maria. 1741-1822. It. P II; GA.

6007. ROCHON, Alexis Marie de. 1741-1817. Fr. P II; GA. Also Astronomy.

6008. SAURI, ... *Abbé*. 1741-1785. Fr. ICB; P II; BLA (SAURY).
1. *Précis de physique*. 2 vols. Paris, 1780.

6009. LICHTENBERG, Georg Christoph. 1742-1799. Ger. DSB; ICB; P I; GA; GB; GC. See also Index.
1. [Trans. of two Latin periodical articles, 1777-78] *Über eine neue Methode, die Natur und die Bewegung der elektrischen Materie zu erforschen*. Leipzig, 1956. (OKEW)
2. *Briefe*. 4 vols in 3. Leipzig, 1901-04. Ed. by A. Leitzmann and C. Schüddekopf. Reprinted, Hildesheim, 1966.
3. *Schriften und Briefe*. 4 vols. Munich, 1967-72. Ed. by W. Promies.
4. *Lichtenberg's visit to England, as described in his letters and diaries*. New York, 1969. Trans. with notes by M.L. Mare and W.H. Quarrell.
5. *Aphorismen, Schriften, Briefe*. Munich, 1974. Ed. by W. Promies.

6010. BREDA, Jacob van. 1743-1818. Holl. P I; GF.

6011. CARRA, Jean Louis. 1743-1793. Fr. P I; GA.

6012. HAÜY, René Just. 1743-1822. Fr. DSB; ICB; P I; BMNH; GA; GB. Also Mineralogy and Crystallography*.
1. *Exposition raisonnée de la théorie de l'électricité et du magnetisme, d'après les principes de M. Aepinus*. Paris, 1787. RM/L.
2. *Traité élémentaire de physique*. 2 vols. Paris, 1803. RM/L. 2nd ed., 1806. 3rd ed., 1821.
2a. ——— *An elementary treatise on natural philosophy*. 2 vols & atlas. London, 1807. Trans. by O. Gregory. RM/L.

6013. IRVINE, William. 1743-1787. Br. P I; GD. Also Chemistry*.

6014. MARAT, Jean Paul. 1743-1793. The revolutionary. ICB; P II; BLA; G.
1. *Recherches physiques sur le feu*. Paris, 1780. RM/L.
2. *Découvertes ... sur la lumière*. Paris, 1780. RM/L.
3. *Recherches physiques sur l'électricité*. Paris, 1782. RM/L.
4. *Notions élémentaires d'optique*. Paris, 1784.
See also 5769/1g.

6015. ERXLEBEN, Johann Christian Polykarp. 1744-1777. Ger. ICB; P I; BMNH; BLA; GA; GC. Also Chemistry and Zoology.
1. *Anfangsgründe der Naturlehre*. Göttingen, 1768. X. Another ed., with additions by G.C. Lichtenberg, 1787.

6016. LENOIR, Etienne. 1744-1832. Fr. P I; GA.

6017. PETETIN, Jacques Henri Désiré. 1744-1808. Fr. P II; BLA.
1. *Nouveau mécanisme de l'électricité, fondé sur les lois de l'équilibre et du mouvement*. Lyons, 1802. RM.

6018. WÜNSCH, Christian Ernst. 1744-1828. Ger. P II; GC. Also Meteorology.

6019. ATWOOD, George. 1745-1807. Br. DSB; P I; GA; GB; GD. Also Mechanics*.
1. *An analysis of a course of lectures on the principles of natural philosophy*. London, 1784. RM.

6020. CUTHBERTSON, John. 1745-1851. Br./Holl. P I.
  1. *Practical electricity and galvanism.* London, 1807. RM/L.

6021. FAULWETTER, Carl Alexander. 1745-1801. Ger. P I.

6022. GIL, Emmanuele Gervasio. 1745-1807. It. P I.

6023. VOLTA, Alessandro. 1745-1827. It. DSB; ICB; P II; GA; GB. Also Chemistry*.
  1. [Trans. of periodical articles, 1792.] *Briefe über thierische Elektricität.* Leipzig, 1900. (OKEW. 114)
  2. [Trans. of periodical articles, 1796-1800.] *Untersuchungen über den Galvanismus.* Leipzig, 1900. (OKEW. 118)
  3. *Epistolario.* 5 vols. Bologna, 1949-55. The national ed.
  4. *Collezione dell'opere.* 3 vols in 5. Florence, 1816. Ed. by V. Antinori. RM/L.
  5. *Le opere.* 7 vols. Milan, 1918-29. The national ed. Reprinted, New York, 1968.
  5a. ——— *Aggiunte alle Opere e all'Epistolario.* Bologna, 1966.
  5b. ——— *Indici delle Opere e dell'Epistolario.* Milan, 1974.
  6. *Opere scelte.* Turin, 1967. Ed. by M. Gliozzi.

6024. CHARLES, Jacques Alexandre César. 1746-1823. Fr. DSB; P I; GA; GB.

6025. MAILLARD, Sebastian von. 1746-1822. Aus. P III; GA; GC.
  1. *Théorie des machines mues par la force de la vapeur de l'eau.* Vienne/Strasbourg, 1784. RM.

6026. MONGE, Gaspard. 1746-1818. Fr. DSB; ICB; P II; GA; GB. Also Mathematics*, Mechanics, and Chemistry. See also Index.
  1. *Encyclopédie méthodique.* [Main entry: 3000] *Physique.* 4 vols & atlas. Paris, 1793-1822. RM/L.

6027. MONTESSON, Jean Louis de. 1746-1802.
  1. *Mémoire sur la vertu répulsive du feu, considérée comme agent principal de la nature, et application de ce principe à la formation des vapeurs....* Le Mans, 1783. RM.

6028. POLI, Giuseppe Saverio. 1746-1825. It. DSB; P II; BMNH; GA. Also Zoology.

6029. SWINDEN, Jan Hendrik van. 1746-1823. Holl. DSB; ICB; P II; GF. Also Earth Sciences.
  1. *Recueil des mémoires sur l'analogie de l'électricité et du magnétisme.* 3 vols. The Hague, 1784. X. Another ed. (variant title), 1785. RM.
  2. *Positiones physicae.* 2 vols. Harderwijk, 1786. A textbook. RM.

6030. VENTURI, Giovanni Battista. 1746-1822. It. ICB; P II.

6031. WOLFF, Franz Ferdinand. 1747-1804. Ger. P II.

6032. KORTUM, Karl von. 1748-1808. Warsaw. P I.

6033. LÜDICKE, August Friedrich. 1748-1822. Ger. P I.

6034. PLAYFAIR, John. 1748-1819. Br. DSB; ICB; P II; BMNH; GA; GB; GD. Also Mathematics and Geology*.

1. *Outlines of natural philosophy*. 2 vols. Edinburgh, 1812-14. RM/L.
2. *Works*. 4 vols. Edinburgh, 1822. Ed. with a biography by James G. Playfair. RM/L.

6035. AMBSCHEL, Anton von. 1749-1821. Aus. P I.

6036. CAVALLO, Tiberius. 1749-1809. Br. DSB; ICB; P I; GA; GB; GD. Also Chemistry*.
1. *A complete treatise of electricity in theory and practice*. London, 1777. RM/L.
2. *An essay on the theory and practice of medical electricity*. London, 1780. RM/L.
3. *A treatise on magnetism, in theory and practice*. London, 1787. RM.
4. *The elements of natural or experimental philosophy*. 4 vols. London, 1803. RM/L.

6037. GOETHE, Johann Wolfgang von. 1749-1832. Ger. The famous writer. DSB; ICB; P I; BMNH; Mort.; G. Other entries: see 3216.
1. *Beyträge zur Optik*. 2 parts. Weimar, 1791-92. RM. Reprinted, Hildesheim, 1964.
2. *Zur Farbenlehre*. 2 vols & atlas. Tübingen, 1810. RM.
2a. ——— *Farbenlehre*. Ravensburg, 1971. Selections ed. by R. Matthai. English trans.: *Goethe's colour theory*, New York, 1971.
2b. ——— *Farbenlehre: Didaktischer Teil*. Cologne, 1974. Selections with introd., etc., by J. Pawlik.
2c. ——— *Theory of colours*. London, 1840. Trans. with notes by C.S. Eastlake. Omits most of the polemic against Newton. Reprinted, London, 1967.
2d. ——— *Le traité des couleurs*. Paris, 1973. With introd. and notes by Rudolf Steiner (1889). French trans. by H. Bideau.

6038. KOHLREIF, Gottfried Albert. 1749-1802. Russ. P I; BLA.

6039. LAPLACE, Pierre Simon de. 1749-1827. Fr. DSB Supp.; ICB; P I; GA. Also Mathematics*, Astronomy*#, and History of Astronomy*. See also Index.

6040. LUDWIG, Christian. 1749-1784. Ger. P I.

6041. VINCE, Samuel. 1749-1821. Br. P II; GD. Also Mathematics*, Astronomy*, and Mechanics*.
1. *The heads of a course of lectures on experimental philosophy*. Cambridge, 1809.

6042. VOGLER, Georg Joseph. 1749-1816. Ger./Swed. P II; GB; GC.

6043. ROULAND, N. late 18th century. Fr. P II.

6044. LANGENBUCHER, Jacob. d. 1791. Ger. P I.

6045. BARLETTI, Carlo. d. 1800. It. P I.
1. *Physica specimina*. Milan, 1772. RM.
2. *Analisi d'un nuovo fenomeno del fulmine, ed osservazioni sopra gli usi medici della elettricita*. Pavia, 1780. RM.

6046. ADAMS, George (second of the name). 1750-1795. Br. P I; BMNH; BLA; GA; GD. Other entries: see 3217.

1. *An essay on electricity*. London, 1784. RM. 2nd ed., 1785. 3rd ed., 1787. Another ed., 1792.
2. *An essay on vision*. London, 1789.
3. *Lectures on natural and experimental philosophy*. 5 vols. London, 1794. RM/L.

6047. BENNET, Abraham. 1750-1799. Br. P I.
1. *New experiments on electricity*. Derby, 1789. RM/L.

6048. MARUM, Martinus van. 1750-1837. Holl. DSB; ICB; P II; BLA; GB. Also Chemistry and Botany.
1. *Sur la théorie de Franklin, suivant lequel les phénomènes électriques sont expliqués par un seul fluide*. [Harlem? 1819] RM.
2. *Life and work*. 6 vols. Harlem/Leiden, 1969-76. Ed. by R.J. Forbes. Includes his original writings as well as secondary material by various authors.

6049. MILNER, Isaac. 1750-1820. Br. ICB; GA; GD.

6050. YOUNG, Matthew. 1750-1800. Br. P II; GD.
1. *An enquiry into the principle* [sic] *phaenomena of sounds and musical strings*. Dublin, 1784.
2. *An analysis of the principles of natural philosophy*. Dublin, 1800. RM.

6051. LANDRIANI, Marsilio. ca. 1751-ca. 1815. Aus./It. DSB; ICB; P I.
1. *Richerche fisiche intorno alla salubrità dell'aria*. Milan, 1775. RM.
2. *Dell'utilità dei conduttori elettrici*. Milan, 1784. RM.

6052. PREVOST, Pierre. 1751-1839. Switz. DSB; ICB; P II; GA; GB.
1. *Recherches physico-mécaniques sur la chaleur*. Geneva, 1792. RM.
2. *Du calorique rayonnant*. Paris, 1809. RM/L.
3. *Exposition élémentaire des principes qui servent de base à la théorie de la chaleur rayonnante*. Geneva, 1832. RM.
See also 15342/1.

6053. TARDY DE LA BROSSY, ... ca. 1751-1831. Fr. P II.

6054. LIBES, Antoine. 1752-1832. Fr. P I; GA. Also History of Physics*.
1. *Traité élémentaire de physique*. 2 vols. Paris, 1801.

6055. MAYER, Johann Tobias. 1752-1830. Ger. P II; GC.
1. *Über die Gesetze und Modificationen des Wärmestoffs*. Erlangen, 1791. RM.

6056. PICTET, Marc Auguste. 1752-1825. Switz. DSB; P II; BMNH; GA. Also Various Fields.

6057. ACHARD, Franz Carl. 1753-1821. Ger. DSB; ICB; P I; GA; GB; GC. Also Chemistry*.
1. *Chymisch-physische Schriften*. Berlin, 1780. X. Reprinted, Hildesheim, 1969.

6058. GEISSLER, Johann. b. 1753. Ger. P I (GEISLER).
1. *Über die Bemühungen der Gelehrten und Künstler, mathematische und astronomische Instrumente einzutheilen*. Dresden, 1792. RM.

2. *Beschreibung und Geschichte der neuesten und vorzüglichsten Instrumente.* Zittau/Leipzig, 1798. RM.

6059. NICHOLSON, William. 1753-1815. Br. DSB; ICB; P II; BLA; GA; GB; GD. Also Science in General* and Chemistry*.
1. *An introduction to natural philosophy.* 2 vols. London, 1782. RM/L. Another ed., 1796.

6060. STANHOPE, Charles. 1753-1816. Br. ICB; P II; GA; GB; GD.
1. *Principles of electricity.* London, 1779. RM.

6061. THOMPSON, Benjamin, *Count* RUMFORD. 1753-1814. Br./Ger./Fr. ICB (RUMFORD); P II,III,IV (RUMFORD); GA; GB; GC; GD; GE. See also 15475/1.
1. *Philosophical papers ... relating to various branches of natural philosophy and mechanics.* Vol. 1 [No more publ.?] London, 1802.
2. *Mémoires sur la chaleur.* Paris, 1804.
3. *The complete works of Count Rumford.* 4 vols in 5. Boston, Mass., 1870-75. Publ. by the American Academy of Arts and Sciences. RM. Reprinted, Ann Arbor, 1960. Another ed., London, 1875-76.
4. *Collected works.* 5 vols. Cambridge, Mass., 1968-70. Ed. by S.C. Brown.

6062. TROUGHTON, Edward. 1753-1836. Br. DSB; P II; GB; GD.

6063. WEBER, Joseph. 1753-1831. Ger. P II; GC.

6064. DOMIN, Joseph Franz von. 1754-1819. Hung. ICB; P I.

6065. DONNDORFF, Johann August. 1754-1837. Ger. P I; BMNH. Also Zoology and History of Science*.

6066. FISCHER, Ernst Gottfried. 1754-1831. Ger. ICB; P I.

6067. KÜHN, Karl Gottlob. 1754-1840. Ger. P I; BLA; GA; GC. See also 159/1 and 220/1.

6068. MORGAN, George Cadogen. 1754-1798. Br. P II; GA; GD.
1. *Lectures on electricity.* 2 vols. Norwich, 1794. RM.

6069. FLAUGERGUES, Honoré. 1755-1835. Fr. P I; GA. Also Astronomy.

6070. HAUCH, Adam Wilhelm von. 1755-1838. Den. P I.

6071. VEAU DE LAUNAY, Claude Jean. 1755-1826.
1. *Manuel de l'électricité.* Paris, 1809. RM.

6072. CHLADNI, Ernst Florenz Friedrich. 1756-1827. Ger. DSB; ICB; P I; BMNH; GA; GC. Also Mineralogy.
1. *Entdeckungen über die Theorie des Klanges.* Leipzig, 1787. RM.
2. *Die Akustik.* Leipzig, 1802. RM.
3. *Neue Beyträge zur Akustik.* Leipzig, 1817. RM.

6073. GATTEY, (Etienne) François. 1756 (or 1753)-1819. Fr. P I; GA.
1. *Eléments du nouveau système métrique, suivis des tables de rapports des anciennes mesures agraires avec les nouvelles.* Paris, 1801. RM.

6074. KELLY, Patrick. 1756-1842. Br. GD.
1. *Metrology; or, An exposition of weights and measures ... comprising tables of comparison and views of various standards.* London, 1816. RM.

6075. SEIFERHELD, Georg Heinrich. 1757-1818. Ger. P II.

6076. BETANCOURT Y MOLINA, Augustin de. 1758-1824. Spain/Fr./Russ. DSB; P I; GA (BETHENCOURT).

6077. CARRADORI, Giachimo. 1758-1818. It. P I; BMNH; BLA; GA. Also Botany.

6078. GERBOIN, Antoine Clément (or Claude). ca. 1758-1827. Fr. P I.
   1. *Recherches expérimentales sur un nouveau mode de l'action électrique*. Strasbourg, 1808. RM.

6079. SOUTHERN, John. 1758-1815. Br. ICB; P II.

6080. WILSON, Patrick. ca. 1758-1788. Br. P II.

6081. BOURGEOIS, Charles Guillaume Alexandre. 1759-1832. Fr. P I; GA.

6082. HELLER, Theodor Aegidius. 1759-1810. Ger. P I.

6083. SPÄTH, Johann Leonhard. 1759-1842. Ger. ICB; P II. Also Astronomy*.

6084. VACCA-BERLINGHIERI, Leopold. fl. 1789-1807. It. P II.

6085. ASH, Edward. d. 1829. Br. P I; BLA.

6086. BARBANÇOIS, Charles Hélion. 1760-1822. Fr. P I; GA.
   1. *Lettre adressée à M. Delamétherie ... contenant un essai sur le fluide électrique*. Paris, 1817. RM.

6087. GREN, (Friedrich Albert) Carl. 1760-1798. Ger. DSB; ICB; P I; BLA; GA; GC. Also Chemistry*.
   1. *Grundriss der Naturlehre*. Halle, 1788. RM.
   2. *Grundriss der Naturlehre in seinem mathematischen und chemischen Theile*. Halle, 1793. RM.

6088. LE BLOND, Auguste Savinien. 1760-1811. Fr. P I.

6089. PERROLLE, Etienne. ca. 1760-1838. Fr. P II; BLA (PÉROLLE). Also Physiology.

6090. WOOD, James. 1760-1839. Br. P II; GD. Also Mathematics* and Mechanics*.
   1. *The elements of optics*. 2nd ed. Cambridge, 1801. 3rd ed., 1811.

6091. FRANCESCONI, Daniele. 1761-1835. It. P I.

6092. PETROV, Vasily Vladimirovich. 1761-1834. Russ. DSB; ICB; P II (PETROFF, Basilius). Also Chemistry.

6093. ROMAGNOSI, Gian Domenico. 1761-1835. It. ICB; P II; GA.

6094. VASSALLI-EANDI, Antonio Maria. 1761-1825. It. P II. Also Meteorology.

6095. ALDINI, Giovanni. 1762-1834. It. DSB; ICB; P I; BLA; GA; GB.
   1. *Memoria intorno all'elettricita*. Bologna, 1794. RM.
   2. *De animale electricitate*. Bologna, 1794. RM.
   3. *An account of the galvanic experiments performed by John Aldini on the body of a malefactor executed at Newgate*. London, 1803. RM.
   4. *An account of the late improvements in galvanism, with a series of ... experiments performed before the commiss-*

ioners of the French National Institute and repeated lately in the anatomical theatres of London. London, 1803. RM.
5. *Essai théorique et expérimental sur le galvanisme.* Paris, 1804. RM.
6. *General views on the application of galvanism to medical purposes.* London, 1819. RM.

See also 14685/1a.

6096. CHAPPE, Claude. 1763-1805. Fr. P I; GA; GB.

6097. JOYCE, Jeremiah. 1763-1816. Br. GA; GD. Also Chemistry*.
1. *Scientific dialogues ... in which the first principles of natural and experimental philosophy are fully explained.* London, 1809. X. Other eds.: 1848, 1855, 1861.

6098. PRIEUR (-DUVERNOIS) (called PRIEUR DE LA CÔTE D'OR), Claude Antoine. 1763-1832. Fr. P II & Supp.; GA; GB.

6099. ROBERTSON, Etienne Gaspard. 1763-1837. Belg./Fr. P II; GA.

6100. SCHRADER, Johann Gottlieb Friedrich. b. 1763. Ger./Russ. P II.

6101. TRALLES, Johann Georg. 1763-1822. Ger./Switz. P II; GC. Also Mathematics and Astronomy.

6102. BEAUFOY, Mark. 1764-1827. Br. P I; GD.
1. *Nautical and hydraulic experiments, with numerous scientific miscellanies.* Vol. 1 [No more publ.] London, 1834. RM/L.

6103. CARPUE. Joseph Constantine. 1764-1846. Br. BLA; GD.
1. *An introduction to electricity and galvanism.* London, 1803. RM.

6104. ERMAN, Paul. 1764-1851. Ger. P I; GA; GB; GC.

6105. MAISTRE, Xavier de. 1764-1852. Fr./Russ. P II; GA; GB.

6106. MARÉCHAUX, Peter Ludwig. b. 1764. Ger. P II.

6107. SILVA TELLES, Vicente Coelho de Seabra. 1764-1804. Also Chemistry*.
1. *Dissertaçao sobre o calor.* Coimbra, 1788. RM.

6108. BRYAN, Margaret. fl. 1797-1815. GD. Also Astronomy*.
1. *Lectures on natural philosophy.* London, 1806. RM.

6109. WILKINSON, Charles Henry. fl. 1798-1814. Br. P II.
1. *Elements of galvanism in theory and practice. With a comprehensive view of its history.* 2 vols. London, 1804. RM/L.

6110. IVORY, James. 1765-1842. Br. DSB; P I; GB; GD. Also Mathematics, Astronomy, Mechanics, and Geodesy.

6111. SCINÀ, Domenico. 1765-1837. It. P II.

6112. BENINI, Stefano. 1766-1794. It. P I.

6113. DALTON, John. 1766-1844. Br. DSB; ICB; P I; GA; GB; GD. Also Chemistry* and Meteorology*. See also Index.

6114. IZARN, Joseph. 1766-after 1834. Fr. P I. Also Astronomy* and Chemistry*.

1. *Manuel du galvanisme; ou, Description et usage des divers appareils galvaniques.* Paris, 1804. RM.

6115. LESLIE, John. 1766-1832. Br. DSB; ICB; P I; GA; GB; GD.
   1. *An experimental inquiry into the nature and propagation of heat.* London, 1804. RM/L.
   2. *A short account of experiments and instruments depending on the relations of air to heat and moisture.* Edinburgh, 1813. RM/L.
   3. *Elements of natural philosophy.* Edinburgh, 1823. RM.
   4. *Treatises on various subjects of natural and chemical philosophy.* Edinburgh, 1838. Includes a biography of Leslie by M. Napier. RM.

6116. PERKINS, Jacob. 1766-1849. U.S.A./Br. ICB; P II & Supp.; GB; GE.

6117. WOLLASTON, William Hyde. 1766-1828. Br. DSB; ICB; P II; BLA; GA; GB; GD. Also Chemistry, Crystallography, and Physiology.

6118. BLEIN, François Ange Alexandre. b. 1767. Fr. P I; GA.

6119. PARROT, Georg Friedrich. 1767-1852. Russ. P II: GA; GC.

6120. SUREMAIN DE MISSERY, Antoine. 1767-1852. Fr. P II & Supp.

6121. FOURIER, Jean Baptiste Joseph. 1768-1830. Fr. DSB; ICB; P I; GA; GB. Also Mathematics*.
   1. *Théorie analytique de la chaleur.* Paris, 1822. RM. Reprinted, Breslau, 1883.
   1a. ——— *The analytical theory of heat.* Cambridge, 1878. Trans. with notes by A. Freeman. RM/L. Reprinted, New York, 1955.
   2. *Oeuvres.* 2 vols. Paris, 1888-90. Ed. by G. Darboux.

6122. GRIMM, Johann Karl Philipp. 1768-1813. Ger. P I.

6123. NEGRO, Salvatore dal. 1768-1839. It. P II.

6124. NICOL, William. 1768-1851. Br. DSB; ICB; P II; GB. Also Geology.

6125. SCHMIDT, Georg Gottlieb. 1768 (not 1789)-1837. Ger. ICB; P II.

6126. EBERMAIER, Johann Erwin Christoph. 1769-1825. Ger. P I; BLA & Supp.
   1. *Versuch einer Geschichte des Lichtes in Rücksicht seines Einflusses auf die gesammte Natur und auf den menschlichen Körper.* Osnabrück, 1799. RM.

6127. GILBERT, Ludwig Wilhelm. 1769-1824. Ger. ICB; P I & Supp.; GA; GC.

6128. HACHETTE, Jean Nicolas Pierre. 1769-1834. Fr. DSB; P I; GA; GB. Also Mathematics and Mechanics.

6129. MARCET, Jane, *Mrs.* 1769-1858. Br. ICB; P II & Supp.(M., Mary); BMNH; GD. Also Chemistry* and Botany*.
   1. *Conversations on natural philosophy.* London, 1819.

6130. REINHOLD, Johann Christoph Leopold. 1769-1809. Ger. P II.

6131. HALDAT DU LYS, Charles Nicolas Alexandre de. 1770-1852. Fr. P I; BLA; GA. Also Physiology.

6132. LA RIVE, Charles Gaspard de. 1770-1834. Switz. DSB; P II (RIVE); GA.

6133. SEEBECK, Thomas (Johann). 1770-1831. Ger. DSB; P II; GC.
1. *Magnetische Polarisation der Metalle und Erze durch Temperatur-Differentz.* [Periodical article, 1822] Leipzig, 1895. (OKEW. 70)

6134. EYCK, Simon Speijert van der. 1771-1837. Holl. ICB; P I.

6135. YELIN, Julius Konrad von. 1771-1826. Ger. ICB; P II.

6136. BACCELLI, Liberato Giovanni. 1772-1835. It. P I.
1. *I fenomeni elettro-magnetici.* Modena, 1821. RM.

6137. MUNCKE, Georg Wilhelm. 1772-1847. Ger. DSB; P II.

6138. FRIES, Jakob Friedrich. 1773-1843. Ger. DSB; P I; GA; GB; GC. Also Mathematics.

6139. FUSINIERI, Ambrogio. 1773-1853. It. P I.

6140. JAEGER, Karl Christoph Friedrich von. 1773-1828. Ger. P I; BLA.

6141. PFAFF, Christian Heinrich. 1773-1852. Ger. P II; BLA; GC. Also Chemistry.

6142. THOMSON, Thomas. 1773-1852. Br. DSB; ICB; P II; BMNH; GB; GD. Also Chemistry*, Mineralogy*, and History of Science*.
1. *An outline of the sciences of heat and electricity.* London, 1830. 2nd ed., 1840.

6143. TRÉMERY, Jean Louis. 1773-1851. Fr. P II.

6144. YOUNG, Thomas. 1773-1829. Br. DSB; ICB; P II; Mort.; BLA; GA; GB; GD. Also Physiology. See also Index.
1. *A syllabus of a course of lectures on natural and experimental philosophy.* London, 1802.
2. *A course of lectures on natural philosophy and the mechanical arts.* 2 vols. London, 1807. Includes an extensive bibliography. RM/L. Reprinted, New York, 1971. Another ed., with notes by P. Kelland, 1845.
3. *Miscellaneous works.* 3 vols. London, 1855. Ed. by G. Peacock and J. Leitch. Reprinted, New York, 1972.

6145. BIOT, Jean Baptiste. 1774-1862. Fr. DSB; ICB; P I,III; GA; GB. Also Mathematics*, Astronomy*, and History of Astronomy*.
1. *Recherches sur les réfractions extraordinaires qui ont lieu près de l'horizon.* Paris, 1810. RM.
2. *Recherches expérimentales et mathématiques sur les mouvements des molécules de la lumière autour de leur centre de gravité.* Paris, 1814. RM/L.
3. *Traité de physique expérimentale et mathématique.* 4 vols. Paris, 1816. RM/L.
4. *Instructions pratiques sur l'observation et la mesure des propriétés optiques appelées rotatoires.* Paris, 1845. RM.
5. *Mélanges scientifiques et littéraires.* 3 vols. Paris, 1858. RM/L.
See also 5769/6.

6146. GREGORY, Olinthus Gilbert. 1774-1841. Br. DSB; P I; GA; GB; GD. Also Astronomy* and Mechanics*. See also 6012/2a.

6147. MOLLWEIDE, Karl Brandan. 1774-1825. Ger. DSB; P II; GA; GC. Also Mathematics and Astronomy.

6148. NEUMANN, Johann Philipp. 1774-1849. Aus. P II,III.

6149. AMPÈRE, André Marie. 1775-1836. Fr. DSB; ICB; P I; GA; GB. Also Science in General*, Mathematics*, and Personal Writings*. See also Index.
1. *Recueil d'observations électro-dynamiques, contenant divers mémoires, etc.* Paris, 1822. RM/L.
2. *Description d'un appareil électro-dynamique.* Paris, 1824. RM.
3. *Théorie mathématique des phénomènes électro-dynamiques uniquement déduite de l'expérience.* Paris, 1826. X. Reprinted, 1883 and 1958.

Correspondence and Collected Works

4. *Journal and correspondance.* Paris, 1872. Ed. by Mme H. Chevreux. cf. 15349/1.
5. *André-Marie Ampère et Jean-Jacques Ampère: Correspondance et souvenirs (de 1805 à 1864).* 2 vols. Paris, 1875. Ed. by Mme H. Chevreux. RM/L.
6. *Correspondance du grand Ampère.* 3 vols. Paris, 1936-43. Ed. by L. de Launay.
7. *Mémoires sur l'électrodynamique.* 2 vols. Paris, 1885-87. Publ. by the Société Française de Physique. X.

6150. EXLEY, Thomas. 1775-1855? Br. GD.
1. *Principles of natural philosophy; or, A new theory of physics, founded on gravitation....* London, 1829. RM.
2. *Physical optics; or, The phenomena of optics explained according to mechanical science.....* London, 1834. RM.

6151. HÄLLSTRÖM, Gustaf Gabriel. 1775-1844. Fin. P I.

6152. HEIDMANN, Johann Anton. 1775-1855. Aus. P III. Also Physiology.
1. *Vollständige auf Versuche und Vernunftschlüsse gegrundete Theorie der Elektricität.* 2 vols. Vienna, 1799. RM.

6153. MALUS, Etienne Louis. 1775-1812. Fr. DSB; P II; GA; GB.

6154. AUGUSTIN, Friedrich Ludwig. 1776-1854. Ger. P I; BLA & Supp. Also Physiology.

6155. AVOGADRO, Amedeo. 1776-1856. It. DSB; ICB; P III; GB. Also Chemistry*#.
1. *Nuove considerazioni sulle affinità de' corpi pel calorico.* Modena, 1822. RM.
2. *Fisica de' corpi ponderabili; ossia, Trattato della costituzione generale de' corpi.* Turin, 1837. RM.

6156. BARLOW, Peter. 1776-1862. Br. DSB; ICB; P I,III; GA; GB; GD. Also Mathematics*.
1. *An essay on magnetic attractions, particularly as regards the deviation of the compass on shipboard.* London, 1820. RM.
2. *Experiments on the transverse strength and other properties of malleable iron.* London, 1835. RM.

6157. BELLANI, Angelo. 1776-1852. It. DSB; P I & Supp. Also Chemistry.

6158. DELEZENNE, Charles Edouard Joseph. 1776-1866. Fr. P III.

6159. LEONELLI, (Giuseppe) Zecchini. 1776-1847. It./Fr. ICB; P I; GA. Also Mathematics.
  1. *Démonstration des causes des phénomènes électriques; ou, Théorie de l'électricité prouvée par l'expérience.* Strasbourg, 1813. RM.

6160. RITTER, Johann Wilhelm. 1776-1810. Ger. DSB; ICB; P II; BLA; GA; GC. Also Chemistry and Physiology.
  1. *Die Begründung der Elektrochemie und Entdeckung der ultravioletten Strahlen: Eine Auswahl aus den Schriften.* Frankfurt, 1968. (OKEW, n.F. 2)
  2. *Fragmente aus dem Nachlass eines jungen Physikers.* Heidelberg, 1969. Ed. by H. Schipperges.
  3. *Briefe eines romantischen Physikers: J.W. Ritter an G.H. Schubert und K. von Hardenberg.* Munich, 1966. Ed. with commentary by F. Klemm and A. Hermann.

6161. ZAMBONI, Giuseppe. 1776-1846. It. P II.

6162. BENZENBERG, Johann Friedrich. 1777-1846. Ger. DSB; ICB; P I; GA; GC. Also Astronomy and Geodesy.

6163. BRANDES, Heinrich Wilhelm. 1777-1834. Ger. DSB; P I; GC. Also Astronomy.

6164. CAGNIARD DE LA TOUR, Charles. 1777-1859. Fr. DSB; ICB; P I, III; Mort.; GA (LATOUR); GB. Also Chemistry.

6165. CUMMING, James. 1777-1861. Br. DSB; P I,III; GD. See also 6248/1a.

6166. GAUSS, Karl Friedrich. 1777-1855. Ger. DSB; ICB; P I,III; GA; GB; GC. Also Mathematics*#, Astronomy*, Mechanics*, and Earth Sciences*. See also Index.
  1. *Dioptrische Untersuchungen.* Göttingen, 1841. RM.
  See 10460/1-2 for works on geomagnetism which were also important for the basic theory of magnetism.

6167. OERSTED, Hans Christian. 1777-1851. Den. DSB; ICB; P II; BLA; GA. Also Science in General*.
  1. *Experimenta circa effectum conflictus electrici in acum magneticam.* Periodical article, 1820. RM.
  1a. ——— Reprinted in *La découverte de l'électromagnetisme, faite en 1820 par H.C. Örsted*, Copenhagen, 1920. Ed. by A. Larsen. Contains a facsimile of the original with translations into Danish, English, French, and German.
  2. *Correspondance avec divers savants.* 2 vols. Copenhagen, 1920. Ed. by M.C. Harding.
  3. *Naturvidenskabelige skrifter. Samlet utgave.* 3 vols. Copenhagen, 1920. Ed. by K. Meyer.
  3a. ——— *Scientific papers. Collected edition.* 3 vols. Copenhagen, 1920. Ed. by K. Meyer.

6168. SCHEIBLER, Johann Heinrich. 1777-1838. Ger. P II; GC.

6169. GAY-LUSSAC, Joseph Louis. 1778-1850. Fr. DSB; ICB; P I; GA; GB. Also Chemistry*.

6170. PRECHTL, Johann Joseph von. 1778-1854. Aus. P II. Also Science in General* and Chemistry*.

6171. BROUGHAM, Henry. 1779-1868. Br. The lawyer and politician. P I,III; GA; GB; GD. Also Science in General*#, Mechanics*, Autobiography*, and Biography*.
1. *Experiments and observations upon the properties of light.* London, 1850.
2. *Tracts, mathematical and physical.* 1860.

6172. CONFIGLIACHI, Pietro. 1779-1844. It. P I & Supp.
1. *L'identità del fluido elettrico col così detto fluido galvanico.* Pavia, 1814. RM.

6173. FARRAR, John. 1779-1853. U.S.A. DSB; P III; GE. Also Astronomy*.
1. *An elementary treatise on optics.* Cambridge, Mass., 1826. RM.

6174. MURHARD, Friedrich Wilhelm August. 1779-1853. Ger. P II; GA. Also Mathematics* and Mechanics*.
1. *Literatur der mechanischen und optischen Wissenschaften.* 3 vols. Leipzig, 1803-05. See 3652/1a.

6175. SCHWEIGGER, Johann Salomon Christoph. 1779-1857. Ger. DSB; ICB; P II; GC. Also Chemistry.

6176. HALLASCHKA, Franz Ignaz Cassian. 1780-1847. Brünn/Prague. P I. Also Astronomy.

6177. HENSCHEL, Karl Anton. 1780-1861. Ger. P I; GC.

6178. ARNIM, (Ludwig) Achim (or Joachim) von. 1781-1831. Ger. ICB; P I; GA; GB; GC.

6179. BREWSTER, David. 1781-1868. Br. DSB; P I,III; GA; GB; GD. Also Science in General* and Biography*. See also Index.
1. *A treatise on new philosophical instruments.* Edinburgh, 1813. RM/L.
2. *A treatise on optics.* London, 1831. RM/L. Other eds: 1835 and 1853.
3. *Letters on natural magic.* London, 1832. RM. 5th ed., 1842. Another ed., 1853.
4. *A treatise on magnetism.* Edinburgh, 1837. RM/L.
5. *The stereoscope: Its history, theory, and construction.* London, 1856. RM/L. Another ed., 1870.
6. *The kaleidoscope: Its history, theory, and construction.* 2nd ed. London, 1858. Another ed., 1870.

6180. BUQUOY, Georg Franz August von. 1781-1851. Prague. P I.

6181. HARE, Robert. 1781-1858. U.S.A. DSB; ICB; P I & Supp.; GA; GE. Also Chemistry*.
1. *A new theory of galvanism.* Philadelphia, 1819. RM.
2. *A brief exposition of the science of mechanical electricity.* Philadelphia, 1840. RM.

6182. PLANA, Giovanni (or Jean). 1781-1864. It. DSB; ICB; P II,III. Also Mathematics, Astronomy, and History of Mechanics*.
1. *Mémoire sur la distribution de l'électricité.* Turin, 1845.

6183. POISSON, Siméon Denis. 1781-1840. Fr. DSB Supp.; ICB; P II; GA; GB. Also Mathematics* and Mechanics*.
1. *Nouvelle théorie de l'action capillaire.* Paris, 1831.

2. *Théorie mathématique de la chaleur.* Paris, 1835. RM/L. Supplement, 1837. RM.
3. *Mémoire sur les déviations de la boussole produits par le fer des vaissaux.* 1838.

6184. NEEFF, Christian Ernst. 1782-1849. Ger. P II; BLA.

6185. STURGEON, William. 1783-1850. Br. DSB; ICB; P II; GD.
1. *Lectures on electricity.* London, 1842. RM/L.
2. *Scientific researches, experimental and theoretical.* Bury, 1850. RM. Another ed., London, 1852.

6186. BARLOCCI, Saverio. 1784-1845. It. P I & Supp.

6187. CHRISTIE, Samuel Hunter. 1784-1865. Br. DSB; P III; GD.

6188. CROSSE, Andrew. 1784-1855. Br. P I & Supp.; GD. See also 15576/1.

6189. HANSTEEN, Christopher. 1784-1873. Nor. DSB; P I,III; GA; GB. Also Astronomy and Geomagnetism*.
1. *Physikalische meddelelser.* Christiania, 1858.

6190. NOBILI, Leopoldo. 1784-1835. It. DSB; P II; GB.

6191. PIANCIANI, Giambattista. 1784-1862. It. P II,III.

6192. ROMERSHAUSEN, Elard. 1784-1857. Ger. ICB; P II.

6193. DULONG, Pierre Louis. 1785-1838. Fr. DSB; ICB; P I; GA; GB. Also Chemistry.

6194. GROTTHUSS, Theodor (Christian Johann Dietrich) von. 1785-1822. Mitau. DSB; ICB; P I (GROTHUSS). Also Chemistry.
1. *Abhandlungen über Elektrizität und Licht.* Leipzig, 1906. (OKEW. 152)

6195. MOLL, Gerard. 1785-1838. Holl. DSB; P II. Also Science in General*.

6196. PELTIER, Jean Charles Athanase. 1785-1845. Fr. DSB; P II; GA; GB.

6197. SCHLEIERMACHER, Ludwig. 1785-1844. Ger. P II.

6198. AMICI, Giovan Battista. 1786-1868. It. DSB; ICB; P I,III; BMNH; Mort.; GA; GB. Also Biology in General* and Microscopy.

6199. ARAGO, (Dominique) François (Jean). 1786-1853. Fr. DSB; ICB; P I; GA; GB. Also Astronomy*#, Earth Sciences*, and Biography*.

6200. CHEVREUL, Michel Eugène. 1786-1889. Fr. DSB; ICB; P I,III,IV; GA; GB. Also Chemistry* and History of Chemistry*.
1. *De la loi du contraste simultané des couleurs et de l'assortiment des objets colorés.* Paris, 1839.
1a. ——— *The principles of harmony and contrast of colours.* London, 1854. Trans. by C. Martel. RM. 2nd ed., 1855. 3rd ed., 1859.
1b. ——— *The laws of contrast of colour.* London, 1857. Trans. by J. Stanton. RM. Another ed., 1883.
2. *Théorie des effets optiques que présentent les étoffes de soie.* Paris, 1846. RM.

6201. ELICE, Ferdinando. b. 1786. It. P I.

6202. MITCHELL, James. ca. 1786-1844. Br. P II; BMNH; GD. Also Chemistry* and Geology*.
1. *The elements of natural philosophy*. London, 1819. RM.
2. *A dictionary of the mathematical and physical sciences*. London, 1823. RM.

6203. RUHLAND, Reinhold Ludwig. 1786-1827. Ger. P II. Also Chemistry.

6204. SEGUIN, Marc (also known as SEGUIN AÎNÉ). 1786-1875. Fr. DSB; ICB; P III.
1. *Origine et propagation de la force*. Paris, 1859. RM.

6205. SINGER, George John. 1786-1817. Br. P II; GD.
1. *Elements of electricity and electrochemistry*. London, 1814. RM/L.

6206. FRAUNHOFER, Joseph. 1787-1826. Ger. DSB; ICB; P I; GA; GB; GC.
1. *Bestimmung des Brechnungs- und Farbenzerstreuungs-Vermögens verschiedener Glasarten*. [Periodical article, 1814] Leipzig, 1906. (OKEW. 150)
2. *Gesammelte Schriften*. Munich, 1888. Ed. by E. Lommel. RM.
3. *Prismatic and diffraction spectra: Memoirs*. New York, 1898. Trans. by J.S. Ames.

6207. NÖRREMBERG, Johann Gottlieb Christian. 1787-1862. Ger. P II, III.

6208. PEREGO, Antonio. 1787-1848. It. P II,III.

6209. ARNOTT, Neil. 1788-1874. Br. BLA & Supp.; GB; GD.
1. *Elements of physics; or, Natural philosophy*. 2 vols. London, 1827-29. RM/L. 5th ed., 1833. 6th ed., 1864-65. 7th ed., 1876.

6210. BECQUEREL, Antoine César. 1788-1878. Fr. DSB; P I,III; GA; GB. Also Chemistry* and Earth Sciences*.
1. *Traité expérimental de l'électricité et du magnétisme*. 7 vols. Paris, 1834-40. RM/L.
2. *Traité de physique considerée dans ses rapports avec la chimie et les sciences naturelles*. 2 vols. Paris, 1842-44. RM/L.
3. *Resumé de l'histoire de l'électricité et du magnétisme, et des applications de ces sciences*. Paris, 1858. With Edmond Becquerel. RM/L.

See also 6450/1.

6211. FRESNEL, Augustin Jean. 1788-1827. Fr. DSB; ICB; P I; GA; GB.
1. *Mémoire sur la diffraction de la lumière*. [Several periodical articles under this title from 1816 onwards.] German trans. Leipzig, 1926. (OKEW. 215)
2. *Mémoire sur la diffraction de la lumière où l'on examine particulièrement le phénomène des franges colorées que présentent les ombres des corps éclairés par un point lumineux*. Paris, n.d. RM.
3. *Oeuvres complètes*. 3 vols. Paris, 1866-70. Ed. by H. de Sénarmont et al. Includes correspondence. RM/L. Reprinted, New York, 1965.

See also 6637/2.

6212. POHL, Georg Friedrich. 1788-1849. Ger. P II.

6213. RONALDS, Francis. 1788-1873. Br. P II,III; GD. Also Meteorology.
   1. *Description of an electric telegraph, and of some other electrical apparatus.* London, 1823. RM.
   2. *Catalogue of books and papers relating to electricity, magnetism, the electric telegraph, etc., including the Ronalds library.* 2 vols. London, 1880. Compiled by Ronalds. Ed. with a biography of him by A.J. Frost. RM/L.

6214. TREDGOLD, Thomas. 1788-1829. Br. P II; GB; GD.

6215. CRAHAY, Jacques Guillaume. 1789-1856. Belg. P I; GF.

6216. OHM, Georg Simon. 1789-1854. Ger. DSB; ICB; P II; GA; GB; GC.
   1. *Das Grundgesetz des elektrischen Stromes. Drei Abhandlungen von G.S. Ohm (1825 und 1826) und G.T. Fechner (1829).* Leipzig, 1938. (OKEW. 244)
   2. *Die galvanische Kette, mathematisch bearbeitet.* Berlin, 1827. RM/L.
   2a. ——— *Théorie mathématique des courants éléctriques.* Paris, 1860. Trans. with notes by J.M. Gaugin.
   3. *Gesammelte Abhandlungen.* Leipzig, 1892. Ed. by E. Lommel.

6217. SCORESBY, William. 1789-1857. Br. P II; BMNH; GB; GD. Also Geomagnetism* and Natural History.
   1. *Magnetical investigations.* London, 1844. RM.

6218. DANIELL, John Frederic. 1790-1845. Br. DSB; ICB; P I; GB; GD. Also Chemistry* and Meteorology*.

6219. DENT, Edward John. 1790-1853. Br. P I; GD. Also Astronomy.
   1. *A treatise on the aneroid, a newly invented portable barometer.* London, 1849.

6220. HERAPATH, John. 1790-1868. Br. DSB; ICB; GD.
   1. *Mathematical physics.* London, 1847. RM. Reprinted with addition of selected papers and ed. with an introd. and bibliography by S.G. Brush, New York, 1972.

6221. MARIANINI, Stefano Giovanni. 1790-1866. It. P II,III.

6222. POUILLET, (Claude Servais) Mathias. 1790-1868. Fr. DSB; ICB; P II,III; GA.
   1. *Eléments de physique expérimentale et de météorologie.* 2 vols in 4. Paris, 1827-30. RM. Another ed., Brussels, 1840. 7th ed., Paris, 1856.
   1a. ——— *Pouillet's "Lehrbuch der Physik und Meteorologie" für deutsche Verhältnisse frei bearbeitet.* 2 vols. Brunswick, 1842-44. X. An adaptation by J.H.J. Müller of the 1837 ed. of Pouillet's textbook. For the subsequent eds of the German adaptation see 6364/1a.

6223. RITCHIE, William. 1790-1837. Br. P II; GD.

6224. SAVART, Nicolas. 1790-1853. Fr. P II.

6225. BELLI, Giuseppe. 1791-1860. It. P I & Supp.; GB.

6226. BOTTO, Giuseppe Domenico. 1791-1865. It. P I,III.

6227. FARADAY, Michael. 1791-1867. Br. DSB; ICB; P I,III; GA; GB; GD. Also Chemistry*. See also Index.
1. *Experimental researches in electricity.* 3 vols. London, 1839-55. A collection of periodical articles from the period 1831-52. RM/L. Reprinted, New York, 1965.
2. *Experimental researches in chemistry and physics.* London, 1859. A collection of periodical articles. RM/L. Reprinted, Brussels, 1969.
3. *A course of six lectures on the various forces of matter and their relations to each other.* London, 1860. Ed. by W. Crookes. RM. 3rd ed., 1861. Another ed., 1890.
4. *Faraday's diary, being the various philosophical notes of experimental investigations made during the years 1820-62.* 7 vols & index. London, 1932-36. Ed. by T. Martin.
5. *Selected correspondence.* 2 vols. Cambridge, 1971. Ed. by L.P. Williams.
6. [Selected texts] *The achievements of Michael Faraday.* New York, 1967. Ed. by L.P. Williams.

German translations of many of his periodical articles are included in OKEW.

6228. HARRIS, William Snow. 1791-1867. Br. P III; GB; GD.
1. *On the nature of thunderstorms, and on the means of protecting buildings and shipping against ... lightning.* London, 1843. RM/L.
2. *Rudimentary magnetism. Being a concise exposition of the general principles of magnetical science.* 3 parts. London, 1850-52. RM.
3. *Rudimentary electricity. Being a concise exposition of the general principles of electrical science.* London, 1851. RM. 4th ed., 1854.
4. *Rudimentary treatise on galvanism, and the general principles of animal and voltaic electricity.* London, 1856. RM.
5. *A treatise on frictional electricity.* London, 1867. Ed. with a memoir of the author by C. Tomlinson. RM/L.

6229. MOSSOTTI, Ottaviano Fabrizio. 1791-1863. It. DSB; ICB; P II, III.

6230. PETIT, Alexis Thérèse. 1791-1820. Fr. DSB; P II; GA.

6231. SAVART, Félix. 1791-1841. Fr. DSB; ICB; P II; BLA; GA.

6232. DESPRETZ, César Mansuète. 1792-1863. Fr. P I,III; GA.

6233. HERSCHEL, John Frederick William. 1792-1871. Br. DSB; ICB; P I,III; GA; GB; GD. Also Science in General* and Astronomy*#.

6234. LOCKE, John. 1792-1856. U.S.A. ICB; P I; GE. Also Geology.

6235. SCHWERD, Friedrich Magnus. 1792-1871. Ger. P II,III; GC. Also Astronomy.
1. *Die Beugungserscheinungen aus den Fundamentalgesetzen der Undulationstheorie analytisch entwickelt und in Bildern dargestellt.* Mannheim, 1835. RM.

6236. BAUMGARTNER, Andreas von. 1793-1865. Aus. P I,III; GA; GC.
1. *Naturlehre nach ihrem gegenwärtigen Zustande.* Vienna, 1836.

6237. BROUWER, Seerp. 1793-1856. Holl. P III; GF.

6238. GREEN, George. 1793-1841. Br. DSB Supp.; ICB; P I; GD. Also Mathematics*#.
1. *An essay on the application of mathematical analysis to the theories of electricity and magnetism.* Nottingham, 1828. X. Reprinted, Gothenburg, 1958.
1a. ——— German trans. Leipzig, 1895. (OKEW. 61)

6239. LARDNER, Dionysius. 1793-1859. Br. ICB; P I & Supp.; BMNH; GA; GB; GD. Other entries: see 3303.
1. *A treatise on heat.* London, 1833.
2. *A manual of electricity, magnetism and meteorology.* London, 1841.
3. *A handbook of natural philosophy and astronomy.* 3 vols. London, 1851-53.
4. *Handbooks of natural philosophy.* 4 vols. London, 1861-63.

6240. PÉCLET, (Jean Claude) Eugène. 1793-1857. Fr. P II; GA.
1. *Traité de l'éclairage.* Paris, 1827.
2. *Traité de la chaleur considerée dans ses applications aux arts et manufactures.* 2 vols. Paris, 1829. X. 3rd ed., 3 vols, 1860-61.

6241. PESCHEL, Karl Friedrich. 1793-1852. Ger. P II.
1. *Lehrbuch der Physik.* ? vols. Dresden, 1842-44. X.
1a. ——— *Elements of physics.* 3 vols. London, 1845-46. Trans. with notes by E. West. 2nd ed., 1854.

6242. WILDE, (Heinrich) Emil. 1793-1859. Ger. P II. Also History of Physics*.

6243. BABINET, Jacques. 1794-1872. Fr. DSB; P I,III; GA. Also Science in General* and Meteorology.

6244. CODDINGTON, Henry. d. 1845. Br. P I; GD.
1. *An elementary treatise on optics.* Cambridge, 1823. 2nd ed., 1825.
2. *A system of optics.* 2 vols. Cambridge, 1829-30.

6245. PARTINGTON, Charles Frederick. d. 1857? Br. BMNH; GD. Also Science in General* and Natural History*.
1. *A manual of natural and experimental philosophy.* 2 vols. London, 1828.

6246. AUGUST, Ernst Ferdinand. 1795-1870. Ger. P I,III. Also Chemistry*.
1. *Handwörterbuch der Chemie und Physik.* ? vols. Berlin, 1842-50. RM.

6247. BLANCHET, Pierre Henry. 1795-1863. Fr. P III.

6248. DEMONFERRAND, Jean Baptiste Firmin. 1795-1844. Fr. P I.
1. *Manuel d'électricité dynamique.* Paris, 1823. RM.
1a. ——— *A manual of electro dynamics.* Cambridge, 1829. Trans. with notes and additions by J. Cumming. RM.

6249. HENRICI, Friedrich Christoph. 1795-1885. Ger. P I,III.
1. *Über die Elektricität der galvanischen Kette.* Göttingen, 1840. RM.

6250. KUNZEK, August. 1795-1865. Lemberg/Aus. P I,III.
1. *Die Lehre vom Lichte.* Lemberg, 1836. A textbook. RM.
2. *Studien aus der höheren Physik.* Vienna, 1856. RM.

6251. LAMÉ, Gabriel. 1795-1870. Fr. DSB; ICB; P I,III; GA. Also Mathematics* and Mechanics*.
   1. *Cours de physique.* 2 vols in 3. Paris, 1836-37. 2nd ed., 3 vols, 1840.
   2. *Leçons sur la théorie analytique de la chaleur.* Ib., 1861.

6252. MAAS, Anton Jacob. b. 1795. Belg. P II,III.

6253. PAMBOUR, François Marie Guyonneau de. b. 1795. Fr. DSB; P II.
   1. *Théorie de la machine à vapeur.* Paris, 1839. RM.
   1a. ——— *The theory of the steam engine.* London, 1839. RM.

6254. WALFERDIN, François Hippolyte. 1795-1880. Fr. P II,III.

6255. APJOHN, James. 1796-1886. Br. P III; BMNH. Also Chemistry* and Mineralogy*.

6256. CARNOT, (Nicolas Léonard) Sadi. 1796-1832. Fr. DSB; ICB; P I, III; GA; GB.
   1. *Réflexions sur la puissance motrice du feu et sur les machines propres à développer cette puissance.* Paris. 1824. RM/L. Reprinted, ib., 1878, and London, 1966.
   1a. ——— *Reflections on the motive power of heat, and on machines fitted to develop this power.* London, 1890. Trans. by R.H. Thurston. RM/L. Reprinted, New York, 1943.
   1b. ——— German trans. Leipzig, 1892. (OKEW. 37)

6257. EKELUND, Adam Wilhelm. 1796-1885. Swed. P I,III.

6258. ETTINGSHAUSEN, Andreas von. 1796-1878. Aus. P I,III. Also Mathematics.
   1. *Anfangsgründe der Physik.* Vienna, 1844. RM.

6259. POGGENDORFF, Johann Christian. 1796-1877. Ger. DSB; ICB; P II, III; GA; GB; GC. Also Biography* and History of Physics*. See also 8655.

6260. POWELL, Baden. 1796-1860. Br. DSB; P II,III; GD. Also Science in General* and History of Science*.
   1. *A general and elementary view of the undulatory theory as applied to the dispersion of light and some other subjects.* London, 1841. RM/L.

6261. QUETELET, (Lambert) Adolphe (Jacques). 1796-1874. Belg. DSB; ICB; P II,III; GA; GB. Other entries: see 3306.
   1. *Positions de physique; ou, Résumé d'un cours de physique générale.* 3 vols. Brussels, 1827-29.

6262. DUHAMEL, Jean Marie Constant. 1797-1872. Fr. DSB; P I,III. Also Mathematics* and Mechanics*.

6263. GASSIOT, John Peter. 1797-1877. Br. DSB; P I; GD.

6264. HENRY, Joseph. 1797-1878. U.S.A. DSB; ICB; P III; GB; GE. See also Index.
   1. *Scientific writings.* 2 vols. Washington, 1886. RM/L.
   2. *The papers.* Washington, 1972-. Ed. by N. Reingold.

6265. POISEUILLE, Jean Léonard Marie. 1797-1869. Fr. DSB; ICB; P II, III; Mort.; BLA. Also Physiology.

6266. REES, Richard van. 1797-1875. Holl. P II,III; GF.

6267. SAIGEY, Jacques Frédéric. 1797-1871. Fr. P II,III.

6268. STREHLKE, Friedrich. 1797-1886. Ger./Danzig. P II,III.

6269. ZANTEDESCHI, Francesco. 1797-1873. It. P II,III.
1. *Saggi dell'elettro-magnetico e magneto-elettrico*. Venice, 1839. RM.

6270. MELLONI, Macedonio. 1798-1854. It. DSB; ICB; P II; GA; GB.
1. *Opere*. Bologna, 1954-. Ed. by G. Polvani and G. Todesco.

6271. METCALFE, Samuel Lytler. 1798-1856. U.S.A. GE.
1. *Caloric, its mechanical, chemical, and vital agencies in the phenomena of nature*. 2 vols. London, 1843.

6272. NEUMANN, Franz Ernst. 1798-1895. Ger. DSB; ICB; P II,III,IV; BMNH; GA; GB; GC. Also Mathematics* and Crystallography.
1. *Theorie der doppelten Strahlenbrechnung, abgeleitet aus den Gleichungen der Mechanik*. [Periodical article, 1832] Leipzig, 1896. (OKEW. 76)
2. *Gesetze der Doppelbrechnung des Lichts*. Berlin, 1842.
3. *Die mathematischen Gesetze der inducirten elektrischen Ströme*. [Periodical article, 1845] Leipzig, 1889. (OKEW. 10)
4. *Über ein allgemeines Princip der mathematischen Theorie inducirter elektrischer Ströme*. [Periodical article, 1847] Leipzig, 1892. (OKEW. 36)
5. *Vorlesungen über mathematische Physik*. 7 vols. Leipzig, 1881-94. Ed. by C. Neumann et al.
6. *Gesammelte Werke*. 3 vols. Leipzig, 1906-28. X.
See also 6569/1.

6273. SOLEIL, Jean Baptiste François. 1798-1878. Fr. DSB; P II,III (SOLEIL, N.).

6274. BARY, Emile Louis François. b. 1799.

6275. CALLAN, Nicholas. 1799-1864. Br. DSB; ICB.

6276. CLAPEYRON, Benoit Pierre Emile. 1799-1864. Fr. DSB; ICB; P I, III.
1. *Mémoire sur la puissance motrice de la chaleur*. [Periodical article, 1834] German trans. Leipzig, 1926. (OKEW. 216)

6277. EISENLOHR, Wilhelm. 1799-1872. Ger. P I,III.

6278. KUPFFER, Adolf Theodor von. 1799-1865. Russ. P I,III; BMNH; GC. Also Crystallography* and Earth Sciences.

6279. MAURICE, George. 1799-1839. Switz. P II.

6280. PETŘINA, Franz Adam. 1799-1855. Prague. P II.

6281. POTTER, Richard. 1799-1886. Br. P II,III; GD. Also Mechanics*.
1. *An elementary treatise on optics*. 2 vols. London, 1847-51. 2nd ed., 1851.
2. *Physical optics*. 2 vols. Cambridge, 1856-59. RM/L.

6282. REICH, Ferdinand. 1799-1882. Ger. P II,III; GC. Also Earth Sciences.

6283. SALM-HORSTMAR, Wilhelm Friedrich Carl August. 1799-1865. Ger. P II,III; BMNH. Also Chemistry.

6284. JEDLIK, (Stephan) Anianus. 1800-1895. Hung. ICB; P III.

6285. LLOYD, Humphrey. 1800-1881. Br. DSB; ICB; P I,III; GD. Also Earth Sciences*.
1. *A treatise on light and vision.* London, 1831.
2. *Lectures on the wave-theory of light.* Dublin, 1841.
3. *The elements of optics.* Dublin, 1849.
4. *An elementary treatise on the wave-theory of light.* Dublin, 1857. X. 2nd ed., London, 1857. 3rd ed., 1873.
5. *A treatise on magnetism, general and terrestrial.* London, 1874.
6. *Miscellaneous papers connected with physical science.* London, 1877.

6286. McGAULEY, James William. ca. 1800-1867. Br. GD.
1. *Lectures on natural philosophy.* Dublin, 1840. RM. 4th ed., 1853. 5th ed., 1857.

6287. RUDBERG, Fredrik. 1800-1839. Swed. ICB; P II.

6288. SPILLER, Philipp. 1800-1879. Ger./Posen. P II,III; GC.

6289. TALBOT, William Henry Fox. 1800-1877. Br. DSB; ICB; P II,III; GB; GD.

6290. AIRY, George Biddell. 1801-1892. Br. DSB; ICB; P I,III,IV; GB; GD1. Also Mathematics*, Astronomy* and Autobiography*.
1. *The undulatory theory of optics.* London, 1866. RM/L. Another ed., 1877.
2. *On sound and atmospheric vibrations. With the mathematical elements of music.* London, 1868. RM/L. 2nd ed., 1871.
3. *A treatise on magnetism.* London, 1870. RM/L.

6291. COMBES, Charles Pierre Mathieu. 1801-1872. Fr. DSB; P I,III. Also Mechanics.
1. *Exposé des principes de la théorie mécanique de la chaleur et de ses applications principales.* Paris, 1863. RM.

6292. FECHNER, Gustav Theodor. 1801-1887. Ger. The psychologist. DSB; ICB; P I,III,IV; BMNH; Mort.; BLA & Supp.; GB; GC. Also Various Fields*.
1. *Massbestimmungen über die galvanische Kette.* Leipzig, 1831. RM.
2. *Repertorium der Experimentalphysik, enthaltend eine vollständige Zusammenstellung der neuern Fortschritte.* 3 vols. Leipzig, 1832. RM.
3. *Über die physikalische und philosophische Atomenlehre.* Leipzig, 1855. RM. 2nd ed., 1864.
See also 6216/1.

6293. FRANKENHEIM, Moritz Ludwig. 1801-1869. Ger. DSB; P I,III; BMNH. Also Crystallography.
1. *Die Lehre von der Kohäsion, umfassend die Elastizität der Gase, die Elastizität und Kohärenz der flussigen und festen Körper, und die Kristallkunde.* Breslau, 1835. RM.

6294. JACOBI, Moritz Hermann von (or Boris Semenovich). 1801-1874. Russ. DSB; ICB; P I,III; GA.

6295. LA RIVE, (Arthur) Auguste de. 1801-1873. Switz. DSB; ICB; P II,III (RIVE); GA; GB.

1. *Traité de l'électricité, théorique et appliquée.* 3 vols. Paris, 1854-58. RM/L.
1a. ——— *A treatise on electricity, in theory and practice.* 3 vols. London, 1853-58. Trans. by C. V. Walker. RM/L.

6296. PERSON, Charles Cléophas. b. 1801. Fr. P II.

6297. PLATEAU, Joseph Antoine Ferdinand. 1801-1883. Belg. DSB; ICB; P II,III,IV; GA; GB.
1. *Statique expérimentale et théorique des liquides soumis aux seules forces moléculaires.* 2 vols. Paris, 1873.

6298. PLÜCKER, Julius. 1801-1868. Ger. DSB; ICB; P II,III; GB; GC. Also Mathematics*#. See also 6501/2.

6299. PORRO, Ignazio. 1801-1875. It./Fr. DSB; P II,III.

6300. STEINHEIL, Karl August. 1801-1870. Ger. DSB; ICB; P II,III; GC. Also Astronomy.

6301. COLLADON, Jean Daniel. 1802-1879. Switz. P I,III,IV. Also Meteorology.

6302. GHERARDI, Silvestro. 1802-1879. It. ICB; P III. Also History of Mathematics and Physics.

6303. MAGNUS, (Heinrich) Gustav. 1802-1870. Ger. DSB; ICB; P II,III; BLA; GB; GC. Also Chemistry.

6304. WHEATSTONE, Charles. 1802-1875. Br. DSB; ICB; P II,III; GB; GD. See also 15600/1.
1. *The scientific papers.* London, 1879. Publ. by the Physical Society of London. RM/L.

6305. WREDE, Fabian Jacob. 1802-1893. Swed. P II,III,IV.

6306. CHALLIS, James. 1803-1882. Br. DSB; P I,III; GD. Also Astronomy*.
1. *An essay on the mathematical principles of physics.* London, 1873.

6307. DOPPLER, (Johann) Christian. 1803-1853. Aus. DSB; ICB; P I; GA; GC. Also Astronomy.
1. *Abhandlungen.* Leipzig, 1907. Ed. with notes by H.A. Lorentz. (OKEW. 161)

6308. DOVE, Heinrich Wilhelm. 1803-1879. Ger. DSB; ICB; P I,III; GC. Also Earth Sciences*.
1. *Untersuchungen im Gebiet der Inductionselektricität.* Berlin, 1842. RM.
2. *Über Wirkungen aus der Ferne.* Berlin, 1845. RM.
3. *Darstellung der Farbenlehre und optische Studien.* Berlin, 1853. RM.

6309. ERICSSON, John. 1803-1889. U.S.A. ICB; P III; GB; GE.

6310. HESSLER, Ferdinand. 1803-1865. Aus. P I,III.

6311. MARCET, François. 1803-1883. Switz. P II,III. Also Chemistry.

6312. SCHMIDT, Johann Karl Eduard. 1803-1832. Ger. P II.

6313. STURM, (Jacques) Charles François. 1803-1855. Fr. DSB; ICB; P II; GA; GB. Also Mathematics*.

6314. BREGUET, Louis François Clément. 1804-1883. Fr. DSB; P I,III.

6315. BROOKE, Charles. 1804-1879. Br. P I,III; BLA; GD.

6316. GANOT, Adolphe. 1804-1887. Fr. P III.
1. *Traité élémentaire de physique expérimentale*. Paris, 1851. X. 4th ed., 1855. 21st ed., 1894.
1a. ——— *Elementary treatise on physics*. London, 1863. Trans. and ed. by E. Atkinson. X. 7th ed., 1875. Another ed., 1877. 10th ed., 1881. 11th ed., 1883. Another ed., 1890. 14th ed., 1893.

6317. LENZ, Emil Khristianovich (or Heinrich Friedrich Emil). 1804-1865. Russ. DSB; ICB; P I,III. Also Earth Sciences.

6318. MAGRINI, Luigi. ca. 1804-1868. It. P III.

6319. MOIGNO, François Napoléon Marie. 1804-1884. Fr. P II,III; GA. Also Mechanics*.
1. *Répertoire d'optique moderne; ou, Analyse complète des travaux modernes relatifs aux phénomènes de la lumière*. 4 vols. Paris, 1847-50.
2. *Traité de télégraphie électrique*. Paris, 1848. RM.

6320. MORREN, Jean François Auguste. 1804-1870. Fr. P III; BMNH. Also Chemistry*.

6321. MUNK AF ROSENSCHÖLD, Peter Samuel. 1804-1860. Swed. P II,III.

6322. PEREIRA, Jonathan. 1804-1853. Br. P II; BMNH; BLA; GD.
1. *Lectures on polarized light*. London, 1843. RM. 2nd ed., 1854.

6323. RIESS, Peter Theophil. 1804-1883. Ger. ICB; P II,III; GC.
1. *Die Lehre von der Reibungselectricität*. 2 vols. Berlin, 1853.
2. *Abhandlungen zur Lehre von der Reibungselektricität*. 2 vols. Berlin, 1867-79.

6324. ROBIDA, Karl (or Lucas). 1804-1877. Aus. P II,III; GC.

6325. VOLPICELLI, Paolo. 1804-1879. It. P III.

6326. WEBER, Wilhelm Eduard. 1804-1891. Ger. DSB; ICB; P II,III,IV; Mort.; GB; GC. See also Index.
1. *Mechanik der menschlichen Gehwerkzeuge*. Göttingen, 1836. With E.F. Weber.
2. *Über die Einführung absoluter elektrischer Masse*. Brunswick, 1968. (OKEW, n.F. 5)
3. *Werke*. 6 vols. Berlin, 1892-94. Publ. by the Gesellschaft der Wissenschaften zu Göttingen.

6327. BANCALARI, Michele Alberto. b. 1805. It. P I,III.

6328. BUFF, Heinrich. 1805-1878. Ger. P I,III; GC. Also Geophysics*.

6329. DELLMANN, Johann Friedrich Georg. 1805-1870. Ger. P I,III. Also Crystallography and Meteorology.

6330. EARNSHAW, Samuel. 1805-1888. Br. ICB; P III.

6331. HAMILTON, William Rowan. 1805-1865. Br. DSB; ICB; P III; GB; GD. Also Mathematics*# and Mechanics. See also 15596/1.

6332. KNOCHENHAUER, Karl Wilhelm. 1805-1875. Ger. P I,III.
1. *Die Undulationstheorie des Lichtes. Eine Beilage zu den Lehrbüchern der Physik.* Berlin, 1839. RM.

6333. KNORR, Ernst. b. 1805. Russ. P I,III.

6334. MOSER, Ludwig Ferdinand. 1805-1880. Ger. P II,III.

6335. MOUSSON, (Joseph Rudolph) Albert. 1805-1890. Switz. P II,III; BMNH. Also Zoology.
1. *Die Physik auf Grundlage der Erfahrung.* 3 vols. Zurich, 1858-63. X. 3rd ed., 1874-80.

6336. SEEBECK, Ludwig Friedrich Wilhelm August. 1805-1849. Ger. P II.

6337. STAMKART, Franciscus Johanes. 1805-1882. Holl. P II,III. Also Earth Sciences.

6338. FRICK, Joseph. 1806-1875. Ger. P I,III.
1. *Die physikalische Technik.* 2nd ed. Brunswick, 1856. X. 4th ed., 1872. 5th ed., 1876. Another ed., 2 vols, 1890-95.

6339. MASSON, Antoine Philibert. 1806-1860. Fr. DSB; ICB; P II,III; BLA.

6340. NOBERT, Friedrich Adolph. 1806-1881. Ger. DSB; ICB; P II,III.

6341. RÖBER, August. 1806-1891. Ger. P II,III.

6342. SILBERMANN, Johann Theobald. 1806-1865. Fr. P II,III.

6343. SVANBERG, Adolph Ferdinand. 1806-1857. Swed. P II.

6344. ALTER, David. 1807-1881. U.S.A. ICB; GE.

6345. DUPREZ, François Joseph Ferdinand. b. 1807. Belg. P I,III.

6346. PACINOTTI, Luigi. 1807-1889. It. ICB; P III.

6347. PETZVAL, Joseph. 1807-1891. Aus. P III. Also Mathematics*.

6348. TATE, Thomas. 1807-1888. Br. P III; GD.

6349. BILLET, Felix. 1808-1882. Fr. P III.
1. *Traité d'optique physique.* 2 vols. Paris, 1858-59.

6350. DUPRÉ, Athanase Louis Victoire. 1808-1869. Fr. DSB; P III. Also Mathematics.
1. *Théorie mécanique de la chaleur.* Paris, 1869. RM.

6351. KELLAND, Philip. 1808-1879. Br. P I,III; GD.
1. *The theory of heat.* Cambridge, 1837. RM/L.
See also 6144/2.

6352. KLEITZ, Charles. b. 1808. Fr. P III.
1. *Etudes sur les forces moléculaires des liquides en mouvement et l'application à l'hydrodynamique.* Paris, 1873.

6353. LABORDE, Edmond. b. 1808. Fr. P III.

6354. LISTING, Johann Benedict. 1808-1882. Ger. P I,III; Mort. Also Physiology*.

6355. TOMLINSON, Charles. 1808-1897. Br. P III,IV; GD. Also Biography*. See also 6228/5.

6356. BARFUSS, Friedrich Wilhelm. b. 1809. Ger. P I,III.
  1. *Theorie der Spiegelmikroscop mit sphärischen Glasspiegeln.* Weimar, 1840. RM.

6357. FLIEDNER, Conrad. 1809-1855. Ger. P I,III.

6358. FORBES, James David. 1809-1868. Br. DSB; P I,III; BMNH; GB; GD. Also Geology*. See also 15488/1.

6359. GREISS, Carl Bernhard. b. 1809. Ger. P I,III.

6360. JOLLY, Philipp Johann Gustav von. 1809-1884. Ger. DSB; P I, III; GC.

6361. KOHLRAUSCH, Rudolph Herrmann Arndt. 1809-1858. Ger. DSB; ICB; P I; GC.

6362. MacCULLAGH, James. 1809-1847. Br. DSB; P II; GB; GD.
  1. *The collected works.* Dublin, 1880. Ed. by J.H. Jellett and S. Haughton.

6363. MAGGI, Pietro. 1809-1854. It. P III.

6364. MÜLLER, Johann Heinrich Jacob. 1809-1875. Ger. DSB; P II,III; GC. Also Crystallography*.
  1. *Pouillet's Lehrbuch der Physik und Meteorologie.* 2 vols. Brunswick, 1842-44. X. See 6222/1a.
  1a. ——— 3rd ed., *Müller-Pouillet's Lehrbuch....* 1847. 4th ed., 1853. 6th ed., 1868. 7th ed., by L. Pfaundler, 3 vols in 4, 1876-81. 8th ed., 1879-82. 9th ed., 1886-98; RM (under Pouillet).
  2. *Grundriss der Physik und Meteorologie.* Brunswick, 1846. X. 9th ed., 1866.
  2a. ——— *Principles of physics and meteorology.* London, 1847. Another ed., Philadelphia, 1848.
  3. *Lehrbuch der kosmischen Physik.* Brunswick, 1856. X. 2nd ed., 2 vols, 1865.

6365. SPASSKII, Mikhail Fedorovich. 1809-1859. Russ. ICB. Also Meteorology.

6366. VORSSELMANN DE HEER, Pieter Otto Conraad. 1809-1841. Holl. P II.

6367. ADIE, Richard. 1810-1880. Br. P III.

6368. ARMSTRONG, William George. 1810-1899. Br. P III,IV; GB; GD1.
  1. *Electrical movement in air and water. With theoretical inferences.* London, 1897. Supplement, 1899.

6369. BACHHOFFNER, George Henry. 1810-1879. Br. GD.
  1. *A popular treatise on voltaic electricity and electro-magnetism.* London, 1838. RM.

6370. EMSMANN, August Hugo. 1810-1889. Ger. P I,III,IV.

6371. LANGBERG, Lorenz Christian. 1810-1857. Nor. P I.

6372. LAROQUE, Frédéric Raymond Noël. 1810-1887. Fr. P III.

6373. QUET, Jean Antoine. 1810-1884. Fr. P II,III,IV.

6374. RADICKE, Gustav. 1810-1883. Ger. P II,III.
  1. *Handbuch der Optik.* 2 vols. Berlin, 1839.

6375. REGNAULT, (Henri) Victor. 1810-1878. Fr. DSB; ICB; P II,III; GA; GB. Also Chemistry*.

6376. ABRIA, Jérémie Joseph Benoit. 1811-1892. Fr. P III,IV; GF.

6377. DRAPER, John William. 1811-1882. U.S.A. DSB; ICB; P I,III; BMNH; BLA & Supp.; GB; GD; GE. Also Chemistry, Physiology*, and History of Science*.
   1. *Scientific memoirs, being experimental contributions to a knowledge of radiant energy.* London, 1878. Reprinted, New York, 1973.

6378. GAUGAIN, Jean Mothée. 1811-1880. Fr. P III. See also 6216/2a.

6379. GROVE, William Robert. 1811-1896. Br. DSB; ICB; P I,III; GA; GB; GD1. Also Chemistry.
   1. *On the correlation of physical forces.* London, 1846. RM/L. 2nd ed., 1850. 4th ed., 1862. 5th ed., *Followed by a discourse on continuity*, 1867. 6th ed., *With other contributions to science* [reprinting of many of his papers], 1874; RM/L.
   See also 6470/1.

6380. HOLTZMANN, Karl Heinrich Alexander. 1811-1865. Ger. P I,III.
   1. *Mechanische Wärme-Theorie.* Stuttgart, 1866.

6381. MATTEUCCI, Carlo. 1811-1868. It. DSB; ICB; P II,III; BMNH; Mort.; BLA; GB. Also Physiology*.

6382. NOUGARÈDE DE FAYET, Auguste. 1811-1853. Fr. P III.
   1. *De l'électricité dans ses rapports avec la lumière, la chaleur et la constitution des corps.* Paris, 1839. RM.

6383. WATERSTON, John James. 1811-1883. Br. DSB; ICB.
   1. *The collected scientific papers.* Edinburgh, 1928. Ed. with a biography by J.S. Haldane.

6384. DESAINS, Edward François. 1812-1865. Fr. P I,III.

6385. PAGE, Charles Grafton. 1812-1868. U.S.A. P II,III; GE.

6386. REUSCH, Friedrich Eduard. 1812-1891. Ger. P II,III.

6387. RIJKE, Pieter Leonhard. 1812-1901. Holl. P II,III,IV.

6388. WALKER, Charles Vincent. 1812-1882. Br. P III; GD. See also 6295/1a and 10519/2a.

6389. WILHELMY, Ludwig Ferdinand. 1812-1864. Ger. DSB; P II,III. Also Chemistry*.

6390. FAVRE, Pierre Antoine. 1813-1880. Fr. DSB; P I,III. Also Chemistry.

6391. FLESCH, J. 1813-1879. Ger. P I,III.

6392. LAURENT, Pierre Alphonse. 1813-1854. Fr. DSB; PI. Also Mathematics.

6393. LOVERING, Joseph. 1813-1892. U.S.A. P I,III,IV; GE.

6394. SMITH, Archibald. 1813-1872. Br. P II,III; GD. See also 10487/2.

6395. ÅNGSTRÖM, Anders Jonas. 1814-1874. Swed. DSB; P I,III; GB. Also Astronomy.

1. *Recherches sur le spectre solaire*. 2 vols. Berlin, 1868-69.

6396. DAGUIN, Pierre Adolphe. 1814-1884. Fr. P III. Also Meteorology.
   1. *Traité élémentaire de physique, théorique et expérimentale, avec les applications à la météorologie et aux arts industrielles*. 3 vols. Toulouse, 1855-60. X. 3rd ed., Paris, 1867-68. RM.

6397. ELLIS, Alexander John. 1814-1890. Br. P III,IV; GB; GD1. See also 14888/1a,b.

6398. HANKEL, Wilhelm Gottlieb. 1814-1899. Ger. DSB; P I,III,IV; GC.

6399. MAYER, Julius Robert von. 1814-1878. Ger. DSB; ICB; P II,III, IV; Mort.; BLA & Supp.; GB; GC. Also Physiology.
   1. *Die Mechanik der Wärme. Zwei Abhandlungen*. Leipzig, 1911. (OKEW. 180)
   2. *Beiträge zur Dynamik des Himmels*. Heilbronn, 1848. X. Reprinted, Leipzig, 1927. (OKEW. 223)
   3. *Die Mechanik der Wärme in gesammelte Schriften*. Stuttgart, 1867. RM.
   4. *Naturwissenschaftliche Vorträge*. Stuttgart, 1871.
   5. *Kleinere Schriften und Briefe*. Stuttgart, 1893. Ed. with a biography by J.J. Weyrauch. RM.

   See also 6470/1.

6400. MELSENS, Louis Henri Frédéric. 1814-1886. Belg. P II,III,IV; BLA. Also Chemistry.

6401. O'BRIEN, Matthew. 1814-1855. Br. P II,III; GD.

6402. CAVALLERI, Giovanni Maria. d. 1874. It. P III.

6403. BIRD, Golding. 1815-1854. Br. P I & Supp.; BLA & Supp.; GD.
   1. *The elements of natural philosophy*. London, 1841. X. 2nd ed., 1844. 4th ed., 1854. 5th ed., 1860. 6th ed., 1867.

6404. COLDING, Ludvig August. 1815-1888. Den. DSB Supp.; ICB; P I, III. Also Meteorology.
   1. *Ludvig Colding and the conservation of energy principle: Experimental and philosophical contributions*. New York, 1972. Introd., trans., and commentary by P.F. Dahl.

6405. DECHARME, Joseph Constantin. 1815-1905. Fr. P III,IV.

6406. DE LA RUE, Warren. 1815-1889. Br. DSB; P III,IV; GB; GD. Also Astronomy* and Chemistry.

6407. GEISSLER, (Johann) Heinrich (Wilhelm). 1815-1879. Ger. DSB; P III; GB.

6408. HIRN, Gustave Adolphe. 1815-1890. Fr. DSB; P I,III,IV.
   1. *Exposition analytique et expérimentale de la théorie mécanique de la chaleur*. Paris, 1862. 2nd ed., 1865. 3rd ed., 2 vols, 1875-76.
   2. *Conséquences philosophiques et métaphysiques de la thermodynamique*. Paris, 1868.

6409. NOAD, Henry Minchin. 1815-1877. Br. P III; GD. Also Chemistry*.

1. *A course of eight lectures on electricity, galvanism, magnetism and electro-magnetism.* London, 1839. X. Another ed., 1844.
2. *Manual of electricity.* 4th ed. 2 vols. London, 1855-57.
3. *The improved induction coil.* London, 1861.
4. *The student's text-book of electricity.* London, 1867.
5. *Textbook of electricity.* London, 1879.

6410. OPPEL, Johann Joseph. 1815-1894. Ger. P II,III.

6411. POPOFF, Alexander Fedorovich. 1815-1879. Russ. P II,III. Also Mathematics.

6412. PROVENZALI, Francesco Saverio. 1815-1894. It. P III,IV.

6413. SAN ROBERTO, Paolo, *Conte* di Ballada. 1815-after 1879. It. P II,III.

6414. SONDHAUSS, Karl Friedrich Julius. 1815-1886. Ger. P II,III.

6415. WERTHEIM, Wilhelm. 1815-1861. Fr. P II.

6416. BAYMA, Joseph. 1816-1892. Br./U.S.A. ICB; P III; GE.
1. *The elements of molecular mechanics.* London, 1866. Another ed., 1886.

6417. CODAZZA, Giovanni. 1816-1877. It. P III.

6418. KUHN, Karl. 1816-1869. Ger. P I,III.

6419. LALLEMAND, Etienne Alexandre. 1816-1886. Fr. P III. Also Chemistry.

6420. SIEMENS, (Ernst) Werner von. 1816-1892. Ger. DSB; ICB; P II, III, IV; GB; GC. Also Autobiography*.
1. *Gesammelte Abhandlungen und Vorträge.* Berlin, 1881. RM.
2. *Wissenschaftliche und technische Arbeiten.* 2 vols. Berlin, 1889-91. X.
2a. ——— *Scientific and technical papers.* 2 vols. London, 1892-95.

6421. BRIOT, Charles Auguste Albert. 1817-1882. Fr. DSB; P III,IV. Also Mathematics*.
1. *Essai sur la théorie mathématique de la lumière.* Paris, 1864. RM.
2. *Théorie mécanique de la chaleur.* Paris, 1869. RM/L. 2nd ed., 1883.

6422. DESAINS, Paul Quentin. 1817-1885. Fr. P I,III.

6423. DUB, (Christoph) Julius. 1817-1873. Ger. P I,III. Also Evolution*.
1. *Der Elektromagnetismus.* Berlin, 1861. RM.
2. *Die Anwendung des Elektromagnetismus.* Berlin, 1863. RM.

6424. FEILITZSCH, Fabian Carl Ottokar. 1817-1885. Ger. P I,III.

6425. JELLETT, John Hewitt. 1817-1888. Br. P III,IV; GD. Also Mathematics* and Chemistry*.
1. *A treatise on the theory of friction.* London/Dublin, 1872.
See also 6362/1.

6426. MARBACH, Christian August Hermann. 1817-1873. Ger. P II,III.

6427. MOON, Robert. 1817-1889. Br. P III.

6428. WARTMANN, Elie François. 1817-1886. Switz. P II,III.

6429. ZAMMINER, Friedrich Georg Karl. 1817-1858. Ger. P II; GC.

6430. BERTIN, Pierre Auguste. 1818-1884. Fr. P I,III,IV.

6431. BROCH, Ole Jacob. 1818-1889. Nor. P I,III,IV.

6432. CANTONI, Giovanni. 1818-1897. It. P III,IV. See also 6580/1.

6433. FORTI, Angiolo. 1818-after 1882. It. P III.

6434. JAMIN, Jules Célestin. 1818-1886. Fr. P I,III,IV.
1. *Cours de physique*. 3 vols. Paris, 1858. X. 2nd ed., 1869-75. 3rd ed., 4 vols, 1879-83. 4th ed., 1885-91. Supplement, by E. Bouty, 2 vols, 1896-99.
2. *Petit traité de physique*. Paris, 1870.

6435. JOULE, James Prescott. 1818-1889. Br. DSB; ICB; P I,III; GB; GD.
1. *The scientific papers*. 2 vols. London, 1884-87. Publ. by the Physical Society of London. RM/L. Reprinted, London, 1963.

6436. LECONTE, John. 1818-1891. U.S.A. DSB; P III; BLA; GE.

6437. LUVINI, Giovanni. 1818-after 1889. It. P III,IV.

6438. SCHELLEN, (Thomas Joseph) Heinrich. 1818-1884. Ger. P II,III.
1. *Die Spectralanalyse in ihrer Anwendung auf die Stoffe der Erde und die Natur der Himmelskörper*. Brunswick, 1870. RM.
1a. ——— *Spectrum analysis in its application to terrestrial substances and the physical constitution of the heavenly bodies*. London, 1872. Trans. from the 2nd enl. and rev. German ed. by J. and C. Lassell. Ed. with notes by W. Huggins. RM/L.
1b. ——— ——— 2nd ed., trans. from the 3rd German ed., 1885.

6439. SWAN, William. 1818-1894. Br. P III; GD.

6440. EDLUND, Erik. 1819-1888. Swed. P I,III,IV.

6441. FELICI, Riccardo. 1819-1902. It. P III,IV & Supp.
1. *Teoria matematica dell'induzione elettro-dinamica*. [Periodical articles, 1854 and 1857] German trans. Leipzig, 1899. (OKEW. 109)

6442. FIZEAU, Armand Hippolyte Louis. 1819-1896. Fr. DSB; ICB; P I, III,IV; GB.

6443. FOUCAULT, (Jean Bernard) Léon. 1819-1868. Fr. DSB; ICB; P I, III; GA; GB. Also Mechanics*.
1. *Recueil des travaux scientifiques*. 2 vols in 1. Paris, 1878. Ed. by C.M. Gabriel. RM/L.

6444. GRAEVELL, Friedrich. 1819-1878. Ger. P I; BLA.

6445. LANE, Jonathen Homer. 1819-1880. U.S.A. DSB; P III.

6446. MONTIGNY, Charles Marie Valentin. 1819-1890. Belg. ICB; P II, III,IV. Also Meteorology.

6447. PIERRE, Victor. 1819-1886. Aus. P II,III,IV.

6448. STOKES, George Gabriel. 1819-1903. Br. DSB; ICB; P II,III,IV; GB; GD2. Also Mechanics.
1. *On light*. London, 1887. Lecture courses. 2nd ed., 1892.
2. *Mathematical and physical papers*. 5 vols. Cambridge, 1880-1905. Vols I-III ed. by Stokes, Vols IV and V by J. Larmor. RM/L. Reprinted with some additions, New York, 1966.
3. *Memoirs and scientific correspondence*. 2 vols. Cambridge, 1907. The memoirs are biographies by various persons. RM/L. Reprinted, New York, 1971.

6449. SZTOCZEK, Joseph. 1819-1890. Hung. P III.

6450. BECQUEREL, (Alexandre) Edmond. 1820-1891. Fr. DSB; P I,III, IV; GB.
1. *Traité d'électricité et de magnétisme*. 3 vols. Paris, 1885-86. With A.C. Becquerel.
2. *La lumière, ses causes et ses effets*. 2 vols. Paris, 1867-68. RM/L.
3. *Des forces physico-chimiques et leur intervention dans la production des phénomènes naturels*. Paris, 1875. RM/L.
See also 6210/3 and 10479/1.

6451. BOUTAN, Augustin. 1820-1900. Fr. P III,IV.

6452. CORNELIUS, Carl Sebastian. 1820-1896. Ger. P I,III,IV; GC. Also Astronomy*.
1. *Grundzüge einer Molecularphysik*. Halle, 1866. RM.

6453. KARSTEN, Gustav. 1820-1900. Ger. P I,III,IV. Also Earth Sciences.

6454. KNOBLAUCH, Karl Hermann. 1820-1895. Ger. P I,III,IV; GC.

6455. MARIÉ-DAVY, Edme Hippolyte. 1820-1893. Fr. P III,IV. Also Meteorology.

6456. RANKINE, William John Macquorn. 1820-1872. Br. DSB; ICB; P II, III; GB; GD.
1. *Miscellaneous scientific papers*. London, 1881. Ed. by W.J. Millar with a biography by P.G. Tait. RM/L.

6457. SAWELJEW, Alexander Stepanovich. 1820-1860. Russ. P II,III.

6458. SCHEFFLER, (August Christian Wilhelm) Hermann. 1820-1903. Ger. P II,III,IV. Also Mathematics and Mechanics.

6459. TYNDALL, John. 1820-1893. Br. DSB; ICB; P II,III,IV; GB; GD. Also Science in General*, Geology*, Microbiology*, and Biography*. See also Index.
1. *Heat considered as a mode of motion*. London, 1863. RM/L. 2nd ed., ib., 1865 and New York, 1867. 5th ed., 1875. 6th ed., 1880. 7th ed., 1887. 9th ed., 1892.
2. *On radiation*. London, 1865.
3. *Sound*. London, 1867. RM/L. 2nd ed., 1869. 3rd ed., 1875; reprinted, New York, 1964. 4th ed., 1883.
4. *Notes of a course of seven lectures on electrical phenomena and theories*. London, 1870. RM/L. Other eds: 1871, 1884.
5. *Notes of a course of nine lectures on light*. London, 1870. 3rd ed., 1871. 5th ed., 1873.
6. *Researches on diamagnetism*. London, 1870. RM/L. Another ed., 1888.

7. *Contributions to molecular physics in the domain of radiant heat.* London, 1872. RM/L.
8. *Six lectures on light.* London, 1873. RM/L. 2nd ed., 1875. 3rd ed., 1882. 4th ed., 1885.
9. *Lessons in electricity.* London, 1876. RM/L. 3rd ed., 1882. 4th ed., 1887.

6460. VALERIUS, Hubert. 1820-1897. Belg. P III; BLA; GF. Also Physiology.

6461. COLNET D'HUART, ... 1821-after 1896. Luxembourg, P III,IV.

6462. DU MONCEL, Théodore Achille Louis. 1821-1884. Fr. DSB. Under MONCEL: P II,III; GA.
1. *Exposé des applications de l'électricité.* 2 vols. Paris, 1853-54. X. 3rd ed., 5 vols, 1872-78.

6463. HELMHOLTZ, Hermann Ludwig Ferdinand von. 1821-1894. Ger. DSB; ICB; P I,III,IV; Mort.; BLA & Supp.; GB; GC. Also Science in General*#, Mechanics*, and Physiology#. See also Index.
1. *Über die Erhaltung der Kraft.* Berlin, 1847. RM. Reprinted, Leipzig, 1889. (OKEW. 1)
2. *Theorie der Luftswingungen in Röhren mit offenen Enden.* [Periodical article, 1859] Leipzig, 1896. (OKEW. 80)
3. *Vorlesungen über theoretische Physik.* 6 vols in 7. Leipzig, 1898-1922. RM/L.
4. *Abhandlungen zur Thermodynamik.* Leipzig, 1902. (OKEW. 124)

6464. LÖSCHE, Eduard G. 1821-1879. Ger. P III.

6465. LOSCHMIDT, Joseph. 1821-1895. Aus. DSB; ICB; P III,IV; GC. Also Chemistry*.

6466. PRIVAT-DESCHANEL, Augustin. 1821-1883. Fr. P III. Also Science in General*.
1. *Traité de physique élémentaire.* Paris, 1869.
1a. ——— *Elementary treatise on natural philosophy.* 4 parts. London, 1870-72. Trans. and ed. with extensive additions by J.D. Everett. Another ed., 1876. 4th ed., 1877. 5th ed., 1880. 6th ed., 1882-83. 8th ed., 1885. 10th ed., 1888-90. 13th ed., 1894. 14th ed., 1897-99.

6467. ROLLMANN, Wilhelm. 1821-after 1872. Ger. ICB; P II,III.

6468. VIARD, Henri Stanislas. 1821-1858. Fr. P II,III.

6469. WEISS, Adam. 1821-after 1855. Ger. P II.

6470. YOUMANS, Edward Livingston. 1821-1887. U.S.A. ICB; GE. See also 15536/1.
1. *The correlation and conservation of forces: A series of expositions by Prof. Grove, Prof. Helmholtz, Dr. Mayer, Dr. Faraday, Prof. Liebig, and Dr. Carpenter. With an introduction and brief biographical notices of the chief promoters of the new views.* New York, 1865. RM.

6471. ALMEIDA, Joseph Charles d'. 1822-1880. Fr. ICB; P III.

6472. BEETZ, Wilhelm von. 1822-1886. Switz./Ger. P I,III,IV; GC.

6473. BERTRAND, Joseph Louis François. 1822-1900. Fr. DSB; ICB;

P I,III,IV; GA. Also Mathematics*, Biography*, and History of Science*.
1. *Thermodynamique.* Paris, 1887.
2. *Leçons sur la théorie mathématique de l'électricité.* Paris, 1890.

6474. CLARK, (Josiah) Latimer. 1822-1898. Br. DSB; P III; GB; GD1.

6475. CLAUSIUS, Rudolph (Julius Emmanuel). 1822-1888. Switz./Ger. DSB; ICB; P I,III,IV; GB; GC.
1. *Über die bewegende Kraft der Wärme und die Gesetze welche sich daraus für die Wärmelehre selbst ableiten lassen.* [Periodical article, 1850] Leipzig, 1898. (OKEW. 99)
2. *Abhandlungen über die mechanische Wärmetheorie.* 2 vols. Brunswick, 1864-67. RM/L.
2a. ——— *The mechanical theory of heat.* London, 1867. Ed. by T.A. Hirst. RM/L.
2b. ——— 2nd ed. *Die mechanische Wärmetheorie.* 3 vols in 2. Brunswick, 1876-79.
2c. ——— ——— *The mechanical theory of heat.* London, 1879. Trans. by W.R. Browne.
3. *Die Potentialfunction und das Potential.* Leipzig, 1864. X. 3rd ed., 1877.
4. *Über den zweiten Hauptsatz der mechanischen Wärmetheorie. Ein Vortrag.* Brunswick, 1867. RM.

6476. GUILLEMIN, Claude Marie. 1822-1890. Fr. P III. Also Plant Physiology.

6477. KRÖNIG, August Karl. 1822-1879. Ger. DSB; P I,III.

6478. LISSAJOUS, Jules Antoine. 1822-1880. Fr. DSB; ICB; P I,III.

6479. THOMSON, James (second of the name). 1822-1892. Br. ICB; P III,IV; GB; GD.
1. *Collected papers in physics and engineering.* Cambridge, 1912. Ed. by J. Larmor and J. Thomson.

6480. WILLIGEN, Volkert Simon Maarten van der. 1822-1878. Holl. P II,III.

6481. WITTWER, Wilhelm Constantin. 1822-1908. Ger. P II,III,IV. Also Biography*.

6482. AMSLER (-LAFFON), Jakob. 1823-1912. Switz. DSB; P I,III,IV.

6483. BRUNNER (VON WATTENWYL), Carl. 1823-after 1906. Switz. P I; BMNH. Also Zoology.

6484. KOOSEN, Johann Heinrich. 1823-after 1887. Ger. P I,III,IV.

6485. PAALZOW (or PAALZOU), (Carl) Adolph. 1823-1908. Ger. P II, III,IV.
1. *Die Bestimmung des elektrischen Leitungswiderstandes von Metalldrähten.* Berlin, 1888.

6486. PYNCHON, Thomas Ruggles. 1823-1904. U.S.A. GE.
1. *Chemical forces, heat, light, electricity.* 1870.

6487. SEGUIN, Jean Marie François. 1823-after 1881. Fr. P III,IV Supp.

6488. SIEMENS, (Charles) William (or Karl Wilhelm). 1823-1883.

Ger./Br. DSB; ICB; P II,III; GB; GC; GD.

1. *The scientific works*. 3 vols. London, 1889. Ed. by E.F. Bamber.
2. *A collection of letters to Sir Charles William Siemens*. London, 1953. With a short biography by W.H. Kennett.

6489. STALLO, John (or Johann) Bernhard. 1823-1900. U.S.A. DSB; P IV; GE. Also Science in General*.

1. *The concepts and theories of modern physics*. London, 1882. 2nd ed., 1882. Reprinted, Cambridge, Mass., 1960.

6490. HOLMGREN, Karl Albert Victor. 1824-1905. Swed. P III,IV.

6491. KERR, John. 1824-1907. Br. DSB; ICB; P III,IV; GD2.

6492. KIRCHHOFF, Gustav Robert. 1824-1887. Ger. DSB; ICB; P I,III, IV; GB; GC. Also Mechanics* and Chemistry*.

1. *Abhandlungen über mechanische Wärmetheorie*. [Three periodical articles, 1858] Leipzig, 1898. (OKEW. 101)
2. *Abhandlungen über Emission und Absorption*. [Three periodical articles, 1859-62] Leipzig, 1898. (OKEW. 100)
3. *Untersuchungen über das Sonnenspectrum und die Spectren der chemischen Elemente*. ? vols. Berlin, 1861-63. RM. Reprinted, Osnabrück, 1972.

3a. ——— *Researches on the solar spectrum, and on the spectra of the chemical elements*. Cambridge, 1862. Trans. by H.E. Roscoe. RM/L.

4. *Gesammelte Abhandlungen*. Leipzig, 1882. Ed. by himself. RM/L. *Nachtrag*, ib., 1891. Ed. by L. Boltzmann.
5. *Vorlesungen über mathematische Physik*. 4 vols in 2. Leipzig, 1891-97. Ed. by K. Hensel and M. Planck. RM/L.

See also 6569/1.

6493. MARTIN, Adolphe Alexandre. 1824-after 1881. Fr. P III.

6494. MEYER, Martin Hermann. 1824-1856. Ger. P II. Also Mathematics.

6495. QUINTUS ICILIUS, E.W. Gustav von. 1824-1885. Ger. P II,III.

6496. THOMSON, William, *Baron* KELVIN. 1824-1907. Br. DSB; ICB (KELVIN); P III,IV; GB; GD2. Also Geology*. See also 15553/1.

1. *The dynamical theory of heat*. [Periodical articles, 1851 and 1852] German trans. Leipzig, 1914. (OKEW. 193)
2. *Treatise on natural philosophy*. Vol. 1. [No more publ. in this ed.] Oxford, 1867. With P.G. Tait. RM/L. 2nd ed., 2 parts, Cambridge, 1879-83. RM/L. Various reissues. Vol. 2, originally intended, was abandoned.
3. *Elements of natural philosophy*. Part 1. [No more publ. in this ed.] Oxford, 1873. With P.G. Tait. 2nd ed., Cambridge, 1879. Another ed., Cambridge, 1894. RM/L.
4. *Reprints of papers on electrostatics and magnetism*. London, 1872. RM/L. 2nd ed., 1884.
5. *Elasticity and heat, being articles contributed to the "Encyclopaedia Britannica."* Edinburgh, 1880. RM/L.
6. *Notes of lectures on molecular dynamics and the wave theory of light*. Baltimore, 1884. Twenty lectures given at the Johns Hopkins University and stenographically reported by A.S. Hathaway. RM/L.

6a. ——— *The Baltimore lectures on molecular dynamics and*

the wave theory of light. London, 1904. Another ed. of item 6 with appendices on allied topics. RM/L.
7. *The molecular tactics of a crystal*. Oxford, 1893. RM.
8. *Popular lectures and addresses*. 3 vols. London, 1889-94. RM/L.
9. *Mathematical and physical papers*. 6 vols. Cambridge, 1882-1911. Vols 1-3 ed. by himself, Vols 4-6 by J. Larmor. RM/L.

6497. VERDET, (Marcel) Emile. 1824-1866. Fr. DSB; P II,III.
1. *Leçons d'optique physique*. 2 vols. Paris, 1869-70.
2. *Oeuvres*. 8 vols in 9. Paris, 1868-72.

6498. BERNARD, Felix. d. 1866. Fr. P III.

6499. ANTOINE, Louis Charles. 1825-after 1892. Fr. P III,IV.

6500. BALMER, Johann Jakob. 1825-1898. Switz. DSB.

6501. BEER, August. 1825-1863. Ger. ICB; P I,III.
1. *Einleitung in die höhere Optik*. Brunswick, 1853. 2nd ed., 1882.
2. *Einleitung in die Elektrostatik, die Lehre vom Magnetismus und die Elektrodynamik*. Brunswick, 1865. Ed. by J. Plücker.
3. *Einleitung in die mathematische Theorie der Elasticität und Capillarität*. Leipzig, 1869. Ed. by A. Giesen.

6502. BOURBOUZE, J.G. 1825-1889. Fr. P III,IV.

6503. GRIPON, Emile. 1825-after 1881. Fr. P III.

6504. MANGIN, Alphonse François Eugène. 1825-1885. ICB.

6505. PUSCHL, Karl. 1825-1912. Aus. P III,IV.

6506. SPOTTISWOODE, William. 1825-1883. Br. P II,III,IV; GB; GD. Also Mathematics.
1. *The polarisation of light*. London, 1874. 3rd ed., 1879. Another ed., 1891.

6507. CHAUTARD, Jules Maria Augustin. 1826-after 1880. Fr. P III. Also Chemistry.

6508. GOVI, Gilberto. 1826-1889. It. ICB; P III,IV.

6509. GRASHOF, Franz. 1826-1893. Ger. DSB; ICB; P III. Also Mechanics.

6510. GUILLEMIN, Amédée Victor. 1826-1893. Fr. P III. Also Astronomy*.
1. *Les phénomènes de la physique*. Paris, 1868. X.
1a. ——— *The forces of nature. A popular introduction to the study of physical phenomena*. London, 1872. Ed. with additions and notes by J.N. Lockyer. X. 2nd ed., 1873.
2. *Les applications de la physique aux sciences, à l'industrie et aux arts*. Paris, 1874. RM/L.
2a. ——— *The applications of physical forces*. London, 1877. Ed. with additions and notes by J.N. Lockyer. RM/L.

3. *Le magnétisme et l'électricité.* Paris, 1890. X.
3a. ——— *Electricity and magnetism.* London, 1891. Trans. and ed. by S.P. Thompson.

6511. RIEMANN, (Georg Friedrich) Bernhard. 1826-1866. Ger. DSB; ICB; P II,III,IV; GB; GC. Also Mathematics*#.
1. *Schwere, Elektricität und Magnetismus.* Hanover, 1876. RM/L. 2nd ed., 1880.

6512. RITTER, (Georg) August (Dietrich). 1826-1908. Ger. P III;IV. Also Mechanics*.

6513. STONEY, George Johnstone. 1826-1911. Br. DSB; P IV; GD2. Also Astronomy.

6514. WIEDEMANN, Gustav Heinrich. 1826-1899. Switz./Ger. DSB; ICB; P II,III,IV; GB; GC.
1. *Über die Wärme-Leitungsfähigkeit der Metalle.* [Periodical article with R. Franz, 1853] Leipzig, 1927. (OKEW. 222)
2. *Die Lehre von Galvanismus und Elektromagnetismus.* 2 vols. Brunswick, 1860-63. X. 2nd ed., 1872-74.
3. *Die Lehre von der Elektricität.* 4 vols in 5. Brunswick, 1882-85. (Also counted as the 3rd ed. of item 2.) 2nd ed. (4th ed. of item 2), 1893-98.

6515. WIENER, (Ludwig) Christian. 1826-1896. Ger. DSB; ICB; P II, III,IV; GC. Also Mathematics*.

6516. FRANZ, Rudolph. 1827-after 1878. Ger. ICB; P I,III. See also 6514/1.

6517. KAHL, Emil. 1827-after 1875. Ger. P I,III.

6518. LIVEING, George Downing. 1827-1924. Br. P III,IV; GD4. Also Chemistry.
1. *Collected papers on spectroscopy.* Cambridge, 1915. With James Dewar.

6519. MARIANINI, Pietro Domenico. 1827-1884. It. P III,IV.

6520. PISKO, Franz Josef. 1827-1888. Aus. P III.
1. *Lehrbuch der Physik.* Vienna, 1855. X. 3rd ed., 1873.

6521. SORET, Jacques Louis. 1827-1890. Switz. ICB; P II,III,IV.

6522. THALÉN, Tobias Robert. 1827-1905. Swed. P II,III,IV.

6523. WATSON, Henry William. 1827-1903. Br. P III,IV. GD2. Also Mechanics*.
1. *A treatise on the kinetic theory of gases.* Oxford, 1876.
2. *The mathematical theory of electricity and magnetism.* 2 vols. Oxford, 1885-89.

6524. ABT, Anton. 1828-after 1901. Hung. P III,IV; BMNH. Also Crystallography.

6525. DELSAULX, Joseph. 1828-1891. Belg. P III,IV.

6526. LERAY, Amand Jean. 1828-after 1902. Fr. P IV.
1. *Essai sur la synthèse des forces physiques.* Paris, 1885.

6527. PETRUSCHEWSKY, Fedor Famitsch. 1828-1904. Russ. P III,IV.

6528. SCHMIDT, Wilibald Gottlob. 1828-1877. Ger. P II,III.

6529. STEWART, Balfour. 1828-1887. Br. DSB; P III,IV; GB; GD.
Also Earth Sciences. See also 2963/2.
1. *An elementary treatise on heat.* Oxford, 1866. X. 2nd ed., 1871. 3rd ed., 1876. 5th ed., 1888.
2. *Lessons in elementary physics.* London, 1870. X. Other eds: 1874, 1879, 1885.
3. *The conservation of energy, being an elementary treatise on energy and its laws.* London, 1873. 2nd ed., 1874; RM. 3rd ed., 1874. 4th ed., 1877. 7th ed., 1887.

6530. WALTENHOFEN, Adalbert Carl von. 1828-1914. Aus. P II,III,IV.

6531. ZECH, Paul Heinrich von. 1828-1893. Ger. P II,III,IV; GC.

6532. ZEUNER, Gustav Anton. 1828-1907. Switz./Ger. DSB; P II,III,IV.
1. *Grundzüge der mechanischen Wärmetheorie.* Freiburg, 1860. RM.
1a. ——— *Théorie mécanique de la chaleur.* Paris, 1869. Trans. by Maurice Arnthal.

6533. ARNDTSEN, Adam Frederik Olaf. 1829-after 1881. Nor. P III.
1. *Physikalske meddelelser.* Christiania, 1858.

6534. FERNET, Emile. 1829-after 1880. Fr. P III.

6535. LORENZ, Ludvig Valentin. 1829-1891. Den. DSB; ICB; P III,IV.
1. *Lehre vom Licht.* Leipzig, 1877.
2. *Oeuvres scientifiques.* 2 vols. Copenhagen, 1898-1904. Ed. with a biography by H. Valentiner. RM/L. Reprinted, New York, 1965.

6536. MOUTIER, Jules. 1829-after 1894. Fr. P III,IV.

6537. TRÈVE, Auguste Hubert Stanislas. 1829-1885. Fr. P III,IV.

6538. FABRI, Ruggiero. 1830?-after 1882. It. P III.

6539. HUGHES, David Edward. 1830-1900. Br. ICB; P III,IV; GB; GD1; GE.

6540. KELLER, Filippo. 1830-1903. It. P IV.

6541. LEDIEU, Alfred Constant Hector. 1830-1891. Fr. P III,IV.

6542. LJUBIMOW, Nicolai Alekseievich. 1830-1897. Russ. P III,IV.

6543. MATTHIESSEN, Heinrich Friedrich Ludwig. 1830-1906. Ger. P II, III,IV. Also Mathematics and Physiology.

6544. REITLINGER, Edmund. ca. 1830-1882. Aus. P III.

6545. ŠUBIC, Simon. 1830-1903. Aus. P III,IV.

6546. BOHN, Johann Conrad. 1831-1897. Ger. P III,IV.

6547. BOSSCHA, Johannes. 1831-1911. Holl. P I,III,IV.

6548. BURBURY, Samuel Hawksley. 1831-1911. Br. P IV & Supp.; GD2.
1. *A treatise on the kinetic theory of gases.* Cambridge, 1899. See also 5679/1.

6549. EISENLOHR, Friedrich. 1831-1904. Ger. P III.

6550. EVERETT, Joseph David. 1831-1904. Br. P III,IV; GD2.
1. *Illustrations of the centimetre-gramme-second (C.G.S.) system of units.* London, 1875.

1a. ——— Another ed. *Units and physical constants.* London, 1879. 2nd ed., 1886. 4th ed., 1891.
2. *Elementary textbook of physics.* London, 1877.
3. *Vibratory motion and sound.* London, 1882.
His translation of Privat-Deschanel's textbook (6466/1a) is sometimes listed under his name.

6551. FERRINI, Rinaldo Eugenio Domenico Tranquillo. 1831-1908. It. P III,IV.

6552. GRINWIS, Cornelius Hubertus Carolus. 1831-1899. Holl. P III, IV.

6553. LANDOLT, Hans Heinrich. 1831-1910. Ger. DSB; P I,III,IV. Also Chemistry*.
1. *Physikalisch-chemische Tabellen.* Berlin, 1883. With R. Börnstein. 2nd ed., 1894.

6554. LORBERG, Hermann. 1831-1906. Ger. P III,IV.

6555. MAXWELL, James Clerk. 1831-1879. Br. DSB; ICB; P III; GB; GD (CLERK-MAXWELL). See also Index.
1. *On the stability of the motions of Saturn's rings.* Cambridge, 1859. A prize essay.
2. *Theory of heat.* London, 1870. 2nd ed., 1872. 4th ed., rev., 1875. 7th ed., 1883. 8th ed., 1885. 11th ed., by Lord Rayleigh, 1894. Another ed., 1899.
3. *A treatise on electricity and magnetism.* 2 vols. Oxford, 1873. RM/L. 2nd ed., by W.D. Niven, 1881. 3rd ed., by J.J. Thomson, 1892.
4. *Matter and motion.* London, 1876. An elementary textbook. RM. Other eds: 1882 and 1888.
5. *An elementary treatise on electricity.* Oxford, 1881. Ed. by W. Garnett. RM/L. 2nd ed., rev., 1888.
6. *The scientific papers.* 2 vols. Cambridge, 1890. Ed. by W.D. Niven. RM/L. Reprinted, New York, 1952.
7. *The origins of Clerk Maxwell's electrical ideas as described in familiar letters to William Thomson.* Cambridge, 1937. Ed. by J. Larmor.
German translations of several of his periodical articles on lines of force are included in OKEW.

6556. ROOD, Ogden Nicholas. 1831-1902. U.S.A. DSB; P III,IV; GE.
1. *Modern chromatics, with applications to art and industry.* London, 1879.

6557. TAIT, Peter Guthrie. 1831-1901. Br. DSB; ICB; P III,IV; GB; GD2. Also Mathematics* and Mechanics*. See also Index.
1. *A sketch of thermodynamics.* Edinburgh, 1868. 2nd ed., 1877.
2. *Lectures on some recent advances in physical science.* London, 1876. RM. 2nd ed., 1876. 3rd ed., 1885.
3. *Light.* Edinburgh, 1884. RM/L. 2nd ed., 1889.
4. *Heat.* London, 1884. RM/L.
5. *Properties of matter.* Edinburgh, 1885. 2nd ed., 1890. 4th ed., 1899.
6. *Scientific papers.* 2 vols. Cambridge, 1898-1900. Ed. by himself. RM/L.
6a. ——— Supplement: *Life and scientific work.* Cambridge, 1911. By C.G. Knott.

6558. TERQUEM, Alfred. 1831-1887. Fr. P III,IV.

6559. CAILLETET, Louis Paul. 1832-1913. Fr. DSB; P III,IV.

6560. CAZIN, Achille Auguste. 1832-1877. Fr. P III.
1. *La chaleur*. Paris, 1866. X.
1a. ——— *The phenomena and laws of heat*. 1868. Trans. from the 2nd French ed. (1868) by E. Rich.

6561. CROOKES, William. 1832-1919. Br. DSB; ICB; P III,IV; GB; GD3. Also Chemistry*. See also Index.
1. *On radiant matter. A lecture*. [London? 1879?] RM.

6562. DUFOUR, Louis. 1832-1892. Switz. P I,III.

6563. FEDDERSEN, Berend Wilhelm. 1832-1918. Ger. DSB; P I,III,IV.
1. *Entladung der Leidener Flasche: Intermittierende, kontinuierliche, oszillatorische Entladung und dabei geltende Gesetze*. [Periodical articles, 1857-66] Leipzig, 1908. (OKEW. 166)

6564. KLEIN, Hermann. 1832-1902. Ger. P III,IV.
1. *Theorie der Elasticität, Akustik und Optik*. Leipzig, 1877.

6565. KÖNIG, (Karl) Rudolph. 1832-1901. Fr. DSB; P III,IV; GB.
1. *Quelques expériences d'acoustique*. Paris, 1882.

6566. LE ROUX, François Pierre. 1832-1907. Fr. P III,IV.

6567. MASSIEU, François Jacques Dominique. 1832-1896. Fr. P III.

6568. MELDE, Franz Emil. 1832-1901. Ger. P II,III,IV.
1. *Die Lehre von den Schwingungscurven*. 2 vols. Leipzig, 1864.
2. *Die Akustik*. Leipzig, 1883.

6569. NEUMANN, Carl Gottfried. 1832-1925. Ger. DSB; P II,III,IV. Also Mathematics*.
1. *Die elektrischen Kräfte: Darlegung und Erweiterung der von A. Ampère, F. Neumann, W. Weber, G. Kirchhoff entwickelten mathematischen Theorien*. 2 vols. Leipzig, 1873-98.
2. *Vorlesungen über die mechanische Theorie der Wärme*. Leipzig, 1875.
See also 6272/5.

6570. CROVA, André Prosper Paul. 1833-1907. Fr. P IV.

6571. DÜHRING, Eugen Karl. 1833-1921. Ger. The philosopher. P III, IV; GB. Also Mechanics*.
1. *Neue Grundgesetze zur rationellen Physik und Chemie*. Leipzig, 1878. RM.
See also 2953/1,1a.

6572. GUTHRIE, Frederick. 1833-1886. Br. P III,IV; GD. Also Chemistry.
1. *The elements of heat and of non-metallic chemistry*. London, 1868.
2. *Magnetism and electricity*. London, 1876.
3. *Practical physics, molecular physics and sound*. London, 1878. Another ed., 1885.

6573. HAGENBACH (-BISCHOFF), (Jacob) Eduard. 1833-1910. Switz. ICB; P III,IV.

6574. ISSALY, Pierre Adolphe. 1833-after 1903. Fr. P IV.

6575. JENKIN, Henry Charles Fleeming. 1833-1885. Br. DSB; P III; GB; GD.
1. *Reports of the Committee on Electrical Standards*. London, 1873.
2. *Electricity and magnetism*. London, 1873. 2nd ed., 1874. 3rd ed., 1876. 4th ed., 1878. 8th ed., 1885.
3. *Papers: literary, scientific, etc*. 2 vols. London, 1887. Ed. by S. Colvin and J.A. Ewing, with a biography by R.L. Stevenson (the well-known writer).

6576. JOCHMANN, Emil Carl Gustav Georg. 1833-1871. Ger. P III.

6577. KREBS, Georg. 1833-1907. Ger. P III.
1. *Die Erhaltung der Energie als Grundlage der neueren Physik*. Munich, 1877. RM.

6578. LENZ, Robert. 1833-1903. Russ. P III,IV.

6579. PLACE, Francis. 1833-after 1896. Ger. P II,III,IV.

6580. ROSSETTI, Francesco. 1833-1885. It. DSB; P III.
1. *Bibliografia italiana di elettricità e magnetismo*. Padua, 1881. With G. Cantoni. Reprinted, Cosenza, 1960.

6581. WILDE, Henry. 1833-1919. Br. P IV. Also Geomagnetism*.

6582. DAHLANDER, Gustav Robert. 1834-1903. Swed. P III,IV.

6583. JOUBERT, Jules. 1834-1910. Fr. P III,IV. See also 6624/2.

6584. LANGLEY, Samuel Pierpoint. 1834-1906. U.S.A. DSB; ICB; P III, IV; GB; GE. Also Astronomy* and Biography*.
1. *Experiments in aerodynamics*. Washington, 1891. RM/L.
2. *The internal work of the wind*. Washington, 1893. RM/L.

6585. LA RIVE, Lucien de. 1834-1924. Switz. P III (RIVE), IV (DE LA RIVE).

6586. MEYER, Oskar Emil. 1834-1909. Ger. ICB; P III,IV.
1. *Die kinetische Theorie der Gase*. Breslau, 1877. RM/L.
1a. ——— *The kinetic theory of gases*. London, 1899. Trans. by R.E. Baynes from the 2nd German ed. RM/L.

6587. PLANTÉ, (Raimond Louis) Gaston. 1834-1889. Fr. P III,IV.
1. *Recherches sur l'électricité*. Paris, 1879.
2. [Trans.?] *The storage of electrical energy*. London, 1887.

6588. QUINCKE, Georg Hermann. 1834-1924. Ger. DSB; P II,III,IV.

6589. TAYLOR, Sedley. 1834-1920. Br. P III.
1. *Sound and music: A non-mathematical treatise*. London, 1873. 3rd ed., 1896.

6590. ANDERSON, William. 1835-1898. Br. GD1.
1. *On the conversion of heat into work*. London, 1887.
2. *The interdependence of abstract science and engineering*. London, 1893. A lecture.

6591. AVENARIUS, Michail Petrovich. 1835-1895. Russ. P III,IV & Supp.

6592. BENEVIDES, Francisco de Fonseca. 1835-after 1880. Port. P III.

6593. FOSTER, George Carey. 1835-1919. Br. P III,IV; GF. Also Chemistry.

6594. KÜLP, Ludwig Georg. 1835-1891. Ger. P III,IV.

6595. KURZ, Johann Philipp August. 1835-after 1903. Ger. P III,IV.

6596. MATHIEU, Emile Léonard. 1835-1890. Fr. DSB; P III,IV. Also Mathematics and Mechanics*.
   1. *Cours de physique mathématique.* Paris, 1873.
   2. *Théorie de la capillarité.* Paris, 1883.
   3. *Théorie du potentiel et ses applications à l'électrostatique et au magnétisme.* 2 vols. Paris, 1885-86.

6597. MENSBRUGGHE, Gustave Léonard van der. 1835-1911. Belg. ICB; P III,IV (under VAN).

6598. RECKNAGEL, Georg Friedrich. 1835-1920. Ger. P III,IV.

6599. STEFAN, Josef. 1835-1893. Aus. DSB; P II,III,IV; GC.

6600. WAND, Theodor von. 1835-1896. Ger. P III,IV.

6601. WÜLLNER, (Friedrich Hugo Anton) Adolph. 1835-1908. Ger. P II, III,IV.
   1. *Lehrbuch der Experimental-Physik.* 4 vols. Leipzig, 1862-65. X. 3rd ed., 1874-75. 4th ed., 1882-86. Another ed., 1896-1907.
   2. *Compendium der Physik.* 2 vols. Leipzig, 1879.

6602. ADAMS, William Grylls. 1836-1915. Br. P III,IV. See also 5034/2.

6603. BLASERNA, Pietro. 1836-1918. It. P III,IV.
   1. [Trans.] *The theory of sound in its relation to music.* London, 1876.

6604. CLIFTON, Robert Bellamy. 1836-1908. Br. P III,IV.

6605. COTTERILL, James Henry. 1836-1922. Br. P IV & Supp. Also Mechanics.
   1. *Notes on the theory of the steam engine.* London, 1871. X.
   1a. ——— 2nd ed. *The steam engine considered as a heat engine.* Ib., 1878. Another ed., rev, 1890.

6606. D'HENRY, Louis. 1836-after 1880. Fr. P III.

6607. HOLTZ, Wilhelm Theodor Bernhard. 1836-1913. Ger. P III,IV.

6608. KETTELER, Eduard. 1836-1900. Ger. P III,IV.

6609. LOCKYER, Joseph Norman. 1836-1920. Br. DSB; ICB; P III,IV; GB; GD3. Also Astronomy* and History of Astronomy*.
   1. *The spectroscope and its applications.* London, 1873. X. 2nd ed., 1873.
   2. *Studies in spectrum analysis.* London, 1878. 4th ed., 1886.
   See also 6510/1a,2a.

6610. LUCAS, Félix. 1836-after 1907. Fr. P III,IV.

6611. MARCO, Felice. 1836-after 1881. It. P III.

6612. MAYER, Alfred Marshall. 1836-1897. U.S.A. DSB; P III,IV; GE.
   1. *Light.* London, 1878. With C. Bernard.
   2. *Sound.* London, 1879. Another ed., 1881.

6613. OETTINGEN, Arthur Joachim von. 1836-1920. Dorpat/Ger. P III, IV.

6614. PAPE, Carl (Johannes Wilhelm Theodor). 1836-1906. Ger. P III, IV.

6615. SCHRÖDER VAN DER KOLK, Heinrich Wilhelm. 1836-1867. Holl. P II,III.

6616. TOEPLER, August Joseph Ignaz. 1836-1912. Ger. P III,IV.
1. *Beobachtungen nach einer neuen optischen Methode.* Bonn, 1864. X. Reprinted, Leipzig, 1906. (OKEW. 157)
2. *Beobachtungen nach der Schlierenmethode.* [Periodical articles, 1866 onwards] Leipzig, 1906. (OKEW. 158)

6617. VILLARI, Emilio. 1836-1904. It. DSB; P III,IV.

6618. WRIGHT, Arthur Williams. 1836-1915. U.S.A. P III,IV.

6619. BAILY, Walter. 1837-after 1892. Br. P III,IV.

6620. BEZOLD, (Johann Friedrich) Wilhelm von. 1837-1907. Ger. P III. Also Meteorology.
1. *Farbenlehre.* Brunswick, 1874. X.
1a. ——— *Theory of colour.* 1876.

6621. CARL, Philipp Franz Heinrich. 1837-1891. Ger. P III; GC.

6622. HANDL, Alois. 1837-1915. Aus. P III,IV.

6623. LOMMEL, Eugen (Cornelius Joseph). 1837-1899. Ger. P III,IV; GC.
1. *Das Wesen des Lichtes.* Leipzig, 1875. X.
1a. ——— *The nature of light, with a general account of physical optics.* London, 1875.
2. *Über die Interferenz des gebeugten Lichtes.* 4 parts in 1. Erlangen, 1875-76.
3. *Lehrbuch der Experimentalphysik.* Leipzig, 1893. X.
3a. ——— *Experimental physics.* London, 1899. Trans. by G.W. Myers.
See also 6206/2 and 6216/3.

6624. MASCART, Eleuthère Elie Nicolas. 1837-1908. Fr. DSB; P III,IV.
1. *Traité de l'électricité statique.* Paris, 1876. RM/L.
2. *Leçons sur l'électricité et le magnétisme.* 2 vols. Paris, 1882-86. With J. Joubert.
2a. ——— *A treatise on electricity and magnetism.* 2 vols. London, 1883-88.
3. *Traité d'optique.* 3 vols & atlas. Paris, 1889-93. RM/L.

6625. WAALS, Johannes Diderik van der. 1837-1923. Holl. DSB; ICB; P III (WAALS), IV (VAN).
1. *Over de continuiteit van den gas- en vloeistoftoestand.* Leiden, 1873.
1a. ——— *La continuité des états gazeux et liquide.* Paris, 1894.

6626. ZINCKE, Hans Friedrich August. 1837-after 1884. Ger. P III.

6627. BOTTOMLEY, James. 1838-after 1902. Br. P III,IV.

6628. GRÜNWALD, Anton Karl. 1838-1920. Prague. P III,IV. Also Mathematics.

6629. LANG, Viktor von. 1838-1921. Aus. P III,IV.

6630. LÉVY, Maurice. 1838-1910. Fr. DSB; P III,IV. Also Mechanics.

6631. LIPPICH, Ferdinand Franz. 1838-1913. Prague. P III,IV.

6632. MACH, Ernst. 1838-1916. Aus./Prague. DSB; ICB; P III,IV. Also Mechanics*, Physiology, and History of Physics*.
  1. *Zwei populäre Vorträge über Optik*. Graz, 1867. RM.
  2. *Grundriss der Physik*. Leipzig, 1894. Ed. by F. Harbordt and M. Fischer.
  3. *Populärwissenschaftliche Vorlesungen*. Leipzig, 1896. X.
  3a. ——— *Popular scientific lectures*. La Salle, Ind., 1895. Trans. by T.J. McCormack. 2nd ed., 1897. 3rd ed., 1898.
  4. *Die Principien der Wärmelehre, historisch-kritisch entwickelt*. Leipzig, 1896. RM/L.
  5. *Ernst Mach Abhandlungen: Die Geschichte und die Wurzel des Satzes von der Erhaltung der Arbeit (1872); Zur Geschichte des Arbeitbegriffes (1873); Kultur und Mechanik (1915)*. Amsterdam, 1969. Ed. by J. Thiele with a bibliography of Mach's writings.

6633. MORLEY, Edward Williams. 1838-1923. U.S.A. DSB; ICB; P III, IV; GE. Also Chemistry*.

6634. SAINT-EDME. Ernest. 1838-after 1889. Fr. P III,IV.

6635. SZILY (VON NAGY-SZIGETH), Coloman (or Kálmán). 1838-1924. Hung. P III,IV.

6636. WERNICKE, Carl Wilhelm. 1838-after 1881. Ger. P III.

6637. ALDIS, William Steadman. 1839-1928. Br. P III,IV. Also Mathematics.
  1. *An elementary treatise on geometrical optics*. Cambridge, 1872.
  2. *A chapter on Fresnel's theory of double refraction*. 2nd ed. Cambridge, 1879.

6638. DITSCHEINER, Leander. 1839-1905. Aus. P IV. Also Crystallography.

6639. GIBBS, (Josiah) Willard. 1839-1903. U.S.A. DSB; ICB; P III, IV & Supp.; GB; GE. Also Mechanics*.
  1. *The scientific papers*. 2 vols. London, 1906. Ed. by H.A. Bumstead and R.G. Van Name. RM/L. Reprinted, New York, 1961.
  2. *The collected works*. 2 vols. London, 1928. Ed. by W.R. Langley and R.G. Van Name. Reprinted, New Haven, Conn., 1948 and 1957.

6640. KUNDT, August Adolph. 1839-1894. Ger. DSB; P III,IV; GB.

6641. PFAUNDLER, Leopold. 1839-1920. Aus. P III,IV. Also Chemistry. See also 6364/1a.

6642. SCHIMKOW, Andrei Petrovich. 1839-after 1904. Russ. P III,IV.

6643. STOLETOV, Aleksandr Grigorievich. 1839-1896. Russ. DSB; P III, IV.

6644. WEBER, Heinrich. 1839-after 1906. Ger. P III,IV.

6645. ABBE, Ernst. 1840-1905. Ger. DSB; ICB; P III,IV; Mort.; BLAF. Also Microscopy.

6646. FERRON, Eugène. 1840-after 1904. Luxembourg. P III,IV.

6647. FUCHS, Friedrich (or Fritz). 1840-1911. Ger. P III,IV; BLA & Supp. Also Physiology.

6648. KOHLRAUSCH, Friedrich Wilhelm Georg. 1840-1910. Ger. DSB; P III, IV. Also Chemistry*.
   1. *Leitfaden der praktischen Physik.* Leipzig, 1870. X. 4th ed., 1880.
   1a. ——— *Introduction to physical measurements.* London, 1873. Trans from the 2nd German ed. by T.H. Waller and H.R. Proctor. 3rd ed. (from the 7th German ed.), 1894.

6649. LAURENT, Léon Louis. 1840-1909. Fr. P III,IV.

6650. MANCE, Henry Christopher. 1840-1926. Br. ICB; GF.

6651. MARANGONI, Carlo Giuseppe Matteo. 1840-1925. It. P III,IV.

6652. POPE, Franklin Leonard. 1840-1895. U.S.A. GE.
   1. *Evolution of the electric incandescent lamp.* Elizabeth, N.J., 1889. RM.

6653. POTIER, Alfred. 1840-1905. Fr. P III,IV.

6654. SCHWEDOFF, Theodor Nikiforovich. 1840-1906. Russ. P III,IV.

6655. AMAGAT, Emile Hilaire. 1841-1915. Fr. DSB; P III,IV.

6656. BAILLE, Jean Baptiste Alexandre. 1841-after 1898. Fr. P III,IV.

6657. BOSANQUET, Robert Holford Macdowall. 1841-1912. Br. P III,IV.

6658. CORNU, (Marie) Alfred. 1841-1902. Fr. DSB; ICB; P III,IV; GB.

6659. GARIEL, Charles Marie. 1841-1924. Fr. P III,IV; BLAF. Also Physiology*.

6660. HOORWEG, Jan Leendert. 1841-1919. Holl. P III,IV. Also Physiology.

6661. KOETTERITZSCH, Ernst Theodor. 1841-after 1885. Ger. P III,IV.
   1. *Lehrbuch der Elektrostatik.* Leipzig, 1872.

6662. LE BON, Gustave. 1841-1931. Fr. P IV. Also Physiology*.

6663. LUNDQUIST, Carl Gustaf. 1841-after 1882. Swed. P III.

6664. MENDENHALL, Thomas Corwin. 1841-1924. U.S.A. ICB; P III,IV; GE. Also History of Physics*.

6665. NACCARI, Andrea. 1841-1926. It. P III,IV.

6666. NEYRENEUF, Vincent. 1841-1903. Fr. P III,IV.

6667. PACINOTTI, Antonio. 1841-1912. It. DSB; ICB; P III,IV.

6668. PINA VIDAL, Adriano Augusto da. 1841-1919. Port. P III.
   1. *Tratado elementar de optica.* Lisbon, 1874.

6669. SCHUMANN, Victor. 1841-1913. Ger. DSB; P IV.

6670. SERRA-CARPI, Giuseppe. 1841-after 1882. It. P III.

6671. VIOLLE, Jules Louis Gabriel. 1841-1923. Fr. DSB; P III,IV.
   1. *Cours de physique.* 4 vols. Paris, 1883-88.

6672. WEINHOLD, Adolf Ferdinand. 1841-1917. Ger. P III,IV.

1. *Vorschule der Experimental-Physik.* Leipzig, 1872. X.
1a. ——— *Introduction to experimental physics.* London, 1875.

6673. BASSO, Giuseppe. 1842-1895. It. P III,IV.

6674. BOUSSINESQ, Joseph Valentin. 1842-1929. Fr. DSB; ICB; P III, IV. Also Mechanics*.

6675. BUDDE, Emil Arnold. 1842-1921. Ger. P III,IV. Also Mechanics*.

6676. DEWAR, James. 1842-1923. Br. DSB; ICB; P III,IV; GB; GD4. Also Chemistry*#. See also 6518/1.

6677. EXNER, Karl Franz Josef. 1842-1914. Aus. P III,IV.

6678. MARTINI, Tito Enrico. 1842-after 1909. It. P III,IV.

6679. MILLER, Andreas. 1842-after 1903. Ger. P IV.

6680. NIVEN, William Davidson. 1842-1917. Br. P III. See also 6555/3,6.

6681. PISATI, Giuseppe. 1842-1891. It. P III,IV.

6682. REYNOLDS, Osborne. 1842-1912. Br. DSB; ICB; P III,IV; GD3. Also Mechanics*.
1. *Papers on mechanical and physical subjects.* 3 vols. Cambridge, 1900-03.

6683. SOHNCKE, Leonhard. 1842-1897. Ger. DSB; P III,IV; BMNH; GC. Also Crystallography* and Meteorology.

6684. STRUTT, John William, *3rd Baron* RAYLEIGH. 1842-1919. Br. DSB; ICB (RAYLEIGH); P III,IV (RAYLEIGH); GB; GD3.
1. *The theory of sound.* 2 vols. London, 1877-78. 2nd ed., 1894-96; reprinted, New York, 1945.

6685. VOLLER, (Karl) August. 1842-1920. Ger. P III,IV.

6686. WAHA, Mathias de. 1842-after 1881. Luxembourg. P III.

6687. ABNEY, William de Wiveleslie. 1843-1920. Br. DSB; P III,IV; GD3. Also Astronomy.
1. *Colour measurement and mixture.* London, 1891. RM/L.
2. *Colour vision.* London, 1895. RM/L.

6688. CHRISTIANI, Arthur. 1843-1887. Ger. P III; BLA & Supp.; GC. Also Physiology.

6689. CHRISTIANSEN, Christian. 1843-1917. Den. DSB Supp.; P III,IV.
1. *Indledning til den matematiske fysik.* 2 vols. Copenhagen, 1887-90. X.
1a. ——— *Elements of theoretical physics.* London, 1897. Trans. by W.F. Magie.

6690. DEPREZ, Marcel. 1843-1918. Fr. DSB; P III,IV.

6691. FEUSSNER. (Friedrich) Wilhelm. 1843-1928. Ger. P III,IV.

6692. FRÖLICH, Oscar. 1843-1909. Ger. P III,IV.

6693. RAYNAUD, Jules François Emmanuel. 1843-1888. Fr. P III,IV.

6694. RÓITI, Antonio. 1843-1921. It. P III,IV.

6695. SARASIN, Edouard. 1843-1917. Switz. P III,IV.

6696. SUNDELL, August Fredrik. 1843-after 1919. Fin. P III,IV.

6697. TROWBRIDGE, John. 1843-1923. U.S.A. DSB; P III,IV; GE.
    1. *What is electricity?* London, 1897.

6698. WEBER, Heinrich Friedrich. 1843-1912. Ger./Switz. P III,IV.

6699. BARRETT, William Fletcher. 1844-1925. Br. P III,IV; GF.

6700. BENOIT, Justin Mirande René. 1844-1922. Fr. DSB; P III,IV (B., Jean René).

6701. BOLTZMANN, Ludwig. 1844-1906. Aus./Ger. DSB; ICB; P III,IV. Also Mechanics*.
    1. *Vorlesungen über Maxwells Theorie der Elektricität und des Lichtes.* 2 vols. Leipzig, 1891-93. RM/L.
    2. *Vorlesungen über Gastheorie.* 2 vols. Leipzig, 1896-98. RM.
    2a. ——— *Lectures on gas theory.* Berkeley, Cal., 1964. Trans. by S.G. Brush with introd., notes, and a bibliography.
    3. *Populäre Schriften.* Leipzig, 1905.
    4. *Wissenschaftliche Abhandlungen.* 3 vols. Leipzig, 1909. Ed. by F. Hasenöhrl. Reprinted, New York, 1968.
    5. *Theoretical physics and philosophical problems. Selected writings.* Dordrecht, 1974. Ed. by B. McGuiness.
    See also 6492/4.

6702. BRANLY, Edouard. 1844-1940. Fr. ICB; P III,IV; GF.

6703. CANTONI, Paolo. 1844-after 1882. It. P III.

6704. CARHART, Henry Smith. 1844-after 1901. U.S.A. P IV.

6705. CROULLEBOIS, Marcel Désiré. 1844-1886. Fr. P III.

6706. HERWIG, Hermann Anton Bernhard. 1844-1881. Ger. P III.

6707. MEES, Rudolf Adriaan. 1844-1886. Holl. P III,IV.

6708. MOUTON, Jean Louis. 1844-1895. Fr. P III.

6709. NARR, Friedrich Gustav. 1844-1893. Ger. P III,IV.

6710. OBERMAYER, (Joseph Vincenz) Albert von. 1844-1915. Aus. P III,IV.

6711. PRESTON, Samuel Tolver. 1844-after 1908. Br. P III,IV.
    1. *Physics of the ether.* London, 1875.

6712. WASSMUTH, Anton. 1844-after 1921. Aus. P III,IV.

6713. WATTS, William Marshall. 1844-1919. Br. P III,IV.
    1. *An index of spectra. With an introduction on the methods of measuring and mapping spectra.* London, 1872. Another ed., Manchester, 1889. Appendix, London, 1891-.

6714. BAUER, Karl Ludwig. 1845-after 1888. Ger. P III,IV.

6715. BICHAT, Ernest Adolphe. 1845-1905. Fr. P IV.

6716. BOTTOMLEY, James Thomson. 1845-1926. Br. P III,IV.

6717. COLLEY, Robert Andreievich 1845-1891. Russ. P IV.

6718. CONROY, John. 1845-1900. Br. P III,IV.

6719. EDELMANN, (Max) Thomas. 1845-1913. Ger. P III,IV.

6720. GAMBERA, Pietro. 1845-after 1882. It. P III.

6721. HENRICHSEN, Sophus. 1845-1928. Nor. DSB.

6722. HESEHUS, Nikolai Aleksandrovich. 1845-1919. Russ. P IV.

6723. KALISCHER, Salomon. 1845-1924. Ger. P III,IV.

6724. LIPPMANN, Gabriel Jonas. 1845-1921. Fr. DSB; ICB; P III,IV.
  1. *Cours de thermodynamique*. Paris, 1889. Ed. by E. Mathias and A. Renault.

6725. LÜDTGE, Franz Hermann Robert. 1845-1880. Ger. ICB; P III.

6726. MACALUSO, Damiano. 1845-1932. It. P III,IV.

6727. MINCHIN, George. 1845-1914. Br. P III. Also Mechanics*.

6728. NIVEN, Charles. 1845-1923. Br. P III,IV.

6729. PERNET, Johann. 1845-1902. Ger./Switz. P III,IV.

6730. PULUJ, Johann. 1845-1918. Aus./Prague. P III,IV.

6731. RIECKE, (Carl Victor) Eduard. 1845-1915. Ger. DSB; P III,IV.

6732. RÖNTGEN, Wilhelm Conrad. 1845-1923. Ger. DSB; ICB; P III, IV; GB.
  1. *Über eine neue Art von Strahlen*. [Periodical article, 1895] Munich, 1972. Ed. by F. Krafft.

6733. SCHULLER, Alois. 1845-after 1916. Hung. P III,IV.

6734. TOMLINSON, Herbert. 1845-1931. Br. ICB; P III,IV.

6735. WRÓBLEWSKI, Zygmunt Florenty von. 1845-1888. Cracow. DSB; P III,IV.

6736. BLEEKRODE, Louis. 1846-1905. Holl. P III,IV.

6737. BOUTY, Edmond (Marie Léopold). 1846-1922. Fr. P III,IV. See also 6434/1.

6738. DOMALÍP, Karl. 1846-1909. Prague. P III,IV.

6739. DONATI, Luigi. 1846-1932. It. P III,IV.

6740. GLAN, Paul. 1846-1898. Ger. P III,IV.

6741. GROSS, Carl Friedrich Theodor. 1846-1913. Ger. P III,IV.

6742. MÜLLER, Johann Jacob. 1846-1875. Ger./Switz. P III. Also Physiology.

6743. OBERBECK, Anton. 1846-1900. Ger. P III,IV.

6744. PICKERING, Edward Charles. 1846-1919. U.S.A. DSB; P III,IV; GE. Also Astronomy.
  1. *Elements of physical manipulation*. 2 vols. Boston, 1873-76. X. Another ed., London, 1886.

6745. PICTET, Raoul Pierre. 1846-1929. Switz./Ger. DSB; P III,IV.

6746. PSCHEIDL, Wenzel. 1846-after 1905. Aus. P III,IV.

6747. QUESNEVILLE, (Gustave) Georges. 1846-after 1903. Fr. P III,IV.

6748. RÜHLMANN, (Moritz) Richard. 1846-1908. Ger. P III,IV. Also Meteorology*.

6749. UMOW, Nikolai Alekseevich. 1846-1915. Russ. ICB; P III,IV.

6750. WARBURG, Emil Gabriel. 1846-1931. Ger. DSB; P III,IV.

6751. AYRTON, William Edward. 1847-1908. Br. P III,IV; GB; GD2.

6752. DUTER, Emile. 1847-after 1889. Fr. P III,IV.

6753. EDISON, Thomas Alva. 1847-1931. U.S.A. The famous inventor. DSB; ICB; P III; GB. See also 15517/1.

6754. FERRARIS, Galileo. 1847-1897. It. DSB; ICB; P III,IV.

6755. GIESE, Wilhelm. 1847-after 1882. Ger. P III.

6756. GRAY, Andrew. 1847-1925. Br. P IV.
   1. *Absolute measurements in electricity and magnetism.* London, 1884.
   2. *The theory and practice of absolute measurements in electricity and magnetism.* 2 vols in 3. London, 1888-93.
   3. *A treatise on Bessel functions and their applications to physics.* London, 1895. With G.B. Mathews.
   4. *A treatise on magnetism and electricity.* Vol. 1. [No more publ.] London, 1898.

6757. GROTRIAN, Otto Natalis August. 1847-1921. Ger. P III,IV.

6758. LÜBECK, Gustav. 1847-after 1881. Ger. P III.

6759. NIPHER, Francis Eugene. 1847-1926. U.S.A. P III,IV; GE.
   1. *The theory of magnetic measurements.* New York, 1886.

6760. STEVENS, Walter Le Conte. 1847-1927. U.S.A. P III,IV.

6761. BELLATI, Manfredo. 1848-1932. It. P III,IV.

6762. BIDWELL, Shelford. 1848-1909. Br. P IV; GD2.

6763. DORN, Friedrich Ernst. 1848-1916. Ger. P III,IV.

6764. DUFET, (Jean Baptiste) Henry. 1848-1905. Fr. P III,IV. Also Crystallography.
   1. *Optique.* 3 vols. Paris, 1898-1900.

6765. DVOŘÁK, Vincenz. 1848-1922. Agram. P III,IV.

6766. EÖTVÖS, Roland (or Loránd) von. 1848-1919. Hung. DSB; ICB; P IV.

6767. FOUSSEREAU, Georges Ernest Marie. 1848-after 1901. Fr. P IV.

6768. HASSELBERG, Clas Bernhard. 1848-1922. Russ./Swed. P III,IV. Also Astrophysics.

6769. HASTINGS, Charles Sheldon. 1848-1932. U.S.A. ICB; P III,IV & Supp.; GE1.

6770. MICHAELIS, Gerrit Jan. 1848-after 1903. Holl. P III,IV.

6771. ROWLAND, Henry Augustus. 1848-1901. U.S.A. DSB; ICB; P III, IV; GB; GE.
   1. *The physical papers.* Baltimore, 1902. RM/L.

6772. RÜCKER, Arthur William. 1848-1915. Br. P III,IV; GF.

6773. SCHILLER, Nikolai Nikolaievich. 1848-1910. Russ. P III,IV.

6774. SLOTTE, Karl Fredrik. 1848-1914. Fin. P III,IV.

6775. STREINTZ, Heinrich. 1848-1892. Aus. P III.

6776. TUCKERMAN, Alfred. 1848-1925. Also Chemistry*.
 1. *Index to the literature of the spectroscope.* Washington, 1888. X.
 1a. ——— *Index to the literature of the spectroscope, 1887-1900.* Washington, 1902.
 2. *Index to the literature of thermodynamics.* Washington, 1890.

6777. WEBER, (Joachim) Leonard. 1848-1919. Ger. P III,IV.

6778. WINKELMANN, Adolph August. 1848-1910. Ger. P III,IV.
 1. *Handbuch der Physik.* 3 vols in 5. Breslau, 1891-96.

6779. WITZ, Aimé Marie Joseph. 1848-1926. Fr. ICB; P III,IV.
 1. *Cours de manipulations de physique.* Paris, 1883.

6780. BAYNES, Robert Edward. 1849-1921. Br. P III,IV.
 1. *Lessons on thermodynamics.* Oxford, 1878.
 See also 6586/1a.

6781. BLONDLOT, René Prosper. 1849-1930. Fr. DSB; P III,IV.

6782. BORGMANN, Ivan Ivanovich. 1849-1914. Russ. P IV.

6783. EXNER, Franz Serafin. 1849-1922. Aus. P III,IV; GF.

6784. FLEMING, (John) Ambrose. 1849-1945. Br. DSB; ICB; P III,IV; GD6. Also Autobiography*.
 1. *Electric lamps and electric lighting.* London, 1894. RM.
 2. *Magnets and electric currents. An elementary treatise.* London, 1898.

6785. FORBES, George. 1849-1936. Br. ICB; P IV. Also Astronomy*.
 1. *A course of lectures on electricity.* London, 1888.

6786. GOURÉ DE VILLEMONTÉE, Louis Anne Gustave Albert. 1849-after 1906. Fr. P IV.

6787. GUÉBHARD, Paul Emile Adrien. 1849-after 1922. Fr. P III,IV. Also Geology.

6788. HOPKINSON, John. 1849-1898. Br. DSB; P III,IV; GB; GD1.
 1. *Original papers.* Cambridge, 1901. Ed. with a memoir by B. Hopkinson. RM/L.

6789. HURION, Louis Alphonse. 1849-after 1901. Fr. P IV.

6790. JEGOROV, Nikolai Grigorievich. 1849-1919. Russ. P IV.

6791. KLEINER, Alfred. 1849-1916. Switz. P IV.

6792. LAMB, Horace. 1849-1934. Br. DSB; ICB; P III,IV; GD5. Also Mechanics* and Geophysics.

6793. MANEUVRIER, Georges. 1849-after 1903. Fr. P IV.

6794. NEESEN, Friedrich. 1849-1923. Ger. P III,IV.

6795. SCHNEEBELI, Heinrich. 1849-1890. Switz. P III,IV.

6796. SEYDLER, August Johann Friedrich. 1849-1891. Prague. P III, IV. Also Astronomy.

6797. THIESEN, Max Ferdinand. 1849-after 1910. Ger. P III,IV.

6798. BERSON, Félix Gustave Adolphe. 1850-after 1902. Fr. P IV.

6799. BRAUN, (Karl) Ferdinand. 1850-1918. Ger. DSB; ICB; P III,IV.

1. *Über physikalische Forschungsart*. Strasbourg, 1899.

6800. CINTOLESI, Filippo. 1850-after 1902. It. P III,IV.

6801. ETTINGSHAUSEN, Albert von. 1850-1932. Aus. P III,IV.

6802. GARBE, Paul. 1850-after 1896. Fr. P III,IV.

6803. GOLDSTEIN, Eugen. 1850-1930. Ger. DSB; P IV.
1. *Canalstrahlen*. [Periodical article, 1886] Leipzig, 1930. (OKEW. 231)

6804. GRAY, Thomas. 1850-1908. U.S.A./Japan. P IV & Supp.

6805. HEAVISIDE, Oliver. 1850-1925. Br. DSB; ICB; P III,IV; GD4.
1. *Electromagnetic waves*. London, 1889.
2. *Electrical papers*. 2 vols. London, 1892. RM/L. Reprinted, New York, 1971.
3. *Electromagnetic theory*. 3 vols. London, 1893-1912. Reprinted, New York, 1971.

6806. PELLAT, (Joseph Solange) Henri. 1850-1909. Fr. ICB; P III,IV.
1. *Leçons sur l'électricité*. Paris, 1890.

6807. PERRY, John. 1850-1920. Br. P III,IV; GF. Also Mechanics*.

6808. RIGHI, Augusto. 1850-1920. It. DSB; ICB; P III,IV.
1. *L'ottica delle oscillazioni elettriche*. Bologna, 1897.

6809. SCHEBUJEW, Georgii Nikolaievich. 1850-1900. Russ. P IV.

6810. SILOW, Petr Alekseievich. 1850-1921. Russ./Warsaw. P III,IV.

6811. STROUHAL, Vincenz. 1850-1922. Prague. P III,IV. See also 6910/1.

6812. VOIGT, Woldemar. 1850-1919. Ger. DSB; P III,IV; BMNH. Also Crystallography.

6813. WEBER, Robert Heinrich. 1850-after 1910. Switz. P III,IV.

6814. BARTOLI, Adolfo Giuseppe. 1851-1896. It. P III,IV.

6815. BONGIOVANNI, Giuseppe. 1851-1918. It. P IV.

6816. CHRYSTAL, George. 1851-1911. Br. DSB; P III,IV; GD2. Also Mathematics*.

6817. FITZGERALD, George Francis. 1851-1901. Br. DSB; ICB; P III, IV; GD2. Also Biography*.
1. *The scientific writings*. Dublin, 1902. Ed. with a historical introd. by J. Larmor. RM/L.

6818. GRASSI, Guido Giovanni. 1851-after 1926. It. P III,IV.

6819. GRIFFITHS, Ernest Howard. 1851-1932. Br. ICB; P IV; GD5.

6820. GROMEKA, Hyppolit. 1851-after 1883. Russ. P III.

6821. GRUNMACH, (Ludwig) Leo. 1851-1923. Ger. P III,IV.

6822. HAGEN, Carl Ernst Bessel. 1851-1923. Ger. P IV.

6823. HEEN, Pierre de. 1851-1915. Belg. ICB; P III,IV (under DE).
1. *Recherches touchant la physique comparée et la théorie des liquides*. Paris, 1888.

6824. HELM, Georg Ferdinand. 1851-1923. Ger. ICB; P III,IV.

1. *Die Energetik, nach ihrer geschichtlichen Entwicklung.* Leipzig, 1898. RM/L.

6825. JULIUS, Victor August. 1851-1902. Holl. P III,IV; GF.

6826. KOLÁČEK, František. 1851-1913. Brünn/Prague. P III,IV.

6827. LODGE, Oliver Joseph. 1851-1940. Br. DSB; ICB; P III,IV & Supp.; GB; GD5. Also Autobiography* and Biography*.
1. *Modern views on electricity.* London, 1889. 2nd ed., 1889. Another ed., 1892.
2. *The work of Hertz and some of his successors.* London, 1894. A lecture. X. 2nd ed., with additions, 1894. Another ed., 1897.

6828. MACÉ DE LÉPINAY, Jules Charles Antonin. 1851-1904. Fr. P III, IV.

6829. MACFARLANE, Alexander. 1851-1913. Br./U.S.A. P III,IV. Also Mathematics*.

6830. MALLOCK, Henry Reginald Arnulph. 1851-1933. Br. ICB.

6831. POLONI, Giuseppe. 1851-1887. It. P III,IV.

6832. PRYTZ, Peter Kristian. 1851-1929. Den. P III,IV.

6833. SCHUSTER, Arthur. 1851-1934. Br. DSB; ICB; P III,IV; GD5. Also Geomagnetism. See also 8028/2.

6834. THOMPSON, Silvanus Philipps. 1851-1916. Br. DSB; ICB; P III, IV; GD3. Also Biography*. See also Index.
1. *Elementary lessons in electricity and magnetism.* London, 1881. X. 5th ed., 1884. Another ed., 1890.
2. *The electromagnet and electromagnetic mechanism.* London, 1887. X. 2nd ed., 1892.
3. *The Cantor lectures on the electromagnet.* London, 1890.
4. *Light visible and invisible.* London, 1897.

6835. BECQUEREL, (Antoine) Henri. 1852-1908. Fr. DSB; ICB; P III, IV; GB.

6836. BÖRNSTEIN, Richard. 1852-1913. Ger. P III,IV. Also Meteorology. See also 6553/1.

6837. BURCH, George James. 1852-1914. Br. P IV.
1. *The capillary electrometer in theory and practice.* London, 1896.

6838. CHISTONI, Giuseppe Ciro Pericle. 1852-1927. It. P IV. Also Earth Sciences.

6839. CHWOLSON, Orest Danilovich. 1852-after 1931. Russ. P III,IV.

6840. DUFOUR, Henri Evert. 1852-1910. Switz. P III,IV.

6841. FROMME, Carl Friedrich Ferdinand. 1852-after 1925. Ger. P III, IV.

6842. GORDON, James Edward Henry. 1852-1893. Br. P III,IV; GD1.
1. *Four lectures on static electric induction.* London, 1879.
2. *A physical treatise on electricity and magnetism.* 2 vols. London, 1880. 2nd ed., 1883.

6843. HAGA, Hermann. 1852-after 1925. Holl. P III,IV.

6844. HIMSTEDT, Franz. 1852-1933. Ger. P III,IV.

6845. HODGES, Nathaniel Dana Carlile. 1852-after 1882. U.S.A. P III.

6846. KOCH, Karl Richard Robert. 1852-1924. Ger. P IV.

6847. MacGREGOR, James Gordon. 1852-1913. Canada/Br. P III,IV.

6848. MICHELSON, Albert Abraham. 1852-1931. U.S.A. DSB; ICB; P III, IV; GE.

6849. MOSER, James. 1852-1908. Aus. P IV.

6850. OBACH, Eugen Friedrich August. 1852-1898. Br. P III,IV.

6851. POYNTING, John Henry. 1852-1914. Br. DSB; P III,IV; GD3.
1. *The mean density of the earth*. London, 1894.
2. *A text-book of physics*. London, 1899. With J.J. Thomson.
3. *Collected scientific papers*. Cambridge, 1920.

6852. WEINSTEIN, Max Bernhard. 1852-1918. Ger. P III,IV.

6853. WIEBE, Hermann Friedrich. 1852-1912. Ger. P IV.

6854. WIEDEMANN, Eilhard Ernst Gustav. 1852-1928. Ger. ICB; P III, IV. Also History of Science*.

6855. WORTHINGTON, Arthur Mason. 1852-1916. Br. P III,IV. Also Mechanics*.
1. *A first course of physical laboratory practice*. London, 1888.

6856. DESLANDRES, Henri Alexander. 1853-1948. Fr. DSB; ICB; P IV. Also Astronomy.

6857. FRÖHLICH, Isidor. 1853-1931. Hung. P III,IV.

6858. GUGLIELMO, Giovanni. 1853-after 1925. It. P IV.

6859. KAMERLINGH ONNES, Heike. 1853-1926. Holl. DSB; ICB (ONNES); P III,IV.

6860. KAYSER, Heinrich Johannes Gustav. 1853-1940. Ger. DSB; ICB; P III,IV.
1. *Lehrbuch der Spektralanalyse*. Berlin, 1883.

6861. KLEMENČIČ, Ignaz. 1853-1901. Aus. P III,IV.

6862. KRÜSS, (Andreas) Hugo. 1853-1925. Ger. P III,IV.

6863. LORENTZ, Hendrik Anton. 1853-1928. Holl. DSB; ICB; P III,IV.
1. *Théorie électromagnétique de Maxwell et son application aux corps mouvants*. Leiden, 1892. RM/L.
2. *Versuch einer Theorie der elektrischen und optischen Erscheinungen in bewegten Körpern*. Leiden, 1895. RM.
3. *Collected papers*. 9 vols. The Hague, 1934-39. Ed. by P. Zeeman and A.D. Fokker.
See also 6307/1.

6864. MAREK, Wenzel. 1853-after 1903. Aus. P IV.

6865. MURANI, Oreste. 1853-after 1929. It. P IV.

6866. MURAOKA, Hanichi. 1853-after 1913. Japan. P III,IV.

6867. THOMSON, Elihu. 1853-1937. U.S.A. DSB; ICB.
1. *Selections from the scientific correspondence*. Cambridge, Mass., 1971. Ed. by H.J. Abrahams and M.B. Savin.

6868. WAITZ, Karl. 1853-1911. Ger. P III,IV.

6869. BASSET, Alfred Barnard. 1854-1930. Br. P IV. Also Mechanics*.
 1. *A treatise on physical optics.* Cambridge, 1892.

6870. BRILLOUIN, Marcel Louis. 1854-1948. Fr. DSB; ICB; P III,IV.

6871. CHAPPUIS, James. 1854-after 1930. Fr. ICB; P IV.
 1. *Leçons de physique générale.* 3 vols. Paris, 1891-92. With A. Berget.

6872. COHN, Emil. 1854-after 1935. Ger. P III,IV.

6873. ELSTER, (Johann Philipp Ludwig) Julius. 1854-1920. Ger. DSB; P III,IV.

6874. FÖPPL, August. 1854-1924. Ger. DSB; ICB; P IV.
 1. *Die Einführing in die Maxwellsche Theorie der Elektricität.* Leipzig, 1894.

6875. GLAZEBROOK, Richard Tetley. 1854-1935. Br. DSB; ICB; P III, IV; GD5. Also Biography*.
 1. *Physical optics.* London, 1883.
 2. *The laws and properties of matter.* London, 1893.

6876. GOUY, (Louis) Georges. 1854-1926. Fr. DSB; ICB; P III,IV.

6877. HOPPE, Edmond. 1854-1928. Ger. ICB; P III,IV. Also History of Physics*.

6878. KRUSCHKOLL, Michel. 1854-after 1902. Fr. P IV.

6879. MAZZOTTO, Domenico. 1854-after 1931. It. P III,IV.

6880. MEBIUS, Claes Albert. 1854-after 1930. Swed. P IV.

6881. NICHOLS, Edward Leamington. 1854-1937. U.S.A. P III,IV.
 1. *Elements of physics.* New York, 1896. With W.S. Franklin.

6882. PEIRCE, Benjamin Osgood. 1854-1914. U.S.A. DSB; P III,IV; GE. Also Mathematics*.

6883. POINCARÉ, (Jules) Henri. 1854-1912. Fr. DSB; ICB; P III,IV. Also Mathematics*#, Astronomy*, and Mechanics*.
 1. *Leçons sur la théorie mathématique de la lumière.* 2 vols. Paris, 1889-92. RM/L.
 2. *Electricité et optique.* 2 vols. Paris, 1890-91. RM/L.
 3. *Thermodynamique.* Paris, 1892.
 4. *Théorie mathématique de la lumière.* Paris, 1892.
 5. *Les oscillations électriques. Leçons....* Paris, 1894. RM/L.
 6. *Théorie analytique de la propagation de la chaleur.* Paris, 1895. RM/L.
 7. *Capillarité.* Paris, 1895.
 8. *Théorie de Maxwell et les oscillations hertziennes.* Paris, 1899.

6884. RYDBERG, Johannes (Janne) Robert. 1854-1919. DSB; P IV.
 1. *Recherches sur la constitution des spectres d'émission des éléments chimiques.* [Periodical article, 1889] German trans. Leipzig, 1922. (OKEW. 196)

6885. SANFORD, Fernando. 1854-after 1922. U.S.A. P IV.

6886. SHAW, William Napier. 1854-1945. Br. DSB; ICB; P III,IV. Also Meteorology.

6887. SLUGINOFF, Nikolai Petrovich. 1854-1897. Russ. P III,IV.

6888. SOKOLOFF, Aleksei Petrovich (second of the name). 1854-after 1904. Russ. P IV.

6889. SORET, Charles. 1854-1904. Switz. P III,IV; BMNH. Also Crystallography*.

6890. WITKOWSKI, August Victor. 1854-1913. Lemberg. P III.

6891. YAMAGAWA, Kenjiro. 1854-1931. Japan. ICB.

6892. ZEHNDER, Ludwig Albert. 1854-1949. Switz./Ger. ICB; P IV.

6893. BOYS, Charles Vernon. 1855-1944. Br. DSB Supp.; ICB; P IV; GD6.
1. *Soap-bubbles and the forces which mould them*. London, 1890. Reprinted, ib., 1959.

6894. CELLÉRIER, Gustave. 1855-1914. Switz. P III,IV.

6895. CHAPPUIS, Pierre. 1855-1916. Fr. ICB; P III,IV.

6896. COOK, Ernest Henry. 1855-after 1899. Br. P III,IV.

6897. EWING, James Alfred. 1855-1935. Br. DSB; ICB; P III,IV; GD5.
1. *Magnetic induction in iron and other metals*. London, 1891.
2. *The steam engine and other heat engines*. Cambridge, 1894.
See also 6575/3.

6898. GEITEL, (Friedrich Karl) Hans. 1855-1923. Ger. DSB; P IV.

6899. HALL, Edwin Herbert. 1855-1938. U.S.A. DSB; ICB; P III,IV & Supp.

6900. JACQUES, William White. 1855-after 1893. U.S.A. P III,IV.

6901. KOHLRAUSCH, Wilhelm Friedrich. 1855-1936. Ger. P III,IV.

6902. LEHMANN, Otto. 1855-1922. Ger. DSB; P III,IV; BMNH. Also Crystallography.
1. *Die Molekularphysik, mit besonderer Berücksichtigung mikroskopischer Untersuchungen*. 2 vols. Leipzig, 1888-89.

6903. OLEARSKI, Kazimierz. 1855-1936. Cracow/Lemberg. P IV.

6904. REIFF, Richard August. 1855-1908. Ger. P IV. Also History of Mathematics*.

6905. ROBIN, Victor Gustave. 1855-1897. Fr. P IV.

6906. STEFANINI, Annibale. 1855-after 1922. It. P IV.

6907. STREINTZ, Franz. 1855-1922. Aus. P III,IV.

6908. AUERBACH, Felix. 1856-1933. Ger. P III,IV.

6909. BAKKER, Gerrit. 1856-after 1928. Holl. P IV.

6910. BARUS, Carl. 1856-1935. U.S.A. DSB; P III,IV; GE1.
1. *Electrical and magnetic properties of the iron carburets*. Washington, 1885. With V. Stronhal.
2. *The mechanism of solid viscosity*. Washington, 1892.

6911. BENOIST, Louis. 1856-after 1925. Fr. P IV.

6912. CARVALLO, Emmanuel. 1856-after 1931. Fr. P IV.

6913. GEIGEL, Robert. 1856-1910. Ger. P IV.

6914. GRAETZ, Leo. 1856-after 1931. Ger. P III,IV.

6915. HEYDWEILLER, Adolf. 1856-1926. Ger. P IV.

6916. HOLMAN, Silas Whitcomb. 1856-1900. U.S.A. P III,IV.

6917. JONES, John Viriamu. 1856-1901. Br. P IV; GD2.

6918. KNOTT, Cargill Gilston. 1856-1922. Br./Japan. DSB; P IV. Also Geophysics. See also Index.

6919. KÖNIG, Arthur. 1856-1901. Ger. DSB; P III,IV; BLAF. Also Physiology.

6920. KOLLERT, Julius August. 1856-after 1928. Ger. P IV.

6921. LECHER, Ernst. 1856-1926. Aus./Prague. P III,IV.

6922. LEDUC, Sylvestre Anatole. 1856-1937. Fr. P IV.

6923. MAGGI, Gian Antonio. 1856-after 1931. It. P III,IV. Also Mechanics.

6924. MARGULES, Max. 1856-1920. Aus. DSB; P III,IV. Also Meteorology.

6925. PAGLIANI, Stefano. 1856-1934. It. P III,IV.

6926. RUNGE, Carl David Tolmé. 1856-1927. Ger. DSB; ICB; P IV. Also Mathematics.

6927. TESLA, Nikola. 1856-1943. U.S.A. DSB; ICB; P IV.

6928. THOMSON, Joseph John. 1856-1940. Br. DSB; ICB; P III,IV; GD5. Also Autobiography*.
   1. *A treatise on the motion of vortex rings*. London, 1883. A prize essay.
   2. *The application of dynamics to physics and chemistry*. London, 1888.
   3. *Notes on recent researches in electricity and magnetism*. Oxford, 1893. RM/L.
   4. *Elements of the mathematical theory of electricity and magnetism*. Cambridge, 1895. RM/L. 2nd ed., 1897.
   5. *The discharge of electricity through gases*. London/New York, 1898. RM/L.

   See also 6555/3 and 6851/2.

6929. TUMLIRZ, Ottokar Anton Alois. 1856-after 1921. Prague/Czernowitz. P III,IV.

6930. VOLKMANN, Paul Oskar Eduard. 1856-1938. Ger. DSB; P III,IV. Also Philosophy of Science and History of Physics.

6931. ÅNGSTRÖM, Knut Johan. 1857-1910. Swed. P IV.

6932. ASCOLI, Moisé. 1857-1921. It. P IV.

6933. BEAULARD, Fernand. 1857-after 1910. Fr. P IV.

6934. BLÜMCKE, (Gustav) Adolf. 1857-after 1905. Ger. P IV.

6935. CANTONE, Michele. 1857-1932. It. P IV.

6936. CZERMAK, Paul. 1857-1912. Aus. P IV.

6937. HALLOCK, William. 1857-1913. U.S.A. P III,IV.

6938. HERTZ, Heinrich Rudolph. 1857-1894. Ger. DSB; ICB; P III,IV; GB; GC. Also Mechanics* and Autobiography*.
1. *Über sehr schnelle elektrische Schwingungen. Vier Arbeiten.* [ca. 1888-90] Leipzig, 1971. (OKEW. 251)
2. *Gesammelte Werke.* 3 vols. Leipzig, 1892-95. Ed. by P. Lenard.
2a. ——— Vol. I. *Schriften vermischten Inhalts.* 1895.
2b. ——— ——— *Miscellaneous papers.* London, 1896. Trans. by D.E. Jones and G.A. Schott. RM/L.
2c. ——— Vol. II. *Untersuchungen über die Ausbreitung der elektrischen Kraft.* 1892.
2d. ——— ——— *Electric waves, being researches on the propagation of electric action with finite velocity through space.* London, 1893. Trans. by D.E. Jones. RM/L. Reprinted, New York, 1962.
2e. ——— Vol. III. *Die Principien der Mechanik in neuem Zusammenhänge.* 1894. See 5757/1.

6939. JOLY, John Swift. 1857-1933. Br. DSB; ICB; P IV; GD5.

6940. KURLBAUM, Ferdinand. 1857-1927. Ger. DSB; P IV.

6941. LARMOR, Joseph. 1857-1942. Br. DSB; ICB; P IV; GD6. See also Index.
1. *Mathematical and physical papers.* 2 vols. Cambridge, 1929.

6942. MACK, Karl Friedrich. 1857-1934. Ger. P IV. Also Meteorology.

6943. MEYER, Georg Franz Julius. 1857-after 1926. Ger. P IV.

6944. OOSTING, Hendrik Jan. 1857-after 1926. Holl. P IV.

6945. PFEIFFER, Emanuel. 1857-1921. Ger. P IV.

6946. PILTSCHIKOFF, Nikolai Dmitrievich. 1857-1909. Russ. P IV & Supp.

6947. RONKAR, Jacques Emile Joseph. 1857-1902. Belg. P IV.

6948. SCHTSCHEGLIAIEFF, Vladimir Sergeievich. 1857-after 1903. Russ. P IV.

6949. WALKER, James. 1857-after 1904. Br. P IV.

6950. WESENDONK, Karl von. 1857-after 1918. Ger. P IV.

6951. BARTON, Edwin Henry. 1858-1925. Br. P IV.

6952. BOSE, Jagadis Chandra. 1858-1937. India. DSB Supp.; ICB; P IV.
1. *Collected physical papers.* London, 1927.

6953. CARDANI, Pietro. 1858-1924. It. ICB; P IV.

6954. DIETERICI, Conrad Heinrich. 1858-1929. Ger. P IV.

6955. DOUMER, Jean Marie Emmanuel. 1858-after 1923. Fr. P IV.

6956. KNIBBS, George Handley. 1858-1929. Australia. P IV; GF.

6957. MAGIE, William Francis. 1858-after 1931. U.S.A. P IV. See also 6689/1a.

6958. NEGREANU, Demetre. 1858-1908. Bucharest. P IV.

6959. PLANCK, Max Karl Ernst Ludwig. 1858-1947. Ger. DSB; ICB; P III,IV. Also Autobiography*.
1. *Die Ableitung des Strahlungsgesetzes. Sieben Abhandlungen.* [1896 onwards] Leipzig, 1923. (OKEW. 206)
2. *Vorlesungen über Thermodynamik.* Leipzig, 1897.
2a. ——— *A treatise on thermodynamics.* London, 1903. Trans. by A. Ogg.
3. *Physikalische Abhandlungen und Vorträge.* 3 vols. Brunswick, 1958.
4. *Planck's original papers in quantum physics.* London, 1972. German and English ed., annotated by H. Kangro, trans. by D. ter Haar and S.G. Brush.
See also 6492/5.

6960. PULFRICH, Carl. 1858-1927. Ger. DSB; P IV.

6961. PUPIN, Michael Idvorsky. 1858-1935. U.S.A. DSB; ICB; GE1.
1. *Thermodynamics of reversible cycles in gases and saturated vapours.* New York, 1894. Ed. by M. Osterberg.

6962. BRACE, De Witt Bristol. 1859-1905. U.S.A. DSB; P IV & Supp.; GE.

6963. CURIE, Pierre. 1859-1906. Fr. DSB; ICB; P IV & Supp.; GB.
1. *Oeuvres.* Paris, 1908. Publ. by the Société Française de Physique. RM/L.

6964. EMO, Angelo. 1859-after 1884. It. P III,IV.

6965. GUMLICH, Ernst Carl Adolph. 1859-1930. Ger. P IV.

6966. HALLWACHS, Wilhelm Ludwig Franz. 1859-1922. Ger. DSB; P IV.

6967. KÖNIG, (Carl Georg) Walter. 1859-1936. Ger. P III,IV.

6968. PIONCHON, Joseph Eugène Napoléon. 1859-after 1912. Fr. P IV.

6969. PRINGSHEIM, Ernst. 1859-1917. Ger. DSB; P IV.

6970. SUTHERLAND, William. 1859-1911. Australia. DSB; ICB; P IV.

Imprint Sequence

6971. FRANCE. Commission temporaire des poids et mesures républicaines.
1. *Instruction abrégée sur les mesures déduites de la grandeur de la terre, uniformes pour toute la République, et sur les calculs relatifs à leur décision décimale.* Paris, [1794]. RM/L.

6972. SOCIETY for the Diffusion of Useful Knowledge.
1. *Natural philosophy.* 4 vols. London, 1829-38. (Library of useful knowledge)

6973. *SCIENTIFIC memoirs selected from the transactions of foreign academies, and from foreign journals. Natural philosophy.* London, 1853. X. Ed. by J. Tyndall and W. Francis. Reprinted, New York, 1966. cf. 3009 and 11267.

6974. SOCIÉTÉ française de physique.
1. *Catalogue de la bibliothèque.* Paris, 1893.

6975. PHILLIPS, Charles Edmund Stanley.
1. *Bibliography of X-ray literature and research, 1896-97.* London, 1899.

3.07 CHEMISTRY

Including biochemistry and medical chemistry.

6976. VATER, Christian. 1651-1732. Ger. P II; BLA; GC. Also Physics.

6977. BOULDUC, Simon. 1652-1729. Fr. ICB; P I; GA.

6978. MANGET, Jean Jacques. 1652-1742. Switz. P II; BLA; GA.

6979. SLEVOGT, Johann Adrian. 1653-1726. Ger. P II; BMNH; BLA. Also Botany.

6980. HOFFMANN, Friedrich. 1660-1742. Ger. DSB; ICB; P I; BMNH; Mort.; BLA; GA; GB; GC. Also Physiology and Part 2.
1. *Opera omnia physico-medica ... Cum vita auctoris.* 6 vols in 3. Geneva, 1740. Another ed., ib., 6 vols in 4, 1748.
1a. ——— *Supplementum.* 2 vols. Ib., 1749. X. 2nd ed., 1754.
1b. ——— *Supplementum secundum.* 3 vols in 2. Ib., 1753-60.

6981. MATTE, Jean. 1660-1742. Fr. P II.

6982. STAHL, Georg Ernst. 1660-1734. Ger. DSB; ICB; P II; Mort.; BLA & Supp.; GA; GB; GC. Also Physiology* and Part 2. See also 15505/1.
1. *Specimen Beccherianum.* Leipzig, 1703.
2. *Fundamenta chymiae dogmaticae et experimentalis.* Nuremberg, 1723. Composed from Stahl's lectures by his students. RM/L.
3. *Fundamenta pharmaciae chymicae manu methodoque Stahliana posita.* Büdingen, 1728. RM.
4. *Experimenta, observationes, animadversiones ... chymicae et physicae.* Berlin, 1731. RM.
5. *Fundamenta chymiae dogmatico-rationalis et experimentalis.* Nuremberg, 1732. (Not the same work as item 2.) RM/L.

6983. FICK, Johann Jacob. 1662-1730. Ger. P I; BMNH; BLA; GA. Also Botany and Anatomy.

6984. POLI, Martino. 1662-1714. It. ICB; P II; GA.

6985. ZANNICELLI, Gian Girolamo. 1662-1729. It. P II; BMNH. Also Botany*.

6986. BOERHAAVE, Hermann. 1668-1738. Holl. DSB; ICB; P I; BMNH; Mort.; BLA & Supp.; GA; GB. Also Botany* and Physiology*. See also Index.
1. *Sermo academicus de chemia.* Leiden, 1718.
2. *Institutiones et experimenta chemiae.* 2 vols. Paris [really Leiden], 1724. An unauthorised publication of his lecture course. X.

2a. ——— *A new method of chemistry ... To which is prefix'd a critical history of chemistry and chemists.* London, 1727. Trans. with notes by P. Shaw and E. Chambers. RM/L.

3. *Elementa chemiae.* 2 vols. Leiden, 1732. His own publication of his lecture course. RM/L. Another ed. 1753.

3a. ——— *Elements of chemistry.* 2 vols. London, 1735. Trans. by T. Dallowe. RM.

3b. ——— *A new method of chemistry. Including the history, theory and practice of the art ... The second edition* [the first being item 2a]. 2 vols. London, 1741. Trans. with notes by P. Shaw. X. 3rd ed. [unchanged re-issue], 1753.

4. *Opuscula omnia.* The Hague, 1738. RM.

5. *Opera omnia medica.* Venice, 1751. RM.

6. *Correspondence.* Leiden, 1962-. Ed. by G.A. Lindeboom.

6987. DEIDIER, Antoine. d. 1746. Fr. P I; BLA; GA.

6988. GHERLI, Fulvio. 1670-1735. It. P I.

6989. VANDI, Andrea Giovanni Domenico. 1670-1763. It. P II.

6990. GEOFFROY, Etienne François. 1672-1731. Fr. DSB; ICB; P I; BMNH; BLA & Supp.; GA; GB.

1. *A treatise of the fossil, vegetable and animal substances that are made use of in physick ... To which is prefixed an enquiry into the constituent principles of mixed bodies.* London, 1736. This English version was the first publication of the work, being translated from a MS copy of the author's lectures (given in Paris) by G. Douglas. Latin and French versions appeared in Paris in 1741 and 1743.

1a. ——— *Tractatus de materia medica.* 3 vols. Paris, 1741. X. Another ed., 2 vols in 3, Venice, 1742-56. RM.

6991. DIPPEL, Johann Conrad. 1673-1734. Ger. P I; BLA & Supp.; GA; GB; GC.

6992. BOULDUC, Gilles (or Egide) François. 1675-1742. Fr. ICB; P I; GA.

6993. FRIEND, John. 1675-1728. Br. DSB; ICB; P I; BLA & Supp.; GA; GB; GD.

1. *Opera omnia.* Venice, 1733. RM.

6994. LAGARAYE, Claude Toussaint Marot de. 1675-1755. Fr. P I; GA.

6995. CARL, Johann Samuel. 1676 (or 1667)-1757. Ger./Den. ICB; P I & Supp.; BLA; GA; GC.

6996. GAROFALO, Biaggio (or CARYOPHILUS, Blasius). 1677-1762. It./Aus. P I; GA.

1. *De antiquis auri, argenti, stanni, aeris, ferri, plumbique fodinis.* Vienna, 1757. RM.

6997. HALES, Stephen. 1677-1761. Br. DSB; ICB; P I; BMNH; Mort.; BLA & Supp.; GA; GB; GD. Also Physiology*, Geology*, and Botany*.

6998. LEMERY, Louis. 1677-1743. Fr. DSB; P I; BLA; GA. Also Anatomy.

1. *Traité des alimens.* Paris, 1702 and later eds. X.

1a. ——— *A treatise of foods in general ... to which are added remarks ... wherein their nature and uses are explained according to the principles of chemistry and mechanism.* London, 1704.

6999. HENCKEL, Johann Friedrich. 1678-1744. Ger. DSB; P I (HENKEL); BMNH (HENKEL); GA; GC. Also Mineralogy.
1. *Pyritologia; oder, Kiess-Historie, als des vornehmsten Minerals, nach dessen Nahmen, Arten....* Leipzig, 1725. X. Another ed., 1754. X.
1a. ——— *Pyritologia; or, A history of the pyrites, the principal body in the mineral kingdom.* London, 1757. RM/L.
2. *Mineralogische, chemische und alchymistische Briefe von ... Gelehrten an ... J.F. Henkel.* Dresden, 1794. RM.

7000. JUNCKER, Johann. 1679-1759. Ger. DSB; P I; BLA (JUNKER); GA. Also Physiology*.
1. *Conspectus chemiae theoretico-practicae.* 2 vols. Halle, 1730-38. X. Another ed., 3 vols, 1749-53; reprinted, Leipzig, 196-?

7001. ROTHE, Gottfried. 1679-1710. Ger. P II.

7002. QUINCY, John. d. 1722. Br. ICB; GA; GD; BLA. Also Science in General*.

7003. ALBERTI, Michael. 1682-1757. Ger. P I; BMNH; BLA; GA; GC. Also Botany.

7004. BECCARI, Jacopo Bartolomeo. 1682-1766. It. ICB; P I; BLA; GA.
1. *De quamplurimis phosphoris nunc primum detectis commentarius.* Bologna, 1744. RM.

7005. BÖTTGER (or BÖTTIGER), Johann Friedrich. 1682-1719. Ger. ICB; P I; GA (BOETTCHER); GC.

7006. ALSTON, Charles. 1683-1760. Br. P I; BMNH; BLA & Supp.; GA; GB; GD. Also Botany.

7007. HENSING, Johann Thomas. 1683-1726. Ger. P I; BLA.

7008. NEUMANN, Caspar. 1683-1737. Ger. DSB; P II; GA (N., Gaspar).
1. *Chymiae medicae dogmatico-experimentalis; oder, Der gründlichen und mit Experimenten erwiesen medicinischen Chymie.* 7 vols in 3. Züllichau, 1749-52.
2. [Trans.] *The chemical works. Abridged and methodized with large additions.* London, 1759. By W. Lewis. RM/L. 2nd ed., 2 vols, 1773.

7009. RÉAUMUR, René Antoine Ferchault de. 1683-1757. Fr. DSB; ICB; P II,III; BMNH; Mort.; GA; GB. Also Physics, Zoology*, and Embryology*.
1. *L'art de convertir le fer forgé en acier, et l'art d'adoucir le fer fondu.* Paris, 1722. X.
1a. ——— *Memoirs on steel and iron.* Chicago, 1956. Trans. by A.G. Sisco, with introd. and notes by C.S. Smith.

7010. VATER, Abraham. 1684-1751. Ger. P II; BMNH; Mort.; BLA; GC. Also Botany and Anatomy.

7011. GEOFFROY, Claude Joseph. 1685-1752. Fr. DSB; ICB; P I; GA.

7012. HARTMANN, Melchior Philipp. 1685-1765. Ger. P I; BLA.

7013. HELLOT, Jean. 1685-1766. Fr. DSB; P I; GA.

7014. ROTH-SCHOLTZ, Friedrich. 1687-1736. Ger. P II.
1. *Bibliotheca chemica; oder, Catalogus von chymischen Bücher*n. Nuremberg, 1719. X. Reprinted, Hildesheim, 1971.

7015. SCHULZE, Johann Heinrich. 1687-1744. Ger. ICB; P II; BMNH; BLA; GA; GC.

7016. FREKE, John. 1688-1756. Br. BLA; GD. Also Physics*.
1. *A treatise on the nature and property of fire*. London, 1752. RM.

7017. ELLER (VON BROCKHAUSEN), Johann Theodor. 1689-1760. Ger. DSB; P I; BLA & Supp.; GA; GC.

7018. BROWN, John (first of the name). d. 1736. Br. P I; GD.

7019. MICKLEBURGH, John. d. 1756. Br. ICB.

7020. FIZES, Antoine. 1690-1765. Fr. ICB; P I; BLA & Supp.; GA.

7021. HUXHAM, John. 1692-1768. Br. P I; Mort.; BLA; GA; GD.
1. *Medical and chemical observations upon antimony*. London, 1756.

7022. POTT, Johann Heinrich. 1692-1777. Ger. DSB; P II; GA; GC.
1. *Exercitationes chymicae*. Berlin, 1738. RM.

7023. GERICKE, Peter. 1693-1750. Ger. P I; BLA & Supp. (GERIKE); GA; GC.

7024. SENAC, Jean Baptiste. ca. 1693-1770. Fr. DSB; ICB; P II; BLA; GA. Also Anatomy.

7025. BRANDT, Georg. 1694-1768. Swed. DSB; ICB; P I; GA.

7026. GESNER, Johann Albrecht. 1694-1760. Ger. P I; BLA; GA.

7027. PEMBERTON, Henry. 1694-1771. Br. DSB; ICB; P II; BLA; GA; GD. Also Mechanics* and Physiology.
1. *A course of chemistry*. London, 1771. Publ. after his death by J. Wilson. RM.

7028. PLATNER, Johann Zacharias. 1694-1747. Ger. P II; BLA; GA; GC.

7029. SHAW, Peter. 1694-1764. Br. DSB; ICB; P II; BLA; GD.
1. *Chemical lectures*. London, [1734?]. RM. 2nd ed., 1755.
See also 6986/2a,3b.

7030. BOURDELIN, Claude Louis. 1696-1777. Fr. P I.

7031. HAUPT, Friedrich Gottlieb. 1696-1742. Ger. P I.

7032. GOULARD, Thomas. 1697-1784. Fr. ICB; BLA & Supp.; GA.
1. *Traité des effets du préparation de plomb*. 1760. X.
1a. ——— *A treatise on the effects of various preparations of lead*. London, 1769. X. Another ed., 1773. Another ed., (with a table exhibiting the differences between English and French weights), 1775.

7033. SCRINCI, Johann Anton Joseph. 1697-1773. Prague. P II; BLA.

7034. PLUMMER, Andrew. ca. 1698-1756. Br. DSB; P II.

7035. RUTTY, John. 1698-1775. Br. ICB; BLA; GA; GD.
1. *A methodical synopsis of mineral waters*. London, 1757.

7036. DENFFER (also called JANSEN), Johann Heinrich. 1700-1770. Mitau. P I.

7037. DUHAMEL DU MONCEAU, Henri Louis. 1700-1782. Fr. DSB; ICB; P I; BMNH; BLA (HAMEL); GA; GB. Also Botany*.

7038. HOFMANN, Gottfried August. 1700-1775. Ger. P I.

7039. HIMSEL, Joachim Gebhard. 1701-1751. Reval. P I.

7040. MALOUIN, Paul Jacques. 1701-1778. Fr. DSB; P II & Supp.; BLA; GA.

7041. GMELIN, Johann Conrad. ca. 1702-1759. Ger. P I.

7042. KÜHNST, Johann Christoph. 1702-1762. Ger. P I.

7043. HEBENSTREIT, (Johann) Ernst. 1703-1757. Ger. P I; BMNH; BLA; GA; GC. Also Natural History*.

7044. ROUELLE, Guillaume François. 1703-1770. Fr. DSB; ICB; P II; GA; GB. Also Geology.

7045. SWAB (or SVAB), Anton. 1703-1768. Swed. P II.

7045A CARTHEUSER, Johann Friedrich. 1704-1777. Ger. P I; BMNH; BLA; GA (CARTHAEUSER). Also Botany.
1. *Elementa chymiae dogmatico-experimentalis*. Halle, 1736.
X. 2nd ed., Frankfurt a.d. Oder, 1753. RM.

7046. BALDASSARRI, Giuseppe. 1705-1785. It. P I; BLA.

7047. GAUB, Hieromymus (or Jerome) David. 1705-1780. Ger./Holl. ICB; P I (GAUBIUS); BLA & Supp. (GAUBIUS); GA; GC.

7048. MEYER, Johann Friedrich 1705-1765. Ger. DSB; P II. See also 7087/1.

7049. JUGEL, Johann Gottfried. 1707-1786. Ger. P I.
1. *Höchstnützliches Berg- und Schmeltz-Buch*. Berlin, 1743. RM.
2. *Philosophische Unterredung zwischen dem fliegenden Mercurium und einem gemeinen Schmeltzer*. Berlin, 1743. RM.
3. *Freyentdeckte Experimentalchymie*. Leipzig, 1766. RM.

7050. LEWIS, William. 1708 (or 1714)-1781. Br. DSB; ICB; P I; BLA; GA; GD.
1. *A course of practical chemistry*. London, 1746. RM.
2. *An experimental history of the materia medica*. London, 1761. X. 2nd ed., 1768. Another ed., 1791.
3. *Commercium philosophico-technicum; or, The philosophical commerce of arts. Designed as an attempt to improve the arts, trades and manufactures*. London, 1763. RM/L.
See also 7008/2.

7051. LUDOLFF, Hieronymus von (second of the name). 1708-1764. Ger. P I; BLA.

7052. MARGGRAF, Andreas Sigismund. 1709-1782. Ger. DSB; ICB; P II; GA; GB; GC.
1. [Trans. of a Latin periodical article of 1743] *Einige*

neue Methoden, den Phosphor im festen Zustande ... darzustellen. Leipzig, 1912. (OKEW. 187)
2. *Chymische Versuche, einen wahren Zucker aus verschiedenen Pflanzen, die in unseren Ländern wachsen, zu ziehen.* [Periodical article, 1747] Leipzig, 1907. (OKEW. 159)

7053. RAHN, Johann Heinrich. 1709-1786. Switz. P II.

7054. WALLERIUS, Johan Gottshalk. 1709-1785. Swed. DSB; ICB; P II; BMNH. Also Mineralogy*. See also 15339/1.
1. *Elementa metallurgiae speciatim chemicae.* Stockholm, 1768.

7055. CLAUSIER, Jean Louis. d. ca. 1750. Fr. P I; BLA.

7056. BOECLER, Johann Philipp. 1710-1759. Strasbourg. P I; BLA; GA.

7057. CRAMER, Johann Andreas. 1710-1770. Ger. DSB Supp.; P I; GA; GC.
1. *Elementa artis docimasticae.* 2 vols. Leiden, 1739. X. Another ed., 1744. RM.
1a. ——— *Elements of the art of assaying metals.* London, 1741. RM/L. 2nd ed., 1746.

7058. CULLEN, William. 1710-1790. Br. DSB; ICB; P I; Mort.; BLA & Supp.; GA; GB; GD. Also Physiology*#. See also Index.

7059. KIRSTEN, Johann Jacob. 1710-1765. Ger. P I; BLA.

7060. SCHEFFER, Henrik Theophilus. 1710-1759. Swed. P II.

7061. SCHWEITZER, Johann Konrad (or Cornelius) Friedrich. 1710-ca. 1776. Ger. P II; BLA.

7062. BECKER, Johann Philipp. 1711-1799. Ger. P I; BLA.

7063. BROWNRIGG, William. 1711-1800. Br. DSB; ICB; P I; BLA; GD.
1. *The art of making common salt, as now practised in most parts of the world.* London, 1748.

7064. LOMONOSOV, Mikhail Vasilievich. 1711-1765. Russ. DSB; ICB; P I; GA; GB. Also Various Fields*# and Physics.
1. [Trans. of Latin and Russian periodical articles] *Physikalisch-chemische Abhandlungen, 1741-42.* Leipzig, 1910. (OKEW. 178)
2. *On the corpuscular theory.* Cambridge, Mass., 1970. Trans. with an introd. by H.M. Leicester.

7065. MODEL, Johann Georg. 1711-1775. Ger./Russ. P II.

7066. BRENDEL, Johann Gottfried. 1712-1758. Ger. P I; BLA & Supp. Also Mathematics.

7067. MAZÉAS, Guillaume. 1712-1776. Fr. P II. Also Physics.

7068. NAVIER, Pierre Toussaint. 1712-1779. Fr. P II; BLA; GA.

7069. GELLERT, Christlieb Ehregott. 1713-1795. Russ./Ger. P I; GA; GC.
1. *Anfangsgründe zur metallurgischen Chymie.* Leipzig, 1751. X. 2nd ed., 1776. X.
1a. ——— *Metallurgic chymistry. Being a system of mineralogy in general, and of all the arts arising from this science.* London, 1776. Trans. by J. Seiferth. RM.

7070. CHARDENON, Jean Pierre. 1714-1769. Fr. DSB.

7071. FABRICIUS, Philipp Konrad. 1714-1774. Ger. P I; BMNH; BLA & Supp.; GA; GC. Also Botany*.

7072. TILLET, Mathieu. 1714-1791. Fr. DSB; P II.

7073. FUCHS, Georg August. d. 1770. Ger. P I.

7074. BARON (D'HÉNOUVILLE), Théodore. 1715-1768. Fr. P I; BLA.

7075. HUNDERTMARK, Karl Friedrich. 1715-1762. Ger. P I; BLA.

7076. LEIDENFROST, Johann Gottlob. 1715-1794. Ger. P I; BLA.

7077. DOSSIE, Robert. 1717-1777. Br. ICB; P I.
1. *The elaboratory laid open; or, The secrets of modern chemistry and pharmacy revealed.* London, 1758. X. 2nd ed., 1768.

7078. LASSONE, Joseph Marie François de. 1717-1788. Fr. ICB; P I; BLA; GA.

7079. MACQUER, Pierre Joseph. 1718-1784. Fr. DSB; ICB; P II; BLA; GA.
1. *Elémens de chymie théorique.* Paris, 1749. RM. Another ed., 1756.
2. *Elémens de chymie pratique, contenant la description des opérations fondamentales de la chymie.* 2 vols. Paris, 1751. RM.
3. [Trans. of items 1 and 2] *Elements of the theory and practice of chymistry.* 2 vols. London, 1758. Trans. by A. Reid. RM. 2nd ed., 1764. 3rd ed., 3 vols, 1768 and 1775.
4. *Dictionnaire de chymie.* 2 vols. Paris, 1766. RM/L. 2nd ed., 4 vols, 1778.
4a. ——— *A dictionary of chemistry.* 2 vols. London, 1771. Trans. by James Keir. RM.

7080. ROUELLE, Hilaire Marin. 1718-1779. Fr. DSB; P II; GA.

7081. LEHMANN, Johann Gottlob. 1719-1767. Ger./Russ. DSB; ICB; P I; BMNH; GA; GB; GC. Also Geology*.
1. *Die Probier-Kunst.* Berlin, 1761. RM.
2. *Physikalisch-chymische Schriften.* Berlin, 1761. RM.

7082. MANGOLD, Christoph Andreas. 1719-1767. Ger. P II; BLA.

7083. ROTHERDAM, John. ca. 1719-1787. Br. P II.

7084. MARTEAU, Pierre Antoine. d. 1772. Fr. P II; BLA.

7085. WIDMER, Georg. d. ca. 1775. Strasbourg. P II.

7086. RÜDIGER, Anton. 1720-1783. Ger. P II.

7087. CRANTZ, Heinrich Johann Nepomuk von. 1722-1799. Aus. BMNH; BLA & Supp. Also Botany*,
1. *Examinis chemici doctrinae Meyerianae de acido pingui, et Blackianae de aere fixo respectu calcis rectificatio.* Leipzig, 1770. Re J.F Meyer (7048) and J. Black (7117).

7088. CRONSTEDT, Axel Frederik. 1722-1765. Swed. DSB; P I; BMNH; GA. Also Mineralogy*.

7089. DAMBOURNEY, Louis Auguste. 1722-1795. Fr. P I; GA. Also Botany.
1. *Recueil de procédés et d'expériences sur les teintures solides que nos végétaux indigènes communiquent aux laines et aux lainages.* Paris, 1786. X. Another ed., 1794. RM.

7090. MAGELLAN, Jean Hyacinthe. 1722-1790. Port./Br. DSB; ICB; P II (MAGELHAENS); GA (MAGELHAENS); GD. Also Physics.

7091. MONTET, Jacques. 1722-1782. Fr. P II; GA.

7092. NICOLAI, Ernst Anton. 1722-1802. Ger. P II; BLA; GA.

7093. SALCHOW, Ulrich Christoph. 1722-1787. Ger./Russ. P II; BLA.

7094. SPIELMANN, Jacques Reinhold. 1722-1783. Strasbourg. ICB; P II; BMNH; BLA; GA; GC. Also Botany.
1. *Institutiones chemiae.* Strasbourg, 1763. X. 2nd ed., 1766. X.
1a. ——— *Instituts de chymie.* Paris, 1770. Trans. from the 2nd Latin ed. by Cadet *le jeune.*

7095. VENEL, Gabriel François. 1723-1775. Fr. DSB; ICB; P II; BLA.

7096. ANDREAE, Johann Gerhard Reinhard. 1724-1793. Ger. P I; GA; GC.

7097. FÜRSTENAU, Johann Friedrich. 1724-1751. Ger. P I; BLA.

7098. ROUSSEAU, Georg Ludwig Claudius. 1724-1794. Ger. ICB; P II; BLA; GA.

7099. VOGEL, Rudolph Augustin. 1724-1774. Ger. P II; BMNH; BLA; GC.

7100. BAYEN, Pierre. 1725-1798. Fr. DSB; ICB; P I; BLA; GA.
1. *Recherches chimiques sur l'étain.* Paris, 1781. With L.M. Charlard. RM.
2. *Opuscules chimiques.* 2 vols. Paris, 1798. Ed. by P. Malatret. RM.

7101. BEWLEY, William. 1725-1783. Br. P I; GD.

7102. DARCET, Jean. 1725-1801. Fr. DSB; ICB; P I; BMNH (ARCET); BLA; GA.
1. *Mémoire sur l'action d'un feu égal, violent, et continué pendant plusieurs jours, sur un grand nombre de terres, de pierres et de chaux métalliques.* Paris, 1771. RM.
2. *Rapport sur la fabrication des savons, sur leurs différentes espèces, suivant la nature des huiles et des alkalis qu'on emploie pour les fabriquer.* Paris, 1795. With Lelièvre and Pelletier. RM.

7103. KURELLA, Ernst Gottfried. 1725-1799. Ger. P I; BLA.

7104. WELL, Johann Jacob von. 1725-1787. Aus. P II; BMNH; BLA. Also Natural History.

7105. LE ROY, Charles. 1726-1779. Fr. DSB; ICB; P I; BLA; GA. Also Physics.

7106. MACBRIDE, David. 1726-1778. Br. DSB; P II; BLA; GA; GD.
1. *Experimental essays on the following subjects. I. On the fermentation of alimentary mixtures. II. On the nature and properties of fixed air. III. On the respective*

powers, and manner of acting, of the different kinds of antiseptics. IV. On the scurvy. V. On the dissolvent power of quicklime. London, 1764.

7107. MARET, Hugues. 1726-1785. Fr. P II; BLA; GA. See also 7150/2.

7108. BUCCI, Antonio. 1727-ca. 1793. It. P I.

7109. DOBSON, Matthew. 1727-1784. Br. ICB; P I.
1. *A medical commentary on fixed air.* Chester, 1779. X. 2nd ed., 1785.

7110. GADD, Peter Adrian. 1727-1797. Fin. ICB; P I; GA.

7111. HARTMANN, Peter Immanuel. 1727-1791. Ger. P I; BLA; GA. Also Botany.

7112. ILSEMANN, Johann Christoph. 1727-1822. Ger. P I.

7113. JACQUIN, Nicolas Joseph von. 1727-1817. Aus. DSB; ICB; P I; BMNH; BLA; GA; GC. Also Botany*.

7114. MONRO, Donald. 1727-1802. Br. P II; BLA; GA; GD.
1. *A treatise on medical and pharmaceutical chymistry.* 3 vols & supp. London, 1788-90.

7115. WOULFE, Peter. 1727?-1803. Br. DSB; ICB; P II; GD.

7116. BAUMÉ, Antoine. 1728-1804. Fr. DSB; ICB; P I; GA; GB.
1. *Elémens de pharmacie, théorique et pratique.* Paris, 1762. X. 8th ed., 1797.
2. *Manuel de chymie; ou, Exposé des opérations et des produits d'un cours de chymie.* Paris, 1763. RM.
2a. ——— *A manual of chemistry; or, A brief account of the operations of chemistry and their products.* Warrington, 1778. Trans. by John Aikin.
3. *Chymie expérimentale et raisonnée.* Paris, 1773. RM.

7117. BLACK, Joseph. 1728-1799. Br. DSB; ICB; P I; BLA & Supp.; GA; GB; GD. Also Physics.
1. *Dissertatio medica inauguralis, de humore acido a cibis orto, et magnesia alba.* Edinburgh, 1754. X.
1a. ——— *On acid humor arising from food, and on white magnesia.* Minneapolis, 1973. Trans. by T. Hanson.
2. *Experiments upon magnesia alba, quicklime, and some other alcaline substances.* [Periodical article, 1756] Edinburgh, 1898 and re-issues. (Alembic Club reprints. 1)
3. *Lectures on chemistry.* Not published during his lifetime; the following is a reproduction of the MS notes of one of his students. *Notes from Doctor Black's lectures on chemistry, 1767/8. By T. Cochrane.* Wilmslow, Cheshire, 1966. Ed. with an introd. by D. McKie.
4. *Lectures on the elements of chemistry.* 2 vols. Edinburgh, 1803. Ed. by J. Robison.
See also 7087/1 and 7149/2.

7118. DEMACHY, Jacques François. 1728-1803. Fr. P I; BLA; GA.

7119. GEOFFROY, Claude François. ca. 1728-1753. Fr. ICB.

7120. JAEGER, Johann Ludolph. ca. 1728-1787. Ger. P I.

1. *Philosophische und physikalische Abhandlungen über verschiedene Materien aus dem Reihe der Natur.* Nuremberg, 1784. RM.

7121. GEORGI, Johann Gottlieb. 1729-1802. Ger./Russ. ICB; P I; BMNH. Also Natural History.

7122. HAHN, Johann David. 1729-1784. Holl. P I; BLA; GA.

7123. BERKENHOUT, John. 1730?-1791. Br. BMNH; BLA & Supp.; GA; GD. Also Natural History*.

7124. DURANDE, Jean François. 1730-1794. Fr. P I; BMNH; BLA; GA. Also Botany. See also 7150/2.

7125. FONTANA, Felice Gaspar Ferdinand. 1730-1805. It. DSB; ICB; P I; BMNH; Mort.; BLA & Supp.; GA. Also Anatomy and Physiology*.
1. *Richerche fisiche sopra l'aria fissa.* Florence, 1775. RM.

7126. GÜNTHER, Christoph. 1730-1790. Den. P I.

7127. SIGAUD DE LAFOND, Joseph Aignan. 1730-1810. Fr. DSB; P II (S., Jean René); BLA; GA. Also Physics*.
1. *Essai sur différentes espèces d'air qu'on désigne sous le nom d'air fixe.* Paris, 1779. RM/L.

7128. WOLLIN, Christian. 1730-1798. Swed. P II.

7129. CADET (DE GASSICOURT), Louis Claude. 1731-1799. Fr. DSB; ICB; P I; BLA; GA.

7130. CARBURI, Marco. 1731-1808. It. P I; GA.

7131. CAVENDISH, Henry. 1731-1810. Br. DSB; ICB; P I; GA; GB; GD. Also Physics*#. See also 15486/1.
1. *Experiments on air.* [Periodical articles, 1784-85] Edinburgh, 1893 and re-issues. (Alembic Club reprints. 3)

7132. HADLEY, John (second of the name). 1731-1764. Br. ICB; P I; GD.
1. *A plan of a course of chemical lectures.* Cambridge, 1758. RM.

7133. PÖRNER, Karl Wilhelm. 1732-1796. Ger. P II; BLA.

7134. WIEGLEB, Johann Christian. 1732-1800. Ger. DSB; P II; GA; GC. Also History of Chemistry*.

7135. WINTERL, Jacob Joseph. 1732-1809. Hung. ICB; P II. Also Botany.

7136. KIRWAN, Richard. 1733?-1812. Br. DSB; ICB; P I; BMNH; GA; GB; GD. Also Mineralogy*, Geology*, and Meteorology*. See also Index.
1. *An essay on phlogiston and the constitution of acids.* London, 1784. X.
1a. ——— *Essai sur la phlogistique et sur la constitution des acides.* Paris, 1788. Trans. by Mme Lavoisier with notes by de Morveau, Lavoisier, Laplace, Monge, Berthollet, and Fourcroy.
1b. ——— *An essay on phlogiston ... A new edition.* London, 1789. With a trans. of the notes added to the

French version (item 1a) by Lavoisier et al., defending the antiphlogistic theory, and with additional remarks and replies by Kirwan. Reprinted, London, 1968.

2. [Trans.] *Physisch-chemische Schriften*. 5 vols in 7. Berlin, 1783-1801. Trans. with introd. and notes by L. Crell.

7137. LAURAGUAIS, Louis Léon Felicité, *Duc* de Brancas, *Comte* de. 1733-1824. Fr. P I; GA (BRANCAS).

7138. PRIESTLEY, Joseph. 1733-1804. Br. DSB; ICB; P II; GA; GB; GD; GE. Also Various Fields*#, Physics*, and Autobiography*.

1. *Directions for impregnating water with fixed air*. London, 1772. RM.
2. *Experiments and observations on different kinds of air*. London, 1774. (Not called Vol. 1.)

2a. ——— Vols 2 and 3. Ib., 1775, 1777.

2b. ——— 2nd ed. Vol. 1, 1775. Vol. 2, 1776. Vol. 3, 1777.

2c. ——— 3rd ed. Vol. 1, 1781. Vol. 2, 1784.

3. *Experiments and observations relating to various branches of natural philosophy. With a continuation of the observations on air*. 3 vols. London, 1779-86. Priestly and others sometimes refer to these three volumes as Vols 4-6 of the foregoing work.
4. *Experiments and observations on different kinds of air, and other branches of natural philosophy connected with the subject*. "In 3 volumes, being the former 6 volumes abridged and methodized, with many additions." Birmingham, 1790. A revision of items 2 and 3. RM/L. Reprinted, 3 vols, New York, 1970.
5. *Considerations on the doctrine of phlogiston and the decomposition of water*. Philadelphia, 1796. X. Another ed., 1797. RM. Reprinted, Princeton, 1929, and New York, 1969.
6. *The doctrine of phlogiston established, and that of the composition of water refuted*. Northumberland (U.S.A.), 1800. RM.

7139. BUCHOLZ, Wilhelm Heinrich Sebastian. 1734-1798. Ger. P I; BLA & Supp.

7140. CARTHEUSER, Friedrich August. 1734-1796. Ger. P I; BMNH; BLA; GA (CARTHAEUSER). Also Geology.

7141. HENRY, Thomas. 1734-1816. Br. DSB; P I; BLA; GD. See also 7182/1a,3.

7142. MONNET, Antoine Grimoald. 1734-1817. Fr. DSB; P II; BMNH; GA. Also Mineralogy* and Geology.

1. *Traité des eaux minérales*. Paris, 1768. RM.

7143. SALUZZO (or SALUCES), Giuseppe Angelo. 1734-1810. It. P II.

7144. BERGMAN, Torbern Olof. 1735-1784. Swed. DSB; ICB; P I & Supp.; BMNH; GA; GB. Also Various Fields and Mineralogy*.

1. [Trans. of a periodical article of 1773] *On acid of air*. Stockholm, 1956. Trans. by S.M. Jonsson.
2. *Commentatio de tubo ferruminatorico, eiusdemque usu in explorandis corporibus praesertim mineralibus*. Vienna, 1779. RM.

3. [Trans. of a Swedish work of 1779] *An essay on the usefulness of chemistry*. London, 1783. Trans. by John Murray. RM.
4. *Opuscula physica et chemica*. Vols 1-3. Uppsala/Leipzig, 1779-83.

4a. ——— Posthumous ed. 6 vols. Leipzig, 1786-92. A reprint of Vols 1-3 together with the first publn of Vols 4-6, the latter ed. by Hebenstreit. X.

4b. ——— *Physical and chemical essays*. 2 vols. London, 1784. Trans. with notes by Edmund Cullen. RM.

4c. ——— ——— Vol. 3. Edinburgh, 1791. Translator anonymous.

4d. ——— ——— Reprint of Vols 1-3, U.S.A., 1979.

5. *Disquisitio de attractionibus electivis*. First publ. in 1775. Included in an extended version in item 4 in 1783.

5a. ——— *A dissertation on elective attractions*. London, 1785. Trans. by T. Beddoes. RM/L. Reprinted, London, 1970.

5b. ——— *A dissertation on elective attractions*. New York, 1968. Trans. with an introd. by J.A. Schufle.

6. *Foreign correspondence*. Stockholm, 1965. Ed. by G. Carlid and J. Nordström.

7145. KEIR, James. 1735-1820. Br. DSB; ICB; P I; GD. See also 7079/4a and 15586/1.

7146. NEBEL, Daniel Wilhelm. 1735-1805. Ger. P II; BLA.

7147. REUSCH, Karl Daniel. 1735-1806. Ger. P II. Also Physics.

7148. ROSE, Valentin (first of the name). 1736-1771. Ger. P II.

7149. WATT, James. 1736-1819. Br. The famous inventor. DSB; ICB; P II; G. See also Index.

1. *Correspondence on his discovery of the theory of the composition of water*. London, 1846. Ed. with introd. and appendix by J.P. Muirhead. RM/L.
2. *Partners in science: Letters of James Watt and Joseph Black*. London, 1970. Ed. with introd. and notes by E. Robinson and D. McKie.

7150. GUYTON DE MORVEAU, Louis Bernard. 1737-1816. Fr. DSB; ICB; P I; BLA & Supp.; GA; GB.

1. *Digressions académiques; ou, Essais sur quelques sujets de physique, de chymie et d'histoire naturelle*. Dijon, 1772. RM.
2. *Elémens de chymie, théorique et pratique*. 3 vols. Dijon, 1777-78. With H. Maret and J.F. Durande. RM.
3. *Encyclopédie méthodique*. [Main entry: 3000] *Chymie, Pharmacie et Métallurgie*. Vol. 1 (in 2 parts). Paris, 1786-89. (The subsequent volumes were written by Fourcroy--item 7262/3.)
4. *Méthode de nomenclature chimique, proposée par MM. de Morveau, Lavoisier, Bertholet [sic] et de Fourcroy. On y a joint un nouveau système de caractères chimiques, adaptés à cette nomenclature, par MM. Hassenfratz et Adet*. Paris, 1787. RM.

4a. ——— *Method of chemical nomenclature*. London, 1788. Trans. by J. St. John. RM.

4b. ——— Another English trans.: see 7240/1.
5. *Traité des moyens de désinfecter l'air, de prévenir la contagion, et d'en arrêter les progrès.* Paris, 1801. RM.
5a. ——— *A treatise on the means of purifying infected air.* London, 1802. Trans. by R. Hall.
See also Index.

7151. NAHUYS, Alexander Peter. 1737-1794. Holl. P II & Supp.; BMNH; BLA.

7152. PARMENTIER, Antoine Augustin. 1737-1813. Fr. DSB; ICB; P II; BLA; GA.

7153. SICKINGEN, Karl Heinrich Joseph von. 1737-1791. Ger. P III; GC.

7154. WATSON, Richard. 1737-1816. Br. DSB; ICB; P II; GA; GB; GD. Also Autobiography*.
1. *Chemical essays.* 5 vols. Cambridge, 1781-87. X. 2nd ed., London, 1782-88. Another ed., 1791-1800. 6th ed., 1793-96. 7th ed., 1800.

7155. HAGEN, Johann Heinrich. 1738-1775. Ger. P I.

7156. PLENK, Joseph Jacob von. 1738-1807. Aus./Hung. P II. Under PLENCK: BMNH; BLA; GA. Also Botany and Physiology*.

7157. TROMMSDORFF, Wilhelm Bernhard. 1738-1782. Ger. P II.

7158. WARLTIRE, John. ca. 1738-1810. Br. ICB.

7159. ZIEGLER, Johann Heinrich. 1738-1818. Switz. P II.

7160. BONVICINO, Costanzo Benedetto. 1739-1812. It. DSB; P I & Supp. (BUONVICINO).

7161. CLOUET, Jean Baptiste Paul Antoine. 1739-1816. ICB.

7162. KELLER, Christian Friedrich. 1739-1797. Ger. P I.

7163. LAFOLIE, Louis Guillaume de. 1739-1780. Fr. GA.

7164. MEYER, Johann Karl Friedrich. 1739-1811. Ger. P II,III.

7165. CARRÈRE, Joseph Barthélemy François. 1740-1802. Fr. P I; BLA; GA.
1. *Catalogue raisonnée des ouvrages qui ont été publiés sur les eaux minérales.* Paris, 1785. RM.

7166. HIGGINS, Bryan. ca. 1740-1818. Br. DSB; ICB; P I; GD.
1. *Experiments and observations made with the view of improving the art of composing and applying calcareous cements and of preparing quick-lime. Theory of the arts.* London, 1780. RM.
2. *Experiments and observations relating to acetous acid, fixable air, dense inflammable air, oils and fuels, the matter of fire and light, metallic reduction, combustion, fermentation, putrefaction, respiration, and other subjects of chemical philosophy.* London, 1786. RM.

7167. PERCIVAL, Thomas. 1740-1804. Br. ICB; P II; BMNH; BLA; GA; GD.
1. *Experiments and observations on water, particularly the hard pump water of Manchester.* London, 1768. X. 2nd ed., 1772.

7168. SAGE, Balthazar Georges. 1740-1824. Fr. DSB; P II; BMNH; GA.
1. *Elémens de minéralogie docimastique*. Paris, 1772. X. 2nd ed., 2 vols, 1777.
2. *Mémoires de chimie*. Paris, 1773. RM/L.
3. *L'art d'essayer l'or et l'argent*. Paris, 1780. RM.

7169. WENZEL, Karl Friedrich. 1740-1793. Ger. ICB; P II; GB; GC.

7170. CHAULNES, Marie Joseph Louis d'Albert d'Ailly, *Duc* de. 1741-1793. Fr. P I; GA.

7171. EHRMANN, Friedrich Ludwig. 1741-1800. Fr. P I.

7172. WEBER, Jacob Andreas. 1741-1792. Ger. P II; BLA; GC.

7173. DUCHANOY, Claude François. 1742-1827. Fr. BLA.
1. *Essais sur l'art d'imiter les eaux minérales*. Paris, 1780. RM.

7174. LEBLANC, Nicolas. 1742-1806. Fr. DSB; ICB; P I & Supp.; GA; GB.

7175. SCHEELE, Carl Wilhelm. 1742-1786. Swed. DSB; ICB; P II; BLA; GA; GB.
1. *Chemische Abhandlungen von der Luft und dem Feuer*. Uppsala /Leipzig, 1777. X. Reprinted, Stockholm, 1970.
1a. ——— *Chemical observations and experiments on air and fire*. London, 1780. Trans. by J.R. Forster with notes by R. Kirwan. RM/L.
1b. ——— Extracts. Edinburgh, 1894 and re-issues. (Alembic Club reprints. 8)
1c. ——— Extracts. Leipzig, 1894. (OKEW. 58)
2. *The chemical essays*. London, 1786. Trans. of periodical articles by T. Beddoes. RM. Reprinted, London, 1901, and New York, 1966.
3. *Sämmtliche physische und chemische Werke*. 2 vols. Berlin, 1793. Trans. and ed. by S.F. Hermbstädt. X. Reprinted, 2 vols in 1, Wiesbaden, 1971.
4. *The collected papers*. London, 1931. Trans. from the Swedish and German originals by L. Dobbin. Reprinted, New York, 1969.
5. *Efterlemnade bref och anteckningar*. Stockholm, 1892. Ed. by A.E. Nordenskiöld.
5a. ——— *Nachgelassene Briefe und Aufzeichnungen*. Ib., 1892. X. Reprinted, Wiesbaden, 1973.
6. *Handschriften, 1756-1777*. Stockholm, 1942. Deciphered by C.W. Oseen.
7. *His life and work*. Vols I-II: *The brown book*. Stockholm, 1968. Ed. and trans. by U. Boklund. (The first two vols of a continuing edition of his works, with English trans. and commentary.)

7176. SENEBIER, Jean. 1742-1809. Switz. DSB; ICB; P II; BMNH; BLA; GA; GB. Also Science in General* and Plant Physiology*. See also Index.
1. *Mémoires physico-chimiques sur l'influence de la lumière solaire pour modifier les êtres des trois règnes de la nature, et sur-tout ceux du règne végétal*. 3 vols. Geneva, 1782. RM.

2. *Recherches sur l'influence de la lumière solaire pour métamorphoser l'air fixe en air pur par la végétation.* Geneva, 1783. RM.
3. *Recherches analytiques sur la nature de l'air inflammable.* Geneva, 1784. RM.

7177. CADET DE VAUX (or CADET LE JEUNE), Antoine Alexis François. 1743-1828. Fr. DSB; P I; BLA; GA. See also 7094/1a.

7178. DEIMAN, Johann Rudolph. 1743-1808. Holl. P I & Supp.; BLA; GA.

7179. IRVINE, William. 1743-1787. Br. P I; GD. Also Physics.
1. *Essays, chiefly on chemical subjects.* London, 1805. RM.

7180. KLAPROTH, Martin Heinrich. 1743-1817. Ger. DSB; ICB; P I; GA; GB; GC.
1. *Beiträge zur chemischen Kenntniss der Mineralkörper.* 6 vols. Posen/Berlin, 1795-1815. RM.
1a. ——— [Trans. of Vols 1 and 2] *Analytical essays towards promoting the chemical knowledge of mineral substances.* 2 vols. London, 1801-04. RM.

7181. LAMÉTHERIE, Jean Claude de. 1743-1817. Fr. DSB; P I; BMNH; BLA (MÉTHERIE); GA. Also Science in General*, Geology*, and Biology in General*. See also Index.
1. *Essai analytique sur l'air pur, et les différentes espèces d'air.* Paris, 1785. RM.

7182. LAVOISIER, Antoine Laurent. 1743-1794. Fr. DSB; ICB; P I; Mort.; GA; GB. Also Physiology. See also Index.
1. *Opuscules physiques et chimiques.* Paris, 1774. RM.
1a. ——— *Essays physical and chemical.* London, 1776. Trans. with notes by T. Henry. X. Reprinted, ib., 1970.
2. *Instruction sur l'établissement des nitrières et sur la fabrication du salpêtre.* Paris, 1777. RM.
3. [Trans. of nine periodical articles] *Essays on the effects produced by various processes on atmospheric air. With a particular view to an investigation of the constitution of the acids.* London, 1783. Trans. by T. Henry. RM.
4. [Trans. of two periodical articles, 1780 and 1784] *Zwei Abhandlungen über die Wärme.* Leipzig, 1892. (OKEW. 40)
5. [Trans. of periodical article, 1781/83] *Das Wasser.* Leipzig, 1930. (OKEW. 230)
6. *Traité élémentaire de chimie.* 2 vols. Paris, 1789. RM. Reprinted, ib., 1937. Another ed., 1793.
6a. ——— *Elements of chemistry.* Edinburgh, 1790. Trans. by R. Kerr. RM. Reprinted, New York, 1965. 2nd ed., Edinburgh, 1793. 3rd ed., 1796. 4th ed., 1799. 5th ed., 2 vols, 1802.
7. *Mémoires de chimie.* 2 vols. [No more publ.] Paris, 1805. Ed. by Mme Lavoisier. The beginning of a proposed complete edition of his memoirs.
8. *Oeuvres.* 6 vols. Paris, 1864-93. Ed. by J.B. Dumas (Vols I-IV) and E. Grimaux (Vols V and VI). RM/L. Reprinted, New York, 1965.
9. *Correspondance.* (*Oeuvres.* Vol. VII) 3 parts. Paris, 1955-64. Ed. by R. Fric. (Incomplete)

7183. NICOLAS, Pierre François. 1743-1816. Fr. P II; BMNH; BLA; GA. Also Zoology*.

7184. TINGRY, Pierre François. 1743-1821. Switz. ICB; P II.

7185. BANCROFT, Edward. 1744-1821. Br. ICB; P I & Supp.; BMNH; GA; GD; GE. Also Natural History*.
   1. *Experimental researches concerning the philosophy of permanent colours, and the best means of producing them, by dyeing, etc.* London, 1794. RM.

7186. CORNETTE, Claude Melchior. 1744-1794. Fr. DSB; P I & Supp.; BLA.

7187. ERXLEBEN, Johann Christian Polykarp. 1744-1777. Ger. ICB; P I; BMNH; BLA; GA; GC. Also Physics* and Zoology.

7188. FALCONER, William. 1744-1824. Br. BMNH; BLA; GA; GD. Also Various Fields.
   1. *An essay on the Bath waters.* 1770.
   2. *Essays and observations ... on the dissolvent power of water impregnated with fixable air.* London, 1776.

7189. HOFFMANN, Friedrich Christian. b. 1744. Ger. P I.

7190. HOPSON, Charles Rivington. 1744-1796. Br. P I; GD.

7191. LAMARCK, Jean Baptiste Pierre Antoine de Monet, *Chevalier* de. 1744-1829. Fr. DSB; ICB; P I; BMNH; GA; GB. Other entries: see 3204.
   1. *Recherches sur les causes des principaux faits physiques, et particulièrement sur celles de la combustion....* Paris, 1794. The title goes on to enumerate a large number of other chemical phenomena. RM.
   2. *Réfutation de la théorie pneumatique, ou de la nouvelle doctrine des chimistes modernes.* Paris, 1796. RM.

7192. MOROZZO (or MOUROUX), Carlo Lodovico. 1744-1804. It. P II.

7193. BRONGNIART, Anton Louis. d. 1804. Fr. P I; GA.

7194. CRELL, Lorenz Florenz Friedrich von. 1745-1816. Ger. DSB; ICB; P I; GC. See also 7136/2.

7195. DEYEUX, Nicolas. 1745-1837. Fr. P I & Supp.

7196. GAHN, Johan Gottlieb. 1745-1818. Swed. DSB; ICB; P I; GA.

7197. LICHTENSTEIN, Georg Rudolph. 1745-1807. Ger. P I; BLA.

7198. OPOIX, Christophe. 1745-1840. Fr. P II; GA.

7199. VOLTA, Alessandro. 1745-1827. It. DSB; ICB; P II; GA; GB. Also Physics*#.
   1. *Lettere sull'aria inflammabile nativa delle paludi.* Milan, 1777. X.
   1a. ——— *Lettres sur l'air inflammable des marais.* Strasbourg, 1778. RM.

7200. BESEKE, Johann Melchior Gottlieb. 1746-1802. Mitau. P I; BMNH. Also Zoology.

7201. BRINKMANN, Johann Peter. 1746-1785. Ger. P I; BLA & Supp.

7202. BUCQUET, Jean Baptiste Michel. 1746-1780. Fr. DSB; P I; BMNH; BLA; GA.

7203. HEYER, Justus Christian Heinrich. 1746-1821. Ger. P I.

7204. HJELM, Petter Jacob. 1746-1813. Swed. P I.

7205. KASTELEYN, Petrus Johannes. 1746-1794. Holl. P I; GF.

7206. LEONHARDI, Johann Gottfried. 1746-1823. Ger. DSB; P I; BLA.

7207. MONGE, Gaspard. 1746-1818. Fr. DSB; ICB; P II; GA; GB. Also Mathematics*, Mechanics, and Physics*. See also Index.

7208. RUSH, Benjamin. 1746-1813. U.S.A. DSB; ICB; P II; Mort.; BLA & Supp.; GA; GB; GE. Also Autobiography*.
1. *Letters*. 2 vols. Princeton, 1951. Ed. by L.H. Butterfield.

7209. VOGLER, Johann Philipp. 1746-1816. Ger. P II; BLA. Also Botany.

7210. BARNEVELD, Willem van. 1747-1826. Holl. P I.

7211. ELLIOT, John. 1747-1787. Br. ICB; P I; BLA (ELLIOTT). See also 10968/1.

7212. GILLET DE LAUMONT, François Pierre Nicolas. 1747-1834. Fr. P I & Supp.

7213. WALL, Martin. 1747-1824. Br. P II; BLA; GD.
1. *A syllabus of a course of lectures in chemistry*. Oxford, 1782. RM.
2. *Dissertations on select subjects in chemistry and medicine*. Oxford, 1783.
3. *Dissertations, letters, etc*. Oxford, 1786.

7214. WERNBERGER, Erasmus Ludwig. 1747-1795. Ger. P II.

7215. ANDRIA, Nicola. 1748-1814. It. P I; BLA; GA. Also Physiology*.

7216. BERTHOLLET, Claude Louis. 1748-1822. Fr. DSB; ICB; P I; BLA; GA; GB.
1. *Eléments de l'art de la teinture*. 2 vols. Paris, 1791. RM/L.
1a. ——— *Elements of the art of dyeing*. 2 vols. London, 1791. Trans. by W. Hamilton. RM/L.
1b. ——— 2nd ed. *Eléments de l'art de la teinture, avec une description du blanchiment par l'acide muriatique oxigéné*. Paris, 1804. RM.
2. *Recherches sur les lois d'affinité*. Paris, 1801. X.
2a. ——— *Researches into the laws of chemical affinity*. London, 1804. Trans. by M. Farrell. Reprinted, New York, 1966.
2b. ——— *Untersuchungen über die Gesetze der Verwandtschaft*. Leipzig, 1896. (OKEW. 74)
3. *Essai de statique chimique*. 2 vols. Paris, 1803. Reprinted, Brussels, 1968, and New York, 1972.
See also 7136/1a, 7150/4, and 7240/1.

7217. BLAGDEN, Charles. 1748-1820. Br. DSB; ICB; P I; GA; GD.
1. German trans. of two periodical articles of 1788. Leipzig, 1894. (OKEW. 56) The articles deal with the cooling of water below its freezing point, and the effect of various substances in lowering its freezing point.

7218. CLOUET, Pierre Romain. 1748-1810. Fr. ICB; P III.

7219. CRAWFORD, Adair. 1748-1795. Br. DSB Supp.; ICB; P I; Mort.;

BLA & Supp.; GA; GD. Also Physiology.
1. *Experiments and observations on animal heat and the inflammation of combustible bodies. Being an attempt to resolve these phenomena into a general law of nature.* London, 1779. RM. 2nd ed., 1788.

7220. GMELIN, Johann Friedrich. 1748-1804. Ger. P I; BMNH; BLA & Supp.; GA; GC. Also Botany and History of Chemistry*.
1. *Grundsätze der technischen Chemie.* Halle, 1786. RM.
2. *Allgemeine Geschichte der Pflanzengifte.* 1803. X. Reprinted, Wiesbaden, 1973.
3. *Allgemeine Geschichte der thierischen und mineralischen Gifte.* Erfurt, 1806. RM.

7221. MINKELERS, Johann Peter. 1748-1824. Belg. P II; GF.

7222. SANGIORGIO, Paolo. 1748-1816. It. P III.

7223. WEIGEL, Christian Ehrenfried. 1748-1831. Ger. DSB; P II; BMNH; BLA; GC. Also Botany.
1. *Observationes chemicae et mineralogicae.* 2 vols. Göttingen, 1771-73. X.
1a. ——— *Chemisch-mineralogische Beobachtungen.* Breslau, 1779. Trans. with many additions by J.T. Pyl. RM.

7224. CAVALLO, Tiberius. 1749-1809. Br. DSB; ICB; P I; GA; GB; GD. Also Physics*.
1. *A treatise on the nature and properties of air and other permanently elastic fluids.* London, 1781. RM.
2. *The history and practice of aerostation.* London, 1785. Re ballooning. RM.
3. *An essay on the medicinal properties of factitious airs, with an appendix on the nature of blood.* London, 1798. RM/L.

7225. DUNDONALD, Archibald COCHRANE, *9th Earl of.* 1749-1831. Br. DSB. Under COCHRANE, A.: P I; GA; GD.

7226. HAGEN, Karl Gottfried. 1749-1829. Ger. P I & Supp.; BMNH; BLA; GA; GC. Also Botany.

7227. HAUSSMANN, Johann Michel. 1749-1824. Strasbourg. P I; GA.

7228. HOCHHEIMER, Karl Friedrich August. 1749-after 1825. Ger. P I.

7229. RUTHERFORD, Daniel. 1749-1819. Br. DSB; ICB; P II; GD.

7230. STORR, Gottlieb Konrad Christian. 1749-1821. Ger. P II.

7231. HARRINGTON, Robert. fl. 1781-1815. Br. GD.

7232. RUPRECHT, Anton von. fl. 1783-1790. Aus. P II.

7233. DEHNE, Johann Christian Conrad. d. 1791. Ger. P I; BLA & Supp.

7234. BINDHEIM, Johann Jacob. 1750-1825. Ger./Russ. P I.

7235. MARUM, Martinus van. 1750-1837. Holl. DSB; ICB; P II; BLA; GB. Also Physics*# and Botany.

7236. CLOUET, Jean François. 1751-1801. Fr. DSB; ICB; P I (C., Louis).

7237. DESCROIZILLES, François Antoine Henri. 1751-1825. Fr. ICB; P I; GA.

7238. ÉTIENNE, André. 1751-1797. ICB.

7239. MORELOT (or MORELLOT). Simon. 1751-1809. Fr. P II.
1. *Histoire naturelle appliquée à la chimie, aux arts....* 2 vols. Paris, 1809. RM.

7240. PEARSON, George. 1751-1828. Br. DSB; P II; BLA; GA; GD.
1. *A translation of the table of chemical nomenclature proposed by De Guyton, ... Lavoisier, Bertholet and De Fourcroy* [item 7150/4], *with additions and alterations. To which are prefixed an alteration of the terms, and some observations on the new system of chemistry.* London, 1794. X.
1a. ——— 2nd ed. *A translation ... To which are subjoined tables of single elective attraction, tables of chemical symbols....* London, 1799.

7241. STRUVE, Henri (or Heinrich). 1751-1826. Switz. P II. Also Geology*.

7242. SUCKOW, Georg Adolph. 1751-1813. Ger. P II; BMNH; GC. Also Botany.

7243. TYCHSEN, Nicolai. 1751-1804. Den. P II.

7244. WESTRUMB, Johann Friedrich. 1751-1819. Ger. P II; BMNH; GC.

7245. HEMPEL, Johann Gottfried. b. 1752. Ger. P I.

7246. PRICE, James. 1752-1783. Br. ICB; P II; GA; GD.

7247. TROOSTWIJK, Adriaan Paets van. 1752-1837. Holl. DSB; P II.

7248. ACHARD, Franz Carl. 1753-1821. Ger. DSB; ICB; P I; GA; GB; GC. Also Physics*.
1. *Chymisch-physische Schriften.* Berlin, 1780. X. Reprinted, Hildesheim, 1969.
2. *Anleitung zum Anbau der zur Zuckerfabrication anwendbaren Runkelrüben.* Breslau, 1809. X. Reprinted, Leipzig, 1907. (OKEW. 159)

7249. AFZELIUS (or ARVIDSON), Johann. 1753-1837. Swed. P I.

7250. DANIEL, Christian Friedrich. 1753-1798. Ger. P I; BLA & Supp.; GA.

7251. DRIESSEN, Petrus. 1753-1828. Holl. P I & Supp.; BLA & Supp.

7252. NICHOLSON, William. 1753-1815. Br. DSB; ICB; P II; BLA; GA; GB; GD. Also Science in General* and Physics*.
1. *The first principles of chemistry.* London, 1790. RM/L. 2nd ed., 1792. 3rd ed., 1796.
2. *A dictionary of chemistry.* 2 vols. London, 1795. RM/L. 2nd ed., 1808.
See also Index.

7253. TIELEBEIN, Christian Friedrich. 1753-1786. Ger. P II.

7254. CLEGHORN, William. 1754-1783. Br. ICB.

7255. ELHUYAR, Juan José d'. 1754-1796. Spain. DSB; ICB.

7256. GALLISCH, Friedrich Anton. 1754-1783. Ger. P I.

7257. MEUSNIER (DE LA PLACE), Jean Baptiste Marie Charles. 1754-1793. Fr. DSB; ICB; P II; GA. Also Mathematics.

7258. PROUST, Joseph Louis. 1754-1826. Fr./Spain. DSB; ICB; P II, IV Supp.; GA; GB.

7259. QUATREMÈRE-DISJONVAL, Denis Bernard. 1754-1830. Fr. P II; BMNH; GA. Also Zoology. See also 5952/1a.

7260. CAVEZZALI, Girolamo. 1755-1830. It. P I.

7261. ELHUYAR, Fausto d'. 1755-1833. Spain/Mexico. DSB; ICB; P I & Supp.; GA.

7262. FOURCROY, Antoine François de. 1755-1809. Fr. DSB; ICB; P I; BMNH; BLA; GA; GB.

1. *Leçons élémentaires d'histoire naturelle et de chimie.* 2 vols. Paris, 1782. RM.
1a. ——— 2nd ed. *Elémens d'histoire naturelle et de chimie.* 4 vols. Paris, 1786. Supplement, 1789.
1b. ——— *Elementary lectures on chemistry and natural history.* 2 vols. Edinburgh, 1785. Trans. of item 1 by T. Elliott, with many additions, notes etc. X.
1c. ——— ——— 2nd ed. *Elements of natural history and chemistry.* 4 vols. London, 1788. Trans. of item 1a by W. Nicholson, with notes.
1d. ——— ——— 3rd ed. 3 vols. London, 1790. Trans. from the 3rd French ed. (1789).
1e. ——— ——— 4th ed. ? vols. Edinburgh, 1796. Trans. from the 4th French ed. (1791) by R. Heron. RM.
1f. ——— ——— 5th ed. 3 vols. Edinburgh, 1800. Trans. with notes by J. Thomson, presumably from the 5th French ed. (1793).
2. *Mémoires et observations de chimie.* Paris, 1784. RM.
3. *Encyclopédie méthodique.* [Main entry: 3000] *Chymie, Pharmacie et Métallurgie.* Vols 2-6. Paris, 1792-1815. (Vol. 1 was written by Guyton de Morveau--7150/3.)
4. *Philosophie chimique; ou, Vérités fondamentales de la chimie moderne.* Paris, 1792. RM. 2nd ed., 1795. X. 3rd ed., 1806.
4a. ——— *The philosophy of chemistry.* London, 1795. Trans. from the 2nd French ed. RM/L.
5. *Tableaux synoptiques de chimie, pour servir de résumé aux leçons données sur cette science.* Paris, 1800. RM/L.
5a. ——— *Synoptic tables of chemistry.* London, 1801. Trans. by W. Nicholson. RM/L.
6. *Système des connaissances chimiques.* 11 vols in 10. Paris, 1801-02. RM/L.

See also Index.

7263. GÖTTLING, Johann Friedrich August. 1755-1809. Ger. ICB; P I; BLA; GA.

7264. HAHNEMANN, Christian Friedrich Samuel. 1755-1843. Ger. DSB; ICB; P I; BLA & Supp.; GA; GB; GC.

7265. HASSENFRATZ, Jean Henri. 1755-1827. Fr. DSB; ICB; P I; GA.

1. *La sidérotechnie; ou, L'art de traiter les minérais de fer pour en obtenir de la fonte, du fer ou d'acier.* 4 vols. Paris, 1812. RM.

2. *Traité théorique et pratique de l'art de calciner la pierre calcaire, et de fabriquer toutes sortes de mortiers, cimens, bétons, etc.* Paris, 1825. RM.

See also 7150/4.

7266. LABILLARDIÈRE, Jacques Julien Houten de. 1755-1834. Fr. P I (HOUTON); BMNH; GA. Also Botany*.

7267. MARTINOVICS, Ignác. 1755-1795. Lemberg. DSB; P II; GA.

7268. SCHERER, Johann Baptist Andreas von. 1755-1844. Aus. P II; BMNH; BLA.

7269. CHAPTAL, Jean Antoine Claude. 1756-1832. Fr. DSB; ICB; P I; GA; GB.

1. *Elémens de chimie.* 3 vols. Montpellier/Paris, 1790. X.

1a. ——— *Elements of chemistry.* 3 vols. London, 1791. Trans. by W. Nicholson. RM/L. 2nd ed., 1795. 3rd ed., 1800.

2. *Essai sur le perfectionnement des arts chimiques en France.* Paris, 1800. RM.

3. *Chimie appliquée aux arts.* 4 vols. Paris, 1807. RM/L.

3a. ——— *Chemistry applied to arts and manufactures.* 4 vols. London, 1807. Trans. by W. Nicholson. RM.

4. *L'art de la teinture du coton en rouge.* Paris, 1807.

5. *Chimie appliquée à l'agriculture.* 2 vols. Paris, 1823. RM.

7270. HOFMANN (or HOFFMANN), Karl August. 1756-1833. Ger. P I.

7271. MARTIUS, Ernst Wilhelm. 1756-1849. Ger. ICB; P II; BMNH. Also Botany*.

7272. PEART, Edward. ca. 1756-1824. Br. P II; BLA; GD.

7273. REUSS, August Christian von. 1756-1824. Ger. P II; BLA.

7274. LOWITZ, (Johann) Tobias. 1757-1804. Russ. DSB (LOVITS); ICB; P I; GA.

7275. ALYON, Pierre Philippe. 1758-1816. Fr. P I & Supp.; BLA.

1. *Cours élémentaire de chimie théorique et pratique d'après les théories modernes.* Paris, 1787. X. 2nd ed., 2 vols, 1798-99. RM.

7276. DANDOLO, Vincenzo. 1758-1819. It. ICB; P I; GA; GB.

1. *Fondamenti della scienza chimico-fisica ... Esposti in due dizionarj che comprehendo il linguaggio nuovo e vecchio, vecchio e nuovo.* Venice, 1795. RM.

7277. FIEDLER, Karl Wilhelm. b. 1758. Ger. P I.

7278. HEBENSTREIT, Ernst Benjamin Gottlieb. 1758 (or 1753)-1803. Ger. P I; BMNH; BLA; GA. Also Botany. See also 7144/4a.

7279. BUTINI, Pierre. b. 1759. Switz. P I.

7280. COOPER, Thomas. 1759-1839. Br./U.S.A. DSB; ICB; P I; GB; GD; GE.

7281. KELS, Heinrich Wilhelm. 1759-1792. Ger. P I; BLA.

7282. LUBBOCK, Richard. 1759?-1808. Br. DSB; ICB; BMNH. Also Zoology.

7283. NOTARIANNI, Francesco Antonio. 1759-1843. It. P II.

7284. REMLER, Johann Christian Wilhelm. b. 1759. Ger. P II.

7285. SCHEIDT, François de Paule. 1759-1807. ICB. "Pioneer of the Lavoisierian theory in Poland."

7286. CRUICKSHANK, William. d. ca. 1810. Br. ICB.

7287. FUCHS, Georg Friedrich Christian. 1760-1813. Ger. P I; BLA. Also History of Chemistry*.

7288. GADOLIN, Johan. 1760-1852. Fin. DSB; P I; BMNH. Also Mineralogy.
1. *Systema fossilium analysibus chemicis.* Berlin, 1825. RM.

7289. GIRTANNER, Christoph. 1760-1800. Ger. DSB; ICB; P I; BLA & Supp.; GA; GC. Also Science in General*.
1. *Anfangsgründe der antiphlogistischen Chemie.* Berlin, 1792. X. 2nd ed., 1795. RM.

7290. GREN, (Friedrich Albert) Carl. 1760-1798. Ger. DSB; ICB; P I; BLA; GA; GC. Also Physics*.
1. *Grundriss der Chemie, nach den neuesten Entdeckungen entworfen.* 2 parts. Halle, 1796-97. X.
1a. ——— *Principles of modern chemistry.* 2 vols. London, 1800. RM.

7291. HERMBSTÄDT, Sigismund Friedrich. 1760-1833. Ger. DSB Supp.; ICB; P I; BMNH; BLA; GA; GC.
1. *Anleitung zu der Kunst ... Zeuge ... zu farben.* Berlin, 1815. RM.
See also 7175/3.

7292. LINCK, Johann Wilhelm. 1760-1805. Ger. P I; BMNH. Also Natural History.

7293. VOGELMANN, Johann Baptist. 1760-1821. Ger. P II.

7294. BRUGNATELLI, Luigi Valentino. 1761-1818. It. P I & Supp.; BLA; GA.

7295. GIOBERT, Giovanni Antonio. 1761-1834. It. P I; GA.

7296. HALL James (first of the name). 1761-1832. Br. DSB; ICB; P I; BMNH; GB; GD. Also Geology.

7297. PARKES, Samuel. 1761-1825. Br. P II; GA; GD.
1. *A chemical catechism.* London, 1806. 5th ed., 1812. 9th ed., 1819. 10th ed., 1822. 13th ed., rev. by Brayley, 1834.
2. *The rudiments of chemistry.* London, 1810. X.
2a. ——— Another ed. *An elementary treatise of chemistry.* London, 1839. X. Another ed., 1852.
3. *Chemical essays on various subjects, principally relating to the arts and manufactures.* 5 vols. London, 1815. 4th ed., 1841.

7298. PELLETIER, Bertrand. 1761-1797. Fr. DSB; P II; GA. See also 7102/2.

7299. PETROV, Vasily Vladimirovich. 1761-1834. Russ. DSB; ICB; P II (PETROFF, Basilius). Also Physics.

7300. TENNANT, Smithson. 1761-1815. Br. DSB; ICB; P II; GA; GB; GD.

7301. DAGOUMER, Thomas. 1762-ca. 1835. Fr. P I; BLA.

7302. RICHTER, Jeremias Benjamin. 1762-1807. Ger. DSB; ICB; P II; GA; GB; GC.
  1. *Anfangsgründe der Stöchyometrie; oder, Messkunst chymischer Elemente.* 4 vols. Breslau, 1792-94. X. Reprinted, 4 vols in 2, Hildesheim, 1968.

7303. ROSE, Valentin (second of the name). 1762-1807. Ger. P II; GB.

7304. SCHRADER, Johann Christian Karl. 1762-1826. Ger. P II.

7305. ADET, Pierre August. 1763-1834. Fr. DSB; P I; GA. See also 7150/4 and 10413/1a.

7306. HÄNLE, Georg Friedrich. 1763-1824. Ger. P I.

7307. HIGGINS, William. 1763-1825. Br. DSB; ICB; P I; GD.
  1. *A comparative view of the phlogistic and antiphlogistic theories.* London, 1789. X.
  2. *Experiments and observations on the atomic theory, and electrical phenomena.* Dublin, 1814. RM/L.
  3. *The life and work of William Higgins ... including reprints of* [items 1 and 2]. Oxford, 1960. By T.S. Wheeler and J.R. Partington.

7308. JOYCE, Jeremiah. 1763-1816. Br. GA; GD. Also Physics*.
  1. *Dialogues in chemistry ... in which the first principles of that science are fully explained.* New edition. 2 vols. London, 1809.

7309. MacNEVEN, William James. 1763-1841. U.S.A. ICB; P III; BLA. Under MacNEVIN: GD; GE.

7310. PIEPENBRING, Georg Heinrich. 1763-1806. Ger. P II; BLA.

7311. SCHILLER, Johann Michael. 1763-1825. Ger. P II,III.

7312. STIPRIAAN-LUISCIUS, Abraham Gerard van. 1763-1829. Holl. P II; BMNH; BLA (LUISCIUS).

7313. VAUQUELIN, Nicolas Louis. 1763-1829. Fr. DSB; ICB; P II; BLA; GA; GB.

7314. WELTER, Jean Joseph. 1763-1852. Fr. P II & Supp.

7315. BOUILLON-LAGRANGE, Edme Jean Baptiste. 1764-1844. Fr. P I; BLA & Supp.; GA.
  1. *Manuel d'un cours de chimie.* 2 vols. Paris, 1799. X.
  1a. ——— *A manual of a course of chemistry.* 2 vols. London, 1800. (Listed under LAGRANGE)

7316. DIZÉ, Michel Jean Jacques. 1764-1852. Fr. P I.

7317. DUBUC, Guillaume. 1764-1837. Fr. P I.

7318. HILDEBRANDT, (Georg) Friedrich. 1764-1816. Ger. DSB; P I; BLA; GA; GC. See also 7377/1.

7319. KIRCHHOF, Konstantin Sigizmundovich. 1764-1833. Russ. DSB; P I (K., Gottlieb Sigismund).

7320. MITCHILL, Samuel Latham. 1764-1831. U.S.A. ICB; P II; BMNH; BLA; GA; GE. Also Natural History. See also 9213/4.

7321. SILVA TELLES, Vicente Coelho de Seabra. 1764-1804. Also Physics*.
1. *Elementos de chimica*. Coimbra, 1788-90. RM.

7322. ARZT, Philipp Edmund Gottlob. d. 1802. Dorpat. P I.

7323. MUSSIN-PUSCHKIN, Apollo von. d. 1806. Russ. P II.

7324. BONDT, Nicolas. 1765-1796. Holl. P I; BMNH; BLA. Also Botany.

7325. CURAUDEAU, François René. 1765-1813. Fr. P I & Supp.; BLA; GA.

7326. HATCHETT, Charles. 1765-1847. Br. DSB; ICB; P I; GD.

7327. HERMANN, Karl Samuel Leberecht. 1765-1846. Ger. P I.

7328. KIELMEYER, Karl Friedrich von. 1765-1844. Ger. DSB; ICB; P I; BMNH; BLA; GA; GC. Other entries: see 3260.

7329. MONS, Jean Baptiste van. 1765-1842. Belg. ICB; P II; GA; BLA.

7330. SCHURER, Friedrich Ludwig. b. ca. 1765. Strasbourg. P II.

7331. SMITHSON, James (Louis Macie). 1765-1829. Br. DSB; ICB; P II, III; GB; GD.
1. *The scientific writings*. Washington, Smithsonian Institution, 1879. RM.

7332. WURZER, Ferdinand. 1765-1844. Ger. ICB; P II; BLA; GC.

7333. BRANCHI, Giuseppe. b. 1766. It. P I.
1. *Sopra alcune proprieta del fosforo*. Pisa, 1813. RM.

7334. DALTON, John. 1766-1844. Br. DSB; ICB; P I; GA; GB; GD. Also Physics and Meteorology*. See also Index.
1. *A new system of chemical philosophy*. Part 1, 1808. Part 2, 1810. Vol. 2, Part 1, 1827. [No more publ. in this ed.] 2 vols in 3. Manchester/London, 1808-27. RM/L. Reprinted, London, 1960?
1a. ——— 2nd ed. Part 1. [No more publ.] London, 1842.
2. *On the phosphates and arseniates, microcosmic salt, acids, bases, and water. With a new and easy method of analysing sugar*. Manchester, 1840-42. RM/L.

7335. GARNETT, Thomas. 1766-1802. Br. DSB; P I; BMNH; BLA; GA; GD. Also Physiology*.

7336. HISINGER, Wilhelm. 1766-1852. Swed. DSB; ICB; P I; BMNH; GA. Also Geology*.

7337. HOPE, Thomas Charles. 1766-1844. Br. DSB; P I; GD.

7338. IZARN, Joseph. 1766-after 1834. Fr. P I. Also Astronomy* and Physics*.
1. *Explication du nouveau langage des chimistes*. Paris, 1803.

7339. JACQUIN, Joseph Franz von. 1766-1839. Aus. P I; BMNH; BLA. Also Botany.
1. *Lehrbuch der allgemeinen und medicinischen Chymie*. 2 vols. Vienna, 1793. X. Another ed., 1798. RM.

7340. RAYMOND, Jean Michel. 1766-1837. Fr. P II & Supp.; GA.

7341. WOLFF, Friedrich Benjamin. 1766-1845. Ger. P II; GC.

7342. WOLLASTON, William Hyde. 1766-1828. Br. DSB; ICB; P II; BLA; GA; GB; GD. Also Physics, Crystallography, and Physiology.

7343. EKEBERG, Anders Gustaf. 1767-1813. Swed. DSB; P I.

7344. SAUSSURE, Nicolas Théodore de. 1767-1845. Switz. DSB; P II; BMNH; GA; GB. Also Plant Physiology.
1. *Recherches chimiques sur la végétation.* Paris, 1804. Reprinted, Paris, 1957.
1a. ——— German trans. 2 vols. Leipzig, 1890. (OKEW)

7345. SÉGUIN, Armand. 1767-1835. Fr. DSB; ICB; P II; Mort.; BLA. Also Physiology.

7346. SIMON, Paul Louis. 1767-1815. Ger. P II.

7347. CARLISLE, Anthony. 1768-1840. Br. DSB; ICB; P I & Supp.; BLA; GD. Also Anatomy.

7348. CELLIER-BLUMENTHAL, Jean Baptiste. 1768-1840. ICB.

7349. SCHREGER, Christian Heinrich Theodor. 1768-1833. Ger. P II; BLA.

7350. SNIADECKI, Jedrzej (or Andreas). 1768-1838. Wilno. ICB; P II; BLA.

7351. ACCUM, Friedrich Christian. 1769-1838. Ger./Br. DSB; ICB; P I; BMNH; BLA; GA; GC; GD. Also Crystallography*.
1. *A system of theoretical and practical chemistry.* 2 vols. London, 1803. RM/L. 2nd ed., enl., 1807.
2. *A practical essay on the analysis of minerals.* London, 1804. RM/L.
3. *A practical treatise on gas-light.* London, 1815. RM.
4. *A practical essay on chemical re-agents or tests.* London, 1816.
5. *Chemical amusement, comprising a series of curious and instructive experiments.* London, 1817. RM. 3rd ed., 1818. 4th ed., 1819.
6. *Description of the process of manufacturing coal gas.* London, 1819. RM.
7. *A treatise on the adulterations of food ... and methods of detecting them.* London, 1820. RM.

7352. CADET DE GASSICOURT, Charles Louis. 1769-1821. Fr. DSB; ICB; P I; BLA & Supp.; GA.
1. *Dictionnaire de chimie.* 4 vols. Paris, 1803. RM/L.

7353. HUMBOLDT, Alexander von. 1769-1859. Ger./Fr. DSB; ICB; P I & Supp.; P III; BMNH; G. Other entries: see 3268. For a chemical investigation see 7415/1.

7354. MARCET, Jane, *Mrs.* 1769-1858. Br. ICB; P II & Supp. (M., Mary); BMNH; GD. Also Physics* and Botany*.
1. *Conversations on chemistry.* 2 vols. London, 1806. X. Another ed., 1809. 5th ed., 1817. 6th ed., 1819. 13th ed., 1837. 16th ed., 1853.

7355. PESCHIER, Jacques. 1769-1832. Switz. P II.

7356. POZZI, Giovanni. 1769-1838. It. P II; BLA.

7357. ALLEN, William. 1770-1843. Br. ICB; P I; GD.

7358. BOURGUET, David Ludwig. 1770-ca. 1805. Ger. P I.

7359. BUCHOLZ, Christian Friedrich. 1770-1818. Ger. DSB; P I; BLA & Supp.; GC.

7360. CLOUD, Joseph. ca. 1770-1845. U.S.A. P I.

7361. LAUGIER, André. 1770-1832. Fr. P I; GA.

7362. MARCET, Alexander John Gaspard. 1770-1822. Br. ICB; P II; BLA; GA; GD.

7363. SCHAUB, Johann. 1770-1819. Ger. P II.

7364. TROMMSDORFF, Johann Bartholomäus. 1770-1837. Ger. DSB; ICB; P II; BMNH; GC.
 1. *Geschichte der Galvanismus, oder der galvanischen Elektricität, vorzüglich in chemischer Hinsicht.* Erfurt, 1803. RM.

7365. WOODHOUSE, James. 1770-1809. U.S.A. ICB; P II; GE.

7366. EIMBKE, Georg. 1771-1843. Ger. P I.

7367. GRUNER, J.L.W. 1771-1849. Ger. P I.

7368. HOPFENGÄRTNER, Philipp Friedrich. 1771-1807. Ger. P I; BLA.

7369. MACLEAN, John. 1771-1814. U.S.A. DSB; ICB; P II; GE.

7370. PACCHIANI, Francesco Giuseppe Maria. 1771-1835. It. P II.

7371. SCHERER, Alexander Nicolaus von. 1771-1824. Russ. P II; GC.
 1. *Grundriss der Chemie.* Tübingen, 1800.

7372. SUERSEN, Johann Friedrich Hermann. b. 1771. Ger. P II.

7373. LAMPADIUS, Wilhelm August. 1772-1842. Ger. ICB; P I; BMNH; GA; GC.
 1. *Handbuch der allgemeinen Hüttenkunde, in theoretischer und praktischer Hinsicht.* 2 vols. Göttingen, 1801-10. RM.
 2. *Systematischer Grundriss der Atmosphärologie.* Freiburg, 1806. RM.
 3. *Die neuern Fortschritte im Gebiete der gesammten Hüttenkunde.* Freiberg, 1839. RM.

7374. MOJON, Giuseppe. 1772-1837. It. P II; BLA.

7375. WERKHOVEN, Pieter van. b. 1772. Holl. P II.

7376. AIKIN, Arthur. 1773-1854. Br. ICB; P I; BMNH; GB; GD. Also Mineralogy*.
 1. *A dictionary of chemistry and mineralogy.* London, 1807. With C.R. Aikin. RM/L.
 2. *An account of the most important recent discoveries and improvements in chemistry and mineralogy. Being an appendix to their "Dictionary of chemistry and mineralogy."* London, 1814. With C.R. Aikin. RM/L.

7377. BOECKMANN, Carl Wilhelm. 1773-1821. Ger. P I.
 1. *Versuche über das Verhalten des Phosphorus in verschiedenen Gasarten.* Erlangen, 1800. Ed. by F. Hildebrandt. RM.

7378. COLLET-DESCOTILS, Hippolyte Victor. 1773-1815. Fr. DSB; P I.

7379. DUPONCHEL, Dénis Joseph. b. 1773. Fr. P I.

7380. MORICHINI, Domenico Lino. 1773-1836. It. DSB; P II.

7381. PFAFF, Christian Heinrich. 1773-1852. Ger. P II; BLA; GC. Also Physics.

7382. THOMSON, Thomas. 1773-1852. Br. DSB; ICB; P II; BMNH; GB; GD. Also Physics*, Mineralogy*, and History of Science*.
   1. *A system of chemistry.* 4 vols. Edinburgh, 1802. 2nd ed., 1804. 6th ed., 1820. 7th ed.: *A system of chemistry of inorganic bodies*, 2 vols, London, 1831.
   2. *The elements of chemistry.* Edinburgh, 1810.
   3. *An attempt to establish the first principles of chemistry by experiment.* 2 vols. London, 1825. RM/L.
   4. *Chemistry of organic bodies. Vegetables.* London, 1838. X.
   5. *Chemistry of animal bodies.* Edinburgh, 1843.

7383. CHENEVIX, Richard. 1774-1830. Br. DSB; ICB; P I; GA; GD.

7384. FUCHS, Johann Nepomuck von. 1774-1856. Ger. DSB; ICB; P I; BMNH; GB; GC. Also Mineralogy*#.

7385. GRISCOM, John. 1774-1852. U.S.A. ICB; GE.

7386. HENRY, William. 1774-1836. Br. DSB; P I & Supp.; GA; GB; GD.
   1. *An epitome of chemistry.* London, 1801. X.
   1a. ——— 6th ed. *The elements of experimental chemistry.* 2 vols, ib., 1810. X.
   1b. ——— ——— 3rd American ed. From the 6th English ed. Ed. with notes by B. Silliman. Boston, 1814. RM.
   1c. ——— 7th ed., 2 vols, London, 1815.
   1d. ——— 8th ed., 2 vols, ib., 1818.
   1e. ——— ——— 2nd [sic] American ed. From the 8th English ed. Ed with additions by R. Hare. Philadelphia, 1823. RM.
   1f. ——— 10th ed. 2 vols, London, 1826.
   1g. ——— 11th ed., 2 vols, ib., 1829.

7387. JUCH, Karl Wilhelm. 1774-1821. Ger. P I; BMNH; BLA.

7388. MICHELOTTI, Vittorio. 1774-1842. It. P II.

7389. SERULLAS, Georges Simon. 1774-1832. Fr. DSB; P II; GA.

7390. STADION, Friedrich Carl Joseph Dismas von. 1774-1821. Ger. P III.

7391. FIGUIER, ... d. 1817. Fr. P I.

7392. GEHLEN, Adolf Ferdinand. 1775-1815. Ger. DSB Supp.; P I; GA; GC.

7393. KINDT, Heinrich Hugo. 1775-1837. Ger. P I.

7394. PEPYS, William Hasledine. 1775-1856. Br. P II; GD.

7395. SALZER, Karl Friedrich. b. 1775. Ger. P II.

7396. AVOGADRO, Amedeo. 1776-1856. It. DSB; ICB; P III; GB. Also Physics*.
   1. *Opere scelte.* Turin, 1911. RM/L.

7397. BELLANI, Angelo. 1776-1852. It. DSB; P I & Supp. Also Physics.

7398. GRINDEL, David Hieronymus. 1776-1836. Dorpat/Riga. P I; BLA & Supp.

7399. PLANCHE, Louis Antoine. 1776-1840. Fr. P II; GA.

7400. RITTER, Johann Wilhelm. 1776-1810. Ger. DSB; ICB; P II; BLA; GA; GC. Also Physics*# and Physiology.

7401. STROMEYER, Friedrich. 1776-1835. Ger. ICB; P II; BLA.

7402. VEST, Lorenz Chrysanth von. 1776-1840. Aus. P II; BLA; GC.

7403. BOULLAY, Pierre François Guillaume. 1777-after 1839. Fr. ICB; P I.

7404. CAGNIARD DE LA TOUR, Charles. 1777-1859. Fr. DSB; ICB; P I, III; Mort.; GA (LATOUR); GB. Also Physics.

7405. CHILDREN, John George. 1777-1852. Br. P I & Supp.; BMNH; GD. Also Zoology.
1. *An essay on chemical analysis.* London, 1819.
See also 7424/6c.

7406. COURTOIS, Bernard. 1777-1838. Fr. DSB; ICB; P I.

7407. DARCET, Jean Pierre Joseph. 1777-1844. Fr. P I; BLA (ARCET); GA.

7408. DESORMES, Charles Bernard. 1777-1862. Fr. DSB; ICB; P I & Supp.

7409. DU MÊNIL, August Peter Julius. 1777-1852. Ger. P II (MÊNIL); BMNH.

7410. SCHUSTER, Johann Nepomuk. 1777-1838. Hung. P II; BLA.

7411. SEMENTINI, Luigi. 1777 (or 1775)-1847. It. P II; BLA.

7412. THENARD, Louis Jacques. 1777-1857. Fr. DSB; ICB; P II; BLA; GA; GB.
1. *Traité de chimie élémentaire, théorique et pratique.* 4 vols. Paris, 1813-16. RM. 2nd ed., 1817-18. 6th ed., 5 vols, 1834-36.
See also 7415/2 and 7540/1.

7413. CLÉMENT, Nicholas. 1778/9-1841. Fr. DSB; ICB. Under CLÉMONT-DÉSORMES: P I; GA.

7414. DAVY, Humphry. 1778-1829. Br. DSB; ICB; P I; GA; GB; GD. Also Autobiography*. See also Index.
1. *Researches chemical and philosophical, chiefly concerning nitrous oxide, or dephlogisticated nitrous air, and its respiration.* London, 1800. X. Reprinted, ib., 1972.
2. *A discourse, introductory to a course of lectures on chemistry.* London, 1802. RM.
3. *A syllabus of a course of lectures on chemistry.* London, 1802.
4. *Outlines of a course of lectures on chemical philosophy.* London, 1804. RM.
5. [Trans. of the Bakerian lectures, 1806 and 1807, publ. as periodical articles] *Electrochemische Untersuchungen.* Leipzig, 1893. (OKEW. 45)
6. *The decomposition of the fixed alkalies and alkaline earths.* [Periodical articles, 1807-08] Edinburgh, 1901 and re-issues. (Alembic Club reprints. 6)
7. *The elementary nature of chlorine.* [Periodical articles, 1809-18] Edinburgh, 1902 and re-issues. (Alembic Club reprints. 9)

8. *A lecture on the plan which it is proposed to adopt for improving the Royal Institution and rendering it permanent.* London, 1810. RM.
9. *Elements of chemical philosophy.* Vol. 1, Part 1. [No more publ.] London, 1812. RM/L.
10. *Elements of agricultural chemistry.* London, 1813. RM/L. 6th ed., 1839. Another ed., 1844.
11. *On the safety lamp for coal miners, with some researches on flame.* London, 1818. RM.
12. *On the safety lamp for preventing explosions in mines, houses lighted by gas, ... with some researches on flame.* 1825. X.

12a. ——— German trans. Leipzig, 1937. (OKEW. 242)
13. *Six discourses delivered before the Royal Society ... preceded by an address to the Society on the progress and prospects of science.* London, 1827. RM/L.
14. *Electro-chemistry, including the Bakerian lectures, and memoirs ... 1806, 1807, and 1808.* London, 1848. RM.
15. *The collected works.* 9 vols. London, 1839-40. Ed. by his brother, John Davy. RM/L. Reprinted, New York, 1972.
16. *Fragmentary remains, literary and scientific.* London, 1858. Ed. by John Davy with a biography and selections from his correspondence. RM/L.

7415. GAY-LUSSAC, Joseph Louis. 1778-1850. Fr. DSB; ICB; P I; GA; GB. Also Physics.

1. [Trans. of periodical articles by A. von Humboldt and J.L. Gay-Lussac, 1805-08] *Das Volumgesetz gasförmiger Verbindungen.* Leipzig, 1893. (OKEW. 42)
2. *Recherches physico-chimiques, faites sur la pile, sur la préparation chimique et les propriétés du potassium et du sodium.* 2 vols. Paris, 1811. With L.J. Thenard. RM/L.
3. [Trans. of periodical article, 1814] *Untersuchungen über das Iod.* Leipzig, 1889. (OKEW. 4)
4. *Recherches sur l'acide prussique.* Paris, 1815.
5. *Cours de chimie, comprenant l'histoire des sels, la chimie végétale et animale.* 2 vols. Paris, 1828.
6. *Instruction sur l'essai des matières d'argent par la voie humide.* Paris, 1832. RM/L.

7416. MEISSNER, Paul Traugott. 1778-1864. Aus. P II,III; GC.

1. *Neues System der Chemie.* 3 vols. Vienna, 1835-41.

7417. MURRAY, John (first of the name). 1778-1820. Br. P II & Supp.; BMNH; BLA; GA; GB; GD. Also Geology*.

1. *A system of chemistry.* 4 vols. Edinburgh, 1806-07. RM. 3rd ed., 1812.

See also 7144/3.

7418. PHILLIPS, Richard. 1778-1851. Br. ICB; P II; GA; GD.

7419. PRECHTL, Johann Joseph von. 1778-1854. Aus. P II. Also Science in General* and Physics.

1. *Grundlehren der Chemie in technischer Beziehung.* Vienna, 1817. RM.

7420. REUSS, Ferdinand Friedrich von. 1778-1852. Russ. P II; BLA; GC.

7421. URE, Andrew. 1778-1857. Br. DSB; ICB; P II; BMNH; GD. Also Geology*.
1. *A dictionary of chemistry*. 2 vols. London, 1821. RM/L. 4th ed.: *A dictionary of chemistry and mineralogy, with their applications*, 1835.
2. *A dictionary of arts, manufactures and mines*. London, 1839. An expansion of item 1. RM. 4th ed., 1853.

7422. VOGEL, Heinrich August. 1778-1867. Ger./Fr. P II,III.

7423. WENDT, Johann Christian Wilhelm. 1778-1838. Den. P II; BMNH; BLA.

7424. BERZELIUS, Jöns Jacob. 1779-1848. Swed. DSB; ICB; P I; BLA & Supp.; GA; GB. Also Autobiography*.
1. *Föreläsningar i djurkemien*. Stockholm, 1806. X.
1a. ——— *A view of the progress and present state of animal chemistry*. London, 1813. Trans. by G. Brunnmark. RM.
2. *Lärbok i kemien*. 3 vols. Stockholm, 1808-18. X.
2a. ——— 2nd ed., 6 vols, 1817-30.
2b. ——— ——— *Lehrbuch der Chemie*. 4 vols in 6. Reutlingen, 1821-32. Trans. from the 2nd Swedish ed. by K.A. Blöde.
2c. ——— ——— *Lehrbuch der Chemie*. 4 vols in 8. Trans. by F. Wöhler. RM/L.
2d. ——— ——— *Traité de chimie*. 8 vols. Paris, 1829-33. Trans. by A.J.L. Jourdan (Vol. 1) and M. Esslinger (Vols 2-8). RM/L.
2e. ——— 3rd ed. *Lehrbuch der Chemie*. 10 vols. Dresden/Leipzig, 1833-41. Trans. from the Swedish MS by F. Wöhler.
2f. ——— 4th ed. *Lehrbuch der Chemie*. 10 vols. Dresden/Leipzig, 1835-41. Trans. by F. Wöhler. X.
2g. ——— ——— *Traité de chimie*. 8 vols. Brussels, 1838-46. Trans. by B. Valerius.
2h. ——— 5th ed. *Lehrbuch der Chemie*. 5 vols. Dresden/Leipzig, 1843-48. Written in German by Berzelius. X. Reprinted, 5 vols, Leipzig, 1856.
2i. ——— ——— *Traité de chimie minérale, végétale et animale*. 2. *éd française*. 6 vols. Paris, 1845-60. Trans. by Esslinger and Hoefer. RM/L.
3. [Trans. of periodical articles, 1811 and 1812] *Versuch, die bestimmten und einfachen Verhältnisse aufzufinden nach welchen die Bestandtheile der inorganischen Natur mit einander verbunden sind*. Leipzig, 1892. (OKEW. 35)
4. *Försök att genom användandet af den elektro-kemiska theorien....* Stockholm, 1814. X.
4a. ——— *An attempt to establish a pure scientific system of mineralogy by the application of the electro-chemical theory and the chemical proportions*. London, 1814. Trans. by J. Black. RM.
4b. ——— *Nouveau système de minéralogie*. Paris, 1819. Trans. under Berzelius' supervision and publ. by him. RM/L.
4c. ——— *Neues chemisches Mineral-system*. Nuremberg 1847.

Trans. by C.F. Rammelsberg. 2nd ed. of the German trans. (the 1st was in a periodical, 1814-15).

5. [Trans. of Vol. III (1818) of item 2] *Essai sur la théorie des proportions chimiques et sur l'influence chimique de l'électricité*. Paris, 1819. Trans. from the Swedish under Berzelius' supervision and publ. by him. RM/L. Reprinted, New York, 1972.

5a. ——— *Versuch über die Theorie der chemischen Proportionen und über die chemischen Wirkungen der Elektricität*. Dresden, 1820. Trans. from the Swedish and French by K.A. Blöde. X. Reprinted, Hildesheim, 1973.

5b. ——— 2nd ed. *Théorie des proportions chimiques, et table synoptique des poids atomiques des corps simples, et de leurs combinaisons les plus importantes*. Paris, 1835. RM.

6. *Om blåsrörets användande i kemien och mineralogien*. Stockholm, 1820. X.

6a. ——— *Von der Anwendung des Löthrohrs in der Chemie und Mineralogie*. Nuremberg, 1821. Trans. by H. Rose. RM.

6b. ——— *De l'emploi du chalumeau dans les analyses chimiques et les déterminations minéralogiques*. Paris, 1821. Trans. by A.J. Fresnel. X.

6c. ——— ——— *The use of the blowpipe in chemical analysis and in the examination of minerals*. London, 1822. Trans. from the French by J.C. Children. RM.

7. [Trans.?] *De l'analyse des corps inorganiques*. Paris, 1827.

7a. ——— *The analysis of inorganic bodies*. London, 1833. Trans. from the French by G.O. Rees. RM.

Correspondence

8. *Berzelius und Liebig, ihre Briefe von 1831-1845*. Munich/Leipzig, 1893. Ed. by J. Carrière. RM. Reprinted, Wiesbaden, 1967.

9. *Zwanzig Briefe gewechselt zwischen J.J. Berzelius und C. Schönbein, 1836-1847*. Basel, 1898. Ed. by G.W.A. Kahlbaum. X.

9a. ——— *The letters of J.J. Berzelius and C.F. Schönbein, 1836-1847*. London, 1900. Trans. by F.V. Darbyshire and N.V. Sidgwick. RM/L.

10. *Aus Jac. Berzelius und Gustav Magnus Briefwechsel in den Jahren 1828-1847*. Brunswick, 1900. Ed. by E. Hjelt.

11. *Briefwechsel zwischen J. Berzelius und F. Wöhler*. 2 vols. Leipzig, 1901. Ed. by O. Wallach. Reprinted, Wiesbaden, 1966.

12. *Bref*. 6 vols. Uppsala, 1912-32. Ed. by H.G. Söderbaum.

12a. ——— Supplement. 3 vols in 1. Ib., 1935-61. Ed. by A. Holmberg.

13. *J.J. Berzelius, G.C.F. Löwenhielm: En brefväxling, 1818-1847*. Lidingö, 1968. Ed. by S. Klemming.

7425. SCHWEIGGER, Johann Salomon Christoph. 1779-1857. Ger. DSB; ICB; P II; GC. Also Physics.

7426. SILLIMAN, Benjamin (first of the name). 1779-1864. U.S.A.

DSB; ICB; P II,III; BMNH; GB; GE. Also Geology*. See also 15510/1.
1. *Elements of chemistry*. 2 vols. New Haven, 1830-31. RM/L.
See also 7386/1b.

7427. ZENNECK, Ludwig Heinrich. 1779-1859. Ger. P II.

7428. ELLIS, Daniel. fl. 1807-1828. Br. BLA.
1. *An inquiry into the changes induced on atmospheric air by the germination of seeds, the vegetation of plants and the respiration of animals*. Edinburgh, 1807. RM/L.
1a. ——— *Further inquiries into the changes induced*.... 1811.

7429. ROLOFF, Johann Christoph Heinrich. d. 1825. Ger. P II.

7430. PLISSON, Auguste Arthur. d. 1832. Fr. P II.

7431. BRACONNOT, Henri. 1780-1855. Fr. DSB; P I,III; BLA & Supp.; GA.

7432. DEROSNE, (Louis) Charles. 1780-1846. Fr. DSB; P I & Supp.; GA.

7433. DÖBEREINER, Johann Wolfgang. 1780-1849. Ger. DSB; ICB; P I; GA; GB; GC.

7434. KAPP, Georg Christian Friedrich. 1780-1806. Ger. P I; BLA.

7435. NASSE, Johann Friedrich Wilhelm. b. 1780. Russ. P II.

7436. ROBIQUET, Pierre Jean. 1780-1840. Fr. DSB; P II; GA.

7437. ZIMMERMANN, Wilhelm Ludwig. 1780-1825. Ger. P II.

7438. BRANDENBURG, Friedrich. 1781-1837. Russ. P I.

7439. GIESE, Johann Emmanuel Ferdinand. 1781-1821. Russ. P I.

7440. HARE, Robert. 1781-1858. U.S.A. DSB; ICB; P I & Supp.; GA; GE. Also Physics*.
1. *A compendium of the course of chemical instruction*.... Philadelphia, 1827. RM.
See also 7386/1e.

7441. MEINECKE, Johann Ludwig Georg. 1781-1823. Ger. P II.

7442. PERETTI, Pietro. b. 1781. It. P II.

7443. VOGEL, Fr. Chr. Max. ca. 1781-1813. Ger. P II.

7444. BERTHIER, Pierre. 1782-1861. Fr. DSB; P I & Supp.

7445. FICINUS, Heinrich David August. 1782-1857. Ger. P I; BMNH; BLA. Also Botany.

7446. FISCHER, Nicolaus Wolfgang. 1782-1850. Ger. DSB; P I.

7447. JOHN, Johann Friedrich. 1782-1847. Ger. P I.

7448. KARSTEN, Karl Johann Bernhard. 1782-1853. Ger. DSB; P I; GA; GB; GC. Also Geology*.
1. *Über Contact-Electricität*. Berlin, 1836. RM.

7449. PELLETAN, Pierre. 1782-1845. Fr. P II; BLA; GA.
1. *Dictionnaire de chimie générale et médicale*. Paris, 1824.

7450. SCHULZE-MONTANUS, Karl August. 1782-1823. Ger. P II. Also Various Fields.

1. *Die chemischen Reagentien und deren Anwendung.* Berlin, 1814. X. 3rd ed., 1820.

7451. BERTHOLLET, Amédée B. ca. 1783-1811. Fr. P I.

7452. BUCHNER, Johann Andreas. 1783-1852. Ger. P I & Supp.; BLA; GA; GC.

7453. GORHAM, John. 1783-1829. U.S.A. P III; GE.
1. *The elements of chemical science.* 2 vols. Boston, 1819-20. RM.

7454. GRANVILLE, Augustus Bozzi. 1783-1872. Br. ICB; P I; BLA; GA; GD. Also Science in General*.

7455. KASTNER, Karl Wilhelm Gottlob. 1783-1857. Ger. P I; GA; GC.

7456. PAGENSTECHER, Johann Samuel Friedrich. 1783-1856. Switz. P II.

7457. PETTENKOFER, Franz Xaver. 1783-1850. Ger. P II.

7458. PORRETT, Robert. 1783-1868. Br. P II,III; GD.

7459. SERTÜRNER, Friedrich Wilhelm Adam Ferdinand. 1783-1841. Ger. DSB; ICB; P II; BLA & Supp.

7460. WUTTIG, Johann Friedrich Christian. 1783-1850. Ger. P II.

7461. COLIN, Jean Jacques. 1784-1865. Fr. P III.

7462. MELANDRI-CONTESSI, Girolamo. 1784-1833. It. P II.

7463. SIGWART, Georg Karl Ludwig. 1784-1864. Ger. ICB; P II,III; BLA; GC.

7464. DAVY, Edmund. 1785-1857. Br. P I & Supp.; GD.

7465. DULONG, Pierre Louis. 1785-1838. Fr. DSB; ICB; P I; GA; GB. Also Physics.

7466. EWELL, Thomas. 1785-1826. U.S.A. GE. Also Physiology.
1. *Plain discourses on the laws or properties of matter, containing the elements or principles of modern chemistry.* New York, 1806.

7467. GEIGER, Philipp Lorenz. 1785-1836. Ger. P I; BLA.

7468. GROTTHUSS, Theodor (Christian Johann Dietrich) von. 1785-1822. Mitau. DSB; ICB; P I (GROTHUSS). Also Physics*.

7469. PARIS, John Ayrton. 1785-1856. Br. P II; BLA; GA; GD. Also Science in General* and Biography*.
1. *The elements of medical chemistry.* London, 1825.

7470. PROUT, William. 1785-1850. Br. DSB; ICB; P II,IV & Supp.; Mort.; BLA; GB; GD.
1. *Chemistry, meteorology, and the function of digestion, considered with reference to natural theology.* London, 1834. (Bridgewater Treatises. 8) RM/L. 2nd ed., 1838. 3rd ed., 1845. 4th ed., 1855.

7471. SCHNAUBERT, Ludwig. b. ca. 1785. Ger./Russ. P II.

7472. STRATINGH, Sibrandus. 1785-1841. Holl. P II; BLA.

7473. CHEVREUL, Michel Eugène. 1786-1889. Fr. DSB; ICB; P I,III,IV; GA; GB. Also Physics* and History of Chemistry*.

1. *Recherches chimiques sur les corps gras d'origine animal.* Paris, 1823. X. Reprinted, Paris, 1889. RM/L.
2. *Considérations générales sur l'analyse organique et sur ses applications.* Paris, 1824. RM/L.
3. *Recherches sur la teinture.* Paris, 1847.

7474. LOW, David. 1786-1859. Br. GD. Also Zoology*.
1. *An inquiry into the nature of the simple bodies of chemistry.* London, 1844. X. 3rd ed., Edinburgh, 1856.

7475. MITCHELL, James. ca. 1786-1844. Br. P II; BMNH; GD. Also Physics* and Geology*.
1. *A dictionary of chemistry, mineralogy and geology.* London, 1823. RM.

7476. MONHEIM, Johann Peter Joseph. 1786-1855. Ger. P II.

7477. RUHLAND, Reinhold Ludwig. 1786-1827. Ger. P II. Also Physics.

7478. SCHOLZ, Benjamin. 1786-1833. Aus. ICB; P II.
1. *Anfangsgründe der Physik als Vorbereitung zum Studium der Chemie.* Vienna, 1816. RM.
2. *Lehrbuch der Chemie.* Vienna, 1824-25. RM.

7479. ENGESTRÖM, Jöns Albin. 1787-1846. Swed. P I.

7480. FALKNER, Johann Ludwig. 1787-ca. 1831. Switz. P I.

7481. ITTNER, Franz von. 1787-1823. Ger. P I; BLA.

7482. NEES VON ESENBECK, Theodor Friedrich Ludwig. 1787-1837. Ger. P II; BMNH; BLA; GA; GC. Also Botany*.

7483. ORFILA, Matthieu Joseph Bonaventura. 1787-1853. Fr. ICB; P II; BLA & Supp.; GA; GB.
1. *Elémens de chimie médicale.* 2 vols. Paris, 1817. X. 6th ed., 1835.

7484. SEFSTRÖM, Nils Gabriel. 1787-1845. Swed. ICB; P II; BLA.

7485. BECQUEREL, Antoine César. 1788-1878. Fr. DSB; P I,III; GA; GB. Also Physics* and Earth Sciences*.
1. *Eléments d'électro-chimie appliquée aux sciences naturelles et aux arts.* Paris, 1843. RM.

1a. ——— *Elemente der Electro-Chemie in ihrer Anwendung auf die Naturwissenschaften und die Künste.* Erfurt, 1845.

7486. BRANDE, William Thomas. 1788-1866. Br. DSB; ICB; P I,III; BMNH; GB; GD. Also Science in General*, Geology*, and History of Chemistry*.
1. *A manual of chemistry.* London, 1819. RM. 2nd ed., 3 vols, 1821. 4th ed., 3 vols, 1836. 6th ed., 2 vols, 1848.
2. *Tables in illustration of the theory of definite proportionals, shewing the prime equivalent numbers of the elementary substances and the volume and weights in which they combine.* London, 1828. RM/L.
3. *A table of chemical equivalents and notation.* London, 1833. RM.
4. *Chemistry.* 1863. With A.S. Taylor.

7487. DULK, Friedrich Philipp. 1788-1851. Ger. P I; GA.

7488. GMELIN, Leopold. 1788-1853. Ger. DSB; ICB; P I; BMNH; BLA & Supp.; GA; GC.
1. *Handbuch der theoretischen Chemie.* 2 vols in 3. Frankfurt, 1817-19. X.
1a. ——— 4th ed. *Handbuch der Chemie.* 5 vols. Heidelberg, 1843-52. After his death this edition was expanded by K. Kraut et al. to 13 vols by 1870.
1b. ——— ——— *Handbook of chemistry.* 19 vols. London, 1848-72. Trans. by H. Watts.
1c. ——— 5th ed. *Handbuch der anorganische Chemie.* 3 vols in 5. Heidelberg, 1871-86. Ed. by K. Kraut.

7489. PELLETIER, Pierre Joseph. 1788-1842. Fr. DSB; ICB; P II; GA.

7490. RAAB, Christian Wilhelm Julius. 1788-1835. Ger. P II.

7491. REICHENBACH, Karl Ludwig. 1788-1869. Aus. DSB; ICB; P II,III; GA; GC. Also Physiology*.

7492. ROTHOFF, Emmanuel. 1788-1831. Swed. P II.

7493. BÉRARD, Jacques Etienne. 1789-1869. Fr. DSB; P I,III.

7494. SIMON, Johann Eduard. 1789-1856. Ger. P II & Supp.

7495. ZEISE, William Christopher. 1789-1847. Den. DSB; P II.

7496. BÜCHNER, Johann August Wilhelm. 1790-1849. Ger. P I.

7497. COOPER, John Thomas. 1790-1854. Br. P I.

7498. DANIELL, John Frederic. 1790-1845. Br. DSB; ICB; P I; GB; GD. Also Physics and Meteorology*.
1. *An introduction to the study of chemical philosophy.* London, 1839. RM. 2nd ed., 1843.

7499. DAVY, John. 1790-1868. Br. DSB; ICB; P I,III; BMNH; BLA & Supp.; GD. Also Natural History, Physiology*, and Biography*. See also Index.

7500. DONOVAN, Michael. 1790-1876. Br. P I (DONAVAN).
1. *A treatise on chemistry.* 3rd ed. London, 1832. No earlier edition is known.

7501. DRIESSEN, Jan Constantin. 1790-1824. Holl. P I & Supp.

7502. GUIBOURT, Nicolas Jean Baptiste Gaston. 1790-1867. Fr. P I, III; BLA; GA.

7503. JULIA DE FONTENELLE, Jean Sébastien Eugène. 1790-1842. Fr. P I; BLA & Supp.; GA.

7504. BAUP, Samuel. 1791-1862. Switz. P I,III.

7504A BIZIO, Bartolommeo. 1791-1862. It. P I,III; BLA.

7505. BONSDORFF, Pehr Adolph von. 1791-1839. Fin. P I.

7506. CANOBBIO, Giambattista. b. 1791. It. P I.

7507. FARADAY, Michael. 1791-1867. Br. DSB; ICB; P I,III; GA; GB; GD. Also Physics*#. See also Index.
1. *The liquefaction of gases.* [Periodical articles, 1823-45] Edinburgh, 1893 and re-issues. (Alembic Club reprints. 12)

2. *Chemical manipulation. Being instructions to students....* London, 1827. RM. 2nd ed., 1830. 3rd ed., rev., 1842. Reprinted, New York, 1974.
3. *The subject matter of a course of six lectures on the non-metallic elements.* London, 1853. Ed. by J. Scoffern. RM/L.
4. *Experimental researches in chemistry and physics.* London, 1859. A collection of periodical articles. RM/L. Reprinted, Brussels, 1969.
5. *A course of six lectures on the chemical history of a candle.* London/New York, 1861. Ed. by W. Crookes. RM/L. Other eds: 1865, 1873, 1886.
6. *The letters of Faraday and Schönbein, 1836-1862.* Basel, 1899. Ed. with notes by G.W.A. Kahlbaum and F.V. Darbishire. RM.
7. *Faraday's diary.* See 6227/4.

7508. PRESL, Jan Swatopluk. 1791-1849. Prague. P II; BMNH. Also Botany.

7509. WEHRLE, Aloys. 1791-1835. Aus./Hung. P II; GC.

7510. WITTSTOCK, Christian. 1791-1867. Ger. P III.

7511. ARFVEDSON, Johann August. 1792-1841. Swed. DSB Supp.; ICB; P I; BMNH. Also Mineralogy.

7512. BACHE, Franklin. 1792-1864. U.S.A. P III; GE.

7513. BISCHOF, (Carl) Gustav (Christoph). 1792-1870. Ger. DSB; ICB; P I,III; BMNH; GA; GC. Also Geology*.

7514. FYFE, Andrew (second of the name). 1792-1861. Br. P I & Supp.; BLA; GD.

7515. GAULTIER DE CLAUBRY, Henri François. 1792-1878. Fr. DSB; P I, III; BLA; GA.

7516. GMELIN, Christian Gottlieb. 1792-1860. Ger. P I & Supp.; BLA & Supp.; GC.
1. *Einleitung in die Chemie.* Tübingen, 1833. RM.

7517. HENSMANS, Pierre Joseph. b. 1792. Belg. P I.

7518. ILISCH, Samuel Friedrich. 1792-1842. Riga. P I.

7519. MEISSNER, Karl Friedrich Wilhelm. 1792-1853. Ger. P II.

7520. TADDEI, Giovacchino. 1792-1860. It. ICB; P II; BLA.

7521. CHEVALLIER, Jean Baptiste Alphonse. 1793-1879. Fr. DSB; P I, III; BLA & Supp.

7522. DANA, James Freeman. 1793-1827. U.S.A. P I & Supp.; BLA & Supp.; GE. Also Geology*.
1. *An epitome of chemical philosophy, being an extended syllabus of the lectures....* Concord, N.H., 1825. RM.

7523. DUPASQUIER, (Gaspard) Alphonse. 1793-1848. Fr. P I; BLA & Supp.; GA.

7524. GRISCHOW, Karl Christoph. 1793-1860. Ger. P I & Supp.

7525. SOUBEIRAN, Eugène. 1793-1858. Fr. ICB; P II; BLA.

7526. TRAUTWEIN, Jacob Bernhard. 1793-1855. Ger. P II.

7527. BUSSY, Antoine Alexandre Brutus. 1794-1882. Fr. ICB; P I,III.

7528. CONNELL, Arthur. 1794-1863. Br. P III.

7529. FERRARI, Girolamo. b. ca. 1794. It. P I.

7530. FONTANA, Francesco. 1794-1867. ICB.

7531. MARSH, James. 1794-1846. Br. ICB; P II; BLA; GA; GD.

7532. MERCK, Heinrich Emmanuel. 1794-1855. Ger. P II.

7533. MITSCHERLICH, Eilhard. 1794-1863. Ger. DSB; ICB; P II; GA; GB; GC. Also Crystallography.
1. [Trans. of a periodical article in Swedish, 1821] *Über das Verhältniss zwischen der chemischen Zusammensetzung und der Krystallform arseniksaurer und phosphorsaurer Salze.* Leipzig, 1898. (OKEW. 94)
2. *Lehrbuch der Chemie.* 2 vols. Berlin, 1829-30. X. 2nd ed., 1831-40. 3rd ed., 1837-45.
2a. ——— *Practical and experimental chemistry, adapted to arts and manufactures.* London, 1838. Trans. by S.L. Hammick. RM.
3. *Über das Benzin und die Verbindungen desselben.* [Periodical article, 1834] Leipzig, 1898. (OKEW. 98)
4. *Gesammelte Schriften. Lebensbild, Briefwechsel und Abhandlungen.* Berlin, 1896. Ed. by A. Mitscherlich. RM.

7534. RUNGE, Friedlieb Ferdinand. 1794-1867. Ger. DSB; ICB; P II, III; GC.

7535. CASASECA, José ... Luis. d. 1863. Cuba. P III.

7536. AUGUST, Ernst Ferdinand. 1795-1870. Ger. P I,III. Also Physics*.
1. *Handwörterbuch der Chemie und Physik.* ? vols. Berlin, 1842-50. RM.

7537. BOON-MESCH, Hendrik Carel van der. 1795-1831. Holl. P I; BMNH; BLA & Supp. Also Botany.

7538. BRANDES, Rudolph. 1795-1842. Ger. P I; BLA; GA; GC.

7539. BRUGNATELLI, Gaspare. 1795-1852. It. P III.

7540. CAVENTOU, Joseph Bienaimé. 1795-1877. Fr. DSB; ICB; P I,III; BLA & Supp.; GA.
1. *Nouvelle nomenclature chimique, d'après la classification adoptée par M. Thenard.* Paris, 1816. RM.

7541. COZZI, Andrea. ca. 1795-1856. It. P I.

7542. DAUBENY, Charles Giles Bridle. 1795-1867. Br. DSB; ICB; P I, III; BMNH; GB; GD. Other entries: see 3305.
1. *An introduction to the atomic theory.* Oxford, 1831. RM/L. Supplement, 1840. 2nd ed., 1850. RM.

7543. LOEWEL, Henri (or Heinrich). 1795-1856. Fr./Ger. P I,IV.

7544. PAYEN, Anselme. 1795-1871. Fr. DSB; P II,III; BMNH; GA. Also Botany*.

7545. ROSE, Heinrich. 1795-1864. Ger. DSB; P II,III; BMNH; GB; GC.

1. *Handbuch der analytischen Chemie*. Berlin, 1829. X.
1a. ——— *A manual of analytical chemistry*. London, 1831. Trans. by J. Griffin.
1b. ——— 4th ed., 2 vols, Berlin, 1838.
1c. ——— ——— *A practical treatise of chemical analysis*. 2 vols. London, 1849. Trans. by A. Normandy.
1d. ——— 5th ed. *Ausführliches Handbuch der analytischen Chemie*. 2 vols. Brunswick, 1851.
See also 7424/6a.

7546. SCHWEIGGER-SEIDEL, Franz Wilhelm. 1795-1838. Ger. P II; BLA; GC.

7547. APJOHN, James. 1796-1886. Br. P III; BMNH. Also Physics and Mineralogy*.
1. *A manual of the metalloids*. London, 1863. X. 2nd ed., 1865.

7548. BOUTRON-CHARLARD, Antoine François. 1796-1878. Fr. P I,III.

7549. BRUNNER, Carl Emanuel. 1796-1867. Switz. P I,III.

7550. CLAUS, Carl Ernst. 1796-1864. Russ. DSB; P I,III; BMNH.

7551. CRUM, Walter. 1796-1867. Br. DSB; P III.

7552. HERAPATH, William. 1796-1867. Br. P I,III; GD.

7553. IOVSKII, Alexander. 1796-1857. Russ. ICB; BLA.

7554. JOHNSTON, James Finlay Weir. 1796-1855. Br. P I; GA; GD.
1. *Lectures on agricultural chemistry and geology*. Edinburgh, 1844. RM. 7th ed., 1856.
2. *The chemistry of common life*. 2 vols. Edinburgh, 1855-56.
See also 7605/1a.

7555. MACAIRE, Isaac François. 1796-1869. Switz. P II,III.

7556. MARTIUS, Theodor Wilhelm Christian. 1796-1863. Ger. P II,III; BLA.

7557. PRELÀ, Benedetto Viale. 1796-1874. It. P III.

7558. ROBINET, Stéphane. 1796-1869. Fr. P II,III; BLA.

7559. TURNER, Edward. 1796-1837. Br. DSB; ICB; P II; GD.
1. *Elements of chemistry, including the recent discoveries and doctrines*. Edinburgh, 1827. RM/L. 5th ed., 1834. 6th ed., 2 vols, 1837-42. 8th ed., by J. Liebig and W. Gregory, 1847.

7560. CHRISTISON, Robert. 1797-1882. Br. P I,III; BLA & Supp.; GB; GD.

7561. DUBRUNFAUT, Augustin Pierre. 1797-1881. Fr. ICB; P III.

7562. FREMERY, Petrus Johannes Isaacus. 1797-1855. Holl. P III (FREMERIJ); BMNH; BLA. Also Zoology.

7563. FROMHERZ, Carl. 1797-1854. Ger. P I; BMNH. Also Geology.

7564. MARTENS, Martin. 1797-1863. Belg. P II & Supp.; BMNH; GF. Also Botany.

7565. MOLDENHAUER, Friedrich. b. 1797. Ger. P II.

7566. MOSANDER, Carl Gustaf. 1797-1858. Swed. DSB; P II.

7567. OSANN, Gottfried Wilhelm. 1797-1866. Ger. P II,III.

7568. SCHUBARTH. Ernst Ludwig. 1797-1868. Ger. P II,III; BLA.
  1. *Lehrbuch der theoretischen Chemie.* Berlin, 1822. X. 5th ed., 1832.

7569. ÅKERMAN, Joachim. 1798-1876. Swed. P I,III.

7570. BOUTIGNY, Pierre Hippolyte. 1798-1884. Fr. P III,IV.

7571. HENRY, Etienne Ossian. 1798-1873. Fr. P I,III; GA.

7572. WACKENRODER, Heinrich Wilhelm Ferdinand. 1798-1854. Ger. DSB; ICB; P II; GC.

7573. HOLGER, Philipp Aloys von. 1799-1866. Aus. P III; BMNH; BLA. Also Geology.

7574. HÜNEFELD, Friedrich Ludwig. 1799-1882. Ger. P I,III; BLA.

7575. MARTIN, Jean Emile. b. 1799. Fr. P II.

7576. PURGOTTI, Sebastiano. 1799-1879. It. P III.

7577. SALM-HORSTMAR, Wilhelm Friedrich Carl August. 1799-1865. Ger. P II,III; BMNH. Also Physics.

7578. SCHLIPPE, Karl Friedrich von. 1799-1867. Ger./Russ. P II,III.

7579. SCHÖNBEIN, Christian Friedrich. 1799-1868. Switz. DSB; ICB; P II,III; BLA; GB; GC. See also Index.

7580. WALCHNER, Friedrich August. 1799-1865. Ger. P II; BMNH; GC. Also Geology.

7581. WETZLAR, Gustav. 1799-1861. Ger. P II.

7582. YORKE, Philip James. 1799-1874. Br. P III; GD.

7583. HENNELL, Henry. d. 1842. Br. ICB; P I.

7584. DESSAIGNES, Victor. 1800-1885. Fr. DSB; P I,III.

7585. DUMAS, Jean Baptiste André. 1800-1884. Fr. DSB; ICB; P I,III; BLA & Supp.; GA; GB.
  1. *Traité de chimie appliquée aux arts.* 9 vols. Paris, 1828-46. RM/L. Another ed., Liége, 1847-48.
  2. *Leçons sur la philosophie chimique.* Paris, 1837. Ed. by Bineau. RM. Reprinted, Brussels, 1972. 2nd ed., 1878.
  3. *De l'action du calorique sur les corps organiques.* Paris, 1838.
  4. *Essai de statique chimique des êtres organisés.* Paris, 1841. With J.B.J.D. Boussingault. X.
  4a. ——— *The chemical and physiological balance of organic nature. An essay.* London, 1844.
  5. *Mémoire sur les équivalents des corps simples.* Paris, 1859. RM.
  See also 7182/8 and 9462/3.

7586. KÜHN, Otto Bernhard. 1800-1863. Ger. P I,III; BLA.

7587. LANGLOIS, Charles. 1800-1880. Fr. P I,III.

7588. LASSAIGNE, Jean Louis. 1800-1859. Fr. P I & Supp.; BLA; GA.

7589. LE CANU, Louis René. 1800-1871. Fr. P I,III; BLA; GA.

7590. MORIN, Antoine. 1800-1879. Switz. P II,III.

7591. PLATTNER, Karl Friedrich. 1800-1858. Ger. DSB; P II; BMNH; GB; GC.
1. *Die Probierkunst mit dem Löthrohre.* Leipzig, 1835. RM.
1a. ——— *The use of the blowpipe in the examination of minerals, ores, furnace products and other metallic combinations.* London, 1845. Trans. with notes by J.S. Muspratt. RM/L. 2nd ed., 1850. 3rd ed., 1854.
2. *Vorlesungen über allgemeine Hüttenkunde.* 2 vols. Freiberg, 1860-63. RM/L.

7592. WÖHLER, Friedrich. 1800-1882. Ger. DSB; ICB; P II,III; BLA; GB; GC.
1. *Untersuchungen über das Radikal der Benzoesäure.* [Periodical article by F. Wöhler and J. Liebig, 1832] Leipzig, 1891. (OKEW. 22)
2. *Grundriss der Chemie.* Comprises the two following works, published separately.
2a. ——— *Grundriss der unorganischen Chemie.* Berlin, 1831. X. 8th ed., 1845. RM. 9th ed., 1846. 11th ed., 1854.
2b. ——— *Grundriss der organischen Chemie.* Berlin, 1840. X. 5th ed., 1854. 10th ed., rev. by R. Fittig, Leipzig, 1877.
2c. ——— ——— *Wöhler's Outlines of organic chemistry.* By R. Fittig. Philadelphia, 1873. Trans. from the 8th German ed., with additions, by I. Remsen. RM.
3. *Practische Übungen in der chemischen Analyse.* Göttingen, 1853. X.
3a. ——— *Hand-book of inorganic analysis.* London, 1854. Trans. and ed. by A.W. Hofmann.
See also Index.

7593. BLEY, Ludwig Franz. b. 1801. Ger. P I,III.

7594. CLARK, Thomas. 1801-1867. Br. DSB; ICB; P III; GB; GD.

7595. BALARD, Antoine Jérome. 1802-1876. Br. DSB; ICB; P I,III; GA.

7596. BOUSSINGAULT, Jean Baptiste Joseph Dieudonné. 1802-1887. Fr./ Colombia. DSB; ICB; P I,III,IV; Mort.; GA; GB. Also Meteorology and Botany. See also 7585/4.

7597. DUFLOS, Adolph Ferdinand. 1802-1889. Ger. P I,III; GC.

7598. ELSNER, Franz Carl Leonhard. 1802-1874. Ger. P I,III.

7599. GRIFFIN, John Joseph. 1802-1877. Br. P I,III; BMNH; GD. Also Crystallography*.
1. *A practical treatise on the use of the blowpipe in chemical and mineral analysis.* Glasgow, 1827. RM/L.
2. *The radical theory in chemistry.* London, 1858.
3. *Chemical handicraft: A classified and descriptive catalogue of chemical apparatus.* London, 1877.
See also 7545/1a.

7600. HESS, Germain Henri. 1802-1850. Russ. DSB; ICB (H., Herman Ivanovich); P I; BLA (H., Hermann Heinrich).
 1. *Thermodynamische Untersuchungen.* [Periodical articles, 1839-42] Leipzig, 1890. (OKEW. 9)

7601. JACQUELAIN, (Victor) Auguste. 1802-1885. Fr. P I,III.

7602. LANDGREBE, Georg. b. 1802. Ger. P I,III.

7603. MAGNUS, (Heinrich) Gustav. 1802-1870. Ger. DSB; ICB; P II, III; BLA; GB; GC. Also Physics. See also 7424/10.

7604. MALAGUTI, Faustino Jovita Mariano. 1802-1878. Fr. P II,III; GA.

7605. MULDER, Gerhardus Johannes. 1802-1880. Holl. DSB; ICB; P II, III; BMNH; BLA & Supp.
 1. *Proeve eener algemeene physiologische scheikunde.* 10 parts. Rotterdam, 1843-50. X.
 1a. ——— *The chemistry of vegetable and animal physiology.* Edinburgh, 1845-49. Trans. by P.F.H. Fromberg, with introd. and notes by J.F.W. Johnston.
 2. *De wijn scheikundig beschouwd.* Rotterdam, 1855. X.
 2a. ——— *The chemistry of wine.* London, 1857. Ed. by H.B. Jones. RM.
 3. [Trans.] *Die Chemie der austrocknenden Oele.* Berlin, 1867. Trans. from the Dutch by J. Müller. RM.

7606. SCHRÖTTER, Anton von. 1802-1875. Aus. DSB; P II,III; GC.
 1. *Die Chemie nach ihrem gegenwärtigen Zustande.* 2 vols. Vienna, 1847-49.

7607. GIRARDIN, Jean Pierre Louis. 1803-1884. Fr. P I,III; BMNH.
 1. *Elémens de minéralogie appliquée aux sciences chimiques.* 2 vols. Paris, 1826. With H. Lecoq. X. Another ed., 1837. RM.
 2. *Leçons de chimie élémentaire.* Paris, 1835. X. 2nd ed., 1839.

7608. GREGORY, William. 1803-1858. Br. DSB; ICB; P I & Supp.; BMNH; GB; GD. See also Index.
 1. *Outlines of chemistry.* 2 vols. London, 1845.
 2. *A handbook of organic chemistry.* London, 1852. Counted as the 3rd ed. of one of the vols of item 1. 4th ed., 1856.
 3. *A handbook of inorganic chemistry.* London, 1853. Counted as the 3rd ed. of one of the vols of item 1. 4th ed., 1857.

7609. KERSTEN, Karl Moritz. 1803-1850. Ger. P I.

7610. KUHLMANN, Charles Frédéric. 1803-1881. Fr. P I,III; GA.
 1. *Recherches scientifiques et publications diverses.* Paris, 1877. RM.

7611. LIEBIG, Justus von. 1803-1873. Ger. DSB; ICB; P I,III; GA; GB; GC. Also Science in General*. See also Index.
 1. *Untersuchungen über das Radikal der Benzoesäure.* [Periodical article by F. Wöhler and J. Liebig, 1832] Leipzig, 1891. (OKEW. 22)

2. *Anleitung zur Analyse organischer Körper*. Brunswick, 1837. X.

2a. ——— *Instructions for the chemical analysis of organic bodies*. Glasgow, 1839. Trans. by W. Gregory. RM/L.

3. *Abhandlung über die Constitution der organischen Säuren*. [Periodical article, 1838] Leipzig, 1891. (OKEW. 26)

4. *Die organische Chemie in ihre Anwendung auf Agricultur und Physiologie*. Brunswick, 1840. X. 5th ed., 1843. 6th ed., 1846. 9th ed., by P. Zöller, 1876. In the later editions the title reads *Die Chemie in ihre*....

4a. ——— *Organic chemistry in its applications to agriculture and physiology*. London, 1840. Trans. by L. Playfair. RM/L.

4b. ——— ——— 2nd ed. *Chemistry in its*.... 1842. RM/L. 3rd ed., 1843. 4th ed., 1847. All eds trans. by L. Playfair, the 4th jointly with W. Gregory.

5. *Die Thierchemie; oder, Die organische Chemie in ihrer Anwendung auf Physiologie und Pathologie*. Brunswick, 1842. RM/L.

5a. ——— *Animal chemistry; or Organic chemistry in its applications to physiology and pathology*. London, 1842. Trans. by W. Gregory. RM/L. Reprinted, New York, 1964. 2nd ed., 1843. 3rd ed., 1846.

5b. ——— [Trans. of an extract] *Chemistry and physics in relation to physiology and pathology*. London, 1846. 2nd ed., 1847.

6. *Chemische Briefe*. Heidelberg, 1844. RM. Many later eds. The Leipzig, 1878 ed. has been reprinted: Hildesheim, 1967.

6a. ——— *Familiar letters on chemistry, and its relation to commerce, physiology and agriculture*. London, 1843. 16 letters. Trans. by J. Gardner. RM/L. 2nd ed., 1844. 3rd ed., 1845. 4th ed., trans. by J. Blyth, 1859.

6b. ——— *Familiar letters on chemistry. Second series. The philosophical principles and general laws of the science*. London, 1844. 11 letters. Trans. by J. Gardner.

7. *Chemische Untersuchung über das Fleisch und seine Zubereitung zum Nahrungsmittel*. Heidelberg, 1847. X.

7a. ——— *Researches on the chemistry of food*. London, 1847. Trans. by W. Gregory. RM/L.

8. *Untersuchungen über einige Ursachen der Säftebewegung im thierischen Organismus*. Brunswick, 1848. X.

8a. ——— *Researches on the motion of the juices in the animal body*. London, 1848. Trans. by W. Gregory. RM/L.

9. *Die Grundsätze der Agricultur-Chemie*. Brunswick, 1855. RM.

9a. ——— *Principles of agricultural chemistry*. London, 1855. Trans. by W. Gregory. RM/L.

10. *Naturwissenschaftliche Briefe über die moderne Landwirthschaft*. Munich, 1859.

10a. ——— *Letters on modern agriculture*. London, 1859. Trans. by J. Blyth. RM.

11. *Reden und Abhandlungen.* Leipzig, 1874. Ed. by J. Carrière. X. Reprinted, Wiesbaden, 1965.

Correspondence

12. *Aus J. Liebig's und F. Wöhler's Briefwechsel in den Jahren 1829-1873.* Brunswick, 1888. Ed. by A.W. Hofmann. RM/L. Reprinted, (place?), 1958.
13. *Berzelius und Liebig: Ihre Briefe von 1831-1845.* Munich/ Leipzig, 1893. Ed. by J. Carrière. X. Reprinted, Wiesbaden, 1967.
14. *J. von Liebig und C.F. Schönbein: Briefwechsel 1853-1868.* Leipzig, 1900. Ed. with notes by G.W.A. Kahlbaum and E. Thon. RM/L. Reprinted, Leipzig, 1970.
15. *J. von Liebig und F. Mohr in ihren Briefen von 1834-1870.* Leipzig, 1904. Ed. by G.W.A. Kahlbaum. X. Reprinted, Leipzig, 1970.

7612. LÖWIG, Karl Jacob. 1803-1890. Ger./Switz. ICB; P I,III; GC.

7613. MARCET, François. 1803-1883. Switz. P II,III. Also Physics.

7614. MARESKA, Joseph Daniel Benoît. 1803-1858. Belg. P II; BLA.

7615. WIGGERS, Heinrich August Ludwig. 1803-1880. Ger. P II,III; BLA; GC.

7616. BERGEMANN, Carl Wilhelm. 1804-1884. Ger. P I,III,IV.

7617. BOON-MESCH, Antonius Hendrik van der. 1804-1874. Holl. P I,III.

7618. ERDMANN, Otto Linné. 1804-1869. Ger. DSB; P I,III; GA; GB; GC.

7619. GAUDIN, Marc Antoine Augustin. 1804-1880. Fr. DSB; ICB; P I, III; GA.

1. *L'architecture du monde des atomes, dévoilant la structure des composés chimiques et leur cristallogénie.* Paris, 1873. RM.

7620. MARQUART, Ludwig Clamor. 1804-1881. Ger. P III; BMNH.

7621. MORREN, Jean François Auguste. 1804-1870. Fr. P III; BMNH. Also Physics.

1. *Recherches sur la rubéfaction des eaux et leur oxygénation par les animalcules et les algues.* Brussels, 1841. With C.F.A. Morren (the botanist).

7622. SIMON, Johann Franz. 1804 (or 1807)-1843. Ger. P II; BLA.

1. *Handbuch der angewandten medicinischen Chemie.* 2 vols. Berlin, 1840-42. X.

1a. ——— *Animal chemistry with reference to the physiology and pathology of man.* 2 vols. London, 1845-46.

7623. BALLING, Carl Joseph Napoleon. 1805-1868. Prague. P I,III.

7624. BECKS, Franz Caspar. 1805-1847. Ger. P I.

7625. GRAHAM, Thomas. 1805-1869. Br. DSB; ICB; P I,III; GB; GD.

1. *Researches on the arseniates, phosphates and modifications of phosphoric acid.* [Periodical article, 1833] Edinburgh, 1904 and re-issues. (Alembic Club reprints. 10)
2. *Elements of chemistry, including the application of the science to the arts.* London, 1842. RM/L. 2nd ed., 2 vols, 1850-58. See also 7663/1.

3. [Trans. of three periodical articles, 1861 and later] *Abhandlungen über Dialyse. (Kolloide).* Leipzig, 1911. (OKEW. 179)
4. *Chemical and physical researches.* Edinburgh, 1876. RM/L.

See also 8656.

7626. HERMANN, Hans Rudolph. 1805-1879. Russ. P I,III; BMNH. Also Mineralogy.

7627. MITSCHERLICH, Karl Gustav. 1805-1871. Ger. ICB; P II,III; BLA.

7628. OPPERMANN, Karl Friedrich. b. 1805. Strasbourg. P II.

7629. PERSOZ, Jean François. 1805-1868. Fr. DSB; P II,III.

7630. PETER, Robert. 1805-1894. U.S.A. ICB; P II; GE.

7631. SVANBERG, Lars (or Lorenz) Fredrik. 1805-1878. P II,III.

7632. BAUDRIMONT, Alexandre Edouard. 1806-1880. Fr. DSB; P III,IV; BMNH. Also Physiology.
1. *Traité de chimie générale et expérimentale, avec les applications....* 2 vols. Paris, 1844-46. RM.

7633. BIBRA, Ernst von. 1806-1878. Ger. P I,III; BLA; GC.

7634. BÖTTGER, Rudolph Christian. 1806-1881. Ger. DSB; ICB; P I, III; GC.

7635. BOUCHARDAT, Apollinaire. 1806-1886. Fr. P III,IV; BMNH; BLA & Supp.; GA.

7636. BOUDET, Felix Henri. 1806-1878. Fr. P I,III.

7637. BOULLAY, Polydore. 1806-1835. Fr. ICB; P I.

7638. ETTLING, Carl Jacob. 1806-1856. Ger. P I.

7639. KRAMER, Anton Johann von. 1806-1853. It. P I.

7640. MOHR, (Carl) Friedrich. 1806-1879. Ger. DSB; ICB; P II,III; GC. See also 7611/15.

7641. MÜLLER, Johannes. b. 1806. Ger. P II. Also Botany.

7642. UNVERDORBEN, Otto. 1806-1873. Ger. ICB; P II,III; GC.

7643. CLEMSON, Thomas Green. 1807-1888. U.S.A. ICB; GE.

7644. DARCET (or d'ARCET), Felix. ca. 1807-1846. Fr. P I.

7645. HUNT, Robert. 1807-1887. Br. P III; GA; GB; GD. Also Science in General*.
1. *Researches on light: An examination of all the phenomena connected with the chemical and molecular changes produced by the influence of the solar rays, embracing all the known photographic processes.* London, 1844. RM.
1a. ——— 2nd ed. *Researches on light in its chemical relations....* London 1854.

7646. LAURENT, Auguste. 1807-1853. Fr. DSB; ICB; P I; GA.
1. *Précis de cristallographie, suivi d'une méthode simple d'analyse au chalumeau.* Paris, 1847.
2. *Méthode de chimie.* Paris, 1854. X.
2a. ——— *Chemical method, notation, classification, and nomenclature.* London, 1855. Trans. by W. Odling. Printed for the Cavendish Society.

7647. MIALHÉ, Louis. 1807-1886. Fr. P II,III; BLA.

7648. PELOUZE, Théophile Jules. 1807-1867. Fr. DSB; ICB; P II,III; BLA; GA; GB.
1. *Traité de chimie générale*. 3 vols & atlas. Paris, 1848-50. With E. Frémy. 3rd ed., 6 vols, 1860-65.

7649. SCHARLING, Edvard August. 1807-1866. Den. P II,III.

7650. WARINGTON, Robert. 1807-1867. Br. P II,III; GD.

7651. BLONDLOT, Nicolas. 1808-1877. Fr. P III; BLA.

7652. FRITZSCHE, Carl Julius. 1808-1871. Russ. DSB; ICB; P I,III.

7653. LEVOL, Alexandre Irénée François. b. 1808. Fr. P I.

7654. LÖWE, Alexander. b. 1808. Aus. P I.

7655. NÖLLNER, Carl. 1808-1877. Ger. P III.

7656. POGGIALE, Antoine Baudin. 1808-1879. Fr. DSB; P II,III; BLA.

7657. FELLENBERG-RIVIER, Ludwig Rudolph von. 1809-1878. Switz. P I, III.

7658. KANE, Robert John. 1809-1890. Br. DSB; P III; BLA; GA; GD.
1. *Elements of chemistry, theoretical and practical*. Dublin, 1849.

7659. KOENE, Corneille Jean. b. 1809. Belg. P I.

7660. KONINCK, Laurent Guillaume de. 1809-1887. Belg. DSB; P I,III, IV; BMNH; GB. Also Geology*.

7661. LANDERER, Xaver. 1809-1885. Greece. P III.

7662. NORMANDY, Alphonse René Le Mire de. 1809-1864. Fr./Br. P II; GD. See also 7545/1c.

7663. OTTO, Friedrich Julius. 1809-1870. Ger. P II,III; GA; GC.
1. *Lehrbuch der Chemie. Zum Theil auf Grundlage von Dr. T. Graham's "Elements of chemistry."* 2 vols in 4. Brunswick, 1844-49.
1a. ——— 2nd ed. *Graham-Otto's ausführliches Lehrbuch der Chemie*. ? vols. Brunswick, 1854-?

7664. REINSCH, Edgar Hugo Emil. 1809-1884. Ger. P III; BMNH.

7665. STENHOUSE, John. 1809-1880. Br. P II,III; GD.

7666. STÖCKHARDT, (Julius) Adolph. 1809-1886. Ger. P II,III; GC.
1. *Die Schule der Chemie*. Brunswick, 1846. Many later eds. X.
1a. ——— *The principles of chemistry*. Cambridge, 1850. Trans. from the 3rd German ed. by C.H. Peirce. X. Other eds, London: 1851 and 1855.
2. *Chemische Feldpredigten für deutsche Landswirthe*. Leipzig, (date?). X.
2a. ——— *Chemical field lectures: A familiar exposition of the chemistry of agriculture addressed to farmers*. London, 1847. Trans. by A. Henfrey. X. Another ed., 1855.

7667. VOSKRESENSKII, Alexandr Abramovich. 1809-1880. Russ. ICB; P II,III (WOSKRESSENSKY).

7668. QUESNEVILLE, Gustave Augustin. 1810-1889. Fr. P II,III,IV.

7669. REDTENBACHER, Joseph. 1810-1870. Prague/Aus. P II,III.

7670. REGNAULT, (Henri) Victor. 1810-1878. Fr. DSB; ICB; P II,III; GA; GB. Also Physics.
   1. *Cours élémentaire de chimie*. 4 vols. Paris, 1847-49. X. 2nd ed., 1850? 3rd ed., 1851. 4th ed., 1854.

7671. SCHRÖDER, Heinrich Georg Friedrich. 1810-1885. Ger. P II,III.

7672. THOMSON, Robert Dundas. 1810-1864. Br. P III; GD.

7673. WALTER, Philippe. 1810-1847. Fr. DSB; ICB (W., Filip Neriusz); P II,III (WALTHER).

7674. WINKELBLECH, Georg Karl. 1810-1865. Ger. P II,III.

7675. WITTSTEIN, Georg Christian. 1810-1887. Ger. P II,III; BMNH. Also History of Chemistry*.
   1. *Anleitung zur chemischen Analyse von Pflanzen und Pflanzentheilen auf ihre organischen Bestandtheile*. Nördlingen, 1868. X.
   1a. ——— *The organic constituents of plants and vegetable substances and their chemical analysis*. Melbourne, 1878. Trans. with additions by J. von Müller (the botanist--12272).
   2. *Taschenbuch der Chemikalienlehre*. Nördlingen, 1879.

7676. BUNSEN, Robert Wilhelm. 1811-1899. Ger. DSB; ICB; P I,III,IV; GA; GB; GC.
   1. *Untersuchungen über die Kakodylreihe*. [Periodical articles, 1837-43] Leipzig, 1891. (OKEW. 27)
   2. *Photochemische Untersuchungen*. [Periodical articles by R. Bunsen and H.E. Roscoe] 2 vols. Leipzig, 1892. (OKEW. 34,38)
   3. *Gasometrische Methoden*. Brunswick, 1857. RM. 2nd ed., 1877.
   3a. ——— *Gasometry. Comprising the leading physical and chemical properties of gases*. (place?), 1857. Trans. by H.E. Roscoe. RM.
   4. *Gesammelte Abhandlungen*. 3 vols. Leipzig, 1904. Ed. by W. Ostwald and M. Bodenstein. RM.
   See also 7885/1.

7677. DRAPER, John William. 1811-1882. U.S.A. DSB; ICB; P I,III; BMNH; BLA & Supp.; GB; GD; GE. Also Physics*, Physiology*, and History of Science*.

7678. FEHLING, Herrmann Christian von. 1811-1885. Ger. P I,III; GB; GC.
   1. *Neues Handwörterbuch der Chemie. Auf Grundlage des von Liebig, Poggendorff und Wöhler....* [see 8655] 10 vols. Brunswick, 1871-1930.

7679. GOBLEY, Nicolas Théodore. 1811-1876. Fr. DSB; P III,IV.

7680. GROVE, William Robert. 1811-1896. Br. DSB; ICB; P I,III; GA; GB; GD1. Also Physics*.

7681. HIMLY, August Friedrich Karl. 1811-1885. Ger. P I,III.

7682. HOEFER, (Jean Chrétien) Ferdinand. 1811-1878. Fr. ICB; P I, III; BMNH; BLA; GA. Also Biography* and History of Science*.
1. *Nomenclature et classifications chimiques, suivies d'un lexique historique et synonymique comprenant les noms anciens, les formules, les noms nouveaux, le nom de l'auteur et la date de la découverte des principaux produits de la chimie.* Paris, 1845.
2. *Dictionnaire de chimie et de physique.* Paris, 1846.
See also 7424/2i.

7683. NENDTWICH, Karl Maximilian. 1811-1892. Hung. P II,III.

7684. PÉLIGOT, Eugène Melchior. 1811-1890. Fr. P II,III,IV; GA.

7685. STEIN, Heinrich Wilhelm. 1811-1889. Ger. P II,III.

7686. TROMMSDORFF, Christian Wilhelm Hermann. 1811-after 1859. Ger. P II.

7687. ULEX, Georg Ludwig, 1811-1883. Ger. P II,III.

7688. ULLGREN, Clemens. 1811-1868. Swed. P II,III.

7689. BERLIN, Nils Johannes. 1812-1891. Swed. P I,III,IV; BLA & Supp.

7690. BINEAU, Amand. 1812-1861. Fr. P I,III. See also 7585/2.

7691. BOLLEY, (Alexander) Pompejus. 1812-1870. Switz. P I,III.

7692. BOYÉ, Martin Hans. 1812-1909. U.S.A. ICB; P III; GE.

7693. DELFFS, (Friedrich) Wilhelm (Herrmann). 1812-1894. Ger. P I, III.
1. *Die organische Chemie in ihren Grundzügen dargestellt.* Kiel, 1840. RM.

7694. HERVÉ DE LA PROVOSTAYE, Frédéric. 1812-1863. Fr. P I,III.

7695. LEHMANN, Karl Gotthelf. 1812-1863. Ger. P I,III; BLA.
1. *Vollständiges Taschenbuch der theoretischen Chemie.* Leipzig, 1840. X. 5th ed., 1851.
2. *Lehrbuch der physiologischen Chemie.* 3 vols. Leipzig, 1842-45. X.
2a. ——— *Physiological chemistry.* 3 vols. London, 1851-54. Trans. from the 2nd German ed. by G.E. Day. Publ. by the Cavendish Society.

7696. MILLON, (Auguste Nicolas) Eugène. 1812-1867. Fr. DSB; P II, III; GA.

7697. PIERRE, Joachim Isidore. 1812-1881. Fr. P II,III.
1. *Sommaire des leçons du cours de chimie appliquée à l'agriculture.* Paris, 1869-70.

7698. SOBRERO, Ascanio. 1812-1888. It. ICB; P II,III.

7699. STEINBERG, Karl. 1812-1852. Ger. P II.

7700. THAULOW, Moritz Christian Julius. 1812-1850. Nor. P II.

7701. VLIET, August Frederik van der. 1812-1862. Holl. P II,III.

7702. WILHELMY, Ludwig Ferdinand. 1812-1864. Ger. DSB; P II,III. Also Physics.
1. *Über das Gesetz nach welchem die Einwirkung der Säuren auf*

den Rohzucker stattfindet. [Periodical article, 1850] Leipzig, 1891. (OKEW. 29)

7703. WILL, Heinrich. 1812-1890. Ger. P II,III.
1. *Anleitung zur qualitativen chemischen Analyse.* Heidelberg, 1846. X.
1a. ——— *Outlines of the course of qualitative analysis followed in the Giessen laboratory.* London, 1846. With a preface by J. Liebig.

7704. ZININ, Nikolai Nikolaievich. 1812-1880. Russ. DSB; ICB; P II, III.

7705. ANDREWS, Thomas. 1813-1885. Br. DSB; P I,III; GB; GD1.
1. *On the continuity of the gaseous and liquid states of matter. On the gaseous state of matter.* [Two periodical articles, 1869 and 1876] German trans. Leipzig, 1902. (OKEW. 132)
2. *The scientific papers.* London, 1889. With a memoir by P.G. Tait and A. Crum Brown. RM/L.

7706. BRAME, Charles Henri Auguste. 1813-ca. 1895. Fr. P III,IV.

7707. BUCHNER, Ludwig Andreas. 1813-1897. Ger. P I,III,IV; BLA & Supp.; GA; GC.

7708. CAHOURS, Auguste André Thomas. 1813-1891. Fr. DSB; P I,III.

7709. FAVRE, Pierre Antoine. 1813-1880. Fr. DSB; P I,III. Also Physics.

7710. HELLER, Johann Florian. 1813-1871. Aus. P I,III; BLA & Supp.

7711. KERCKHOFF, Petrus Johannes van. 1813-1876. Holl. P I,III.

7712. LE BLANC, Félix. 1813-1886. Fr. P I,III.

7713. MARCHAND, Richard Felix. 1813-1850. Ger. DSB; P II; BLA; GC.

7714. MÜLLER, Ferdinand Albrecht Louis. 1813-after 1852. Ger. P II.

7715. RAMMELSBERG, Karl Friedrich. 1813-1899. Ger. DSB; ICB; P II, III,IV; BMNH; GB.
1. *Handwörterbuch des chemischen Theils der Mineralogie.* Berlin, 1841. Supplements, 5 parts in 4, 1843-53.
2. *Leitfaden für die qualitative chemische Analyse.* Berlin, 1847. X. 5th ed., 1867.
3. *Lehrbuch der chemischen Metallurgie.* Berlin, 1850. 2nd ed., 1865.
4. *Lehrbuch der Krystallkunde.* Berlin, 1852.
5. *Handbuch der krystallographischen Chemie.* Berlin, 1855.
6. *Die neuesten Forschungen in der krystallographischen Chemie.* Leipzig, 1857.
7. *Handbuch der Mineralchemie.* Berlin, 1860. 2nd ed., Leipzig, 1875. Supplements, 1886 and 1895.
8. *Handbuch der krystallographisch-physikalischen Chemie.* 2 vols. Leipzig, 1881-82.
See also 7424/4c.

7716. SCHEERER, (Karl Johann August) Theodor. 1813-1875. Ger./Nor. P II,III; BMNH. Also Mineralogy*.

7717. SCHIEL, Jacob Heinrich Wilhelm. 1813-after 1879. Ger. P III.

7718. SCHNEIDER, Franz Coelestin von. 1813-1897. Aus. P II,III,IV; BLA; GC.

7719. SCHOEDLER, Friedrich Karl Ludwig. 1813-1884. Ger. P II,III, IV. Also Science in General*.

7720. STAS, Jean Servais. 1813-1891. Belg. DSB; ICB; P II,III,IV; BLA; GB.
1. *Nouvelles recherches sur les lois des proportions chimiques, sur les poids atomiques et leurs rapports mutuels.* Brussels, 1865.
1a. ——— German trans. Leipzig, 1867.
2. *Oeuvres complètes.* 3 vols. Brussels, 1894. RM/L.

7721. WELTZIEN, Karl. 1813-1870. Ger. ICB; P II,III; BMNH.

7722. BECQUEREL, (Louis) Alfred. 1814-1862. Fr. BMNH; BLA & Supp.
1. *Traité de chimie pathologique, appliquée à la médecine pratique.* Paris, 1854. With M.A. Rodier.
1a. ——— *Pathological chemistry.* London, 1857.

7723. BIEWEND, Eduard. 1814-after 1850. Ger. P I.

7724. BLYTH, John. 1814-1871. Br. P I,III. See also 7611/6a,10a.

7725. DÖPPING, Otto. 1814-after 1857. Russ. P I.

7726. EBELMEN, Jacques Joseph. 1814-1852. Fr. P I; GA.

7727. FILHOL. Edouard. 1814-1883. Fr. P III,IV; BMNH.

7728. FRÉMY, Edmond. 1814-1894. Fr. DSB; P I,III,IV; GA; GB. See also 7648/1 and 8659.

7729. JEANNEL, Julien François. 1814-1896. Fr. P III.

7730. JONES, Henry Bence. 1814-1873. Br. ICB; P I,III; BMNH; BLA; GB (BENCE-JONES); GD. Also Biography* and History of Science*.
1. *On animal chemistry.* London, 1850.
2. *On animal electricity. (Abstract of the discoveries of E. Du Bois-Reymond).* London, 1852.
3. *The Croonian lectures on matter and force.* London, 1868.
See also 7605/2a and 7743/1.

7731. KNAPP, Friedrich Ludwig. 1814-1904. Ger. P I,III,IV.
1. *Lehrbuch der chemischen Technologie.* 2 vols. Brunswick, 1847. X.
1a. ——— *Chemical technology; or, Chemistry applied to the arts and to manufactures.* 2 vols. London, 1848. Ed. with notes by E. Ronalds and T. Richardson.

7732. LAWES, John Bennet. 1814-1900. Br. DSB; ICB; GB; GD1.
1. *On the sources of the nitrogen of vegetation.* London, 1860. With J.H. Gilbert and E. Pugh. X. Another ed., 1862.

7733. MELSENS, Louis Henri Frédéric. 1814-1886. Belg. P II,III,IV; BLA. Also Physics.

7734. SCHERER, Johann Joseph. 1814-1869. Ger. P II,III; BLA; GC.

7735. WALZ, Georg Friedrich. ca. 1814-1862. Ger. P III.

7736. ZWENGER, Constantin. 1814-1884. Ger. P II,III; BLS; GC.

7737. MARGUERITTE, Frédéric. fl. 1843-1865. Fr. P III.

7738. BAHR, Johann Friedrich. 1815-1875. Swed. P I,III.

7739. BARFOED, Christen Thomsen. 1815-1889. Den. P III,IV.

7740. BOEDEKER, Carl Heinrich Detlev. 1815-1895. Ger. P I,III.

7741. DAY, George Edward. 1815-1872. Br. BMNH; BLA; GD.
1. *Chemistry in its relations to physiology and medicine.* London, 1860.
See also 7695/2a and 8656.

7742. DE LA RUE, Warren. 1815-1889. Br. DSB; P III,IV; GD. Also Astronomy* and Physics.

7743. FOWNES, George. 1815-1849. Br. DSB; ICB; P I; GD.
1. *A manual of elementary chemistry, theoretical and practical.* London, 1844. X. 4th ed., by H. Bence Jones and A.W. Hofmann, 1852. 7th ed., 1858. 9th ed., 1863. 11th ed., 1873. 12th ed., by H. Watts, 2 vols, 1877. 13th ed., by H. Watts and W.A. Tilden, 2 vols, 1883-86. 14th (last) ed., by W.A. Tilden, 1889.

7744. GÉLIS, Amédée. 1815-after 1880. Fr. P I,III.

7745. GOTTLIEB, Johann. 1815-1875. Aus. P I,III.

7746. HAGEN, Robert Herrmann Heinrich. 1815-1858. Ger. P I,III.

7747. MORIN, Pyrame Louis. 1815-1864. Switz. P II,III.

7748. NOAD, Henry Minchin. 1815-1877. Br. P III; GD. Also Physics*.
1. *Lectures on chemistry, including its applications in the arts.* London, 1843. RM/L.
2. *Chemical manipulation and analysis, qualitative and quantitative.* London, 1848. RM. Another ed., 1852.
3. *A manual of chemical analysis, qualitative and quantitative.* London, 1864. Another ed., 1870.

7749. PIRIA, Rafaelle. 1815-1865. It. ICB; P III.

7750. SCHULZE, Franz Ferdinand. 1815-1873. Ger. DSB; P II,III; BMNH; GC. Also Microbiology.

7751. SIMPSON, Maxwell. 1815-1902. Br. ICB; P II,III; GD2.

7752. VARRENTRAPP, Franz. 1815-1877. Ger. P II,III.

7753. WATTS. Henry. 1815-1884. Br. P III; GD.
1. *A dictionary of chemistry and the allied branches of other sciences.* 5 vols. London, 1863-68 and re-issues. 1st supp., 1872. 2nd supp., 1875. 3rd supp., 2 parts, 1879-81.
1a. ——— 2nd ed. 4 vols. 1888-94. Rev. by H.F. Morley and M.M.P. Muir. Re-issued, 4 vols, 1894-99.
See also 7488/1b and 7743/1.

7754. WERTHER, (August Friedrich) Gustav. 1815-1869. Ger. P II,III.
1. *Die unorganische Chemie.* 2 parts. Berlin, 1850-52.

7755. WOODS, Thomas. 1815-after 1881. Br. P III.

7756. BÉCHAMP, (Pierre Jacques) Antoine. 1816-1908. Fr. DSB Supp.; ICB; P III,IV; BMNH; BLA & Supp. Also Microbiology.

7757. FORDOS, Mathurin Joseph. 1816-1878. Fr. DSB; P I,III.

7758. GERHARDT, Charles Frédéric. 1816-1856. Fr. DSB; ICB; P I; GA; GB. See also 15525/2.
  1. *Précis de chimie organique.* 2 vols. Paris, 1844-45. RM.
  2. *Traité de chimie organique.* 4 vols. Paris, 1853-56.
  2a. ——— *Lehrbuch der organischen Chemie.* Leipzig, 1854. Ed. by Gerhardt and R. Wagner. RM.
  3. *Correspondance.* 2 vols. Paris, 1918-25. Ed. by M. Tiffeneau. Incomplete.

7759. LALLEMAND, Etienne Alexandre. 1816-1886. Fr. P III. Also Physics.

7760. LERCH, Josef Udo. 1816-1892. Prague. P III,IV.

7761. MARCHAND, Jean Eugène Augustin. 1816-1895. Fr. P III.

7762. MARSSON, Theodor Friedrich. 1816-1892. Ger. P II; BMNH; GC. Also Botany.

7763. PENNY, Frederick. 1816-1869. Br. DSB; P III.

7764. PERSONNE, Jacques. 1816-1880. Fr. DSB; P III.

7765. SCHAFFGOTSCH, Franz G.J.J.K.M. 1816-1864. Ger. P II,III.

7766. SILLIMAN, Benjamin (second of the name). 1816-1885. U.S.A. DSB; P II,III; BMNH; BLA; GE. Also Geology.
  1. *First principles of chemistry.* Philadelphia, 1847. X. 48th ed., 1859.

7767. BARRESWIL, Charles Louis. 1817-1870. Fr. DSB; P I,III.

7768. BENSCH, August. ca. 1817-after 1848. Ger. P I.

7769. BRODIE, Benjamin Collins (second of the name). 1817-1880. Br. DSB; P III; GD.
  1. *The calculus of chemical operations.* Periodical articles, 1866 and 1877.
  1a. ——— *Le calcul des opérations chimiques.* Paris, 1879. Trans. by A. Naquet.

7770. BROWN, Samuel Morison. 1817-1856. Br. P III; GB; GD.
  1. *Lectures on the atomic theory, and essays scientific and literary.* Edinburgh, 1858. RM.

7771. CLOËZ, François Stanislas. 1817-1883. Fr. P III.

7772. FRANCIS, William. 1817-1904. Br. P I,III; BMNH. See also 6973.

7773. GILBERT, Joseph Henry. 1817-1901. Br. DSB (under LAWS, J.B.); ICB; P III,IV; GB; GD2. See also 7732/1.

7774. GORUP-BESANEZ, Eugen Franz Cajetan von. 1817-1878. Ger. P I, III; BLA; GC.
  1. *Anleitung zur qualitativen und quantitativen zoochemischen Analyse.* Nuremberg, 1850. X. 3rd ed., Brunswick, 1871.
  2. *Lehrbuch der Chemie.* 2 vols. Brunswick, 1859-60. X. Another ed., 3 vols, 1876-81. The individual parts were also publ. separately, as follows. (All at Brunswick.)
  3. *Lehrbuch der anorganischen Chemie.* 1861. X. Another ed., 1873.
  4. *Lehrbuch der organischen Chemie.* 1862. X. 6th ed., 1881.

5. *Lehrbuch der physiologischen Chemie.* 1863. X. 2nd ed., 1867. 4th ed., 1878.

7775. HEINTZ, Wilhelm Heinrich. 1817-1880. Ger. P I,III.

7776. JELLETT, John Hewitt. 1817-1888. Br. P III,IV; GD. Also Mathematics* and Physics*.
   1. [Trans. of periodical articles, 1860, 1863, and 1873] *Chemisch-optische Untersuchungen.* Leipzig, 1908. (OKEW. 163)

7777. KNOP, Johann August Ludwig Wilhelm. 1817-1891. Ger. P I,III,IV.

7778. KOPP, Emil. 1817-1875. Strasbourg. P III.

7779. KOPP, Hermann. 1817-1892. Ger. DSB; ICB; P I,III,IV; GB; GC. Also History of Chemistry*.

7780. MARIGNAC, Jean Charles Galissard de. 1817-1894. Switz. DSB; ICB; P II,III,IV; GB.

7781. MILLER, William Allen. 1817-1870. Br. DSB; ICB; P II,III; GA; GD. Also Astronomy.
   1. *On the importance of chemistry to medicine.* London, 1845. Two lectures.
   2. *Elements of chemistry, theoretical and practical.* 3 vols. London, 1855-57. X. 2nd ed., 1860-62. 4th ed., 1867-69. 5th ed., rev. by H. McLeod, 1872-80. 6th ed., 1877-80.
   3. *Introduction to the study of inorganic chemistry.* London, 1870. X. Another ed., 1871.

7782. POSSELT, Louis. 1817-after 1860. Ger. P II.

7783. ROWNEY, Thomas Henry. 1817-1894. Br. P III.

7784. SELMI, Francesco. 1817-1881. It. ICB; P II,III; BLA.

7785. SMITH, Robert Angus. 1817-1884. Br. DSB; P III,IV; BLA; GD. Also Biography* and History of Science*.
   1. *Air and rain: The beginnings of a chemical climatology.* London, 1872. Re atmospheric and water pollution. RM/L.

7786. VOGEL, August. 1817-1889. Ger. P II,III,IV.

7787. WURTZ, Charles Adolphe. 1817-1884. Fr. DSB; ICB; P II,III,IV (WÜRTZ, Karl A.); BLA; GB. Also History of Chemistry*.
   1. *Mémoire sur les glycols ou alcoöls diatoniques.* [Periodical article, 1859] German trans. Leipzig, 1909. (OKEW. 170)
   2. *Leçons de philosophie chimique.* Paris, 1864.
   3. *Leçons élémentaires de chimie moderne.* Paris, 1866 and later eds. X.
   3a. ——— *Elements of modern chemistry.* Philadelphia, 1880. Trans. by W.H. Greene. X. 5th American ed., 1895.
   4. *Dictionnaire de chimie pure et appliquée.* 3 vols in 5. Paris, 1869-78 (re-issued 1882). RM/L. 1st supp., 2 vols, 1880-86. RM/L. 2nd supp., ed. by Friedel and Chabrié, 7 vols, 1892-1908. RM.
   5. *La théorie atomique.* Paris, 1879. RM/L. 3rd ed., 1880.
   5a. ——— *The atomic theory.* London, 1880. Trans. by E. Cleminshaw. RM. 2nd ed., 1880. 5th ed., 1888. 6th ed., 1892.

6. *Traité de chimie biologique.* Paris, 1885.

7788. ARPPE, Adolph Eduard. 1818-1894. Fin. P I,III,IV.

7789. BABO, (Clemens Heinrich) Lambert von. 1818-1899. Ger. P I, III,IV; GC.

7790. BAUMERT, Friedrich Moritz. 1818-1865. Ger. P I,III; BMNH; BLA.

7791. BLEY, Heinrich. 1818-1858. Ger. P III.

7792. BOUQUET, Jean Pierre. 1818-after 1858. Fr. P I,III.

7793. BUCKTON, George Bowdler. 1818-1905. Br. P III; BMNH; GD2. Also Zoology.

7794. CHODNEW, Alexius. 1818-after ca. 1880. Russ. P I,III.

7795. DEVILLE, Henri Etienne Sainte-Claire. 1818-1881. Fr. DSB. Under SAINTE-CLAIRE-DEVILLE: ICB; P II,III; GA; GB.
1. *De l'aluminium: Ses propriétés, sa fabrication et ses applications.* Paris, 1859. RM.

7796. FRESCENIUS, Carl Remigius. 1818-1897. Ger. DSB; P I,III,IV; GC.
1. *Anleitung zur qualitativen chemischen Analyse.* Bonn, 1841. X. Another ed., 1846. 15th ed., 1885.
1a. ——— *Instruction in chemical analysis (qualitative).* London, 1841. Trans. by J.L. Bullock. With a preface by J. Liebig. X. 2nd ed., 1843. 3rd ed., 1850. 4th ed.: *A system of instruction in ....* 1855. 5th ed., 1859. 6th ed., 1864. 9th ed.: *Qualitative chemical analysis*, trans. from the 14th German ed. by A. Vacher, 1876. 10th ed., trans. from the 15th German ed. by C.E. Graves, 1887.
2. *Anleitung zur quantitativen chemischen Analyse.* Brunswick, 1846. X. 6th ed., 1887.
2a. ——— *Instruction in chemical analysis (quantitative).* London, 1846. Trans. by J.L. Bullock. 3rd ed.: *A system of instruction in ....* 1860. 4th ed., 1865. 6th ed.: *Quantitative chemical analysis*, trans. from the 5th German ed. by A. Vacher, 1872. 7th ed., trans. from the 6th German ed. by A. Vacher, 2 vols, 1876.

7797. GÖRGEY, Arthur. 1818-1916. Hung./Aus. DSB; P I; GA.

7798. GROSHANS, John Antony. 1818-1904. Holl. P I,III,IV.

7799. HOFMANN, August Wilhelm von. 1818-1892. Ger./Br. DSB; ICB; P I,III,IV; GB; GC. Also Biography* and History of Chemistry*. See also Index.
1. *Contributions to the history of the phosphorus bases.* London, 1861.
1a. ——— French trans. Paris, 1862.
2. *Introduction to modern chemistry, experimental and theoretic.* London, 1865.

7800. HORSFORD, Ebenezer Norton. 1818-1893. U.S.A. DSB; ICB; P I, III; GE.

7801. KHODNEV, Aleksei Ivanovich. 1818-1883. Russ. ICB.

7802. KOLBE, (Adolph Wilhelm) Hermann. 1818-1884. Ger. DSB; P I, III,IV; GB; GC.
1. [Trans. of periodical articles, 1845-68] *The electrolysis of organic compounds*. Edinburgh, 1893 and re-issues. (Alembic Club reprints. 15)
2. *Über den naturlichen Zusammenhang der organischen mit den unorganischen Verbindungen*. [Periodical article, 1859] Leipzig, 1897. (OKEW. 92)
3. *Kurzes Lehrbuch der Chemie*. 2 parts. Brunswick, 1877-79. The two parts were also publ. individually (at Brunswick), as follows.
4. *Kurzes Lehrbuch der anorganischen Chemie*. 1877.
4a. ——— *A short text-book of inorganic chemistry*. London, 1884. Trans. by T.S. Humpidge. 2nd ed., 1888. 3rd ed., 1892.
5. *Kurzes Lehrbuch der organischen Chemie*. 1879. X.

7803. LEWY, Bernhard Karl. 1818-after 1858. Fr. P I,III.

7804. LOUYET, Paulin Laurent Charles Evalery. 1818-1850. Belg. P I; GF.
1. *Cours élémentaire de chimie générale*. 3 vols. Brussels? 1841-44.

7805. MAUMENÉ, Edme Jules. 1818-after 1895. Fr. P II,III,IV.

7806. PETTENKOFER, Max Josef von. 1818-1901. Ger. DSB; ICB; P II, III,IV; Mort.; BLA & Supp.; GB. Also Physiology.

7807. PLAYFAIR, Lyon. 1818-1898. Br. DSB; ICB; P II,III,IV; BLA; GB; GD1. See also 7611/4a,4b and 15539/1.

7808. REISET, Jules. 1818-after 1891. Fr. P II,III,IV.

7809. SCHNEDERMANN, Georg Heinrich Eberhard. 1818-1881. Ger. P II, III.

7810. SCHWEIZER, Matthias Eduard. 1818-1860. Switz. P II.

7811. SMEE, Alfred. 1818-1877. Br. P II,III; BMNH; BLA; GD. Also Physiology*.
1. *Elements of electro-metallurgy*. London, 1841. RM. 2nd ed., 1843. 3rd ed., 1851.

7812. SMITH, John Lawrence. 1818-1883. U.S.A. ICB; P III; BMNH; BLA; GE. Also Mineralogy*.
1. *Mineralogy and chemistry*. Louisville, Ky., 1873. RM.

7813. WEBER, Thomas Reinhardt. 1818-1876. Ger. P II,III.

7814. WILSON, George. 1818-1859. Br. P II; GD. Also Biography*.
1. *Chemistry*. London, 1850. X. Another ed., 1856.
2. *Religio chemici: Essays*. London, 1862. RM/L.
3. *Inorganic chemistry*. London, 1866. X. Another ed., rev. and enl. by H.G. Madan, 1871.

7815. WOLFF, Emil Theodor von. 1818-1896. Ger. P II,III; GC.
1. *Quellen-Literatur der theoretisch-organischen Chemie*. Halle, 1845. X. Reprinted?

7816. ANDERSON, Thomas. 1819-1874. Br. DSB; P I,III; BLA; GD.

7817. BOILLOT, Alexis. 1819-after 1883. Fr. P III.

7818. BOWMAN, John Eddowes (second of the name). 1819-1854. Br. P III; GD.
    1. *Introduction to practical chemistry.* London, 1848. X. 6th ed., 1871.
    2. *Practical handbook of medical chemistry.* London, 1850. X. 4th ed., 1862.

7819. CRACE-CALVERT, Frederick. 1819-1873. Fr./Br. P III. Under CALVERT: ICB; GB; GD.

7820. HAUER, Carl von. 1819-1880. Aus. P III.

7821. LEFORT, Jules. 1819-1896. Fr. P III.

7822. LUDWIG, Johann Friedrich Herrmann. 1819-1873. Ger. P I,III.

7823. MANSFIELD, Charles Blackford. 1819-1855. Br. ICB; P II; GD.
    1. *A theory of salts: A treatise on the constitution of bipolar (two-membered) chemical compounds.* London, 1865. Ed. with a preface by N. Story Maskelyne.

7824. RIECKHER, Theodor. ca. 1819-after 1858. Ger. P II.

7825. ROCHLEDER, Friedrich. 1819-1874. Prague/Aus. P II,III; GC.

7826. SACC, Frédéric. 1819-after 1884. Fr. P II,III,IV; BMNH.

7827. SCHLOSSBERGER, Julius Eugen. 1819-1860. Ger. ICB; P II; BLA; GC.

7828. SONNENSCHEIN, Franz Leopold. 1819-1879. Ger. P II,III.

7829. THENARD, Arnould Paul Edward. 1819-1884. Fr. P III.

7830. UNGER, Julius Bodo. ca. 1819-after 1852. Ger. P II.

7831. VOELKEL, Friedrich Karl. 1819-1880. Switz. P II,III.

7832. TISSIER, Charles. d. 1864. Fr. P III.

7833. BAUMHAUER, Eduard Henrik van. 1820-1885. Holl. DSB; P I,III, IV.

7834. BROMEIS, Johann Conrad. 1820-1862. Ger. P I & Supp.

7835. CASSELMANN, Wilhelm Theodor Oscar. 1820-1872. Ger. P I,III.

7836. CORENWINDER, Benjamin. 1820-1884. Fr. P I,III.

7837. DEXTER, William Prescott. 1820-after 1882. U.S.A. P III.

7838. GERDING, Theodor. 1820-after 1875. Ger. P I,III. Also History of Chemistry*.

7839. HERAPATH, William Bird. 1820-1868. Br. DSB; P I,III.

7840. LAMY, Claude Auguste. 1820-1878. Fr. P III.

7841. LUCA, Sebastiano de. 1820-1880. It. P III.

7842. MORFITT, Campbell. 1820-1897. U.S.A. P II; GE.

7843. NICKLÈS, (François) Joseph Jérôme. 1820-1869. Fr. P II,III. See also 8807/1a.

7844. REGNAULD, Jules Antoine. 1820-1895. Fr. P III,IV.

7845. SCHUNCK, (Henry) Edward. 1820-1903. Br. DSB; P II,III,IV; GD2.

7846. WERTHEIM, Theodor. 1820-1864. Aus./Hung. P II,III; GC.

7847. BAUDRIMONT, Marie Victor Ernest. 1821-1885. Fr. P III.

7848. KERNDT, Karl Huldreich Theodor. 1821-after 1857. Ger. P I.

7849. KEYSER, Carl Johan. 1821-after 1880. Swed. P III.

7850. LOSCHMIDT, Joseph. 1821-1895. Aus. DSB; ICB; P III,IV; GC. Also Physics.
1. *Chemische Studien.* Part I. [No more publ.] Vienna, 1861. X.
1a. ——— [Extract] *Konstitutions-Formeln der organischen Chemie in graphischer Darstellung.* Leipzig, 1913. (OKEW. 190)

7851. MUSPRATT, (James) Sheridan. 1821-1871. Br. ICB; P II,III; GD.
1. *Chemistry, theoretical, practical and analytical, as applied and relating to the arts and manufactures.* 2 vols. London, 1860.
See also 7591/1a.

7852. NATTERER, Johann August. 1821-1901. Aus. P II,III,IV.

7853. POLECK, Thomas August Theodor. 1821-1906. Ger. P III,IV.

7854. QUADRAT, Bernhard. 1821-1895. Brünn. P II,III.

7855. ROBIN, Charles Philippe. 1821-1885. Fr. DSB; ICB; BMNH; BLA & Supp.; GA. Also Microscopy and Anatomy*.
1. *Traité de chimie anatomique et physiologique, normale et pathologique.* Paris, 1853.

7856. STAEDELER, Georg Andreas Karl. 1821-1871. Ger./Switz. P II, III; GC.

7857. BECK, Wilhelm von. 1822-after 1890. Russ. P III; BMNH.

7858. BOUIS, Jules. 1822-1886. Fr. P III,IV.

7859. CHANCEL, Gustav Charles Bonaventure. 1822-1890. Fr. DSB; P III,IV.

7860. DEACON, Henry. 1822-1876. Br. ICB; P III.

7861. DONNY, François Marie Louis. 1822-after 1878. Belg. P I,III.

7862. GIBBS, (Oliver) Wolcott. 1822-1908. U.S.A. DSB; ICB; P I,III, IV; GB; GE.

7863. PASTEUR, Louis. 1822-1895. Fr. DSB; ICB; P II,III; BMNH; Mort.; BLA; G. Also Microbiology*#. See also 15566/1.
1. *La dissymétrie moléculaire des produits organiques naturels.* Periodical article, 1860.
1a. ——— German trans. Leipzig, 1891. (OKEW. 28)
1b. ——— English trans. Edinburgh, 1905 and re-issues. (Alembic Club reprints. 14) RM/L.

7864. PODWYSSOTZKY, Valerian Ossipovich. 1822-1892. Russ. P III; BLA.

7865. PORTER, John Addison. 1822-1866. U.S.A. P III; GE.
1. *Principles of chemistry.* New York, 1857. RM.

7866. ROBIQUET, Henri Edme. 1822-1860. Fr. P II.

7867. SCHMIDT, Karl Ernst Heinrich. 1822-1894. Dorpat. P II,III,IV. BMNH; Mort.; BLA. See also 14832/1.

7868. SPIRGATIS, Johann Julius Hermann. 1822-1899. Ger. P III,IV.

7869. STRECKER, Adolph (Friedrich Ludwig). 1822-1871. Ger./Nor. P II,III; GC.
 1. *Kurzes Lehrbuch der organischen Chemie.* Brunswick, 1851. X. 5th ed., 1868. Another ed., 1874.
 1a. ——— *Short text-book of organic chemistry.* London, 1881. Rev. by J. Wislicenus. Trans. and ed., with additions, by W.H. Hodgkinson. 2nd ed., 1885.

7870. STRUVE, Heinrich Wilhelm. 1822-1908. Russ. P II,III,IV; BMNH.

7871. WICKE, Johann Anton Wilhelm. 1822-1871. Ger. P II,III.

7872. BERNAYS, Albert James. 1823-1892. Br. GD1.
 1. *Household chemistry; or Rudiments of the science applied to every-day life.* London, 1852. X. 7th ed., entitled *Student's chemistry*, 1869.

7873. BIZIO, Giovanni. 1823-1881? It. P III.

7874. ENGELBACH, Theophil. 1823-1872. Ger. P III.

7875. HELDT, Wilhelm. 1823-1865. Ger. P I,III.

7876. LEA, Matthew Carey. 1823-1897. U.S.A. ICB; P III,IV; GE.

7877. LÖWE, Julius Friedrich Ferdinand Franz. 1823-after 1883. Ger. P I,III.

7878. SULLIVAN, William Kirby. 1823-1890. Br. P III.

7879. VOHL, Eduard Hermann Ludwig. 1823-after 1878. Ger. P II,III.

7880. WAGNER, (Johann) Rudolph von. 1823-1880. Ger. P II,III; GC. See also 7758/2a.

7881. CZYRNIANSKI, Emil. 1824-1884. Cracow. P III,IV.

7882. DEBUS, Henry. 1824-1916. Ger./Br. P I,III,IV.

7883. HITTORF, Johann Wilhelm. 1824-1914. Ger. DSB; ICB; P I,III,IV.
 1. *Über die Wanderungen der Ionen während der Elektrolyse.* [Periodical articles, 1853-59] 2 vols. Leipzig, 1891. (OKEW. 21,23)

7884. KESSLER, Friedrich Christian Ludwig. 1824-1896. Ger. P I,III, IV.

7885. KIRCHHOFF, Gustav Robert. 1824-1887. Ger. DSB; ICB; P I,III, IV; GB; GC. Also Mechanics* and Physics*#.
 1. *Chemische Analyse durch Spectralbeobachtungen.* [Periodical article by G. Kirchhoff and R. Bunsen, 1860] Leipzig, 1895. (OKEW. 72)

7886. LIST, Karl Georg Ernst. 1824-after 1880. Ger. P I,III.

7887. SCHLOESING, Jean Jacques Théophile. 1824-1919. Fr. P III,IV.

7888. SCHWARZ, Karl Leonhard Heinrich. 1824-1890. Ger./Aus. P II, III,IV.

7889. WARREN, Cyrus Moors. 1824-1891. U.S.A. P III,IV; GE.

7890. WILLIAMSON, Alexander William. 1824-1904. Br. DSB; P II,III, IV; GB; GD2.
 1. *Papers on etherification and the constitution of salts,*

1850-1856. Edinburgh, 1902 and re-issues. (Alembic club reprints. 16)
2. *Chemistry for students*. Oxford, 1865. 2nd ed., 1868. 3rd ed., 1873. Another ed., 1883.

7891. BECHI, Emilio. fl. 1851-1879. It. P III.

7892. ERLENMEYER, (Richard August Carl) Emil. 1825-1909. Ger. DSB; P III,IV.

7893. ETTI, Carl. 1825-1890. Aus. P IV.

7894. FRANKLAND, Edward. 1825-1899. Br. DSB; ICB; P I,III,IV; GB; GD1.
1. *On recent chemical researches in the Royal Institution*. London, 1864.
2. *Lecture notes for chemical students*. 2 vols. London, 1866-70. Another ed., 1876-82.
3. *Inorganic chemistry*. London, 1870. X. Another ed., rev. by F.R. Japp, 1884.
4. *Experimental researches in pure, applied and physical chemistry*. London, 1877. Collected papers, ed. by himself. (The text of the originals is sometimes modified.)

7895. HENNEBERG, Johann Wilhelm Julius. 1825-1890. Ger. P I,III.

7896. HLASIWETZ, Heinrich Hermann. 1825-1875. Aus. P I,III; GC.

7897. HOPPE-SEYLER, (Ernst) Felix (Immanuel). 1825-1895. Ger. DSB; P I,III,IV (HOPPE, E.F.I.); Mort.; BLA; GC.
1. *Anleitung zur pathologisch-chemischen Analyse*. Berlin, 1858. X.
1a. ——— Another ed. *Handbuch der physiologisch-chemischen Analyse*. Berlin, 1875. 5th ed., 1883.
2. *Medicinisch-chemische Untersuchungen*. Berlin, 1866-71.
3. *Physiologische Chemie*. 4 parts. Berlin, 1877-81.

7898. POHL, Joseph Johann. 1825-after 1882. Aus. P II,III; BMNH.

7899. SCHLIEPER, Adolf. 1825-1887. Ger. GC.

7900. SCHNEIDER, Ernst Robert. 1825-1900. Ger. P II,III,IV.

7901. TROOST, Louis Joseph. 1825-1911. Fr. DSB; ICB; P II,III.

7902. WETHERILL, Charles Mayer. 1825-1871. U.S.A. ICB; P III; GE.

7903. BLOMSTRAND, Christian Wilhelm. 1826-1897. Swed. DSB; P I,III, IV.

7904. CANNIZZARO, Stanislao. 1826-1910. It. DSB; ICB; P IV; GB.
1. *Sunto di un corso di filosofia chimica*. Periodical article, 1858.
1a. ——— German trans. Leipzig, 1891. (OKEW. 30)
1b. ——— English trans. Edinburgh, 1910 and re-issues. (Alembic Club reprints. 18)
2. [Trans. of a periodical article of 1871] *Historische Notizen und Betrachtungen über die Anwendung der Atomtheorie in der Chemie, und über die Systeme der Konstitutionsformeln von Verbindungen*. Stuttgart, 1913. Trans. from the Italian by B.L. Vanzetti and M. Speter.

7905. CHAUTARD, Jules Maria Augustin. 1826-after 1880. Fr. P III. Also Physics.

7906. DAVY, Edmund William. 1826-after 1896. Br. P III,IV.

7907. FIELD, Frederick. 1826-1885. Br. GD.

7908. GORE, George. 1826-1908. Br. DSB; P III,IV; GD2. Also Science in General*.
  1. *Practical chemistry, including the theory and practice of electro-deposition.* London, 1856.
  2. *The art of electro-metallurgy, including all known processes of electro-deposition.* London, 1870. X. Another ed., New York, 1877. 2nd ed., 1884.

7909. HINTERBERGER, Friedrich. 1826-1875. Aus. P I,III. See also 8067/1.

7910. HUNT, Thomas Sterry. 1826-1892. U.S.A./Canada. DSB; ICB; P III,IV; BMNH; GB; GE. Also Mineralogy* and Geology*.
  1. *A new basis for chemistry: A chemical philosophy.* Boston, 1887. X. 4th ed., New York, 1892.

7911. KLETZINSKY, Vincenz. 1826-1882. Aus. P III; BLA.
  1. *Compendium der Biochemie.* Vienna, 1858.

7912. PEBAL, Leopold von. 1826-1887. Lemberg/Aus. P II,III.

7913. RITTHAUSEN, Karl Heinrich Leopold. 1826-1912. Ger. P II,III, IV.

7914. SOKOLOFF, Nikolai Nikolaievich. 1826-1877. Russ. P II,III.

7915. THOMSEN, (Hans Peter Jörgen) Julius. 1826-1909. Den. DSB; P II,III,IV; GB.
  1. *Thermochemische Untersuchungen.* 4 vols. Leipzig, 1882-86.

7916. TRAUBE, Moritz. 1826-1894. Ger. DSB; ICB; P II,III,IV; Mort.

7917. WORMLEY, Theodor George. 1826-1897. U.S.A. P III,IV; BLA; GE.

7918. ABEL, Frederick Augustus. 1827-1902. Br. P III,IV; GB; GD2.
  1. *Hand-book of chemistry, theoretical, practical and technical.* London, 1854. With C.L. Bloxam.

7919. BEKETOV, Nikolai Nikolaievich. 1827-1911. Russ. DSB; ICB; P IV.

7920. BERTAGNINI, Cesare Pietro T. 1827-1857. It. P III.

7921. BERTHELOT, (Pierre Eugène) Marcellin. 1827-1907. Fr. DSB; ICB; P I,III,IV; GB. Also Science in General* and History of Chemistry*.
  1. *Chimie organique fondée sur la synthèse.* 2 vols. Paris, 1860. RM/L. Reprinted, Brussels, 1966.
  2. *Recherches sur les affinités. De la formation et de la décomposition des éthers.* Paris, 1862. With P. de Saint-Gilles.
  2a. ——— German trans. Leipzig, 1910. (OKEW. 173)
  3. *Leçons sur les méthodes générales de synthèse en chimie organique.* Paris, 1864. RM/L.
  4. *Sur la force de la poudre et des matières explosives.* Paris, 1871. RM/L.
  5. *Traité élémentaire de chimie organique.* Paris, 1872. RM/L. Another ed., 2 vols, 1886.
  6. *La synthèse chimique.* Paris, 1876. RM.

7. *Essai de mécanique chimique fondée sur la thermochimie.* 2 vols. Paris, 1879. RM/L.
8. *Traité pratique de calorimétrie chimique.* Paris, 1893. RM.
9. *Thermochimie: Données et lois numériques.* 2 vols. Paris, 1897. RM/L.
10. *Chaleur animale: Principes chimiques de la production de la chaleur chez les êtres vivants.* 2 vols. Paris, 1899. RM.
11. *Chimie végétale et agricole.* 2 vols. Paris, 1899. RM.
12. *Les carbures d'hydrogène, 1851-1901: Recherches expérimentales.* 3 vols. Paris, 1901.
13. *Traité pratique de l'analyse des gaz.* Paris, 1906. RM.

7922. BOTHE, Friedrich Ferdinand. 1827-after 1881. Ger. P III.

7923. COOKE, Josiah Parsons. 1827-1894. U.S.A. DSB; P III,IV; GE. Also Science in General*.
1. *Elements of chemical physics.* Boston, 1860. 4th ed., London, 1886.
2. *First principles of chemical philosophy.* Boston, 1869. Another ed., London, 1882.
3. *The new chemistry.* London, 1873. X. 2nd ed., 1874. 4th ed., 1876. 5th ed., 1879. 8th ed., 1884. 9th ed., 1887.

7924. DEBRAY, Henri Jules. 1827-1888. Fr. DSB; ICB; P III,IV.
1. *Cours élémentaire de chimie.* Paris, 1863.

7925. GLADSTONE, John Hall. 1827-1902. Br. DSB; P III,IV; GB; GD2. Also Biography*.

7926. GOESSMANN, Charles Anthony. 1827-1910. U.S.A. P III; GE.

7927. GÜNSBERG, Hermann Rudolph. 1827-after 1878. Lemberg. P III.

7928. GUNNING, Jan Willem. 1827-1901. Holl. P III,IV.

7929. KREMERS, Peter. 1827-after 1870. Ger. P I,III.

7930. LIMPRICHT, Heinrich. 1827-1909. Ger. P I,III,IV.

7931. LIVEING, George Downing. 1827-1924. Br. P III,IV; GD4. Also Physics*.

7932. MAYER, Wilhelm Karl Heinrich. 1827-after 1867. Ger. P III.

7933. MÈNE, Charles. 1827-1876. Fr. P III.

7934. REICHARDT, Eduard. 1827-1891. Ger. P II,III.

7935. SCHEIBLER, Carl Bernhard Wilhelm. 1827-1899. P III,IV.

7936. ARENDT, Rudolf Friedrich Eugen. 1828-1902. Ger. P III.

7937. BUFF, Heinrich Ludwig. 1828-1872. Ger./Prague. P III.

7938. BUTLEROV, Aleksandr Mikhailovich. 1828-1886. Russ. DSB; ICB; P I,III,IV.

7939. DUPPA, Baldwin Francis. 1828-1873. Br. P III.

7940. FLÜCKIGER, Friedrich August. 1828-1894. Switz./Ger. ICB; P III,IV; BMNH; BLA & Supp.

7941. FUNKE, Otto. 1828-1879. Ger. P I,III; BMNH; Mort.; BLA & Supp. Also Physiology*.

1. *Atlas der physiologischen Chemie.* Leipzig, 1853. X. 2nd ed., 1858.
1a. ——— *Atlas of physiological chemistry.* London, 1853.

7942. HIRZEL, (Christoph) Heinrich. 1828-1908. Ger. P I,III.

7943. JACQUEMIN, Eugène Théodore. 1828-after 1877.

7944. MÜLLER, Karl Alexander. 1828-after 1880. Ger./Swed. P II,III.

7945. PUGH, Evan. 1828-1864. U.S.A. P III; GE. See also 7732/1.

7946. ROSING, Anton. 1828-1867. Nor. P II,III.

7947. SCHWANERT, Franz Hugo. 1828-1902. Ger. P III,IV.

7948. TERREIL, (Claire) Auguste. 1828-after 1892. Fr. P III,IV.

7949. USLAR, Julius Wilhelm Louis von. 1828-after 1861. Ger. P II.

7950. WESELSKY, Philipp. 1828-1889. Aus. P III,IV.

7951. ABASHEV, Dmitri Nikolaievich. 1829-1880. ICB.

7952. CARIUS, Ludwig. 1829-1875. Ger. P III.

7953. EBERMAYER, Ernst Wilhelm Ferdinand. 1829-1908. Ger. P III,IV. Also Meteorology.
1. *Die physiologische Chemie der Pflanzen. Zugleich Lehrbuch der organischen Chemie und Agrikulturchemie.* Berlin, 1882.

7954. GRIESS, (Johann) Peter. 1829-1888. Br. DSB; ICB; P III,IV; GC.

7955. GUIGNET, Charles Ernest. 1829-after 1899. Fr. P III,IV.

7956. HOUZEAU, (Jean) Auguste. 1829-after 1893. Fr. P I,III,IV.

7957. KEKULÉ (VON STRADONITZ), (Friedrich) August. 1829-1896. Ger./Belg. DSB; ICB; P I,III,IV; GB; GC.
1. *Über die Konstitution und die Metamorphosen der chemischen Verbindungen und über die chemische Natur des Kohlenstoffs* [Periodical article, 1858]. *Untersuchungen über aromatische Verbindungen* [Periodical articles, 1865 onwards]. Leipzig, 1904. (OKEW. 145)
2. *Lehrbuch der organischen Chemie oder der Chemie der Kohlenstoffverbindungen.* Erlangen, 1859. X. Another ed., 2 vols, 1861-66.
3. *Chemie der Benzolderivate oder der aromatischen Substanzen.* Erlangen, 1867-87. RM.
4. *Cassirte Kapitel aus der Abhandlung "Über die Carboxytartronsäure und die Constitution des Benzols."* Weinheim, 1965. Facsimile of an unpublished MS.

7958. KRAUT, Karl Johann. 1829-1912. Ger. P III,IV. See also 7488/1a,1c.

7959. MARCET, William. 1829-1900. Mort.

7960. MUSCULUS, Frédéric Alphonse. 1829-1888. Fr. P III.

7961. ODLING, William. 1829-1921. Br. DSB; P III.
1. *A course of practical chemistry.* London, 1854. X. 2nd ed., 2 parts, 1863-65.
2. *A manual of chemistry, descriptive and theoretical.* Part 1. [No more publ.] London, 1861.

3. *Lectures on animal chemistry.* London, 1866.
4. *A course of six lectures on the chemical changes of carbon.* London, 1869. Ed. with notes by W. Crookes.
5. *Outlines of chemistry; or Brief notes of chemical facts.* London, 1870.

See also 7646/2a.

7962. REBOUL, Pierre Edmond. 1829-1902. Fr. P III,IV.

7963. REYNOSO, Alvaro. 1829-1889. Cuba. P III.

7964. RICHE, (Jean Baptiste Léopold) Alfred. 1829-1908. Fr. P II, III,IV.

7965. SCHÜTZENBERGER, Paul. 1829-1897. Fr. P III,IV; GB.
1. *Les fermentations.* Paris, 1875. X.
1a. ——— *On fermentation.* London, 1876.

7966. THUDICHUM, Johann Ludwig Wilhelm. 1829-1901. Ger./Br. ICB; P III; BLA.
1. *A manual of chemical physiology.* London, 1872.
2. *Annals of chemical medicine.* 2 vols. London, 1879-81.
3. *A treatise on the chemical constitution of the brain.* London, 1884. X. Reprinted, Hamden, Conn., 1962.

7967. VALENTIN, William George. 1829-1879. Br. P III.
1. *Introduction to inorganic chemistry.* London, 1872. 3rd ed., 1876.

7968. WEBER, Friedrich Rudolph. 1829-1894. Ger. P II,III,IV.

7969. WILLIAMS, Charles Greville. 1829-1910. Br. P III; GD2.
1. *Handbook of chemical manipulation.* London, 1857.

7970. BEMMELEN, Jakob Maarten van. 1830-1911. Holl. P III,IV.

7971. FROEHDE, Karl Friedrich August. 1830-after 1867. Ger. P III.

7972. GIRARD, Aimé. 1830-1898. Fr. P III,IV.

7973. HERAPATH, Thornton John. 1830-1858. Br. P I & Supp.

7974. JOHNSON, Samuel William. 1830-1909. U.S.A. P III,IV; GE.
1. *From the letter files of S.W. Johnson, professor of agricultural chemistry in Yale University, 1856-1896.* New Haven, 1913. Ed. by his daughter, E.A. Osborne.

7975. LORIN, Jacques Esther. 1830-after 1881. Fr. P III.

7976. MEYER, (Julius) Lothar. 1830-1895. Ger. DSB; ICB; P II,III, IV; BLA; GB; GC.
1. *Die modernen Theorien der Chemie.* Breslau, 1864. X. 2nd ed., 1872. 4th ed., 1880. 5th ed., 1884.
1a. ——— *Modern theories of chemistry.* London, 1888. Trans. from the 5th German ed. by P.P. Bedson and W.C. Williams. RM/L.
2. *Grundzüge der theoretischen Chemie.* Leipzig, 1890. X.
2a. ——— *Outlines of theoretical chemistry.* London, 1892. Trans. by P.P. Bedson and W.C. Williams. RM/L.

7977. NEUBAUER, Carl Theodor Ludwig. 1830-1879. Ger. P II,III.

7978. RAOULT, François Marie. 1830-1901. Fr. DSB; ICB; P III,IV; GB.

7979. RUSSELL, William James. 1830-1909. Br. ICB; P III,IV; GD2.

7980. SCHLAGDENHAUFFEN, Charles Frédéric. 1830-1907. Fr. P II,III, IV; BMNH. Also Botany.

7981. SCHMITT, Rudolf Wilhelm. 1830-1898. Ger. P III,IV.

7982. SCHWARZENBACH, Valentin. 1830-1890. Switz. P II,III,IV.

7983. SHISHKOV, Leon (or Liew) Nikolaievich. 1830-1908. Russ. ICB; P II,III (SCHISCHOFF).

7984. BLOXAM, Charles Loudon. 1831-1887. Br.
1. *Chemistry, inorganic and organic.* London, 1867. 2nd ed., 1872. 3rd ed., 1875. Other eds, 1883 and 1890.
2. *Laboratory teaching.* London, 1869. 2nd ed., 1871.
3. *Metals: Their properties and treatment.* London, 1870. X. 2nd ed., 1871. Other eds, 1882 and 1885.
See also 7918/1.

7985. CASAMAJOR, Paul. 1831-after 1881. U.S.A. P III.

7986. CLERMONT, Philippe Henri Arnout de. 1831-1921. Fr. P III,IV.

7987. COUPER, Archibald Scott. 1831-1892. Br. DSB; ICB.
1. *On a new chemical theory.* [Periodical article, 1858] German trans. Leipzig, 1911. (OKEW. 183)
2. *On a new chemical theory. Researches on salicylic acid.* [Periodical articles, 1858] Edinburgh, 1933 and re-issues. (Alembic Club reprints. 21)

7988. GERLAND, Balthasar William. 1831-after 1900. Ger./Br. P III, IV.

7989. HVOSLEF, Hans Henrik. 1831-after 1869. Nor. P III.

7990. LANDOLT, Hans Heinrich. 1831-1910. Ger. DSB; P I,III,IV. Also Physics*.
1. *Das optische Drehungsvermögen organischer Substanzen und dessen praktische Anwendungen.* Brunswick, 1879.
1a. ——— *Handbook of the polariscope and its applications.* London, 1882. Adapted from the German ed. by D.C. Robb and V.H. Veley.
2. *Physikalisch-chemische Tabellen.* Berlin, 1883. With R. Börnstein. 2nd ed., 1894.

7991. LUBOLDT, Rudolph August. 1831-after 1860. Ger. P III.

7992. MAISCH, John Michael. 1831-1893. U.S.A. P III; GE.

7993. MATTHIESSEN, Augustus. 1831-1870. Br. DSB; P II,III; GD.

7994. OUDEMANS, Anthonie Cornelis (first of the name). 1831-1895. Holl. P III,IV.

7995. PISANI, Félix. 1831-after 1882. Fr. P III; BMNH. Also Mineralogy*.

7996. STAHLSCHMIDT, Johann Karl Friedrich. 1831-1902. Ger. P III,IV.

7997. BINZ, Carl. 1832-1913. Ger. P III,IV; BLA & Supp.

7998. CROOKES, William. 1832-1919. Br. DSB; ICB; P III,IV; GB; GD3. Also Physics*. See also Index.
1. *Select methods in chemical analysis (chiefly inorganic).* London, 1871. RM/L. 2nd ed., 1886.

7999. ENGELHARDT, Alexander Nikolaievich. 1832-1893. Russ. P I, III; ICB.

8000. FRIEDEL, Charles. 1832-1898. Fr. P III,IV; GB. Also Mineralogy*. See also 7787/4.

8001. GÄNGE, Christian. 1832-1909. Ger. P IV.

8002. HUPPERT, (Karl) Hugo. 1832-1904. Prague. P III,IV; BLA.

8003. MALLET, John William. 1832-1912. Br./U.S.A. ICB; P III,IV; BLA; GE; GF. See also 9584/1.

8004. MULDER, Eduard. 1832-1924. Holl. P III,IV.

8005. NATANSON, Jacob. 1832-1884. Warsaw. P III.

8006. PÉAN DE SAINT-GILLES, Léon. 1832-1862. Fr. DSB; P III.

8007. POLLACCI, Egidio. 1832-1913. It. ICB; P III,IV.

8008. RÜDORFF, Friedrich. 1832-1902. Ger. P III,IV.

8009. STOHMANN, Friedrich Carl Adolf. 1832-1897. Ger. P III,IV; GC.

8010. STORER, Francis Humphreys. 1832-1914. U.S.A. P III,IV; GE.
    1. *First outline of a dictionary of solubilities of chemical substances.* Cambridge, Mass., 1864.
    2. *A cyclopedia of quantitative chemical analysis.* Boston, 1870-73.

8011. ZAENGERLE, Max. 1832-1903. Ger. P IV.
    1. *Lehrbuch der Chemie.* 1870.

8012. BORODIN, Aleksandr Porfirievich. 1833-1887. Russ. The composer. DSB; ICB; P III,IV; GB.

8013. BUCHNER, Max. 1833-1899. Aus. P III,IV.

8014. BURG, Eduard Alexander van der. 1833-after 1881. Holl. P III.

8015. COSSA, Alfonso. 1833-1902. It. P III,IV & Supp.; BMNH.

8016. DITTMAR, William. 1833-1892. Br. DSB; P III,IV.
    1. *Analytical chemistry.* London, 1879. X. Another ed., 1886.

8017. GEUTHER, Anton. 1833-1889. Ger. DSB; P III,IV.

8018. GORGEU, Alexandre. 1833-after 1892. Fr. P III,IV.

8019. GUTHRIE, Frederick. 1833-1886. Br. P III,IV; GD. Also Physics*.

8020. HUSEMANN, August. 1833-1877. Switz. P III.

8021. LANG, Johan Robert Tobias. 1833-1902. Swed. P III,IV.

8022. MOITESSIER, Albert. 1833-1889. Fr. P III.

8023. MÜLLER, Hugo Heinrich Wilhelm. 1833-after 1881. Ger./Br. P III.

8024. OPPENHEIM, (Friedrich Ludwig) Alphons. 1833-1877. Ger. P III; GC.

8025. OSER, Johann. 1833-1912. Aus. P IV.

8026. PERROT, Adolphe. 1833-1887. Switz. P III.

8027. PHIPSON, Thomas Lamb. 1833-1908. Br./Fr. P II,III,IV.
1. *Phosphorescence; or, The emission of light by minerals, plants and animals*. London, 1862. RM/L. Another ed., 1870.
See also 5105/3a.

8028. ROSCOE, Henry Enfield. 1833-1915. Br. DSB; ICB; P II,III,IV; GB; GD3. Also Autobiography* and Biography*. See also Index.
1. *Lessons in elementary chemistry, inorganic and organic*. London, 1867. Other eds, 1871 and 1877.
2. *Spectrum analysis*. London, 1869. 2nd ed., 1870. 3rd ed., 1873. 4th ed., with A. Schuster, 1885.
3. *Chemistry*. London, 1872. (In the series *Science primers*, ed. by Huxley, Roscoe, and Stewart; see 2963/2.) X. Another ed., 1883.
4. *A treatise on chemistry*. 3 vols in 8. London, 1877-92. With C. Schorlemmer.
5. *Inorganic chemistry for advanced students*. London, 1899. With A. Harden.

8029. SCHEURER-KESTNER, Auguste. 1833-1899. Fr. P II,III,IV.

8030. SIMLER, Rudolph Theodor. 1833-1874. Switz. P II (SIMMLER),III.

8031. WAAGE, Peter. 1833-1900. Nor. DSB; P III,IV. See also 8074/1.

8032. WILLM, Edmond. 1833-1910. Fr. P III,IV.

8033. ZULKOWSKY, Karl. 1833-1908. Brünn/Prague. P III,IV.

8034. CARO, Heinrich. 1834-1910. Ger./Br. DSB.

8035. CHURCH, Arthur Herbert. 1834-1915. Br. P III,IV.

8036. ELIOT, Charles William. 1834-1926. U.S.A. President of Harvard, 1869-1909. ICB; GB; GE.
1. *Manual of inorganic chemistry*. Boston, 1867. X. 2nd ed., New York, 1868.

8037. FINKENER, Rudolph Heinrich. 1834-1902. Ger. P III,IV.

8038. FUDAKOWSKI, Hermann Boleslaw. 1834-1878. Warsaw. ICB; P III; BLA.

8039. GERNEZ, Désiré Jean Baptiste. 1834-1910. Fr. P III,IV.

8040. HALLWACHS, Wilhelm. 1834-1881. Ger. P III.

8041. HARCOURT, Augustus George Vernon. 1834-1919. Br. DSB; P IV (VERNON-HARCOURT); GD3.

8042. HENRY, Louis. 1834-1913. Belg. P III,IV.

8043. LUGININ, Vladimir Fedorovich. 1834-1911. Russ./Fr. DSB; P III, IV.

8044. MENDELEEV, Dmitry Ivanovich. 1834-1907. Russ. DSB; ICB; P III, IV (MENDELEJEFF); GB.
1. [In Russian. *Principles of chemistry*.] 2 vols. St. Petersburg, 1870 and later eds. X.
1a. ——— *Grundlagen der Chemie*. St. Petersburg, 1890. Trans. from the 5th Russian ed. by L. Jawein. RM.
1b. ——— *Principles of chemistry*. 2 vols. London, 1891. Trans. from the 5th Russian ed. Another ed., from the 6th Russian, 1897.

8045. NAQUET, Alfred Joseph. 1834-1916. Fr. ICB; P III; GB.
1. *Principes de chimie, fondés sur les théories modernes.* Paris, 1865. X.
1a. ——— *Principles of chemistry, founded on modern theory.* London, 1868. Trans. from the 2nd French ed. by W. Cortis. Rev. by T. Stevenson.
See also 7769/1a.

8046. SCHIFF, Hugo Josef. 1834-1915. Switz./It. DSB; ICB; P II,III, IV.

8047. SCHORLEMMER, Carl. 1834-1892. Br. DSB; ICB; P III,IV; GD. Also History of Chemistry*.
1. *A manual of the chemistry of the carbon compounds; or, Organic chemistry.* London, 1874.
See also 8028/4.

8048. SPRENGEL, Hermann Johann Philipp. 1834-1906. Br. P III; GD2.

8049. THAN, Károly. 1834-1908. Hung. DSB; P III,IV (T., Anton Karl von); BLA.

8050. VILLIERS-MORIAMÉ, Antoine. 1834-1932. Fr. ICB.

8051. VOGEL, Hermann Wilhelm. 1834-1898. Ger. P II,III,IV.

8052. VOLHARD, Jacob. 1834-1910. Ger. ICB; P III,IV.

8053. WANKLYN, James Alfred. 1834-1906. Br. DSB; P II,III,IV; GD2.

8054. ZINNO, Silvestre. 1834-after 1882. It. P III.

8055. ATTFIELD, John. 1835-1911. Br. P III.
1. *A manual of chemistry, general, medical and pharmaceutical.* London, 1867. X. 15th ed. (with variant title), 1893.

8056. BAEYER, Adolf (Johann Friedrich Wilhelm) von. 1835-1917. Ger. DSB; ICB; P III,IV; BLAF; GB.
1. *Gesammelte Werke.* 2 vols. Brunswick, 1905. RM/L.

8057. BARKER, George Frederick. 1835-1910. U.S.A. P III,IV; GE.

8058. DUPRÉ, August. 1835-1907. Br. P III; GD2.

8059. FITTIG, Rudolph. 1835-1910. Ger. DSB; P III,IV; GB. See also 7592/2b,2c.

8060. FOSTER, George Carey. 1835-1919. Br. P III,IV; GF. Also Physics*.

8061. FREUND, August. 1835-1892. Lemberg. P IV.

8062. GRIMAUX, (Louis) Edouard. 1835-1900. Fr. P III,IV. Also Biography*.
1. *Chimie organique élémentaire.* Paris, 1874. X. 2nd ed., 1878.
See also 7182/8.

8063. HESSE, (Julius) Oswald. 1835-1917. Ger. P III,IV.

8064. LORSCHEID, J. 1835-1884. Belg. P III.

8065. WILDE, Prosper de. 1835-after 1885. Belg. P III.

8066. WISLICENUS, Johannes. 1835-1902. Switz./Ger. DSB; P II,III, IV; GB. See also 7869/1a and 8466/1c.

8067. BAUER, Alexander Anton Emil. 1836-1921. Aus. ICB; P III,IV.
1. *Lehrbuch der chemischen Technik.* Vienna, 1859. With F. Hinterberger.

8068. BOURGOIN, Edme Alfred. 1836-1897. Fr. P III,IV.

8069. CARSTANJEN, Ernst. 1836-1884. Ger. P III.

8070. CHANDLER, Charles Frederick. 1836-1925. U.S.A. ICB; P III,IV; GE.

8071. CHYDENIUS, Johan Jacob. 1836-1890. Fin. P III.

8072. COQUILLION, Jacques J. 1836-after 1896. Fr. P III,IV.

8073. DRAGENDORFF, (Johann) Georg (Noël). 1836-1898. Dorpat. P III, IV; BMNH; BLA & Supp.; GC.

8074. GULDBERG, Cato Maximilian. 1836-1902. Nor. DSB; P III,IV.
1. [Trans. of periodical articles in Norwegian by C.M. Guldberg and P. Waage, 1864, 1867, 1879] *Untersuchungen über die chemischen Affinitäten.* Leipzig, 1899. (OKEW. 104)
2. [Trans. of three periodical articles in Norwegian, 1867-72] *Thermodynamische Abhandlungen über Molekulartheorie und chemische Gleichgewichte.* Leipzig, 1903. (OKEW. 139)

8075. HAUTEFEUILLE, Paul Gabriel. 1836-1902. Fr. DSB; ICB; P III,IV.

8076. HINRICHS, Gustavus Detlef. 1836-1923. Den./U.S.A. P III,IV & Supp.; BMNH; BLA. Also Various Fields.
1. *The true atomic weights of the chemical elements and the unity of matter.* St. Louis, 1894. RM.

8077. ISAMBERT, Nicolas Ferdinand Irénée. 1836-1890. Fr. P III,IV.

8078. LAUTH, Charles. 1836-1913. Fr. P IV.

8079. LIEBEN, Adolf. 1836-1914. It./Prague/Aus. ICB; P III,IV.

8080. MICHAELSON, Karl August. 1836-ca. 1875. Swed./Fr. P III.

8081. MITSCHERLICH, Alexander. 1836-1918. Ger. P III,IV. See also 7533/4.

8082. MORTON, Henry. 1836-1902. U.S.A. P III; GE.

8083. PETERSEN, Carl Theodor. 1836-1919. Ger. P III,IV.

8084. SCHÖNN, Johann Ludwig. 1836-1894. Ger. P III,IV.

8085. TICHBORNE, Charles R.C. 1836-after 1883. Br. P III.

8086. WHEELER, Charles Gilbert. 1836-after 1883. U.S.A. P III.

8087. BRÜNING, Adolf von. 1837-1884. Ger. P IV; GC.

8088. DIVERS, Edward. 1837-1912. Br./Japan. ICB; P III,IV.

8089. FERGUSON, John. 1837-1916. Br. ICB.

8090. GAUTIER, Armand (Emile Justin). 1837-1920. Fr. DSB; ICB; P III,IV; BLAF.
1. *Chimie appliquée à la physiologie, à la pathologie et à la hygiène.* 2 vols. Paris, 1874.

8091. GIRARD, Charles. 1837-1918. Fr. P III,IV.

8092. GOPPELSRÖDER, Friedrich. 1837-1919. Switz./Ger. ICB; P III,IV.

8093. HÜBNER, Hans. 1837-1884. Ger. P III,IV.

8094. JÖRGENSEN, Sophus Mads. 1837-1914. Den. DSB; ICB; P III,IV.

8095. KÜHNE, Wilhelm Friedrich. 1837-1900. Ger. DSB; P IV; BMNH; Mort.; BLA; GB. Also Physiology.
1. *Lehrbuch der physiologischen Chemie*. Leipzig, 1868.

8096. LATSCHINOFF, Pavel Alexandrovich. 1837-1902. Russ. P III,IV.

8097. LENSSEN, Ernst. 1837-after 1870. Ger. P III.

8098. MARKOVNIKOV, Vladimir Vasilevich. 1837-1904. Russ. DSB; ICB; P III,IV.

8099. NAUMANN, Alexander Nicolaus Franz. 1837-1922. Ger. DSB; P III, IV.
1. *Lehr- und Handbuch der Thermochemie*. Brunswick, 1882.

8100. NEWLANDS, John Alexander Reina. 1837-1898. Br. DSB; ICB; GB.
1. *On the discovery of the periodic law, and on relations among the atomic weights*. London, 1884. RM/L.

8101. OTTO, Friedrich Wilhelm Robert. 1837-1907. Ger. P III,IV.

8102. SILVA, Roberto Duarte da. 1837-after 1882. Fr. P III.

8103. BARBAGLIA, Giovanni Angelo. 1838-1891. It. P III,IV.

8104. BEILSTEIN, Friedrich Konrad. 1838-1906. Russ. DSB; P III,IV. Also History of Chemistry*.
1. *Handbuch der organischen Chemie*. 2 vols. Leipzig, 1881-83. X. 2nd ed., 3 vols, Hamburg, 1886-90.

8105. BOISBAUDRAN, Paul Emile Lecoq de (*called* François). 1838-1912. Fr. DSB; P III,IV (LECOQ).
1. *Spectres lumineux*. 2 vols. Paris, 1874. On spectrum analysis.
His works are commonly listed under LECOQ DE BOISBAUDRAN.

8106. BROWN, Alexander Crum. 1838-1922. Br. DSB; P III,IV. Also Physiology. See also 7705/2.

8107. CLAUS, Adolf (Carl Ludwig). 1838-1900. Ger. DSB; P III,IV.

8108. DANILEVSKY, Aleksandr Yakovlevich. 1838-1923. Russ. ICB; Mort.

8109. DELAFONTAINE, Marc Abraham. 1838-after 1882. Switz./U.S.A. P III.

8110. DIBBITS, Hendrik Cornelis. 1838-1903. Holl. P III.

8111. LIPPMANN, Eduard. 1838-1919. Aus. P IV.

8112. LOSSEN, Wilhelm Clemens. 1838-1906. Ger. P III,IV.

8113. MORLEY, Edward Williams. 1838-1923. U.S.A. DSB; ICB; P III, IV; GE. Also Physics.
1. *Outlines of organic chemistry*. 1886.
2. *On the densities of oxygen and hydrogen and on the ratio of their atomic weights*. Washington, 1895.

8114. PERKIN, William Henry. 1838-1907. Br. DSB; ICB; P III,IV; GB; GD2.

8115. RADZISZEWSKI, Bronislaus. 1838-1914. Lemberg. P III,IV.

8116. RIBAN, (Alexandre) Joseph. 1838-1917. Fr. P III,IV.

8117. SCHMIEDEBERG, (Johann Ernst) Oswald. 1838-1921. Ger. P III, IV; Mort.; BLA.

8118. SCHÖNE, Hermann Emil. 1838-1896. Russ. P III,IV.

8119. WINKLER, Clemens (Alexander). 1838-1904. Ger. DSB; P III,IV.

8120. BARTH VON BARTHENAU, Ludwig. 1839-1890. Aus. P III,IV.

8121. BAUMSTARK, Ferdinand. 1839-1889. Ger. P III,IV.

8122. BETTENDORFF, Anton Joseph Hubert Maria. 1839-1902. Ger. P III,IV.

8123. BIRNBAUM, Carl. 1839-1887. Ger. P III,IV.

8124. BLAS, Charles. 1839-after 1901. Belg. P III,IV.

8125. CARNOT, (Marie) Adolphe. 1839-1920. Fr. P III,IV.

8126. CRAFTS, James Mason. 1839-1917. U.S.A./Fr. P III,IV; GE.

8127. ESSON, William. 1839-1916. Br. DSB; P IV.

8128. GAL, Henri. 1839-after 1886. Fr. P III,IV.

8129. HILGER, Albert. 1839-1905. Ger. P III,IV.

8130. HIORTDAHL, Thornstein Hallager. 1839-1925. Nor. P III,IV; BMNH.

8131. JUNGFLEISCH, Emile Clément. 1839-1916. Fr. ICB; P III,IV.

8132. KLIMENKO, Jefim Filimonovich. 1839-after 1903. Russ. P III (K., Euthyme), IV.

8133. KÖRNER, Wilhelm. 1839-1925. It. ICB.
   1. [Trans. of four periodical articles in French and Italian, 1867 onwards] *Über die Bestimmung des chemischen Ortes bei den aromatischen Substanzen.* Leipzig, 1910. (OKEW)

8134. LIEBREICH, Matthias Eugen Oscar. 1839-1908. Ger. P III,IV; BLA.

8135. LUNGE, Georg. 1839-1923. Ger./Switz. DSB; P III,IV.

8136. MALY, (Leo) Richard. 1839-1891. Aus. P III,IV; BLA; BLAF.

8137. MERZ, Victor. 1839-1893. Switz. P III,IV.

8138. MOND, Ludwig. 1839-1909. Br. DSB; ICB; P IV; GB; GD2.

8139. MÜLLER-ERZBACH, Wilhelm. 1839-1914. Ger. P III,IV.

8140. NASSE, Otto Johann Friedrich. 1839-1903. Ger. P III, IV & Supp.; BMNH; BLA.

8141. PFAUNDLER, Leopold. 1839-1920. Aus. P III,IV. Also Physics.

8142. ROSENSTIEHL, (Daniel) Auguste. 1839-1916. Fr. P III,IV.

8143. SESTINI, Fausto Alessandro. 1839-1904. It. P III,IV.

8144. ŠTOLBA, Franz. 1839-1910. Prague. P III,IV.

8145. SWARTS, Théodore. 1839-1911. Belg. P IV.

8146. URBAIN, Victor. 1839-after 1897. Fr. P III,IV.

8147. VINCENT, Camille Philippe. 1839-1910. Fr. P III,IV.

8148. WAHLFORSS, Henrik Alfred. 1839-1899. Fin. P III,IV.

8149. WALKER, John Francis. 1839-after 1901. Br. P III,IV. Also Palaeontology.

8150. AMATO, Domenico. d. 1896. It. P IV.

8151. ALEXEJEFF, Peter. 1840-after 1882. Russ. P III.

8151A ČECH, Carl Otokar. ca. 1840-after 1885. Ger./Russ. P IV.

8152. CLEVE, Per Teodor. 1840-1905. Swed. DSB; P III,IV; BMNH. Also Oceanography.

8153. DUCLAUX, (Pierre) Emile. 1840-1904. Fr. DSB; P III,IV & Supp.; BMNH; BLAF. Also Microbiology*.

8154. HÜFNER, (Carl) Gustav von. 1840-1908. Ger. P III,IV; Mort.; BLAF.

8155. JACOBSEN, Oscar Georg Friedrich. 1840-1889. Ger. P III,IV.

8156. KAEMMERER, Hermann. 1840-1898. Ger. P III,IV.

8157. KOHLRAUSCH, Friedrich Wilhelm Georg. 1840-1910. Ger. DSB; P III,IV. Also Physics*#
   1. *Das Leitvermögen der Elektrolyte insbesondere der wässrigen Lösungen: Methoden, Resultate und chemische Anwendungen.* Leipzig, 1898. With L. Holborn.

8158. MILLS, Edmund James. 1840-1921. Br. P III,IV.
   1. *Destructive distillation.* London, 1886.

8159. NILSON, Lars (or Lorenz) Frederik. 1840-1899. Swed. ICB; P III,IV.

8160. PICCARD, Jules. 1840-1933. Switz. P III,IV.

8161. POPOV, Aleksandr Nikiforivich. 1840?-1881. Russ./Warsaw. DSB; P III (POPOFF).

8162. RATHKE, (Heinrich) Bernhard. 1840-1923. Ger. P III,IV.

8163. SACHSSE, Georg Robert. 1840-1895. Ger. P III,IV.

8164. SCHULZE, Ernst August. 1840-1912. Ger./Switz. P III,IV.

8165. SCHWEITZER, Johann Paul. 1840-1911. U.S.A. P III,IV.

8166. TRIBE, Alfred. 1840-1885. Br. P III,IV.

8167. WIBEL, Ferdinand. 1840-1902. Ger. P IV.

8168. ARONSTEIN, Ludwig. 1841-1913. Holl. P III,IV.

8169. COPPET, Louis Casimir de. 1841-after 1902. Switz. P III,IV.

8170. GABBA, Luigi. 1841-after 1911. It. P III.

8171. GAMGEE, Arthur. 1841-1909. Br. BMNH; BLA & Supp.; BLAF; GD2. Also Physiology.
   1. *A textbook of the physiological chemistry of the human body.* 2 vols. London, 1880-93.

8172. GLASER, Karl Andreas. 1841-after 1926. Ger. P III,IV.

8173. GRAEBE, Karl James Peter. 1841-1927. Ger./Switz. DSB; ICB; P III,IV.

8174. GROVES, Charles Edward. 1841-1920. Br. P III.

8175. HABERMANN, Josef. 1841-1914. Brünn. P IV.

8176. HAMMARSTEN, Olof. 1841-1932. Swed. P III,IV; Mort.; BLAF.
1. *Lehrbuch der physiologischen Chemie*. Wiesbaden, 1891. Trans. from the 2nd Swedish ed. X. 3rd ed., 1895.
1a. ——— *Text-book of physiological chemistry*. New York, 1893. Trans. by J.A. Mandl.

8177. HAMPE, Johann Friedrich Wilhelm. 1841-1899. Ger. P III,IV.

8178. JAFFÉ, Max. 1841-1911. Ger. P IV; Mort.; BLAF.

8179. JANNASCH, Paul Ehrhardt. 1841-1921. Ger. P III,IV.

8180. LANGLEY, John Williams. 1841-1918. U.S.A. P III,IV; GE.

8181. LEMOINE, (Clément) Georges. 1841-1922. Fr. P III,IV.

8182. LINNEMANN, Eduard. 1841-1886. Prague. P III,IV.

8183. McLEOD, Herbert. 1841-1923. Br. P III,IV.

8184. PRUNIER, Léon Louis Adolphe. 1841-1906. Fr. P III,IV; BLAF.

8185. RICHTER, Victor von. 1841-1891. Ger. P III,IV.
1. *Kurzes Lehrbuch der anorganischen Chemie*. Bonn, 1875. X.
1a. ——— *Text-book of inorganic chemistry*. Philadelphia, 1890. Another ed., 1893.
2. *Kurzes Lehrbuch der organischen Chemie*. Bonn, 1875. X.
2a. ——— 5th ed. *Chemie der Kohlenstoffverbindungen; oder, Organische Chemie*. Bonn, 1888.

8186. SAYTZEFF, Alexander Michailovich. 1841-1910. Russ. P III,IV.

8187. SENHOFER, Karl. 1841-1904. Aus. P III.

8188. THOMSEN, Thomas Gottfried. 1841-1901. Den. P III,IV.

8189. TOLLENS, Bernhard Christian Gottfried. 1841-1918. Ger. P III, IV.

8190. ULLIK, Franz. ca. 1841-after 1881. Aus. P III.

8191. WILLGERODT, Conrad Heinrich Christoph. 1841-after 1913. Ger. P III,IV.

8192. WREDEN, Felix Romanovich. 1841-1878. Russ. P III.

8193. BAUBIGNY, Henri. 1842-1912. Fr. P IV.

8194. BEHRENS, Theodor Heinrich. 1842-1905. Holl. P III,IV; BMNH; GF. Also Geology.

8195. BOUCHARDAT, Gustave. 1842-1918. Fr. P III,IV; BLAF.

8196. BUNGE, Nicolaus. 1842-1915. Russ. P III,IV.

8197. DEWAR, James. 1842-1923. Br. DSB; ICB; P III,IV; GB; GD4. Also Physics.
1. *Collected papers*. 2 vols. Cambridge, 1927. Ed. by Lady Dewar et al.

8198. DUBOIS, Edouard. 1842-after 1882. Belg. P III.

8199. EMMERLING, Adolf. 1842-1906. Ger. P III.

8200. ENDEMANN, (Samuel Theodor) Hermann (Carl). 1842-after 1905. Ger./U.S.A. P III,IV.

8201. ENGLER, Karl Oswald Viktor. 1842-1925. Ger. ICB; P III,IV.

8202. FITZ, Albert. 1842-1885. Ger. P IV.

8203. FREBAULT, Aristide. 1842-after 1905. Fr. P IV.

8204. GUSTAVSON, Gabriel. 1842-1908. Russ. P IV.

8205. HOFMANN, Karl Berthold. 1842-1922. Aus. P III,IV.

8206. HORSTMANN, August Friedrich. 1842-1929. Ger. DSB; P III,IV.
1. *Abhandlungen zur Thermodynamik chemischer Vorgänge.* [Periodical articles, 1868 onwards] Leipzig, 1903. (OKEW)

8207. KRAEMER, Gustav Wilhelm. 1842-1915. Ger. P IV.

8208. LADENBURG, Albert. 1842-1911. Ger. DSB; ICB; P III,IV. Also History of Chemistry*.

8209. LIEBERMANN, Carl Theodore. 1842-1914. Ger. DSB; P III,IV.

8210. LUDWIG, Ernst. 1842-1915. Aus. P III,IV; BLAF.

8211. MENSHUTKIN, Nikolai Aleksandrovich. 1842-1907. Russ. DSB; P III,IV.

8212. PINNER, Adolf. 1842-1909. Ger. P III,IV.
1. *Repetitorium der anorganischen Chemie.* Berlin, 1873. X. Another ed., 1887.

8213. SABANEJEFF, Alexander Pavlovich. 1842-after 1899. Russ. P III,IV.

8214. SCHAER, Eduard. 1842-1913. Switz./Ger. P IV. See also 15558/1.

8215. SELL, Eugen. 1842-1896. Ger. P III.

8216. TILDEN, William Augustus. 1842-1926. Br. DSB; P III,IV. Also History of Chemistry*.
1. *Introduction to the study of chemical philosophy: The principles of theoretical and systematic chemistry.* London, 1876. 4th ed., 1884. Other eds: 1882, 1888, 1895.
2. *A manual of chemistry, theoretical and practical.* London, 1897.
See also 7743/1.

8217. TOPSÖE, Haldor Frederic Axel. 1842-1935. Den. P III,IV.

8218. WICHELHAUS, (Carl) Hermann. 1842-after 1921. Ger. P III,IV.

8219. ANNAHEIM, Joseph. 1843-after 1881. Switz. P III.

8220. BOLTON, Henry Carrington. 1843-1903. U.S.A. P III,IV; GE. Also History of Science*.

8221. CLASSEN, Alexander. 1843-1934. Ger. DSB Supp.; P III,IV.

8222. DITTE, Alfred. 1843-1908. Fr. ICB; P III,IV.

8223. DRECHSEL, (Heinrich Ferdinand) Edmund. 1843-1897. Ger. P III; Mort.; GC.

8224. FLEISCHER, Emil. 1843-after 1873. Ger. P III.

8225. GINTL, Wilhelm Friedrich. 1843-1908. Prague. P III,IV.

8226. LAGERMARCK, Berndt Herman. 1843-after 1901. Fin./Russ. P III, IV.

8227. LEEDS, Albert Ripley. 1843-1902. U.S.A. P III,IV.

8228. PURDIE, Thomas. 1843-1916. Br. P IV.

8229. STAEDEL, Wilhelm. 1843-1919. Ger. P III,IV.

8230. TIDY, Charles Meymott. 1843(or 1849)-1892. Br. BLA; GD.
 1. *Handbook of modern chemistry, inorganic and organic.* London, 1878. 2nd ed., 1887.

8231. WEDDIGE, Anton Johann Julius. 1843-after 1904. Ger. P III,IV.

8232. WYRUBOFF, Grigory Nikolaievich. 1843-1913. Fr. ICB; P IV; BMNH. Also Crystallography.

8233. ZINCKE, Ernst Carl Theodor. 1843-after 1922. Ger. P III,IV.

8234. BALLÓ, Mathias. 1844-1930. Hung. P III,IV.

8235. BOEHM, Rudolf Albert Martin. 1844-1926. Ger. P III,IV; BLAF.

8236. BUNGE, Gustav von. 1844-1920. Dorpat/Switz. DSB; ICB; P III, IV; Mort.; BLAF. Also Physiology.
 1. *Lehrbuch der physiologischen und pathologischen Chemie.* Leipzig, 1887. X. 2nd ed., 1889.
 1a. ——— *Text-book of physiological and pathological chemistry.* London, 1890. Trans. from the 2nd German ed. by L.C. Wooldridge.

8237. FRANCHIMONT, Antoine Paul Nicolas. 1844-1919. Holl. P III,IV.

8238. GLUTZ, Joseph Anton Ludwig. 1844-1873. Ger. P III.

8239. KONINCK, Lucien Louis de. 1844-1921. Belg. P III,IV.

8240. LENGYEL, Béla. 1844-1913. Hung. P IV.

8241. LOEW, Oscar Carl Benedikt. 1844-after 1930. Ger./Japan. P III,IV; BMNH.

8242. MIESCHER, (Johann) Friedrich (second of the name). 1844-1895. Switz. DSB; ICB; Mort.; BLAF. Also Physiology*.

8243. O'SULLIVAN, Cornelius. 1844-1907. Br. GD2.

8244. PERGER, Hugo von. 1844-1901. Aus. P III,IV.

8245. PLOSZ, Pál. 1844-1902. Hung. BLAF.

8246. REYNOLDS, James Emerson. 1844-1920. Br. P III,IV; GD3.

8247. (Entry cancelled)

8248. SALET, Pierre Gabriel Georges. 1844-1894. Fr. P III,IV.

8249. SALKOWSKI, Ernst Leopold. 1844-1923. Ger. P II,III; BLAF.

8250. SCHULTZ-SELLACK, Carl Heinrich Theodor. 1844-1879. Ger. P III.

8251. SOMMARUGA, Erwin von. 1844-1897. Aus. P III.

8252. URECH, Friedrich Wilhelm Karl. 1844-after 1904. Ger. P IV.

8253. WARTHA, Vincze. 1844-1914. Hung./Switz. P III,IV.

8254. WEITH, Wilhelm. 1844 (or 1846)-1881. Switz. P III; GC.

8255. WILEY, Harvey Washington. 1844-1930. U.S.A. DSB; P III,IV; GE.

8256. WRIGHT, Charles Romley Alder. 1844-after 1894. Br. P III,IV.
1. *The threshold of science: A variety of simple and amusing experiments illustrating some of the chief physical and chemical properties of surrounding objects, and the effects on them of light and heat.* London, 1891.

8257. ADOR, Emile. 1845-1920. Switz. P III,IV.

8258. BARSILOWSKY, Jakov Nikolaievich. 1845-after 1900. Russ. P III,IV.

8259. BASAROW, Alexander. 1845-after 1883. Russ. P III.

8260. BENDER, Carl. 1845-after 1902. Ger. P III,IV.

8261. BIEDERMANN, Rudolf. 1845-1929. Ger. P III,IV.

8262. CAVAZZI, Alfredo. 1845-after 1923. It. P IV.

8263. CHAPMAN, Ernest Theophron. ca. 1845-1872. Br. P III.

8264. FLEISCHER, Anton. 1845-1877. Klausenburg. P III.

8265. GAYON, Ulysse. 1845-1929. Fr. P III,IV.

8266. JOLY, Alphons (or Eugène) Alexandre. 1845-1897. Fr. ICB; P IV.

8267. JONES, Francis. 1845-1925. Br. P III,IV.

8268. LJUBAVIN, Nicolai Nicolaievich. 1845-after 1903. Russ. P III, IV.

8269. ROSENFELD, Maximilian. 1845-1932. Teschen. P III,IV.

8270. SAYTZEFF, Michail Michailovich. 1845-after 1877. Russ. P III.

8271. SCHMIDT, Ernst Albert. 1845-1921. Ger. P III,IV.

8272. SMITH, Watson. 1845-1920. Br. P III,IV.

8273. THORPE, (Thomas) Edward. 1845-1925. Br. DSB; P III,IV; GD4. Also Biography* and History of Chemistry*.
1. *A manual of inorganic chemistry.* 2 vols. London, 1874. X. Another ed., 1887-96.
2. *A dictionary of applied chemistry.* 3 vols. London, 1890-93. See also 5968/2.

8274. WILM, Theodor Eduard. 1845-1893. Russ. P III,IV.

8275. ATTERBERG, Albert. 1846-after 1882. Swed. P III.

8276. BAUMANN, Eugen. 1846-1896. Ger. P IV; Mort.; BLAF.

8277. BERGLUND, Emil. 1846-1887. Swed. P III,IV.

8278. BURCKER, Emile Eugène. 1846-after 1901. Fr. P IV.

8279. DARMSTAEDTER, Ludwig. 1846-1927. Ger. P IV.

8280. EKSTRAND, Åke Gerhard. 1846-after 1919. Swed. P III,IV.

8281. HAMONET, Jules Léandre. 1846-after 1918. Fr. P IV.

8282. HARTLEY, Walter Noel. 1846-1913. Br. P IV.
1. *Air and its relation to life.* London, 1875. 2nd ed., 1876.

8283. HECHT, Otto. 1846-after 1903. Ger. P III,IV.

8284. LERMONTOV, Iulia Vsevolodna. 1846-1919. Russ. ICB; P III.

8285. MEYER, Richard Emil. 1846-1926. Switz./Ger. P III,IV.

8286. MIXTER, William Gilbert. 1846-1936. U.S.A. P III,IV.

8287. MONSELISE, Giulio. 1846-after 1878. It. P III.
1. *La chimica moderna*. 2 vols. Mantua, 1878.

8288. MÜNTZ, (Charles) Achille. 1846-1916. Fr. P III,IV.

8289. OLIVERI, Vincenzo. 1846-1918. It. P IV.

8290. OLSZEWSKI, Karol Stanislaw. 1846-1915. Cracow. DSB; ICB; P IV.

8291. POST, Julius. 1846-after 1883. Ger. P III.

8292. PREIS, Karl. 1846-after 1880. Prague. P III.

8293. REMSEN, Ira. 1846-1927. U.S.A. DSB; ICB; P III,IV; GE.
1. *Principles of theoretical chemistry*. Philadelphia, 1877.
X. 3rd ed., 1887.
2. *An introduction to the study of the compounds of carbon; or, Organic chemistry*. Boston/London, 1885. Another ed., 1889.
3. *An introduction to the study of chemistry*. New York, 1886.
4. *Elements of chemistry*. New York/London, 1887.
5. *Text-book of inorganic chemistry*. New York/London, 1889.
See also 7592/2c.

8294. RENARD, (Guillaume) Adolphe. 1846-after 1895. Fr. P III,IV.
1. *Traité de chimie appliquée à l'industrie*. Paris, 1890.

8295. SALKOWSKI, Heinrich Otto. 1846-after 1916. Ger. P III,IV.

8296. TAWILDAROW, Nikolai Ivanovich. 1846-after 1904. Russ. P III,IV.

8297. WALZ, Isidor. 1846-1877. U.S.A. P III.

8298. BERTONI, Giacomo. 1847-1921. It. P IV.

8299. BREMER, Gustav Jacob Wilhelm. 1847-1907. Holl. P IV.

8300. BRUNNER, Heinrich. 1847-1910. Switz. P IV.

8301. CHASTAING, Paul Louis. 1847-after 1888. Fr. P III,IV.

8302. CLARKE, Frank Wigglesworth. 1847-1931. U.S.A. DSB; P III,IV; BMNH; GE1.
1. *The constants of nature*. 3 parts. Washington, 1888-96.
1a. ——— Part V. *A recalculation of the atomic weights*. Ib., 1897.

8303. DORP, Willem Anne van. 1847-1914. Holl. P III,IV.

8304. ELTEKOFF, Alexander. 1847-after 1883. Russ. P III.

8305. FRESENIUS, (Remigius) Heinrich. 1847-1920. Ger. P IV.

8306. GUARESCHI, Icilio. 1847-1918. It. P III,IV; BLAF.

8307. GUTHZEIT, Max Adolf. 1847-1915. Ger. P IV.

8308. HOOGEWERFF, Sebastian. 1847-1933. Holl. P III,IV.

8309. JACKSON, Charles Loring. 1847-after 1926. U.S.A. P III,IV.

8310. LE BEL, Joseph Achille. 1847-1930. Fr. DSB; ICB; P III,IV.

8311. LIVERSIDGE, Archibald. 1847-1927. Australia. P III,IV; BMNH; GF.

8312. MEDICUS, Ludwig. 1847-1915. Ger. P III,IV; BLAF.

8313. MEYER, Ernst (Sigismund Christian) von. 1847-1916. Ger. P III,IV. Also History of Chemistry*.

8314. MICHAELIS, Karl Arnold August. 1847-1916. Ger. P III,IV.

8315. NENCKI, Marceli. 1847-1901. Switz. DSB; P III,IV; BLAF.

8316. NIETZKI, Rudolf Hugo. 1847-1917. Ger./Switz. P III,IV.

8317. PATERNÒ, Emanuele. 1847-1935. It. ICB; P III,IV.

8318. PLUGGE, Pieter Cornelis. 1847-1897. Holl. P III,IV; BLAF.

8319. PŘIBRAM, Richard. 1847-1928. Prague. P III,IV.

8320. SADTLER, Samuel Philip. 1847-1923. U.S.A. P III; GE.

8321. SELL, William James. 1847-1915. Br. P IV.

8322. TANRET, Charles Joseph. 1847-1917. Fr. P III,IV.

8323. WALLACH. Otto. 1847-1931. Ger. DSB; ICB; P III,IV. See also 7424/11.

8324. ARMSTRONG, Henry Edward. 1848-1937. Br. DSB; ICB; P IV; GD5. Also Science in General*.
   1. *An introduction to the study of organic chemistry.* London, 1874. 5th ed., 1886.
   2. *The chemistry of carbon and its compounds.* 2nd ed., London, 1880.

8325. BARBIER, (François Antoine) Philippe. 1848-1922. Fr. P III,IV.

8326. BLOCHMANN, (Georg Rudolf) Reinhard. 1848-1920. Ger. P IV.

8327. BROWN, Horace T. 1848-1925. Br. P III,IV.

8328. CHATARD, Thomas Marean. 1848-1947. U.S.A. P IV.

8329. CLOWES, Frank. 1848-1923. Br. P IV.
   1. *An elementary treatise on practical chemistry and qualitative inorganic analysis.* London, 1874. X. 7th ed., 1899.

8330. CONRAD, Max. 1848-1920. Ger. P III,IV.

8331. DONATH, Eduard. 1848-1932. Aus./Brünn. P III,IV.

8332. FLAWITZKY, Flavian Mikailovich. 1848-1917. Russ. P III,IV.

8333. GREEN, Joseph Reynolds. 1848-1914. Br. BMNH. Also Botany*.
   1. *The soluble ferments and fermentation.* Cambridge, 1899.

8334. GRETE, Ernst August. 1848-after 1878. Switz. P III.

8335. HERRMANN, Carl Felix. 1848-1912. Ger. P IV.

8336. JAPP, Francis Robert. 1848-1925. Br. P III,IV. See also 7894/3.

8337. KLASON (before 1886: CLAËSSON), Johan Peter. 1848-after 1933. Swed. P III (under C), IV (under K).

8338. LANDAUER, John. 1848-after 1903. Ger. P III,IV.
  1. *Die Spectralanalyse.* Brunswick, 1896. X.
  1a. ——— *Spectrum analysis.* New York, 1898. Trans. by J.B. Tingle.

8339. LESCOEUR, Jean Joseph Charles Henri. 1848-1935. Fr. P III,IV.

8340. MEYER, Victor. 1848-1897. Switz./Ger. DSB; ICB; P III,IV; GC.

8341. MILLER, Heinrich Oswald. 1848-after 1911. Russ. P III,IV.

8342. MILLER, Wilhelm von. 1848-1899. Ger. P III,IV.

8343. MORSE, Harmon Northrop. 1848-1920. U.S.A. P IV; GE.

8344. MÜLLER, Friedrich Carl Georg. 1848-1931. Ger. P III,IV.

8345. MUIR, Matthew Moncrieff Pattison. 1848-1931. Br. DSB; P III, IV. Also History of Chemistry*. See also Index.
  1. *A treatise on the principles of chemistry.* Cambridge, 1884. 2nd ed., 1889.

8346. PETTERSSON, Sven Otto. 1848-after 1930. Swed. P III,IV.

8347. ROUSSEAU, Henri Gustave. 1848-1897. Fr. P IV.

8348. SOROKIN, Vassily Ivanovich. 1848-after 1904. Russ. P IV; BMNH.

8349. SPRING, Walthère Victor. 1848-1911. Belg. DSB; ICB; P III,IV.

8350. TIEMANN, Johann Carl Wilhelm Ferdinand. 1848-1899. Ger. DSB; P III,IV.

8351. TOMMASI, Donato. 1848-1907. Fr. P III,IV.
  1. *Traité théorique et pratique d'électrochimie.* Paris, 1889.

8352. TUCKERMAN, Alfred. 1848-1925. Also Physics*.
  1. *Bibliography of the chemical influence of light.* Washington, 1891.

8353. ARATA, Pedro Narciso. 1849-1923. Argentina. P IV.

8354. BRIEGER, Ludwig. 1849-1919. Ger. Mort.; BLAF.

8355. CHRUSTSCHOFF, Pavel Dmitrievich. 1849-after 1902. Russ. P IV.

8356. FABINYI, Rudolf. 1849-1920. Hung. P III,IV.

8357. FRISWELL, Richard John. 1849-1908. Br. P IV.

8358. HALLER, Albin. 1849-1925. Fr. ICB; P IV.

8359. HELL, Carl Magnus. 1849-1926. Ger. P IV.

8360. HILL, Henry Barker. 1849-1903. U.S.A. P III,IV; GE.

8361. KEHRER, Eduard Alexander. 1849-1906. Ger. P IV.

8362. KJELDAHL, Johann Gustav Christoffer. 1849-1900. Den. DSB.

8363. LLOYD, John Uri. 1849-1936. U.S.A. DSB; ICB.

8364. MELDOLA, Raphael. 1849-1915. Br. P III,IV; GF.

8365. MERMET, Achille. 1849-after 1907. Fr. P III,IV.

8366. MUNROE, Charles Edward. 1849-1938. U.S.A. ICB; P III.

8367. REVERDIN, Frédéric. 1849-1931. Switz. P III,IV.

8368. SALOMON, Friedrich Ludwig Rudolph. 1849-1898. Ger. P III,IV.

8369. TANATAR, Sebastian. 1849-after 1914. Russ. P III,IV.

8370. THOMSON, John Millar. 1849-1933. Br. ICB; P IV.

8371. VAGNER, Egor Egorovich. 1849-1903. Russ./Warsaw. DSB; P IV (WAGNER, J.J.).

8372. VILLE, Jules Joseph Mathieu. 1849-1915. Fr. P IV.

8373. WEIDEL, Hugo. 1849-1899. Aus. P IV.

8374. COMSTOCK, William J. fl. 1879-1898. U.S.A. P IV.

8375. ARONHEIM, B. d. 1881. Ger. P III.

8376. BRÜHL, Julius Wilhelm. 1850-1911. Lemberg/Ger. P III,IV.

8377. BRUYLANTS, Gustave J.M. 1850-after 1879. Belg. P III.

8378. DELACHANAL, Antoine Benedict. 1850-after 1900. Fr. P III,IV.

8379. DOEBNER, Oscar Gustav. 1850-1907. Ger. P III,IV.

8380. ENGEL, Rodolphe Charles. 1850-1916. Fr. P III,IV & Supp.

8381. FAHLBERG, Constantin. 1850-1910. Ger./U.S.A. P III,IV.

8382. FITTICA, Friedrich. 1850-1912. Ger. P III,IV.

8383. GOLDSCHMIEDT, Guido. 1850-1915. Aus./Prague. P III,IV.

8384. HENNINGER, Arthur Rodolphe Marie. 1850-1884. Fr. P III,IV.

8385. HEUMANN, Karl. 1850-1893. Switz. P III,IV.

8386. HOFMEISTER, Franz. 1850-1922. Prague/Ger. P III,IV; BLAF.

8387. ISTRATI, Constantin. 1850-1918. Bucharest. P IV; BLAF.

8388. JANOVSKY, Jaroslav. 1850-1907. Aus. P IV.

8389. KUCHEROV, Mikhail Grigorievich. 1850-1911. Russ. ICB.

8390. LE CHATELIER, Henri Louis. 1850-1936. Fr. DSB; ICB; P III,IV.

8391. MABERY, Charles Frederic. 1850-1927. U.S.A. P III,IV; GE.

8392. MAGNIER DE LA SOURCE, Louis. 1850-after 1904. Fr. P III,IV.

8393. MAZZARA, Girolamo. 1850-ca. 1903. It. P IV.

8394. MELIKOFF, Peter Grigorievich. 1850-1927. Russ. P IV.

8395. PECHMANN, Hans von. 1850-1902. Ger. P III,IV.

8396. POEHL, Alexander Vassilievich. 1850-1908. Russ. P III,IV; Mort.; BLAF.

8397. RÜGHEIMER, Leopold. 1850-1917. Ger. P III,IV.

8398. SCHRÖDER, Woldemar von. 1850-1898. Ger. P III,IV; BLAF.

8399. SHENSTONE, William Ashwell. 1850-1908. Br. P IV; GD2. Also Biography*.

8400. SKRAUP, Zdenko Hans. 1850-1910. Aus. DSB; P III,IV.

8401. VIGNON. Léo. 1850-after 1921. Fr. P IV.

8402. WILLIAMS, William Carleton. 1850-after 1904. Br. P III,IV. See also 7976/1a,2a.

8403. BERTRAM, Julius. 1851-1925. Ger. P IV.

8404. BOETTINGER, Carl Conrad. 1851-1901. Ger. P IV.

8405. BOURQUELOT, (Elie) Emile. 1851-1921. Fr. ICB; P IV.

8406. BOUTROUX, Louis Désiré Léon. 1851-after 1900. Fr. P III,IV.

8407. CANZONERI, Francesco. 1851-1930. Argentina. P IV.

8408. CHRISTENSEN, Odin Tidemand. 1851-1914. Den. P III,IV.

8409. CLAISEN, Ludwig. 1851-1930. Ger. DSB; P III,IV.

8410. COUNCLER, Constantin Alfred. 1851-1910. Ger. P III,IV.

8411. DEGENER, Paul. 1851-1901. Ger. P III,IV.

8412. DUVILLIER, Edouard. 1851-1904. Fr. P III,IV.

8413. EYKMAN, Johan Frederik. 1851-1915. Holl./Japan. ICB; P IV.

8414. FABRE, Charles. 1851-after 1919. Fr. P IV.

8415. FILETI, Michele. 1851-1914. It. P III,IV.

8416. GABRIEL, Siegmund. 1851-1924. Ger. DSB; P III,IV.

8417. GRIMSHAW, Harry. 1851-after 1902. Br. P III,IV.

8418. HEMPEL, Walther Mathias. 1851-1916. Ger. P III,IV.

8419. HODGKINSON, William Richard. 1851-after 1903. Br. P IV.

8420. ILOSVAY DE NAGY-ILOSVA, Ludwig. 1851-1936. Hung. P IV.

8421. KÖNIGS, Wilhem. 1851-1906. Ger. P IV.

8422. LETTS, Edmund Albert. 1851-1918. Br. P III,IV.

8423. NOELTING, (Domingo) Emilio. 1851-1922. Ger. P III,IV.

8424. OECHSNER DE CONINCK, William François. 1851-1917. Fr. P IV.

8425. SCHULTZ, Gustav Theodor August Otto. 1851-1928. Ger. P III.
   1. *Die Chemie des Steinkohltheers, mit besonderer Berücksichtigung der künstlichen organischen Farbstoffe*. Brunswick, 1882. X. 2nd ed., 2 vols, 1886-90.

8426. SEUBERT, Karl Friedrich Otto. 1851-after 1921. Ger. P III,IV.

8427. VAUGHAN, Victor Clarence. 1851-1929. U.S.A. Mort.; BLAF; GE.

8428. VYSHNEGRADSKY, Aleksei. 1851-1880. Russ. P III (WYSCHNE...).

8429. ANSCHÜTZ, Richard. 1852-1937. Ger. DSB; P III,IV.

8430. AUSTEN, Peter Townsend. 1852-1907. U.S.A. P III,IV; GE.

8431. BAILEY, George Herbert. 1852-1924. Br. P IV.

8432. BALBIANO, Luigi. 1852-1917. It. P III,IV.

8433. BAUMERT, Georg Paul Gustav. 1852-1927. Ger. P III,IV.

8434. BENEDIKT, Rudolf. 1852-1896. Aus. P III,IV.

8435. CARNELLEY, Thomas. 1852-1890. Br. P III,IV.

8436. CAZENEUVE, Paul. 1852-after 1921. Fr. P IV & Supp.

8437. DEMARÇAY, Eugène Anatole. 1852-1904. Fr. P III,IV.

8438. DENNSTEDT, Max Eugen Hermann. 1852-1931. Ger. P IV.

8439. DIXON, Harold Baily. 1852-1930. Br. DSB; P IV.

8440. DOBBIE, James Johnston. 1852-1924. Br. P IV.

8441. ÉTARD, Alexandre Léon. 1852-1910. Fr. P III,IV; BLAF.

8442. FISCHER, Emil. 1852-1919. Ger. DSB; ICB; P III,IV; Mort.; BLAF. Also Autobiography*.
   1. *Gesammelte Werke.* 8 vols. Berlin, 1906-24.

8443. FISCHER, (Philipp) Otto. 1852-1932. Ger. P III,IV.

8444. GENVRESSE, Pierre. 1852-after 1904. Fr. P IV.

8445. GIESEL, Friedrich Oskar. 1852-1927. Ger. DSB; P IV.

8446. GNEHM, Robert. 1852-1926. Switz. P III,IV.

8447. GOOCH, Frank Austin. 1852-1929. U.S.A. P III,IV.

8448. HARNACK, Friedrich Moritz Erich. 1852-1915. Ger. P III,IV.

8449. HÖNIG, Max. 1852-after 1932. Brünn. P IV.

8450. ILES, Malvern Wells. 1852-after 1882. U.S.A. P III.

8451. KINGZETT, Charles Thomas. 1852-1935. Br. P III,IV.
   1. *The history, products and processes of the alkali trade.* London, 1877.
   2. *Animal chemistry; or, The relations of chemistry to physiology and pathology.* London, 1878.

8452. KRAFFT, Wilhelm Ludwig Friedrich Emil. 1852-1923. Ger. P III, IV.

8453. LENZ, Wilhelm Georg Leberecht. 1852-1916. Ger. P IV.

8454. LIEBERMANN, Leo. 1852-1926. Aus./Hung. P III,IV; BLAF.

8455. MAUTHNER, Julius. 1852-1917. Aus. P III,IV; BLAF.

8456. MOISSAN, Henri. 1852-1907. Fr. DSB; ICB; P III,IV; GB.
   1. *Recherches sur l'isolement du flor.* Paris, 1887.
   2. *Le nickel.* Paris, 1896. With L. Ouvrard.

8457. MORAWSKI, Theodor Rudolf. 1852-after 1903. Aus. P III,IV.

8458. OST, (Friedrich) Hermann (Theodor). 1852-1931. Ger. P III,IV.

8459. PABST, Jean Albert. 1852-after 1901. Fr. P III,IV.

8460. PESCI, Leone. 1852-1917. It. P III,IV.

8461. PRECHT, Heinrich. 1852-1924. Ger. P III,IV.

8462. RAMSAY, William. 1852-1916. Br. DSB; ICB; P III,IV; GB; GD3.
   1. *A system of inorganic chemistry.* London, 1891.
   2. *The gases of the atmosphere: The history of their discovery.* London, 1896. RM/L.

8463. RAÝMAN, Bohuslav. 1852-1910. Prague. P IV.

8464. RIMBACH, Friedrich Eberhard. 1852-1933. Ger. P IV.

8465. SCHRAMM, Julian. 1852-after 1904. Lemberg/Cracow. P IV.

8466. VAN'T HOFF, Jacobus Henricus. 1852-1911. Holl./Ger. DSB;

ICB (under H); P III (under H), IV (under V); GB.
1. *La chimie dans l'espace.* Rotterdam, 1875. X.
1a. ——— 2nd ed., entitled *Dix années dans l'histoire d'une théorie.* Rotterdam, 1887. RM/L.
1b. ——— ——— *Chemistry in space.* Oxford, 1891. Trans. from *Dix années...* and ed. by J.E. Marsh. RM/L.
1c. ——— *Die Lagerung der Atome im Raume.* Brunswick, 1877. Trans. by F. Hermann with a preface by J. Wislicenus. X. 2nd ed., rev., 1894. X.
1d. ——— ——— *The arrangement of atoms in space.* London, 1898. Trans. from the 2nd German ed. by A. Eiloart.
2. *De verbeeldingskracht in de wetenschap.* Rotterdam, 1878. X.
2a. ——— *Imagination in science.* New York, 1967. Trans. with introd. and notes by G.F. Springer.
3. *Ansichten über die organische Chemie.* 2 vols. Brunswick, 1878-81.
4. *Etudes de dynamique chimique.* Amsterdam, 1884. RM.
4a. ——— *Studien zur chemischen Dynamik.* Amsterdam, 1896. Rev. and enlarged by E. Cohen.
4b. ——— ——— *Studies in chemical dynamics.* London, 1896. Trans. from the German ed. by T. Ewan.
5. *L'équilibre chimique dans les systèmes gazeux, ou dissous à l'état dilué.* [Periodical article, 1886] German trans. Leipzig, 1900. (OKEW. 110)
6. *Vorlesungen über theoretische und physikalische Chemie.* 3 vols. Brunswick, 1898-1901. RM/L.
6a. ——— *Leçons de chimie physique.* 3 vols. Paris, 1898-1901. Trans. by Corvisy.
6b. ——— *Lectures on theoretical and physical chemistry.* 3 vols. London, 1899-1900. RM/L.

8467. VONGERICHTEN, Eduard. 1852-after 1922. Ger. P IV.

8468. WIDMAN, (Karl) Oskar. 1852-after 1919. Swed. P III,IV.

8469. ARNAUD, Albert Léon. 1853-ca. 1915. Fr. P IV.

8470. ARTH, Georges Marie Florent. 1853-1909. Fr. P IV.

8471. BANDROWSKI, Ernst Titus von. 1853-1920. Cracow. P IV.

8472. BECKMANN, Ernest Otto. 1853-1923. Ger. DSB; P IV.

8473. BOGUSKI, Jozef Jery von. 1853-after 1902. Warsaw. P IV.

8474. COLSON, Albert Jules. 1853-after 1926. Fr. P IV.

8475. ELION, Hartog. 1853-1930. Holl. P IV.

8476. EMMERLING, Oscar. 1853-1933. Ger. P III,IV.

8477. GREENE, William Houston. 1853-1918. U.S.A. P III,IV. See also 7787/3a.

8478. HAEUSSERMANN, Carl. 1853-1918. Ger. P IV.

8479. HERZIG, Josef. 1853-1924. Aus. P IV.

8480. JAHN, Hans Max. 1853-1906. Ger. DSB; P IV.

8481. KAHLBAUM, Georg Wilhelm August. 1853-1905. Ger./Switz. P III, IV; BLAF. Also Biography* and History of Chemistry. See also Index.

8482. KLINGER, Heinrich Konrad. 1853-after 1922. Ger. P IV.

8483. KOSSEL, (Karl Martin Leonhard) Albrecht. 1853-1927. Ger. DSB; ICB; P IV; Mort.; BLAF. Also Physiology.

8484. LAAR, Peter Conrad. 1853-1929. Ger. P III,IV.

8485. MAQUENNE, Léon Gervais Marie. 1853-1925. Fr. P IV.

8486. MICHAEL, Arthur. 1853-1942. U.S.A. DSB; P III,IV.

8487. OGIER, Jules François. 1853-1913. Fr. P III,IV.

8488. OSTWALD, (Friedrich) Wilhelm. 1853-1932. Riga/Ger. DSB Supp.; ICB; P III,IV; BLAF.
1. *Volumchemische Studien über Affinität, und volumchemische und optisch-chemische Studien.* [Periodical articles, 1877 and later] Leipzig, 1966. (OKEW. 250)
2. *Über Katalyse.* [Periodical articles, 1883 and later] Leipzig, 1923. (OKEW. 200)
3. *Lehrbuch der allgemeinen Chemie.* 2 vols. Leipzig, 1885-87. X. 2nd ed., 1891-1902.
3a. ——— [Trans. of an extract] *Solutions.* London, 1891. Trans. from the 2nd German ed. by M.M.P. Muir.
4. *Grundriss der allgemeinen Chemie.* Leipzig, 1889. 3rd ed., 1899.
4a. ——— *Outlines of general chemistry.* London, 1890. Trans. by J. Walker.
5. *Hand- und Hilfsbuch zur Ausführung physiko-chemischer Messungen.* Leipzig, 1893. X.
5a. ——— *Manual of physico-chemical measurements.* London, 1894. Trans. by J. Walker.
6. *Die wissenschaftlichen Grundlagen der analytischen Chemie.* Leipzig, 1894. X.
6a. ——— *The scientific foundations of analytical chemistry.* London, 1895. Trans. by McGowan.
7. *Elektrochemie. Ihre Geschichte und Lehre.* Leipzig, 1896. RM/L.
8. *Abhandlungen und Vorträge allgemeinen Inhaltes, 1887-1903.* Leipzig, 1904.
Later works not included.
9. *Aus dem wissenschaftlichen Briefwechsel.* 2 parts. Berlin, 1961-69. Ed. by H.G. Körber.

8489. POWER, Frederick Belding. 1853-1927. U.S.A. DSB; GE.

8490. SCHOTTEN, Carl Ludwig Johannes. 1853-1910. Ger. P III,IV.

8491. SCOTT, Alexander. 1853-1947. Br. ICB; P IV.
1. *Introduction to chemical theory.* London, 1891.

8492. SEELIG, Eduard. 1853-after 1904. Ger. P III,IV.

8493. SENIER, Alfred. 1853-after 1918. Br. P IV.

8494. SMOLKA, Alois. 1853-after 1904. Aus./Prague. P IV.

8495. WENGHÖFFER, Ludwig Johann. 1853-1914. Ger. P III.
1. *Lehrbuch der Chemie der Kohlenstoffverbindungen.* Stuttgart, 1882.

8496. WITT, Otto Nicolaus. 1853-1915. Ger. P III,IV.

8497. ALLAIN-LE CANU, (Jacques François Louis) Jules. 1854-1919. Fr. P IV.

8498. BAYLEY, Thomas. 1854-after 1886. Br. P III,IV.

8499. BÖRNSTEIN, Ernst Gustav. 1854-1932. Ger. P III,IV.

8500. FUNARO, Angiolo. 1854-after 1914. It. P IV.

8501. GARZAROLLI VON THURNLACKH, Karl. 1854-1906. Prague. P III,IV.

8502. GÖTTIG, Heinrich Julius Christian. 1854-after 1903. Ger. P IV.

8503. HANRIOT, (Adrien Armand) Maurice. 1854-1933. Fr. P IV.

8504. HORBACZEWSKI, Jan. 1854-1942. Aus./Prague. DSB; P III,IV.

8505. JONES, Henry Chapman. 1854-1932. Br. P IV.

8506. KANONNIKOFF, Inokentii Ivanovich. 1854-1901. Russ. P IV.

8507. LA COSTE, Wilhelm. 1854-1885. Ger. P IV.

8508. LEUCKART, Rudolf. 1854-1889. Ger. P III,IV.

8509. LOUÏSE, Emile Alphonse Camille. 1854-after 1910. Fr. P IV.

8510. MYLIUS, Franz Benno. 1854-1931. Ger. P IV.

8511. NASINI, Raffaello. 1854-1931. It. P IV.

8512. PICCINI, Augusto. 1854-1905. It. ICB; P IV.

8513. REYCHLER, Albert Marie Joseph. 1854-1938. Belg. ICB; P IV.
1. *Théories physico-chimiques*. Brussels, 1897. X.
1a. ——— *Outlines of physical chemistry*. London, 1899. Trans. by J. McCrae.

8514. RIEGLER, Emanuel. 1854-1929. Jassy. P IV; BLAF.

8515. ROOZEBOOM, Hendrik Willem Bakhuis. 1854-1907. Holl. DSB; P IV.

8516. SABATIER, Paul. 1854-1941. Fr. DSB; ICB; P III,IV.

8517. SCHIFF, Robert. 1854-after 1909. Ger./It. P III,IV.

8518. SMITH, Edgar Fahs. 1854-1928. U.S.A. DSB; ICB; P III,IV; GE.

8519. SPICA (-MERCATAIO), Pietro. 1854-1929. It. P IV; BLAF.

8520. VILLIERS, (Charles) Antoine (Théodore). 1854-after 1919. Fr. P III,IV.

8521. VORTMANN, Georg. 1854-after 1920. Aus. P III,IV.

8522. WILL, (Carl) Wilhelm. 1854-1919. Ger. P IV.

8523. WURSTER, Casimir. 1854-1913. Ger. P IV.

8524. ZEISEL, Simon Josef Maria. 1854-after 1913. Aus. P III,IV.

8525. HENTSCHEL, Willibald. fl. 1884-. Ger. P IV.

8526. LOVÉN, Johan Martin. d. 1920. Swed. P IV.

8527. BECKURTS, Heinrich August. 1855-1929. Ger. P III,IV.

8528. BERNTHSEN, (Heinrich) August. 1855-1931. Ger. DSB; P III,IV.
1. *Kurzes Lehrbuch der organischen Chemie*. Brunswick, 1887. X.
1a. ——— *Text-book of organic chemistry*. London, 1889. Trans. by G. M'Gowan. 2nd ed., 1894.

8529. BINET, Paul. 1855-1896. Switz. BLAF.

8530. BISCHOFF, Carl Adam. 1855-1903. Ger./Riga. P III,IV.
    1. *Handbuch der Stereochemie*. Frankfurt, 1894. With P. Walden.

8531. BODLÄNDER, Guido. 1855-1904. Ger. P IV.

8532. BRAUNER, Bohuslav. 1855-1935. Prague. DSB; ICB; P III,IV.

8533. BREDT, Conrad Julius. 1855-after 1931. Ger. P IV.

8534. BRUNNER, Karl. 1855-after 1931. Aus./Prague. P III,IV.

8535. CROSS, Charles Frederick. 1855-1935. Br. ICB; P IV; GD5.

8536. DACCOMO, Girolamo. 1855-1925. It. P IV; BLAF.

8537. EDER, Josef Maria. 1855-1944. Aus. DSB; ICB; P III,IV.

8538. GOLDBERG, Alwin Heinrich. 1855-after 1930. Ger. P IV.

8539. HANNAY, James Ballantyne. 1855-after 1915. Br. ICB; P III,IV.

8540. HJELT, Edvard Immanuel. 1855-1921. Fin. P III,IV. See also 7424/10.

8541. KILIANI, Heinrich. 1855-1945. Ger. ICB; P III,IV.

8542. LEIDIÉ, Emile Jules. 1855-1904. Fr. P IV.

8543. LIPP, Andreas. 1855-1916. Ger. P III,IV.

8544. MORLEY, Henry Forster. 1855-after 1901. Br. P III,IV.

8545. MÜLLER, Joseph Auguste. 1855-after 1900. Algeria. P IV & Supp.

8546. ROMBURGH, Pieter van. 1855-after 1931. Holl./Java. P IV (under VAN).

8547. TEIXEIRA MENDES, Raymundo. 1855-1927.
    1. *La philosophie chimique d'après Auguste Comte*. Rio de Janeiro, 1887.

8548. WELLS, Horace L. 1855-1924. U.S.A. P IV.

8549. ADAM, Paul Gabriel. 1856-ca. 1915. Fr. P IV.

8550. ANDRÉ, Gustave. 1856-1927. Fr. ICB; P IV.

8551. ANDREWS, Launcelot Winchester. 1856-ca. 1930. U.S.A. P IV.

8552. BEHREND, Anton Friedrich Robert. 1856-1926. Ger. DSB Supp.; P IV.

8553. BLADIN, Johan Adolf. 1856-1902. Swed. P IV.

8554. BOURGEOIS, Léon Zéphrin. 1856-after 1901. Fr. P III,IV; BMNH.
    1. *Reproduction artificielle des minéraux*. Paris, 1884.

8555. BUCHKA, Karl Heinrich von. 1856-1917. Ger. ICB; P III,IV.

8556. BUISINE, Alphonse Jean-Baptiste Aimable. 1856-after 1903. Fr. P IV.

8557. CHITTENDEN, Russell Henry. 1856-1943. U.S.A. DSB; P III,IV; Mort.; BLAF.
    1. *Studies in physiological chemistry, being reprints ... 1897-1900*. New York, 1901.

8558. FOCK, Andreas Ludwig. 1856-after 1923. Ger. PIII,IV; BMNH. Also Crystallography*.

8559. FORCRAND, (Robert) Hippolyte de. 1856-1933. Fr. P IV.

8560. FRESCENIUS, Theodor Wilhelm. 1856-1936. Ger. P IV.

8561. KONOVALOV, Dmitry Petrovich. 1856-1929. Russ. DSB; ICB; P IV.

8562. LACHOWICZ, Bronislaus. 1856-1903. Lemberg. P IV.

8563. LELLMANN, Eugen Karl. 1856-1893. Ger. P III,IV.

8564. LONG, John Harper. 1856-1918. U.S.A. P III,IV; GE.

8565. MEUNIER, Jean Alexis. 1856-after 1921. Fr. P IV.

8566. PETERSEN, Christian Ulrick Emil. 1856-1907. Den. P IV.

8567. PLIMPTON, Richard Tayler. 1856-after 1897. Br. P III,IV.

8568. REIMER, Carl Ludwig. 1856-1921. Ger. P III,IV.

8569. RETGERS, Jan Willem. 1856-1896. Holl./Java. P IV. Also Mineralogy.

8570. RÖHMANN, Franz. 1856-1919. Ger. P IV; BLAF.

8571. SCHALL, (Johann Friedrich) Carl. 1856-after 1922. Switz. P III,IV.

8572. SCHLOESING, Alphonse Théophile. 1856-after 1920. Fr. P IV.

8573. SCHRÖTTER, Hugo. 1856-1911. Aus. P III,IV.

8574. SCHULTÉN, August Benjamin af. 1856-1912. Fin. P III,IV.

8575. SENDERENS, Jean Baptiste. 1856-after 1922. Fr. P IV.

8576. VELEY, Victor Herbert. 1856-1933. Br. ICB; P III,IV. See also 7990/1a.

8577. VENABLE, Francis Preston. 1856-1934. U.S.A. P IV; GE. Also History of Chemistry*.

8578. VERNEUIL, Auguste Victor Louis. 1856-1913. Fr. P IV.

8579. ANDREASCH, Rudolf. 1857-1928. Aus. P IV.

8580. BAMBERGER, Eugen. 1857-1932. Ger./Switz. DSB; P III,IV.

8581. BARTHE, Joseph Paul Léonce. 1857-after 1926. Fr. P IV.

8582. CIAMICIAN, Giacomo Luigi. 1857-1922. It. DSB; P IV; BLAF.

8583. CURTIUS, Theodor. 1857-1928. DSB; P IV.

8584. EINHORN, Alfred. 1857-1917. Ger. P IV.

8585. ERDMANN, Ernst Immanuel. 1857-1925. Ger. P IV.

8586. FISCHER, Bernhard (second of the name). 1857-1905. Ger. BLAF.

8587. FRIEDLÄNDER, Paul. 1857-1923. Aus. P IV.

8588. GARROD, Archibald Edward. 1857-1936. Br. ICB; BLAF; GD5.

8589. GOLDSCHMIDT, Heinrich Jacob. 1857-after 1931. Switz./Ger. P III,IV.

8590. GRÖGER, Maximilian. 1857-after 1919. Aus. P IV.

8591. HANTZSCH, Arthur Rudolf. 1857-1935. Switz./Ger. DSB; P III,IV.
1. *Grundriss der Stereochemie*. Leipzig, 1893. X.
1a. ——— *Précis de stéréochimie*. Paris, 1896. Trans. by Guye and Gautier. X.
1b. ——— ——— *Elements of stereochemistry*. Easton, Pa., 1901. Trans. from the French ed. by C.G.L. Wolf.

8592. HINSBERG, Oscar Heinrich Daniel. 1857-after 1931. Ger./Switz. P IV.

8593. JOANNIS, (Jean) Alexandre. 1857-after 1919. Fr. ICB; P IV.

8594. KABLUKOV, Ivan Alexsevich. 1857-1942. Russ. DSB; P IV Supp.

8595. KONDAKOV, Ivan Lavrentievich. 1857-1931. Warsaw. DSB; P IV.

8596. LINDET, Léon Gaston Aimé. 1857-1929. Fr. P IV.

8597. LIPPMANN, Edmund Oskar von. 1857-1940. Ger. ICB; P III,IV. Also History of Chemistry.

8598. LOBRY DE BRUYN, Cornelis Adriaan. 1857-1904. Holl. P IV.

8599. LÜPKE, Robert Theodor Wilhelm. 1857-1903. Ger. P IV.
1. *Grundzüge der Elektrochemie*. Berlin, 1895. X.
1a. ——— *The elements of electro-chemistry*. London, 1897. Trans. from the 2nd German ed. by M.M.P. Muir.

8600. MASSOL, Gustave. 1857-after 1937. Fr. P IV.

8601. MÖHLAU, (Bernard Julius) Richard. 1857-after 1910. Ger. P III,IV.

8602. NOYES, William Albert. 1857-1941. U.S.A. DSB; P IV.

8603. PICTET, Amé. 1857-1937. Switz. ICB; P III,IV. Also Autobiography*.

8604. PIUTTI, Arnaldo Teofilo Pietro. 1857-1928. It. P IV.

8605. REICHER, Lodewijk Theodorus. 1857-after 1933. Holl. P IV.

8606. STOEHR, Karl. 1857-after 1904. Ger. P IV.

8607. TREADWELL, Frederick Pearson. 1857-1918. Switz. P IV.

8608. WAGNER, Julius Eugen. 1857-1924. Ger. P III,IV.

8609. YOUNG, Sydney. 1857-1937. Br. DSB; ICB; P IV; GD5.

8610. CRISMER, Léon. 1858-1944. Belg. ICB; P IV.

8611. DOBBIN, Leonard. 1858-after 1931. Br. P IV.
1. *Chemical theory for beginners*. London, 1896. With J. Walker.

8612. ELBS, Karl. 1858-1933. Ger. P IV.

8613. FRANKLAND, Percy Faraday. 1858-1946. Br. DSB; P IV; GD6. Also Bacteriology and Biography*.

8614. FRIEDHEIM, Carl. 1858-1909. Switz. P IV.

8615. KASSNER, Georg Max Julius. 1858-1929. Ger. P IV.

8616. LASSAR-COHN. 1858-1922. Ger. DSB (COHN, Lassar); P IV.
1. *Die Chemie im täglichen Leben: Gemeinverständliche Vorträge*. Leipzig, 1896. X.

1a. ——— *Chemistry in daily life: Popular lectures.* London, 1896. Trans. by M.M.P. Muir.

8617. MALBOT, Hippolyte. 1858-1900. Algeria. P IV.

8618. MASSON, David Orme. 1858-1937. Australia. ICB; P IV; GD5.

8619. PELLIZZARI, Guido. 1858-after 1929. It. P IV.

8620. PICKERING, Percival Spencer Umfreville. 1858-1920. Br. ICB; P IV.

8621. PONSOT, Auguste. 1858-1907. Fr. P IV.

8622. POULSSON, Edvard. 1858-1935. Nor. BMNH; BLAF.

8623. RÖSSING, Adelbert. 1858-after 1903. Ger. P IV.

8624. ROSER, Wilhelm. 1858-1923. Ger. P IV.

8625. SAKURAI, Joji. 1858-1939. Japan. P IV & Supp.

8626. STOKLASA, Julius. 1858-1936. Aus./Prague. ICB; P IV.

8627. ARRHENIUS, Svante August. 1859-1927. Swed. DSB; ICB; P IV; GB.
    1. *Recherches sur la conductibilité galvanique des électrolytes.* Stockholm, 1884. X.
    1a. ——— German trans. Leipzig, 1907. (OKEW. 160) RM/L.

8628. BÉHAL, Auguste. 1859-1941. Fr. ICB; P IV.

8629. BÜLOW, (Theodor) Carl (Heinrich Ernst). 1859-1933. Ger. P IV.

8630. COHEN, Julius Berend. 1859-1935. Br. ICB; P IV.

8631. COLLIE, John Norman. 1859-1942. Br. DSB; ICB; P IV & Supp.; GD6.

8632. DENIGÈS, Georges. 1859-after 1931. Fr. P IV; BLAF.

8633. FRENTZEL, Johannes. 1859-1902. Ger. P IV.

8634. FRITSCH, Paul Ernst Moritz. 1859-1913. Ger. P IV.

8635. GEORGIEVICS, Georg Cornelius Theodor von. 1859-after 1929. Aus. P IV.

8636. GUNTZ, Antoine Nicolas. 1859-1935. Fr. P IV (GÜNTZ).

8637. HEYCOCK, Charles Thomas. 1859-1931. Br. P IV.

8638. HOLLEMAN, Arnold Frederik. 1859-after 1930. Holl. P IV.

8639. HOWE, James Lewis. 1859-1955. U.S.A. DSB; P IV.

8640. JACOBSON, Paul Heinrich. 1859-1923. Ger. P III,IV.

8641. KNORR, Ludwig. 1859-1921. Ger. P IV.

8642. KNORRE, Georg Karl von. 1859-1910. Ger. P III,IV.

8643. KRÜSS, (Alexander) Gerhard. 1859-1895. Ger. P IV; GC.

8644. OSBORNE, Thomas Burr. 1859-1929. U.S.A. DSB; ICB; Mort.; GE.

8645. RUHEMANN, Siegfried. 1859-after 1931. Ger./Br. P IV.

8646. SELIWANOFF, Feodor Feodorovich. 1859-after 1921. Russ. P IV.

8647. STOKES, Henry Newlin. 1859-after 1909. U.S.A. P IV.

8648. THOMS, Hermann Friedrich Maria. 1859-1931. Ger. P IV; BLAF.

8649. WEGSCHEIDER, Rudolf Franz Johann. 1859-after 1922. Aus. P IV.

8650. WOLFF, Ludwig. 1859-1919. Ger. P IV.

Imprint Sequence

8651. *ART de la verrerie, de Neri, Merret et Kunckel.* Paris, 1752. Trans. by P.H.T. d'Holbach--one of a number of his translations of German writings on various aspects of chemical technology. It is chiefly a trans. of Kunckel's *Ars vitraria* (2391/2) which incorporates the works of Neri and Merrett.

8652. ACADÉMIE des Sciences. Paris.
1. *Recueil de mémoires et d'observations sur la formation et sur la fabrication du salpêtre.* Paris, 1776. Partly by Lavoisier. RM.

8653. DICKSON, Stephen.
1. *An essay on chemical nomenclature ... in which are comprised observations on the same subject by R. Kirwan.* London, 1796. RM.

8654. *EXPLANATORY dictionary of the apparatus and instruments employed in the various operations of philosophical and experimental chemistry.* London, 1824.

8655. *HANDWÖRTERBUCH der reinen und angewandten Chemie.* 9 vols & supp. Brunswick, 1842-64. Ed. by J. Liebig, J.C. Poggendorff and F. Wöhler. RM/L. For a later version see 7678/1.

8656. *CHEMICAL reports and memoirs.* London, 1848. Trans. by G.E. Day. Ed. by T. Graham. Printed for the Cavendish Society. RM/L (under Graham).

8657. ZUCHOLD, Ernst Amandus.
1. *Bibliotheca chemica, 1840-58: Verzeichniss der auf dem Gebiete der reinen, pharmaceutischen und technischen Chemie ... 1840-1858 ... erschienenen Schriften.* Göttingen/Leipzig, 1859.

8658. REIMANN, M.
1. *Technologie des Anilins.* Berlin, 1866. X.
1a. ——— *On aniline and its derivatives: A treatise upon the manufacture of aniline and aniline colours, by M. Reimann. To which is added in an appendix, "The report of the colouring matters derived from coal, shown at the French Exhibition, 1867", by A.W. Hofmann.* London, 1868. Ed. by W. Crookes. RM.

8659. *ENCYCLOPÉDIE chimique.* 10 vols. Paris, 1881-1905. Ed. by E. Frémy.

8660. CHEMICAL Society of London.
1. *A catalogue of the library, arranged according to subjects, with indexes.* London, 1886.
2. *Jubilee. 1841-1891.* London, 1896.
3. *A catalogue of the library, arranged according to authors, with a subject index.* London, 1903.

3.08 MINERALOGY AND CRYSTALLOGRAPHY

Apart from crystallography, the section is confined to mineralogy in roughly the modern sense of the word. As is discussed at the beginning of the following section, the word was also used in the eighteenth and early nineteenth centuries to denote what would now be called geology.

Books dealing with both mineralogy and geology are included in the following section, not in this section.

8661. GIMMA, Giacinto. 1668-1735. It. GA.

8662. HENCKEL, Johann Friedrich. 1678-1744. Ger. DSB; P I (HENKEL); BMNH (HENKEL); GA; GC. Also Chemistry*.

8663. CAPPELER, Moritz Anton. 1685-1769. Switz. ICB; P I; BMNH; GA.

8664. BARRÈRE, Pierre. 1690?-1755. Fr. BMNH; BLA; GA. Also Natural History*.
   1. *Observations sur l'origine et la formation des pierres figurées, et sur celles qui, tant extérieurement qu'intérieurement, ont une figure regulière et déterminée.* Paris, 1746. RM.

8665. HILL, John. 1707?-1775. Br. DSB; ICB; P I; BMNH; BLA; GA; GB; GD. Other entries: see 3105.
   1. *Theophrastus's "History of stones." With an English version and critical and philosophical notes.* London, 1746. X. 2nd ed., enlarged, 1774.
   2. *Spatogenesia. The origin and nature of spar, its qualities and uses. With a description and history of eighty-nine species ... A specimen of a general distribution of fossils.* London, 1772. RM.

8666. WALLERIUS, Johan Gottshalk. 1709-1785. Swed. DSB; ICB; P II; BMNH. Also Chemistry*. See also 15339/1.
   1. *Mineralogia; eller, Mineral-riket.* Stockholm, 1747. X.
   1a. ——— *Mineralogie; oder, Mineralreich.* Berlin, 1750. Trans. by J.D. Denso. X.
   1b. ——— ——— *Minéralogie; ou, Description générale des substances du règne minéral.* 2 vols. Paris, 1753. Trans. from the German by P.H.D. Holbach. RM.

8667. DAUBENTON, Louis Jean Marie. 1716-1800. Fr. DSB Supp.; ICB; P I; BMNH; BLA & Supp.; GA; GB. Also Natural History, Botany, and Zoology. See also Index.

8668. SCHMIDEL (or SCHMIEDEL), Casimir Christoph. 1718-1792. Ger. DSB; P II; BMNH; BLA. Also Botany.

1. *Fossilium, metalla et res metallicas concernentium, glebae, suis coloribus expressae.* Nuremberg, 1753.

8669. CRONSTEDT, Axel Frederik. 1722-1765. Swed. DSB; P I; BMNH; GA. Also Chemistry.
1. *Försök till mineralogie; eller, Mineral-rikets upställning.* Stockholm, 1758. X.
1a. ——— *An essay towards a system of mineralogy. Translated ... with notes, by G. von Engeström. To which is added A treatise on the pocket-laboratory, containing an easy method ... for trying mineral bodies, written by the translator. The whole revised and corrected by E. Mendes de Costa.* London, 1770. RM/L. 2nd ed., 1772. cf. 8680/1.
See also 8718/1.

8670. SCOPOLI, Giovanni Antonio. 1723-1788. It./Hung. ICB; P II; BMNH; GA. Also Geology, Natural History*, Botany*, and Zoology*.
1. *Principia mineralogiae systematicae et practicae.* Prague, 1772. RM.
2. *Crystallographia Hungarica.* Prague, 1776. RM.

8671. BRÜCKMANN, Urban Friedrich Benedict. 1728-1812. Ger. P I; BMNH.

8672. FORSTER, Johann Reinhold. 1729-1798. Ger./Br. DSB; ICB; P I & Supp.; BMNH; GA; GC. Also Geology*, Natural History*, Botany*, and Zoology*. See also Index.
1. *An easy method of assaying and classing mineral substances.* London, 1772.

8673. RASHLEIGH, Philip. 1729-1811. Br. BMNH; GD.
1. *Specimens of British minerals.* 2 vols. London, 1797-1802.

8674. VALMONT DE BOMARE, Jacques Christophe. 1731-1807. Fr. DSB; P II; BMNH; GA. Also Geology and Natural History*.
1. *Minéralogie.* 2 vols. Paris, 1762. RM.

8675. KIRWAN, Richard. 1733?-1812. Br. DSB; ICB; P I; BMNH; GA; GB; GD. Also Chemistry*, Geology*, and Meteorology*. See also Index.
1. *Elements of mineralogy.* London, 1784. 2nd ed., 2 vols, 1794-96.

8676. ADELUNG, Johann Christoph. 1734-1806. Ger. P I; GA; GB; GC.

8677. MONNET, Antoine Grimoald. 1734-1817. Fr. DSB; P II; BMNH; GA. Also Chemistry* and Geology.
1. *Traité de l'exploitation des mines.* Paris, 1773. RM.
2. *Nouveau système de minéralogie.* Bouillon, 1779.

8678. BERGMAN, Torbern Olof. 1735-1784. Swed. DSB; ICB; P I & Supp.; BMNH; GA; GB. Also Various Fields and Chemistry*#.
1. *Sciagraphia regni mineralis.* Leipzig/Dresden, 1782. RM.
1a. ——— *Outlines of mineralogy.* Birmingham, 1783. Trans. by W. Withering. RM/L.
1b. ——— *Manuel du minéralogiste.* Paris, 1784. Trans. by Mongez *le jeune*. Another ed., enlarged by J.C. Delamétherie, 1792.
2. *Meditationes de systemate fossilium naturali.* Florence, 1784. RM.

8679. ROMÉ DE L'ISLE, Jean Baptiste Louis. 1736-1790. Fr. DSB; P II; BMNH; GB. Also Geology* and History of Science*.
1. *Essai de cristallographie*. Paris, 1772. RM.
2. *Description méthodique d'une collection de minéraux*. Paris, 1773. RM.
3. *Cristallographie*. 4 vols. Paris, 1783. X. 2nd ed., 1783.
See also 10973/1.

8680. ENGESTRÖM, Gustav von. 1738-1813. Swed. P I; BMNH; GA.
1. [Trans.] *Beschreibung eines mineralogischen Taschen-Laboratoriums und inbesondere des Nutzens des Blaser-rohrs in der Mineralogie*. Greifswald, 1782. RM. The original appeared in English in 1770: see 8669/1a.

8681. GESNER, Johann Augustin Philipp. 1738-1801. Ger. P I; BLA.

8682. GOLITSYN, Dmitri Alekseievich. 1738-1803. Holl./Ger. P I (GALLITZIN); BMNH (GHOLITZUIN); GA (GALITZIN).
1. *Traité ou description abrégée et méthodique des minéraux*. Maastricht, 1792. RM.
2. *Recueil de noms par ordre alphabétique apropriés en minér-alogie aux terres et pierres, aux métaux et demi métaux, et aux bitumes. Avec un précis de leur histoire-naturelle et leurs synonymies*. Brunswick, 1802. RM.

8683. BLANK, Joseph Bonavita. 1740-1827. Ger. P I.

8684. MÜLLER (VON REICHENSTEIN), Franz Joseph. 1740-1825. Hung./Aus. DSB; ICB; P II; BMNH.

8685. TREBRA, Friedrich Wilhelm Heinrich von. 1740-1819. Ger. DSB; ICB; P II; BMNH; GC. Also Geology*.
1. *Mineraliencabinett*. Clausthal, 1795. RM.

8686. VELTHEIM, August Ferdinand von. 1741-1801. Ger. P II; BMNH; GC. Also Geology.

8687. CARANGEOT, Arnould. 1742-1806. Fr. DSB; BMNH. Also Zoology.

8688. HAÜY, René Just. 1743-1822. Fr. DSB; ICB; P I; BMNH; GA; GB. Also Physics*.
1. *Essai d'une théorie sur la structure des crystaux*. Paris, 1784. RM.
2. *Traité de minéralogie*. 4 vols & atlas. Paris, 1801. RM/L. Reprinted, Brussels, 1968. 2nd ed., 1822.
2a. ——— *Lehrbuch der Mineralogie*. Paris, 1804-10. Trans. with notes by D.L.G. Karsten.
3. *Tableau comparatif des résultats de la cristallographie et de l'analyse chimique, relativement à la classification des minéraux*. Paris, 1809. RM/L. Reprinted, Brussels, 1968.
4. *Traité de cristallographie, suivi d'une application des principes de cette science à la détermination des espèces minérales*. 2 vols & atlas. Paris, 1822. RM. Reprinted, Brussels, 1968.
See also 8720/1.

8689. MACQUART, Louis Charles Henri. 1745-1808. Fr. ICB; P II; BMNH; BLA.
1. *Essais; ou, Recueil de mémoires sur plusieurs points de minéralogie*. Paris, 1789. RM.

8690. LAUMONT, François Pierre Nicolas Gillet de. 1747-1834. Fr. BMNH; GA; GB.

8691. STÜTZ, Andreas (or Anton). 1747-1806. Aus. P II.

8692. WERNER, Abraham Gottlob. 1749-1817. Ger. DSB; ICB; P II; BMNH; GA; GB; GC. Also Geology*.
1. *Von den äusserlichen Kennzeichen der Fossilien*. Leipzig, 1774. RM. Reprinted, Amsterdam, 1965.
1a. ——— *A treatise on the external characters of fossils*. Dublin, 1805. Trans. by T. Weaver. RM/L.
1b. ——— *On the external characters of minerals*. Urbana, Ill., 1962. Trans. by A.V. Carozzi of a copy annotated by Werner for a 2nd ed. (which was never published).
2. *Letztes Mineral-System*. Freiberg, 1817. Posthumous. Ed. with notes. RM.
See also 8708/1, 8714/1, and 8726/1.

8693. DOLOMIEU, Dieudonné (or Déodat) de Gratet de. 1750-1801. Fr. DSB; ICB; P I; BMNH; GA; GB. Also Geology*.
1. *Sur la philosophie minéralogique et sur l'espèce minéralogique*. Paris, 1801. RM.

8694. BOURNON, Jacques Louis de. 1751-1825. Fr./Br. P I; BMNH; GA.
1. *Traité complet de la chaux carbonatée et de l'arragonite*. 3 vols. London, 1808.
2. *Traité de minéralogie*. 3 vols. London, 1808. RM.

8695. LUDWIG, Christian Friedrich. 1751-1823. Ger. P I; BMNH; BLA.

8696. LELIÈVRE, Claude Hugues. 1752-1835. Fr. P I; GA. See also 7102/2.

8697. BORCH, Michael Johann, *Graf* von der. 1753-1811. Livonia/It. P I; BMNH; GA.
1. *Minéralogie Sicilienne docimastique et métallurgique; ou, Connaissance de tous les minéraux que produit l'île de Sicilie*. Turin, 1780. RM.

8698. WIDMANNSTÄTTEN (or WIDMANSTETTER), Aloys Joseph Beck von. 1754-1849. Aus. DSB; ICB; P II,III.

8699. BABINGTON, William. 1756-1833. Br. DSB; P I & Supp.; BMNH; BLA; GD.
1. *A systematic arrangement of minerals, founded on the joint consideration of their chemical, physical, and external characters*. London, 1795. RM.
2. *A new system of mineralogy in the form of a catalogue*. London, 1799.

8700. CHLADNI, Ernst Florenz Freidrich. 1756-1827. Ger. DSB; ICB; P I; BMNH; GA; GC. Also Physics*.

8701. SCHÖNBAUER, Joseph Anton von. 1757-1807. Hung. P II; BMNH; Also Zoology.

8702. SCHUMACHER, (Heinrich) Christian Friedrich. 1757-1830. Den. P II; BMNH; BLA. Also Botany and Anatomy.
1. *Versuch eines Verzeichnisses der in den dänisch-nordischen Staaten sich findenden einfachen Mineralien*. Copenhagen, 1801. RM.

8703. SOWERBY, James. 1757-1822. Br. DSB; P II; BMNH; GB; GD.
Also Geology*, Botany*, and Zoology*.
1. *British mineralogy.* 5 vols. London, 1804-17. RM/L.
2. *Exotic mineralogy ... A supplement to "British mineralogy."* 2 vols. London, 1811-17.

8704. MONTEIRO, Joao Antonio. b. 1758. P II.

8704A GADOLIN, Johan. 1760-1852. Fin. DSB; P I; BMNH. Also Chemistry*.

8705. HOFFMANN, Christian August. 1760-1813. Ger. P I; BMNH.

8706. GIESECKE, Carl Ludwig (or Charles Lewis). 1761-1833. Den./Br. ICB; P I; BMNH.

8707. GREGOR, William. 1761-1817. Br. P I; GD.

8708. REUSS, Franz Ambrosius. 1761-1830. Bohemia. DSB; ICB; P II; BMNH; BLA. Also Geology*.
1. *Neues mineralogisches Wörterbuch; oder, Verzeichniss aller Wörter welche auf Oryctognosie bezughaben. Mit Angabe ihrer wahren Bedeutung nach des ... Werners neuester Nomenclatur.* Hof, 1798. RM.
2. *Lehrbuch der Mineralogie nach des Herrn D.L.G. Karsten mineralogischer Tabelle.* 8 vols. Leipzig, 1801-06. RM.

8709. THOMSON, William. 1761-ca. 1806. Br. ICB.

8710. GISMONDI, Carlo Giuseppe. 1762-1824. It. P II; GA.

8711. BRUNNER, Joseph. 1764-1807. Ger. P I; BMNH.
1. *Handbuch der mineralogischen Diagnosis.* Leipzig, 1804. RM.

8712. HABERLE, Karl Constantin. 1764-1832. Ger./Hung. P I; BMNH.
Also Botany.

8713. MAWE, John. 1764-1829. Br. P II; BMNH; GA; GD. Also Zoology*.
1. *The mineralogy and geology of Derbyshire.* London, 1802.

8714. RÍO, Andrés Manuel del. 1764-1849. Mexico. DSB; ICB (DEL RIO); P II; BMNH.
1. *Elementos de orictognosia; ó, Del conocimiento de los fósiles, dispuestos segun los principios de A.G. Werner.* 2 vols. Mexico City, 1795-1805. RM.

8715. WIDENMANN, Johann Friedrich Wilhelm. 1864-1798. Ger. P II; BMNH. Also Geology.
1. *Handbuch des oryktognostischen Theils der Mineralogie.* Leipzig, 1794. RM.

8716. EMMERLING, Ludwig August. 1765-1842. Ger. P I; BMNH; GC.
1. *Lehrbuch der Mineralogie.* 3 vols. Giessen, 1793-97. RM.

8717. WOLLASTON, William Hyde. 1766-1828. Br. DSB; ICB; P II; BLA; GA; GB; GD. Also Physics, Chemistry, and Physiology.

8718. SCHMEISSER, Johann Gottfried. 1767-1837. Ger./Br. P II; BMNH.
1. *A system of mineralogy, formed chiefly on the plan of Cronstedt.* London, 1795. RM.

8719. BAKEWELL, Robert (second of the name). 1768-1843. Br. DSB; P I; BMNH; GB; GD. Also Geology*.
1. *An introduction to mineralogy.* London, 1819. RM.

8720. ACCUM, Friedrich Christian. 1769-1838. Ger./Br. DSB; ICB; P I; BMNH; BLA; GA; GC; GD. Also Chemistry*.
1. *Elements of crystallography, after the method of Haüy.* London, 1813. RM/L.

8721. CLARKE, Edward Daniel. 1769-1822. Br. DSB; P I; BMNH; GA; GB; GD. Also Geology. See also 15425/1.

8722. TOWNSON, Robert. fl. 1795-1798. Br. ICB; BMNH; GD. Also Natural History*.
1. *Philosophy of mineralogy.* London, 1798. RM.

8723. BROOKE, Henry James. 1771-1857. Br. P I & Supp.; P III; BMNH; GD.
1. *A familiar introduction to crystallography, including an explanation of the principle and use of the goniometer.* London, 1823.

8724. JORDAN, Johann Ludwig. 1771-1853. Ger. P I; BMNH. Also Geology.
1. *Mineralogische und chemische Beobachtungen und Erfahrungen.* Göttingen, 1800. RM.

8725. BROCCHI, Giovanni Battista. 1772-1826. It. DSB; ICB; P I; BMNH; GA; GB. Also Geology*.
1. *Trattato mineralogico e chimico sulle miniere di fero del dipartimento del Mella. Con l'esposizione della costituzione fisica della montagna metallifere della Val-Trompia.* Brescia, 1807-08. RM.

8726. BROCHANT DE VILLIERS, André Jean François Marie. 1772-1840. Fr. DSB; P I; BMNH; GA; GB. Also Geology.
1. *Traité de minéralogie, suivant les principes du professeur Werner.* 2 vols. Paris, 1801-02. 2nd ed., 1808.

8727. MIELICHHOFER, Mathias. 1772-1847. Aus. P III. Also Botany.

8728. STRUVE, Heinrich Christoph Gottfried. 1772-1851. Ger. P II; BMNH.

8729. AIKIN, Arthur. 1773-1854. Br. ICB; P I; BMNH; GB; GD. Also Chemistry*.
1. *A manual of mineralogy.* London, 1814. RM. 2nd ed., 1815.

8730. MOHS, Friedrich. 1773-1839. Ger./Aus. DSB; P II; BMNH; GA; GC.
1. *Die Charaktere der Klassen, Ordnungen, Geschlechter und Arten, oder die Charakteristik der naturhistorischen Mineral-Systems.* Dresden, 1820. RM.
1a. ——— *The characters of the classes, orders, genera and species, or the characteristic of the natural history system of mineralogy.* Edinburgh, 1820. RM/L.
2. *Grundriss der Mineralogie.* 2 vols. Dresden, 1822-24. RM.
2a. ——— *A treatise on mineralogy.* 3 vols. Edinburgh, 1825. With additions by W. Haidinger. RM.
3. *Leichtfassliche Anfangsgründe der Naturgeschichte des Mineralsreiches.* 2 vols. Dresden, 1832-38. X. 2nd ed., Vienna, 1836-39.
4. *Anleitung zum Schürfen.* Vienna, 1838. RM.

8731. SEYBERT, Adam. 1773-1825. U.S.A. ICB; P II; BLA; GE.

8732. THOMSON, Thomas. 1773-1852. Br. DSB; ICB; P II; BMNH; GB; GD. Also Physics*, Chemistry*, and History of Science*.
1. *Outlines of mineralogy, geology and mineral analysis.* 2 vols. London, 1836.

8733. BERNHARDI, Johann Jacob. 1774-1850. Ger. P I & Supp.; BMNH; BLA & Supp.; GA; GC. Also Botany.

8734. FUCHS, Johann Nepomuck von. 1774-1856. Ger. DSB; ICB; P I; BMNH; GB; GC. Also Chemistry.
1. *Über die Theorien der Erde, den Amorphismus fester Körper und den gegenseitigen Einfluss der Chemie und Mineralogie.* Munich, 1844. RM.
2. *Gesammelte Schriften.* Munich, 1856. Ed. with an obituary by C.G. Kaiser. RM.

8735. JAMESON, Robert. 1774-1854. Br. DSB; ICB; P I; BMNH; GA; GB; GD. Also Geology*.
1. *An outline of the mineralogy of the Shetland Islands and of the island of Arran.* Edinburgh, 1798. RM.
2. *Mineralogy of the Scottish isles.* 2 vols. Ib., 1800. RM.
3. *A system of mineralogy.* 3 vols. Ib., 1804-08. (Vol. 3 is subtitled *Elements of geognosy*; see 9238/1.) RM/L. 2nd ed., 1816. 3rd ed., 1820.
4. *A treatise on the external characters of minerals.* Ib., 1805. RM. 2nd ed., entitled *A treatise on the external, chemical and physical characters....* 1816.
5. *Mineralogical travels through the Hebrides, Orkney and Shetland Islands, and mainland of Scotland.* Ib., 1813. RM.
6. *A manual of mineralogy.* Ib., 1821. RM.
7. *Mineralogy according to the natural history system.* Ib., 1837. RM.

8736. KÖNIG, Karl Dietrich Eberhardt (or KONIG, Charles). 1774-1851. Ger./Br. P I; BMNH; GB; GD.

8737. SCHNEIDER, Johann Georg. b. 1774. Ger. P II.

8738. KICKX, Jean (first of the name). 1775-1831. Belg. P I (KICKS); BMNH; GA. Also Botany.

8739. KIDD, John. 1775-1851. Br. DSB; P I; BMNH; GB; GD. Also Science in General* and Various Fields.
1. *Outlines of mineralogy.* Oxford, 1809. RM.

8740. PHILLIPS, William (first of the name). 1775-1828. Br. DSB; P II; BMNH; GA; GB; GD. Also Geology.
1. *An elementary introduction to the knowledge of mineralogy.* London, 1816. X. 3rd ed., 1823. Another ed., 1852.

8741. GIBBS, George (first of the name). 1776-1833. U.S.A. P I,III; GE.

8742. NEERGAARD, Tönnes Christian Brunn. 1776-1824. Den. P II.

8743. NÜSSLEIN, Franz Anton. 1776-1832. Ger. P II. Also Zoology.

8744. ALLAN, Thomas. 1777-1833. Br. P I; BMNH; GA; GD. Also Geology.
1. *Mineralogical nomenclature.* Edinburgh, 1814. RM/L.

8745. CORDIER, Pierre Louis Antoine. 1777-1861. Fr. DSB; P I & Supp.; BMNH; GA. Also Geology.

8746. KOPP, Johann Heinrich. 1777-1858. Ger. P I; BMNH; BLA. See also 8747/1.

8747. LEONHARD, Karl Cäsar von. 1779-1862. Ger. DSB; P I,III; BMNH; GA; GC. Also Geology*.
   1. *Systematisch-tabellarische Übersicht und Charakteristik der Mineralkörper. In oryktognostischer und orologischer Hinsicht aufgestellt.* Frankfurt, 1806. With K.F. Merz and J.H. Kopp. RM.
   2. *Handbuch der Oryktognosie.* Heidelberg, 1821. RM.

8748. SCHWABE, Johann Friedrich Heinrich. 1779-1834. Ger. P II; BMNH.

8749. CLEAVELAND, Parker. 1780-1858. U.S.A. DSB; P III; BMNH; GE. Also Geology*.

8750. GÖSSEL, Johann Heinrich Gottlieb. 1780-1846. Ger. P I & Supp.; BMNH.

8751. LUCAS, Jean André Henri. 1780-1825. Fr. P I; BMNH; GA.

8752. WEISS, Christian Samuel. 1780-1856. Ger. DSB; ICB; P II; BMNH; GC.

8753. HAUSMANN, Johann Friedrich Ludwig. 1782-1859. Ger. P I & Supp.; BMNH; GA; GB; GC. Also Geology.

8754. TROLLE-WACHTMEISTER, Hans Gabriel. 1782-1871. Swed. P II,III. Also Zoology.

8755. ZIPSER, Christian Andreas von. 1783-1864. Hung. P II,III; BMNH; GC.

8756. RAU, Ambrosius. 1784-1830. Ger. P II; BMNH. Also Botany.

8757. GERMAR, Ernst Friedrich. 1786-1853. Ger. P I; BMNH; GC. Also Geology* and Zoology.
   1. *Lehrbuch der gesammten Mineralogie.* Halle, 1824. With J.L.H. Meinecke. X. 2nd ed., 1837.
   2. *Grundriss der Krystallkunde.* Halle, 1830. RM.

8758. BEUDANT, François Sulpice. 1787-1850. Fr. DSB; P I; BMNH; GA; GB. Also Geology*.
   1. *Traité élémentaire de minéralogie.* Paris, 1824. RM. 7th ed., 1857.

8759. SOKOLOV, Dmietri Ivanovich. 1788-1852. Russ. DSB. Also Geology.

8760. TESCHEMACHER, James Englebert. 1790-1853. U.S.A. P II.

8761. ZINCKEN, Johann Karl Ludwig. 1790-1862. Ger. P II; BMNH.

8762. BREITHAUPT, Johann Friedrich August. 1791-1873. Ger. DSB; P I,III; BMNH; GA; GC.
   1. *Über die Ächtheit der Kristalle.* Freiberg, 1815. RM.
   2. *Vollständige Charakteristik des Mineral-Systems.* 2nd ed. Dresden, 1823. RM.
   3. *Vollständiges Handbuch der Mineralogie.* 3 vols. Dresden/ Leipzig, 1836-47.
   4. *Die Paragenesis der Mineralen, mineralogisch, geognostisch und chemisch beleuchtet.* Freiberg, 1849. RM.

8763. ZIPPE, Franz Xaver Maximilian. 1791-1863. Prague/Aus. P II, III; BMNH; GC.

8764. ARFVEDSON, Johann August. 1792-1841. Swed. DSB Supp.; ICB; P I; BMNH. Also Chemistry.

8765. DUFRÉNOY, (Ours) Pierre Armand. 1792-1857. Fr. DSB; P I; BMNH; GA; GB. Also Geology*.
1. *Voyage métallurgique en Angleterre; ou, Recueil de mémoires sur le gisement, l'exploitation et le traitment des minérais d'étain, de cuivre, de plomb, de zinc et de fer dans la Grande-Bretagne.* Paris, 1827. With Elie de Beaumont. RM.
2. *Traité de minéralogie.* 4 vols. Paris, 1844-47. RM/L. 2nd ed., 5 vols, 1856-59.

8766. NORDENSKJÖLD, Nils Gustaf. 1792-1866. Fin. ICB; P II,III; BMNH.

8767. GLOCKER, Ernst Friedrich. 1793-1858. Ger. P I & Supp.; BMNH; GA; GC.

8768. MARX, Karl Michael. 1794-1864. Ger. P II & Supp.; BMNH. Also History of Crystallography*.

8769. MITSCHERLICH, Eilhard. 1794-1863. Ger. DSB; ICB; P II; GA; GB; GC. Also Chemistry*#.

8770. WHEWELL, William. 1794-1866. Br. DSB; ICB; P II,III; BMNH; GB; GD. Other entries: see 3304.

8771. HAIDINGER. Wilhelm Karl von. 1795-1871. Aus. DSB; P I,III; BMNH; GA; GB; GC. Also Geology*.
1. *Handbuch der bestimmende Mineralogie.* Vienna, 1845. RM.
2. *Naturwissenschaftliche Abhandlungen.* 4 vols. Vienna, 1847-51.

8772. LÉVY, Serve-Dieu Abailard (called Armand). 1795-1841. Fr. DSB; P I & Supp. (L., Armand); BMNH (L., Armand).

8773. SORET, Frédéric Jacob. 1795-1865. Switz./Ger. P II,III; GC.

8774. APJOHN, James. 1796-1886. Br. P III; BMNH. Also Physics and Chemistry*.
1. *A descriptive catalogue of the simple minerals in the systematic collection of Trinity College.* Dublin, 1850.

8775. DELAFOSSE, Gabriel. 1796-1878. Fr. DSB Supp.; P III; BMNH.

8776. HESSEL, Johann Friedrich Christian. 1796-1872. Ger. DSB; P I,III,IV; BMNH.
1. *Krystallometrie.* Leipzig, 1831. X. Reprinted, 2 vols, Leipzig, 1897. (OKEW. 88,89)

8777. NAUMANN, (Georg Amadeus) Karl Friedrich. 1797-1873. Ger. DSB; P II,III; BMNH; GA; GB; GC. Also Geology*.
1. *Lehrbuch der reinen und angewandten Krystallographie.* 2 vols & atlas. Leipzig, 1830.
2. *Elemente der Mineralogie.* Leipzig, 1846. X. 6th ed., 1864. 8th ed., 1871. 9th ed., 1874. 10th ed., rev. by F. Zirkel, 1877. 12th ed., 1885. 14th ed., 1901.

8778. KURR, Johann Gottlob von. 1798-1870. Ger. P I,III; BNMH; GC.

Also Geology.
1. *Das Mineralreich in Bildern.* Stuttgart, 1858. X.
1a. ——— *The mineral kingdom.* 1859.

8779. NEUMANN, Franz Ernst. 1798-1895. Ger. DSB; ICB; P II,III,IV; BMNH; GA; GB; GC. Also Mathematics* and Physics*.

8780. ROSE, Gustav. 1798-1873. Ger. DSB; P II,III; BMNH; GB; GC.
1. *Elemente der Krystallographie.* Berlin, 1833.
2. *Das krystallo-chemische Mineralsystem.* Leipzig, 1852.

8781. KUPFFER, Adolf Theodor von. 1799-1865. Russ. P I,III; BMNH; GC. Also Physics and Earth Sciences.
1. *Handbuch der rechnenden Krystallonomie.* St. Petersburg, 1831.

8782. HARTWALL, Victor. 1800-1857. Fin. P I.

8783. FRANKENHEIM, Moritz Ludwig. 1801-1869. Ger. DSB; P I,III; BMNH. Also Physics*.

8784. MILLER, William Hallowes. 1801-1880. Br. DSB; ICB; P II,III; BMNH; GA; GB; GD.
1. *A treatise on crystallography.* Cambridge, 1838. RM. Other eds, 1839 and 1863.

8785. SEYBERT, Henry. 1801-1883. U.S.A. GE.

8786. BLUM, Johann Reinhard. 1802-1883. Ger. P I,III; BMNH; GC.
1. *Die Pseudomorphosen des Mineralreichs.* 5 vols in 3. Stuttgart, 1843.

8787. GRIFFIN, John Joseph. 1802-1877. Br. P I,III; BMNH; GD. Also Chemistry*.
1. *A system of crystallography, with its application to mineralogy.* Glasgow, 1841. RM/L.

8788. WISER, David Friedrich. 1802-1878 (or 1868). Switz. P II, III; GC.

8789. KAYSER, Gustav Eduard. b. 1803. Ger. P I; BMNH.

8790. KOBELL, (Wolfgang Xaver) Franz von. 1803-1882. Ger. P I,III; BMNH; GB; GC. Also History of Mineralogy*.
1. *Tafeln zur Bestimmung der Mineralien mittelst chemischer Versuche.* Munich, 1833 and later eds. X.
1a. ——— *Instructions for the discrimination of minerals by simple chemical experiments.* Glasgow, 1841. Trans. by R.C. Campbell.
2. *Skizzen aus dem Steinreich.* Munich, 1850. X.
2a. ——— *Popular sketches from the mineral kingdom.* London, 1852. Trans. and ed. by A. Henfrey.

8791. SUCKOW, Gustav. 1803-1867. Ger. P II,III; BMNH; GC.

8792. SHEPARD, Charles Upham. 1804-1886. U.S.A. P II,III,IV; BMNH; GE.
1. *A treatise on mineralogy.* New Haven, Conn., 1832. RM.

8793. BURHENNE, Georg Heinrich. 1805-1876. Ger. P I,III.

8794. DELLMANN, Johann Friedrich Georg. 1805-1870. Ger. P I,III. Also Physics and Meteorology.

8795. FRÖBEL, Julius. 1805-1893. Switz. P I,III; BMNH.

8796. HERMANN, Hans Rudolph. 1805-1879. Russ. P I,III; BMNH. Also Chemistry.

8797. JACKSON, Charles Thomas. 1805-1880. U.S.A. DSB; ICB; P I,III; BMNH; BLA; GA; GE. Also Geology.

8798. KÖHLER, Friedrich Wilhelm. 1805-1871. Ger. P I,III; BMNH.

8799. ALLAN, Robert. 1806-1863. Br. P III; BMNH.
1. *A manual of mineralogy.* Edinburgh, 1834. RM/L.

8800. ALGER, Francis. 1807-1863. U.S.A. P III; BMNH. See also 9508/1.

8801. DAMOUR, Augustin Alexis. 1808-1902. Fr. P I,III,IV; BMNH.

8802. JUNGHANN, Gustav Julius. b. 1808. Ger. P III.

8803. SÉNARMONT, Henri Hureau de. 1808-1862. Fr. DSB; P II,III; BMNH; GA; GB. See also 6211/3.

8804. TENNANT, James. 1808-1881. Br. BMNH; GD. Also Geology*.
1. *Mineralogy and crystallography.* London, 1856.

8805. BURAT, Amédée. 1809-1883. Fr. P III; BMNH. Also Geology*.
1. *Minéralogie appliquée.* Paris, 1864. RM/L.

8806. KRANTZ, August. 1809-1872. Ger. P III.

8807. MÜLLER, Johann Heinrich Jacob. 1809-1875. Ger. DSB; P II,III; GC. Also Physics*.
1. *Grundzüge der Krystallographie.* Brunswick, 1845. X.
1a. ——— *Eléments de cristallographie.* Paris, 1847. Trans. by J. Nicklès.

8808. QUENSTEDT, Friedrich August von. 1809-1889. Ger. DSB; P II, III; BMNH; GB; GC. Also Geology*.

8809. HESSENBERG, Friedrich. 1810-1874. Ger. P I,III; BMNH.

8810. LEYDOLT, Franz. 1810-1859. Aus. P I & Supp.; BMNH; GC.

8811. NICOL. James. 1810-1879. Br. P III; BMNH; GB; GD. Also Geology*.
1. *A manual of mineralogy.* Edinburgh, 1849.
2. *Elements of mineralogy.* Ib., 1858. 2nd ed., 1873.

8812. BRAVAIS, Auguste. 1811-1863. Fr. DSB; ICB; P I,III; BMNH; GA. Also Various Fields.
1. *Sur les propriétés géométriques des assemblages de points régulièrement distribués dans l'espace.* [Periodical article, 1848] German trans. Leipzig, 1897. (OKEW. 90)
2. *Sur les polyèdres de forme symétrique.* [Periodical article, 1849] German trans. Leipzig, 1890. (OKEW. 17)

8813. GÜMBEL, Wilhelm Theodor von. 1812-1858. Ger. P III; BMNH. Also Botany.

8814. SCHARFF, Friedrich A. 1812-1881. Ger. P III; BMNH.

8815. DANA, James Dwight. 1813-1895. U.S.A. DSB; ICB; P I,III,IV; BMNH; GB; GE. Also Geology* and Zoology*. See also Index.
1. *A system of mineralogy.* New Haven, Conn., 1837. RM. 2nd

ed., New York, 1844. 3rd ed., ib., 1850. 4th ed., London, 1855. 5th ed., London/New York, 1868; many re-issues.

1a. ——— 6th ed. *The system of mineralogy of J.D. Dana, 1837-68. Descriptive mineralogy.* London, 1892; re-issued 1898. Ed. by E.S. Dana.

1b. ——— ——— Appendices to the 6th ed. New York/London, 1899-. By E.S. Dana and W.E. Ford.

2. *Manual of mineralogy, including observations on mines, rocks, reduction of ores, and the applications of the science to the arts.* New Haven, Conn., 1848. X. Many later editions with varying titles published at London and New York.

8816. KOLENATI, Friedrich. 1813-1864. Prague/Brünn. P I,III; BMNH. Also Natural History.

8817. SCHEERER, (Karl Johann August) Theodor. 1813-1875. Ger./Nor. P II,III; BMNH. Also Chemistry.

1. *Löthrohrbuch: Eine Anleitung zum Gebrauch des Löthrohrs.* Brunswich, 185-? X.

1a. ——— *An introduction to the use of the mouth-blowpipe.* London, 1856. Trans. by H.F. Blanford.

8818. AICHORN, Sigmund Johann Nepomuck. 1814-1892. BMNH.

8819. BESNARD, Anton Franz. 1814-1885. Ger. P III; BMNH. Also Botany.

8820. RAMSAY, Andrew Crombie. 1814-1891. Br. DSB; P III; BMNH; GB; GD. Also Geology*. See also 15524/2.

1. *Rudiments of mineralogy.* 2nd ed. London, 1874.

8821. FELLOECKER, Sigismundus. 1816-1887. Aus. P III; BMNH.

8822. BRISTOW, Henry William. 1817-1889. Br. P III,IV; BMNH; GB; GD1. Also Geology.

1. *Glossary of mineralogy.* London, 1861.

8823. DES CLOIZEAUX, Alfred Louis Olivier Legrand. 1817-1897. Fr. DSB; P I,III,IV; BMNH; GB.

1. *Manuel de minéralogie.* 2 vols in 3. Paris, 1862-93.

2. *Nouvelles recherches sur les propriétés optiques des cristaux.* Paris, 1867.

8824. FISCHER, Leopold Heinrich. 1817-1886. Ger. P I; BMNH. Also Zoology.

8825. KENNGOTT, (Johann Gustav) Adolf. 1818-1897. Ger./Aus./Switz. P I,III,IV; BMNH; GB. Also Geology*. See also 3023.

8826. KOKSCHAROW, Nicolai Ivanovich. 1818-1893. Russ. P I,III,IV; BMNH; GB.

8827. SMITH, John Lawrence. 1818-1883. U.S.A. ICB; P III; BMNH; BLA; GE. Also Chemistry*.

1. *Mineralogy and chemistry.* Louisville, Ky., 1873. RM.

8828. GENTH, Frederick Augustus. 1820-1893. U.S.A. DSB; P III,IV; BMNH; GE.

8829. PURGOLD, Alfred. 1820-1892. It./Ger. P III,IV.

8830. CHAPMAN, Edward John. 1821-1904. Br./Canada. P I,III,IV;

GD2. Also Geology.
1. *Practical mineralogy; or, A compendium of the distinguishing characters of minerals.* London, 1843.

8831. IGELSTRÖM, Lars Johan. 1822-1897. Swed. P III,IV; BMNH.

8832. DAUBER, Hermann. 1823-1861. Ger. P I & Supp.

8833. MASKELYNE, (Mervyn Herbert) Nevil Story. 1823-1911. Br. P III, IV; BMNH; GD2 (STORY-MASKELYNE).
1. *Crystallography.* Oxford, 1895.
See also 7823/1 and 8956/1,2.

8834. LUDLAM, Henry. 1824-1880. Br. BMNH; GD.

8835. WEBSKY, Christian Friedrich Martin. 1824-1886. Ger. P II,III, IV; BMNH; GC.

8836. KREJČI, Jan. 1825-1887. Prague. P I,III; BMNH. Also Geology.

8837. SCHABUS, Jacob (or Joseph). 1825-1867. Aus. P II,III; BMNH.

8838. SOWERBY, Henry. 1825-1891. Br. BMNH; GD.
1. *Popular mineralogy.* London, 1850.

8839. GREG, Robert Philips. 1826-1906. Br. P IV; BMNH.
1. *A manual of the mineralogy of Great Britain and Ireland.* London, 1858. With W.G. Lettsom. RM.

8840. HEUSSER, Jacob Christian. 1826-after 1867. Switz./Argentina. P I,III; BMNH.

8841. HUNT, Thomas Sterry. 1826-1892. U.S.A./Canada. DSB; ICB; P III,IV; BMNH; GB; GE. Also Chemistry* and Geology*.
1. *Systematic mineralogy, based on a natural classification.* New York, 1891. RM/L. 2nd ed., 1892.

8842. RAIMONDI, Antonio. 1826-1890. Peru. ICB; P III; BMNH.

8843. SCHRÖDER, Friedrich Heinrich. 1826-after 1856. Ger. P II; BMNH.

8844. SELLA, Quintino. 1827-1884. It. ICB; P II,III; BMNH; GB.
1. *Memorie di cristallografia.* Rome, 1885. X.
1a. ——— German trans. Leipzig, 1906. (OKEW. 155)

8845. ABT, Anton. 1828-after 1901. Hung. P III,IV; BMNH. Also Physics.

8846. BUCHNER, Christian Ludwig Otto. 1828-1897. Ger. P III,IV; BMNH.
1. *Die Meteoriten in Sammlungen, ihre Geschichte, mineralogische und chemische Beschaffenheit.* Leipzig, 1863.

8847. FOUQUÉ, Ferdinand André. 1828-1904. Fr. DSB; P III,IV & Supp.; BMNH; GB. Also Geology*.
1. *Minéralogie micrographique: Roches éruptives françaises.* 2 vols. Paris, 1879. With A.M. Levy. RM/L.
2. *Synthèse des minéraux et des roches.* Paris, 1882. With A.M. Levy.

8848. GADOLIN, Axel Wilhelm. 1828-1893. Russ. P III,IV; BMNH.
1. *Abhandlung über die Herleitung aller krystallographischer Systeme mit ihren Unterabtheilungen aus einem einzigen Prinzipe.* [Trans. of a periodical article of 1867] Leipzig, 1896. (OKEW. 75)

8849. HEDDLE, Matthew Forster. 1828-1897. Br. P III,IV; BMNH; GB.

8850. HUGO, Léopold Armand. 1828-after 1889. Fr. P III,IV; BMNH.

8851. KNOP, Adolf. 1828-1893. Ger. P I,III,IV; BMNH.

8852. GRAILICH, Wilhelm Joseph. 1829-1859. Aus. P I,III; BMNH.

8853. JENZSCH, Gustav Julius Siegmund. 1830-1877. Ger. P I,III; BMNH.

8854. JEREMEJEW, Pavel Vladimirovich. 1830-1899. Russ. P IV; BMNH (EREMYEEV).

8855. RATH, Gerhard vom. 1830-1888. Ger. P II,III,IV; BMNH; GB; GC. Also Geology.

8856. SÖCHTING, Johann Wilhelm Edmund. 1830-after 1866. Ger. P II, III.

8857. STRENG, Johann August. 1830-1897. Ger. P II,III,IV; BMNH.

8858. ZEPHAROVICH, Victor Leopold von. 1830-1890. Aus./Prague. P II,III,IV; BMNH; GC.

8859. BRUSH, George Jarvis. 1831-1912. U.S.A. P III,IV; BMNH; GE.
1. *A manual of determinative mineralogy.* New York/London, 1875. X. 6th ed., 1884. 15th ed., 1898.

8860. PISANI, Félix. 1831-after 1882. Fr. P III; BMNH. Also Chemistry.
1. *Traité élémentaire de minéralogie.* Paris, 1875. 2nd ed., 1883.

8861. EGLESTON, Thomas. 1832-1900. U.S.A. P IV; BMNH; GE. See also 8957/1.

8862. FRIEDEL, Charles. 1832-1898. Fr. P III,IV; GB. Also Chemistry.
1. *Cours de minéralogie.* Paris, 1893.

8863. JANNETTAZ, (Pierre Michel) Edouard. 1832-1899. Fr. P III, IV; BMNH. Also Geology*.

8864. BOMBICCI-PORTA, Luigi. 1833-after 1903. It. P III,IV; BMNH.
1. *Corso di mineralogia generale.* Bologna, 1862. X. 2nd ed., 2 vols in 3, 1873-75.
2. *Mineralogia e cristallografia.* 1868-70.

8865. MALLARD, (François) Ernest. 1833-1894. Fr. DSB; P III,IV; BMNH.
1. *Traité de cristallographie géometrique et physique.* 3 vols. Paris, 1879-84.

8866. WEISBACH, (Julius) Albin. 1833-1901. Ger. P III,IV; BMNH.
1. *Tabellen zur Bestimmung der Mineralien mittels äusserer Kennzeichen.* Leipzig, 1866. X. 4th ed., 1892.

8867. BAUERMAN, Hilary. 1834-1909. Br./Canada. BMNH; GD2. Also Geology.
1. *Text-book of systematic mineralogy.* London, 1881.
2. *Text-book of descriptive mineralogy.* London, 1884.

8868. LASPEYRES, (Ernst Adolf) Hugo. 1836-1913. Ger. P III,IV; BMNH. Also Geology.

8869. ROSENBUSCH, Harry (or Karl Heinrich Ferdinand). 1836-1914.

Ger. DSB; P III,IV; BMNH. Also Geology.
1. *Mikroskopische Physiographie der petrographisch wichtigen Mineralien.* Stuttgart, 1873. X.
1a. ——— *Microscopic physiography of the rock-forming minerals.* London/New York, 1888. Trans. by J.P. Iddings. 3rd ed., 1895.
2. *Mikroskopische Physiographie der massigen Gesteine.* Stuttgart, 1877.
3. *Mikroskopische Physiographie Mineralien und Gesteine.* 2 vols. Stuttgart, 1892-96.

8870. TSCHERMAK, Gustav von. 1836-1927. Aus. DSB; P III,IV; BMNH.
1. *Lehrbuch der Mineralogie.* Vienna, 1881. 3rd ed., 1888. 5th ed., 1897.

8871. DAVIES, Thomas. 1837-1891. Br. BMNH; GD1.

8872. FUCHS, Carl Wilhelm C. 1837-1886. Ger. P III,IV; BMNH. Also Geology.
1. *Tafeln zur Bestimmung der Mineralien durch das Löthrohr.* Heidelberg, 1867. X.
1a. ——— *Practical guide to the determination of minerals by the blowpipe.* London, 1868. Trans. and ed. by T.W. Danby.
2. *Die künstlich dargestellten Mineralien.* 1872.

8873. HOFFMANN, George Christian. 1837-after 1902. Br./Australia/Canada. P IV.

8874. MINNIGERODE, Ludwig Bernhard. 1837-1896. Ger. P III,IV; BMNH.

8875. SCHRAUF, Albrecht. 1837-1897. Aus. P II,III,IV; BMNH.

8876. DITSCHEINER, Leander. 1839-1905. Aus. P IV. Also Physics.

8877. HAUSHOFER, Karl. 1839-1895. Ger. P III,IV; BMNH.

8878. KRENNER, Joseph Alexander. 1839-1920. Hung. P III,IV; BMNH.

8879. LAPPARENT, Albert Auguste Cochon de. 1839-1908. Fr. DSB; P III,IV; BMNH; GB. Also Geology*.
1. *Cours de minéralogie.* Paris, 1884.

8880. LASAULX, Arnold Contantin Peter Franz von. 1839-1886. Ger. P III,IV; BMNH; GB; GC. Also Geology.

8881. WERNER, Gotthilf. 1839-1881. Ger. P III.

8882. BOŘICKY, Emanuel. 1840-1881. Prague. P III,IV; BMNH.

8883. COLLINS, Joseph Henry. 1841-1916. BMNH. Also Geology.
1. *The general principles of mineralogy.* 2 vols. London, 1878-83.

8884. FLIGHT, Walter. 1841-1885. Br. P III; BMNH; GD.
1. *A chapter in the history of meteorites.* London, 1887. Posthumous. Ed. with an obituary by L. Fletcher.

8885. RUMPF, Johann. 1841-1923. Aus. P IV.

8886. COHEN, Emil William. 1842-1905. Ger. P III,IV; BMNH. Also Geology.
1. *Meteoritenkunde.* Stuttgart, 1894-1905.

8887. FRENZEL, (Friedrich) August. 1842-1902. Ger. DSB; BMNH.

8888. KLEIN, Johann Friedrich Carl. 1842-1907. Ger. P III,IV; BMNH.

8889. LEMBERG, Johann Theodor. 1842-1902. Dorpat. P III,IV.

8890. RUTLEY, Frank. 1842-1904. Br. P III,IV; BMNH; GB. Also Geology*.
1. *Elements of mineralogy.* London, 1874. X. 3rd ed., 1876.
2. *Rock-forming minerals.* London, 1888.

8891. SOHNCKE, Leonhard. 1842-1897. Ger. DSB; P III,IV; BMNH; GC. Also Physics and Meteorology.
1. *Entwickelung einer Theorie der Krystallstruktur.* Leipzig, 1879. RM.

8892. SPEZIA, Giorgio. 1842-1911. It. P IV; BMNH.

8893. STRÜVER, Johannes Karl Theodor. 1842-1915. It. P III,IV; BMNH.

8894. CHESTER, Albert Huntington. 1843-1903. U.S.A. P III,IV; BMNH.
1. *A dictionary of the names of minerals, including their history and etymology.* New York, 1896.

8895. GROTH, Paul Heinrich. 1843-1927. Ger. DSB; P III,IV; BMNH.
1. *Tabellarische Übersicht der einfachen Mineralien, nach ihren krystallographisch-chemischen Beziehungen.* Brunswick, 1874. 4th ed., 1898.
2. *Physikalische Krystallographie.* Leipzig, 1876. 2nd ed., 1885. 3rd ed., 1895.
3. *Grundriss der Edelsteinkunde.* Leipzig, 1887.

8896. LEUCHTENBERG, Nicolaus von, *Prince* ROMANOVSKI. 1843-1891. Russ. P III.

8897. MEUNIER, (Etienne) Stanislas. 1843-1925. Fr. P III,IV; BMNH. Also Geology*.
1. *Les méthodes de synthèse en minéralogie.* Paris, 1891.

8898. SADEBECK, Alexander. 1843-1879. Ger. P III; BMNH.

8899. SARASIN, Emile Edmond. 1843-1890. Switz. P IV.

8900. WYRUBOFF, Grigory Nikolaievich. 1843-1913. Fr. ICB; P IV; BMNH. Also Chemistry.

8901. BAUER, Max Hermann. 1844-1917. Ger. P III,IV; BMNH. Also Geology.
1. *Edelsteinkunde.* Leipzig, 1896.

8902. KÖNIG, George Augustus. 1844-1913. U.S.A. P III,IV; GE.

8903. MICHEL-LÉVY, Auguste. 1844-1911. Fr. DSB; P IV (LÉVY); BMNH (LÉVY). Also Geology*.
1. *Les minéraux des roches.* Paris, 1888. With A. Lacroix.
2. *Tableaux des minéraux des roches: Résumé de leurs propriétés optiques, crystallographiques et chimiques.* Paris, 1889. With A. Lacroix.

His books are generally listed under LÉVY, A.M.
See also 8847/1,2.

8904. BARLOW, William. 1845-1934. Br. DSB; ICB; P IV.

8905. HALL, Townshend Monckton. 1845-1899. Br. P III; BMNH. Also Geology.

1. *The mineralogist's directory; or, A guide to the principal mineral localities in the United Kingdom.* London, 1868.

8906. HIRSCHWALD, Julius. 1845-1928. Ger. P III,IV; BMNH.

8907. VRBA, Karl. 1845-1922. Czernowitz/Prague. P III,IV; BMNH.

8908. ARZRUNI, Andreas. 1847-1898. Ger. P IV; BMNH.
1. *Physikalische Chemie der Krystalle.* Brunswick, 1893.

8909. GURNEY, Henry Palin. 1847-1904. Br. BMNH; GD2.
1. *Crystallography.* London, 1873. A textbook.

8910. KLOCKE, Friedrich. 1847-1884. Ger. P III.

8911. LEWIS, William James. 1847-1926. Br. P III,IV; BMNH.
1. *A treatise on crystallography.* Cambridge, 1899.

8912. BAUMHAUER, Heinrich Adolf. 1848-1926. Ger./Switz. DSB; P III, IV; BMNH.
1. *Darstellung der 32 möglichen Krystallklassen.* Leipzig, 1899.

8913. BREZINA, (Maria) Aristides (Severin Ferdinand). 1848-1909. Aus. P III,IV; BMNH.

8914. DUFET, (Jean Baptiste) Henry. 1848-1905. Fr. P III,IV. Also Physics*.

8915. PANEBIANCO, Ruggiero. 1848-1930. It. P III,IV; BMNH.

8916. CESÀRO, Giuseppe. 1849-1939. Belg. ICB; P IV; BMNH.

8917. DANA, Edward Salisbury. 1849-1935. U.S.A. P III,IV; BMNH; GE1.
1. *A text-book of mineralogy ... on the plan and with the co-operation of ... J.D. Dana.* New York, 1877. 2nd ed., 1878. Another ed., 1898.
2. *Minerals and how to study them.* New York, 1895. X. 2nd ed., 1895.

See also 8815/1a,1b.

8918. FLINK, Gustaf. 1849-1931. Swed. P IV; BMNH.

8919. SELIGMANN, Gustav. 1849-after 1904. Ger. P IV.

8920. BERWERTH, Friedrich Martin. 1850-1918. Aus. P III,IV; BMNH.

8921. DOELTER (Y CISTERICH), Cornelio August Severinus. 1850-1930. Aus. DSB; P III,IV; BMNH.
1. *Allgemeine chemische Mineralogie.* Leipzig, 1890.
2. *Edelsteinkunde.* Leipzig, 1893.

8922. FOULLON-NORBECK, Heinrich von. 1850-1896. Aus. P IV. Also Geology.

8923. VOIGT, Woldemar. 1850-1919. Ger. DSB; P III,IV; BMNH. Also Physics.

8924. ENDLICH, Frederic Miller. 1851-1899. U.S.A. P III; BMNH. Also Geology.
1. *Manual of qualitative blowpipe analysis and determinative mineralogy.* New York, 1892. X. 2nd ed., 1895.

8925. HINTZE, Carl Adolf Ferdinand. 1851-1917. Ger. P III,IV; BMNH.
1. *Handbuch der Mineralogie.* ? parts. Leipzig, 1889-.

8926. LUEDECKE, Otto Paul. 1851-1910. Ger. P III,IV; BMNH.

8927. TRECHMANN, Charles Otto. 1851-after 1907. Br. P III,IV.

8928. LIEBISCH, Theodor. 1852-1922. Ger. P IV; BMNH.
1. *Physikalische Krystallographie*. Leipzig, 1891.

8929. FYODOROV, Evgraf Stepanovich. 1853-1919. Russ. DSB; ICB (FEDOROV); P IV (FEDOROW, J.S.); BMNH (FEDOROV). Also Geology.

8930. GOLDSCHMIDT, Victor. 1853-1933. Ger. DSB; P III,IV; BMNH.
1. *Krystallographische Winkeltabellen*. Berlin, 1897.

8931. LEWIS, Henry Carvill. 1853-1888. U.S.A. P III; BMNH; GB. Also Geology.
1. *Papers and notes on the genesis and matrix of the diamond*. London, 1897. Ed. by T.G. Bonney.

8932. SCHOENFLIES, Arthur Moritz. 1853-1928. Ger. DSB; P III,IV; BMNH. Also Mathematics.

8933. FLETCHER, Lazarus. 1854-1921. Br. P III,IV; BMNH; GF.
1. *Optical indicatrix and the transmission of light in crystals*. London, 1892.
See also 8884/1.

8934. SORET, Charles. 1854-1904. Switz. P III,IV; BMNH. Also Physics.
1. *Eléments de cristallographie physique*. Geneva, 1893.

8935. BECKE, Friedrich Johann Karl. 1855-1931. Prague/Aus. DSB; P IV; BMNH.

8936. BECKENKAMP, Jacob. 1855-1931. Ger. P IV.

8937. LEHMANN, Otto. 1855-1922. Ger. DSB; P III,IV; BMNH. Also Physics*.

8938. MÜLLER, Paul Friedrich Wilhelm. 1855-after 1902. Ger. P IV; BMNH.

8939. VIOLA, Carlo Maria. 1855-after 1920. It. P IV; BMNH.

8940. FOCK, Andreas Ludwig. 1856-after 1923. Ger. P III,IV; BMNH. Also Chemistry.
1. *Einleitung in die chemische Krystallographie*. Leipzig, 1888. X.
1a. ——— *Introduction to chemical crystallography*. Oxford, 1895. Trans. and ed. by W.J. Pope.

8941. HUSSAK, Eugen. 1856-1911. Aus./Ger./Brazil. P III,IV; BMNH.
1. *Anleitung zum Bestimmung der gesteinbildenden Mineralien*. Leipzig, 1885.
1a. ——— *Determination of rock-forming minerals*. New York, 1886. Trans. by E.G. Smith.

8942. KUNZ, George Frederick. 1856-1932. U.S.A. ICB; P IV; BMNH; GE1.

8943. PENFIELD, Samuel Lewis. 1856-1906. U.S.A. P IV; BMNH.

8944. RETGERS, Jan Willem. 1856-1896. Holl./Java. P IV. Also Chemistry.

8945. SCHUSTER, Maximilian Joseph. 1856-1887. Aus. P III,IV.

8946. WEIBULL, (Kristian Oskar) Mats. 1856-1923. Swed. P IV; BMNH.

8947. WILLIAMS, George Huntington. 1856-1894. U.S.A. BMNH; GE. Also Geology.
1. *Elements of crystallography*. New York, 1890. X. 2nd ed., 1890. 3rd ed., 1899.

8948. LINCK, Gottlob Eduard. 1858-after 1930. Ger. P IV; BMNH.

8949. MIERS, Henry Alexander. 1858-1942. Br. DSB; ICB; P IV; BMNH; GD6.

8950. MÜGGE, (Johannes) Otto (Conrad). 1858-1932. Ger. P III,IV; BMNH.

8951. WALLERANT, Frédéric Félix Auguste. 1858-1936. Fr. P IV; BMNH. Also Geology.
1. *Groupements cristallins*. Paris, 1899.

8952. MOSES, Alfred Joseph. 1859-1920. U.S.A. P IV; BMNH.
1. *Characters of crystals: An introduction to physical crystallography*. New York, 1899.

8953. SCHARIZER, Rudolf. 1859-after 1920. Czernowitz. P IV.

Imprint Sequence

8954. K.K. HOF-MINERALIEN-KABINET. Vienna.
1. *Kurze Übersicht der ... zur Schau gestellten acht Sammlungen*. Vienna, 1843. Ed. by P. Partsch. X. 2nd ed., 1855.
2. *Katalog der Bibliothek*. Vienna, 1851. Ed. by P. Partsch.

8955. WERNERIAN Club. (Founded 1844) BMNH.
1. *An analysis of the natural system and its application to the mineral kingdom*. London, 1846-47.

8956. BRITISH Museum (Natural History). Department of Mineralogy.
1. *Guide to the collection of minerals*. London, 1862. By M.H.N. Story-Maskelyne. Other eds, 1869 and 1870.
2. *Catalogue of minerals*. London, 1863. By M.H.N. Story-Maskelyne.
3. *Index to the collection of minerals*. London, 1868.
4. *Introduction to the study of minerals*. London, 1889. Another ed., 1894.

8957. SMITHSONIAN Institution. Washington.
1. *Catalogue of minerals, with their formulas, etc*. Washington, 1863. By T. Egleston.

3.09 GEOLOGY

Including palaeontology, petrology, and physical geography.

In the eighteenth and early nineteenth centuries the distinction from mineralogy (in the modern sense) is difficult because the word 'mineralogy' was then used for what would now be called geology, as well as for mineralogy in the modern sense. (The abandonment of the old usage of the word is illustrated by the difference between the titles of items 9213/1 and 1a.)

In the second half of the nineteenth century a large number of titles of very specialised monographs has been omitted.

8958. ROBINSON, Thomas (of Ousby). d. 1719. Br. BMNH; GD. Also Part 2*.
  1. *An essay towards a natural history of Westmorland and Cumberland. Wherein an account is given of their several mineral and surface productions ... To which is annexed a vindication of the philosophical and theological paraphrase of the Mosaick account of the Creation.* London, 1709. RM.

8959. MAILLET, Benoît de. 1656-1738. Fr. DSB; ICB; P II; GA.
  1. *Telliamid; ou, Entretiens d'un philosophe indien avec un missionaire françois sur la diminution de la mer, la formation de la terre, l'origine de l'homme, etc.* Amsterdam, 1748. RM.
  1a. ——— *Telliamid; or, Discourses between an Indian philosopher and a French missionary on the diminution of the sea, the formation of the earth, the origin of men and animals, etc.* London, 1750. RM.
  1b. ——— *Telliamid; or, Conversations between....* Urbana, Ill., 1968. Trans. and ed. by A.V. Carozzi.

8960. MARSIGLI, Luigi Ferdinando. 1658-1730. It. DSB (MARSILI); ICB; P II; BMNH; GA; GB. Also Oceanography* and Part 2.

8961. VALLISNIERI, Antonio. 1661-1730. It. DSB; ICB; P II; BMNH; Mort.; BLA; GA. Also Various Fields* and Embryology.
  1. *Lezione accademica intorno all'origine delle fontane.* Venice, 1715. RM.
  2. *De corpi marini che sui monte si trovano: Della loro origine, e dello stato del mondo avanti'l diluvio, nel diluvio, e dopo il diluvio.* Venice, 1721. RM.

8962. WOODWARD, John. 1665-1728. Br. DSB; P II; BMNH; BLA; GB; GD. Also Botany and Part 2*.
  1. *Naturalis historia telluris, illustrata et aucta.* London, 1714. X.

1a. ——— *The natural history of the earth, illustrated, inlarged and defended.* London, 1726. Trans. by B. Holloway. RM.
2. *Fossils of all kinds, digested into a method suitable to their mutual relation and affinity.* London, 1728. RM.
3. *An attempt towards a natural history of the fossils of England.* 2 vols. London, 1729. RM/L.

8963. BERINGER (or BEHRINGER), Johann Bartholomaeus Adam. ca. 1667-1738. Ger. DSB; ICB; P I; BLA; GA. Also Natural History.
1. *Lithographiae Wirceburgensis ... specimen primum.* Wurzburg, 1726. X. 2nd ed., Frankfurt/Leipzig, 1767. RM.
1a. ——— *The lying stones of Dr. Beringer, being his "Lithographiae Wirceburgensis", translated and annotated.* Berkeley, Cal., 1963. By M.E. Jahn and D.J. Woolf.

8964. LANG, Carl Nicolaus. 1670-1741. Switz. DSB; ICB; P I (LANGE); BMNH (LANGIUS); GA.
1. *Historia lapidum figuratorum Helvetiae, ejusque viciniae.* Venice, 1708. RM. Appendix, 1735. RM.
2. *Tractatus de origine lapidum figuratorum.* Lucerne, 1709. RM.
3. *Methodus nova et facilis testacea marina pleraque ... in suas ... classes, genera et species distribuendi.* Lucerne, 1722. RM.

8965. STRACHEY, John. 1671-1743. Br. DSB; GD1 (under S., Henry).

8966. SCHEUCHZER, Johann Jacob. 1672-1733. Switz. DSB; ICB; P II; BMNH; BLA; GA; GB; GC. Also Natural History* and Zoology*.
1. *Specimen lithographiae Helvetiae curiosae, quo lapides ex figuratis Helveticis ... describuntur.* Zurich, 1702. RM.
2. *Museum diluvianum quod possidet J.J. Scheuchzer.* Zurich, 1716. RM.

8967. LEUPOLD, Jacob. 1674-1727. Ger. ICB; P I; BMNH; GA. Also Physics.

8968. MYLIUS, Gottlieb Friedrich. 1675-1726. Ger. P II; BMNH.

8969. BAIER, Johann Jacob. 1677-1735. Ger. DSB; P I; BMNH; BLA; GA; GC.
1. *Oryctographia Norica; sive, Rerum fossilium ... in territorio Norimbergensi eiusque vicinia observatarum succincta descriptio.* Nuremberg, 1708. RM.
2. *Sciagraphia musei sui.* Nuremberg, 1730. RM.
3. *Monumenta rerum petrificatarum praecipua, "Oryctographiae Noricae" supplementi.* Nuremberg, 1757. Ed. by his son, F.J. Baier. RM.
4. *Epistolae ad viros eruditos eorumque responsiones.* Frankfurt, 1760. Ed. by F.J. Baier. RM.
See also 10918/1.

8970. HALES, Stephen. 1677-1761. Br. DSB; ICB; P I; BMNH; Mort.; BLA & Supp.; GA; GB; GD. Also Chemistry, Physiology*, and Botany*.
1. *Some considerations on the causes of earthquakes.* London, 1750. RM/L.

8971. BOURGUET, Louis. 1678-1742. Switz. DSB Supp.; ICB; P I; BMNH; GA.
1. *Lettres philosophiques sur la formation des sels et des crystaux, et sur la génération et le mechanisme organique des plantes et des animaux ... Avec un mémoire sur la théorie de la terre.* Amsterdam, 1729. RM.
2. *Traité des pétrifications.* Paris, 1742. RM.
2a. ——— Another ed., entitled *Mémoires pour servir à l'histoire naturelle des pétrifications.* The Hague, 1742. RM.

8972. BROMELL, Magnus von. 1679-1731. Swed. DSB; P I; BMNH; BLA.

8973. ARGENVILLE, Antoine Joseph Dezallier d'. 1680-1765. Fr. DSB. Under DEZALLIER: P I; BMNH; GA. Also Natural History*.
1. *Enumerationis fossilium, quae in omnibus Galliae provinciis reperiuntur, tentamina.* Paris, 1751. RM.

8974. HOTTINGER, Johann Heinrich. 1680-1756. Switz. ICB; P I; GA.

8975. RITTER, Albrecht. 1684-1755. Ger. P II; BMNH.

8976. KLEIN, Jacob Theodor. 1685-1759. Danzig. DSB; P I; BMNH; GA; GC. Also Zoology*.
1. *Specimen descriptionis petrefactorum Gedanensium ... Probe einer Beschreibung ... der in der Danziger ... Gegend befindlichen Versteinerungen.* Nuremberg, 1770. RM.

8977. CALVÖR, Henning. 1686-1766. Ger. P I & Supp.; GA.
1. *Acta historico-chronologico-mechanica circa metallurgiam in Hercynia Superiori; oder, Historisch-chronologische Nachricht und ... Beschreibung ... der Hülfsmittel bey dem Bergbau auf dem Oberharz.* 2 vols. Brunswick, 1763. RM.

8978. MORO, Antonio Lazzaro. 1687-1764. It. DSB; P II; BMNH.
1. *Dei crostacei e degli altri corpi marini che si trovano sui monti.* 2 vols. Venice, 1740. RM.

8979. ROSINUS, Michael Reinhold. 1687-1725. BMNH.

8980. KÜHN, Heinrich. 1690-1769. P I. Danzig. Also Various Fields.
1. *Meditationes de origine fontium et aquae putealis.* Bordeaux, 1741. A prize essay. X.
1a. ——— *Vernünftige Gedanken von dem Ursprung der Quellen und des Grundwassers.* Berlin, 1746. RM.

8981. LESSER, Friedrich Christian. 1692-1754. Ger. P I; BMNH. Also Zoology.
1. *Lithotheologie; das ist, Natürliche Historie und geistliche Betrachtung derer Steine....* Hamburg, 1735. RM.
2. *Nachrichtliche Beschreibung des ... neu entdeckten Muschel-Marmors.* Nordhausen, 1752. RM.
3. *Testaceo-Theologia; oder, Gründlicher Beweis des Daseyns ... eines göttlichen Wesens aus ... Betrachtung der Schnecken und Keuscheln.* Leipzig, 1756. RM.

8982. EDWARDS, George. 1694-1773. Br. ICB; P I; BMNH; GA; GB; GD. Also Zoology*.
1. *Elements of fossilogy; or, An arrangement of fossils into classes, orders, genera and species, with their characteristics.* London, 1776. RM.

8983. SCHÜTTE, Johann Heinrich. 1694-1774. Ger. P II; BMNH; BLA.

8984. ALTMANN, Johann Georg. 1697-1758. Switz. P I; GA; GC.
1. *Versuch einer historischen und physischen Beschreibung der helvetischen Eisbergen.* Zurich, 1751. RM.

8985. BONO, Michele del. 1697-after 1753. It. P I.

8986. BRÜCKMANN, Franz Ernst. 1697-1753. Ger. P I; BMNH; BLA; GA; GB. Also Natural History.

8987. LANGE, Johann Joachim. ca. 1698-1765. Ger. P I.

8988. EHRHART, (Johann) Balthasar. 1700-1756. Ger. P I; BMNH; BLA; GA. Also Botany.
1. *De belemnitis Suevicis.* Leiden, 1724. X. Another ed., Augsburg, 1727. RM.

8989. EVANS, Lewis. 1700-1756. U.S.A. DSB; ICB; GE.

8990. ROUELLE, Guillaume François. 1703-1770. Fr. DSB; ICB; P II; GA; GB. Also Chemistry.

8991. KNORR, Georg Wolfgang. 1705-1761. Ger. DSB; P I; BMNH; GA; GC. Also Botany*.
1. *Sammlung von Merckwürdigkeiten der Natur und Alterthümern des Erdbodens, welche petrificirte Cörper enthält.* 4 vols. Nuremberg, 1755-74. After Knorr's death the work was continued by J.E.I. Walch. RM.
1a. ——— [Part of item 1] *Die Naturgeschichte der Versteinerungen, zur Erläuterung der Knorrischen "Sammlung von Merckwürdigkeiten der Natur."* Nuremberg, 1768-73. Ed. by J.E.I. Walch.

8992. BUFFON, Georges Louis Leclerc, *Comte* de. 1707-1788. Fr. DSB; ICB; P I; BMNH; GA; GB. Also Natural History*#. See also Index.
1. *Les époques de la nature.* Paris, 1779. X. Reprinted 1971.
1a. ——— *Edition critique, avec le manuscript, une introduction, et des notes.* Paris, 1965. By J. Roger.
His *Histoire et théorie de la terre* forms part of his *Histoire naturelle* (10954/1).

8993. GESSNER, Johannes. 1709-1790. Switz. DSB; BLA & Supp.; GC. Under GESNER: P I & Supp.; BMNH; GA. Also Physics and Botany. See also 14659/9.
1. *Tractatus physicus de petrificatis.* Leiden, 1758. RM.

8994. SAUVAGES DE LA CROIX, Pierre Augustin Boissier de. 1710-1795. Fr. P II; BMNH.

8995. BERTRAND, Elie. 1712-1777 (or ca. 1790). Switz. P I; BMNH; GA.
1. *Mémoires sur la structure intérieure de la terre.* Zurich, 1752. RM.
2. *Essai sur les usages des montagnes.* Zurich, 1754. RM.
3. *Mémoires historiques et physiques sur les tremblemens de terre.* The Hague, 1757. RM.
4. *Dictionnaire universel des fossiles propres et des fossiles accidentels.* Avignon, 1763. RM.
5. *Recueil de divers traités sur l'histoire naturelle de la terre et des fossils.* Avignon, 1766. RM.

8996. TARGIONE-TOZZETTI, Giovanni. 1712-1783. It. DSB; ICB; P II; BMNH; BLA; GA. Also Natural History*, Biography*, and History of Science*.

8997. TILAS, Daniel. 1712-1772. Swed. DSB; P II.

8998. (Entry cancelled)

8999. WHITEHURST, John. 1713-1788. Br. DSB; P II; BMNH; GA; GD.
1. *An enquiry into the original state and formation of the earth*. London, 1778. RM/L. 2nd ed., 1786. 3rd ed., 1792.
2. *Works. With memoirs of his life and writings*. London, 1792. Ed. by Charles Hutton. RM/L.

9000. ARDUINO (or ARDUINI), Giovanni. 1714-1795. It. DSB; ICB; P I; BMNH.
1. *Raccolta di memorie chimico-mineralogiche, metallurgiche e orittografiche*. Venice, 1775. RM.
2. *Memoria epistolare sopra varie produzioni vulcaniche, minerali e fossili*. Venice, 1782. RM.

9001. FRISCH, Jodocus (or Joseph) Leopold. 1714-1787. Ger. P I; BMNH; GA.

9002. BERTRAND, Bernard Nicolas. 1715-1780. Fr. P I; BMNH.
1. *Elémens d'oryctologie; ou, Distribution méthodique des fossiles*. Neuchâtel, 1773. RM.

9003. GUETTARD, Jean Etienne. 1715-1786. Fr. DSB; P I; BMNH; GA; GB. Also Botany*.

9004. KRÜGER, Johann Gottlob. 1715-1759. Ger. P I; BLA; GA. Also Physics.

9005. COSTA, Emanuel Mendes da. 1717-1791. Br. P I; BMNH; GD. Also Zoology*.
1. *Natural history of fossils*. Vol. 1, Part 1. [No more publ.] London, 1757. RM/L. (Generally listed under MENDES DA COSTA)
See also 8669/1a.

9006. GRUNER, Gottlieb Siegmund. 1717-1778. Switz. P I; BMNH; GA; GB; GC.

9007. BAUMER, Johann Wilhelm. 1719-1788. Ger. P I; BMNH; BLA & Supp.; GA; GC.

9008. LEHMANN, Johann Gottlob. 1719-1767. Ger./Russ. DSB; ICB; P I; BMNH; GA; GB; GC. Also Chemistry*.
1. *Versuch einer Geschichte von Flötz-Gebürgen*. Berlin, 1756. RM.

9009. STADEL, Eberhard Friedrich. d. 1755. Ger. P II.

9010. SALERNE, François. d. 1760. Fr. P II; BLA. Also Zoology.

9011. JUSTI, Johann Heinrich Gottlob von. 1720-1771. Ger./Aus. DSB; P I; BMNH; GA; GC.
1. *Grundriss der gesamten Mineralreiches*. Göttingen, 1757. RM.
2. *Geschichte des Erd-Cörpers, aus sienen äusserlichen und unterirdischen Beschaffenheiten hergeleitet und erwiesen*. Berlin, 1771. RM.

9012. RECUPERO, Giuseppe. 1720-1778. It. P II; BMNH; GA.

9013. WILKENS, Christian Friedrich. 1721-1784. BMNH.

9014. WOLTERSDORF, Johann Lucas. 1721-1772. Ger. P II; GC.

9015. BOULLANGER, Nicolas Antoine. 1722-1759. Fr. DSB.

9016. FÜCHSEL, Georg Christian. 1722-1773. Ger. DSB; ICB.

9017. MEINECKE, Johann Christoph. 1722-1790. Ger. P II.

9018. NICOLIS DI ROBILANTE, Spirito Benedetto. 1722-1801. It. P II; GA (ROBILANT).

9019. MERIAN, Hans (or Johann) Bernhard. 1723-1807. Switz./Ger. P II; GA; GC.

9020. PODA VON NEUHAUS, Nicolaus. 1723-1798. Aus. P II. Also Zoology.

9021. SCOPOLI, Giovanni Antonio. 1723-1788. It./Hung. ICB; P II; BMNH; GA. Also Mineralogy*, Natural History*, Botany*, and Zoology*.

9022. TATA, Domenico. b. 1723. It. P II.

9023. BÜTTNER, David Sigismund August. 1724-1768. Ger. P I; BMNH; BLA & Supp.; GA (BUTTNER). Also Botany.
1. *Rudera diluvii testes; i.e. Zeichen und Zeugen der Sündfluth.* Leipzig, 1710. RM.
2. *Coralliographia subterranea; seu, Dissertatio de coralliis fossilibus.* Leipzig, 1714. RM.
The dates of these two works (also cited thus in P and BMNH) are incompatible with his birth date. Possibly the attribution is incorrect.

9024. MICHELL, John. 1724?-1793. Br. DSB; ICB; P II; GB; GD. Also Astronomy* and Physics*.
1. *Conjectures concerning the cause, and observations upon the phenomena, of earthquakes.* London, 1760. RM.

9025. PFENNIG, Johann Christoph. 1724-1804. Ger. P II.

9026. DESMAREST, Nicolas. 1725-1815. Fr. DSB; P I; GB.
1. *L'ancienne jonction de l'Angleterre à la France; ou, Le détroit de Calais, sa formation par la rupture de l'isthme....* Amiens, 1753. X. Reprinted, Paris, 1875. RM.
2. *Encyclopédie méthodique.* [Main entry: 3000] *Géographie physique.* 5 vols & atlas. Paris, 1794-1828. RM/L.
See also 5780/1a.

9027. PRYCE, William. 1725?-1790. Br. BMNH; GA; GD.
1. *Mineralogia Cornubiensis: A treatise on minerals, mines, and mining, containing the theory and natural history of strata, fissures, and lodes.* London, 1778. RM/L.

9028. WALCH, Johann Ernst Immanuel. 1725-1778. Ger. DSB; P II; BMNH; GB; GC. See also 8991/1,1a.

9029. FUCHS, Johann Christoph. 1726-1795. Ger. P I; GA.

9030. GABRINI, Tommaso Maria. 1726-1808. It. P I.

9031. HUTTON, James. 1726-1797. Br. DSB; ICB; P I; BMNH; GA; GB; GD. Also Science in General* and Physics*.
1. *Theory of the earth*. 2 vols. Edinburgh, 1795. RM. Reprinted, Weinheim, 1960. (HNC. 1)
1a. ——— Vol. 3. London, 1899. MS discovered in 1895. Ed. by A. Geikie.
1b. ——— *The lost drawings*. Edinburgh, 1978. Ed. by G.Y. Craig.
2. *System of the earth, 1785. Theory of the earth, 1788. Observations on granite, 1794. Together with Playfair's biography of Hutton*. Darien, Conn., 1970. Facsimiles of a pamphlet (1785) and three periodical articles. Ed. by V.A. Eyles.

9032. OLAFSSON, Eggert. 1726-1768. Den. ICB; P II; BMNH; GA (OLAFSEN).
1. *Reise durch Island*. Copenhagen/Leipzig, 1774. (Mainly physical geography) RM.
See also 10984/1.

9033. CHABAUD LA TOUR, Antoine de. 1727-1791. Fr. P I; GA.

9034. COLLINI, Cosimo Alessandro. 1727-1806. It/Ger. P I; BMNH; GA.
1. *Journal d'un voyage, qui contient différentes observations minéralogiques, particulièrement sur les agates et le basalte*. Mannheim, 1776. RM.

9035. DELUC, Jean André (first of the name). 1727-1817. Switz./Br./Ger. DSB; P I; BMNH; GA; GB; GD. Also Physics* and Meteorology*.
1. *Lettres physiques et morales, sur les montagnes et sur l'histoire de la terre et de l'homme*. ? vols. En Suisse, 1778. RM. Another ed., 5 vols in 6, Paris, 1779-80. RM/L.
2. *Lettres sur l'histoire physique de la terre*. Paris, 1798. RM.
2a. ——— *Letters on the physical history of the earth*. London, 1831. Trans. by H. de la Fitte. RM/L.
3. *Traité élémentaire de géologie*. Paris, 1809. RM.
3a. ——— *An elementary treatise on geology*. London, 1809. Trans. by H. de la Fitte. RM/L.
4. *Geological travels in the north of Europe and in England*. 3 vols. London, 1810-11. Trans. from the French MS. RM.

9036. ANDRÉ, Noël (or CHRYSOLOGUE, *Père*). 1728-1808. Fr. Under CHRYSOLOGUE, Noël André: P II; GA. Also Astronomy.
1. *Théorie de la surface actuelle de la terre*. Paris, 1806. RM (under CHRYSOLOGUE, *Père*)/L.

9037. BENVENUTI, Giuseppe. 1728-1770? (or 1789?). It. P I; BMNH; BLA & Supp.

9038. DELIUS, Christoph Traugott. 1728-1779. Aus. P I; GA; GC.
1. *Anleitung zu der Bergbaukunst nach ihrer Theorie und Ausubung*. Vienna, 1773. RM.

9039. HUNTER, John. 1728-1793. Br. DSB; ICB; P I; BMNH; Mort.; BLA & Supp.; GA; GB; GD. Other entries: see 3162.
1. *Observations and reflections on geology. Intended to serve as an introduction to the catalogue of his collection of extraneous fossils*. London, 1859.

9040. DELUC, Guillaume Antoine. 1729-1812. Switz. P I.

9041. FORSTER, Johann Reinhold. 1729-1798. Ger./Br. DSB; ICB; P I & Supp.; BMNH; GA; GC. Also Mineralogy*, Natural History*, Botany*, and Zoology. See also Index.
1. *Beobachtungen und Wahrheiten, nebst einigen Lehrsätzen ... als Stoff zur künstigen Entwerfung einer Theorie der Erde.* Leipzig, 1798. RM/L.

9042. BERTRAND, Philippe. 1730-1811. Fr. P I; GA.

9043. HAMILTON, William. 1730-1803. Br./It. DSB; P I; BMNH; GA; GB; GD.
1. *Observations on Mount Vesuvius, Mount Etna, and other volcanos.* London, 1772. RM.
2. *Campi phlegrai: Observations on the volcanos of the Two Sicilies.* Naples, 1776.
2a. ——— *Supplement. Being an account of the great eruption of Mount Vesuvius, in ... August, 1779.* Ib., 1779.

9044. SCHULTZE, Christian Friedrich. 1730-1775. Ger. P II (SCHULZE); BMNH; BLA (SCHULZE).
1. *Kurtze Betrachtung derer versteinerten Höltzer.* Dresden/Leipzig, 1754. RM.
2. *Betrachtung der versteinerten Seesterne und ihrer Theile.* Warsaw/Dresden, 1760. RM.

9045. WILLIAMS, John. 1730-1797. Br. P II; BMNH.
1. *The natural history of the mineral kingdom.* 2 vols. Edinburgh, 1789. X. 2nd ed., 1810.

9046. BUC'HOZ, Pierre Joseph. 1731-1807. Fr. P I; BMNH; BLA; GA. Also Natural History*, Botany, and Zoology*.
1. *Dictionnaire minéralogique et hydrologique de la France.* Paris, 1772-76. RM.

9047. KESSLER VON SPRENGSEISEN, Christian Friedrich. ca. 1731-1809. Ger. P I.

9048. LEYSSER, Friedrich Wilhelm von. 1731-1815. Ger. ICB; P I; BMNH. Also Botany.

9049. LIMBOURG, Robert de. 1731-1792. Belg. ICB; BLA.

9050. PIGNATARI, Filippo Jacopo. 1731-1827. It. P III.

9051. RIMROD, Friedrich August. 1731-1809. Ger. P II.

9052. SCHWAB, Johann. 1731-1795. Ger. P II.

9053. VALMONT DE BOMARE, Jacques Christophe. 1731-1807. Fr. DSB; P II; BMNH; GA. Also Mineralogy* and Natural History*.

9054. WALKER, John (first of the name). 1731-1803. Br. DSB; BMNH; GD. Also Natural History*.
1. *Lectures on geology.* Chicago, 1966. First publication of the MS, written ca. 1780. Ed. by H.W. Scott.

9055. ESPER, Johann Friedrich. 1732-1781. Ger. P I; BMNH; GA; GC.
1. *Ausführliche Nachricht von neuentdeckten Zoolithen unbekannter vierfüssiger Thiere.* Nuremberg, 1774. RM. Reprinted, Wiesbaden, 1978.

9056. FICHTEL, Johann Ehrenreich von. 1732-1795. Aus./Hung. P I; BMNH; GA; GC.
1. *Beytrag zur Mineralgeschichte von Siebenbürgen.* Nuremberg, 1780. RM.
2. *Mineralogische Bemerkungen von den Karpathen.* Vienna, 1791. RM.

9057. FOUGEROUX DE BONDAROY, Auguste Denis. 1732-1789. Fr. ICB; P I; BLA; GA.

9058. GEHLER, Johann Karl. 1732-1796. Ger. P I; BMNH; BLA & Supp.; GA.

9059. PÖTSCH, Christian Gottlieb. 1732-1805. Ger. P II; BMNH.

9060. STRANGE, John. 1732-1799. Br./It. P II; BMNH; GD.
1. *De monti colonnari e d'altri fenomeni vulcani dello stato Veneto.* Milan, 1778. RM.

9061. CALINDRI, Serafino. 1733-1811. It. P I.

9062. KIRWAN, Richard. 1733?-1812. Br. DSB; ICB; P I; BMNH; GA; GB; GD. Also Chemistry*, Mineralogy*, and Meteorology*. See also Index.
1. *Geological essays.* London, 1799.
2. *Observations on the proofs of the Huttonian theory of the earth, adduced by Sir James Hall.* Dublin, 1800. RM.

9063. CARTHEUSER, Friedrich August. 1734-1796. Ger. P I; BMNH; BLA; GA (CARTHAEUSER). Also Chemistry.

9064. MONNET, Antoine Grimoald. 1734-1817. Fr. DSB; P II; BMNH; GA. Also Chemistry* and Mineralogy*.

9065. FELLER, François Xavier de. 1735-1802. Belg. P I; GA; GB; GC.
1. *Examen impartiel des "Epoques de la nature" de Mr. le comte de Buffon.* Luxemburg, 1780. RM.

9066. SCHRÖTER, Johann Samuel. 1735-1808. Ger. P II; BMNH; GC. Also Zoology*.
1. *Lithologisches Real- und Verballexikon.* 8 vols. Frankfurt, 1779-80. RM.

9067. ORDINAIRE, Claude Nicolas. 1736-1809. Fr. P II; BMNH; BLA.
1. *Histoire naturelle des volcans.* Paris, 1802. RM.
1a. ——— *The natural history of volcanoes.* London, 1801. Trans. from the original French MS by R.C. Dallas. RM.

9068. ROMÉ DE L'ISLE, Jean Baptiste Louis. 1736-1790. Fr. DSB; P II; BMNH; GB. Also Crystallography* and History of Science*.
1. *L'action du feu central bannie de la surface de la terre, et le soleil rétabli dans ses droits.* Paris, 1778. X.
1a. ——— 2nd ed. *L'action du feu central démontrée nulle à la surface du globe, contre les assertions de MM. le comte de Buffon, Bailly, de Mairan, etc.* Ib., 1781. RM/L.
See also 10973/1.

9069. SOLDANI, Ambrogio (or Baldo Maria). 1736-1808. It. DSB; P II; BMNH; GA.
1. *Saggio orittografico; offero, Osservazioni sopra le terre*

nautilitiche ed ammonitiche della Toscana. Siena, 1780. RM.
2. *Testaceographia ac zoophitographia parva et microscopica*. 2 vols in 4. Siena, 1789-98.

9070. BRÜNNICH, Morten Thrane. 1737-1827. Den./Nor. P I; BMNH. Also Zoology*.

9071. GOBET, Nicolas. 1737-ca. 1781. Fr. P I; BMNH; GA. See also 1473/2.

9072. LAXMANN, Eric. 1737-1796. Russ. P I; BMNH.

9073. RASPE, Rudolph Erich. 1737-1794. Ger./Br. DSB; ICB; P II; BMNH; GA; GB; GD.
1. *Specimen historiae naturalis globi terraquei, praecipue de novis e mari natis insulis, et ... Hookiana telluris hypothesi*. Amsterdam/Leipzig, 1763. RM.
1a. ——— *An introduction to the natural history of the terrestrial sphere, principally concerning new islands born from the sea, and Hooke's hypothesis of the earth*. New York, 1970. Trans. and ed. by A.N. Iverson and A.V. Carozzi.
2. *An account of some German volcanos and their productions. With a new hypothesis of prismatical basaltes*. London, 1776. RM/L.
See also 9098/1a,2a and 9101/1a.

9074. CANCRIN, Franz Ludwig von. 1738-1812 (or 1816). Ger./Russ. DSB; P I; GA; GB; GC.
1. *Beschreibung der vorzüglichsten Bergwerke*. Frankfurt, 1767. RM.
2. *Abhandlung von der Natur und Einrichtung einer Bergbelehnung*. Giessen, 1788. RM.

9075. CHARPENTIER, Johann Friedrich Wilhelm Toussaint von. 1738-1805. Ger. P I; GA; GC.
1. *Mineralogische Geographie der chursächsischen Lande*. Leipzig, 1778. RM.
2. *Beobachtungen über die Lagerstätte der Erze, hauptsächliche aus den sächsischen Gebirgen. Ein Beytrag zur Geognosie*. Leipzig, 1799. RM.
3. *Beytrag zur geognostischen Kenntnis des Riesengebirges schlesischen Antheils*. Leipzig, 1804. RM.

9076. GERHARD, Carl Abraham. 1738-1821. Ger. P I; BMNH; BLA.

9077. BOURRIT, Marc Théodore. 1739-1819. Switz. P I; BMNH; GA; GB.

9078. ESTNER, Franz Joseph Anton. 1739-1803. Aus. P I; BMNH.

9079. HACQUET, Balthasar. 1739-1815. Aus./Lemberg. P I; BMNH; BLA; GA; GC.

9080. MODEER, Adolf. 1739-1799. Swed. P II & Supp.; BMNH. Also Zoology.

9081. PINI, Ermenegildo. 1739-1825. It. P II; BMNH; GA.
1. *De venarum metallicarum excoctione*. Milan, 1779-80. RM.

9082. SCHREBER, Johann Christian Daniel von. 1739-1810. Ger. ICB; P II; BMNH; BLA; GC. Also Botany and Zoology*. See also 9098/3 and 10958/10,11a.

9083. TOWNSEND, Joseph. 1739-1816. Br. DSB; BMNH; GD.
1. *Geological and mineralogical researches ... wherein the effects of the Deluge are traced and the veracity of the Mosaic account is established.* Bath, 1824. RM.

9084. KERN, Johann Gottlieb. d. ca. 1776. Ger. P I; BMNH.
1. *Vom Schnekensteine oder dem sächsischen Topasfelsen.* Prague, 1776. Posthumous. Ed. with notes by I. von Born. RM.

9085. BEROLDINGEN, Franz Cölestin von. 1740-1798. Ger. P I; BMNH; GA.

9086. FRIDVALSZKI, Johann. 1740-1784. Hung. P I.

9087. LAUNAY, Louis de. 1740-1805. Belg. Under DELAUNAY: P I; BMNH; GA.
1. *Traités sur l'histoire naturelle et la minéralogie.* London, 1780. RM.
2. *Essai sur l'histoire naturelle des roches, précédé d'un exposé systématique des terres et des pierres.* Brussels, 1786. RM.

9088. RICHARDSON, William (first of the name). 1740-1820. Br. GA; GD.

9089. SAUSSURE, Horace Bénédict de. 1740-1799. Switz. DSB; ICB; P II; BMNH; GA; GB. Also Physics*.
1. *Voyages dans les Alpes, précédés d'un essai sur l'histoire naturelle des environs de Genève.* 4 vols. Geneva, 1779-96. RM/L.

9090. TREBRA, Friedrich Wilhelm Heinrich von. 1740-1819. Ger. DSB; ICB; P II; BMNH; GC. Also Mineralogy*.
1. *Erfahrungen vom Innern der Gebirge.* Dessau/Leipzig, 1785. RM.

9091. FAUJAS DE SAINT-FOND, Barthélemy. 1741-1819. Fr. DSB; P I; BMNH; GA; GB.
1. *Recherches sur les volcans éteints du Vivarais et du Velay. Avec un discours sur les volcans brulans.* Grenoble, 1778. RM.
2. *Minéralogie des volcans; ou, Description de toutes les substances produites ou rejetées par les feux souterrains.* Paris, 1784. RM.
3. *Essai sur l'histoire naturelle des roches de trapp.* Paris, 1788. RM.
4. *Essai sur le gaudron du charbon de terre ... Précédé de recherches sur l'origine et les différentes sortes de charbon de terre.* Paris, 1790. RM.
5. *Voyage en Angleterre....* 2 vols. Paris, 1797. X.
5a. ——— *Travels in England, Scotland, and the Hebrides.* 2 vols. London, 1799. RM.
5b. ——— *A journey through England and Scotland to the Hebrides in 1784.* 2 vols. Glasgow, 1907. Rev. ed. of item 5a, with notes and a memoir of the author, by A. Geikie. RM.
6. *Histoire naturelle de la montagne de Saint-Pierre de Maestricht.* Paris, 1799. RM.

7. *Essai de géologie; ou, Mémoires pour servir à l'histoire naturelle du globe.* 2 vols. Paris, 1809. RM.
See also 1473/2.

9092. FORTIS, Giovanni Battista (called Alberto). 1741-1803. It. P I; BMNH; GA.

9093. HEIM, Johann Ludwig. 1741-1819. Ger. P I; BMNH; GC.

9094. MERCK, Johann Heinrich. 1741-1791. Ger. ICB; P II; BMNH; GA; GB; GC.

9095. PALLAS, Peter Simon. 1741-1811. Russ. DSB; ICB; P II; BMNH; BLA; GA; GB; GC. Also Natural History*, Botany*, and Zoology*.
1. *Observations sur la formation des montagnes et les changemens arrivés à notre globe.* St. Petersburg, 1782.

9096. VELTHEIM, August Ferdinand von. 1741-1801. Ger. P II; BMNH; GC. Also Mineralogy*.

9097. BARRAL, Pierre. 1742-1826. Fr. P I & Supp.; BMNH.

9098. BORN, Ignaz von. 1742-1791. Aus. DSB; P I; BMNH; GA; GB; GC. Also Zoology*.
1. *Briefe über mineralogische Gegenstände, auf seine Reise durch das Temeswarer Bannat, Siebenbürgen, Ober- und Nieder-Hungarn.* Frankfurt, 1774. Ed. by J.J. Ferber. RM.
1a. ——— *Travels through the Bannat of Temeswar, Transylvania, and Hungary in the year 1770.* London, 1777. Trans. by R.E. Raspe. RM.
2. *Über das Anquicken der gold- und silberhältigen Erze, Rohsteine, Schwarzkupfer, und Hüttenspeise.* Vienna, 1786. RM.
2a. ——— *New process of amalgamation of gold and silver ores, and other metallic mixtures.* London, 1791. Trans. by R.E. Raspe. RM.
3. [In Czech. *Letters to D.G. and J.C.D. Schreber.*] Prague, 1971. The letters are in German and there is a German summary of the introd.
See also 9084/1 and 9101/1a.

9099. PATRIN, Eugène Louis Melchior. 1742-1815. Fr. P II; BMNH; GA.

9100. BURTIN, François Xavier de. 1743-1818. Belg. P I,III; BMNH; BLA; GA; GF.
1. *Oryctographie de Bruxelles; ou, Description des fossiles, tant naturels qu'accidentels, découverts jusqu'à ce jour dans les environs de cette ville.* Brussels, 1784. RM.

9101. FERBER, Johann Jakob. 1743-1790. Mitau/Russ./Ger. P I; BMNH; GA; GC.
1. *Briefe aus Wälschland über natürliche Merkwürdigkeiten dieses Landes.* Prague, 1773. X.
1a. ——— *Travels through Italy, in the years 1771 and 1772, described in a series of letters to Baron Born, on the natural history, particularly the mountains and volcanos of that country.* London, 1776. Trans. with notes by R.E. Raspe. RM.
2. *Beyträge zu der Mineral-Geschichte von Böhmen.* Berlin, 1774. RM.

2a. ——— *Mineralogical history of Bohemia.* Appended to 9098/1a.
3. *Beschreibung des Quecksilber-Bergwerks zu Idiria im Mittel-Crayn.* Berlin, 1774. RM.
4. *Physikalisch-metallurgische Abhandlungen über die Gebirge und Bergwerke in Ungarn.* Berlin, 1780. RM.
5. *Nachricht von dem Anquicken der gold- und silberhältigen Erze, Kupfersteine und Speisen in Ungarn und Böhmen.* Berlin, 1787. RM.
6. *Briefe an Friedrich Nicolai, aus Mitau und St. Petersburg.* Berlin, 1974. Ed. by H. Ischreyt.

See also 9098/1.

9102. LAMÉTHERIE, Jean Claude de. 1743-1817. Fr. DSB; P I; BMNH; BLA (MÉTHERIE); GA. Also Science in General*, Chemistry*, and Biology in General*. See also Index.
1. *Théorie de la terre.* Paris, 1795. RM.

9103. CAROSI, Johann Philipp von. 1744-1801. Warsaw. P I.

9104. HERMELIN, Samuel Gustaf. 1744-1820. Swed. P I; GA.

9105. LAMARCK, Jean Baptiste Pierre Antoine de Monet, *Chevalier* de. 1744-1829. Fr. DSB; ICB; P I; BMNH; GA; GB. Other entries: see 3204.
1. *Hydrogéologie; ou, Recherches sur l'influence qu'ont les eaux sur la surface du globe terrestre.* Paris, 1802. RM.

1a. ——— *Hydrogeology.* Urbana, Ill., 1964. Trans. by A.V. Carozzi.

9106. TITIUS, Karl Heinrich. 1744-1813. Ger. P II.

9107. HABEL, Christian Friedrich. d. 1814. Ger. P I.

9108. ANSCHÜTZ, Johann Mattäus. 1745-1802. Ger. P I.
1. *Über die Gebirgs- und Steinarten des chursächsischen Hennebergs.* Leipzig, 1788. RM.

9109. SANTI, Giorgio. 1746-1822. It. P II; BLA.
1. [Trans.] *Voyage au Montamiata et dans le Siennois, contenant des observations nouvelles sur la formation des volcans, l'histoire géologique, minéralogique et botanique de cette partie de l'Italie.* 2 vols. Lyons, 1802. Trans. from the Italian by Bodard. RM.

9110. SCHREIBER, Johann Gottfried. 1746-1827. Ger./Fr. P II.

9111. GIOENI, Giuseppe. 1747-1822. It. P I; BMNH; GA.

9112. KLIPSTEIN, Philipp Engel. 1747-1808. Ger. P I; BMNH.

9113. BREISLAK, Scipione. 1748-1826. It. DSB; P I & Supp.; BMNH; GA; GB.
1. *Memoria sulla fabbricazione e raffinazione de' nitri.* Milan, 1802. RM.
2. *Introduzione alla geologia.* Milan, 1811. RM.
3. *Descrizione geologica della provincia di Milano.* Milan, 1822. RM.

9114. DIETRICH, Philipp Friedrich von. 1748-1793. Strasbourg. P I; BMNH; GA; GC.

9115. LENZ, Johann Georg. 1748-1832. Ger. P I; BMNH; GC.

9116. MAIRONI DA PONTE, Giovanni. 1748-1833. It. P II; BA.

9117. MÖLLER, Johann Wilhelm. 1748-1807. Ger. P II.

9118. PLAYFIELD, John. 1748-1819. Br. DSB; ICB; P II; BMNH; GA; GB; GD. Also Mathematics and Physics*#.
1. *Illustrations of the Huttonian theory of the earth.* Edinburgh, 1802. RM. Reprinted, Urbana, Ill., 1956, and New York, 1964.
See also 9259/1 and 9439/1.

9119. DAUDIN, Jean Antoine. 1749-1832. Fr. P III.

9120. GLAESER, Friedrich Gottlob. 1749-1804. Ger. P I.

9121. GOETHE, Johann Wolfgang von. 1749-1832. Ger. The famous writer. DSB; ICB; P I; BMNH; Mort.; G. Other entries: see 3216.

9122. WERNER, Abraham Gottlob. 1749-1817. Ger. DSB; ICB; P II; BMNH; GA; GB; GC. Also Mineralogy*.
1. *Kurze Klassification und Beschreibung der verschiedenen Gebirgsarten.* Dresden, 1787. RM. Reprinted, Freiberg, 1967.
1a. ——— *Short classification and description of the various rocks.* New York, 1971. Trans. with introd. and notes by A.M. Ospovat. With a facsimile of the original.
2. *Neue Theorie von den Entstehung der Gänge.* Freiburg, 1791. RM.
2a. ——— *Nouvelle théorie de la formation des filons.* Paris, 1802. Trans. by J.F. Daubuisson and augmented with many notes, some of which are by Werner.
2b. ——— *New theory of the formation of veins.* Edinburgh, 1809. RM/L.
3. *Bestandsübersicht des handschriftlichen wissenschaftlichen Werner-Nachlasses.* Freiberg, Bergsakademie, 1967. By K.F. Zillmann.

9123. DOLLER, Johann Lorenz. 1750-1820. Ger. P I.

9124. DOLOMIEU, Dieudonné (or Déodat) de Gratet de. 1750-1801. Fr. DSB; ICB; P I; BMNH; GA; GB. Also Mineralogy*.
1. *Voyage aux îles de Lipari, fait en 1781; ou, Notices sur les Isles Aeoliennes, pour servir à l'histoire des volcans.* Paris, 1783. RM.
2. *Mémoire sur les tremblements de terre de la Calabre pendant l'année 1783.* Rome, 1784. RM.
3. *Mémoire sur les îles Ponces, et catalogue raisonné des produits de l'Etna, pour servir à l'histoire des volcans.* Paris, 1788. RM.
See also 9135/2.

9125. KOLLATAJ, Hugo. 1750-1812. Cracow. ICB.

9125A MELOGRANI, Giuseppe. It. b. 1750. P II.

9126. NAPIONE, Carlo Antonio Galeani. 1750-1814. It. P II; GA.

9127. STRUVE, Henri (or Heinrich). 1751-1826. Switz. P II. Also Chemistry.
1. *Méthode analytique des fossiles, fondée sur leurs caractères*

extérieurs. Lausanne, 1797. X. 2nd ed., Paris, 1798. RM.

9128. ASSMAN, Christian Gottfried. 1752-1822. Ger. P I; BMNH.
1. *Reise im Riesengebirge. Ein geologischer Versuch*. Leipzig, 1789. RM.

9129. BECHER, Johann Philipp. 1752-1831. Ger. P I; BMNH.

9130. CUMBERLAND, George. 1752-1848. Br. P I; BMNH.

9131. LASIUS, Georg Siegmund Otto. 1752-1833. Ger. P I; BMNH.

9132. SOULAVIE, Jean Louis Giraud. 1752-1813. Fr. DSB; P II; GA.

9133. VOIGT, Johann Carl Wilhelm. 1752-1821. Ger. DSB; P II; BMNH; GC.

9134. DOUGLAS, James. 1753-1819. Br. BMNH; GD.
1. *A dissertation on the antiquity of the earth*. London, 1785. RM/L.

9135. NOSE, Karl Wilhelm. 1753-1835. Ger. P II; BMNH; BLA.
1. *Orographische Briefe über das Siebengebirge und die benachbarten zum Theil vulkanischen Gegenden*. 3 vols. Frankfurt, 1789-91. RM.
2. *Beschreibung einer Sammlung von meist vulkanisirten Fossilien, die Deodat Dolomieu im Jahre 1791 von Maltha aus nach Augsburg und Berlin versandte*. Frankfurt, 1797. RM.

9136. JIRASEK (or IRASEK), Johann. 1754-1797. Aus. P I; BMNH.

9137. LACOSTE, Pierre François. 1754-1826. Fr. P I; BMNH.
1. *Lettres minéralogiques et géologiques sur les volcans de l'Auvergne*. Clermont, 1805. RM.

9138. TOULMIN, George Hoggart. 1754-1817. Br. DSB.

9139. COQUEBERT DE MONTBRET, Charles Etienne. 1755-1831. Fr. P I; BMNH; GA.

9140. CRAMER, Ludwig Wilhelm. 1755-1832. Ger. P I; BMNH.
1. *Vollständige Nachricht von dem Hollerter Zuge, einem wichtigen Eisensteinwerke*. Freiberg, 1793. RM.

9141. DA-RIO, Nicolo. 1755-1845. It. P I & Supp.

9142. HERMANN, Benedict Franz Johann von. 1755-1815. Aus./Russ. P I; BMNH.
1. *Versuch einer mineralogischen Beschreibung des Uralischen Erzgebirges*. Berlin/Stettin, 1789. RM.
2. *Über die Entstehung der Gebirge, und ihre gegenwärtige Beschaffenheit*. Leipzig, 1797. RM.

9143. LANGER, Johann Heinrich Siegmund. 1755-1788. Ger. P I.

9144. MONTLOSIER, François Dominique Reynaud de. 1755-1838. Fr. P II; BMNH; GA; GB.

9145. PARKINSON, James. 1755-1824. Br. DSB; ICB; BMNH; BLA; GB; GD.
1. *Organic remains of a former world. An examination of the mineralised remains of the vegetables and animals of the antediluvian world, generally termed extraneous fossils*. 3 vols. London, 1804-11. RM/L. 2nd ed., 1833.
2. *Outlines of oryctology: An introduction to the study of*

fossil organic remains, especially of those found in the British strata. London, 1822. 3rd ed., Leicester, n.d. [1835?]. Another ed., 184-?

9146. RAMOND DE CARBONNIÈRES, Louis François Elizabeth. 1755-1827. Fr. DSB; P II; BMNH; GA. Also Botany.

9147. SCHMIDT, Friedrich Christian. 1755-1830. Ger. P II; GC.

9148. SELB, Carl Joseph. ca. 1755-1827. Ger. P II; BMNH.

9149. STASZIC, Stanisław Wawrzyniec. 1755-1826. Poland. DSB Supp.; ICB. Also Science in General*.

9150. WAD, Gregers. 1755-1832. Den. P II; BMNH.

9151. BALLENSTEDT, Johann Georg Justus. 1756-1840. Ger. P I & Supp.; BMNH; GC.

9152. FLURL, Mathias von. 1756-1823. Ger. P I; GC.
 1. *Über die Gebirgsformationen in den dermaligen Churpfalzbaierischen Staaten*. Munich, 1805. RM.

9153. HAIDINGER, Karl. 1756-1797. Aus. P I.
 1. *Systematische Eintheilung der Gebirgsarten*. Vienna, 1787. RM.

9154. SCHROLL, Kaspar Melchior Balthasar. 1756-1829. Aus. P II,III.

9155. DETHIER, Laurence François. 1757-1843. Belg. BMNH; GF.

9156. MITTROWSKY, Johann Nepomuk von. 1757-1799. Brünn. P II.

9157. SOWERBY, James. 1757-1822. Br. DSB; P II; BMNH; GB; GD. Also Mineralogy*, Botany*, and Zoology*.
 1. *The mineral conchology of Great Britain; or, Coloured figures and descriptions of those remains of testaceous animals or shells which have been preserved at various times and depths in the earth*. 7 vols. London, 1812-46. From Vol. 4 the work was continued by J.D.C. Sowerby. RM/L.
 1a. ——— *Systematical, stratigraphical, and alphabetical indexes to the first six volumes*. London, 1834. By J.D.C. Sowerby. With a brief biography. RM.

9158. VOLNEY, Constantin François Chasse-Boeuf, *Comte* de. 1757-1820. Fr. DSB; ICB; P II; BMNH; GA; GB.
 1. *Tableau du climat et du sol des Etats-Unis d'Amérique*. 1803. X.
 1a. ——— *A view of the soil and climate of the United States of America*. London/Philadelphia, 1804. Trans. by C.B. Brown. X. "The earliest connected account in English of the geology of the United States." Reprinted, New York, 1968.

9159. BORSON, Stefano. 1758-1832. It. P I; BMNH.

9160. DEFRANCE, Jacques Louis Marin. 1758-1850. Fr. P I; BMNH.
 1. *Tableau des corps organisés fossiles, précédé de remarques sur leur pétrification*. Paris, 1824. RM/L.

9161. PINKERTON, John. 1758-1826. Br. P II; BMNH; GA; GB; GD.
 1. *Petrology: A treatise on rocks*. 2 vols. London, 1811. RM/L.

9162. RAMONDINI, Vincenzo. 1758-1811. It. P II.

9163. GATTERER, Christoph Wilhelm Jakob. 1759-1838. Ger. P I; BMNH.
1. *Allgemeines Repertorium der mineralogischen, Bergwerks- und Salzwerkswissenschaftlichen Literatur.* 2 vols. Giessen, 1798-99. RM.

9164. HAILSTONE, John. 1759-1847. Br. P I; GD.

9165. KAPF, Georg Friedrich. 1759-1797. Ger. P I.

9166. MONTICELLI, Teodoro. 1759-1846. It. P II; BMNH.

9167. VOITH, Ignaz von. 1759-1848. Ger. P II; BMNH; GC.

9168. RAZUMOWSKY, Grigorii. d. 1837. Russ./Switz. P II; BMNH.

9169. MOLL, Karl Ehrenheit von. 1760-1838. Aus./Ger. P II & Supp.; BMNH; GC.

9170. WATSON, White. 1760-1835. Br. BMNH.
1. *A delineation of the strata of Derbyshire ... together with a description of the fossils found in these strata.* Sheffield, 1811. X. Reprinted with an introd. by T.D. Ford, Buxton, Derbyshire, 1973.

9171. FLEURIAU DE BELLEVUE, ... ca. 1761-1852. Fr. P I.

9172. HALL, James (first of the name). 1761-1832. Br. DSB; ICB; P I; BMNH; GB; GD. Also Chemistry. See also 9062/2.

9173. KNEIFL (or KNEIFEL), Reginald. 1761-1826. Aus. P I.

9174. PENN, Granville. 1761-1844. Br. BMNH; GD.
1. *A comparative view of the mineral and Mosaical geologies.* London, 1822. RM/L.
2. *Conversations on geology. Comprising a familiar explanation of the Huttonian and Wernerian systems, the Mosaic geology as explained by G. Penn, and the late discoveries of Prof. Buckland, Humboldt, Dr. Macculloch, and others.* London, 1828. RM.

9175. REUSS, Franz Ambrosius. 1761-1830. Bohemia. DSB; ICB; P II; BMNH; BLA. Also Mineralogy*.
1. *Orographie des nordwestlichen Mittelgebirges in Böhmen: Ein Beitrag zur Beantwortung der Frage, ist der Basalt vulkanisch oder nicht?* Dresden, 1790. RM.
2. *Mineralogische Geographie von Böhmen.* 2 vols. Dresden, 1793-97. RM.
3. *Sammlung naturhistorischer Aufsätze, mit vorzüglicher Hinsicht auf die Mineralgeschichte Böhmens.* Prague, 1796. RM.

9176. STERNBERG, Kaspar Maria von. 1761-1838. Prague/Ger. DSB; ICB; P II; BMNH; GC. Also Botany.
1. *Versuch einer geognostich-botanischen Darstellung der Flora der Vorwelt.* Leipzig, 1838.

9177. TONSO, Alessandro. 1761-1820. It. ICB.

9178. KUHN, Bernhard Friedrich. 1762-1825. Switz. ICB; GC.

9179. RHODE, Johann Gottlieb. 1762-1827. Ger. P II; BMNH; GC. Also Various Fields.
1. *Beiträge zur Pflanzenkunde der Vorwelt.* Breslau, 1820.

9180. SAVARESI, Andrea. 1762-1810. It. P II.

9181. TONDI, Matteo. 1762-1835. It. P II.

9182. ANDRÉ, Christian Carl. 1763-1831. Brünn. P I.

9183. DELUC, Jean André (second of the name). 1763-1847. Switz. P I.
1. *Mémoire sur le phénomène des grandes pierres primitives alpines distribuées par groupes dans le bassin du lac de Genève et dans les vallées de l'Arve*. Geneva, 1827. RM.

9184. DUBUISSON, François René André. 1763-1836. Fr. P I; GA.

9185. ESMARCH (or ESMARK), Jens. 1763-1839. Nor. ICB; P I; GA.

9186. MACLURE, William. 1763-1840. U.S.A. DSB; ICB; P II; GB; GE.
1. *Observations on the geology of the United States of America*. Philadelphia, 1817. RM.
2. *An essay on the formation of rocks*. Philadelphia, 1818. RM.

9187. REBOUL, Henri Paul Irenée. 1763-1839. Fr. P III; BMNH; GA.

9188. STUCKE, Kaspar Heinrich. b. 1763. Ger. P II.
1. *Chemische Untersuchungen einiger niederrheinischen Fossilien, eines Vesuvians, und des Wassers in Basalt*. Frankfurt, 1793. RM.

9189. EBEL, Johann Gottfried. 1764-1830. Ger./Switz. DSB; P I; BMNH; BLA; GA; GB; GC.
1. *Über den Bau der Erde in dem Alpen-Gebirge*. 2 vols. Zurich, 1808. RM.

9190. PETZL, Joseph. 1764-1817. Ger. P II.

9191. RENGGER, (Johann) Albert. 1764-1835. Switz. P II; GC.

9192. SPADONI, Paolo. 1764-1826. It. P II; BMNH.

9193. WIDENMANN, Johann Friedrich Wilhelm. 1764-1798. Ger. P II; BMNH. Also Mineralogy*.

9194. ARNDTS, Anton Wilhelm Stephan. 1765-ca. 1830. Ger. P I.

9195. DURINI, Giuseppe Nicola. 1765-1845. It. P I.

9196. HERRGEN, J. Christian. 1765-1816. Spain. P III.

9197. RICCI, Vito Procaccini. 1765-1845. It. P II.

9198. RITTER, Christian Wilhelm Jonathan. 1765-1819. Ger. P II; BMNH.
1. *Beschreibung merkwürdiger Berge, Felsen und Vulkane. Ein Beytrag zur physikalischen Geschichte der Erde*. Posen, 1806. RM.

9199. SCHLOTHEIM, Ernst Friedrich von. 1765-1832. Ger. DSB; P II; BMNH; GB; GC.
1. *Die Petrefactenkunde auf ihrem jetzigen Standpunkte*. Gotha, 1820. RM. *Nachträge*, 1822-23. RM.

9200. SEVERGIN, Vassily Mikhaylovich. 1765-1826. Russ. DSB; P II.

9201. FAREY, John. 1766-1826. Br. DSB; GB; GD.

9202. HISINGER, Wilhelm. 1766-1852. Swed. DSB; ICB; P I; BMNH; GA. Also Chemistry.

1. *Samling till en minerografi öfver Sverige*. Stockholm, 1790. X.
1a. ——— *Versuch einer mineralogischen Geographie von Schweden*. Freiberg, 1819. Trans. with notes and additions by K.A. Blöde. RM.
2. *Lethaea Svecica; seu, Petrifacta Sveciae, iconibus et characteribus illustrata*. Stockholm, 1837.

9203. ESCHER VON DER LINTH, Hans Conrad. 1767-1823. Switz. DSB; P I; GA; GC.

9204. FERRARA, Francesco. 1767-1850. It. P I,III; BMNH.

9205. MARTIN, William. 1767-1810. Br. BMNH; GA; GB; GD.
1. *Outlines of an attempt to establish a knowledge of extraneous fossils on scientific principles*. Macclesfield, 1809. RM.
2. *Petrefacta Derbiensia; or, Figures and descriptions of petrifactions collected in Derbyshire*. Vol. 1. [No more publ.] Wigan, 1809. EM/L.

9206. SALMON, Urbain Pierre. 1767-1805. Fr. P II; BLA.

9207. BAKEWELL, Robert (second of the name). 1768-1843. Br. DSB; P I; BMNH; GB; GD. Also Mineralogy*.
1. *An introduction to geology, illustrative of the general structure of the earth*. London, 1813. RM/L. 2nd ed., 1815. 3rd ed., 1828. 4th ed., 1833. 5th ed., 1838.

9208. KARSTEN, Diedrich Ludwig Gustav. 1768-1810. Ger. ICB; P I; BMNH; GA; GC. See also 8688/2a and 8708/2.

9209. NICOL, William. 1768-1851. Br. DSB; ICB; P II; GB. Also Physics.

9210. SIEMSSEN, Adolph Christian. 1768-1833. Ger. P II; BMNH. Also Zoology.

9211. AUBUISSON DE VOISINS, Jean François d'. 1769-1841. Fr. DSB; P I; BMNH. Also Mechanics*.
1. *Des mines de Freiberg en Saxe et de leur exploitation*. Leipzig, 1802. RM.
2. *Mémoire sur les basaltes de la Saxe, accompagné d'observations sur l'origine des basaltes en général*. Paris, 1803. RM.
2a. ——— *An account of the basalts of Saxony, with observations on the origin of basalt in general*. Edinburgh, 1814. Trans. by P. Neill. RM.
3. *Traité de géognosie; ou, Exposé des connaissances actuelles sur la constitution physique et minérale du globe terrestre*. 2 vols. Strasbourg, 1819. RM. Another ed., 1821.
See also 9122/2a.

9212. CLARKE, Edward Daniel. 1769-1822. Br. DSB; P I; BMNH; GA; GB; GD. Also Mineralogy*. See also 15425/1.

9213. CUVIER, Georges. 1769-1832. Fr. DSB; ICB; P I; BMNH; GA; GB. Also Science in General*, Zoology*, Biography*, and History of Science*. See also Index.
1. *Essai sur la géographie minéralogique des environs de Paris, avec une carte géognostique et des coupes de terrain*. Paris, 1811. With A. Brogniart. RM.

1a. ——— *Description géologique des environs de Paris. Nouvelle édition.* Paris, 1822. (Much enlarged) RM. 3rd ed., 1835.
2. *Recherches sur les ossements fossiles de quadrupèdes, où l'on rétablit les caractères de plusieurs espèces d'animaux que les révolutions du globe paroissent avoir détruits.* 4 vols. Paris, 1812. RM. 2nd ed., 7 vols, 1821-24. 3rd ed., 2 parts in 7 vols, 1825. 4th ed., 12 vols, 1834-36.
3. *Discours sur les révolutions de la surface du globe, et sur les changemens qu'elles ont produit dans le règne animal.* Separate publication of the *Discours préliminaire* of item 2. Date of first separate publication not known. Many later editions.
3a. ——— 3rd ed., Paris, 1825. X. Reprinted, Brussels, 1969. Another ed., 1826.
3b. ——— *A discourse on the revolutions of the surface of the globe, and the changes thereby produced in the animal kingdom.* London, 1829.
4. *An essay on the theory of the earth, translated from the French* [i.e. from the *Discours préliminaire* of item 2] *by Robert Kerr, with mineralogical notes and an account of Cuvier's geological discoveries by Professor Jameson.* Edinburgh, 1813. RM. Reprinted, Farnborough, 1971. 3rd ed., London, 1817. American ed., by S.L. Mitchell, New York, 1818. RM. 4th ed., with additions, Edinburgh, 1822. 5th ed., trans. from the last French ed., with additions by the author and translator, ib., 1827.

9214. GAUTIERI, Giuseppe. 1769-1833. It. P I; BMNH; BLA; GA.

9215. HAYDEN, Horace H. 1769-1844. U.S.A. GE.
1. *Geological essays; or, An enquiry into some of the geological phenomena to be found in various parts of America and elsewhere.* Baltimore, 1820. RM.

9216. HUMBOLDT, Alexander von. 1769-1859. Ger./Fr. DSB; ICB; PI,III; BMNH; G. Other entries: see 3268.
1. *Mineralogische Beobachtungen über einige Basalte am Rhein.* Brunswick, 1790. RM.
2. *Über die unterirdischen Gasarten und die Mittel ihren Nachtheil zu vermindern. Ein Beytrag zur Physik der praktischen Bergbaukunde.* Brunswick, 1799.
3. *Essai géognostique sur le gisement des roches dans les deux hemisphères.* Paris, 1823. RM. 2nd ed., 1826.
3a. ——— *A geognostical essay on the superposition of rocks in both hemispheres.* London, 1823. RM/L.

9217. SMITH, William. 1769-1839. Br. DSB; ICB; P II; BMNH; GB; GD. See also 15454/1.
1. *A delineation of the strata of England and Wales, with part of Scotland.* London, 1815. RM/L.
1a. ——— *A memoir to the map and delineation....* 1815. RM/L.
2. *Strata identified by organized fossils.* Ib., 1816. RM/L.
3. *Stratigraphical systems of organized fossils.* Ib., 1817. RM.
4. *Geological section from London to Snowdon, showing the variety of the strata and the correct altitudes of the hills.* Ib., 1817. RM.

5. *Geological view and section in Essex and Hertfordshire. Geological view and section of the country between London and Cambridge.* Ib., 1819. RM.
6. *Geological view and section of Norfolk. Geological view and section through Suffolk to Ely.* Ib., 1819. RM.
7. *Geological view and section through Dorsetshire and Somersetshire to Taunton.* Ib., 1819. RM.
8. *Section of the strata through Hampshire and Wiltshire to Bath.* Ib., 1819. RM.
9. *Vertical section of the strata in Surrey dipping northward. Section of strata in Sussex dipping southward.* Ib., 1819. RM.

9218. BRONGNIART, Alexandre. 1770-1847. Fr. DSB; ICB; P I; BMNH; GA; GB.
1. *Histoire naturelle des crustacés fossiles, sous les rapports zoologiques et géologiques.* Paris, 1822. With A.G. Desmarest. RM/L.
2. *Sur les caractères zoologiques des formations, avec l'application de ces caractères à la détermination de quelques terrains de craie.* Paris, 1822.
3. *Mémoire sur les terrains de sédiments supérieurs calcaréotrappéens du Vincentin, et sur quelques terrains d'Italie, de France, d'Allemagne, etc.* Paris, 1823. RM.
4. *Classification et caractères minéralogiques des roches homogènes et hétérogènes.* Paris, 1827. RM.

See also 9213/1.

9219. HOSER, Joseph Karl Eduard. 1770-1848. Prague. P I; BMNH; BLA.

9220. REICHETZER, Franz. b. 1770. Aus. P II.

9221. VARGAS-BEDEMAR, Eduard Romeo. 1770-1847. Den. P II; BMNH.

9222. FISCHER (VON WALDHEIM), (Johann) Gotthelf. 1771-1853. Russ. ICB; P I; BMNH; BLA; GA; GC. Also Natural History*.
1. *Prodromus petromatognosiae animalium systematicae continens bibliographiam animalium fossilium.* 2 parts. Moscow, 1829-32. RM.

9223. HOFF, Karl Ernst Adolf von. 1771-1837. Ger. DSB; ICB; P I; BMNH; GA.

9224. JORDAN, Johann Ludwig. 1771-1853. Ger. P I; BMNH. Also Mineralogy*.

9225. ROSENMÜLLER, Johann Christian. 1771-1820. Ger. ICB; P II; BMNH; BLA; GC. Also Anatomy amd Embryology.

9226. ULLMANN, Johann Christoph. 1771-1821. Ger. P II; BMNH; GC.

9227. ANKER, Matthias Joseph. 1772-1843. Aus. P I; BLA.

9228. BROCCHI, Giovanni Battista. 1772-1826. It. DSB; ICB; P I; BMNH; GA; GB. Also Mineralogy*.
1. *Conchiologia fossile sub-Appennina.* 2 vols. Milan, 1814.
2. *Catologo ragionato di una raccolta di rocce, disposito con ordine geografico per servire alla geognosia dell'Italia.* Milan, 1817. RM.

9229. BROCHANT DE VILLIERS, André Jean François Marie. 1772-1840. Fr. DSB; P I; BMNH; GA; GB. Also Mineralogy*. See also 9358/1,2.

9230. FORSTER, Westgarth. 1772-1835. Br. BMNH.
1. *A treatise on a section of the strata from Newcastle-upon-Tyne to Cross-Foll, with remarks on mineral veins.* Newcastle, 1809. X. 2nd ed., Alston, Cumberland, 1821.

9231. MacCULLOCH, John. 1773-1835. Br. DSB; ICB; P II; BMNH; BLA; GA; GB; GD.
1. *A description of the western islands of Scotland ... comprising an account of their geological structure.* 2 vols & atlas. London, 1819.
2. *A geological classification of rocks ... comprising the elements of practical geology.* London, 1821. RM/L.
3. *A system of geology, with a theory of the earth and an explanation of its connection with the sacred records.* 2 vols. London, 1831. RM/L.

See also 9174/2.

9232. STEFFENS, Henrik (or Heinrich). 1773-1845. Den./Ger. ICB; P II; BMNH; GA; GB; GC.
1. *Beyträge zur innern Naturgeschichte der Erde.* Freiberg, 1801. X. Reprinted, Amsterdam, 1973.
2. *Geognostich-geologische Aufsäze als Vorbereitung zu einer innern Naturgeschichte der Erde.* Hamburg, 1810. RM.
3. *Vollständiges Handbuch der Oryktognosie.* Halle, 1811-14. RM.
4. *Schriften, alt und neu.* Breslau, 1821.

9233. WEAVER, Thomas. 1773-1855. Br. P II; BMNH; GD. See also 8692/1a.

9234. WEBSTER, Thomas. 1773-1844. Br. DSB; ICB; P II; GB; GD. See also 3003.

9235. BUCH, (Christian) Leopold von. 1774-1853. Ger. DSB; ICB; P I; BMNH; GA; GB; GC.
1. *Versuch einer mineralogischen Beschreibung von Landeck.* Breslau, 1797. RM.

1a. ——— *A mineralogical description of the environs of Landeck.* Edinburgh, 1810. Trans. from the French with notes by C. Anderson. RM.

2. *Geognostische Beobachtungen auf Reisen durch Deutschland und Italien.* 2 vols. Berlin, 1802-09. RM.
3. *Reise durch Norwegen und Lappland.* 2 vols. Berlin, 1810. RM.

3a. ——— *Travels through Norway and Lapland.* London, 1813. Trans. by John Black. With notes and an account of the author by R. Jameson. RM.

4. *Über den Jura in Deutschland.* Berlin, 1839. RM.
5. *Gesammelte Schriften.* 4 vols. Berlin, 1867-85. Ed. by J. Ewald et al. RM.

9236. FREIESLEBEN, Johann Karl. 1774-1846. Ger. DSB; P I; BMNH; GC.

9237. GAILLARDOT, Claude (or Charles) Antoine. 1774-1833. Fr. BMNH; BLA; GA.

9238. JAMESON, Robert. 1774-1854. Br. DSB; ICB; P I; BMNH; GA; GB; GD. Also Mineralogy*.
1. *Elements of geognosy.* Edinburgh, 1808. Vol. 3 of his *System of mineralogy*: see 8735/3.

1a. ——— Reprint entitled *The Wernerian theory of the Neptunian origin of rocks*. New York, 1976.
See also 9213/4 and 9235/3a.

9239. MARASCHINI, Pietro. 1774-1825. It. P II; BMNH.

9240. SARTORIUS, Georg Christian. 1774-1838. Ger. P II.
1. *Geognostische Beobachtungen und Erfahrungen, vorzüglich in Hinsicht des Basaltes*. Eisenach, 1821. RM.

9241. WANGER, Andreas. 1774-1836. Switz. P II.

9242. BECKER, Wilhelm Gottlob Ernst. ca. 1775-1836. Ger./Kielze. P I.
1. *Über die Flotzgebirge in südlichen Polen, besonders in Hinsicht auf Steinsalz und Soole*. Freiburg, 1830. RM.

9243. MÉNARD DE LA GROYE, François Jean Baptiste. 1775-1827. Fr. P II. Also Zoology.

9244. PHILLIPS, William (first of the name). 1775-1828. Br. DSB; P II; BMNH; GA; GB; GD. Also Mineralogy*.

9245. BERTRAND DE DOUE, Jacques Mathieu. 1776-1862. BMNH.
1. *Description géognostique des environs du Puy en Velay*. Paris, 1823.

9246. BIGOT DE MOROGUES, Pierre Marie Sébastien de. 1776-1840. Fr. P I; BMNH.

9247. EATON, Amos. 1776-1842. U.S.A. DSB; ICB; P I; BMNH; GE. Also Botany.
1. *Index to the geology of the northern states*. Leicester, Mass., 1818. RM.
2. *A geological and agricultural survey of Rensselaer county, in the state of New York*. Albany, N.Y., 1822. RM.
3. *A geological and agricultural survey of the district adjoining the Erie canal in the state of New York*. Albany, N.Y., 1824. RM.
4. *A geological nomenclature for North America*. Albany, N.Y., 1828. RM.
5. *A geological text-book, prepared for popular lectures on North American geology*. Albany, N.Y., 1830. RM.

9248. HÉRICART DE THURY, Louis Etienne François. 1776-1854. Fr. P I; BMNH; GA.

9249. MÜNSTER, Georg zu. 1776-1844. Ger. P II; BMNH; GB; GC.
1. *Beiträge zur Petrefactenkunde*. 7 parts. Bayreuth, 1839-46.
See also 9282/2.

9250. NEIL, Patrick. 1776-1851. Br. P II; BMNH; GD. See also 9211/2a.

9251. TROOST, Gerard. 1776-1850. Holl./U.S.A. DSB; P II; BMNH; GE.

9252. ALLAN, Thomas. 1777-1833. Br. P I; BMNH; GA; GD. Also Mineralogy*.

9253. CORDIER, Pierre Louis Antoine. 1777-1861. Fr. DSB; P I & Supp.; BMNH; GA. Also Mineralogy.

9254. ESCHWEGE, Wilhelm Ludwig von. 1777-1855. Ger./Port./Brazil. P I; BMNH; GC.

9255. HACK, Maria. 1777-1844. Br. BMNH; GD.
1. *Geological sketches and glimpses of the ancient earth.* London, 1832.

9256. LEHMANN, Ernst Johann Traugott. 1777-1847. Ger. P I; BMNH,

9257. YOUNG, George. 1777-1848. Br. BMNH; GD.

9258. GREENOUGH, George Bellas. 1778-1855. Br. DSB; ICB; P I; BMNH; GB; GD.
1. *A critical examination of the first principles of geology.* London, 1819. RM/L.

9259. MURRAY, John (first of the name). 1778-1820. Br. P II & Supp.; BMNH; BLA; GA; GB; GD. Also Chemistry*.
1. *A comparative view of the Huttonian and Neptunian systems of geology, in answer to the "Illustrations of the Huttonian theory of the earth" by Prof. Playfair.* Edinburgh, 1802. RM.

9260. SCHMIEDER, Karl Christoph. 1778-1850. Ger. ICB; P II; BMNH.
1. *Topographische Mineralogie der Gegend um Halle in Sachsen; oder, Beschreibung der sich um Halle findenden Mineralien und Fossilien.* Halle, 1797. RM.
2. *Versuch einer Lithurgik; oder, Ökonomischen Mineralogie.* 2 vols. Leipzig, 1803-04. RM.

9261. URE, Andrew. 1778-1857. Br. DSB; ICB; P II; BMNH; GD. Also Chemistry*.
1. *A new system of geology, in which the great revolutions of the earth and animated nature are reconciled at once to modern science and sacred history.* London, 1829. RM/L.

9262. ENGELHARDT, Moritz von. 1779-1842. Dorpat. P I; BMNH; GC.
1. *Geognostiche Versuche.* Berlin, 1815. With K. von Raumer. RM.
2. *Geognostiche Umrisse von Frankreich, Grossbrittanien, einem Theile Teutschlands, und Italiens.* Berlin, 1816. With K. von Raumer. RM.

9263. LEONHARD, Karl Cäsar von. 1779-1862. Ger. DSB; P I,III; BMNH; GA; GC. Also Mineralogy*.
1. *Charakteristik der Felsarten.* 3 parts. Heidelberg, 1823-24. RM.
2. *Geologie, oder Naturgeschichte der Erde.* 5 vols. Stuttgart, 1836-44. RM.

9264. MARZARI-PENCATI, Giuseppe. 1779-1836. It. P II; BMNH; GA. Also Botany.

9265. SILLIMAN, Benjamin (first of the name). 1779-1864. U.S.A. DSB; ICB; P II,III; BMNH; GB; GE. Also Chemistry*. See also 15510/1.
1. *Outline of the course of geological lectures given in Yale College.* New Haven, 1829. RM.

9266. WITHAM, Henry Thornton Maire. 1779-1844. Br. DSB; BMNH.
1. *Observations on fossil vegetables, accompanied by representations of their internal structure as seen through the microscope.* Edinburgh, 1831. RM.
2. *The internal structure of fossil vegetables found in the*

carboniferous and oolitic deposits of Great Britain. Edinburgh, 1833. RM/L.

9267. BROWN, John (second of the name). 1780-1859. Br. BMNH; GD1.

9268. CLEAVELAND, Parker. 1780-1858. U.S.A. DSB; P III; BMNH; GE. Also Mineralogy.
1. *An elementary treatise on mineralogy and geology.* Boston, 1816. RM. 2nd ed., 2 vols, 1822.

9269. FEATHERSTONHAUGH, George William. 1780-1866. U.S.A. DSB Supp.; P III; BMNH.
1. *Geological report of an examination, made in 1834, of the elevated country between the Missouri and Red rivers.* Washington, 1835. RM.
2. *Report of a geological reconnaissance ... to the Coteau de Prairie, an elevated ridge dividing the Missouri from the St. Peter's river.* Washington, 1836. RM.

9270. FITTON, William Henry. 1780-1861. Br. DSB; ICB; P III; BMNH; GB; GD.
1. *A geological sketch of the vicinity of Hastings.* London, 1833.

9271. LARDY, Charles. 1780-1858. Switz. P I.

9272. MACKENZIE, George Steuart. 1780-1848. Br. BMNH; GD.

9273. SCHUBERT, Gotthilf Heinrich von. 1780-1860. Ger. ICB; P II; BMNH; BLA; GC. Other entries: see 3285.
1. *Handbuch der Geognosie und Bergbaukunde.* Nuremberg, 1813. RM.

9274. SERRES (DE MESPLÈS), Marcel (Pierre Toussaint) de. 1780-1862. Fr. DSB; ICB; P II,III; BMNH. Also Zoology.
1. *Observations pour servir à l'histoire des volcans éteints du Département de l'Herault.* Montpellier, 1808. RM.
2. *Géognosie des terrains tertiaires; ou, Tableau des principaux animaux invertébrés des terrains marins tertiaires du midi de la France.* Paris/Montpellier, 1829. RM.

9275. SOMERVILLE, Mary Fairfax Greig. 1780-1872. Br. DSB; ICB; P II,III; GB; GD. Also Science in General*, Astronomy* and Autobiography*.
1. *Physical geography.* London, 1848. X. 2nd ed., 1849. 3rd ed., 1851. X. 4th ed., 1858. 5th ed., 1862. 6th ed., rev. by H.W. Bates, 1870. 7th ed., 1877.

9276. STIFFT, Christian Ernst. 1780-1855. Ger./Holl. P II; BMNH.
1. *Geognostische Beschreibung des Herzogthums Nassau.* Wiesbaden, 1831. RM.

9277. BONNARD, Augustin Henri de. 1781-1857. Fr. P I,III; BMNH.

9278. JASCHE, Christoph Friedrich. b. 1781. Ger. P I; BMNH.

9279. ACKNER, Johann Michael. 1782-1862. Hermannstadt. P III; BMNH.

9280. CARNE, Joseph. 1782-1858. Br. GD.

9281. CATULLO, Tommaso Antonio. 1782-1869. It. P I,III; BMNH.
1. *Saggio di zoologia fossile.* Padua, 1827. RM.

9282. GOLDFUSS, Georg August. 1782-1848. Ger. P I; BMNH; GA; GB;

GC. Also Zoology.
1. *Physikalisch-statistische Beschreibung des Fichtelgebirges.* Nuremberg, 1817. With G. Bischof. RM.
2. *Petrefacta Germaniae ... Abbildungen und Beschreibungen der Petrefacten Deutschlands und der angränzenden Länder.* 3 vols. Düsseldorf, 1826/33-44. With G. zu Münster. RM/L. 2nd ed., Leipzig, 1863. *Repertorium*, Leipzig, 1866.

9283. GRATELOUP, Jean Pierre Sylvestre de. 1782-1861. Fr. BMNH; BLA. Also Zoology.

9284. HAUSMANN, Johann Friedrich Ludwig. 1782-1859. Ger. P I & Supp.; BMNH; GA; GB; GC. Also Mineralogy.

9285. HIBBERT (later HIBBERT-WARE), Samuel. 1782-1848. Br. BMNH; GD.
1. *History of the extinct volcanoes of the basin of Neuwied, on the lower Rhine.* Edinburgh/London, 1832.

9286. KARSTEN, Karl Johann Bernhard. 1782-1853. Ger. DSB; P I; GA; GB; GC. Also Chemistry*.
1. *Handbuch der Eisenhüttenkunde.* 2 vols. Halle, 1816. X. 3rd ed., 5 vols & atlas, Berlin, 1841.
2. *Metallurgische Reise durch einen Theil von Baiern und durch die süddeutschen Provinzen Österreichs.* Halle, 1821. RM.
3. *Untersuchungen über die kohligen Substanzen des Mineralreichs.* Berlin, 1826. RM.
4. *System der Metallurgie, geschichtlich, statistisch, theoretisch und technisch.* 5 vols. Berlin, 1831-32. RM/L.

9287. MacLAREN, Charles. 1782-1866. Br. BMNH; GB; GD.
1. *Sketch of the geology of Fife and the Lothians.* Edinburgh, 1839.

9288. POHL, Johann Emmanuel. 1782-1834. Aus. ICB; P II; BMNH; GA; GC. Also Botany.

9289. SMITH, James. 1782-1867. Br. P III; BMNH; GD.

9290. HILDRETH, Samuel Prescott. 1783-1863. U.S.A. ICB; P III; GE. Also Zoology.

9291. KÜHN, Karl Amandus. 1783-1848. Ger. P I & Supp.

9292. OMALIUS D'HALLOY, Jean Baptiste Julien d'. 1783-1875. Belg. DSB; P II,III; BMNH; BLA; GA; GB.

9293. RAUMER, Karl Georg von. 1783-1865. Ger. P II,III; GA; GC. See also 9262/1,2.

9294. BUCKLAND, William. 1784-1856. Br. DSB; ICB; P I; BMNH; GA; GB; GD.
1. *Vindiciae geologicae; or, The connexion of geology with religion.* Oxford, 1820. RM/L.
2. *Reliquiae diluvianae; or, Observations on the organic remains ... and on other geological phenomena attesting the action of an universal deluge.* London, 1823. RM/L.
3. *Geology and mineralogy considered with reference to natural theology.* 2 vols. London, 1836. (Bridgewater Treatises. 6) RM/L. 2nd ed., 1837. 3rd ed., with additions by R. Owen et al. and a biography, ed. by F.T. Buckland, 1858.

See also 9174/2 and 15603/1.

9295. DEWEY, Chester. 1784-1867. U.S.A. P III; BMNH; GE. Also Botany.

9296. GRIFFITH, Richard John. 1784-1878. Br. DSB; P III; BMNH; GB; GD.
1. *Notice respecting the fossils of the mountain limestone of Ireland as compared with those of Great Britain, and also with the Devonian system.* Dublin, 1842.
2. *A synopsis of the Silurian fossils of Ireland.* Dublin, 1846. With F. McCoy.

9297. KEFERSTEIN, Christian. 1784-1866. Ger. P I; BMNH; GA; GC. Also History of Geology*.
1. *Die Naturgeschichte des Erdkörpers in ihren ersten Grundzügen.* 2 vols. Leipzig, 1834. RM.

9298. STRÖM, Hans Christian. 1784-1836. Nor. P II.

9299. BOWMAN, John Eddowes (first of the name). 1785-1841. Br. P I; GD.

9300. BROWN, Thomas (third of the name). 1785-1862.
1. *Illustrations of the fossil conchology of Great Britain and Ireland.* London, 1849.

9301. FLEMING, John. 1785-1857. Br. DSB; P III; BMNH; GD. Also Zoology*.

9302. HORNER, Leonard. 1785-1864. Br. DSB; P III; GB; GD. See also 2901/3b.

9303. JAEGER, Georg Friedrich von. 1785-1866. Ger. DSB; BMNH; BLA; GC.
1. *Über die Pflanzenversteinerungen welche in dem Bausandstein von Stuttgart vorkommen.* Stuttgart, 1827.

9304. SEDGWICK, Adam (first of the name). 1785-1873. Br. DSB; P II, III; BMNH; GB; GD. Also Science in General*. See also 15519/1.
1. *A synopsis of the classification of the British Palaeozoic rocks.* Cambridge, 1854. With F. McCoy.

9305. VELTHEIM, Franz Wilhelm Werner von. 1785-1839. Ger. P II; GC.

9306. VOLTZ, Philippe Louis. 1785-1840. Fr. DSB; P II; BMNH.

9307. WARBURTON, Henry. ca. 1785-1858. Br. P II; GD.

9308. ZIETEN, Karl Hartwig von. 1785-1846. Ger. P II; BMNH; GC.
1. *Die Versteinerungen Württembergs.* 10 parts. Stuttgart, 1830-33.

9309. BABEY, Claude Marie Philibert. 1786-1848. BMNH.

9310. CHARPENTIER, Johann (or Jean) de. 1786-1855. Switz. DSB; P I; BMNH; GC.
1. *Essai sur la constitution géognostique des Pyrénées.* Paris, 1823. RM.
2. *Essai sur les glaciers et sur le terrain erratique du bassin du Rhône.* Lausanne, 1841. RM/L.

9311. DUNIN-BORKOWSKI, Stanislaus. b. 1786. Lemberg. P I.

9312. FÉRUSSAC, André Etienne (Justin Pascal Joseph François) d'Audebard, *Baron* de. 1786-1836. Fr. ICB; P I; BMNH; GA. Also Zoology*.

9313. GERMAR, Ernst Friedrich. 1786-1853. Ger. P I; BMNH; GC. Also Mineralogy* and Zoology.
1. *Die Versteinerungen des Steinkohlengebirges von Wettin und Löbejün im Saalkreise.* 8 parts. Halle, 1844-53.

9314. KLÖDEN, Karl Friedrich von. 1786-1856. Ger. P I; BMNH; GC. Also Autobiography*.
1. *Die Versteinerungen der Mark Brandenburg.* Berlin, 1834.

9315. KNIGHT, William. 1786-1844. Br. GD.
1. *Facts and observations towards forming a new theory of the earth.* Edinburgh, 1818. RM.

9316. MARTIN, Peter John. 1786-1860. Br. BMNH; GD.
1. *A geological memoir on a part of western Sussex, with some observations upon chalk basins, the Weald denudation, and outliers by protrusion.* London, 1828.

9317. MICHELIN, (Jean Louis) Hardouin. 1786-1867. Fr. BMNH. Also Zoology.
1. *Iconographie zoophytologique: Description ... des polypiers fossiles de France.* Text & atlas. Paris, 1840-47.

9318. MITCHELL, James. ca. 1786-1844. Br. P II; BMNH; GD. Also Physics* and Chemistry*.
1. *A dictionary of chemistry, mineralogy and geology.* London, 1823. RM.

9319. NECKER (DE SAUSSURE), Louis Albert. 1786-1861. Switz. DSB; ICB; P II & Supp.; BMNH. Also Zoology.
1. *Etudes géologiques dans les Alpes.* Paris, 1841.

9320. ALLIES, Jabez. 1787-1856. Br. BMNH; GD.

9321. BEUDANT, François Sulpice. 1787-1850. Fr. DSB; P I; BMNH; GA; GB. Also Mineralogy*.
1. *Voyage minéralogique et géologique en Hongrie ... 1818.* 3 vols. Paris, 1822. RM.
1a. ——— *Travels in Hungary in 1818.* London, 1823. RM.
2. *Cours élémentaire de minéralogie et de géologie.* 2 parts. Paris, 1841. X. 17th ed., 1886.

9322. BLACK, James. ca. 1787-1867. Br. DSB; BMNH; BLA & Supp.; GD.

9323. CONYBEARE, William Daniel. 1787-1857. Br. DSB; P III; BMNH; GB; GD.
1. *Outlines of the geology of England and Wales. Part 1. With an introductory compendium of the general principles of that science and comparative views of the structure of foreign countries.* [No more publ.] London, 1822. RM/L.

9324. GEMMELLARO, Carlo. 1787-1866. It. P III. Also Zoology.

9325. GODEFFROY, Karl. 1787-1848. Ger. P I.

9326. JONAS, Joseph. 1787-1821. Hung. P I.
1. *Ungerns Mineralreich orycto-geognostich und topographisch dargestellt.* Pest, 1820. RM.

9327. PRÉVOST, (Louis) Constant. 1787-1856. Fr. DSB; P II; BMNH; GA; GB.
1. *Documents pour l'histoire des terrains tertiaires.* Paris, n.d. [1842?]. RM.

9328. SOWERBY, James De Carle. 1787-1871. Br. BMNH; GD. Also Zoology. See also 9157/1,1a.

9329. BRANDE, William Thomas. 1788-1866. Br. DSB; ICB; P I,III; BMNH; GB; GD. Also Science in General*, Chemistry*, and History of Chemistry*.
  1. *A descriptive catalogue of the British specimens deposited in the geological collection of the Royal Institution.* London, 1816.
  2. *Outlines of geology.* London, 1817. RM/L. Another ed., 1829.

9330. BRARD, Cyprien Prosper. 1788-1838. Fr. P I,III; BMNH; GA.
  1. *Manuel du minéralogiste et du géologiste voyageur.* Paris, 1805.

9331. MENGE, Johann. 1788-1852. Ger. P II; GC,

9332. NÖGGERATH, Johann Jacob. 1788-1877. Ger. P II,III; BMNH; GC.

9333. SOKOLOV, Dmitry Ivanovich. 1788-1852. Russ. DSB. Also Mineralogy.

9334. VENETZ, Ignatz. 1788-1859. Switz. DSB; P II.

9335. ZIMMERMANN, Johann Christian. 1788-1853. Ger. P II.

9336. ARTIS, Edmund Tyrell. 1789-1847. Br. BMNH.
  1. *Antediluvian phytology, illustrated by a collection of the fossil remains of plants peculiar to the coal formations of Great Britain.* London, 1825. Another ed., with additions, 1838.

9337. HARCOURT, William Venables Vernon. 1789-1871. Br. P III; GB; GD.

9338. KÄMMERER, August Alexander. 1789-1858. Russ. P I.

9339. REDFIELD, William C. 1789-1857. U.S.A. DSB; P II; GE. Also Meteorology.

9340. TAYLOR, Richard Cowling. 1789-1851. Br./U.S.A. P II; BMNH; GD; GE.

9341. YATES, James. 1789-1871. Br. P III; GD.

9342. BREDSDORFF, Jacob Hornemann. 1790-1841. Den. P I; BMNH; GA.

9343. COVELLI, Niccola. 1790-1829. It. P I; BMNH; GA.

9344. HUOT, Jean Jacques Nicolas. 1790-1845. Fr. P I; BMNH; GA.
  1. *Cours élémentaire de géologie.* Paris, 1837. X.
  1a. ——— *Nouveau cours....* 2 vols & atlas. Ib., 1837-39.

9345. MANTELL, Gideon Algernon. 1790-1852. Br. DSB; ICB; P II; BMNH; GA; GB; GD. Also Microscopy*.
  1. *The fossils of the South Downs; or, Illustrations of the geology of Sussex.* London, 1822. RM/L.
  2. *Illustrations of the geology of Sussex, containing a general view of the geological relations of the south-eastern part of England.* Ib., 1827. RM/L.
  3. *The geology of the south-east of England.* Ib., 1833. RM/L.
  4. *Thoughts on a pebble; or A first lesson in geology.* Ib., 1837. X. 8th ed., 1849.

5. *The wonders of geology; or, A familiar exposition of geological phenomena.* 2 vols. Ib., 1838. X. 3rd ed., 1839. 6th ed., 1848. 7th ed., 1857-58.
6. *The medals of creation; or, First lessons in geology and the study of organic remains.* 2 vols. Ib., 1844. RM/L. 2nd ed., 1854.
7. *Geological excursions around the Isle of Wight and along the adjacent coast of Dorsetshire.* Ib., 1847. RM/L. 2nd ed., 1851. 3rd ed., 1854.
8. *A pictorial atlas of fossil remains.* Ib., 1850. RM/L.
9. *Petrifactions and their teachings; or, A handbook to the gallery of organic remains of the British Museum.* Ib., 1851. RM/L.
10. *The journal ... covering the years 1818-1852.* Ib., 1940. Ed. with introd. and notes by E.C. Curwen.

9346. PUSCH, Georg Gottlieb. 1790-1846. Warsaw. P II; BMNH.
1. *Polens Paläontologie; oder, Abbildung und Beschreibung der ... Petrefacten und den Gebirgsformationen in Polen.* Stuttgart, 1837.

9347. ROSE, Caleb Burrell. 1790-1872. Br. BMNH; GD.

9348. WOODWARD, Samuel. 1790-1838. Br. BMNH; GB; GD.

9349. ALTHAUS, August Heinrich Jacob. b. 1791. Ger. P III.

9350. FIEDLER, Karl Gustav. 1791-1853. Ger. P I; BMNH; GC.

9351. MENKE, Karl Theodor. 1791-1861. Ger. P III; BMNH; BLA. Also Zoology.

9352. PARTSCH, Paul Maria. 1791-1856. Aus. P II; BMNH. See also 8954/1,2.

9353. SCHMERLING, Philippe Charles. 1791-1836. Belg. DSB; ICB; P II; BMNH.

9354. BABBAGE, Charles. 1792-1871. Br. DSB; ICB; P I,III; GA; GB; GD. Also Science in General*, Mathematics*, and Autobiography*.
1. *Observations on the Temple of Serapis at Pozzuoli near Naples.* London, 1847.

9355. BIGSBY, John Jeremiah. 1792-1881. Canada/Br. P III; BMNH; GB; GD.
1. *Thesaurus Siluricus: The flora and fauna of the Silurian period.* London, 1868. RM/L.
2. *Thesaurus Devonico-Carboniferous: The flora and fauna of the Devonian and Carboniferous periods.* Ib., 1878. RM/L.

9356. BISCHOF, (Carl) Gustav (Christoph). 1792-1870. Ger. DSB; ICB; P I,III; BMNH; GA; GC. Also Chemistry.
1. *Die Wärmelehre des Innern unseres Erdkörpers.* Leipzig, 1837. RM.
1a. ——— *Physical, chemical, and geological researches on the internal heat of the globe.* Vol. 1. [No more publ.] London, 1841.
2. *Lehrbuch der chemischen und physikalischen Geologie.* 2 vols in 3. Bonn, 1847-54.
2a. ——— *Elements of chemical and physical geology.* 3 vols. London, 1854-59. Trans. by Drummond. RM/L.

See also 9282/1.

9357. BOBLAYE, Emil le Puillon de. 1792-1843. Fr. P I; BMNH (LE PUILLON); GA (LE PUILLON).

9358. DUFRÉNOY, (Ours) Pierre Armand. 1792-1857. Fr. DSB; P I; BMNH; GA; GB. Also Mineralogy*.
1. *Mémoires pour servir à une description géologique de la France.* 4 vols. Paris, 1830-38. With Elie de Beaumont. Under the direction of Brochant de Villiers. RM.
2. *Carte géologique de la France. Explication de la carte....* 3 vols. Paris, 1841. With Elie de Beaumont. Under the direction of Brochant de Villiers.

9359. LEA, Isaac. 1792-1886. U.S.A. DSB; P I,III; BMNH; GE. Also Zoology.
1. *Contributions to geology.* Philadelphia, 1833. RM/L.

9360. LOCKE, John. 1792-1856. U.S.A. ICB; P I; GE. Also Physics.

9361. MURCHISON, Roderick Impey. 1792-1871. Br. DSB; P II,III; BMNH; GA; GB; GD.
1. *Outline of the geology of the neighbourhood of Cheltenham.* Cheltenham, 1834. X. Another ed., augmented and rev. by J. Buckman and H.E. Strickland, London, 1845.
2. *The Silurian system.* 2 vols. London, 1839. RM/L.
3. *The geology of Russia in Europe, and the Ural Mountains.* 2 vols. London, 1845. With E. de Verneuil and A. von Keyserling. RM/L.
4. *Notes illustrative of the geological map of Europe.* Edinburgh, 1854. With J. Nicol.
5. *Siluria: The history of the oldest known rocks containing organic remains.* London, 1854. RM/L. 3rd ed., 1859. 4th ed., 1867. 5th ed., 2 vols, 1872.
6. *First sketch of a new geological map of Scotland.* Edinburgh, 1862. With A. Geikie.
See also 15524/1.

9362. PASSY, Antoine François. 1792-1873. Fr. P III; BMNH. Also Botany.

9363. SEGATO, Gerolamo. 1792-1836. It. ICB.

9364. VANUXEM, Lardner. 1792-1848. U.S.A. DSB; P II; BMNH; GE. Also Science in General*.
1. *Observations on the geology and organic remains of the secondary, tertiary and alluvial formations of the Atlantic coast of the United States.* Philadelphia, 1828. With S.G. Morton. RM.

9365. BEINERT, Carl Christian. 1793-1868. Ger. P I,III; BMNH.

9366. DANA, James Freeman. 1793-1827. U.S.A. P I & Supp.; BLA Supp.; GE. Also Chemistry*.
1. *Outlines of the mineralogy and geology of Boston and its vicinity.* Boston, 1818. With S.L. Dana. RM.

9367. HITCHCOCK, Edward. 1793-1864. U.S.A. DSB; ICB; P I,III; BMNH; GA; GB; GE.
1. *Elementary geology.* New York, 1840. X. 2nd ed., 1841.
2. *The religion of geology and its connected sciences.* Boston, 1851. RM/L. Another ed., Glasgow, 1866.
3. *Outline of the geology of the globe, and of the United*

States in particular. Boston, 1853. RM.
4. *Illustrations of surface geology.* Washington, 1857. RM.

9368. HOPKINS, William. 1793-1866. Br. DSB; P III; GB; GD.

9369. LARDNER, Dionysius. 1793-1859. Br. ICB; P I & Supp.; BMNH; GA; GB; GD. Other entries: see 3303.
1. *Popular geology.* London, 1856.

9370. MITCHELL, Elisha. 1793-1857. U.S.A. DSB; P II; BMNH; GE.

9371. AUSTIN, Thomas. 1794?-1881. Br. BMNH.

9372. BOUÉ, Aimé. 1794-1881. Fr./Aus. DSB; ICB; P I,III; BMNH; GA; GB; GC.
1. *Essai géologique sur l'Ecosse.* Paris, 1820. RM/L.
2. *Mémoires géologiques et paléontologiques.* Vol. 1. [No more publ.] Paris, 1832. RM.
3. *Essai d'une carte géologique du globe terrestre.* Paris, 1845. RM.

9373. EUDES-DESLONGCHAMPS, Jacques Armand. 1794-1867. Fr. BMNH; GB (DESLONGCHAMPS). Also Zoology.

9374. FORCHHAMMER, Johan Georg. 1794-1865. Den. DSB; ICB; P I,III; BMNH; GB. Also Oceanography.

9375. LONSDALE, William. 1794-1871. Br. DSB; ICB; P III; BMNH; GB; GD.

9376. PANDER, Christian Heinrich von. 1794-1865. Russ. DSB; ICB; P II,III; BMNH; Mort.; BLA; GC. Also Zoology and Embryology.
1. *Beiträge zur Geognosie des russischen Reiches.* St. Petersburg, 1830.

9377. PORTLOCK, Joseph Ellison. 1794-1864. Br. P III; BMNH; GB; GD. Also Biography*.
1. *Report on the geology of the county of Londonderry.* 2 vols. Dublin, 1843.
2. *A rudimentary treatise on geology.* London, 1849. X. Another ed., 1854.

9378. PROVANA DI COLLEGNO, Giacinto. 1794-1856. It./Fr. ICB; P II; BMNH (COLLEGNO).

9379. STEININGER, Johann. 1794-1874. Ger. P II,III; BMNH.
1. *Geognostische Studien am Mittelrheine.* Mainz, 1819. RM.
2. *Die erloschenen Vulkane in der Eifel und am Niederrheine.* Mainz, 1820. RM.
3. *Neue Beiträge zur Geschichte der rheinischen Vulkane.* Mainz, 1821. RM.

9380. STUDER, Bernhard. 1794-1887. Switz. DSB; ICB; P II,III; BMNH; GB; GC.
1. *Geologie der westlichen Schweizer-Alpen.* Heidelberg, 1834. RM.

9381. ALBERTI, Friedrich August von. 1795-1878. Ger. DSB; P I,III; BMNH.
1. *Über die Gebürge des Königreiches Württemberg, in besonderer Beziehung auf Halurgie.* Stuttgart, 1826. RM.
2. *Beiträge zu einer Monographie des bunten Sandsteins, Musch-*

elkalks und Keupers, und der Verbindung dieser Gebilde zu einer Formation. Stuttgart, 1834. RM.
3. *Überblick über die Trias, mit Berücksichtigung ihres Vorkommens in den Alpen.* Stuttgart, 1864. RM.

9382. BERTRAND, Alexandre Jacques François. 1795-1831. Fr. P I; BLA.
1. *Lettres sur les révolutions du globe.* Paris, 1824. X. 4th ed., 1833.

9383. CAUCHY, François Philippe. 1795-1842. Belg. P I,III; BMNH; GF.

9384. DAUBENY, Charles Giles Bridle. 1795-1867. Br. DSB; ICB; P I, III; BMNH; GB; GD. Other entries: see 3305.
1. *A description of active and extinct volcanos.* London, 1826. RM/L. 2nd ed., 1848.

9385. EHRENBERG, Christian Gottfried. 1795-1876. Ger. DSB; ICB; P I,III; BMNH; Mort.; BLA & Supp.; GA; GB; GC. Also Zoology and Microbiology*.
1. *Über die Natur und Bildung der Coralleninseln und Corallenbäuke im Rothen Meere.* Berlin, 1834. RM.
2. *Die Bildung der europäischen, libyschen und arabischen Kreidefelsen und des Kreidemergels aus mikroskopischen Organismen.* Berlin, 1839. RM.
3. *Mikrogeologie. Das Erden und Felsen schaffende Wirken des unsichtbar kleinen selbstständigen Lebens auf der Erde.* Leipzig, 1854. RM.

9386. EICHWALD, Karl Eduard Ivanovich. 1795-1876. Ger./Russ. DSB; P I,III; BMNH; BLA; GA; GB. Also Zoology.
1. *Über das silurische Schichtensystem in Esthland.* St. Petersburg, 1840.
2. *Lethaea Rossica; ou, Paléontologie de la Russie.* 3 vols in 5. Stuttgart, 1852-68.

9387. FALLOU, Friedrich Albert. 1795-1877. Ger. P III.

9388. HAIDINGER, Wilhelm Karl von. 1795-1871. Aus. DSB; P I,III; BMNH; GA; GB; GC. Also Mineralogy*.
1. *Naturwissenschaftliche Abhandlungen.* 4 vols. Vienna, 1847-51.

9389. MERIAN (-THURNEYSEN), Peter. 1795-1883. Switz. P II,III; GC.

9390. OEYNHAUSEN, Karl von. 1795-1871. Ger. P II,III; BMNH; GC.
1. *Geognostische Umrisse der Rheinländer zwischen Basel und Mainz, mit besonderer Rücksicht auf das Vorkommen des Steinsalzes.* 2 vols. Essen, 1825.

9391. PERCIVAL, James Gates. 1795-1856. U.S.A. P II; BMNH; GB; GE.

9392. TRIMMER, Joshua. 1795-1857. Br. P II; BMNH; GB; GD.
1. *Practical geology and mineralogy.* London, 1841.

9393. CURIONI, Giulio. 1796-1878. It. P III; BMNH.

9394. DE LA BECHE, Henry Thomas. 1796-1855. Br. DSB; ICB; P I (BECHE); BMNH; GB; GD (BECHE).
1. *A selection of the geological memoirs contained in the "Annales des mines", together with a synoptical table of equivalent formations. Translated with notes.* London, 1824. RM/L. Another ed., 1836.

2. *Sections and views illustrative of geological phaenomena.* Ib., 1830. Another ed., 1839.
3. *Geological notes.* Ib., 1830.
4. *A geological manual.* Ib., 1831. X. 2nd ed., 1832. 3rd ed., 1833.
5. *Researches in theoretical geology.* Ib., 1834. RM/L.
6. *Report on the geology of Cornwall, Devon, and West Somerset.* Ib., 1839. RM/L.
7. *The geological observer.* Ib., 1851. RM/L. 2nd ed., 1853.

9395. HARTMANN, Carl Friedrich Alexander. 1796-1863. Ger. DSB; P I, III; BMNH.
1. *Die Wunder der Erdrinde; oder, Gemeinfassliche Darstellung der Mineralogie und Geologie.* Stuttgart, 1838. RM.
2. *Handwörterbuch der Berg-, Hütten- und Salzwerkskunde, der Mineralogie und Geognosie.* 3 vols & atlas. Weimar, 1859-60.

9396. MACGILLIVRAY, William. 1796-1852. Br. BMNH; GA; GB; GD. Also Botany*, Zoology*, and Biography*.
1. *A manual of geology.* London, 1840.

9397. RICHARDSON, George Fleming. 1796?-1848. Br. P III (R., George Frederick); BMNH; GD.
1. *Geology for beginners.* London, 1842. X.
1a. ——— 4th ed.: *An introduction to geology and its associate sciences.* Ib., 1851. Rev. and enlarged by T. Wright. Another ed., 1869.

9398. ROSTHORN, Franz von. 1796-1877. Aus. P III.

9399. ROULIN, François Désiré. 1796-1874. Fr. ICB; P II,III; BMNH; BLA & Supp. Also Natural History.

9400. STRZELECKI, Paul Edmund. 1796-1873. Australia. ICB; P III; BMNH; GD.
1. *A physical description of New South Wales and Van Diemen's Land, accompanied by a geological map ... and figures of organic remains.* London, 1845.

9401. ZIMMERMANN, Karl Gottfried. 1796-1876. Ger. P II; BLA; GC.

9402. ADHÉMAR, Joseph Alfonse. 1797-1862. Fr. P I,III; BMNH.
1. *Révolutions de la mer, déluges périodiques.* Paris, 1842. RM. 2nd ed., 1860.

9403. BAIN, Andrew Geddes. 1797-1864. Br./South Africa. GB.

9404. BOWERBANK, James Scott. 1797-1877. Br. BMNH; GB; GD. Also Natural History.
1. *A history of the fossil fruits and seeds of the London clay.* London, 1840.

9405. DESHAYES, Gerard Paul. 1797-1875. Fr. DSB; BMNH; GB. Also Zoology.
1. *Description des coquilles fossiles des environs de Paris.* 2 vols & atlas. Paris, 1824. RM/L.
2. *Description de coquilles caractéristiques des terrains.* Paris, 1831. RM.
3. *Traité élémentaire de conchyliologie, avec les applications de cette science à la géologie.* 3 vols. Paris, 1839-57.

4. *Description des animaux sans vertèbres découverts dans le bassin de Paris.* 3 vols & atlas of 2 vols. Paris, 1857. RM/L.

9406. DOLLFUS-AUSSET, Daniel. 1797-1870. Fr. P III; GC.
1. *Matériaux pour servir à l'étude des glaciers.* 13 parts. Paris, 1864-71.

9407. FROMHERZ, Carl. 1797-1854. Ger. P I; BMNH. Also Chemistry.

9408. GESNER, Abraham. 1797-1864. Canada. P III; BMNH; GB.

9409. HAGENOW, (Karl) Friedrich von. 1797-1865. Ger. P III; BMNH; GC.

9410. HOFFMANN, Friedrich. 1797-1836. Ger. P I; BMNH; GC.

9411. HUMBLE, William. 1797-1878. Br. BMNH.
1. *A dictionary of geology and mineralogy.* London, 1840. X. 3rd ed., 1860.

9412. KEILHAU, Baltazar Mathias. 1797-1858. Nor. P I; BMNH.

9413. LYELL, Charles. 1797-1875. Br. DSB; ICB; P I,III; BMNH; GA; GB; GD. Also Evolution*. See also Index.
1. *Principles of geology. Being an attempt to explain the former changes of the earth's surface by reference to causes now in operation.* 3 vols. London, 1830-33. RM/L. Reprinted, New York, 1969.
1a. ——— 2nd ed. Vols I and II only (these were published before Vol. III of the 1st ed.), 1832-33.
1b. ——— 3rd ed., 4 vols, 1834-35. 4th ed., 4 vols, 1835. X. 5th ed., 4 vols, 1837. X. 6th ed., 3 vols, 1840. 7th ed., 1 vol., 1847. 8th ed., 1 vol., 1850. 9th ed., 1 vol., 1853. 10th ed., 2 vols, 1867-68. 11th ed., 2 vols, 1872.
1c. ——— 12th ed.: *Principles of geology; or, The modern changes of the earth and its inhabitants considered as illustrative of geology.* 2 vols. 1875. RM/L.
2. *Elements of geology.* London, 1838. RM/L. 2nd ed., 2 vols, 1841.
2a. ——— 3rd ed.: *A manual of elementary geology*, 1851. 4th ed., 1852. 5th ed., 1855. 6th ed., 1865.
3. *Travels in North America in the years 1841-2, with geological observations on the United States, Canada and Nova Scotia.* 2 vols. London/New York, 1845. RM/L.
4. *A second visit to the United States of North America.* 2 vols. London/New York, 1849. RM. 2nd ed., 1850.
5. *The geological evidences of the antiquity of man, with remarks on theories of the origen of species by variation.* London/Philadelphia, 1863. RM/L. 4th ed., 1873; reprinted, New York, 1973.
6. *The student's elements of geology.* London, 1871. 2nd ed., 1874. 3rd ed., 1878. 4th ed., 1885. Another ed.: *The student's Lyell: A manual of elementary geology*, 1896.

9414. NAUMANN, (Georg Amadeus) Karl Friedrich. 1797-1873. Ger. DSB; P II,III; BMNH; GA; GB; GC. Also Mineralogy*.
1. *Lehrbuch der Geognosie.* 3 vols. Leipzig, 1850-54. 2nd ed., 1858-62.

9415. SCROPE, George Julius Poulett. 1797-1876. Br. DSB; P II,III; BMNH; GB; GD.
1. *Considerations on volcanos ... leading to the establishment of a new theory of the earth.* London, 1825. RM/L.
1a. ——— 2nd ed. *Volcanos: The character of their phenomena, their share in the structure and composition of the surface of the globe, and their relation to its internal forces.* Ib., 1862. RM/L. Re-issued with a new preface and new title-page dated 1872.
2. *Memoir of the geology of central France, including the volcanic formations of Auvergne, the Velay, and the Vivarais.* London, 1827. RM.
2a. ——— 2nd ed. *The geology and extinct volcanos of central France.* Ib., 1858. RM/L.

9416. STIEHLER, August Wilhelm. 1797-1878. Ger. BMNH; GC.

9417. TERQUEM, Olry (second of the name). 1797-1887. Fr. BMNH.

9418. TREVELYAN, Walter Calverley. 1797-1879. Br. P III; BMNH; GD.

9419. WAGNER, (Andreas) Johann. 1797-1861. Ger. P II; BMNH; GC. Also Zoology.

9420. ANDERSON, Henry James. 1798-1875. U.S.A. P III; BMNH.

9421. BURKART, Herrmann Joseph. 1798-1874. Ger./Mexico. P I,III; BMNH.

9422. CLARKE, William Branwhite. 1798-1878. Br./Australia. DSB; P III; BMNH; GB; GD.
1. *On the transmutation of rocks in Australia.* Sydney, 1866.
2. *On marine fossiliferous secondary formations in Australia.* London, 1866.
3. *Remarks on the sedimentary formations of New South Wales.* Sydney, 1867. 4th ed., 1878.

9423. DUBOIS (DE MONTPÉREUX), Frédéric. 1798-1850. Switz./Russ. P I & Supp.; BMNH; GA.

9424. ÉLIE DE BEAUMONT, (Jean Baptiste Armand Louis) Léonce. 1798-1874. Fr. DSB; ICB (BEAUMONT); P I,III; BMNH; GA; GB.
1. *Coup d'oeil sur les mines.* Paris, 1824. RM.
2. *Observations géologiques sur les différentes formations qui, dans le système des Vosges, séparent la formation houillière de celle du lias.* n.p., 1828. RM.
3. *Recherches sur quelques-unes des révolutions de la surface du globe.* Paris, 1829. RM.
4. *Notice sur les systèmes de montagnes.* 3 vols. Paris, 1852. RM/L.
5. *Rapport sur les progrès de la stratigraphie.* Paris, 1869. RM.

His main works were written with Dufrénoy: see 8765/1 and 9358/1,2.

9425. GUTBIER, (Christian) August von. 1798-1866. Ger. P I,III; BMNH; GC.
1. *Geognostische Beschreibung des Zwickauer Schwarzkohlengebirges und seiner Umgebungen.* Zwickau, 1834.

9426. HUTTON, William. 1798-1860. Br. BMNH; GD. See also 11964/7.

9427. KURR, Johann Gottlob von. 1798-1870. Ger. P I,III; BMNH; GC. Also Mineralogy*.

9428. LOGAN, William Edmond. 1798-1875. Br./Canada. DSB; ICB; P III; BMNH; GB; GD.

9429. ROZET, Claude Antoine. 1798-1858. Fr. P II; BMNH.

9430. WOOD, Searles Valentine (first of the name). 1798-1880. Br. BMNH; GB; GD.

9431. ANNING, Mary. 1799-1847. Br. ICB; P I; GB; GD1.

9432. BARRANDE, Joachim von. 1799-1883. Prague. DSB; P III,IV; BMNH; GB.
   1. *Système silurien du centre de Bohême.* 8 parts in 30 vols. Prague, 1852-1902.
   2. *Parallèle entre les dépôts siluriens de Bohême et de Scandinavie.* Prague, 1856.

9433. BOASE, Henry Samuel. 1799-1883. Br. BMNH; GB; GD. Also Science in General*.
   1. *A treatise on primary geology, being an examination, both practical and theoretical, of the older formations.* London, 1834.

9434. BOUILLETT, Jean Baptiste. 1799-1878. Fr. P I,III; BMNH; GA.
   1. *Topographie minéralogique du Département du Puy-de-Dôme.* Clermont-Ferrand, 1829.
   See also 9485/1.

9435. DIXON, Frederick. 1799-1849. Br. BMNH.
   1. *The geology and fossils of the Tertiary and Cretaceous formations of Sussex.* London, 1850. RM/L. Another ed., Brighton, 1878.

9436. EDWARDS, Frederic Erasmus. 1799-1875. Br. BMNH.

9437. EMMONS, Ebenezer. 1799-1863. U.S.A. DSB; P III; BMNH; GB; GE.
   1. *The Taconic system.* Albany, N.Y., 1844. RM.
   2. *Manual of geology.* New York, 1859. X. 2nd ed., 1860.

9438. ENGELSBACH-LARIVIÈRE, Auguste. 1799-1831. Belg. P III; BMNH; GA (ENGELSPACH).

9439. GREENWOOD, George. 1799-1875.
   1. *Rain and rivers; or, Hutton and Playfair against Lyell and all comers.* London, 1857. RM/L. 2nd ed., 1866.
   2. *River terraces. Letters on geological and other subjects.* London, 1877. RM/L.

9440. HOLGER, Philipp Aloys von. 1799-1866. Aus. P III; BMNH; BLA. Also Chemistry.

9441. LEVALLOIS, J.B. Jules. 1799-1877. Fr. P I,III; BMNH.

9442. SIMPSON, Martin. 1799-1892. Br. BMNH.

9443. WALCHNER, Friedrich August. 1799-1865. Ger. P II; BMNH; GC. Also Chemistry.

9444. ZENKER, Jonathan Karl. 1799-1837. Ger. P II; BMNH; BLA; GC.
   1. *Beiträge zur Naturgeschichte der Urwelt.* Jena, 1833.

9445. RHIND, William. fl. 1833-1867. Br. BMNH. Also Botany*.

1. *The age of the earth considered geologically and historically.* Edinburgh, 1838.

9446. ROBERTS, George. d. 1860. Br. P II; GD.

9447. PARETO, Lorenzo. d. 1865. It. P III; BMNH.

9448. PASINI, Ludovico. d. 1870. It. P III.

9449. COOLEY, William Desborough. d. 1883. Br. BMNH; GD.
1. *Physical geography.* London, 1876.

9450. BASTEROT, Barthélemy. 1800-1887. Fr. BMNH.
1. *Description géologique du bassin tertiaire du sud-ouest de la France.* Paris, 1825.

9451. BRAUN, Karl Friedrich Wilhelm. 1800-1864. Ger. BMNH; GC.

9452. BRONN, Heinrich Georg. 1800-1862. Ger. DSB; ICB; P I & Supp.; BMNH; Mort.; GA; GB; GC. Also Zoology*.
1. *System der urweltlichen Konchylien.* Heidelberg, 1824.
2. *System der urweltlichen Pflanzenthiere.* Ib., 1825.
3. *Italiens Tertiär-Gebilde und deren organische Einschlüsse.* Ib., 1831. RM.
4. *Lethaea geognostica; oder, Abbildungen und Beschreibungen der für die Gebirgs-Formationen bezeichnendsten Versteinerungen.* 2 vols. Stuttgart, 1835-38. RM. 2nd ed., 1837. 3rd ed., 3 vols & atlas, 1850-56. For the continuation of this work see 9700/3.
5. *Handbuch einer Geschichte der Natur.* 4 vols. Stuttgart, 1841-49.
6. *Index palaeontologicus; oder, Übersicht der bis jetzt bekannten fossilen Organismen.* 2 vols. Stuttgart, 1848-49. RM/L.
7. *Beiträge zur triasischen Fauna und Flora der bituminösen Schiefer von Raibl.* Stuttgart, 1858. RM.

9453. DAVREUX, Charles Joseph. 1800-1863. Belg. P I,III; BMNH; GF.

9454. DECHEN, (Ernst) Heinrich (Carl) von. 1800-1889. Ger. DSB; P I,III,IV; BMNH; GB; GC.
1. *Geognostische Beschreibung des Siebengebirges am Rhein.* Bonn, 1852. RM.
2. *Die nutzbaren Mineralien und Gebirgsarten im deutschen Reiche, nebst einer physiographischen und geognostischen Übersicht des Gebietes.* Berlin, 1873. RM.

9455. DESNOYERS, Jules Pierre François Stanislas. 1800-1887. Fr. P I,III; BMNH; GA; GB.

9456. GÖPPERT, Heinrich Robert. 1800-1884. Ger. DSB; P I,III; BMNH; BLA; GC. Also Botany*.
1. *Die fossilen Farnkräuter.* Breslau, 1836.
2. *Die Gattungen der fossilen Pflanzen, vergliechen mit denen der Jetztwelt.* 6 parts. Bonn, 1841-46. Text in French as well as German.
3. *Die fossilen Coniferen, mit steter Berücksichtigung der lebenden.* Leiden, 1850.
4. *Beiträge zur Tertiärflora Schlesiens.* Cassel, 1852.

9457. PEACH, Charles William. 1800-1886. Br. GB; GD. Also Zoology.

9458. PHILLIPS, John. 1800-1874. Br. DSB; ICB; P II,III; BMNH; GA; GB; GD. Also Biography*.
1. *Illustrations of the geology of Yorkshire.* 2 parts. York, 1829-36. RM/L. 2nd ed., London, 1835.
2. *A guide to geology.* London, 1834. X. 3rd ed., 1836.
3. *A treatise on geology.* 2 vols. London, 1837-39. X. Other eds: 1839-41, 1842-46, 1852.
4. *Figures and descriptions of the Palaeozoic fossils of Cornwall, Devon, and West Somerset.* London, 1841. RM/L.
5. *Manual of geology, theoretical and practical.* London, 1855. RM. Another ed., by R. Ethridge and H.G. Seeley, 1885.
6. *Life on the earth, its origin and succession.* Cambridge, 1860. RM/L.
7. *Vesuvius.* Oxford, 1869.
8. *Geology of Oxford and the valley of the Thames.* Oxford, 1871. RM.

9459. UNGER, Franz von. 1800-1870. Aus. DSB; ICB; P III; BMNH; GC. Also Botany*.
1. *Synopsis plantarum fossilium.* Leipzig, 1845.
2. *Chloris protogaea: Beiträge zur Flora der Vorwelt.* Ib., 1847.
3. *Genera et species plantarum fossilium.* Vienna, 1850.
4. *Die Urwelt in ihren verschiedenen Bildungsperioden.* Vienna, 1851. X.
4a. ——— *Ideal views of the primitive world in its geological and palaeontological phases.* London, 1863.

9460. VIRLET D'AOUST, Pierre Théodore. 1800-1894? Fr. P III; BMNH.

9461. VOLBORTH, Alexander von. 1800-1876. Russ. P II,III; BMNH.

9462. BRONGNIART, Adolphe Théodore. 1801-1876. Fr. DSB; ICB; P I, III; BMNH; GA; GB. Also Botany.
1. *Prodrome d'une histoire des végétaux fossiles.* Paris, 1828. RM.
2. *Histoire des végétaux fossiles.* 2 vols. Paris, 1828-37. RM/L. Reprinted, Amsterdam, 1965.
3. *Recherches sur les graines fossiles silicifiées.* Paris, 1881. With an obituary by J.B. Dumas. RM.

9463. CLAY, Charles. 1801-1893. Br. BMNH; BLA & Supp.; GB; GD1.
1. *Geology.* London, 1839.

9464. DUMORTIER, Vincent Eugène. 1801-1876. Fr. P III; BMNH.

9465. EDMONDS, Richard. 1801-1886. Br. P III; BMNH; GD.

9466. ESMARK, Hans Morten Thrane. 1801-1882. Nor. ICB; P I,III.

9467. FOURNET, Joseph Jean Baptiste Xavier. 1801-1869. Fr. ICB; P III; BMNH; GB.

9468. GUMPRECHT, Thaddäus Eduard. 1801-1856. Ger. P I; BMNH.

9469. HOFMANN, Ernst von (first of the name). 1801-1871. Russ. P I, III; BMNH.

9470. KLIPSTEIN, August von. 1801-1894. Ger. P I; BMNH.

9471. LARTET, Edouard Amant Isidore Hippolyte. 1801-1871. Fr. DSB; BMNH; GB.

9472. LEYMERIE, Alexandre Félix Gustave Achille. 1801-1878. Fr. P III; BMNH.

9473. MEYER, (Christian Erich) Hermann von. 1801-1869. Ger. DSB; P II,III (M., Christian Friedrich Hermann); BMNH; GB; GC.

9474. SCHMIDT, Carl Joseph. 1801-1862. Brünn. P III.

9475. ARCHIAC, Etienne Jules Adolphe Desmier de Saint-Simon, *Vicomte* d'. 1802-1868. Fr. DSB; P I,III; BMNH; GA; GB.
1. *Histoire des progrès de la géologie de 1834 à 1859*. 8 vols in 9. Paris, 1847-60. RM/L.
2. *Description des animaux fossiles du groupe nummulitique de l'Inde*. Paris, 1853. With J. Haime.
3. *Introduction à l'étude de la paléontologie stratigraphique*. 2 vols. Paris, 1864.
4. *Paléontologie de la France*. Paris, 1868. RM.

9476. BAUZÁ Y RÁVARA, Felipe. 1802-1875. Spain. BMNH.

9477. BILLY, Edouard de. 1802-1874. Fr. P I,III; BMNH.

9478. CAUMONT, Arcisse de. 1802-1873. Fr. P III; BMNH.

9479. CAUTLEY, Proby Thomas. 1802-1871. Br./India. BMNH; GB; GD.

9480. CHAMBERS, Robert. 1802-1871. Br. DSB; ICB; P III; BMNH; GA; GB; GD. Also Evolution*.
1. *Ancient sea-margins as memorials of change in the relative level of sea and land*. Edinburgh, 1848. RM/L.

9481. CHRISTOL (or CRISTOL), Jules de. 1802-1861. Fr. DSB.

9482. DEICKE, J. Carl. 1802-1869. Switz. P III.

9483. DOMEYKO, Ignacio. 1802-1889. Chile. ICB; P III,IV; BMNH.

9484. FUCHS, Wilhelm. 1802-1853. Aus./Hung. P I; BMNH; GC.

9485. LECOQ, Henri. 1802-1871. Fr. P I,III; BMNH; GA. Also Botany*.
1. *Vues et coupes des principales formations géologiques du ... Puy-de-Dôme*. 2 vols. Paris, 1830. With J.B. Bouillet. RM/L.
2. *Des glaciers et des climats; ou, Des causes atmosphériques en géologie*. Paris, 1847. RM.

See also 7607/1.

9486. MILLER, Hugh. 1802-1856. Br. DSB; ICB; P II; BMNH; GA; GB; GD. Also Autobiography*. Other biographies: see Index.
1. *The old red sandstone*. Edinburgh, 1841. RM. 3rd ed., 1847. 4th ed., 1850. 6th ed., 1854. 7th ed., 1857. Other eds: 1861, 1865, 1869. 13th ed., 1870. 20th ed., 1875.
2. *The foot-prints of the Creator; or, The Asterolepis of Stromness*. Ib., 1847. X. 3rd ed., with a memoir of the author, by L. Agassiz, Boston, 1859. RM. Another ed., Edinburgh, 1861; reprinted, Farnborough, 1968. 13th ed., Edinburgh, 1871. 16th ed., London, 1874. Another ed., 1881. 22nd ed., 1883.
3. *The testimony of the rocks; or, Geology in its bearing on the two theologies, natural and revealed*. Ib., 1857. RM/L. Other eds: 1860, 1869, 1873, 1874, 1881.

4. *The cruise of the "Betsey"; or, A summer ramble among the fossiliferous deposits of the Hebrides.* Ib., 1858. RM/L. Other eds: 1861, 1869, 1879. 11th ed., 1874. 13th ed., 1883.
5. *Sketch-book of popular geology.* Ib., 1859. RM/L. 3rd ed., 1869. 7th ed., 1874. Other eds: 1880, 1889.
6. *Edinburgh and its neighbourhood, geological and historical.* Ib., 1863. Another ed., 1864; RM/L. 3rd ed., 1869. 5th ed., 1873. 6th ed., 1875. Another ed., 1879.

9487. MÖLLER, Nikolai Benjamin. 1802-1860. Nor. P II,III.

9488. ORBIGNY, Alcide Dessalines d'. 1802-1857. Fr. DSB; P II; BMNH; GA; GB. Also Natural History*.
1. *Paléontologie française: Description zoologique et géologique de tous les animaux mollusques et rayonnés fossiles de la France.* 40 vols. Paris, 1840-94. Continued after his death under the auspices of the Société Géologique de France.
2. *Foraminifères fossiles du bassin tertiaire de Vienne (Autriche).* Paris, 1846. RM/L.
3. *Prodrome de paléontologie stratigraphique universelle des animaux mollusques et rayonnés.* 3 vols in 1. Paris, 1850-52.
4. *Géologie appliquée aux arts et à l'agriculture.* Paris, 1851. RM.

See also 12923/3.

9489. RUSSEGGER, Joseph von. 1802-1863. Aus. P II; BMNH; GC.

9490. CONRAD, Timothy Abbott. 1803-1877. U.S.A. DSB; BMNH.

9491. HELMERSEN, Grigory Petrovich. 1803-1885. Russ. DSB; P I,III; BMNH; GA; GB.

9492. LONGUEMAR, Alphonse le Touzé de. 1803-1881. Fr. P III; BMNH; GA.

9493. SCHAFHÄUTL, Karl Franz Emil. 1803-1890. Ger. P II,III; BMNH; GC.

9494. SCHIMPER, Karl Friedrich. 1803-1867. Ger. DSB; ICB; BMNH; GC. Also Botany.

9495. SHCHUROVSKII, Grigory Efimovich. 1803-1884. Russ. P III (SCHTSCHUROWSKI); BMNH.

9496. SIMMS, Frederick Walter. 1803-1865. Br. P III; GD.

9497. SOPWITH, Thomas. 1803-1879. Br. P III; BMNH; GD.
1. *Description of a series of geological models illustrating the nature of stratification, valleys of denudation....* Newcastle-upon-Tyne, 1841.

9498. BÖBERT, Carl Friedrich. 1804-1869. Nor. P I,III; BMNH.

9499. CARNALL, Rudolf von. 1804-1874. Ger. DSB; P III; GC.

9500. FUHLROTT, (Johann) Karl. 1804-1877. Ger. DSB; ICB; P IV; BMNH.

9501. MATHER, William Williams. 1804-1859. U.S.A. DSB; P II; BMNH; GE.

9502. OWEN, Richard (first of the name). 1804-1892. Br. DSB; ICB; BMNH; Mort.; BLA; GA; GB; GD. Also Natural History* and Zoology*. See also Index.
Unless otherwise indicated, the following works were published at London.
1. *Description of the skeleton of an extinct gigantic sloth, Mylodon robustus.* 1842. RM/L.
2. *A descriptive catalogue of the fossil Mammalia and Aves in the museum of the Royal College of Surgeons.* 1845.
3. *A history of British fossil mammals and birds.* 1846. RM/L.
4. *A monograph on the Reptilia of the London clay.* 2 vols. 1848-56. With J. Bell.
5. *A history of British fossil reptiles.* 4 vols. 1849-84. RM/L.
6. *A monograph on the fossil Reptilia of the Cretaceous formations.* 4 vols. 1851-64.
7. *A monograph of the fossil Reptilia of the Wealden and Purbeck formations.* 1853-89.
8. *The geology and inhabitants of the ancient world.* 1854.
9. *Key to the geology of the globe. An essay.* Boston, 1857. RM.
10. *Paleontology; or, A systematic summary of extinct animals and their geological relations.* Edinburgh, 1860. RM/L. 2nd ed., 1861.
11. *A monograph on the Reptilia of the Kimmeridge clay and Portland stone.* 3 parts. 1861-89.
12. *A monograph of the fossil Reptilia of the Liassic formations.* 3 parts. 1861-81.
13. *A monograph on the British fossil Cetacea from the Red Crag.* 1870.
14. *A monograph of the fossil Mammalia of the Mesozoic formations.* 1871.
15. *A Cuvierian principle in palaeontology, tested by evidences of an extinct leonine marsupial.* 1871.
16. *A monograph on the fossil Reptilia of the Mesozoic formations.* 1874-89.
17. *A descriptive and illustrated catalogue of the fossil Reptilia of South Africa in the collection of the British Museum.* 1876.
18. *Researches on the fossil remains of the extinct mammals of Australia.* 2 vols. 1877.
19. *On the extinct animals of the colonies of Great Britain.* 1879.
20. *A memoir on the extinct wingless birds of New Zealand.* 2 vols. 1879.

9503. ROGERS, William Barton. 1804-1882. U.S.A. DSB; P III; BMNH; GE.

9504. THURMANN, Julius. 1804-1855. Fr./Switz. P II; BMNH.

9505. ANGELIN, Nils Peter. 1805-1876. Swed. BMNH.

9506. HAMILTON. William John. 1805-1867. Br. P III; BMNH; GD.

9507. HENWOOD, William Jory. 1805-1875. Br. P III; BMNH; GB; GD.

9508. JACKSON, Charles Thomas. 1805-1880. U.S.A. DSB; ICB; P I,III; BMNH; BLA; GA; GE. Also Mineralogy*.

1. *Remarks on the mineralogy and geology of the peninsula of Nova Scotia.* Cambridge, Mass., 1832. With F. Alger. RM.
2. *Report on the geological and agricultural survey of the state of Rhode Island.* Providence, 1840. RM.

9509. MILNE-HOME, David. 1805-1890. Br. P III.

9510. PILLA, Leopoldo. 1805-1848. It. ICB; P II; BMNH.
   1. *Studii di geologia.* Naples, 1841.

9511. PONZI, Giuseppe. 1805-1885. It. P II,III,IV; BMNH.

9512. RIVIÈRE, Alphonse Ennemond Auguste. 1805-1877. Fr. BMNH.

9513. TATE, George. 1805-1871. Br. BMNH; GD.

9514. TUOMEY, Michael. 1805-1857. U.S.A. P III; BMNH.

9515. VERNEUIL, (Philippe) Edouard (Poulletier) de. 1805-1873. Fr. DSB; P II,III; BMNH; GB. See also 9361/3.

9516. YANDELL, Lunsford Pitts. 1805-1878. U.S.A. GE.

9517. ZEJSZNER, Ludwik. 1805-1871. Cracow. DSB Supp.; P III (ZEUSCHNER); BMNH.

9518. ABICH, (Otto) Hermann Wilhelm. 1806-1886. Russ. DSB; ICB; P I,III; BMNH; GB; GC.

9519. BEUST, Friedrich Constantin von. 1806-1891. Ger./Aus. P I,III, IV; BMNH.
   1. *Geognostische Skizze der wichtigsten Porphyrgebilde zwischen Freyberg ... und Nossen.* Freiberg, 1835.

9520. BOUBÉE, Nérée. 1806-1863. Fr. P I,III; BMNH.

9521. BRYCE, James. 1806-1877. Br. P III; BMNH; GD.

9522. CZJZEK (or CZIZEK), Johann. 1806-1855. Aus. P I & Supp.
   1. *Erläuterungen zur geognostischen Karte der Umgebung Wiens.* Vienna, 1849.

9523. GRAS, Joseph Scipion. 1806-ca. 1875. Fr. P III; BMNH.

9524. GREY-EGERTON, Philip de Malpas. 1806-1881. Br. Under EGERTON: BMNH; GB; GD.

9525. LESQUEREUX, (Charles) Leo. 1806-1889. Switz./U.S.A. DSB; ICB; P III; BMNH; GE. Also Botany.

9526. MARTINS, Charles Frédéric. 1806-1889. Fr. P II,III; BMNH; BLA; GA. Also Mineralogy and Botany.

9527. ORBIGNY, (Alcide) Charles (Victor Dessalines) d'. 1806-1876. Fr. P II,III; BMNH; GA. See also 11265.

9528. REYNAUD, Jean Ernest. 1806-1863. Fr. P II,III; GA.

9529. SAWKINS, James Gay. 1806-1878. Several countries. P III; BMNH.

9530. SHARPE, Daniel. 1806-1856. Br. P II; BMNH; GB; GD.

9531. AGASSIZ, (Jean) Louis (Rodolphe). 1807-1873. Switz./U.S.A. DSB; ICB; P I,III; BMNH; BLA; GA; GB; GE. Also Natural History* and Zoology*. See also Index.
   1. *Recherches sur les poissons fossiles.* 5 vols in 2 and atlas. Neuchâtel, 1833-43.

2. *Discours prononcé à l'ouverture des séances de la Société Helvétique des Sciences Naturelles, à Neuchâtel, le 24 juillet 1837.* X.
2a. ——— *The discourse of Neuchâtel.* New York, 1967. Trans. and ed. by A.V. Carozzi.
3. *Etudes sur les glaciers.* 2 vols. Neuchâtel, 1840. RM/L. Reprinted, London, 1966.
3a. ——— *Studies on glaciers.* New York, 1967. Trans. and ed. by A.V. Carozzi.
4. *Etudes critiques sur les mollusques fossiles ... du Jura et de la craie suisse.* 4 vols. Neuchâtel, 1840-45. RM/L.
5. *Monographie des poissons fossiles du vieux grès rouge, ou système Devonien des Iles Britanniques et de Russie.* Text & atlas. Neuchâtel, 1844-45.
6. *Iconographie des coquilles tertiaires reputées identique avec les espèces vivantes.* Neuchâtel, 1845.
7. *Nouvelles études et expériences sur les glaciers actuels, leur structure, leur progression et leur action physique sur le sol.* Text & atlas. Paris, 1847. RM/L.
8. *Bibliographia zoologiae et geologiae: A general catalogue of all books, tracts and memoirs on zoology and geology.* 4 vols. London, 1848-54. (Ray Society publns. 13,18,22, 26) Corr., enl., and ed. by H.E. Strickland and W. Jardine. RM/L. Reprinted, New York, 1968.
9. *Geological sketches.* Boston, 1866. RM/L. Second series, ib., 1876. RM.

9532. BEIMA, Elte Martens. 1807-1873. Holl. P I,III.

9533. BURMEISTER, (Carl) Hermann (Conrad). 1807-1892. Ger./Argentina. ICB; P I,III; BMNH; GA; GC. Also Zoology*.
1. *Die Organisation der Trilobiten.* Berlin, 1843.
1a. ——— *The organization of trilobites, deduced from their living affinities.* London, 1846. (Ray Society publn. 8) Trans. by Bell and E. Forbes.

9534. ESCHER VON DER LINTH, Arnold. 1807-1872. Switz. P I,III; BMNH; GB; GC.

9535. FISCHER-OOSTER, Carl von. 1807-1875. Switz. P III; BMNH.

9536. GUYOT, Arnold Henry. 1807-1884. U.S.A. DSB; ICB; P I,III,IV; BMNH; GB; GE.

9537. HOHENEGGER, Ludwig. 1807-1864. Ger./Aus. P III; BMNH.

9538. MATHERON, Pierre Philippe Emile. 1807-1899. Fr. BMNH.

9539. OWEN, David Dale. 1807-1860. U.S.A. DSB; ICB; P III; BMNH; GB; GE.
1. *Mineral lands of the United States.* Washington, 1845. RM.

9540. PALMIERI, Luigi. 1807-1896. It. P II,III,IV; BMNH; GF. Also Other Earth Sciences.
1. *Incendio Vesuviano del ... 1872.* Turin/Berlin, 1872. X.
1a. ——— *The eruption of Vesuvius in 1872.* London, 1873. Trans. with introd. and notes by R. Mallet.

9541. SISMONDA, Angelo. 1807-1879. It. ICB; P II,III; BMNH.

9542. COTTA, (Carl) Bernhard von. 1808-1879. Ger. DSB; P I,III;

BMNH; GA; GB; GC.
1. *Geologische Briefe aus den Alpen*. Leipzig, 1850. RM.
2. *Geognostische Karten unseres Jahrhunderts*. Freiberg, 1850. RM.
3. *Der innere Bau der Gebirge*. Freiberg, 1851. RM.
4. *Geologische Bilder*. Leipzig, 1852. RM.
5. *Deutschlands Boden, sein geologischer Bau und dessen Einwirkung auf das Leben der Menschen*. Leipzig, 1854. X. 2nd ed., 1858. RM.
6. *Die Gesteinlehre*. Freiberg, 1855. RM/L. 2nd ed., 1862. X.
6a. ——— *Rocks classified and described. A treatise on lithology*. London, 1866. Trans. by P.H. Lawrence. RM/L. Another ed., 1893.
7. *Die Lehre von den Erzlagerstätten*. Freiberg, 1855. RM. 2nd ed., 2 vols, 1859-61. X.
7a. ——— *A treatise on ore deposits*. New York, 1870. Trans. from the 2nd German ed. by F. Prime. RM/L.
8. *Die Lehre von den Flötzformationen*. Freiberg, 1856. RM.
9. *Geologische Fragen*. Freiberg, 1858. RM.
9a. ——— *Geology and history*. 1865.
10. *Die Geologie der Gegenwart*. Leipzig, 1866. RM. 5th ed., 1878.
11. *Geologisches Repertorium*. Leipzig, 1877. (*Beiträge zur Geschichte der Geologie*. Abt. 1) RM.

9543. EHRLICH, Franz Carl. 1808-1886. Aus. P III; BMNH.

9544. FALCONER, Hugh. 1808-1865. Br./India. DSB; BMNH; GB; GD.
1. *Palaeontological memoirs and notes*. 2 vols. London, 1868. Ed. with a biographical sketch by Charles Murchison.

9545. GODWIN-AUSTEN, Robert Alfred Cloyne. 1808-1884. Br. DSB; P III; BMNH; GB; GD. See also 13235/2.

9546. MOORE, Francis. 1808-1864. U.S.A. ICB.

9547. PERREY, Alexis. ca. 1808-1882. Fr. P III; BMNH.

9548. PHILIPPI, Rudolph Amandus. 1808-1904. Ger./Chile. P II,III, IV & Supp.; BMNH. Also Botany amd Zoology.

9549. ROGERS, Henry Darwin. 1808-1866. U.S.A./Br. DSB; P III; BMNH; GB; GE.

9550. SCHIMPER, Wilhelm Philipp. 1808-1880. Strasbourg. DSB; BMNH; GC. Also Botany.
1. *Monographie des plantes fossiles du grès bigarré de la chaîne des Vosges*. Leipzig, 1844. With A. Mougeot.
2. *Traité de paléontologie végétale*. 3 vols & atlas. Paris, 1869-74.

9551. SEXE, Sjurd Aamundsen. 1808-1888. Nor. P II,III,IV.

9552. STARING, Winand Carel Hugo. 1808-1877. Holl. BMNH; GF.

9553. STREFFLEUR, Valentin. 1808-1870. Aus. P II,III; GC.

9554. TENNANT, James. 1808-1881. Br. BMNH; GD. Also Mineralogy*.
1. *A stratigraphical list of British fossils*. London, 1847.

9555. WHITTLESEY, Charles. 1808-1886. U.S.A. ICB; P III,IV; BMNH.

9556. BAUDIN, Desiré Pierre. 1809-1870. Fr. P III; BMNH.

9557. BIANCONI, Gian Giuseppe. 1809-1878. It. P III; BMNH. Also Zoology.

9558. BOEHTLINGK, Wilhelm. 1809-1841. Russ. P I.

9559. BURAT, Amédée. 1809-1883. Fr. P III; BMNH. Also Mineralogy*.
1. *Description des terrains volcaniques de la France centrale.* Paris, 1833. RM.

9560. CORDA, August Carl Joseph. 1809-1849. Ger. BMNH; GC. Also Botany*.
1. *Beiträge zur Flora der Vorwelt.* Prague, 1845.

9561. COSTA, Francisco Antonio Pereira da. 1809-1889. Port. P III; BMNH (PEREIRA DA COSTA).

9562. CREDNER, Carl Friedrich Heinrich. 1809-1876. Ger. P III; BMNH; GB; GC.

9563. DARWIN, Charles Robert. 1809-1882. Br. DSB; ICB; P III; BMNH; G. Also Natural History*, Botany*, Zoology*, Evolution*#, and Autobiography*. See also Index.
1. *The structure and distribution of coral reefs. Being the first part of "The geology of the voyage of the 'Beagle' ... during the years 1832 to 1836."* London, 1842. RM/L. Another ed., by J.W. Williams, 1842. 2nd ed., 1874. Another ed., with an introd. by J.W. Williams, [188-?]. 3rd ed., with an appendix by T.G. Bonney, 1889.
2. *Geological observations on the volcanic islands visited during the voyage of H.M.S. "Beagle", together with some brief notes on the geology of Australia and the Cape of Good Hope. Being the second part of "The geology of the voyage of the 'Beagle' ... during the years 1832-1836."* London, 1844. RM/L.
3. *Geological observations on South America. Being the third part of "The geology of the voyage of the 'Beagle' ... during the years 1832 to 1836."* London, 1846. RM/L.
3a. ——— 2nd ed. *Geological observations on the volcanic islands and parts of South America visited during the voyage of the "Beagle."* London, 1876. 3rd ed., New York, 1897.
4. *Geological observations on coral reefs, volcanic islands, and on South America. Being the geology of the voyage of the "Beagle" ... 1832 to 1836.* 3 parts in 1 vol. London, 1851. A re-issue, with a covering title, of items 1-3.
4a. ——— Another ed. With a critical introduction to each work by J.W. Judd. London, 1890.
4b. ——— Reprint of Part 1. *The structure and distribution of coral reefs.* Berkeley, Cal., 1962 and 1977.

9564. DUMONT, André Hubert. 1809-1857. Belg. ICB; P I; BMNH; GB.

9565. DUNKER, Wilhelm Bernhard Rudolph Hadrian. 1809-1885. Ger. P I,III; BMNH. Also Zoology.
1. *Monographie der norddeutschen Wealdenbildung. Ein Beitrag zur Geognosie und Naturgeschichte der Vorwelt.* Brunswick, 1846.

9566. FORBES, James David. 1809-1868. Br. DSB; P I,III; BMNH; GB; GD. Also Physics. See also 15488/1.

1. *Travels through the Alps of Savoy ... with observations on the phenomena of glaciers.* Edinburgh, 1843. RM/L. 2nd ed., 1845.
2. *Norway and its glaciers.* Ib., 1853. RM.
3. *Occasional papers on the theory of glaciers.* Ib., 1859. RM/L.
4. *An index to the correspondence and papers.* St. Andrews, 1968. Compiled at St. Andrews University Library.

9567. GIBBES, Robert Wilson. 1809-1866. U.S.A. BMNH; BLA; GE.

9568. GRUNER, Emmanuel Louis. 1809-1883. Fr. P I,III.

9569. HEER, Oswald. 1809-1883. Switz. DSB; P I,III; BMNH; GB; GC. Also Botany and Zoology.

9570. HOUGHTON, Douglass. 1809-1845. U.S.A. DSB; P I; BMNH; GE.

9571. KING, William. 1809-1886. Br. BMNH; GD.
1. *A monograph of the Permian fossils of England.* London, 1850.

9572. KONINCK, Laurent Guillaume. 1809-1887. Belg. DSB; P I,III,IV; BMNH; GB. Also Chemistry.
1. *Description des animaux fossiles qui se trouvent dans le terrain carbonifère de Belgique.* 2 vols. Liége, 1842-44.
2. *Recherches sur les fossiles paléozoiques de la Nouvelle Galles du Sud (Australie).* Brussels, 1876-77.

9573. LE HON, Henri Sébastien. 1809-1872. Belg. BMNH; GF.
1. *Périodicité des grands déluges ... Théorie prouvée par les faits géologiques.* Brussels, 1858. X. 2nd ed., 1861.

9574. PICTET (DE LA RIVE), François Jules. 1809-1872. Switz. BMNH; GA; GB.
1. *Traité élémentaire de paléontologie.* 4 vols in 2. Geneva/Paris, 1844-46. 2nd ed., 4 vols & atlas, 1853-57.

9575. QUENSTEDT, Friedrich August von. 1809-1889. Ger. DSB; P II, III; BMNH; GB; GC. Also Mineralogy.
1. *Handbuch der Petrefaktenkunde.* 2 vols. Tübingen, 1852.

9576. RÖMER, Friedrich Adolph. 1809-1869. Ger. DSB; P II,III; BMNH; GB; GC.
1. *Die Versteinerungen des norddeutschen Oolithen-Gebirges.* Hanover, 1836.
2. *Die Versteinerungen des norddeutschen Kreidegebirges.* Ib., 1841.
3. *Die Versteinerungen des Harzgebirges.* Ib., 1843.

9577. SARTORIUS VON WALTERSHAUSEN, Wolfgang. 1809-1876. Ger. P II, III; BMNH; GB (WALTERSHAUSEN); GC.

9578. STROMBECK, August von. 1809-1900. Ger. P II,III,IV.

9579. WRIGHT, Thomas. 1809-1884. Br. P III; BMNH; GB; GD. See also 9397/1a.

9580. GORINI, Paolo. d. 1881. It. P III; BMNH.

9581. BELGRAND, Marie François Eugène. 1810-1878. Fr. P III; BMNH.

9582. HAWKINS, Thomas. 1810-1889. Br. BMNH; GD.
1. *The book of the great sea dragons, Ichthyosauri and Plesiosauri, extinct monsters of the ancient earth.* London, 1840.

9583. KRUG VON NIDDA, Otto Ludwig. 1810-1885. Ger. P III,IV; GC (NIDDA).

9584. MALLET, Robert. 1810-1881. Br. DSB; P III; BMNH; GB; GD.
1. *Earthquake catalogue of the British Association.* London, 1858. With J.W. Mallet.
2. *The great Neapolitan earthquake of 1857.* 2 vols. Ib., 1862.
See also 9540/1a.

9585. MORRIS, John. 1810-1886. Br. BMNH; GB; GD.
1. *Catalogue of British fossils.* London, 1843. X. 2nd ed., 1854.
2. *A new geological chart, showing at one view the order of succession of the stratified rocks.* Ib., 1859.

9586. NICOL, James. 1810-1879. Br. P III; BMNH; GB; GD. Also Mineralogy*. See also 9361/4.
1. *Guide to the geology of Scotland.* Edinburgh, 1854.

9587. ORMEROD, George Wareing. 1810-1891. Br. P III,IV; BMNH; GD.

9588. OWEN, Richard (second of the name). 1810-1890. U.S.A. P III, IV; BMNH; GB.

9589. PETZHOLDT, (George Paul) Alexander. 1810-1889. Ger./Dorpat. P II,III; BMNH.
1. *Uber Calamiten und Steinkohlenbildung.* Dresden/Leipzig, 1841.

9590. SCACCHI, Arcangelo. 1810-1893. It. P II,III,IV; BMNH. Also Zoology.

9591. SENFT, (Christian Carl Friedrich) Ferdinand. 1810-1893. Ger. P II,III; BMNH.
1. *Classification und Beschreibung der Felsarten.* Breslau, 1857.
2. *Die krystallinischen Felsgemengtheile.* Berlin, 1868.

9592. THEOBALD, Gottfried Ludwig. 1810-1869. Ger./Switz. P III; BMNH; GC. Also Natural History.

9593. BOUCHEPORN, René Charles Félix Bertrand de. 1811-1857.
1. *Etudes sur l'histoire de la terre et sur les causes des révolutions de sa surface.* Paris, 1844. RM.

9594. DESOR, (Pierre Jean) Edouard. 1811-1882. Switz./U.S.A. DSB; ICB; P III; BMNH; GB.
1. *Excursions et séjours dans les glaciers et les hautes régions des Alpes de M. Agassiz et ses compagnons.* Neuchâtel, 1844. RM.
2. *Nouvelles excursions et ... de M. Agassiz....* Ib., 1845. RM.
3. *Le paysage morainique, son origine glaciaire et ses rapports avec les formations pliocènes d'Italie.* Paris, 1875. RM.

9595. EWALD, Julius Wilhelm. 1811-1891. Ger. P I,III; BMNH; GC. See also 9235/5.

9596. HALL, James (second of the name). 1811-1898. U.S.A. DSB; ICB; P III,IV; BMNH; GB; GE.
1. *Geology of New York.* 4 vols. Albany, 1842-43.
2. *Palaeontology of New York.* 8 vols in 13. Ib., 1847-94.

9597. JUKES, Joseph Beete. 1811-1869. Br. DSB; P III; BMNH; GB; GD.
1. *The student's manual of geology.* Edinburgh, 1857. 2nd ed., 1862. 3rd ed., by A. Geikie, 1872.
2. *Letters and extracts from the addresses and occasional writings.* London, 1871. Ed. by his sister.

9598. MENEGHINI, Giuseppe. 1811-1889. It. P III; BMNH. Also Botany.

9599. REUSS, August Emmanuel von. 1811-1873. Prague/Aus. DSB; P II, III; BMNH; BLA; GB; GC.
1. *Die Versteinerungen der böhmischen Kreideformation.* 2 parts. Stuttgart, 1845-46.

9600. STRICKLAND, Hugh Edwin. 1811-1853. Br. P II; BMNH; GB; GD. Also Zoology*. See also Index.

9601. WALSER, Franz Xaver. 1811-after 1859. Ger. P III.

9602. WATELET, Adolphe. 1811-1879. Fr. BMNH.
1. *Description des plantes fossiles du bassin de Paris.* Paris, 1866.

9603. BACH, Karl Philipp Heinrich. 1812-1870. Ger. P III; BMNH.

9604. BINNEY, Edward William. 1812-1881. Br. P III; BMNH; GB; GD.
1. *Observations on fossil plants found in the Carboniferous strata.* 4 vols. London, 1868-75.

9605. BISCHOF, Karl. 1812-1884. Ger. P III; BMNH.

9606. CASTELNAU, Francis de. 1812-1880. Fr. P III; BMNH. Also Zoology*.
1. *Essai sur le système silurien de l'Amérique septentrionale.* Paris, 1843.

9607. CHIKHACHEV, Pierre de. 1812-1890. Fr. P II,III (TSCHICHATSCHEW). Under TCHIHATCHEFF: BMNH; GB. Also Natural History.

9608. EVANS, John. 1812-1861. U.S.A. ICB; P III.

9609. HÉBERT, Edmond. 1812-1890. Fr. P III,IV; BMNH; GB.

9610. KUTORGA, Stepan Semenovich. 1812-1861. Russ. P III; BMNH. Also Microbiology*.
1. *Beiträge zur Geognosie und Palaeontologie Dorpats.* St. Petersburg, 1835.

9611. LUDWIG, Rudolph August Birminhold Sebastian. 1812-1880. Ger. P I,III; BMNH.

9612. MICHELOTTI, Giovanni. 1812-1889 (or 1814-1898). It. P II; BMNH.

9613. PENGELLY, William. 1812-1894. Br. P III; BMNH; GB; GD.

9614. PISSIS, (Pierre Joseph) Aimé. 1812-1889. Fr./Chile. P II; BMNH.

9615. PRESTWICH, Joseph. 1812-1896. Br. DSB; P III,IV; BMNH; GB; GD1. See also 15520/1.
1. *A geological inquiry respecting the water-bearing strata of the country around London.* London, 1851. RM/L. 2nd ed., 1895.
2. *The past and future of geology.* Ib., 1875.

3. *Geology; chemical, physical and stratigraphical.* 2 vols. Oxford, 1886-88. RM/L.
4. *On certain phenomena belonging to the close of the last geological period, and on their bearing upon the tradition of the flood.* London, 1895. RM/L.
5. *Collected papers on some controverted questions of geology.* London, 1895.

9616. SANTAGATA, Domenico. 1812-after 1880. It. P III.

9617. BOSQUET, Joseph Augustin Hubert. 1813-1880. Belg./Holl. BMNH.

9618. COQUAND, Henri. 1813-1881. Fr. P III; BMNH.
1. *Traité des roches considérées au point de vue de leur origine, de leur composition, de leur gisement et de leurs applications.* Paris, 1857.

9619. DANA, James Dwight. 1813-1895. U.S.A. DSB; ICB; P I,III,IV; BMNH; GB; GE. Also Mineralogy* and Zoology*. See also Index.
1. *On coral reefs and islands.* New York, 1853. RM.
2. *Manual of geology.* Philadelphia/London, 1863. 2nd ed., 1865 and re-issues. 3rd ed., 1879. 4th ed., (date?).
2a. ——— Abridgment: *A text-book of geology.* New York, 1871. 3rd ed., 1877. 4th ed., 1883. 5th ed., 1897. Title varies.
3. *Corals and coral islands.* London, 1872. 2nd ed., 1875. Another ed., 1885.
4. *The geological story briefly told.* New York, 1875.
5. *Characteristics of volcanoes.* New York/London, 1890. RM/L.

9620. GUTBERLET, Wilhelm Karl Julius. 1813-1864. Ger. P III; BMNH; GC.

9621. LEICHHARDT, (Friedrich Wilhelm) Ludwig. 1813-1848. Australia. ICB; P III; BMNH; GC; GD.
1. *Beiträge zur Geologie von Australien.* Halle, 1855.
2. *The letters.* 3 vols. London, 1968. Trans. and ed. by M. Aurousseau.

9622. NYST, (Henri Joseph) Pierre. 1813-1880. Belg. BMNH; GF.
1. *Description des coquilles et des polypiers fossiles des terrains tertiaires de la Belgique.* Brussels, 1843.

9623. OZERSKY, Aleksandr Dmitrievich. 1813-1880. Russ. DSB; BMNH.

9624. RIBEIRO SANTOS, Carlos. 1813-1882. Port. DSB; P III; BMNH.

9625. RICHTER, Reinhard. 1813-1884. Ger. P II,III; BMNH; GC.

9626. WORTHEN, Amos Henry. 1813-1888. U.S.A. P III; BMNH; GE.

9627. ZIGNO, Achille de. 1813-1892. It. P III; BMNH.
1. *Flora fossilis formationis oolithicae. Le plante fossili dell'oolite.* 2 vols. Padua, 1856-?

9628. ADAMS, Charles Baker. 1814-1853. U.S.A. P III; BMNH; GE. Also Zoology.

9629. ANSTED, David Thomas. 1814-1880. Br. ICB; P III; BMNH; GB; GD.
1. *Geology; introductory, descriptive and practical.* 2 vols. London, 1844.
2. *A geologist's text-book.* Ib., 1845.

3. *The ancient world.* Ib., 1847. 2nd ed., 1848.
4. *The gold-seeker's manual.* Ib., 1849. RM.
5. *Elementary course of geology, mineralogy and physical geography.* Ib., 1850. X. 2nd ed., 1856.
6. *Scenary, science and art, being extracts from the notebook of a geologist and mining engineer.* Ib., 1854. RM.
7. *The natural history of inanimate creation.* Ib., 1856. Another ed., 3 vols, 1860.
8. *Geological gossip.* Ib., 1860.
9. *The great stone book of nature.* Ib., 1863.
10. *Applications of geology to the arts and manufactures.* Ib., 1865. RM/L.
11. *Physical geography.* Ib., 1867. 5th ed., 1871.
12. *The world we live in.* Ib., 1870.

9630. ATHERSTONE, William Guybon. 1814-1898. Br./South Africa. ICB; GB; GF.

9631. DAUBRÉE, (Gabriel) Auguste. 1814-1896. Fr. DSB; P I,III,IV; BMNH; GB.
1. *Etudes et expériences synthétiques sur le métamorphisme et sur la formation des roches cristallines.* Paris, 1859.
2. *Substances minérales.* Paris, 1867.
3. *Etudes synthétiques de géologie expérimentale.* 2 vols. Paris, 1879. RM/L.
4. *Les météorites et la constitution géologique du globe terrestre.* Paris, 1886. RM.
5. *Les eaux souterraines à l'époque actuelle.* Paris, 1887. RM.
6. *Les eaux souterraines aux époques anciennes.* Paris, 1887. RM.

9632. DAVIES, William. 1814-1891. Br. BMNH; GD1.

9633. DEVILLE, Charles Joseph Sainte-Claire. 1814-1876. Fr. Under SAINTE-CLAIRE-DEVILLE: P II,III; BMNH; GB. Also Meteorology.

9634. ERDMANN, Axel Joakim. 1814-1869. Swed. P I,III; BMNH.

9635. GALEOTTI, Henri Guillaume. 1814-1858. Belg. P I; BMNH; GF. Also Botany.

9636. GEINITZ, Hanns Bruno. 1814-1900. Ger. P I,III,IV; BMNH; GB.
1. *Grundriss der Versteinerungskunde.* 2 vols. Dresden/Leipzig, 1845-46. 2nd ed., 1856.
2. *Die Versteinerungen der Steinkohlenformation in Sachsen.* Leipzig, 1855.
3. *Geognostische Darstellung der Steinkohlenformation in Sachsen mit besonderer Berücksichtigung des Rothliegenden.* Text & atlas. Leipzig, 1856.
4. *Die Leitpflanzen des Rothliegenden und des Zechsteingebirges, oder der Permischen Formation in Sachsen.* Dresden, 1858.

9637. GIRARD, (Carl Adolph) Heinrich. 1814-1878. Ger. P I,III; BMNH.

9638. GRESSLY, Amanz. 1814-1865. Switz. DSB; P III; BMNH; GC.

9639. LECKENBY, John. 1814-1877. Br. P III.

9640. PAGE, David. 1814-1879. Br. P III; BMNH; GD.
1. *Rudiments of geology.* 1851.
2. *Introductory text-book of geology.* Edinburgh/London, 1854.

X. 6th ed., 1864. 11th ed., 1877.
3. *Advanced text-book of geology.* Ib., 1856. X. 3rd ed., 1861. 6th ed., 1876.
4. *Handbook of geological terms.* Ib., 1859. 2nd ed , 1865.
5. *The past and present life of the globe.* Ib., 1861.
6. *The philosophy of geology.* Ib., 1863.
7. *Advanced text-book of physical geography.* Ib., 1864.
8. *Geology for general readers. A series of popular sketches.* Ib., 1866. X. 2nd ed., 1866. 3rd ed., 1870.
9. *Chips and chapters ... for geologists.* Ib., 1869.
10. *Economic geology; or, Geology in its relations to the arts and manufactures.* Ib., 1874.

9641. RAMSAY, Andrew Crombie. 1814-1891. Br. DSB; P III; BMNH; GB; GD. Also Mineralogy*. See also 15524/2.
1. *Old glaciers of Switzerland and North Wales.* London, 1860.
2. *The physical geology and geography of Great Britain.* Ib., 1863. X. 3rd ed., 1872. 5th ed., 1878.

9642. ROUILLIER, Karl Frantsovich. 1814-1858. Russ. DSB. Also Zoology.

9643. AUERBACH, Johann Alexander. 1815-1867. Russ. P III; BMNH.

9644. BEYRICH, Heinrich Ernst. 1815-1896. Ger. DSB; P I,III,IV; BMNH; GB; GC.

9645. BRODIE, Peter Bellinger. 1815-1897. Br. BMNH; GB.
1. *A history of fossil insects in the secondary rocks of England.* London, 1845. RM/L.

9646. EMMRICH, Hermann Friedrich. 1815-1879. Ger. P I,III; BMNH.

9647. FAVRE, (Jean) Alphonse. 1815-1890. Switz. P III,IV; BMNH; GB.

9648. FORBES, Edward. 1815-1854. Br. DSB; ICB; P III; BMNH; GB; GD. Also Oceanography and Zoology*. See also Index.

9649. FOSTER, John Wells. 1815-1873. U.S.A. P III; BMNH.

9650. FRAPOLI, Louis. 1815-after 1848. Fr. P III.

9651. GIBBS, George (second of the name). 1815-1873. U.S.A. ICB; P III; BMNH; GE. Also Zoology.

9652. HÖRNES, Moritz. 1815-1868. Aus. P I,III; BMNH; GB (HORNES); GC.

9653. HOLMES, Francis Simmons. 1815-after 1882. U.S.A. P III; BMNH. Also Zoology.

9654. HÓRBYE, Jens Carl. 1815-after 1881. Nor. P III.

9655. KEYSERLING, Alexander Andreevich (or Alexander F.M.L.A. von). 1815-1891. Russ. DSB; P III,IV; BMNH. Also Botany. See also 9361/3.

9656. MOORE, Charles (first of the name). 1815-1881. Br. BMNH; GD.

9657. RAULIN, Félix Victor. 1815-1905. Fr. P II,III,IV; BMNH. Also Meteorology.

9658. SCHMID, Ernst Erhard Friedrich Wilhelm. 1815-1885. Ger. P II, III; BMNH; GC.

9659. SHARP, Samuel. 1815-1882. Br. P III; BMNH; GD.

9660. SISMONDA, Eugenio. 1815-1870. It. BMNH.

9661. VUKOTINOVIĆ, Ljudevit. 1815-1893. Agram. P III; BMNH. Also Botany.

9662. WISSMANN, Heinrich Ludolf. 1815-after 1842. Ger. P II; BMNH.

9663. ALLPORT, Samuel. 1816-1897. Br. BMNH; GB.

9664. BUCHMAN, James. 1816-1884. Br. BMNH; GD. Also Botany.
1. *The ancient straits of Malvern. An essay on the former marine conditions which separated England and Wales.* London, [1849?].
See also 9361/1.

9665. HARKNESS, Robert. 1816-1878. Br. P III; GB; GD.

9666. LEONHARD, Gustav von. 1816-1878. Ger. P I,III; BMNH.

9667. MYLNE, Robert William. 1816-1890. Br. BMNH; GD.
1. *Map of the geology and contours of London and its environs.* London, 1856.

9668. OLDHAM, Thomas B. 1816-1878. Br. DSB; P III; BMNH; GB; GD.
1. *Geological glossary for the use of students.* London, 1879. Ed. by R.D. Oldham.

9669. SILLIMAN, Benjamin (second of the name). 1816-1885. U.S.A. DSB; P II,III; BMNH; BLA; GE. Also Chemistry*.

9670. SONKLAR (VON INNSTÄDTEN), Karl Albrecht. 1816-1885. Aus. P III; BMNH; GC.
1. *Allgemeine Orographie. Die Lehre von den Relief-Formen der Erdoberfläche.* Vienna, 1873. RM.

9671. WILLIAMSON, William Crawford. 1816-1895. Br. DSB; BMNH; GB; GD. Also Botany.

9672. ZURCHER, Frédéric. 1816-1890. BMNH.
1. *Volcans et tremblements de terre.* Paris, 1866. With E. Margollé. X.
1a. ——— *Volcanoes and earthquakes.* London, 1868. Trans. by Mrs. N. Lockyer. Another ed., 1874.

9673. ANDRAE, Carl Justus. 1817-1885. Ger. P III; BMNH.
1. *Vorweltliche Pflanzen aus dem Steinkohlengebirge der preussischen Rheinlande und Westphalens.* 3 parts. Bonn, 1865-69.

9674. BOLL, Ernst (Friedrich August). 1817-1868. Ger. P III; BMNH. Also Natural History.

9675. BRISTOW, Henry William. 1817-1889. Br. P III,IV; BMNH; GB; GD1. Also Mineralogy*. See also 9708/1a.

9676. DAVIDSON, Thomas. 1817-1885. Br. BMNH; GB; GD.

9677. DEFFNER, Carl. 1817-1877. Ger. P III; BMNH.

9678. DELESSE, Achille Ernest Oscar Joseph. 1817-1881. Fr. P I,III; BMNH; GB.

9679. DUROCHER, Joseph Marie Elisabeth. 1817-1860. Fr. P I,III; BMNH; GB.

9680. FISHER, Osmond. 1817-1914. Br. P III,IV; BMNH.
1. *Physics of the earth's crust*. London, 1881. 2nd ed., 1889.

9681. HONEYMAN, David. 1817-after 1884. Canada. P III,IV.

9682. JACQUOT, André Eugène. 1817-after 1891. Fr. P I,III,IV; BMNH.

9683. MEEK, Fielding Bradford. 1817-1876. U.S.A. DSB; P III; BMNH; GB; GE.

9684. SMYTH,Warington Wilkinson. 1817-1890. Br. P III; BMNH; GB; GD.

9685. SWALLOW, George Clinton. 1817-1899. U.S.A. BMNH; GE.

9686. TRAUTSCHOLD, Hermann von. 1817-1902. Russ. P II,III,IV; BMNH.

9687. VOGT, Carl Christoph. 1817-1895. Ger./Switz. DSB; P II,III; BMNH; BLA; GB; GC. Also Zoology*.
1. *Lehrbuch der Geologie und Petrefactenkunde*. 2 vols. Brunswick, 1846-47. 2nd ed., 1854. 4th ed., 1879.

9688. WIES, Nicolas. 1817-1879. Luxembourg. P III; BMNH.

9689. ZADDACH, Ernst Gustav. 1817-1881. Ger. P II,III; BMNH. Also Zoology.

9690. BALL, John. 1818-1889. Br. ICB; P III,IV; BMNH; GB; GD1. Also Botany.

9691. BELLARDI, Luigi. 1818-1889. It. DSB; BMNH. Also Zoology.

9692. COOK, George Hammell. 1818-1889. U.S.A. P III,IV; BMNH; GE.

9693. COTTEAU, (Honoré) Gustave. 1818-1894. Fr. P III,IV; BMNH.

9694. GASTALDI, Bartolomeo. 1818-1879. It. P III; BMNH.

9695. GREEN, William Lowthian. 1818?-1890. Br. BMNH.
1. *Vestiges of the molten globe, as exhibited in the figure of the earth, volcanic action, and physiography*. 2 vols. London, 1875-87.

9696. HINGENAU, Otto Bernhard von. 1818-1872. Aus. P I; P III (HINGEMAN); GC.

9697. HOLMBERG, Henrik Johan. 1818-1864. Fin. P III; BMNH.

9698. JOHNSTRUP, Johannes Frederik. 1818-1894. Den. P III; BMNH.

9699. KENNGOTT, (Johann Gustav) Adolf. 1818-1897. Ger./Aus./Switz. P I,III,IV; BMNH; GB. Also Mineralogy. See also 3023.
1. *Handwörterbuch der Mineralogie, Geologie und Palaeontologie*. 3 vols. Breslau, 1882-87.

9700. RÖMER, (Carl) Ferdinand von. 1818-1891. Ger. DSB; P II,III; BMNH; GB; GC; GE.
1. *Das rheinische Übergangsgebirge. Eine paläontologisch-geognostische Darstellung*. Hanover, 1844.
2. *Die Kreidebildungen von Texas und ihre organischen Einschlüsse*. Bonn, 1852. RM.
3. *Lethaea geognostica; oder, Beschreibung und Abbildung der für die Gebirgs-Formationen bezeichnendsten Versteinerungen*. Stuttgart, 1876-. Continued after his death by F. Frech et al. For the predecessor of this work see 9452/4.

4. *Die Knochenhöhlen von Ojcow in Polen.* Cassel, 1883. X.
4a. ——— *The bone caves of Ojcow in Poland.* London, 1884. Trans. by J.E. Lee.

9701. ROTH, Justus Ludwig Adolph. 1818-1892. Ger. DSB; P II,III,IV; BMNH; GB; GC.
1. *Allgemeine und chemische Geologie.* 3 vols. Berlin, 1879-83.

9702. SYMONDS, William Samuel. 1818-1887. Br. P III; BMNH; GB; GD.
1. *Old bones; or, Notes for young naturalists.* London, 1861.
2. *Records of the rocks; or, Notes on the geology ... of North and South Wales, Devon and Cornwall.* London, 1872.

9703. ZERRENNER, Karl Michael. 1818-1878. Ger./Russ./Aus. P II, III; BMNH; GC.

9704. ALTH, (F.) Alois von. 1819-1886. Cracow. P III; BMNH.

9705. BAILY, William Hellier. 1819-1888. Br. BMNH; GB.

9706. ETHERIDGE, Robert (first of the name). 1819-1903. Br. BMNH; GB; GD2.
1. *Fossils of the British Islands, stratigraphically and zoologically arranged.* Vol. 1. [No more publ.] Oxford, 1888.
See also 9458/5 and 9788/1.

9707. FEISTMANTEL, Karl. 1819-1885. Prague. P III,IV; BMNH.

9708. FIGUIER, Louis Guillaume. 1819-1894. Fr. P I,III; BMNH; GA.
Other entries: see 3332.
1. *La terre avant le déluge.* Paris, 1863. X. 6th ed., 1872.
1a. ——— *The world before the deluge.* London, 1865. Other eds: 1866, 1867, 1869 (rev., with additions, by H.W. Bristow), 1882, 1891.
2. *La terre et les mers; ou, Description physique du globe.* Paris, 1864. X. 3rd ed., 1866. 4th ed., 1872.
2a. ——— *The earth and sea.* 1870.

9709. GREWINGK, Constantin Caspar Andreas. 1819-1887. Russ. P III, IV; BMNH; GC.

9710. JONES, Thomas Rupert. 1819-1911. Br. BMNH; GB; GD2.

9711. KINGSLEY, Charles. 1819-1875. Br. The writer. ICB; BMNH; GA; GB; GD. Also Science in General* and Zoology*.
1. *Town geology.* London, 1873.

9712. KUDERNATSCH, Johann. 1819-1856. Aus. P I & Supp.; BMNH.

9713. LESLEY, (Joseph) Peter. 1819-1903. U.S.A. DSB; P III; BMNH; GB; GE.

9714. MÜLLER, Albrecht. 1819-1890. Switz. P II,III; BMNH.

9715. NAUCK, Ernst Friedrich. 1819-1875. Ger./Riga. P III.

9716. PICHLER (VON RAUTENKAR), Adolph. 1819-1900. Aus. P II,III,IV; BMNH.

9717. PILLET, Louis. 1819-1894. Fr. P III; BMNH.

9718. WHITNEY, Josiah Dwight. 1819-1896. U.S.A. DSB; P II,III; BMNH; GB; GE.

1. *Climatic changes of later geological times.* Cambridge, Mass., 1882.

9719. VILLE, Ludovic. d. 1877. Algeria. P III; BMNH.

9720. ANDREWS, Ebenezer Baldwin. 1820-1880. U.S.A. P III; BMNH.

9721. BÉGUYER DE CHANCOURTOIS, Alexandre Emile. 1820-1886. Fr. DSB. Under CHANCOURTOIS: P III; BMNH.

9722. BILLINGS, Elkanah. 1820-1876. Canada. DSB; P III; BMNH.
1. *Palaeozoic fossils.* 3 vols. Montreal, 1865-84. With J.F. Whiteaves.

9723. CAMPBELL, John L. 1820-1886. U.S.A. P III,IV.

9724. CASTILLO, Antonio del. 1820-1895. Mexico. BMNH.

9725. DAWSON, (John) William. 1820-1899. Canada. DSB; P III,IV; BMNH. Also Science in General* and Evolution*.
Unless otherwise stated, the following works were published at London (and many of them simultaneously at Montreal as well.)
1. *Acadian geology: An account of the geological structure and mineral resources of Nova Scotia.* Edinburgh, 1855. RM. 2nd ed., 1868. Supplement, 1878.
2. *The story of the earth and man.* 1873. X. 7th ed., 1882. 9th ed., 1887. Another ed., 1898.
3. *Life's dawn on earth. Being the history of the oldest known fossil remains....* 1875. RM/L. Another ed., 1885.
4. *The origin of the world, according to Revelation and science.* 1877. 5th ed., 1888.
5. *The chain of life in geological times.* 1880. X. 3rd ed., 1888.
6. *Fossil men and their modern representatives.* 1880. RM. 3rd ed., 1888.
7. *The geological history of plants.* 1888.
8. *Modern ideas of evolution as related to Revelation and science.* 1890.
9. *Some salient facts in the science of the earth.* 1893. RM/L.
10. *The meeting-place of geology and history.* 1894.
11. *The Canadian ice age.* Montreal, 1893. RM. Another ed., New York, 1894.
12. *Eden lost and won.* 1895.
13. *Relics of primeval life.* 1897. RM/L.

9726. GIEBEL, Christoph Gottfried Andreas. 1820-1881. P I,III; BMNH; GB; GC. Also Zoology*.

9727. MEYN, (Claus Christian) Ludwig. 1820-1878. Ger. P II,III; BMNH; GC.

9728. MORLOT, Charles Adolphe. 1820-1867. Aus./Switz. P III; BMNH; GC.

9729. ROMINGER, Carl. 1820-1907. Ger./U.S.A. P III; BMNH.

9730. SALTER, John William. 1820-1869. Br. BMNH; GD.

9731. SHUMARD, Benjamin Franklin. 1820-1869. U.S.A. P III; BMNH.

9732. TYNDALL, John. 1820-1893. Br. DSB; ICB; P II,III,IV; GB; GD.

Also Science in General*, Physics*, Microbiology*, and Biography*. See also Index.
1. *The glaciers of the Alps*. London, 1860. RM/L. Another ed., 1896.
2. *Hours of exercise in the Alps*. New York, 1871.
3. *Forms of water in clouds and rivers, ice and glaciers*. London, 1872. RM/L. 3rd ed., 1873. 5th ed., 1875. 9th ed., 1885. 10th ed., 1889.

9733. WINKLER, Gustav Georg. 1820-1896. Ger. P III; BMNH; GC.

9734. CHAPMAN, Edward John. 1821-1904. Br./Canada. P I,III,IV; BMNH; GD2. Also Mineralogy*.

9735. CROLL, James. 1821-1890. Br. DSB; P III,IV; BMNH; GB; GD1. Also Evolution* and Autobiography*.
1. *Climate and time in their geological relations: A theory of secular changes in the earth's climate*. London, 1875. RM/L.
2. *Discussions on climate and cosmology*. Edinburgh, 1885.
3. *Stellar evolution and its relations to geological time*. London, 1889. RM/L.

9736. DUNCAN, Peter Martin. 1821 (or 1824)-1891. Br. P III,IV; BMNH; GB; GD1. Also Zoology* and Biography*.

9737. HAUGHTON, Samuel. 1821-1897. Br. P III,IV; BMNH; Mort.; GB; GD1. Other entries: see 3336.
1. *Manual of geology*. London, 1865. Another ed., 1871. 4th ed., 1876.
2. *Six lectures on physical geography*. Dublin, 1880.

9738. HOWSE, Richard. 1821-1901. Br. BMNH.

9739. POMEL, Nicolas Auguste. 1821-1898. Fr./Algeria. P III,IV; BMNH.

9740. SANDBERGER, B.E.F. Guido. 1821-1869. Ger. P II,III; BMNH. Also Zoology.

9741. VÉZIAN, Jacques Marie Alexandre. 1821-after 1903. Fr. P III,IV.

9742. WEISS, Friedrich. 1821-1868. Ger. P II,III; BMNH; GC.

9743. WOODWARD, Samuel Pickworth. 1821-1865. Br. BMNH; GD. Also Zoology*.

9744. CAMPBELL, John Francis. 1822-1885. Br. ICB; P III; BMNH; GB; GD.
1. *Frost and fire. Natural engines, tool-marks and chips*. 2 vols. Edinburgh, 1865.

9745. CLOSE, Maxwell Henry. 1822-1903. Br. GB; GD2.

9746. CONDON, Thomas. 1822-1907. U.S.A. GE.

9747. GAUDIN, Charles Théodore. 1822-1866. Switz. P III; BMNH.

9748. HAUER, Franz von. 1822-1899. Aus. P I,III,IV; BMNH; GB; GC.
1. *Geologische Übersicht der Bergbau der österreichischen Monarchie*. Vienna, 1855.
2. *Die Geologie ... der österreichisch-ungarischen Monarchie*. Vienna, 1875.

9749. NEWBERRY, John Strong. 1822-1892. U.S.A. DSB; ICB; P III,IV; BMNH; GB; GE.

9750. PHILLIPS, John Arthur. 1822-1887. Br. P III; BMNH; GD.

9751. SAFFORD, James Merrill. 1822-1907. U.S.A. P III,IV & Supp.; BMNH; GE.

9752. SJÖGREN, Carl Anton Hjalmar. 1822-1893. Swed. P II,III,IV; BMNH.

9753. SZABÓ (VON SZENTMIKLÓS), Joseph. 1822-1894. Hung. P III,IV; BMNH; GB.

9754. VILANOVA Y PIERA, Juan. 1822-1893. Spain. BMNH.

9755. VOLGER, Georg Heinrich Otto. 1822-1897. Ger./Switz. P II,III, IV; BMNH.

9756. BOTELLA Y DE HORNOS, Federigo de. 1823-1899. Spain. BMNH.

9757. EBRAY, Charles Henri Théophile. 1823-1879. Fr. P III; BMNH.

9758. FÖTTERLE, Franz. 1823-1876. Aus. P III; BMNH.

9759. GÜMBEL, (Carl) Wilhelm von. 1823-1898. Ger. P III,IV; BMNH; GB; GC.

9760. HIND, Henry Youle. 1823-1908. Canada. P III; BMNH; GD2.

9761. LE CONTE, Joseph. 1823-1901. U.S.A. DSB; P III,IV; BMNH; BLA & Supp.; GB; GE. Also Evolution*, Physiology*, and Autobiography*.
   1. *Elements of geology. A text-book.* New York, 1878. 3rd ed., 1891.

9762. LEIDY, Joseph. 1823-1891. U.S.A. DSB; ICB; P III; BMNH; Mort.; BLA & Supp.; GB; GE. Also Zoology and Anatomy.

9763. LORY, Charles. 1823-1889. Fr. P III,IV; BMNH; GF.

9764. McCOY, Frederick. 1823-1899. Br./Australia. BMNH; GB; GD1. Also Zoology.
   1. *A synopsis of the characters of the Carboniferous limestones of Ireland.* Dublin, 1844.
   2. *A synopsis of the Silurian fossils of Ireland.* Dublin, 1846. Another ed., London, 1862.
   3. *Contributions to British palaeontology.* Cambridge, 1854.
   4. *The glacial action of ice and nature of the glacial period.* 1872.

   See also 9296/2 and 9304/1.

9765. MARTIN, Jules Jean Baptiste. 1823-after 1885. Fr. P III,IV; BMNH.

9766. PROBST, Joseph. 1823-1905. Ger. P III,IV; BMNH.

9767. ROUVILLE, Paul Gervais de. 1823-1907. Fr. P IV; BMNH.

9768. SAPORTA, (Louis Charles Joseph) Gaston de. 1823-1896. Fr. DSB; P III; BMNH. See also 13873/16.

9769. THOMSON, James (third of the name). 1823-1900. Br. P III; BMNH.

9770. CONTEJEAN, Charles Louis. 1824-1897. Fr. P III,IV; BMNH. Also Natural History.

9771. FRAAS, Oscar Friedrich von. 1824-1897. Ger. P III,IV; BMNH; GC.

9772. FROMENTEL, Louis Edouard Gourdan de. 1824-after 1891. Fr. BMNH.

9773. HAAST, (Johann Franz) Julius von. 1824-1887. New Zealand. DSB; ICB; P III; BMNH; GB; GD.

9774. HAIME, Jules. 1824-1856. Fr. P III; BMNH. Also Zoology. See also 9475/2.

9775. MARCOU, Jules Belknap. 1824-1898. Fr./Switz./U.S.A. DSB; P III,IV; BMNH; GB; GE. Also Science in General* and Biography*.
1. *Lettres sur les roches du Jura et leur distribution géographique dans les deux hemisphères.* 2 vols. Paris, 1857-60. RM/L.
2. *Geology of North America.* Zurich, 1858. RM/L.

9776. SELWYN, Alfred Richard Cecil. 1824-1902. Br./Australia/Canada. DSB; P IV; BMNH; GB; GD2: GF.

9777. STODDART, William Walter. 1824-1880. Br. P III; BMNH.

9778. STOPPANI, Antonio. 1824-1891. It. P III; BMNH; GB.
1. *Note ad un corso annuali di geologia.* 3 vols. Milan, 1866-70.

9779. THOMSON, William, *Baron* KELVIN. 1824-1907. Br. DSB; ICB (KELVIN); P III,IV; GB; GD2. Also Physics*#. See also 15553/1.
1. *The age of the earth.* London, 1897. Reprinted?

9780. TYLOR, Alfred. 1824-1884. Br. P III; BMNH; GD. Also Natural History*.

9781. WINCHELL, Alexander. 1824-1891. U.S.A. DSB; P III,IV; BMNH; GE.
1. *Sketches of creation.* New York, 1870.
2. *Pre-Adamites.* Chicago, 1880. Another ed., 1890.
3. *World-life; or, Comparative geology.* Chicago, 1883.
4. *Shall we teach geology? A discussion of the proper place of geology in modern education.* Chicago, 1889.
5. *Geological studies; or, Elements of geology.* Chicago, 1889.

9782. BLAKE, William Phipps. 1825-1910. U.S.A. P III; BMNH; GE.

9783. BRIART, Alphonse. 1825-1898. Belg. P III; BMNH.

9784. DAHLL, Tellef. 1825-1893. Nor. P III,IV; BMNH.

9785. FERNANDEZ DE CASTRO Y SUERO, Manuel. 1825-1895. Spain. BMNH.

9786. FORBES, David. 1825-1876. Br./Nor. P III; BMNH; GB; GD.

9787. HOSIUS, August. 1825-1896. Ger. P III,IV; BMNH.

9788. HUXLEY, Thomas Henry. 1825-1895. Br. DSB; ICB; BMNH; Mort.; BLA; GB; GD1. Other entries: see 3340.
1. *A catalogue of the collection of fossils in the Museum of Practical Geology, with an explanatory introduction.* London, 1865. With R. Etheridge.
2. *On a piece of chalk.* [A lecture given in 1868] New York, 1967. Ed. with introd. and notes by L. Eisley.
3. *Physiography. An introduction to the study of nature.* London, 1877. 2nd ed., 1878; re-issued 1882. 3rd ed., 1883. Physical geography.

9789. KARRER, Felix. 1825-1903. Aus. P IV; BMNH; GB.

9790. KJERULF, Theodor. 1825-1888. Nor. P I,III,IV; BMNH; GB.

9791. KREJČI, Jan. 1825-1887. Prague. P I,III; BMNH. Also Mineralogy.

9792. MEGLITZKY, Nicolai Gavrilovich. 1825-1857. Russ. P III.

9793. PETERS, Karl Ferdinand. 1825-1881. Aus. P III; BMNH.

9794. PFAFF, Alexius Burkhard Immanuel Friedrich. 1825-1886. Ger. P II,III,IV; BMNH; GC.

9795. RÜTIMEYER, (Karl) Ludwig. 1825-1895. Switz. DSB; ICB; BMNH; Mort.; BLA; GC. Also Zoology.

9796. SCHLUMBERGER, Charles. 1825-1905. Fr. DSB.

9797. SCHMIDT, (Johann Friedrich) Julius. 1825-1884. Ger./Greece. DSB; ICB; P II,III,IV; BMNH; GC. Also Astronomy*.

9798. WOLF, Heinrich Wilhelm. 1825-1882. Aus. P III; GC.

9799. BOSCOWITZ, Arnold. b. 1826. BMNH.
   1. *Les tremblements de terre*. Paris, 1885.
   1a. ——— *Earthquakes*. London, 1890. Trans. by C.B. Pitman.

9800. CROSSKEY, Henry William. 1826-1893. Br. P III; BMNH; GB.

9801. DEWALQUE, (Georges Jean) Gustave. 1826-1905. Belg. P III,IV; BMNH.

9802. GILLIÉRON, Victor. 1826-1899 (or 1890). Switz. P III,IV; BMNH.

9803. GÖBEL, Friedemann Adolph. 1826-1895. Russ. P I,III; BMNH.

9804. HUNT, Thomas Sterry. 1826-1892. U.S.A./Canada. DSB; ICB; P III,IV; BMNH; GB; GE. Also Chemistry* and Mineralogy*.
   1. *Report on the chemistry of the earth*. Washington, 1871.
   2. *Chemical and geological essays*. Boston, 1875. RM/L. 2nd ed., London, 1879. 3rd ed., New York, 1890.
   3. *Mineral physiology and physiography. A second series of chemical and geological essays*. Boston, 1886. 2nd ed., New York, 1891.

9805. JOKÉLY, Johann von. 1826-1862. Aus. P III.

9806. MAYER-EYMAR, Karl David Wilhelm. 1826-1907. Switz. DSB; P II, III,IV; BNMH.

9807. MORTON, George Highfield. 1826-1900. Br. P III,IV; BMNH; GD1.

9808. PERRY, John B. ca. 1826-1872. U.S.A. P III.

9809. PETTERSEN, Karl Johan. 1826-1890. Nor. P III,IV.

9810. SANDBERGER, (Karl Ludwig) Fridolin von. 1826-1898. Ger. P II, III,IV; BMNH; GB; GC.

9811. SORBY, Henry Clifton. 1826-1908. Br. DSB; ICB; P III,IV; BMNH; GB; GD2. Also Microscopy.
   1. *Sorby on sedimentology: A collection of papers from 1851-1908*. Miami, 1976. Ed. by C.H. Summerson.

9812. WHITE, Charles Abithar. 1826-1910. U.S.A. P III,IV; BMNH; GE.
   1. *On Mesozoic fossils*. Washington, 1884.

9813. ADAMS, Andrew Leith. 1827-1882. Br. P III; BMNH; GB; GD.
Also Zoology.

9814. BRAUNS, David August. 1827-1893. Ger. P III,IV; BMNH.

9815. BROADHEAD, Garland Carr. 1827-1912. U.S.A. P III,IV; BMNH; GE.

9816. COCCHI, Igino. 1827-1913. It. DSB; P III; BMNH.

9817. DAVIES, David Christopher. 1827-1885. Br. P III; BMNH; GD.

9818. GAUDRY, Albert Jean. 1827-1908. Fr. DSB; P III,IV; BMNH; GB.
1. *Matériaux pour l'histoire des temps quaternaires.* 4 parts. Paris, 1876-92.
2. *Les enchaînements du monde animal dans les temps géologiques.* 3 vols. Paris, 1878-90.
3. *Les ancêtres de nos animaux dans les temps géologiques.* Paris, 1888.

9819. GRÜNEWALDT, (Johann Georg) Moritz von. 1827-1873. Russ. P I, III; BMNH.

9820. HANTKEN, Maximilian von Prudnik. 1827-1893. Hung. P III; BMNH.

9821. KERR, Washington Caruthers. 1827-1885. U.S.A. P III; BMNH; GE.

9822. KOCH, Carl Jacob Wilhelm Ludwig. 1827-1882. Ger. P III; BMNH; GC. Also Zoology.

9823. LA VALLÉE-POUSSIN, Charles Louis Xavier Joseph de. 1827-1903. Belg. P III,IV (under DE); BMNH.

9824. MOESCH, Casimir. 1827-1898. Switz. P III,IV; BMNH. Also Zoology.

9825. MURPHY, Joseph John. 1827-1894. Br. P III. Also Science in General*.

9826. PIETTE, (Louis) Eduard (Stanislas). 1827-1906. Fr. DSB; BMNH.

9827. PLATZ, Philipp. 1827-after 1893. Ger. P III,IV; BMNH.

9828. ROLLE, Friedrich. 1827-1887. Ger./Aus. P III; BMNH; GC. Also Evolution*.

9829. SALMON, Henry Curwen. 1827-1873. Br. P III; BMNH.

9830. SPEYER, Oscar Wilhelm Karl. 1827-1881. Ger. P III; BMNH.

9831. STUR, Dionys Rudolf Josef. 1827-1893. Aus. P III; BMNH.

9832. COLLETT, John. 1828-after 1882. U.S.A. P III; BMNH.

9833. FOUQUÉ, Ferdinand André. 1828-1904. Fr. DSB; P III,IV & Supp.; BMNH; GB. Also Mineralogy*.
1. *Santorin et ses éruptions.* Paris, 1879.

9834. HUDLESTON (formerly SIMPSON), Wilfred Hudleston. 1828-1909. Br. BMNH; GD2.

9835. LAWRENCE, Philip Henry. 1828-1895. BMNH.
1. *Lithology; or, Classification of rocks.* London, 1865.
See also 9542/6a.

9836. LEWAKOWSKI, Ivan Fedorovich. 1828-1893. Russ. P III,IV.

9837. LIEBE, Karl Theodor. 1828-1894. Ger. P IV; BMNH; GC.

9838. LORIOL, Perceval de. 1828-after 1905. Switz. ICB; P III,IV; BMNH.

9839. TORELL, Otto Martin. 1828-1900. Swed. P III,IV; BMNH; GB.

9840. WHITFIELD, Robert Parr. 1828-1910. U.S.A. DSB; BMNH; GE.

9841. HAYDEN, Ferdinand Vandiveer. 1829-1887. U.S.A. DSB; ICB; P III; BMNH; GB; GE.

9842. HOCHSTETTER, Ferdinand Christian von. 1829-1884. Aus. P I, III,IV; BMNH; GB; GC; GF.

9843. HULL, Edward. 1829-1917. Br. P III,IV; BMNH.
1. *A sketch of geological history.* 1887.
2. *Volcanoes past and present.* London, 1892.

9844. JAMIESON, Thomas Francis. 1829-after 1901. Br. P III,IV.

9845. KINAHAN, George Henry. 1829-1908. Br. P III,IV; BMNH; GD2.
1. *Handy-book of rock names, with brief descriptions of the rocks.* London, 1873.
2. *Valleys and their relation to fissures, fractures and faults.* Ib., 1875.
3. *Manual of the geology of Ireland.* Ib., 1878.

9846. LINDSTRÖM, Gustav. 1829-1901. Swed. BMNH; GB.

9847. MEDLICOTT, Henry Benedict. 1829-1905. Br./India. P III,IV; BMNH; GD2; GF.

9848. OMBONI, Giovanni. 1829-1910. It. P III,IV; BMNH.
1. *Sullo stato geologico dell'Italia.* Milan, 1856.

9849. O'REILLY, Joseph Patrick. 1829-1905. Br. P III,IV; BMNH.

9850. ORTON, Edward Francis Baxter. 1829-1899. U.S.A. ICB; P III, IV; BMNH; GE.

9851. SCHLAGINTWEIT, Adolph von. 1829-1857. Ger. P II; BMNH; GC.
1. *Untersuchungen über die physicalische Geographie der Alpen.* 2 vols. Leipzig, 1850. With H. Schlagenweit.
2. *Neue Untersuchungen über die physicalische Geographie und die Geologie der Alpen.* 2 vols. Leipzig, 1854. With H. Schlagenweit.

9852. WACHSMUTH, Charles. 1829-1896. U.S.A. BMNH; GB; GE.

9853. KAUFMANN, Franz Joseph. fl. 1860-1887. Switz. BMNH.

9854. WINKLER, Tiberius Cornelius. d. 1898. Holl. BMNH.

9855. DIEULAFAIT, Louis. 1830-after 1885. Fr. P III,IV; BMNH.

9856. HULKE, John Whitaker. 1830-1895. Br. BMNH; BLA; GB; GD1.

9857. JEITTELES, Ludwig Heinrich. 1830-1883. Hung. P III; BMNH. Also Zoology.

9858. KLUGE, Carl Emil. 1830-1864. Ger. P III; BMNH.
1. *Über die Ursachen der in den Jahren 1850 bis 1857 stattgefundenen Erd-Erschütterungen und die Beziehungen zu den Vulkanen und zur Atmosphäre.* Stuttgart, 1861.

9859. LA HARPE, Philippe George de. 1830-1882. Switz. P III (HARPE); BMNH.

9860. LAUGHTON, John Knox. 1830-1915. Br. P III; GD3. Also Meteorology*.

9861. McMAHON, Charles Alexander. 1830-1904. Br. GD2; GF.

9862. OCHSENIUS, Carl Christian. 1830-1906. Ger./Chile. DSB; ICB; P III,IV; BMNH.

9863. RATH, Gerhard vom. 1830-1888. Ger. P II,III,IV; BMNH; GB; GC. Also Mineralogy.

9864. SIMONIN, Louis Laurent. 1830-1886. Fr. P III; BMNH.

9865. TRIBOLET, Georges de. 1830-1873. Switz. P III; BMNH.

9866. ULRICH, Georg Heinrich Friedrich. 1830-1900. Australia/New Zealand. P III,IV; BMNH.

9867. WOOD, Searles Valentine (second of the name). 1830-1884. Br. BMNH; GD.

9868. BARBOT DE MARNI, Nikolai Pavlovich. 1831-1877. Russ. P III (MARNY, N.P.); BMNH.

9869. BORNEMANN, Johann Georg. 1831-1896. Ger. BMNH.

9870. COMPTER, Gustav. 1831-1922. Ger. ICB; BMNH.

9871. EVANS, Caleb. 1831-1886. Br. BMNH; GD.

9872. HUGHES, Thomas McKenny. 1831-1917. Br. P IV; BMNH; GF.

9873. MARSH, Othniel Charles. 1831-1899. U.S.A. DSB; ICB; P III, IV; BMNH; GB; GE.

9874. OPPEL, (Carl) Albert. 1831-1865. Ger. DSB; P III; BMNH; GB; GC.

9875. RENEVIER, Eugène. 1831-1906. Switz. DSB; P III,IV; BMNH; GB.

9876. ROBERTS, George Edward. 1831-1865. Br. BMNH; GD.

9877. SUESS, Edward. 1831-1914. Aus. DSB; ICB; P III,IV; BMNH; GB.
1. *Die Entstehung der Alpen.* Vienna, 1875.
2. *Das Antlitz der Erde.* 3 vols in 4. Ib., 1883-1909.
2a. ——— *La face de la terre.* 3 vols in 8. Paris, 1897-1918. Trans. with notes under the direction of E. de Margerie.
2b. ——— *The face of the earth.* 5 vols. Oxford, 1904-24. Trans. under the direction of W.J. Sollas.

9878. VOSE, George Leonard. 1831-1910. U.S.A. P III; GE.
1. *Orographic geology; or, The origin and structure of mountains. A review.* Boston, 1866. X. Another ed., 1886.

9879. BELT, Thomas. 1832-1878. Br. P III; BMNH; GB; GD.

9880. BLANFORD, William Thomas. 1832-1905. Br./India. P III,IV; BMNH; GB; GD2; GF. Also Zoology.

9881. FRIČ, Antonín Jan. 1832-1916? Prague. P IV (FRITSCH); BMNH. Also Zoology.

9882. GOSSELET, Jules Auguste. 1832-1916. Fr. DSB; P III,IV; BMNH.

9883. GREEN, Alexander Henry. 1832-1896. Br. P III; BMNH; GB; GD1.
1. *Geology for students and general readers. Part 1. Physical*

geology. [No more publ.] London. 1876. 3rd ed., 1882. Other eds, 1892 and 1898.
2. *The birth and growth of worlds*. London, 1890.

9884. JANNETTAZ, (Pierre Michel) Edouard. 1832-1899. Fr. P III,IV; BMNH. Also Mineralogy.
1. *Les roches. Description de leurs éléments, méthode de détermination, etc.* Paris, 1874.
1a. ——— *Guide to the determination of rocks, being an introduction to lithology*. New York, 1877. Trans. by G.W. Plympton. X. 2nd ed., 1883.

9885. MAW, George. 1832-after 1886. Br. P III; BMNH. Also Botany.

9886. MÖHL, Heinrich. 1832-1903. Ger. P III; BMNH.

9887. NORDENSKIÖLD, (Nils) Adolf Erik. 1832-1901. Swed. DSB; ICB; P II,III,IV; BMNH; GB. Also History of Cartography*. See also 7175/5,5a.

9888. O'KELLY, Joseph. 1832-1883. Br. BMNH; GD.

9889. PELLAT, Edmond. 1832-after 1901. Fr. P III,IV; BMNH.

9890. RAMES, Jean Baptiste. 1832-1894. Fr. DSB.

9891. READE, Thomas Mellard. 1832-1909. Br. P III,IV; BMNH; GD2.
1. *Chemical denudation in relation to geological time*. London, 1879.
2. *The origin of mountain ranges*. London, 1886.

9892. SCHMIDT, Friedrich. 1832-1908. Russ. P III,IV; BMNH.

9893. WEISS, Christian Ernst. 1832-1901 (or 1890). Ger. P III; BMNH; GC.

9894. WOODS, Julian Edmond Tenison-. 1832-1889. Australia. BMNH; GD.

9895. WOODWARD, Henry. 1832-1921. Br. BMNH. See also 10358/1,3.

9896. BONNEY, Thomas George. 1833-1923. Br. DSB; P III,IV; BMNH; GB; GD4. Also Biography*.
1. *The story of our planet*. London, 1893. Another ed., 1896.
2. *Ice-work, present and past*. Ib., 1896.
3. *Volcanoes, their structure and significance*. Ib., 1899.
See also 8931/1 and 9563/1.

9897. CAPELLINI, Giovanni. 1833-1922. It. P III,IV; BMNH.

9898. CZEKANOWSKI, Aleksander Piotr. 1833-1876. Russ. DSB; P III,IV.

9899. DWIGHT, William Buck. 1833-1906. U.S.A. P IV.

9900. FALSAN, Albert. 1833-after 1889. Fr. P III; BMNH.

9901. HILGARD, Eugen Woldemar. 1833-1916. U.S.A. P III,IV; BMNH; GE.

9902. JACCARD, Auguste. 1833-1895. Switz. DSB; P III,IV; BMNH.

9903. JESPERSEN, Johan Peter Magnus. 1833-after 1882. Den. P III.

9904. MAKOWSKY, Alexander. 1833-after 1897. Brünn. P IV; BMNH.

9905. RICHTHOFEN, Ferdinand Paul Wilhelm von. 1833-1905. Ger. DSB; P II,III,IV; BMNH; GB.
1. *Führer fur Forschungsreisende: Anleitung zu Beobachtungen über Gegenstände der physischen Geographie und Geologie*. Berlin, 1886.

9906. SCOTT, Robert Henry. 1833-1916. Br. P III,IV; BMNH. Also Meteorology* and History of Science*.

9907. SEGUENZA, Giuseppe. 1833-1889. It. P III,IV; BMNH.

9908. STACHE, (Karl Heinrich Hector) Guido. 1833-1921. Aus. P III, IV; BMNH.

9909. BAUERMAN, Hilary. 1834-1909. Br./Canada. BMNH; GD2. Also Mineralogy*.

9910. BLANFORD, Henry Francis. 1834-1893. India. P IV; BMNH; GD1; GF. Also Meteorology. See also 8817/1a.

9911. CORNET, François Léopold. 1834-1887. Belg. P III,IV; BMNH.

9912. FOOTE, Robert Bruce. 1834-after 1895. India. P III,IV; BMNH.

9913. GODWIN-AUSTEN, Henry Haversham. 1834-1923. Br./India. BMNH; GD4. Also Zoology.

9914. GOLOVKINSKY, Nikolai Alexievich. 1834-1897. Russ. ICB; P III; BMNH (GHOLOVKINSKII).

9915. HECTOR, James. 1834-1907. Canada/New Zealand. BMNH; GD2; GF. Also Natural History.

9916. LOBLEY, James Logan. 1834-1913. Br. BMNH.
1. *Mount Vesuvius. A descriptive, historical and geological account of the volcano.* London, 1868. X. Another ed., 1889.

9917. LUBBOCK, John, *1st Baron* AVEBURY. 1834-1913. Br. DSB; ICB; BMNH; GB; GD3. Other entries: see 3351.
1. *The scenery of Switzerland and the causes to which it is due.* London, 1896. 3rd ed., 1898.

9918. MALAISE, Constantin Henri Gérard Louis. 1834-after 1880. Belg. P III; BMNH.

9919. PÉRON, (Pierre) Alphonse. 1834-1908. Fr./Algeria. BMNH.

9920. POWELL, John Wesley. 1834-1902. U.S.A. DSB; ICB; P III,IV; BMNH; GB; GE.
1. *Report on the exploration of the Colorado River of the West and its tributaries.* Washington, 1875. X.
1a. ——— *Canyons of Colorado.* Meadville, Pa., 1895. A re-issue of Chapters 1-9 of item 1, with six additional chapters. RM.

9921. ROSSI, Michele Stefano de. 1834-1898. It. P III,IV.

9922. WOLDŘICH, Johann Nepomuk. 1834-1906. Aus./Prague. P IV; BMNH.

9923. ANDRIAN-WERBURG, Ferdinand von. 1835-after 1895. Aus. P III,IV.

9924. CLAYPOLE, Edward Waller. 1835-1901. U.S.A. BMNH; GE.

9925. DELGADO, Joaquim Filippe Nery. 1835-1908. Port. BMNH.

9926. FONTAINE, William Morris. 1835-1913. U.S.A. P III,IV; BMNH.

9927. GARRIGOU, Joseph Louis Félix. 1835-1920. Fr. ICB; P III,IV; BMNH; BLA & Supp.

9928. GEIKIE, Archibald. 1835-1924. Br. DSB; ICB; P III,IV; BMNH; GB; GD4. Also Autobiography* and Biography*. See also Index.

1. *The story of a boulder; or, Gleanings from the note-book of a field geologist.* Edinburgh, 1858.
2. *The scenery of Scotland viewed in connexion with its physical geology.* London, 1865. RM.
3. *Physical geography.* London, 1874. X. Rev. ed., 1886.
4. *Outlines of field geology.* London, 1876. X. 5th ed., 1896.
5. *Geological sketches at home and abroad.* London, 1882. RM/L.
6. *Text-book of geology.* London, 1882. RM/L. 2nd ed., 1885. 3rd ed., 1893.
7. *Class-book of geology.* London, 1886. 2nd ed., 1890.
8. *Elementary lessons in physical geography.* London, 1890.
9. *The ancient volcanoes of Great Britain.* 2 vols. London, 1897. RM/L.

9929. LYMAN, Benjamin Smith. 1835-1920. U.S.A. DSB; P III,IV; BMNH; GE.

9930. SCHLÜTER, Clemens August Joseph. 1835-1906. Ger. P III,IV; BMNH.

9931. SILVESTRI, Orazio. 1835-1890. It. P III,IV; BMNH.

9932. STUEBEL, Moritz Alphons. 1835-1904. Ger. BMNH.

9933. WHITEAVES, Joseph Frederick. 1835-1909. Canada. BMNH; GB.
1. *Mesozoic fossils.* Montreal, 1876-79.
2. *Palaeozoic fossils.* Ib., 1897.
See also 9722/1.

9934. YOUNG, John. 1835-1902. Br. BMNH.
1. *Physical geography.* London, 1874.

9935. BERENDT, Gottlieb Michael. 1836-after 1899. Ger. P III,IV; BMNH.

9936. BROOKS, Thomas Benton. 1836-1900. U.S.A. BMNH; GE.

9937. DREW, Frederick. 1836-1891. Br. GD1; GF.

9938. FLICHE, Paul. 1836-1908. Fr. P IV; BMNH.

9939. GLENN, John Wright. 1836-1892. U.S.A. ICB.

9940. HITCHCOCK, Charles Henry. 1836-1919. U.S.A. P III,IV; BMNH; GE.

9941. HUTTON, Frederick Wollaston. 1836-1905. New Zealand. P III,IV; BMNH; GD2. Also Zoology* and Evolution*.

9942. KINKELIN, Georg Friedrich. 1836-1913. Ger. P III,IV; BMNH.

9943. LASPEYRES, Ernst Adolf Hugo. 1836-1913. Ger. P III,IV; BMNH. Also Mineralogy.

9944. LORETZ, Martin Friedrich Heinrich Hermann. 1836-after 1901. Ger. P IV.

9945. MELLO, John Magens. 1836-after 1893. Br. P III; BMNH.

9946. MILLER, Samuel A. 1836?-1897. U.S.A. BMNH.

9947. POŠEPNÝ, Ferencz (or Franz). 1836-1895. Příbram. DSB; ICB; P III,IV; BMNH.

9948. RAND, Theodore Dehon. 1836-1903. U.S.A. P IV & Supp.

9949. ROSENBUSCH, Harry (or Karl Heinrich Ferdinand). 1836-1914. Ger. DSB; P III,IV; BMNH. Also Mineralogy*.

9950. SIDENBLADH, Per Elis. 1836-after 1880. Swed. P III; BMNH.

9951. WHITAKER, William. 1836-1925. Br. P III,IV; BMNH.

9952. BACHMANN, Isidor. 1837-1885. Switz. P III; BMNH.

9953. BARRETT, Lucas. 1837-1862. Br./West Indies. BMNH; GB; GD.

9954. ECK, Heinrich Adolf von. 1837-1925. Ger. P III,IV; BMNH.

9955. FUCHS, Carl Wilhelm C. 1837-1886. Ger. P III,IV; BMNH. Also Mineralogy*.

9956. FUCHS, (Ph. J.) Edmond. 1837-1889. Fr. P III,IV; BMNH.

9957. GRODDECK, Albrecht von. 1837-1887. Ger. DSB; P III,IV; BMNH.

9958. HICKS, Henry. 1837-1899. Br. GB; GD1.

9959. KOENEN, Adolph von. 1837-after 1915. Ger. P III,IV; BMNH.

9960. MATTHEW, George Frederic. 1837-after 1901. Canada. P III,IV; BMNH.

9961. PUMPELLY, Raphael. 1837-1923. U.S.A. DSB; P III,IV; BMNH; GE.

9962. TAYLOR, John Ellor. 1837-1895. Br. BMNH; GD. See also 15394.

9963. BENECKE, Ernst Wilhelm. 1838-1917. Ger. P III,IV; BMNH.

9964. BLEICHER, Marie Gustave. 1838-1901. Fr. P III,IV; BMNH.

9965. BRADLEY, Frank Howe. 1838-1879. U.S.A. P III; BMNH; GE.

9966. DAWKINS, William Boyd. 1838-1929. Br. P III,IV; BMNH; GB; GD4.

9967. FRITSCH, Karl Wilhelm Georg von. 1838-after 1900. Ger. P III; BMNH.

9968. GOODYEAR, Watson Andrews. 1838-after 1882. U.S.A. P III; BMNH.

9969. NEGRI, Gaetano de. 1838-1902. It. P III,IV; BMNH.

9970. NILES, William Harmon. 1838-1910. U.S.A. P III,IV.

9971. PAUL, Carl Maria. 1838-1900. Aus. P III,IV; BMNH.

9972. PENNING, William Henry. 1838-1902. Br. BMNH.
1. *A text book of field geology.* London, 1876. With A.J. Jukes-Brown. 2nd ed., 1879.

9973. REISS, (Johann) Wilhelm. 1838-1908. Ger. ICB; P III; BMNH.

9974. STOLICZKA, Ferdinand. 1838-1874. Aus./India. ICB; P III; BMNH; GB; GC; GF.

9975. TÖRNEBOHM, Alfred Elis. 1838-1911. Swed. P III,IV; BMNH.

9976. VOGELSANG, Hermann Peter. 1838-1874. Ger./Holl. P III; BMNH; GC.
1. *Philosophie der Geologie und mikroscopische Gesteinsstudien.* Bonn, 1867. RM/L.

9977. WRIGHT, George Frederick. 1838-1921. U.S.A. DSB; P IV; BMNH; GE.
1. *The Ice Age in North America and its bearings on the antiquity of man.* New York, 1889. Another ed., London, 1890.
2. *Man and the glacial period.* London, 1892. 2nd ed., 1893.

9978. ZIRKEL, Ferdinand. 1838-1912. Ger./Lemberg. DSB; P III,IV; BMNH.
1. *Lehrbuch der Petrographie.* Bonn, 1866. X. 2nd ed., 3 vols, Leipzig, 1893-94.
2. *Untersuchungen über die mikroscopische Zusammensetzung und Structur der Basaltgesteine.* Bonn, 1870.
3. *Die mikroscopische Beschaffenheit der Mineralien und Gesteine.* Leipzig, 1873.
See also 8777/2.

9979. ACHIARDI, Antonio d'. 1839-1902. It. P III,IV; BMNH.

9980. BAILEY, Loring Woart. 1839-1925. Canada. DSB; ICB; P III,IV; BMNH.

9981. ENGELHARDT, Hermann. 1839-after 1902. Ger. P III,IV; BMNH.

9982. FONTANNES, Charles Francisque. 1839-1886. Fr. BMNH.

9983. GABB, William More. 1839-1878. U.S.A. DSB; P III; BMNH; GE.

9984. GEIKIE, James. 1839-1915. Br. DSB; P III,IV; BMNH; GB.
1. *The great Ice Age and its relation to the antiquity of man.* London, 1874. 2nd ed., 1877. 3rd ed., 1894.
2. *Prehistoric Europe. A geological sketch.* Ib., 1880. RM/L.
3. *Outlines of geology.* Ib., 1886. 2nd ed., 1888.
4. *Fragments of earth lore ... geological and geographical.* Edinburgh, 1893. RM/L.
5. *Earth structure; or, The origin of landforms.* London, 1898.

9985. GRAND'EURY, (François) Cyrille. 1839-1917. Fr. DSB; BMNH.

9986. HINDE, George Jennings. 1839-1918. Br. BMNH.

9987. HOFMANN, Karl. 1839-1891. Hung. P III; BMNH.

9988. LAPPARENT, Albert Auguste Cochon de. 1839-1908. Fr. DSB; P III,IV; BMNH; GB. Also Mineralogy*.
1. *Traité de géologie.* Paris, 1882. X. 3rd ed., 1893.
2. *Leçons de géographie physique.* Paris, 1896.

9989. LASAULX, Arnold Contantin Peter Franz von. 1839-1886. Ger. P III,IV; BMNH; GB; GC. Also Mineralogy.

9990. LAUBE, Gustav Karl. 1839-1923. Aus./Prague. P III,IV; BMNH.

9991. LINDSTRÖM, Axel Fredrik. b. 1839. Swed. BMNH.

9992. MACPHERSON, Joseph. 1839-1902. Spain. BMNH.

9993. MOJSISOVICS VON MOJSVAR, (Johann August Georg) Edmund. 1839-1907. Aus. P III,IV; BMNH; GB.

9994. NIES, Friedrich, 1839-1895. Ger. P III,IV; GC.

9995. REMELÉ, Adolf. 1839-1915. Ger. P III,IV; BMNH.

9996. SEEBACH, Karl Albert Ludwig von. 1839-1880. Ger. P III,IV; BMNH; GC.

9997. SEELEY, Harry Govier. 1839-1909. Br. BMNH; GD2. Also Zoology.
1. *The story of the earth in past ages.* London, 1895.
See also 9458/5.

9998. WALKER, John Francis. 1839-after 1901. Br. P III,IV. Also Chemistry.

9999. WIIK, Fredrik Johan. 1839-1909. Fin. P III,IV & Supp.; BMNH.

10000. WINCHELL, Newton Horace. 1839-1914. U.S.A. DSB; P III,IV; BMNH; GE.

10001. ZITTEL, Karl Alfred von. 1839-1904. Ger. DSB; P III,IV; BMNH; GB. Also History of Geology*.
1. *Aus der Urzeit. Bilder aus der Schöpfungsgeschichte.* 2 vols. Munich, 1871-72.
2. *Handbuch der Paläontologie.* 5 vols. Munich, 1876-93.
2a. ——— *Traité de paléontologie.* 4 vols. Paris, 1883-94. Trans. by C. Barrois.
2b. ——— *Textbook of palaeontology.* 3 vols. London, 1896; re-issued 1900-03. Trans. by C.R. Eastman.
3. *Grundzüge der Paläontologie.* 2 vols. Munich, 1895.

10002. AREITIO Y LARRINAGA, Alfonso de. d. 1884. Spain. BMNH.

10003. MANZONI, Angelo. d. 1895. It. BMNH.

10004. BOECKH, János. 1840-1909. Hung. BMNH.

10005. CALVIN, Samuel. 1840-1911. Br./U.S.A. BMNH; GE.

10006. DOLLFUS, Auguste. 1840-1869. Fr. P III; BMNH.

10007. ERDMANN, Edvard. 1840-1923. Swed. BMNH.

10008. HAGUE, Arnold. 1840-1917. U.S.A. DSB; P III,IV; BMNH; GE.

10009. HARTT, Charles Frederick. 1840-1878. U.S.A. ICB; P III (HARTL); BMNH.
1. *The Thayer expedition. Scientific results of a journey in Brazil by Louis Agassiz and his travelling companions. Geology and physical geography of Brazil.* Boston, 1870.

10010. JUDD, John Wesley. 1840-1916. Br. P III,IV; BMNH; GF.
1. *Volcanoes. What they are and what they teach.* London, 1881. 4th ed., 1888. 5th ed., 1893.
See also 9563/4a.

10011. JULIEN, Alexis Anastay. 1840-1919. U.S.A. P III,IV; BMNH.

10012. LARTET, Louis. 1840-1899. Fr. DSB; BMNH.

10013. MELLER, Valerian Ivanovich. 1840-after 1883. Russ. P III (MOELLER); BMNH.

10014. MÜHLBERG, Fritz. 1840-1915. Switz. P III,IV; BMNH.

10015. NEWTON, Edwin Tulley. 1840-1930. Br. DSB; ICB; P IV; BMNH.

10016. RUDLER, Frederick William. 1840-1915. Br. BMNH.

10017. SMITH, Stephenson Percy. 1840-1922. New Zealand. BMNH; GF.

10018. STELZNER, Alfred Wilhelm. 1840-1895. Ger. P III,IV; BMNH; GC.

10019. TATE, Ralph. 1840-1901. Br./Australia. BMNH; GB; GF. Also Botany.
1. *Rudimentary treatise on geology.* London, 1871. 2nd ed., 2 vols, 1874-75.
2. *Class-book of geology, physical and historical.* Ib., 1872.

10020. TÖRNQVIST, Sven Leonhard. 1840-1920. Swed. P III,IV; BMNH.

10021. TRAQUAIR, Ramsay Heatley. 1840-1912. Br. BMNH.

10022. BELL, Robert. 1841-1917. Canada. DSB; P IV & Supp.; BMNH.

10023. BRIGHAM, William Tufts. 1841-after 1891. U.S.A. P III,IV; BMNH. Also Botany.

10024. CALKER, Friedrich Julius Peter van. 1841-1913. Holl. P III, IV; BMNH.

10025. COLLINS, Joseph Henry. 1841-1916. BMNH. Also Mineralogy.

10026. CREDNER, (Karl) Hermann Georg. 1841-1913. Ger. DSB; P III, IV; BMNH.
1. *Elemente der Geologie*. Leipzig, 1872. X. 2nd ed., 1872. 6th ed., 1887.

10027. DUPONT, Edouard François. 1841-1911. Belg. P III,IV; BMNH.

10028. DUTTON, Clarence Edward. 1841-1912. U.S.A. DSB; P III,IV; BMNH; GE.

10029. EMMONS, Samuel Franklin. 1841-1911. U.S.A. DSB; P III,IV; BMNH; GE.

10030. FOSTER, Clement Le Neve. 1841-1904. Br. P III,IV; BMNH; GB; GD2.

10031. GALLI, Ignazio. 1841-after 1913. It. P IV.

10032. LINNARSSON, Jonas Gustaf Oskar. 1841-1881. Swed. P III; BMNH.

10033. LOSSEN, Karl August. 1841-1893. Ger. DSB; P III,IV; BMNH.

10034. MALLADA, Lúcas. 1841-after 1909. Spain. BMNH.

10035. SCHLÖNBACH, Georg Justin Karl Urban. 1841-1870. Aus. P III; BMNH; GC.

10036. SHALER, Nathaniel Southgate. 1841-1906. U.S.A. DSB; ICB; P III,IV; BMNH; GE. Also Autobiography*.
1. *Illustrations of the earth's surface. Glaciers*. Boston, 1881. With W.M. Davis.
2. *Aspects of the earth. A popular account of some geological phenomena*. London, 1890.
3. *Outlines of the earth's history*. London, 1898.

10037. SMITH, Eugene Allen. 1841-1927. U.S.A. P III,IV; BMNH; GE.

10038. STERZEL, Johann Traugott. 1841-1914. Ger. BMNH.

10039. STEVENSON, John James. 1841-1924. U.S.A. P III,IV; BMNH; GE.

10040. STONE, George Hapgood. 1841-1917. U.S.A. P IV; BMNH.

10041. TOPLEY, William. 1841-1894. Br. BMNH; GD.

10042. WAAGEN, Wilhelm Heinrich. 1841-1900. Aus./Prague/India. P III,IV; BMNH; GB.

10043. WARD, Lester Frank. 1841-1913. U.S.A. ICB; BMNH; GE.

10044. BALTZER, (R.) Armin. 1842-1913. Switz. P III,IV; BMNH.

10045. BEHRENS, Theodor Heinrich. 1842-1905. Holl. P III,IV; BMNH; GF. Also Chemistry.

10046. BROWN, Robert (of Campster). 1842-1895. Br. P III; BMNH; GD1. Also Science in General*, Botany*, and Zoology.

1. *Our earth and its story. A popular treatise on physical geography.* 3 vols. London, 1887-89.

10047. COHEN, Emil William. 1842-1905. Ger. P III,IV; BMNH. Also Mineralogy*.

10048. ENGEL, Karl Theodor. 1842-after 1911. Ger. P IV; BMNH.

10049. FUCHS, Theodor. 1842-1925. Aus. P III,IV; BMNH.

10050. GIORGI, Cosimo de. 1842-after 1880. It. P III (under DE); BMNH.

10051. HOWORTH, Henry Hoyle. 1842-1923. Br. P IV; BMNH; GF.

10052. ISSEL, Arturo. 1842-1922. It. DSB; P III,IV; BMNH.

10053. KING, Clarence Rivers. 1842-1901. U.S.A. DSB; ICB; P III, IV; BMNH; GB; GE.

10054. KLOOS, Johan Hermann. 1842-1901. Ger./U.S.A. P IV; BMNH.

10055. KROPOTKIN, Petr Alekseevich. 1842-1921. Russ. DSB; ICB; GB.

10056. LAPWORTH, Charles. 1842-1920. Br. DSB; BMNH; GB.
1. *An intermediate text-book of geology.* Edinburgh, 1899.

10057. LOVISATO, Domenico. 1842-after 1903. It. P IV; BMNH.

10058. MARTIN, David Salomon. 1842-after 1902. Fr. P IV.

10059. RENARD, Alphonse François. 1842-1903. Belg. DSB; P III,IV; BMNH; GB.

10060. RUTLEY, Frank. 1842-1904. Br. P III,IV; BMNH; GB. Also Mineralogy*.
1. *The study of rocks. An elementary textbook of petrology.* London, 1879. 2nd ed., 1881. 3rd ed., 1884. 4th ed., 1888. 5th ed., 1891. Another ed., 1899.

10061. SAUVAGE, Henri Emile. 1842-1917. Fr. BMNH.

10062. VIDAL, Luis Mariano. 1842-1922. Spain. BMNH.

10063. BALL, Valentine. 1843-1895. Br./India. BMNH; GF.

10064. BÖLSCHE, Wilhelm. 1843-1893. Ger. P III; BMNH.

10065. CHAMBERLIN, Thomas Chrowder. 1843-1928. U.S.A. DSB; ICB; P IV; BMNH; GE.

10066. DAMES, Wilhelm Barnim. 1843-1898. Ger. P IV; BMNH.

10067. EMERSON, Benjamin Kendall. 1843-1932. U.S.A. DSB; P III,IV; BMNH; GE1.

10068. FILHOL, (Antoine Pierre) Henri. 1843-1902. Fr. BMNH.

10069. GILBERT, Grove Karl. 1843-1918. U.S.A. DSB; P III,IV; BMNH; GB; GE.

10070. HÖFER, Hans. 1843-1924. Aus. P III,IV; BMNH.

10071. INOSTRANZEFF, Alexander Alexandrovich. 1843-after 1900. Russ. P III,IV; BMNH.

10072. KOCH, Anton. 1843-after 1903. Hung. P IV; BMNH.

10073. LUNDGREN, Sven Anders Bernhard. 1843-1897. Swed. P III,IV; BMNH.

10074. MEUNIER, (Etienne) Stanislas. 1843-1925. Fr. P III,IV; BMNH. Also Mineralogy*.
1. *Cours de géologie comparée*. Paris, 1874.

10075. MUNIER-CHALMAS, Ernest Charles Philippe Auguste. 1843-1903. Fr. DSB; BMNH.

10076. THOMAS, Philippe Etienne. 1843-1910. Fr. P IV; BMNH.

10077. THOULET, (Marie) Julien Olivier. 1843-after 1921. Fr. P III, IV; BMNH. Also Oceanography.

10078. VOGDES, Anthony Wayne. 1843-1923. U.S.A. P IV; BMNH.

10079. WARD, James Clifton. 1843-1880. Br. P III; BMNH; GD.

10080. WILKINSON, Charles Smith. 1843-1891. Australia. P III; BMNH; GD.

10081. BAUER, Max Hermann. 1844-1917. Ger. P III,IV; BMNH. Also Mineralogy*.

10082. BLOMBERG, Albert. b. 1844. Swed. BMNH.

10083. BRANCO, (Carl) Wilhelm (Franz). 1844-after 1907. Ger. P III, IV; BMNH.

10084. BROWN, Henry Yorke Lyell. 1844-1928. Canada/Australia. BMNH; GF.

10085. FRAZER, Persifor. 1844-1909. U.S.A. P III,IV; BMNH; GE.

10086. KREUTZ, Felix Franz Xaver. 1844-after 1896. Lemberg/Cracow. P III,IV.

10087. MICHEL-LÉVY, Auguste. 1844-1911. Fr. DSB; P IV (LÉVY); BMNH (LÉVY). Also Mineralogy*.
1. *Structures et classification des roches éruptives*. Paris, 1889.
2. *Etude sur la détermination des feldspaths dans les plaques minces, au point de vue de la classification des roches*. Paris, 1894.
His works are generally listed under LÉVY, A.M.

10088. NICHOLSON, Henry Alleyne. 1844-1899. Br. BMNH; GB; GD1. Also Biology in General*, Zoology*, Biography*, and History of Science*.
1. *A manual of palaeontology*. Edinburgh, 1872. 2nd ed., 1879. 3rd ed., 2 vols, 1889.
2. *The ancient life history of the earth*. Ib., 1877.

10089. PANTANELLI, Dante. 1844-1913. It. P III,IV. Also Zoology.

10090. PERKINS, George Henry. 1844-1933. U.S.A. BMNH; GE.

10091. STUCKENBERG, Alexander Antonovich. 1844-1905. Russ. P III, IV; BMNH (SHTUKENBERGH).

10092. BATZEVICH, L. fl. 1873-1887. Russ. BMNH.

10093. (Entry cancelled)

10094. LANDERER, José J. fl. 1873-1907. Spain. BMNH.

10095. SINTZOV, Ivan Theodorovich. fl. 1871-1898. Russ. BMNH.

10096. CORTÁZAR, Daniel de. b. 1845. Spain. BMNH.

10097. CZERSKI, Jan. 1845-1892. Russ. DSB. Also Zoology.

10098. ELLS, Robert Wheelock. 1845-after 1909. Canada. P IV; BMNH.

10099. FAVRE, Ernest. 1845-after 1887. Switz. P III; BMNH.

10100. GORCEIX, Henri. ca. 1845-after 1887. Fr./Brazil. P III,IV; BMNH.

10101. GUROW (or GOUROV), Alexander Vassilevich. 1845-after 1883. Russ. P III.

10102. HALL, Townshend Monckton. 1845-1899. Br. P III; BMNH. Also Mineralogy*.

10103. HARDMAN, Edward Townley. 1845-1887. Br./Australia. P III; BMNH; GD.

10104. JACK, Robert Logan. 1845-1921. Australia. BMNH; GF. See also 10122/1.

10105. KAYSER, (Friedrich Heinrich) Emmanuel. 1845-1927. Ger. P III; IV; BMNH.
1. *Die Fauna der ältesten Devon-Ablagerungen des Harzes.* 2 vols. Berlin, 1878.
2. *Lehrbuch der Geologie.* 2 parts. Stuttgart, 1891-93. X.
2a. ——— Trans. of Part 2: *Text-book of comparative geology.* London, 1893. Trans. and ed. by P. Lake. 2nd ed., 1895.

10106. MOURLON, Michel Félix. 1845-1915. Belg. P III,IV; BMNH.
1. *Recherches sur l'origine des phénomènes volcaniques.* Brussels, 1867.
2. *Géologie de la Belgique.* 2 vols. Paris, 1880-81.

10107. NEUMAYR, Melchior. 1845-1890. Aus. DSB; P III,IV; BMNH; GB. Also Zoology*.
1. *Erdgeschichte.* 2 vols. Leipzig, 1886-87. X. Other eds, 1890 and 1895.

10108. NIEDZWIEDZKI, Julian. 1845-1918. Lemberg. P III,IV; BMNH.

10109. RICE, William North. 1845-1928. U.S.A. BMNH; GE. Also Science in General*.

10110. ROHON, Joseph Victor. 1845-1923. Aus./Russ. BMNH.

10111. STUART-MENTEATH, Patrick William. 1845-after 1903. Br./Fr. P IV; BMNH.

10112. STUDER, Theophil. 1845-1922. Switz. P III,IV; BMNH; BLAF. Also Zoology.

10113. TARAMELLI, Torquato. 1845-1922. It. ICB; P III,IV; BMNH.

10114. TIETZE, Emil Ernst August. 1845-1931. Aus. P III,IV; BMNH.

10115. TOULA, Franz. 1845-after 1915. Aus. P III,IV; BMNH.
1. *Mineralogische und petrographische Tabellen.* Prague, 1886.

10116. VÉLAIN, Charles. 1845-1925. Fr. P IV; BMNH.
1. *Conférences de pétrographie.* Paris, 1889.

10117. VERBEEK, Rogier Diederik Marius. 1845-1926. Holl./East Indies. P III,IV; BMNH.
1. *Krakatau.* 2 vols. Batavia, 1885-86.

10118. COLLOT, Louis. 1846-after 1905. Fr. P IV.

10119. DAVIS, James William. 1846-1893. Br. BMNH.

10120. DOKUCHAEV, Vasily Vasilievich. 1846-1903. Russ. DSB; BMNH.

10121. DOUVILLÉ, (Joseph) Henri (Ferdinand). 1846-1937. Fr. P IV; BMNH.

10122. ETHERIDGE, Robert (second of the name). 1846-1920. Australia. BMNH; GF.
   1. *Catalogue of works, papers, reports, and maps of the geology, palaeontology, mineralogy, mining, and metallurgy, etc., of the Australian continent and Tasmania.* London, 1881. With R.L. Jack.

10123. HAY, Oliver Perry. 1846-1930. U.S.A. BMNH; GE.

10124. HELLAND, Amund Theodor. 1846-after 1903. Nor. P IV; BMNH.

10125. HOLMES, William Henry. 1846-1933. U.S.A. P III; BMNH; GE1.

10126. HOLST, Nils Olof. 1846-after 1915. Swed. P III,IV; BMNH.

10127. LAHUSEN, Iosif Ivanovich. 1846-after 1901. Russ. P III,IV; BMNH (LAGHUZEN).

10128. LANG, Heinrich Otto. 1846-after 1896. Ger. P III,IV; BMNH.

10129. TODD, James Edward. 1846-1922. U.S.A. P IV; BMNH.

10130. BECKER, George Ferdinand. 1847-1919. U.S.A. DSB; P III,IV; BMNH; GE.

10131. BERTRAND, Marcel Alexandre. 1847-1907. Fr. DSB; ICB; P IV; BMNH.

10132. DE RANCE, Charles Eugene. 1847-1906. Br. BMNH.

10133. GRIESBACH, Charles Ludolf. 1847-after 1902. India. P III,IV; BMNH.

10134. IRVING, Roland Duer. 1847-1888. U.S.A. P III,IV; BMNH; GE.

10135. KARPINSKY, Alexander Petrovich. 1847-1936. Russ. DSB; ICB; P III,IV; BMNH.

10136. RICHTER, Eduard. 1847-1905. Aus. P IV; BMNH; GF.

10137. RUTOT, Aimé Louis. 1847-1933. Belg. P IV; BMNH.

10138. SIMONOVICH, Spiridon Eghorovich. 1847-after 1897. Russ. BMNH.

10139. SVEDMARK, (Lennert) Eugène. 1847-1922. Swed. BMNH.

10140. WADSWORTH, Marsham Edward. 1847-1921. U.S.A. P III,IV; BMNH.

10141. WILLIAMS, Henry Shaler. 1847-1918. U.S.A. DSB; P IV; BMNH; GE.
   1. *Geological biology. An introduction to the geological history of organisms.* New York, 1895.

10142. DATHE, Johann Friedrich Ernst. 1848-after 1913. Ger. P IV; BMNH.

10143. FEISTMANTEL, Ottokar. 1848-1891. Prague/India. ICB; P III, IV; BMNH.
   1. *Palaeontologische Beiträge.* Cassel, 1876-79.

10144. FLETCHER, Hugh. 1848-1909. Canada. P IV & Supp.; BMNH.

10145. HARRINGTON, Bernard James. 1848-1907. Canada. BMNH; GF.

10146. HAWES, George Wesson. 1848-1882. U.S.A. P III; BMNH.

10147. LENZ, (Heinrich) Oscar. 1848-1925. Aus. P III; BMNH.

10148. ROUSSEL, Joseph. 1848-after 1904. Fr. P IV; BMNH.

10149. SPRINGER, Frank. 1848-1927. U.S.A. BMNH; GE.

10150. THOMPSON, Beeby. 1848-1931. Br. P IV; BMNH.

10151. VAČEK, Michael. 1848-after 1904. Aus. P III,IV; BMNH.

10152. WHITE, Israel Charles. 1848-1927. U.S.A. DSB; BMNH; GE.

10153. WOODWARD, Horace Bolingbroke. 1848-1914. Br. P III,IV; BMNH; GB.
1. *The geology of England and Wales. A concise account.* London, 1876. 2nd ed., 1887.
See also 8884/1.

10154. BOURGEAT, (François) Emilien. b. 1849. BMNH.

10155. CHOFFAT, Paul. 1849-1919. Port. BMNH.

10156. COMSTOCK, Theodore Bryant. 1849-after 1901. U.S.A. P III, IV; BMNH.

10157. DAWSON, George Mercer. 1849-1901. Canada. ICB; P III,IV; BMNH; GD2; GF.

10158. GUÉBHARD, Paul Emile Adrien. 1849-after 1922. Fr. P III,IV. Also Physics.

10159. HEIM, Albert. 1849-1937. Switz. DSB; ICB; P III,IV; BMNH; GB.
1. *Untersuchungen über den Mechanismus der Gebirgsbildung.* 3 vols. Basel, 1878.

10160. LOKHTIN, Vladimir Mikhaylovich. 1849-1919. Russ. DSB.

10161. OEHLERT, Daniel Victor. 1849-1920. Fr. P IV; BMNH.

10162. PEALE, Albert Charles. 1849-after 1893. U.S.A. P III; BMNH.

10163. REYER, Eduard. 1849-1914. Aus. P III,IV; BMNH.
1. *Beitrag zur Physik der Eruptionen und der Eruptiv-Gesteine.* Vienna, 1877.
2. *Die Euganeen. Bau und Geschichte eines Vulcans.* Ib., 1877.
3. *Theoretische Geologie.* Stuttgart, 1888.
4. *Geologische und geographische Experimente.* Leipzig, 1892-94.

10164. SEUNES, Jean. 1849-after 1903. Fr. P IV; BMNH.

10165. SOLLAS, William Johnson. 1849-1936. Br. DSB; ICB; P III,IV; BMNH; GD5. See also 9877/2b.

10166. TEALL, Jethro Justinian Harris. 1849-1924. Br. DSB; P IV; BMNH; GD4.
1. *British petrography.* London, 1888. RM/L.

10167. WILLIAMS, Edward Higginson. 1849-1933. U.S.A. BMNH.
1. *Manual of lithology.* New York, 1886. 2nd ed., 1899.

10168. GONNARD, Ferdinand. fl. 1870-1908. Fr. P IV; BMNH.

10169. PRENDEL, Romul Aleksandrovich. fl. 1875-1893. Russ. BMNH.

10170. KROTOV, Petr Ivanovich. fl. 1877-1912. Russ. BMNH.

10171. CHRUSTSCHOFF, Konstantin Dmitrievich. fl. 1878-1894. Russ. P IV; BMNH.

10172. PALACIOS, Pedro. fl. 1879-1919. Spain. BMNH.

10173. AMMON, Johann Georg Friedrich Ludwig von. 1850-1922. Ger. P IV; BMNH.

10174. BITTNER, Alexander. 1850-1902. Aus. P III,IV; BMNH.

10175. BRANNER, John Casper. 1850-1922. U.S.A. P IV; BMNH; GE.

10176. COSSMANN, Alexandre Edouard Maurice. 1850-1924. Fr. BMNH.

10177. CROSBY, William Otis. 1850-1925. U.S.A. P III,IV; BMNH; GE.
1. *Common minerals and rocks.* Boston, 1881.

10178. DAVIS, William Morris. 1850-1934. U.S.A. DSB; ICB; P IV & Supp.; BMNH; GE1. Also Other Earth Sciences*.
1. *Physical geography.* Boston, 1898.
See also 10036/1.

10179. DILLER, Joseph Silas. 1850-1928. U.S.A. P IV & Supp.; BMNH; GE1.

10180. DOLLFUS, Gustave Frédéric. 1850-1912? Fr. BMNH.

10181. DRASCHE, Richard von. 1850-after 1886. Aus. P III; BMNH. Also Zoology.

10182. FAIRCHILD, Herman Leroy. 1850-1943. U.S.A. ICB; P IV. Also History of Science*.

10183. FOULLON-NORBECK, Heinrich von. 1850-1896. Aus. P IV. Also Mineralogy.

10184. GRUBENMANN, (Johann) Ulrich. 1850-1924. Switz. DSB; P IV; BMNH.

10185. HOERNES, Rudolf. 1850-1912. Aus. P III,IV; BMNH.
1. *Elemente der Palaeontologie.* Leipzig, 1884.

10186. JENTZSCH, Karl Alfred. 1850-after 1919. Ger. P III,IV; BMNH.

10187. KLEBS, Richard Hermann Erdmann. 1850-1911. Ger. P III,IV; BMNH.

10188. MILNE, John. 1850-1913. Br. DSB; BMNH; GD3.
1. *Earthquakes and other earth movements.* London, 1886. 4th ed., 1898.
2. *Seismology.* London, 1896. Another ed., 1898.

10189. MOURET, (Ernest Jean) Georges. 1850-after 1917. Fr. P IV; BMNH.

10190. MUSHKETOV, Ivan Vasilievich. 1850-1902. Russ. DSB; P III, IV (MUSCHKETOW); BMNH.

10191. NATHORST, Alfred Gabriel. 1850-1921. Swed. DSB; P III,IV; BMNH.

10192. SKERTCHLY, Sydney Barber Josiah. 1850-after 1908. Br./Australia. BMNH.
1. *Elements of physical geography.* London, 1877.
2. *The physical system of the universe. An outline of physiography.* Ib., 1878.

10193. UPHAM, Warren. 1850-1934. U.S.A. P III,IV; BMNH; GE.

10194. WALCOTT, Charles Doolittle. 1850-1927. U.S.A. DSB; ICB; P III,IV; BMNH; GE.
1. *The fauna of the lower Cambrian, or Olenellus zone.* Washington, 1890.

10195. BARROIS, Charles Eugène. 1851-1939. Fr. DSB; ICB; P IV; BMNH.

10196. BLAAS, Josef. 1851-after 1926. Aus. P III,IV; BMNH.

10197. BROECK, Ernest van den. 1851-after 1909. Belg. BMNH.

10198. BRÖGGER, Waldemar Christopher. 1851-1940. Nor. DSB; ICB; P III,IV; BMNH; GB.

10199. BÜCKING, (Ferdinand Carl Bertram) Hugo. 1851-1932. Ger. P III,IV; BMNH.

10200. DERBY, Orville Adelbert. 1851-1915. U.S.A./Brazil. P III, IV; BMNH.

10201. ENDLICH, Frederic Miller. 1851-1899. U.S.A. P III; BMNH. Also Mineralogy*.

10202. JUKES-BROWNE, Alfred Joseph (or Alfred John). 1851-1914. Br. BMNH; GF.
1. *The student's handbook of historical geology.* London, 1886.
2. *The building of the British Isles. A study in geographical evolution.* Ib., 1888. 2nd ed., 1892.
3. *Geology. An elementary textbook.* Ib., 1893.
See also 9972/1.

10203. KALKOWSKY, (Louis) Ernst. 1851-after 1919. Ger. P IV; BMNH.

10204. LEHMANN, Johannes Georg. 1851-1925. Ger. P III,IV; BMNH.

10205. LEPSIUS, (Carl George) Richard. 1851-1917. Ger. P III,IV; BMNH.

10206. MARTIN, (Johann) Karl (Ludwig). 1851-after 1931. Holl. P III, IV; BMNH.

10207. MONTESSUS DE BALLORE, Fernand de. 1851-1923. Fr. ICB; P IV; BMNH; GF (BALLORE).

10208. NIKITIN, Sergei Nikolaievich. 1851-1909. Russ. DSB; P III, IV; BMNH.

10209. RICHE, Attale. 1851-after 1921. Fr. P IV; BMNH.

10210. ROBERTS, Robert Davies. 1851-1911. Br. BMNH; GD2.
1. *The earth's history. An introduction to modern geology.* London, 1893.

10211. SPENCER, (Joseph) William Winthrop. 1851-1921. U.S.A. P III, IV; BMNH.

10212. STEFANI, Carlo de. 1851-after 1912. It. P III (under S), IV (under DE); BMNH.

10213. WAHNSCHAFFE, Gustaf Albert Bruno Felix. 1851-1914. Ger. P III,IV; BMNH.

10214. WICHMANN, (Carl Ernst) Arthur. 1851-after 1922. Holl. P III, IV; BMNH.

10215. BRÉON, (Charles) René. 1852-after 1902. Fr. P IV; BMNH.

10216. COLEMAN, Arthur Philemon. 1852-1939. Canada. ICB.

10217. FRANCO, Pasquale. 1852-after 1902. It. P IV.

10218. FRÜH, Johann Jacob. 1852-after 1904. Switz. P IV; BMNH.

10219. HIBSCH, Josef Emanuel. 1852-after 1931. Aus. P III,IV; BMNH.

10220. JOHN VON JOHNESBERG, Conrad Heinrich. 1852-1918. Aus. P III, IV; BMNH.

10221. KOBY, Frédéric Louis. 1852-after 1904. Switz. BMNH.

10222. LAGORIO, Alexander Karl Leo. 1852-after 1897. Warsaw. P III, IV; BMNH.

10223. RATHBUN, 1852-1918. U.S.A. P III; BMNH; GE. Also Zoology.

10224. REUSCH, Hans Henrik. 1852-1922. Nor. P III,IV; BMNH; GB.

10225. ROLLAND, Georges François Joseph. 1852-after 1901. Fr. P IV; BMNH.

10226. RUSSELL, Israel Cook. 1852-1906. U.S.A. ICB: P III,IV; BMNH; GB; GE.

10227. STRAHAN, Aubrey. 1852-1928. Br. P IV; BMNH.

10228. SVENONIUS, Fredrik Vilhelm. 1852-1928. Swed. BMNH.

10229. TELLER, Friedrich Joseph. 1852-1913. Aus. P III,IV; BMNH.

10230. TRIBOLET (-HARDY), (Frédéric) Maurice de. 1852-1929. Switz. ICB; P III; BMNH.

10231. TULLBERG, Sven Axel Theodor. 1852-1886. Swed. BMNH.

10232. BARROW, George. 1853-1932. Br. BMNH.

10233. BASSANI, Francesco. 1853-1916. It. DSB; P IV; BMNH.

10234. BERGERON, Pierre Joseph Jules. 1853-after 1902. Fr. P IV.

10235. CATHREIN, Alois. 1853-after 1902. Aus. P IV.

10236. FYODOROV, Evgraf Stepanovich. 1853-1919. Russ. DSB; ICB (FEDOROV); P IV (FEDOROW, J.S.); BMNH (FEDOROV). Also Mineralogy and Crystallography.

10237. HEILPRIN, Angelo. 1853-1907. U.S.A. ICB; BMNH; GE. Also Zoology.
1. *The earth and its story. A first book of geology.* New York, 1896.

10238. HILBER, Vincenz. 1853-1931. Aus. P IV; BMNH.

10239. HILLEBRAND, William Francis. 1853-1925. U.S.A. ICB; P IV; BMNH; GE.

10240. HOLM, (Edvard Johan) Gerhard. 1853-after 1918. Swed. BMNH.

10241. HOLZAPFEL, (Gustav Hermann) Eduard. 1853-1913. Ger. P IV; BMNH.

10242. LEWIS, Henry Carvill. 1853-1888. U.S.A. P III; BMNH; GB. Also Mineralogy*.

10243. McGEE, William John. 1853-1912. U.S.A. BMNH; GE.

10244. OEBBEKE, Konrad Josef Ludwig. 1853-1932. Ger. P IV; BMNH.

10245. QUIROGA Y RODRÍGUEZ, Francisco. 1853-1894. Spain. BMNH.

10246. RAUFF, Carl Friedrich Hermann. 1853-after 1896. Ger. P III, IV; BMNH.

10247. REID, Clement. 1853-1916. Br. BMNH.

10248. ROTHPLETZ, (Friedrich) August. 1853-1918. Ger. P III,IV; BMNH.
1. *Geotektonische Probleme*. Stuttgart, 1894.

10249. ZURCHER, Philippe Elie Frédéric. 1853-after 1904. Fr. P IV (ZÜRCHER); BMNH.

10250. AMEGHINO, Florentino. 1854-1911. Argentina. DSB; ICB; BMNH.

10251. ASHBURNER, Charles Albert. 1854-1889. U.S.A. P III,IV; BMNH; GE.

10252. BÖHM, Georg. 1854-1913. Ger. P III; BMNH.

10253. CALDERÓN Y ARANA, Salvador. 1854-after 1910. Spain. BMNH.

10254. CAREZ, Léon Louis Hippolyte. 1854-after 1902. Fr. P IV; BMNH.

10255. CROSS, (Charles) Whitman. 1854-1949. U.S.A. DSB; P IV; BMNH.

10256. DEPÉRET, Charles Jean Julien. 1854-1929. Fr. DSB; P IV; BMNH.

10257. FICHEUR, (Louis) Emile. 1854-after 1902. Algeria. P IV; BMNH.

10258. GEINITZ, Franz Eugen. 1854-1925. Ger. P III,IV; BMNH.

10259. KITTL, Ernst Anton Leopold. 1854-1913. Aus. P III; BMNH.

10260. LANGENBECK, Rudolf August Justus. 1854-after 1922. Ger. BMNH.
1. *Die Theorien über die Entstehung der Koralleninseln und Korallenriffe und ihre Bedeutung für geophysische Fragen*. Leipzig, 1890.

10261. MERRILL, George Perkins. 1854-1929. U.S.A. DSB; ICB; BMNH; GE.
1. *A treatise on rocks, rock-weathering and soils*. New York, 1897.

10262. MOBERG, Johan Christian. 1854-1915. Swed. BMNH.

10263. NAUMANN, Edmund. 1854-1927. Ger./Japan. P III,IV; BMNH,

10264. PAVLOV, Aleksei Petrovich. 1854-1929. Russ. DSB; P IV; BMNH.

10265. SÁNCHEZ LOZANO, Rafael. b. 1854. Spain. BMNH.

10266. BODENBENDER, Wilhelm (or Guillermo). fl. 1885-1922. Argentina. P IV.

10267. GRAEFF, Franz Friedrich. 1855-1902. Ger. P IV; BMNH.

10268. HAAS, Hippolyte Julius. 1855-1913. Ger. BMNH.

10269. MIKHALSKII, Aleksandr Oktavianovich. 1855-1904. Russ. ICB (MICHALSKI); BMNH.

10270. PARONA, Carlo Fabrizio. 1855-after 1924. It. P III,IV; BMNH.

10271. POHLIG, Hans. 1855-after 1911. Ger. P IV; BMNH.

10272. RZEHAK, Anton. 1855-1923. Brünn. P IV.

10273. BEECHER, Charles Emerson. 1856-1904. U.S.A. P IV; BMNH; GB; GE. Also Evolution*.

10274. CHERNYSHEV, Feodosy Nikolaievich. 1856-1914. Russ. DSB; BMNH (CHERNUISHEV, T.N.).

10275. CRICK, George Charles. 1856-1917. Br. BMNH.

10276. HUTCHINSON, Henry Neville. 1856-1927. Br. BMNH.
1. *The autobiography of earth. A popular account of geological history.* London, 1890.
2. *Extinct monsters. A popular account.* Ib., 1892. 4th ed., 1896.
3. *Creatures of other days.* Ib., 1894.

10277. KENDALL, Percy Fry. 1856-1936. Br. ICB; BMNH.

10278. KOTO, Bunjiro. 1856-1935. Japan. DSB; BMNH.

10279. LÖWL, Ferdinand. 1856-1908. Prague/Czernowitz. P IV; BMNH.

10280. SJÖGREN, Sten Anders Hjalmar. 1856-1922. Swed. BMNH.

10281. SOKOLOV, Nikolai. 1856-1907. Russ. P IV; BMNH.

10282. STEINMANN, (Johann Heinrich Conrad Gottfried) Gustav. 1856-1929. Ger. ICB; P III,IV; BMNH.
1. *Elemente der Paläontologie.* Leipzig, 1890. With L. Döderlein.

10283. WILLIAMS, George Huntington. 1856-1894. U.S.A. BMNH; GE. Also Crystallography*.

10284. ZAITZEV, Aleksyei Mikhailovich. 1856-after 1902. Russ. BMNH.

10285. ZHUIOVIĆ (or ZUJOVIĆ), Iovan M. 1856-1936. Belgrade. BMNH.

10286. CHELIUS, Carl Robert Ludwig. 1857-after 1902. Ger. P IV.

10287. CLARKE, John Mason. 1857-1925. U.S.A. P IV; BMNH; GE.

10288. EBERT, Theodor. 1857-1899. Ger. P III,IV; BMNH.

10289. FALLOT, (Jean) Emmanuel. 1857-after 1902. Fr. P IV; BMNH.

10290. HÖGBOM, Arvid Gustaf. 1857-after 1936. Swed. P IV; BMNH.

10291. HOLLICK, (Charles) Arthur. 1857-1933. U.S.A. BMNH; GE1.

10292. IDDINGS, Joseph Paxson. 1857-1920. U.S.A. DSB; BMNH; GE. See also 8869/1a.

10293. KLVAŇA, Josef. 1857-after 1903. Prague. P III,IV; BMNH.

10294. LOHEST, Marie Joseph Maximin. 1857-1926. Belg. DSB; P IV; BMNH.

10295. MARR, John Edward. 1857-1933. Br. ICB; P IV & Supp.; BMNH; GD5.
1. *Principles of stratigraphical geology.* Cambridge, 1898.

10296. MIKLUKHO-MAKLAY, Mikhail Nikolaevich. 1857-1927. Russ. DSB Supp.

10297. MORGAN, Jacques Jean Marie de. 1857-1924. Fr. ICB; BMNH.

10298. NOETLING, Friedrich Wilhelm (or Fritz). 1857-after 1913. Ger./India. P IV; BMNH.

10299. STOCKBRIDGE, Horace Edward. 1857-1930. U.S.A. BMNH; GE.
1. *Rocks and soils, their origin, composition and characteristics.* New York, 1888.

10300. UHLIG, Viktor Karl. 1857-1911. Aus. P IV; BMNH.

10301. ULRICH, Edward Oscar. 1857-1944. U.S.A. DSB; BMNH.

10302. VAN HISE, Charles Richard. 1857-1918. U.S.A. DSB; ICB; P IV; BMNH; GE.

10303. WILLIS, Bailey. 1857-1949. U.S.A. DSB; P IV; BMNH.

10304. WOLFF, John Eliot. 1857-1940. U.S.A. P IV; BMNH.

10305. AGAMENNONE, Giovanni. 1858-after 1931. It. P IV.

10306. BECK, Carl Richard. 1858-1919. Ger. P III,IV; BMNH.

10307. DAVID, (Tannatt William) Edgeworth. 1858-1934. Australia. DSB; ICB; P IV; BMNH; GD5.

10308. DAVISON, Charles. 1858-after 1925. Br. P IV.

10309. DELVEAUX, Eugène. 1858-after 1896. Belg. P IV.

10310. GEER, Gerhard Jakob de. 1858-1943. Swed. DSB; ICB; P IV (under DE); BMNH.

10311. HAYES, Charles Willand. 1858-1916. U.S.A. BMNH; GE.

10312. HILL, Robert Thomas. 1858-after 1900. U.S.A. P IV; BMNH.

10313. KEILHACK, (Friedrich Ludwig Heinrich) Konrad. 1858-after 1937. Ger. P IV; BMNH.
1. *Lehrbuch der praktischen Geologie. Arbeits- und Untersuchungsmethoden.* Stuttgart, 1896.

10314. KLOCKMANN, Friedrich Ferdinand Hermann. 1858-after 1923. Ger. P IV; BMNH.

10315. OLDHAM, Richard Dixon. 1858-1936. India. DSB; ICB; P IV; BMNH. See also 9668/1.

10316. PENCK, (Friedrich Carl) Albrecht. 1858-1945. Ger./Aus. DSB; P III,IV; BMNH.
1. *Morphologie der Erdoberfläche.* Stuttgart, 1894.

10317. SALISBURY, Rollin Daniel. 1858-1922. U.S.A. DSB; BMNH; GE.

10318. SCHARDT, Hans. 1858-1931. Switz. DSB; P IV; BMNH.

10319. SCHUCHERT, Charles. 1858-1942. U.S.A. DSB; P IV; BMNH.

10320. SCOTT, William Berryman. 1858-1947. U.S.A. DSB; ICB; BMNH.
1. *Introduction to geology.* New York, 1898.

10321. SIEMIRADZKI, Josef. 1858-1933. Lemberg. P III,IV; BMNH.

10322. TAUSCH VON GLÖCKELSTHURN, Leopold. 1858-1899. P IV; BMNH.

10323. TOLL, Eduard Vasilievich von. 1858-1903? Russ. BMNH.

10324. TYRRELL, Joseph Burr. 1858-1957. Canada. DSB; ICB; P IV; BMNH.

10325. VOGT, Johan Hermann Lie. 1858-1932. Nor. DSB; P IV; BMNH.

10326. WALLERANT, Frédéric Félix Auguste. 1858-1936. Fr. P IV; BMNH. Also Crystallography*.

10327. WELSCH, Jules Augustin. 1858-1929. Algeria/Fr. ICB; P IV.

10328. WENTZEL, Josef. 1858-after 1904. Prague. P IV.

10329. ZUBER, Rudolf. 1858-after 1904. Lemberg. P IV; BMNH.

10330. ADAMS, Frank Dawson. 1859-1942. Canada. DSB; ICB; P IV; BMNH.

10331. ANDREAE, Achilles. 1859-after 1903. Ger. P IV; BMNH.

10332. COLE, Grenville Arthur James. 1859-1924. Br. P IV; BMNH.

10333. FELIX, Johannes Paul. 1859-after 1912. Ger. P III,IV; BMNH.

10334. GOURRET, Paul Gabriel Marie. 1859-after 1906. Fr. P IV; BMNH. Also Zoology.

10335. GÜRICH, Georg Julius Ernst. 1859-after 1933. Ger. P IV; BMNH.

10336. HARKER, Alfred. 1859-1939. Br. DSB; ICB; P IV; BMNH; GD5.
1. *Petrology for students*. Cambridge, 1895. X. 2nd ed., 1897.

10337. HERRMANN, Max Otto. 1859-after 1903. Ger. P IV; BMNH.

10338. JOHNSON, Willard Drake. 1859-1917. U.S.A. DSB.

10339. KEMP, James Furman. 1859-1926. U.S.A. ICB; P IV & Supp.; BMNH; GE.
1. *The ore deposits of the United States*. New York, 1893. RM.

10340. LAMPLUGH, George William. 1859-1926. Br. P IV; BMNH.

10341. LEPPLA, August. 1859-1924. Ger. P IV; BMNH.

10342. LEVERETT, Frank. 1859-after 1931. U.S.A. P IV; BMNH.

10343. MIDDLEMISS, Charles Stewart. 1859-1945. Br./India. ICB; BMNH.

10344. NICKLÈS, Toussaint Joseph René. 1859-1917. Spain. BMNH.

10345. OSANN, Carl Alfred. 1859-1923. Ger. P IV; BMNH.

10346. POČTA, Philipp (or Filip). 1859-1924. Prague. P IV; BMNH.

10347. REID, Harry Fielding. 1859-1944. U.S.A. DSB.

10348. ROLLIER, Henri Louis. 1859-1932. Switz. BMNH.

10349. TERMIER, Pierre. 1859-1930. Fr. DSB; ICB; P IV; BMNH.

10350. THÜRACH, Hans. 1859-after 1912. Ger. P IV; BMNH.

10351. UDDEN, Johan August. 1859-1932. U.S.A. ICB; BMNH; GE.

10352. VATER, Heinrich August. 1859-after 1904. Ger. P IV; BMNH.

Imprint Sequence

10353. ACCADEMIA delle Scienze. Naples.
1. [Trans.] *The natural history of Mount Vesuvius, with the explanation of the various phenomena that usually attend the eruptions of this celebrated volcano*. London, 1743.

10354. KLOTZSCH, Johann Friedrich.
1. *Gedanken von der Erfindung des Bergwerkes zu Freyberg*. Chemnitz, 1763. RM.
2. *Ursprung der Bergwerke in Sachsen, aus der Geschichte mittler Zeiten untersucht*. Ib., 1764. RM.
3. *Vom Gegenbuche. Ein Beytrag zur sächsischen Bergwerksgeschichte*. Ib., 1780. RM.

10355. PALAEONTOGRAPHICAL Society. London. (Founded 1847)
BMNH gives a list of the numerous monographs published by the Society from 1848 onwards.

10356. REYNOLDS, James. (Publisher) BMNH.
1. *Reynold's geological atlas of Great Britain*. London, 1860.

10357. MUSEUM of Practical Geology. London.
1. *Catalogue of the library*. London, 1878.

10358. BRITISH Museum (Natural History). Department of Geology.
1. *A guide to the exhibition galleries*. London, 1881. By H. Woodward. X. 4th ed., 1886.
2. *Catalogue of the fossil fishes*. Ib., 1889-91. By A.S. Woodward.
3. *A guide to the fossil invertebrates and plants*. Ib., 1897. By H. Woodward.

10359. GEOLOGICAL Society. London.
1. *Catalogue of the library*. London, 1881.

10360. ROYAL Society of London. Krakatoa Committee.
1. *The eruption of Krakatoa and subsequent phenomena*. London, 1888. Ed. by G.J. Symons.

3.10 OTHER EARTH SCIENCES

Comprising geodesy, geomagnetism, geophysics, meteorology (including atmospheric physics), climatology, oceanography, and limnology.

Oceanographic expeditions are included in the list of scientific expeditions given in Section 3.121 (notably entries 11299-11302, 11307, and 11309).

10361. BARLOW (or BOOTH), Edward. 1639-1719. Br. GD. Meteorology. (Also Part 2)
   1. *Meteorological essays, concerning the origin of springs, generation of rain, and production of wind.* London, 1715. RM/L.

10362. LA HIRE, Philippe de. 1640-1718. Fr. DSB; ICB; P I; GA. Geodesy. (Also Mathematics, Astronomy*, Physics, and Part 2*)

10363. HALLEY, Edmond. 1656?-1743. Br. DSB; ICB; P I; GA; GB; GD. Geomagnetism. (Also Various Fields*, Astronomy*#, and Part 2)

10364. MARSIGLI, Luigi Ferdinando. 1658-1730. It. DSB (MARSILI); ICB; P II; BMNH; GA; GB. Oceanography. (Also Geology and Part 2)
   1. *Histoire physique de la mer.* Amsterdam, 1725. With a preface by H. Boerhaave.

10365. MARALDI, Giacomo Filippo. 1665-1729. Fr. DSB; P II; GA. Geodesy. (Also Astronomy)

10366. MÜLLER, Johann Christoph. 1673-1721. Aus. P II. Geodesy. Cartography.

10367. DELISLE, Guillaume. 1675-1726. Fr. DSB; ICB; P I. Cartography.

10368. LE MONNIER, Pierre. 1675-1757. Fr. P I; GA. Geodesy.

10369. CASSINI, Jacques. 1677-1756. Fr. DSB; P I; GA; GB. Geodesy. (Also Astronomy*)

10370. MAIRAN, Jean Jacques d'Ortus de. 1678-1771. Fr. DSB; ICB; P II; GA. Meteorology. (Also Mechanics*, Physics*, and Biography*)
   1. *Traité physique et historique de l'aurore boréale.* Paris, 1733. RM.

10371. TRAUTMANN, Christian. 1678-1740. Ger. P II. Meteorology.

10372. GRISCHOW, Augustin. 1683-1749. Ger. P I; GA. Meteorology.

10373. SPIDBERG, Jens Christian. 1684-1762. Nor. ICB; P II. Meteorology.

10374. HADLEY, George. 1685-1768. Br. GD. Meteorology.

10375. BARBIER, Edmond Jean François. 1689-1771. Fr. ICB. Meteorology.

10376. MICHELI DU CREST, Jacques Barthélemi. 1690-1766. Switz. ICB; P II; GA. Meteorology.

10377. SHORT, Thomas. 1690?-1772. Br. P II; BMNH; BLA; GA. Meteorology.

10378. OUTHIER, Réginald. 1694-1774. Fr. DSB; P II; GA. Geodesy and cartography. (Also Astronomy) See also 10382/1.

10379. PEYSSONNEL, Jean André. 1694-1759. Fr. DSB. Oceanography. (Also Zoology)

10380. ANVILLE, Jean Baptiste Bourguignon d'. 1697-1782. Fr. DSB; ICB; P I; GA; GB; GF. Geodesy and cartography.

10381. BOUGUER, Pierre. 1698-1758. Fr. DSB; ICB; P I; GA; GB. Geodesy. (Also Astronomy*, Mechanics*, and Physics*)
1. *La figure de la terre, déterminée par les observations de Messieurs De la Condamine et Bouguer ... envoyés par ordre du roi au Pérou pour observer aux environs de l'équateur.* Paris, 1749. RM/L.

10382. MAUPERTUIS, Pierre Louis Moreau de. 1698-1759. Fr./Ger. DSB; ICB; P II; GA; GB; GC. Geodesy. (Other entries: see 3089)
1. *La figure de la terre, déterminée par les observations de Messieurs de Maupertuis, Clairaut, Camus, Le Monnier et de M. l'abbé Outhier, accompagnés de M. Celsius ... au cercle polaire.* Paris, 1738. RM.
1a. ——— *The figure of the earth, determined from the observations, made by order of the French King, at the polar circle, by Messrs de Maupertuis....* London, 1738. RM/L.
2. *Examen désintéressé des différens ouvrages qui ont été faits pour déterminer la figure de la terre.* Oldenburg, 1738. RM.
3. *Examen des trois dissertations que Monsieur Desaguliers a publiées sur la figure de la terre.* Oldenburg, 1738. RM.
4. *Elémens de géographie.* Paris, 1740. Anonymous.

10383. DODSON, James. d. 1757. Br. P I; GA; GD. Geomagnetism.

10384. BUACHE, Philippe. 1700-1773. Fr. DSB; P I; GA. Physical geography and cartography.

10385. CELSIUS, Anders. 1701-1744. Swed. DSB; ICB; P I; GA; GB. Geodesy. (Also Astronomy)
1. *De observationibus pro figura telluris determinanda in Gallia habitis disquisitio.* Upsala, 1738. RM.
See also 10382/1.

10386. LA CONDAMINE, Charles Marie. 1701-1774. Fr. DSB Supp.; ICB; P I (CONDAMINE); BMNH; BLA (CONDAMINE); GA; GB. Geodesy. (Also Natural History)
1. *Mesure des trois premiers degrés du méridien dans l'hémisphère austral.* Paris, 1751.
See also 10381/1 and 10415/2.

10387. WERENBERG, Johann Georg. 1702-1780. Ger. P II. Meteorology.

10388. WINKLER, Johann Heinrich. 1703-1770. Ger. P II; GC (WINCKLER). Atmospheric physics. (Also Physics*)
1. *De vi luminis borealis in commovenda acu magnetica.* Leipzig, 1768. RM.

10389. FRANKLIN, Benjamin. 1706-1790. U.S.A. The famous statesman. DSB; ICB; P I; G. Oceanography and meteorology. (Also Physics* and Autobiography*. See also Index.)

10390. WALLERIUS, Nils. 1706-1764. Swed. P II. Meteorology.

10391. EULER, Leonhard. 1707-1783. Ger./Russ. DSB; ICB; P I; GA; GB; GC. Geodesy and cartography. (Also Science in General*, Mathematics*#, Astronomy*, Mechanics*, and Physics*. See also Index.)
1. *Tabula geographica haemisphaerii australis.* Berlin, 1753.
2. *Drei Abhandlungen über Kartenprojection.* [1777] Leipzig, 1898. (OKEW. 93)

10392. MEAD, Joseph. 1707-1799. Br. P II. Oceanography.

10393. MARALDI, Giovanni Domenico. 1709-1788. Fr. DSB; P II; GA. Geodesy. (Also Astronomy)

10394. BOSCOVICH, Ruggiero Giuseppe. 1711-1787. It./Fr. DSB (BOŠKOVIĆ); ICB; P I; GA; GB. Geodesy and meteorology. (Other entries: see 3114)
1. *Sopra il turbine che ...* [in June 1749] *danneggio una gran parte di Roma.* Rome, 1749. RM.

10395. CLAIRAUT, Alexis Claude. 1713-1765. Fr. DSB; ICB; P I; GA; GB. Geodesy. (Also Mathematics*, Astronomy*, Mechanics, and Physics. See also Index.)
1. *Théorie de la figure de la terre, tirée des principes de l'hydrostatique.* Paris, 1743. RM. Reprinted, ib., 1808.
1a. ——— German trans. Leipzig, 1913. (OKEW. 189)

10396. JUAN Y SANTACILLA, Jorge. 1713-1773. Spain/Peru. DSB; P I; GA. Geodesy.
1. *Observaciones astronomicas y fisicas hechas de orden de S.M. en los reynos del Peru ... de las quales se deduce la figura y magnitud de la tierra.* Madrid, 1748. With A. de Ulloa. RM/L.

10397. LACAILLE, Nicolas Louis de. 1713-1762. Fr. DSB; ICB; P I; GA (CAILLE); GB (CAILLE). Geodesy. (Also Astronomy*)

10398. CASSINI DE THURY, César François. 1714-1784. Fr. DEB; ICB; P I; GA; GB. Geodesy and cartography. (Also Astronomy)
1. *La méridienne de l'Observatoire royal de Paris vérifée ... par de nouvelles observations.* Paris, 1744. RM.
2. *Rélation d'un voyage en Allemagne qui comprend les opérations relatives à la figure de la terre.* Ib., 1775. RM.
3. *Description géométrique de la France.* Ib., 1783. RM.

10399. LE MONNIER, Pierre Charles. 1715-1799. Fr. DSB; ICB; P I; GA; GB. Geomagnetism. (Also Astronomy*)
1. *Loix du magnétisme comparées aux observations et aux expériences dans les différentes parties du globe terrestre*

pour perfectionner la théorie générale de l'aimant et indiquer par-là les courbes magnétiques qu'on cherche à la mer, sur les cartes réduites. Paris, 1776. RM.

See also 2278/1a and 10382/1.

10400. LIESGANIG, Joseph Xaver. 1719-1799. Aus. DSB; ICB; P I. Geodesy.

10401. TOALDO, Giuseppe. 1719-1797. It. P II; GA. Meteorology.
1. *Saggio meteorologico della vera influenza degli astri sulli stagioni e mutazione del tempo.* Padua, 1770. X.
1a. ——— *Essai météorologique sur la véritable influence des astres, des saisons et changemens de tems.* Chambéry, 1784. Trans. by Daquin. RM.
2. *Dei conduttori per preservare gli edifizi da' fulmine.* Venice, 1778. RM.

10402. BARKER, Thomas. 1722-1809. Br. P I; GD. Meteorology. (Also Astronomy)

10403. FELBIGER, Johann Ignatz von. 1724-1788. Ger./Aus. P I; GC. Meteorology.

10404. MOUNTAINE, William. d. 1779. Br. ICB; P II. Geomagnetism.

10405. ROY, William. 1726-1790. Br. P II; GB; GD. Geodesy.

10406. DELUC, Jean André (first of the name). 1727-1817. Switz./Br./Ger. DSB; P I; BMNH; GA; GB; GD. Meteorology. (Also Physics* and Geology*)
1. *Recherches sur les modifications de l'atmosphère. Contenant l'histoire critique du baromètre et du thermomètre, un traité sur la construction de ces instrumens, des expériences relatives à leurs usages....* 2 vols. Geneva, 1772. RM.

10407. GERLACH, Friedrich Wilhelm Anton. 1728-1802. Aus. P I. Geodesy.
1. *Die Bestimmung der Gestalt und Grösse der Erde, wie auch der Vorrückung der Nachtgleichen, Schwankung der Erdaxe, etc.* Vienna, 1782. RM.

10408. LAMBERT, Johann Heinrich. 1728-1777. Switz./Ger. DSB; ICB; P I; GA; GB; GC. Cartography. (Also Mathematics*, Astronomy* and Physics*)
1. *Anmerkungen und Zusätze zur Entwerfung der Land- und Himmelscharten.* [Periodical article, 1772] Leipzig, 1894. (OKEW. 54)
1a. ——— *Notes and comments on the composition of terrestrial and celestial maps.* Ann Arbor, Mich., 1972. Trans. with an introd. by W.R. Tobler.

10409. MASON, Charles. 1728-1786. Br. DSB; ICB; P II; GA; GD. Geodesy. (Also Astronomy)

10410. LORIMER, John. 1732-1795. Br. P I. Geomagnetism.

10411. BORDA, Jean Charles. 1733-1799. Fr. DSB; ICB; P I; GA; GB. Geodesy. (Also Mechanics)
1. *Description et usage du cercle de réflexion.* Paris, 1787. X. Another ed., 1810. A surveying instrument which proved to be valuable in 18th-century geodesy.

2. *Mémoire sur les expériences faites par Borda sur la longeur du pendule.* [Periodical article by J.D. Cassini, 1792] German trans. Leipzig, 1911. (OKEW. 181)

10412. HEMMER, Johann Jacob. 1733-1790. Ger. P I. Meteorology.

10413. KIRWAN, Richard. 1733?-1812. Br. DSB; ICB; P I; BMNH; GA; GB; GD. Meteorology. (Also Chemistry*, Mineralogy*, and Geology*. See also Index.)
1. *An estimate of the temperature of different latitudes.* London, 1787.
1a. ——— *Estimation de la température de différens degrés de latitude.* Paris, 1790. Trans. by P.A. Adet.

10414. CRAMER, Joseph Anton. 1737-1794. Ger. P I. Meteorology.

10415. DAVID, Jean Pierre. 1737-1784. Fr. P III; BLA; GA. Geodesy.
1. *Dissertation sur la figure de la terre, ou l'on tâche de prouver ... d'après les expériences mêmes faites au Pérou et au cercle polaire que cette planète est allongée par les poles.* The Hague, 1769. RM.
2. *Réplique à la lettre de M. de la Condamine.* Ib., 1769. RM.

10416. DUCARLA-BONIFACE, Marcellin. 1738-1816. Fr. P I; GA. Physical geography, meteorology, etc.

10417. CHANGEUX, Pierre Nicolas. 1740-1800. Fr. P I; GA (C., Pierre Jacques). Meteorology.

10418. COTTE (or COTTA), Louis. 1740-1815. Fr. DSB; P I; GA. Meteorology.
1. *Traité de météorologie.* Paris, 1774. RM.

10419. CHIMINELLO, Vincenzo. 1741-1815. It. P I. Meteorology. (Also Astronomy)

10420. INOCHODZOW, Peter. ca. 1741-1806. Russ. P I. Meteorology and geomagnetism.

10421. GRONAU, Karl Ludwig. 1742-1826. Ger. P I. Meteorology.

10422. RENNELL, James. 1742-1830. Br./India. DSB; ICB; P II; GA; GB; GD. Physical geography and oceanography.

10423. CAPPER, James. 1743-1825. Br. P I; GA; GD. Meteorology.
1. *Observations on the winds and monsoons.* London, 1801. RM.
2. *Meteorological and miscellaneous tracts.* Cardiff, 1810. RM.

10424. KRAFFT, Wolfgang Ludwig. 1743-1814. Russ. P I; GC (KRAFT). Geomagnetism. (Also Astronomy)

10425. OTTO, Johann Friedrich Wilhelm. 1743-1814. Ger. P II & Supp. Physical geography and oceanography.

10426. DALBY, Isaac. 1744-1824. Br. P I & Supp.; GD. Geodesy. See also 10444/1.

10427. MÉCHAIN, Pierre François André. 1744-1804. Fr. DSB; ICB; P II; GA. Geodesy. (Also Astronomy) See also 10432/1 and 10433/2.

10428. WÜNSCH, Christian Ernst. 1744-1828. Ger. P II; GC. Meteorology. (Also Physics)

1. *Neue Theorie von der Atmosphäre und Höhenmessung mit Barometern*. Leipzig, 1782. RM.

10429. ROSENTHAL, Gottfried Erich. 1745-1814. Ger. P II. Meteorology.

10430. SWINDEN, Jan Hendrick van. 1746-1823. Holl. DSB; ICB; P II; GF. Meteorology and geomagnetism. (Also Physics*)

10431. STRNAD, Anton. 1747-1799. Prague. ICB; P II; GC. Meteorology.

10432. CASSINI, Jean Dominique. 1748-1845. Fr. DSB; ICB; P I; GA; GB. Geodesy. (Also Astronomy, Autobiography*, and History of Astronomy*. See also Index.)
1. *Exposé des opérations faites en France en 1787 pour la jonction des observatoires de Paris et de Greenwich, par MM. Cassini, Méchain, et Le Gendre*. Paris, 1792. RM.

10433. DELAMBRE, Jean Baptiste Joseph. 1749-1822. Fr. DSB; ICB; P I; GA; GB. Geodesy. (Also Mathematics*, Astronomy*, and History of Astronomy*)
1. *Méthodes analytiques pour la détermination d'un arc de méridien*. Paris, 1799. RM.
2. *Base du système métrique décimal; ou, Mesure de l'arc du méridien compris entre les parallèles de Dunkerque et Barcelone, executée en 1792 et années suivantes par MM. Méchain et Delambre*. 3 vols. Paris, 1806-10.
2a. ——— German trans. of selected extracts. Leipzig, 1911. (OKEW. 181)

10434. LA COUDRAYE, François Célestin de Loynes-Barraud, *Chevalier* de. ca. 1750-1815. Fr./Den./Russ. P I. Oceanography.
1. *Théories des vents et des ondes*. Copenhagen, 1796. RM.

10435. LEGENDRE, Adrien Marie. 1752-1833. Fr. DSB; ICB; P I; GA; GB. Geodesy. (Also Mathematics* and Astronomy*)
1. *Méthode pour déterminer la longeur exacte du quart du méridien, d'après les observations faites pour la mesure de l'arc compris entre Dunkerque et Barcelone*. Paris, 1799. RM.
See also 10432/1.

10436. SCHIEGG, Ulrich. 1752-1810. Ger. ICB; P II; GC. Geodesy.

10437. EHRENHEIM, Frederik Wilhelm von. ca. 1753-1828. Swed. ICB; P I. Meteorology.

10438. ZACH, Franz Xaver von. 1754-1832. Aus./Ger. DSB; ICB; P II; GA; GB; GC. Geodesy. (Also Astronomy*)
1. *Mémoire sur le degré du méridien, mesuré en Piédmont par le P. Beccaria*. Turin, 1811. RM.
2. *L'attraction des montagnes, et ses effets sur les fils à plomb ou sur les niveaux des instrumens d'astronomie, constatés et determinés par des observations astronomiques et géodesiques*. Avignon, 1814.

10439. WELLS, William Charles. 1757-1817. Br./U.S.A. DSB; ICB; P II; BMNH; BLA; GD; GE. Meteorology. (Other entries: see 3243)
1. *An essay on dew, and several appearances connected with it.*

London, 1814. RM. Another ed., 1818 (see 3243/1). Reprinted, London, 1866 and 1877.

10440. ELLINGER, Anselm. 1758-1816. Ger. P I. Meteorology.
1. *Von den bisherigen Versuchen über längere Voraussicht der Witterung. Eine geschichtliche Skizze mit Bemerkungen.* Munich, 1815. RM.

10441. KRAYENHOFF, Cornelis Rudolphus Theodorus. 1758-1840. Holl. DSB; P I. Geodesy.

10442. DITTMAR (or DIETMAR), Siegmund Gottfried. 1759-1834. Ger. P I. Meteorology.

10443. VASSALLI-EANDI, Antonio Maria. 1761-1825. It. P II. Meteorology. (Also Physics)

10444. MUDGE, William. 1762-1821. Br. P II; GA; GD. Geodesy.
1. *An account of the ... trigonometrical survey of England and Wales.* London, 1799. With Isaac Dalby.

10445. ZYLIUS, Johann Diedrich Otto. 1764-1820. Ger. P II. Meteorology.

10446. IVORY, James. 1765-1842. Br. DSB; P I; GB; GD. Geodesy. (Also Mathematics, Astronomy, Mechanics, and Physics)

10447. LACROIX, Sylvestre François. 1765-1843. Fr. DSB; ICB; P I; GA. Mathematical and physical geography. (Also Mathematics*)
1. *Introduction à la géographie mathématique et critique, et à la géographie physique.* Paris, 1811. X. Another ed., 1849.

10448. DALTON, John. 1766-1844. Br. DSB; ICB; P I; GA; GB; GD. Meteorology. (Also Physics and Chemistry*. See also Index.)
1. *Meteorological observations and essays.* London, 1793. RM/L. 2nd ed., Manchester, 1834.

10449. STEINHÄUSER, Johann Gottfried. 1768-1825. Ger. P II; GC. Geomagnetism.

10450. HUMBOLDT, Alexander von. 1769-1859. Ger./Fr. DSB; ICB; P I, III; BMNH; G. The earth sciences generally. (Other entries: see 3268)
1. *Fragments de géologie et de climatologie asiatiques.* Paris, 1831. RM.
2. *Asie centrale. Recherches sur les chaînes de montagnes et la climatologie comparée.* 3 vols. Paris, 1843.
2a. ——— *Central Asien. Untersuchungen über die Gebirgsketten und die vergleichende Klimatologie.* Berlin, 1844. Trans. with notes by W. Mahlmann. RM/L.

10451. PUISSANT, Louis. 1769-1843. Fr. P II; GA. Geodesy and cartography.
1. *Traité de géodesie; ou, Exposition des méthodes astronomiques et trigonométriques....* Paris, 1805. RM/L. 2nd ed., 2 vols, 1819.

10452. HASSLER, Ferdinand Rudolph. 1770-1843. U.S.A. DSB; ICB; P I; GE. Geodesy.

10453. KRUSENSTERN, Adam Johann von. 1770-1846. Russ. P I; BMNH;

GA; GB; GC. "The founder of Russian oceanography."

10454. SCHÖN, Johann. 1771-1839. Ger. P II; GC. Meteorology. (Also Mathematics)

10455. SVANBERG, Jöns. 1771-1851. Swed. P II. Geodesy. (Also Mathematics)
1. *Exposition des opérations faites en Lapponie pour la détermination d'un arc du méridien en 1801-1803 par Ms. Ofverbom et al. Redigé par J. Svanberg.* Stockholm, 1805. RM.

10456. HOWARD, Luke. 1772-1864. Br. ICB; P I,III; GD. Meteorology.
1. *The climate of London.* 2 vols. London, 1818-20. X. Another ed., 3 vols, 1842.
2. *Papers on meteorology.* London, 1850-54.
3. *His correspondence with Goethe and his Continental journey of 1816.* York, 1976. Ed. by D.F.S. Scott.

10457. BAIN, William. 1775 (or 1771)-1853. Br. P I; GF. Geomagnetism.
1. *An essay on the variation of the compass.* Edinburgh, 1817. RM/L.

10458. SOLDNER, Johann Georg von. 1776-1833. Ger. DSB; ICB; P II; GC. Geodesy. (Also Astronomy)
1. *Theorie der Landesvermessung.* 1810. X. Reprinted, Leipzig, 1911. (OKEW. 184)

10459. BENZENBERG, Johann Friedrich. 1777-1846. Ger. DSB; ICB; P I; GA; GC. Geodesy. (Also Astronomy and Physics)

10460. GAUSS, Karl Friedrich. 1777-1855. Ger. DSB; ICB; P I,III; GA; GB; GC. Geodesy and geomagnetism. (Also Mathematics*#, Astronomy*, Mechanics*, and Physics*. See also Index.)
1. *Intensitas vis magneticae terrestris ad mensuram absolutam revocata.* Göttingen, 1833. RM.
1a. ——— *Die Intensität der erdmagnetischen Kraft auf absolutes Maass zuruckgeführt.* Leipzig, 1894. (OKEW. 53)
2. *Resultate aus den Beobachtungen des magnetischen Vereins, 1836-41.* 6 vols & atlas of 3 vols. Göttingen, 1837-43. With W.E. Weber.
3. *Untersuchungen über Gegenstände der höheren Geodäsie.* [Periodical articles, 1845 and 1847] Leipzig, 1910. (OKEW. 177)

10461. KATER, Henry. 1777-1835. Br. DSB; P I; GA; GB; GD. Geodesy. (Also Mechanics*)

10462. MACKENZIE, George. 1777-1856. Br. GD. Meteorology.

10463. MUDIE, Robert. 1777-1842. Br. BMNH; GA; GD. Meteorology. (Also Various Fields)
1. *The air. A popular account of the atmospheric fluid in its composition, action, phenomena, and uses in the economy of nature.* London, 1835. X. Another ed., 1838.
2. *The earth.* Ib., 1835. Another ed., 1837.
3. *The sea.* Ib., 1835. X. Another ed., 1837.

10464. HOPKINS, Thomas. 1780-1864. Br. P IV. Meteorology.
1. *On the atmospheric changes which produce rain and wind,*

and the fluctations of the barometer. London, 1844. X. 2nd ed., with additions, 1854.
2. *On winds and storms.* London, 1860.

10465. SCHUMACHER, Heinrich Christian. 1780-1850. Den. DSB; ICB; P II; GB; GC. Geodesy. (Also Astronomy)

10466. FARQUHARSON, James. 1781-1843. Br. P I; GD. Meteorology.

10467. MURPHY, Patrick. 1782-1847. Br. P II; GD. Meteorology.

10468. OLTMANNS, Jabbo. 1783-1833. Ger. P II; GA. Astronomical geography.

10469. BESSEL, Friedrich Wilhelm. 1784-1846. Ger. DSB; ICB; P I; GA; GB; GC. Geodesy and geophysics. (Also Mathematics and Astronomy*#. See also Index.)
1. *Gradmessung in Ostpreussen.* Berlin, 1838. With J.J. von Baeyer.
2. *Untersuchungen über die Länge des einfachen Secundenpendels.* [Periodical article, 1826] Leipzig, 1889. (OKEW. 7)

10470. COLBY, Thomas Frederick. 1784-1852. Br. P I; GB; GD. (See also 15443/1) Geodesy.

10471. CRONSTRAND, Simon Anders. 1784-1850. Swed. P I. Geodesy. (Also Astronomy)

10472. HANSTEEN, Christopher. 1784-1873. Nor. DSB; P I,III; GA; GB. Geomagnetism. (Also Astronomy and Physics*)
1. *Untersuchungen über den Magnetismus der Erde.* Christiania, 1819.

10473. ESPY, James Pollard. 1785-1860. U.S.A. DSB; ICB; P I & Supp.; GE. Meteorology.
1. *The philosophy of storms.* Boston, 1841.

10474. ARAGO, (Dominique) François (Jean). 1786-1853. Fr. DSB; ICB; P I; GA; GB. Geodesy and meteorology. (Also Astronomy*#, Physics, and Biography*)
1. [Trans.] *Meteorological essays.* London, 1855. Trans. under the superintendence of E. Sabine. RM/L.

10475. DUPERREY, Louis Isidore. 1786-1865. Fr. DSB; P I,III; BMNH; GA. Oceanography and geomagnetism. See also 11279/1.

10476. GALBRAITH, William. 1786-1850. Br. P I. Geodesy.

10477. NICOLLET, Joseph Nicolas. 1786-1843. Fr./U.S.A. ICB; P II & Supp.; GA; GE. Geodesy, etc.
1. *The journals.* St. Paul, Minn., 1970. Trans. from the French by A. Ferty. Ed. by M.C. Bray.

10478. SCHÜBLER, Gustav. 1787-1834. Ger. P II; BMNH; BLA. Meteorology. (Also Botany)

10479. BECQUEREL, Antoine César. 1788-1878. Fr. DSB; P I,III; GA; GB. Meteorology. (Also Physics* and Chemistry*)
1. *Eléments de physique terrestre et de météorologie.* Paris, 1847. With E. Becquerel. RM.

10480. RONALDS, Francis. 1788-1873. Br. P II,III; GD. Meteorology. (Also Physics*)

10481. SABINE, Edward. 1788-1883. Br. DSB; P II,III; GB; GD. Geomagnetism, etc.
1. *An account of experiments to determine the figure of the earth by means of the pendulum vibrating seconds in different latitudes.* London, 1825. RM/L.
2. *Report on the variations of the magnetic intensity observed at different points of the earth's surface.* Ib., 1838. RM.
3. *Observations on days of unusual magnetic disturbances, made at the British colonial magnetic observatories.* 2 parts. Ib., 1843-51.

See also 10474/1 and 10504/1.

10482. FORSTER, Thomas Ignatius Maria. 1789-1860. Br. P I; BMNH; GA; GD. Meteorology. (Also Zoology)
1. *Researches about atmospheric phaenomena.* London, 1813. RM. 2nd ed., 1815. 3rd ed., 1823.
2. *Illustrations of the atmospherical origin of epidemic disorders of health.* Chelmsford, 1829.
3. *Essai sur l'influence des comètes sur les phénomènes de la terre. Seconde édition, augmentée d'un autre sur les étoiles filantes.* Bruges, 1843. RM.
4. *Mémoire sur les étoiles filantes ainsi que sur les météores en général, par rapport à leurs causes déterminantes.* Bruges, 1846. RM.

10483. FOX, Robert Were. 1789-1877. Br. P I,III; GB; GD. Geophysics.

10484. REDFIELD, William C. 1789-1857. U.S.A. DSB; P II; GE. Meteorology. (Also Geology)

10485. SCHOUW, Joakim Frederik. 1789-1852. Den. DSB; P II; BMNH. Climatology. (Also Botany*)

10486. SCHUBERT, Theodor Friedrich von. 1789-1865. Russ. P III; GC. Geodesy.

10487. SCORESBY, William. 1789-1857. Br. P II; BMNH; GB; GD. Geomagnetism. (Also Physics* and Natural History)
1. *Magnetical investigations.* 2 vols. London, 1844-52.
2. *Journal of a voyage to Australia and round the world for magnetic research.* Ib., 1859. Ed. with an introd. by Archibald Smith.

10488. DANIELL, John Frederic. 1790-1845. Br. DSB; ICB; P I; GB; GD. Meteorology. (Also Physics and Chemistry*)
1. *Meteorological essays and observations.* London, 1823. RM. 2nd ed., 1827. 3rd ed., entitled *Elements of meteorology,* 2 vols, 1845.

10489. JACKSON, Julian. 1790-1853. Br./Russ. P I; BMNH; GD. Limnology.
1. *Observations on lakes, being an attempt to explain the laws of nature regarding them.* London, 1833. RM.

10490. SYKES, William Henry. 1790-1872. Br./India. P II,III; GD; GF. Meteorology.

10491. LARTIGUE, Joseph. 1791-1876. Fr. P III; GA. Meteorology.

10492. MORIN, Pierre Etienne. 1791-1848. Fr. P II & Supp. Meteorology.

10493. REID, William. 1791-1858. Br. P II; GA; GB; GD. Meteorology.
1. *An attempt to develop the law of storms.* London, 1838. RM/L. 3rd ed., 1850.
2. *The progress of the development of the law of storms and of the variable winds.* London, 1849.

10494. KOLLER, Marian. 1792-1866. Aus. P I,III. Meteorology and geomagnetism. (Also Astronomy)

10495. STAMPFER, Simon. 1792-1864. Aus. P II,III,IV; GC. Geodesy. (Also Astronomy)

10496. STRUVE, Friedrich Georg Wilhelm. 1793-1864. Russ. DSB; ICB; P II,III; GA; GB; GC. Geodesy. (Also Astronomy*)

10497. BABINET, Jacques. 1794-1872. Fr. DSB; P I,III; GA. Meteorology. (Also Science in General* and Physics)

10498. BAEYER, Johann Jacob von. 1794-1885. Ger. P I,III,IV. Geodesy. See also 10469/1.

10499. FISHER, George. 1794-1873. Br. P III; GD. Geomagnetism, atmospheric physics, and geodesy.

10500. FORCHHAMMER, Johan Georg. 1794-1865. Den. DSB; ICB; P I,III; BMNH; GB. Oceanography. (Also Geology)

10501. JOHNSON, Edward John. 1794-1853. Br. P I. Geomagnetism.
1. *Practical illustrations of the necessity for ascertaining the deviations of the compass. With ... notes on magnetism, etc.* London, 1847. RM.

10502. MACLEAR, Thomas. 1794-1879. Br./Cape Obs. DSB; ICB; P III; GD. Geodesy. (Also Astronony)
1. *The verification and extension of La Caille's arc of meridian at the Cape of Good Hope.* 2 vols. London, 1866.

10503. THORSTENSEN, Jón. 1794-1855. Iceland. P II,III; BLA (THORSTEINSON). Meteorology.

10504. WRANGEL (or VRANGEL), Ferdinand Petrovich (or F. Ludwig). 1794-1870. Russ. P II,III,IV; BMNH; GC. Earth sciences generally.
1. [Trans.] *Narrative of an expedition to the Polar Sea, 1820-23, commanded by F. von Wrangell.* New York, 1841. Trans. from a German version of the then unpublished Russian MS. Ed. by E. Sabine.

10505. DAUBENY, Charles Giles Bridle. 1795-1867. Br. DSB; ICB; P I, III; BMNH; GB; GD. Climatology. (Other entries: see 3305)
1. *Climate. An inquiry into the causes of its differences, and into its influence on vegetable life.* Oxford/London, 1863.

10506. RÖDER, Georg Wilhelm. 1795-1872. Ger./Switz. P III. Meteorology.

10507. QUETELET, (Lambert) Adolphe (Jacques). 1796-1874. Belg. DSB; ICB; P II,III; GA; GB. Meteorology and geomagnetism. (Other entries: see 3306)
1. *Sur le climat de la Belgique.* 2 vols. Brussels, 1845-53.
2. *Sur la physique du globe.* Brussels, 1861.

10508. FOSTER, Henry. 1797-1831. Br. DSB; P I; GA; GD. Geomagnetism and geophysics.

10509. PIDDINGTON, Henry. 1797-1858. Br./India. GD. Meteorology.

10510. KREIL, Karl. 1798-1862. Prague/Aus. P I,III; GA. Meteorology and geomagnetism.

10511. WILKES, Charles W. 1798-1877. U.S.A. ICB; P III; BMNH; GB; GE. Meteorology.
1. *Theory of the winds.* Philadelphia, 1856.
See also 11291/1 and 12633/1.

10512. KUPFFER, Adolf Theodor von. 1799-1865. Russ. P I,III; BMNH; GC. Meteorology and geomagnetism. (Also Physics and Crystallography*)

10513. REICH, Ferdinand. 1799-1882. Ger. P II,III; GC. Geomagnetism and meteorology. (Also Physics)

10514. CALDECOTT, John. 1800-1849. Br./India. P I; GD. Meteorology and geomagnetism. (Also Astronomy)

10515. JENYNS (after 1871 BLOMEFIELD), Leonard. 1800-1893. Br. BMNH; GD1 (BLOMEFIELD). Meteorology. (Also Zoology* and Biography*)
1. *Observations in meteorology.* London, 1858.

10516. LACHMANN, Heinrich Wilhelm Ludolph. 1800-1861. Ger. P III; GC. Meteorology.

10517. LLOYD, Humphrey. 1800-1881. Br. DSB; ICB; P I,III; GD. Geomagnetism. (Also Physics*)
1. *An account of the magnetic observatory of Dublin.* Dublin, 1842.
2. *Observations made at the magnetic and meteorological observatory at Trinity College, Dublin.* 2 vols. Ib., 1865-69.
3. *Treatise on magnetism, general and terrestrial.* London, 1874.

10518. ROSS, James Clark. 1800-1862. Br. DSB; P II; BMNH; GB; GD. Geomagnetism. (Also Natural History*)

10519. KÄMTZ, Ludwig Friedrich. 1801-1867. Ger./Russ. P I,III. Meteorology.
1. *Lehrbuch der Meteorologie.* 3 vols. Halle, 1831-36. RM.
2. *Vorlesungen über Meteorologie.* Halle, 1840.
2a. ——— *A complete course of meteorology.* London, 1845. Trans. with additions by C.V. Walker. RM/L.

10520. BOUSSINGAULT, Jean Baptiste Joseph Dieudonné. 1802-1887. Fr./Colombia. DSB; ICB; P I,III,IV; Mort.; GA; GB. Meteorology. (Also Chemistry and Botany)

10521. COLLADON, Jean Daniel. 1802-1876. Switz. P I,III,IV. Meteorology. (Also Physics)

10522. SVANBERG, Gustav. 1802-1882. Swed. P II,III,IV. Geomagnetism, meteorology, and geodesy.

10523. DOVE, Heinrich Wilhelm. 1803-1879. Aus. DSB; ICB; P I,III; GC. Meteorology and geomagnetism. (Also Physics*)
1. *Correspondirende Beobachtungen über die regelmässigen*

stündlichen Veränderungen und über Perturbationen der magnetischen Abweichung im mitteleren und östlichen Europa. Leipzig, 1830. RM.
2. *Meteorologische Untersuchungen.* Berlin, 1837. RM.
3. *Über den Zusammenhang der Wärmeveränderungen der Atmosphäre mit der Entwickelung der Pflanzen.* Berlin, 1846.
4. *Die Verbreitung der Wärme auf der Oberfläche der Erde, erläutert durch Isothermen, thermische Isanomeln, und Temperaturcurven.* Berlin, 1852. X.
4a. ——— *The distribution of heat over the surface of the globe, illustrated by isothermal, thermic isabnormal, and other curves of temperature.* London, 1853. RM/L.
5. *Das Gesetz der Stürme in Beziehung zu den allgemeinen Bewegungen in der Atmosphäre.* Berlin, 1857. X.
5a. ——— *The law of storms considered in connection with the ordinary movements of the atmosphere.* London, 1862. Trans. from the 2nd German ed. by R.H. Scott. RM/L.

10524. JAMES, Henry. 1803-1877. Br. P III; GD. Geodesy and meteorology.
1. *Instructions for taking meteorological observations.* London, 1860.
See also 10832/1.

10525. KELLER, Franz Anton Eduard. 1803-1874. Fr. P III. Meteorology and oceanography.

10526. BIRT, William Radcliff. 1804-1881. Br. DSB; P III. Meteorology. (Also Astronomy)

10527. BREMIKER, Carl. 1804-1877. Ger. DSB; P I,III. Geodesy. (Also Astronomy)

10528. LENZ, Emil Khristianovich (or Heinrich Friedrich Emil). 1804-1865. Russ. DSB; ICB; P I,III. Oceanography and meteorology. (Also Physics)

10529. NERENBURGER, Adolph Wilhelm. 1804-1869. Belg. P II,III; GF. Geodesy.

10530. ROWELL, George Augustus. 1804-1892. Br. GD. Meteorology.
1. *An essay on the cause of rain and its allied phenomena.* Oxford, 1859. RM.
2. *Electric meteorology.* Ib., 1886? A collection of periodical articles.

10531. BUFF, Heinrich. 1805-1878. Ger. P I,III; GC. Geophysics. (Also Physics)
1. *Zur Physik der Erde.* Brunswick, 1850. Lectures. X.
1a. ——— *Familiar letters on the physics of the earth. Treating of the chief movements of the land, the water, and the air.* London, 1851. Trans. by A.W. Hofmann.

10532. DELLMANN, Johann Friedrich Georg. 1805-1870. Ger. P I,III. Meteorology. (Also Physics and Crystallography)

10533. FITZROY, Robert. 1805-1865. Br. DSB; ICB; P III; BMNH; GB; GD. Meteorology.

1. *The weather book. A manual of practical meteorology.* London, 1863. RM. 2nd ed., 1863.
See also 11283/1.

10534. LAMONT, Johann von. 1805-1879. Ger./Br. DSB; ICB; P I,III; GB; GC; GD. Geomagnetism and meteorology. (Also Astronomy)
1. *Handbuch des Erdmagnetismus.* Berlin, 1849.
2. *Magnetische Karten von Deutschland und Bayern.* Munich, 1854.
3. *Magnetische Ortsbestimmungen ausgeführt an verschiedenen Punkten Bayerns.* 2 vols in 1. Munich, 1854-56.
4. *Handbuch des Magnetismus.* Leipzig, 1867.

10535. NERVANDER, Johann Jacob. 1805-1848. Fin. P II. Meteorology and geomagnetism.

10536. STAMKART, Franciscus Johanes. 1805-1882. Holl. P II,III. Meteorology and geomagnetism. (Also Physics)

10537. BACHE, Alexander Dallas. 1806-1867. U.S.A. DSB; ICB; P I, III; GB; GE. Geodesy, meteorology, oceanography, and geomagnetism. (Also Various Fields)

10538. BUTLER, Thomas Belden. 1806-1873. U.S.A. P III; GE. Meteorology.

10539. COFFIN, James Henry. 1806-1873. U.S.A. P III; GE. Meteorology.
1. *Winds of the globe; or, Laws of atmospheric circulation.* Washington, 1875.

10540. EISENLOHR, Otto. 1806-1853. Ger. P I. Meteorology.

10541. ERMAN, (Georg) Adolph. 1806-1877. Ger. DSB; P I,III; GA; GC. Earth sciences generally.
1. *Die Grundlagen der Gaussischen Theorie und die Erscheinungen des Erdmagnetismus im Jahre 1829.* Berlin, 1874. With H. Petersen. RM.

10542. MARTINS, Charles Frédéric. 1806-1889. Fr. P II,III; BMNH; BLA; GA. Meteorology. (Also Geology and Botany)

10543. MAURY, Matthew Fontaine. 1806-1873. U.S.A. DSB; ICB; P II, III; GA; GB; GE. Oceanography and meteorology. (Also Astronomy)
1. *The physical geography of the sea.* New York/London, 1855. RM. Reprinted, Cambridge, Mass., 1963. 3rd ed., 1855. Other eds: 1856, 1857, 1859. 9th ed., rev. and enl., 1860. 10th ed., 1861. 14th ed., 1869 and 1870. 15th ed., 1872. 20th ed., 1886.

10544. GOLDSCHMIDT, Carl Wolfgang Benjamin. 1807-1851. Ger. P I. Geomagnetism. (Also Astronomy)

10545. LADAME, Henri. 1807-1870. Switz. P I,III. Meteorology.

10546. NOWÁK, Alois F.P. 1807-1880. Prague. P III. Meteorology.

10547. PALMIERI, Luigi. 1807-1896. It. P II,III,IV; BMNH; GF. Meteorology and geomagnetism. (Also Geology*)

10548. LOOF, Friedrich Wilhelm. 1808-1889. Ger. P III; GC. Meteorology.

10549. OSLER, Abraham Follett. 1808-1903. Br. GD2. Meteorology.

10550. RESLHUBER, (Augustin) Wolfgang. 1808-1875. Aus. P II,III; GC. Meteorology and geomagnetism. (Also Astronomy)

10551. DREW, John. 1809-1857. Br. P III; GD. Meteorology. (Also Astronomy)
1. *Practical meteorology*. London, 1855. 2nd ed., 1860.

10552. GLAISHER, James. 1809-1903. Br. DSB; P III,IV; GB; GD2. Meteorology.

10553. PRESTEL, Michael August Friedrich. 1809-1880. Ger. P II,III. Meteorology.

10554. SPASSKII, Mikhail Fedorovich. 1809-1859. Russ. ICB. Meteorology. (Also Physics)

10555. PRETTNER, Johann. d. 1875. Aus. P III. Meteorology.

10556. MEISTER, Franz Xaver. 1810-1872. Ger. P II,III. Meteorology.

10557. MÜHRY, (Adalbert) Adolf. 1810-1888. Ger. P III,IV; BLA. Meteorology.

10558. ARNDT, Johann Albert. 1811-1882. Ger. P I,III. Meteorology.

10559. BROCKLESBY, John. 1811-1889. U.S.A. P III. Meteorology and physical geography.
1. *Elements of physical geography*. Philadelphia, 1868. X. Another ed., 1875.

10560. GILLISS, James Melville. 1811-1865. U.S.A. P I,III; GE. Meteorology and geomagnetism. (Also Astronomy)
1. *Magnetical and meteorological observations made at Washington*. Washington, 1845. RM.

10561. LE VERRIER, Urbain Jean Joseph. 1811-1877. Fr. DSB; ICB; P I,III; GA; GB. Meteorology. (Also Astronomy*)

10562. LOOMIS, Elias. 1811-1889. U.S.A. DSB; P I,III,IV; GE. Meteorology and geomagnetism. (Also Astronomy)
1. *Treatise on meteorology*. New York, 1868. X. Re-issue, 1875.

10563. LOSE, Ludwig. 1811-1879. Ger. P III. Meteorology.

10564. ANDRAE, Carl Christopher George. 1812-1893. Den. P III. Geodesy.

10565. DINES, George. 1812-1888. Br. P III. Meteorology.

10566. FRITSCH, Karl. 1812-1879. Prague/Aus. P I,III; BMNH. Meteorology. (Also Botany*)

10567. KRECKE, Friedrich Wilhelm Christian. 1812-1882. Holl. P I, III. Meteorology.

10568. MAHLMANN, Wilhelm. 1812-1848. Ger. P II. Meteorology. See also 10450/2a.

10569. SIMONY, Friedrich. 1812-1896. Aus. P III; BMNH. Physical geography and meteorology.

10570. DAGUIN, Pierre Adolphe. 1814-1884. Fr. P III. Meteorology. (Also Physics*)

10571. DENZLER, Hans Heinrich. 1814-1876. Switz. P I,III. Meteorology.

10572. DEVILLE, Charles Joseph Sainte-Claire. 1814-1876. Fr. Under SAINTE-CLAIRE-DEVILLE: P II,III; BMNH; GB. Meteorology. (Also Geology)

10573. FAYE, Hervé. 1814-1902. Fr. DSB; P I,III,IV. Geodesy and meteorology. (Also Astronomy)

10574. BAXENDELL, Joseph. 1815-1887. Br. P III,IV; GD1. Meteorology. (Also Astronomy)

10575. COLDING, Ludvig August. 1815-1888. Den. DSB Supp.; ICB; P I, III. Meteorology. (Also Physics*)

10576. EVANS, Frederick John Owen. 1815-1885. Br. DSB; P III; GD. Geomagnetism.

10577. FORBES, Edward. 1815-1854. Br. DSB; ICB; P III; BMNH; GB; GD. Oceanography. (Also Geology and Zoology*. See also Index.)

10578. PETERSEN, Heinrich Jacob Reinhold. 1815-after 1878. Ger. P III. Geomagnetism. See also 10541/1.

10579. PLANTAMOUR, Emile. 1815-1882. Switz. P II,III. Meteorology and geodesy. (Also Astronomy)

10580. RAULIN, Félix Victor. 1815-1905. Fr. P II,III,IV; BMNH. Meteorology. (Also Geology)

10581. RENOU, Emilien Jean. 1815-1902. Fr./Algeria. P III,IV. Meteorology.

10582. WÜLLERSTORF-URBAIR, Bernhard von. 1816-1883. Aus. P III; GC. Meteorology.

10583. BÖRSCH, Carl Cäsar Ludwig Otto Haubold. 1817-1890. Ger. P IV. Geodesy.

10584. BROUN, John Allan. 1817-1879. Br./India. P III; GD. Geomagnetism and meteorology.

10585. BUYS-BALLOT, Christoph Hendrik Diederik. 1817-1890. Holl. DSB; ICB; P I,III,IV. Meteorology.
1. *Sur la marche annuelle du thermomètre et du baromètre en Néerlande et en divers lieux de l'Europe, 1849-59.* Amsterdam, 1861.

10586. FERREL, William. 1817-1891. U.S.A. DSB; P III,IV; GE. Meteorology and geophysics.
1. *A popular treatise on the winds.* New York, 1889. X. 2nd ed., 1893.

10587. LEFROY, John Henry. 1817-1890. Br./Canada. ICB; GD. Geomagnetism.

10588. MANN, Robert James. 1817-1886. Br./South Africa. P III; GD. Meteorology.

10589. MARGUET, Etienne Jean Jules. 1817-after 1903. Switz. P III, IV. Meteorology.

10590. NEGRETTI, Henry. 1817-1879. Br. P III; GD. Meteorology.

1. *Treatise on meteorological instruments.* London, 1864.

10591. RIDDELL, Charles James Buchanan. 1817-1903. Br. GD2. Geomagnetism.

10592. STRACHEY, Richard. 1817-1908. Br./India. P IV; BMNH; GB; GD2. Meteorology and physical geography.

10593. BLASIUS, William. 1818-1899. U.S.A. ICB. Meteorology.
1. *Storms: Their nature, classification and laws. With the means of predicting them.* Philadelphia, 1875.

10594. HOLTEN, Carl Valentin. 1818-1886. Den. P III. Meteorology.

10595. MARSH, Benjamin. 1818-after 1882. U.S.A. P III. Meteorology.

10596. STEVENSON, Thomas. 1818-1887. Br. P III; GD. Meteorology.

10597. LAUSSEDAT, Aimé. 1819-1907. Fr. P III. Geodesy.

10598. MONTIGNY, Charles Marie Valentin. 1819-1890. Belg. ICB; P II, III,IV. Meteorology. (Also Physics)

10599. SILBERMANN, Ignaz Joseph. 1819-ca. 1875. Fr. P II,III. Meteorology.

10600. SMYTH, (Charles) Piazzi. 1819-1900. Br. DSB; P II,III,IV; GB; GD1. Meteorology. (Also Astronomy*)
1. *Madeira meteorologic.* Edinburgh, 1882. RM.

10601. STRUVE, Otto Wilhelm. 1819-1905. Russ. DSB; P II,III,IV. Geodesy. (Also Astronomy*)

10602. TOYNBEE, Henry. 1819-1909. Br. P III,IV; BMNH. Meteorology and oceanography.

10603. WESSELOVSKY, Constantin Stepanovich. 1819-1901. Russ. P II, IV. Meteorology.

10604. BENT, Silas. 1820-1887. U.S.A. GE. Oceanography.

10605. KARSTEN, Gustav. 1820-1900. Ger. P I,III,IV. Meteorology and oceanography. (Also Physics)

10606. MARIÉ-DAVY, Edme Hippolyte. 1820-1893. Fr. P III,IV. Meteorology. (Also Physics)

10607. RAGONA (-SCINÀ), Domenico. 1820-1892. It. P III,IV. Meteorology.

10608. ROCHE, Edouard Albert. 1820-1883. Fr. DSB; P II,III. Geophysics and meteorology. (Also Astronomy)

10609. VETTIN, Ulrich Franz Friedrich. 1820-1905. Ger. P II,III,IV. Meteorology.

10610. HAUGHTON, Samuel. 1821-1897. Br. P III,IV; BMNH; Mort.; GB; GD1. Climatology. (Other entries: see 3336)
1. *New researches on sun-heat, terrestrial radiation, etc.* Dublin, 1886.

10611. MELDRUM, Charles. 1821-1901. Mauritius. P III,IV; GD2. Meteorology.

10612. MORITZ, Paul Heinrich Arnold. 1821-1902. Russ. P II,III,IV. Meteorology.

10613. GALTON, Francis. 1822-1911. Br. DSB; ICB; P III,IV; BMNH; Mort.; BLA Supp.; BLAF; GB; GD2. Meteorology. (Also Heredity* and Autobiography*)
1. *Meteorologica; or, Methods of mapping the weather.* London, 1863.

10614. JELINEK, Karl. 1822-1876. Prague/Aus. P I,III. Meteorology.

10615. MENDEL, (Johann) Gregor. 1822-1884. Brünn. DSB; ICB; BLA. Meteorology. (Also Heredity*)

10616. BLODGET, Lorin. 1823-1901. U.S.A. GE. Meteorology and climatology.

10617. GARIBALDI, Pietro Maria. 1823-1902. It. P III,IV & Supp. Geomagnetism.

10618. MEUCCI, Ferdinando. 1823-after 1878. It. P III. Meteorology.

10619. POURTALÈS, Louis François de. 1823-1880. U.S.A. DSB; BMNH; GE. Oceanography and marine biology.

10620. SCHENZL, Guido Johann Jeremias Max. 1823-1890. Hung. P III, IV. Meteorology and geomagnetism.

10621. WELSH, John. 1824-1859. Br. P II; GD. Meteorology and geomagnetism.

10622. HILGARD, Julius Erasmus. 1825-1891. U.S.A. P III; GE. Geodesy.

10623. IBÁÑEZ E IBÁÑEZ DE IBERO, Carlos. 1825-1891. Spain. DSB; ICB; P III,IV. Geodesy.

10624. LOWE, Edward Joseph. 1825-1900. Br. P III; BMNH. Meteorology. (Also Botany)
1. *Treatise on atmospheric phenomena.* London, 1846.

10625. POINCARÉ, Nicolas Antoine Hélène. 1825-after 1903. Fr. P IV. Meteorology.

10626. QUETELET, Ernst Adolphe François. 1825-1878. Belg. ICB; P III; GF. Meteorology and geomagnetism. (Also Astronomy)

10627. BERTELLI, (Leopoldo) Timoteo. 1826-1905. It. P III,IV. Meteorology and seismology.

10628. LIAIS, Emmanuel. 1826-1900. Fr./Brazil. P III,IV. Meteorology and geodesy. (Also Astronomy)
1. *Traité d'astronomie appliquée et de géodésie pratique.* Paris, 1867.

10629. NEUMAYER, Georg Balthasar. 1826-1909. Ger./Australia. P III, IV; BMNH. Meteorology, etc.
1. *Anleitung zu wissenschaftlichen Beobachtungen auf Reisen. Mit besonderer Rücksicht auf die Bedürfnisse der kaiserlichen Marine.* Berlin, 1875. Compiled by P. Ascherson et al. and edited by Neumayer. 2nd ed., 2 vols, 1888.

10630. SCHLAGINTWEIT, Hermann Rudolph Alfred von. 1826-1882. Ger. P II,III; BMNH; GC. Physical geography (see 9851/1,2), meteorology, and geomagnetism.

10631. SCHOTT, Charles Anthony. 1826-1901. U.S.A. DSB; P III,IV; GE. Geodesy, meteorology, and geomagnetism.

10632. BOGUSLAWSKI, Georg Heinrich von. 1827-1884. Ger. P I,III. Meteorology and oceanography. (Also Astronomy)
1. *Handbuch der Ozeanographie*. 2 vols. Stuttgart, 1884-87. With O. Krümmel.

10633. BOUQUET DE LA GRYE, Jean Jacques Anatole. 1827-1909. Fr. P III,IV. Oceanography.

10634. LIEBIG, Georg von. 1827-1903. Ger. P III,IV. Meteorology. (Also Physiology)

10635. OUDEMANS, Jean Abraham Chrétien. 1827-1906. Holl./Java. P II,III,IV; GF. Geodesy. (Also Astronomy)

10636. STUART, Lewis Cohen. 1827-1878. Holl. P III. Geodesy.

10637. CLARKE, Alexander Ross. 1828-1914. Br. P III. Geodesy.
1. *Comparisons of the standards of length of England, France, Belgium, Prussia, Russia, India, Australia*. London, 1856. Another ed., 1866.
2. *Geodesy*. Oxford, 1880.
See also 10832/1.

10638. ELLIS, William. 1828-1916. Br. DSB; P III,IV. Geomagnetism and meteorology. (Also Astronomy)

10639. STEWART, Balfour. 1828-1887. Br. DSB; P III,IV; GB; GD. Meteorology and geomagnetism. (Also Physics*. See also 2963/2.)

10640. BUCHAN, Alexander. 1829-1907. Br. P IV & Supp.; BMNH; GD2. Meteorology.
1. *Handy book of meteorology*. Edinburgh, 1867. X. 2nd ed., 1868.
2. *Atlas of meteorology*. Ib., 1899.

10641. EBERMAYER, Ernst Wilhelm Ferdinand. 1829-1908. Ger. P III, IV. Meteorology. (Also Chemistry*)

10642. RUBENSON, Robert. 1829-1902. Swed. P III,IV. Meteorology.

10643. BERGSMA, Pieter Adrian. d. 1882. Java. P III. Meteorology and geomagnetism.

10644. FRITZ, Hermann. 1830-1893. Switz. P III,IV. Meteorology.

10645. HIRSCH, Adolph. 1830-1901. Switz. P III,IV. Geodesy and meteorology. (Also Astronomy)

10646. KARLINSKI, Michael Franciszek. 1830-1906. Cracow. P I,III. Meteorology. (Also Astronomy)

10647. LAUGHTON, John Knox. 1830-1915. Br. P III; GD3. Meteorology. (Also Geology)
1. *Physical geography in its relation to the prevailing winds and currents*. London, 1870. Another ed., 1873.

10648. POEY, André. 1830?-after 1882. Cuba. P III. Meteorology.

10649. THOMSON, Charles Wyville. 1830-1882. Br. DSB; P III; BMNH; GB; GD. Oceanography. (Also Zoology)
1. *The depths of the sea. An account of the general results of the dredging cruises of H.M.SS. "Porcupine" and "Lightning" during 1868-70, under the scientific*

direction of Dr. Carpenter ... J.G. Jeffreys ... and W. Thomson. London, 1873. RM/L. 2nd ed., 1874.
See also 11302/1,2.

10650. ZENGER, Karel Václay. 1830-1908. Prague. ICB; P IV. Meteorology. (Also Astronomy)

10651. BERGER, Joseph. 1831-after 1870. Ger. P III. Meteorology.

10652. BRITO-CAPELLO, João Carlos de. 1831-after 1879. Port. P III. Meteorology and geomagnetism.

10653. CORNELISSEN, Jan Eduard. 1831-1876. Holl. P III. Meteorology.

10654. VERNON, George Venables. 1831-1878. Br. P III. Meteorology.

10655. LINDENKOHL, Adolf. 1833-1904. U.S.A. GE. Oceanography.

10656. MÜTTRICH, (Gottlieb) Anton. 1833-1904. Ger. P III,IV. Meteorology.

10657. PAULSEN, Adam Frederik Wivet. 1833-1907. Den. P IV. Meteorology and geomagnetism.

10658. SCHÜCK, (Karl Wilhelm) Albert. 1833-after 1913. Ger. P III, IV. Meteorology, oceanography, and geomagnetism.

10659. SCOTT, Robert Henry. 1833-1916. Br. P III,IV; BMNH. Meteorology. (Also Geology and History of Science*)
1. *Weather charts and storm warnings.* London, 1876.
2. *Elementary meteorology.* Ib., 1883. Another ed., 1887.
See also 10523/5a.

10660. WILD, Heinrich. 1833-1902. Switz./Russ. DSB; P II,III,IV. Meteorology and geomagnetism.

10661. WILDE, Henry. 1833-1919. Br. P IV. Geomagnetism. (Also Physics)
1. *On the causes of the phenomena of terrestrial magnetism.* [London? 1891?]

10662. BLANFORD, Henry Francis. 1834-1893. India. P IV; BMNH; GD1; GF. Meteorology. (Also Geology)

10663. CHAMBERS, Charles. 1834-1896. Br./India. P III,IV. Geomagnetism and meteorology.

10664. DENZA, Francesco. 1834-1894. It. P III,IV. Meteorology and geomagnetism.

10665. EATON, Henry Storks. 1834-after 1880. Br. P III. Meteorology.

10666. PERRIER, François. 1834-1888. Fr. ICB; P III. Geodesy.

10667. AGASSIZ, Alexander. 1835-1910. U.S.A. DSB; ICB; BMNH; BLAF; GB; GE. Oceanography. (Also Zoology* and Autobiography*)
1. *A contribution to American thalassography. Three cruises of the U.S.S. "Blake" ... 1877-1880.* 2 vols. Cambridge, Mass., 1888.
2. *The coral reefs of the Hawaiian Islands.* Ib., 1889.
3. *The elevated reefs of Florida.* Ib., 1896.

10668. HENSEN, (Christian Andreas) Victor. 1835-1924. Ger. DSB; P IV; BMNH; BLA & Supp. Oceanography (plankton studies). (Also Physiology. See also 11309/1.)

10669. MOHN, Henrik. 1835-1916. Nor. DSB; P III,IV. Meteorology and oceanography.

10670. STRACHAN, Richard. 1835-after 1881. Br. P III. Meteorology.

10671. FISCHER, Amandus. 1836-1894. Ger. P IV; GC. Geodesy.

10672. HOFFMEYER, Niels. 1836-1884. Den. P III. Meteorology.

10673. RUSSELL, Henry Chamberlaine. 1836-1907. Australia. DSB Supp.; P III,IV; GD2. Meteorology. (Also Astronomy*)

10674. SCHIO, Almerico Alvise Cassiano da. 1836-after 1882. It. P III. Meteorology.

10675. SCHODER, Hugo von. 1836-1884. Ger. P III,IV. Meteorology.

10676. WEYHER, Charles Louis. 1836-1916. ICB. Meteorology.

10677. BEZOLD, (Johann Friedrich) Wilhelm von. 1837-1907. Ger. P III. Meteorology. (Also Physics*)

10678. DINKLAGE, Ludwig Eduard. 1837-1903. Ger. P III,IV. Oceanography.

10679. KOLDEWEY, Karl. 1837-1908. Ger. P III,IV; BMNH. Oceanography and meteorology.

10680. NEUMANN, Franz. 1837-after 1865. Aus. P II,III. Meteorology.

10681. SASS, Arthur Ferdinand von. 1837-ca. 1870. Oesel. P III. Oceanography and meteorology.

10682. ABBE, Cleveland. 1838-1916. U.S.A. DSB; ICB; P III,IV & Supp.; GE. Meteorology.
1. *The mechanics of the earth's atmosphere. A collection of translations.* 2 vols. Washington, 1891-1910. RM/L.

10683. HILDEBRAND-HILDEBRANDSSON, Hugo. 1838-1925. Swed. P III,IV. Meteorology.

10684. LEMSTRÖM, Karl Selim. 1838-1904. Fin. P III,IV. Meteorology and geomagnetism.

10685. SYMONS, George James. 1838-1900. Br. P III,IV; GB; GD1. Meteorology. (See also 10360/1)

10686. TACCHINI, Pietro. 1838-1905. It. DSB; P III,IV. Meteorology, geomagnetism, and seismology. (Also Astronomy)

10687. WEYPRECHT, Karl. 1838-1881. Aus. ICB; P III; GC. Meteorology and geomagnetism.

10688. AITKEN, John. 1839-1919. Br. ICB; P III,IV; GF. Meteorology and oceanography.
1. *Collected scientific papers.* Cambridge, 1923. Ed. with a memoir by C.G. Knott.

10689. BRAULT, Louis Désiré Léon. 1839-1885. Fr. P III. Meteorology.

10690. COLONGUE, Jean Alexander Heinrich Clapier de (or Ivan Petrovich). 1939-1901. Russ. P IV. Geomagnetism.

10691. ELIOT, John. 1839-1908. Br./India. P III; GD2. Meteorology.

10692. FRITSCHE, Hermann Peter Heinrich. 1839-1913. Russ./China. P III,IV. Geomagnetism. (Also History of Science*)

10693. HANN, Julius (Ferdinand) von. 1839-1921. Aus. DSB; ICB; P III,IV; GF. Meteorology and climatology.
1. *Handbuch der Klimatologie*. Stuttgart, 1883. RM/L.

10694. PEIRCE, Charles (Santiago) Sanders. 1839-1914. U.S.A. DSB; ICB; P III,IV; GE. Geodesy. (Also Science in General*# and Mathematics*)

10695. TILLO, Aleksey Andreevich. 1839-1900. Russ. DSB; P III,IV; BMNH. Geodesy and earth sciences generally.

10696. TIZARD, Thomas Henry. 1839-1924. Br. GD4. Oceanography.

10697. CLEVE, Per Teodor. 1840-1905. Swed. DSB; P III,IV; BMNH. Oceanography (plankton studies). (Also Chemistry)

10698. FRIESENHOF, Gregor von. 1840-after 1906. Hung. P III,IV. Meteorology.

10699. GRONEMAN, Hendrik Jan Herman. 1840-after 1903. Holl. P III, IV. Atmospheric physics.

10700. LEY, William Clement. 1840-1896. Br. P III,IV. Meteorology.
1. *Laws of the winds prevailing in western Europe*. London, 1872.

10701. ROTH, Friedrich. 1840-1891. Ger. P III,IV. Geophysics.

10702. RYKACHEV, Michail Alexandrovich. 1840-1919. Russ. ICB; P III, IV (RYKATSCHEW). Meteorology.

10703. SCHMIDT, Carl August von. 1840-1929. Ger. DSB; P III,IV. Geophysics. (Also Astrophysics)

10704. SNELLEN, Maurits. 1840-1907. Holl. P III,IV. Meteorology.

10705. ZIEGLER, Julius Joseph Ernst Friedrich. 1840-1902. P III,IV. Meteorology.

10706. BEBBER, Wilhelm Jakob van. 1841-after 1902. Ger. P III,IV. Meteorology.

10707. CHAMBERS, George Frederick. 1841-after 1881. Br. P III. Meteorology. (Also Astronomy*)
1. *The story of the weather*. London, 1897.

10708. DRAPER, Daniel. 1841-1931. U.S.A. P III,IV. Meteorology.

10709. FOREL, François Alphonse. 1841-1912. Switz. DSB Supp.; ICB; P III,IV; BMNH; BLAF. "The founder of limnology." (Also Zoology)

10710. LEVÄNEN, Sakari. 1841-1898. Fin. P III,IV. Meteorology. (Also Mathematics)

10711. MIELBERG, Johann. 1841-1894. Russ. P III,IV. Meteorology and geomagnetism.

10712. MURRAY, John (third of the name). 1841-1914. Br. DSB; P IV; BMNH; GB; GD3. Oceanography. See also 11301/2.

10713. OPPOLZER, Theodor von. 1841-1886. Aus. DSB; ICB; P III,IV; GC. Geodesy. (Also Astronomy*)

10714. SLUDSKY, Fedor Alekseevich. 1841-1897. Russ. P III,IV. Geophysics and geodesy. (Also Mechanics)

10715. WEIRAUCH, Johann Karl Friedrich. 1841-1891. Dorpat. P III, IV. Geophysics and meteorology.

10716. ABERCROMBY, Ralph. 1842-1897. Br. P IV. Meteorology.
1. *Principles of forecasting by means of weather charts.* London, 1885.
2. *Weather. A popular exposition.* Ib., 1887. 2nd ed., 1888.
3. *Instructions for observing clouds on land and sea.* Ib., 1888.

10717. ANDRÉ, Charles (Louis François). 1842-1912. Fr. DSB; P III, IV. Meteorology. (Also Astronomy and History of Astronomy*)

10718. MOUREAUX, Théodule. 1842-1919. Fr. P IV. Geomagnetism.

10719. SOHNCKE, Leonhard. 1842-1897. Ger. DSB; P III,IV; BMNH; GC. Meteorology. (Also Physics and Crystallography*)

10720. VOEYKOV, Aleksandr Ivanovich. 1842-1916. Russ. DSB; P III, IV (WOJEJKOF). Meteorology and climatology.

10721. WHIPPLE, George Mathews. 1842-1893. Br. P III,IV; GD. Meteorology.

10722. WITTE, Emil. 1842-after 1904. Ger. P III,IV. Oceanography and meteorology.

10723. ALBRECHT, Carl Theodor. 1843-1915. Ger. DSB; P III,IV. Geodesy.

10724. BLYTT, Axel Gudbrand. 1843-1898. Nor. P IV; BMNH. Meteorology. (Also Botany)

10725. BOERGEN, Carl Nikolaus Jensen. 1843-1909. Ger. P III,IV. Oceanography, meteorology, and geomagnetism.

10726. HELMERT, Friedrich Robert. 1843-1917. Ger. DSB; P III,IV. Geodesy.
1. *Die mathematischen und physikalischen Theorien der höheren Geodäsie.* 2 vols. Leipzig, 1880-84. X. Reprinted, New York, 1962.

10727. LORENZONI, Giuseppe. 1843-1914. It. DSB; P III,IV. Geodesy. (Also Astronomy)

10728. THOULET, Marie Julien Olivier. 1843-after 1921. Fr. P III, IV; BMNH. Oceanography. (Also Geology)

10729. TISSANDIER, Gaston. 1843-1899. Fr. P III,IV. Meteorology.

10730. WEILENMANN, August. 1843-1906. Switz. P III,IV. Meteorology.

10731. BUCHANAN, John Young. 1844-1925. Br. DSB; P III,IV; BMNH. Oceanography.
1. *Scientific papers.* Cambridge, 1913.
2. *Comptes rendus of observation and reasoning.* Ib., 1917. A collection of papers and essays reprinted from various sources.
3. *Accounts rendered of work done and things seen.* Ib., 1919.

10732. KLEIN, Hermann Joseph. 1844-1914. Ger. DSB; P III,IV. Meteorology. (Also Astronomy*)

10733. KNIPPING, Erwin Rudolph Theobald. 1844-1922. Ger./Japan. P III,IV; BMNH. Meteorology and oceanography.

10734. RICCÒ, Annibale. 1844-1919. It. DSB; P III,IV. Geophysics. (Also Astrophysics)

10735. WRANGEL, Ferdinand. 1844-1919. Russ. P III,IV. Oceanography.

10736. ASSMANN, Richard. 1845-1918. Ger. P IV. Meteorology.

10737. BROCARD, Pierre René Jean Baptiste Henri. 1845-1922. Fr./Algeria. DSB; P III,IV. Meteorology. (Also Mathematics)

10738. DARWIN, George Howard. 1845-1912. Br. DSB; ICB; P III,IV; GD3. Geophysics. (Also Astronomy*#)

10739. DECHEVRENS, Marc. 1845-1923. Fr./China. P IV. Meteorology.

10740. LAIS, Giuseppe. 1845-1921. It. P IV. Meteorology. (Also Astronomy)

10741. ABELS, Hermann Ferdinand. 1846-1929. Russ. P IV. Meteorology.

10742. AUGUSTIN, Franz. 1846-1908. Prague. P III,IV. Meteorology.

10743. CHAVANNE, Josef. 1846-after 1883. Aus. P III. Meteorology.

10744. KLOSSOVSKII, A.V. 1846-1917. Russ. ICB. Geophysics.

10745. KÖPPEN, Wladimir Peter. 1846-1940. Ger. ICB; P III,IV. Meteorology and oceanography.

10746. LA COUR, Paul. 1846-1908. Den. P III,IV (COUR). Meteorology.

10747. PILLSBURY, John Elliott. 1846-1919. U.S.A. GE. Oceanography.

10748. RIJSSELBERGHE, François van. 1846-1893. Belg. ICB; P III (RYSSELBERGHE). Meteorology.

10749. RÜHLMANN, (Moritz) Richard. 1846-1908. Ger. P III,IV. Meteorology. (Also Physics)
   1. *Die barometrischen Höhenmessungen und ihre Bedeutung für die Physik der Atmosphäre.* Leipzig, 1870.

10750. HAMBERG, Hugo Emanuel. 1847-after 1922. Swed. P III,IV. Meteorology.

10751. MILLOT, Charles. 1847-after 1903. Fr. P IV. Meteorology.

10752. ALBERT I, *Prince of Monaco* (GRIMALDI, Honoré Charles). 1848-1922. Fr. DSB; ICB; P IV (MONACO); BMNH; (ALBERT HONORÉ CHARLES); GF. Oceanography. See also 11307.

10753. ANGOT, (Charles) Alfred. 1848-1924. Fr. P III,IV; GF. Meteorology.
   1. *Les aurores polaires.* Paris, 1895. X.
   1a. ——— *The aurora borealis.* London, 1896.

10754. EKHOLM, Nils. 1848-1923. Swed. P IV. Meteorology.

10755. GUENTHER, (Adam Wilhelm) Siegmund. 1848-1923. Ger. DSB; P III,IV & Supp. Meteorology, geophysics, and geography. (Also Mathematics and History of Mathematics*)
   1. *Lehrbuch der Geophysik und physikalischen Geographie.* 2 vols. Stuttgart, 1884-85. X. 2nd ed., *Handbuch der Geophysik*, 2 vols, ib., 1897-99.
   2. *Handbuch der mathematischen Geographie.* Ib., 1890.
   3. *Handbuch der physikalischen Geographie.* Ib., 1891.

10756. HARRINGTON, Mark Walrod. 1848-1926. U.S.A. P III; GE. Meteorology.

10757. MARRIOTT, William. 1848-1916. Br. P III,IV. Meteorology.

10758. MERRIMAN, Mansfield. 1848-1925. U.S.A. P III,IV. Geodesy. (Also Mechanics*)
1. *The figure of the earth. An introduction to geodesy.* New York, 1881.

10759. PERNTER, Josef Maria. 1848-1908. Aus. P III,IV. Meteorology.

10760. SCHREIBER, Carl Adolph Paul. 1848-1924. Ger. P III,IV. Meteorology.

10761. SPRUNG, Adolf Friedrich Wichard. 1848-1909. Ger. DSB; P III, IV. Meteorology.

10762. BILLWILLER, Robert August. 1849-1905. Switz. P III,IV; BMNH. Meteorology.

10763. HAZEN, Henry Allen. 1849-1900. U.S.A. P III,IV; GE. Meteorology.

10764. LAMB, Horace. 1849-1934. Br. DSB; ICB; P III,IV; GD5. Geophysics. (Also Mechanics* and Physics)

10765. LANCASTER, Albert Benoît Marie. 1849-1908. Belg. ICB; P III. Meteorology. (Also Astronomy*)

10766. LANG, Karl. 1849-1893. Ger. P III,IV. Meteorology.

10767. MAKAROV, Stepan Osipovich. 1849-1904. Russ. DSB; P IV & Supp. Oceanography.

10768. SCHOLS, Charles Mathieu. 1849-1897. Holl. P III,IV; GF. Geodesy.

10769. WIJKANDER, Erik Anders Gustaf August. 1849-1913. Swed. P III, IV. Meteorology and geomagnetism.

10770. WOODWARD, Robert Simpson. 1849-1924. U.S.A. DSB; ICB; P IV; GE. Geophysics.

10771. BERGMANN, Reinhold von. 1850-1913. Russ. P IV. Meteorology.

10772. DAVIS, William Morris. 1850-1934. U.S.A. DSB; ICB; P IV & Supp.; BMNH; GE1. Geography and meteorology. (Also Geology*)
1. *Elementary meteorology.* Boston, 1894.

10773. HESS, Clemens. 1850-1918. Switz. P III,IV. Meteorology.

10774. STELLING, Eduard Reinhold. 1850-1922. Russ. P III,IV. Meteorology and geomagnetism.

10775. ARCHIBALD, Edmund Douglas. 1851-after 1898. Br. P III,IV. Meteorology.
1. *The story of the earth's atmosphere.* London, 1897.

10776. BIGELOW, Frank Hagar. 1851-1924. U.S.A. P IV & Supp.; GE. Meteorology and geomagnetism.
1. *Abstract of a report on solar and terrestrial magnetism in their relations to meteorology.* Washington, 1898.

10777. HILL, Samuel Alexander. 1851-1890. India. P III,IV. Meteorology.

10778. LAGRANGE, Charles Henri. 1851-1932. Belg. P III,IV. Geomagnetism. (Also Astronomy)

10779. PRESTON, Erasamus Darwin. 1851-after 1902. U.S.A. P IV. Geodesy.

10780. SCHUSTER, Arthur. 1851-1934. Br. DSB; ICB; P III,IV; GD5. Geomagnetism. (Also Physics)

10781. STOK, Johannes Paulus van der. 1851-after 1918. Holl./East Indies. P III (under S), IV (under VAN). Meteorology and geomagnetism.

10782. TROMHOLT, Sophus. 1851-1896. Nor. P III,IV. Meteorology.

10783. BÖRNSTEIN, Richard. 1852-1913. Ger. P III,IV. Meteorology. (Also Physics)

10784. BROUNOW, Peter Ivanovich. 1852-after 1902. Russ. P III,IV. Meteorology.

10785. CHISTONI, Giuseppe Ciro Pericle. 1852-1927. It. P IV. Geomagnetism and meteorology. (Also Physics)

10786. LEYST, Ernst. 1852-1918. Russ. P IV. Meteorology.

10787. LIZNAR, Josef. 1852-1932. Aus. P III,IV. Meteorology and geomagnetism.

10788. MARCHAND, Charles Emile Honoré. 1852-after 1913. Fr. P IV. Meteorology.

10789. WRAGGE, Clement Lindley. 1852-1922. Australia. GF. Meteorology.

10790. SATKE, Wladislaw. 1853-1904. Tarnopol. P IV. Meteorology.

10791. SCHÖNROCK, Alexander. 1853-after 1922. Russ. P IV. Meteorology.

10792. ANDRÉE, Salomon August. 1854-1897. Swed. ICB; P IV; GB. Meteorology.

10793. BÖRSCH, (Franz) Anton (Carl Cäsar). 1854-1920. Ger. P III, IV. Geodesy.

10794. HELLMANN, Johann Georg Gustav. 1854-after 1929. Ger. P III, IV. Meteorology.

10795. MÖLLER, Max Emil Karl. 1854-1935. Ger. P III,IV. Meteorology.

10796. PROHASKA, Karl. 1854-1937. Aus. P IV. Meteorology.

10797. RAJNA, Michele. 1854-1920. It. P III,IV. Geodesy and geomagnetism.

10798. RIGGENBACH, (Germann) Albert. 1854-1921. Switz. P IV. Meteorology.

10799. SCHERING, Karl Julius Eduard. 1854-1925. Ger. P III,IV. Geomagnetism.

10800. SHAW, William Napier. 1854-1945. Br. DSB; ICB; P III,IV. Meteorology. (Also Physics)

10801. SMITH, Charles Michie. 1854-1922. India. P III,IV. Meteorology.

10802. VALLOT, Joseph. 1854-after 1921. Fr. P IV. Meteorology.

10803. DANCKELMANN, Alexander von. 1855-1919. Ger. P III,IV. Meteorology.

10804. DINES, William Henry. 1855-1927. Br. ICB; GD4. Meteorology.

10805. GROSSMANN, Louis Adolph. 1855-1917. Ger. P III,IV. Meteorology.

10806. LESS, Emil. 1855-after 1930. Ger. P IV. Meteorology.

10807. MEYER, Carl Friedrich Louis Hugo. 1855-1936. Ger. P IV. Meteorology.

10808. TEISSERENC DE BORT, Léon Philippe. 1855-1913. Fr. P IV. Meteorology.

10809. BIESE, Franz Carl Otto August Ernst. 1856-1926. Fin. P IV. Meteorology and geomagnetism.

10810. FOLGHERAITER, Giuseppe. 1856-1913. It. P IV. Geomagnetism.

10811. GORE, James Howard. 1856-1939. U.S.A. P IV. Geodesy. (Also History of Geodesy)
 1. *Elements of geodesy*. New York, 1886. X. 2nd ed., 1889. Other eds: Boston, 1891; London, 1891.

10812. KNOTT, Cargill Gilston. 1856-1922. Br./Japan. DSB; P IV. Geophysics, especially seismology. (Also Physics. See also Index.)

10813. MARGULES, Max. 1856-1920. Aus. DSB; P III,IV. Meteorology. (Also Physics)

10814. MÜLLER, Paul Alfred. 1856-1926. Russ. P IV. Meteorology and geomagnetism.

10815. SHOKALSKY, Yuly Mikhaylovich. 1856-1940. Russ. DSB; ICB. Oceanography, meteorology, and physical geography.

10816. ERK, Fritz. 1857-1909. Ger. P IV. Meteorology.

10817. LACHMANN, Georg Friedrich Otto Alexander. 1857-1913. Ger. P IV. Meteorology.

10818. MACK, Karl Friedrich. 1857-1934. Ger. P IV. Meteorology. (Also Physics)

10819. MARCHI, Luigi de. 1857-after 1932. It. P IV (under DE). Meteorology.

10820. MAURER, Julius Maximilian. 1857-1938. Switz. DSB; P III,IV. Meteorology.

10821. SRESNEVSKY, Boris Ismailovich. 1857-after 1902. Russ. P IV. Meteorology.

10822. WALDO, Frank. 1857-1920. Meteorology.
 1. *Modern meteorology. An outline of the growth and present condition of some of its phases*. London, 1893.

10823. EKAMA, Henri. 1858-after 1902. Holl. P IV. Meteorology.

10824. ESCHENHAGEN, Johann Friedrich August Max. 1858-1901. Ger. P IV. Geomagnetism.

10825. HAYDEN, Edward Everett. 1858-1932. U.S.A. GE1. Meteorology.

10826. KREBS, Christian Ludwig Wilhelm. 1858-1924. Ger. P IV. Meteorology.

10827. KREMSER, Victor. 1858-1909. Ger. P IV. Meteorology.

10828. STRUVE, (Gustav Wilhelm) Ludwig. 1858-1920. Russ. DSB; P III, IV. Geodesy. (Also Astronomy)

10829. CARLHEIM-GYLLENSKÖLD, Vilhelm. 1859-after 1916. Swed. P IV. Geomagnetism and geodesy.

10830. SCHUBERT, Johannes Oscar. 1859-after 1922. Ger. P IV. Meteorology.

Imprint Sequence

10831. ROYAL Society of London. Committee of Physics and Meteorology.
1. *Report relative to the observations to be made in the Antarctic Expedition and in the magnetic observatories.* London, 1840.

10832. UNITED Kingdom. Ordnance Board.
1. *Ordnance trigonometrical survey. Account of the observations and calculations of the principal triangulation, and of the figure, dimensions, and mean specific gravity of the earth as derived therefrom.* 2 vols. London, 1858. Drawn up by A.R. Clarke under the direction of H. James.

10833. ROYAL Geographical Society. London.
1. *Catalogue of the library.* London, 1865.

10834. METEOROLOGICAL Society. (From 1883: Royal Meteorological Society.) London.
1. *Modern meteorology.* London, 1879.
2. *Catalogue of the library.* Ib., 1891.

10835. UNITED States. Naval Academy.
1. *Mathematical theory of the deviations of the compass.* 1879.

10836. GREENWICH. Royal Observatory.
1. *Diagrams representing the diurnal change in magnitude and direction of the magnetic forces in the horizontal plane, 1841-76.* London, 1886.

3.11 BIOLOGY IN GENERAL

10837. BONNET, Charles. 1720-1793. Switz. DSB; ICB; P I; BMNH; Mort.; BLA; GA; GB. Also Botany*, Zoology, and Embryology.
1. *Essai de psychologie*. Leiden, 1754. X. Another ed., London, 1755.
2. *Essai analytique sur les facultés de l'âme*. Copenhagen, 1760. Reprinted, Hildesheim, 1973.
3. *Considérations sur les corps organisés, où l'on traite de leur origine, de leur développement, de leur reproduction*. 2 vols. Amsterdam, 1762. RM.
4. *Contemplation de la nature*. 2 vols. Amsterdam, 1764. X. Another ed., ib., 1766. 2nd ed., 1769.
4a. ——— *The contemplation of nature*. 2 vols. London, 1766.
5. *La palingénésie philosophique; ou, Idées sur l'état passé et sur l'état futur des êtres vivans*. Geneva, 1769. RM.
6. *Oeuvres d'histoire naturelle et de philosophie*. 9 vols. Neuchâtel, 1779-83. RM/L.
7. *Lettres à M. l'Abbé Spallanzani*. Milan, 1971. Critical ed. with introd. and notes (in Italian) by C. Castellani.

10838. DARWIN, Erasmus. 1731-1802. Br. DSB; ICB; P I; BMNH; BLA & Supp.; GA; GB; GD. Also Botany*. See also Index.
1. *Zoonomia; or, The laws of organic life*. 2 vols. London, 1794-96. RM/L. Reprinted, New York, 1974. 2nd ed., 1796. 3rd ed., 4 vols, 1801.
2. *The essential writings*. London, 1968. Ed. with commentary by D. King-Hele.

10839. LAMÉTHERIE, Jean Claude de. 1743-1817. Fr. DSB; P I; BMNH; BLA (MÉTHERIE); GA. Also Science in General*, Chemistry*, and Geology*. See also Index.
1. *Vues physiologiques sur l'organisation animale et végétale*. Amsterdam, 1780. RM.

10840. TREVIRANUS, Gottfried Reinhold. 1776-1837. Ger. DSB; ICB; P II; BMNH; BLA; GB; GC. Also Zoology and Physiology.
1. *Biologie; oder, Philosophie der lebenden Natur*. 6 vols. Göttingen, 1802-22.

10841. BROWN, Thomas (first of the name). 1778-1820. Br. BMNH; GA; GB; GD.
1. *Observations on the "Zoonomia" of Erasmus Darwin*. Edinburgh, 1798. RM.

10842. KIESER, Dietrich Georg. 1779-1862. Ger. BLA; BMNH; GA. 'Naturphilosophie' and general biological theory. (Also Botany)

10843. OKEN (or OKENFUSS), Lorenz. 1779-1851. Ger./Switz. DSB; ICB;

P II; BMNH; Mort.; BLA & Supp.; GA; GB; GC. Also Natural History* and Zoology. See also Index.
1. *Die Zeugung*. Bamberg, 1805. RM.

10844. AMICI, Giovan Battista. 1786-1868. It. DSB; ICB; P I,III; BMNH; Mort.; GA; GB. Also Physics and Microscopy.
1. *Collezione di alcune memorie e lettre*. Modena, 1825. RM.

10845. LOTZE, Hermann Rudolph. 1817-1881. Ger. DSB; ICB; BLA; GA; GB; GC.

10846. GRINDON, Leopold Hartley. 1818-1904. Br. BMNH. Also Botany.
1. *Life: Its nature, varieties and phenomena*. London, 1856. X. 4th ed., 1875.

10847. SPENCER, Herbert. 1820-1903. Br. The philosopher. DSB; ICB; BMNH; GB; GD2. Also Science in General*, Evolution*, and Autobiography*.
1. *The principles of biology*. 2 vols. (Vols 2 and 3 of his *System of synthetic philosophy*) London, 1864; re-issued 1867. Other eds: 1880, 1884, 1898-99.

10848. STIRLING, James Hutchison. 1820-1909. Br. The philosopher. BMNH; GB; GD2. Also Evolution*.
1. *As regards protoplasm, in relation to Professor Huxley's essay "On the physical basis of life."* Edinburgh, 1869. Another ed., London, 1872.

10849. HINTON, James. 1822-1875. Br. ICB; BLA; GB; GD. Also Physiology*.
1. *Life in nature*. London, 1872. X. A collection of articles on various biological topics. 2nd ed., 1875.

10850. HUXLEY, Thomas Henry. 1825-1895. Br. DSB; ICB; BMNH; Mort.; BLA; GB; GD1. Other entries: see 3340.
1. *A course of practical instruction in elementary biology*. London, 1875. With H.N. Martin. X. 2nd ed., 1876. Other eds: 1883, 1889, 1892.

10851. BEALE, Lionel Smith. 1828-1906. Br. DSB; ICB; BMNH; BLA & Supp.; GD2. Also Microscopy* and Histology*.
1. *Protoplasm; or, Life, force and matter*. London, 1870. X. 2nd ed., 1870. 3rd ed., 1874. (Title varies)
2. *Life theories: Their influence upon religious thought*. London, 1871.
3. *Bioplasm: An introduction to the study of physiology and medicine*. London, 1872.
4. *On life and on vital action, in health and disease*. London, 1875.

10852. LETOURNEAU, Charles Jean Marie. 1831-1902. BMNH.
1. *La biologie*. Paris, 1876.
1a. ——— *Biology*. London, 1878. Trans. by W. Maccall. Another ed., 1890.

10853. NICHOLSON, Henry Alleyne. 1844-1899. Br. BMNH; GB; GD1. Also Geology*, Zoology*, Biography*, and History of Science*.
1. *Introduction to the study of biology*. Edinburgh, 1872.

10854. HOWES, (Thomas) George Bond. 1853-1905. Br. BMNH; GD2.

10855. DELAGE, Yves. 1854-1920. Fr. DSB; BMNH. Also Zoology*.
1. *La structure du protoplasma, les théories sur l'hérédité, et les grands problèmes de la biologie générale.* Paris, 1895.

Imprint Sequence

Bibliographies and Dictionaries

Most of these are primarily concerned with medicine but they are included because of their coverage of many of the biological sciences.

10856. ENGELMANN, Wilhelm. 1808-1878. Ger. Bibliographer and publisher. BMNH; BLA & Supp. Also Zoology*
1. *Bibliotheca medico-chirurgica et anatomico-physiologica: Alphabetisches Verzeichniss der medizinischen ... anatomischen und physiologischen Bücher welche vom Jahre 1750 bis zu ... 1847 in Deutschland erschienen sind, mit einem vollständigen Materien-Register.* Leipzig, 1848. X.
1a. ——— 6th ed., bound with *Supplement enthaltend die Literatur vom Jahre 1848 bis ... 1867.* Ib., 1867. X. Reprinted, Hildesheim, 1965.

10857. BIBLIOTHÈQUE Nationale. Paris.
1. *Catalogue des sciences médicales.* 2 vols. Paris, 1857-73.

10858. YALE Medical Library.
1. *The library of the Medical Institution of Yale College and its catalogue of 1865.* New Haven, Conn., 1960. By F.G. Kilgour.

10859. UNITED States. Surgeon-General's Office.
1. *Catalogue of the library.* Washington, 1872. Another ed., 3 vols, 1873-74.
2. *Index catalogue of the library. Authors and subjects.* 16 vols. Washington, 1880-95.

10860. PAULY, Alphonse. 1830-1909. Fr. BLA.
1. *Bibliographie des sciences médicales.* Paris, 1874. X. Reprinted, London, 1954.

10861. NEALE, Richard. 1827-1900.
1. *Medical digest. Being a means of ready reference to the principal contributions to medical science during the last thirty years.* London, New Sydenham Society, 1877. 3rd ed., 1891.

10862. ROYAL Medical and Chirurgical Society. London.
1. *Catalogue of the library.* 3 vols. London, 1879.

10863. UNIVERSITY College. London.
1. *Catalogue of the books in the medical and biological libraries.* London, 1887.

10864. DUNMAN, Thomas. 1849-1882.
1. *A glossary of anatomical, physiological and biological terms.* London, 1889.

10865. GOULD, George Milbury. 1848-1922. U.S.A. BLAF; GE.
1. *An illustrated dictionary of medicine, biology and allied sciences.* London, 1894.

10866. ROYAL Medical Society. Edinburgh.
1. *Catalogue of the library.* Edinburgh, 1896.

3.111 Microscopy

10867. JOBLOT, Louis. 1645-1723. Fr. DSB; ICB. Also Physics.
1. *Descriptions et usages de plusieurs nouveaux microscopes, tant simples que composez. Avec de nouvelles observations.* Paris, 1718. X.
1a. ——— 2nd ed. *Observations d'histoire naturelle faites avec le microscope ... Avec la description et les usages des différens microscopes.* 2 vols. Paris, 1754-55. RM.
See also 10872/1.

10868. BAKER, Henry. 1698-1774. Br. DSB; ICB; P I; BMNH; Mort.; BLA & Supp.; GA; GB; GD. Also Zoology*.
1. *The microscope made easy; or, The nature, uses and magnifying powers of the best kinds of microscopes described.* London, 1742. X. 2nd ed., 1743. 3rd ed., 1744. 4th ed., 1754. 5th ed., 1769.
1a. ——— *Le microscope à la portée de tout le monde.* Paris, 1754.
2. *Employment for the microscope.* London, 1753. 2nd ed., 1764.

10869. NEEDHAM, John Turberville. 1713-1781. Br./Belg. DSB; ICB; P II; BMNH; BLA; GA; GD. Also Embryology*.
1. *An account of some new microscopical discoveries, founded on an examination of the calamary and its wonderful milt-vessels.* London, 1745. RM.
2. *New microscopical discoveries, containing observations: I. On the calamary and its milt-vessels....* Ib., 1745.

10870. GLEICHEN (-RUSSWORM), Wilhelm Friedrich von. 1717-1783. Ger. DSB; ICB; P I; BMNH; BLA; GA; GC.

10871. LEDERMÜLLER, Martin Frobenius. 1719-1769. Ger. ICB; P I; BMNH; BLA.

10872. ADAMS, George (first of the name). 1720-1773. Br. P I; BMNH; GA; GD. Also Astronomy*.
1. *Micrographia illustrata; or, The knowledge of the microscope explain'd ... To which is added a translation of Mr. Joblot's observations on the animalcula that are found in ... infusions, and a ... particular account of ... the fresh water polype, translated from the French treatise of Mr. Trembley.* London, 1746. RM. 4th ed., 1771.

10873. COLOMBO, Michele. 1747-1838. BMNH.

10874. ADAMS, George (second of the name). 1750-1795. Br. P I; BMNH; BLA; GA; GD. Other entries: see 3217.
1. *Essays on the microscope.* London, 1787. RM/L. 2nd ed., 1798.

10875. AMICI, Giovan Battista. 1786-1868. It. DSB; ICB; P I,III; BMNH; Mort.; GA; GB. Also Physics and Biology in General*.

10876. LISTER, Joseph Jackson. 1786-1869. Br. DSB; Mort.; GD.

10877. MANTELL, Gideon Algernon. 1790-1852. Br. DSB; ICB; P II; BMNH; GA; GB; GD. Also Geology*.
1. *Thoughts on animalcules; or, A glimpse of the invisible world revealed by the microscope.* London, 1846.

10878. GORING, C.R. 1792-1840. Br. P I; BMNH.
1. *Microscopic illustrations of a few ... living objects.* London, 1830. With A. Pritchard. X. 3rd ed., 1845.
2. *Micrographia.* 1837. With A. Pritchard.

10879. RASPAIL, François Vincent. 1794-1878. Fr. DSB; ICB; P II, III; BMNH; BLA; GA. Also Botany*.

10880. PRITCHARD, Andrew. 1804-1882. Br. P III; BMNH; GD. Also Microbiology*.
1. *The microscopic cabinet.* London, 1832. X.
2. *Notes on natural history, selected from "The microscopic cabinet."* London, 1844. X. 2nd ed., 1849.
See also 10878/1,2.

10881. MOHL, Hugo von. 1805-1872. Ger. DSB; ICB; P III; BMNH; GA; GB; GC. Also Botany*.
1. *Mikrographie.* Tübingen, 1846.

10882. HARTING, Pieter. 1812-1885. Holl. DSB; P I,III; BMNH; Mort.; BLA. Also Zoology and History of Microscopy*.

10883. CARPENTER, William Benjamin. 1813-1885. Br. DSB; ICB; P III; BMNH; BLA & Supp.; GB; GD. Other entries: see 3326.
1. *The microscope and its revelations.* London, 1856. 2nd ed., 1857. 5th ed., 1875. 6th ed., 1881. 7th ed., enl. by W.H. Dallinger, 1891.

10884. HANNOVER, Adolph. 1814-1894. Den. ICB; P I; BMNH; BLA & Supp. Also Histology.

10885. LANKESTER, Edwin. 1814-1874. Br. BMNH; BLA; GD. Also Zoology* and Biography*. See also Index.
1. *Half-hours with the microscope.* London, 1859. X. Another ed., 1877.

10886. QUEKETT, John Thomas. 1815-1861. Br. BMNH; BLA; GD. Also Histology*.
1. *A practical treatise on the use of the microscope.* London, 1848. X. 2nd ed., 1852. 3rd ed., 1855.

10887. HOGG, Jabez. 1817-1899. Br. BMNH; BLA; GD1.
1. *The microscope. Its history, construction and application.* London, 1854. 2nd ed., 1855. 4th ed., 1859. 5th ed., 1861. 6th ed., 1867. 15th ed., 1898.

10888. NÄGELI, Karl Wilhelm von. 1817-1891. Ger./Switz. DSB; ICB; P III; BMNH; GB; GC. Also Botany* and Evolution*.
1. *Das Mikroscop. Theorie und Anwendung desselben.* 2 vols. Leipzig, 1867. With S. Schwendener. 2nd ed., 1877.
1a. ——— *The microscope in theory and practice.* London, 1887. 2nd ed., 1892.

10889. GRIFFITH, John William. 1819?-1901. Br. BMNH.
1. *The micrographic dictionary. A guide to the examination and investigation of the structure and nature of microscopic objects.* London, 1856. With A. Henfy. 2nd ed., 1860. 3rd ed., 1875. 4th ed., 1883.

10890. SMITH, Hamilton Lamphere. 1819-1903. U.S.A. P III,IV; BMNH.

10891. ROYSTON-PIGOTT, George West. 1820-1889. Br. P III.

10892. ROBIN, Charles Philippe. 1821-1885. Fr. DSB; ICB; BMNH; BLA & Supp.; GA. Also Chemistry* and Anatomy*.

10893. FREY, (Johann Friedrich) Heinrich (Konrad). 1822-1890. Switz. BMNH; BLA & Supp.; GC. Also Zoology* and Histology*.
1. *Das Mikroscop und die mikroscopische Technik.* Leipzig, 1863. X. 5th ed., 1873. 7th ed., 1881.
1a. ——— *The microscope and microscopal technology.* New York, 1872. Trans. and ed. by G.R. Cutler. 2nd ed., 1880.

10894. SCHULTZE, Max Johann Sigismund. 1825-1874. Ger. DSB; ICB; P III; BMNH; Mort.; BLA; GB; GC. Also Zoology, Histology, Embryology, and Cytology.

10895. SORBY, Henry Clifton. 1826-1908. Br. DSB; ICB; P III,IV; BMNH; GB; GD2. Also Geology*.

10896. DIPPEL, Leopold. 1827-1914. BMNH. Also Botany*.
1. *Das Mikroscop und seine Anwendung.* 2 vols. Brunswick, 1869-72. 2nd ed., 1882-83.
2. *Grundzüge der allgemeinen Mikroskopie.* Ib., 1885.

10897. BEALE, Lionel Smith. 1828-1906. Br. DSB; ICB; BMNH; BLA & Supp.; GD2. Also Biology in General* and Histology*.
1. *How to work with the microscope.* London, 1857. X. 2nd ed., 1861. 3rd ed., 1865. 4th ed., 1878. 5th ed., 1880.

10898. ARCHER, William. 1830-1897. Br. BMNH; GD1. Also Microbiology.

10899. WOODWARD, Joseph Janvier. 1833-1884. U.S.A. P III; BLA; GE.

10900. KLEBS, (Theodor Albrecht) Edwin. 1834-1913. Ger./Switz. ICB; Mort.; BLA. Also Microbiology.

10901. RICHARDSON, Joseph Gibbons. 1836-1886. U.S.A. BLA.

10902. WARD, Richard Halstead. 1837-1917. U.S.A. GE.

10903. HEURCK, Henri Ferdinand van. 1838-1909. Belg. ICB; BMNH; GF (under VAN). Also Botany.
1. *Le microscope. Sa construction, son maniement et son application.* Paris, 1865 and later eds. X.
1a. ——— *The microscope.* 1893.

10904. STERNBERG, George Miller. 1838-1915. U.S.A. ICB; BMNH; BLA; BLAF; GE. Also Microbiology*.
1. *Photo-micrographs and how to make them.* Boston, 1883.

10905. ABBE, Ernst. 1840-1905. Ger. DSB; ICB; P III,IV; Mort.; BLAF. Also Physics.

10906. DALLINGER, William Henry. 1842-1909. Br. BMNH; GD2. Also Microbiology. See also 10883/1.

10907. MERKEL, Friedrich Siegismund. 1845-1919. Ger. BMNH; Mort.; BLAF. Also Anatomy and Embryology.
   1. *Das Mikroscop und seine Anwendung.* 1875.

10908. FRIEDLÄNDER, Carl. 1847-1887. Ger. BMNH; BLA; BLAF; GC. Also Microbiology.
   1. *Microscopische Technik.* Berlin, 1882. X. Another ed., 1889.

10909. LANKESTER, Edwin Ray. 1847-1929. Br. DSB; BMNH; BLAF; GD4. Also Science in General*, Zoology*, and Evolution*. See also Index.

10910. MAYER, Paul. 1848-1923. Ger. BMNH; BLAF. Also Zoology.

10911. LEE, Arthur Bolles. 1849-1927. BMNH.
   1. *The microtomist's vade-mecum. A handbook of the methods of microscopic anatomy.* London, 1885.

10912. GAGE, Simon Henry. 1851-1944. U.S.A. BMNH; BLAF. Also Zoology.

10913. SPITTA, Edmund Johnson. 1853-1921. Br. BMNH. Also Microbiology.
   1. *Photo-micrography.* London, 1899.

10914. AMBRONN, Hermann. 1856-1927. Ger. P IV; BMNH.

3.12 NATURAL HISTORY

This section includes books dealing with all three, or any two, of the "three kingdoms of Nature", and also books whose titles do not specify their scope beyond indicating that they deal with natural history (some of these may, in fact, be confined to zoology).

As well as some major authors, the section includes many minor ones who wrote on both botany and zoology, together or separately. Accounts of travels or explorations are included only if they contain significant material relating to the natural history sciences.

10915. SLOANE, Hans. 1660-1753. Br. DSB; ICB; P II; BMNH; BLA; GA; GB; GD. Also Botany* and Personal Writings*.

10916. MOLYNEUX, Thomas. 1661-1733. Br. P II; BMNH; GA; GD.

10917. LE LONG, Jacques. 1665-1709. Fr. BMNH.

10918. SEBA, Albert. 1665-1736. Holl. ICB; P II; BMNH; GA.
1. *Locupletissimi rerum naturalium thesauri accurata descriptio.* 2 vols of text & 2 vols of plates. Amsterdam, 1734-38. The half-title reads: *Description exacte des principales curiositez naturelles du ... cabinet d'A. Seba.* Vol. 1 contains a preface by H. Boerhaave and Vol. 2 one by J.J. Baier.

10919. HARRIS, John. ca. 1666-1719. Br. DSB; ICB; P I; GA; GB; GD. Also Science in General*, Mathematics*, and Astronomy*.
1. *Navigantium atque itinerantium bibliotheca; or, A compleat collection of voyages and travels.* London, 1705. X. Another ed., 2 vols, 1744-48.

10920. HELWING, Georg Andreas. 1666-1748. Ger. P I (HELWIG); BMNH; GA (HELWIG).

10921. BERINGER (or BEHRINGER), Johann Bartholomaeus Adam. ca. 1667-1738. Ger. DSB; ICB; P I; BLA; GA. Also Geology*.

10922. WALDSCHMIDT, Wilhelm Hulderich. 1669-1731. Ger. P II; BMNH; BLA; GC.

10923. DIÈREVILLE, ... *Sieur* de. ca. 1670-after 1703. Fr. ICB; BMNH; GA.

10924. HERMANN, Leonard David. 1670-1736. Ger. P I; BMNH.

10925. MORTON, John. 1670-1726. Br. DSB; BMNH; GD.

10926. SCHEUCHZER, Johann Jacob. 1672-1733. Switz. DSB; ICB; P II; BMNH; BLA; GA; GB; GC. Also Geology* and Zoology*.
1. *[Oresiphoites] Helveticus; sive, Itineris alpini descriptio physico-medica.* Zurich, 1702. X.

1a. ——— Another ed. ... *sive, Itinera alpina tria. In quibus incolae, animalia, plantae ... exponitur.* London, 1708. RM/L.

1b. ——— Another ed. ... *sive, Itinera per Helveticae alpinas regiones facta annis* [1702-07, 1709-11]. 4 vols. Leiden, 1723. RM.

2. *Helvetiae historia naturalis; oder, Natur-Historie des Schweitzerlands.* Zurich, 1716. RM.

10927. DOUGLAS, James. 1675-1742. Br. DSB; BMNH; Mort.; BLA & Supp.; GD. Also Anatomy and History of Anatomy*.

10928. ARGENVILLE, Antoine Joseph Dezallier d'. 1680-1765. Fr. DSB. Under DEZALLIER: P I; BMNH; GA. Also Geology*.

1. *L'histoire naturelle éclaircie dans deux de ses parties principales, la lithologie et la conchyliologie.* Paris, 1742. RM (under DEZALLIER)/L.

1a. ——— *L'histoire naturelle éclaircie dans une de ses parties principales, l'oryctologie.* Ib., 1755.

1b. ——— *L'histoire naturelle éclaircie dans une de ses parties principales, la conchyliologie ... Nouvelle édition.* Ib., 1757.

1c. ——— ——— *La conchyliologie; ou, Histoire naturelle des coquilles ... Troisième édition.* 2 vols & atlas. Ib., 1780. Ed. with a biography by Favanne de Montcerville.

10929. CATESBY, Mark. 1683-1749. Br./U.S.A. DSB; ICB; BMNH; GA; GD; GE.

1. *The natural history of Carolina, Florida, and the Bahama Islands.* London, 1731-43, 1729-47. X. Reprinted, Savannah, Ga., 1974.

10930. ASTRUC, Jean. 1684-1766. Fr. DSB; ICB; P I; BMNH; BLA & Supp.; GA; GB; GF. Also Various Fields.

1. *Mémoires pour servir à l'histoire naturelle de la province de Languedoc.* Paris, 1737. RM.

10931. KUNDMANN, Johann Christian. 1684-1751. Breslau. P I; BMNH; BLA; GA.

10932. EGEDE, Hans Poulsen. 1686-1758. Den. ICB; P I; BMNH; GA; GB.

10933. PLUCHE, Noël Antoine. 1688-1761. Fr. DSB; ICB; BMNH; GA. Also Astronomy*.

1. *Spectacle de la nature; ou, Entretiens sur les particularités de l'histoire naturelle.* 8 vols in 9. Paris, 1732-50. Many later eds. X.

1a. ——— *Spectacle de la nature; or, Nature displayed. Being discourses....* 3 vols. London, 1736-37. Trans. by S. Humphreys. 2nd ed., 7 vols? 1749. Other eds, 1757-63 and 1770.

10934. RICHTER, Johann Christoph. 1689-1751. Ger. P II.

10935. BROOKES, Richard. fl. 1721-1762. Br. BMNH; BLA & Supp.; GA; GD.

10936. RZACZYNSKI, Gabriel. d. 1737. Poland. BMNH.

10937. BARRÈRE, Pierre. 1690?-1755. Fr. BMNH; BLA; GA. Also

Mineralogy*.
1. *Essai sur l'histoire naturelle de la France équinoxiale.* Paris, 1741. Deals with various French colonies.

10938. FISCHER, Christian Gabriel. 1690?-1751. Ger. ICB; P I; BMNH; GA.

10939. GARSAULT, François Alexandre Pierre de. 1691-1778. Fr. BMNH; GA.

10940. COLLINSON, Peter. 1693/4-1768. Br. DSB; ICB; P I; BMNH; GA; GD. Important as an "intelligencer", especially in his contacts with American scientists. See also 5857/1b.

10941. BORLASE, William. 1695-1772. Br. P I (BORLACE); BMNH; GA; GB; GD.
1. *The natural history of Cornwall.* Oxford, 1758. RM/L. Reprinted, London, 1970.

10942. BÜCHNER, Johann Gottfried. 1695-1749. Ger. P I; BMNH.

10943. BRÜCKMANN, Franz Ernst. 1697-1753. Ger. P I; BMNH; BLA; GA; GB. Also Geology.

10944. PONTOPPIDAN, Erik Ludvigsen. 1698-1764. Den. P II; BMNH; GA; GB.
1. [Trans.] *The natural history of Norway.* London, 1755. Trans. from the Danish. RM/L.

10945. TORRUBIA, José. 1698-1761. Spain. BMNH.

10946. ALLÉON DULAC, Jean Louis. ca. 1700-1768.
1. *Mémoires pour servir à l'histoire naturelle des provinces de Lyonnois, Forez, et Beaujolois.* Lyons, 1765. RM.
2. *Mélanges d'histoire naturelle.* Lyons, 1765. RM.

10947. LA CONDAMINE, Charles Marie. 1701-1774. Fr. DSB Supp.; ICB; P I (CONDAMINE); BMNH; BLA (CONDAMINE); GA; GB. Also Geodesy*.

10948. RAPPOLT, Karl Heinrich. 1702-1753. Ger. ICB; P II.

10949. ARDERON, William. 1703-1767. Br. GD.

10950. HEBENSTREIT, (Johann) Ernst. 1703-1757. Ger. P I; BMNH; BLA; GA; GC. Also Chemistry.
1. *Museum Richterianum, continens fossilia animalia, vegetabilia marina, illustrata iconibus et commentariis.* Leipzig, 1743.

10951. BERGEN, Carl August von. 1704 (or 1709)-1759. Ger. P I; BMNH; BLA & Supp.; GA; GC. Also Anatomy.

10952. JUSSIEU, Joseph de. 1704-1779. Fr./South America. DSB; ICB; BMNH; BLA; GA; GB.

10953. BOWLES, William. 1705-1780. Br./Spain. BMNH; GA; GD.

10954. BUFFON, Georges Louis Leclerc, *Comte* de. 1707-1788. Fr. DSB; ICB; P I; BMNH; GA; GB. Also Geology*. See also Index.
1. *Histoire naturelle, générale et particulière, avec la description du Cabinet du Roi.* 44 vols. Paris, 1749-1804. RM/L. Some sections were written by Daubenton. The volumes completed after Buffon's death were by Lacépède. For an analysis of the work see BMNH.

1a. ——— Another ed. *Histoire naturelle par Buffon.* 76 vols. Ib., 1799-1809. Ed. by Lacépède.
1b. ——— Another ed. *Histoire naturelle ... Nouvelle édition ... rédigé par C.S. Sonnini* [de Manoncourt]. 64 vols. Ib., 1799-1805.

English Translations of the *Histoire naturelle* or sections of it.

2. *A natural history, general and particular, containing the history and theory of the earth, a general history of man, the brute creation, vegetables, minerals, etc. etc.* 9 vols. London, 1781. Trans. by W. Smellie.
2a. ——— *A natural history, general and particular. The history of man and quadrupeds. Translated with notes and observations by W. Smellie ... A new edition ... enlarged ... by W. Wood.* 20 vols. London, 1812. RM/L.
3. *Barr's Buffon. Buffon's natural history, containing a theory of the earth, a general history of man, of the brute creation, and of vegetables, minerals, etc.* 10 vols. London, 1792. The publisher was J.S. Barr. Another ed., 16 vols, 1797-1808.

Collected Works and Correspondence

4. *Oeuvres complètes.* 33 vols. Paris, 1820-22. Ed. by Lacépède. X. Other eds, 1825-28 and 1828-33.
5. *Oeuvres complètes, avec des extraits de Daubenton et la classification de Cuvier.* 6 vols. Paris, 1837-39.
6. *Oeuvres complètes, avec la nomenclature linnéenne et la classification de Cuvier.* 12 vols. Paris, 1853-55. Revised and annotated by Flourens.
7. *Oeuvres complètes. Edition annotée, suivie de la correspondance générale.* 14 vols. Paris, 1884-85. By J.L. Lanessan. X.
8. *Correspondance générale. Recueillie et annotée par H. Nadault de Buffon.* Paris, 1885. X. Reprinted, Geneva, 1971.

Selections

9. *Oeuvres philosophiques.* Paris, 1954. Critical ed. by J. Piveteau.

There are many editions of *Oeuvres choisies* and of extracts from the *Histoire naturelle.*

10955. DARLUC, Michel. 1707-1783. Fr. P I; BMNH; BLA; GA.

10956. HILL, John. 1707?-1775. Br. DSB; ICB; P I; BMNH; BLA; GA; GD. Other entries: see 3105.
1. *A general natural history; or, New and accurate descriptions of the animals, vegetables and minerals of the world.* 3 vols. London, 1748-52. RM/L.
2. *Essays in natural history and philosophy. Containing a series of discoveries, by the assistance of microscopes.* London, 1752. RM/L.

10957. HUGHES, Griffith. b. ca. 1707. West Indies. BMNH; GA; GD.
1. *The natural history of Barbados.* London, 1750.

10958. LINNAEUS (later VON LINNÉ), Carl. 1707-1778. Swed. DSB; ICB; P I; BMNH. Under LINNÉ: BLA & Supp.; GA; GB. Also Botany*, Zoology*, and Personal Writings*. See also Index.

1. *Systema naturae; sive, Regna tria naturae systematice proposita per classes, ordines, genera, et species.* Leiden, 1735. Reprinted, Stockholm, 1907. Another reprint, with the first English trans. of the *Observationes*, Nieuwkoop, 1964. (Dutch classics on history of science. 8)

1a. ——— 2nd ed., Stockholm, 1740. X. 3rd ed., Halle, 1740. X. 4th ed., Paris, 1744. X. 5th ed., Halle, 1747. X.

1b. ——— 6th ed., *Emendata et aucta*, Stockholm, 1748. X. 7th ed., Leipzig, 1748. 8th ed., Swedish trans., Stockholm, 1753. X.

1c. ——— 9th ed., *Multo auctior et emandatior*, Leiden, 1756.

1d. ——— 10th ed., *Reformata*, 2 vols, Stockholm, 1758-59. RM/L. Reprint of Vol. 1 (*Animalia*), Leipzig, 1894. Another reprint of Vol. 1, London, 1956. Reprint of Vol. 2 (*Vegetabilia*), Weinheim, 1964. (HNC. 34)

1e. ——— 11th ed., Leipzig, 1762. Unchanged. Very rare. X.

1f. ——— 12th ed., *Reformata*, 3 vols in 4, Stockholm, 1766-68.

1g. ——— ——— *Appendices zoologicae*. Ib., 1768-71. X. Reprinted, The Hague, 1935.

1h. ——— ——— Reprint, purporting to be the 13th ed., 3 vols in 4, Vienna, 1767-70.

1i. ——— 13th ed., *Aucta, reformata, cura J.F. Gmelin*, 3 vols in 7, Leipzig, 1788-93.

1j. ——— ——— Reprint, 3 vols, Lyons, 1789-96.

Translations of the *Systema naturae*

2. *Vollständiges Natursystem.* 6 vols in 8. Nuremberg, 1773-75. Trans. from the 12th ed., with notes, by P.L.S. Müller. *Supplement und Register*, 3 vols, 1776.

3. *A genuine and universal system of natural history.* 14 vols. London, 1794-1810. Trans. from the 13th ed., with additions, by the editors of the *Encyclopaedia Londinensis*.

4. *A general system of nature.* 7 vols. London, 1802-06. Trans. from the 13th ed., with notes and additions, by W. Turton. RM. Another ed., 1806.

Linnaeus' Travels

5. *Öländska och Gothländska resa ... 1741.* Stockholm, 1745. X.

5a. ——— *Linnaeus' Oland and Gotland journey, 1741.* London, 1973.

6. *Lachesis Lapponica; or, A tour in Lapland. Now first published from the original manuscript journal of the celebrated Linnaeus. By J.E. Smith.* 2 vols. London, 1811. (The MS dates from 1732.) RM. Reprinted, New York, 1971.

7. *Skånska resa förrättad 1749.* Stockholm, 1751. Reprinted, ib., 1973.

8. *Linné i Lappland: Utdrag ur C. Linnaeus' dagbok från resan till Lappland, 1732.* Stockholm, 1969.

9. *Linné på Öland: Utdrag ur C. Linnaeus' dagbokmanuskript från Öländska resan, 1741*. Stockholm, 1970.

Students' Theses

10. *Amoenitates academicae; seu, Dissertationes variae physicae, medicae botanicae antehac seorsim editae, nunc collectae et auctae*. 10 vols. Stockholm, Erlangen, etc., 1749-90. A collection of 186 theses (originally published individually) of Linnaeus' students at the University of Uppsala in the period 1743-76. Vols 1-7 of the collection were ed. by Linnaeus and publ. at Stockholm, 1749-69. Vols 8-10 were ed. by J.C.D. von Schreber, one of his students, and publ. at Erlangen, 1785-90, together with a new ed. of Vols. 1-7. There are various later editions, including a pirated one publ. at Amsterdam and Leiden.

10a. ——— *Miscellaneous tracts relating to natural history, husbandry, and physik*. London, 1759. Trans. of some extracts by B. Stillingfleet. See item 11402/1.

Other Works

11. *Materia medica*. 2 vols. Stockholm, 1749-63. X.

11a. ——— 2nd ed. *Materia medica per regna tria naturae*. Leipzig/Erlangen, 1772. Ed. by J.C.D. von Schreber. Another ed., Vienna, 1773.

12. *Museum S.R.M. Adolphi Friderici Regis*. Stockholm, 1754. See item 12663/2.

12a. ——— Trans. of preface: *Reflections on the study of nature*. London, 1785. Trans. by J.E. Smith.

Correspondence

13. *Collectio epistolarum, quas ad viros illustres et clarissimos scripsit. Accedunt opuscula* ... [see item 15339/1]. Hamburg, 1792. Ed. by D.H. Stoever. RM.

14. *A collection of the correspondence of Linnaeus and other naturalists. From the original manuscripts*. 2 vols. London, 1821. By J.E. Smith. RM/L.

15. *Brefvexling*. Stockholm, 1885. A catalogue of all the letters known to have been written by and to Linnaeus. Ed. by J.E. Ährling. Reprinted, Zug, 1967.

16. *Bref och skrifvelser af och till Carl von Linné*. 8 vols. Stockholm, etc., 1908-22. Ed. for the University of Uppsala.

Collected Works and Selections

17. *Skrifter*. 5 vols. Uppsala, 1905-13. The Latin works trans. into Swedish. RM.

18. *L'équilibre de la nature*. Paris, 1972. Trans. by B. Jasmin. Introd. and notes by C. Limoges.

10959. SENCKENBERG, Johann Christian. 1707-1772. Ger. ICB; BMNH; BLA & Supp.; GC. Founder of the Frankfurt museum.

10960. THOMAS, Pierre. 1708-1781. Switz. ICB.

10961. GMELIN, Johann Georg. 1709-1755. Ger./Russ. DSB; ICB; P I;

BMNH; BLA; GA; GC. Also Botany*
1. *Reise durch Sibirien, von 1733 bis 1743.* 4 vols. Göttingen, 1751-52.

10962. STELLER, Georg Wilhelm. 1709-1746. Russ. DSB; ICB; P III; BMNH; GC.

10963. BRICKELL, John. 1710?-1745. BMNH.
1. *The natural history of North Carolina.* Dublin, 1737. X. Reprinted, New York, 1969.

10964. LIGNAC, Joseph Adrien le Large de. 1710-1762. Fr. BMNH; GA.

10965. MOEHRING, Paul Heinrich Gerhard. 1710-1792. Ger. BMNH; BLA.

10966. BASTER, Job. 1711-1775. Holl. ICB; BMNH; BLA & Supp.

10967. KRASHENINNIKOV, Stepan Petrovich. 1711-1755. Russ. DSB; ICB; BMNH.

10968. FOTHERGILL, John. 1712-1780. Br. ICB; P I; BMNH; BLA & Supp.; GA; GB; GD.
1. *A complete collection of the medical and philosophical works.* London, 1781. Ed. with notes and a biography by John Elliot. RM.
See also 11343/1a.

10969. HORREBOW, Niels. 1712-1760. Den. P I; BMNH; GA.
1. *Tilforladelige efterretninger om Island.* Copenhagen, 1752. RM.
1a. ——— *The natural history of Iceland.* London, 1758. RM/L.

10970. RYCHKOV, Petrov Ivanovich. 1712-1777. Russ. ICB.

10971. TARGIONI-TOZZETTI, Giovanni. 1712-1783. It. DSB; ICB; P II; BMNH; BLA; GA. Also Geology, Biography*, and History of Science*.
1. *Relazione d'alcuni viaggi fatti in diverse parti della Toscana.* 6 vols. Florence, 1751-54. X.
1a. ——— *Voyage minéralogique, philosophique et historique en Toscane.* Paris, 1792. RM.

10972. ALLAMAND, Jean Nicolas Sébastien. 1713-1787. Holl. P I; BMNH; GA.

10973. DAVILA, Pedro Francisco. 1713-1785. Spain. BMNH; GA.
1. *Catalogue systématique et raisonné des curiosités de la nature et de l'art, qui composent le cabinet de M. Davila.* 3 vols. Paris, 1767. Compiled by Davila with the assistance of Romé de l'Isle.

10974. DONATI, Vitaliano. 1713 (or 1717)-1763. It. P I; BMNH; BLA; GA.
1. *Della storia naturale marina dell'Adriatico.* Venice, 1750. RM.
1a. ——— *Essai sur l'histoire naturelle de la Mer Adriatique.* The Hague, 1758. RM/L.

10975. BATTARRA, Giovanni Antonio. 1714-1789. BMNH.

10976. HAWKESWORTH, John. 1715?-1773. Br. BMNH; GA; GB; GD.
1. *An account of the voyages ... in the southern hemisphere*

... drawn up from the journals which were kept by the several commanders, and from the papers of Joseph Banks. 3 vols. London, 1773. X.

1a. ——— *Geschichte der See-Reisen und Entdeckungen im Süd-Meer aus den Tagebüchern und Handschriften Joseph Banks.* 3 vols. Berlin, 1774.

10977. RUSSELL, Alexander. ca. 1715-1768. Br. P II; BMNH; BLA; GD.
1. *The natural history of Aleppo and parts adjacent.* London, 1756. Through a misprint the date on the title page reads 1856.

10978. BOCK, Friedrich Samuel. 1716-1786. Ger. P I; BMNH; GA.

10979. DAUBENTON, Louis Jean Marie. 1716-1800. Fr. DSB Supp.; ICB; P I; BMNH; BLA & Supp.; GA; GB. Also Mineralogy, Botany, and Zoology. See also Index.

10980. ULLOA (Y DE LA TORRE GIRAL), Antonio de. 1716-1795. Spain. DSB; P II; BMNH; GA; GE. See also 10396/1.

10981. GUNNERUS, Johan Ernst. 1718-1773. Nor. P I; BMNH.
1. *Brevekslíng, 1761-1772.* Trondheim, 1976. Correspondence with Linnaeus. Ed. by L. Amundsen.

10982. LA FAILLE, Clément de. 1718-1782. Fr. ICB; BMNH; GA.

10983. MEUSCHEN, Friedrich Christian. 1719-1790. ICB; BMNH.

10984. PÁLSSON, Bjarne. 1719-1779. Den. P II (PAULSEN); BMNH (POVELSON).
1. [Trans.] *Reise durch Island.* 2 vols. Copenhagen/Leipzig, 1774-75. With E. Olafsen. Trans. from the Danish by J.M. Geuss.

10985. BROWNE, Patrick. ca. 1720-1790. Br. BMNH; BLA & Supp.; GA; GD.

10986. FERMIN, Philippe. 1720-1790 (or 1729-1813). Holl./East Indies. BMNH; BLA & Supp.; GA; GF.

10987. GOECKEL, Philipp Caspar. 1720-1759. Ger. BMNH; BLA.

10988. HAMMER, Christopher. 1720-1804. Nor. P I; BMNH.

10989. HOUTTUYN, Martinus. 1720-1798. Holl. ICB; BMNH.

10990. WHITE, Gilbert. 1720-1793. Br. DSB; ICB; BMNH; GB; GD.
1. *The natural history and antiquities of Selborne.* London, 1789. RM/L. Other eds, 1813 and 1822; later eds not included.
2. *A naturalist's calendar, with observations in various branches of natural history.* London, 1795. RM.
3. *The works in natural history. Comprising "The natural history of Selborne", "The naturalist's calendar", and miscellaneous observations extracted from his papers.* 2 vols. London, 1802.
4. *A nature calendar.* London, 1911. Ed. with an introd. by W.M. Webb.
5. *Journals.* London, 1931. Ed. by W. Johnson. X. Reprinted, Newton Abbot, 1970.
6. *Writings.* 2 vols. London, 1938. Selected and ed. by H.J. Massingham.

10991. TURGOT, Etienne François. 1721-1788. Fr. DSB; P II; BMNH; GA.
1. *Mémoire instructif sur la manière de rassembler, de préparer, de conserver, et d'envoyer les diverses curiosités d'histoire naturelle.* Lyons, 1758. RM.

10992. HASSELQUIST, Frederik. 1722-1752. Swed. ICB; P I; BMNH; BLA; GA; GB.
1. *Iter Palaestinum; eller, Resa til Heliga Landet ... 1749 til 1752.* Stockholm, 1757. Ed. by Linnaeus. RM. Reprinted, Stockholm, 1969.
1a. ——— *Voyages and travels in the Levant in the years 1749-52. Containing observations in natural history, etc.* London, 1766. RM/L.

10993. ASCANIUS, Peder. 1723-1803. Den. BMNH; GA.

10994. MANETTI, Xaverio. 1723-1784. It. BMNH; GA.

10995. MONTIN, Lars. 1723-1785. Swed. ICB; BMNH; BLA.

10996. OSBECK, Pehr. 1723-1805. Swed. ICB; BMNH; GA.

10997. SCOPOLI, Giovanni Antonio. 1723-1788. It./Hung. ICB; P II; BMNH; GA. Also Mineralogy*, Geology, Botany*, and Zoology*.
1. *Annus I (-V) historico-naturalis.* 5 vols in 1. Leipzig, 1769 (-1772).
2. *Introductio ad historiam naturalem, sistens genera lapidum, plantarum et animalium....* Prague, 1777.
3. *Deliciae florae et faunae Insubricae.* Pavia, 1786. Natural history of Insubria Austriaca, now part of northern Italy.

10998. WELL, Johann Jacob von. 1725-1787. Aus. P II; BMNH; BLA. Also Chemistry.

10999. HÜPSCH, Johann Wilhelm Karl Adolph von. 1726-1805. Ger. P I; BMNH; GC.

11000. PENNANT, Thomas. 1726-1798. Br. DSB; ICB; BMNH; GA; GB; GD. Also Zoology* and Autobiography*.
1. *Tour on the Continent, 1765.* London, 1948. Ed. from the unpubl. MS by G.P. De Beer. (Ray Society publn. 132)
2. *A tour in Scotland, 1769.* Chester, 1771. 2nd ed., London, 1772. 4th ed., 1776. 5th ed., 1790.
3. *A tour in Scotland and voyage to the Hebrides, 1772.* 2 vols. Chester, 1774-76. 2nd ed., London, 1776. Another ed., 1790.
4. *A tour in Wales.* 2 vols. London, 1778-81. X. Another ed., 1784.
5. *The history of the parishes of Whiteford and Holywell.* London, 1796.
6. *Outlines of the globe.* 4 vols. London, 1798-1800.
See also 11034/2.

11001. COMMERSON, Philibert. 1727-1773. Fr. DSB; GA. See also 15482/1.

11002. RUSSELL, Patrick. 1727-1805. Br./India. P II (RUSSEL); BMNH; BLA; GD; GF. See also 11603/1.

11003. GARDEN, Alexander. 1728-1791. Br./U.S.A. ICB; P I; BLA & Supp.; GD; GE.

11004. KÖNIG, Johann Gerhard. 1728-1785. Den./East Indies. ICB; P I; BMNH; BLA & Supp.; GA; GF.

11005. WULFFEN (or WÜLFEN, etc.), Franz Xaver von. 1728-1805. Aus. P II; BMNH (WULFEN).

11006. BOUGAINVILLE, Louis Antoine de. 1729-1811. Fr. The explorer. DSB; ICB; P I; GA; GB. Also Mathematics*.
1. *Voyage autour du monde ... en 1766-1769*. Paris, 1771. X.
1a. ——— *A voyage around the world ... in 1766-1769*. London, 1772. Trans. by J.R. Forster.

11007. FORSTER, Johann Reinhold. 1729-1798. Ger./Br. DSB; ICB; P I & Supp.; BMNH; GA; GC. Also Mineralogy*, Geology*, Botany* and Zoology*. See also Index.
1. *Observations made during a voyage around the world, on physical geography, natural history and ethic philosophy*. London, 1778. RM/L.
1a. ——— *Bemerkungen über Gegenstände ... auf seine Reise um die Welt gesammelt*. Berlin, 1783. Trans. with notes by his son, Georg Forster.
2. *Tagebuch einer Entdeckungsreise nach der Südsee in den Jahren 1776 bis 1780, unter Anführung der Capitains Cook, Clerke, Gore und King*. Berlin, 1781.
3. *Geschichte der Entdeckungen und Schiffahrten im Norden*. Frankfurt on Oder, 1784. X.
3a. ——— *History of the voyages and discoveries made in the north*. London, 1786.
4. *Enchiridion historiae naturali inserviens*. Halle, 1788. Another ed., Edinburgh, 1794.
5. *Magazin von merkwürdigen neuen Reisebeschreibungen*. 25 vols. Berlin, 1790-1800. Another ed., Vienna, 1792-1801.
6. *Die neuesten Reisen nach Botany-Bay und Port-Jackson*. 3 vols. Berlin, 1794.

11008. GEORGI, Johann Gottlieb. 1729-1802. Ger./Russ. ICB; P I; BMNH. Also Chemistry.

11009. LEVER, Ashton. 1729-1788. Br. BMNH; GD.
1. *A companion to the museum (late Sir Ashton Lever's)*. London, 1790.
2. *The Leverian museum*. [180-?]
See also 11060/2.

11010. SPALLANZANI, Lazzaro. 1729-1799. It. DSB; ICB; P II; BMNH; Mort.; BLA; GA; GB. Other entries: see 3167.
1. *Viaggi alle due Sicilie e in alcune parti dell'Appennino*. 6 vols. Pavia, 1792-97. X.
1a. ——— *Travels in the two Sicilies and some parts of the Apennines*. 4 vols. London, 1798. RM.

11011. TITIUS (or TIETZ), Johann Daniel. 1729-1796. Ger. DSB; P II; BMNH; GC. Also Astronomy and Physics.

11012. BERKENHOUT, John. 1730?-1791. Br. BMNH; BLA & Supp.; GA; GD. Also Chemistry*.
1. *Outlines of the natural history of Great Britain and Ireland*. 3 vols in 1. London, 1769-72. 2nd ed., *Synopsis of....* 2 vols, 1789. 3rd ed., 1795.

11013. BRUCE, James. 1730-1794. Br. DSB; ICB; P I; BMNH; GA; GD.
1. *Travels to discover the source of the Nile in the years 1768-1773.* 5 vols. Edinburgh/London, 1790.

11014. GOLDSMITH, Oliver. 1730 (or 1728)-1774. Br. The well known writer. ICB; BMNH; G. Also Science in General*.
1. *A history of the earth and of animated nature.* 8 vols. London, 1774. RM/L. 2nd ed., 1779. Another ed., with corrections and additions by W. Turton, 6 vols, 1816. Later eds not included.

11015. GRONOVIUS, Laurentius Theodorus. 1730-1777. Holl. ICB; P I & Supp.; BMNH; GA. Also Zoology*.
1. *Bibliotheca regni animalis atque lapidei; seu, Recensio auctorum et librorum qui de regno animali et lapidei ... tractant.* Leiden, 1760. RM/L.

11016. BUC'HOZ, Pierre Joseph. 1731-1807. Fr. P I; BMNH; BLA; GA. Also Geology*, Botany, and Zoology*.
1. *Première (-Seconde) centurie de planches ... représentant au naturel ce qui se trouve de plus intéressant et de plus curieux parmi les animaux, les végétaux et les minéraux.* 2 vols. Paris/Amsterdam, 1775-81.

11017. VALMONT DE BOMARE, Jacques Christophe. 1731-1807. Fr. DSB; P II; BMNH; GA. Also Geology and Mineralogy*.
1. *Dictionnaire raisonné universel d'histoire naturelle.* 5 vols. Paris, 1765. X. 2nd ed., 12 vols, Yverdon, 1768-69. RM. 4th ed., 12 vols, "Suisse", 1780. Another ed., 15 vols, Lyons, 1800.

11018. WALKER, John (first of the name). 1731-1803. Br. DSB; BMNH; GD. Also Geology*.
1. *Institutes of natural history.* Edinburgh, 1792. A lecture course.

11019. FALK, Johan Peter. 1733-1774. Swed./Russ. BMNH; BLA (FALCK); GA.

11020. CALONNE, Charles Alexandre de. 1734-1802. Fr. BMNH; GA; GB.
1. *Museum Calonnianum: Specification of the various articles which comprise the magnificent museum of natural history ....* London, 1797. A sale catalogue.

11021. MITTEPACHER (VON MITTERBURG), Ludwig. 1734-1814. Aus./Hung. P II; BMNH. Also Astronomy*.

11022. PETAGNA, Vincenzo. 1734-1810. It. BMNH; GA.

11023. LINCK, Johann Heinrich (second of the name). 1735-1807. Ger. P I.

11024. ROBINET, Jean Baptiste René. 1735-1820. Fr. DSB; ICB; BMNH; GA.
1. *Vue philosophique de la gradation naturelle des formes de l'être; ou, Les essais de la nature qui apprend à faire l'homme.* Amsterdam, 1768. Also publ. as *Considérations philosophiques de la gradation....*

11025. VANDELLI, Domenico. 1735-1816. It./Port. ICB; BMNH; BLA. Also Physiology.

11026. ZUECKERT, Johann Friedrich. 1737-1778. Ger. BMNH; BLA.

11027. DEMIDOFF, Pavel Grigorievich. 1738-1821. Russ. BMNH; GA.

11028. BARTRAM, William. 1739-1823. U.S.A. DSB; ICB; BMNH; GE.
1. *Travels through North and South Carolina, Georgia, East and West Florida....* Philadelphia, 1791. X. Another ed., Dublin, 1793.
2. *Botanical and zoological drawings, 1756-88.* Philadelphia, 1968. Ed. with introd. and commentary by J. Ewan.
3. *John and William Bartram's America. Selections from the writings of the early American naturalists.* New York, 1961. Ed. with introd. by H.G. Cruickshank.

11029. DAVIES, Hugh. ca. 1739-1821. Br. BMNH; GD.

11030. SEPP, Jan Christiaan. 1739-1811. Holl. BMNH.

11031. LEPEKHIN, Ivan Ivanovich. 1740-1802. Russ. DSB; ICB; P I; BMNH.

11032. MOLINA, Juan Ignacio. 1740-1829. Chile/It. DSB; ICB; P II; BMNH; GA.

11033. SMELLIE, William. 1740-1795. Br. BMNH; GD. See also 15420/1.
1. *The philosophy of natural history.* 2 vols. Edinburgh, 1790-99. RM/L.
See also 10954/2,2a.

11034. PALLAS, Peter Simon. 1741-1811. Russ. DSB; ICB; P II; BMNH; BLA; GA; GB; GC. Also Geology*, Botany*, and Zoology*.
1. *Bemerkungen auf einer Reise in die südlichen Statthalterschaften des russischen Reichs in den Jahren 1793 und 1794.* 2 vols. Leipzig, 1799. RM.
1a. ——— *Travels through the southern provinces of the Russian Empire in the years 1793 and 1794.* London, 1802-03. RM.
2. *A naturalist in Russia: Letters from P.S. Pallas to Thomas Pennant.* Minneapolis, 1967. Ed. by C. Urness.

11035. ASSO Y DEL RIO, Ignacio Jordan de. 1742-1814. Spain. BMNH; GA.

11036. AZARA, Félix de. 1742-1821. Spain/South America. DSB; ICB; P I; BMNH; GA.

11037. RETZIUS, Anders Jahn. 1742-1821. Swed. P II & Supp.; BMNH; GA. See also 12663/1a.

11038. BANKS, Joseph. 1743-1820. Br. DSB; ICB; P I; BMNH; GA; GB; GD. Also Botany*. See also Index.
1. *Catalogus bibliothecae historico-naturalis Josephi Banks.* 5 vols. London, 1796-1800. By J. Dryander. RM/L. Reprinted, New York, 1966.

Journals of Voyages

2. *A journal of a voyage round the world in H.M. ship "Endeavour."* London, 1771. Anonymous; attribution to Banks doubtful. Another ed., Dublin, 1772.
3. *Journal of ... Sir Joseph Banks during Captain Cook's first voyage ... in 1768-71.* London, 1896. Ed. by J.D. Hooker.

4. *The "Endeavour" journal of Joseph Banks, 1768-1771.* 2 vols. Sydney, 1962. Ed. by J.C. Beaglehole.
5. *Joseph Banks in Newfoundland and Labrador, 1766: His diary, manuscripts, and collections.* Berkeley, Cal., 1970. Ed. by M.A. Lysaght.

Correspondence

6. *The Banks letters: A calendar of the manuscript correspondence of Sir Joseph Banks preserved in the British Museum, etc.* London, 1958. Ed. by W.R. Dawson. Supplement, 1962.

11039. JEFFERSON, Thomas. 1743-1826. U.S.A. President of the United States. DSB; ICB; GA; GB; GE.

11040. BANCROFT, Edward. 1744-1821. Br. ICB; P I & Supp.; BMNH; GA; GD; GE. Also Chemistry*.
1. *An essay on the natural history of Guiana, in South America.* London, 1769. RM/L.

11041. GMELIN, Samuel Gottlieb. 1744-1774. Russ. P I; BMNH; BLA & Supp.; GA; GC. Also Botany. See also 11426/1.

11042. LAMARCK, Jean Baptiste Pierre Antoine de Monet, *Chevalier* de. 1744-1829. Fr. DSB; ICB; P I; BMNH; GA; GB. Other entries: see 3204.
1. *Extrait de quelques articles du "Nouveau dictionnaire d'histoire naturelle", édition (an 1818) de Déterville* [entry 11260], *précédés de considérations préliminaires données comme bases des connoissances qui sont le plus dans l'interêt de l'homme.* [n.p., 1818?] RM.

11043. LETTSOM, John Coakley. 1744-1815. Br. ICB; BMNH; BLA; GA; GD.
1. *The naturalist's and traveller's companion, containing instructions for discovering and preserving objects of natural history.* London, 1772. X. 2nd ed., 1774. 3rd ed., 1779.

11044. PICOT DE LA PEYROUSE, Philippe Isidore. 1744-1818. Fr. P II; BMNH (LAPEYROUSE); GA (LA PEROUSE).

11045. RACKNITZ, Joseph Friedrich von. 1744-1818. Ger. P II; GC (RACKWITZ).

11046. PALASSOU, Pierre Bernard. 1745-1830. Fr. P II; BMNH.
1. *Mémoires pour servir à l'histoire naturelle des Pyrénées et des pays adjacents.* Pau, 1815-21. RM.

11047. PARKINSON, Sydney. ca. 1745-1771. Br. DSB; BMNH; GD.

11048. MAJOLI, Cesare. 1746-1823. It. GA.

11049. TROIL, Uno von. 1746-1803. Swed. BMNH; GA.
1. [Trans.] *Letters on Iceland, containing observations on the ... natural history ... made during a voyage undertaken in the year 1772 by Joseph Banks, assisted by Dr. Solander, Dr. J. Lind, Dr. Uno von Troil.* London. 1780. Trans. from the German version (of J.G.P. Möller) by Mrs. S.D. Dixon. RM. The Swedish original appeared in 1777.

11050. AIKIN, John. 1747-1822. Br. ICB; BMNH; BLA; GA; GB; GD. See also 7116/2a.

11051. BEXON, Gabriel Léopold Charles Amé. 1747-1784. Fr. DSB; BMNH; GA.

11052. CUBIÈRES, Simon Louis Pierre de. 1747-1821. Fr. BMNH; GA.

11053. LOW, George. 1747-1795. Br. BMNH; GA; GD.

11054. ANDERSON, William. 1748-1778. Br. ICB; BMNH; GA; GD.

11055. GRAY, Edward Whittaker. 1748-1806. Br. P I; BMNH; GD.

11056. SONNERAT, Pierre. 1748-1814. Fr. DSB; P II; BMNH; GA.

11057. WALCOTT, John. fl. 1778-1822. Br. BMNH.

11058. GÉRARDIN, Sébastien. 1751-1816. Fr. BMNH.

11059. LESKE, Nathanael Gottfried. 1751-1786. Ger. P I; BMNH; BLA.

11060. SHAW, George. 1751-1813. Br. P II; BMNH; GA; GD. Also Zoology*. See also 3004/1.
1. *The naturalist's miscellany; or, Coloured figures of natural objects drawn and described immediately from nature.* 24 vols. London, 1789/90-1813. The figures were drawn by F.P. Nodder.
2. *Museum Leverianum. Containing select specimens from the museum of ... Sir A. Lever, with descriptions.* 6 parts. London, 1792-96. cf. items 11009/1,2.
3. *Zoology and botany of New Holland.* London, [1794?] With J.E. Smith. The illustrations by James Sowerby.
4. *Cimelia physica: Figures of rare and curious quadrupeds, birds, etc., together with several of the most elegant plants.* London, 1796. Engravings by J.F. Miller with descriptions by Shaw.

11061. SONNINI DE MANONCOURT, Charles Nicolas Sigisbert. 1751-1812. Fr. ICB; BMNH; GA. See also 10954/1b.

11062. BLUMENBACH, Johann Friedrich. 1752-1840. Ger. DSB; ICB; P I; BMNH; Mort.; BLA & Supp.; GA; GB; GC. Also Zoology*, Embryology*, and Physiology*.
1. *Handbuch der Naturgeschichte.* 2 vols. Göttingen, 1779-80. 7th ed., 1803.
1a. ——— *A manual of the elements of natural history.* London, 1825. Trans. from the 10th German ed. by R.T. Gore.

11063. SCHÖPF, Johann David. 1752-1800. Ger./U.S.A. ICB; P II; BMNH; Mort.; BLA; GC; GE.

11064. DELEUZE, Joseph Philippe François. 1753-1835. Fr. BMNH; BLA; GA.
1. *Histoire et description du Muséum Royal d'Histoire Naturelle.* 2 vols. Paris, Le Muséum, 1823. RM/L.
1a. ——— *History and description of the Royal Museum of Natural History.* Paris, The Museum, 1823.

11065. FORSTER, (Johann) Georg (Adam). 1754-1794. Ger./Br. DSB; ICB; P I; BMNH; GA; GB; GD. Also Botany*. See also Index.
1. *A voyage round the world in His Britannic Majesty's sloop "Resolution", commanded by Captain J. Cook, during the years 1772-75.* 2 vols. London, 1777. Reprinted, Berlin, 1968.

1a. ——— *Johann Reinhold Forster's Reise um die Welt während den Jahren 1772 bis 1775 in dem ... durch den Capitain Cook geführten Schiffe "Resolution".* 2 vols. Berlin, 1778-80. Trans. from the English by the author. (He had accompanied his father, J.R. Forster, on the voyage.) Other eds, 1780 and 1784.
2. *Reply to Mr. Wales's remarks.* London, 1778. Re remarks on item 1 by William Wales, published in 1778.
3. *A journey from Bengal to England.* 2 vols. London, 1798.

Correspondence and Collected Works

4. *Briefwechsel.* 2 parts. Leipzig, 1829. Ed. by T. Haber, with biographical material.
5. *Briefwechsel mit S.T. Sömmerring.* Brunswick, 1877. Ed. by H. Hettner.
6. *Verzeichnis der Korrespondenten von G. Forster: Forster-ausgabe.* Berlin, Akademie der Wissenschaften, 1958.
7. *Kleine Schriften.* 6 vols. Leipzig/Berlin, 1789-97.
8. *Tagebücher.* Berlin, 1914. Ed. by P. Zincke .
9. *Werke: Sämtliche Schriften, Tagebücher, Briefe.* Berlin, Akademie der Wissenschaften, 1958-.

11066. LAICHARTING, Johann Nepomuk von. 1754-1797. Aus. P I; BMNH.

11067. MAYER, Johann. 1754-1807. Prague. P II; BMNH.

11068. FIBIG, Johann. d. 1792. Ger. P I; BMNH.

11069. BECHSTEIN, Johann Matthäus. 1757-1822. Ger. BMNH; GA; GC.
1. *Gemeinnützige Naturgeschichte Deutschlands nach allen drey Reichen.* 4 vols. Leipzig, 1789-95. 2nd ed., 1801-09.

11070. KITAIBEL, Pál (or Paul). 1757-1817. Hung. DSB; ICB; P I; BMNH. Also Botany.

11071. CASTEL, René Louis Richard. 1758-1832. Fr. BMNH; GA. See also 12685/4.

11072. HÖPFNER, Johann Georg Albrecht. 1759-1813. Switz. P I; BLA.

11073. MILLIN (DE GRANDMAISON), Aubin Louis. 1759-1818. Fr. BMNH; GA.
1. *Eléments d'histoire naturelle.* 3rd ed. Paris, 1802.

11074. RENIER, Stefano Andrea. 1759-1830. It. P II; BMNH; GA.

11075. BORKHAUSEN, Moritz Balthasar. 1760-1806. Ger. BMNH; GA; GC.

11076. LINCK, Johann Wilhelm. 1760-1805. Ger. P I; BMNH. Also Chemistry.

11077. SUE, Jean Joseph (second of the name). 1760-1830. Fr. BMNH; BLA (S., Jean Baptiste); GA.
1. *Essai sur la physiognomie des corps vivans, considérée depuis l'homme jusqu'à la plante.* Paris, 1796. RM.

11078. BATSCH, August Johann Georg Karl. 1761-1802. Ger. ICB; P I; BMNH; BLA; GA; GC. Also Botany*.

11079. MERREM, Blasius. 1761-1824. Ger. P II; BMNH.

11080. BUCHANAN (later HAMILTON), Francis. 1762-1829. India. ICB; BMNH; GD.

11081. FLOERKE, Heinrich Gustav. 1764-1835. Ger. P I; BMNH. Also Botany*.

11082. MEIGEN, Johann Wilhelm. 1764-1845. ICB; BMNH.

11083. MITCHILL, Samuel Latham. 1764-1831. U.S.A. ICB; P II; BMNH; BLA; GA (MITCHELL); GE. Also Chemistry. See also 9213/4.

11084. NEMNICH, Philipp Andreas. 1764-1822. BMNH.
1. *Allgemeines Polyglotten-Lexicon der Naturgeschichte.* 4 vols. Hamburg, 1793-95. RM/L.

11085. TRATTINICK, Leopold. 1764-1849. Aus. BMNH; GC.

11086. FORSTER, Edward. 1765-1849. Br. BMNH; GD.

11087. WOLF, Johann. 1765-1824. Ger. BMNH; GC.

11088. GUILLEMEAU, Jean Louis Marie. 1766-1850. Fr. BMNH; BLA; GA.

11089. NAU, Bernhard Sebastian von. 1766-1845. Ger. P II; GC.

11090. CLINTON, De Witt. 1769-1828. U.S.A. GB; GE.
1. *An introductory discourse, delivered before the Literary and Philosophical Society of New York....* New York, 1815. "An excellent summary of the state of scientific knowledge in the U.S." RM.
2. *Letters on the natural history and internal resources of the State of New York.* New York, 1822. RM.

11091. HUMBOLDT, Alexander von. 1769-1859. Ger./Fr. DSB; ICB; P I, III; BMNH; G. Other entries: see 3268.
1. *Voyage aux régions équinoxiales du Nouveau Continent, fait en 1799-1804 par Al. de Humboldt et A. Bonpland.* 34 vols (text & atlases). Paris, 1805-34. Contents: I. Relation historique (7 vols). II. Zoologie (2 vols). III. Essai politique (3 vols). IV. Astronomie (3 vols). V. Essai sur la géographie des plantes (1 vol.) VI. Botanique (18 vols). Reprinted, Hildesheim, 1971-.
1a. ——— Part I. *Relation historique du voyage aux....* Reprinted, 2 vols, Stuttgart, 1970.
1b. ——— ——— *Personal narrative of travels to the equinoctial regions of the New Continent during 1799-1804.* 4 vols in 3. London, 1814-19. Trans. by H.M. Williams. RM. Another ed., 3 vols, 1852-53.

11092. MIKAN, Johann Christian. 1769-1844. Prague. ICB; P II; BMNH.

11093. MOUTON-FONTENILLE DE LA CLOTTE, Marie Jacques Philippe. 1769-1837. Fr. BMNH; GA.

11094. OLIVI, Giuseppe. 1769-1795. It. P II; BMNH.

11095. TOWNSON, Robert. fl. 1795-1798. Br. ICB; BMNH; GD. Also Mineralogy*.
1. *Tracts and observations in natural history and physiology.* London, 1799. RM/L.

11096. DVIGHUBSKII, Ivan Aleksyeevich. 1771-1839. Russ. BMNH.

11097. FISCHER (VON WALDHEIM), (Johann) Gotthelf. 1771-1853. Russ. ICB; P I; BMNH; BLA; GA; GC. Also Geology*.
1. *Das Nationalmuseum der Naturgeschichte zu Paris. Von seinem ersten Ursprunge bis zu seinem jetzigen Glanze.* 2 vols. Frankfurt, 1802-03.

11098. STURM, Jacob. 1771-1848. Ger. BMNH; GC.

11099. BILLBERG, Gustav Johan. b. 1772. Swed. BMNH.

11100. KIRCKS, Jean. 1772-1831. Belg. GA.

11101. SCHILLING, Peter Samuel. 1773-1852. BMNH.

11102. KIELSEN, Frederik Christian. 1774-1850. Den. BMNH; GA.

11103. LANGSDORFF, Georg Heinrich von. 1774-1852. Ger. P I; BMNH; BLA; GA; GC.

11104. MATON, William George. 1774-1835. Br. ICB; BMNH; GD.

11105. WOOD, William. 1774-1857. Br. ICB; BMNH; GD. Also Zoology*.
1. *Catalogue of an extensive and valuable collection of the best works on natural history.* London, 1824. X. Another ed., 1832.
See also 10954/2a.

11106. DESCOURTILZ, Michel Etienne. 1775-1836. Fr. DSB; ICB; BMNH; BLA.

11107. ILLIGER, Johann Karl Wilhelm. 1775-1813. Ger. ICB; BMNH; GC. Also Zoology*.
1. *Versuch einer systematischen vollständigen Terminologie für das Thierreich und Pflanzenreich.* Helmstädt, 1800. RM.

11108. ANSLIJN, Nicolaas. 1777-1838. Holl. BMNH; GF.

11109. RISSO, (Giovanni) Antonio. 1777-1845. Fr. ICB; BMNH.

11110. BORY DE SAINT-VINCENT, Jean Baptiste Georges Marie. 1778-1846. Fr. DSB; ICB; P I; BMNH; GA; GB.
1. *Expédition scientifique de Morée. Section des sciences physiques.* 6 vols. Paris, 1836.
See also 11262.

11111. FREYCINET, Claude Louis Desaulses de. 1779-1842. Fr. P I; BMNH; GA; GB.

11112. LHERMINIER, Félix Louis. 1779-1833. Fr. GA.

11113. OKEN (or OKENFUSS), Lorenz. 1779-1851. Ger./Switz. DSB; ICB; P II; BMNH; Mort.; BLA & Supp.; GA; GB; GC. Also Biology in General* and Zoology. See also Index.
1. *Lehrbuch der Naturgeschichte.* 3 parts in 6 vols. Leipzig/Jena, 1813-26.
2. *Allgemeine Naturgeschichte für alle Stände.* 7 vols in 13. Stuttgart, 1833-41. RM/L.
2a. ——— Trans. of an extract: *Elements of physiophilosophy.* London, 1847. Trans. by A. Tulk. (Ray Society publns. 10) RM/L.

11114. JESSE, Edward. 1780-1868. Br. BMNH; GB; GD.
1. *Lectures on natural history.* London, 1861. X. 2nd ed., enlarged, 1863.

11115. MOHR, Daniel Matthias Heinrich. 1780-1808. Ger. BMNH; GC.

11116. SCHUBERT, Gotthilf Heinrich von. 1780-1860. Ger. ICB; P II; BMNH; BLA; GC. Other entries: 3285.
1. *Allgemeine Naturgeschichte*. Erlangen, 1826. RM.

11117. RAFFLES, Thomas Stamford Bingley. 1781-1826. Br./East Indies, etc. The colonial administrator. DSB; ICB; P II; BMNH; G.

11118. VOIGT, Friedrich Siegmund. 1781-1850. Ger. BMNH; GC. See also 12818/3b.

11119. WEBER, Friedrich. 1781-1823. Ger. BMNH; BLA; GC.

11120. BURCHELL, William John. ca. 1782-1863. Br. BMNH; GD.

11121. GAILLON, (François) Benjamin. 1782-1839. Fr. BMNH; GA.

11122. WIED, (Alexander Philip) Maximilian zu. 1782-1867. Ger. DSB; BMNH (WIED-NEUWIED). Under NEUWIED: ICB; GA; GC.

11123. DRUMMOND, James Lawson. 1783-1853. Br. BMNH; BLA & Supp.; GD. Also Botany*.

11124. RAFINESQUE, Constantine Samuel. 1783-1840. It./U.S.A. DSB; ICB; BMNH; GE. Also Botany*, Zoology*, and Autobiography*.
1. *Principes fondamentaux de somniologie; ou, Les lois de la nomenclature et de la classification de l'empire organique, ou des animaux et des végétaux*. Palermo, 1814.
2. *Analyse de la nature; ou, Tableau de l'univers et des corps organisés*. Palermo, 1815.
3. *Circular address on botany and zoology*. Philadelphia, 1816.

11125. SCHWEIGGER, August Friedrich. 1783-1821. Ger. BMNH; GC.

11126. ZETTERSTEDT, Johan Wilhelm. 1785-1874. Swed. BMNH.

11127. MURRAY, John (second of the name). ca. 1786-1851. Br. BMNH; GD.

11128. NUTTALL, Thomas. 1786-1859. Br./U.S.A. DSB; BMNH; GB; GD; GE. Also Botany* and Zoology.
1. *A journal of travels into the Arkansas Territory during the year 1819*. Philadelphia, 1821. RM.

11129. BOITARD, Pierre. 1787-1859. Fr. BMNH; GA.
1. *Le Jardin des Plantes. Description et moeurs des mammifères de la ménagerie et du Muséum d'Histoire Naturelle. Précédé d'une introduction historique, descriptive et pittoresque par J. Janin*. Paris, 1842.

11130. DELALANDE, Pierre Antoine. 1787-1823. Fr. BMNH; GA.

11131. RICHARDSON, John. 1787-1865. Br. P II,III; BMNH; GB; GD. Also Zoology*. See also 15585/1.
1. *The museum of natural history. With an introductory essay on the natural history of the primeval world*. 4 vols. London, 1859-62.

11132. CAP (or GRATACAP), Paul Antoine. 1788-1877. Fr. P III; BMNH; GA. Also Various Fields. See also 1473/3.
1. *Le Muséum d'Histoire Naturelle*. Paris, 1854.

11133. RUTHE, Johann Friedrich. 1788-1859. Ger. BMNH; GC. See also 13077/1.

11134. THOMAS, (Abraham Louis) Emmanuel. 1788-1859. Switz. ICB; BMNH.

11135. SCORESBY, William. 1789-1857. Br. P II; BMNH; GB; GD. Also Physics* and Geomagnetism*.

11136. ARAGO, Jacques Etienne Victor. 1790-1855. Fr. GA.
1. *Promenade autour du monde pendant les années 1817-1820 sur les corvettes du roi "l'Uranie" et la "Physicienne", commandées par M. Freycinet.* 2 vols. Paris, 1822.
1a. ——— *Narrative of a voyage round the world ... on a scientific expedition....* London, 1823.

11137. BERENDT, Georg Carl. 1790-1850. Danzig. BMNH; BLA.

11138. DAVY, John. 1790-1868. Br. DSB; ICB; P I,III; BMNH; BLA & Supp.; GD. Also Chemistry, Physiology*, and Biography*. See also Index.

11139. DUMONT D'URVILLE, Jules Sébastien César. 1790-1842. Fr. ICB; P I; BMNH; GA; GB. See also 11282/1 and 11290/1.

11140. PARRY, William Edward. 1790-1855. U.S.A. The Arctic explorer. ICB; P II; BMNH; GA; GD; GE.

11141. BOWDICH (later LEE), Sarah, *Mrs.* 1791-1856. Br. BMNH; GD (LEE). Also Biography*.
1. *Elements of natural history.* London, 1844. X. Another ed., 1853.

11142. MEYER, Johann Rudolph. 1791-1833. Switz. P II; GC.

11143. DE KAY, James Ellsworth. 1792-1851. U.S.A. BMNH; GE.
1. *Anniversary address on the progress of the natural sciences in the United States. Delivered before the Lyceum of Natural History of New York.* New York, 1826. X. Reprinted, ib., 1970.

11144. KUNZE, Gustav. 1793-1851. Ger. BMNH; BLA; GC.

11145. BERTHELOT, Sabin. 1794-1880. BMNH.

11146. EVERSMANN, Eduard Friedrich. 1794-1860. Ger./Russ. BMNH; BLA.

11147. GODMAN, John Davidson. 1794-1830. U.S.A. ICB; BMNH; BLA & Supp.; GE. Also Anatomy.

11148. PARTINGTON, Charles Frederick. d. 1857? Br. BMNH; GD. Also Science in General* and Physics*.
1. *The British cyclopaedia of natural history.* 3 vols. London, 1835-37.

11149. HUGI, Franz Joseph. 1796-1855. Switz. P I; GA.

11150. ROULIN, François Désiré. 1796-1874. Fr. ICB; P II,III; BMNH; BLA & Supp. Also Geology.

11151. BOWERBANK, James Scott. 1797-1877. Br. BMNH; GB; GD. Also Geology*.

11152. BECK, Lewis Caleb. 1798-1853. U.S.A. ICB; P I; BMNH; BLA & Supp.

11153. PÖPPIG, Eduard Friedrich. 1798-1868. Ger. P II,III; BMNH; GA; GC.

1. *Tropenvegetation und Tropenmenschen. Zwei Vorträge.* Leipzig, 1965. (OKEW. 249)

11154. AUTENRIETH, Hermann Friedrich. 1799-1874. Ger. BMNH; BLA.

11155. KITTLITZ, Friedrich Heinrich von. 1799-1874. Ger. BMNH; GC.

11156. MANDT, Martin Wilhelm. 1799-1858. Ger./Russ. ICB; BMNH; BLA; GC.

11157. GAY, Claude. 1800-1873. Fr./Chile. P I,III; BMNH; GA.

11158. HOGG, John. 1800-1869. Br. BMNH; GD.

11159. POUCHET, Félix Archimède. 1800-1872. Fr. DSB; ICB; BMNH; BLA; GA. Also Zoology*, Embryology*, Microbiology*, and History of Science*.
1. *L'univers.* Paris, 1865. X.
1a. ——— *The universe; or, The infinitely great and the infinitely little.* London, 1870. Another ed., 1871. RM/L. 7th ed., 1883.

11160. ROSS, James Clark. 1800-1862. Br. DSB; P II; BMNH; GB; GD. Also Geomagnetism.
1. *A voyage of discovery and research in the southern and Antarctic regions during the years 1839-43.* 2 vols. London, 1847. Reprinted, New York, 1969.

11161. BLOXAM, Andrew. 1801-1878. Br. BMNH; GD.
1. *Diary of Andrew Bloxam, naturalist of the "Blonde", on her trip from England to the Hawaiian Islands, 1824-25.* Honolulu, 1925. Publ. by the Bernice P. Bishop Museum.

11162. JACQUEMONT, Victor. 1801-1832. Fr. ICB; P I; BMNH; GA.
1. *Correspondance ... pendant son voyage dans l'Inde, 1828-32.* Paris, 1833.
2. *Voyage dans l'Inde, 1828 à 1832.* 4 vols & atlas of 2 vols. Paris, 1841-44. X.

11163. LA SAGRA, Ramon de. 1801-1871. Cuba. P II,III (SAGRA); BMNH (SAGRA).

11164. NEWMAN, Edward. 1801-1876. Br. BMNH; GD.

11165. RATZEBURG, Julius Theodor Christian. 1801-1871. Ger. P II, III; BMNH; BLA.

11166. COMTE, (Joseph) Achille. 1802-1866. Fr. BMNH; GA.

11167. LEUNIS, Johannes. 1802-1873. Ger. BMNH; GC.
1. *Synopsis der drei Naturreiche.* 3 vols. Hanover, 1844-53.

11168. LOWE, Richard Thomas. 1802-1874. Br. BMNH; GD.

11169. NOULET, Jean Baptiste. 1802-1890. Fr. BMNH.

11170. ORBIGNY, Alcide Dessalines d'. 1802-1857. Fr. DSB; P II; BMNH; GA; GB. Also Geology*.
1. *Voyage dans l'Amérique méridionale ... 1826-33.* 7 vols & atlas of 2 vols. Paris, 1835-47. RM.
See also 12923/3.

11171. BAIRD, William. 1803-1872. Br. BMNH; GD. Also Zoology.
1. *A cyclopaedia of the natural sciences.* London, 1858. RM/L.

11172. BENNETT, George. 1804-1893. Australia. BMNH; GF.
1. *Gatherings of a naturalist in Australasia.* London, 1860.

11173. BERTOLONI, Giuseppe. 1804-1878. It. BMNH.

11174. OWEN, Richard (first of the name). 1804-1892. Br. DSB; ICB; BMNH; Mort.; BLA; GA; GB; GD. Also Geology* and Zoology*. See also Index.
1. *On the extent and aims of a national museum of natural history.* London, 1862. X. 2nd ed., 1862.

11175. SCHOMBURGK, Robert Hermann. 1804-1865. Several countries. P III; BMNH; GB; GC; GD.

11176. BADHAM, Charles David. 1806-1857. Br. BMNH; GD.

11177. AGASSIZ, (Jean) Louis (Rodolphe). 1807-1873. Switz./U.S.A. DSB; ICB; P I,III; BMNH; BLA; GA; GB; GE. Also Geology* and Zoology*. See also Index.
1. *An introduction to the study of natural history.* New York, 1847.
2. *Lake Superior. Its physical character, vegetation, and animals.* Boston, 1850. RM. Reprinted, Huntington, N.Y., 1974.
3. *Contributions to the natural history of the United States of America.* 4 vols. Boston, 1857-62. RM/L.
4. *An essay on classification.* London, 1859. First publ. as part of item 3. Reprinted, Cambridge, Mass., 1962.
5. *Methods of study in natural history.* Boston, 1863. RM/L. Reprinted, New York, 1970.
6. *A journey in Brazil.* Boston/London, 1868. With Elizabeth C.C. Agassiz.
7. *The intelligence of Louis Agassiz: A specimen book of scientific writings.* Boston, 1963. Selected, with an introd. and notes, by G. Davenport.

11178. CATLOW, Agnes. 1807?-1889. Br. BMNH.

11179. FIEBER, Franz Xaver. 1807-1872. Prague. BMNH; GC.

11180. GODRON, Dominique Alexandre. 1807-1880. Fr. BMNH. See also 12070/1.

11181. BACH, Michael. 1808-1878. Ger. BMNH.

11182. MENGE, Franz Anton. 1808-1880. Danzig. BMNH.

11183. DARWIN, Charles Robert. 1809-1882. Br. DSB; ICB; P III; BMNH; G. Also Geology*, Botany*, Zoology*, Evolution*#, and Autobiography*. See also Index.
1. *Journal of researches into the geology and natural history of the various countries visited by H.M.S. "Beagle" ... from 1832 to 1836.* London, 1839. (Also issued as Vol. 3 of *Narrative of the surveying voyages of H.M.S. "Adventure" and "Beagle"....* see 11283/1c.) RM/L. 2nd ed., 1845. Many reprintings.
2. *Diary of the voyage of H.M.S. "Beagle."* Cambridge, 1933. Ed. from the MS by Nora Barlow. Reprinted, New York, 1969.
3. *Charles Darwin and the voyage of the "Beagle."* London. 1945. Letters and notebooks, ed. by Nora Barlow.

11184. HINDS, Richard Brinsley. fl. 1844. Br. BMNH.

11185. FOCKE, Gustav Woldemar. 1810-1877. Ger. BMNH; BLA; GC.

11186. SHUTTLEWORTH, Robert James. 1810-1874. Br./Switz. BMNH; GD.

11187. THEOBALD, Gottfried Ludwig. 1810-1869. Ger./Switz. P III; BMNH; GC. Also Geology.

11188. BERGE, (Karl) Friedrich. 1811-1883. ICB; BMNH.
1. [Trans.] *Berge's complete natural history of the animal, mineral, and vegetable kingdoms.* London, 1890. Ed. by R.F. Crawford.

11189. DIEFFENBACH, Ernst. 1811-1855. Ger. P I; BMNH; GA; GC.
1. *Travels in New Zealand, with contributions to the geography, geology, botany and natural history of that country.* London, 1843. RM.

11190. LAPHAM, Increase Allen. 1811-1875. U.S.A. P III; BMNH; GE.

11191. ALLMAN, George James. 1812-1898. Br. BMNH; GB; GD1.

11192. CHIKHACHEV, Pierre de. 1812-1890. Fr. P II,III (TSCHICHATSCHEW). Under TCHIHATCHEFF: BMNH; GB. Also Geology. See also 12135/2a.

11193. DUPUY, Dominique. 1812-1885. Fr. BMNH.

11194. HARTWIG, Georg Ludwig. 1813-1880. Ger./Belg. P III; BMNH.
1. *Die Tropenwelt in Thier- und Pflanzenleben dargestellt.* Wiesbaden, 1860. X.
1a. ——— *The tropical world. A popular scientific account of the natural history....* London, 1863. X. Another ed., 1873.
2. *Die Unterwelt, mit ihren Schätzen und Wundern.* Wiesbaden, 1863. X.
2a. ——— *The subterranean world.* London, 1871. 3rd ed., 1875. Another ed., 1885.
3. *Gott in der Natur; oder, Die Einheit der Schöpfung.* Wiesbaden, 1864. X.
3a. ——— *The harmonies of nature; or, The unity of creation.* London, 1866.
4. *Das Leben des Luftmeeres. Populäre Streifzüge in das atmosphärische Reich.* Wiesbaden, 1872. X.
4a. ——— *The aerial world. A popular account of the phenomena and life of the atmosphere.* London, 1874. Other eds, 1877 and 1886.

11195. HEHN, Victor. 1813-1890. Russ./Ger. BMNH; GC.
1. *Kulturpflanzen und Hausthiere in ihrem Übergang aus Asien nach Griechenland und Italien sowie in das übrige Europa.* Berlin, 1870. X.
1a. ——— *The wanderings of plants and animals from their first home.* London, 1885. Ed. by J.S. Stallybras. Another ed. (variant title), 1891.

11196. KOLENATI, Friedrich. 1813-1864. Prague/Brünn. P I,III; BMNH. Also Crystallography.

11197. WAGNER, Moritz Friedrich. 1813-1887. Ger. P III; BMNH; GC.

11198. MIDDENDORF, Aleksandr Fedorovich. 1815-1894. Russ. DSB; P II,III; BMNH; BLA; GC. Biogeography and ecology.

11199. BOLL, Ernst (Friedrich August). 1817-1868. Ger. P III; BMNH. Also Geology.

11200. THOREAU, Henry David. 1817-1862. U.S.A. The writer and philosopher. ICB; GB; GE.

11201. HILL, Thomas. 1818-1891. U.S.A. P III; GE. Also Mathematics.

11202. CALCARA, Pietro. 1819-1854. It. BMNH.

11202A ADAMS, Arthur. 1820-1878. Br. BMNH. Also Zoology*.
1. *A manual of natural history, for the use of travellers.* London, 1854. With W.B. Baikie and C. Barron.

11203. FERGUSON, William. 1820-1887. Br./Ceylon. BMNH; GD.

11204. PARFITT, Edward. 1820-1893. Br. BMNH; GD.

11205. BABINGTON, Churchhill. 1821-1889. Br. BMNH; GB; GD1.

11206. HAUGHTON, Samuel. 1821-1897. Br. P III,IV; BMNH; Mort.; GB; GD1. Other entries: see 3336.
1. *The three kingdoms of nature briefly described.* London, 1868.

11207. KARSCH, Anton. 1822-1892. Ger. BMNH; GC.

11208. MACGILLIVRAY, John. 1822-1867. Br./Australia. BMNH; GD.
1. *Narrative of the voyage of H.M.S. "Rattlesnake", commanded by the late Captain Owen Stanley, during the years 1846-50, including discoveries and surveys in New Guinea, etc.* 2 vols. London, 1852. Reprinted, Adelaide, 1967.

11209. TRISTRAM, Henry Baker. 1822-1906. Br. ICB; BMNH; GD2.

11210. LESSONA, Michele. 1823-1894. It. BMNH.
1. *Naturalisti italiani.* Rome, 1884. RM.

11211. WALLACE, Alfred Russel. 1823-1913. Br. DSB; ICB; P III; BMNH; GB; GD3. Also Zoology*. Evolution*, Autobiography*, and History of Science*.
1. *A narrative of travels on the Amazon and Rio Negro, with ... observations on the climate, geology and natural history of the Amazon Valley.* London, 1853. RM.
2. *The Malay Archipeligo ... A narrative of travel, with studies of man and nature.* 2 vols. London, 1869. RM/L. Another ed., 1890.
3. *Tropical nature, and other essays.* London, 1878. RM/L.
4. *Island life; or, The phenomena and causes of insular faunas and floras.* London, 1880. RM/L. 2nd ed., 1892. 3rd ed., 1902.
See also 12185/1.

11212. CONTEJEAN, Charles Louis. 1824-1897. Fr. P III,IV; BMNH. Also Geology.

11213. THOMAS, Jean Louis. 1824-1886. Switz. ICB.

11214. TYLOR, Alfred. 1824-1884. Br. P III; BMNH; GD. Also Geology.
1. *Colouration in animals and plants.* 1886.

11215. HUXLEY, Thomas Henry. 1825-1895. Br. DSB; ICB; BMNH; Mort.; BLA; GB; GD1. Other entries: see 3340.
 1. *Diary of the voyage of H.M.S. "Rattlesnake."* London, 1935. Ed. from the unpublished MS by Julian Huxley. Reprinted, New York, 1972.

11216. SIRODOT, Simon. 1825-1903. Fr. BMNH.

11217. COLLINGWOOD, Cuthbert. 1826-1908. Br. P III; BMNH; GD2.

11218. SEVERTSOV, Nikolai Alekseevich. 1827-1885. Russ. ICB; BMNH (SYEVERTZOV).

11219. BEDDOME, Richard Henry. 1830-1911. India. ICB; BMNH.

11220. COOPER, James Graham. 1830-1902. U.S.A. P III; BMNH; GE.

11221. MORE, Alexander Goodman. 1830-1895. Br. BMNH.

11222. ORTON, James. 1830-1877. U.S.A. DSB; ICB; P III; BMNH; GE.
 1. *The Andes and the Amazon; or, Across the continent of South America.* London, 1870. RM.

11223. PALACKÝ, Jan. 1830-1908. Prague. P III,IV; BMNH.

11224. PALMER, Edward. 1831-1911. U.S.A. DSB; ICB.

11225. RADDE, Gustav Ferdinand Richard. 1831-1903. Russ. ICB; P III, IV; BMNH.

11226. KIRK, John. 1832-1922. Br./East Africa. ICB; GB; GD4.

11227. DECKEN, Carl Claus von der. 1833-1865. Ger. P III; BMNH.

11228. GODMAN, Frederick Ducane. 1834-1919. Br. BMNH; GF.
 1. *Natural history of the Azores.* London, 1870. RM.

11229. HECTOR, James. 1834-1907. Canada/New Zealand. BMNH; GD2; GF. Also Geology.

11230. ROCHEBRUNE, Alphonse Trémeau de. 1834-after 1904. Fr. BMNH.

11231. WARD, Henry Augustus. 1834-1906. U.S.A. ICB; BMNH; GE.

11232. WRIGHT, Edward Perceval. 1834-1910. Br. BMNH; GD2.

11233. POTANIN, Grigory Nikolaevich. 1835-1920. Russ. DSB; BMNH.

11234. ÄHRLING, (Johan Erik) Ewald. 1837-1888. Swed. ICB; BMNH. See also 10958/15.

11235. BURROUGHS, John. 1837-1921. U.S.A. ICB; GB; GE.
 1. *John Burroughs' "America." Selections.* Garden City, N.Y., 1961. Ed. with an introd. by F.A. Wiley.

11236. YATES, Lorenzo Gordin. 1837-1909. U.S.A. ICB; BMNH.

11237. MUIR, John. 1838-1914. U.S.A. ICB; BMNH; GE. Also Autobiography*.

11238. PRZHEVALSKY, Nikolai Mikhailovich. 1839-1888. Russ. DSB; ICB; BMNH; GB (PRJEVALSKY).

11239. SPRY, William James Joseph. d. 1906. BMNH.
 1. *The cruise of H.M.S. "Challenger." Voyages over many seas, scenes in many lands.* London, 1876. RM. Another ed., New York, 1877.

11240. SWEDERUS, Magnus Bernhard. 1840-1911. Swed. ICB; BMNH. See also 11418/24.

11241. ALBERTUS, Luigi Maria d'. 1841-1901. BMNH.

11242. WOOD, Horatio Charles. 1841-1920. U.S.A. A well known physician. DSB; ICB; BMNH; BLAF; GE.

11243. WHITE, Francis Buchanan. 1842-1894. Br. BMNH; GD.

11244. BECCARI, Odoardo. 1843-1920. It. ICB; BMNH.

11245. MAJOR, (Charles Immanuel) Forsyth. 1843-1923. It. ICB; BMNH.

11246. FEDCHENKO, Aleksei Pavlovich. 1844-1873. Russ. BMNH; GB.

11247. MOSELEY, Henry Nottidge. 1844-1891. Br. BMNH; GD.
1. *Notes of a naturalist on the "Challenger." Being an account of various observations made during the voyage of H.M.S. "Challenger" round the world, 1872-1876.* London, 1879. RM/L. Another ed., with a brief memoir of the author, by G.C. Bourne, 1892.

11248. DALL, William Healey. 1845-1927. U.S.A. DSB; ICB; P III,IV; BMNH; GE. Also Zoology.
1. *Nomenclature in zoology and botany.* Salem, Mass., 1877.

11249. ELWES, Henry John. 1846-1922. Br. ICB; BMNH; GD4.

11250. MIKLUKHO-MAKLAY, Nikolai Nikolaievich. 1846-1888. Russ. P III (MIKLUCHO); BMNH.

11251. AUSTEN, Nathaniel Laurence. 1847-1874. BMNH.

11252. McALPINE, Daniel. 1848-1932. Australia. BMNH; GF.
1. *Biological atlas. A guide to the practical study of plants and animals.* Edinburgh, 1880. With A. McAlpine.
2. *Zoological atlas.* 2 vols. Edinburgh, 1881.
3. *Life histories of plants.* London, 1886.

11253. SWIRE, Herbert. d. 1934. Br. ICB.
1. *The voyage of the "Challenger." A personal narrative of the historic circumnavigation of the globe, 1872-1876.* 2 vols. London, 1938.

11254. BETTANY, George Thomas. 1850-1891. BMNH. Also Biography*. See also 13330/1.

11255. DALLA-TORRE, Karl Wilhelm von. 1850-1928. BMNH.

11256. FORBES, Henry Ogg. 1851-1932. Br. BMNH.
1. *A naturalist's wanderings in the Eastern Archipeligo ... 1878 to 1883.* London, 1885. X. 2nd ed., 1885.

11257. DENIKER, Joseph. 1852-1918. Fr. BMNH. Also Science in General*.

11258. FLETCHER, James. 1853-1908. Br./Canada. BMNH; GD2; GF.

11259. JOHNSTON, Henry Hamilton. 1858-1927. Br. ICB; BMNH; GD4.

Imprint Sequence

11260. *NOUVEAU dictionnaire d'histoire naturelle, appliquée aux arts, principalement à l'agriculture*. 24 vols. Paris, 1803-04. 2nd ed., 36 vols, 1816-19.

11261. *DICTIONNAIRE des sciences naturelles*. 60 vols & 12 vols of plates. Paris, 1804-05, 1816-30. (Publn suspended between 1805 and 1816.) Ed. by F. Cuvier. Prospectus by G. Cuvier. Introd. by Fourcroy.

11262. *DICTIONNAIRE classique d'histoire naturelle*. 18 vols. Paris, 1822-31. Ed. by Bory de Saint-Vincent.

11263. *DICTIONNAIRE pittoresque d'histoire naturelle et des phénomènes de la nature*. 9 vols. Paris, 1833-39. Ed. by F. Guérin-Méneville.

11264. *SUITES à Buffon*.
1. *Suites* [in some editions *Collection des suites*, in others *Nouvelles suites*] *à Buffon, formant avec les oeuvres de cet auteur un cours complet d'histoire naturelle*. Many vols. Paris, 1834-74. For a list of authors and subjects see BMNH under Buffon. In the present catalogue the individual works in this series are entered under their respective authors.

11265. *DICTIONNAIRE universel d'histoire naturelle*. 13 vols & 3 vols of plates. Paris, 1839-49; re-issued 1861. Ed. by C. d'Orbigny. 2nd ed., 17 vols, 1867-69.

11266. RAY Society. London. (Founded 1844)
Publications from 1845 to the present; for a list see BMNH. The individual items in this series which fall within the limits of the present catalogue are entered under their respective authors.

11267. *SCIENTIFIC memoirs, selected from the transactions of foreign academies of science and from foreign journals. Natural history*. London, 1853. X. Ed. by A. Henfrey and T.H. Huxley. Reprinted, New York, 1966. cf. 3009 and 6973.

11268. MUSÉUM National d'Histoire Naturelle. Paris.
1. *Guide des étrangers dans le Muséum*. Paris, 1855.

11269. KONINKLIJKE Natuurkundige Vereeniging in Nederlandsch Indie. Batavia.
1. *Catalogus van de bibliotheek*. 1861.

11270. McNICOLL, David Hudson. BMNH.
1. *Dictionary of natural history terms*. London, 1863.

11271. LINNEAN Society. London.
1. *Catalogue of the natural history library*. London, 1866.

11272. BRITISH Museum (Natural History). London.
1. *A general guide to the Museum*. London, 1887.
2. *List of the natural history publications*. Ib., 1897.

11273. KÖNIGLICH Friedrich-Wilhelms-Universität in Berlin.
1. *Das Museum für Naturkunde*. Berlin, 1889.

3.121 Scientific Expeditions

The section is confined to seagoing expeditions. Expeditions on land are included in various other sections under the names of the scientists concerned. (The seagoing expeditions are bibliographically distinct because they can be listed under the names of the ships involved.)

For a valuable bibliography see BMNH under the names of the chief countries.

11274. ENDEAVOUR, H.M.S. (British ship) Voyage, 1768-71. BMNH. See also 11038/2-4 and 11560/1.

11275. RESOLUTION, H.M.S. (British ship) Voyages 1772-75 and 1776-80. BMNH. See also 4663/1 and 11065/1.

11276. ADVENTURE, H.M.S. (British ship) Voyage, 1772-75. BMNH. See also 11283 and 4663/1.

11277. URANIE. (French ship) Voyage, 1817-20. BMNH. See also Index.

11278. PHYSICIENNE. (French ship) Voyage, 1817-20. BMNH. See also Index.

11279. COQUILLE. (French ship) Voyage, 1822-25. BMNH.
1. *Voyage autour du monde ... sur la corvette ... la "Coquille" ... 1822-25.* 6 vols & atlas of 5 vols. Paris, 1826-30. By L.I. Duperry. Contents: Histoire du voyage; Zoologie; Botanique; Hydrographie.
See also 12992/1.

11280. BLONDE, H.M.S. (British ship) Voyage, 1824-25. BMNH (under BLOXAM, A.). See also 11161/1.

11281. BLOSSOM, H.M.S. (British ship) Voyage, 1825-28. BMNH. See also 11848/9 and 12942/2.

11282. ASTROLABE. (French ship) Voyage, 1826-29. BMNH. Also 11290. See also Index.
1. *Voyage de la corvette "l'Astrolabe" ... 1826-29, sous le commandement de M.J. Dumont d'Urville.* 14 vols in 15 and atlas of 7 vols. Paris, 1830-35. Contents: Histoire du voyage; Botanique; Zoologie; Faune entomologique de l'Océan Pacifique; etc.

11283. ADVENTURE, H.M.S. (British ship) Voyage 1826-36. BMNH. See also 11276 and 4663/1.
1. *Narrative of the surveying voyages of H.M.S. "Adventure" and "Beagle" between the years 1826 and 1836, describing their examination of the southern shores of South America, and the "Beagle's" circumnavigation of the globe.* 3 vols in 4. London, 1839. RM (under FITZROY, R.)/L.
1a. ——— Vol. 1. *Proceedings of the first expedition, 1826-30.* By Capt. P.P. King. Ed. by Capt. R. Fitzroy.
1b. ——— Vol. 2. *Proceedings of the second expedition, 1831-36.* By Capt. R. Fitzroy.
1c. ——— Vol. 3. *Journal and remarks, 1832-36.* By C. Darwin. Also issued as *Journal of researches....* See 11183/1.

11284. SENYAVIN. (Russian ship) Voyage, 1826-29. BMNH.

11285. BEAGLE, H.M.S. (British ship) Voyage, 1832-36. BMNH. Also 11283/1. See also Index.

11286. RECHERCHE. (French ship) Voyages, 1835-36 and 1838-40. BMNH.

11287. BONITÉ. (French ship) Voyage, 1836-37. BMNH.
1. *Voyage autour du monde ... 1836 et 1837 sur la corvette la "Bonité."* 15 vols in 11 & atlas of 3 vols. Paris, 1840-66. Contents: Relation du voyage; Zoologie; Botanique; Géologie et minéralogie; Physique.
See also 11806/2 and 13070/1.

11288. VÉNUS. (French ship) Voyage, 1836-39. BMNH. See also 12053/2.

11289. SULPHUR, H.M.S. (British ship) Voyage, 1836-42. BMNH. See also 11971/4.

11290. ASTROLABE. (French ship) Voyage, 1837-40. BMNH. Also 11282. See also Index.
1. *Voyage au Pôle Sud et dans l'Océanie sur les corvettes "l'Astrolabe" et la "Zélée" ... 1837-40, sous le commandement de M.J. Dumont d'Urville.* 23 vols in 22 & atlas of 7 vols in 5. Paris, 1841-54. Contents: Histoire du voyage; Anthropologie; Zoologie; Botanique; Géologie; Physique; Hydrographie.

11291. WILKES Expedition, 1838-42. (An American naval expedition to the Pacific, Antarctica, and around the world.)
1. *United States exploring expedition during the years 1838-42, under the command of Charles Wilkes. Narrative, by C. Wilkes.* 5 vols & atlas. Philadelphia, 1845.
See also 12633.

11292. EREBUS, H.M.S. (British ship) Voyage, 1839-43. BMNH. See also Index.

11293. TERROR, H.M.S. (British ship) Voyage, 1839-43. BMNH. See also Index.

11294. SAMARANG, H.M.S. (British ship) Voyage, 1843-46. BMNH. See also 13292/1.

11295. HERALD, H.M.S. (British ship) Voyage, 1845-51. BMNH. See also Index.

11296. RATTLESNAKE, H.M.S. (British ship) Voyage, 1846-50. BMNH. See also Index.

11297. NOVARA. (Austrian ship) Voyage, 1857-59. BMNH.
1. *Reise der österreichischen Fregatte "Novara" um die Erde in den Jahren 1857-59....* 12 vols in 17. Vienna, 1861-75. Contents: Beschreibender Theil; Botanischer Theil; Zoologischer Theil; etc.
1a. ——— Supplement. *Physikalische und geognostiche Erinnerungen.* By A. von Humboldt.

11298. MAGENTA. (Italian ship) Voyage, 1865-68. BMNH.

11299. PORCUPINE, H.M.S. (British ship) Dredging cruise, 1868-70. BMNH. See also 10649/1.

11300. LIGHTNING, H.M.S. (British ship) Dredging cruise, 1868-70. BMNH. See also 10649/1.

11301. BLAKE. (U.S. ship) Oceanographic cruises, 1868-80. BMNH. See also 10667/1.

11302. CHALLENGER, H.M.S. (British ship) Voyage, 1873-76. BMNH.
 1. *The voyage of the "Challenger." The Atlantic: A preliminary account of the general results of the exploring voyage of H.M.S. "Challenger" during the year 1873 and the early part of the year 1876.* London, 1877. By C.W. Thomson. RM/L.
 2. *Report on the scientific results of the voyage of H.M.S. "Challenger" during the years 1873-76.* 40 vols in 50. Edinburgh/London, 1880-95. Ed. by C.W. Thomson and later by J. Murray. Reprinted, New York, 1966. Contents: I. Narrative (3 vols). II. Physics and chemistry [of ocean water, etc.] (2 vols). III. Deep-sea deposits (1 vol.). IV. Botany (2 vols). V. Zoology (40 vols). VI. Summary (2 vols).

 See also Index.

11303. GAZELLE. (German ship) Voyage, 1874-76. BMNH.
 1. *Die Forschungsreise S.M.S. "Gazelle" in den Jahren 1874 bis 1876.* 5 vols. Berlin, 1889-90. Contents: Reisebericht; Physik und Chemie; Zoologie und Geologie; Botanik; Meteorologie.

11304. WILLEM BARENTS. (Dutch ship) Voyages, 1878, 1879, 1880-84. BMNH.

11305. ALERT, H.M.S. (British ship) Voyage, 1881-82. BMNH. See also 13855/6.

11306. VETTOR PISANI. (Italian ship) Voyage, 1882-85. BMNH.

11307. HIRONDELLE. (Ship belonging to Prince Albert I of Monaco) Oceanographic cruises, 1885 onwards. BMNH.

11308. RAMBLER, H.M.S. (British ship) Dredging cruise, 1888. BMNH.

11309. HUMBOLDT-Stiftung.
 1. *Ergebnisse der in dem Atlantischen Ocean ... 1889 ausgeführten Plankton-Expedition.* 48 vols. Kiel, 1892-1911. Ed. by V. Hensen. X. Reprinted, New York, ca. 1970.

3.13 BOTANY

Including plant anatomy and plant physiology.

In the second half of the nineteenth century a large number of titles of very specialised monographs has been omitted.

Minor authors who wrote on both botany and zoology are listed under Natural History.

11310. HILLER, Mattheus. 1646-1725. Ger. BMNH; GA; GC.
1. *Hierophyton; sive, Commentarius in loca Scripturae Sanctae quae plantarum faciunt mentionem.* Utrecht, 1725. RM.

11311. PLUMIER, Charles. 1646-1704. Fr. DSB; BMNH; GA. Also Part 2.
1. *Traité des fougères de l'Amérique.* Paris, 1705. RM.
See also 11411/3.

11312. MERIAN, Maria Sibylla. 1647-1717. Holl. ICB (MARIAN); BMNH; GA; GC; GF. Also Zoology* and Part 2*.

11313. NISSOLE, Guillaume. 1647-1734. Fr. GA.

11314. EYSEL, Johann Philipp. 1651-1717. Ger. P I; BMNH; BLA; GA.

11315. KAEMPFER, Engelbert. 1651-1716. Ger. DSB; ICB; P I; BMNH; BLA & Supp.; GA; GB; GC. Also Part 2.
1. *Amoenitatum exoticarum politico-physico-medicarum, fasciculi V.* Lemgo, 1712. RM/L.

Posthumous

2. *Geschichte und Beschreibung von Japan.* 2 vols. Lemgo, 1777-79. Ed. from the MS by C.W. Dohm. X. Reprinted, Stuttgart, 1964.
2a. ——— *The history of Japan.* 2 vols. London, 1727. Trans. from the MS in the British Museum by J.G. Scheuchzer. Thus the work first appeared as the English version. X.
2b. ——— ——— 2nd ed. Ed. by T. Woodward and C. Davis. London, 1728.
3. *Icones selectae plantarum quas in Japonica collegit et delineavit.* London, 1791. Ed. by Joseph Banks from the original drawings in the British Museum.

11316. SLEVOGT, Johann Adrian. 1653-1726. Ger. P II; BMNH; BLA. Also Chemistry.

11317. ALEXANDRE, Nicolas. 1654-1728. Fr. P I; BMNH.

11318. KNAUT, Christian. 1654-1716. BMNH.

11319. TILLI, Michael Angelus. 1655-1740. It. BMNH; GA.

11320. TOURNEFORT, Joseph Pitton de. 1656-1708. Fr. DSB; ICB; P II; BMNH; GA; GB. Also Part 2*.
1. *Relation d'un voyage du Levant ... enrichie de descriptions ... de plantes rares*. 2 vols. Paris, 1717. RM. The voyage was made in 1700-02.
1a. ——— *A voyage into the Levant*. 2 vols. London, 1718. Trans. by J. Ozell with a biography. X. Another ed., 3 vols, 1741. RM/L.
See also 11351/1.

11321. SAVASTANO, Francesco Eulalio. 1657-1717. It. BMNH.

11322. BUDDLE, Adam. d. 1715. Br. GD.

11323. FEUILLÉE, Louis Econches. 1660-1732. Fr. DSB; P I; BMNH; GA (FEUILLET). Also Astronomy.
1. *Journal des observations physiques, mathématiques et botaniques faites par ordre du roy sur les côtes orientales de l'Amérique Méridionale et dans les Indes Occidentales, 1707-1712*. Paris, 1714. RM.

11324. SLOANE, Hans. 1660-1753. Br. DSB; ICB; P II; BMNH; BLA; GA; GB; GD. Also Natural History and Personal Writings*.
1. *Catalogus plantarum quae in insula Jamaica sponte proveniunt vel vulgo coluntur*. London, 1696. Wing S3998. X.
2. *A voyage to the islands Madeira, Barbadoes ... and Jamaica, with the natural history of the herbs and trees ... of the last of these islands*. 2 vols. London, 1707-25. X. Reprinted, Weinheim, 196-?

11325. CAMELLUS, Georgius Josephus. 1661-1706. Moravia. BMNH; GA.

11326. FICK, Johann Jacob. 1662-1730. Ger. P I; BMNH; BLA; GA. Also Chemistry and Anatomy.

11327. LOCHNER (VON HUMMELSTEIN), Michael Friedrich. 1662-1720. Ger. P I; BMNH; BLA; GA.

11328. ZANNICELLI, Gian Girolamo. 1662-1729. It. P II; BMNH. Also Chemistry.
1. *Istoria delle piante che nascono ne' lidi intorno a Venezia*. Venice, 1735. RM.

11329. RICHARDSON, Richard. 1663-1741. Br. BMNH; GD.

11330. NEBEL, Daniel. 1664-1733. Ger. BMNH; BLA; GA.

11331. WOODWARD, John. 1665-1728. Br. DSB; P II; BMNH; BLA; GB; GD. Also Geology* and Part 2*.

11332. FILIUS, Marcus Mappus. 1666-1736. Fr. ICB.

11333. COMMELIN, Caspar. 1667-1731. Holl. BMNH; GA.
1. *Praeludia botanica ad publicas plantarum exoticarum demonstrationes*. Leiden, 1703.
2. *Horti medici Amstelaedamensis plantae rariores et exoticae ad vivum aeri incisae*. Ib., 1706. X. Another ed., 1715.

11334. FISCHER, Johann Andreas. 1667-1729. Ger. P I; BMNH; BLA.

11335. BOERHAAVE, Hermann. 1668-1738. Holl. DSB; ICB; P I; BMNH; Mort.; BLA & Supp.; GA; GB. Also Chemistry*# and Physiology*. See also Index.

1. *Sermo academicus quem habuit quum ... botanicam at chemicam professionem publice poneret.* Leiden, 1729.

11336. VAILLANT, Sébastien. 1669-1722. Fr. DSB; ICB; BMNH; GA.

11337. CUNNINGHAM, James. d. ca. 1709. Br. BLA; GA; GD.

11338. RAND, Isaac. d. 1743. Br. BMNH; GD.

11339. BARHAM, Henry. ca. 1670-1726. Br. BMNH; GD.

11340. CELSIUS, Olof. 1670-1756. Swed. ICB; GA.

11341. CHOMEL, Pierre Jean Baptiste. 1671-1740 (or 1748). Fr. P I; BMNH; BLA; GA.

11342. AGRICOLA, Georg Andreas. 1672-1738. Ger. BMNH; BLA; GA; GC.

11343. LOGAN, James. 1674-1751. U.S.A. DSB; ICB; P I & Supp.; BMNH; GD; GE. Also Various Fields.
1. *Experimenta et meletemata de plantarum generatione.* Leiden, 1739. X.
1a. ——— *Experiments and considerations on the generation of plants.* London, 1747. Trans. by J. Fothergill. Parallel text in Latin and English. RM.

11344. WEDEL, Johann Adolf. 1675-1747. Ger. BMNH (WEDELIUS); BLA.

11345. BURCKHARD, Johann Heinrich. 1676-1738. BMNH.

11346. ERNDTEL (or ERNDL), Christian Heinrich. 1676-1734. Ger. BMNH; GA.

11347. LINDER, Johan. 1676-1723. Swed. BMNH; BLA.

11348. THRELKELD, Caleb. 1676-1728. Br. BMNH; GD.
1. *Synopsis stirpium Hibernicarum ... Being a short treatise of native plants.* Dublin, 1727.

11349. HALES, Stephen. 1677-1761. Br. DSB; ICB; P I; BMNH; Mort.; BLA & Supp.; GA; GB; GD. Also Chemistry, Geology*, and Physiology*.
1. *Vegetable staticks; or, An account of some statical experiments on the sap in vegetables ... Also a specimen of an attempt to analyse the air by a great variety of chymio-statical experiments.* London, 1727. RM/L. Reprinted,ib., 1961.
1a. ——— *Statical essays. Containing "Vegetable staticks."* Ib., 1731. The revised 2nd ed. X.
1b. ——— *Statical essays.* 2 vols. Ib., 1733. Vol. I is the 3rd ed. of item 1, and Vol. II is the 1st ed. of *Haemastaticks*, for which see 14642/1b. Another ed., 1769.
1c. ——— ——— *Le statique des végétaux et celle des animaux.* 2 vols. Paris, 1779-80.

11350. HESSELIUS, Andreas. 1677-1733. Swed. ICB; BMNH.

11351. MICHELI, Pier Antonio. 1679-1737. It. DSB; BMNH; GA (MICHIELI). See also 15406/1.
1. *Nova plantarum genera juxta Tournefortii methodum disposita.* Florence, 1729.

11352. BLAIR, Patrick. d. 1728. Br. DSB; BMNH; GA; GD; BLA.

11353. BREYNE, Johann Philipp. 1680 (or 1690)-1764. Danzig. P I; BMNH (BREYNIUS); GA.
1. *Icones rariorum plantarum.* Leiden, 1700.

11354. KEOGH, John. 1681?-1754. Br. BMNH; GD.
1. *Botanologia universalis Hibernica; or, A general Irish herbal ... giving an account of the herbs, shrubs and trees....* Cork, 1735.

11355. ALBERTI, Michael. 1682-1757. Ger. P I; BMNH; BLA; GA; GC. Also Chemistry.

11356. LINDERN, Franz Balthasar von. 1682-1755. Strasbourg. BMNH; BLA.

11357. MONTI, Giuseppe. 1682-1760. It. P II; BMNH; GA.

11358. ALSTON, Charles. 1683-1760. Br. P I; BMNH; BLA & Supp.; GA; GB; GD. Also Chemistry.

11359. HEISTER, Lorenz. 1683-1758. Ger. DSB; ICB; BMNH; BLA & Supp.; GA; GC. Also Anatomy*.

11360. WEINMANN, Johann Wilhelm. b. 1683. BMNH.

11361. BORELL, Jean. 1684-1747. Ger. P I; BLA & Supp.

11362. SCHEUCHZER, Johann. 1684-1738. Switz. P II; BMNH; BLA; GA; GC.
1. *Agrostographia; sive, Graminum, Juncorum, Cyperorum, Cyperoidum, iisque affinium historia.* Zurich, 1719.

11363. VATER, Abraham. 1684-1751. Ger. P II; BMNH; Mort.; BLA; GC. Also Chemistry and Anatomy.

11364. BREWER, Samuel. d. 1743. Br. BMNH; GA; GD.

11365. ALSTRÖMER, Johan (or Jonas). 1685-1761. Swed./Br. ICB; P I; GA; GB.

11366. CLAYTON, John (second of the name). ca. 1685-1773. Br./U.S.A. ICB; BMNH; BLA; GA; GD; GE.

11367. JUSSIEU, Antoine de. 1686-1758. Fr. DSB; ICB; P I; BMNH; BLA; GA; GB.

11368. SIEGESBECK, Johann Georg. 1686-1755. Ger./Russ. P II & Supp.; BMNH; BLA; GC. Also Astronomy.

11369. DILLENIUS, Johann Jacob. 1687-1747. Ger./Br; DSB; ICB; BMNH; BLA & Supp. (DILLEN); GA; GB; GC; GD.
1. *Catalogus plantarum circa Gissam sponte nascentium.* Frankfurt, 1718. RM.
2. *Hortus Elthamensis; seu, Plantarum rariorum quas in horto suo Elthami in Cantio coluit ... J. Sherard ... delineationes et descriptiones.* 2 vols. London, 1732. RM/L. Another ed., Leiden, 1774.
3. *Historia Muscorum.* Oxford, 1741.
3a. ——— *Historia muscorum. A general history of land and water, etc., mosses and corals.* London, 1763.

11370. MARQUET, François Nicolas. 1687-1759. Fr. BMNH; BLA; GA.

11371. BRADLEY, Richard. 1688-1732. Br. DSB; ICB; P I; BMNH; BLA & Supp.; GA; GD.

1. *New improvements of planting and gardening, both philosophical and practical, explaining the motion of the sap and generation of plants.* London, 1717. X. 7th ed., 3 parts in 1 vol., 1738-39.
2. *A philosophical account of the works of nature, endeavouring to set forth the several graduations remarkable in the mineral, vegetable and animal parts of the creation ... To which is added an account of the state of gardening.* London, 1721. RM/L.

11372. COLDEN, Cadwallader. 1688-1776. Br./U.S.A. DSB; ICB; P I; BLA & Supp.; GA; GB; GD; GE. Also Mechanics.

11373. PONTEDERA, Giulio. 1688-1757. It. DSB; BMNH; GA.

11374. ROHR, Julius Bernhard von. 1688-1742. Ger. P II; BMNH; GC.

11375. RUPPIUS, Heinrich Bernhard. 1688-1719. Ger. BMNH.

11376. WALTHER, Augustin Friedrich. 1688-1746. Ger. BMNH; BLA; GC. Also Anatomy.

11377. ARDÈNE, Jean Paul de Rome d'. 1689-1769. BMNH.

11378. DOBBS, Arthur. 1689-1765. Br./U.S.A. ICB; GD; GE.

11379. MILLER, Joseph. fl. 1722-1748. Br. BMNH.
1. *Botanicum officinale; or, A compendious herbal, giving an account of all such plants as are now used in the practice of physick.* London, 1722.

11380. GRONOVIUS, Joannes Fredericus. 1690-1760. Holl. ICB; BMNH; GA.
1. *Flora Virginica.* Leiden, 1739. RM.
2. *Flora orientalis; sive, Recensio plantarum quas ... L. Rauwolffus annis 1573-75 ... observavit et collegit ... Has methodo sexuali disposuit.* Leiden, 1755. RM.

11381. DUVERNOY, Johann Georg. 1691-1759. Ger./Russ. BMNH; BLA. Also Anatomy.

11382. MILLER, Philip. 1691-1771. Br. DSB; ICB; BMNH; GD.
1. *The gardener's dictionary.* Dublin, 1731. X. 4th ed., London, 1743.
1a. ——— *The gardener's dictionary abridg'd.* 3 vols in 1. London, 1754. X. Reprinted, Weinheim, 1969. (HNC. 72)
1b. ——— Another ed. *The gardener's and botanist's dictionary.* 2 vols in 4. London, 1807. Ed. by T. Martyn.
2. *The gardener's kalendar.* Dublin, 1732. X. 15th ed., London, 1769.

11383. KNOWLTON, Thomas. 1692-1782. Br. BMNH; GA; GD.

11384. BUXBAUM, Joannes Christianus. 1693-1730. Ger. BMNH; GA; GC.
1. *Plantarum minus cognitarum centuriae, complectans plantas circa Byzantium et in Oriente observatas.* 5 parts. St. Petersburg, 1728-40.

11385. BORETIUS, Matthias Ernestus. 1694-1738. Ger. BMNH; BLA.

11386. DEERING, George Charles. ca. 1695-1749. Ger./Br. BMNH; BLA (D., Karl); GA; GD.

11387. HOUSTOUN, William. 1695-1733. Br. BMNH; GA; GD.

11388. QUER Y MARTINEZ, José. 1695-1764. Spain. BMNH; GA.

11389. STAEHELI, Benedict. 1695-1750. Switz. P II; GA (STÄHELIN). Also Physics.

11390. TREW, Christoph Jacob. 1695-1769. Ger. ICB; BMNH; BLA; GA; GC. Also Anatomy.

11391. CAMERARIUS, Alexander. 1696-1736. Ger. P I; BMNH; BLA & Supp.

11392. BARTRAM, John. 1699-1777. U.S.A. DSB; ICB; BMNH; GA; GE.
1. *John and William Bartram's America: Selections from the writings of the early American naturalists.* New York, 1961. Ed. with an introd. by H.G. Cruickshank.
See also 15433/1.

11393. JUSSIEU, Bernard de. 1699-1777. Fr. DSB; ICB; BMNH; BLA; GA; GB.

11394. LA CHESNAYE-DESBOIS, François Alexandre Aubert de. 1699-1784. Fr. BMNH (CHESNAYE-D.); GA (DESBOIS). Also Zoology*.
1. *Dictionnaire universel d'agriculture et de jardinage.* 2 vols in 3. Paris, 1751. RM.

11395. MARTYN, John. 1699-1768. Br. ICB; P II; BMNH; GA; GB; GD.
1. *Historia plantarum rariorum.* London, 1728. X.
1a. ——— *Historia plantarum rariorum ... Beschreibung seltener Pflanzen....* Nuremberg, 1752. Ed. by J.D. Meyer. Parallel text in Latin and German.
See also 2997/1c,1e and 2998/1.

11396. WILSON, John. d. 1751. Br. BMNH; GD.

11397. BLACKSTONE, John. d. 1753. Br. BMNH; GA; GD.

11398. BLACKWELL, Elizabeth. ca. 1700-after 1747. Br. BMNH; GD; GF.
1. *A curious herbal ... of the most useful plants ... in the practice of physic.* 2 vols. London, 1737. X.
1a. ——— *Sammlung der Gewächse die zum Arzney-Gebrauch....* 2 vols. Nuremberg, 1750-65.

11399. DUHAMEL DU MONCEAU, Henri Louis. 1700-1782. Fr. DSB; ICB; P I; BMNH; BLA (HAMEL); GA; GB. Also Chemistry.
1. *La physique des arbres, où il est traité de l'anatomie des plantes et de l'économie végétale.* 2 vols. Paris, 1758. RM/L.
2. *Des semis et plantations des arbres, et leur culture.* Paris, 1760. RM.
3. *De l'exploitation des bois.* 2 vols. Paris, 1764. RM.
4. *Du transport, de la conservation et de la force des bois.* Paris, 1767. RM.
5. *Traité des arbres fruitiers.* 2 vols. Paris, 1768. RM.

11400. EHRHART, (Johann) Balthasar. 1700-1756. Ger. P I; BMNH; BLA; GA. Also Geology*.

11401. KNOOP, Johann Hermann. 1700?-1769? Holl. BMNH; GF.
1. *Pomologia; dat is, Beschryvingen en afbeeldingen van de beste zoorten van appels en peeren....* Leeuwarden, 1758. RM.

2. *Dendrologia; of, Beschryving der plantagiegewassen.* Ib., 1760. RM.
3. *Fructologia; of, Beschryving der vrugtbomen en vrugten.* Ib., 1763. RM.

11402. STILLINGFLEET, Benjamin. 1702-1771. Br. BMNH; GA; GD.
1. *Miscellaneous tracts relating to natural history, husbandry and physik. Translated from the Latin, with notes.* London, 1759. A trans. of six essays from Linnaeus' *Amoenitates academicae* (item 10958/10) with a preface. The preface is said to be "the first fundamental treatise on the principles of Linnaeus published in England." 2nd ed., 1762. RM. 4th ed., 1791.

11403. WACHENDORFF, Everardus Jacobus van. 1702-1758. Holl. BMNH; BLA. Also Anatomy.

11404. DALIBARD, Thomas François. 1703-1779. Fr. DSB; P III; BMNH.
1. *Florae Parisiensis prodromus; ou, Catalogue des plantes qui naissent dans les environs de Paris.* Paris 1749.

11405. SÉGUIER, Jean François. 1703-1784. Fr. P II; BMNH; GA.
1. *Bibliotheca botanica; sive, Catalogus auctorum et librorum qui de re botanica, de medicamentis ex vegetabilibus paratis, de re rustica et de horticultura tractant.* The Hague, 1740. RM.

11406. CARTHEUSER, Johann Friedrich. 1704-1777. Ger. P I; BMNH; BLA; GA (CARTHAEUSER). Also Chemistry*.

11407. HAEN, Antonius de. 1704-1776. Holl. ICB; BMNH; BLA & Supp.; GA; GC.

11408. KNIPHOFF, Johannes Hieronymus. 1704-1762 (or 1765). Ger. BMNH; BLA.

11409. ROYEN, Adrianus van. 1704-1779. Holl. BMNH; BLA; GF. See also 11455/1.

11410. KNORR, Georg Wolfgang. 1705-1761. Ger. DSB; P I; BMNH; GA; GC. Also Geology*.
1. *Thesaurus rei herbariae hortensisque universalis ... Allgemeines Blumen-, Kräuter-, Frucht- und Garten-Buch.* 2 vols in 3. Nuremberg, 1750-72.

11411. BURMAN, Johannes. 1706-1779. Holl. ICB; BMNH; BLA & Supp.; GA.
1. *Thesaurus Zeylanicus, exhibens plantas in insula Zeylana nascentes.* Amsterdam, 1737. (Flora of Ceylon) Reprinted, New York, 1964.
2. *Rariorum Africanarum plantarum....* Amsterdam, 1738-39.
3. *Plantarum Americanarum ... continens plantas quas olim C. Plumierius ... detexit ... Has primum in lucem edidit ....* 10 parts. Amsterdam, 1755-60.

11412. SAUVAGES (DE LA CROIX), François Bossier de. 1706-1767. Fr. ICB; BMNH; BLA; GA.

11413. AMMANN, Johann. 1707-1741. Russ. BMNH; GA; GC.

11414. BROWALLIUS, Johannes. 1707-1755. Swed. ICB; P I; BMNH; GA.

11415. HECKER, Johann Julius. 1707-1768. Ger. BMNH; GC.

11416. HILL, John. 1707?-1775. Br. DSB; ICB; P I; BMNH; BLA; GA; GB; GD. Other entries: see 3105.

1. *The British herbal. An history of plants and trees, natives of Britain, cultivated for use or raised for beauty.* London, 1756.
2. *The vegetable system; or, A series of experiments and observations tending to explain the internal structure and the life of plants.* 26 vols. London, 1759-75. X. 2nd ed., 1773-75.
3. *Flora Britannica.* London, 1760.
4. *Hortus Kewensis.* London, 1768. X. 2nd ed., 1769.
5. *The construction of timber, from its early growth, explained by the microscope and proved from experiments.* London, 1770. RM.

11417. HUBER, Johann Jacob. 1707-1778. Ger. DSB; BLA; GA; GC. Also Anatomy.

11418. LINNAEUS (later VON LINNÉ), Carl. 1707-1778. Swed. DSB; ICB; P I; BMNH. Under LINNÉ: BLA & Supp.; GA; GB. Also Natural History*#, Zoology*, and Personal Writings*. See also Index.

1. *Fundamenta botanica, quae majorum operum prodromi instar theoriam scientiae botanices per breves aphorismos tradunt.* Amsterdam, 1736. X. Issued both as a separate work and with items 2,5, and 8.
2. *Bibliotheca botanica recensens libros plus mille de plantis huc usque editos, secundum systema auctorum naturale ... Fundamentorum botanicorum pars prima.* Amsterdam, 1736. X. Reprinted, Munich, 1968. Other eds: Halle, 1747; Amsterdam, 1751.
3. *Musa Cliffortiana.* Leiden, 1736.
4. *Hortus Cliffortianus.* Amsterdam, 1737. RM. Reprinted, Weinheim, 1965. (HNC. 63)
5. *Critica botanica in quo nomina plantorum generica, specifica et variantia examini subjiciuntur ... seu Fundamentorum botanicorum pars IV.* Leiden, 1737.

5a. ——— *The "Critica botanica" of Linnaeus.* London, 1938. Trans. by A. Hort. (Ray Society publn. 124)

6. *Flora Lapponica.* Amsterdam, 1737. RM. Another ed., London, 1792.
7. *Genera plantarum.* Leiden, 1737. 2nd ed., ib., 1742.

7a. ——— 5th ed., Stockholm, 1754. Reprinted, Weinheim, 1960. (HNC. 3)

7b. ——— 6th ed., Stockholm, 1764. Reprinted, Vienna, 1767.

7c. ——— 9th ed., 2 vols, Göttingen, 1830-31. Ed. by C. Sprengel.

8. *Classes plantarum ... Fundamentorum botanicorum pars II.* Leiden, 1738.
9. *Flora Svecica.* Stockholm, 1745. 2nd ed., 1755.
10. *Flora Zeylonica, sistens plantas indicas Zeylonae insulae.* Stockholm, 1747. (Flora of Ceylon)
11. *Hortus Upsaliensis, exhibens plantas exoticas horto Upsaliensis academiae.* Vol. 1. [No more publ.] Stockholm, 1748. RM/L.
12. *Philosophia botanica.* Stockholm, 1751. RM. Reprinted, Weinheim, 1966. (HNC. 48) For an English adaptation see 11438/1. Other eds: Vienna, 1755, 1763, 1770.

2nd ed., Berlin, 1780. Ed. by J.G. Gleditsch. Another so-called 2nd ed., Vienna, 1783.

13. *Species plantarum.* 2 vols. Stockholm, 1753. RM/L. Reprinted, London, 1957-59. (Ray Society publns. 140,142) 2nd ed., Stockholm, 1762-63. 3rd ed., Vienna, 1764. 4th ed., *Adjectis vegetabilibus hucusque cognitis*, 6 vols in 20, Berlin, 1797-1830. Ed. by C.L. Willdenow.

14. *Disquisitio de questione ... sexum plantarum.* St. Petersburg, 1760. A prize essay. X.

14a. ——— *A dissertation on the sexes of plants.* London, 1786. Trans. by J.E. Smith. RM.

15. *Mantissa plantarum "Generum" editionis VI* [1764; item 7b] *et "Specierum" editionis II* [1762-63; item 13]. Stockholm, 1767. A separately published appendix to Vol. 2 of the 12th ed. of the *Systema naturae* (item 10958/1f). X. Another ed., ib., 1771.

15a. ——— Reprint of both the 1767 and the 1771 eds. Weinheim, 1961. (HNC. 7)

16. *Systema vegetabilium ... Editio decima tertia.* Göttingen, 1774. Ed. by J.A. Murray (the new additions being written by Linnaeus). A new ed. of the botanical part (i.e. Vol. 2) of the *Systema naturae*; counted after the 12th ed. of that work.

16a. ——— *A system of vegetables.* 2 vols. London, 1783. Trans. by Erasmus Darwin et al. RM.

16b. ——— 14th ed., Göttingen, 1784. Ed. by J.A. Murray.

16c. ——— 15th ed., Göttingen, 1797. Ed. by C.H. Persoon.

16d. ——— ——— *Editio nova. Specibus inde ab editione 15 detectis aucta et locupletata.* 7 vols in 8. Stuttgart, 1817-30. Ed. by J.J. Roemer and J.A. Schultes.

16e. ——— ——— ——— *Mantissa.* 3 vols, ib., 1822-27.

16f. ——— 16th ed., 5 vols, Göttingen, 1825-28. Ed. by C. Sprengel.

Students' Theses

For the collected edition of the theses, *Amoenitates academicae*, see 10958/10.

17. *Flora Jamaicensis, quam ... praeside ... C. Linnaeo ... submittit C.G. Sandmark.* Uppsala, 1759.

18. *Auctores botanici. In dissertatione propositi, quam sub praesidio C. Linnaei ... defert A. Loo.* Uppsala, 1759. A biographical catalogue of some 350 botanical writers. X. Reprinted, Stockholm, 1970.

19. *Termini botanici, quos ... praeside ... C. Linnaeo ... examinandos sistit J. Elmgren.* Uppsala, 1762. X.

19a. ——— *Caroli a Linné ... Termini botanici.* Hamburg, 1781. By P.D. Giseke. Compiled chiefly from the above thesis and partly from various works of Linnaeus; gives the German equivalents of the terms. X. Another ed., 1787; gives the French and English, as well as the German equivalents.

20. *Dissertatio botanico-medico sistens rariora Norvegiae quam ... praeside C. a Linné ... offert H. Tonning.* Uppsala, 1768.

21. *Pandora et flora Rybyensis, quam dissertatione academica ... praeside ... C. a Linné ... offert D.H. Söderberg.* Uppsala, 1771.

Posthumous Works

22. *Praelectiones in ordines naturales plantarum.* Hamburg, 1792. Ed. by P.D. Giseke. RM.
23. *Hortus Uplandicus; sive, Enumeratio plantarum quae in variis hortis Uplandiae....* Uppsala, 1899. Ed. by T.M. Fries from a MS dating from 1730.
24. *Vorlesungen über die Cultur der Pflanzen.* Uppsala, 1907. Ed. by M.B. Swederus.

Collections and Selections

25. *Systema plantarum Europae.* 7 vols. Geneva, 1785-87. Ed. by J.E. Gilibert.
26. *The families of plants, with their natural characters ... Translated from ... "Genera plantarum" ... "Mantissa plantarum."* 2 vols. London, 1787. By Erasmus Darwin et al.
27. *Opera. Editio prima critica ... Vol. secundum, systema vegetabilium libros diagnostico-botanicos continens.* [No more publ.] Leipzig, 1835.

11419. CUNO, Johann Christian. 1708-1783. Ger. BMNH; GA; GC.

11420. HALLER, Albrecht von. 1708-1777. Switz./Ger. DSB; ICB; P I; BMNH; Mort.; BLA & Supp.; GA; GB; GC. Also Anatomy*, Physiology*#, and Personal Writings*.
1. *Opuscula sua botanica prius edita. Recensuit, retractavit, auxit coniuncta, edidit A. Hallerus.* Göttingen, 1749.
2. *Historia stirpium indigenarum Helvetiae inchoata.* 3 vols. Bern, 1768.
2a. ——— *Icones plantarum Helvetiae, ex ipsius "Historia stirpium Helveticarum" denuo recusae. Cum descriptionibus auctoris, ejusque praefatione.* Bern, 1795. Ed. with notes by J.S. Wyttenbach. RM.
3. *Bibliotheca botanica, qua scripta ad rem herbariam facientia a rerum initiis recensentur.* 2 vols. Zurich, 1771-72. X. Reprinted, Hildesheim, 1969.

11421. PLAZ, Anton Wilhelm. 1708-1784. Ger. BMNH; BLA.

11422. WEDEL, Johann Wolfgang. 1708-1757. Ger. BMNH; BLA.

11423. BARBEU-DUBOURG, Jacques. 1709-1779. Fr. BMNH; BLA & Supp.; GA.

11424. ERNSTING, Arthur Conrad. 1709-1768. Ger. P I; BMNH; BLA; GA.

11425. GESSNER, Johannes. 1709-1790. Switz. DSB; BLA & Supp.; GC. Under GESNER: P I & Supp.; BMNH; GA. Also Physics and Geology*. See also 14659/9.

11426. GMELIN, Johann Georg. 1709-1755. Ger./Russ. DSB; ICB; P I; BMNH; BLA; GA; GC. Also Natural History*.
1. *Flora Sibirica; sive, Historia plantarum Sibiriae.* 4 vols. St. Petersburg, 1747-69. Vols III and IV were ed. by S.G. Gmelin.

11427. LUDWIG, Christian Gottlieb. 1709-1773. Ger. P I; BMNH; BLA; GA; GC.

11428. GERBER, Traugott. d. 1743. Ger./Russ. BMNH; BLA.

11429. BAZIN, Gilles Augustin. d. 1754. Fr. P I; BMNH; BLA; GA.

11430. EHRET, Georg Dionysius. 1710-1770. Ger./Br. BMNH; GA; GD; GF. An illustrator.

11431. ELLIS, John. ca. 1710-1776. Br. ICB; BMNH; GA; GD. Also Zoology*.

11432. DIETRICH, Adam. 1711-1782. ICB.

11433. ROUSSEAU, Jean Jacques. 1712-1778. Fr. The famous philosophe. ICB; BMNH; G.
1. *Essais élémentaires sur la botanique.* Paris, 1771. X.
1a. ——— *La botanique de J.J. Rousseau.* Paris, 1805.
2. [Trans.] *Letters on the elements of botany ... Translated, with notes and twenty-four additional letters, fully explaining the system of Linnaeus.* London, 1785. By T. Martyn. RM. 5th ed., 1796. Another ed., 1815.

11434. FABRICIUS, Philipp Konrad. 1714-1774. Ger. P I; BMNH; BLA & Supp.; GA; GC. Also Chemistry.

11435. GLEDITSCH, Johann Gottlieb. 1714-1786. Ger. P I; BMNH; BLA; GA; GC.
1. *Pflanzenverzeichnis zum Nutzen und Vergnügen der Lust- und Baumgärtner.* Berlin, 1773. RM.
See also 11418/12.

11436. ROSEN, Eberhard. 1714-1796. Swed. BMNH; BLA.

11437. GUETTARD, Jean Etienne. 1715-1786. Fr. DSB; P I; BMNH; GA; GB. Also Geology.
1. *Observations sur les plantes.* Paris, 1747. RM.

11438. LEE, James. 1715-1795. Br. ICB; BMNH; GD.
1. *An introduction to botany, containing an explanation of the theory of that science, and an interpretation of its technical terms. Extracted from the works of Linnaeus.* London, 1760. A large part of the work is a free trans. of Linnaeus' *Philosophia botanica.* 4th ed., 1810.; includes a biography by Lee's son.

11439. LOUREIRO, João de. 1715-1796. BMNH.
1. *Flora Cochin-Chinensis.* 2 vols. Lisbon, 1790. Another ed., with notes by C.L. Willdenow, Berlin, 1793.

11440. MUELLER, Johann Sebastian (later MILLER, John). 1715?-1790? Br. BMNH; GA; GC.
1. *Illustratio systematis sexualis Linnaei ... An illustration of the sexual system of Linnaeus.* 20 parts in 1 vol. London, 1775-77. Parallel text in Latin and English. Another ed., 1794.

11441. WATSON, William. 1715-1787. Br. DSB; ICB; P II; BLA; GA; GD. Also Physics*.

11442. DAUBENTON, Louis Jean Marie. 1716-1800. Fr. DSB Supp.; ICB; P I; BMNH; BLA & Supp.; GA; GB. Also Mineralogy, Natural History, and Zoology. See also Index.

11443. KALM, Pehr. 1716-1779. Fin. DSB; ICB; P I; BMNH; GA.

11444. GAUTIER D'AGOTY, Jacques Fabian. 1717-1785. Fr. P I (GAUTHIER); BMNH; Mort.; BLA & Supp.; GA (GAUTHIER). Also Physics* and Anatomy.

11445. GORTER, David de. 1717-1783. Belg. BMNH; BLA & Supp.

11446. LE MONNIER, Louis Guillaume. 1717-1799. Fr. DSB; ICB; P I; BMNH; BLA; GA. Also Physics.

11447. SCHAEFFER, Jacob Christian. 1718-1790. Ger. P II; BMNH; GA; GC. Also Zoology*.
1. *Erleichterte Arzneykräuterwissenschaft*. Regensburg, 1759. RM.
2. *Fungorum qui in Bavaria et Palatinatu circa Ratisbonam nascuntur icones*. 4 vols. Regensburg, 1762-74. X. Another ed., with notes by C.H. Persoon, Erlangen, 1800.

11448. SCHMIDEL (or SCHMIEDEL), Casimir Christoph. 1718-1792. Ger. DSB; P II; BMNH; BLA. Also Mineralogy*.

11449. WILLICH, Christianus Ludovicus. 1718-1773. Ger. BMNH.

11450. AGNETHLER, Michael Gottlieb. 1719-1752. Ger. BMNH; GC.

11451. POIVRE, Pierre. 1719-1786. Fr. DSB; GA.

11452. AUBLET, Jean Baptiste Christophe Fusée. 1720-1778. Fr. BMNH; GA.
1. *Histoire des plantes de la Guiane françoise*. 4 vols. Paris, 1775. Reprinted, Weinheim, 1977. (HNC. 100)

11453. BONNET, Charles. 1720-1793. Switz. DSB; ICB; P I; BMNH; Mort.; BLA; GA; GB. Also Biology in General*#, Zoology, and Embryology.
1. *Recherches sur l'usage des feuilles dans les plantes*. Göttingen, 1754. RM/L.

11454. ROLANDER, Daniel. 1720-1774. Swed. GA.

11455. GMELIN, Philipp Friedrich. 1721-1768. Ger. P I; BMNH; BLA & Supp.; GA; GC.
1. *Otia botanica ... reddidit "Prodromum florae Leydensis" Adriani van Royen*. Tübingen, 1760. RM.

11456. MALESHERBES, Crétien Guillaume de Lamoignon de. 1721-1794. Fr. The governmental official. DSB; ICB; BMNH; GA; GB.
1. *Observations sur "l'Histoire naturelle" ... de Buffon et Daubenton*. 2 vols. Paris, 1798. X. Reprinted, 2 vols in 1, Geneva, 1971.

11457. CRANTZ, Heinrich Johann Nepomuk von. 1722-1799. Aus. BMNH; BLA & Supp. Also Chemistry*.
1. *Stirpium Austriarum....* 3 parts in 1. Vienna, 1762-67. X. 2nd ed., 1769. RM.

11458. DAMBOURNEY, Louis Auguste. 1722-1795. Fr. P I; GA. Also Chemistry*.

11459. MARSHALL, Humphry. 1722-1801. U.S.A. ICB; BMNH; GE. See also 15433/1.
1. *Arbustum Americanum ... or, An alphabetical catalogue of*

forest trees and shrubs, natives of the American United States. Philadelphia, 1785. X. Reprinted (together with the French trans., below), New York, 1967.

1a. ——— *Catalogue alphabétique des arbres....* Paris, 1788. Trans. with notes by Lézermes. X.

11460. RAMSPECK, Jacob Christophe. b. 1722. Switz. ICB.

11461. SPIELMANN, Jacques Reinhold. 1722-1783. Strasbourg. ICB; P II; BMNH; BLA; GA; GC. Also Chemistry*.

11462. BERGIUS (or BERG), Benedict (or Bengt). 1723-1784. Swed. ICB; BMNH; GA.

11463. BOEHMER, Georg Rudolph. 1723-1803. Ger. P I; BMNH; BLA; GA.
1. *Flora Lipsiae indigena*. Leipzig, 1750.

11464. BOSE, Ernestus Gottlob. 1723-1788. Ger. BMNH; BLA.

11465. MEESE, David. b. 1723. Holl. BMNH.

11466. SCOPOLI, Giovanni Antonio. 1723-1788. It./Hung. ICB; P II; BMNH; GA. Also Mineralogy*, Geology, Natural History*, and Zoology*.
1. *Flora Carniolica*. Vienna, 1760. RM.

11467. BÜTTNER, David Sigismund August. 1724-1768. Ger. P I; BMNH; BLA & Supp.; GA (BUTTNER). Also Geology*.

11468. ALLIONI, Carlo. 1725-1804. It. P I & Supp.; BMNH; BLA; GA.
1. *Flora Pedemontana*. 3 vols. Turin, 1785.

11469. HOPE, John. 1725-1786. Br. BMNH; GD.

11470. BOSSECK, Heinrich Otto. 1726-1776. BMNH.

11471. MARTIN, Roland. 1726-1788. Swed. ICB; BMNH; BLA.

11472. ADANSON, Michel. 1727-1806. Fr. DSB; ICB; P I; BMNH; GA; GB. Also Zoology*.
1. *Familles des plantes*. 2 vols. Paris, 1763-64. RM. Reprinted, Weinheim, 1966. (HNC. 46)

11473. HALLE, Johann Samuel. 1727-1810. Ger. ICB; P I; BMNH.
1. *Die deutschen Giftpflanzen*. Berlin, 1784.

11474. HARTMANN, Peter Immanuel. 1727-1791. Ger. P I; BLA; GA. Also Chemistry.

11475. JACQUIN, Nicolas Joseph von. 1727-1817. Aus. DSB; ICB; P I; BMNH; BLA; GA; GC. Also Chemistry.
1. *Enumeratio systematica plantarum quas in insulis Caribaeis vicinaque Americes continente detexit*. Leiden, 1760. X. Reprinted, Zug, 1967.
2. *Selectarum stirpium Americanarum historia*. 2 vols. Vienna, 1763. RM/L.
3. *Observationum botanicarum....* 4 parts. Vienna, 1764-71.
4. *Anleitung zur Pflanzenkenntniss nach Linné's Methode*. Vienna, 1785. RM.
5. *Oxalis monographia*. Vienna, 1794.

11476. LEERS, Joannes Daniel. 1727-1774. Ger. BMNH.

11477. REICHEL, Georg Christian. 1727 (or 1717)-1771. Ger. ICB; BMNH; BLA.

11478. ROTTBOELL, Christian Friis. 1727-1797. Den. BMNH; BLA.
1. *Cyperaceae. Descriptionum et iconum....* Copenhagen, 1773.

11479. ZINN, Johann Gottfried. 1727-1759. Ger. P II; BMNH; Mort.; BLA; GC. Also Anatomy.

11480. ARDUINO, Pietro. 1728-1805. It. BMNH.

11481. OEDER, Georg Christian. 1728-1791. Den. ICB; BNMH; BLA; GA; GC.
1. *Elementa botanicae.* Copenhagen, 1764-66. RM.
2. *Einleitung zu der Kräuterkenntniss.* Ib., 1764-66. RM.

11482. AUGUSTIN, Samuel. 1729-1792. BMNH.

11483. BERKHEY, Jan le Francq van. 1729-1812. Holl. BMNH; BLA & Supp.; GA.
1. *Expositio characteristica structurae florum qui dicuntur Compositi.* Leiden, 1760. RM.

11484. CORTI, Bonaventura. 1729-1813. It. DSB; ICB; P I; BMNH. Also Physics.

11485. FORSTER, Johann Reinhold. 1729-1798. Ger./Br. DSB; ICB; P I & Supp.; BMNH; GA; GC. Also Mineralogy*, Geology*, Natural History*, and Zoology*. See also Index.
1. *Characteres generum plantarum quas in itinere ad insulas maris australis collegerunt ... 1772-1775.* London, 1776. With his son, Georg (i.e. J.G.A.) Forster.
1a. ——— *Beschreibungen der Gattungen von Pflanzen auf einer Reise nach den Inseln der Süd-See gesammelt ... 1772 bis 1775.* Stuttgart, 1779. Trans. by J.S. Kerner.

11486. LA TOURETTE, Marc Antoine Louis Claret de Fleurieu de. 1729-1793. Fr. BMNH; GA. See also 15548/1.

11487. LÖFLING, Pehr. 1729-1756. Swed. ICB; BMNH; GA.

11488. NECKER, Noel Joseph de. 1729-1793. Belg. BMNH; GA; GF.

11489. MARATTI, Giovanni Francesco. d. 1777. It. BMNH.
1. *De plantis zoophytis et lithophytis in Mari Mediterraneo viventibus.* Rome, 1776. RM.

11490. BERGIUS, Petter Jonas. 1730-1790. Swed. ICB; P I; BMNH; BLA & Supp.; GA.

11491. DURANDE, Jean François. 1730-1794. Fr. P I; BMNH; BLA; GA. Also Chemistry.

11492. HEDWIG, Johann. 1730-1799. Ger. DSB; BMNH; BLA; GA; GC.
1. *Fundamentum historiae naturalis muscorum frondosorum.* 2 parts. Leipzig, 1782. RM.
2. *Theoria generationis et fructificationis plantarum Cryptogamicarum Linnaei.* St. Petersburg, 1784. RM. Another ed., Leipzig, 1798.
3. *Descriptio et adumbratio microscopico-analytica muscorum frondosorum, nec non aliorum vegetantium e classe Cryptogamica Linnaei.* 4 vols. Leipzig, 1787-97. RM.
4. *Species muscorum frondosorum.* Leipzig, 1801. Ed. by F. Schwaegrichen. RM/L. Reprinted, Weinheim, 1960. (HNC.

16) Supplements: 1st, 2 vols, 1811-16, RM/L; 2nd, 2 vols, 1823-27, RM/L; 3rd, 2 vols, 1827-30; 4th, 1842.

11493. INGENHOUSZ, Jan. 1730-1799. Holl./Br. DSB; ICB; P I; BMNH; BLA; GA; GD. Also Physics*.
1. *Experiments upon vegetables, discovering their great power of purifying the common air in the sun-shine, and of injuring it in the shade and at night.* London, 1779. RM.
2. *An essay on the food of plants and the renovation of soils.* London, 1796. RM.

11494. MÜLLER, Otto Frederik. 1730-1784. Den. DSB; ICB; BMNH; BLA; GA (MULLER). Also Zoology* and Microbiology*.

11495. AITON, William. 1731-1793. Br. DSB; BMNH; GA; GB; GD.
1. *Hortus Kewensis; or, A catalogue of the plants cultivated in the Royal Botanic Garden at Kew.* 3 vols. London, 1789. 2nd ed., by W.T. Aiton, 5 vols, 1810-13. RM/L.

11496. BUC'HOZ, Pierre Joseph. 1731-1807. Fr. P I; BMNH; BLA; GA. Also Geology*, Natural History*, and Zoology*.

11497. DARWIN, Erasmus. 1731-1802. Br. DSB; ICB; P I; BMNH; BLA & Supp.; GA; GB; GD. Also Biology in General*#. See also Index.
1. *The botanic garden. Poem in two parts, with philosophical notes.* London, 1791. Reprinted, Menston, 1973. The two parts were also published separately, as follows.
1a. ——— Part I. *The economy of vegetation.* London, 1791. 4th ed., 1799.
1b. ——— Part II. *The loves of the plants.* London, 1789. X. 2nd ed., 1790. 3rd ed., 1791. 4th ed., 1794.
2. *Phytologia; or, The philosophy of agriculture and gardening.* London, 1800. RM/L.

11498. KOELPIN, Alexander Bernhard. 1731-1801. Ger. BMNH; BLA & Supp.

11499. LEYSSER, Friedrich Wilhelm von. 1731-1815. Ger. ICB; P I; BMNH. Also Geology.

11500. MIEG, Achilles. 1731-1799. Switz. ICB; BMNH; BLA.

11501. FORSSKÅL, Peter. 1732-1763. Den. DSB; ICB. Under FORSKÅL: P I; BMNH; GA; GB. Also Zoology*.
1. *Flora Aegypto-Arabica; sive, Descriptiones plantarum quas per Aegyptum Inferiorem et Arabiam Felicem detexit.* Copenhagen, 1775. Ed. by C. Niebuhr. RM/L.

11502. GAERTNER, Joseph. 1732-1791. Ger. DSB; BMNH; BLA; GA; GC.
1. *De fructibus et seminibus plantarum.* 3 vols. Stuttgart, 1788-1807. Another ed., Leipzig, 1801-07.

11503. HOLMSKIOLD, Theodor von. 1732-1793. Den. BMNH (HOLM); GA.

11504. JÖRLIN, Engelbert. 1732-1810. Swed. ICB; BMNH.

11505. MUTIS (Y BOSSIO), José Celestino Bruno. 1732-1808. South America. DSB Supp.; ICB; P II; BMNH; BLA; GA.

11506. WINTERL, Jacob Joseph. 1732-1809. Hung. ICB; P II. Also Chemistry.

11507. GERARD, Louis. 1733-1819. Fr. BMNH; GA.

11508. GOUAN, Antoine. 1733-1821. Fr. ICB; BMNH; GA. Also Zoology*.
1. *Flora Monspeliaca.* Lyons, 1765. RM.
2. *Illustrationes et observationes botanicae ad speciarum historiam facientes.* Zurich, 1773. RM.

11509. HUDSON, William. 1733-1793. Br. DSB; BMNH; GA; GD.
1. *Flora Anglica.* London, 1762. 3rd ed., 1798.

11510. KOELREUTER, (Joseph) Gottlieb. 1733-1806. Ger. DSB; ICB; BMNH; BLA; GA; GC.
1. *Vorläufige Nachricht von einigen das Geschlecht der Pflanzen betreffenden Versuchen und Beobachtungen, nebst Fortsetzungen 1,2 und 3.* Leipzig, 1761-66. RM. Reprinted, Leipzig, 1893. (OKEW. 41)

11511. SOLANDER, Daniel Carl. 1733-1782. Swed./Br. DSB; ICB; P II; BMNH; GA; GD. Also Zoology. See also Index.

11512. TODE, Heinrich Julius. 1733-1797. Ger. ICB; BMNH.

11513. WESTON, Richard. 1733-1806. Br. BMNH; GD.
1. *The English flora.* London, 1775. RM. Supplement, 1780. RM.

11514. BURMAN, Nicolaas Laurens. 1734-1793. Holl. ICB; BMNH; BLA Supp.
1. *Flora Indica.* Leiden, 1768.

11515. MARTELLIUS, Nicolaus. 1734?-1829. It. BMNH.

11516. MATTUSCHKA, Heinrich Gottfried von. 1734-1779. Ger. P II; BMNH; GA; GC. Also Astronomy.

11517. WIRSING, Adam Ludwig. 1734-1797. BMNH.

11518. LIGHTFOOT, John. 1735-1788. Br. ICB; BMNH; GA; GD.

11519. MARTYN, Thomas (first of the name). 1735-1825. Br. ICB; BMNH; GA; GD.
1. *Thirty-eight plates, with explanations, intended to illustrate Linnaeus' system of vegetables.* London, 1788. X. Another ed., 1799.
2. *The language of botany, being a dictionary of the terms made use of ... principally by Linnaeus. With ... an attempt to establish significant English terms.* London, 1793. X. 2nd ed., 1796. 3rd ed., 1807.
See also 11382/1b and 11433/2.

11520. WILLEMET, Pierre Remi. 1735-1807. Fr. BMNH; GA.

11521. LA CHENAL, Werner de. 1736-1800. Switz. ICB; BMNH.

11522. MEDICUS, Friedrich Casimir. 1736-1808. Ger. DSB; ICB; P II; BLA; GA.

11523. MERLET DE LA BOULAYE, Gabriel Eléonor. 1736-1807. Fr. BMNH; GA.

11524. ROUCEL, François Antoine. 1736-1831. Belg. BMNH; GF.

11525. BALDINGER, Ernst Gottfried. 1738-1804. Ger. ICB; P I; BMNH; BLA & Supp. Also Biography.

11526. BOLOTOV, Andrei Timofeevich. 1738-1833. Russ. DSB.

11527. DICKSON, James. 1738-1822. Br. BMNH; GA; GD.

11528. LODDIGES, Conrad. 1738?-1826. Br. BMNH.
1. *The botanical cabinet, consisting of coloured delineations of plants from all countries, with a short account of each.* 20 vols. London, 1817-33. The plates by G. Cook.

11529. PLENK, Joseph Jacob von. 1738-1807. Aus./Hung. P II. Under PLENCK: BMNH; BLA; GA. Also Chemistry and Physiology*.

11530. CIRILLO, Domenico. 1739-1799. It. BMNH; BLA & Supp.; GA; GB.

11531. KLUK, Christoph. 1739-1796. Poland. ICB; BMNH; GA.

11532. LYONS, Israel. 1739-1775. Br. P I; GA; GD. Also Mathematics* and Astronomy.

11533. MOSCATI, Pietro. 1739-1824. It. ICB; P II; BMNH; BLA; GA.
1. *Dissertazione sopra una gramigna che nella Lombardia infesta la segale.* Milan, 1772.

11534. SCHREBER, Johann Christian Daniel von. 1739-1810. Ger. ICB; P II; BMNH; BLA; GC. Also Geology and Zoology*. See also 10958/10,11a.

11535. UCRIA, Bernardinus ab. 1739-1796. BMNH.
1. *Hortus regius Panhormitanus.* Palermo, 1789.

11535A AMBODIK, Nestor Maksimovich. 1740-1812. Russ. BMNH.

11536. CELS, Jacques Philippe Martin. 1740-1806. Fr. DSB; ICB; BMNH; GA.

11537. MURRAY, Johan Anders. 1740-1791. Ger. BMNH; BLA; GA. See also 11418/16,16b.

11538. ORTEGA, Casimiro Gomez. 1740-1818. Spain. BMNH; BLA.

11539. THOMAS, Abraham. 1740-1824. Switz. ICB.

11540. WALTER, Thomas. ca. 1740-1789. U.S.A. BMNH; GE.

11541. AMOREUX, Pierre Joseph. 1741-1824. Fr. BMNH; BLA; GA. Also Zoology*.

11542. BELLARDI, Carl Antonio Ludovico. 1741-1826. It. ICB; BMNH; GA.

11543. COYTE, William Beeston. ca. 1741-1810. Br. BMNH; GD.

11544. GILIBERT, Jean Emmanuel. 1741-1814. Fr. ICB; BMNH; BLA; GA.
1. *Exercitia phytologica, quibus omnes plantae Europeae ... ex typo naturae describuntur.* 2 vols. Lyons, 1792.
See also 11418/25.

11545. LINNÉ, Carl von (the younger). 1741-1783. Swed. The son of the great naturalist. ICB (LINNAEUS, Carl, the younger); BMNH; BLA; GA.
1. *Supplementum plantarum "Systematis vegetabilium" editionis decimae tertiae, "Generum plantarum" editionis sextae et "Specierum plantarum" editionis secundae.* Brunswick, 1781. RM/L.

11546. MASSON, Francis. 1741-1805. Br. BMNH; GA; GD.

11547. PALLAS, Peter Simon. 1741-1811. Russ. DSB; ICB; P II; BMNH; BLA; GA; GB; GC. Also Geology*, Natural History*, and Zoology*.
1. *Charakteristik der Thierpflanzen.* Nuremberg, 1787. RM.

11548. ROBSON, Stephen. 1741-1779. Br. BMNH; GD.

11549. SCHKUHR, Christian. 1741-1811. Ger. BMNH.

11550. WITHERING, William (first of the name). 1741-1799. Br. DSB; ICB; P II; BMNH; BLA; GD. See also 8678/1a.
1. *A botanical arrangement of all the vegetables growing in Great Britain.* 2 vols. Birmingham, 1776. 5th ed., 4 vols, 1812. 6th ed., 1818. 7th ed., 1830. Title varies.
1a. ——— Another ed. *A systematic arrangement of British plants ... corrected and condensed.* London, 1830. By W. MacGillivray. X. 6th ed., 1845.

11551. BULLIARD, Pierre. 1742-1793. Fr. BMNH; BLA; GA.
1. *Flora Parisiensis; ou, Descriptions et figures des plantes qui croissent aux environs de Paris.* 6 vols. Paris, 1776-83. RM.
2. *Dictionnaire élémentaire de botanique.* Paris, 1783. RM.

11552. CUTLER, Manasseh. 1742-1823. U.S.A. ICB; GB; GE.

11553. DOMBEY, Joseph. 1742-1794. Fr. DSB; GA.

11554. EHRHART, Friedrich. 1742-1795. Switz. ICB; P I; BMNH; BLA; GA; GC.

11555. ESPER, Eugen Johann Christoph. 1742-1810. Ger. P I; BMNH; GA; GC. Also Zoology*.
1. *Icones fucorum cum characteribus systematicis.* 2 vols. Nuremberg, 1797-1802. X. Reprinted, Weinheim, 1965. (HNC. 50)

11556. HIRSCHFELD, Christian Cajus Lorenz. 1742-1792. Ger. GA; GC.
1. *Theorie der Gartenkunst.* 5 vols. Leipzig, 1779-85. Reprinted, Hildesheim, 1971.

11557. SENEBIER, Jean. 1742-1809. Switz. DSB; ICB; P II; BMNH; BLA; GA; GB. Also Science in General* and Chemistry*. See also Index.
1. *Physiologie végétale.* 5 vols in 2. Geneva, 1800. RM.

11558. STACKHOUSE, John. 1742-1819. Br. BMNH; GD.

11559. VELLOZO, José Mariano da Conceição. 1742-1811. Brazil. DSB; BMNH.

11560. BANKS, Joseph. 1743-1820. Br. DSB; ICB; P I; BMNH; GA; GB; GD. Also Natural History*#. See also Index.
1. *Illustrations of the botany of Captain Cook's voyage round the world in H.M.S. "Endeavour" in 1768-71.* 3 vols. London, British Museum, 1900-05. With D. Solander.

11561. MILNE, Colin. 1743?-1815. Br. BMNH; GD.
1. *A botanical dictionary; or, Elements of systematic and philosophical botany.* London, 1770. 3rd ed., 1805.

11562. REICHARD, Johann Joseph. 1743-1782. Ger. BMNH; BLA; GA; GC.

11563. THUNBERG, Carl Pehr. 1743-1828. Swed. DSB; ICB; P II; BMNH; BLA; GA; GB.
1. *Flora Japonica.* Leipzig, 1784.
2. *Resa uti Europa, Africa, Asia.* 4 vols. Uppsala, 1793-95. X.
2a. ——— [Trans. of an extract] *Le Japon du XVIIIe siècle vu par un botaniste suédois.* Paris, 1966. Trans. by L. Langlès.
3. *Icones plantarum Japonicarum.* 5 parts. Uppsala, 1794-1805.

11564. AVELLAR BROTERO, Felix de. 1744-1828. Port. BMNH; GA (BROTERO).

11565. GMELIN, Samuel Gottlieb. 1744-1774. Russ. P I; BMNH; BLA & Supp.; GA; GC. Also Natural History. See also 11426/1.

11566. LAMARCK, Jean Baptiste Pierre Antoine de Monet, *Chevalier* de. 1744-1829. Fr. DSB; ICB; P I; BMNH; GA; GB. Other entries: see 3204.
1. *Flore françoise.* 3 vols. Paris, 1778. X. 2nd ed., 1795. 3rd ed.: see 11790/2.
2. *Extrait de la "Flore française."* 2 parts. Paris, 1792. RM.
3. *Histoire naturelle des végétaux.* 15 vols. Paris, 1803. RM.

11567. MOENCH, Conrad. 1744-1805. Ger. DSB; P II; BMNH.
1. *Methodus plantas horti botanici et agri Marburgensis a staminum situ describendi.* Marburg, 1794. X. Supplement, 1802. X. Reprinted, 2 vols, Koenigstein, 1966.

11568. WEISS, Friedrich Wilhelm. b. 1744. Ger. BMNH.

11569. BOLTON, James. fl. 1775-1795. Br. ICB; BMNH; GD.

11570. CAVANILLES, Antonio José. 1745-1804. Spain. DSB; BMNH; GA; GB.
1. *Monadelphiae classis dissertationes.* 10 vols in 3. Paris, 1785-89.
2. *Icones et descriptiones plantarum quae ... in Hispania crescunt.* 6 vols in 2. Madrid, 1791-1801. Reprinted, Weinheim, 1965. (HNC. 42)

11571. VILLARS, Dominique. 1745-1814. Fr. P II,III; BMNH; BLA; GA.

11572. VOGEL, Benedict Christian. 1745-1825. Ger. BMNH; BLA.

11573. CURTIS, William. 1746-1799. Br. ICB; BMNH; GA; GD.
1. *Flora Londinensis; or, Plates and descriptions of such plants as grow wild in the environs of London.* 3 vols. London, 1777-87. RM/L.
2. *Lectures on botany.* 3 vols. London, 1803-04. Ed. by S. Curtis.

11574. L'HÉRITIER DE BRUTELLE, Charles Louis. 1746-1800. Fr. DSB; ICB; BMNH; GA (HÉRITIER).
1. *Stirpes novae aut minus cognitae.* 9 parts. Paris, 1784-85 [publ. 1785-1805]. X. Reprinted, Weinheim, 196-?
2. *Sertum Anglicum; seu, Plantae rariores quae in hortis juxta Londinium....* 4 parts. Paris, 1788 [publ. 1789-

92]. X. Reprinted, Pittsburgh, 1963.
3. *Cornis. Specimen botanicum sistens descriptiones et icones Corni minus cognitarum.* Paris, 1788.

11575. MICHAUX, André. 1746-1802. Fr./U.S.A. DSB; ICB; BMNH; GA; GB; GE.
1. *Flora Boreali-Americana.* Paris, 1803. RM. Reprinted, New York, 1973.

11576. POHL, Johann Ehrenfried. 1746-1800. Ger. P II; BMNH; BLA.

11577. VOGLER, Johann Philipp. 1746-1816. Ger. P II; BLA. Also Chemistry.

11578. DUCHESNE, Antoine Nicolas. 1747-1827. Fr. P I; BMNH; GA.

11579. KOHLHAAS, Johann Jacob. 1747-1811. Ger. P I; BLA.

11580. MAYER, Johann Christoph Andreas. 1747-1801. Ger. BMNH; BLA & Supp.; GA. Also Anatomy.
1. *Einheimische Giftgewächse.* 2 vols. Berlin, 1798-1800.

11581. SCHRANK, Franz von Paula von. 1747-1835. Ger. P II; BMNH; GC. Also Zoology*.

11582. THOUIN, André. 1747-1824. Fr. DSB; BMNH; GA.
1. *Monographie des greffes.* [Paris, 1821?]

11583. DRYANDER, Jonas. 1748-1810. Swed./Br. BMNH; GA; GB; GD. See also 11038/1.

11584. GMELIN, Johann Friedrich. 1748-1804. Ger. P I; BMNH; BLA & Supp.; GA; GC. Also Chemistry* and History of Chemistry*. See also 10958/1i.

11585. JUSSIEU, Antoine Laurent de. 1748-1836. Fr. DSB; ICB; BMNH; BLA; GA; GB.
1. *Genera plantarum, secundum ordines naturales disposita.* Paris, 1789. RM/L. Reprinted, Weinheim, 1964. (HNC. 35)
See also 11652/7.

11586. SAINT-AMANS, (Jean) Florimond Boudon de. 1748-1831. Fr. ICB; BMNH; GA.

11587. TRENTEPOHL, Johann Friedrich. 1748-1806. Ger. BMNH; GC.

11588. WEIGEL, Christian Ehrenfried. 1748-1831. Ger. DSB; P II; BMNH; BLA; GC. Also Chemistry*.

11589. GOETHE, Johann Wolfgang von. 1749-1832. Ger. The famous writer. DSB; ICB; P I; BMNH; Mort.; G. Other entries: see 3216.
1. *Versuch die Metamorphose der Pflanzen zu erklären.* Gotha, 1790. RM.
1a. ——— *Saggio sulla metamorfosi delle piante.* Milan, 1842. Trans. by P. Robiati.
1b. ——— *Goethe's botany: "The metamorphosis of plants" (1790) and Tobler's "Ode to nature" (1782).* Waltham, Mass., 1946. Trans. with an introd. by A. Arber.
1c. ——— *La métamorphose des plantes. Introduction, commentaires, notes par Rudolf Steiner (1884).* Paris, 1975. Trans. by H. Bideau.

2. [Trans.] *Botanical writings*. Honolulu, 1952. Trans. by B. Mueller.

11590. HAGEN, Karl Gottfried. 1749-1829. Ger. P I & Supp.; BMNH; BLA; GA; GC. Also Chemistry.

11591. VAHL, Martin Hendriksen. 1749-1804. Den. BMNH; GA.
1. *Ecologae Americanae; seu, Descriptiones plantarum praesertim Americae meridionalis nondum cognitarum*. 3 parts. Copenhagen, 1796-1807.
2. *Enumeratio plantarum, vel ab aliis, vel ab ipso observatarum, cum ... descriptionibus succinctis*. 2 vols. Ib., 1804-06. Another ed., Göttingen, 1827.

11592. HELLENIUS, Carl Niclas. fl. 1776-1795. Swed. BMNH.

11593. BROUGHTON, Arthur. d. 1796. Br. BMNH; GD.

11594. JOLY-CLERC, Nicolas. d. 1817. Fr. BMNH; GA.

11595. AFZELIUS, Adam. 1750-1837. Swed. ICB; BMNH; BLA; GA; GB.
1. *Genera plantarum Guineensium revisa et aucta*. Uppsala, 1804. With N.W. Elgenstierna.
2. [Trans.] *Sierra Leone journal, 1795-1796*. Stockholm, 1967. Ed. by A.P. Kup.

11596. CORREA DA SERRA, José Francisco. 1750-1823. Port. ICB; P I & Supp.; GA; GB.

11597. DESFONTAINES, René Louiche. 1750-1833. Fr. ICB; P I & Supp.; BMNH; GA; GB. Also Zoology*.
1. *Flora Atlantica; sive, Historia plantarum quae in Atlante, agro Tunetano et Algeriensi crescunt*. 2 vols. Paris, 1798-1800.

11598. HUBER, François. 1750-1831. Switz. ICB; P I; BMNH; GA; GB. Also Zoology.
1. *Mémoires sur l'influence de l'air et de diverses substances gazeuses dans la germination de différentes graines*. Geneva, 1801. With J. Senebier.

11599. LUMNITZER, Stephen. 1750-1806. BMNH.
1. *Flora Posoniensis*. Leipzig, 1791. Flora of Posen.

11600. MARUM, Martinus van. 1750-1837. Holl. DSB; ICB; P II; BLA; GB. Also Physics*# and Chemistry.

11601. SPRENGEL, Christian Konrad. 1750-1816. Ger. DSB; ICB; BMNH; GC.
1. *Das entdeckte Geheimniss der Natur im Bau und in der Befruchtung der Blumen*. Berlin, 1793. X. Reprinted in 4 parts, Leipzig, 1894. (OKEW. 48-51) Another reprint, Weinheim, 1971. (HNC. 97)

11602. BERGERET, Jean Pierre. 1751-1813. Fr. BMNH; BLA; GA.

11603. ROXBURGH, William. 1751-1815. Br./India. P II; BMNH; GD.
1. *Plants of the coast of Coromandel*. 3 vols. London, 1795-1819. Ed. by Patrick Russell.
2. *Flora Indica*. 2 vols. Serampore, 1820-24. Ed. by W. Carey. X. Another ed., 3 vols, ib., 1832; reprinted, New Delhi, 1971.

11604. SESSÉ (Y LACASTA), Martin de. 1751?-1808. Mexico. DSB; BMNH.

11605. SUCKOW, Georg Adolph. 1751-1813. Ger. P II; BMNH; GC. Also Chemistry.

11606. WAKEFIELD, *Mrs.* Priscilla (*née* BELL). 1751-1832. Br. BMNH; GD.
1. *An introduction to botany, in a series of familiar letters.* Dublin, 1796. 2nd ed., (place?), 1798. 7th ed., London, 1816.

11607. DUBOIS, François Noel Alexandre. 1752-1824. Fr. BMNH; GA.

11608. PALISOT DE BEAUVOIS, Ambroise Marie François Joseph. 1752-1820. Fr. ICB (BEAUVOIS); BMNH; GA.
1. *Flore d'Oware et de Benin en Afrique.* Paris, 1804-07. X. Reprinted, Weinheim, 196-?
2. *Essai d'une nouvelle agrostographie; ou, Nouveaux genres des Graminées.* Text & atlas. Paris, 1812.

11609. WEBER, Georg Heinrich. 1752-1828. Ger. BMNH; BLA; GC.

11610. WOODVILLE, William. 1752-1805. Br. BMNH; BLA; GA; GD.
1. *Medical botany.* 4 vols. London, 1790-94. 2nd ed., 3 vols, 1810. 3rd ed., rev. by W.J. Hooker and G. Spratt, 5 vols, 1832.

11611. LERCHENFELD, Joseph Raditschnigg von. 1753-1812. Aus. GC.

11612. MUHLENBERG, Henry (or MUEHLENBERG, Gotthilf Heinrich Ernst). 1753-1815. U.S.A. ICB; BMNH; GE. See also 11796/1.

11613. WESTRING, Johann Peter. 1753-1833. Swed. P II; BMNH.

11614. BARON, P. Alexis. b. 1754. Fr. BMNH.

11615. FORSTER, (Johann) Georg (Adam). 1754-1794. Ger. DSB; ICB; P I; BMNH; GA; GB; GD. Also Natural History*#. See also Index.
1. *Dissertatio inauguralis botanico-medica de plantis esculantis oceani australis.* Halle, 1786.
1a. ——— *De plantis esculantis insularum oceani australis commentatio botanica.* Berlin, 1786.
2. *Florulae insularum australium prodromus.* Göttingen, 1786.

11616. LEFEBURE, Louis Henri. 1754-1839. Fr. BMNH; GA.

11617. MENZIES, Archibald. 1754-1842. Br. ICB; BMNH; GD.

11618. PAVÓN (Y JIMÉNEZ), José Antonio. 1754-1840. Spain/South America. DSB Supp.; ICB; BMNH. See also 11620/1,2.

11619. RICHARD, (Louis) Claude (Marie). 1754-1821. Fr. BMNH; GA.
1. *Démonstrations botaniques; ou, Analyse du fruit considéré en général.* Paris, 1808. X.
1a.——— *Observations on the structure of fruits and seeds.* London, 1819. Trans. with notes by J. Lindley.
2. *De Musaceis commentatio botanica.* Bonn, 1831. Completed and published by his son, A. Richard.

11620. RUIZ (LOPEZ), Hipólito. 1754-1816. Spain. DSB; ICB; BMNH.
1. *Florae Peruvianae et Chilensis prodromus.* Madrid, 1794. With José Pavon. Reprinted, Weinheim, 1965. (HNC. 43)
2. *Flora Peruviana et Chilensis.* 4 vols. Madris, 1798-1802. With J. Pavon. Reprinted, Weinheim, 1965. (HNC. 43)

11621. CERVANTES, Vincente. 1755-1829. Mexico. DSB; ICB; BMNH.
  1. *Discurso pronunciado en el Real Jardin Botanico*. Mexico, 1794. RM.

11622. DANCER, Thomas. ca. 1755-1811. Br./West Indies. BMNH; BLA & Supp.; GD.

11623. FIELD, Henry. 1755-1837. Br. BMNH; GD. Also History of Botany*.

11624. KERNER, Johann Simon von. 1755-1830. Ger. BMNH. See also 11485/1a.

11625. LABILLARDIÈRE, Jacques Julien Houten de. 1755-1834. Fr. P I (HOUTON); BMNH; GA. Also Chemistry.
  1. *Icones plantarum Syriae rariorum*. Paris, 1791-1812. X. Reprinted, Weinheim, 1968. (HNC. 60)
  2. *Novae Hollandiae plantarum specimen*. 2 vols. Paris, 1804-06. Reprinted, Weinheim, 1966. (HNC. 45)
  3. *Sertum Austro-Caledonicum*. 2 parts. Paris, 1824-25. Reprinted, Weinheim, 1968. (HNC. 59)

11626. POIRET, Jean Louis Marie. 1755-1834. Fr. BMNH; GA. See also 12628/1a,1b.

11627. RAMOND DE CARBONNIÈRES, Louis François Elizabeth. 1755-1827. Fr. DSB; P II; BMNH; GA. Also Geology.

11628. TARGIONI-TOZZETTI, Ottaviano. 1755-1826. It. P II; BMNH; BLA.

11629. CASTIGLIONI, Luigi. 1756-1832. It. ICB; BMNH.

11630. MARTIUS, Ernst Wilhelm. 1756-1849. Ger. ICB; P II; BMNH. Also Chemistry.
  1. *Neueste Anweisung Pflanzen nach dem Leben abzudrucken*. Wezlar, 1784. Reprinted, Marburg, 1977.

11631. ACHARIUS, Erik. 1757-1819. Swed. DSB; BMNH; BLA; GA; GB.
  1. *Methodus qua omnes detectos Lichenes secundum organa carpomorpha ad genera, species et varietates redigere*. Stockholm, 1803.
  2. *Lichenographia universalis*. Göttingen, 1810.
  3. *Synopsis methodica Lichenum*. Lund, 1814. RM/L.

11632. BOUCHER (DE CRÈVECOEUR), Jules Armand Guillaume. 1757-1844. Fr. BMNH.

11633. IBBETSON, Agnes, *Mrs*. 1757-1823. Br. GA; GD. Plant anatomy and physiology.

11634. KITAIBEL, Pál (or Paul). 1757-1817. Hung. DSB; ICB; P I; BMNH. Also Natural History.

11635. MOCIÑO (or MOZIÑO), José Mariano. 1757-1820. Mexico/Spain. DSB; BMNH.

11636. RÖHLING, Johann Christoph. 1757-1813. BMNH.
  1. *Deutschlands Flora*. Bremen, 1796. X. 2nd ed., 3 parts, Frankfurt, 1812-13. X. Another ed., 5 parts, rev. and enlarged by F.C. Mertens and W.D.J. Koch, 1823-39. RM.

11637. ROTH, Albrecht Wilhelm. 1757-1834. Ger. BMNH; BLA.
  1. *Novae plantarum species, praesertim Indiae orientalis, ex*

collectione ... B. Heynii. Halberstadt, 1821. From the collection of Benjamin Heyne (d. 1819).

11638. SCHUMACHER, (Heinrich) Christian Friedrich. 1757-1830. Den. P II; BMNH; BLA. Also Mineralogy* and Anatomy.

11639. SOWERBY, James. 1757-1822. Br. DSB; P II; BMNH; GB; GD. Also Mineralogy*, Geology*, and Zoology*.
1. *English botany; or, Coloured figures of British plants, with their essential characters*. 36 vols. London, 1790-1814. With notes by J.E. Smith. RM/L.
1a. ——— *Supplement*. 5 vols. Ib., 1831-63. By W.J. Hooker et al.
1b. ——— 3rd ed.: see 12244/1.

11640. THUILLIER, Jean Louis. 1757-1822. Fr. BMNH; GA.

11641. VENTENANT, Etienne Pierre. 1757-1808. Fr. ICB; BMNH; GA.

11642. BAUER, Franz Andreas. 1758-1840. Br. DSB; ICB; BMNH.
1. *Delineations of exotic plants cultivated in the Royal Garden at Kew*. London, 1796. Ed. by W.T. Aiton.

11643. CARRADORI, Gioachimo. 1758-1818. It. P I; BMNH; BLA; GA. Also Physics.

11644. COLLADON, Jean Antoine. 1758-1830. Switz. P I; ICB.

11645. DONN, James. 1758-1813. Br. BMNH; GD.
1. *Hortus Cantabrigiensis; or, A catalogue of plants ... cultivated in the Walkerian Botanic Garden, Cambridge*. Cambridge, 1796. X.
1a. ——— 13th ed. (variant title), London, 1845. Enlarged by P.N. Don.

11646. DUPETIT-THOUARS, (Louis Marie) Aubert. 1758-1831. Fr. BMNH; GA. See also 15478/1.

11647. HEBENSTREIT, Ernst Benjamin Gottlieb. 1758 (or 1753)-1803. Ger. P I; BMNH; BLA; GA. Also Chemistry.

11648. NOCCA, Domenico. 1758-1841. It. BMNH.

11649. SIBTHORP, John. 1758-1796. Br. BMNH; GA; GB; GD.
1. *Flora Graecae prodromus*. 2 vols. London, 1806-13. With J.E. Smith.

11650. KNIGHT, Thomas Andrew. 1759-1838. Gr. DSB; BMNH; GD.
1. [German trans. of six periodical articles (1803-12) on plant physiology] Leipzig, 1895. (OKEW. 62)

11651. REDOUTÉ, Pierre Joseph. 1759-1840. Fr. DSB; ICB; BMNH; GA.
1. *Les roses*. 2 parts in 3 vols. Paris, 1817-24. The text by C.A. Thory. X. 3rd ed., by M. Pirolle, 1828-29.

11652. SMITH, James Edward. 1759-1828. Br. DSB; ICB; BMNH; GA; GD. See also Index.
1. *Plantarum icones hactenus ineditae, plerumque ad plantas in Herbario Linnaeano conservatas delineatae*. 3 vols. London, 1789-91.
2. *A specimen of the botany of New Holland*. Vol. 1. [No more publ.] London, 1793.
3. *Tracts relating to natural history*. London, 1798. RM/L.

4. *Flora Britannica*. 3 vols. London, 1800-04. RM. Another ed., by J.J. Roemer, with notes, 3 vols, Zurich, 1804-05.
5. *Exotic botany ... new, beautiful or rare plants as are worthy of cultivation*. 2 vols. London, 1804-08.
6. *An introduction to physiological and systematic botany*. London, 1807. X. 4th ed., 1819. 5th ed., 1825. 7th ed., by W.J. Hooker, 1833. Another ed., by W. Macgillivray, 1836.
7. *A grammer of botany, illustrative of artificial as well as natural classification, with an explanation of Jussieu's system*. London, 1821. RM. 2nd ed., 1826.
8. *The English flora*. 4 vols. London, 1824-28. X. 2nd ed., 5 vols in 6, 1824-36.
9. *Memoir and correspondence*. London, 1832. Ed. by Lady Smith. RM.

11653. VIBORG, Erik Nissen. 1759-1822. Den. P II; BMNH; BLA; GC.

11654. GRAY, Samuel Frederick. fl. 1780-1836. Br. BMNH; BLA & Supp., GD.
1. *A natural arrangement of British plants ... With an introduction to botany*. 2 vols. London, 1821.

11655. BAUER, Ferdinand Lucas. 1760-1826. Aus./Br. DSB; ICB; BMNH; GA; GC.
1. *Illustrationes florae Novae Hollandiae*. London, 1813. Illustrations of plants described in R. Brown's *Prodromus* (item 11754/1).

See also 12012/1.

11656. HOPPE, David Heinrich. 1760-1846. Ger. BMNH.

11657. SWARTZ, Olof. 1760-1818. Swed. DSB; BMNH; GA; GB.
1. *Nova genera et species plantarum ... quae sub itinere in Indiam occidentalem annis 1783-87 digessit*. Stockholm, 1788. X. Reprinted, Weinheim, 1962. (HNC. 25)
2. *Observationes botanicae, quibus plantae Indiae occidentalis....* Erlangen, 1791.
3. *Flora Indiae occidentalis*. 3 vols. Erlangen, 1797-1806.
4. *Synopsis Filicum*. Kiel, 1806.

11658. ABBOT, Charles. 1761?-1817. Br. BMNH; GD.

11659. BATSCH, August Johann Georg Karl. 1761-1802. Ger. ICB; P I; BMNH; BLA; GA; GC. Also Natural History.
1. *Elenchus Fungorum*. 3 vols. Halle, 1783-89. RM/L.

11660. BRIDEL (-BRIDERI), Samuel Elisée de. 1761-1828. Switz. BMNH; GA.
1. *Bryologia universa*. 2 vols. Leipzig, 1826-27.

11661. BROUSSONET, Pierre Marie Auguste. 1761-1807. Fr. DSB; P I; BMNH; BLA & Supp.; GA; GB. Also Zoology*.

11662. CAREY, William. 1761-1834. India. ICB; BMNH; GA; GB; GD. See also 11603/2.

11663. FORSTER, Thomas Furly. 1761-1825. Br. BMNH; GD.

11664. HAENKE, Thaddaeus. 1761-1817. Prague/South America. ICB; BMNH; GA; GC. See also 11909/1.

11665. HOFFMANN, Georg Franz. 1761-1826 (or 1766-1821). Ger. BMNH; BLA.
1. *Enumeratio Lichenum*. Erlangen, 1784.

11666. HOST, Nicolaus Thomas. 1761-1834. Aus. BMNH; GA; GC.
1. *Flora Austriaca*. 2 vols. Vienna, 1827-31.

11667. HULL, John. 1761-1843. Br. BMNH; GD.

11668. LAMBERT, Aylmer Bourke. 1761-1842. Br. BMNH; GD.
1. *A description of the genus Cinchona*. London, 1797.
2. *A description of the genus Pinus*. 2 vols. London, 1803-04. X. 2nd ed., 1828. Another ed., 1832.

11669. PERSOON, Christiaan Hendrik. 1761-1836. Ger./Fr. DSB; ICB; BMNH; GA.
1. *Observationes mycologicae*. 2 parts. Leipzig, 1796-99. X. Reprinted, East Ardsley (U.K.), 1967.
2. *Icones et descriptiones Fungorum minus cognitorum*. 2 parts. Leipzig, 1798-1800.
3. *Synopsis methodica Fungorum*. Göttingen, 1801. X. Reprinted, New York, 1952.
4. *Synopsis plantarum*. 2 vols. Paris, 1805-07.
5. *Mycologia Europaea*. 3 vols. Erlangen, 1822-28.
See also 11418/16c and 11447/2.

11670. SALISBURY, Richard Anthony. 1761-1829. Br. BMNH; GD.

11671. STERNBERG, Kaspar Maria von. 1761-1838. Prague/Ger. DSB; ICB; P II; BMNH; GC. Also Geology*.

11672. GMELIN, Karl Christian. 1762-1837. Ger. BMNH; GC. Also Zoology.
1. *Flora Badensis Alsatica*. 4 vols. Karlsruhe, 1805-26. Flora of the middle Rhineland.

11673. THORE, Jean. 1762-1823. Fr. ICB; BMNH.

11674. CALEY, George. 1763-1831 (or 1775-1829). Br./Australia. ICB; BMNH; GF.

11675. HAYNE, Friedrich Gottlob. 1763-1832. Ger. BMNH; GA.
1. *Termini botanici ... oder, Botanische Kunstsprache*. 2 vols. Berlin, 1799-1817.

11676. RE, Filippo. 1763-1817. It. ICB; P II; BMNH; GA.

11677. ROEMER, Johann Jakob. 1763-1819. Switz. BMNH; BLA; GC. Also Zoology*. See also 11418/16d and 11652/4.

11678. RUDGE, Edward. 1763-1846. Br. BMNH; GD.

11679. SAGERET, Augustin. 1763-1851. Fr. ICB; BMNH.

11680. VAUCHER, Jean Pierre Etienne. 1763-1841. Switz. DSB; P II; BMNH.

11681. FLOERKE, Heinrich Gustav. 1764-1835. Ger. P I; BMNH. Also Natural History.
1. *De Cladoniis, difficillimo Lichenum genere*. Rostock, 1828.

11682. HABERLE, Karl Constantin. 1764-1832. Ger./Hung. P I; BMNH. Also Mineralogy.

11683. PETIF, Carl. 1764-1845. Ger. ICB; BMNH.

11684. WERNEKINK, Franz. 1764-1839. Ger. GC.

11685. SALISBURY, William. d. 1823. Br. BMNH; GD.

11686. WADE, Walter. d. 1825. Br. BMNH; GD.

11687. BALBIS, Giovanni Battista. 1765-1831. It. BMNH; GA.

11688. BAUMGARTEN, Johann Christian Gottlob. 1765-1843. Ger. BMNH.

11689. BONDT, Nicolas. 1765-1796. Holl. P I; BMNH; BLA. Also Chemistry.

11690. KER, John Bellenden. ca. 1765-1842. Br. BMNH; GD.

11691. KOPS, Jan. 1765-1849. Holl. BMNH; GF.

11692. THÉIS, Alexandre Etienne Guillaume de. 1765-1842. Fr. BMNH; GA.
1. *Glossaire de botanique.* Paris, 1810. RM.

11693. WILLDENOW, Karl Ludwig. 1765-1812. Ger. DSB; BMNH; GA; GC.
1. *Flora Berolinensis prodromus.* Berlin, 1787.
2. *Grundriss der Kräuterkunde.* Berlin, 1792. X.
2a. ——— *The principles of botany and of vegetable physiology.* Edinburgh, 1805.
See also 11418/13 and 11439/1.

11694. AITON, William Townsend. 1766-1849. Br. DSB; BMNH; GD. See also 11495/1 and 11642/1.

11695. BARTON, Benjamin Smith. 1766-1815. U.S.A. DSB; ICB; P I; BMNH; BLA & Supp.; GA; GB; GE. Also Zoology*.

11696. COLLA, Luigi. 1766-1848. It. BMNH.

11697. GAUDIN, Jean François Gottlieb Philippe. 1766-1833. Switz. BMNH.

11698. HOFFMANNSEGG, Johann Centurius. 1766-1849. Ger. GA.

11699. JACQUIN, Joseph Franz von. 1766-1839. Aus. P I; BMNH; BLA. Also Chemistry*.

11700. MOLDENHAWER, Johann Jacob Paul. 1766-1827. Ger. DSB; BMNH.

11701. RUSSELL, John, *6th Duke of Bedford.* 1766-1839. Br. BMNH; GB; GD.

11702. SPRENGEL, Kurt Polycarp Joachim. 1766-1833. Ger. DSB; BMNH; BLA & Supp.; GA; GB; GC. Also History of Botany*.
1. *Florae Halensis tentamen novum.* Halle, 1806. RM/L.
2. *Mantissa florae Halensis.* Halle, 1807. RM.
3. *Observationes botanicae in floram Halensem. Mantissa secunda.* Halle, 1811. RM.
4. *Plantarum minus cognitarum....* 2 parts. Halle, 1813-15.

11703. GEUNS, Stephan Jan van. 1767-1795. Holl. BMNH; BLA; GA.

11704. KNAPP, John Leonard. 1767-1845. Br. BMNH; GD.
1. *The journal of a naturalist.* London, 1829. "A botanical companion to White's *Selborne*." 3rd ed., 1830. 4th ed., 1838.

11705. LINK, Heinrich Friedrich. 1767-1851. Ger. DSB; P I; BMNH; BLA; GA; GC. Also Various Fields.

1. *Enumeratio plantarum Horti Regii Botanici Berolinensis altera*. 2 parts. Berlin, 1821. The first *Enumeratio* was by C.L. Willdenow in 1809.
1a. ——— Another ed. *Hortus Regius Botanicus Berolinensis descriptus*. Berlin, 1827-33.

11706. SAUSSURE, Nicolas Théodore de. 1767-1845. Switz. DSB; P II; BMNH; GA; GB. Also Chemistry (plant biochemistry)*.

11707. SCHRADER, Heinrich Adolph. 1767-1836. Ger. BMNH; GC.

11708. CALDAS, Francisco José de. 1768-1816. South America. DSB; P III; BMNH; GA.

11709. HAWORTH, Adrian Hardy. 1768-1833. Br. DSB; BMNH; GA; GD. Also Zoology*.
1. *Synopsis plantarum succulentarum*. London, 1812. Supplement, 1819.

11710. MARSCHALL VON BIEBERSTEIN, Friedrich August. 1768-1826. Russ. P II; BMNH; GA (BIEBERSTEIN).

11711. THORNTON, Robert John. 1768?-1837. Br. ICB; BMNH; BLA; GD.
1. *A new illustration of the sexual system of C. von Linnaeus*. 3 parts in 2 vols. London, 1799-1807. Another ed., 1807-10.
2. *Botanical extracts; or, Philosophy of botany*. 3 vols. London, 1810.
3. *The British flora*. 5 vols. London, 1812. RM/L.
4. *Elements of botany*. 2 vols. London, 1812.

11712. USTERI, Paul. 1768-1831. Switz. ICB; BMNH; BLA.

11713. ALBERTINI, Johann Baptist von. 1769-1831. BMNH.

11714. DUNCAN, John Shute. 1769-1844. Br. BMNH; GD (under D.; Philip Bury).
1. *Botano-theology*. Oxford, 1825.
2. *Analogies of organized beings*. Oxford, 1831.

11715. GALPINE, John. 1769?-1806. Br. BMNH; GD.
1. *A synoptical compend of British botany*. London, 1806. X. 2nd ed., 1819.

11716. GOMES, Bernardino Antonio. 1769-1823. Port. Mort.; BLA; GA.
1. *Plantas medicinais do Brazil*. São Paulo, 1972. Facsimile reprints of his works, with a biographical introd. and bibliography.

11717. HOSACK, David. 1769-1835. U.S.A. DSB; ICB; P I; BMNH; BLA; GA.

11718. HUMBOLDT, Alexander von. 1769-1859. Ger./Fr. DSB; ICB; P I, III; BMNH; G. Other entries: see 3268.
1. *Aphorismi ex doctrina physiologiae chemicae plantarum*. Berlin, 1793. X.
1a. ——— *Aphorismen aus der chemischen Physiologie der Pflanzen*. Leipzig, 1794. Trans. by G. Fischer. RM.
2. *Ideen zu einer Physiognomik der Gewächse*. Tübingen, 1806. X. Reprinted, Leipzig, 1959. (OKEW. 247)
3. *Essai sur la géographie des plantes*. Paris, 1805. (Part

V of *Voyage aux régions équinoxiales....* See 11091/1) X. Reprinted, London, 1959.
3a. ——— *Ideen zu einer Geographie der Pflanzen.* Tübingen, 1807. Reprinted, Leipzig, 1960. (OKEW. 248) Another reprint, Darmstadt, 1974.
4. *Plantes équinoxiales....* 2 vols. Paris, 1808-17. (Section 1 of Part VI of *Voyage aux régions équinoxiales....* See 11091/1)
5. *Nova genera et species plantarum....* 7 vols. Paris, 1815-25. (Section 3 of Part VI of *Voyage aux régions équinoxiales....* See 11091/1) Reprinted, Weinheim, 1963.
6. *De distributione geographica plantarum.* Paris, 1817. RM.

11719. KEITH, Patrick. 1769-1840. Br. BMNH.
1. *A system of physiological botany.* 2 vols. London, 1816. RM.

11720. MARCET, Jane, *Mrs.* 1769-1858. Br. ICB; P II & Supp. (M., Mary); BMNH; GD. Also Physics* and Chemistry*.
1. *Conversations on vegetable physiology.* London, 1829. X. 3rd ed., 1830.
See also 12629/1.

11721. RAFN, Carl Gottlob. 1769-1808. Den. P II; BMNH.
1. [Trans.] *Entwurf einer Pflanzenphysiologie, auf die neuern Theorien der Physik und Chemie gegründet.* Copenhagen/Leipzig, 1798. Trans. by J.A. Markussen from the Danish original of 1796. RM.

11722. SAVI, Gaetano. 1769-1844. It. BMNH.

11723-11732. (Numbers cancelled)

11733. WINCH, Nathaniel John. 1769-1838. Br. P II; BMNH; GD.

11734. ANDREWS, Henry C. fl. 1799-1828. Br. BMNH; GA; GD.
1. *The botanist's repository, for new and rare plants.* 10 vols. London, 1797-1811.

11735. GRAUMÜLLER, Johann Christian Friedrich. d. 1824. Ger. BMNH.

11736. DUPPA, Richard. 1770-1831. Br. BMNH; GA; GD.
1. *Elements of the science of botany, as established by Linnaeus.* 2 vols. London, 1809. X. 2nd ed., 1809.
2. *The classes and orders of the Linnean system of botany.* 3 vols. London, 1816. RM/L.

11737. HORNEMANN, Jens Wilken. 1770-1841. Den. BMNH; GA.
1. *Hortus Regius Botanicus Hafniensis.* 2 vols. Copenhagen, 1813-15.

11738. MICHAUX, François André. 1770-1855. Fr. ICB; BMNH; GA; GE.
1. *Histoire des arbres forestiers de l'Amérique septentrionale.* 3 vols. Paris, 1810-13. RM.
1a. ——— *The North American "Sylva"; or, Description of the forest trees of the United States, Canada and Nova Scotia.* Paris, 1819. RM.

11739. RADDI, Giuseppe. 1770-1829. It. ICB; BMNH; GA.

11740. ELLIOTT, Stephen. 1771-1830. U.S.A. ICB; BMNH; GE.

1. *A sketch of the botany of South Carolina and Georgia.* 2 vols. Charleston, 1821-24. RM. Reprinted, New York, 1971.

11741. FUNCK, Heinrich Christian. 1771-1839. Ger. BMNH; GC.
1. *Cryptogamische Gewächse des Fichtelgebirgs.* 42 parts. Leipzig, 1800-38.

11742. KOCH, Wilhelm Daniel Joseph. 1771-1849. Ger. BMNH; GA; GC.
1. *Synopsis florae Germanicae et Helveticae.* 2 vols. Frankfurt, 1836-37. X. 2nd ed., 3 vols, Leipzig, 1843-45.
See also 11636/1.

11743. RUDOLPHI, Karl Asmund. 1771-1832. Ger. DSB; BMNH; BLA & Supp.; GA; GC. Also Zoology and Physiology.

11744. WIEGMANN, A.J.Fr. 1771-1853. Ger. P II; BMNH.

11745. BILLERBECK, (Heinrich Ludwig) Julius. 1772-1838. BMNH.
1. *Flora classica.* Leipzig, 1824.

11746. BIROLI, Giovanni. 1772-1825. It. BMNH; GA.

11747. GAERTNER, Karl Friedrich von. 1772-1850. Ger. DSB; BMNH; BLA & Supp.; GC.

11748. HEDWIG, Romanus Adolf. 1772-1806. Ger. BMNH; BLA.

11749. JAUME SAINT-HILAIRE, Jean Henri. 1772-1845. Fr. BMNH; GA.
1. *Plantes de la France.* 10 vols. Paris, 1805-22. RM.

11750. MIELICHHOFER, Mathias. 1772-1847. Aus. P III. Also Mineralogy.

11751. ROQUES, Joseph. 1772-1850. Fr. BMNH; BLA.

11752. VIVIANI, Domenico. 1772-1840. It. P II; BMNH.

11753. BONPLAND, Aimé Jacques Alexandre Goujaud. 1773-1858. Fr. ICB; BMNH; GA; GB. See also 11091/1 and 11866/1.

11754. BROWN, Robert (of the British Museum). 1773-1858. Br. DSB; ICB; P III; BMNH; GA; GB; GD.
1. *Prodromus florae Novae Hollandiae.* Vol. 1. [No more publ.] London, 1810. Reprinted, Leipzig, 1821. Another reprint, Weinheim, 1960. (HNC. 6)
1a. ——— 2nd ed., Nuremberg, 1827. Ed. by C.G. Nees von Esenbeck.
1b. ——— *Supplementum primum.* London, 1830. Reprinted, Weinheim, 1960. (HNC. 6)
2. *General remarks, geographical and systematical, on the botany of Terra Australis.* London, 1814.
3. *Vermischte botanische Schriften.* 5 vols. Nuremberg, 1825-34. Trans. and ed. by C.G. Nees von Esenbeck et al.
4. *The miscellaneous botanical works.* 3 vols. London, 1866-68. (Ray Society publns. 39,43,44) RM/L.
See also 11655/1.

11755. DELESSERT, Benjamin. 1773-1847. Fr. BMNH.
1. *Icones selectae plantarum.* 5 vols. Paris, 1820-46. The plant descriptions are by A.P. de Candolle. RM (under CANDOLLE)/L.
2. *Voyages dans les deux océanes, Atlantique et Pacifique, 1844-47.* Paris, 1848.

3. *Musée botanique de M. Benjamin Delessert. Notices sur les collections de plantes et la bibliothèque qui le composent.* Paris, 1845. By Antoine Lasèque. Reprinted, Weinheim, 1970. (HNC. 81) Listed under LASÈQUE.

See also 12746/4b.

11756. FORBES, James. 1773-1861. Br. BMNH.

11757. HOOPER, Robert. 1773-1835. Br. BMNH; BLA; GD. Also Anatomy*.

11758. HORSFIELD, Thomas. 1773-1859. Br./Java. P III; BMNH; GD; GE. Also Zoology*.

11759. LESCHENAULT DE LA TOUR, (Jean Baptiste) Louis Théodor. 1773-1826. Fr. P I; BMNH; GA.

11760. RE, Giovanni Francesco. 1773-1833. It. BMNH; BLA; GA.

11761. REINWARDT, Kaspar Georg Karl. 1773-1854. Holl. P II; BMNH; GC.

11762. SALM-REIFFERSCHEID-DYCK, Joseph. 1773-1861. Ger. BMNH; GC.

11763. BERNHARDI, Johann Jacob. 1774-1850. Ger. P I & Supp.; BMNH; BLA & Supp.; GA; GC. Also Mineralogy.

11764. BINGLEY, William. 1774-1823. Br. BMNH; GA; GD. Also Zoology*.

1. *A practical introduction to botany.* London, 1817. X. 3rd ed., 1831.

11765. BIVONA-BERNARDI, Antonio. 1774-1837. It. BMNH.

1. *Sicularum plantarum....* 2 vols. Palermo, 1806-07.

11766. LOISELEUR-DESLONGCHAMPS, Jean Louis Auguste. 1774-1849. Fr. BMNH; BLA; GA.

11767. PURSH, Frederick. 1774-1820. U.S.A. DSB; ICB; BMNH; GE.

1. *Flora Americae septentrionalis; or, A systematic arrangement and description of the plants of North America.* 2 vols. London, 1814. RM/L. 2nd ed., 1816.
2. *Journal of a botanical excursion in the northeastern parts of the states of Pennsylvania and New York during 1807.* Philadelphia, 1869. X. Reprinted, Port Washington, N.Y., 1969.

11768. WENDEROTH, Georg Wilhelm Franz. 1774-1861. Ger. BMNH; GC.

11769. BERTOLONI, Antonio. 1775-1869. It. BMNH; BLA.

11770. CHAUMETON, François Pierre. 1775-1819. Fr. BMNH; BLA & Supp.

1. *Flore médicale.* 7 vols. Paris, 1814-20. X. Another ed., 1828-35.

11771. GALLIZIOLI, Filippo. 1775-1844. BMNH.

1. *Elementi botanico-agrari.* 4 vols. Florence, 1809-12. RM.

11772. KICKX, Jean (first of the name). 1775-1831. Belg. P I (KICKS); BMNH; GA. Also Mineralogy.

11773. MACKAY, James Townsend. ca. 1775-1862. Br. BMNH; GD.

11774. PHILLIPS, Henry. 1775-1838. Br. BMNH; GD.

11775. SCHWÄGRICHEN, (Christian) Friedrich. 1775-1853. ICB; BMNH.

See also 11492/4.

11776. TURNER, Dawson. 1775-1858. Br. ICB; BMNH; GD.

11777. TURPIN, Pierre Jean François. 1775-1840. Fr. DSB; BMNH; GA.

11778. VROLIK, Gerard. 1775-1859. Holl. P II; BMNH; BLA.

11779. WITHERING, William (second of the name). 1775-1832. Br. BMNH.

11780. ALLMAN, William. 1776-1846. Br. P III; BMNH; GD.

11781. DUTROCHET, (René Joaquim) Henri. 1776-1847. Fr. DSB; ICB; P I; BMNH; Mort.; BLA; GA; GB. Also Physiology.
1. *Recherches anatomiques et physiologiques sur la structure intime des animaux et des végétaux, et sur leur motilité*. Paris, 1824. RM.
1a. ——— German trans. Leipzig, 1906. (OKEW. 154)
2. *L'agent immédiat du mouvement vital dévoilé dans sa nature et dans son mode d'action, chez les végétaux et chez les animaux*. Paris, 1826. RM.
3. *Nouvelles recherches sur l'endosmose et l'exosmose, suivies de l'application expérimentale de ces actions physiques à la solution du problème de l'irritabilité végétale et à la détermination de la cause de l'ascension des tiges et de la descente des racines*. Paris, 1828. RM/L.
4. *Mémoires pour servir à l'histoire anatomique et physiologique des végétaux et des animaux*. 2 vols & atlas. Paris, 1837. RM.

11782. EATON, Amos. 1776-1842. U.S.A. DSB; ICB; P I; BMNH; GE. Also Geology*.

11783. LAMOUROUX, Jean Vincent Félix. 1776-1825. Fr. DSB; ICB; P I & Supp.; BMNH; GA.

11784. MIRBEL, Charles François Brisseau de. 1776-1854. Fr. DSB; BMNH; GA.
1. *Traité d'anatomie et de physiologie végétales*. Paris, 1802. (Suites à Buffon) RM.
2. *Histoire naturelle, générale et particulière, des plantes*. 18 vols. Paris, 1802-06. By Mirbel et al. (Suites à Buffon) RM.
3. *Elemens de physiologie végétale et de botanique*. 3 vols. Paris, 1815. RM.

11785. NEES VON ESENBECK, Christian Gottfried (Daniel). 1776-1858. Ger. DSB; ICB; BMNH; BLA; GA; GC. See also Index.

11786. SERINGE, Nicolas Charles. 1776-1858. Switz. BMNH.

11787. WOODS, Joseph. 1776-1864. Br. BMNH; GD.

11788. MARQUIS, Alexandre Louis. 1777-1828. Fr. BMNH; GA.

11789. BERTANI, Pellegrino. 1778-1822. It. BMNH.

11790. CANDOLLE, Augustin Pyramus de. 1778-1841. Switz. DSB; ICB; P I (DECANDOLLE); BMNH; GA; GB. Also Autobiography*.
1. *Plantarun succulentarum historia; ou, Histoire naturelle des plantes grasses*. 4 vols. Paris, 1799-1829.
2. *Synopsis plantarum in "Flora gallica" descriptarum. Auctoribus J.B. de Lamarck et A.P. de Candolle*. 4 vols. Paris, 1805. A complete revision of Lamarck's *Flore française* (item 11566/1); sometimes listed as its 3rd ed. RM/L. 2nd ed., 5 vols, 1815.

2a. ——— Another ed. *Botanicon gallicum; seu, Synopsis plantarum in "Flora gallica" descriptarum.* 2 vols. Paris, 1828-30. Ed. by J.E. Duby. RM/L. (Generally listed under Candolle, but under Lamarck in RM.)
3. *Théorie élémentaire de la botanique.* Paris, 1813. RM.
4. *Regni vegetabilis systema naturale.* 2 vols. Paris, 1818-21. RM/L.
5. *Grundzüge der wissenschaftlichen Pflanzenkunde.* 1820. With K. Sprengel. X.
5a. ——— *Elements of the philosophy of plants, containing the principles of scientific botany.* Edinburgh, 1821. RM/L.
6. *Notices sur les plantes rares cultivées dans le Jardin Botanique de Genève.* 10 parts in 1 vol. Geneva, 1823-47. With his son, Alphonse de Candolle.
7. *Prodromus systematis naturalis regni vegetabilis.* 17 parts in 21 vols. Paris, 1824-73. Continued after his death by Alphonse de Candolle et al. For a successor to this work see 12039/5. For an index see 11924/1. RM/L.
8. *Mémoires sur la famille des Légumineuses.* Paris, 1825. RM/L. Reprinted, Weinheim, 1966. (HNC. 44)
9. *Revue de la famille des Lythraires.* Geneva, 1826.
10. *Organographie végétale; ou, Description raisonnée des organes des plantes. (Cours de botanique, lère partie)* 2 vols. Paris, 1827. RM/L.
10a. ——— *Vegetable organography; or, An analytical description of the organs of plants.* 2 vols. London, 1839-40. Trans. by B. Kingdom. 2nd ed., 1841.
11. *Collection de mémoires pour servir à l'histoire du règne végétal.* Paris, 1828-38. RM. Reprinted, Weinheim, 1971. (HNC. 88)
12. *Histoire de la botanique genevoise.* Geneva, 1830. A lecture. RM.
13. *Physiologie vegetale. (Cours de botanique, 2e partie)* 3 vols. Paris, 1832. RM/L.
14. *Mémoire sur la famille des Myrtacées.* Geneva, 1842.

See also 11755/1 and 11924/1.

11791. HERBERT, William. 1778-1847. Br. DSB; ICB; BMNH; GD.
1. *Amaryllidaceae. Preceded by an attempt to arrange the monocotyledonous orders, and followed by a treatise on cross-bred vegetables, and Supplement.* London, 1837. Reprinted, Weinheim, 1970.

11792. RAFFENEAU-DELILE, Alire. 1778-1850. Fr. DSB (DELILE); BMNH; BLA; GA.

11793. SCHELVER, Friedrich Joseph. 1778-1832. Ger. P II; BMNH; BLA.

11794. THOMSON, Anthony Todd. 1778-1849. Br. P III; BMNH; BLA; GD.
1. *Lectures on the elements of botany.* Part 1. [No more publ.] London, 1822. Another ed., 1827.

11795. TRINIUS, Carl Bernhard. 1778-1844. Russ. BMNH; BLA; GC.
1. *Species Graminum.* 3 vols. St. Petersburg, 1828-36. Reprinted, Weinheim, 1970.

11796. BALDWIN, William. 1779-1819. U.S.A. BMNH; GE.

1. *Reliquiae Baldwinianae. Selections from the correspondence.* Philadelphia, 1843. Compiled by W. Darlington. Consists mainly of Baldwin's correspondence with H. Muhlenberg and W. Darlington. X. Reprinted, New York, 1969. (Classica botanica Americana. Supp. 2)

11797. CURTIS, Samuel. 1779-1860. Br. BMNH; GD. See also 11573/2.

11798. KIESER, Dietrich Georg. 1779-1862. Ger. BLA; BMNH; GA. Also Biology in General.

11799. LANDSBOROUGH, David. 1779-1854. Br. BMNH; GD.

11800. LEJEUNE, Alexandre Louis Simon. 1779-1858. Belg. BMNH; GF.
1. *Compendium florae Belgicae.* 3 vols. Liege, 1828-36. With R. Courtois.

11801. MARZARI-PENCATI, Giuseppe. 1779-1836. It. P II; BMNH; GA. Also Geology.

11802. SAINT-HILAIRE, Auguste (François César Prouvençal) de. 1779-1853. Fr. DSB; ICB; BMNH. Under PROUVENÇAL DE S.-H.: GA; GB.

11803. TREVIRANUS, Ludolph Christian. 1779-1864. Ger. DSB; BMNH; GC.
1. *Physiologie der Gewächse.* 2 vols. Bonn, 1835-38. RM.

11804. KAULFUSS, Georg Friedrich. d. 1830. BMNH.

11805. FÉBURIER, C. Romain. fl. 1812-1825. Fr. BMNH.

11806. GAUDICHAUD-BEAUPRÉ, Charles. 1780-1854. Fr. BMNH; GA; GB.
1. *Voyage autour du monde ... sur les corvettes "l'Uranie" et la "Physicienne" ... 1817-20. Botanique.* Text & atlas. Paris, 1826-30.
2. *Voyage autour du monde ... 1836 et 1837 sur la corvette la "Bonité" ... Botanique.* Paris, 1844-46.

11807. MERAT (DE VAUMARTOISE), François Victor. 1780-1851. Fr. BMNH; BLA; GA.

11808. REUM, Johann Adam. 1780-1839. Ger. BMNH; GC.

11809. SCHWEINITZ, Lewis David von. 1780-1834. U.S.A. BMNH; GE.
1. *Synopsis Fungorum in America Boreali media degentium.* [Periodical article, 1834] Weinheim, 1962. (HNC. 24)

11810. TENORE, Michele. 1780-1861. It. BMNH.

11811. WAHLENBERG, Göran (or George). 1780-1851. Swed. DSB; P II; BMNH; BLA.
1. *Flora Lapponica.* Berlin, 1812. RM.
2. *Flora Carpatorum principalium.* Göttingen, 1814. RM.

11812. AUNIER, Jean Juste Noel Antoine. 1781-1859. BMNH.

11813. CASSINI, (Alexandre) Henri (Gabriel) de. 1781-1832. Fr. BMNH; GA.
1. *Cassini on Compositae: Collected from the "Dictionnaire des sciences naturelles."* 3 vols. New York, 1975. Ed. with introd. and index by R.M. King and H.W. Dawson.

11814. CHAMISSO, Adelbert (Louis Charles Adélaïde) von. 1781-1838. Ger. DSB Supp.; ICB; P I; BMNH; GA; GB; GC.

11815. LINDENBERG, Johann Bernhard Wilhelm. 1781-1851. Ger. BMNH; GC.
1. *Synopsis Hepaticarum Europaeum.* Bonn, 1829.
2. *Species Hepaticarum.* 3 vols in 1. Bonn, 1839-51. With C.M. Gottsche.
See also 12069/1.

11816. STEVEN, Christian von. 1781-1864. Russ. ICB; BMNH.

11817. DARLINGTON, William. 1782-1863. U.S.A. DSB; ICB; BMNH; BLA & Supp.; GE. Also Biography*. See also 11796/1.

11818. FICINUS, Heinrich David August. 1782-1857. Ger. P I; BMNH; BLA. Also Chemistry.

11819. FISCHER, Friedrich Ernst Ludwig von. 1782-1854. Russ. BMNH.
1. *Sertum Petropolitanum; seu, Icones et descriptiones plantarum quae in Horto Botanico Imperiali Petropolitano floruerunt.* St. Petersburg, 1846. With C.A. Meyer.

11820. HOBSON, Edward. 1782-1830. Br. GD.

11821. LIBERT, Marie Anne. 1782-1865. Belg. GF.

11822. MEYER, Georg Friedrich Wilhelm. 1782-1856. Ger. P II; BMNH.

11823. MORETTI, Giuseppe. 1782-1853. It. P II; BMNH.

11824. POHL, Johann Emmanuel. 1782-1834. Aus. ICB; P II; BMNH; GA; GC. Also Geology.

11825. POLLINI, Ciro. 1782-1833. It. BMNH.
1. *Flora Veronensis.* 3 vols. Verona, 1822-24.

11826. ROHDE, Michael. 1782-1812. Ger. BMNH; GC.

11827. SEBASTIANI, Antonio. 1782-1821. It. BMNH.
1. *Florae Romanae prodromus.* Rome, 1818.

11828. BERTA, Tommaso Luigi. 1783-1845. It. ICB; BMNH.

11829. CASSEL, Franz Peter. 1783-1821. Ger. BMNH; GA.

11830. DRUMMOND, James Lawson. 1783-1853. Br. BMNH; BLA & Supp.; GD. Also Natural History.
1. *First steps to botany.* London, 1823. X. 2nd ed., 1826.

11831. HUNDESHAGEN, Johann Christian. 1783-1834. Ger. P I; BMNH; GA; GC.

11832. LOUDON, John Claudius. 1783-1843. Br. BMNH; GA; GD.
1. *An encyclopaedia of gardening.* London, 1822. 2nd ed., 1830. 3rd ed., 1835.
2. *An encyclopaedia of plants.* Ib., 1829. 1st supplement, 1841. 2nd ed., 1841. Other eds: 1855, 1866, 1872, 1880.
3. *Hortus Britannicus. A catalogue of all the plants ... in ... Britain. Part I. The Linnean arrangement ... Part II. The Jussieuean arrangement.* Ib., 1830. X. Another ed., 1850.
4. *Arboretum et fruticetum Britannicum; or, The trees and shrubs of Britain, native and foreign.* 8 vols. Ib., 1838. Another ed., 1844. 2nd ed., 1854.

11833. OTTO, Christoph Friedrich. 1783-1856. Ger. BMNH.

11834. RAFINESQUE, Constantine Samuel. 1783-1840. It./U.S.A. DSB; ICB; BMNH; GE. Also Natural History*, Zoology*, and Autobiography*.
1. *Florula Ludoviciana; or, A flora of the State of Louisiana. Translated, revised and improved from the French of C.C. Robin*. New York, 1817. X. Reprinted, ib., 1967. (Classica botanica Americana. 5)
2. *Medical flora; or, Manual of the medical botany of the United States*. Philadelphia, 1828. RM.
3. *New flora and botany of North America*. 4 parts. Philadelphia, 1836. RM.

11835. STEUDEL, Ernst Gottlieb. 1783 (or 1780) -1856. Ger. BMNH; BLA.
1. *Nomenclator botanicus, enumerans ordine alphabetico nomina atque synonyma, tum generica tum specifica, et a Linnaeo et recentioribus de re botanica scriptoribus plantis phanerogamis imposita*. Stuttgart, 1821. RM/L.
1a. ——— 2nd ed. *Nomenclator botanicus; seu, Synonymia plantarum universalis*. 2 parts. Ib., 1840-41.

11836. SWEET, Robert. 1783-1835. Br. BMNH; GD.
1. *Geraniaceae: The natural order of Gerania*. 5 vols. London, 1820-30.
2. *Cistineae: The natural order of Cistus, or rock-rose*. Ib., 1825-30.
3. *Hortus Britannicus; or, A catalogue of plants cultivated in the gardens of Great Britain*. Ib., 1826-27. X. 2nd ed., 1830. 3rd ed., by G. Don, 1839.
4. *Flora Australasica; or, A selection of ... plants, natives of New Holland and the South Sea islands*. Ib., 1827-28.

11837. BASTARD (or BATARD), Toussaint. 1784-1846. Fr. BMNH.

11838. BEHLEN, Stephan. 1784-1847. Ger. BMNH.

11839. BERLESE, Lorenzo. 1784-1863. BMNH.
1. *Monographie du genre Camellia*. Paris, 1837. X.
1a. ——— *Monograph of the genus Camellia*. Boston, 1838. Trans. by H.A.S. Dearborn.

11840. BESSER, Willibald von. 1784-1842. ICB; BMNH.

11841. DESVAUX, Auguste Nicaise. 1784-1856. Fr. BMNH.

11842. DEWEY, Chester. 1784-1867. U.S.A. P III; BMNH; GE. Also Geology.

11843. MONTAGNE, (Jean François) Camille. 1784-1866. Fr. BMNH; GA.
1. *Voyage au Pôle Sud et dans l'Océanie sur les corvettes "l'Astrolabe" et la "Zélée" ... 1837-40 ... Botanique. Plantes cellulaires*. Paris, 1845.

11844. RAU, Ambrosius. 1784-1830. Ger. P II; BMNH. Also Mineralogy.

11845. TAYLOR, Thomas. d. 1848. Br. BMNH; GD.

11846. AFZELIUS, Arvid August. 1785-1871. Swed. BMNH; GA; GB.

11847. AGARDH, Carl Adolf. 1785-1859. Swed. DSB; ICB; P III; BMNH; BLA; GA.
1. *Synopsis Algarum Scandinaviae*. Lund, 1817. RM.
2. *Icones Algarum ineditae*. 2 vols. Lund, 1820-21. Another ed., 1846.

3. *Species Algarum.* 2 vols. Greifswald, 1823-28. X. Reprinted, Amsterdam, 1969.
4. *Systema Algarum.* Lund, 1824.
5. *Essai de réduire la physiologie végétale à des principes fondamentaux.* Lund, 1828.

11848. HOOKER, William Jackson. 1785-1865. Br. DSB; ICB; BMNH; GA; GB; GD. See also Index.

1. *British Jungermanniae.* London, 1816.
2. *Musci exotici.* 2 vols. Ib., 1818-20. In English.
3. *Muscologia Britannica.* Ib., 1818. In English. X. 2nd ed., 1827.
4. *Flora Scotia; or, A description of Scottish plants.* 2 vols. Ib., 1821. RM/L.
5. *Exotic flora.* 3 vols. Edinburgh, 1823-27.
6. (Entry cancelled)
7. *Flora boreali-Americana; or, The botany of the northern parts of British America.* 2 vols. London, 1829-40. RM/L. Reprinted, Weinheim, 1960. (HNC. 5)
8. *The British flora.* Ib., 1830. 6th ed., 1850. 8th ed., 1860.
9. *The botany of Captain Beechey's voyage ... to the Pacific and Bering's Strait performed in H.M.S. "Blossom" ... 1825-28.* 9 parts. Ib., 1831. With G.A.W. Arnott. Reprinted, Weinheim, 1965. (HNC. 39)
10. *Icones plantarum; or, Figures ... of new or rare plants.* Series 1 and 2. Vols 1-10 in 2 vols. Ib., 1837-54. Reprinted, Weinheim, 1965. (HNC)

10a. ——— Series 3. Vols 11-20 in 2 vols. Ib., 1867-91. Reprinted, Weinheim, 1965. (HNC)

11. *Genera Filicum.* Ib., 1842. In English.
12. *Notes on the botany of the Antarctic voyage ... in H.M. Discovery Ships "Erebus" and "Terror".* Ib., 1843. cf. item 12179/2.
13. *Species Filicum.* 5 vols. Ib., 1846-64. Reprinted, Weinheim, 1970. (HNC. 82)
14. *Niger flora; or, An enumeration of the plants of western tropical Africa.* Ib., 1849. Reprinted, Weinheim, 1966. (HNC. 51)
15. *A manual of scientific enquiry, prepared for the use of Her Majesty's Navy and ... travellers in general. Botany.* Ib., 1849. X. 4th ed., rev. by J.D. Hooker, 1871. X. Other eds : 1878,1881,1894,1897. cf. items 2919/2,2a.
16. *A century of ferns.* Ib., 1854. A re-issue of Vol. X of *Icones plantarum* (item 10).
17. *Filices exoticae.* Ib., 1859. In English.
18. *A second century of ferns.* Ib., 1861.
19. *The British ferns.* Ib., 1861. RM.
20. *Synopsis filicum.* Ib., 1868. In English. 2nd ed., 1874.

11849. HOPKIRK, Thomas. 1785-1841. Br. BMNH; GD.

11850. LEDEBOUR, Karl Friedrich von. 1785-1851. Ger./Dorpat. P I; BMNH.

1. *Icones plantarum novarum ... floram Rossicam....* 5 vols in 1. Riga, 1829-34. X. Reprinted, Weinheim, 1965. (HNC. 70)

11851. SMITH, Christian. 1785-1816. Nor. P II; BMNH.

11852. BIGELOW, Jacob. 1786-1879. U.S.A. BMNH; Mort.; BLA & Supp.; GE.
1. *Florula Bostoniensis. A collection of plants of Boston and its environs.* Boston, 1814. RM.
2. *American medical botany.* 3 vols in 1. Ib., 1817-20. RM.
See also 15475/2.

11853. GAY, Jacques. 1786-1864. Fr. BMNH.

11854. NUTTALL, Thomas. 1786-1859. Br./U.S.A. DSB; BMNH; GB; GD; GE. Also Natural History* and Zoology.
1. *The genera of North American plants, and a catalogue of the species.* 2 vols. Philadelphia, 1818. RM. Reprinted, New York, 1971.
2. *An introduction to systematic and physiological botany.* Cambridge, Mass., 1827. RM.

11855. SINCLAIR, George. 1786-1834. Br. BMNH; GD.

11856. WALLICH, Nathanael. 1786-1854. India. BMNH; GD; GF.

11857. BARTON, William Paul Crillon. 1787-1856. U.S.A. ICB; BMNH; Mort.; BLA & Supp.; GE.
1. *Compendium florae Philadelphicae.* Philadelphia, 1818. In English. RM.
2. *A flora of North America.* 3 vols. Ib., 1821-23. RM.

11858. BAXTER, William. 1787-1871. Br. BMNH; GD.

11859. HOCHSTETTER, Christian Ferdinand. 1787-1860. BMNH.

11860. LASCH, Wilhelm Gottfried. 1787-1863. Ger. GC.

11861. NEES VON ESENBECK, Theodor Friedrich Ludwig. 1787-1837. Ger. P II; BMNH; BLA; GA; GC. Also Chemistry.
1. *Genera plantarum florae Germanicae.* 7 vols in 4. Bonn, 1835-54.

11862. OPIZ, Philipp Maximilian. 1787-1858. Prague. ICB; BMNH; GC.

11863. SCHÜBLER, Gustav. 1787-1834. Ger. P II; BMNH; BLA. Also Meteorology.

11864. SPORLEDER, Friedrich Wilhelm. 1787-1875. Ger. GC.

11865. DIERBACH, Johann Heinrich. 1788-1845. Ger. BMNH; BLA & Supp. Also History of Botany*.

11866. KUNTH, Karl Sigismund. 1788-1850. Ger. DSB Supp.; ICB; BMNH; GA; GC.
1. *Synopsis plantarum quas, in itinere ad plagam aequinoctialem orbis novi, collegerunt Al. de Humboldt et Am. Bonpland.* 4 vols. Paris, 1822-25. RM.
2. *Enumeratio plantarum omnium hucusque cognitarum, secundum familias naturales.* 5 vols & supp. Stuttgart, 1833-50. RM/L.

11867. MARTENS, Georg Matthias von. 1788-1872. Ger. BMNH; GC.

11868. BARTON, John. 1789-1852. Br. BMNH.
1. *A lecture on the geography of plants.* London, 1827. RM/L.

11869. BLYTT, Mathias Numsen. 1789-1862. Nor. BMNH.

11870. COMSTOCK, John Lee. 1789-1858. U.S.A. P III; BMNH.

11871. DUNAL, (Michel) Félix. 1789-1856. Fr. BMNH.
1. *Histoire naturelle, médicale et économique, des Solanum.* Paris, 1813.

11872. FÉE, Antoine Laurent Apollinaire. 1789-1874. Fr. DSB; BMNH; GA.
1. *Essai sur les cryptogames des écorces exotiques officinales.* Paris, 1824.
1a. ——— Part 2. *Supplément et révision.* Strasbourg, 1837.
2. *Mémoires sur la famille des fougères.* 11 parts in 2 vols. Strasbourg, 1844-66. X. Reprinted, Weinheim, 1966. (HNC. 52)

11873. HEGETSCHWEILER, Johann Heinrich. 1789-1839. Switz. BMNH; BLA; GA; GC.

11874. MIERS, John. 1789-1879. Br. BMNH; GD.

11875. SCHÖNHEIT, Friedrich Christian Heinrich. 1789-1870. Ger. BMNH; GC.

11876. SCHOUW, Joakim Frederik. 1789-1852. Den. DSB; P II; BMNH. Also Climatology.
1. [Trans.] *The earth, plants and man.* London, 1852. Another ed., 1859.

11877. SIEBER, Franz Wilhelm. 1789 (or 1785) -1844. Several countries. ICB; BMNH; BLA; GC.

11878. WIKSTRÖM, Johan Emanuel. 1789-1856. Swed. BMNH.
1. *Conspectus litteraturae botanicae in Suecia ... notis bibliographicis et biographicis auctorum adjectis.* Stockholm. 1831. RM.

11879. GREEN, Jacob. 1790-1841. U.S.A. DSB; ICB; P III; BMNH; GE. Also Various Fields.
1. *An address on the botany of the United States.* Albany, N.Y., 1814. RM.

11880. HARTMAN, Carl Johan. 1790-1849. Swed. BMNH; BLA; GA.

11881. HENSCHEL, August Wilhelm Eduard Theodor. 1790-1856. Ger. BMNH; BLA. Also Biography* and History of Botany.

11882. MAUND, Benjamin. 1790-1863. Br. BMNH; GD.
1. *The botanic garden. Consisting of ... representations of hardy ornamental flowering plants.* 13 vols. London, 1825-51.

11883. ALSCHINGER, Andreas. 1791-1863. BMNH.

11884. ASPEGREN, Gustav Casten. 1791-1828. Swed. ICB; BMNH.

11885. CUNNINGHAM, Allan. 1791-1839. Australia. ICB; BMNH; GD. See also 15573/1.

11886. HERBICH, Franz. 1791-1865. Cracow. BMNH; BLA & Supp.

11887. JOHNSON, Charles. 1791-1880. Br. BMNH; GD.

11888. MEYER, Ernst Heinrich Friedrich. 1791-1858. Ger. P II; BMNH; GC. Also History of Botany*.

11889. NOLTE, Ernst Ferdinand. 1791-1875. Ger. BMNH; GC.

11890. PRESL, Jan Swatopluk. 1791-1849. Prague. P II; BMNH. Also Chemistry.

11891. SOYER-WILLEMET, Hubert Félix. 1791-1861. Fr. BMNH.

11892. WARD, Nathaniel Bagshaw. 1791-1868. Br. BMNH; GD.

11893. BOOTT, Francis. 1792-1863. Br. BMNH; BLA; GD.

11894. COLLADON, Louis Théodore Frédéric. 1792-1862. Fr. P III; BMNH.

11895. LEHMANN, Johann Georg Christian. 1792-1860. Ger. BMNH; GC.
1. *Monographia generis Primularum.* Leipzig, 1817.
2. *Plantae e familia Asperfoliarum nuciferae.* 2 vols. Berlin, 1818.
3. *Generis Nicotianarum historia.* Hamburg, 1818.
4. *Novarum et minus cognitarum stirpium pugillus.* 10 parts. Hamburg, 1832-57. Reprinted, Weinheim, 196-?
5. *Plantae Preissianae; sive, Enumeratio plantarum quas in Australasia occidentali ... annis 1838-41 collegit Ludovicus Preiss.* 2 vols. Hamburg, 1844-47.
6. *Revisio Potentillarum.* Breslau, 1856.

11896. PASSY, Antoine François. 1792-1873. Fr. P III; BMNH. Also Geology.

11897. WALLROTH, Carl Friedrich Wilhelm. 1792-1857. Ger. BMNH; BLA; GC.

11898. WENDLAND, Heinrich Ludolph. 1792-1869. BMNH.

11899. BAYRHOFFER, Johann David Wilhelm. 1793-1868. BMNH.

11900. BEILSCHMIED, Carl Traugott. 1793-1848. BMNH.

11901. IRVINE, Alexander. 1793-1873. Br. BMNH; GD.

11902. REICHENBACH, Heinrich Gottlieb Ludwig. 1793-1879. Ger. ICB; BMNH; GA; GC. Also Zoology*.
1. *Iconographia botanica ... Abbildungen seltener und weniger genau bekannter Gewächse.* 10 vols. Leipzig, 1823-32.
2. *Iconographia botanica exotica.* 3 vols. Leipzig, 1827-30. In German.
3. *Icones florae Germanicae et Helveticae.* 22 vols. Leipzig, 1834-70. With several collaborators.
4. *Handbuch der natürlichen Pflanzensystems.* Dresden, 1837. RM.

11903. WEBB, Philipp Barker. 1793-1854. Br. P II; BMNH; GD.

11904. BAINES, Henry. 1794?-1878. Br. BMNH.

11905. DURAND, Elias. 1794-1873. Fr./U.S.A. BMNH; GE.

11906. FRIES, Elias Magnus. 1794-1878. Swed. DSB; ICB; BMNH; GB.
1. *Systema mycologicum.* 3 vols. Lund, 1821-32. Reprinted, New York, 1952, and Weinheim, 1960.
2. *Elenchus Fungorum.* 2 vols. Greifswald, 1828. Reprinted, New York, 1952, and Weinheim, 1960.
3. *Lichenographia Europaea reformata.* Lund, 1831.
4. *Epicrisis systematis mycologici; seu, Synopsis Hymenomycetum.* 2 vols. Uppsala, 1836-38. X. Reprinted, New York, 1965.

4a. ——— Another ed. *Hymenomycetes Europaei*. Ib., 1874. X. Reprinted, Leipzig, 1937, and Amsterdam, 1963.
5. *Summa vegetabilium Scandinaviae*. 2 parts. Stockholm, 1846-49.
6. *Monographia Hymenomycetum Sueciae*. 2 vols. Uppsala, 1857-63. X. Reprinted, Amsterdam, 1963.

11907. GREVILLE, Robert Kaye. 1794-1866. Br. BMNH; GD.
1. *Scottish cryptogamic flora*. 6 vols. Edinburgh, 1823-28.

11908. MARTIUS, Karl Friedrich Philipp von. 1794-1868. Ger. DSB; P II,III; BMNH; GA; GB; GC.
1. *Historia naturalis Palmarum*. 3 vols. Leipzig, 1823-50.
2. *Icones selectae plantarum cryptogamicarum quas in itinere per Brasiliam annis 1817-20 ... collegit et descripsit*. 4 parts. Munich, 1828-34.
3. *Die Eriocauleae, als selbständige Pflanzen-Familie*. Breslau, 1835.
4. *Flora Brasiliensis*. 15 vols in 40. Munich, 1840-1906. Contd after his death by numerous collaborators.
See also 12895/2.

11909. PRESL, Karel Boriwoj. 1794-1852. Prague. DSB; P II; BMNH; GC.
1. *Reliquae Haenkeanae; seu, Descriptiones et icones plantarum quas in America meridionali et boreali, in insulis Philippinis et Marianis collegit Thaddaeus Haenke*. 2 vols, Prague, 1830-31. X. Reprinted, Amsterdam, 1973.

11910. RASPAIL, François Vincent. 1794-1878. Fr. DSB; ICB; P II,III; BMNH; BLA; GA. Also Microscopy.
1. *Nouveau système de physiologie végétale et de botanique*. Brussels, 1837.

11911. RICHARD, Achille. 1794-1852. Fr. P II; BMNH; BLA; GA.
1. *Nouveaux élémens de botanique et de physiologie végétale*. Paris, 1819. X. 5th ed., 1833.
1a. ——— *Elements of botany and vegetable physiology*. Edinburgh, 1831. Trans. from the 4th French ed., with notes, by W. MacGillivray.
2. *Monographie du genre Hydrocotyle, de la famille des Ombellifères*. Brussels, 1820.
See also 11619/2 and 12032/1.

11912. SCHLECHTENDAL, Dietrich Franz Leonhard von. 1794-1866. Ger. BMNH; GC.
1. *Flora von Deutschland*. 22 vols. Jena, 1841-69. With C.E. Langethal and E. Schenk.

11913. SCHOTT, Heinrich Wilhelm. 1794-1865. Aus. BMNH; GC. See also 12012/4.

11914. SHORT, Charles Wilkins. 1794-1863. U.S.A. ICB; BMNH; GE.
1. *Scientific publications*. New York, 1978. Ed. with an introd. by R.L. Stuckey.

11915. TOMMASINI, Mutius von. 1794-1879. Trieste. GC.

11916. BOON-MESCH, Hendrik Carel van der. 1795-1831. Holl. P I; BMNH; BLA & Supp. Also Chemistry.

11917. DAUBENY, Charles Giles Bridle. 1795-1867. Br. DSB; ICB; P I, III; BMNH; GB; GD. Other entries: see 3305.

11918. DIETRICH, Albert. 1795-1856. Ger. BMNH; GA.

11919. ECKLON, Christian Friedrich. 1795-1868. BMNH.
1. *Enumeratio plantarum Africae australis extratropicae.* 3 parts. Hamburg, 1834-37. With C. Zeyher.

11920. HAMPE, Georg Ernst Ludwig. 1795-1880. Ger. BMNH.
1. *Icones Muscorum novorum vel minus cognitorum.* Bonn, 1844.

11921. MEYER, Carl Anton. 1795-1855. Russ. BMNH. See also 11819/1.

11922. PAYEN, Anselme. 1795-1871. Fr. DSB; P II,III; BMNH; GA. Also Chemistry.
1. *Mémoires sur les développements des végétaux.* Paris, 1842.

11923. BLUME, Karel Lodewijk. 1796-1862. Holl./East Indies. BMNH.
1. *Enumeratio plantarum Javae et insularum adjacentium.* Leiden, 1827. Reprinted, Amsterdam, 1968.
2. *Flora Javae nec non insularum adjacentium.* 3 vols. Brussels, 1828. With J.B. Fischer. *Nova series.* Leiden, 1858.
3. *Rumphia; sive, Commentationes botanicae inprimis de plantis Indiae Orientalis.* 4 vols. Leiden, 1836-48.
4. *Museum botanicum Lugdano-Batavum; sive, Stirpium exoticarum novarum vel minus cognitarum brevis expositio et descriptio.* 2 vols. Leiden, 1849-51.

11924. BUEK, Heinrich Wilhelm. 1796-1879. Ger. BMNH; BLA.
1. *Genera, species et synonyma Candolleana, alphabetico ordine disposita; seu, Index generalis et specialis ad A.P. de Candolle "Prodromum systematis naturalis regni vegetabilis."* 4 parts. Hamburg, 1842-74. An index to item 11790/7.

11925. DESMAZIÈRES, Jean Baptiste Henri Joseph. 1796-1862. Fr. BMNH.

11926. GORIANINOV, Pavel Feodorovich. 1796-1865. Russ. ICB; BMNH (GHORYANINOV); BLA. Also Science in General.

11927. GUILLEMIN, (Jean Baptiste) Antoine. 1796-1842. Fr. BMNH; GA.
1. *Icones lithographicae plantarum Australasiae rariorum.* Paris, 1827.
2. *Enumération des plantes découvertes par les voyageurs dans les Isles de la Société, principalement dans celle de Taiti.* Paris, 1837.

11928. HENSLOW, John Stevens. 1796-1861. Br. DSB; BMNH; GB; GD. See also Index.
1. *The principles of descriptive and physiological botany.* London, 1835.
2. *A dictionary of botanical terms.* London, 1857.

11929. JAMESON, William. 1796-1873. Br./South America. BMNH; GD.

11930. LÉVEILLÉ, Joseph Henri. 1796-1870. Fr. BMNH.

11931. MACGILLIVRAY, William. 1796-1852. Br. BMNH; GA; GB; GD. Also Geology*, Zoology*, and Biography*.
1. *A manual of botany, comprising vegetable anatomy and physiology.* London, 1840.
See also Index.

11932. MORIS, Giuseppe Giacinto. 1796-1869. It. BMNH.

11933. SIEBOLD, Philipp Franz von. 1796-1866. Holl./Japan. ICB; P III; BMNH; BLA; GB; GC.
1. *Flora Japonica*. 2 vols. Leiden, 1835-70. With J.G. Zuccarini.

11934. TORREY, John. 1796-1873. U.S.A. DSB; ICB; P III; BMNH; GB; GE.
1. *A flora of North America*. 2 vols. New York, 1838-43. With A. Gray. Reprinted, ib., 1969. (Classica botanica Americana. 4)
2. *Plantae Fremontianae; or, Descriptions of plants collected by Col. J.C. Frémont in California*. Washington, 1853. RM.
3. *Phanerogamia of Pacific North America*. Philadelphia, 1874. (U.S. Exploring Expedition [1838-42], Vol. 17) RM.

11935. WIGHT, Robert. 1796-1872. India. ICB; BMNH; GD.
1. *Prodromus florae peninsulae Indiae Orientalis*. London, 1834. With G.A.W. Arnott.
2. *Contributions to the botany of India*. London, 1834.
3. *Illustrations of Indian botany*. 2 vols. Madras, 1840-50.
4. *Icones plantarum Indiae Orientalis; or, Figures of Indian plants*. 6 vols. Madras, 1840-56. Reprinted, 6 vols in 3, Weinheim, 1963. (HNC. 31)

11936. BISCHOFF, Gottlieb Wilhelm. 1797-1854. Ger. DSB; BMNH.
1. *Handbuch der botanischen Terminologie und System-Kunde*. 3 vols. Nuremberg, 1833-44.

11937. BOHLER, John. 1797-1872. Br. BMNH; GD.

11938. DES MOULINS, Charles. 1797-1875. Fr. BMNH.

11939. DUMORTIER, Barthélemy Charles Joseph. 1797-1878. Belg. BMNH; GA. Also Zoology.
1. *Analyse des familles des plantes*. Tournai, 1829.

11940. JAMES, Edwin P. 1797-1861. U.S.A. P III; BMNH; GE.

11941. JUSSIEU, Adrien (Henri Laurent) de. 1797-1853. Fr. DSB; ICB; BMNH; BLA; GA; GB.
1. *De Euphorbiacearum generibus*. Paris, 1824.
2. *La botanique*. Paris, 1842. X.
2a. ——— *The elements of botany*. London, 1849. Trans. by J.H. Wilson. Another ed., enl., 1858.
3. *Monographie des Malpighiacées*. 2 parts. Paris, 1843.

11942. LESTIBOUDOIS, Thémistocle Gaspard. 1797-1876. Fr. BMNH; GA.

11943. MALY, Joseph Karl. 1797-1866. Aus. BMNH.

11944. MARTENS, Martin. 1797-1863. Belg. P II & Supp.; BMNH; GF. Also Chemistry.

11945. ZUCCARINI, Joseph Gerhard. 1797-1848. Ger. BMNH; GC. See also 11933/1.

11946. BARTLING, Friedrich Gottlieb. 1798-1875. Ger. BMNH; GC.
1. *Ordines naturales plantarum, eorumque characteres et affinitates*. Göttingen, 1830.

11947. BREBISSON, Louis Alphonse de. 1798-1872. Fr. BMNH.

11948. DON, George. 1798-1856. Br. BMNH; GD.
1. *A general system of gardening and botany, containing a complete enumeration and description of all plants hitherto known*. 4 vols. London, 1831-37. X.
1a. ——— Another ed. *A general history of the dichlamydeous plants*. 4 vols. Ib., 1831-38.
See also 11836/3.

11949. DUBY, Jean Etienne. 1798-1885. Switz. BMNH. See also 11790/2a.

11950. JAUBERT, Hippolyte François. 1798-1874. Fr. BMNH; GA.
1. *Illustrationes plantarum orientalium; ou, Choix de plantes nouvelles, ou peu connues, de l'Asie occidentale*. 5 vols. Paris, 1842-57. With E. Spach.
2. *La botanique à l'Exposition Universelle de 1855*. Ib., 1855.

11951. KITTEL, Martin Balduin. 1798-1885. Ger. P III; BMNH.

11952. LAURER, Johann Friedrich. 1798-1873. Ger. BMNH; BLA; GC.

11953. PIEPER, Philipp Anton. 1798-1851. Ger. P II; BMNH; BLA.

11954. SCHULTZ (-SCHULTZENSTEIN), Karl Heinrich. 1798-1871. Ger. P II,III; BMNH; BLA; GC. Also Physiology*.

11955. SMITH, John. 1798-1888. Br. BMNH.

11956. SPENNER, Fridolin Karl Leopold. 1798-1841. Ger. BMNH; GC.

11957. ARNOTT, George Arnott Walker. 1799-1868. Br. BMNH; GD. See also 11848/9 and 11935/1.

11958. CAMBESSÈDES, Jacques. 1799-1863. Fr. BMNH.

11959. CHOISY, Jacques Denys. 1799-1859. Switz. BMNH.

11960. DON, David, 1799-1841. Br. BMNH; GD.
1. *Prodromus florae Nepalensis*. London, 1825. With (D.D.) F. Hamilton.

11961. DOUGLAS, David. 1799-1834. Br. ICB; BMNH; GD.

11962. EKART, Tobias Philipp. 1799-1877. Ger. BMNH.
1. *Synopsis Jungermanniarum in Germania vicinisque terris*. Coburg, 1832.

11963. GRIGOLATO, Gaetano. 1799-1884. It. ICB; BMNH.

11964. LINDLEY, John. 1799-1865. Br. DSB; ICB; BMNH; GA; GB; GD.
1. *Rosarum monographia; or, A botanical history of roses*. London, 1820.
2. *Collectanea botanica; or, Figures and botanical illustrations of ... exotic plants*. 8 parts. Ib., 1821.
3. *A synopsis of the British flora*. Ib., 1829. 3rd ed., 1841.
4. *An outline of the first principles of botany*. Ib., 1830. X. 4th ed., *Elements of botany*, 1841. 5th ed., 1847 and 1849. 6th ed., 1861.
5. *An introduction to the natural system of botany*. Ib., 1830. X. 2nd ed., 1836.
6. *The genera and species of orchidaceous plants*. 7 parts. Ib., 1830-40.
7. *The fossil flora of Great Britain*. 3 vols. Ib., 1831-37. With W. Hutton. RM/L.

8. *An introduction to botany.* Ib., 1832. 2nd ed., 1835. 3rd ed., 1839. 4th ed., 2 vols, 1848.
9. *Ladies' botany; or, A familiar introduction to ... the natural system of botany.* 2 vols. Ib., 1834-37. X. Another ed., 1847.
10. *Sertum orchidaceum. A wreath of the most beautiful orchidaceous flowers.* 10 parts. Ib., 1837-41.
11. *Flora medica. A botanical account of all the more important plants used in medicine.* Ib., 1838.
12. *School botany; or, The rudiments of botanical science.* Ib., 1839. X. Other eds, 1845 and 1856. 12th ed., 1862.
13. *The vegetable kingdom; or, The structure, classification, and uses of plants.* Ib., 1846. RM. 2nd ed., 1847. 3rd ed., 1853.
14. *Medical and oeconomical botany.* Ib., 1849.
15. *Folia orchidaceae.* 9 parts. Ib., 1852-59.
16. *Descriptive botany; or, The art of describing plants correctly in scientific language.* Ib., 1858. X. 7th ed., n.d. 9th ed., n.d.
17. *The treasury of botany. A popular dictionary of the vegetable kingdom.* 2 vols. Ib., 1866. With T. Moore. Other eds: 1870, 1874, 1876, 1889, 1899.

See also 11619/1a.

11965. ROYLE, John Forbes. 1799-1858. Br./India. P II; BMNH; BLA; GB; GD.
1. *Illustrations of the botany ... of the Himalayan Mountains and of the flora of Cashmere.* 2 vols & atlas. London, 1833-40.

11966. SCHEIDWEILER, Michel Joseph François. 1799-1861. Belg. BMNH; GF.

11967. WEIHE, Karl Ernst August. 1799-1834. Ger. ICB; BMNH.

11968. WILSON, William. 1799-1871. Br. BMNH; GD.

11969. RHIND, William. fl. 1833-1867. Br. BMNH. Also Geology*.
1. *A history of the vegetable kingdom, embracing the physiology, classification and the culture of plants.* Glasgow, 1840? Other eds: 1855, 1868, 1874? 1877.

11970. HÜBENER, J.W.P. d. 1847. Ger. BMNH.

11971. BENTHAM, George. 1800-1884. Br. DSB; ICB; BMNH; GB; GD.
1. *Labiatarum genera et species.* London, 1832-36.
2. *Commentationes de Leguminosarum generibus.* Vienna, 1837.
3. *Plantas Hartwegianas imprimis Mexicanas ... enumerat....* London, 1839-57. Reprinted, New York, 1970. (HNC. 80)
4. *The botany of the voyage of H.M.S. "Sulphur" ... 1836-42.* London, 1844. Reprinted, New York, 1968. (HNC. 61)
5. *Handbook of the British flora.* London, 1858. 2nd ed., 2 vols, 1863-65. Another ed., 1866. 5th ed., by J.D. Hooker, 1887.
6. *Genera plantarum ad exemplaria imprimis in Herbariis Kewensis servata definata.* 3 vols. Ib., 1862-83. With J.D. Hooker. Reprinted, Weinheim, 1965. (HNC. 84). Index: see 12573/1.

6a. ——— *Supplemental papers.* Ib., 1860-81. X. Reprinted, Weinheim, 1970.

7. *Flora Australiensis*. 7 vols. Ib., 1863-78. With F. Mueller. Reprinted, (place?), 1963.
See also 12012/4.

11972. BURNETT, Gilbert Thomas. 1800-1835. Br. BMNH; GD.
1. *Outlines of botany*. London, 1835.

11973. DIETRICH, David Nathanael Friedrich. 1800-1888. Ger. BMNH; GC.
1. *Das Wichtigste aus dem Pflanzenreiche*. 22 parts. Jena, 1831-38.
2. *Synopsis plantarum*. Weimar, 1839. RM.
3. *Encyclopädie der Pflanzen*. 2 vols. Jena, 1841-46.

11974. FRANCIS, George William. 1800-1865. Br./Australia. BMNH; GD.

11975. GÖPPERT, Heinrich Robert. 1800-1884. Ger. DSB; P I,III; BMNH; BLA; GC. Also Geology*.

11976. HOWITT, Godfrey. 1800-1873. Br. BMNH.

11977. LEES, Edwin. 1800-1887. Br. BMNH; GD.

11978. LE MAOUT, Emmanuel. 1800-1877. Fr. P I & Supp.; BMNH.
1. *Les trois règnes de la nature. Règne végétal ... Histoire naturelle des familles végétales*. Paris, 1851. X. Another ed., 1884.
2. *Traité général de botanique descriptive et analytique*. 2 vols. Paris, 1868. With J. Decaisne. X. 2nd ed., 1876.
2a. ——— *A general system of botany, descriptive and analytical*. 2 parts. London, 1873. Trans. by Mrs. Hooker. With additions by J.D. Hooker. 2nd ed., 1876. RM/L.

11979. MEISSNER, Carl Friedrich. 1800-1874. Switz. BMNH (MEISNER); GC.
1. *Monographia generis Polygoni prodromus*. Geneva, 1826.
2. *Plantarum vascularum genera*. 2 parts. Leipzig, 1836-43.

11980. POLLENDER, (Franz) Aloys (Antoine). 1800-1879. Ger. DSB; ICB; BMNH; BLA. Also Microbiology.

11981. SHAW, Henry. 1800-1889. U.S.A. ICB; GE.

11982. UNGER, Franz von. 1800-1870. Aus. DSB; ICB; P III; BMNH; GC. Also Geology*.
1. *Versuch einer Geschichte der Pflanzenwelt*. Vienna, 1852. RM.
2. *Botanische Briefe*. Vienna, 1852. RM.
3. *Anatomie und Physiologie der Pflanzen*. Vienna, 1855.
4. *Briefwechsel zwischen F. Unger und S. Endlicher*. Berlin, 1899. Ed. with commentary by G. Haberlandt. RM.
See also 12012/9.

11983. WYDLER, Heinrich. 1800-1883. Switz. BMNH; GC.

11984. BAUTIER, Alexandre. b. 1801. Fr. BMNH.

11985. BENNETT, John Joseph. 1801-1876. Br. BMNH; GD.

11986. BONORDEN, Hermann Friedrich. b. 1801. Ger. BMNH; BLA & Supp.

11987. BROMFIELD, William Arnold. 1801-1851. Br. BMNH; GD.

11988. BRONGNIART, Adolphe Théodore. 1801-1876. Fr. DSB; ICB; P I, III; BMNH; GA; GB. Also Geology*.

11989. FLEISCHER, Franz von. 1801-1879. Ger. P III; BMNH; GC.

11990. HALL, Hermann Christian van. 1801-1874. Holl. BMNH; BLA.

11991. LEMAIRE, Charles Antoine. 1801-1871. Fr. BMNH.

11992. PAXTON, Joseph. 1801-1865. Br. ICB; BMNH; GA; GB; GD.
1. *A pocket botanical dictionary.* London, 1840. X. Another ed., 1849. 2nd ed., 1868.

11993. RAINEY, George. 1801-1884. Br. BMNH; Mort.; GD. Also Anatomy.
1. *An experimental inquiry into the cause of the ascent and descent of the sap, with some observations on the nutrition of plants and the cause of endosmose and exosmose.* London, 1847.

11994. RÖPER, Johannes August Christian. 1801-1885. Ger. BMNH; GC.

11995. VISIANI, Roberto de. 1801-1878. It. BMNH.

11996. WESTERHOFF, Rembertus. 1801-1874. Holl. BMNH; GF.

11997. BAYER, Johann N. 1802-1870. Aus. BMNH.

11998. BOUSSINGAULT, Jean Baptiste Joseph Dieudonné. 1802-1887. Fr./Colombia. DSB; ICB; P I,III,IV; Mort.; GA; GB. Also Chemistry and Meteorology.

11999. BRANDT, Johann Friedrich. 1802-1879. Ger./Russ. DSB; BMNH; BLA; GC. Also Zoology.

12000. LECOQ, Henri. 1802-1871. Fr. P I,III; BMNH; GA. Also Geology*.
1. *Dictionnaire raisonné des termes de botanique et des familles naturelles.* Paris, 1831. With J. Juillet. RM.

12001. BEER, Joseph Georg. 1803-1873. BMNH.

12002. BERKELEY, Miles Joseph. 1803-1889. Br. DSB; ICB; BMNH; GD1.
1. *Introduction to cryptogamic botany.* London, 1857.
2. *Notices of British fungi.* Weinheim, 196-? (HNC) Collected papers.

12003. BOREAU, Alexandre. b. 1803. Fr. BMNH.

12004. BUNGE, Alexander von. 1803-1890. Russ. BMNH; BLA; GA; GC.

12005. FISCHER VON WALDHEIM, Alexander. 1803-1884. Russ. BMNH; BLA.

12006. JAMES, Thomas Potts. 1803-1882. U.S.A. BMNH; GE.

12007. KICKX, Jean (second of the name). 1803-1864. Belg. BMNH; GF.

12008. NEILREICH, August. 1803-1871. Aus. BMNH; GC.

12008A SCHIMPER, Karl Friedrich. 1803-1867. Ger. DSB; ICB; BMNH; GC. Also Geology.

12009. SULLIVANT, William Starling. 1803-1873. U.S.A. ICB; BMNH; GE.
1. *Icones muscorum; or, Figures and descriptions of ... mosses peculiar to eastern North America.* 1864. X. Supplement, 1874. X.
1a. ——— Reprinted, 2 vols, Amsterdam/New York, 1969.

12010. WIMMER, Christian Friedrich Heinrich. 1803-1868. Ger. BMNH; GC.

12011. CASTLE, Thomas. ca. 1804-1838. Br. BMNH; GD.

12012. ENDLICHER, István László (or Stephen Ladislaus). 1804-1849. Aus. BMNH; BLA; GA; GC.
1. *Prodromus florae Norfolkicae; sive, Catalogus stirpium quae in Insula Norfolk 1804 et 1805 a F. Bauer collectae et depictae.* Vienna, 1833. Re Ferdinand Bauer.
2. *Ataka botanika. Nova genera et species plantarum descripta.* Ib., 1833.
3. *Genera plantarum secundum ordines naturales disposita.* [with] *Supplementum I.* 2 vols. Ib., 1836-40. RM/L.
3a. ——— *Mantissa botanica sistens "Genera plantarum" supplementum II.* Ib., 1842. RM/L.
3b. ——— *Mantissa botanica altera sistens ... supplementum III.* Ib., 1843.
3c. ——— *Supplementum IV.* Ib., 1847.
3d. ——— *Supplementum V.* Ib., 1850.
4. *Enumeratio plantarum quas in Novae Hollandiae ... collegit ... Baron de Hügel.* Ib., 1837. With G. Bentham, E. Fenzl, and H.W. Schott.
5. *Stirpium Australasicarum herbarii Hügeliani....* Ib., 1838.
6. *Iconographia generum plantarum.* Ib., 1838.
7. *Novarum stirpium decades 1-10.* Ib., 1839. With E. Fenzl.
8. *Enchiridion botanicum, exhibens classes et ordines plantarum.* Leipzig, 1841. RM/L.
9. *Grundzüge der Botanik.* Vienna, 1843. With F. Unger. RM/L.
10. *Synopsis Coniferarum.* Sangalli, 1847. RM/L.
See also 11982/4.

12013. GASPARRINI, Guglielmo. 1804-1866. It. BMNH.

12014. KIRSCHLEGER, Frédéric. 1804-1869. Fr. BMNH.

12015. LÜBEN, August Heinrich Philipp. 1804-1874. Ger. BMNH; GC.

12016. MEYEN, Franz Julius Ferdinand. 1804-1840. Ger. DSB; P II,III; BMNH; GC.
1. *Grundriss der Pflanzengeographie.* Berlin, 1836. X.
1a. ——— *Outlines of the geography of plants.* London, 1846. (Ray Society publn. 7)

12017. MOQUIN-TANDON, (Christian Horace Bénédict) Alfred. 1804-1863. Fr. ICB; BMNH; BLA; GA.
1. *Eléments de tératologie végétale.* Paris, 1841.

12018. PHOEBUS, Philipp. 1804-1880. Ger. P II,III; BMNH; BLA; GC.

12019. SCHLEIDEN, Matthias Jacob. 1804-1881. Ger. DSB; ICB; BMNH; Mort.; BLA & Supp.; GB; GC. Also Science in General* and Cell Theory. See also Index.
1. *Grundzüge der wissenschaftlichen Botanik.* 2 vols. Leipzig, 1842-43. X.
1a. ——— 2nd ed. *Grundzüge ... Botanik, nebst einer methodologischen Einleitung.* Ib., 1845-46. RM.
1b. ——— 3rd ed. *Die Botanik als inductive Wissenschaft. Grundzüge....* Ib., 1849. X.

1c. ——— ——— *Principles of scientific botany; or, Botany as an inductive science.* London, 1849. Trans. by Edwin Lankester. RM/L. Reprinted, New York, 1969.

2. *Die Pflanze und ihr Leben. Populäre Vorträge.* Leipzig, 1848. X. 2nd ed., 1850. 3rd ed., 1852. RM.
2a. ——— *The plant. A biography.* London, 1848. Trans. by A. Henfrey. RM/L. 2nd ed., 1853.
2b. ——— *Poetry of the vegetable world. A popular exposition of the science of botany and its relations to man.* Cincinnati, 1853. 1st American ed., from the London ed.
3. *Studien. Populäre Vorträge.* Leipzig, 1855. RM.

12020. SCHULTZ, Friedrich Wilhelm. 1804-1876. Ger. BMNH; BLA; GC.

12021. STEETZ, Joachim. 1804-1862. Ger. BMNH.

12022. WATSON, Hewett Cottrell. 1804-1881. Br. DSB; BMNH; GD. Also Evolution.
1. *The new botanist's guide to the localities of the rarer plants of Britain.* 2 vols. London, 1835-37.

12023. ZANARDINI, Giovanni Antonio Maria. 1804-1878. It. BMNH.

12024. FIELDING, Henry Barron. d. 1851. Br. BMNH; GD.

12025. DEAKIN, Richard. d. 1873. Br. BMNH.

12026. BLUFF, Mathias Joseph. 1805-1837. Ger. BMNH; BLA; GA (BLUF).

12027. BRAUN, Alexander Carl Heinrich. 1805-1877. Ger. DSB; BMNH; GC.
1. *Betrachtungen über die Erscheinung der Verjüngung in der Natur, inbesondere in der Lebens- und Bildungsgeschichte der Pflanze.* Freiburg i.B., 1849-50. RM.
2. *Über Parthenogenesis bei Pflanzen.* Berlin, 1857.

12028. GAROVAGLIO, Santo. 1805-1882. It. BMNH.

12029. HARTIG, Theodor. 1805-1880. Ger. DSB; BMNH.
1. *Entwickelungsgeschichte des Pflanzenkeims.* Leipzig, 1858.

12030. KLOTZSCH, Johann Friedrich. 1805-1860. Ger. BMNH; GC.

12031. LEIGHTON, William Allport. 1805-1889. Br. BMNH; GD.

12032. LESSON, Pierre Adolphe. b. 1805. Fr. BMNH.
1. *Voyage de découvertes de "l'Astrolabe" ... Botanique.* 2 vols & plates. Paris, 1832-34. With A. Richard.

12033. McCLELLAND, John. 1805-1875. India. BMNH. See also 12094/1.

12034. MOHL, Hugo von. 1805-1872. Ger. DSB; ICB; P III; BMNH; GA; GB; GC. Also Microscopy*.
1. *Grundzüge der Anatomie und Physiologie der vegetabilischen Zelle.* Brunswick, 1851. X.
1a. ——— *Principles of the anatomy and physiology of the vegetable cell.* London, 1852. Trans. by A. Henfrey.
2. *Vermischte Schriften, botanischen Inhalts.* Tübingen, 1845. RM.

12035. NOTARIS, Giuseppe de. 1805-1877. It. BMNH.

12036. PFEIFFER, Ludwig Georg Carl. 1805-1877. Ger. BMNH; BLA; GA. Also Zoology.
1. *Nomenclator botanicus.* 2 vols. Cassel, 1873-74.

12037. PICKERING, Charles. 1805-1878. U.S.A. BMNH; GB; GE. Also History of Botany*.

12038. SCHULTZ, Karl Heinrich. 1805-1867. Ger. BMNH; BLA; GC.

12039. CANDOLLE, Alphonse (Louis Pierre Pyramus) de. 1806-1893. Switz. DSB; ICB; P I,III (DECANDOLLE); BMNH. Also History of Science*.
1. *Monographie des Campanulées.* Paris, 1830.
2. *Introduction à l'étude de la botanique.* 2 vols. Ib., 1835.
3. *Géographie botanique raisonnée.* 2 vols. Ib., 1855.
4. *Lois de la nomenclature botanique adoptés par le Congrès International de Botanique ... en ... 1867.* Geneva, 1867. X.
4a. ——— *Laws of botanical nomenclature adopted ... in ... 1867.* London, 1868. Trans. by H.A. Weddell.
4b. ——— *Nouvelles remarques sur la nomenclature botanique ... Supplément au "Commentaire ..." qui accompagnait le texte des lois.* Geneva, 1883.
5. *Monographiae Phanerogamarum: "Prodromi" nunc continuatio, nunc revisio....* 9 vols in 10. Paris, 1878-96. With A.C.P. de Candolle. A continuation of item 11790/7.
6. *La phytographie; ou, L'art de décrire les végétaux considérés sous différents points de vue.* Paris, 1880.
7. *Origine des plantes cultivées.* Paris, 1882. X.
7a. ——— *Origin of cultivated plants.* London, 1884. 2nd ed., 1886; reprinted, New York, 1959.
See also 11790/6,7.

12040. CESATI, Vincenzo. 1806-1883. It. BMNH.

12041. COURTOIS, Richard Joseph. 1806-1835. Belg. BMNH; GF. See also 11800/1.

12042. KALCHBRENNER, Károly. 1806?-1886. Hung. BMNH.

12043. LANGETHAL, Christian Eduard. 1806-1878. Ger. BMNH; GC. See also 11912/1.

12044. LESQUEREUX, (Charles) Leo. 1806-1889. Switz./U.S.A. DSB; ICB; P III; BMNH; GE. Also Geology.

12045. MARTINS, Charles Frédéric. 1806-1889. Fr. P II,III; BMNH; BLA; GA. Also Geology and Meteorology.

12046. MORITZI, Alexander. 1806-1850. Switz. ICB; BMNH. Also Evolution*.

12047. MÜLLER, Johannes. b. 1806. Ger. P II. Also Chemistry.

12048. PETERMANN, Wilhelm Ludwig. 1806-1855. Ger. BMNH.
1. *Deutschlands Flora.* Leipzig, 1849.

12049. PRATT, Anne. 1806-1893. Br. GD.

12050. RABENHORST, Gottlob Ludwig. 1806-1881. Ger. BMNH; GC.

12051. WELWITSCH, Friedrich Martin Josef. 1806-1872. Aus./Port./Br. BMNH; GC; GD.

12052. WIRTGEN, Philipp Wilhelm. 1806-1870. Ger. BMNH; GC.

12053. DECAISNE, Joseph. 1807-1882. Fr. BMNH; GA.
1. *Herbarii Timorensis descriptio.* Paris, 1835.
2. *Voyage autour du monde sur la frégate la "Vénus" ... 1836-1839 ... Botanique.* Atlas. Ib., 1846. Text. Ib., 1864.
3. *Voyage au Pôle du Sud et dans l'Océanie sur les corvettes "l'Astrolabe" et la "Zélée"... 1837-1840 ... Botanique. Description des plantes vasculaires.* Text & atlas. Ib., 1852-53.
See also 11978/2.

12054. DOZY, Franz. 1807-1856. Holl. BMNH; GF.

12055. KÜTZING, Friedrich Traugott. 1807-1893. Ger. DSB; ICB; P I, III; BMNH; GA; GC.
1. *Synopsis diatomearum.* Halle, 1834. In German.
2. *Phycologia generalis; oder Anatomie, Physiologie und Systemkunde der Tange.* Leipzig, 1843.
3. *Tabulae Phycologicae; oder, Abbildungen der Tange.* 20 parts. Nordhausen, 1845-71.
4. *Species Algarum.* Leipzig, 1849. Reprinted, Amsterdam, 1969.

12056. MOORE (formerly MUIR), David. 1807-1879. Br. BMNH; GD.

12057. MORREN, (Charles François) Antoine. 1807-1858. Belg. ICB; P II; BMNH; GA. See also 7621/1.

12058. RALFS, John. 1807-1890. Br. BMNH; GD.
1. *The British Desmidieae.* London, 1848. RM. Reprinted, Weinheim, 1962. (HNC. 18)

12059. REICHENBACH, Anton Benedict. 1807-1880. BMNH.
1. *Allgemeine Pflanzenkunde; oder, Einleitung in die Botanik.* Leipzig, 1837. RM.

12060. RIDDELL, John Leonard. 1807-1865. U.S.A. ICB; P III; BMNH; BLA; GE.
1. *A synopsis of the flora of the western states.* Cincinnati, 1835. RM.

12061. VRIESE, Willem Hendrik de. 1807-1862. Holl. BMNH; BLA.

12062. ANGREVILLE, J.E. d'. 1808-1867. Switz. BMNH.

12063. BABINGTON, Charles Cardale. 1808-1895. Br. DSB; BMNH; GD1.
1. *Memorials, journal and botanical correspondence.* Cambridge, 1897. Ed. by A.M. Babington. RM/L.

12064. BALFOUR, John Hutton. 1808-1884. Br. DSB; BMNH; BLA Supp.; GD.
1. *A manual of botany.* Edinburgh, 1849. X. 2nd ed., London, 1851. Another ed., 1860. 5th ed., Edinburgh, 1875.
2. *Phyto-theology; or, Botanical sketches intended to illustrate the works of God.* Edinburgh, 1851. X. 2nd ed., London, 1851.
3. *Class book of botany.* Edinburgh, 1854. 3rd ed., 1871.
4. *The elements of botany.* Ib., 1869. X. 3rd ed., 1876.
5. *Introduction to the study of palaeontological botany.* Ib., 1872.

12065. CURTIS, Moses Ashley. 1808-1872. U.S.A. ICB; BMNH; GE.

12066. DÖLL, Johann Christoph. 1808-1885. Ger. BMNH; GC.

12067. FENZL, Eduard. 1808-1879. Aus. BMNH; GC. See also 12012/4,7.

12068. FRESENIUS, Johann Baptist Georg Wolfgang. 1808-1866. Ger. BMNH; BLA & Supp.; GC.

12069. GOTTSCHE, Carl Moritz. 1808-1892. Ger. BMNH; BLA Supp.; GC.
1. *Synopsis Hepaticarum*. 5 parts. Hamburg, 1844-47. With J.B.G. Lindenberg and C.G. Nees von Esenbeck. Reprinted, Weinheim, 1967. (HNC. 56)
See also 11815/2.

12070. GRENIER, Jean Charles Marie. 1808-1875. Fr. BMNH.
1. *Flore de France*. 3 vols. Paris, 1848-56. With D.A. Godron.

12071. KORTHALS, Pieter Willem. 1808?-1892. East Indies. BMNH.
1. *Botanie van Nederlandsch Indiae*. 2 vols. Leiden, 1839-42.

12072. PHILIPPI, Rudolph Amandus. 1808-1904. Ger./Chile. P II,III, IV & Supp.; BMNH. Also Geology and Zoology.

12073. QUEKETT, Edwin John. 1808-1847. Br. BMNH; GD (under Q., John Thomas).

12074. RICHTER, Hermann Friedrich Eberhard. 1808-1876. Ger. BMNH; BLA; GA; GC.

12075. SCHIMPER, Wilhelm Philipp. 1808-1880. Strasbourg. DSB; BMNH; GC. Also Geology*.

12076. CHAPMAN, Alvan Wentworth. 1809-1899. U.S.A. DSB; BMNH; GE.
1. *Flora of the southern United States*. New York, 1860. Another ed., 1872.

12077. CORDA, August Carl Joseph. 1809-1849. Ger. BMNH; GC. Also Geology*.
1. *Icones Fungorum*. 6 parts. Prague, 1837-54. In German. Reprinted in 1 vol., Weinheim, 1963. (HNC. 33)

12078. DARWIN, Charles Robert. 1809-1882. Br. DSB; ICB; P III; BMNH; G. Also Geology*, Natural History*, Zoology*, Evolution*#, and Autobiography*. See also Index.
1. *On the various contrivances by which ... orchids are fertilised by insects, and on the good effects of intercrossing*. London, 1862. 2nd ed., 1877; re-issued 1885, 1888, and 1890.
2. *On the movements and habits of climbing plants*. Ib., 1865. Reprinted, Brussels, 1969. 2nd ed., 1875. RM/L. Re-issues of 2nd ed.: 1876, 1882, 1885.
3. *Insectivorous plants*. Ib., 1875. RM/L. 2nd ed., rev. by Francis Darwin, 1888.
4. *The effects of cross and self fertilisation in the vegetable kingdom*. Ib., 1876. RM/L. 2nd ed., 1878; re-issued 1888.
5. *The different forms of flowers on plants of the same species*. Ib., 1877. RM/L. Re-issued 1884 and 1888. Reprinted, Brussels, 1969. 2nd ed., 1892.
6. *The power of movement in plants*. Ib., 1880. With Francis Darwin.

12079. ENGELMANN, George. 1809-1884. Ger./U.S.A. DSB Supp.; BMNH; BLA & Supp.; GC; GE.
1. *The botanical works.* Cambridge, Mass., 1887. Ed. by W. Trelease and A. Gray. RM/L.

12080. HEER, Oswald. 1809-1883. Switz. DSB; P I,III; BMNH; GB; GC. Also Geology and Zoology.

12081. JUNGHUHN, Franz Wilhelm. 1809-1864. East Indies. ICB; P I, III; BMNH; BLA; GA; GC.

12082. KLINGGRÄFF, Karl Julius Meyer von. 1809-1879. Ger. BMNH; GC.

12083. KOCH, Karl Heinrich Emil. 1809-1879. Ger. ICB; P III; BMNH; GA; GC.

12084. LESSING, Christian Friedrich. 1809-1862 (or 1810-1880). Ger. BMNH; GC.

12085. TRAUTVETTER, Ernst Rudolph von. 1809-1889. Russ. BMNH.

12086. TURCHANINOV, Nikolai Stepanovich. d. 1864. Russ. BMNH.

12087. BOISSIER, Edmond Pierre. 1810-1885. Switz. BMNH.
1. *Flora orientalis ... a Graecia et Aegypto ad Indiae fines.* Geneva, 1867-84. Supplement, ed. by R. Buser, ib., 1888.

12088. BOSCH, Roelof Benjamin van den. 1810-1862. Holl. BMNH.

12089. BRACKENRIDGE, William D. 1810-1893. U.S.A. BMNH; GE.

12090. DUVAL-JOUVE, Joseph. 1810-1883. Fr. BMNH.

12091. FERMOND, Charles. 1810-after 1867. BMNH.
1. *Essai de phytomorphie; ou, Etudes des causes qui déterminent les principales formes végétales.* 2 vols. Paris, 1864-68.

12092. GOLDMANN, Ignaz. 1810-1856 (or 1848). Ger. P I; BMNH.

12093. GRAY, Asa. 1810-1888. U.S.A. DSB; ICB; BMNH; GA; GB; GE. Also Evolution* and Biography*. See also Index.
1. *Elements of botany.* New York, 1836. RM.
1a. ——— 5th ed. *Introduction to structural and systematic botany, and vegetable physiology.* Ib., 1857. 6th ed., 2 vols, 1879-85.
2. *A manual of the botany of the northern United States.* Boston, 1848. RM.
3. *Genera florae Americae ... The genera of the plants of the United States.* 2 vols. Boston, 1848-49. RM/L.
4. *Lessons in botany and vegetable physiology.* New York, 1857. X. Reprinted, ib., 1971. Another ed., 1868.
5. *Synoptical flora of North America.* 2 vols. New York, 1878-97. With S. Watson. Contd. by B.L. Robinson.
6. *Letters.* 2 vols. Boston, 1893. Ed. by J.L. Gray. RM/L. Reprinted, New York, 1973.
7. *Scientific papers.* 2 vols. London, 1889. Selected by C.S. Sargent. RM/L.

12094. GRIFFITH, William. 1810-1845. India. DSB; BMNH; GD.
1. *Posthumous papers. Journals of travels in Assam, Burma, Bootan, Afghanistan and the neighbouring countries.* 2 vols. Calcutta, 1847-48. Ed. by J. McClelland who also

edited the three following appendices.
1a. ——— *Notulae ad plantas asiaticas*. 4 parts. Ib., 1847-54.
1b. ——— *Icones plantarum asiaticarum*. 4 parts. Ib., 1847-54.
1c. ——— *Palms of British East India*. Ib., 1850.

12095. STEINHEIL, Adolphe. 1810-1839. Ger./Fr. BMNH; GC.

12096. WOOD, Alphonso. 1810-1881. U.S.A. ICB; BMNH.

12097. ARESCHOUG, Johan Erhard. 1811-1887. Swed. BMNH.

12098. ARRHENIUS, Johan Pehr. 1811-1889. Swed. BMNH.

12099. BATEMAN, James. 1811-1897. Br. BMNH; GD1.

12100. DUCHARTRE, Pierre Etienne Simon. 1811-1894. Fr. BMNH.

12101. HARVEY, William Henry. 1811-1866. Br. DSB; BMNH; GD. Also Autobiography*.
1. *The genera of South African plants*. Cape Town, 1838. 2nd ed., by J.D. Hooker, ib., 1868.
2. *A manual of the British marine algae*. London, 1841. X. 2nd ed., 1849.
3. *Phycologia Britannica; or, A history of British seaweeds*. 4 vols. London, 1846-51.
4. *Nereis australis; or, Algae of the Southern Ocean*. London, 1847. Reprinted, Weinheim, 1965. (HNC. 40)

12102. HASSKARL, Justus Karl. 1811-1894. East Indies. BMNH; GA (HASZKARL); GC.
1. *Plantae Javanicae rariores*. Berlin, 1848.

12103. JOHNS, Charles Alexander. 1811-1874. Br. BMNH; GD.

12104. MENEGHINI, Giuseppe. 1811-1889. It. P III; BMNH. Also Geology.

12105. MIQUEL, Frederik Anton Willem. 1811-1871. Holl. DSB; BMNH; BLA.
1. *Commentarii phytographici*. 3 parts. Leiden, 1838-40.
2. *Monographia Cycadearum*. Utrecht, 1842.
3. *Systema Piperacearum*. Rotterdam, 1843.
4. *Revisio critica Casuarinarum*. Amsterdam, 1848.

12106. SCHOMBURGK, (Moritz) Richard. 1811-1891. Ger./Australia. BMNH; GD (under S., Robert H.)

12107. THWAITES, George Henry Kendrick. 1811-1882. Br./Ceylon. BMNH; GD.

12108. WRIGHT, Charles. 1811-1885. U.S.A. ICB; BMNH; GE.

12109. BASKERVILLE, Thomas. 1812-1840. Br. BMNH; GD.
1. *Affinities of plants*. London, 1839.

12110. DELPONTE. Giovanni Battista. 1812-1884. It. BMNH.

12111. EDGEWORTH, Michael Pakenham. 1812-1881. India. BMNH; GD.
1. *Pollen*. London, 1877. 2nd ed., 1879.

12112. FRITSCH, Karl. 1812-1879. Prague/Aus. P I,III; BMNH. Also Meteorology.

1. *Resultate mehrjährigen Beobachtungen über jene Pflanzen deren Blumenkronen sich täglich periodisch öffnen und schliessen.* Prague, 1851.

12113. GARDNER, George. 1812-1849. Br. BMNH; GD.

12114. GARREAU, Lazare. 1812-1892. Fr. DSB; BMNH.

12115. GÜMBEL, Wilhelm Theodor von. 1812-1858. Ger. P III; BMNH. Also Crystallography.

12116. SONDER, Otto Wilhelm. 1812-1881. Ger. BMNH; GC.

12117. VOGEL, Julius Rudolf Theodor. 1812-1841. Ger. BMNH; GC.

12118. AGARDH, Jacob Georg. 1813-1901. Swed. DSB; BMNH.
1. *Recensio specierum generis Pteridis.* Lund, 1839.
2. *Species, genera et ordines Algarum.* 3 vols in 6. Lund, 1848-80. RM/L.
3. *Theoria systematis plantarum.* 2 vols. Lund, 1858. RM/L.

12119. CARPENTER, William Benjamin. 1813-1885. Br. DSB; ICB; P III; BMNH; BLA & Supp.; GB; GD. Other entries: see 3326.
1. *Vegetable physiology and botany.* London, 1844. Other eds, 1858 and 1865.

12120. CHATIN, (Gaspard) Adolphe. 1813-1901. Fr. BMNH.

12121. CLARKE, Benjamin. 1813-1890. Br. BMNH.
1. *A new arrangement of phanerogamous plants.* London, 1866.

12122. DICKIE, George. 1813-1882. Br. BMNH; GD.

12123. DREJER, Salomon Thomas Nicolai. 1813-1842. Den. BMNH.

12124. KELLOGG, Albert. 1813-1887. U.S.A. DSB; BMNH; GE.

12125. KOTSCHY, Theodor. 1813-1866. Aus. BMNH.

12126. SCHAUER, Johann Conrad. 1813-1848. Ger. BMNH; GC.

12127. SENDTNER, Otto. 1813-1859. Ger. BMNH; GC.

12128. TORNABENE, Francesca. 1813-1897. It. BMNH.

12129. ÅNGSTRÖM, Johan. 1814-1879. Swed. BMNH.

12130. BELLYNCK, Auguste Alexis Adolphe Alexandre. 1814-1877. Belg. BMNH.

12131. BERGER, Ernst Friedrich. 1814-1853. Ger. BMNH.

12132. BESNARD, Anton Franz. 1814-1885. Ger. P III; BMNH. Also Mineralogy.

12133. DURAND, Pierre Bernard. 1814-1853. Fr. GA.

12134. GALEOTTI, Henri Guillaume. 1814-1858. Belg. P I; BMNH; GF. Also Geology.

12135. GRISEBACH, August Heinrich Rudolph. 1814-1879. Ger. DSB; P I,III; BMNH; BLA Supp.; GC.
1. *Flora of the British West Indian islands.* London, 1859-64. X. Reprinted, Weinheim, 1963. (HNC. 30)
2. *Die Vegetation der Erde nach ihrer klimatischen Anordnung. Ein Abriss der vergleichenden Geographie der Pflanzen.* 2 vols. Leipzig, 1872. X.

2a. ——— *La végétation du globe d'après sa disposition suivant les climats.* 2 vols. Paris, 1877-78. Trans. with notes by P. de Tchihatchef.

12136. JORDAN, (Claude Thomas) Alexis. 1814-1897. Fr. DSB; BMNH.

12137. PANCHIĆ, Josif. 1814-1888. Belgrade. BMNH.

12138. RAVENEL, Henry William. 1814-1887. U.S.A. ICB; GE.

12139. RUPRECHT, Franz Joseph. 1814-1870. Russ. BMNH; GC.

12140. SCHACHT, Hermann. 1814-1864. Ger. BMNH; GC.
1. *Das Mikroscop und seine Anwendung, inbesondere für Pflanzen-Anatomie und Physiologie.* Berlin, 1851. X.
1a. ——— *The microscope and its application to vegetable anatomy and physiology.* 2nd ed., London, 1855. Ed. by F. Currey.

12141. SCHNIZLEIN, Adalbert Carl Friedrich Hellwig Conrad. 1814-1868. Ger. BMNH; GC.
1. *Iconographia familiarum naturalium regni vegetabilis delineata.* 4 vol. Bonn, 1843-70.

12142. THEDENIUS, Knut Fredrik. 1814-1894. Swed. BMNH.

12143. WOOLLS, William. 1814-1893. Australia. BMNH; GF.
1. *A contribution to the flora of Australia.* Sydney, 1867.
2. *Lectures on the vegetable kingdom, with special reference to the flora of Australia.* Ib., 1879. Another ed., 1889.

12144. ANTOINE, Franz. 1815-1886. Aus. BMNH.

12145. KEYSERLING, Alexander Andreevich (or Alexander F.M.L.A. von). 1815-1891. Russ. DSB; P III,IV; BMNH. Also Geology.

12146. LANTZIUS-BÉNINGA, Bojung Scato Georg. 1815-1871. Holl./Ger. BMNH; GC.

12147. MOUGEOT, Antoine. 1815?-1889. Fr. BMNH. See also 9550/1.

12148. NAUDIN, Charles Victor. 1815-1899. Fr. DSB; ICB; BMNH.

12149. PRITZEL, Georg August. 1815-1874. Ger. ICB; BMNH; GC.
1. *Thesaurus literaturae botanicae.* Leipzig, 1851.
1a. ——— Editio nova reformata. Ib., 1872-77. Reprinted, Milan, 1950, and Königstein, 1972.
2. *Iconum botanicarum index ... Die Abbildungen ... aus der botanischen und Gartenliteratur des XVIII. und XIX. Jahrhunderts in alphabetischer Folge.* Vol.I. Berlin, 1855. RM/L. 2nd ed., 1861. Vol. II. Ib., 1866.

12150. REGEL, Eduard August von. 1815-1892. Ger./Russ. ICB; BMNH; GC.

12151. SANDE LACOSTE, Cornelius Marinus van der. 1815 (or 1821)-1887. Holl. BMNH; BLA.

12152. SCHENK, (Joseph) August von. 1815-1891. Ger. BMNH; GC.
1. *Handbuch der Botanik.* 4 vols in 5. Breslau, 1881-90. (*Encyclopaedie der Naturwissenschaften.* Abt. I. Th. 1)

12153. TULASNE, Louis René. 1815-1885. Fr. DSB; ICB; BMNH; GA.

12154. VUKOTINOVIĆ, Ljudevit. 1815-1893. Agram. P III; BMNH. Also Geology.

12155. ABERLE, Karl. 1816-1892. Aus. BMNH; BLA Supp.

12156. BUCKMAN, James. 1816-1884. Br. BMNH; GD. Also Geology*.

12157. COLEMAN, William Higgins. ca. 1816-1863. Br. BMNH; GD.

12158. COLMEIRO Y PENIDO, Miguel. 1816-1901. Spain. BMNH.
1. *La botánica y los botánicos de la península Hispano-Lusitana. Estudios bibliograficos y biograficos.* Madrid, 1858.

12159. FUSS, Michael. 1816-1883. Ger. BMNH; GC.

12160. IBBOTSON, Henry. ca. 1816-1886. Br. BMNH; GD.

12161. IRMISCH, Johann Friedrich Thilo. 1816-1879. Ger. BMNH; GC.

12162. KESSLER, Hermann Friedrich. 1816-1897. Ger. BMNH.

12163. MARSSON, Theodor Friedrich. 1816-1892. Ger. P II; BMNH; GC. Also Chemistry.

12164. MOLKENBOER, Julian Hendrik. 1816-1854. Holl. BMNH.

12165. ÖRSTED, Anders Sandöe. 1816-1872. Den. BMNH.

12166. PARLATORE, Filippo. 1816-1877. It. ICB; BMNH; BLA.
1. *Flora italiana.* 10 vols. Florence, 1848-94. Contd. after his death by T. Caruel.
2. *Les collections botaniques du Musée Royal de Physique et d'Histoire Naturelle de Florence.* Florence, 1874.

12167. SCHRENCK, Alexander Gustav von. 1816-1876. Russ. P III.

12168. TULASNE, Charles. 1816-1884. Fr. ICB; BMNH.

12169. VILMORIN, Pierre Louis François Leveque de. 1816-1860. Fr. DSB; BMNH.

12170. WALPERS, Wilhelm Gerhard. 1816-1853. Ger. BMNH; GC.

12171. WILLIAMSON, William Crawford. 1816-1895. Br. DSB; BMNH; GB; GD. Also Geology.

12172. BARLA, Josef Hieronymus Jean Baptiste. 1817-1896. Fr. BMNH.

12173. BASINER, Theodor Friedrich Julius. 1817-1862. Russ. BMNH.

12174. COSTA, Antonio Cipriano. 1817-1886. Spain. BMNH.

12175. DALZELL, Nicol Alexander. 1817-1878. Br./India. BMNH; GD.

12176. FITCH, Walter Hood. 1817-1892. Br. BMNH.
1. *Illustrations of the British flora.* London, 1880. With W.G. Smith. X. Another ed., 1887.

12177. HASSALL, Arthur Hill. 1817-1894. Br. BMNH; Mort.; BLA. Also Anatomy*.

12178. HEUFLER, Ludwig von. 1817-1885. Aus. BMNH.

12179. HOOKER, Joseph Dalton. 1817-1911. Br. DSB; ICB; BMNH; GA; GB; GD2. See also Index.
1. *Contributions towards a flora of Van Diemen's Land.* London, 1840.
2. *The botany of the Antarctic voyage of H.M. Discovery Ships "Erebus" and "Terror" ... 1838-1843.* 3 vols in 6. Ib., 1844-60. RM/L. Reprinted in 4 vols, Weinheim, 1963. cf. item 11848/12.

3. *Notes of a tour on the plain of India, the Himala and Borneo. Being extracts from the private letters of Dr. Hooker.* Ib., 1849.
4. *The Rhododendrons of Sikkim-Himalaya.* Ib., 1849. Ed. by W.J. Hooker. 2nd ed., 1849.
5. *Himalayan Journals.* 2 vols. Ib., 1854. RM/L. Another ed., rev. and condensed, 2 vols, 1855. Other eds, 187-? and 1891.
6. *Illustrations of Himalayan plants.* Ib., 1855.
7. *Flora Indica.* Vol. 1. [No more publ.?] Ib., 1855. With T. Thomson. In English.
8. *Handbook of the New Zealand flora.* 2 vols. Ib., 1864-67.
9. *The student's flora of the British Isles.* Ib., 1870. X. 2nd ed., 1878.
10. *The flora of British India.* 7 vols. Ib., 1875-97.
11. *Journal of a tour in Marocco.* Ib., 1878.
12. *A sketch of the flora of British India.* Ib., 1904.

12180. KARSTEN, (Carl Wilhelm Gustav) Hermann. 1817-after 1886. Ger./Aus. P I,III; BMNH.

12181. KÖRBER, Gustav Wilhelm. 1817-1885. Ger. BMNH.

12182. MORIÈRE, Jules. 1817-1888. Fr. BMNH.

12183. MÜLLER, Carl. 1817-1870. Ger. BMNH.

12184. NÄGELI, Karl Wilhelm von. 1817-1891. Ger./Switz. DSB; ICB; P III; BMNH; GB; GC. Also Microscopy* and Evolution*.

1. *Beiträge zur wissenschaftlichen Botanik.* 4 parts in 1 vol. Leipzig, 1858-68. RM.
2. *Die niederen Pilze in ihren Beziehungen zu den Infectionskrankheiten.* Munich, 1877. RM.
3. *Untersuchungen über niedere Pilze.* Munich, 1882. With H. Buchner. RM.
4. *Pflanzenphysiologische Untersuchungen.* 4 parts. Zurich, 1855-58. With C. Cramer. X.

4a. ——— Extracts: *Die Micellartheorie.* Leipzig, 1928. (OKEW. 227)

12185. SPRUCE, Richard. 1817-1893. Br./South America. DSB; ICB; BMNH; GD.

1. *Notes of a botanist on the Amazon and Andes. Being records of travel ... during the years 1849-1864.* London, 1908. Ed. by A.R. Wallace, with a biography. RM.

12186. THOMSON, Thomas (second of the name). 1817-1878. Br./India. BMNH; GD.

12187. THURET, Gustave Adolphe. 1817-1875. Fr. DSB; GB.

12188. TUCKERMAN, Edward. 1817-1886. U.S.A. BMNH; GE.

1. *The collected lichenological papers.* 2 vols. Weinheim, 196-? Ed. by W.L. Culberson. (HNC. 9)

12189. WICHURA, Max Ernst. 1817-1866. Ger. BMNH; GC.

12190. AUERSWALD, Bernhard. 1818-1870. BMNH.

12191. BALL, John. 1818-1889. Br. ICB; P III,IV; BMNH; GB; GD1. Also Geology.

12192. (Entry cancelled)

12193. CARRIÈRE, Elie Abel. 1818-1896. Fr. BMNH.

12194. CASPARY, Johann Xaver Robert. 1818-1887. Ger. BMNH; GC.

12195. GIBSON, George Stacey. 1818-1883. Br. BMNH; GD.

12196. GRINDON, Leopold Hartley. 1818-1904. Br. BMNH. Also Biology in General*.

12197. HAZSLINSKY, Frigyes. 1818-1896. Hung. BMNH.

12198. LANGE, Johan Martin Christian. 1818-1898. Den. BMNH.

12199. MÜLLER, Johann Karl August. 1818-1899. Ger. BMNH.

12200. MUNRO, William. 1818-1880. Br./India. BMNH; GD.

12201. PAYER, Jean Baptiste. 1818-1860. Fr. BMNH.
1. *La botanique cryptogamique; ou, Histoire des familles naturelles des plantes inférieures.* Paris, 1850. 2nd ed., 1868.
2. *Traité d'organogénie comparée de la fleur.* Paris, 1857. X. Reprinted, Weinheim, 1966. (HNC. 47)
See also 12701/2.

12202. SEUBERT, Moritz August. 1818-1878. Ger. BMNH; GC.

12203. TODARO, Agostino. 1818-1892. It. BMNH.

12204. TRÉCUL, Auguste Adolphe Lucien. 1818-1896. Fr. BMNH.

12205. TREVISAN DI SAN LEON, Vittore Benedetto Antonio. 1818-1897. It. BMNH.

12206. ZOLLINGER, Heinrich. 1818-1859. Switz./East Indies. P II; BMNH; GC.

12207. ARDOINO, Honoré Jean Baptiste. 1819-1871. Fr. BMNH.

12208. BRÜCKE, Ernst Wilhelm von. 1819-1892. Ger./Aus. DSB; P I, III,IV; Mort.; BLA & Supp.; GC. Also Anatomy and Physiology*.
1. *Pflanzenphysiologische Abhandlungen.* [Periodical articles, 1844-62] Leipzig, 1898. (OKEW. 95)

12209. COSSON, Ernst Saint-Charles. 1819-1889. Fr. BMNH.
1. *Flore ... des environs de Paris.* Text & atlas. Paris, 1845. With E. Germain de Saint Pierre.
2. *Compendium florae Atlanticae ... ou, Flore des états barbaresques Algérie, Tunisie et Maroc.* 2 vols. Paris, 1881-84.

12210. CURREY, Frederick. 1819-1881. Br. BMNH; GD. See also Index.

12211. FIGUIER, Louis Guillaume. 1819-1894. Fr. P I,III; BMNH; GA. Other entries: see 3332.
1. *Histoire des plantes.* Paris, 1865. 2nd ed., 1874.
1a. ——— *The vegetable world. Being a history of plants.* London, 1869.

12212. GARCKE, Friedrich August. 1819-1904. Ger. BMNH.
1. *Flora von Nord- und Mittel-Deutschland.* Berlin, 1849. X.
1a. ——— 15th ed. *Flora von Deutschland.* Ib., 1885. 17th ed., 1895. 18th ed., 1898.

12213. HENFRY, Arthur. 1819-1859. Br. DSB; BMNH; GD. See also Index.
1. *Outlines of structural and physiological botany.* London, 1846.
2. *An elementary course of botany, structural, physiological and systematic.* London, 1857. X. 2nd ed., rev. by M.T. Masters, 1870.

12214. HOFFMANN, (Heinrich Karl) Hermann. 1819-1891. Ger. ICB; P III, IV; BMNH; GC. Also Microbiology.

12215. MERCKLIN, Carl Eugen von. 1819-1904. Russ. BMNH.

12216. MERGET, Antoine Eugène. 1819-after 1883. Fr. P III.

12217. WEDDELL, Hugh Algernon. 1819-1877. Fr. BMNH. See also 12039/4a.

12218. GERMAIN DE SAINT-PIERRE, Ernest. d. 1882. Fr. BMNH. See also 12209/1.

12219. CLEGHORN, Hugh Francis Clarke. 1820-1895. India. BMNH; GF.

12220. LAMOTTE, Martial. 1820-1883. Fr. BMNH.

12221. MOORE, Charles (second of the name). 1820-1905. Australia. BMNH.

12222. NYLANDER, Fredrik. 1820-1880. Fin. DSB; BMNH.

12223. NYMAN, Carl Fredrik. 1820-1893. Swed. BMNH.

12224. PASQUALE, Giuseppe Antonio. 1820-1893. It. BMNH.

12225. AMBROSI, Francesco. 1821-1897. It. BMNH.

12226. ANDERSSON, Nils Johan. 1821-1880. Swed. BMNH.

12227. BENTLY, Robert. 1821-1893. Br. BMNH; BLA & Supp.; GD1.
1. *A manual of botany.* London, 1861. 2nd ed., 1870. 4th ed., 1882. 5th ed., 1887.
2. *Student's guide to structural, morphological and physiological botany.* London, 1883.

12228. CLOS, Dominique. 1821-1908. Fr. BMNH.

12229. DIETRICH, Amalie, *Frau*. 1821-1891. ICB; BMNH.

12230. FUCKEL, (Karl Wilhelm Gottlieb) Leopold. 1821-1876. Ger. BMNH.
1. *Symbolae mycologicae. Beiträge zur Kenntniss der rheinischen Pilze.* Wiesbaden, 1869. X. Reprinted, Weinheim, 1966. (HNC)

12231. HILLEBRAND, Wilhelm. 1821-1886. Ger. ICB; BMNH; GC.

12232. JESSEN, Karl Friedrich Wilhelm. 1821-1889. Ger. ICB; BMNH; GC. Also History of Botany.

12233. MOORE, Thomas. 1821-1887. Br. BMNH; GD. See also 11964/17.

12234. THURBER, George. 1821-1890. U.S.A. BMNH; GE.

12235. VAUPELL, Christian Theodor. 1821-1862. Den. BMNH.

12236. WIGAND, (Julius Wilhelm) Albert. 1821-1886. Ger. DSB; BMNH; GC. Also Evolution*.

12237. WILLKOMM, (Heinrich) Moritz. 1821-1895. Ger./Dorpat/Prague. P III,IV; BMNH; GC.

12238. GUILLEMIN, Claude Marie. 1822-1890. Fr. P III. Also Physics.

12239. HANSTEIN, Johannes Ludwig Emil Robert von. 1822-1880. Ger. BMNH; GC.

12240. NYLANDER, William. 1822-1899. Fin./Fr. DSB; BMNH; BLA.
1. *Collected lichenological papers, 1853-1900*. 2 vols. Weinheim, 196-?

12241. PHILLIPS, William (second of the name). 1822-1905. Br. BMNH; GD2.

12242. PORTER, Thomas Conrad. 1822-1901. U.S.A. BMNH; GE.

12243. POST, Hampus Adolf von. 1822-after 1878. Swed. P III; BMNH.

12244. SYME (later BOSWELL), John Thomas Irvine. 1822-1888. Br. BMNH.
1. *English botany; or, Coloured figures of British plants*. 12 vols. London, 1863-86. The 3rd ed. of item 11639/1, rev. and enlarged by Syme and N.E. Brown.
1a. ——— Supplement. 4 vols. Ib., 1891-92. By N.E. Brown.

12245. THURY, Jean Marc Antoine. 1822-1905. Switz. P II,III,IV; BMNH.

12246. VASEY, George. 1822-1893. U.S.A. BMNH; GE.

12247. DUPUIS, Augustin Noël Aristide. 1823-1883. Fr. BMNH.

12248. LE JOLIS, Auguste François. 1823-1904. Fr. BMNH.

12249. METTENIUS, Georg Heinrich. 1823-1866. Ger. DSB; BMNH; BLA.

12250. NORMAN, Johannes Musaeus. 1823-1903. Nor. BMNH.

12251. PARRY, Charles Christopher. 1823-1890. U.S.A. BMNH; GE.

12252. PLANCHON, Jules Emile. 1823-1888. Fr. BMNH.

12253. PRINGSHEIM, Nathanael. 1823-1894. Ger. DSB; BMNH; GB; GC.

12254. SCHMIDT, Johann Anton. 1823-1905. Ger. BMNH.

12255. BEKETOV, Andrei Nikolaievich. 1824-1902. Russ. BMNH; BLA & Supp.

12256. HOFMEISTER, Wilhelm Friedrich Benedict. 1824-1877. Ger. DSB; ICB; P I,III; BMNH; GB; GC.
1. *Vergleichende Untersuchungen der Keimung, Entfaltung und Fruchtbildung höherer Kryptogamen ... und der Samenbildung der Coniferen*. Leipzig, 1851. X.
1a. ——— *On the germination, development and fructification of the higher Cryptogamia, and on the fructification of the Coniferae*. London, 1862. Trans. by F. Currey. (Ray Society publn. 33) RM/L.
2. *Die Lehre von der Pflanzenzelle*. Leipzig, 1867.

12257. KRELAGE, Jacob Heinrich. 1824-1901. Holl. BMNH; GF.

12258. MASSALONGO, Abramo Bartolommeo. 1824-1860. It. ICB; BMNH.

12259. MILDE, Carl August Julius. 1824-1871. Ger. BMNH; GC.

12260. MOHR, Charles Theodore. 1824-1901. U.S.A. GE.

12261. REICHENBACH, Heinrich Gustav. 1824-1889. Ger. BMNH; GA; GC.

12262. WOOSTER, David. 1824-1888. Br. BMNH.

12263. ARRONDEAU, E. Théodore. d. 1874. Fr. BMNH.

12264. ALLAN, James. 1825-1866. Br. P III; BMNH.

12265. COOKE, Mordecai Cubitt. 1825-1914. Br. BMNH.
1. *A manual of botanic terms*. London, 1862. 2nd ed., 1873.

12266. HANBURY, Daniel. 1825-1875. Br. P III; BMNH; GD.
1. *Science papers, chiefly pharmacological and botanical*. London, 1876. Ed. with a biography by J. Ince. RM/L.

12267. HOLLE, Georg von. 1825-1893. Ger. BMNH; GC.

12268. KLATT, Friedrich Wilhelm. 1825-1897. Ger. BMNH.

12269. KÜHN, Julius Gotthelf. 1825-1910. Ger. ICB; BMNH.

12270. LOWE, Edward Joseph. 1825-1900. Br. P III; BMNH. Also Meteorology*.

12271. MEJER, Ludwig. 1825-1895. Ger. BMNH; GC.

12272. MUELLER, Ferdinand Jakob Heinrich von. 1825-1896. Ger./ Australia. BMNH; GB; GC.
1. *Fragmenta phytographiae Australiae*. 12 vols. Melbourne, 1858-82. Reprinted, Amsterdam, 1974.
2. *Systematic census of Australian plants*. Part I. Vasculares. Melbourne, 1882. Supplement, 1884.
3. *Key to the system of Victorian plants*. 2 vols. Melbourne, 1885-88.
See also 7675/1a and 11971/7.

12273. OUDEMANS, Corneille Antoine Jean Abram. 1825-1906. Holl. DSB; BMNH; BLA; GF.

12274. SAINT-LAGER, Jean Baptiste. 1825-1912. Fr. BMNH. Also History of Botany.

12275. SEEMANN, Berthold Carl. 1825-1871. Ger./Br. ICB; BMNH; GC; GD.
1. *The botany of the voyage of H.M.S. "Herald" ... 1845-51*. London, 1852-57.
2. *Flora Vitiensis*. London, 1865-73. X. Reprinted, Weinheim, 1977. (HNC. 103)

12276. SOWERBY, John Edward. 1825-1870. Br. BMNH; GD.

12277. BAHRDT, Heinrich August. 1826-after 1869. Ger. P III; BMNH.

12278. BRUN, Jacques. 1826-after 1900. Switz. P III,IV; BMNH.

12279. ETTINGSHAUSEN, Contantin von. 1826-1897. Aus. BMNH; GB; GC.
1. *Die wissenschaftliche Anwendung des Naturselbstdruckes zur graphischen Darstellung von Pflanzen*. Vienna, 1856. With A. Pokorny.

12280. HABERLANDT, Friedrich von. 1826-1878. Aus. BMNH; GC.

12281. MEEHAN, Thomas. 1826-1901. U.S.A. BMNH; GE.

12282. POKORNY, Alajos (or Alois). 1826-1886. Aus. BMNH. See also 12279/1.

12283. REMY, Jules. 1826-1893. Fr. BMNH; GA.

12284. STENZEL, Karl Gustav Wilhelm. 1826-1905. Ger. BMNH.

12285. TRIANA, José Gerónimo (or Jerónimo). 1826-1890. Colombia. DSB; BMNH.

12286. WATSON, Sereno. 1826-1892. U.S.A. DSB; BMNH; GE. See also 12093/5.

12287. BAILEY, Frederick Manson. 1827-1915. Br./Australia. BMNH; GF.

12288. BAILLON, Henri Ernest. 1827-1895. Fr. BMNH.
1. *Histoire des plantes*. 13 vols. Paris, 1867-95.
1a. ——— *The natural history of plants*. 8 vols. London, 1871-88. Trans. by M.M. Hartog.
2. *Dictionnaire de botanique*. 4 vols. Paris, 1876-92.

12289. CAUVET, Désiré. 1827-1890. Fr. BMNH.

12290. DIPPEL, Leopold. 1827-1914. BMNH. Also Microscopy*.
1. *Beiträge zur vegetablischen Zellbildung*. Leipzig, 1858.

12291. GAUTHIER, Vincent. 1827-1903. Belg. ICB.

12292. HANCE, Henry Fletcher. 1827-1886. Br./China. BMNH; GD.

12293. HELLBOM, Pehr Johan. 1827-1903. Swed. BMNH.

12294. LAWSON, George. 1827-1895. Br./Canada. BMNH; GF.

12295. MAKSIMOVICH, Karl Ivanovich. 1827-1891. Russ. BMNH.

12296. PIRÉ, Louis Alexandre Henri Joseph. 1827-1887. Belg. BMNH; GF.

12297. STIZENBERGER, Ernst. 1827-1895. Switz. BMNH; BLA; GC.

12298. ARNOLD, Ferdinand Christian Gustav. 1828-1901. BMNH.
1. *Gesammelte lichenologische Schriften*. 1858-97. 3 vols. Weinheim, 1970-74. (HNC. 86)

12299. BESCHERELLE. Emile. 1828-1903. Fr. BMNH.

12300. BOLL, Jacob. 1828-1880. Switz./U.S.A. ICB; BMNH; GE.

12301. BORNET, (Jean Baptiste) Edouard. 1828-1911. Fr. BMNH.

12302. BOUDIER, (Jean Louis) Emile. 1828-1920. Fr. BMNH.

12303. BREWER, William Henry. 1828-1910. U.S.A. ICB; P III,IV; BMNH; GE.

12304. BRISSON, Théodore Polycarpe. 1828-after 1893. Fr. BMNH.

12305. COHN, Ferdinand Julius. 1828-1898. Ger. DSB; ICB; BMNH; BLA & Supp.; GB; GC. Also Microbiology.
1. *Beiträge zur Biologie der Pflanzen*. 8 vols. Breslau, 1870-98.

12306. HERDER, Ferdinand Godofried Theobald Maximilian von. 1828-1896. Russ. BMNH.

12307. MÜLLER, Jean (or Johann). 1828-1896. Switz. BMNH; GC.
1. *Gesammelte lichenologische Schriften*. 2 vols. Weinheim, 1965-67. (HNC)

12308. ROUMEGUÈRE, Casimir. 1828-1892. Fr. BMNH.

12309. ZETTERSTEDT, Johan Emanuel. 1828-1880. Swed. BMNH.

12310. ELLIS, Job Bicknell. 1829-1905. U.S.A. BMNH; GE.

12311. LINDSAY, William Lauder. 1829-1880. Br. BMNH; BLA; GD. Also Zoology*.
1. *The flora of Iceland.* Edinburgh, 1861.

12312. MÜLLER, Hermann. 1829-1883. Ger. ICB (MÜLLER-LIPPSTADT); BMNH; GC.
1. *Die Befruchtung der Blumen durch Insekten.* Leipzig, 1873. RM.
1a. ——— *The fertilization of flowers.* London, 1883. Trans. and ed. by D'A.W. Thompson, with a preface by C. Darwin. RM/L.

12313. RADLKOFER, Ludwig Adolph Timotheus. 1829-1927. Ger. BMNH.

12314. SALOMON, Carl. 1829-1899. Ger. BMNH.

12315. SCHWENDENER, Simon. 1829-1919. Switz./Ger. DSB; ICB; P III, IV; BMNH. See also 10888/1.

12316. DÉSÉGLISE, Alfred. d. 1883. Fr. BMNH.

12317. ARESCHOUG, Frederik Wilhelm Christian. 1830-1908. Swed. BMNH.

12318. BUREAU, (Louis) Edouard. 1830-1918. Fr. BMNH.

12319. CARRUTHERS, William. 1830-1922. Br. BMNH; GF.

12320. CARUEL, Teodoro. 1830-1898. It. BMNH.

12321. CRÉPIN, François. 1830-1903. Belg. ICB; BMNH.

12322. DEHÉRAIN, Pierre Paul. 1830-1902. Fr. P III,IV.

12323. LISTER, Arthur. 1830-1908. Br. BMNH; GD2.

12324. MOTELAY, Léonce. 1830-1917. Fr. ICB; BMNH.

12325. OLIVER, Daniel. 1830-1916. Br. BMNH.

12326. SCHLAGDENHAUFFEN, Charles Frédéric. 1830-1907. Fr. P II,III, IV; BMNH. Also Chemistry.

12327. BARTHÉLEMY, Aimé François Prosper. 1831-1885. Fr. P III,IV.

12328. BOEHM, Josef Anton. 1831-1893. Aus. GC.

12329. BRANTH, Jakob Severin Deichmann. 1831-1917. Den. BMNH.

12330. BUCHENAU, Franz. 1831-1906. Ger. ICB; BMNH.

12331. CRAMER, Karl Eduard. 1831-1901. Switz. BMNH. See also 12184/4.

12332. DE BARY, (Heinrich) Anton. 1831-1888. Ger. DSB; GB. Under BARY: ICB; BMNH; BLA & Supp; GC. Also Microbiology*.
1. *Morphologie und Physiologie der Pilze, Flechten und Myxomyceten.* Leipzig, 1866.
2. *Vergleichende Anatomie der Vegetationsorgane der Phanerogamen und Farne.* Leipzig, 1877. RM (under BARY)/L.
2a. ——— *Comparative anatomy of the vegetative organs of the phanerogams and ferns.* Oxford, 1884. Trans. with notes by F.O. Bower and D.H. Scott.
3. *Vergleichende Morphologie und Biologie der Pilze, Mycetozoen und Bacterien.* Leipzig, 1884. X.

3a. ——— *Comparative morphology and biology of the fungi, mycetozoa and bacteria.* Oxford, 1887. Trans. by H.E.F. Gurnsey. Rev. by I.B. Balfour. RM. Reprinted, New York, 1966.

12333. GIBELLI, Giuseppe. 1831-1898. BMNH.

12334. HALLIER, Ernst Hans. 1831-1904. Ger. DSB; P III,IV; BMNH; BLA. Also Microbiology and History of Science*.
1. *Gährungserscheinungen. Untersuchungen über Gährung, Fäulniss und Verwesung.* Leipzig, 1867.
2. *Parasitologische Untersuchungen bezüglich auf die pflanzlichen Organismen.* Leipzig, 1868.
3. *Die Parasiten der Infectionskrankheiten ... I. Die Plastiden der niederen Pflanzen.* Leipzig, 1878.

12335. HELLRIEGEL, Hermann. 1831-1895. Ger. DSB; GC.

12336. KERNER VON MARILAUN, Anton. 1831-1898. Aus. BMNH; GC.
1. *Das Pflanzleben der Donauländer.* Innsbruck, 1863. X.
1a. ——— *The background of plant ecology.* Ames, Iowa, 1951. "A translation from the German of *The plant life of the Danube basin.*"
2. [Trans.] *Flowers and their unbidden guests.* London, 1878. Ed. by W. Ogle. With a preface by Charles Darwin.
3. *Pflanzenleben.* 2 vols. Leipzig, 1887-91. X.
3a. ——— *The natural history of plants.* 2 vols. London, 1894-95. Trans. by F.W. Oliver et al.

12337. KILLEBREW, Joseph Buckner. 1831-1906. U.S.A. BMNH.

12338. RIETMANN, Othmar. 1831-1869. Switz. P III.

12339. WAWRA (VON FERNSEE), Heinrich. 1831-1887. Aus. BMNH; BLA; GC.

12340. ANDERSON, Thomas. 1832-1870. India. BMNH; GD; GF.

12341. CLARKE, Charles Baron. 1832 (or 1834)-1906. India. BMNH; GD2.

12342. FRIES, Theodor Magnus. 1832-1913. Swed. BMNH. See also 11418/23.

12343. KINDBERG, Nils Conrad. 1832-1910. Swed. BMNH.
1. *Svensk flora.* Lindkoping, 1877.

12344. LEMMON, John Gill. 1832-1908. U.S.A. GE.

12345. MACOUN, John. 1832-1920. Canada. BMNH; GF.

12346. MAW, George. 1832-after 1886. Br. P III; BMNH. Also Geology.

12347. QUÉLET, Lucien. 1832-1889. Fr. BMNH.

12348. SACHS, (Ferdinand Gustav) Julius von. 1832-1897. Ger. DSB; ICB; BMNH; GB; GC. Also History of Botany*.
1. *Lehrbuch der Botanik.* Leipzig, 1868. X. 4th ed., 1874.
1a. ——— *Text-book of botany.* Oxford, 1875. Trans. with notes by A.W. Bennett. 2nd ed., 1882.
2. *Vorlesungen über Pflanzen-Physiologie.* Leipzig, 1882. RM. 2nd ed., 1887.
2a. ——— *Lectures on the physiology of plants.* Oxford, 1887. Trans. by H.M. Ward.
3. *Gesammelte Abhandlungen über Pflanzen-Physiologie.* 2 vols. Leipzig, 1892-93.

12349. SANIO, Karl Gustav. 1832-1891. Ger. DSB; BMNH; GC.

12350. SURINGAR, Willem Frederik Reinier. 1832-1898. Holl. DSB; BMNH; GF.

12351. ALCOCK, Randal Hibbert. 1833-1885. Br. BMNH.

12352. BAIL, (Carl Adolph Emmo) Theodor. 1833-1922. Danzig. BMNH.

12353. BEAL, William James. 1833-1924. U.S.A. GE.

12354. BEBB, Michael Schuck. 1833-1895. U.S.A. ICB; BMNH.

12355. BENNETT, Alfred William. 1833-1902. Br. BMNH; GD2.
1. *A handbook of cryptogamic botany*. London, 1889. With G.R.M. Murray.
See also 12348/1a and 12434/1a.

12356. BRETSCHNEIDER, Emil. 1833-1901. Russ./China. BMNH; BLAF. Also History of Science*.
1. *Botanicon Sinicum. Notes on Chinese botany*. [Periodical articles, 1881-95] 3 vols in 2. Nendeln, 1967.

12357. CHRIST, Hermann. 1833-1933. Switz. BMNH.

12358. CROMBIE, James Morrison. 1833-1906. Br. BMNH.

12359. DELPINO, Giacomo Giuseppe Federico. 1833-1905. BMNH.
1. *Teoria generale della fillotassi*. 1883.

12360. GREMLI, August. 1833-1899. Switz. BMNH.

12361. KURZ, Sulpiz. ca. 1833-1878. Ger./Burma. BMNH; GD; GF.

12362. MARCHAND, Nestor Léon. 1833-1911. Fr. BMNH.

12363. MASTERS, Maxwell Tylden. 1833-1907. Br. BMNH; GD2.
1. *Vegetable teratology. An account of the principal deviations from the usual construction of plants*. London, 1869. (Ray Society publn. 45)
See also 12213/2.

12364. MORREN, (Charles Jacques) Edouard. 1833-1886. Belg. BMNH.

12365. PECK, Charles Horton. 1833-1917. U.S.A. BMNH; GE.

12366. SEYNES, Jules de. 1833-1912. Fr. BMNH.

12367. ASCHERSON, Paul Friedrich August. 1834-1913. Ger. BMNH.
See also 10629/1.

12368. BAKER, John Gilbert. 1834-1920. Br. DSB; BMNH.
1. *Elementary lessons in botanical geography*. London, 1875.

12369. BERNOULLI, Karl Gustav. 1834-1878. Switz. BMNH; GC.

12370. BOCQUILLON, Henri. 1834-1883. Fr. BMNH.

12371. BOLUS, Harry. 1834-1911. South Africa. BMNH.

12372. ČELAKOVSKY, Ladislav. 1834-1902. Prague. ICB; BMNH.

12373. EATON, Daniel Cady. 1834-1895. U.S.A. BMNH; GE.

12374. FOCKE, Wilhelm Olbers. 1834-1922. Ger. BMNH.
1. *Die Pflanzen-Mischlinge. Ein Beitrag zur Biologie der Gewächse*. Berlin, 1881.

12375. FOURNIER, Pierre Nicolas Eugène. 1834-1884. Fr. BMNH.

12376. FRANCHET, Adrien. 1834-1900. Fr. BMNH.

12377. KARSTEN, Petter Adolf. 1834-1917. Fin. BMNH.
1. *Mycologia Fennica*. 4 vols. 1871-78. X. Reprinted, Weinheim, 1966. (HNC)

12378. LUBBOCK, John, *1st Baron* AVEBURY. 1834-1913. Br. DSB; ICB; BMNH; GB; GD3. Other entries: see 3351.
1. *On British wild flowers considered in relation to insects*. London, 1875. 4th ed., 1882.
2. *Scientific lectures*. Ib., 1879.
3. *Flowers, fruits and leaves*. Ib., 1886. Another ed., 1888.
4. *A contribution to our knowledge of seedlings*. 2 vols. Ib., 1892. Another ed., 1896.
5. *On buds and stipules*. Ib., 1899.

12379. NITSCHKE, Theodor Rudolf Josef. 1834-1883. Ger. BMNH; GC.

12380. SCHUMACHER, Wilhelm. 1834-1888. Ger. P III; BMNH.

12381. AITCHISON, James Edward Tierny. 1835-1898. Br. BMNH.

12382. BORSHCHOV, Ilya Grigorevich. 1835?-1879. Russ. BMNH.

12383. FAMINTZUIN, Andrei Sergeivich. 1835-1918. Russ. BMNH.

12384. HENSLOW, George. 1835-1925. Br. BMNH.
1. *The origin of floral structures through insects and other agencies*. London, 1888.
2. *The making of flowers*. Ib., 1891.
3. *The origin of plant structures by self-adaptation to the environment*. Ib., 1895.

12385. HILDEBRAND, Friedrich Hermann Gustav. 1835-1915. Ger. BMNH.

12386. KABSCH, (Albert Walter) Wilhelm. 1835-1864. Ger. GC.

12387. LEITGEB, Hubert. 1835-1888. Aus. BMNH; GC.
1. *Untersuchungen über die Lebermoose*. Jena/Graz, 1874-81. X. Reprinted, Weinheim, 1968.

12388. LINDBERG, Sextus Otto. 1835-1889. Fin. BMNH.

12389. LORENTZ, Paul Günther. 1835-1881. Ger./Argentina. BMNH; GC.

12390. PEYRITSCH, Johann Josef. 1835-1889. Aus. BMNH; GC.

12391. REICHARDT, Heinrich Wilhelm. 1835-1885. Aus. BMNH; GC.

12392. SMITH, Worthington George. 1835-1917. Br. BMNH.

12393. CANDOLLE, (Anne) Casimir (Pyramus) de. 1836-1918. Switz. ICB; BMNH.

12394. DICKSON, Alexander. 1836-1887. Br. GD.

12395. RAULIN, Jules. 1836-1896. Fr. DSB.

12396. REINSCH, Paul Friedrich. b. 1836. Ger. BMNH.

12397. RENAULT, Bernhard. 1836-1904. Fr. DSB; BMNH.

12398. SCHEUTZ, Nils Johann Wilhelm. 1836-1889. Swed. BMNH.

12399. SCHWEINFURTH, Georg August. 1836-1925. Ger. ICB; P II; BMNH; GB.

12400. WOSSIDLO, Paul. 1836-after 1895. Ger. BMNH. Also Zoology*.

12401. ALLEN, Timothy Field. 1837-1902. U.S.A. BMNH; BLA & Supp.; BLAF; GE.

12402. ARDISSONE, Francesco. 1837-1910. It. BMNH.

12403. BERGGREN, Sven. b. 1837. Swed. BMNH.

12404. BOULAY, (Jean) Nicolas. 1837-1905. Fr. BMNH.

12405. JACKSON, John Reader. 1837-1920. Br. BMNH; Also History of Botany*.

12406. MARTINS, Arthur. 1837-1871. Belg. BMNH; GF.

12407. RENAULD, Ferdinand. 1837-1910. Fr. BMNH.

12408. SCHRÖTER, Josef. 1837 (or 1835)-1894. Ger. BMNH; BLA; GC. Also Microbiology.

12409. TERRACCIANO, Nicola. 1837-after 1901. It. BMNH.

12410. WEISS, (Gustav) Adolph. 1837-1891. Aus./Prague. P II,III; BMNH; GC.

12411. HEURCK, Henri Ferdinand van. 1838-1909. Belg. ICB; BMNH; GF (under VAN HEURCK). Also Microscopy*.

12412. MILLARDET, Pierre Marie Alexis. 1838-1902. Fr. BMNH.

12413. NORDSTEDT, Carl Frederik Otto. 1838-1924. Swed. BMNH.

12414. POST, George Edward. 1838-1909. U.S.A. ICB; BMNH; GE.

12415. VECHTRITZ, Rudolf Karl Friedrich von. 1838-1886. Ger. GC.

12416. VORONIN, Mikhail Stepanovich. 1838-1903. Russ. DSB; BMNH.

12417. WIESNER, Julius von. 1838-1916. Aus. DSB; P II,III,IV; BMNH.

12418. BREFELD, (Julius) Oscar. 1839-1925. Ger. DSB; BMNH; BLAF.

12419. BURRILL, Thomas Jonathan. 1839-1916. U.S.A. GE.

12420. EICHLER, August Wilhelm. 1839-1887. Ger. DSB; BMNH; GC.
1. *Blüthendiagramme construirt und erläutert*. 2 vols. Leipzig, 1875-78. RM/L.

12421. FRANK, Albert Bernhard. 1839-1900. Ger. BMNH.
1. *Beiträge zur Pflanzenphysiologie*. Leipzig, 1868.
2. *Die natürliche wagerechte Richtung von Pflanzentheilen und ihre Abhängigkeit vom Lichte und von der Gravitation*. Leipzig, 1870.

12422. HIERN, William Philip. 1839-after 1906. Br. P III; BMNH.

12423. PRYOR, Alfred Reginald. 1839-1881. Br. BMNH; GD.

12424. ROTHROCK, Joseph Trimble. 1839-1922. U.S.A. BMNH; GE.

12425. SORAUER, Paul Carl Moritz. 1839-1916. Ger. BMNH.
1. *Populäre Pflanzenphysiologie für Gärtner, etc.* Stuttgart, 1891. X.
1a. ——— *A popular treatise on the physiology of plants for the use of gardeners, etc.* London, 1895. Trans. by F.E. Weiss.

12426. THÜMEN, Felix Karl Albert Ernst Joachim von. 1839-1892. Aus. BMNH; GC.

12427. VAN TIEGHEM, Philippe Edouard Léon. 1839-1914. Fr. DSB (under T); BMNH (under T).
1. *Traité de botanique.* 2 vols. Paris, 1884.

12428. WITTROCK, Veit Brecher. 1839-1914. Swed. BMNH.

12429. WÜNSCHE, Friedrich Otto. 1839-1905. Ger. BMNH.

12430. DUSS, Antoine. 1840-1924. Fr. ICB; BMNH.

12431. HUSNOT, (Pierre) Tranquille. 1840-1929. Fr. BMNH.

12432. KING, George. 1840-1909. India. BMNH; GD2.

12433. TATE, Ralph. 1840-1901. Br./Australia. BMNH; GB; GF. Also Geology*.

12434. THOMÉ, Otto Wilhelm. 1840-after 1900. Ger. BMNH.
1. *Lehrbuch der Botanik.* Brunswick, 1869. X.
1a. ——— *Textbook of structural and physiological botany.* London, 1877. Trans. and ed. by A.W. Bennett. 5th ed., 1885. 6th ed., 1887. 8th ed., 1897.

12435. VINES, Sydney Howard. 1840 (or 1849)-1934. Br. ICB; BMNH; GD5.
1. *Lectures on the physiology of plants.* Cambridge, 1886.
See also 12519/1a.

12436. WAGNER, Hermann. 1840-1894. Ger. BMNH. Also Zoology.

12437. ARVET-TOUVET, (Jean Maurice) Casimir. 1841-1913. Fr. BMNH.

12438. BRIGHAM, William Tufts. 1841-after 1891. U.S.A. P III,IV; BMNH. Also Geology.

12439. COGNIAUX, (Celestin) Alfred. 1841-1916. Belg. BMNH.

12440. JUST, (Johann) Leopold. 1841-1891. Ger. P III; BMNH; GC.

12441. KRAUS, Gregor. 1841-1915. Ger. BMNH.

12442. RUSSOW, Edmund August Friedrich. 1841-1897. Dorpat. BMNH.

12443. SARGENT, Charles Sprague. 1841-1927. U.S.A. BMNH; GE. See also 12093/7.

12444. WARMING, (Johannes) Eugenius (Bülow). 1841-1924. Den. DSB; BMNH; GF.
1. *Haandbog i den systematiske botanik.* Copenhagen, 1879. X.
1a. ——— *A handbook of systematic botany.* London, 1895. Trans. and ed., with notes, by M.C. Potter.
2. *Plantesamfund. Grundträk af den ökologiske plantegeografi.* Copenhagen, 1895. X.
2a. ——— *Lehrbuch der ökologischen Pflanzengeographie. Eine Einführung in die Kenntnis der Pflanzenvereine.* Berlin, 1896. Ed. by E. Knoblauch. 2nd ed., rev. by P. Graebner, 1902.
2b. ——— *Oecology of plants. An introduction to the study of plant communities.* Oxford, 1909. Trans. and ed. by P. Groom and I.B. Balfour.

12445. BARBEY, William. 1842-1914. BMNH.

12446. BROWN, Robert (of Campster). 1842-1895. Br. P III; BMNH; GD1. Also Science in General*, Geology*, and Zoology.
1. *Manual of botany, anatomical and physiological.* London/ Edinburgh, 1874.

12447. HANSEN, Emil Christian. 1842-1909. Den. DSB; BMNH. Also Microbiology.

12448. HARZ, Karl Otto. 1842-1906. Ger. BMNH; BLAF.

12449. HOWELL, Thomas Jefferson. 1842-1912. U.S.A. ICB; GE.

12450. KUHN, (Friedrich Adalbert) Maximilian. 1842-1894. Ger. BMNH; GC.

12451. RIMPAU, Wilhelm. 1842-1903. Ger. BMNH.

12452. SOLMS-LAUBACH, Hermann M.C.L.F. 1842-1915. Ger. BMNH.
1. *Einleitung in die Paläophytologie vom botanischen Standpunkt.* Leipzig, 1887. X.
1a. ——— *Fossil botany. Being an introduction to palaeophytology from the standpoint of the botanist.* Oxford, 1891. Trans. by H.E.F. Garnsey. Rev. by I.B. Balfour.

12453. STAUB, Móricz. 1842-1904. Hung. BMNH.

12454. BLYTT, Axel Gudbrand. 1843-1898. Nor. P IV; BMNH. Also Meteorology.

12455. BRANDEGEE, Townshend Stith. 1843-1925. U.S.A. ICB; BMNH; GE.

12456. CORNU, Maxime. 1843-1901. Fr. BMNH.

12457. DODEL (-PORT), Arnold. 1843-1908. Switz. BMNH.

12458. GREENE, Edward Lee. 1843-1915. U.S.A. ICB; BMNH; GE.

12459. HECKEL, Edouard Marie. 1843-1916. Fr. ICB; BMNH.

12460. HEMSLEY, William Botting. 1843-1924. Br. BMNH.

12461. KANITZ, Agost. 1843-1896. Hung. BMNH.

12462. KUNTZE, (Carl Ernst) Otto. 1843-1907. Ger. DSB Supp.; BMNH.
1. *Revisio generum plantarum vascularium omnium atque cellularium multarum secundum leges nomenclaturae internationales.* 3 vols. Leipzig, 1891-93.

12463. LUERSSEN, Christian. 1843-1916. Ger. BMNH.

12464. THISELTON-DYER, William Turner. 1843-1928. Br. DSB; BMNH (DYER); GD4.

12465. TIMIRYAZEV, Kliment Arkadievich. 1843-1920. Russ. DSB; ICB.

12466. TRIMEN, Henry. 1843-1896. Br./Ceylon. ICB; BMNH; GD.

12467. BORBÁS, Vincze. 1844-1905. Hung. BMNH.

12468. ENGLER, (Heinrich Gustav) Adolf. 1844-1930. Ger. DSB Supp.; ICB; BMNH.
1. *Versuch einer Entwicklungsgeschichte der Pflanzenwelt.* 2 vols. Leipzig, 1879-82. Reprinted, Weinheim, 1972. (HNC. 91)
2. *Die natürlichen Pflanzenfamilien.* Many vols. Leipzig, 1887-1915. With K. Prantl.

3. *Syllabus der Vorlesungen über specielle und medicinisch-pharmaceutische Botanik.* Berlin, 1892. 2nd ed., 1898.
4. *Das Pflanzenreich.* Many vols. Leipzig, 1900-. Reprinted, Weinheim, 1956-.

12469. FARLOW, William Gilson. 1844-1919. U.S.A. ICB; BMNH; GE.

12470. KLEIN, Gyula. b. 1844. Hung. BMNH.

12471. McNAB, William Ramsay. 1844-1889. Br. BMNH; GD.
1. *Botany. Outlines of morphology and physiology.* London, 1878. X. 2nd ed., 1879. 3rd ed., 1880. 5th ed., 1886. Another ed., 1888.
2. *Botany. Outlines of classification of plants.* London, 1878. Another ed., 1879. 4th ed., 1886. Another ed., 1889.

12472. SCHEFFER, Rudolph Herman Christian Carel. 1844-1880. Holl./East Indies. BMNH.

12473. STRASBURGER, Eduard Adolf. 1844-1912. Ger. DSB; ICB; BMNH; Mort. Also Cytology*.
1. *Wirkung des Lichtes und der Wärme auf Schwärmsporen.* Jena, 1878.
2. *Die Angiospermen und die Gymnospermen.* Jena, 1879.
3. *Das botanische Prakticum. Anleitung zum Selbstudium der mikroskopischen Botanik.* Jena, 1884. X. 2nd ed., 1887.

3a. ——— *Handbook of practical botany.* London, 1887. Trans. with notes by W. Hillhouse. 2nd ed., 1889. 3rd ed., 1893.

4. *Neue Untersuchungen über den Befruchtungsvorgang bei den Phanerogamen als Grundlage für eine Theorie der Zeugung.* Jena, 1884.
5. *Lehrbuch der Botanik.* Jena, 1894. With F. Noll, H. Schenck, and A.F.W. Schimper. X.

5a. ——— *A text-book of botany.* London, 1898. Trans. by H.C. Porter.

12474. WOLKOW, Alexander. 1844-after 1875. Russ. P III.

12475. GHOBI, Khristofor Yakovlevich. fl. 1874-1886. Russ. BMNH.

12476. RISHAVI, Ludvig Albertovich. fl. 1873-1885. Russ. BMNH.

12477. RODRÍGUEZ (Y FEMENÍAS), Juan Joaquin. fl. 1874-1894. Spain. BMNH.

12478. SHELL, Yulian Karlovich. fl. 1872-1883. Russ. BMNH.

12479. SOROKIN, Nikolai Vasilevich. fl. 1870-1892. Russ. BMNH.

12480. LICOPOLI, Gaetano. d. 1897. It. BMNH.

12481. ALVERSON, Andrew. 1845-1916. U.S.A. ICB.

12482. BESSEY, Charles Edwin. 1845-1915. U.S.A. DSB; BMNH; GE.

12483. DUTHIE, John Firminger. 1845-1922. India. BMNH.

12484. HAUCK, Ferdinand. 1845-1889. Aus. BMNH; GC.

12485. PFEFFER, Wilhelm Friedrich Philipp. 1845-1920. Ger. DSB; P III,IV; BMNH; Mort.
1. *Pflanzenphysiologie. Ein Handbuch des Stoffwechsels und Kraftwechsels in der Pflanze.* 2 vols. Leipzig, 1881. X.

1a. ——— *The physiology of plants. A treatise upon the metabolism and sources of energy in plants.* 3 vols. Oxford, 1900-06. Trans. and ed. by A.J. Ewart.

12486. REESS, Max Ferdinand Friedrich. 1845-1901. Ger. BMNH.

12487. SACCARDO, Pietro Andrea. 1845-1920. It. BMNH.

12488. BRITTEN, James. 1846-1924. Br. DSB; BMNH.
1. *A biographical index of British and Irish botanists.* London, 1893. With G.S. Boulger.
See also 1451/3a and 11560/1.

12489. DINGLER, Hermann. 1846-1935. Ger. BMNH.

12490. HIERONYMUS, Georg Hans Emo Wolfgang. 1846-1921. BMNH.

12491. JACKSON, Benjamin Daydon. 1846-1927. Br. BMNH. Also Biography*.
1. *Guide to the literature of botany.* London, 1881. Reprinted, New York, 1964.
2. *Index Kewensis. An enumeration of the genera and species of flowering plants.* 2 vols. Oxford, 1893-95. "Compiled at the expense of the late C.R. Darwin, under the direction of J.D. Hooker, by B.D. Jackson." Reprinted, ib., 1960. Supplements, 1901 and later.

12492. KJELLMAN, Frans Reinhold. 1846-1907. Swed. BMNH.

12493. MARION, Antoine Fortuné. 1846-1900. Fr. DSB; BMNH. Also Zoology.

12494. PFITZER, Ernst Hugo Heinrich. 1846-1906. Ger. BMNH.

12495. ROHRBACH, Paul. 1846-1871. Ger. BMNH; GC.

12496. ZOPF, Wilhelm Friedrich. 1846-1909. Ger. P IV; BMNH; BLAF. Also Microbiology.

12497. BATALIN, Aleksander Fedorovich. 1847-1896. Russ. ICB.

12498. BURBIDGE, Frederick William. 1847-1905. Br. BMNH; GD2.

12499. LUNDSTRÖM, Axel Nicolaus. 1847-1906. Swed. BMNH.
1. *Pflanzenbiologische Studien.* 2 vols. Uppsala, 1884-87.

12500. MASSEE, George Edward. 1847-1917. Br. BMNH. Also Evolution*.

12501. NICHOLSON, George. 1847-1908. Br. BMNH; GD2.

12502. SAGORSKI, Ernst Adolf. 1847-after 1910. Ger. P IV; BMNH.

12503. VÖCHTING, Hermann. 1847-1917. Ger. BMNH.
1. *Über Organbildung im Pflanzenreich. Physiologische Untersuchungen.* Bonn, 1878-84.

12504. ZEILLER, René Charles. 1847-1915. Fr. DSB; BMNH.

12505. ALLEN, Grant. 1848-1899. Br. BMNH (A., Charles Grant); GB; GD1. Also Evolution* and Biography*.
1. *The colours of flowers.* London, 1882.
2. *The story of the plants.* London, 1895. 2nd ed., 1896.

12506. ARNELL, Hampus Wilhelm. b. 1848. Swed. BMNH.

12507. COLLINS, Frank Shipley. 1848-1920. U.S.A. ICB; GE.

12508. COMES, Orazio. 1848-1917. It. BMNH.

12509. DARWIN, Francis. 1848-1925. Br. DSB; BMNH; GD4. See also Index.
1. *Practical physiology of plants*. Cambridge, 1894. With E.H. Acton. 2nd ed., 1895.
2. *The elements of botany*. Ib., 1895. X. 2nd ed., 1896. 3rd ed., 1899.

12510. GREEN, Joseph Reynolds. 1848-1914. Br. BMNH. Also Chemistry*.
1. *A manual of botany*. 2 vols. London, 1895-96.

12511. KOEHNE, (Bernhard Adalbert) Emil. 1848-1918. Ger. BMNH.

12512. URBAN, Ignatz. 1848-1931. Ger. BMNH.

12513. VRIES, Hugo de. 1848-1935. Holl. DSB; ICB; P IV (under DE); BMNH. Also Heredity.
1. *De invloed der temperatuur op de levensverschijnselen der planten*. The Hague, 1870. RM.
2. *De voeding der planten*. Haarlem, 1876. RM.
3. *Het leven der bloem*. Haarlem, n.d. RM.
4. *Opera e periodicis collata*. 7 vols. Utrecht, 1917-20.

12514. WINTER, Heinrich Georg. 1848-1887. Ger. GC.

12515. BROTHERUS, Victor Ferdinand. 1849-1929. Fin. BMNH.

12516. BURBANK, Luther. 1849-1926. U.S.A. ICB; GE.

12517. DUDLEY, William Russel. 1849-1911. U.S.A. BMNH; GE.

12518. JÖNSSON, Bengt. b. 1849. Swed. BMNH.

12519. PRANTL, Karl Anton Eugen. 1849-1893. Ger. BMNH; GC.
1. *Lehrbuch der Botanik*. Leipzig, 1874. X.
1a. ——— *An elementary text-book of botany*. London, 1880. Ed. by S.H. Vines. X. 4th ed., 1885. 5th ed., 1890.
See also 12468/2.

12520. REINKE, Johannes. 1849-1931. Ger. P IV; BMNH.

12521. SCHMALHAUSEN, Johannes Theodor (or Ivan Fedorovich). 1849-1894. Russ. BMNH.

12522. SPALDING, Volney Morgan. 1849-1918. U.S.A. GE.

12523. BRITZELMAYR, Max. fl. 1879-1897. Ger. BMNH.

12524. STEPHANI, Franz. fl. 1879-1914. BMNH.

12525. YATABE, Ryokichi. d. 1899. Japan. BMNH.

12526. CRIÉ, Louis. 1850-after 1900. Fr. BMNH.

12527. DETMER, Wilhelm Alexander. 1850-1930. BMNH.
1. *Das pflanzenphysiologische Praktikum*. Jena, 1888. X. 2nd ed., 1895. X.
1a. ——— *Practical plant physiology*. London, 1898.

12528. DRUCE, George Claridge. 1850-1932. Br. ICB; BMNH; GD5.

12529. GANDOGER, Michel. 1850-1926. Fr. BMNH.
1. *Flora Europae terrarumque adjacentium*. 27 vols. Paris, 1883-91.

12530. KOCH, Ludwig Konrad Albert. 1850-1937. Ger. BMNH.

12531. KRUILOV (or KRYLOV), Porfirii Nikitich. b. 1850. Russ. BMNH.

12532. ROSTAFINSKI, Josef Tomasz. 1850-1928. Poland. BMNH.

12533. SCHMITZ, (Carl Johann) Friedrich. 1850-1895. Ger. BMNH; GC.

12534. BEIJERINCK, Martinus Willem. 1851-1931. Holl. DSB Supp.; ICB; BMNH. Also Microbiology*.
1. *Verzamelde geschriften*. 5 vols. The Hague, 1921-22.

12535. BERTRAND, Charles Eugène (or Charles Egmont). 1851-1917. Fr. DSB; BMNH.

12536. COULTER, John Merle. 1851-1928. U.S.A. ICB; BMNH; GE.

12537. HANSEN, Adolph. 1851-1920. Ger. BMNH.

12538. KIRCHNER, (Emil Otto) Oskar. 1851-1925. Ger. BMNH.

12539. LUDWIG, Friedrich. 1851-1918. Ger. BMNH.

12540. ROUY, Georges. 1851-1924. Fr. BMNH.

12541. SCHUMANN, Karl Moritz. 1851-1904. Ger. BMNH.

12542. SYDOW, Paul. 1851-1925. Ger. BMNH.

12543. TRAIL, James William Helenus. 1851-1919. Br. ICB; BMNH.

12544. TREUB, Melchior. 1851-1910. Holl./Java. DSB; ICB; BMNH.

12545. BORZI, Antonino. 1852-after 1915. It. BMNH.

12546. DRUDE, (Carl George) Oscar. 1852-1933. Ger. P IV; BMNH.
1. *Atlas der Pflanzenverbreitung*. Gotha, 1887.
2. *Handbuch der Pflanzengeographie*. Stuttgart, 1890.

12547. FLAHAULT, Charles Henri Marie. 1852-1935. Fr. BMNH.

12548. GUIGNARD, (Jean Louis) Léon. 1852-1928. Fr. DSB.

12549. HOLWAY, E.W.D. 1852-1923. U.S.A. BMNH; GF.

12550. MANGIN, Louis Alexandre. 1852-1937. Fr. DSB; BMNH.

12551. WESTERMAYER, Maximilian. 1852-1903. Ger./Switz. ICB; BMNH.

12552. BALFOUR, Isaac Bayley. 1853-1922. Br. DSB; BMNH; GD4. See also Index.

12553. BONNIER, Gaston. 1853-1922. Fr. DSB; ICB; BMNH.

12554. TRABUT, Louis. 1853-1929. Algeria. BMNH.

12555. UNDERWOOD, Lucien Marcus. 1853-1907. U.S.A. BMNH; GE.

12556. WAINIO, Edvard August. 1853-1929. Fin. BMNH.

12557. ATKINSON, George Francis. 1854-1918. U.S.A. BMNH; GE.

12558. BEHRENS, Wilhelm Julius. 1854-1903. BMNH.
1. *Methodisches Lehrbuch der allgemeinen Botanik*. Brunswick, 1880. X. 2nd ed., 1882. X.
1a. ——— *Text-book of general botany*. Edinburgh, 1885. Rev. by P. Geddes.

12559. BERTHOLD, Gottfried Dietrich Wilhelm. 1854-1937. Ger. BMNH.

12560. ELFVING, Fredrik Emil Wolmar. 1854-1942. Fin. ICB; BMNH.

12561. GEDDES, Patrick. 1854-1932. Br. ICB; BMNH; GD5. Also Evolution* and Cytology*.
1. *Chapters in modern botany*. 1893. 2nd ed., 1899.
See also 12558/1a.

12562. HABERLANDT, Gottlieb Friedrich Johann. 1854-1945. Aus./Ger. DSB; ICB; BMNH.
1. *Eine botanische Tropenreise. Indo-Malayische Vegetationsbilder und Reiseskizzen*. Leipzig, 1893.
See also 11982/4.

12563. HANSGIRG, Antonín. 1854-1917. Prague. BMNH.

12564. KNUTH, Paul Erich Otto Wilhelm. 1854-1900. Ger. DSB; BMNH; GC.

12565. MILLSPAUGH, Charles Frederick. 1854-1923. U.S.A. GE.

12566. SCOTT, Dukinfield Henry. 1854-1934. Br. DSB; ICB; BMNH; GD5.
1. *An introduction to structural botany*. 2 vols. London, 1894-96. 2nd ed., 1894-97.
See also 12332/2a.

12567. SMITH, Annie Lorrain, *Miss*. 1854-1937. Br. BMNH; GF.

12568. WARD, Harry Marshall. 1854-1906. Br. BMNH; GD2. See also 12348/2a.

12569. LAGERHEIM, (Nils) Gustaf. fl. 1884-1903. Swed. BMNH.

12570. ROTHERT, Władysław. fl. 1885-1909. Russ. BMNH.

12571. BOWER, Frederick Orpen. 1855-1948. Br. DSB; ICB (POWER); BMNH; GD6. See also 12332/2a.

12572. COCKAYNE, Leonard. 1855-1934. New Zealand. ICB.

12573. DURAND, Théophile. 1855-1912. Belg. BMNH.
1. *Index generum phanerogamorum ... promulgatorum in Benthami et Hookeri "Genera plantarum."* Brussels, 1888. An index to 11971/6.

12574. FOSLIE, Mikael Heggelund. 1855-1909. Nor. BMNH.

12575. GOEBEL, Karl. 1855-1932. Ger. DSB; ICB; BMNH.
1. *Grundzüge der Systematik und speciellen Pflanzenmorphologie*. Leipzig, 1882. RM.
1a. ——— *Outlines of classification and special morphology of plants*. Oxford, 1887. Trans. by H.E.F. Garnsey. Rev. by I.B. Balfour.
2. *Pflanzenbiologische Schilderungen*. 2 vols in 3. Marburg, 1889-93. RM.
3. *Organographie der Pflanzen*. Jena, 1898-1901.

12576. KOHL, (Georg) Friedrich. 1855-1910. Ger. BMNH.

12577. KRABBE, (Heinrich) Gustav. 1855-1895. Ger. BMNH; GC.

12578. LIGNIER, Elie Antoine Octave. 1855-1916. Fr. DSB.

12579. McALPINE, Archibald N. 1855-1924. Br. BMNH. See also 11252/1.

12580. MacFARLANE, John Muirhead. 1855-1943. Br./U.S.A. ICB.

12581. VELENOVSKÝ, Josef. 1855-1949. Prague. BMNH.

12582. VOLKENS, Georg. 1855-1917. Ger. BMNH.
1. *Die Flora der aegyptisch-arabischen Wüste*. Berlin, 1887.

12583. BOKORNY, Thomas. 1856-1933. Ger. P IV.

12584. HEINRICHER, Emil. 1856-1934. Aus. BMNH.

12585. LECOMTE, (Paul) Henri. 1856-1934. Fr. ICB; BMNH.

12586. LINDMAN, Carl Axel Magnus. 1856-1928. Swed. BMNH.

12587. MATSUMURA, Jinzo. b. 1856. Japan. BMNH.

12588. MOLISCH, Hans. 1856-1937. ICB; BMNH.

12589. PENZIG, (Alberto Giulio) Ottone. 1856-1929. It. BMNH.

12590. SCHIMPER, Andreas Franz Wilhelm. 1856-1901. Ger. DSB; BMNH.
1. *Pflanzen-Geographie auf physiologischer Grundlage*. Jena, 1898.
1a. ——— *Plant geography upon a physiological basis*. Oxford, 1903. Trans. by W. Fisher. Rev. and ed. by I.B. Balfour. Reprinted, Weinheim, 1960. (HNC. 2)
See also 12473/5.

12591. TSCHIRCH, (Wilhelm Oswald) Alexander. 1856-1939. Ger./Switz. ICB; P IV; BMNH; BLAF.

12592. COSTANTIN, Julien Noël. 1857-1936. Fr. DSB; BMNH.

12593. GRAVIS, Auguste. 1857-1937. BMNH.
1. *De l'influence de la lumière sur la végétation*. Brussels, 1880.

12594. HULT, Ragnar. 1857-1899. Fin. BMNH.

12595. JOHANNSEN, Wilhelm Ludvig. 1857-1927. Den. DSB; BLAF. Also Heredity.

12596. KLEBS, Georg Albrecht. 1857-1918. Ger./Switz. DSB; BMNH.

12597. NAVASHIN, Sergy Gavrilovich. 1857-1930. Russ. DSB; BMNH.

12598. NIEDENZU, Franz. 1857-1937. Ger. BMNH.
1. *Handbuch für botanische Bestimmungsübungen*. Leipzig, 1895.

12599. POTONIÉ, Henry. 1857-1913. Ger. BMNH.

12600. PRAIN, David. 1857-1944. Br./India. ICB; BMNH; GD6.

12601. REINITZER, Friedrich. 1857-1927. Prague/Aus. P IV.

12602. STAPF, Otto. 1857-1933. Aus./Br. ICB; BMNH.

12603. TANFILEV, Gavril Ivanovich. 1857-1928. Russ. DSB.

12604. TRELEASE, William. 1857-1945. U.S.A. DSB; ICB; BMNH. See also 12079/1.

12605. BAILEY, Liberty Hyde. 1858-1954. U.S.A. DSB; ICB; BMNH. Also Evolution*.
1. *Plant breeding*. New York, 1895.

12606. BARNES, Charles Reid. 1858-1910. U.S.A. BMNH; GE.

12607. ERRERA, Léo Abram. 1858-1905. Belg. DSB; ICB; BMNH.

12608. FISCHER, Alfred. 1858-1913. BMNH. Also Microbiology and Cytology*.

12609. KIHLMAN (or KAIRAMO), Alfred Oswald. b. 1858. Fin. BMNH.

12610. MURRAY, George Robert Milne. 1858-1911. Br. DSB; BMNH; GD2. See also 12355/1.

12611. PAX, Ferdinand (Albin). 1858-1942. Ger. BMNH.

12612. POST, Tomas Erik von. 1858-1912. Swed. BMNH.

12613. SCHINZ, Hans. 1858-1942. Switz. ICB; BMNH. See also 12573/2.

12614. SPEGAZZINI, Carlos. 1858-1926. Argentina. BMNH.

12615. THAXTER, Roland. 1858-1932. U.S.A. DSB; ICB; BMNH; GE.

12616. WEHMER, Carl Friedrich Wilhelm. 1858-after 1919. Ger. P IV; BMNH.

12617. WILLE, (Johan) Nordal (Fischer). 1858-1924. Nor. BMNH.

12618. BRITTON, Nathaniel Lord. 1859-1934. U.S.A. DSB; BMNH; GE1.
1. *An illustrated flora of the northern United States, Canada, [etc.]* 3 vols. New York, 1896-98. With A. Brown. Reprinted, ib., 1970.

12619. CAMPBELL, Douglas Houghton. 1859-1953. U.S.A. DSB; ICB; BMNH. Also Evolution*.

12620. ELFSTRAND, Mårten. 1859-after 1905. Swed. BMNH.

12621. GARDINER, Walter. 1859-1941. Br. ICB.

12622. IHNE, Egon. 1859-after 1929. Ger. P IV.

12623. MAIDEN, Joseph Henry. 1859-1925. Australia. BMNH; GF.

12624. MÖBIUS, Martin August Johannes. 1859-after 1937. BMNH.

12625. PALLADIN, Vladimir Ivanovich. 1859-1922. Russ. DSB; ICB; BMNH.

12626. SCHUETT, Franz. 1859-1921. Ger. BMNH.

12627. WARBURG, Otto. 1859-1938. Ger. BMNH; GF.

Imprint Sequence

12628. *ENCYCLOPÉDIE méthodique.* [Main entry: 3000]
1. *Encyclopédie méthodique ... Botanique.* 16 vols & 4 vols of plates. Paris, 1789-1823. RM/L. Contents as follows.
1a. ——— Botanique. Par M. de Lamarck (continuée par J.L.M. Poiret). 8 vols & supp. of 5 vols. 1789-1817.
1b. ——— Tableau encyclopédique et méthodique des trois règnes de la nature. Botanique. Illustrations des genres.... Par M. de Lamarck (continuée par J.L.M. Poiret). 3 vols. 1793, 1823.
1c. ——— Forêts et bois ... Arbres et arbustes.... Par M. L.M. Blanquart de Septfontaines. Ces deux parties précédées de la Physiologie végétale, par M. J. Senebier. 2 parts in 1 vol. 1791-1815.

12629. FITTON, Elizabeth. ICB; BMNH.
1. *Conversations on botany.* London, 1817. (This anonymous

work is also attributed to Jane Marcet.) X. Other eds, 1825 and 1828. 9th ed., 1840.

12630. SOCIETY for the Promotion of Popular Instruction. London.
1. *Popular cyclopaedia of natural sciences. Vegetable physiology.* London, 1841.
1a. ——— American ed. *Popular treatise on vegetable physiology.* Philadelphia, 1842.

12631. NEW YORK [State]. Natural History Survey.
1. *Natural history of New York: Botany. A flora of the State of New York.* 2 vols. Albany, 1843. By J. Torrey. RM.

12632. RAY SOCIETY. London. (Main entry: 11266)
1. *Reports on the progress of zoology and botany, 1841-1842.* London, 1845. (Ray Society publn. 1)
2. *Reports and papers on botany.* Ib., 1846. (Ray Society publn. 6)
3. *Reports and papers on botany.* Ib., 1849. (Ray Society publn. 16) Ed. by A. Henfrey.
4. *Botanical and physiological memoirs.* Ib., 1853. (Ray Society publn. 24) Ed. by A. Henfrey. RM/L.
No further items of this kind were published by the Society.

12633. WILKES Expedition. (Main entry: 11291)
1. *United States exploring expedition during the years 1838-42, under the command of Charles Wilkes.* [Botany] Vols 14-17. Philadelphia, 1854-74. Reprinted, Weinheim, 1971. (HNC. 87)

12634. DULAU and Co. London.
1. *Catalogue of botanical works (containing over 10,000 titles) offered for sale.* London, 1891-92.

12635. ROYAL Botanic Gardens. Kew.
1. *Catalogue of the library.* London, 1899.

3.14 ZOOLOGY

Including comparative anatomy.

In the second half of the nineteenth century a large number of titles of very specialised monographs has been omitted.

Minor authors who wrote on both zoology and botany are listed under Natural History.

12636. MERIAN, Maria Sibylla. 1647-1717. Holl. ICB (MARIAN); BMNH; GA; GC; GF. Also Botany and Part 2*.
1. *Metamorphosis insectorum Surinamensium.* Amsterdam, 1705.
1a. ——— *Die schönesten Tafeln aus dem grossen Buch der Schmetterlinge und Pflanzen, "Metamorphosis insectorum Surinamensium."* Hamburg, 1964. Ed. with introd. and notes by G. Nebel.
2. *De europische insecten.* Amsterdam, 1730.
3. *The wondrous transformations of caterpillars: Fifty engravings selected from "Erucarum ortus" (1718).* London, 1978. Introd. by W.T. Stearn.

12637. VINCENT, Levinus. 1658-1727. Holl. BMNH; GF.

12638. FRISCH, Johann (or Jodocus) Leonhard. 1666-1743. Ger. P I; BMNH; GA; GC. Also History of Science.
1. *Beschreibung von allerley Insecten in Teutsch-Land.* Berlin, 1721. RM.

12639. SCHEUCHZER, Johann Jacob. 1672-1733. Switz. DSB; ICB; P II; BMNH; BLA; GA; GB; GC. Also Geology* and Natural History*.
1. *Piscium querelae et vindiciae.* Zurich, 1708. RM.

12640. LINCK, Johann Heinrich (first of the name). 1674-1734. Ger. P I; BMNH; GA; GC.

12641. RÉAMUR, René Antoine Ferchault de. 1683-1757. Fr. DSB; ICB; P II,III; BMNH; Mort.; GA; GB. Alsp Physics, Chemistry*, and Embryology*.
1. *Mémoires pour servir à l'histoire des insectes.* 6 vols. Paris, 1734-42.
2. [Trans.] *The natural history of ants, from an unpublished manuscript in the archives of the Academy of Sciences in Paris.* London, 1926. Trans. with notes by W.M. Wheeler.

12642. KLEIN, Jacob Theodor. 1685-1759. Danzig. DSB; P I; BMNH; GA; GC. Also Geology*.
1. *Descriptiones tubulorum marinorum.* Danzig, 1731. RM/L.
2. *Naturalis dispositio echinodermatum.* Ib., 1734.
3. *Historiae piscium naturalis.* 5 parts. Ib., 1740-49.
4. *Quadrupedum dispositio brevisque historia naturalis.* Leipzig, 1751. RM.

12643. GUALTIERI, Niccola. 1688-1747. It. P I; BMNH; BLA.
1. *Index testarum conchyliorum*. 5 parts. Florence, 1742.

12644. KADE, David. 1688-1763. BMNH.

12645. ALBIN, Eleazar. fl. 1713-1759. Br. ICB; BMNH; GA; GD.
1. *A natural history of English insects*. London, 1720. RM. Another ed., with notes by W. Derham, 1735.

12646. BOUGEANT, Guillaume Hyacinthe. 1690-1743. Fr. P I; GA. Also Science in General*.
1. *Amusement philosophique sur le langage des bêtes*. Paris, 1739. X. Critical ed. by H. Hastings, Geneva, 1954.

12647. GINANNI, Giuseppe. 1692-1753. It. BMNH; GA.
1. *Delle uova e dei nidi degli uccelli*. Venice, 1737. RM.
2. *Opere postume*. 2 vols. Venice, 1755-57. RM.

12648. LESSER, Friedrich Christian. 1692-1754. Ger. P I; BMNH. Also Geology*.

12649. BIANCHI, (Simon) Giovanni (pseudonym PLANCUS, Janus). 1693-1775. It. ICB; BMNH (PLANCUS); BLA & Supp.; GA.
1. *Jani Planci ... De conchis minus notis liber*. Venice, 1739. RM.

12650. EDWARDS, George. 1694-1773. Br. ICB; P I; BMNH; GA; GB; GD. Also Geology*.
1. *A natural history of uncommon birds and of some other rare and undescribed animals*. 4 parts. London, 1743-51.
2. *Gleanings of natural history*. 3 parts. Ib., 1758-64.
3. *Essays upon natural history*. Ib., 1770. X. Reprinted, Chicheley, 1972.

12651. PEYSSONNEL, Jean André. 1694-1759. Fr. DSB. Also Oceanography.

12652. REIMARUS, Hermann Samuel. 1694-1765. Ger. BMNH; GA; GB; GC.
1. *Allegemeine Betrachtungen über die Triebe der Thiere*. Hamburg, 1760. X. 4th ed., 1798.

12653. BAKER, Henry. 1698-1774. Br. DSB; ICB; P I; BMNH; Mort.; BLA & Supp.; GA; GB; GD. Also Microscopy*
1. *An attempt towards a natural history of the polype*. London, 1743. RM/L.

12654. LA CHESNAYE-DESBOIS, François Alexandre Aubert de. 1699-1784. Fr. BMNH (CHESNAYE-D.); GA (DESBOIS). Also Botany*.
1. *Dictionnaire raisonné et universel des animaux*. 4 vols. Paris, 1759. RM/L.

12655. SELLIUS, Godofredus. d. 1767. BMNH.

12656. L'ADMIRAL, Jacob. 1700-1770. Holl. BMNH.

12657. ARTEDI, Peter. 1705-1735. Swed. DSB; ICB; BMNH; BLA & Supp; GA; GB.
1. *Ichthyologia; sive, Opera omnia de piscibus*. Leiden, 1738. Ed. with a biography by C. Linnaeus. RM. Reprinted, Weinheim, 1962. (HNC. 15)
1a. ——— Another ed., 2 vols, Greifswald, 1788-93. Rev. and ed. by J.J. Wahlbaum.

1a. ——— Extract: *Genera piscium.* 1792. Rev. and ed. by J.J. Wahlbaum. X. Reprinted, Weinheim, 1967. (HNC. 53)

12658. HUTH, Georg Leonhardt. 1705-1761. Ger. BMNH; BLA; GA.

12659. ROESEL VON ROSENHOF, August Johann. 1705-1759. Ger. DSB; BMNH.
1. *Historia naturalis ranarum nostratium ... Die natürliche Historie der Frösche hiesigen Landes.* Nuremberg, 1758.

12660. LYONET (or LYONNET), Pierre (or Pieter). 1706-1789. Holl. DSB; ICB; P I; BMNH; GA; GF.
1. *Traité anatomique de la chenille qui ronge le Bois de Saule.* The Hague, 1760. X. Another ed., enl., 1762.

12661. VOET, Jan Eusebius. 1706-1778. Holl. BMNH; GF.

12662. HILL, John. 1707?-1775. Br. DSB; ICB; P I; BMNH; BLA; GA; GB; GD. Other entries: see 3105.
1. *An history of animals.* London, 1752. (Vol. 3 of his *General natural history*) RM.

12663. LINNAEUS (later VON LINNÉ), Carl. 1707-1778. Swed. DSB; ICB; P I; BMNH. Under LINNÉ: BLA & Supp.; GA; GB. Also Natural History*#, Botany*, and Personal Writings*. See also Index.
1. *Fauna Svecica.* Stockholm, 1746. RM.
1a. ——— *Faunae Svecicae inchoatae pars prima.* Leipzig, 1800. Rev. and ed. By A.J. Retzius.
2. *Museum S.R.M. Adolphi Friderici Regis ... in quo animalia rariora imprimis et exotica ... describuntur et determinantur.* Tom. I. Stockholm, 1754. X.
2a. ——— *Tomi secundi prodromus.* Ib., 1764.
3. *Museum S.R.M. Ludovicae Ulricae Reginae ... in quo animalia rariora, exotica, imprimis insecta et conchilia, describuntur et determinantur.* Stockholm, 1764.

Extracts from the *Systema naturae* (10958/1)

4. *Institutions of entomology. Being a translation of Linnaeus's "Ordines et genera insectorum."* London, 1773. Trans. from the 12th ed. by T.P. Yeats.
5. *Entomologia, faunae Svecicae descriptionibus, aucta.* 4 vols & atlas. Lyons, 1789. Ed. and enlarged by C. de Villiers. From the 12th ed.
6. *The animal kingdom.* London, 1792. Trans. from the 13th ed., with additions, by R. Kerr.

Students' Theses

For the collected ed. of the theses, *Amoenitates academicae*, see 10958/10.
7. *Dissertatio corallia Baltica adumbrans, quam ... praeside ... C. Linnaeo ... submittit H. Fougt.* Uppsala, 1745.

Posthumous Works

8. *Methodus avium Svecicarum.* Uppsala, 1907. Ed. by E. Lönnberg. (The MS dates from 1731.)

12664. CLERCK, Carl Alexander. 1709-1765. Swed. DSB; BMNH; GA.

12665. OWEN, Charles. d. 1746. Br. ICB; BMNH; GD.

12666. ELLIS, John. ca. 1710-1776. Br. ICB; BMNH; GA; GD. Also Botany.
1. *An essay towards a natural history of the corallines and other marine productions*. London, 1755. RM/L.
2. *The natural history of many curious and uncommon zoophytes collected ... by the late J. Ellis ... Systematically arranged and described by the late D. Solander*. London, 1786. RM/L.

12667. TREMBLEY, Abraham. 1710-1784. Switz/Holl. DSB; ICB; P II; BMNH; Mort.; GA.
1. *Mémoires pour servir à l'histoire d'un genre de polypes d'eau douce*. Leiden, 1744. RM/L.
See also 10872/1 and 14479/1a.

12668. HÉRISSANT, François David. 1714-1771. Fr. BLA; GA.

12669. BROGIANI, Dominico. b. 1716. It. BMNH; BLA.
1. *De veneno animantium naturali et adquisitio*. Florence, 1752.

12670. DAUBENTON, Louis Jean Marie. 1716-1800. Fr. DSB Supp; ICB; P I; BMNH; BLA & Supp.; GA; GB. Also Mineralogy, Natural History, and Botany. See also Index.

12671. SEVERINI, János. 1716-1789. Hung. BMNH.

12672. COSTA, Emanuel Mendes da. 1717-1791. Br. P I; BMNH; GD. Also Geology*.
1. *Elements of conchology*. London, 1776. RM (under MENDES DA COSTA)/L.
2. *Historia naturalis testaceorum Britanniae; or, The British conchology*. London, 1778.

12673. GRISELINI, Francesco. 1717-1783. It. ICB.

12674. SCHAEFFER, Jacob Christian. 1718-1790. Ger. P II; BMNH; GA; GC. Also Botany*.
1. *Die Armpolypen in den süssen Wassern um Regensburg*. Regensburg, 1754. RM.
2. *Erstere und fernere Versuche mit Schnecken*. Ib., 1770. RM.
3. *Icones insectorum circa Ratisbonam indigenorum*. Ib., 1779. RM.

12675. SALERNE, François. d. 1760. Fr. P II; BLA. Also Geology.

12676. BONNET, Charles. 1720-1793. Switz. DSB; ICB; P I; BMNH; Mort.; BLA; GA; GB. Also Biology in General*#, Botany*, and Embryology.

12677. GEER, Carl de. 1720-1778. Swed. DSB; ICB (under DE); BMNH; GA.
1. *Genera et species insectorum*. Leipzig, 1783.

12678. GUÉNEAU DE MONTBÉLIARD, Philibert. 1720-1785. Fr. ICB; P I; BMNH; GA.

12679. SELIGMANN, Johann Michael. 1720-1762. BMNH.

12680. VOSMAER, Arnout. 1720-1799. Holl. BMNH.

12681. DODD, James Solas. 1721-1805. Br. BMNH; GD.
1. *An essay towards a natural history of the herring*. London, 1752.

12682. NOZEMAN, Cornelius. 1721-1786. Holl. BMNH; GF.

12683. CAMPER, Peter. 1722-1789. Holl. DSB; ICB; P I; BMNH; Mort.; BLA & Supp.; GA; GB. Also Anatomy and Physiology*.
1. *Oeuvres qui ont pour l'objet l'histoire naturelle, la physiologie et l'anatomie comparée.* 3 vols & atlas. Paris, 1803. Ed. by H.J. Jansen.

12684. HUBER, Jean. 1722-1786. Switz. P I; BMNH; GA.

12685. BLOCH, Marcus Eliezer. 1723-1799. Ger. ICB; BMNH; BLA & Supp.; GA; GB; GC.
1. *Oeconomische Naturgeschichte der Fische Deutschlands.* 3 vols. Berlin, 1782-85.
2. *Naturgeschichte der ausländischen Fische.* 9 vols in 2. Berlin, 1785-95.
3. *Ichthyologie; ou, Histoire naturelle, générale et particulière, des poissons.* 12 vols. Paris, 1785-97. Another ed., 12 vols, Berlin, 1795-97.
4. *Histoire naturelle des poissons ... par Bloch. Ouvrage classé d'après le système de Linné ... par R.R. Castel.* 10 vols. Paris, 1801. X. 3rd ed., 1837.
5. *Systema ichthyologiae.* 1801. Ed. by J.G. Schneider. X. Reprinted, Weinheim, 1967. (HNC. 55)

12686. BRISSON, Mathurin Jacques. 1723-1806. Fr. DSB; P I; BMNH; GA; GB. Also Physics*.
1. *Ornithologia.* 6 vols. Paris, 1760. Parallel text in Latin and French. RM/L. Supplement, 1760. RM.

12687. PODA VON NEUHAUS, Nicolaus. 1723-1798. Aus. P II. Also Geology.

12688. SCOPOLI, Giovanni Antonio. 1723-1788. It./Hung. ICB; P II; BMNH; GA. Also Mineralogy*, Geology, Natural History*, and Botany*.
1. *Entomologia Carniolica.* Vienna, 1763. RM/L.
2. *Ornithological papers from his "Deliciae florae et faunae Insubricae."* [See 10997/3] London, 1882. (Willughby Society publn.) Ed. by A. Newton.

12689. BOHADSCH, Joannes Baptista. 1724-1772. Prague. ICB (BOHÁČ, J.K.); P I; BMNH; BLA; GA.

12690. STUBBS, George. 1724-1806. Br. ICB; BMNH; GD.
1. *The anatomy of the horse.* London, 1766. Reprinted, ib., 1965.
2. *The anatomical works.* Ib., 1974. Ed. by T. Doherty.

12691. WALBAUM, Johann Julius. 1724-1799. Ger. BMNH; BLA.

12692. BAKEWELL, Robert (first of the name). 1725-1795. Br. ICB; GA; GB; GD.

12693. DRURY, Dru. 1725-1803. Br. ICB; BMNH; GD.
1. *Illustrations of natural history, wherein are exhibited ... figures of exotic insects according to their different genera, with a particular description of each insect.* 3 vols. London, 1770-82.

12694. FAVART D'HERBIGNY, Christophe Elisabeth. 1725-1793. BMNH.

1. *Dictionnaire d'histoire naturelle, qui concerne les testacées, ou les coquillages de mer, de terre et d'eau douce*. 3 vols. Paris, 1775.

12695. GEOFFROY, Etienne Louis. 1725-1810. Fr. DSB; BMNH; Mort.; BLA; GA.
1. *Histoire abrégée des insectes qui se trouvent aux environs de Paris*. 2 vols. Paris, 1762. RM.
2. *Traité sommaire des coquilles, tant fluviatiles que terrestres, qui se trouvent aux environs de Paris*. Paris, 1767. RM.
3. *Entomologia Parisiensis*. 2 vols. Paris, 1785. Ed. by A.F. de Fourcroy and sometimes attributed to him. RM.

12696. MÜLLER, Philipp Ludwig Statius. 1725-1776. BMNH. See also 10958/2.

12697. SPALOWSKY, Joachim Johann Nepomuk Anton. 1725-1797. BMNH.

12698. CETTI, Francesco. 1726-1778. It. ICB; P III; BMNH.

12699. GÜNTHER, Friedrich Christian. 1726-1774. BMNH.

12700. PENNANT, Thomas. 1726-1798. Br. DSB; ICB; BMNH; GA; GB; GD. Also Natural History* and Autobiography*.
1. *The British zoology*. London, 1766. X. Another ed., 4 vols, 1768-70. RM. 4th ed., 4 vols, 1776-77.
2. *Indian zoology*. London, 1769. X. 2nd ed., 1790.
3. *Synopsis of quadrupeds*. Chester, 1771. RM.
3a. ——— 2nd ed. *History of quadrupeds*. 2 vols. London, 1781. 3rd ed., 1793.
4. *Genera of birds*. Edinburgh, 1773. RM. Another ed., London, 1781.
5. *Arctic zoology*. 2 vols. London, 1784-85. Supplement, 1787.

12701. ADANSON, Michel. 1727-1806. Fr. DSB; ICB; P I; BMNH; GA; GB. Also Botany*.
1. *Histoire naturelle du Sénégal. Coquillages*. Paris, 1757. RM/L.
1a. ——— *A voyage to Senegal*. London, 1759. RM.
2. *Cours d'histoire naturelle fait en 1772*. 2 vols. Paris, 1845. Introd. and notes by J. Payer.

12702. ENGRAMELLE, Marie Dominique Joseph. 1727-1805. Fr. ICB; BMNH; GA.

12703. HUNTER, John. 1728-1793. Br. DSB; ICB; P I; BMNH; Mort.; BLA & Supp.; GA; GB; GD. Also Various Fields*#, Geology*, and Anatomy.
1. *Observations on certain parts of the animal anatomy*. London, 1786. X. 2nd ed., 1792. Another ed., with notes by R. Owen, 1837.

12704. FORSTER, Johann Reinhold. 1729-1798. Ger./Br. DSB; ICB; P I & Supp.; BMNH; GA; GC. Also Mineralogy*, Geology*, Natural History*, and Botany*. See also Index.
1. *A catalogue of the animals of North America*. London, 1771. RM.
2. *Animals of Hudson's Bay*. [Periodical article, 1772] London, 1882. (Willughby Society publn.)

3. *Descriptiones animalium quae in itinere ad maris australis terras per annos 1772-1774 suscepto collegit.* Berlin, 1844. Ed. by H. Lichtenstein. Reprinted, London, 1882. (Willughby Society publn.)

12705. MARTINI, Freidrich Heinrich Wilhelm. 1729-1778. Ger. P II; BMNH; BLA; GA; GC.
1. *Neues systematisches Conchylien-Cabinet.* 12 vols. Nuremberg,1769-1829. Vols. 4-11 by J.H. Chemnitz and Vol. 12 by G.H. Schubert and J.A. Wagner.

12706. BODDAERT, Pieter. 1730-ca. 1796. Holl. ICB; BMNH; BLA & Supp.; GA.

12707. CHEMNITZ, Johann Hieronymus. 1730-1800. Ger./Den. P I; BMNH.
1. *Von einem Geschlechte vielschalichter Conchylien mit sichtbaren Gelenken, welche beym Linné Chitons Heissen.* Nuremberg, 1784.
See also 12705/1.

12708. GRONOVIUS, Laurentius Theodorus. 1730-1777. Holl. ICB; P I & Supp.; BMNH; GA. Also Natural History*.
1. *Zoophylacium Gronovianum, exhibens animalia ... quae in museo suo ... systematice disposuit et descripsit.* 3 parts. Leiden, 1763-81.

12709. MÜLLER, Otto Frederik. 1730-1784. Den. DSB; ICB; BMNH; BLA; GA (MULLER). Also Botany and Microbiology*.
1. *Fauna insectorum Friedrichsdalina.* Copenhagen, 1764. RM.
2. *Von Würmern des süssen und salzigen Wassers.* Ib., 1771. RM.
3. *Vermium terrestrium et fluviatilium.* 2 vols. Ib., 1773-74. RM/L.
4. *Zoologiae Danicae prodromus.* Ib., 1776. RM/L.
5. *Hydrachnae, quas in aquis Daniae palustribus detexit.* Leipzig, 1781. RM.
6. *Entomostraca, seu insecta testacea, quae in aquis Daniae et Norvegiae reperit.* Leipzig, 1785. RM.
7. *Animalcula infusoria et marina.* Copenhagen, 1786. RM.

12710. BUC'HOZ, Pierre Joseph. 1731-1807. Fr. P I; BMNH; BLA; GA. Also Geology*, Natural History*, and Botany.
1. *Les dons merveilleux ... de la nature dans le règne animal.* 2 vols. Paris, 1782.
2. *Histoire des insectes nuisibles.* Paris, 1781. RM.

12711. GIRARDI, Michele. 1731-1797. It. ICB; BLA; GA.

12712. GOEZE, Johann August Ephraim. 1731-1793. Ger. BMNH; BLA & Supp. Also Various Fields*.
1. *Versuch einer Naturgeschichte der Eingeweidewürmer thierischen Körper.* Blakenburg, 1782. RM/L. Supplement, 1800.

12713. HARRIS, Moses. ca. 1731-ca. 1785. Br. BMNH; GD.
1. *An exposition of English insects.* London, 1776. X. Another ed., 1782.

12714. BERGSTRÄSSER, Johann Andreas Benignus. 1732-1812. Ger. P I; BMNH; GA; GC.

12715. FORSSKÅL, Peter. 1732-1763. Den. DSB; ICB. Under FORSKÅL: P I; BMNH; GA; GB. Also Botany*.
1. *Descriptiones animalium ... quae in itinere orientali observavit.* Copenhagen, 1775. Ed. by C. Niebuhr. RM/L.

12716. GOUAN, Antoine. 1733-1821. Fr. ICB; BMNH; GA. Also Botany*.
1. *Historia piscium ... Histoire des poissons.* Strasbourg, 1770. Parallel text in Latin and French. RM/L.

12717. MONRO, Alexander (second of the name). 1733-1817. Br. DSB; ICB; P II; BMNH; Mort.; BLA; GA; GD. Also Anatomy*.
1. *The structure and physiology of fishes explained and compared with those of man and other animals.* Edinburgh, 1785. RM.

12718. SOLANDER, Daniel Carl. 1733-1782. Swed./Br. DSB; ICB; P II; BMNH; GA; GD. Also Botany. See also Index.

12719. ENGRAMELLE, Jacques Louis Florentin. 1734-1814. ICB.

12720. FOUCHER D'OBSONVILLE, ... 1734-1802. Fr. BMNH; GA.
1. *Essais philosophiques sur les moeurs de divers animaux étrangers.* Paris, 1783.

12721. KLEEMAN, Christian Friedrich Karl. 1735-1789. Ger. BMNH; GA.

12722. SCHRÖTER, Johann Samuel. 1736-1808. Ger. P II; BMNH; GC. Also Geology*.
1. *Über den innern Bau der See- und einiger ausländischen Erd- und Flusschnecken.* Frankfurt, 1783. RM.

12723. SULZER, Johann Heinrich. 1735-1813. BMNH.
1. *Die Kennzeichen der Insekten, nach Anleitung des ... Linnaeus.* Zurich, 1761. RM.

12724. BRÜNNICH, Morten Thrane. 1737-1827. Den./Nor. P I; BMNH. Also Geology.
1. *Zoologiae fundamenta, praelectionibus academicis accomodata.* Copenhagen, 1772.

12725. HERMANN, Johann (or Jean). 1738-1800. Fr. ICB; P I; BMNH; BLA; GA.
1. *Affinitatum animalium tabulam....* Strasbourg, 1777. The thesis of one of his students, G.C. Würtz. X.
1a. ——— Another ed. *Tabula affinitatum animalium ... Nunc uberiore commentario.* Ib., 1783.

12726. LOSANA, Matteo. 1738-1833. It. BMNH; GA.

12727. MODEER, Adolf. 1739-1799. Swed. P II & Supp.; BMNH. Also Geology.

12728. SCHREBER, Johann Christian Daniel von. 1739-1810. Ger. ICB; P II; BMNH; BLA; GC. Also Geology and Botany.
1. *Die Säugethiere in Abbildungen ... mit Beschreibungen.* Erlangen, 1774-1810. X.
1a. ——— Another ed., ib., 7 vols, 1774 (or rather 1817?) -1846? Ed. by G.A. Goldfuss and later by J.A. Wagner. Supplement, 5 parts, 1840-55.
See also 10958/10,11a.

12729. LATHAM, John. 1740-1837. Br. ICB; BMNH; GA; GD.
1. *A general synopsis of birds.* 3 vols. London, 1781-85.

Supplement, 1787. 2nd supplement, 1802.
2. *Index ornithologicus; sive, Systema ornithologiae.* London, 1790.
3. *A general history of birds.* 10 vols. Winchester, 1821-24. Index, 1828.

12730. SCHNELLENBERG, Johann Rudolf. 1740-1806. Switz. BMNH; GC.

12731. AMOREUX, Pierre Joseph. 1841-1824. Fr. BMNH; BLA; GA. Also Botany.
1. *Notice des insectes de la France, réputés venimeux.* Paris, 1789. RM.

12732. PALLAS, Peter Simon. 1741-1811. Russ. DSB; ICB; P II; BMNH; BLA; GA; GB; GC. Also Geology*, Natural History*, and Botany*.
1. *Miscellanea zoologica.* The Hague, 1766. RM/L.
2. *Spicilegia zoologica.* 2 vols. Berlin, 1767-80.

12733. ALESSANDRI, Innocenzio. b. ca. 1742. BMNH.

12734. BORN, Ignaz von. 1742-1791. Aus. DSB; P I; BMNH; GA; GB; GC. Also Geology*.
1. *Index rerum naturalium Musei Caesarei Vindobonensis. I. Testacea.* Vienna, 1778. Parallel text in Latin and German. X.
1a.——— Another ed. *Testacea Musei Caesarei Vindobonensis.* Ib., 1780.

12735. CARANGEOT, Arnould. 1742-1806. Fr. DSB; BMNH. Also Crystallography.

12736. ESPER, Eugen Johann Christoph. 1742-1810. Ger. P I; BMNH; GA; GC. Also Botany*.
1. *Die Schmetterlinge in Abbildungen ... mit Beschreibungen.* 11 vols. Erlangen, 1777-1830.
2. *Die Pflanzenthiere in Abbildungen ... nebst Beschreibungen.* 3 vols. Nuremberg, 1788-1830.

12737. FENN, Eleanor, *Lady*. 1743-1813. Br. BMNH; GD.
1. *A short history of insects.* London, 1797.

12738. FUESSLI, Johann Caspar. 1743-1786. Switz. ICB; BMNH.
1. *Archiv der Insectengeschichte.* 8 parts. Winterthur, 1781-86. X.
1a. ——— *Archives de l'histoire des insectes.* Ib., 1794.

12739. HELLWIG, Johann Christian Ludwig. 1743-1831. Ger. P I; BMNH; GC.

12740. HERBST, Johann Friedrich Wilhelm. 1743-1807. Ger. BMNH; GA.
1. *Versuch einer Naturgeschichte der Krabben und Krebse.* 3 vols & atlas. Berlin, 1782-1804.

12741. NICOLAS, Pierre François. 1743-1816. Fr. P II; BMNH; BLA; GA. Also Chemistry.
1. *Méthode de préparer et conserver les animaux ... pour les cabinets d'histoire naturelle.* Paris, 1801.

12742. TUNSTALL, Marmaduke. 1743-1790. Br. BMNH; GD.
1. *Ornithologia Britannica.* London, 1771. X. Reprinted, ib., 1880. (Willughby Society publn.)

12743. BEILBY, Ralph. 1744-1817. Br. BMNH; GD.

12744. ERXLEBEN, Johann Christian Polykarp. 1744-1777. Ger. ICB; P I; BMNH; BLA; GA; GC. Also Physics* and Chemistry.

12745. FABRICIUS, Otto. 1744-1822. Nor. P I; BMNH.
1. *Fauna Groenlandica*. Copenhagen/Leipzig, 1780. RM.

12746. LAMARCK, Jean Baptiste Pierre Antoine de Monet, *Chevalier* de. 1744-1829. Fr. DSB; ICB; P I; BMNH; GA; GB. Other entries: see 3204.
1. *Système des animaux sans vertèbres*. Paris, 1801.
2. *Philosophie zoologique*. 2 vols. Paris, 1809. RM. Reprinted, ib., 1873 and 1968; also Weinheim, 1960. (HNC)
2a. ——— *Zoological philosophy*. London, 1914. Trans. and introd. by H. Elliot. Reprinted, New York, 1963.
3. *Extrait du cours de zoologie du Muséum d'Histoire Naturelle sur les animaux sans vertèbres*. Paris, 1812.
4. *Histoire naturelle des animaux sans vertèbres*. 7 vols. Paris, 1815-22. RM/L. 2nd ed., rev. by G.P. Deshayes and H. Milne-Edwards, 11 vols, 1835-45.
4a. ——— *An epitome of Lamarck's arrangement of Testacea, being a free translation of that part of* [item 4]. London, 1823. By C. Dubois. Another ed., 1825.
4b. ——— *Recueil de coquilles décrites par Lamarck dans* [item 4] *et non encore figurées*. Paris, 1841. By B. Delessert.

12747. NAUMANN, Johann Andreas. 1744 (or 1747)-1826. Ger. BMNH; GA; GC.

12748. STOLL, Caspar. d. 1795. Holl. BMNH.

12749. FABRICIUS, Johann Christian. 1745-1808. Den. DSB; ICB; P I & Supp.; BMNH; GA; GB; GC.
1. *Systema entomologiae*. Flensburg, 1775.
2. *Genera insectorum*. Kiel, 1777.
3. *Philosophia entomologica*. Hamburg, 1778.
4. *Species insectorum*. 2 vols. Hamburg, 1781.
5. *Mantissa insectorum*. 2 vols. Copenhagen, 1787.
6. *Entomologia systematica*. 4 vols. Copenhagen, 1792-94. Index alphabeticus, 1796. Supplement, 1798; index, 1799.
6a. ——— *Nomenclator entomologicus, enumerans insecta omnia in J.C. Fabricii "Entomologia systematica."* Manchester, 1796.
7. *Systema Eleutheratorum*. 2 vols. Kiel, 1801.
8. *Systema Rhyngotorum*. Brunswick, 1803.
9. *Systema Piezatorum*. Brunswick, 1804.
10. *Systema Antliatorum*. Brunswick, 1805.

12750. GÜLDENSTAEDT, Anton Johann von. 1745-1781. Russ. ICB; P I; BMNH; BLA; GA; GC.

12751. HARWOOD, Busick. 1745?-1814. Br. BMNH; BLA; GA; GD.
1. *A system of comparative anatomy and physiology*. Cambridge, 1796.

12752. BESEKE, Johann Melchior Gottlieb. 1746-1802. Mitau. P I; BMNH. Also Chemistry.

12753. POLI, Giuseppe Saverio. 1746-1825. It. DSB; P II; BMNH; GA. Also Physics.

12754. SCHRANK, Franz von Paula von. 1747-1835. Ger. P II; BMNH; GC. Also Botany.
1. *Enumeratio insectorum Austriae indigenorum.* Augsburg, 1781. RM.

12755. VICQ D'AZYR, Félix. 1748-1794. Fr. DSB; ICB; BMNH; Mort.; BLA; GA. See also 13849/3b.

12756. VIEILLOT, Louis Jean Pierre. 1748-1831? Fr. BMNH.
1. *Analyse d'une nouvelle ornithologie élémentaire.* Paris, 1816. X. Reprinted, London, 1883. (Willughby Society publn.)

12757. GOETHE, Johann Wolfgang von. 1749-1832. Ger. The famous writer. DSB; ICB; P I; BMNH; Mort.; G. Other entries: see 3216.

12758. MARTYN, Thomas (second of the name). fl. 1760-1816. Br. BMNH; GD.
1. *The universal conchologist, exhibiting the figure of every known shell.* 4 vols. London, 1784-92.

12759. BRUGUIÈRES, Jean Guillaume. 1750-1799. Fr. BMNH; BLA; GA.

12760. DESFONTAINES, René Louiche. 1750-1833. Fr. ICB; P I & Supp.; BMNH; GA; GB. Also Botany*.
1. *Mémoire sur quelques nouvelles espèces d'oiseaux des côtes de Barbarie.* [Periodical article, 1787] London, 1880. (Willughby Society publn.)

12761. HUBER, François. 1750-1831. Switz. ICB; P I; BMNH; GA; GB. Also Botany*.

12762. MEIDINGER, Carl von. 1750-1820. Aus. P II; BMNH. Also Various Fields.
1. *Icones piscium Austriae indigenorum.* 5 vols. Vienna, 1785-94.

12763. SCHNEIDER, Johann Gottlob. 1750-1822. Ger. P II; BMNH; GA; GB; GC. Also History of Science.
1. *Historiae amphibiorum naturales et literariae.* 2 parts in 1. Jena, 1799-1801. Reprinted, Amsterdam, 1968.
See also 12685/5.

12764. JURINE, Louis. 1751-1819. Switz. P I; BMNH; BLA; GA.
1. *Histoire des monocles qui se trouvent aux environs de Genève.* Geneva, 1820. RM.

12765. SHAW, George. 1751-1813. Br. P II; BMNH; GA; GD. Also Natural History*. See also 3004/1.
1. *Zoology of New Holland.* London, 1794.
2. *General zoology; or, Systematic natural history.* 14 vols in 28. London, 1800-26. Vols IX-XIV by J.F. Stephens.
3. *Zoological lectures delivered at the Royal Institution.* London, 1809. RM/L.

12766. BLUMENBACH, Johann Friedrich. 1752-1840. Ger. DSB; ICB; P I; BMNH; Mort.; BLA & Supp.; GA; GB; GC. Also Natural History*, Embryology*, and Physiology*.
1. *Handbuch der vergleichenden Anatomie.* Göttingen, 1805. RM.
1a. ——— *A short system of comparative anatomy.* London, 1807. Trans. by W. Lawrence. RM. Another ed., 1827.

12767. BONNATERRE, Joseph P. ca. 1752-1804. Fr. BMNH.

12768. GYLLENHAAL, Leonhard. 1752-1840. Swed. DSB; BMNH; GA.

12769. BEWICK, Thomas. 1753-1828. Br. ICB; BMNH; GA; GB; GD.
1. *A general history of quadrupeds*. Newcastle-upon-Tyne, 1790. 2nd ed., 1791. 3rd ed., 1792. 4th ed., 1800. 6th ed., 1811. 7th ed., 1820. 8th ed., 1824.
2. *A history of British birds*. 2 vols. Ib., 1797-1804. X. Other eds, 1832 and 1847.
3. *Works. Memorial edition*. 5 vols. London, 1885-87.

12770. COQUEBERT DE MONTBRET, Antoine Jean. 1753-1825. BMNH.
1. *Illustratio iconographica insectorum*. 3 parts. Paris, 1799-1804.

12771. LEVAILLANT, François. 1753-1824. Fr. DSB; BMNH; GA.

12772. LICHTENSTEIN, Anton August Heinrich. 1753-1816. Ger. BMNH.
1. *Catalogus rerum naturalium rarissimarum ... Verzeichnis von höchstseltenen Naturalien*. 3 parts. Hamburg, 1793-96. X.
1a. ——— Reprint of Part 1 (Mammalia et Aves). London, 1882. (Willughby Society publn.)

12773. DONNDORFF, Johann August. 1754-1837. Ger. P I; BMNH. Also Physics and History of Science*.

12774. QUATREMÈRE-DISJONVAL, Denis Bernard. 1754-1830. Fr. P II; BMNH; GA. Also Chemistry.

12775. MONTAGU, George. 1755 (or 1751)-1815. Br. ICB; BMNH; GA; GD.
1. *An ornithological dictionary*. 2 vols. London, 1802. Supplement, Exeter, 1813.

12776. PANZER, Georg Wolfgang Franz. 1755-1829. Ger. BMNH; BLA.
1. *Faunae insectorum Germanicae initia; oder, Deutschlands Insecten*. 110 parts. Nuremberg, 1792-1823.
2. *Deutschlands Insectenfauna; oder, Entomologisches Taschenbuch für ... 1795*. Nuremberg, 1795.

12777. CAVOLINI, Filippo. 1756-1810. It. BMNH; GA.
1. *Memorie per servire alla storia de' polipi marini*. Naples, 1785.
2. *Opere*. Naples, 1910. Ed. by the Società di Naturalisti in Napoli. Includes a biography.

12778. HOME, Everard. 1756-1832. Br. DSB; P I; BMNH; Mort.; BLA; GA; GD.
1. *Lectures on comparative anatomy*. 4 vols. London, 1814-23.
1a. ——— Supplement. Vols V and VI. Ib., 1828.

12779. JABLONSKY, Carl Gustav. 1756-1787. Ger. BMNH; GC.

12780. LACÉPÈDE, Bernard Germain Etienne de la Ville-sur-Illon, *Comte* de. 1756-1825. Fr. DSB; ICB; P I; BMNH; GA; GB.
1. *Histoire naturelle des quadrupèdes ovipares*. Paris, 1788. X.
2. *Histoire naturelle des serpents*. Paris, 1789. X.
3. *Histoire naturelle des poissons*. 5 vols. Paris, 1798-1804.
4. *Histoire naturelle des cétacés*. Paris, 1804. X.
These four items are part of Buffon's *Histoire naturelle*. See also item 8, below.

5. *Discours d'ouverture et de clôture du cours d'histoire naturelle donné dans le Muséum National d'Histoire Naturelle*. Paris, 1798. RM.
5a. ——— ... *donné ... l'an VII....* Paris, 1799. RM.
5b. ——— ... *donné ... l'an VIII....* Paris, 1800. RM.
5c. ——— ... *donné ... l'an IX....* Paris, 1801. RM.
6. *Les ages de la nature et histoire de l'espèce humaine*. 2 vols. Paris, 1830. RM.
7. *Histoire naturelle de l'homme*. Paris, 1827. RM.
8. *Oeuvres du compte de Lacépède, comprenant l'histoire naturelle des quadrupèdes ovipares, des serpents, des poissons et des cétacés*. 12 vols. Paris, 1830-33. (Part of *Oeuvres complètes de Buffon*, nouvelle édition, suite.) RM. Other eds, 1833-36 and 1844.

See also 10954/1,1a,4.

12781. OLIVIER, Guillaume Antoine. 1756-1814. Fr. BMNH; GA.

12782. PAYKULL, Gustaf. 1757-1826. Swed. DSB; BMNH; GA.

12783. SCHÖNBAUER, Joseph Anton von. 1757-1807. Hung. P II; BMNH. Also Mineralogy.

12784. SOWERBY, James. 1757-1822. Br. DSB; P II; BMNH; GB; GD. Also Mineralogy*, Geology*, and Botany*.
1. *The British miscellany; or, Coloured figures of new, rare, or little known animal subjects*. 12 parts in 1 vol. London, 1804-06.
2. *The genera of recent and fossil shells*. 2 vols. London, 1820-34. Completed by G.B. Sowerby (first of the name).

12785. AUDEBERT, Jean Baptiste. 1759-1800. Fr. BMNH; GA; GB.

12786. BOSC (D'ANTIC), Louis Augustin Guillaume. 1759-1828. Fr. DSB; P I; BMNH; GA; GB.
1. *Histoire naturelle des coquilles*. 5 vols. Paris, 1802. X. 3rd ed., enlarged, 1836.

12787. CAMPER, Adrian Gilles. 1759-1820. Holl. BMNH; GF. Son of Peter Camper (12683) and editor of many of his works.

12788. KIRBY, William. 1759-1850. Br. BMNH; GA; GB; GD.
1. *Monographia Apum Angliae; or, An attempt to divide into their natural genera and families such species of the Linnean genus Apis as have been discovered in England*. 2 vols. Ipswich, 1802. RM.
2. *An introduction to entomology*. 3 vols. London, 1815-26. With W. Spence. 6th ed., 2 vols, 1843. 7th ed., 1856. Other eds, 1857 and 1870.
3. *On the power, wisdom and goodness of God, as manifested in the creation of animals and in their history, habits and instinct*. 2 vols. London, 1835. (Bridgewater treatises. 7). Another ed., Philadelphia, 1836. RM. Other London eds, 1838 and 1852.

12789. LUBBOCK, Richard. 1759?-1808. Br. DSB; ICB; BMNH. Also Chemistry.

12790. GRAVES, George. fl. 1777-1834. Br. BMNH.
1. *Ovarium Britannicum. Being a correct delineation of the eggs of birds ... in Great Britain*. London, 1816. RM.

12791. STEWART, Charles. fl. 1787-1817. Br. BMNH.
1. *Elements of the natural history of the animal kingdom.* 2 vols. Edinburgh, 1801. X. 2nd ed., 1817.

12792. ABBOT, John. fl. 1791-1802. ICB; BMNH.

12793. WILHELM, Gottlieb Tobias. d. 1811. BMNH.
1. *Unterhaltungen aus der Naturgeschichte der Säugethiere.* Augsburg, 1792.
2. *Unterhaltungen aus der Naturgeschichte der Vögel.* 2 vols. Ib., 1795.
3. *Unterhaltungen aus der Naturgeschichte der Insecten.* 3 vols. Ib., 1799-1800.
4. *Unterhaltungen aus der Naturgeschichte der Fische.* 2 vols. Ib., 1799-1800.

12794. MARSHAM, Thomas. d. 1819. Br. BMNH; GD.
1. *Entomologia Britannica.* Vol. 1. [No more publ.] London, 1802. RM.

12795. BROOKES, Joshua. 1761-1833. Br. BMNH; BLA; GA; GD. Also Anatomy.

12796. BROUSSONET, Pierre Marie Auguste. 1761-1807. Fr. DSB; P I; BMNH; BLA & Supp.; GA; GB. Also Botany.
1. *Ichthyologia.* London/Paris, 1782.

12797. HÜBNER, Jacob. 1761-1826. Ger. ICB; BMNH.

12798. BONSDORFF, Gabriel. 1762-1831. Fin. BMNH; BLA.

12799. GMELIN, Karl Christian. 1762-1827. Ger. BMNH; GC. Also Botany*.

12800. LATREILLE, Pierre André. 1762-1833. Fr. DSB; ICB; BMNH; GA.
1. *Genera crustaceorum et insectorum.* 4 vols. Paris, 1806-09.
2. *Considérations générales sur l'ordre naturel des animaux composant les classes des crustacés, des arachnides et des insectes.* Paris, 1810. Reprinted, Farnborough,1970.
3. *Familles naturelles du règne animal.* Paris, 1825. X.
3a. ——— *Natürliche Familien des Thierreichs.* Weimar, 1827. Trans. with notes by A.A. Berthold.

12801. TURTON, William. 1762-1835. Br. BMNH; GD. See also 10958/4 and 11014/1.

12802. PECK, William Danbridge. 1763-1822. U.S.A. GE.

12803. ROEMER, Johann Jakob. 1763-1819. Switz. BMNH; BLA; GC. Also Botany.
1. *Genera insectorum Linnaei et Fabricii.* Winterthur, 1789.

12804. FALLÉN, Carl Frederic. 1764-1835. Swed. BMNH.

12805. MAWE, John. 1764-1829. Br. P II; BMNH; GA; GD. Also Mineralogy*.
1. *The Linnean system of conchology.* London, 1823.

12806. CHAVANNES, Daniel Alexander. 1765-1846. Switz. P I.

12807. KIELMEYER, Karl Friedrich von. 1765-1844. Ger. DSB; ICB; P I; BMNH; BLA; GA; GC. Other entries; see 3260.

12808. NOËL DE LA MORINIÈRE, Simon Barthélemy Joseph. 1765-1822. Fr. GA.

12809. BARTON, Benjamin Smith. 1766-1815. U.S.A. DSB; ICB; BMNH; BLA & Supp.; GA; GB; GE. Also Botany.
1. *Notes on the animals of North America*. New York, 1974. Facsimile of the MS, written in 1792. Ed. with an introd. by K.B. Sterling.
2. *Fragments of the natural history of Pennsylvania*. Part I. Philadelphia, 1799. X. Reprinted, London, 1883. (Willughby Society publn.)

12810. WILSON, Alexander (second of the name). 1766-1813. U.S.A. DSB; ICB; BMNH; GA; GD; GE.

12811. BREMSER, Johann Gottfried. 1767-1827. Aus. BMNH; BLA & Supp.; GA.

12812. MACLEAY, Alexander. 1767-1845. Br./Australia. GD; GF.

12813. DENYS DE MONTFORT, Pierre. ca. 1768-1820. Fr. BMNH (MONTFORT).
1. *Conchyliologie systématique*. 2 vols. Paris, 1808-10.

12814. DONOVAN, Edward. 1768-1837. Br. BMNH; GB; GD.
1. *The natural history of British birds*. 10 vols in 5. London, 1794-1819?
2. *The natural history of British shells*. 5 vols. Ib., 1799-1804.
3. *An epitome of the natural history of the insects of India*. Ib., 1800. X. Another ed., rev., 1842.
4. *An epitome of the natural history of the insects of New Holland, New Zealand ... and other islands in the Indian, Southern, and Pacific Oceans*. Ib., 1805.

12815. HAWORTH, Adrian Hardy. 1768-1833. Br. DSB; BMNH; GA; GD. Also Botany*.
1. *Lepidoptera Britannica*. 4 parts. London, 1803-28.

12816. MEYER, Friedrich Albrecht Anton. 1768-1795. Ger. P II; BMNH; BLA.
1. *Systematisch-summarische Übersicht der neuesten zoologischen Entdeckungen in Neuholland und Afrika*. Leipzig, 1793.

12817. SIEMSSEN, Adolph Christian. 1768-1833. Ger. P II; BMNH. Also Geology.

12818. CUVIER, Georges. 1769-1832. Fr. DSB; ICB; P I; BMNH; GA; GB. Also Science in General*, Geology*, Biography*, and History of Science*. See also Index.
1. *Tableau élémentaire de l'histoire naturelle des animaux*. Paris, 1798.
2. *Leçons d'anatomie comparée*. 5 vols. Paris, 1800-05. Ed. by C. Duméril and G.L. Duvernoy. RM/L. Reprinted, Brussels, 1969.
2a. ——— *Lectures on comparative anatomy*. London, 1802. Trans. by W. Ross.
3. *Le règne animal*. 4 vols. Paris, 1817. RM/L. 2nd ed., enlarged, 5 vols, 1829-30. 3rd ed., ... accompagnée de planches gravées, text & atlas, 17 vols in 20, 1839-49.
3a. ——— *Das Thierreich*. 4 vols. Stuttgart, 1821-25. Trans. with additions by H.R. Schinz.
3b. ——— *Das Thierreich*. 6 vols. Leipzig, 1831-43. Trans. from the 2nd French ed., with additions, by F.S. Voigt.

3c. ——— *The animal kingdom.* 16 vols. London, 1827-35. Trans. with additions by E. Griffith. RM/L.
3d. ——— *Cuvier's animal kingdom.* 4 vols. London, 1834-36. Trans. and abridged for the use of students by H. McMurtie.
3e. ——— ——— Another ed., trans. from the 2nd French ed., 4 vols in 8, ib., 1833-37.
3f. ——— *The animal kingdom.* London, 1849. Trans. and adapted, with additions, by W.B. Carpenter and J.O. Westwood. Other eds: 1851,1854,1863,1890.
4. *Histoire naturelle des poissons.* 25 vols. Paris, 1828-49. With A. Valenciennes.
5. *Anatomie comparée. Recueil de planches de myologie.* 2 vols. Paris,1850. Ed. by Laurillard et al.

12819. TILESIUS VON TILENAU, Wilhelm Gottlieb von. 1769-1857. Ger. P II; BMNH; BLA.
1. [Trans. of periodical article] *On the mammoth or fossil elephant found in the ice ... in Siberia.* London, 1819.

12820. LEWIN, John William. fl. 1805-1814. Br./Australia. BMNH; GD (under L., William).

12821. JAUFFRET, Louis François. 1770-ca. 1850. Fr. BMNH; GA.
1. *Zoographie des divers régions.* Paris, 1799.

12822. LE PELETIER DE SAINT-FARGEAU, Amédée Louis Michel. 1770-1845. Fr. BMNH.

12823. MACARTNEY, James. 1770-1843. Br. BMNH; BLA; GD1. Also Anatomy.

12824. SUCKOW, Friedrich Wilhelm Ludwig. 1770-1838. Ger. BMNH; GC.

12825. TEMMINCK, Coenraad Jacob. 1770 (or 1778)-1858. Holl. ICB; BMNH.

12826. WIEDEMANN, Christian Rudolph Wilhelm. 1770-1840. Ger. BMNH; BLA; GC.

12827. BAKKER, Gerbrand. 1771-1828. Holl. BMNH; BLA & Supp.

12828. BREZ, Jacques. 1771-1798. It. BMNH; GA.
1. *La flore des insectophiles.* Utrecht, 1791. RM.

12829. RUDOLPHI, Karl Asmund. 1771-1832. Ger. DSB; BMNH; BLA & Supp.; GA; GC. Also Botany and Physiology.

12830. WALCKENAER, Charles Athanase. 1771-1852. Fr. P II; BMNH; GA.

12831. ALBERS, Johann Abraham. 1772-1821. Ger. P I; BMNH; BLA; GA; GC.

12832. ALTON, Josef Wilhelm Eduard d'. 1772-1840. Ger. BMNH; GA; GC.

12833. AUTENRIETH, Johann Heinrich Ferdinand von. 1772-1835. Ger. P I & Supp.; BMNH; BLA & Supp.; GA; GC. Also Physiology.

12834. DRAPARNAUD, Jacques Philippe Raymond. 1772-1804. Fr. DSB; ICB; P I; BMNH; GA.

12835. GEOFFROY SAINT-HILAIRE, Etienne. 1772-1844. Fr. DSB; ICB; P I; BMNH; Mort.; BLA; GA; GB. See also 15458/1.
1. *Philosophie anatomique. Des organes respiratoires sous le*

rapport de la détermination et de l'identité de leurs pièces osseuses. 2 vols. Paris, 1818-22. RM/L. Reprinted, Brussels, 1968.
2. *Histoire naturelle des mammifères.* 4 vols. Paris, 1819-42. With Frédéric Cuvier.
3. *Leçons sur l'histoire naturelle des mammifères.* Vol. I. Paris, 1828. RM.

12836. SCHÖNHERR, Carl Johan. 1772-1848. Swed. DSB; BMNH.

12837. CUVIER, Frédéric. 1773-1838. Fr. DSB; BMNH; GA. See also Index.
1. *Des dents des mammifères, considerées comme caractères zoologiques.* Strasbourg, 1825. RM/L.
2. *De l'histoire naturelle des cétacés.* Paris, 1836. RM.

12838. HORSFIELD, Thomas. 1773-1859. Br./Java. P III; BMNH; GD; GE. Also Botany.
1. *Zoological researches in Java and the neighbouring islands.* London, 1821-24.

12839. ROLANDO, Luigi. 1773-1831. It. DSB; Mort.; BLA. Also Anatomy and Physiology.

12840. BINGLEY, William. 1774-1823. Br. BMNH; GA; GD. Also Botany*.
1. *Animal biography; or, Authentic anecdotes of the lives, manners and economy of the animal creation.* 3 vols. London, 1802-08. X. 2nd ed., 1804. 4th ed., 1813. 6th ed., 4 vols in 2, 1824.

12841. DAUDIN, François Marie. 1774-1804. Fr. ICB; BMNH; GA.
1. *Histoire naturelle des rainettes, des grenouilles et des crapauds.* Paris, 1802. RM.

12842. DUMÉRIL, (André Marie) Constant. 1774-1860. Fr. DSB Supp.; BMNH; BLA & Supp.; GA.
1. *Traité élémentaire d'histoire naturelle.* Paris, 1804. X.
1a. ——— 3rd ed. *Elémens des sciences naturelles.* 2 vols. ib., 1825. X. 5th ed., 1846.
2. *Zoologie analytique; ou, Méthode naturelle de classification des animaux.* Paris, 1806. RM/L.
2a. ——— *Analytische Zoologie.* Weimar, 1806. Trans. by L.F. Froriep.
3. *Erpétologie générale; ou, Histoire naturelle complète des reptiles.* 9 vols in 10 & atlas. Paris, 1834-54. With G. Bibron. RM/L.
See also 12818/2.

12843. DUPONCHEL, Philogène Auguste Joseph. 1774-1846. Fr. BMNH.

12844. WOOD, William. 1774-1857. Br. ICB; BMNH; GD. Also Natural History*.
1. *Zoography; or, The beauties of nature displayed.* 3 vols. London, 1807.
2. *Index testaceologicus; or, A catalogue of shells.* London, 1818. X. Another ed., 1824-25.
3. *Index entomologicus; or, A complete illustrated catalogue ... of the lepidopterous insects of Great Britain.* London, 1839. RM.

12845. GEBAUER, Johann Jacob. d. 1849. BMNH.

1. *Systematisches Verzeichniss der Seesterne, Seeigel, Conchylien und Pflanzenthiere.* Halle, 1802.

12846. AUDINET-SERVILLE, Jean Guillaume. 1775-1858. Fr. BMNH. See also 13032/1.

12847. CLIFT, William. 1775-1849. Br. DSB; ICB; BMNH; BLA & Supp.; GD.

12848. DALYELL, John Graham. 1775-1851. Br. BMNH; GD.
1. *The powers of the Creator displayed in the creation; or, Observations on life amidst the various forms of the humbler tribes of animated nature.* 3 vols. London, 1851-58.
2. *Observations on some interesting phenomena in animal physiology exhibited by several species of Planariae.* Edinburgh, 1814. RM.
See also 14677/2b.

12849. GODART, Jean Baptiste. 1775-1825. Fr. BMNH; GA.

12850. ILLIGER, Johann Karl Wilhelm. 1775-1813. Ger. ICB; BMNH; GC. Also Natural History*.
1. *Prodromus systematis mammalium et avium.* Berlin, 1811.

12851. KLUG, Johann Christoph Friedrich. 1775-1856. Ger. BMNH; BLA.

12852. MÉNARD DE LA GROYE, François Jean Baptiste. 1775-1827. Fr. P II. Also Geology.

12853. NEERGAARD, Jens Veibel. 1775-1856. Den. BMNH; BLA.

12854. PÉRON, François. 1775-1810. Fr. DSB; ICB; P II; BMNH; GA.

12855. RANZANI, Camillo. 1775-1841. It. BMNH; GA.

12856. SCHREIBERS, Karl Franz Anton von. 1775-1852. Aus. DSB; P II; BMNH; GC.

12857. VIREY, Jules (or Julien) Joseph. 1775-1846. Fr. DSB; P III; BMNH; BLA. Also Physiology.
1. *Histoire naturelle du genre humain.* Paris, 1801.
2. *Histoire des moeurs et de l'instinct des animaux.* 2 vols. Paris, 1822. X. Reprinted, ib., 1882.

12858. BOJANUS, Ludwig Heinrich. 1776-1827. Wilno. BMNH; BLA.

12859. NÜSSLEIN, Franz Anton. 1776-1832. Ger. P II. Also Mineralogy.

12860. REINHARDT, Johannes Christopher Hagemann. 1776-1845. Den. BMNH.

12861. RUSCONI, Mauro. 1776-1849. It. ICB; BMNH; BLA. Also Embryology.

12862. SMITH, Charles Hamilton. 1776-1859. Br. BMNH; GD.
Three works on mammals in the Naturalist's Library (see 13852).

12863. TREITSCHKE, (Georg) Friedrich. 1776-1842. Ger. BMNH; GC.

12864. TREVIRANUS, Gottfried Reinhold. 1776-1837. Ger. DSB; ICB; P II; BMNH; BLA; GB; GC. Also Biology in General* and Physiology.

12865. BLAINVILLE, Henri Marie Ducrotay de. 1777-1850. Fr. DSB; P I;

BMNH; BLA; GA; GB. Also History of Biology*.
1. *Mémoire sur les bélemnites, considerées zoologiquement et géologiquement.* Paris, 1827.
2. *Ostéographie; ou, Description iconographique comparée du squelette et du système dentaire des cinq classes d'animaux vertébrés, récents et fossiles.* 4 vols & atlas of 4 vols. Paris, 1839-64.

12866. CHILDREN, John George. 1777-1852. Br. P I & Supp.; BMNH; GD. Also Chemistry*.

12867. DUVERNOY, Georges Louis. 1777-1855. Fr. BMNH; BLA; GA. Also Biography*. See also 12818/2.

12868. GRAVENHORST, Johann Ludwig. 1777-1857. Ger. P I; BMNH.

12869. SAVIGNY, (Marie) Jules (César Lelorgne de). 1777-1851. Fr. DSB; BMNH.
1. *Mémoires sur les animaux sans vertèbres.* 2 vols. Paris, 1816.

12870. SCHINZ, Heinrich Rudolf. 1777-1861. Switz. BMNH; BLA; GC. See also 12818/3a and 12888/2.

12871. DILLWYN, Lewis Weston. 1778-1855. Br. BMNH; GD.
1. *A descriptive catalogue of recent shells.* 2 vols. London, 1817.
2. *An index to the "Historia conchyliorum" of Lister, with the names of the species to which each figure belongs.* Oxford, 1823. Re Martin Lister: see 2492/2.

12872. KOCH, Carl Ludwig. 1778-1857. Ger. BMNH; GC.

12873. LESUEUR, Charles Alexandre. 1778-1846. Fr./U.S.A. DSB; ICB; BMNH; GA; GE.

12874. MACQUART, (Pierre) Justin (Marie). 1778-1855. Fr. BMNH.
1. *Diptères exotiques.* Paris, 1838.
2. *Facultés intérieures des animaux invertébrés.* Lille, 1850. RM.

12875. PEALE, Rembrandt. 1778-1860. U.S.A. DSB Supp.; BMNH; GB; GE.
1. *An historical disquisition on the mammoth ... whose fossil remains have been found in North America.* London, 1803. RM.

12876. WOLFF, Johann Friedrich. 1778-1806. BMNH.
1. *Icones Cimicum descriptionibus illustratae.* 5 parts. Erlangen, 1800-11. X.
1a. ——— *Abbildungen der Wanzen....* 5 parts. Ib., 1800-11.

12877. FRORIEP, Ludwig Friedrich von. 1779-1847. Ger. P I; BMNH; BLA. See also 12842/2a.

12878. JACOPI, Giuseppe. 1779-1813. It. BLA.

12879. OKEN (or OKENFUSS), Laurenz. 1779-1851. Ger./Switz. DSB; ICB; P II; BMNH; Mort.; BLA & Supp.; GA; GB; GC. Also Biology in General* and Natural History*. See also Index.

12880. ROSENTHAL, Friedrich Christian. 1779-1829. Ger. BMNH; BLA; GC. Also Anatomy.

12881. SCHEITLIN, Peter S. 1779-1848. Switz. BMNH; GC.

1. *Versuch einer vollständigen Thierseelenkunde.* Stuttgart, 1840. RM.

12882. THOMPSON, John Vaughan. 1779-1847. Br. DSB; BMNH; GD.
1. *Zoological researches and illustrations.* 5 parts. Cork, 1828-34. X. Reprinted in 1 vol., London, 1968.

12883. WILBRAND, Johann Bernhard. 1779-1846. Ger. DSB; BMNH; BLA; GC. Also Physiology.

12884. CHARPENTIER, Toussaint von. 1780-1847. Ger. P I; BMNH.

12885. DEJEAN, Pierre François Marie Auguste. 1780-1845. BMNH.
1. *Catalogue des coléoptères.* Paris, 1821. 3rd ed., 1837.

12886. DUFOUR, (Jean Marie) Léon. 1780-1865. Fr. ICB; BMNH.

12887. LICHTENSTEIN, (Martin) Heinrich (Carl). 1780-1857. Ger. BMNH; BLA; GA (LIECHTENSTEIN); GC. See also 12704/3 and 12888/1.

12888. NAUMANN, Johann Friedrich. 1780-1857. ICB; BMNH.
1. *Briefwechsel mit H. Lichtenstein, 1818-1856.* Copenhagen, 1954. Ed. by E. Stressemann and P. Thomsen.
2. *Die ornithologische Korrespondenz zwischen J.F. Naumann und H.R. Schinz in den Jahren 1815 bis 1835.* Odense, 1969. Ed. by E. Stressemann and L. Baege.

12889. NODIER, Charles. 1780-1844. Fr. ICB; BLA; GA; GB.

12890. SERRES (DE MESPLÈS), Marcel Pierre Toussaint de. 1780-1862. Fr. DSB; ICB; P II,III; BMNH. Also Geology*.

12891. SPINOLA, Massimiliano. 1780-1857. It. BMNH.
1. *Essai monographique sur les clérites, insectes coléoptères.* 2 vols. Genoa, 1844.

12892. JACOB, Nicholas Henri. 1781-1871. BMNH.
1. *Storia naturale delle scimie.* Milan, 1812.
See also 14131/1.

12893. KNIP, Pauline, *Madame.* 1781-1851. Fr. ICB; BMNH.

12894. MECKEL, Johann Friedrich (second of the name). 1781-1833. Ger. DSB; ICB; BMNH; Mort.; BLA & Supp.; GA; GC. Also Anatomy* and Embryology.
1. *System der vergleichenden Anatomie.* 7 vols. Halle, 1821-33.
2. *Ornithorhynci paradoxi descriptio anatomica.* Leipzig, 1826. X. Another ed., 1827.

12895. SPIX, Johann Baptist von. 1781-1826. Ger. DSB; BMNH; GC.
1. *Selecta genera et species piscium quos in itinere per Brasiliam ... collegit.* Munich, 1829. RM.
2. *Reise in Brasilien ... in der Jahren 1817-1820.* 3 vols & atlas. Munich, 1823-31. With K.F.P. Martius. "One of the most important scientific expeditions of the nineteenth century." X. Reprinted, Stuttgart, 1966-67.

12896. TIEDEMANN, Friedrich. 1781-1861. Ger. DSB; ICB; P III; BMNH; Mort.; BLA; GA; GB; GC. Also Embryology* and Physiology*.

12897. GEBLER, Friedrich August von. 1782-1850. Ger./Russ. BMNH; BLA.

12898. GOLDFUSS, Georg August. 1782-1848. Ger. P I; BMNH; GA; GB; GC. Also Geology*. See also 12728/1a.

12899. GRATELOUP, Jean Pierre Sylvestre de. 1782-1861. Fr. BMNH; BLA. Also Geology.

12900. NITZSCH, Christian Ludwig. 1782-1837. Ger. BMNH; BLA.

12901. TROLLE-WACHTMEISTER, Hans Gabriel. 1782-1871. Swed. P II,III. Also Mineralogy.

12902. WATERTON, Charles. 1782-1865. Br. DSB; ICB; BMNH; GB; GD. See also 15444/1.
1. *Essays on natural history, chiefly ornithology. With an autobiography of the author.* London, 1838. X. *Second series. With a continuation of the autobiography of the author.* Ib., 1844. X. *Third series.* Ib., 1857.
1a. ——— Another ed., 3 vols, ib., 1858-66. Another ed., by N. Moore, with a biography, 1870.
2. *Letters.* London, 1955. Ed. by R.A. Irwin.

12903. BRESCHET, Gilbert. 1783-1845. Fr. DSB; P I & Supp.; BMNH; BLA & Supp.; GA. Also Anatomy.

12904. HILDRETH, Samuel Prescott. 1783-1863. U.S.A. ICB; P III; GE. Also Geology.

12905. JACOBSON, Ludvig Levin. 1783-1843. Den. ICB; BNMH; Mort.; BLA & Supp.; GA. Also Anatomy.

12906. LAURILLARD, Charles Léopold. 1783-1853. Fr. BMNH. See also 12818/5.

12907. LAWRENCE, William. 1783-1867. Br. DSB; ICB; BMNH; BLA; GA; GD. Also Anatomy and Physiology.
1. *An introduction to comparative anatomy and physiology.* London, 1816. Another ed., 1823.
2. *Lectures on physiology, zoology and the natural history of man.* London, 1819. X. Another ed., 1822. 6th ed., 1834. 9th ed., 1848.
See also 12766/1a.

12908. RAFINESQUE, Constantine Samuel. 1783-1840. It./U.S.A. DSB; ICB; BMNH; GE. Also Natural History*, Botany*, and Autobiography*.
1. *Ichthyologia Ohiensis; or, Natural history of the fishes inhabiting the River Ohio and its tributary streams.* Lexington, Ky., 1820. RM.
2. *The complete writings on recent and fossil conchology.* New York, 1864. Ed. by W.G. Binney and G.W. Tryon. RM.

12909. SPENCE, William. 1783-1860. Br. BMNH; GD. See also 12788/2.

12910. BONELLI, Franco Andrea. 1784-1830. It. ICB; GA.

12911. DESMAREST, Anselme Gaëtan. 1784-1838. Fr. P I; BMNH.
1. *Histoire naturelle des Tangaras, des Manakins et des Todiers.* Paris, 1805-07.
2. *Considérations générales sur la classe des Crustacés.* Paris, 1825. RM/L.
See also 9218/1.

12912. LAURENT, Jean Louis Maurice. 1784-1854. Fr. BMNH; BLA; GA. Also Anatomy.

12913. LECONTE, John Eatton. 1784-1860. ICB.

12914. WALTON, John. 1784-1862. Br. BMNH.

12915. YARRELL, William. 1784-1856. Br. ICB; BMNH; GB; GD.

12916. BROWN, Thomas, *Captain*. fl. 1816-1849. Br. BMNH.

12917. HAHN, Carl Wilhelm. d. 1836. Ger. BMNH.

12918. AUDUBON, John James. 1785-1851. U.S.A. DSB; ICB; BMNH; GA; GB; GE. Also Autobiography*.
1. *Ornithological biography; or, An account of the habits of birds*. 5 vols. Edinburgh, 1831-39. RM.
2. *A synopsis of the birds of America*. Edinburgh, 1839. X. Reprinted, Amsterdam, 1972.
3. *The birds of America*. 7 vols. New York, 1840-44. X. Reprinted, ib., 1967.
4. *The viviparous quadrupeds of North America*. 3 vols & atlas of 3 vols. New York, 1845-54. With J. Bachman. Reprinted, Maplewood, N.J., 1967.
5. *Audubon and his journals*. 2 vols. New York, 1898. By Maria R. Audubon. Ed. by E. Coues. Reprinted, ib., 1960.
6. *Journal made during his trip to New Orleans in 1820-21*. Boston, 1929. Ed. by H. Corning.
7. *The 1826 journal*. Norman, Okla., 1967. Ed. by A. Ford.

12919. BURROW, Edward John. 1785-1861. Br. BMNH; GD.
1. *Elements of conchology*. London, 1815. RM/L. 2nd ed., 1818. Another ed., 1825.

12920. FLEMING, John. 1785-1857. Br. DSB; P III; BMNH; GD. Also Geology.
1. *The philosophy of zoology; or, A general view of the structure, functions, and classification of animals*. Edinburgh, 1822.
2. *A history of British animals*. Edinburgh, 1828.

12921. PANIZZA, Bartolomeo. 1785-1867. It. BMNH; Mort.; BLA. Also Physiology.

12922. VIGORS, Nicholas Aylward. 1785-1840. Br. BMNH; GD.

12923. FÉRUSSAC, André Etienne (J.P.J.F.) d'Audebard, *Baron* de. 1786-1836. Fr. ICB; P I; BMNH; GA. Also Geology.
1. *Histoire naturelle générale et particulière des mollusques terrestres et fluviatiles*. 4 vols. Paris, 1819-51. With G.P. Deshayes.
2. *Tableaux systématiques des animaux mollusques*. Ib., 1821.
3. *Histoire naturelle générale et particulière des céphalopods Acétabulifères, vivants et fossiles*. Text & atlas. Ib., 1834-48. With A. d'Orbigny.

12924. GERMAR, Ernst Friedrich. 1786-1853. Ger. P I; BMNH; GC. Also Mineralogy* and Geology*.

12925. LOW, David. 1786-1859. Br. GD. Also Chemistry*.
1. *On the domesticated animals of the British Islands*. London, 1845.

12926. MICHELIN, (Jean Louis) Hardouin. 1786-1867. Fr. BMNH. Also Geology*.

12927. NECKER (DE SAUSSURE), Louis Albert. 1786-1861. Switz. DSB; ICB; P II & Supp.; BMNH. Also Geology*.

12928. NUTTALL, Thomas. 1786-1859. Br./U.S.A. DSB; BMNH; GB; GD; GE. Also Natural History* and Botany*.

12929. OTTO, Adolph Wilhelm. 1786-1845. Ger. ICB; BMNH; BLA. See also 12950/2.

12930. SCHUMMEL, Theodor Emil. 1786-1848. Ger. BMNH.
1. *Beiträge zur Entomologie*. Breslau, 1832.

12931. SERRES, (Antoine) Etienne Reynaud Augustin. 1786-1868. Fr. DSB; ICB; BMNH; BLA; GA. Also Embryology.

12932. ATKINSON, John. 1787-1828. Br. BMNH.

12933. BREHM, Christian Ludwig. 1787-1864. Ger. BMNH; GA.

12934. CAILLIAUD, Frédéric. 1787-1869. Fr. P III; BMNH; GA.

12935. CLOQUET, Hippolyte. 1787-1840. Fr. BMNH; BLA; GA. Also Anatomy*. See also 13849/3a-d.

12936. COSTA, Oronzio Gabriele. 1787-1867. It. BMNH.

12937. DALMAN, Johan Wilhelm. 1787-1828. Swed. BMNH; BLA & Supp.

12938. GEMMELLARO, Carlo. 1787-1866. It. P III. Also Geology.

12939. MAYER, August Franz Carl. 1787-1865. Ger. BMNH; BLA.

12940. NILSSON, Sven. 1787-1883. Swed. ICB; BMNH.

12941. RENNIE, James. 1787-1867. Br. BMNH; GD.
1. *Insect transformations*. London, 1830.

12942. RICHARDSON, John. 1787-1865. Br. P II,III; BMNH; GB; GD. Alao Natural History*. See also 15585/1.
1. *Fauna Boreali-Americana; or, The zoology of the northern parts of British America*. 4 parts. London, 1829-37.
2. *The zoology of Captain Beechey's voyage ... to the Pacific and Behring's Straits performed in H.M.S. "Blossom" ... 1825-28*. London, 1839.
3. *The zoology of the voyage of H.M.S. "Erebus" and "Terror" ... 1839-43*. 2 vols. London, 1844-75. With J.E. Gray.
See also 13235/1.

12943. SAY, Thomas. 1787-1834. U.S.A. DSB; ICB; BMNH; GE.
1. *An account of the crustacea of the United States*. [Periodical article, 1817] Weinheim, 1969. (HNC. 73)
2. *American entomology*. 3 vols. Philadelphia, 1824-28. RM.
3. *The complete writings on the entomology of North America*. 2 vols. New York, 1959. Ed. by J.L. Le Conte, with a biography by G. Ord.

12944. SOWERBY, James De Carle. 1787-1871. Br. BMNH; GD. Also Geology.

12945. CREPLIN, Friedrich Heinrich Christian. 1788-1863. Ger. BMNH; BLA; GC.

12946. ROUSSEAU, Louis François Emanuel. 1788-1868. Fr. BMNH; BLA.

12947. SELBY, Prideaux John. 1788-1867. Br. BMNH; GD.
1. *A catalogue of the generic and sub-generic types of the class Aves*. Newcastle, 1840.
Also two ornithological works in the Naturalist's Library. (See 13852).

12948. SOWERBY, George Brettingham (first of the name). 1788-1854. Br. ICB; BMNH; GD. See also 12784.

12949. BRODERIP, William John. 1789-1859. Br. BMNH; GB; GD.
1. *Zoological recreations.* London, 1847.

12950. CARUS, Carl Gustav. 1789-1869. Ger. ICB; BMNH; BLA & Supp.; GA; GB; GC. Also Physiology*.
1. *Lehrbuch der Zootomie.* Text & atlas. Leipzig, 1818. X.
1a. ——— *An introduction to the comparative anatomy of animals.* London, 1827. Trans. by R.T. Gore.
1b. ——— 2nd ed. *Lehrbuch der vergleichenden Zootomie.* 2 vols & atlas. Leipzig, 1834.
1c. ——— ——— *Traité élémentaire d'anatomie comparée.* 3 vols & atlas. Paris, 1835. Trans. by A.J.L. Jourdan.
2. *Erläuterungstafeln zur vergleichenden Anatomie.* 9 parts. [Leipzig? ca. 1826-55] With A.W. Otto and E. d'Alton. X.
2a. ——— *Tables synoptiques de l'anatomie comparée.* Leipzig, 1826-55. Trans. by E. Martini.

12951. COUCH, Jonathan. 1789-1870. Br. ICB; BMNH; Mort.; GD.

12952. FORSTER, Thomas Ignatius Maria. 1789-1860. Br. P I; BMNH; GA; GD. Also Meteorology*.

12953. LA MARMORA, Alberto Ferrero de. 1789-1863. It. ICB; P I,III; BMNH (MARMORA); GA.

12954. SWAINSON, William. 1789-1855. Br./New Zealand. DSB; BMNH; GD.
1. *Zoological illustrations; or, Original figures and descriptions of new, rare, or interesting animals.* 3 vols. London, 1820-23. Second series, 3 vols, ib., 1829-33.
2. *A preliminary discourse on the study of natural history.* Ib., 1834.
3. *A treatise on the geography and classification of animals.* Ib., 1835.
4. *The natural history and classification of quadrupeds.* Ib., 1835.
5. *The natural history and classification of birds.* 2 vols. Ib., 1836-37.
6. *Animals in menageries.* Ib., 1838.
7. *The natural history of fishes, amphibians and reptiles, or monocardian animals.* 2 vols. Ib., 1838-39.
8. *Taxidermy.* Ib., 1840.
9. *A treatise on malacology.* Ib., 1840.
10. *On the habits and instincts of animals.* Ib., 1840.
11. *On the history and natural arrangement of insects.* Ib., 1840. With W.E. Shuckard.

Also two ornithological works in the Naturalist's Library (see 13852).

12955. LIPPI, Regolo. d. 1854. It. ICB; BMNH; BLA.

12956. BACHMANN, John. 1790-1874. U.S.A. BMNH; GE. See also 12918/4.

12957. BLACKWALL, John. 1790-1881. Br. BMNH; GD.
1. *Researches in zoology, illustrative of the structure, habits, and economy of animals.* London, 1834. X. 2nd ed., 1873.

12958. CLOQUET, Jules Germain. 1790-1883. Fr. BMNH; Mort.; BLA & Supp.; GA. Also Anatomy*.

12959. GRIFFITH, Edward. 1790-1858. Br. BMNH; GD.
1. *General and particular descriptions of the vertebrated animals ... order Carnivora ... order Quadrumana.* 2 vols. London, 1821.

12960. HECKEL, Johann Jakob. 1790-1857. Aus. BMNH; GC.

12961. LEACH, William Elford. 1790-1836. Br. BMNH; GD.
1. *The zoological miscellany, being descriptions of new or interesting animals.* 3 vols. London, 1814-17.
2. *Systematic catalogue of the specimens of the indigenous mammalia and birds that are preserved in the British Museum.* London, 1816. X. Reprinted, ib., 1882. (Willughby Society publn.)

12962. QUOY, Jean René Constant. 1790-1869. Fr. DSB; ICB; P II,III; BMNH; BLA.
1. *Voyage autour du monde ... sur ... "l'Uranie" et la "Physicienne" ... 1817-20: Zoologie.* Text & atlas. Paris, 1824-26. With J.P. Gaimard.
2. *Voyage de ... "l'Astrolabe" ... 1826-29: Zoologie.* 4 vols & atlas of 2 vols. Paris, 1830-35. With J.P. Gaimard.

12963. BOWDICH, Thomas Edward. 1791-1824. Br. P III; BMNH; GA; GD.

12964. BREMI-WOLF, Johann Jakob. 1791-1857. Switz. BMNH.

12965. CURTIS, John. 1791-1862. Br. BMNH; GD1.

12966. JAN, Georg. 1791-1866. It. BMNH.
1. *Iconographie générale des Ophidiens.* 3 vols. Milan, 1860-81. With F. Sordelli. X. Reprinted, 3 vols in 1, Weinheim, 1961.

12967. KNIGHT, Charles. 1791-1873. Br. BMNH; GA; GB; GD.

12968. MENKE, Karl Theodor. 1791-1861. Ger. P III; BMNH; BLA. Also Geology.

12969. SWAN, Joseph. 1791-1874. Br. BMNH; GD. Also Anatomy*.

12970. ALDER, Joshua. 1792-1867. Br. BMNH; GD.

12971. BELL, Thomas. 1792-1880. Br. BMNH; GD.
1. *A history of the British quadrupeds.* London, 1837.
2. *A history of the British reptiles.* Ib., 1839. X. 2nd ed., 1849.

12972. HEUSINGER (VON WALDEGG), Karl Friedrich. 1792-1883. Ger. BMNH; BLA; GA; GC. Also Physiology.

12973. LEA, Isaac. 1792-1886. U.S.A. DSB; P I,III; BMNH; GE. Also Geology*.

12974. MACLEAY, William Sharp. 1792-1865. Br./Australia. BMNH; GD.
1. *Horae entomologicae; or, Essays on the annulose animals.* Vol. I, Parts 1 & 2. [No more publ.] London, 1819-21. X. Another ed., 1829.

12975. STEPHENS, James Francis. 1792-1853. Br. BMNH; GD.
1. *Illustrations of British entomology.* 11 vols. London, 1827-45. Supplement, 1846.

2. *A systematic catalogue of British insects.* 2 parts. Ib., 1829. RM/L.
3. *The nomenclature of British insects.* Ib., 1829.
4. *A manual of British coleoptera, or beetles.* Ib., 1839.
See also 12765/2.

12976. THON, Theodor. 1792-1838. Ger. BMNH; GC.

12977. ESCHSCHOLZ, Johann Friedrich. 1793-1831. Dorpat. DSB; ICB; P I; BMNH; BLA; GA; GB.
1. *Entomographien.* Berlin, 1824.
2. *Zoologischer Atlas, enthaltend Abbildungen und Beschreibungen neuer Thierarten, während des Flottcapitans von Kotzebue zweiter Reise um die Welt ... in ... 1823-26, beobachtet von F. Eschscholtz.* 5 parts. Ib., 1829-33.

12978. GRANT, Robert Edmund. 1793-1874. Br. ICB; BMNH; BLA; GD.
1. *Outlines of comparative anatomy.* London, 1841.

12979. HEYDEN, Karl Heinrich Georg von. 1793-1866. Ger. BMNH; GC.

12980. KIRTLAND, Jared Potter. 1793-1877. U.S.A. ICB; BMNH; BLA; GE.

12981. RATHKE, (Martin) Heinrich. 1793-1860. Ger. DSB; BMNH; Mort.; BLA; GC. Also Embryology*.
1. *Beiträge zur Geschichte der Thierwelt.* 4 parts. Halle, 1820-27.
2. *Beiträge zur vergleichenden Anatomie und Physiologie.* Danzig, 1842.

12982. REICHENBACH, Heinrich Gottlieb Ludwig. 1793-1879. Ger. ICB; BMNH; GA; GC. Also Botany*.
1. *Avium systema naturale ... Das natürliches System der Vögel ... Ornithologie méthodique.* Berlin, 1849-52.

12983. THIENEMANN, Friedrich August Ludwig. 1793-1858. Ger. P II; GC.

12984. CHIAJE, Stefano delle. 1794-1860. It. BMNH; BLA.
1. *Descrizione et anatomia degli animali invertebrati della Sicilia.* 8 vols. Naples, 1841-44.

12985. DESMOULINS, (Louis) Antoine. 1794-1828. Fr. BMNH; BLA; GA.

12986. EUDES-DESLONGCHAMPS, Jacques Armand. 1794-1867. Fr. BMNH; GB (DESLONGCHAMPS). Also Geology.

12987. FLOURENS, (Marie Jean) Pierre. 1794-1867. Fr. DSB; ICB; BMNH; Mort.; BLA; GA; GB. Also Evolution*, Embryology*, Physiology*, and Biography*.
1. *De l'instinct et de l'intelligence des animaux. Résumé des observations de Frédéric Cuvier.* Paris, 1841. X. Another ed., 1870.
2. *De la vie et de l'intelligence.* Paris, 1858.
3. *Ontologie naturelle; ou, Etude philosophique des êtres.* Paris, 1861. X. 3rd ed., 1864.

12988. FOHMANN, Vincent. 1794-1837. Ger./Belg. BMNH; BLA; GF. Also Anatomy.

12989. GARNOT, Prosper. 1794-1838. Fr. DSB; BMNH; BLA. See also 12992/1.

12990. HOLBROOK, John Edwards. 1794-1871. U.S.A. DSB Supp.; BMNH; GA; GE.

12991. KELLNER, August. 1794-1883. Ger. BMNH; GC.

12992. LESSON, René Primevère. 1794-1849. Fr. DSB; ICB; BMNH; GA.
1. *Voyage autour du monde ... sur la corvette la "Coquille" ... 1822-25: Zoologie.* 2 vols & atlas. Paris, 1826-31. With P. Garnot. Another ed., 1838.
2. *Manuel de mammalogie; ou, Histoire naturelle des mammifères.* Ib., 1827.
3. *Histoire naturelle générale et particulière des mammifères et des oiseaux découverts depuis 1788.* 10 vols. Ib., 1828-37. (Complément des oeuvres de Buffon). 2nd ed., 1838.
4. *Histoire naturelle des oiseaux-mouches.* Ib., 1829-30.
5. *Centurie zoologique; ou, Choix d'animaux rares....* Ib., 1830-32.
6. *Histoire naturelle des colibris.* Ib., 1830-32.
7. *Traité d'ornithologie.* Ib., 1830-31.
8. *Les trochilidées; ou, Les colibris et les oiseaux-mouches.* Ib., 1832-33.
9. *Illustrations de zoologie; ou, Recueil de figures d'animaux.* Ib., 1832-35.
10. *Nouveau tableau du règne animal ... mammifères.* Ib., 1842.
11. *Histoire naturelle des zoophytes. Acalèphes.* Ib., 1843.
12. *Articles d'ornithologie parus ... 1842-45.* Ib., 1913. Ed. by A. Menegaux.

12993. LEUCKART, Friedrich Sigismund. 1794-1843. Ger. BMNH; BLA.

12994. PANDER, Christian Heinrich von. 1794-1865. Russ. DSB; ICB; P II,III; BMNH; Mort.; BLA; GC. Also Geology* and Embryology.

12995. RAPP, Wilhelm Ludwig von. 1794-1868. Ger. BMNH; BLA; GC.

12996. RÜPPEL, Wilhelm Peter Eduard Simon. 1794-1884. Ger. P II, III; BMNH; GC.

12997. VALENCIENNES, Achille. 1794-1865. Fr. DSB; BMNH. See also 12818/4.

12998. ALBERS, Johann Christian (or J. Christoph). 1795-1857. Ger. BMNH; BLA & Supp.; GC.

12999. EHRENBERG, Christian Gottfried. 1795-1876. Ger. DSB; ICB; P I,III; BMNH; Mort.; BLA & Supp.; GA; GB; GC. Also Geology* and Microbiology*.

13000. EICHWALD, Karl Eduard Ivanovich. 1795-1876. Ger./Russ. DSB; P I,III; BMNH; BLA; GA; GB. Also Geology*.

13001. GAEDE, Heinrich Moritz. 1795-1834. Ger. GA; GF.

13002. HARRIS, Thaddeus William. 1795-1856. U.S.A. BMNH; GE.

13003. RENGGER, Johann Rudolph. 1795-1832. Switz. ICB; P II; BMNH; GC.

13004. WILSON, James. 1795-1856. Br. BMNH; GD. See also 15476/1.

13005. BOHEMAN, Carl Henrik. 1796-1868. Swed. BMNH.

13006. GAIMARD, (Joseph) Paul. 1796-1858. Fr. DSB; BMNH; BLA. See also 12962/1,2.

13007. HARLAND, Richard. 1796-1843. U.S.A. DSB; BMNH; Mort.; GE.

1. *Fauna Americana, being a description of the mammiferous animals inhabiting North America.* Philadelphia, 1825. RM.
2. *Medical and physical researches; or, Original memoirs in ... zoology and comparative anatomy.* Ib., 1835. RM.

13008. MACGILLIVRAY, William. 1796-1852. Br. BMNH; GA; GB; GD. Also Geology*, Botany*, and Biography*.
1. *The conchologist's text-book.* Edinburgh, 1845. The 6th ed. of a work originally authored by T. Brown.
Also a work on British quadrupeds in the Naturalist's Library (see 13852).

13009. RETZIUS, Anders Adolf. 1796-1860. Swed. DSB; ICB; BMNH; Mort.; BLA; GA. Also Anatomy. See also 14813/3.

13010. AUDOUIN, Jean Victor. 1797-1841. Fr. DSB; ICB; P I; BMNH; BLA; GA; GB.

13011. BENNETT, Edward Turner. 1797-1836. Br. BMNH; GD.

13012. DESHAYES, Gerard Paul. 1797-1875. Fr. DSB; BMNH; GB. Also Geology*. See also 12746/4 and 12923/1.

13013. DUGÈS, Antoine Louis. 1797-1838. Fr. BMNH; BLA & Supp.; GA. Also Physiology.

13014. DUMORTIER, Barthélemy Charles Joseph. 1797-1878. Belg. BMNH; GA. Also Botany*.

13015. FREMERY, Petrus Johannes Isaacus. 1797-1855. Holl. P III (FREMERIJ); BMNH; BLA. Also Chemistry.

13016. HASSELT, Jan Conrad van. 1797-1823. Holl. BMNH; GF.

13017. HOPE, Frederick William. 1797-1862. Br. BMNH; GD.
1. *The coleopterist's manual.* 3 vols. London, 1837-40.
2. *A catalogue of Hemiptera.* 2 vols. Ib., 1837-42.

13018. HUSCHKE, Emil. 1797-1858. Ger. DSB; BMNH; BLA; GC. Also Anatomy, Embryology, and Physiology.
1. *Schädel, Hirn und Seele des Menschen und der Thiere nach Alter, Geschlecht und Race.* Jena, 1854.

13019. JOHNSTON, George. 1797-1855. Br. BMNH; GA; GD.
1. *An introduction to conchology.* London, 1850.

13020. KOLLAR, Vincenz. 1797-1860. Aus. BMNH; GC.

13021. LEURET, François. 1797-1851. Fr. DSB; BLA; GA.
1. *Anatomie comparée du système nerveux considéré dans ces rapports avec l'intelligence.* 2 vols & atlas. Paris, 1839-57. With P. Gratiolet.

13022. LINDEN, Pierre Léon van der. 1797-1831. Belg. BMNH; GF (VAN).

13023. MUHAUT, Etienne. b. 1797. Fr. GA.

13024. MULSANT, (Martial) Etienne. 1797-1880. Fr. BMNH.
1. *Histoire naturelle des coléoptères de France.* 31 parts. Paris/Lyons, 1839-77.

13025. PERCHERON, Achille Remy. 1797-1869. BMNH.
1. *Monographie des Passales.* Paris, 1835.
2. *Bibliographie entomologique.* Ib., 1837. RM.
See also 13042/1.

13026. SCHROEDER VAN DER KOLK, Jacobus Ludovicus Conradus. 1797-1862. Holl. DSB. Under KOLK: BMNH; BLA & Supp.; GF. Also Anatomy and Physiology.

13027. SMITH, Andrew. 1797-1872. Br./South Africa. ICB; BMNH; BLA & Supp.; GD.
1. *Miscellaneous ornithological papers.* London, 1880. Ed. by O. Salvin. (Willughby Society publn.)

13028. WAGNER, (Andreas) Johann. 1797-1861. Ger. P II; BMNH; GC. Also Geology.

13029. BARKOW, Hans Carl Leopold. 1798-1873. Ger. BMNH; BLA. Also Physiology.

13030. MARTIN, William Charles Linnaeus. 1798-1864. Br. BMNH; GD.
1. *A general introduction to the natural history of mammiferous animals, with a particular view of the physical history of man and the more closely allied genera.* London, 1841. RM.

13031. SAVI, Paolo. 1798-1871. It. ICB; P III; BMNH; GB.

13032. AMYOT, Charles Jean Baptiste. 1799-1866. Fr. BMNH.
1. *Histoire naturelle des insectes. Hémiptères.* 2 vols. Paris, 1843. With J.G. Audinet-Serville.

13033. BECK, Henrich Henrichsen. 1799-1863. Den. P III; BMNH.

13034. CHEVROLAT, Louis Alexandre Auguste. 1799-1884. Fr. BMNH.

13035. GUÉRIN-MÉNEVILLE, Félix Edouard. 1799-1874. Fr. BMNH; GA.
1. *Iconographie du "Règne animal" de G. Cuvier.* 3 vols. Paris, 1829-44. RM/L.
See also 11263.

13036. HERRICH-SCHAEFFER, Gottlieb August Wilhelm. 1799-1874. Ger. BMNH; BLA; GA.

13037. KROEYER, Henrick Nicolas. b. 1799. Den. GA.

13038. LENZ, Harald Othmar. 1799-1870. Ger. BMNH. Also History of Zoology and Botany*.
1. *Schlangenkunde.* Gotha, 1832.

13039. PEALE, Titian Ramsay. 1799-1885. U.S.A. DSB; ICB; BMNH; GE.

13040. POEY, Felipe. 1799-1891. Cuba. BMNH.

13041. FRASER, Louis. fl. 1841-1866. Br. BMNH; GD.

13042. GORY, Hippolyte Louis. d. 1852. BMNH.
1. *Monographie des Cétoines et genres voisins.* Paris, 1833. With A. Percheron.

13043. ASMUSS, Hermann Martin. d. 1860. Dorpat. BMNH.

13044. LA FRESNAYE, Frédéric de. d. 1861. Fr. BMNH.

13045. VERANY, Jean Baptiste (or Giovanni Battista). d. 1865. BMNH.

13046. ASSMANN, Friedrich Wilhelm. b. 1800. BMNH.

13047. BRONN, Heinrich Georg. 1800-1862. Ger. DSB; ICB; P I & Supp.; BMNH; Mort.; GA; GB; GC. Also Geology*.
1. *Die Klassen und Ordnungen des Tier-Reichs.* Many vols. Leipzig, 1859-. Contd. after his death by W. Keferstein et al.

13048. DIESING, Carl Moritz. 1800-1867. Aus. BMNH; BLA.
1. *Systema Helminthum.* 2 vols. Vienna, 1850-51. Reprinted, 2 vols in 1, Weinheim, 1960. (HNC. 11)

13049. GRAY, John Edward. 1800-1875. Br. ICB; BMNH; GA; GB; GD.
1. *The zoological miscellany.* 4 parts. London, 1831-44.
See also 12942/3 and 13855/1.

13050. JARDINE, William. 1800-1874. Br. BMNH; GD. Also Biography*.
Twelve works in the Naturalist's Library, of which he was the editor (see 13852). See also Index.

13051. JENYNS (after 1871 BLOMEFIELD), Leonard. 1800-1893. Br. BMNH; GD1 (BLOMEFIELD). Also Meteorology* and Biography*.
1. *Observations in natural history. With an introduction on habits of observing.* London, 1846.

13052. MILNE-EDWARDS, Henri. 1800-1885. Fr. DSB; ICB; BMNH; BLA (EDWARDS); GA; GB.
1. *Zoologie. (Cours élémentaire d'histoire naturelle)* Paris, 1841. X.
1a. ——— *A manual of zoology.* London, 1856. Trans. by R. Knox. Another ed., 1863.
2. *Recherches anatomiques et zoologiques faites pendant un voyage sur les côtes de la Sicilie.* 3 parts. Paris, 1845-49. With A. de Quatrefages and E. Blanchard.
3. *Leçons sur la physiologie et l'anatomie comparée de l'homme et des animaux.* 14 vols. Paris, 1857-81.
4. *Recherches pour servir à l'histoire naturelle des mammifères.* Text & atlas. Paris, 1868-74.
See also 12746/4.

13053. PEACH, Charles William. 1800-1886. Br. GB; GD. Also Geology.

13054. POUCHET, Félix Archimède. 1800-1872. Fr. DSB; ICB; BMNH; BLA; GA. Also Natural History*, Embryology*, Microbiology*, and History of Science*.
1. *Traité élémentaire de zoologie.* Rouen, 1832. RM.
1a. ——— 2nd ed. *Zoologie classique; ou, Histoire naturelle du règne animal.* 2 vols & atlas. Paris, 1841.

13055. WAGLER, Johann Georg. 1800-1832. Ger. BMNH; GC.
1. *Systema Avium.* Part I. Stuttgart, 1827.
2. *Natürliches System der Amphibien.* Munich, 1830.
3. *Six ornithological memoirs* [1829-32]. London, 1884. Ed. by P.L. Sclater. (Willughby Society publn.)

13056. BOISDUVAL, Jean (Baptiste) Alphonse (Déchauffour de). 1801-1879. Fr. BMNH; ICB; GA.
1. *Voyage de découvertes de "l'Astrolabe" ... 1826-1829 ... Faune entomologique de l'Océan Pacifique.* Paris, 1832-35.
2. *Histoire naturelle des insectes. Species général des Lépidoptères.* 2 vols. Paris, 1836-57. With A. Guénée.

13057. CANTRAINE, François Joseph. 1801-1863. Belg. BMNH.

13058. DUJARDIN, Félix. 1801-1860. Fr. DSB; ICB; BMNH; BLA & Supp.; GA.
1. *Histoire naturelle des zoophytes. Infusoires.* Paris, 1841. RM.
2. *Histoire naturelle des Helminthes, ou vers intestinaux.* Paris, 1845. RM.

13059. HAAN, Willem de. 1801-1855. Holl. ICB; BMNH.

13060. HOEVEN, Jan van der. 1801-1868. Holl. DSB; ICB; BMNH; BLA; GA.
1. *Handboek der dierkunde.* 2 vols. Rotterdam, 1828-33. X.
1a. ——— *Handbook of zoology.* Cambridge, 1856-58. Trans. from the 2nd Dutch ed. by W. Clark.

13061. LACORDAIRE, (Jean) Théodore. 1801-1870. Belg. ICB; BMNH; GA.
1. *Introduction à l'entomologie.* 2 vols. Paris, 1834-38.

13062. MÜLLER, Johannes (Peter). 1801-1858. Ger. DSB; ICB; P II; BMNH; Mort.; BLA & Supp.; GB; GC. Also Anatomy, Embryology, and Physiology*.

13063. RAMBUR, Jules Pierre. 1801-1870. Fr. BMNH.

13064. SUNDEVALL, Carl Jacob. 1801-1875. Swed. BMNH. Also Biography*.

13065. VROLIK, Willem. 1801-1863. Holl. BMNH; BLA; GF. Also Embryology*.

13066. ANDREWS, William. 1802-1880. Br. BMNH; GD.

13067. AUBÉ, Charles. 1802-1869. Fr. BMNH.
1. *Monographia Pselaphiorum.* Paris, 1834.

13068. BALL, Robert. 1802-1857. Br. BMNH; GD.

13069. BRANDT, Johann Friedrich. 1802-1879. Ger./Russ. DSB; BMNH; BLA; GC. Also Botany.

13070. EYDOUX, (Joseph) Fortuné (Théodore). 1802-1841. Fr. BMNH.
1. *Voyage autour du monde ... 1836-1837 sur la "Bonité" ... Zoologie.* 2 vols & atlas. Paris, 1841-52. With F.L.A. Souleyet.

13071. FITZINGER, Leopold Joseph Franz Johann. 1802-1884. Aus. BMNH.

13072. MÉNÉTRIÉS, Eduard. 1802-1861. Russ. BMNH.

13073. NARDO, Giovanni Domenico. 1802-1877. It. BMNH.

13074. PATTERSON, Robert. 1802-1872. Br. BMNH; GD.
1. *First steps to zoology.* London, 1849. X. 2nd ed., 2 vols., 1850.

13075. SHUCKARD, William Edward. 1802-1868. Br. BMNH; GD. See also 12954/11 and 13122/1a.

13076. SICHEL, Julius (or Frédéric Jules). 1802-1868. Ger./Fr. BMNH; BLA; GA; GC.

13077. WIEGMANN, Arend Friedrich August. 1802-1841. BMNH.
1. *Handbuch der Zoologie.* Berlin, 1832. With J.F. Ruthe. X. Another ed., 1843.

13078. BAIRD, William. 1803-1872. Br. BMNH; GD. Also Natural History*.

13079. BERCE, Jean Etienne. 1803-1879. Fr. BMNH.

13080. BINNEY, Amos. 1803-1847. U.S.A. BMNH; GE.

13081. BONAPARTE, Charles Lucien (Jules Laurent). 1803-1857. Fr. DSB; BMNH; GA; GB.

13082. DENNY, Henry. 1803-1871. Br. BMNH; GD.

13083. DOHRN, Karl August. 1803-1892. Ger. BMNH; GC.

13084. HERBST, (Ernst Friedrich) Gustav. 1803-1893. Ger. BMNH; BLA.

13085. KAUP, Johann Jacob. 1803-1873. Ger. P I; BMNH; GB; GC.
1. *Skizzirte Entwickelungs-Geschichte und natürliches System der europäischen Thierwelt ... Erster Theil.* [No more publ.] Darmstadt, 1829.
2. *Das Thierreich in seinen Hauptformen systematisch beschrieben.* 3 vols. Darmstadt, 1835-37.

13086. MARTIN SAINT-ANGE, Gaspard Joseph. 1803-1888. Fr. BMNH; BLA; GA.
1. *Etude de l'appareil reproducteur dans les cinq classes d'animaux vertébrés, au point de vue anatomique, physiologique et zoologique.* Paris, 1854.

13087. NEWPORT, George. 1803-1854. Br. DSB; BMNH; GD. Also Embryology.
1. *On the respiration of insects.* London, 1836.

13088. ANTHONY, John Gould. 1804-1877. U.S.A. GE.

13089. BENOIT, Luigi. 1804-1890. It. BMNH.

13090. DES MURS, Marc Athanese Parfait Oeillet. b. 1804. Fr. BMNH.

13091. DUBOIS, Charles Frédéric. 1804-1867. Belg. BMNH.

13092. DUNCAN, James. 1804-1861. Br. BMNH.
1. *Entomology.* 6 vols. London, 1845-46. A collection of six works first published in 1835-42 in the Naturalist's Library (see 13852).

13093. EGGER, Johann Georg. 1804-1866. Aus. BMNH; GC.

13094. GOULD, John. 1804-1881. Br./Australia. DSB; BMNH; GA; GD. See also 15542/1.
1. *The birds of Australia.* 7 vols. London, 1840-48. Reprinted, Melbourne, 1972.
2. *An introduction to the mammals of Australia.* London, 1863.
3. *An analytical index to the works of ... John Gould.* London, 1893. By R.B. Sharpe.

13095. LEREBOULLET, Dominique Auguste. 1804-1865. Fr. DSB; BMNH; BLA. Also Embryology.

13096. MARSCHALL, August Friedrich von. 1804-1887. Aus. BMNH.
1. *Nomenclator zoologicus continens nomina systematica generum animalium, tam viventium quam fossilium.* Vienna, 1873.

13097. OWEN, Richard (first of the name). 1804-1892. Br. DSB; ICB; BMNH; Mort.; BLA; GA; GB; GD. Also Geology* and Natural History*. See also Index.
1. *Memoir on the pearly nautilus.* London, 1832.
2. *Odontography; or, A treatise on the comparative anatomy of the teeth.* 2 vols & atlas. Ib., 1840-45.
3. *Lectures on the comparative anatomy and physiology of the invertebrate animals.* Ib., 1843. RM/L. 2nd ed., 1855.
4. *Lectures on the comparative anatomy and physiology of the vertebrate animals.* Ib., 1846.

5. *On the archetype and homologies of the vertebrate skeleton.* Ib., 1848. RM/L.
6. *On the nature of limbs. A discourse.* Ib., 1849.
7. *On parthenogenesis ... A discourse.* Ib., 1849.
8. *The principal forms of the skeleton and the teeth, as a basis for a system of natural history and comparative anatomy.* Ib., 1854. Re-issued, 1860.
9. *On the classification and geographical distribution of the mammalia.* Ib., 1859. RM/L.
10. *Memoir on the megatherium, or giant ground-sloth of America.* Ib., 1861.
11. *Monograph on the aye-aye.* Ib., 1862.
12. *Instances of the power of God as manifested in His animal creation.* Ib., 1863.
13. *Memoir on the gorilla.* Ib., 1865. RM/L.
14. *On the anatomy of the vertebrates.* 3 vols. Ib., 1866-68. Reprinted, New York, 1973.

13098. SAVAGE, Thomas Staughton. 1804-1880. U.S.A. BMNH; GE.

13099. SCHLEGEL, Hermann. 1804-1884. Holl. BMNH.

13100. SIEBOLD, Carl Theodor Ernst von. 1804-1885. Ger. DSB; BMNH; BLA; GB; GC. Also Embryology*.
1. *Handbuch der Zootomie.* 2 vols. Berlin, 1845-46. With F.H. Stannius. X. 2nd ed., 1854-56.
2. *Lehrbuch der vergleichenden Anatomie der wirbellosen Thiere.* 2 vols. Berlin, 1845-48. With F.H. Stannius.

2a. ——— *Comparative anatomy of the invertebrata.* Vol. 1. [No more publ.] Trans.by W.I. Burnett.

13101. GEOFFROY SAINT-HILAIRE, Isidore. 1805-1861. Fr. DSB; BMNH; BLA; GA; GB. Also Science in General* and Biography*.
1. *Histoire générale et particulière des anomalies de l'organisation chez l'homme et les animaux ... ou, Traité de tératologie.* 3 vols & atlas. Paris, 1832-36. RM/L.
2. *Essais de zoologie générale.* Paris, 1841. RM.
3. *Domestication et naturalisation des animaux utiles.* Paris, 1849. X. 4th ed., 1861.
4. *Histoire naturelle générale des règnes organiques.* 3 vols. Paris, 1854-62. RM.

13102. GOULD, Augustus Addison. 1805-1866. U.S.A. DSB; ICB; BMNH; BLA & Supp.; GB; GE. See also 13120/5.

13103. MARMOCCHI, Francesco Constantino. 1805-1858. It. ICB; BMNH.

13104. MOUSSON, (Joseph Rudolph) Albert. 1805-1890. Switz. P II, III; BMNH. Also Physics*.

13105. NORDMANN, Alexander Danilevsky. 1805-1866. Russ. ICB; BMNH.

13106. PFEIFFER, Ludwig Georg Carl. 1805-1877. Ger. BMNH; BLA; GA. Also Botany*.

13107. SARS, Michael. 1805-1869. Nor. DSB; P III; BMNH.

13108. SMITH, Frederick. 1805-1879. Br. BMNH.

13109. SUFFRIAN, Christian Wilhelm Ludwig Eduard. 1805-1876. Ger. GC.

13110. THOMPSON, William. 1805-1852. Br. BMNH; GD.

13111. WAGNER, Rudolph. 1805-1864. Ger. DSB; BMNH; Mort.; BLA; GB; GC. Also Embryology*, Physiology*, and Biography*.
1. *Lehrbuch der vergleichenden Anatomie*. 2 parts. Leipzig, 1834-35. X.
1a. ——— *Elements of the comparative anatomy of the vertebrate animals*. London, 1845. Trans. by A. Tulk.
2. *Icones zootomicae. Handatlas zur vergleichenden Anatomie*. Leipzig, 1841.

13112. WESTWOOD, John Obadiah. 1805-1893. Br. BMNH; GD.
1. *The entomologist's text book*. London, 1838. RM/L.
2. *An introduction to the modern classification of insects*. 2 vols. London, 1839-40. RM/L.
3. *Arcana entomologica; or, Illustrations of new, rare and interesting insects*. 2 vols. London, 1841-45.
See also 12818/3f.

13113. BENDZ, Henrik Carl Bang. 1806-1882. Den. BMNH; BLA.

13114. BIBRON, Gabriel. 1806-1848. Fr. BMNH; GA. See also 12842/3.

13115. DAHLBOM, Anders Gustav. 1806-1859. Swed. BMNH.

13116. HANCOCK, Albany. 1806-1873. Br. BMNH; GD.

13117. HEWITSON, William Chapman. 1806-1878. Br. BMNH; GD.

13118. LAWRENCE, George Newbold. 1806-1895. U.S.A. GE.

13119. ROSSMAESSLER, Emil Adolf. 1806-1867. Ger. BMNH; GC.

13120. AGASSIZ, (Jean) Louis (Rodolphe). 1807-1873. Switz./U.S.A. DSB; ICB; P I,III; BMNH; BLA; GA; GB; GE. Also Geology* and Natural History*. See also Index.
1. *Selecta genera et species piscium quas in itinere per Brasiliam annis 1817-1820....* Munich, 1829. RM.
2. *Monographie d'échinodermes vivans et fossiles*. 4 vols. Neuchâtel, 1838-42. RM/L.
3. *Histoire naturelle des poissons d'eau douce de l'Europe centrale*. 2 vols. Neuchâtel, 1839-42. RM/L.
4. *Nomenclator zoologicus, continens nomina systematica generum animalium tam viventium quam fossilium*. 12 parts. Solothurn, 1842-46. With several collaborators. RM/L.
4a. ——— *Nomenclatoris zoologici index universalis*. Ib., 1846. X. Another ed., 1848.
5. *Principles of zoology. Part I. Comparative physiology*. [No more publ. in this ed.] Boston, 1848. With A.A. Gould. X. Reprinted, New York, 1970.
5a. ——— Another ed. *Outlines of comparative physiology*. London, 1851. Ed. by T. Wright. Other eds: 1855, 1870, 1878.
6. *Bibliographia zoologiae et geologiae. A general catalogue ....* See 9531/8.

13121. BISCHOFF, Theodor Ludwig Wilhelm. 1807-1882. Ger. DSB; BMNH; Mort.; BLA & Supp.; GA; GC. Also Embryology* and Physiology.

13122. BURMEISTER, (Carl) Hermann (Conrad). 1807-1892. Ger./Argentina. ICB; P I,III; BMNH; GA; GC. Also Geology*.
1. *Handbuch der Entomologie*. 5 vols in 6. Berlin, 1832-55.

1a. ——— *A manual of entomology.* London, 1836. Trans. by W.E. Shuckard.
2. *Genera quaedam insectorum.* 10 parts. Berlin, 1838-46.

13123. BUSK, George. 1807-1886. Br. DSB; BMNH; BLA & Supp.; GB; GD1. See also Index.

13124. HALIDAY, Alexander Henry. 1807-1870. Br. BMNH.
1. *Hymenoptera Britannica.* 2 parts. London, 1839.

13125. HAWKINS, Benjamin Waterhouse. 1807-1889. Br. BMNH.
1. *A comparative view of the human and animal frame.* London, 1860.

13126. KÜSTER, Heinrich Carl. 1807-1876. Ger. BMNH.

13127. LOEW, Hermann. 1807-1879. Ger. BMNH.

13128. VERREAUX, Jules. 1807-1873. Fr. BMNH.

13129. CHENU, Jean Charles. 1808-1879. Fr. ICB; BMNH; BLA; GA.
1. *Leçons élémentaires d'histoire naturelle, comprenant un aperçu sur toute la zoologie.* 2 vols. Paris, 1847.
2. *Encyclopédie d'histoire naturelle; ou, Traité complet de cette science.* 22 vols & 9 vols of plates. Paris, 1850-61. Several re-issues.

13130. DOUBLEDAY, Henry. 1808-1875. Br. BMNH; GD.

13131. ENGELMANN, Wilhelm. 1808-1878. Ger. Bibliographer and publisher. BMNH; BLA. Also Biology in General*.
1. *Bibliotheca historico-naturalis. Verzeichniss der Bücher über Naturgeschichte welche in ... 1700-1846 erschienen sind.* Vol. 1. [No more publ.] Leipzig, 1846. RM/L. Reprinted, Weinheim, 1960. (HNC. 14)
1a. ——— Continued as *Bibliotheca zoologica. Verzeichniss ....* See 13323/1.

13132. GRAY, George Robert. 1808-1872. Br. BMNH; GA; GD.
1. *The entomology of Australia ... Part 1 ... the genus Phasma.* London, 1833.
2. *The genera of birds.* 3 vols. London, 1844-49.
3. *Notices of insects that are known to form the bases of fungoid parasites.* Hampstead, 1858.
See also 13855/3.

13133. HANF, Blasius. 1808-1892. Aus. GC.

13134. JAY, John Clarkson. 1808-1891. U.S.A. BMNH.
1. *A catalogue of recent shells in the cabinet of J.C. Jay.* New York, 1835. X. 3rd ed., 1839. 4th ed., 1850. Supplement, 1852.

13135. KIRCHENPAUER, Gustav Heinrich. 1808-1887. Ger. BMNH; GC.

13136. PHILIPPI, Rudolph Amandus. 1808-1904. Ger./Chile. P II,III, IV & Supp.; BMNH. Also Geology and Botany.

13137. STANNIUS, Hermann (Friedrich). 1808-1883. Ger. DSB; BMNH; Mort.; BLA; GC. Also Physiology.
1. *Das peripherische Nervensystem der Fische.* Rostock, 1849.
See also 13100/1,2.

13138. WALSH, Benjamin Daniel. 1808-1869. U.S.A. BMNH; GE.

13139. BENEDEN, Pierre Joseph van. 1809-1894. Belg. DSB; ICB; BMNH; BLA & Supp.

13140. BIANCONI, Gian Giuseppe. 1809-1878. It. P III; BMNH. Also Geology.

13141. BLAND, Thomas. 1809-1885. Br./U.S.A. BMNH; GE.

13142. BLASIUS, Johann Heinrich. 1809-1870. Ger. P I,III; BMNH.

13143. BRULLÉ, (Gaspard) Auguste. 1809-1873. Fr. BMNH.

13144. CANTOR, Theodore Edward. b. 1809. Malaya. BMNH.

13145. DARWIN, Charles Robert. 1809-1882. Br. DSB; ICB; P III; BMNH; G. Also Geology*, Natural History*, Botany*, Evolution*#, and Autobiography*. See also Index.
1. *The zoology of the voyage of H.M.S. "Beagle" ... 1832-1836.* 5 parts. London, 1839-43.
2. *A monograph on the sub-class Cirripedia.* 2 parts. Ib., 1851-54. (Ray Society publns 21,25) Reprinted, 2 vols in 1, Weinheim, 1964. (HNC. 38)
3. *A monograph on the fossil cirripedes of Great Britain.* Ib., 1851-54.
4. *A monograph on the fossil Lepadidae, or pedunculated cirripedes, of Great Britain.* Ib., 1851.
5. *A monograph on the fossil Balanidae and Verrucidae of Great Britain.* Ib., 1854.
6. *The expression of emotions in man and animals.* Ib., 1872. RM/L. Reprinted, Brussels, 1969. 2nd ed., 1892.
7. *The formation of vegetable mould through the action of worms, with observations on their habits.* Ib., 1881. RM/L. Reprinted, ib., 1961. Other eds: 1883,1888,1892.
8. *Ornithological notes.* Ib., 1963. Ed. with introd. and notes by N. Barlow.
9. *On the routes of male bumble bees.* Ib., 1968. Ed. by R.B. Freeman.

13146. DUNKER, Wilhelm Bernhard Rudolph Hadrian. 1809-1885. Ger. P I,III; BMNH. Also Geology*.

13147. ERICHSON, Wilhelm Ferdinand. 1809-1849. Ger. BMNH.

13148. EYTON, Thomas Campbell. 1809-1880. Br. BMNH; GD.

13149. FITCH, Asa. 1809-1879. U.S.A. BMNH; GE.

13150. GUÉNÉE, Achille. 1809-1880. Fr. BMNH. See also 13056/2.

13151. HEER, Oswald. 1809-1883. Switz. DSB; P I,III; BMNH; GB; GC. Also Geology and Botany.

13152. HOMEYER, Eugen Ferdinand von. 1809-1889. Ger. BMNH; GC.

13153. JEFFREYS, John Gwyn. 1809-1885. Br. DSB; BMNH; GD. See also 10649/1.

13154. LOVÉN, Sven Ludvig. 1809-1895. Swed. DSB; BMNH.
1. *On Pourtalesia, a genus of Echinoidea.* Stockholm, 1833.
2. *Etudes sur les échinoidées.* Ib., 1874.

13155. MORELET, Pierre Marie Arthur. 1809-1892. Fr. BMNH.

13156. PUTZEYS, Jules Antoine Adolphe Henri. 1809-1882. Belg. BMNH; GF.

13157. SAUNDERS, William Wilson. 1809-1879. Br. BMNH; GD.

13158. TOWNSEND, John Kirk. 1809-1851. U.S.A. BMNH; GE.

13159. WALKER, Francis. 1809-1874. BMNH.
1. *Monographia Chalciditum.* London, 1839.

13160. MOCHULSKII, Victor Ivanovich. fl. 1839-1860. Russ. BMNH.

13161. DEGLAND, Côme Damien. d. 1856. Fr. BMNH.

13162. BLYTH, Edward. 1810-1873. Br./India. DSB; ICB; BMNH; GD.

13163. BONSDORFF, Evert Julius. 1810-1898. Fin. BMNH; BLA & Supp.

13164. BUSHNAN, John Stevenson. 1810-1884. Br. BMNH; BLA & Supp.; GD.
1. *The natural history of fishes.* Edinburgh, 1840. (Naturalist's Library; see 13852)

13165. GOSSE, Philip Henry. 1810-1888. Br./Canada. ICB; BMNH; GA; GB; GD. Also Evolution*. See also 15547/1.
1. *The aquarium. An unveiling of the wonders of the deep sea.* London, 1854. X. 2nd ed., 1856.
2. *A manual of marine zoology for the British Isles.* 2 parts. Ib., 1855-56.
3. *Life in its lower, intermediate and higher forms.* Ib., 1857.
4. *The romance of natural history.* Ib., 1860. Second series, 1861.
5. *Land and sea.* Ib., 1865.

13166. GUNDLACH, Johannes. 1810-1896. Ger./Cuba. BMNH; GC.

13167. HUMPHREYS, Henry Noel. 1810-1879. Br. BMNH; GD.
1. *Ocean gardens. The history of the marine aquarium, and the best methods now adopted for its establishment and preservation.* London, 1857.

13168. HYRTL, (Carl) Joseph. 1810-1894. Prague/Aus. DSB; ICB; BMNH; Mort.; BLA. Also Anatomy and History of Anatomy*.
1. *Beiträge zur vergleichenden Angiologie.* Vienna, 1849.

13169. JONES, Thomas Rymer. 1810-1880. Br. BMNH; GA; GD.
1. *A general outline of the animal kingdom, and manual of comparative anatomy.* London, 1838-41. 2nd ed., 1855. 3rd ed., 1861. 4th ed., 1871.
2. *The natural history of animals.* 2 vols. Ib., 1845-52.

13170. KNER, Rudolf. 1810-1869. Aus. BMNH; GC.

13171. MORRIS, Francis Orpen. 1810-1893. Br. ICB; BMNH; GD.
1. *A history of British birds.* 6 vols. London, 1851-57.

13172. QUATREFAGES DE BRÉAU, (Jean Louis) Armand de. 1810-1892. Fr. DSB; ICB; BMNH; Mort.; BLA; GA; GB. Also Autobiography* and Biography*.
1. *Physiologie comparée. Metamorphoses de l'homme et des animaux.* Paris, 1862.
1a. ——— *Metamorphoses of man and the lower animals.* London, 1864. Trans. by H. Lawson.
See also 13052/2.

13173. RODD, Edward Hearle. 1810-1880. Br. BMNH; GD.

13174. SAPPEY, Marie Philibert Constant. 1810-1896. Fr. BMNH; Mort.; BLA. Also Anatomy.

13175. SCACCHI, Arcangelo. 1810-1893. It. P II,III,IV; BMNH. Also Geology.

13176. TROSCHEL, Franz Hermann. 1810-1882. BMNH.

13177. WATERHOUSE, George Robert. 1810-1888. Br. BMNH; GD.
1. *A natural history of the Mammalia*. 2 vols. London, 1846-48.
Also a work on marsupials in the Naturalist's Library (see 13852).

13178. ANTINORI, Orazio. 1811-1882. It. BMNH.

13179. BREE, Charles Robert. 1811-1886. Br. BMNH. Also Evolution*.

13180. DOUBLEDAY, Edward. 1811-1849. Br. BMNH; GD.

13181. DOYÈRE, Louis. 1811-1863. Fr. BMNH; BLA.

13182. HERRICK, Edward Claudius. 1811-1862. U.S.A. P III; BMNH; GE. Also Astronomy.

13183. JERDON, Thomas Claverhill. 1811-1872. India. BMNH; GD.

13184. LONGET, François Achille. 1811-1871. Fr. BMNH; BLA; GA. Also Physiology*.

13185. REICHERT, Karl Bogislaus. 1811-1883. Ger./Dorpat. DSB; ICB; BMNH; Mort.; BLA; GC. Also Anatomy and Embryology.

13186. SOULEYET, Louis François Auguste. 1811-1852. Fr. DSB; BMNH; BLA. See also 13070/1.

13187. STRICKLAND, Hugh Edwin. 1811-1853. Br. P II; BMNH; GB; GD. Also Geology. See also Index.
1. *Ornithological synonyms*. Vol. 1. [No more publ.] London, 1855. Ed. by Mrs. H.E. Strickland and W. Jardine. RM.
2. *Rules for zoological nomenclature*. Edinburgh, 1863. X. Another ed., by P.L. Sclater, London, 1878.

13188. BALDAMUS, August Carl Eduard. 1812-1893. Ger. BMNH.

13189. CASTLENAU, Francis (de Laporte), *Comte* de. 1812-1880. Fr. P III; BMNH. Also Geology*.
1. *Histoire naturelle et iconographie des insectes coléoptères*. 4 vols in 5. Paris, 1837-41.

13190. DAVAINE, Casimir Joseph. 1812-1882. Fr. DSB; ICB; BMNH; BLA & Supp. Also Microbiology.
1. *L'oeuvre*. Paris, 1889. RM.

13191. DUMERIL, Auguste Henri André. 1812-1870. Fr. BMNH; BLA.

13192. GRUBE, Adolph Eduard. 1812-1880. Ger. BMNH; GC.

13193. HALDEMAN, Samuel Stehman. 1812-1880. U.S.A. BMNH; GB; GE.
1. *A monograph of the Limniades*. Philadelphia, 1840-42. RM.

13194. HARTING, Pieter. 1812-1885. Holl. DSB; P I,III; BMNH; Mort.; BLA. Also Microscopy and History of Microscopy*.

13195. JOLY, Nicolas. 1812-1885. Fr. BMNH.

13196. KIRSCHBAUM, Karl Ludwig. 1812-1880. Ger. GC.

13197. KRAUSS, Christian Ferdinand Friedrich. 1812-1890. BMNH.

13198. LEAR, Edward. 1812-1888. Br. The poet and painter. ICB; BMNH; GB; GD.

13199. LETZNER, Karl Wilhelm. 1812-1889. Ger. GC.

13200. MARSEUL, Silvin Augustin de. 1812-1890. Fr. BMNH.
1. *Catalogue des coléoptères d'Europe*. Paris, 1857. 2nd ed., 1863.

13201. MURRAY, Andrew. 1812-1878. Br. BMNH; GD.
1. *The geographical distribution of mammals*. London, 1866.

13202. SOWERBY, George Brettingham (second of the name). 1812-1884. Br. BMNH; GD.
1. *Conchological illustrations*. London, 1832-41.
2. *A conchological manual*. Ib., 1839. 4th ed., 1852.
3. *Thesaurus conchyliorum; or, Monographs of genera of shells*. 5 vols. Ib., 1842-87.
4. *Illustrated index of British shells*. Ib., 1859.

13203. ADAMS, Henry. 1813-1877. BMNH.

13204. CARPENTER, William Benjamin. 1813-1885. Br. DSB; ICB; P III; BMNH; BLA & Supp.; GB; GD. Other entries: see 3326.
1. *Zoology. A systematic account*. 2 vols. London, 1845. Another ed., 1857.
2. *Introduction to the study of the Foraminifera*. Ib., 1862. (Ray Society publn. 32)

13205. CARTER, Henry John. 1813-1895. India. BMNH.

13206. CASSIN, John. 1813-1869. U.S.A. BMNH; GE.

13207. DANA, James Dwight. 1813-1895. U.S.A. DSB; ICB; P I,III,IV; BMNH; GB; GE. Also Mineralogy* and Geology*. See also Index.
1. *On the classification and geographical distribution of Crustacea*. Philadelphia, 1853.

13208. GLOVER, Townend. 1813-1883. Br./U.S.A. BMNH; GE.

13209. HALLMANN, Eduard. 1813-1855. Ger. ICB; P I; BMNH; BLA. Also Physiology.

13210. PASCOE, Francis Polkinghorne. 1813-1893. Br. BMNH; GD.
1. *Zoological classification. A handy book of reference*. London, 1877. 2nd ed., 1880.

13211. SELYS-LONGCHAMPS, Michel Edmond de. 1813-1900. Belg. BMNH; GF.

13212. STEENSTRUP, (Johannes) Japetus Smith. 1813-1897. Den. DSB; BMNH; Mort.; BLA.
1. [Trans.] *On the alternation of generations; or, The propagation and development of animals through alternate generations*. London, 1845. (Ray Society publn. 4) Trans. by G. Busk from the German version (by C.H. Lorenzen) of the Danish original of 1842.

13213. WISSMANN, Otto Ludwig. 1813-1877. Ger. GC.

13214. ADAMS, Charles Baker. 1814-1853. U.S.A. P III; BMNH; GE. Also Geology.

13215. ATKINSON, John Christopher. 1814-1900. Br. BMNH; GD1.

13216. BODINUS, (Karl August) Heinrich. 1814-1884. Ger. GC.

13217. BREWER, Thomas Mayo. 1814-1880. U.S.A. BMNH; GE.

13218. BUCKLER, William. 1814-1884. Br. BMNH; GD.

13219. DOUGLAS, John William. 1814-1905. Br. BMNH.

13220. FELDER, Cajetan von. 1814-1894. Aus. BMNH; GA; GC.

13221. FILIPPI, Filippo de'. 1814-1867. It. BMNH; GA.

13222. GÄTKE, Heinrich. 1814-1897. Ger. BMNH; GC.
1. *Die Vogelwarte Helgoland.* Brunswick, 1891. RM.

13223. GOODSIR, John. 1814-1867. Br. DSB; ICB; BMNH; BLA & Supp.; GB; GD. Also Anatomy*.

13224. GRUBER, Wenzel Leopold. 1814-1890. Russ. BMNH; BLA & Supp. Also Anatomy.
1. *Beobachtungen aus der menschlichen und vergleichenden Anatomie.* Berlin, 1879-89.

13225. HARTLAUB, Carl Johann Gustav. 1814-1900. Ger. ICB; BMNH.

13226. LANKESTER, Edwin. 1814-1874. Br. BMNH; BLA; GD. Also Microscopy* and Biography*. See also Index.
1. *Aquavivarium, fresh and marine.* London, 1856.

13227. NUHN, Anton. 1814-1889. Ger. BMNH; BLA. Also Anatomy.
1. *Lehrbuch der vergleichenden Anatomie.* 2 parts. Heidelberg, 1875-78. X. 2nd ed., 1886.

13228. REDTENBACHER, Ludwig. 1814-1876. Aus. BMNH.

13229. REEVE, Lovell Augustus. 1814-1865. Br. BMNH; GD.

13230. ROUILLIER, Karl Frantsovich. 1814-1858. Russ. DSB. Also Geology.

13231. WYMAN, Jeffries. 1814-1874. U.S.A. DSB; ICB; BMNH; Mort.: GE. Also Physiology.

13232. SCHMARDA, Ludwig Karl. fl. 1846-1872. Aus. BMNH.

13233. DANIELSSEN, Daniel Cornelius. 1815-1894. Nor. BMNH; BLA & Supp.

13234. ERDL, Michael Pius. 1815-1848. Ger. BMNH; BLA; GA. Also Embryology.

13235. FORBES, Edward. 1815-1854. Br. DSB; ICB; P III; BMNH; GB; GD. Also Geology and Oceanography. See also Index.
1. *The zoology of the voyage of H.M.S. "Herald" ... 1845-51 ... Vertebrates, including fossil mammals.* London, 1852-54. With J. Richardson.
2. *Outlines of the natural history of Europe. The natural history of the European seas.* London, 1859. Completed by R.A.C. Godwin-Austen.

13236. GIBBS, George (second of the name). 1815-1873. U.S.A. ICB; P III; BMNH; GE. Also Geology.

13237. GRATIOLET, (Louis) Pierre. 1815-1865. Fr. DSB; BMNH; BLA. See also 13021/1.

13238. HARDY, James. 1815-1898. Br. ICB; BMNH.

13239. HOLMES, Francis Simmons. 1815-after 1882. U.S.A. P III; BMNH. Also Geology.

13240. KESSLER, Karl Theodorovich. 1815-1881. Russ. BMNH.

13241. LUCAS, Pierre Hippolyte. 1815-1899. Fr. BMNH.

13242. PETERS, Wilhelm Karl Hartwig. 1815-1883. Ger. ICB; BMNH; BLA; GC.

13243. ROTH, Johannes Rudolf. 1815-1858. Ger. BMNH; GC.

13244. SCHIÖDTE, Jörgen Matthias Christian. 1815-1884. Den. BMNH.

13245. TOMES, John. 1815-1895. Br. ICB; BLA; GD.

13246. WALLICH, George Charles. 1815-1899. Br. DSB; BMNH; GD.

13247. WILL, Johann Georg Friedrich. 1815-1868. Ger. BMNH; BLA; GC.

13248. DESMAREST, Eugène. 1816-1889. BMNH.

13249. ECKER, Alexander. 1816-1887. Switz. BMNH; BLA & Supp.; GC. Also Anatomy, Physiology*, and Biography*.
1. *Die Anatomie des Frosches. Ein Handbuch.* 3 parts. Brunswick, 1864-82. X.
1a. ——— *The anatomy of the frog.* 1887. Trans. with notes by G. Haslam. X. Reprinted, Amsterdam, 1971.

13250. GASSIES, Jean Baptiste. 1816-1883. Fr. BMNH; GA.

13251. GERVAIS, (François Louis) Paul. 1816-1879. Fr. BMNH; BLA; GB.
1. *Zoologie et palaeontologie française. Nouvelles recherches sur les animaux vertébrés.* Paris, 1848-52. X. 2nd ed., 1859.
2. *Ostéographie des monotrèmes vivants et fossiles.* Text & atlas. Paris, 1877-78.

13252. HALL, John Charles. 1816-1876. Br. BMNH; BLA.

13253. LILLJEBORG, Wilhelm. 1816-1908. Swed. ICB; BMNH.

13254. REINHARDT, Johannes Theodor. 1816-1882. Den. BMNH.

13255. VOLLENHOVEN, Samuel Constant Snellen van. 1816-1880. Holl. BMNH.

13256. BALY, Joseph Sugar. 1817-1890. Br. BMNH.

13257. CHAUDOIR, Maximilien de. 1817-1881. BMNH.

13258. COOTE, Holmes. 1817-1872. Br. BMNH; BLA; GD.
1. *The homologies of the human skeleton.* London, 1849.

13259. FISCHER, Leopold Heinrich. 1817-1886. Ger. P I; BMNH. Also Mineralogy.

13260. FUSS, Karl Adolf. 1817-1874. Ger. GC.

13261. HAGEN, Hermann August. 1817-1893. Ger./U.S.A. ICB; BMNH; GE.
1. *Bibliotheca entomologica. Die Litteratur über das ganze Gebiet der Entomologie.* 2 vols. Leipzig, 1862-63. RM/L. Reprinted, 2 vols in 1, Weinheim, 1960. (HNC. 13)

13262. KOELLIKER, (Rudolf) Albert von. 1817-1905. Ger. DSB; ICB; BMNH; Mort.; BLA & Supp.; GA; GB. Also Anatomy*, Embryology*, Physiology, and Cytology.

13263. LEWES, George Henry. 1817-1878. Br. ICB; BMNH; GA; GB; GD. Also Physiology* and Biography*.
1. *Studies in animal life.* London, 1862.

13264. VOGT, Carl Christopher. 1817-1895. Ger./Switz. DSB; P II,III; BMNH; BLA; GB; GC. Also Geology*.
1. *Lehrbuch der praktischen vergleichenden Anatomie.* Brunswick, 1855. X. Another ed., 2 vols, 1885-94.
2. *Les mammifères. Edition française originale.* Paris, 1884.
2a. ——— *Natural history of animals. Mammalia.* London, 1887. Trans. with additions by G.G. Chisholm.

13265. WHITE, Adam. 1817-1879. Br. BMNH; GD.

13266. ZADDACH, Ernst Gustav. 1817-1881. Ger. P II,III; BMNH. Also Geology.

13267. BATE, Charles Spence. 1818-1889. Br. BMNH; GD.

13268. BELLARDI, Luigi. 1818-1889. It. DSB; BMNH. Also Geology*.

13269. BIGOT, Jacques M.F. 1818-1893. Fr. BMNH.

13270. BUCKTON, George Bowdler. 1818-1905. Br. P III; BMNH; GD2. Also Chemistry.

13271. FOLIN, Alexandre Guillaume Léopold de. 1818-ca. 1895. Fr. BMNH.

13272. GLASER, Ludwig. 1818-1898. Ger. BMNH; GC.

13273. GRAELLS Y DE LA AGÜERA, Mariano de la Paz. 1818?-1898. Spain. BMNH.

13274. HINCKS, Thomas. 1818-1899. Br. BMNH; GD1.

13275. KÖSTLIN, Otto. 1818-1884. Ger. BMNH; BLA; GC.

13276. LORD, John Keast. 1818-1872. Br. BMNH; GD.

13277. STEIN, (Samuel) Friedrich (Nathaniel) von. 1818-1885. BMNH. Also Microbiology*.

13278. TASCHENBERG, Ernst Ludwig. 1818-1898. Ger. BMNH.

13279. TSCHUDI, Johann Jacob von. 1818-1889. Aus. P III; BMNH; GC.

13280. WULP, F.M. van der. 1818?-1899. Holl. BMNH.

13281. BLANCHARD, (Charles) Emile. 1819-1900. Fr. BMNH; GA.
1. *L'organisation du règne animal.* 5 vols. Paris, 1851-64. See also 13052/2.

13282. BLEEKER, Pieter. 1819-1878. Holl./East Indies. BMNH; BLA.

13283. BOCOURT, Firmin. b. 1819. Fr. BMNH.

13284. BRUCH, Karl Wilhelm Ludwig. 1819-1884. Ger. BMNH; Mort.; BLA.

13285. CARPENTER, Philip Pearsall. 1819-1877. Br./U.S.A. BMNH; GD.

13286. FIGUIER, Louis (Guillaume). 1819-1894. Fr. P I,III; BMNH; GA. Other entries: see 3332.
1. *La vie et les moeurs des animaux ... zoophytes et mollusques.* Paris, 1866.
2. [Trans.] *The ocean world. Being a description of the sea and some of its inhabitants.* London, 1868. Trans. from various works. Another ed., 187-?

13287. FISCHER, Johann Gustav. 1819-1889? Ger. BMNH.

13288. HANLEY, Sylvanus Charles Thorp. 1819-1899. Br. BMNH.
1. *An illustrated and descriptive catalogue of recent bivalve shells.* London, 1842-56.
2. *Ipsa Linnaei conchylia. The shells of Linnaeus, determined from his manuscripts and collection.* London, 1855.

13289. KINGSLEY, Charles. 1819-1875. Br. The writer. ICB; BMNH; GA; GB; GD. Also Science in General* and Geology*.
1. *Glaucus; or, The wonders of the shore.* Cambridge, 1855. Another ed., 1858. 4th ed., 1859. 5th ed., 1873. 6th ed., 1878. Another ed., 1890.

13290. SCHAUM, Hermann Rudolph. 1819-1865. Ger. BMNH.

13291. VILLA, Antonio. d. 1885. It. BMNH.

13292. ADAMS, Arthur. 1820-1878. Br. BMNH. Also Natural History*.
1. *The zoology of the voyage of H.M.S. "Samarang" ... 1843-46.* London, 1850.

13293. BRUEHL, Carl Bernhard. 1820-1899. Aus. BMNH; BLA & Supp.
1. *Zootomie aller Thierklassen.* 40 parts. Vienna, 1874-88.

13294. FAIRMAIRE, Léon Marc Herminie. 1820-1906. Fr. BMNH.

13295. GEMMINGER, Max. 1820-1887. Ger. BMNH.

13296. GIEBEL, Christoph Gottfried Andreas. 1820-1881. Ger. P I,III; BMNH; GB; GC. Also Geology.
1. *Thesaurus ornithologiae. Repertorium der gesammten ornithologischen Literatur und Nomenclator sämmtlicher Gattungen und Arten der Vögel.* 3 vols. Leipzig, 1872-77.

13297. HERKLOTS, Janus Adrian. 1820-1872. Holl. BMNH.

13298. LEISERING, August Gottlob Theodor. 1820-1892. BMNH.
1. *Handbuch der vergleichenden Anatomie der Haus-Säugethiere.* Berlin, 1890.

13299. ROFFIAEN, François Xavier. 1820-1898. Belg. GF.

13300. DUNCAN, Peter Martin. 1821 (or 1824)-1891. Br. P III,IV; BMNH; GB; GD1. Also Palaeontology and Biography*.
1. *The transformations, or metamorphoses, of insects.* London, 1882.
See also 13860.

13301. LACAZE-DUTHIERS, (Félix Joseph) Henri de. 1821-1901. Fr. DSB; BMNH.

13302. LEYDIG, Franz von. 1821-1908. Ger. DSB; BMNH; Mort.; BLA. Also Anatomy.
1. *Lehrbuch der Histologie des Menschen und der Thiere.* Hamm, 1857.
2. *Tafeln zur vergleichenden Anatomie.* Tübingen, 1864.
3. *Untersuchungen zur Anatomie und Histologie der Thiere.* Bonn, 1883.
4. *Zelle und Gewebe. Neue Beiträge zur Histologie der Thierkörpers.* Bonn, 1885.

13303. MALM, August Wilhelm. 1821-1882. Swed. BMNH.

13304. SANDBERGER, B.E.F. Guido. 1821-1869. Ger. P II,III; BMNH. Also Geology.

13305. WOODWARD, Samuel Pickworth. 1821-1865. Br. BMNH; GD. Also Palaeontology.
1. *A manual of the Mollusca.* 3 parts. London, 1851-56. X. 2nd ed., 1866, re-issued 1868. 4th ed., 2 vols, 1880.

13306. ANGAS, George French. 1822-1886. Br./Australia. BMNH; GD1.

13307. BAILLY, Jean Baptiste. b. 1822. Fr. BMNH.

13308. BETTA, Francesco Edoardo de. 1822-1896. It. BMNH.

13309. DARESTE, (G.M.) Camille. 1822-1899. Fr. ICB; BMNH; BLA & Supp.

13310. EDWARDS, William Henry. 1822-1909. U.S.A. BMNH; GE.

13311. FREY, (Johann Friedrich) Heinrich (Konrad). 1822-1890. Switz. BMNH; BLA & Supp.; GC. Also Microscopy* and Histology*.
1. *Beiträge zur Kenntniss wirbelloser Thiere.* Brunswick, 1847. With R. Leuckart. RM.

13312. GIRARD, Charles Frédéric. 1822-1895. U.S.A. BMNH; GE.

13313. LEUCKART, (Carl Georg Friedrich) Rudolf. 1822-1898. Ger. DSB; BMNH; BLA & Supp.; GC.
1. *Zoologische Untersuchungen.* 3 parts. Giessen, 1853-54.
2. *Bau und Entwicklungsgeschichte der Pentastomen.* Leipzig, 1860.
See also 13311/1.

13314. LINTNER, Joseph Albert. 1822-1898. U.S.A. GE.

13315. MÜLLER, Fritz (or Johann Friedrich Theodor). 1822-1897. Ger./Brazil. DSB; ICB; BMNH; Mort. (M., J.F.T.); GC (M., J.F.T.). Also Evolution*.

13316. ROWLEY, George Dawson. 1822-1878. Br. BMNH; GF.

13317. STAINTON, Henry Tibbats. 1822-1892. Br. BMNH; GD.

13318. WAGENER, Guido Richard. 1822-1896. Ger. BMNH; BLA; GC.

13319. WOLLASTON, Thomas Vernon. 1822-1878. Br. BMNH; GD.
1. *On the variation of species, with especial reference to the Insecta. Followed by an inquiry into the nature of genera.* London, 1856. RM.

13320. BAIRD, Spencer Fullerton. 1823-1887. U.S.A. DSB; ICB; BMNH; GB; GE.
1. *The mammals of North America.* Philadelphia, 1857.
2. *Correspondence between S.F. Baird and L. Agassiz.* Washington, 1963. Ed. by E.C. Herber.

13321. BARBOSA DU BOCAGE, José Vivente. 1823-1908. Port. BMNH.

13322. BRUNNER (VON WATTENWYL), Carl. 1823-after 1906. Switz. P I; BMNH. Also Physics.

13323. CARUS, Julius Victor. 1823-1903. Ger. DSB Supp.; BMNH; BLA & Supp. Also History of Zoology*.
1. *Bibliotheca zoologica. Verzeichniss der Schriften über Zoologie welche in den periodischen Werken enthalten und vom Jahre 1846-1860 selbständig erschienen sind.* 2 vols.

Leipzig, 1861. With W. Engelmann. Continuation of a work begun by Engelmann: see 13131/1.
1a. ——— Continued as *Bibliotheca zoologica II. Verzeichniss ....* See 13762/1.
2. *Handbuch der Zoologie*. 2 vols. Leipzig, 1863-75. With C.E.A. Gerstaecker.
See also 13873/19.

13324. CLARK, Hamlet. 1823?-1867. Br. BMNH.

13325. COSTA, Achille. 1823-1898. It. BMNH.

13326. EICHHOFF, Wilhelm Josef. 1823-1893. Ger. BMNH; GC.

13327. FABRE, Jean Henri. 1823-1915. Fr. DSB; ICB; BMNH.
1. [Trans.] *The life of the spider*. London, 1912. Trans. by A. Teixeira de Mattos. X. Reprinted, New York, 1971.
2. [Trans.] *Social life in the insect world*. London, 1912. Trans. by B. Miall. X. Reprinted, Freeport, N.Y., 1971.

13328. LEIDY, Joseph. 1823-1891. U.S.A. DSB; ICB; P III; BMNH; Mort.; BLA & Supp.; GB; GE. Also Palaeontology and Anatomy.

13329. McCOY, Frederick. 1823-1899. Br./Australia. BMNH; GB; GD1. Also Palaeontology*.

13330. PARKER, William Kitchen. 1823-1890. Br. BMNH; GD. Also Evolution*.
1. *The morphology of the skull*. London, 1877. With G.T. Bettany.

13331. SCHMIDT, (Eduard) Oscar. 1823-1886. BMNH. Also Evolution*.
1. *Handbuch der vergleichenden Anatomie*. Jena, 1849. X. Another ed., 1852. 8th ed., 1882.
2. *Hand-Atlas der vergleichenden Anatomie*. Jena, 1852-53.
3. *Die niederen Thiere*. Leipzig, 1884.

13332. TARGIONI-TOZZETTI, Adolfo. 1823-1902. It. BMNH.

13333. WALLACE, Alfred Russel. 1823-1913. Br. DSB; ICB; P III; BMNH; GB; GD3. Also Natural History*, Evolution*, Autobiography*, and History of Science*.
1. *The geographical distribution of animals, with a study of the relations of living and extinct faunas as elucidating the past changes of the earth's surface*. 2 vols. London, 1876. RM/L. Reprinted, New York, 1962.

13334. WALLENGREN, Hans Daniel Johan. 1823-1894. Swed. BMNH.

13335. ALTUM, Bernard. 1824-1900. BMNH.
1. *Lehrbuch der Zoologie*. 4th ed. Freiburg, 1878. With H. Landois.

13336. BENNECKE, Berthold. 1824-1886. Ger. BLA Supp.; GC.

13337. BERGH, Ludwig Sophus Rudolf. 1824-1909. Den. BMNH; BLA & Supp.

13338. CHAPUIS, Félicien. 1824-1879. Belg. BMNH.

13339. CORNALIA, Emilio. 1824-1882. It. BMNH.

13340. DALLAS, William Sweetland. 1824-1890. Br. BMNH.
1. *A natural history of the animal kingdom*. London, 1856.
See also 13888/1a and 14513/1a.

13341. HAIME, Jules. 1824-1856. Fr. P III; BMNH. Also Geology.

13342. HALFORD, George Britton. 1824-1910. Australia. BMNH; BLA; GF.
1. *Lines of demarcation between man, gorilla and macaque.* Melbourne, 1864.

13343. HAY, Arthur. 1824-1878. Br. BMNH; GD.

13344. HEUGLIN, (Martin) Theodor von. 1824-1876. Aus. P III; BMNH; GB; GC.

13345. HOLDER, Joseph Bassett. 1824-1888. U.S.A. GE.

13346. McILWRAITH, Thomas. 1824-1903. Canada. GF.

13347. THOMSON, Carl Gustav. 1824-1899. Swed. BMNH.

13348. UDEKEM, (Gérard) Jules (Marie Ghislain) d'. 1824-1864. Belg. ICB; BMNH.

13349. BAIKIE, William Balfour. 1825-1864. Br. ICB; P III; BMNH; GB; GD. See also 11202A/1.

13350. BATES, Henry Walter. 1825-1892. Br. DSB; ICB; P III,IV; Mort.; BMNH; GB; GD1.
1. *The naturalist on the River Amazon.* 2 vols. London, 1863. "One of the finest scientific travel books of the nineteenth century." RM. Reprinted, ib., 1969.
2. *The principal contributions of H.W. Bates to a knowledge of the butterflies and longicorn beetles of the Amazon valley.* New York, 1978. Ed. by E.G. Lindsley.
See also 9275/1.

13351. BILHARZ, Theodor. 1825-1862. Ger./Egypt. DSB; ICB; P III; BMNH; BLA & Supp. Also Anatomy.

13352. GRAY, Robert. 1825-1887. Br. BMNH; GD.

13353. HUXLEY, Thomas Henry. 1825-1895. Br. DSB; ICB; BMNH; Mort.; BLA; GB; GD1. Other entries: see 3340.
1. *The oceanic hydrozoa ... observed during the voyage of H.M.S. "Rattlesnake" in the years 1846-50.* London, 1859. (Ray Society publn. 30)
2. *Lectures on the elements of comparative anatomy.* Ib., 1864. RM/L.
3. *An elementary atlas of comparative osteology.* Ib., 1864.
4. *An introduction to the classification of animals.* Ib., 1869. RM/L.
5. *A manual of the anatomy of invertebrated animals.* Ib., 1877. RM/L. Another ed., 1882.
6. *The crayfish. An introduction to the study of zoology.* Ib., 1880. RM/L. Reprinted, Cambridge, Mass., 1974. 3rd ed., 1881. Another ed., 1884. 6th ed., 1896.
7. *The scientific memoirs.* 4 vols. Ib., 1898-1902. Ed. by M. Foster and E.R. Lankester. RM/L. Supplement, 1903.

13354. KOCH, Ludwig. 1825-1908. Ger. BMNH.

13355. LE CONTE, John Lawrence. 1825-1883. U.S.A. BMNH; GE.

13356. LE MOINE, James MacPherson. 1825-1912. Canada. BMNH; GF.

13357. MÖBIUS, Karl August. 1825-1908. Ger. DSB; BMNH.

13358. OLPHE-GALLIARD, (Victor Aimé) Leon. 1825?-1893. BMNH.
1. *Quelques remarques sur les règles de la nomenclature zoologique appliquées à toutes les branches de l'histoire naturelle.* Bulle, 1872.

13359. PAGENSTECHER, Heinrich Alexander. 1825-1899. Ger. BMNH; GC.
1. *Allgemeine Zoologie.* 4 vols. Berlin, 1875-81.

13360. PELZELN, August von. 1825-1891. Aus. BMNH.

13361. RÜTIMEYER, (Karl) Ludwig. 1825-1895. Switz. DSB; ICB; BMNH; Mort.; BLA; GC. Also Geology.

13362. SCHULTZE, Max Johann Sigismund. 1825-1874. Ger. DSB; ICB; P III; BMNH; Mort.; BLA; GB; GC. Also Microscopy, Histology, Embryology, and Cytology.

13363. THOMAS, Cyrus. 1825-1910. U.S.A. BMNH; GE.

13364. XÁNTUS, János. 1825-1894. Hung./U.S.A. ICB; P III; BMNH; GE.

13365. BUCKLAND, Francis Trevelyan. 1826-1880. Br. BMNH; GB; GD.
1. *Curiosities of natural history.* [First series] London, 1858. X. 9th ed., 1866. 11th ed., 1871.
1a. ——— Second series. Ib., 1860. 5th ed., 1865. 6th ed., 1871.
1b. ——— New [i.e. third] series. Ib., 1866. X. 2nd ed., 1868.
1c. ——— Series 1-4. 4 vols. Ib., 1878-79. Other eds: 1888-90 and 1891-93.
1d. ——— *Buckland's Curiosities of natural history. A selection.* Ib., 1948. Ed. by L.R. Brightwell.
See also 9294/3 and 15598/1.

13366. CLARK, Henry James. 1826-1873. U.S.A. BMNH; GE.
1. *Mind in nature; or, The origin of life and the mode of development of animals.* New York, 1865.

13367. CROSSE, Joseph Charles Hippolyte. 1826-1898. Fr. BMNH.

13368. DAVID, Armand. 1826-1900. Fr./China. ICB; BMNH.

13369. DUGÈS, Alfred Auguste Delsescautz. 1826-1910. Mexico. ICB; BMNH.

13370. GEGENBAUER, Karl. 1826-1903. Ger. DSB Supp.; ICB; BMNH; Mort.; BLA & Supp.; GB. Also Anatomy and Embryology.
1. *Grundzüge der vergleichenden Anatomie.* Leipzig, 1859. X. 2nd ed., 1870.
2. *Untersuchungen zur vergleichenden Anatomie der Wirbelsäule bei Amphibien und Reptilien.* Ib., 1862.
3. *Untersuchungen zur vergleichenden Anatomie der Wirbelthiere.* 3 parts in 1 vol. Ib., 1864-72.
4. *Grundriss der vergleichenden Anatomie.* Ib., 1874. 2nd ed., 1878.
4a. ——— *Elements of comparative anatomy.* London, 1878. Trans. from the 2nd German ed. by F.J. Bell.

13371. GRÜNDLER, Emil Otto. 1826-1893. Ger. BLA Supp.; GC.

13372. SCHRENCK, Peter Leopold von. 1826-1894. Russ. P III,IV; BMNH.

13373. ADAMS, Andrew Leith. 1827-1882. Br. P III; BMNH; GB; GD. Also Geology.

13374. BIELZ, Eduard Albert. 1827-1898. Ger. BMNH; GC.

13375. CANDÈZE, Ernest Charles Auguste. 1827-1898. Belg. BMNH.

13376. CHAUVEAU, (Jean Baptiste) Auguste. 1827-1917. Fr. DSB; BMNH; Mort.; BLA & Supp. Also Physiology.
1. *Traité d'anatomie comparée des animaux domestiques.* Paris, 1857.
1a. ——— *The comparative anatomy of domesticated animals.* London, 1873. Trans. from the 2nd French ed. by G. Fleming.

13377. HÉMENT, Félix. 1827-1891. BMNH.
1. *De l'instinct et de l'intelligence.* Paris, 1880.

13378. KOCH, Carl Jacob Wilhelm Ludwig. 1827-1882. Ger. P III; BMNH; GC. Also Geology.

13379. LÜTKEN, Christian Frederik. 1827-1901. Den. BMNH.

13380. MIVART, St. George Jackson. 1827-1900. Br. DSB; ICB; BMNH; GB; GD1. Also Science in General* and Evolution*.
1. *Dogs, jackals, wolves and foxes. A monograph of the Canidae.* London, 1890.
2. *Types of animal life.* London, 1893.

13381. MOESCH, Casimir. 1827-1898. Switz. P III,IV; BMNH. Also Geology.

13382. OVSYANNIKOV, Filipp Vasilevich. 1827-1906. Russ. BMNH.

13383. PULS, Jacques Charles. 1827-1889. Belg. BMNH; GF.

13384. STEARNS, Robert Edwards Carter. 1827-1909. U.S.A. BMNH; GE.

13385. TROUVELOT, Etienne Léopold. 1827-1895. Fr./U.S.A. DSB; P III, IV. Also Astronomy.

13386. WOOD, John George. 1827-1889. Br. BMNH; GB; GD.
1. *The illustrated natural history.* 3 vols. London, 1851-53. X. Other eds: 1861-63, 1872-75, 1883.
2. *Man and beast, here and hereafter.* 2 vols. Ib., 1874.

13387. BERNSTEIN, Heinrich Agathon. 1828-1865. Java. P III; BMNH.

13388. COBBOLD, Thomas Spencer. 1828-1886. Br. ICB; BMNH; BLA & Supp.; GB; GD.

13389. GERSTÄCKER, Karl Eduard Adolph. 1828-1895. Ger. BMNH; GC. See also 13323/2.

13390. HUDSON, Charles Thomas. 1828-1903. Br. BMNH; GD2.

13391. MÖRCH, Otto Andreas Lowson. 1828-1878. Den. BMNH.

13392. OSTEN-SACKEN, Carl Robert Romanovich von der. 1828-1906. Russ./U.S.A. BMNH; GE.

13393. THOMSON, James (fourth of the name). 1828-1897. Fr. BMNH.

13394. WRAXALL, Frederick Charles Lascelles. 1828-1865. Br. BMNH; GD.
1. *Life in the sea; or, The nature and habits of marine animals.* London, 1860.

13395. BAMBEKE, Charles Eugène van. 1829-1918. Belg. ICB; BMNH. Under VAN BAMBEKE : BLA; GF.

13396. BOURGUIGNAT, Jules René. 1829-1892. Fr. BMNH.

13397. BREHM, Alfred Edmund. 1829-1884. Ger. ICB; P III; BMNH; GC.
1. *Illustriertes Thierleben*. Hildburghausen, 1863-69. X.
1a. ——— 2nd ed. *Thierleben. Allgemeine Kunde des Thierreichs*. 10 vols. Leipzig, 1876-80. Another ed., 1882-85. 3rd ed., 1890-93.

13398. DAY, Francis. 1829-1889. Br./India. BMNH; GD1; GF.

13399. DROUET, Henri. 1829-after 1892. Fr. BMNH.

13400. GREEF, Richard. 1829-1892. Ger. BMNH; BLA Supp; GC.

13401. HOLMGREN, August Emil. 1829-1888. Swed. BMNH.

13402. LINDSAY, William Lauder. 1829-1880. Br. BMNH; BLA; GD. Also Botany*.
1. *Mind in the lower animals, in health and disease*. 2 vols. London, 1879.

13403. NEWTON, Alfred. 1829-1907. Br. ICB; BMNH; GB; GD2. See also 12688/2.

13404. ROLLESTON, George. 1829-1881. Br. DSB; BMNH; BLA; GD.
1. *Forms of animal life*. Oxford, 1870. 2nd ed., rev. by W.H. Jackson, 1888.
2. *Scientific papers and addresses*. 2 vols. Oxford, 1884. Ed. by W. Turner with a biography by E.B. Tylor. RM/L.

13405. SAUSSURE, Henri (Louis Frédéric) de. 1829-1905. Switz. P III; BMNH.
1. *Etude sur la famille des Vespides*. Paris, 1852-58. X.
1a. ——— Part II. *Monographie des guêpes sociales; ou, De la tribu des Vespiens*. Text & atlas. Paris, 1853-58.

13406. SCLATER, Philip Lutley. 1829-1913. Br. DSB; BMNH. See also Index.

13407. THEOBALD, William. 1829-1908. India. BMNH.

13408. VAGNER, Nikolai Petrovich. 1829-1907. Russ. ICB; BMNH (WAGNER).

13409. WEINLAND, (Christoph) David Friedrich. 1829-1915. Ger. BMNH.

13410. DODERLEIN, Pietro. d. 1895. It. BMNH.

13411. BROT, Auguste Louis. d. 1896. Switz. BMNH.

13412. EUDES-DESLONGCHAMPS, Eugène. 1830-1889. Fr. BMNH.

13413. GÜNTHER, Albert Carl Ludwig Gotthilf. 1830-1914. Ger./Br. BMNH; GD3.
1. *Handbuch der Ichthyologie*. Vienna, 1886. RM.
2. *Report on the ... fishes collected by H.M.S. "Challenger."* Text & atlas. London, 1880-89. X. Reprinted, Weinheim, 1963. (HNC. 28)
3. *Andrew Garret's "Fische der Südsee." Three periodical articles, 1873-1916*. Weinheim, 1966. (HNC. 49)

13414. JEITTELES, Ludwig Heinrich. 1830-1883. Hung. P III; BMNH. Also Geology.

13415. KREFFT, Johann Ludwig Gerhard. 1830-1881. Ger./Australia. BMNH; GC; GF.

13416. MAYR, Gustav L. 1830-1908. Aus. BMNH.

13417. MOORE, Frederic. 1830-1907. Br. BMNH.

13418. MUYBRIDGE, Eadweard. 1830-1904. Br./U.S.A. ICB; BMNH; Mort.; GD2; GE.
1. *Animals in motion. An electro-photographic investigation.* London, 1899.

13419. RIESENTHAL, (Julius Adolf) Oskar von. 1830-1898. Ger. BMNH; GC.

13420. STAUDINGER, Otto. 1830-1900. Ger. BMNH.

13421. THOMSON, Charles Wyville. 1830-1882. Br. DSB; P III; BMNH; GB; GD. Also Oceanography*. See also 11302/1,2.

13422. THORELL, Tord Tamerlan Theodor. 1830-1901. Swed. BMNH.

13423. BÖCKING, Adolf. 1831-1898. Ger./U.S.A. BMNH; GC.

13424. CASEY, Thomas Lincoln. 1831-1896. U.S.A. ICB; BMNH; GE1.

13425. FLOWER, William Henry. 1831-1899. Br. DSB; BMNH; BLA & Supp.; GB; GD1.
1. *An introduction to the osteology of the Mammalia.* London, 1870. 3rd ed., 1885.
2. *The horse. A study in natural history.* Ib., 1891.
3. *An introduction to the study of mammals, living and extinct.* Ib., 1891. With R. Lydekker.
4. *Essays on museums and other subjects connected with natural history.* Ib., 1898.

13426. KRAATZ, (Ernst) Gustav. 1831-1909. Ger. BMNH.

13427. MARTENS, Eduard Carl. 1831-1904. BMNH.

13428. MÖSCHLER, Heinrich Benno. 1831-1888. Ger. BMNH; GC.

13429. NORMAN, Alfred Merle. 1831-1918. Br. BMNH; GF.

13430. ROGENHOFER, Alois Friedrich. 1831-1897. Aus. BMNH; GC.

13431. SCHNEIDER, Anton Friedrich. 1831-1890. Ger. DSB; BMNH. Also Cytology.
1. *Zoologische Beiträge.* 2 vols. Breslau, 1883-90.

13432. SOUTHWELL, Thomas. 1831-1909. Br. BMNH; GD2.

13433. BLAKISTON, Thomas Wright. 1832-1891. Br. GD1.

13434. BLANFORD, William Thomas. 1832-1905. Br./India. P III,IV; BMNH; GB; GD2; GF. Also Geology.

13435. BRADY, George Stewardson. 1832-1921. Br. BMNH.

13436. BRAUER, Friedrich Moritz. 1832-1904. Aus. BMNH.

13437. CLAPARÈDE, (Jean Louis René Antoine) Edouard. 1832-1871. Switz. P III; BMNH; BLA; GB.
1. *Beobachtungen über Anatomie und Entwicklungsgeschichte wirbelloser Thiere.* Leipzig, 1863.
2. *Lettres.* Basel, 1971. Ed. by G. de Morsier.

13438. FRIČ, Antonín Jan. 1832-1916? Prague. P IV (FRITSCH); BMNH. Also Geology.

13439. GEMMELLARO, Gaetano Giorgio. 1832-1904. It. P III; BMNH.

13440. JÄGER, Gustav. 1832-1917. Ger. ICB; BMNH; BLAF.
1. *Lehrbuch der allgemeiner Zoologie.* 3 vols in 2. 1871-85.
2. *Zoologische Briefe.* Vienna, 1876.
3. *Handwörterbuch der Zoologie, Anthropologie und Ethnologie.* 8 vols. Breslau, 1880-1900.
4. *Problems of nature. Researches and discoveries of Gustav Jaeger, selected from his published writings.* London, 1897. Trans. and ed. by H.G. Schlichter.
See also 3023.

13441. NOLL, Friedrich Karl. 1832-1893. Ger. BMNH; GC.

13442. ROSS, Alexander Milton. 1832-1897. Canada. BMNH; GF.

13443. RÜDINGER, Nicholaus. 1832-1896. Ger. BMNH; BLA; GC. Also Anatomy.

13444. RYE, Edward Caldwell. 1832-1885. Br. BMNH; GD.

13445. SAALMÜLLER, Max. 1832-1890. Ger. BMNH; GC.

13446. SEEBOHM, Henry. 1832-1895. Br. BMNH; GD.
1. *Classification of birds.* London, 1890. (Vol. 4 of his *History of British birds*) X. Another ed., 1895.

13447. SEMPER, Carl Gottfried. 1832-1893. Ger. DSB; BMNH; GC.
1. *Die Verwandtschaftsbeziehungen der gegliederten Thiere.* 3 parts in 1 vol. Wurzburg/Hamburg, 1875-76.
2. *Die natürlichen Existenzbedingungen der Thiere.* 2 vols. Leipzig, 1880. "The first textbook on animal ecology."
2a. ——— *The natural conditions of existence as they affect animal life.* London, 1881.

13448. STIMPSON, William. 1832-1872. U.S.A. DSB; BMNH; GE.

13449. STRAUCH, Alexander. 1832-1893. Russ. BMNH.

13450. TURNER, William. 1832-1916. Br. DSB; BMNH; BLA; GD3. Also Anatomy. See also 13404/2.

13451. ANDERSON, John. 1833-1900. India. BMNH; GD1; GF.

13452. BINNEY, William Greene. b. 1833. U.S.A. BMNH. See also 12908/2.

13453. CLARK, John Willis. 1833-1910. Br. BMNH; GD2. Also Biography*.
1. *Illustrations of comparative anatomy.* Cambridge, 1875.

13454. KEFERSTEIN, Wilhelm Moritz. 1833-1870. Ger. P I,III; BMNH.
1. *Untersuchungen über niedere Seethiere.* Leipzig, 1862.
2. [Papers on zoology] Berlin, 1862-78.
See also 13047/1.

13455. LYMAN, Theodore. 1833-1897. U.S.A. BMNH; GE.

13456. MEINERT, Frederik Vilhelm August. 1833-1912. Den. BMNH.

13457. PANCERI, Paolo. 1833-1877. It. BMNH; BLA.

13458. PÉREZ, Jean. 1833-1914. Fr. BMNH.

13459. POUCHET, (Henri Charles) Georges. 1833-1894. Fr. ICB; BMNH; BLA; BLAF. Also History of Science*.

13460. POWYS, Thomas Littleton. 1833-1896. Br. BMNH; GD.

13461. RUSS, Karl Friedrich Otto. 1833-1899. Ger. BMNH; GC.

13462. SCHOCH, Gustav. 1833-1899. Switz. BMNH.

13463. STÅL, Carl. 1833-1878. Swed. BMNH.

13464. BAUDELOT, Emile. 1834-1875. Fr. BMNH.
1. *Recherches sur le système nerveux des poissons*. Paris, 1883. RM.

13465. BOGDANOV, Anatoli Petrovich. 1834-1896. Russ. ICB; BMNH (BOGHDANOV).

13466. GODWIN-AUSTEN, Henry Haversham. 1834-1923. Br./India. BMNH; GD4. Also Geology.

13467. HAECKEL, Ernst Heinrich Philipp August. 1834-1919. Ger. DSB; ICB; BMNH; Mort.; BLAF; GB. Also Science in General*, Evolution*#, Embryology*, and Personal Writings*.
1. *Die Radiolarien*. Text & atlas. Berlin, 1862. RM/L.
2. *Beiträge zur Naturgeschichte der Hydromedusen*. Leipzig, 1865.
3. *Die Kalkschwämme*. 2 vols & atlas. Berlin, 1872.
4. *Arabische Korallen. Ein Ausflug nach den Korallenbänken des Rothen Meeres ... Populäre Vorlesung*. Berlin, 1876. RM/L.
5. *Studien zur Gastraea-Theorie*. Jena, 1877.
6. *Das Protistenreich. Eine populäre Übersicht über das Formengebiet der niedersten Lebwesen*. Leipzig, 1878. RM.
7. *Das System der Medusen*. Text & atlas. Jena, 1879-80.
8. *Report on the Radiolaria collected by H.M.S. "Challenger" during ... 1873-76*. 3 vols. London, 1887.

13468. KLUNZINGER, Carl Benjamin. 1834-1914. Ger. ICB; P III,IV; BMNH; BLAF.
1. *Synopsis der Fische des Rothen Meeres*. 2 parts. Frankfurt, 1870-71. X. Reprinted, 2 parts in 1 vol., Weinheim, 1964. (HNC. 36)

13469. LUBBOCK, John, *1st Baron* AVEBURY. 1834-1913. Br. DSB; ICB; BMNH; GB; GD3. Other entries: see 3351.
1. *On the origin and metamorphoses of insects*. London, 1874. Another ed., 1883.
2. *On the senses, instincts and intelligence of animals, with special reference to insects*. London, 1888. 3rd ed., 1889. Another ed., 1899.
3. *Scientific lectures*. London, 1879.

13470. ORD, William Miller. 1834-1902. Br. BMNH; GD2; BLA.
1. *Notes on comparative anatomy. A syllabus of a course of lectures*. London, 1871.

13471. SABATIER, (Charles Paul Dieudonné) Armand. 1834-1910. Fr. DSB; BMNH.

13472. STEINDACHNER, Franz. 1834-1919. Aus. BMNH.

13473. VAILLANT, Léon Louis. 1834-1914. Fr. DSB; BMNH.

13474. WEISMANN, August Friedrich Leopold. 1834-1914. Ger. DSB; ICB; BMNH; Mort.; BLAF; GB. Also Evolution and Heredity*.

13475. WESTERLUND, Carl Agardh. fl. 1861-1901. Swed. BMNH.

13476. AGASSIZ, Alexander. 1835-1910. U.S.A. DSB; ICB; BMNH; BLAF; GB; GE. Also Oceanography* and Autobiography*.
1. *Embryology of the starfish.* Boston, 1864.

13477. BRADY, Henry Bowman. 1835-1891. Br. BMNH; GD1.

13478. CANESTRINI, Giovanni. 1835-1900. It. BMNH; BLA & Supp.

13479. CLAUS, Carl Friedrich Wilhelm. 1835-1899. Ger./Aus. BMNH; GC.
1. *Grundzüge der Zoologie.* Marburg, 1866-68. X. 4th ed., 2 vols, 1880-82.
1a. ——— *Traité de zoologie.* Paris, 1878. Trans., with notes, from the 3rd German ed. by G. Moquin-Tandon. 2nd ed., trans. from the 4th German ed., 1884.
1b. ——— *Elementary text-book of zoology.* 2 vols, London, 1884-85. Trans. and ed. by A. Sedgwick and F.G. Heathcote.
2. *Kleines Lehrbuch der Zoologie.* Marburg, 1880.
2a. ——— 2nd ed. *Lehrbuch der Zoologie.* Ib., 1883.

13480. EHLERS, Ernst Heinrich. 1835-1925. Ger. BMNH.

13481. ELLIOT, Daniel Giraud. 1835-1915. U.S.A. BMNH; GE.

13482. FISCHER, Paul Henri. 1835-1893. Fr. BMNH.

13483. KENNICOTT, Robert. 1835-1866. U.S.A. BMNH; GE.

13484. LANDOIS, Hermann. 1835-1905. Ger. P III,IV; BMNH; BLAF. See also 13335/1.

13485. MILNE-EDWARDS, Alphonse. 1835-1900. Fr. Under EDWARDS: ICB; BMNH; BLA Supp.
1. *Notes sur quelques crustacés nouveaux ou peu connus.* Paris, 1871.
2. *Recherches sur la faune des régions australes.* 3 parts. Paris, 1879-82.

13486. SALVADORI (ADLARD), Tommaso. 1835-1923. It. BMNH.

13487. SALVIN, Osbert. 1835-1898. Br. BMNH; GD1. See also 13027/1.

13488. SAUNDERS, Howard. 1835-1907. Br. BMNH; GD2.

13489. STEBBING, Thomas Roscoe Rede. 1835-1926. Br. DSB; BMNH. Also Evolution*.

13490. UHLER, Phillip Reese. 1835-1913. U.S.A. BMNH; GE.

13491. BENDIRE, Charles Emil. 1836-1897. U.S.A. BMNH.

13492. BOLAU, (Cornelius Carl) Heinrich. 1836-1920. Ger. BMNH.

13493. GRANDIDIER, Alfred. 1836-1921. Fr. BMNH.

13494. HUTTON, Frederick Wollaston. 1836-1905. New Zealand. P III, IV; BMNH; GD2. Also Geology and Evolution*.

13495. IRBY, Leonard Howard Lloyd. 1836-1905. Br. BMNH; GD2.

13496. LORTET, Louis. 1836-1909. Fr. BMNH; BLA.

13497. MICHAEL, Albert Davidson. 1836-1927. Br. BMNH.

13498. PENNETIER, Georges. 1836-1923. Fr. ICB; BMNH.

13499. SAMUELS, Edward Augustus. 1836-1908. U.S.A. BMNH; GE.

13500. SAXBY, Henry Linckmyer. 1836-1873. Br. BMNH; GD.

13501. WOSSIDLO, Paul. 1836-after 1895. BMNH. Also Botany.
1. *Lehrbuch der Zoologie*. Berlin, 1886.

13502. BOSGOED, Dirk Mulder. 1837-1880. Holl. BMNH.
1. *Bibliotheca ichthyologica et piscatoria*. Haarlem, 1873.

13503. BUCHHOLZ, Reinhold Wilhelm. 1837-1876. Ger. P III; BMNH.

13504. GILL, Theodore Nicholas. 1837-1914. U.S.A. DSB; BMNH; GE.

13505. HOFMANN, Ernst (second of the name). 1837-1892. Ger. BMNH.

13506. McCOOK, Henry Christopher. 1837-1911. U.S.A. BMNH; GB; GE.

13507. McLACHLAN, Robert. 1837-1904. Br. BMNH; GD2.

13508. SARS, Georg Ossian. 1837-1927. Nor. ICB; BMNH.

13509. SCUDDER, Samuel Hubbard. 1837-1911. U.S.A. DSB; BMNH; GE. Also History of Science*.
1. *Nomenclator zoologicus. An alphabetical list of all the generic names employed ... for recent and fossil animals from the earliest times to ... 1879*. Washington, 1882.

13510. STIEDA, (Christian Hermann) Ludwig. 1837-1918. Dorpat/Ger. BMNH; BLA. Also History of Neurology*.

13511. ADAMI, Giambattista. 1838-1887. ICB.

13512. ALLEN, Joel Asaph. 1838-1921. U.S.A. BMNH; GE.

13513. BULLER, Walter Lawry. 1838-1906. New Zealand. BMNH; GD2.

13514. CAPELLO, Felix (Antonio) de Brito. b. 1838. Port. BMNH.

13515. CRESSON, Ezra Townsend. 1838-1926. U.S.A. BMNH; GE.

13516. FERNALD, Charles Henry. 1838-1921. U.S.A. BMNH; GE.

13517. FRITSCH, Gustav Theodor. 1838-1927. Ger. DSB; ICB; P III,IV; BMNH; Mort.; BLA & Supp. Also Anatomy and Physiology.

13518. HYATT, Alpheus. 1838-1902. U.S.A. DSB; ICB; P III,IV; BMNH; GB; GE.

13519. M'INTOSH, William Carmichael. 1838-1931. Br. DSB; BMNH; GD5.

13520. MORSE, Edward Sylvester. 1838-1925. U.S.A. ICB; BMNH; GE.

13521. OLSSON, Petter. 1838-1923. Swed. BMNH.

13522. TRYON, George Washington. 1838-1888. U.S.A. BMNH; GE.
1. *Structural and systematic conchology. An introduction to the study of the Mollusca*. 3 vols. Philadelphia, 1882-84.
See also 12908/2.

13523. BOUCARD, Adolphe. 1839-1904. Fr./Br. ICB; BMNH.

13524. DUBOIS, Alphonse. 1839-after 1912. Belg. BMNH.

13525. FINSCH, (Friedrich Hermann) Otto. 1839-1917. BMNH.

13526. PACKARD, Alpheus Spring. 1839-1905. U.S.A. DSB; P III,IV; BMNH; GE. Also Embryology*.
1. *Zoology*. New York, 1879. 4th ed., 1883.

13527. SEELEY, Harry Govier. 1839-1909. Br. BMNH; GD2. Also Geology*.

13528. SHMANKEVICH, Vladimir Ivanovich. 1839-1880. Russ. ICB; BMNH.

13529. SMITT, Fredrik Adam. 1839-1904. Swed. BMNH.

13530. VERRILL, Addison Emery. 1839-1926. U.S.A. DSB; ICB; BMNH; GE.

13531. MARTÍNEZ Y SAEZ, Francisco de Paula. d. 1908. Spain. BMNH.

13532. CUNNINGHAM, Robert Oliver. d. 1918. BMNH.
1. *Notes on the reptiles, amphibia, fishes, mollusca and crustacea obtained during the voyage of H.M.S. "Nassau" in the years 1866-69*. London, Linnean Society, 1870?

13533. COPE, Edward Drinker. 1840-1897. U.S.A. DSB Supp.; ICB; P III,IV; BMNH; GB; GE. Also Evolution*.

13534. DOHRN, (Felix) Anton. 1840-1909. Ger. DSB Supp.; BMNH; Mort.
1. *Studien zur Embryologie der Arthropoden*. Leipzig, 1868.
2. *Untersuchungen über Bau und Entwicklung der Arthropoden*. Leipzig, 1870.
3. *Der Ursprung der Wirbelthiere und das Princip des Functionwechsels*. Leipzig, 1875.

13535. FAUVEL, Albert. 1840-1921. Fr. BMNH.

13536. GOETTE, Albert Wilhelm. 1840-1922. Ger. DSB; BMNH; BLAF. Also Embryology*.

13537. HORN, George Henry. 1840-1897. U.S.A. DSB; BMNH; GE.

13538. KOBELT, Wilhelm. 1840-1916. Ger. BMNH.

13539. MEYER, Adolf Bernhard. 1840-1911. Ger. BMNH.

13540. SCHULZE, Franz Eilhard. 1840-1921. Ger. BMNH.

13541. SEIDLITZ, Georg Carl Maria von. 1840-1917. Dorpat. BMNH. Also Evolution.

13542. SENNETT, George Burritt. 1840-1900. U.S.A. GE.

13543. SHARP, David. 1840-1922. Br. BMNH.

13544. SHELLEY, George Ernest. 1840-1910. Br. BMNH.

13545. SNOW, Francis Huntington. 1840-1908. U.S.A. BMNH; GE.

13546. WAGNER, Hermann. 1840-1894. Ger. BMNH. Also Botany*.

13547. YARROW, Henry Crécy. 1840-1929. U.S.A. BMNH.

13548. ALLEN, Harrison. 1841-1897. U.S.A. BMNH; BLA & Supp.; GE. Also Anatomy.

13549. BOGDANOV, Modest Nikolaevich. 1841-1888. Russ. ICB; BMNH (BOGHDANOV).

13550. CROTCH, George Robert. 1841?-1874. Br. BMNH.

13551. FOREL, François Alphonse. 1841-1912. Switz. DSB Supp.; ICB; P III,IV; BMNH; BLAF. Also Limnology.

13552. GROTE, Augustus Radcliff. 1841-1903. U.S.A./Ger. DSB; BMNH; GE.

13553. HASSE, (Johannes) Carl (Franz). 1841-1922. Ger. BMNH; BLAF.
1. *Anatomische Studien*. Vol. 1. [No more publ.] Leipzig, 1873.
2. *Die vergleichende Morphologie und Histologie des häutigen Gehörorganes des Wirbelthiere*. Leipzig, 1873.

13554. HOFFMANN, Christian Karl. 1841-1903. Holl. BMNH; BLA. Also Embryology.

13555. HUDSON, William Henry. 1841-1922. Argentina. ICB; BMNH; GD4.
1. *Collected works*. 24 vols. London, 1922-23.

13556. LOCARD, Arnould. 1841-1904. Fr. BMNH.

13557. PLATEAU, Félix Auguste Joseph. 1841-1911. Belg. ICB; P III; BMNH. Also Embryology* and Physiology.

13558. WALKER, Francis Augustus. 1841-1905. Br. BMNH.

13559. BROWN, Robert (of Campster). 1842-1895. Br. P III; BMNH; GD1. Also Science in General*, Geology*, and Botany*.

13560. COLLETT, Robert. 1842-1913. Nor. BMNH.

13561. COUES, Elliott. 1842-1899. U.S.A. DSB; BMNH; GB; GE.
1. *Key to North American birds*. Salem, Mass., 1872. RM.
See also 12918/5.

13562. KLEINENBERG, Nicolaus. 1842-1897. It. DSB; BMNH. Also Embryology.

13563. KOVALEVSKY, Vladimir Onufrievich. 1842-1883. Russ. DSB; ICB; BMNH. Also Evolution.

13564. LINSTOW, Otto (Friedrich Bernhard) von. 1842-1916. Ger. BMNH.

13565. MAUPAS, (François) Emile. 1842-1916. Fr. DSB; ICB.

13566. MIALL, Louis Compton. 1842-1921. Br. BMNH; GF. Also Biography*.
1. *Studies in comparative anatomy*. 3 vols. 1878-86.

13567. RAMSAY, Edward Pierson. 1842-1917. Australia. BMNH; GF.

13568. RETZIUS, (Magnus) Gustaf. 1842-1919. Swed. DSB; BMNH; Mort.; BLAF. Also Anatomy and Embryology.

13569. ROSENBERG, Emil Woldemar. 1842-1925. Dorpat/Holl. BMNH; BLAF. Also Anatomy and Embryology.

13570. SELENKA, Emil. 1842-1902. Ger. BMNH; BLAF.

13571. TROUESSART, Edouard Louis. 1842-1927. Fr. BMNH. Also Microbiology*.

13572. TULLBERG, Tycho Fredrik Hugo. 1842-1920. Swed. BMNH.

13573. WHITMAN, Charles Otis. 1842-1910. U.S.A. DSB; BMNH; Mort.; BLAF; GE.
1. *Methods of research in microscopical anatomy and embryology*. Boston, 1880.
2. *Biological lectures*. Woods Hole, Mass., 1894.

13574. ABBOTT, Charles Conrad. 1843-1919. U.S.A. ICB; BMNH; GE.

13575. ADOLPH, Georg Ernst. 1843-after 1920. Ger. BMNH.

13576. BENECKE, Berthold Adolph. 1843-1886. Ger. BMNH; BLA.

13577. BERG, (Federico Guillermo) Carlos. 1843-1902. Argentina. BMNH.

13578. BOBRETZKII, Nikolai Vasilevich. b. 1843. Russ. BMNH.

13579. EIMER, Gustav Heinrich Theodor. 1843-1898. Switz. BMNH; Mort.; GC. Also Evolution*.

13580. GARMAN, Samuel. 1843-1927. U.S.A. BMNH; GE.

13581. RILEY, Charles Valentine. 1843-1895. U.S.A. BMNH; GE.

13582. SMITH, Sidney Irving. 1843-1926. U.S.A. DSB; BMNH.

13583. WATERHOUSE, Charles Owen. 1843-1917. Br. BMNH.

13584. ANDRÉ, Edmond. 1844?-1891. Fr. BMNH.

13585. BOETTGER, Oskar. 1844-1910. Ger. BMNH.

13586. BUTLER, Arthur Gardiner. 1844-1925. Br. BMNH.

13587. ESPINAS, Alfred Victor. 1844-1922. ICB; BMNH.
1. *Des sociétés animales. Etude de psychologie comparée.* Paris, 1877.

13588. FORBES, Stephen Alfred. 1844-1930. U.S.A. DSB; BMNH; GE.

13589. GRABER, Bitus (or Veit). 1844-1892. Aus. BMNH; GC.

13590. KIRBY, William Forsell. 1844-1912. Br. BMNH.
1. *Elementary text book of entomology.* London, 1885. X. 2nd ed., 1892.

13591. MACALISTER, Alexander. 1844-1919. Br. BMNH; BLAF; GF. Also Anatomy.
1. *An introduction to animal morphology and systematic zoology.* 2 vols. London, 1876.
2. *An introduction to the systematic zoology and morphology of vertebrate animals.* Dublin, 1878.
3. *Zoology of the invertebrate animals.* London, 1878. Other eds: 1879, 1885, 1889, 1890.

13592. NICHOLSON, Henry Alleyne. 1844-1899. Br. BMNH; GB; GD1. Also Geology*, Biography*, and History of Science*.
1. *A manual of zoology.* 2 vols. London, 1870. 4th ed., Edinburgh, 1875. 5th ed., 1878. 7th ed., 1887.
2. *Advanced text-book of zoology.* London, 1870. 3rd ed., Edinburgh, 1878.
3. *Synopsis of the classification of the animal kingdom.* Edinburgh, 1882.
4. *Textbook of zoology.* Edinburgh, 1886.

13593. OUSTALET, Emile. 1844-1905. Fr. BMNH.

13594. PANTANELLI, Dante. 1844-1913. It. P III,IV. Also Geology.

13595. PERRIER, (Jean Octave) Edmond. 1844-1921. Fr. DSB; BMNH; BLAF. Also History of Zoology*.

13596. SCHÖYEN, Wilhelm Maribo. 1844-1918. Nor. BMNH.

13597. SCHWARZ, Eugen Amandus. 1844-1928. Ger./U.S.A. BMNH; GE.

13598. CLESSIN, Stephan. fl. 1873-1890. BMNH.

13599. GERVAIS, Henri Frédéric Paul. fl. 1876-1891. Fr. BMNH.
1. *Les poissons.* 3 vols. Paris, 1876-77. With R. Boulart. RM.

13600. GESTRO, Raffaello. fl. 1874-1910. It. BMNH.

13601. MONTEROSATO, Tommaso Allery di. fl. 1872-1893. It. BMNH.

13602. UHAGÓN, Serafín de. fl. 1872-1904. Spain. BMNH.

13603. BROCCHI, Paul. d. 1898. Fr. BMNH.

13604. CUNÍ Y MARTORELLI, Miguel. d. 1909. Spain. BMNH.

13605. ALSTON, Edward Richard. 1845-1881. Br. BMNH; GD.

13606. BRUSINA, Spiridion. 1845-1908. Adriatic region. BMNH.

13607. CARLET, Gaston. 1845-1892. Fr. BMNH.

13608. CZERSKI, Jan. 1845-1892. Russ. DSB. Also Geology.

13609. DALL, William Healey. 1845-1927. U.S.A. DSB; ICB; P III,IV; BMNH; GE. Also Natural History*.

13610. DISTANT, William Lucas. 1845-1922. Br. BMNH.

13611. DOHIĆ, Lazar. 1845-1893. Belgrade. BLAF.

13612. FOL, Hermann. 1845-1892. Switz. DSB; ICB; BMNH. Also Embryology and Cytology.
1. *Lehrbuch der vergleichenden mikroskopischen Anatomie.* Leipzig, 1884-96.

13613. GIGLIOLI, Enrico Hillyer. 1845-1909. It. BMNH.

13614. MARSHALL, William Adolf Ludwig. 1845-1907. Ger. BMNH.

13615. MAYNARD, Charles Johnson. 1845-1929. U.S.A. ICB; BMNH; GE.

13616. METCHNIKOFF, Elie. 1845-1916. Russ./Fr. DSB; ICB (MECHNIKOV); BMNH (MECHNIKOV); BLAF (METSCHNIKOW). Also Embryology and Microbiology.

13617. NEHRING, (Carl Wilhelm) Alfred. 1845-1904. Ger. DSB; P III, IV; BMNH.

13618. NEUMAYR, Melchior. 1845-1890. Aus. DSB; P III,IV; BMNH; GB. Also Geology*.
1. *Die Stämme des Thierreiches ... Wirbellose Thiere.* Vol. 1. [No more publ.] Vienna, 1889.

13619. OBERTHÜR, Charles. 1845-1924. Fr. BMNH.

13620. PALMÉN, Johan Axel. 1845-1919. Fin. BMNH.

13621. PECKHAM, George Williams. 1845-1914. U.S.A. BMNH; GE.

13622. REITTER, Edmund. 1845-1920. BMNH.

13623. SAHLBERG, John Reinhold. 1845-1920. Fin. BMNH.

13624. STUDER, Theophil. 1845-1922. Switz. P III,IV; BMNH; BLAF. Also Geology.

13625. WATERHOUSE, Frederick Herschel. 1845-1919. Br. BMNH. Also Biography*.

13626. ARLOING, Saturnin. 1846-1911. Fr. BMNH; BLAF; GF. Also Microbiology.

13627. BEAN, Tarleton Hoffmann. 1846-1916. U.S.A. BMNH; GE.

13628. BENEDEN, Edouard van. 1846-1910. Belg. DSB; ICB; BMNH; Mort,; BLA & Supp. Also Embryology*.

13629. FÜRBRINGER, Max. 1846-1920. Ger. BMNH; BLAF. Also Anatomy and Embryology.

13630. GARROD, Alfred Henry. 1846-1879. Br. BMNH; BLA; GD.
1. *The collected scientific papers.* London, 1881. Ed. with a biography by W.A. Forbes.

13631. GIARD, Alfred Mathieu. 1846-1908. Fr. DSB; ICB; BMNH; BLAF. Also Embryology.

13632. MARION, Antoine Fortuné. 1846-1900. Fr. DSB; BMNH. Also Botany.

13633. SIMPSON, Charles Torrey. 1846-1932. U.S.A. GE1.

13634. STERKI, Victor. 1846-1933. U.S.A. GE1.

13635. TOMES, Charles Sissmore. 1846-1928. Br. BMNH.

13636. AUBUSSON, Louis Magaud d'. b. 1847. Fr. BMNH.

13637. BESSELS, Emil. 1847-1888. Ger./U.S.A. P III; BMNH; GC.

13638. CAMERON, Peter. 1847-1912. Br. BMNH.

13639. CHATIN, Joannès. 1847-1912. Fr. BMNH.
1. *Les organes des sens dans la série animale. Leçons d'anatomie et de physiologie comparées.* Paris, 1880.

13640. EISEN, August Gustav. 1847-1940. Swed. BMNH; GF.

13641. EMERTON, James Henry. 1847-1930. U.S.A. BMNH; GE1.

13642. GASKELL, Walter Holbrook. 1847-1914. Br. DSB; Mort.; GD3. Also Physiology.

13643. HOLUB, Emil. 1847-1902. Aus. P III,IV; BMNH; GB.

13644. LANKESTER, Edwin Ray. 1847-1929. Br. DSB; BMNH; BLAF; GD4. Also Science in General*, Microscopy, and Evolution*. See also Index.
1. *On comparative longevity in man and the lower animals.* London, 1870.
2. *Spolia maris.* London, 1889.
3. *Zoological articles contributed to the "Encyclopaedia Britannica."* London, 1891.

13645. REICHENOW, Anton. 1847-1915. Ger. BMNH.

13646. SHARPE, Richard Bowdler. 1847-1909. Br. BMNH; GD2. See also 13094/3 and 13855/4.

13647. SMITH, Edgar Albert. 1847-1916. Br. BMNH.

13648. ZALENSKII, Vladimir Vladimirovich. 1847-1918. Russ. BMNH. Also Embryology.

13649. BRIDGE, Thomas William. 1848-1909. Br. GD2.

13650. BROOKS, William Keith. 1848-1908. U.S.A. DSB; ICB; BMNH; Mort.; BLAF; GE. Also Embryology.
1. *Handbook of invertebrate zoology.* Boston, 1882.
2. *Foundations of zoology.* New York, 1899.

13651. BÜTSCHLI, (Johann Adam) Otto. 1848-1920. Ger. DSB; ICB; P IV; BMNH; BLAF. Also Cytology*.

13652. BURGESS, Edward. 1848-1891. U.S.A. BMNH; GE.

13653. DOBSON, George Edward. 1848-1895. Br./India. BMNH; BLA; GD.

13654. FOREL, Auguste Henri. 1848-1931. Switz. DSB; ICB; BMNH; Mort.; BLAF. Also Anatomy.
1. *Briefe. Correspondance. 1864-1927.* Bern, 1967. Ed. by H.H. Walser.

13655. HOLLAND, William Jacob. 1848-1932. U.S.A. BMNH; GE1.

13656. KELLER, Conrad. 1848-1930. Switz. ICB; BMNH.

13657. KRAEPELIN, Karl Mathias Friedrich. 1848-1915. Ger. BMNH.

13658. MAYER, Paul. 1848-1923. Ger. BMNH; BLAF. Also Microscopy.

13659. PORTSCHINSKY, J.A. 1848-1916. Russ. BLAF.

13660. SAUNDERS, Edward. 1848-1910. Br. BMNH; GD2.

13661. SIMON, Eugène Louis. 1848-1924. Fr. BMNH.

13662. THÉEL, (Johan) Hjalmar. 1848-1937. Swed. BMNH.

13663. VETTER, Benjamin. 1848-1893. Ger. BMNH; GC.

13664. WIEDERSHEIM, Robert (Ernst Eduard). 1848-1923. Ger. DSB; BMNH; Mort.; BLAF.
1. *Grundriss der vergleichenden Anatomie der Wirbelthiere.* Jena, 1884. X.
1a. ——— *Elements of the comparative anatomy of vertebrates.* London, 1886. Trans. and adapted by W.N. Parker. 2nd ed., 1897.
2. *Der Bau des Menschen als Zeugniss fur seine Vergangenheit.* Friburg i.B., 1887. X.
2a. ——— *The structure of man: An index to his past history.* London, 1895. Trans. by H.M. Bernard.

13665. BERTKAU, Philipp. 1849-1895. BMNH.
1. *Bericht über die wissenschaftlichen Leistungen im Gebiete der Entomologie.* Berlin, 1883.

13666. BOLL, Franz Christian. 1849-1878. Ger./It. P IV; BMNH; Mort.; BLA & Supp. Also Anatomy and Physiology.

13667. BOULART, Raoul A. b. 1849. Fr. BMNH. See also 13599/1.

13668. BOVALLIUS, Carl Erik Alexander. 1849-1907. Swed. BMNH.

13669. COMSTOCK, John Henry. 1849-1931. U.S.A. ICB; BMNH; GE.

13670. HERTWIG, (Wilhelm August) Oscar. 1849-1922. Ger. DSB; ICB; BMNH; Mort.; BLAF. Also Embryology* and Cytology*.
1. *Das Nervensystem und die Sinnesorgane der Medusen.* Leipzig, 1878. With C.W.T.R. Hertwig.

13671. JANET, Charles. 1849-1932. Fr. BMNH.

1. *Etudes sur les fourmis, les gûepes et les abeilles.* 5 parts. Beauvais, 1895-97.
2. *Notice sur les traveaux scientifiques présentés par ... C. Janet à l'Académie des Sciences au concours de 1896.* Lille, 1896.

13672. KOCH, Gottlieb von. 1849-after 1905. Ger. BMNH.
1. *Grundriss der Zoologie.* Jena, 1876.

13673. KOSSMANN, Robby August. 1849-1907. Ger. BMNH; BLAF.
1. *Elemente der wissenschaftlichen Zoologie.* Munich, 1878.

13674. LYDEKKER, Richard. 1849-1915. Br. BMNH; GF.
1. *Phases of animal life.* 1892.
2. *The royal natural history.* 6 vols. London, 1893-96.
3. *Life and rock.* 1894.
4. *A hand-book to the Marsupialia and Monotremata.* London, 1894.
5. *Marsupials.* London, 1894?
6. *A geographical history of animals.* Cambridge, 1896.
See also 13425/3.

13675. SLADEN, Walter Percy. 1849-1900. Br. BMNH.

13676. THAYER, Abbott Handerson. 1849-1921. U.S.A. ICB; BMNH; GE.

13677. VEJDOVSKÝ, František. 1849-1939. Prague. DSB Supp.; ICB; BMNH. Also Embryology* and Cytology.

13678. BOSCÁ (Y CASANOVES), Eduardo. fl. 1877-1915. Spain. BMNH.

13679. MENZBIR, Mikhail Aleksandrovich. fl. 1880-1916. Russ. BMNH.

13680. SERVAIN, Georges. fl. 1880-1891. Fr. BMNH.

13681. KRAMER, Paul. d. 1898. Ger. BMNH.

13682. BOLÍVAR Y URRUTIA, Ignacio. b. 1850. Spain. BMNH.

13683. BRAUN, Maximilian Gustav Christian Carl. 1850-1930. Ger. ICB; BMNH; BLAF.

13684. CAVANNA, Guelfo. 1850-after 1914. It. BMNH.
1. *Elementi per una bibliografia italiana intorno all'idro-fauna agli allevamenti degli animali acquatici e alla pesce.* Florence, 1880.

13685. CUNNINGHAM, Daniel John. 1850-1909. Br. BMNH; BLAF; GD2. Also Anatomy*.

13686. DRASCHE, Richard von. 1850-after 1886. Aus. P III; BMNH. Also Geology.

13687. FEWKES, Jesse Walter. 1850-1930. U.S.A. ICB; BMNH; GE.

13688. HENSHAW, Henry Wetherbee. 1850-1930. U.S.A. BMNH; GE.

13689. HERTWIG, (Karl Wilhelm Theodor) Richard von. 1850-1937. Ger. DSB; ICB; BMNH; Mort.; BLAF. Also Embryology and Cytology. See also Index.
1. *Lehrbuch der Zoologie.* Jena, 1891-92. X.
1a. ——— *Manual of zoology.* London, 1903. Trans. from the 5th German ed. by J.S. Kingsley.

13690. HUBBARD, Henry Guernsey. 1850-1899. U.S.A. GE.

13691. JHERING, Hermann von. b. 1850. Ger. BMNH.

13692. LECHE, Wilhelm. 1850-1927. Swed. BMNH.

13693. MAN, Johannes Govertus de. 1850-1930. Holl. BMNH.

13694. PARKER, Thomas Jeffery. 1850-1897. New Zealand. BMNH.
1. *A course of instruction in zootomy (vertebrata).* London, 1884.
2. *A text-book of zoology.* 2 vols. Ib., 1897. With W.A. Haswell.
3. *A manual of zoology.* 1899.

13695. PAULY, August. 1850-1914. Ger. DSB; BMNH.

13696. REUTER, Odo Morannal. 1850-1913. Fin. BMNH.

13697. RIDGWAY, Robert. 1850-1929. U.S.A. DSB; BMNH; GE.

13698. ROUGEMONT, Philippe Albert de. 1850-1881. Switz. P III; BMNH.

13699. SHUFELDT, Robert Wilson. 1850-1934. U.S.A. BMNH.

13700. TIKHOMIROV, Aleksandr Andeevich. b. 1850. Russ. BMNH.

13701. ALBRECHT, (Carl Martin) Paul. 1851-1894. Ger./Belg. BMNH; BLA & Supp.; GC. Also Anatomy.

13702. BEAUREGARD, Henri. 1851-1900. Fr. BMNH.

13703. BIRGE, Edward Asahel. 1851-1950. U.S.A. DSB; BMNH.

13704. BREWSTER, William. 1851-1919. U.S.A. GE.

13705. EWART, James Cossar. 1851-1933. Br. ICB; BMNH.

13706. FRANCOTTE, (Charles Joseph) Polydore. 1851-1916. Belg. BMNH.

13707. GAGE, Simon Henry. 1851-1944. U.S.A. BMNH; BLAF. Also Microscopy.

13708. GOODE, George Brown. 1851-1896. U.S.A. ICB; BMNH; GE. Also History of Science*. See also 15523/1.

13709. GRAFF, Ludwig von. 1851-1924. BMNH.

13710. HARTOG, Marcus Manuel. 1851-after 1913. Br. BMNH. See also 12288/1a.

13711. HOEK, Paulus Peronius Cato. 1851-1914. Holl. BMNH.

13712. HOLDER, Charles Frederick. 1851-1915. U.S.A. BMNH; GE. Also Biography*.

13713. JORDAN, David Starr. 1851-1931. U.S.A. DSB; ICB; BMNH; GE. Also Evolution*.

13714. MIERS, Edward John. b. 1851. Br. BMNH.

13715. PERAGALLO, Hippolyte. b. 1851. Fr. ICB; BMNH.

13716. PEREYASLAVTZEVA, Sofya Mikhailovna. 1851-1903. Russ. BMNH.

13717. SIMROTH, Heinrich Rudolf. 1851-after 1913. Ger. BMNH.

13718. STEJNEGER, Leonhard Hess. 1851-1943. U.S.A. DSB; BMNH.

13719. WILLISTON, Samuel Wendell. 1851-1918. U.S.A. DSB; BMNH; BLAF; GE.

13720. ZOGHRAF (or TSOGRAF), Nikolai Yurevich. 1851-after 1904. Russ. BMNH.

13721. BARROIS, Jules. 1852-after 1934. Fr. BMNH.

13722. CARPENTER, Philip Herbert. 1852-1891. Br. BMNH; GD1.

13723. CHUN, Karl. 1852-1914. Ger. BMNH.

13724. DOGIEL, Alexander Stanislavovich. 1852-1922. Russ. BMNH (DOGHEL); Mort.; BLAF. Also Anatomy.

13725. INGERSOLL, Ernest. 1852-1946. U.S.A. ICB; BMNH.

13726. LUCAS, Frederick August. 1852-1929. U.S.A. BMNH; GE.

13727. LUDWIG, Hubert (Jacob). 1852-1913. Ger. BMNH.
1. *Morphologische Studien an Echinodermen*. 2 vols. Leipzig, 1877-82.

13728. MARSHALL, Arthur Milnes. 1852-1893. Br. BMNH; GD1. Also Evolution* and Embryology*.
1. *Biological lectures and addresses*. 1894.

13729. MORGAN, Conwy Lloyd. 1852-1936. Br. DSB; ICB; BMNH; GD5. See also 2988/4.
1. *Animal biology. An elementary text-book*. London, 1887. X. 3rd ed., 1899.
2. *Animal life and intelligence*. London, 1890-91.
3. *An introduction to comparative psychology*. London, 1895.
4. *Habit and instinct*. London, 1896.

13730. PHISALIX, Césaire Auguste. 1852-1906. Fr. BLAF.

13731. RATHBUN, Richard. 1852-1918. U.S.A. P III; BMNH; GE. Also Geology.

13732. RUGE, Georg Hermann. 1852-1919. Ger./Holl./Switz. BMNH; BLAF. Also Anatomy.

13733. WEBER, Max Wilhelm Carl. 1852-1937. Holl. DSB; ICB; BMNH.
1. *Zoologische Ergebnisse einer Reise in niederländisch Ost-Indien*. 4 vols. Leiden, 1890-1907.

13734. WILSON, Andrew. 1852-1912. Br. BMNH. Also Evolution*.

13735. AURIVILLIUS, Per Olof Christopher. 1853-1928. Swed. BMNH.

13736. BALASHEVA, Mariya Dmitrievna. b. 1853. Russ. BMNH.

13737. BERNARD, Henry Meyners. 1853-after 1911. Br. BMNH. See also 13664/2a and 13780/1a.

13738. BUCHINSKII, Petr Nikolaievich. b. 1853. Russ. BMNH.

13739. CLARKE, William Eagle. 1853-1938. Br. BMNH; GF.

13740. EVERMANN, Barton Warren. 1853-1932. U.S.A. BMNH; GE1.

13741. HEILPRIN, Angelo. 1853-1907. U.S.A. ICB; BMNH; GE. Also Geology*.
1. *The geographical and geological distribution of animals*. London, 1887. 2nd ed., 1894.

13742. HUBRECHT, (Ambrosius Arnold) Willem. 1853-1915. Holl. DSB; BMNH; BLAF. Also Evolution* and embryology.

13743. MERRILL, James Cushing. 1853-1902. U.S.A. GE.

13744. SOVINSKII, Vasilii Karlovich. 1853-1917. Russ. ICB.

13745. ARRUDA FURTADO, Francisco d'. 1854-1887. Port. BMNH.

13746. AURIVILLIUS, Carl Vilhelm Samuel. b. 1854. Swed. BMNH.

13747. BEDRYAGHA, Yakov Vladimirovich. 1854-after 1912. Russ. BMNH.

13748. BRANDT, Karl. 1854-1931. Ger. BMNH.

13749. CARRIÈRE, Justus Wilhelm Johannes. 1854-1893. Ger. BMNH.
1. *Die Sehorgane der Thiere vergleichend-anatomisch dargestellt*. Munich, 1885.

13750. DELAGE, Yves. 1854-1920. Fr. DSB; BMNH. Also Biology in General*.
1. *Traité de zoologie concrète*. 8 vols. Paris, 1896-1903. With E. Hérouard.

13751. DISSELHORST, Rudolf. 1854-1930. Ger. BLAF.

13752. GRASSI, Giovanni Battista. 1854-1925. It. DSB; ICB; BMNH; BLAF.

13753. GULDBERG, Gustav Adolph. 1854-1908. Nor. BMNH; BLAF.

13754. HASWELL, William Aitcheson. 1854-1925. Australia. BMNH; GF. See also 13694/2.

13755. HATSCHEK, Berthold. 1854-1941. Aus. DSB; BMNH.
1. *Studien über die Entwicklung des Amphioxus*. Vienna, 1881.
1a. ——— *The Amphioxus and its development*. London, 1893.
2. *Lehrbuch der Zoologie*. 3 parts. Jena, 1888-91.

13756. KINGSLEY, John Sterling. 1854-1929. U.S.A. BMNH.
1. *The standard natural history*. 6 vols. Boston, 1884-85.
See also 13689/1a.

13757. MEYRICK, Edward. 1854-1938. Br. BMNH; ICB (MAYRICK); GD5.

13758. NOMAN, Dirk van Haren. 1854-1896. Holl. BMNH; GF.

13759. PFEFFER, Georg Johann. 1854-1931. Ger. BMNH.

13760. RÜCKERT, Johannes. 1854-1923. Ger. BMNH; BLAF.

13761. SEDGWICK, Adam (second of the name). 1854-1913. Br. BMNH; GD3.
1. *Studies from the Morphological Laboratory in the University of Cambridge*. London, 1884.
2. *A student's text-book of zoology*. 3 vols. Ib., 1898-1909.
See also Index.

13762. TASCHENBERG, (Ernst) Otto (Wilhelm). 1854-1922. Ger. BMNH.

13763. VAYSSIÈRE, Albert Jean Baptiste Marie. 1854-after 1937. Fr. BMNH.
1. *Atlas d'anatomie comparée des invertébrés*. Paris, 1888-90.

13764. WELTNER, Wilhelm. 1854-1917. Ger. BMNH.

13765. YUNG, Emile Jean Jacques. 1854-1918. Switz. ICB; BMNH.

13766. BAYER, František. fl. 1884-1908. Prague. BMNH.

13767. BERGENDAL, David. fl. 1882-1903. Swed. BMNH.

13768. GANGLBAUER, Ludwig. fl. 1882-1894. Aus. BMNH.

13769. HALLER, Béla. fl. 1882-1897. Aus. BMNH.

13770. ASHMEAD, William Harris. 1855-1908. U.S.A. BMNH; GE.

13771. BELL, Francis Jeffrey. 1855-1924. BMNH.
1. *Comparative anatomy and physiology.* London, 1885.
See also 13370/4a.

13772. BLAND-SUTTON, John. 1855-1936. Br. ICB; BLAF; GD5. Also Evolution*.

13773. DADAY, Jenö. 1855-1920. Hung. BMNH.

13774. FORBES, William Alexander. 1855-1883. Br. BMNH; GD.
1. *The collected scientific papers.* London, 1885. Ed. by F.E. Beddard. Preface by P.S. Sclater. RM/L.
See also 13630/1.

13775. GADOW, Hans Friedrich. 1855-1928. Br. BMNH. See also 13900/8.

13776. HAACKE, (Johann) Wilhelm. 1855-1912. Ger. BMNH.

13777. HADDON, Alfred Cort. 1855-1940. Br. ICB; BMNH; GD5. Also Embryology*.

13778. HANSEN, Hans Jacob. 1855-1936. Den. BMNH.

13779. KOLBE, Hermann Julius. b. 1855. Ger. BMNH.

13780. LANG, Arnold. 1855-1914. Switz. DSB; BMNH.
1. *Lehrbuch der vergleichenden Anatomie der wirbellosen Thiere.* 4 parts. Jena, 1888-94. X.
1a. ——— *Text-book of comparative anatomy.* 2 parts. London, 1891-96. Trans. by H.M. Bernard.

13781. MERRIAM, Clinton Hart. 1855-1942. U.S.A. DSB; BMNH.
1. *Selected works.* New York, 1974. Introd. by K.B. Sterling.

13782. NASONOV, Nikolai Viktorovich. 1855-1939. Russ. BMNH.

13783. NELSON, Edward William. 1855-1934. U.S.A. BMNH; GE1.

13784. BOUVIER, (Louis) Eugène. 1856-1944. Fr. BMNH.

13785. COQUILLETT, Daniel William. 1856-1911. U.S.A. BMNH; GE.

13786. FERTON, Charles. 1856-1921. Fr. ICB.

13787. HAUSER, Gustav. 1856-1935. Ger. BMNH; BLAF. Also Microbiology.

13788. KULTSCHITZKY, Nikolai. 1856-1925. Russ. BMNH; Mort. Also Anatomy.

13789. MORTENSEN, Hans Christian Cornelius. 1856-1921. ICB.

13790. POULTON, Edward Bagnall. 1856-1943. Br. ICB; BMNH; GD6. Also Evolution* and Biography*.
1. *The colours of animals, their meaning and use.* London, 1890.

13791. SARASIN, Paul Benedict. 1856-after 1922. Switz. P IV; BMNH.

13792. VIALLANES, Henri. 1856-1893. Fr. BMNH.

13793. APPELLÖF, (Jakob Johan) Adolf. 1857-1921. Nor. BMNH.

13794. BERLEPSCH, Hans von. 1857-1933. ICB.

13795. BLANCHARD, Raphael. 1857-1919. Fr. ICB; BMNH; BLAF.

13796. BROOK, George. 1857-1893. Br. BMNH.

13797. CORY, Charles Barney. 1857-1921. U.S.A. BMNH; GE.

13798. DOLLO, Louis Antoine Marie Joseph. 1857-1931. Belg. DSB; ICB; BMNH.

13799. FRAIPONT, Julien Jean Joseph. 1857-1910. Belg. DSB; ICB; BMNH.

13800. HAMANN, Otto. 1857-1925. BMNH. Also Evolution*.

13801. HOWARD, Leland Ossian. 1857-1950. U.S.A. DSB; ICB; BMNH.

13802. LOCY, William Albert. 1857-1924. U.S.A. ICB; BMNH; GE.

13803. MITROFANOV, Pavel Ilich. 1857-1920. Warsaw. BMNH; BLAF MITROPHANOW). Also Embryology.

13804. OSBORN, Henry Fairfield. 1857-1935. U.S.A. DSB; ICB; BMNH; GE1. Also History of Science*.

13805. RAWITZ, Bernhard. 1857-after 1919. Ger. BMNH; BLAF.

13806. ROSA, Daniele. 1857-1944. It. DSB; BMNH. Also Evolution*.

13807. SMETS, Gérard. b. 1857. Belg. BMNH.

13808. TUCKERMAN, Frederick. 1857-1929. U.S.A. GE.

13809. WINGE, (Adolf) Herluf. 1857-1923. Den. ICB; BMNH.

13810. BEDDARD, Frank Evers. 1858-1925. Br. BMNH.
1. *Animal coloration. An account of the principal facts and theories*. London, 1892. Another ed., 1895.
2. *Text-book of zoogeography*. Cambridge, 1895.
3. *The structure and classification of birds*. London, 1898.
See also 13774/1.

13811. BLOCHMANN, Friedrich. 1858-after 1924. Ger. BMNH. Also Embryology and Microbiology.

13812. BOULENGER, George Albert. 1858-1937. Br. ICB; BMNH.

13813. CHOLODKOWSKY, N.A. 1858-1921. Russ. BMNH; BLAF.

13814. CORNISH, Charles John. 1858-1906. Br. GD2.
1. *Animals at work and play, their activities and emotions*. London, 1896. X. 2nd ed., 1897.

13815. FORBUSH, Edward Howe. 1858-1929. U.S.A. GE.

13816. FRENZEL, Johannes. 1858?-1897. BMNH.

13817. HEATHCOTE (later SINCLAIR), Frederick Granville. b. 1858. BMNH. See also 13479/1b.

13818. HENKING, Hermann. 1858-1942. Ger. DSB; BMNH. Also Cytology.

13819. HERDMAN, William Abbott. 1858-1924. Br. BMNH; GD4.

13820. HERRICK, Clarence Luther. 1858-1904. U.S.A. DSB; BMNH.

13821. LENDENFELD, Robert von. 1858-1913. Ger./Australia. BMNH.

13822. NUTTING, Charles Cleveland. 1858-1927. U.S.A. BMNH; GE.

13823. OSTROUMOV, Aleksei Aleksandrovich. b. 1858. Russ. BMNH.

13824. OUDEMANS, Anthonie Cornelius (second of the name). 1858-1943. Holl. ICB; BMNH.

13825. PLESKE, Thedor (or Fedor) Dmitrievich. 1858-1932. Russ. BMNH.

13826. RUSSELL, Herbrand Arthur, *11th Duke of* BEDFORD. 1858-1940. Br. ICB; BMNH; GD5.

13827. SCHARFF, Robert Francis. 1858-1934. Br. BMNH.
1. *The history of the European fauna.* London, 1899.

13828. SMITH, John Bernard. 1858-1912. U.S.A. ICB; BMNH; GE.

13829. THOMAS, (Michael Rogers) Oldfield. 1858-1929. Br. BMNH. See also 13855/7.

13830. TRUE, Frederick William. 1858-1914. U.S.A. BMNH; GE.

13831. VANHOEFFEN, Ernst. 1858-1918. BMNH.

13832. ZIEGLER, Heinrich Ernst. 1858-1925. Ger. BMNH; BLAF.

13833. ALCOCK, Alfred William. 1859-1933. Br. ICB; BMNH.

13834. BALLOWITZ, Emil. 1859-1936. Ger. BLAF.

13835. BARROIS, Theodore. b. 1859. Fr. BMNH.

13836. BÉRANECK, Edmond. 1859-1920. Switz. BLAF.

13837. BOURNE, Alfred Gibbs. 1859-1940. Br./India. ICB.

13838. BRONGNIART, Charles J.E. 1859-1899. Fr. BMNH.

13839. ETZOLD, Franz. 1859-1928. Ger. P IV.

13840. GILBERT, Charles Henry. 1859-1928. U.S.A. BMNH; GE.

13841. GOURRET, Paul Gabriel Marie. 1859-after 1906. Fr. P IV; BMNH. Also Geology.

13842. HICKSON, Sydney John. 1859-1940. Br. ICB; BMNH; GF.
1. *The fauna of the deep sea.* London, 1894.

13843. JAYNE, Horace Fort. 1859-1913. U.S.A. GE.

13844. NEWSTEAD, Robert. 1859-1947. Br. BMNH; BLAF.

13845. NUSBAUM-HILAROWICZ, Josef. 1859-1917. Lemberg. BMNH (NUSBAUM, Osip I.); BLAF.

13846. SARASIN, Fritz (or Carl Friedrich). 1859-1942. Switz. P IV; BMNH.

13847. SEMON, Richard Wolfgang. 1859-1918. Ger. DSB; BMNH; BLAF.
1. *Zoologische Forschungsreisen in Australien und dem malayischen Archipel.* Jena, 1893-1913.

13848. WASMANN, Erich. 1859-1931. ICB; BMNH.
1. *Instinct und Intelligenz im Thierreich.* Freiburg i.B., 1897. X. 2nd ed., 1899.

Imprint Sequence

13849. *ENCYCLOPÉDIE méthodique*. [Main entry: 3000]
1. *Encyclopédie méthodique ... Histoire naturelle*. 10 vols of text & 10 vols of plates. Paris, 1782-1827. Entirely zoological. Contains sections by many authors.
2. *Encyclopédie méthodique ... Histoire naturelle des vers*. 4 vols of text & 3 vols of plates. Paris, 1792-1832. Contains sections by several authors.
3. *Encyclopédie méthodique ... Système anatomique*. 4 vols of text & 1 vol. of plates. Paris, 1792-1830. Contents as follows.
3a. ——— Vol. 1. Dictionnaire raisonné des termes d'anatomie et de physiologie. Par H. Cloquet. 1823.
3b. ——— Vol. 2. Quadrupèdes. Par F. Vicq-Dazyr. 1792.
3c. ——— Vol. 3. Mammifères et oiseaux. Commencé par F. Vicq-Dazyr et continué par H. Cloquet. 1819.
3d. ——— Vol. 4. Reptiles, poissons, mollusques, etc. Par H. Cloquet. 1830.

13850. MARTYN, William Frederick.
1. *A new dictionary of natural history*. 2 vols. London, 1785.

13851. ZOOLOGICAL Society of London.
1. *The gardens and menagerie of the Zoological Society delineated*. 2 vols. London, 1830-31. Ed. by E.T. Bennett.
2. *Catalogue of the library*. London, 1854. X. 4th ed., 1887. 5th ed., 1902.

13852. *THE NATURALIST'S Library*.
1. *The naturalist's library*. 40 vols. Edinburgh, 1833-43. Ed. by W. Jardine. Re-issued with different volume numbering in 1845-46; there were further re-issues in 1848 and later. Each volume consists of a monograph on some aspect of zoology together with a biography of a distinguished zoologist. See BMNH for a list of the authors and the varying volume numbering. In the present catalogue the individual volumes are noted under the names of their respective authors.
2. *Lives of eminent naturalists*. Edinburgh, 1841. A collection of the biographies included in the 40 volumes of the *Naturalist's library*.

13853. RAY SOCIETY. London. (Main entry: 11266)
1. *Reports on the progress of zoology and botany, 1841-1842*. London, 1845. (Ray Society publn. 1)
2. *Reports on zoology, 1843, 1844*. London, 1847. (Ray Society publn. 11) Trans. from the German by G. Busk, A. Tulk, and A.H. Halliday.

No further items of this kind were published by the Society.

13854. MUSÉUM National d'Histoire Naturelle. Paris.
1. *Catalogue de la collection entomologique. Classe des insectes. Ordre des coléoptères*. 2 vols. Paris, 1850.

13855. BRITISH Museum (Natural History). Department of Zoology.
1. *Catalogue of the ... Mammalia in the ... Museum*. 3 parts. London, 1850-52. By J.E. Gray.

2. *Catalogue of marine Polyzoa in the collection.* Ib., 1852-75. By G. Busk. X. Reprinted, New York, 1966.
3. *Hand-list of genera and species of birds, distinguishing those contained in the British Museum.* 3 vols. Ib., 1869-71. By G.R. Gray.
4. *Catalogue of the birds in the British Museum.* 27 vols. Ib., 1874-75. By R. Bowdler Sharpe et al.
5. *Catalogue of the books in the Department of Zoology.* Ib., 1884.
6. *Report on the zoological collections made in the Indo-Pacific Ocean during the voyage of H.M.S. "Alert", 1881-2.* Ib., 1884.
7. *Catalogue of the Marsupialia and Monotremata in the collection.* Ib., 1888. By O. Thomas.

13856. ENTOMOLOGISCHER Verein. Stettin.
1. *Catalogus Coleopterum Europae.* Stettin, 1856. X. 7th ed., 1858.

13857. RIJKSMUSEUM van Natuurlijke Historie. Leiden.
1. *Revue méthodique et critique des collections.* 10 vols. Leiden, 1862-80. A catalogue. (Listed under Muséum d'Histoire Naturelle des Pays-Bas)

13858. HAYEK, Gustav von. BMNH.
1. *Handbuch der Zoologie.* 4 vols. Vienna, 1877-93.

13859. WILLUGHBY Society. London.
The Society published twelve reprints of important zoological works in the period 1880-84. For a list see BMNH. In the present catalogue they are entered under the names of their respective authors.

13860. *CASSELL'S Natural history.* 6 vols. London, 1883. Ed. by P.M. Duncan. For contents see BMNH.

13861. *ZOOLOGISCHES Adressbuch. Namen und Adressen der lebenden Zoologen, Anatomen, Physiologen und Zoopaläontologen.* 2 parts. Berlin, 1895-1901. By the publishers, R. Friedländer und Sohn. X.

13862. *THE CAMBRIDGE Natural history.* 10 vols. London, 1895-1910. Ed. by S.F. Harmer and A.E. Shipley. Reprinted, Codicote (U.K.), 1958-59.

3.15 EVOLUTION AND HEREDITY

In this section, unlike most other sections, the classification is based very largely on the book titles. The section represents not so much a separate field as a new dimension to botany and zoology.

13863. WHITE, Charles. 1728-1813. Br. DSB; ICB; BMNH; BLA; GD.
1. *An account of the regular gradation in man and in different animals and vegetables, and from the former to the latter.* London, 1799.
See also 13997/1.

13864. MARTINET, Johannes Florentinus. 1729-1795. ICB. Heredity.

13865. LAMARCK, Jean Baptiste Pierre Antoine de Monet, *Chevalier* de. 1744-1829. Fr. DSB; ICB; P I; BMNH; GA; GB. Other entries; see 3204.
Titles of works containing his evolutionary ideas are listed under Zoology (entry 12746).

13866. MATTHEW, Patrick. 1790-1874. Br. ICB.

13867. FLOURENS, (Marie Jean) Pierre. 1794-1867. Fr. DSB; ICB; BMNH; Mort.; BLA; GA; GB. Also Zoology*, Embryology*, Physiology*, and Biography*.
1. *Examen du livre de M. Darwin sur l'origine des espèces.* Paris, 1864.

13868. HODGE, Charles. 1797-1878. U.S.A. GB; GE.
1. *What is Darwinism?* New York, 1874.

13869. LYELL, Charles. 1797-1875. Br. DSB; ICB; P I,III; BMNH; GA; GB; GD. Also Geology*. See also Index.
1. *Scientific journals on the species question.* New Haven, 1972. First publication of the journals. Ed. by L.G. Wilson.

13870. CHAMBERS, Robert. 1802-1871. Br. DSB; ICB; P III; BMNH; GA; GB; GD. Also Geology*.
1. *Vestiges of the natural history of creation.* London, 1844. Published anonymously. RM. Reprinted, Leicester, 1969.
1a. ——— 6th ed. 1847.
1b. ——— 10th ed. With extensive additions and emendations. 1853.
1c. ——— 11th ed. The last ed. revised by Chambers. 1860.
1d. ——— 12th ed. Introd. by A. Ireland. 1884.
1e. ——— Another ed. Introd. by H. Morley. 1887.
2. *Explanations. A sequel to Vestiges....* London, 1845. RM. 2nd ed., 1846.

13871. WATSON, Hewett Cottrell. 1804-1881. Br. DSB; BMNH; GD. Also Botany*.

13872. MORITZI, Alexander. 1806-1850. Switz. ICB; BMNH. Also Botany.
1. *Reflexions sur l'espèce en histoire naturelle*. 1842. X.
1a. ——— Reprint. *Mit einer biographischen Einleitung nach J. Bloch ... und einer Würdigung Moritizis als Vorläufer Charles Darwins, von A. Lang*. Aarau, 1834.

13873. DARWIN, Charles Robert. 1809-1882. Br. DSB; ICB; BMNH; G. Also Geology*, Natural History*, Botany*, Zoology*, and Autobiography*. See also Index.
1. *On the origin of species by means of natural selection; or, The preservation of favoured races in the struggle for life*. London, 1859. RM/L. Reprinted, ib., 1950, and Cambridge, Mass., 1964. 2nd ed., 1860. RM/L. 3rd ed., 1861. 4th ed.,1866. 5th ed., 1869. 6th ed., 1872; many reprintings.
1a. ——— Variorum ed. Philadelphia, 1959. By M. Peckham.
2. *The variation of animals and plants under domestication*. 2 vols. London, 1868. RM/L. 2nd ed., 1875; several reprintings.
3. *The descent of man, and selection in relation to sex*. 2 vols. London, 1871. RM (three issues)/L. 2nd ed., 1874; many reprintings.

Correspondence and Posthumous Publications

4. *Extracts from letters addressed to Professor Henslow by C. Darwin, read at a meeting of the Cambridge Philosophical Society, 16 Nov., 1835*. Cambridge, 1835. RM. Reprinted, ib., 1960.
5. *Life and letters, including an autobiographical chapter*. 3 vols. London, 1887-88. Ed. by his son, Francis Darwin. Reprinted, New York, 1969.
6. *More letters of Charles Darwin. A record of his work in a series of hitherto unpublished letters*. 2 vols. London, 1903. Ed. by Francis Darwin. Reprinted, New York, 1972.
7. *The foundations of the "Origin of species": Two essays written in 1842 and 1844*. Cambridge, 1909. Ed. by Francis Darwin. RM/L. Reprinted, New York, 1970.
8. *Autobiography, with two appendices containing a chapter of reminiscences and a statement of* [his] *religious views by his son, Sir Francis Darwin*. London, 1929.
9. *The autobiography, with original omissions restored*. London, 1958. Ed., with appendix and notes, by his granddaughter, Nora Barlow.
10. *Catalogue of the letters of C.R. Darwin to Asa Gray*. Boston, Mass., 1939. Introd. by B.J. Loewenberg. Reprinted, Wilmington, Del., 1973.
11. *Journal*. London, 1959. Ed. by G. de Beer.
12. *Notebooks on transmutation of species*. Several parts. London, 1960-. Ed. with introd. and notes by G. de Beer.
13. *Darwin and Henslow: The growth of an idea. Letters 1831-1860*. London, 1967. Ed. by Nora Barlow.
14. *Questions about the breeding of animals*. London, 1968. Facsimile of an unpublished pamphlet of 1840. Ed. with an introd. by G. de Beer.

15. *A letter about preparations for the voyage of the "Beagle", 1831.* Philadelphia, 1971.
16. *Correspondance entre Charles Darwin et Gaston de Saporta.* Paris, 1972. Ed. by Y. Conry.
17. *Darwin on man. A psychological study of scientific creativity, by H.E. Gruber, together with Darwin's early and unpublished notebooks, transcribed and annotated by P.H. Barrett.* New York, 1974.
18. *Charles Darwin's "Natural selection", being the second part of his big species book written from 1856 to 1858.* London/New York, 1975. Ed. from the MS by R.C. Stauffer.

Collections

19. *Gesammelte Werke.* 13 vols. Stuttgart, 1875-81. Trans. by J.V. Carus.
20. *Handlist of Darwin papers at the University Library, Cambridge.* Cambridge, 1960.
21. *The collected papers.* 2 vols. Chicago, 1977. Collection of periodical articles. Ed. by P.H. Barrett.

Selections

22. *The student's Darwin.* London, 1881. Ed. by E.B. Aveling.
23. *Darwinism stated by Darwin himself. Characteristic passages from* [his] *writings.* New York, 1884. Selected by N. Sheppard.
24. *The Darwin-Wallace celebration, held on 1st July, 1908, by the Linnean Society of London.* London, 1908. Includes reprints of papers by Darwin and Wallace originally published in the Society's *Journal.*

Later selections not included.

13874. GOSSE, Philip Henry. 1810-1888. Br./Canada. ICB; BMNH; GA; GB; GD. Also Zoology*. See also 15547/1.
1. *Omphalos. An attempt to untie the geological knot.* London, 1857.

13875. GRAY, Asa. 1810-1888. U.S.A. DSB; ICB; BMNH; GA; GB; GE. Also Botany*# and Biography*. See also Index.
1. *Darwiniana. Essays and reviews pertaining to Darwinism.* New York, 1876. RM/L. Reprinted, Cambridge, Mass., 1963.

13876. BREE, Charles Robert. 1811-1886. Br. BMNH. Also Zoology.
1. *Species not transmutable nor the results of secondary causes. Being a critical examination of Mr Darwin's work entitled "Origin and variation of species."* London, 1860. RM/L.
2. *An exposition of fallacies in the hypothesis of Mr Darwin.* London, 1872. RM/L.

13877. FREKE, Henry. 1813-1899. ICB; BMNH.
1. *On the origin of species.* London, 1861.

13878. DUB, (Christoph) Julius. 1817-1873. Ger. P I,III. Also Physics*.
1. *Kurze Darstellung der Lehre Darwin's über die Entstehung der Arten der Organismen, mit erläuternden Bemerkungen.* Stuttgart, 1870. RM.

13879. NÄGELI, Karl Wilhelm von. 1817-1891. Ger./Switz. DSB; ICB;

P III; BMNH; GB; GC. Also Microscopy* and Botany*.
1. *Mechanisch-physiologische Theorie der Abstammungslehre.* Munich, 1884. RM/L.

13880. DAWSON, (John) William. 1820-1899. Canada. DSB; P III,IV; BMNH. Also Science in General* and Geology*.
1. *Modern ideas in evolution as related to Revelation and science.* London, 1890. X. Reprinted, New York, 1977.

13881. MILTON, John Laws. 1820-after 1887. Br. BLA.
1. *The stream of life on our globe. Its archives, traditions and laws, as revealed by modern discoveries in geology and palaeontology. A sketch in untechnical language of the beginning and growth of life.* London, 1864.

13882. SPENCER, Herbert. 1820-1903. Br. The philosopher. DSB; ICB; BMNH; GB; GD2. Also Science in General*, Biology in General*, and Autobiography*.
1. *The factors of organic evolution.* London, 1887.
2. *The inadequacy of "natural selection."* London, 1893.

13883. STIRLING, James Hutchison. 1820-1909. Br. The philosopher. BMNH; GB; GD2. Also Biology in General*.
1. *Darwinism. Workmen and work.* Edinburgh, 1894.

13884. CROLL, James. 1821-1890. Br. DSB; P III,IV; BMNH; GB; GD1. Also Geology*, Climatology*, and Autobiography*.
1. *The philosophical basis of evolution.* London, 1890. RM/L.

13885. WIGAND, (Julius Wilhelm) Albert. 1821-1886. Ger. DSB; BMNH; GC. Also Botany.
1. *Der Darwinismus und die Naturforschung Newtons und Cuviers. Beiträge zur Methodik der Naturforschung und zur Speciesfrage.* 3 vols. Brunswick, 1874-77.

13886. GALTON, Francis. 1822-1911. Br. DSB; ICB; P III,IV; BMNH; Mort.; BLA Supp.; BLAF; GB; GD2. Also Meteorology* and Autobiography*.
1. *Hereditary genius. An inquiry into its laws and consequences.* London, 1869. RM/L. 2nd ed., 1892. Reprinted, ib., 1950.
2. *English men of science. Their nature and nurture.* Ib., 1874. RM/L. Reprinted, ib., 1970.
3. *Inquiries into human faculty and its development.* Ib., 1883. RM/L.
4. *Record of family faculties.* Ib., 1884.
5. *Natural inheritance.* Ib., 1889. RM/L.
6. *Finger prints.* Ib., 1892. RM/L. Reprinted, New York, 1965.
7. *A list of papers and correspondence of Sir Francis Galton held in the Library, University College, London.* Ib., 1976.

13887. MENDEL, (Johann) Gregor. 1822-1884. Brünn. DSB; ICB; BLA. Also Meteorology.
1. *Versuche über Pflanzenhybriden.* Periodical article, 1865.
1a. ——— Reprint (also including the periodical article of 1869). Leipzig, 1901. (OKEW. 121) RM/L.
1b. ——— Another reprint. Weinheim, 1960. (HNC. 4)

1c. ——— *Experiments in plant hybridisation*. Edinburgh, 1965. Trans. with commentary by R.A. Fisher, together with a reprint of W. Bateson's biography of Mendel. Ed. by J.H. Bennett.

13888. MÜLLER, Fritz (or Johann Friedrich Theodor). 1822-1897. Ger. /Brazil. DSB; ICB; BMNH; Mort. (M., J.F.T.); GC (M., J.F.T.). Also Zoology.
1. *Für Darwin*. Leipzig, 1864. X.
1a. ——— *Facts and arguments for Darwin*. London, 1869. Trans. by W.S. Dallas. RM/L. Reprinted, Farnborough, 1968.

13889. LE CONTE, Joseph. 1823-1901. U.S.A. DSB; P III,IV; BMNH; BLA & Supp.; GB; GE. Also Geology*, Physiology*, and Autobiography*.
1. *Evolution and its relation to religious thought*. London, 1888.
1a. ——— 2nd ed. *Evolution. Its nature, its evidences and its relation....* Ib., 1898. Reprinted, New York, 1970.

13890. PARKER, William Kitchen. 1823-1890. Br. BMNH; GD. Also Zoology*.
1. *On mammalian descent*. London, 1885. RM/L.

13891. SCHMIDT, (Eduard) Oscar. 1823-1886. BMNH. Also Zoology*.
1. *Descendenzlehre und Darwinismus*. Leipzig, 1873. RM/L.
1a. ——— *The doctrine of descent and Darwinism*. London, 1874. X. 2nd ed., 1875. RM/L. 3rd ed., 1876. 4th ed., 1881. 7th ed., 1887.
2. *Die Säugethiere in ihrem Verhältniss zur Vorwelt*. Leipzig, 1884.
2a. ——— *The Mammalia in their relation to primeval times*. London, 1885.

13892. WALLACE, Alfred Russel. 1823-1913. Br. DSB; ICB; P III; BMNH; GB; GD3. Also Natural History*, Zoology*, Autobiography*, and History of Science*.
1. *Contributions to the theory of natural selection*. London, 1870. RM/L. Reprinted, New York, 1973. 2nd ed., 1871. Another ed., 1875.
2. *Darwinism. An exposition of the theory of natural selection, with some of its applications*. London, 1889. RM/L. 2nd ed., 1889 and re-issues; reprinted, ib., 1975.
3. *Natural selection and tropical nature. Essays on descriptive and theoretical biology*. London, 1891. Reprinted, Farnborough, 1969. Another ed., 1895.
See also 13873/24.

13893. HUXLEY, Thomas Henry. 1825-1895. Br. DSB; ICB; BMNH; Mort.; BLA; GB; GD1. Other entries: see 3340.
1. *Evidence as to man's place in nature*. London, 1863. RM/L.
2. *On our knowledge of the causes of the phenomena of organic nature*. Ib., 1863. RM/L.
3. *More criticisms on Darwin, and administrative nihilism*. New York, 1872.
4. *Darwiniana*. (*Collected essays*. Vol. 2) See 2963/8.

5. *Man's place in nature, and other anthropological essays. (Collected essays.* Vol. 7) See 2963/8.
6. *Evolution and ethics.* (*Collected essays.* Vol. 9) See 2963/8.
7. *On the origin of species; or, The causes of the phenomena of organic nature.* Ann Arbor, Mich., 1968. A reprint.

13894. BONAVIA, Emmanuel. 1826?-1908.
1. *Studies in the evolution of animals.* London, 1895.

13895. FAIVRE, Ernest. 1827-1879. Fr. BMNH; Mort. Also Anatomy.

13896. MIVART, St. George Jackson. 1827-1900. Br. DSB; ICB; BMNH; GB; GD1. Also Science in General* and Zoology*.
1. *On the genesis of species.* London, 1871. X. 2nd ed., 1871. RM/L.
2. *Man and apes. An exposition of structural resemblances and differences bearing upon questions of affinity and origin.* London, 1873.
3. *Contemporary evolution. An essay upon some recent social changes.* London, 1876.

13897. ROLLE, Friedrich. 1827-1887. Ger./Aus. P III; BMNH; GC. Also Geology.
1. *Der Mensch, seine Abstammung und Gesittung im Lichte der Darwin'schen Lehre.* Prague, 1865. X. 2nd ed., 1870.

13898. WRIGHT, Chauncy. 1830-1875. U.S.A. The philosopher. ICB; BMNH; GB; GE.

13899. GULICK, John Thomas. 1832-1913. U.S.A. ICB; GE.

13900. HAECKEL, Ernst Heinrich Philipp August. 1834-1919. Ger. DSB; ICB; BMNH; Mort.; BLAF; GB. Also Science in General*, Zoology*, Embryology*, and Personal Writings*.
1. *Generelle Morphologie der Organismen. Allgemeine Grundzüge der organischen Formen-Wissenschaft, mechanisch begründet durch die von Charles Darwin reformirte Descendenz-Theorie.* 2 vols. Berlin, 1866. RM/L.
2. *Natürliche Schöpfungsgeschichte. Gemeinverständliche wissenschaftliche Vorträge über die Entwickelungslehre.* Berlin, 1868. X. 2nd ed., 1870. 5th ed., 1874. 8th ed., 1889.

2a. ——— *The history of creation; or, The development of the earth and its inhabitants by the action of natural causes. A popular exposition of the doctrine of evolution.* 2nd ed., 2 vols. London, 1876. Trans. & rev. by E.R. Lankester. 3rd ed., 1883. Another ed., 1899.

3. *Anthropogenie; oder, Entwickelungsgeschichte des Menschen ... Keims- und Stammesgeschichte.* Leipzig, 1874. X.

3a. ——— *The evolution of man. A popular exposition of the principal points of human ontogeny and phylogeny.* 2 vols. London, 1879. Another ed., 1883.

4. *Gesammelte populäre Vorträge aus dem Gebiete der Entwickelungslehre.* 2 vols. Bonn, 1878-79. RM/L.

4a. ——— *The pedigree of man, and other essays.* London, 1883. Trans. by E.R. Aveling. RM.

5. *Die Naturanschauung von Darwin, Goethe und Lamarck.* Jena, 1882. A lecture. RM.

6. *Systematische Phylogenie. Entwurf eines natürlichen Systems der Organismen auf Grund ihrer Stammesgeschichte.* 3 vols. Berlin, 1894-96.
7. *Über unsere gegenwärtige Kenntniss vom Ursprung des Menschen.* Bonn, 1898. A lecture. X.
7a. ——— *The last link. Our present knowledge of the descent of man.* London, 1898. Trans. with notes by H. Gadow. RM/L.
8. *Der Kampf um den Entwickelungs-Gedanken. Drei Vorträge.* Berlin, 1905. RM.

13901. WEISMANN, August Friedrich Leopold. 1834-1914. Ger. DSB; ICB; BMNH; Mort.; BLAF; GB. Also Zoology.
1. *Über den Einfluss der Isolirung auf die Artbildung.* Leipzig, 1872.
2. *Studien zur Descendenz-Theorie.* 2 vols. Ib., 1875-76. RM.
2a. ——— *Studies in the theory of descent.* 2 vols. London, 1882. Trans. with notes by R. Meldola. Preface by C. Darwin. RM/L.
3. *Über die Vererbung. Ein Vortrag.* Jena, 1883.
4. *Zur Annahme einer Continuität des Keimplasma's.* Freiburg, 1886.
5. *Die Bedeutung der sexuellen Fortpflanzung für die Selektions-Theorie.* Jena, 1886. RM/L.
6. *Über den Rückschritt in der Natur.* Freiburg, 1886.
7. *Über die Hypothese einer Vererbung von Verletzungen.* Jena, 1889. A lecture.
8. *Das Keimplasma. Eine Theorie der Vererbung.* Jena, 1892. X.
8a. ——— *The germ-plasm. A theory of heredity.* London, 1893. Trans. by W.N. Parker and H. Rönnfeldt. RM/L.
9. *Aufsätze über Vererbung und verwandte biologische Frage.* Jena, 1892. X.
9a. ——— *Essays upon heredity and kindred biological problems.* 2 vols. Oxford, 1889-92. Trans. by E.B. Poulton et al. RM/L.
10. *The effect of external influences upon development.* London, 1894. Romanes lecture, 1894.
11. *Über Germinal-Selection, eine Quelle bestimmt gerichteter Variation.* Jena, 1896. RM.
11a. ——— *On germinal selection as a source of definite variation.* Chicago, 1896. Trans. by T.J. MacCormack.
12. *Vorträge über Descendenztheorie.* Jena, 1902. RM/L.
12a. ——— *The evolution theory.* 2 vols. London, 1904. Trans. by J.A. Thomson and M.R. Thomson. RM/L.

See also 13916/6.

13902. BUTLER, Samuel. 1835-1902. Br. ICB; Mort.; GB; GD2.
1. *Life and habit.* London, 1877. X. 2nd ed., 1878.
2. *Evolution, old and new; or, The theories of Buffon, Dr Erasmus Darwin and Lamarck, as compared with that of Mr Charles Darwin.* Ib., 1879. 2nd ed., 1890.
3. *Unconscious memory.* Ib., 1880. Re the "psycho-physical theory" of heredity of Ewald Hering. (14956)
4. *Luck or cunning as the main means of organic modification. An attempt to throw additional light upon the late Mr*

Charles Darwin's theory of natural selection. Ib., 1887.
5. *Selections from previous works. With remarks upon Mr G.J. Romanes' "Mental evolution in animals."* Ib., 1884.

13903. STEBBING, Thomas Roscoe Rede. 1835-1926. Br. DSB; BMNH. Also Zoology.
1. *Essays on Darwinism.* London, 1871.

13904. HUTTON, Frederick Wollaston. 1836-1905. New Zealand. P III, IV; BMNH; GD2. Also Geology and Zoology*.
1. *Darwinism and Lamarckism, old and new.* London, 1899.
2. *The lesson of evolution.* Ib., 1902.

13905. KRAUSE, Ernst Ludwig. 1839-1903. Ger. DSB; BMNH. Also Biography*.

13906. CLODD, Edward. 1840-1930. Br. GD4.
1. *The story of creation. A plain account of evolution.* London, 1888. 2nd ed., 1894. Another ed., 1898.
2. *Pioneers of evolution from Thales to Huxley.* Ib., 1897.
See also 2963/10.

13907. COPE, Edward Drinker. 1840-1897. U.S.A. DSB Supp.; ICB; P III,IV; BMNH; GB; GE. Also Zoology.
1. *The origin of the fittest. Essays on evolution.* London, 1887. RM/L.
2. *Primary factors of organic evolution.* Chicago, 1896.

13908. SEIDLITZ, Georg Carl Maria von. 1840-1917. Dorpat. BMNH. Also Zoology.

13909. HARTMANN, (Carl Robert) Eduard von. 1842-1906. Ger. BMNH; GB.
1. *Darwinismus und Thierproduktion.* Munich, 1876. RM/L.

13910. KOVALEVSKY, Vladimir Onufrievich. 1842-1883. Russ. DSB; ICB; BMNH. Also Zoology.

13911. EIMER, Gustav Heinrich Theodor. 1843-1898. Switz. BMNH; Mort.; GC. Also Zoology.
1. *Die Entstehung der Arten auf Grund von Vererben erworbener Eigenschaften nach den Gesetzen organischen Wachsens.* 3 parts. Jena, 1888-1901. RM/L.
1a. ——— Trans. of Part 1. *Organic evolution as the result of the inheritance of acquired characters according to the laws of organic growth.* London, 1890. Trans. by J.T. Cunningham. RM/L.
2. [Trans.] *On orthogenesis and the impotence of natural selection in species-formation.* Chicago, 1898. Trans. by J.T. McCormack.

13912. CHAPMAN, Henry Cadwalader. 1845-1909. U.S.A. BMNH; BLAF; GE. Also Physiology*.
1. *Evolution of life.* Philadelphia, 1873. X. 2nd ed., 1873.

13913. LANKESTER, Edwin Ray. 1847-1929. Br. DSB; BMNH; BLAF; GD4. Also Science in General*, Microscopy, and Zoology*. See also Index.
1. *Degeneration. A chapter in Darwinism.* London, 1880.

13914. MASSEE, George Edward. 1847-1917. Br. BMNH. Also Botany.
1. *The evolution of plant life. Lower forms.* London, 1891.

13915. ALLEN, Grant. 1848-1899. Br. BMNH (A., Charles Grant); GB; GD1. Also Botany* and Biography*.
1. *The evolutionist at large*. London, 1881. "A collection of popular scientific articles, the value and accuracy of which is attested by letters from Darwin and Huxley." -GD1. X. 2nd ed., rev., 1884.

13916. ROMANES, George John. 1848-1894. Br. DSB; ICB; BMNH; Mort.; GD. Also Science in General*, Physiology, and Biography*. See also 15607/1.
1. *The scientific evidences of organic evolution*. London, 1882.
2. *Animal intelligence*. Ib., 1882. Reprinted, Farnborough, 1970. Another ed., 1886.
3. *Mental evolution in animals. With a posthumous essay on instinct by Charles Darwin*. Ib., 1883 and re-issues. RM/L. Reprinted, Farnborough, 1970.
4. *Mental evolution in man*. Ib., 1888. X. Reprinted, Farnborough, 1970.
5. *Darwin and after Darwin. An exposition of the Darwinian theory and a discussion of post-Darwinian questions*. 3 vols. Ib., 1892-97. RM/L.
6. *An examination of Weismannism*. Ib., 1893. RM/L. 2nd ed., 1899.
See also 13902/5.

13917. VRIES, Hugo de. 1848-1935. Holl. DSB; ICB; P IV (under DE); BMNH. Also Botany*#.

13918. JORDAN, David Starr. 1851-1931. U.S.A. DSB; ICB; BMNH; GE. Also Zoology.
1. *Factors in organic evolution*. Boston, 1895.

13919. MARSHALL, Arthur Milnes. 1852-1893. Br. BMNH; GD1. Also Zoology* and Embryology*.
1. *Lectures on the Darwinian theory*. London, 1894. RM/L.

13920. WILSON, Andrew. 1852-1912. Br. BMNH. Also Zoology.
1. *Chapters on evolution*. London, 1883.

13921. HUBRECHT, (Ambrosius Arnold) Willem. 1853-1915. Holl. DSB; BMNH; BLAF. Also Zoology and Embryology.
1. *The descent of the primates*. New York, 1897.

13922. GEDDES, Patrick. 1854-1932. Br. ICB; BMNH; GD5. Also Botany* and Cytology*.
1. *The evolution of sex*. London, 1889. With J.A. Thomson.

13923. BLAND-SUTTON, John. 1855-1936. Br. ICB; BLAF; GD5. Also Zoology.
1. *Evolution and disease*. London, 1890.

13924. VARIGNY, Henry Crosnier de. 1855-after 1927. Fr. BMNH. Also Physiology.
1. [Trans.?] *Experimental evolution*. London, 1892.

13925. BEECHER, Charles Emerson. 1856-1904. U.S.A. P IV; BMNH; GB; GE. Also Palaeontology.
1. *Studies in evolution, mainly reprints*. New York, 1901.

13926. POULTON, Edward Bagnall. 1856-1943. Br. ICB; BMNH; GD6.

Also Zoology* and Biography*.
1. *Essays on evolution, 1889-1907.* Oxford, 1908. RM/L.
See also 13901/9a.

13927. HAMANN, Otto. 1857-1925. BMNH. Also Zoology.
1. *Entwicklungslehre und Darwinismus.* 1882.

13928. JOHANNSEN, Wilhelm Ludvig. 1857-1927. Den. DSB; BLAF. Also Botany.

13929. PEARSON, Karl. 1857-1936. Br. DSB; ICB; P IV; GD5. Also Science in General*, Mathematics*, and Biography*.
1. *The chances of death, and other studies in evolution.* London, 1897.

13930. ROSA, Daniele. 1857-1944. It. DSB; BMNH. Also Zoology.
1. *La riduzione progressiva della variabilità ed i suoi rapporti coll'estinzione e l'origine della specie.* Turin, 1899. X.
1a. ——— *Die progressive Reduktion der Variabilität und ihre Beziehungen zum Aussterben und zur Entstehung der Arten.* Jena, 1903. Trans. by H. Bosshard.

13931. BAILEY, Liberty Hyde. 1858-1954. U.S.A. DSB; ICB; BMNH. Also Botany*.
1. *Survival of the unlike. A collection of essays suggested by the study of domestic plants.* New York, 1896.

13932. SHUTE, Daniel Kerfoot. 1858-1935. U.S.A. BLAF.
1. *A first book in organic evolution.* Chicago, 1899.

13933. CAMPBELL, Douglas Houghton. 1859-1953. U.S.A. DSB; ICB; BMNH. Also Botany.
1. *Lectures on the evolution of plants.* New York, 1899.

3.16 HUMAN ANATOMY

Including histology.

Comparative anatomy is included in the Zoology section.

13934. PALFYN, Jean. 1650-1730. Belg./Fr. ICB (PALFIJN); BLA & Supp. (PALFIJN); GA; GF.

13935. LANCISI, Giovanni Maria. 1654-1720. It. DSB; ICB; P I; BMNH; Mort.; BLA & Supp.; GA. See also 1438/1.

13936. FICK, Johann Jacob. 1662-1730. Ger. P I; BMNH; BLA; GA. Also Chemistry and Botany.

13937. POURFOUR DU PETIT, François. 1664-1741. Fr. DSB; ICB; BMNH (DU PETIT); Mort. Under PETIT: P II; BLA; GA. Also Physiology.

13938. PACCHIONI, Antonio. 1665-1726. It. DSB; ICB; Mort.; BLA; GA.

13939. VALSALVA, Anton Maria. 1666-1723. It. DSB; ICB; Mort.; BLA & Supp.; GA.

13940. DRAKE, James. 1667-1707. Br. BLA; GA; GD.
1. *Anthropologia nova; or, A new system of anatomy.* London, 1707. X. 3rd ed., 1750.

13941. WINSLOW, Jacques Bénigne. 1669-1760. Fr. DSB; ICB; BMNH; Mort.; BLA & Supp.; GA.
1. *Exposition anatomique de la structure du corps humain.* Paris, 1732. X. Other eds (in 4 vols), 1743 and 1776.
1a. ——— *An anatomical exposition of the structure of the human body.* 2 vols in 1. London, 1733. Trans. by G. Douglas.

13942. GOELICKE, Andreas Ottomar. 1671-1744. Ger. BMNH; BLA; GA.
1. *Introductio in historiam literariam anatomiae.* Frankfurt, 1738.

13943. KEILL, James. 1673-1719. Br. DSB; P I; BLA; GA; GD. Also Physiology.
1. *The anatomy of the humane body, abridged.* London, 1698. Wing K131. X. 15th ed., 1771. X.
See also 2573/1c,1d.

13944. PETIT, Jean Louis. 1674-1750. Fr. BLA.

13945. SCHACHER, Polycarp Gottlieb. 1674-1737. Ger. BLA.

13946. DOUGLAS, James. 1675-1742. Br. DSB; BMNH; Mort.; BLA & Supp.; GD. Also Natural History and History of Anatomy*.

13947. FANTONI, Giovanni. 1675-1758. It. BLA & Supp.; GA.

13948. HOVIUS, Jacob. b. ca. 1675. Holl. BLA.

13949. NABOTH, Martin. 1675-1721. Ger. Mort.; BLA.

13950. LEMERY, Louis. 1677-1743. Fr. DSB; P I; BLA; GA. Also Chemistry*.

13951. ASSALTI, Pietro. 1680-1728. It. DSB; BMNH; BLA & Supp.

13952. BIANCHI, Giovanni Battista. 1681-1761. It. BLA; GA.

13953. SANTORINI, Giovanni Domenico. 1681-1737. It. DSB; Mort.; BLA; GA.

13954. MORGAGNI, Giambattista. 1682-1771. It. DSB; ICB; P II; BMNH; BLA & Supp.; GA; GB.
1. *Adversaria anatomica omnia.* Lyons, 1711. X. Other eds, Leiden, 1723 and 1741.
2. *De sedibus et causis morborum per anatomen indagatis.* Venice, 1761. X.
2a. ——— *The seats and causes of diseases investigated by anatomy.* 3 vols. London, 1769. X. Reprinted, New York, 1960.
3. *Opera postuma.* 2 vols. Rome, 1964-65.
See also 2661/3 and 14659/6.

13955. HEISTER, Lorenz. 1683-1758. Ger. DSB; ICB; BMNH; BLA & Supp.; GA; GC. Also Botany.
1. *Compendium anatomicum.* Altorf/Nuremberg, 1717. X. 5th ed., 1741.

13956. VATER, Abraham. 1684-1751. Ger. P II; BMNH; Mort.; BLA; GC. Also Chemistry and Botany.

13957. HELVÉTIUS, Jean Claude Adrien. 1685-1755. Fr. ICB; GA.

13958. GALEAZZI, Domenico Gusmano. 1686-1775. It. DSB; ICB; P I (GALEATI); BLA & Supp.

13959. CHESELDEN, William. 1688-1752. Br. ICB; BMNH; Mort.; BLA & Supp.; GA; GB; GD.
1. *The anatomy of the humane body.* London, 1713. X. 2nd ed., 1722. 3rd ed., 1726. 4th ed., 1732. 5th ed., 1740. 6th ed., 1741. 10th ed., 1773. 12th ed., 1784.
2. *Osteographia; or, The anatomy of the bones.* London, 1733. Another ed., n.d.

13960. WALTHER, Augustin Friedrich. 1688-1746. Ger. BMNH; BLA; GC. Also Botany.

13961. DUDDELL, Benedict. early 18th century. Br. Mort.; BLA & Supp.

13962. DUVERNOY, Johann Georg. 1691-1759. Ger./Russ. BMNH; BLA. Also Botany.

13963. FERREIN, Antoine. 1693-1769. Fr. DSB; P I; BLA; GA.

13964. SENAC, Jean Baptiste. ca. 1693-1770. Fr. DSB; ICB; P II; BLA; GA. Also Chemistry*.

13965. NESBITT (or NISBET), Robert. d. 1761. Br. ICB; GD.

13966. TREW, Christoph Jacob. 1695-1769. Ger. ICB; BMNH; BLA; GA; GC. Also Botany.

13967. MAUCHART, Burkhard David. 1696-1751. Ger. BLA.

13968. ALBINUS, Bernhard Siegfried. 1697-1770. Holl. DSB Supp.; ICB; P I; BMNH; Mort.; BLA & Supp.; GA; GB.
1. *De ossibus corporis humani.* Leiden, 1726. RM.
2. *Explicatio tabularum anatomicarum B. Eustachii.* Leiden, 1744. RM/L.
3. *Tabulae sceleti et musculorum corporis humani.* Leiden, 1747. X.
3a. ——— *Tables of the skeleton and muscles of the human body.* London, 1749. X. Another ed., Edinburgh, 1777.
3b. ——— *The explanation of Albinus' anatomical figures of the human skeleton and muscles, with an historical account of the work.* London, 1754. RM/L.
See also 1497/4.

13969. MONRO, Alexander (first of the name). 1697-1767. Br. DSB; ICB; BMNH; BLA; GA; GD.
1. *The anatomy of the humane bones.* Edinburgh, 1726. X. 4th ed., 1746.

13970. MORAND, Sauveur François. 1697-1773. Fr. P II; BLA; GA.

13971. ALBINUS, Christian Bernhard. 1698/9-1752. Holl. DSB Supp.; BLA & Supp; GA.

13972. CASSEBOHM, Johann Friedrich. 1699?-1743. Ger. Mort.; BLA & Supp.

13973. NICHOLLS, Frank. 1699-1778. Br. P II; BLA; GD.

13974. LE CAT, Claude Nicolas. 1700-1768. Fr. DSB; ICB; P I; BMNH; BLA & Supp.; GA. Also Physiology*.

13975. HUNAULD, François Joseph. 1701-1742. Fr. BLA; GA.

13976. SANTUCCI, Bernardo. 1701-1764. It./Port. ICB.

13977. DEMOURS, Pierre. 1702-1795. Fr. BMNH; BLA & Supp.; GA.

13978. WACHENDORFF, Everardus Jacobus van. 1702-1758. Holl. BMNH; BLA. Also Botany.

13979. WEITBRECHT, Josias. 1702-1747. Russ. P II; BLA; GC. Also Physics.
1. *Syndesmologia; sive, Historia ligamentorum corporis humani.* St. Petersburg, 1742. X.
1a. ——— *Syndesmology; or, A description of the ligaments of the human body.* Philadelphia, 1969. Trans. by E.B. Kaplan.

13980. LIEUTAUD, Joseph. 1703-1780. Fr. DSB; ICB; Mort.; BLA. Also Physiology*.

13981. BERGEN, Carl August von. 1704 (or 1709)-1759. Ger. P I; BMNH; BLA & Supp.; GA; GC. Also Natural History.

13982. CHOVET, Abraham. 1704-1790. Br./U.S.A. ICB; Mort.; GE.

13983. MENGHINI, Vincenzo Antonio. 1704-1759. It. DSB; Mort.

13984. HUBER, Johann Jacob. 1707-1778. Ger. DSB; BLA; GA; GC. Also Botany.

13985. HALLER, Albrecht von. 1708-1777. Switz./Ger. DSB; ICB; P I; BMNH; Mort.; BLA & Supp.; GA; GB; GC. Also Botany*, Physio-ogy*#, and Personal Writings*.
1. *Iconum anatomicarum, quibus praecipuae partes corporis humani delineatae continentur....* 8 parts in 1 vol. Göttingen, 1743-56. RM/L.
2. *Bibliotheca anatomica, qua scripta ad anatomen et physiologiam facientia a rerum initiis recensentur.* 2 vols. Zurich, 1774-77. X. Reprinted, Hildesheim, 1969.
3. *De partium corporis humani praecipuum fabrica et functionibus.* 8 vols in 6. Berne/Lausanne, 1777-78. RM/L.

13986. SUE, Jean Joseph (first of the name). 1710-1792. Fr. BLA; GA.

13987. BOEHMER, Philipp Adolph. 1711 (or 1717)-1789. Ger. BLA; GA; GC.

13988. LALLOUETTE, Pierre. 1711-1792. Fr. BLA.

13989. LIEBERKÜHN, Johannes Nathanael. 1711-1756. Ger. DSB; ICB; P I; Mort.; BLA & Supp.; GA; GC.

13990. BERTIN, Exupère Joseph. 1712-1781. Fr. BLA & Supp.; GA.

13991. GUNZ, Justus Gottfried. 1714-1755. Ger. GA; GC.

13992. ALBINUS, Frederik Bernard. 1715-1778. Holl. DSB Supp.; BLA & Supp.

13993. ACREL, Olof. 1717-1806. Swed. P I; BLA & Supp.; GA.

13994. BOUVARD, Michel Philippe. 1717-1787. Fr. BLA (BOUVART); GA.

13995. GAUTIER D'AGOTY, Jacques Fabian. 1717-1785. Fr. P I (GAUTHIER); BMNH; Mort.; BLA & Supp.; GA (GAUTHIER). Also Physics* and Botany.

13996. TANDON, Antoine. 1717-1806. Fr. BLA. Also Physiology.

13997. HUNTER, William. 1718-1783. Br. DSB; ICB; Mort.; BLA & Supp.; GA; GB; GD.
1. *Lectures of anatomy. Facsimile of manuscript notes, ca. 1752.* Amsterdam, 1972. Lecture notes, taken by C. White, of a course given by Hunter in Manchester, ca. 1752.

13998. PETIT, Antoine. 1718-1794. Fr. ICB; BLA; GA. Also Physiology.

13999. HAARSTRICK, Johann Michael. 1719-after 1786. Ger. ICB.

14000. HENSING, Friedrich Wilhelm. 1719-1745. Ger. BLA.

14001. SCHAARSCHMIDT, August. 1720-1791. Ger. BLA; GC.

14002. BORDEU, Théophile de . 1722-1776. Fr. DSB; ICB; Mort.; BLA & Supp.; GA. Also Physiology.

14003. CAMPER, Peter. 1722-1789. Holl. DSB; ICB; P I; BMNH; Mort.; BLA & Supp.; GA; GB. Also Zoology*# and Physiology*.

14004. MECKEL, Johann Friedrich (first of the name). 1724-1774. Ger. Mort.; BLA; GA; GC.

14005. TENON, Jacques René. 1724-1816. Fr. ICB; Mort.; BLA; GA.

14006. CALDANI, (Leopoldo) Marco Antonio. 1725-1813. It. DSB; ICB; P I; BMNH; BLA & Supp.; GA; GB. Also Physiology.

14007. TARIN, Pierre. 1725-1761. Fr. BLA; GA.

14008. BELL, Andrew. 1726-1809. Br. GD.
1. *Anatomica Britannica. A system of anatomy.* 3 parts. Edinburgh, 1798.

14009. LEBER, Ferdinand Joseph. 1727-1808. Aus. ICB; BLA; GC.
1. *Praelectiones anatomicae.* Vienna, (date?). X. See 14068/1.

14010. ZINN, Johann Gottfried. 1727-1759. Ger. P II; BMNH; Mort.; BLA; GC. Also Botany.

14011. HUNTER, John. 1728-1793. Br. DSB; ICB; P I; BMNH; Mort.; BLA & Supp.; GA; GB; GD. Also Various Fields*#, Geology*, and Zoology*.

14012. ANDRE, William. fl. 1751-1807. Br. ICB; BLA & Supp.

14013. GASSER, Johann Ludwig (or Laurentius). fl. 1757-1765. Aus. ICB; BLA & Supp.

14014. FONTANA, Felice Gaspar Ferdinand. 1730-1805. It. DSB; ICB; P I; BMNH; Mort.; BLA & Supp.; GA. Also Chemistry* and Physiology*.

14015. JANIN DE COMBE-BLANCHE, Jean. 1731-ca. 1799. Fr. P I & Supp.; BLA; GA.

14016. ANDERSCH, Karl Samuel. 1732-1777. Ger. BLA & Supp.

14017. DESCEMET, Jean. 1732-1810. Fr. BMNH; BLA.

14018. SABATIER, Raphaël Bienvenu. 1832-1811. Fr. BLA; GA.

14019. MONRO, Alexander (second of the name). 1733-1817. Br. DSB; ICB; P II; BMNH; Mort.; BLA; GA; GD. Also Zoology*.
1. *Observations on the structure and functions of the nervous system.* Edinburgh, 1783.

14020. WALTER, Johann Gottlieb. 1734-1818. Ger. BLA; GC.

14021. HIRSCH, Anton Balthasar Raymund. late 18th century. Aus. ICB; Mort.

14022. COTUGNO, Domenico Felice Antonio. 1736-1822. It. DSB; ICB; P I; Mort.; BLA & Supp.; GA. Also Physiology.
1. *De aquaductibus auris humanae internae.* Naples, 1761. X. Another ed., Vienna, 1774.
2. *De ischiade nervosa.* Naples, 1764. X. Another ed., Vienna, 1770.

14023. LOBSTEIN, Jean Frédéric. 1736-1784. Fr. BLA & Supp.; GA.

14024. GALVANI, Luigi. 1737-1798. It. DSB; ICB; P I; Mort.; BLA & Supp.; GA; GB. Also Physiology*#.

14025. HEWSON, William. 1739-1774. Br. DSB; ICB; Mort.; BLA & Supp.; GA; GD.
1. *An experimental inquiry into the properties of the blood.* London, 1771. RM.
2. *An experimental inquiry into the figure and composition of the red particles of the blood.* Ib., 1773. RM.
3. *Experimental inquiries. Part 2 ... the lymphatic system.* Ib., 1774.

4. *The works*. Ib., 1846. Ed. with introd. and notes by G. Gulliver for the Sydenham Society. RM/L.

14026. INNES, John. 1739-1777. Br. GD.

14027. WRISBERG, Heinrich August. 1739-1808. Ger. ICB; BMNH; Mort.; BLA; GC. Also Microbiology.

14028. AITKEN, John. d. 1790. Br. BLA & Supp.; GA; GD (AITKIN).
1. *A system of anatomical tables, with explanations*. London, 1786.

14029. EHRENRITTER, Johannes. d. ca. 1790. Aus. BLA & Supp.

14030. AZZOGUIDI, Germano. 1740-1817. It. BLA; GA.

14031. LEVELING, Heinrich Palmatius von. 1742-1798. Ger. ICB; BLA.

14032. NEUBAUER, Johann Ernst. 1742-1777. Ger. BLA.

14033. PORTAL, Antoine. 1742-1832. Fr. DSB; ICB; BLA; GA. Also History of Anatomy.

14034. SANDIFORT, Eduard. 1742-1814. Holl. ICB; BMNH; BLA; GF.
1. *Observationes anatomico-pathologicae*. 4 vols. Leiden, 1777-81.
2. *Opuscula anatomica*. Ib., 1780-84.
3. *Exercitationes academicae*. 2 vols. Ib., 1783-85.
4. *Museum anatomicum Academiae Lugduno-Batavae descriptum*. 4 vols. Ib., 1793-1835.

14035. CHOPART, François. 1743-1795. Fr. BLA & Supp.

14036. MALACARNE, (Michele) Vincenzo (Giacinto) 1744-1816. It. ICB; P II; BLA; GA.

14037. CRUIKSHANK, William Cumberland. 1745-1800. Br. DSB; P I (CRUICKSHANKS); Mort.; BLA & Supp.; GA; GD. Also Embryology and Physiology.

14038. CHAUSSIER, François. 1746-1828. Fr. ICB; P I; BMNH; BLA & Supp.; GA.

14039. MAYER, Johann Christoph Andreas. 1747-1801. Ger. BMNH; BLA & Supp.; GA. Also Botany*.

14040. LAUMONIER, Jean Baptiste Philippe Nicolas René. 1749-1818. Fr. ICB; BLA.

14041. PROCHÁSKA, Georgius. 1749-1820. Prague/Aus. DSB; ICB; P II; Mort.; BLA; GC. Also Embryology and Physiology*.

14042. MURRAY, Adolph. 1751-1803. Swed. ICB; P III; BMNH; BLA.

14043. GENNARI, Francesco. 1752-1797. It. ICB; Mort.

14044. SCARPA, Antonio. 1752-1832. It. DSB; ICB; P II; BMNH; Mort.; BLA & Supp.; GA.
1. *Tabulae neurologicae*. Pavia, 1794.

14045. GAVARD, Hyacinth. 1753-1802. Fr. BMNH; BLA; GA.

14046. LODER, Justus Christian von. 1753-1832. Ger./Russ. ICB; BMNH; BLA; GA; GC.

14047. FYFE, Andrew (first of the name). 1754-1824. Br. BMNH; BLA & Supp.; GD.

1. *A compendium of anatomy, human and comparative.* Edinburgh, 1815. X. Another ed., 1826.

14048. RAMSAY, Alexander. ca. 1754-1824. Br./U.S.A. GE.

14049. FLAXMAN, John. 1755-1826. Br. Mort.; GA; GB; GD.

14050. MASCAGNI, Paolo. 1755-1815. It. DSB; ICB; P II; Mort.; BLA; GA.

14051. SÖMMERRING, Samuel Thomas von. 1755-1830. Ger. DSB; ICB; P II; BMNH; Mort.; BLA; GA; GC.
1. *Vom Baue des menschlichen Körpers.* 5 vols. Frankfurt, 1791-96. X.
1a. ——— Another ed. 8 vols. Leipzig, 1839-45. Vol. 1 contains a biography of the author by Rudolph Wagner. X.
1b. ——— ——— Trans. of Vols 2-8. *Encyclopédie anatomique.* 8 vols. Paris, 1843-47. Trans. by A.J.L. Jourdan.
2. *Über das Organ der Seele.* Königsberg, 1796. X. Reprinted, Amsterdam, 1966.
See also 11065/5.

14052. BOYER, Alexis. 1757-1833. Fr. BLA & Supp.; GA; GB.

14053. CANOVA, Antonio. 1757-1822. It. GA; GB.
1. *Disegni anatomici.* Rome, 1949. Facsimile of a MS. Ed. by M. Pantaleoni.

14054. SCHUMACHER, (Heinrich) Christian Friedrich. 1757-1830. Den. P II; BMNH; BLA. Also Mineralogy* and Botany.

14055. BARCLAY, John. 1758-1826. Br. DSB; BMNH; BLA & Supp.; GD. Also Physiology*.
1. *Introductory lectures to a course of anatomy.* Edinburgh, 1827. With a biography by G. Ballingall. RM.

14056. DUVAL, Jacques René. 1758-1854. Fr. ICB; BLA & Supp.; GA.

14057. GALL, Franz Josef. 1758-1828. Aus./Fr. DSB; ICB; BMNH; Mort.; BLA & Supp.; GA; GB; GC.
1. *Recherches sur le système nerveux en général et sur celui du cerveau en particulier.* Paris, 1809. With G. Spurzheim. X. Reprinted, Amsterdam, 1967.
2. *Anatomie et physiologie du système nerveux en général et du cerveau en particulier, avec observations sur la possibilité de reconnoître plusieurs dispositions intellectuelles et morales de l'homme et des animaux par la configuration de leurs têtes.* 4 vols & atlas. Paris, 1810-19. With J.G. Spurzheim (Vols 1 and 2 only).
See also 14064/2 and 14088/1.

14058. LAUTH, Thomas. 1758-1826. Ger./Fr. ICB; BLA; GC. Also History of Anatomy.

14059. HESSELBACH, Franz Kaspar. 1759-1816. Ger. Mort.; BLA; GA; GC.

14060. REIL, (Johann) Christian. 1759-1813. Ger. DSB; ICB; P II; BMNH; Mort.; BLA; GA; GC. Also Physiology.

14061. BROOKES, Joshua. 1761-1833. Br. BMNH; BLA; GA; GD. Also Zoology.

14062. WISTAR, Caspar. 1761-1818. U.S.A. DSB; P II; BLA; GE.

14063. BELL, John. 1762-1820. Br. ICB; BLA & Supp.; GA; GB; GD.
1. *The anatomy of the human body.* 4 vols. Edinburgh/London, 1797-1804. With Charles Bell. X. Another ed., 1802-04.
1a. ——— 6th ed. *The anatomy and physiology of....* 3 vols. London, 1826. Rev. by Charles Bell. 7th ed., 1829.

14064. ABERNETHY, John. 1764-1831. Br. P I; BMNH; BLA & Supp.; GA; GB; GD. Also Physiology*.
1. *The surgical works.* 2 vols. London, 1816. X.
1a. ——— Another ed. *The surgical and physiological works.* Ib., 1825. Another ed., 4 vols, 1830.
2. *Reflections on Gall and Spurzheim's system of physiognomy and phrenology.* Ib., 1821. X. Also included in item 1a.
3. *Lectures on anatomy, surgery and pathology.* Ib., 1828. RM.

14065. ACKERMANN, Jacob Fidelis. 1765-1815. Ger. P I; BMNH; BLA & Supp.

14066. RIBES, François. 1765-1845. Fr./It. BLA; GA.

14067. VETTER, Alois Rudolf. 1765-1806. Aus. BLA; GC.

14068. VAUGHAN, Walter. 1766-1828.
1. *An exposition of the principles of anatomy and physiology ... containing the "Praelectiones anatomicae" of F. Leber translated from the original.* 2 vols. London, 1791. cf. item 14009/1.

14069. BOI, Francesco. 1767-1860. It. ICB.

14070. CHEVALIER, Thomas. 1767-1824. Br. BLA; GA; GD.
1. *Lectures on the general structure of the human body, and on the anatomy and functions of the skin.* London, 1823.

14071. CARLISLE, Anthony. 1768-1840. Br. DSB; ICB; P I & Supp.; BLA; GD. Also Chemistry.

14072. COOPER, Astley Paston. 1768-1841. Br. The famous surgeon. Mort.; BLA & Supp.; GA; GB; GD.

14073. WENZEL, Joseph. 1768-1808. Ger. BMNH; BLA; GC.

14074. RIEMER, Pieter de. 1769-1831. Holl. Mort.; BLA.

14075. MACARTNEY, James. 1770-1843. Br. BMNH; BLA; GD1. Also Zoology.

14076. BICHAT, Xavier. 1771-1802. Fr. DSB; ICB; Mort.; BLA & Supp.; GA; GB. Also Physiology*.
1. *Traité des membranes en général, et de diverses membranes en particulier.* Paris, 1800. X. Another ed., with a biography by Husson, 1802.
2. *Anatomie générale, appliquée à la physiologie et à la médecine.* 4 vols. Paris, 1801. X. Other eds: 1812, 1821, 1830.
3. *Traité d'anatomie descriptive.* 5 vols. Paris, 1801-03. Completed by M.F.R. Buisson amd P.J. Roux. RM/L.

14077. ISENFLAMM, Heinrich Friedrich. 1771-1828. Ger./Russ. BLA; GC.

14078. ROSENMÜLLER, Johann Christian. 1771-1820. Ger. ICB; P II; BMNH; BLA; GC. Also Geology and Embryology.

14079. CALDANI, Floriano. 1772-1836. It. ICB; P I; BMNH; BLA Supp.

14080. COLLES, Abraham. 1773-1843. Br. BLA & Supp; GD.

14081. HOOPER, Robert. 1773-1835. Br. BMNH; BLA; GD. Also Botany.
1. *The anatomist's vade-mecum, containing the anatomy and physiology of the human body.* Boston, 1801.
See also 14687/1a.

14082. MONRO, Alexander (third of the name). 1773-1859. Br. ICB; BLA; GA; GD.

14083. REISSEISEN, François Daniel. 1773-1828. Fr./Ger. BMNH; Mort.; BLA.

14084. ROLANDO, Luigi. 1773-1831. It. DSB; Mort.; BLA. Also Zoology and Physiology.

14085. BELL, Charles. 1774-1842. Br. DSB; ICB; Mort.; BLA & Supp.; GA; GB; GD. See also Index.
1. *A system of dissections, explaining the anatomy of the human body.* 2 vols in 1. Edinburgh, 1798-1803. 2nd ed., 1799-1805.
2. *Engravings of the arteries.* London, 1801. RM. 4th ed., 1824.
3. *The anatomy of the brain.* London, 1802. RM.
4. *A series of engravings explaining the course of the nerves.* London, 1803. RM/L.
5. *Essays on the anatomy of expression in painting.* London, 1806. RM.
5a. ——— 3rd ed. *The anatomy and physiology of expression.* Ib., 1844. RM. 6th ed., 1872. 7th ed., 1877.
6. *Idea of a new anatomy of the brain.* London, 1811. X. Reprinted, ib., 1966.
7. *An exposition of the natural system of the nerves of the human body.* London, 1824. RM.
8. *The nervous system of the human body.* London, 1830. 3rd ed., 1844.
9. *The hand. Its mechanism and vital endowments as evincing design.* London, 1833. (Bridgewater treatises. 4) RM/L. 2nd ed., 1833. 3rd ed., 1834. 4th ed., 1837. American ed., New York, 1840. 5th ed., 1852. 6th ed., 1854. 7th ed., 1865. 9th ed., 1874.
10. *Practical essays.* Edinburgh, 1841.
11. *Letters.* 1870.
12. [Manuscript of drawings of the arteries] New York, 197-?

14086. BURDACH, Karl Friedrich. 1776-1847. Ger. DSB; ICB; BMNH; Mort.; BLA & Supp.; GA; GC. Also Physiology.

14087. LANGENBECK, Conrad Johann Martin. 1776-1851. Ger. ICB; BLA; GA; GC.

14088. SPURZHEIM, Johann Christoph (or Johann Gaspar). 1776-1832. Aus/Fr./Br. DSB; ICB; BMNH; Mort.; BLA; GA; GB; GC.
1. *The physiognomical system of Drs Gall and Spurzheim.* London, 1815. X. 2nd ed., 1815.

2. [Trans.] *The anatomy of the brain, with a general view of the nervous system.* London, 1826. Trans. by R. Willis from the unpublished French MS.

See also 14057/1,2 and 14064/2.

14089. DUPUYTREN, Guillaume. 1777-1835. Fr. ICB; BLA & Supp.; GA.

14090. LOBSTEIN, Jean Georges Crétien Frédéric Martin. 1777-1835. Fr. ICB; Mort.; BLA.

14091. ROSENTHAL, Friedrich Christian. 1779-1829. Ger. BMNH; BLA; GC. Also Zoology.

14092. SANDIFORT, Gerard. 1779-1848. Holl. BMNH; BLA.

14093. SHRAPNELL, Henry Jones. d. 1834. Br. ICB; Mort.

14094. BURNS, Allan. 1781-1813. Br. Mort.; BLA; GD.

14095. MECKEL, Johann Friedrich (second of the name). 1781-1833. Ger. DSB; ICB; BMNH; Mort.; BLA & Supp.; GA; GC. Also Zoology* and Embryology.

1. *Handbuch der menschlichen Anatomie.* 4 vols. Halle, 1815-20. X.

1a. ——— *Manual of general anatomy.* London, 1837. Trans. by A.S. Doane from a French version.

14096. BOCK, (Karl) August. 1782-1833. Ger. BLA; GA.

14097. KROMBHOLZ, Julius Vincenz von. 1782-1843. Prague. BMNH; GC.

14098. BRESCHET, Gilbert. 1783-1845. Fr. DSB; P I & Supp.; BMNH; BLA & Supp.; GA. Also Zoology.

14099. JACOBSON, Ludvig Levin. 1783-1843. Den. ICB; BMNH; Mort.; BLA & Supp.; GA. Also Zoology.

14100. LAWRENCE, William. 1783-1867. Br. DSB; ICB; BMNH; BLA; GA; GD. Also Zoology* and Physiology.

14101. LAURENT, Jean Louis Maurice. 1784-1854. Fr. BMNH; BLA; GA. Also Zoology.

14102. BÉCLARD, Pierre Augustin. 1785-1825. Fr. ICB; BLA & Supp.; GA.

14103. GUTHRIE, George James. 1785-1856. Br. BLA & Supp.; GD.

14104. MÜNZ, Martin. 1785-1848. Ger. ICB; BLA.

14105. GORDON, John. 1786-1818. Br. BLA.

14106. CLOQUET, Hippolyte. 1787-1840. Fr. BMNH; BLA; GA. Also Zoology.

1. *Traité d'anatomie descriptive.* 4th ed. Paris, 1828. X.

1a. ——— *System of human anatomy.* Edinburgh, 1828. Trans. from the 4th French ed. by R. Knox.

See also 13849/3a-d.

14107. LIZARS, John. 1787?-1860. Br. BLA; GD.

1. *A system of anatomical plates of the human body, accompanied with descriptions.* Edinburgh, 1822-33.

14108. PURKYNĚ, Jan Evangelista. 1787-1869. Prague/Aus. DSB; ICB; BMNH. Under PURKINJE: P II,III; Mort.; BLA & Supp.; GC. Also Embryology and Physiology*#.

14109. BUIAL'SKII, Ilya (or Elias). 1789-1864. Russ. BLA.

14110. CLOQUET, Jules Germain. 1790-1883. Fr. BMNH; Mort.; BLA & Supp.; GA. Also Zoology.
1. *Anatomie de l'homme.* 5 vols in 3. Paris, 1821-31.
2. *Manuel d'anatomie descriptive du corps humain.* 2 vols. Paris, 1825.

14111. JACOB, Arthur. 1790-1874. Br. Mort.; BLA & Supp.; GD.

14112. CRUVEILHIER, Jean. 1791-1874. Fr. DSB; ICB; BLA & Supp.; GB.

14113. SWAN, Joseph. 1791-1874. Br. BMNH; GD. Also Zoology.
1. *A demonstration of the nerves of the human body.* London, 1834.
2. *The brain in its relation to the mind.* London, 1854.

14114. EHRMANN, Charles Henri. 1792-1878. Fr. BLA (E., Karl Heinrich).

14115. CRAIGIE, David. 1793-1866. Br. Mort.; BLA; GD.

14116. HORNER, William Edmonds. 1793-1853. U.S.A. ICB; Mort.; GE.

14117. JUNG, Karl Gustav. 1793-1864. Switz. BLA.

14118. KNOX, Robert. 1793-1862. Br. DSB; ICB; BMNH; Mort.; BLA & Supp.; GD. See also Index.

14119. LEE, Robert. 1793-1877. Br. BLA; GD.

14120. STANLEY, Edward. 1793-1862. Br. ICB; BLA; GD.

14121. FOHMANN, Vincent. 1794-1837. Ger./Belg. BMNH; BLA; GF. Also Zoology.

14122. GODMAN, John Davidson. 1794-1830. U.S.A. ICB; BMNH; BLA & Supp.; GE. Also Natural History.

14123. QUAIN, Jones. 1795-1865. Br. Mort.; BMNH; BLA; GD.

14124. SCHLEMM, Friedrich. 1795-1858. Ger. BLA; GC.

14125. VELPEAU, Alfred Armand Louis Marie. 1795-1867. Fr. BLA & Supp.; GA.

14126. WEBER, Ernst Heinrich. 1795-1878. Ger. DSB; ICB; P II,III; BMNH; Mort.; BLA; GC. Also Physiology*.

14127. WEBER, Moritz Ignatz. 1795-1875. Ger. BMNH; BLA; GC.

14128. AMUSSAT, Jean Zuléma. 1796-1856. Fr. BLA & Supp.; GA.

14129. MAYO, Herbert. 1796-1852. Br. DSB; Mort.; BLA; GA; GD. Also Physiology*.

14130. RETZIUS, Anders Adolf. 1796-1860. Swed. DSB; ICB; BMNH; Mort.; BLA; GA. Also Zoology. See also 14813/3.

14131. BOURGERY, Marc Jean. 1797-1849. Fr. BLA; GA.
1. *Traité complet de l'anatomie de l'homme.* 8 vols. Paris? 1831-54. With N.H. Jacob.

14132. GERDY, Pierre Nicholas. 1797-1856. Fr. BLA & Supp.; GA. Also Physiology.

14133. HUSCHKE, Emil. 1797-1858. Ger. DSB; BMNH; BLA; GC. Also Zoology*, Embryology, and Physiology.

14134. KRAUSE, Karl Friedrich Theodor. 1797-1868. Ger. P I,III; BLA; GC.

14135. SCHROEDER VAN DER KOLK, Jacobus Ludovicus Conradus. 1797-1862. Holl. DSB. Under KOLK: BMNH; BLA & Supp.; GF. Also Zoology and Physiology.

14136. BLANDIN, Philippe Frédéric. 1798-1849. Fr. BLA & Supp.; GA.

14137. MICHAELIS, Gustav Adolf. 1798-1848. Ger. ICB; BLA; GC.

14138. AMMON, Friedrich August von. 1799-1861. Ger. BMNH; BLA & Supp. Also Embryology.

14139. FOVILLE, Achille Louis. 1799-1878. Fr. BMNH; BLA & Supp.; GA.

14140. FLEISCHMANN, Friedrich Ludwig. fl. 1832-1841. Ger. BMNH; BLA.

14141. BERGMANN, Gottlieb Heinrich. d. 1861. Ger. BLA & Supp.

14142. FLOOD, Valentine. 1800-1847. Br. BLA & Supp.; GD.

14143. BOCHDALEK, Vincenz (not Victor) Alexander. 1801-1883. Prague. BLA & Supp.

14144. BURDACH, Ernst. 1801-1876. Ger. BMNH; Mort.; BLA.

14145. FIELDING, George Hunsley. 1801-1871. Br. Mort.; BLA.

14146. GORGONE, Giovanni. 1801-1868. It. ICB; BMNH.

14147. GRAINGER, Richard Dugard. 1801-1865. Br. BMNH; BLA; GD.
1. *Elements of general anatomy, containing an outline of the organization of the human body.* London, 1829.
2. *Observations on the structure and functions of the spinal cord.* London, 1837.

14148. MÜLLER, Johannes (Peter). 1801-1858. Ger. DSB; ICB; P II; BMNH; Mort.; BLA & Supp.; GB; GC. Also Zoology, Embryology, and Physiology*.

14149. RAINEY, George. 1801-1884. Br. BMNH; Mort.; GD. Also Botany*.

14150. THEILE, Friedrich Wilhelm. 1801-1879. Ger./Switz. BLA; GC.

14151. BARRY, Martin. 1802-1855. Br. DSB; BLA; GD. Also Embryology.

14152. BÉRARD, Auguste. 1802-1846. Fr. BLA & Supp.; GA.

14153. HOUSTON, John. 1802-1845. Br. BLA; GD.

14154. HUECK, Alexander Friedrich. 1802-1842. Dorpat. BLA & Supp.

14155. SHARPEY, William. 1802-1880. Br. DSB; ICB; Mort.; BLA; GD. Also Physiology.

14156. TOURTUAL, Kaspar Theobald. 1802-1865. Ger. P II; BLA; GC. Also Physiology.

14157. ALTON, (Johann Samuel) Eduard d'. 1803-1854. Ger. BMNH; BLA & Supp.; GC. See also 12950/2.

14158. ARNOLD, Friedrich. 1803-1890. Ger. BMNH; BLA & Supp.

14159. DALRYMPLE, John. 1803-1852. Br. Mort.; BLA & Supp.; GD.
1. *Anatomy of the human eye.* London, 1834.

14160. LAUTH, Ernest Alexandre. 1803-1837. Fr. BMNH; BLA; GA.

14161. FRORIEP, Robert. 1804-1861. Ger. BMNH; BLA.

14162. GULLIVER, George. 1804-1882. Br. BMNH; BLA; GD. Also Physiology. See also 14025/4.

14163. HUGUIER, Pierre Charles. 1804-1874. Fr. BLA; GA.

14164. KOBELT, Georg Ludwig. 1804-1857. Ger. BLA.

14165. CHASSAIGNAC, (Charles Marie) Edouard. 1805-1879. Fr. BLA & Supp.; GA.

14166. CIVININI, Filippo Antonio Romolo. 1805-1844. It. ICB; BLA.

14167. HILTON, John. 1805-1878. Br. ICB; BLA; GB; GD.

14168. SOLLY, Samuel. 1805-1871. Br. BLA; GD.
1. *The human brain. Its configuration, structure, development and physiology.* London, 1836.

14169. MALGAIGNE, Joseph François. 1806-1865. Fr. BLA; GA.

14170. BUCK, Gordon. 1807-1877. U.S.A. BLA; GE.

14171. CALORI, Luigi. 1807-1896. It. BMNH; BLA & Supp.

14172. ZEIS, Eduard. 1807-1868. Ger. BMNH; BLA; GC.

14173. DENONVILLIERS, Charles Pierre. 1808-1872. Fr. BLA & Supp.

14174. GIRALDÈS, Joachim Albin Cardozo. 1808-1875. Port./Fr. BMNH; BLA & Supp.

14175. JONES, Thomas Wharton. 1808-1891. Br. BMNH; BLA.

14176. BOCK, Carl Ernst. 1809-1874. Ger. BLA & Supp.

14177. BONNET, Amédée. 1809-1852. Fr. BLA & Supp.

14178. BUROW, (Karl) August von. 1809-1874. Ger. BMNH; BLA & Supp.

14179. FÄSEBECK, Georg Matthias Ferdinand. 1809-1900. Ger. BLA Supp.; GC.

14180. HENLE, (Friedrich Gustav) Jakob. 1809-1885. Ger. DSB; ICB; BMNH; Mort.; BLA & Supp.; GA; GB; GC.
1. *Über die pacinischen Körperchen an den Nerven des Menschen und der Säugethiere.* Zurich, 1844. With A. Kölliker. RM.
2. *Der Briefwechsel zwischen Jakob Henle und Karl Pfeufer, 1843-69.* Wiesbaden, 1970. Ed. by H. Hoepke.
See also 14277/1.

14181. TODD, Robert Bentley. 1809-1860. Br. BMNH; BLA; GD.
1. *Cyclopaedia of anatomy and physiology.* 5 vols in 6. London, 1835-59.
2. *The physiological anatomy and physiology of man.* Vol. 1. London, 1845. With W. Bowman. 2nd ed., 2 vols (first publn of Vol. 2), 1856. Another ed., by L.S. Beale, 1866.
3. *The descriptive and physiological anatomy of the brain, spinal cord, and ganglions.* London, 1845.

14182. VOIGT, Christian August. 1809-1890. Aus. BMNH; BLA.

14183. WILSON, William James Erasmus. 1809-1884. Br. ICB; BLA; GD.

14184. BIDDER, Friedrich Heinrich. 1810-1894. Dorpat. DSB; BMNH; Mort.; BLA & Supp.; GC. Also Physiology*.

14185. HYRTL, (Carl) Joseph. 1810-1894. Prague/Aus. DSB; ICB; BMNH; Mort; BLA. Also Zoology* and History of Anatomy*.

14186. PIROGOV, Nikolay Ivanovich. 1810-1881. Russ. DSB; ICB; Mort.; BLA & Supp; GB.

14187. SAPPEY, Marie Philibert Constant. 1810-1896. Fr. BMNH; Mort.; BLA. Also Zoology.

14188. STILLING, Benedict. 1810-1879. Ger./Aus. BMNH; BLA; GC.

14189. VALENTIN, Gabriel Gustav. 1810-1883. Ger./Switz. DSB; ICB; P II,III; BMNH; BLA; GC. Also Embryology, Physiology*, and Cytology.

14190. KOHLRAUSCH, Otto Ludwig Bernhard. 1811-1854. Ger. BLA.

14191. MERCIER, Louis Auguste. 1811-1882. Fr. BLA.

14192. MIESCHER, (Johann) Friedrich (first of the name). 1811-1887. Ger./Switz. BMNH; BLA. Also Physiology.

14193. REICHERT, Karl Bogislaus. 1811-1883. Ger./Dorpat. DSB; ICB; BMNH; Mort.; BLA; GC. Also Zoology and Embryology.

14194. ARLT, (Carl) Ferdinand von. 1812-1887. Prague/Aus. ICB; BLA & Supp.; GC.

14195. ELLIS, George Viner. 1812-1900. Br. BLA & Supp.

14196. MANDL, Louis. 1812-1881. Fr. BLA.
1. *Anatomie microscopique*. 2 vols. 1838-57.

14197. MERKEL, Carl Ludwig. 1812-1876. Ger. Mort.; BLA. Also Physiology.

14198. PACINI, Filippo. 1812-1883. It. DSB; ICB; BMNH; Mort.; BLA.

14199. DAVEY, James George. 1813-1895. Br. BLA.
1. *The ganglionic nervous system*. London, 1858.

14200. FICK, (Franz) Ludwig. 1813-1858. Ger. P I & Supp.; BLA & Supp. Also Physiology.

14201. TIGRI, Atto Ferdinando Romolo. 1813-1875. It. ICB.

14202. GOODSIR, John. 1814-1867. Br. DSB; ICB; BMNH; BLA & Supp.; GB; GD. Also Zoology.
1. *Anatomical memoirs*. 2 vols. Edinburgh, 1868.

14203. GRUBER, Wenzel Leopold. 1814-1890. Russ. BMNH; BLA & Supp. Also Zoology*.

14204. HANNOVER, Adolph. 1814-1894. Den. ICB; P I; BMNH; BLA & Supp. Also Microscopy.

14205. LUCAE, Johann Christian Gustav. 1814-1885. Ger. BMNH; BLA; GC.

14206. NUHN, Anton. 1814-1889. Ger. BMNH; BLA. Also Zoology*.

14207. PAGET, James. 1814-1899. Br. ICB; BMNH; BLA; GB; GD1.

14208. SIBSON, Francis. 1814-1876. Br. Mort.; BLA; GD.

14209. BAILLARGER, (Jules Gabriel) François. 1815 (or 1806)-1890. Fr. ICB; BLA & Supp.

14210. DITTEL, Leopold von. 1815-1898. Aus. BLA & Supp.; GC.

14211. JARJAVAY, Jean François. 1815-1868. Fr. BLA.

14212. MEYER, Georg Hermann von. 1815-1892. Switz. P II,III. Also Physiology.

14213. PASSAVANT, (Philipp) Gustav. 1815-1893. Ger. BLA.

14214. QUEKETT, John Thomas. 1815-1861. Br. BMNH; BLA; GD. Also Microscopy*.
1. *Lectures on histology.* 2 vols. London, 1852-54.

14215. REMAK, Robert. 1815-1865. Ger. DSB; ICB; Mort.; BLA & Supp.; GC. Also Embryology, Physiology, and Cytology.

14216. TOYNBEE, Joseph. 1815-1866. Br. BLA; GD. Also Physiology.

14217. WEDL, Karl. 1815-1891. Aus. BLA; GC.

14218. BOWMAN, William. 1816-1892. Br. DSB; ICB; Mort.; BLA & Supp.; GD1. Also Physiology. See also 14181/2.

14219. DENNIS, James Blatch Piggott. 1816-1861. Br. GD.

14220. ECKER, Alexander. 1816-1887. Switz. BMNH; BLA & Supp.; GC. Also Zoology*, Physiology*, and Biography*.

14221. GUERIN, Alphonse François Marie. 1816-1895. Fr. ICB; BLA & Supp.

14222. HIRSCHFELD, Ludwik Maurycz. 1816-1876. Warsaw. BLA.

14223. KRUKENBERG, Adolph. 1816-1877. Ger. BLA.

14224. RICHET, Louis Dominique Alfred. 1816-1891. Fr. BLA.

14225. WALLER, Augustus Volney. 1816-1870. Br. DSB; ICB; P III; Mort.; BLA; GD. Also Physiology.

14226. CLARKE, (Jacob Augustus) Lockhart. 1817-1880. Br. Mort.; BLA & Supp.; GD.

14227. HASSALL, Arthur Hill. 1817-1894. Br. BMNH; Mort.; BLA. Also Botany.
1. *The microscopic anatomy of the human body.* 2 vols. London, 1849. Another ed., New York, 1855.

14228. KOELLIKER, (Rudolf) Albert von. 1817-1905. Ger. DSB; ICB; BMNH; Mort.; BLA & Supp.; GA; GB. Also Zoology, Embryology*, Physiology, and Cytology.
1. *Mikroskopische Anatomie; oder, Gewebelehre des Menschen.* 2 vols. Leipzig, 1850-54. RM/L.
2. *Handbuch der Gewebelehre des Menschen.* Leipzig, 1852. X. 6th ed., 3 vols, 1889-99.
2a. ——— *Manual of human histology.* 2 vols. London, Sydenham Society, 1853-54. Trans. and ed. by G. Busk and T. Huxley. RM/L.
2b. ——— ——— Another ed. *Manual of human microscopic anatomy.* Ib.,1860. Rev. by G. Buchanan.
See also 14180/1.

14229. ZIMMERMANN, Gustav Heinrich Eduard. 1817-1866. Ger. BLA; GC.

14230. BIGELOW, Henry Jacob. 1818-1890. U.S.A. BLA & Supp.; GE.

14231. LENHOSSEK, Joseph. 1818-1888. Hung. BMNH; BLA.

14232. MARSHALL, John. 1818-1891. Br. BMNH; BLA; GB; GD. Also Physiology*.
1. *A description of the human body, its structure and functions.* London, 1860. X. 3rd ed., 1875.

14233. BRÜCKE, Ernst Wilhelm von. 1819-1892. Ger./Aus. DSB; P I, III,IV; Mort.; BLA & Supp.; GC. Also Botany* and Physiology*.

14234. HASNER, Joseph, *Ritter* von Artha. 1819-1892. Prague. BLA; GC.

14235. LANGER (VON EDENBERG), Karl. 1819-1887. Hung./Aus. BMNH; BLA; GC.

14236. GERLACH, Joseph von. 1820-1896. Ger. ICB; BMNH; Mort.; BLA & Supp.; GC.

14237. HUMPHRY, George Murray. 1820-1896. Br. ICB; Mort.; BLA; GD1.

14238. LUSCHKA, Hubert von. 1820-1875. Ger. BMNH; Mort.; BLA; GC.

14239. MÜLLER, Heinrich. 1820-1864. Ger. BMNH; Mort.; BLA. Also Physiology.

14240. LEYDIG, Franz von. 1821-1908. Ger. DSB; BMNH; Mort.; BLA. Also Zoology*.

14241. M'DOWEL, Benjamin George. 1821 (or 1829)-1885. Br. BLA.

14242. RANDACIO, Francesco. 1821-1903. It. BLA.

14243. ROBIN, Charles Philippe. 1821-1885. Fr. DSB; ICB; BMNH; BLA & Supp.; GA. Also Chemistry* and Microscopy.
1. *Du microscope et des injections dans leurs applications à l'anatomie et à la pathologie.* Paris, 1849.

14244. CORTI, Alfonso Giacomo Gaspare. 1822-1876. It. DSB; ICB; Mort.; BLA & Supp.

14245. FOLTZ, Jean Charles Eugène. 1822-1876. Fr. BLA.

14246. FREY, (Johann Friedrich) Heinrich (Konrad). 1822-1890. Switz. BMNH; BLA & Supp.; GC. Also Microscopy* and Zoology*.
1. *Histologie und Histochemie des Menschen.* Leipzig, 1859. X. 5th ed., 1876.
1a. ——— *Histology and histochemistry of man.* London, 1874. Trans. by A.E.J. Barker from the 4th German ed.
2. *Grundzüge der Histologie.* Leipzig, 1875. X. 2nd ed., 1879.
2a. ——— *Compendium of histology.* London, 1876. Trans. by G.R. Cutler.

14247. KUSSMAUL, Adolf. 1822-1902. Ger. ICB; Mort.; BLA & Supp.

14248. RICHARD, (Félix) Adolphe. 1822-1872. Fr. ICB; BLA.

14249. FOLLIN, François Anthime Eugène. 1823-1867. Fr. BLA.

14250. LEDWICH, Thomas Hawkesworth. 1823-1858. Br. BLA; GD.

14251. LEIDY, Joseph. 1823-1891. U.S.A. DSB; ICB; P III; BMNH; Mort.; BLA & Supp.; GB; GE. Also Palaeontology and Zoology.

14252. STRUTHERS, John. 1823-1899. Br. ICB; BLA; GD1. Also History of Anatomy*.

14253. TEICHMANN, Ludwik Karol. 1823-1895. Cracow/Ger. DSB; BLA.

14254. BROCA, (Pierre) Paul. 1824-1880. Fr. DSB; ICB; BMNH; Mort.; BLA & Supp.; GB.

14255. CIACCIO, Giuseppe Vincenzo. 1824-1901. It. BLA & Supp.

14256. GUDDEN, (Johann) Bernhard Aloys von. 1824-1886. Ger. DSB; ICB; BMNH; Mort.; BLA & Supp.; GC.

14257. MAIER, Rudolf Robert. 1824-1888. Ger. BLA.

14258. REISSNER, Ernst. 1824-1878. Dorpat/Ger. Mort.; BMNH; BLA.

14259. ROUGET, Charles Marie Benjamin. 1824-1904. Fr. DSB. Also Physiology.

14260. BILHARZ, Theodor. 1825-1862. Ger./Egypt. DSB; ICB; P III; BMNH; BLA & Supp. Also Zoology.

14261. GRAY, Henry. 1825-1861. Br. DSB; ICB; Mort.; BLA & Supp.
1. *Anatomy, descriptive and surgical.* London, 1859. Many editions, continuing to the present day.

14262. SCHULTZE, Max Johann Sigismund. 1825-1874. Ger. DSB; ICB; P III; BMNH; Mort.; BLA; GB; GC. Also Microscopy, Zoology, Embryology, and Cytology.

14263. WOOD, John. 1825 (or 1827)-1891. Br. BLA; GD.

14264. GEGENBAUR, Karl. 1826-1903. Ger. DSB Supp.; ICB; BMNH; Mort.; BLA & Supp.; GB. Also Zoology* and Embryology.

14265. ALBINI, Giuseppe. 1827-1911. It. BLA & Supp.

14266. BURNETT, Waldo Irving. 1827-1854. U.S.A. ICB; BLA. Also Cytology. See also 13100/2a.

14267. FAIVRE, Ernest. 1827-1879. Fr. BMNH; Mort. Also Evolution.

14268. OEHL, Eusebio. 1827-1903. It. BLA.

14269. AUERBACH, Leopold (first of the name). 1828-1897. Ger. ICB; Mort.; BLA & Supp.; GC. Also Physiology.

14270. BEALE, Lionel Smith. 1828-1906. Br. DSB; ICB; BMNH; BLA & Supp.; GD2. Also Biology in General* and Microscopy*.
1. *On the structure of the simple tissues of the human body.* London, 1861.
See also 14181/2.

14271. CZERMAK, Johann Nepomuk. 1828-1873. Prague/Ger. DSB; ICB; P III; BMNH; Mort.; BLA & Supp.; GC. Also Physiology*#.

14272. LUYS, (Jules) Bernard. 1828-1897. Fr. ICB; Mort.; BLA. Also Physiology.

14273. BILLROTH, (Christian Albert) Theodor. 1829-1894. Switz./Aus. DSB; ICB; BMNH; BLA & Supp.; GB; GC; GF.

14274. GOLL, Friedrich. 1829-1903. Switz. Mort.; BLA & Supp.

14275. KUPFFER, Carl Wilhelm von. 1829-1902. Ger. BMNH; Mort.; BLA. Also Embryology.

14276. LE FORT, Léon Clément. 1829-1893. Fr. ICB; BLA.

14277. MEISSNER, Georg. 1829-1905. Ger. DSB; ICB; P III; Mort.; BLA. Also Physiology.
1. *Briefe an Jacob Henle, 1855-1878*. Göttingen 1975. Ed. by H.H. Eulner and H. Hoepke.

14278. BEAUNIS, Henri Etienne. 1830-1921. Fr. BLA & Supp.
1. *Nouveaux éléments d'anatomie descriptive et d'embryologie*. Paris, 1868. With H.A. Bouchard. X. 2nd ed., 1873.

14279. BLESSIG, Robert. 1830-1878. Russ. BLA & Supp.

14280. OLLIER, (Louis Xavier Edouard) Léopold. 1830-1900. Fr. ICB; Mort.; BLA.

14281. BÖTTCHER, Arthur. 1831-1889. Dorpat. BMNH; BLA & Supp.

14282. BRAUNE, Christian Wilhelm. 1831-1892. Ger. BMNH; Mort.; BLA & Supp.; GC. Also Physiology.

14283. FROMMANN, Carl Friedrich Wilhelm. 1831-1892. Ger. BMNH; BLA & Supp.; GC. Also Cytology.

14284. GUYON, Jean Casimir Félix. 1831-1920. Fr. ICB; BLA & Supp.

14285. HIS, Wilhelm (first of the name). 1831-1904. Switz./Ger. DSB; ICB; BMNH; Mort.; BLA & Supp. Also Embryology*.
1. *Über die Aufgaben und Zielpunkte der wissenschaftliche Anatomie*. Leipzig, 1872. A lecture.
2. *Lebenserinnerungen und ausgewählte Schriften*. Bern, 1965.

14286. GOTTSTEIN, Jacob. 1832-1895. Ger. BLA & Supp.; GC.

14287. KEY, (Ernst) Axel (Henrik). 1832-1901. Swed. ICB; BMNH; BLA.

14288. MOLL, Jacob Antonius. 1832-1914. Holl. BLAF.

14289. RÜDINGER, Nicholaus. 1832-1896. Ger. BMNH; BLA; GC. Also Zoology.

14290. TURNER, William. 1832-1916. Br. DSB; BMNH; BLA; GD3. Also Zoology.

14291. KRAUSE, Wilhelm. 1833-1910. Ger. BMNH; BLA.
1. *Allgemeine und microscopische Anatomie*. Hanover, 1876.
2. *Specielle und macroscopische Anatomie*. Hanover, 1879.

14292. MANZ, Wilhelm. 1833-1911. Ger. BLA.

14293. MEYNERT, Theodor Hermann. 1833-1892. Aus. ICB; Mort.; BLA & Supp.

14294. RECKLINGHAUSEN, Friedrich Daniel von. 1833-1910. Ger. Mort.; BLA & Supp.

14295. SOUSA, Manuel Bento de. 1833-1899. Port. ICB.

14296. WESTPHAL, Carl Friedrich Otto. 1833-1890. Ger. Mort.; BLA.

14297. ZAUFAHL, Emmanuel. 1833 (or 1837)-1910. Prague. BLA.

14298. BETS (or BETZ), Vladimir Aleksandrovich. 1834-1894. Russ. Mort.; BLAF.

14299. DEITERS, Otto Friedrich Carl. 1834-1863. Ger. BMNH; Mort.; BLA & Supp.

14300. HEIDENHAIN, Rudolf (Peter Heinrich). 1834-1897. Ger. DSB; P III,IV; BMNH; Mort.; BLA; GC. Also Physiology*.

14301. HENKE, (Phillip Jacob) Wilhelm. 1834-1896. Ger. BLA & Supp.; GC.

14302. HOYER, Heinrich. 1834-1907. Warsaw. BLA.

14303. KOLLMANN, Julius Konstantin Ernst. 1834-1918. Ger./Switz. BMNH; Mort.; BLA; BLAF.

14304. NEUMANN, Ernst Francis Christian. 1834-1918. Ger. BLA.

14305. PETTIGREW, James Bell. 1834-1908. Br. BMNH; BLA & Supp.; GD2. Also Physiology*.

14306. SCHWEIGGER-SEIDEL, Franz. 1834-1871. Ger. Mort.; BLA. Also Cytology.

14307. STRICKER, Salomon. 1834-1898. Aus. BMNH; Mort.; BLA; GC. Also Physiology*.
1. *Handbuch der Lehre von den Geweben des Menschen und der Thiere.* 2 vols. Leipzig, 1869-72.
1a. ——— *Manual of human and comparative histology.* 3 vols. London, 1870-73. Trans. by H. Power.

14308. FRANKENHÄUSER, Ferdinand. d. 1894. Ger./Switz. BLA Supp.

14309. AEBY, Christoph Theodor. 1835-1885. Switz./Prague. BMNH; BLA & Supp.

14310. ARNOLD, Julius. 1835-1915. Ger. BLA & Supp.

14311. BOCHDALEK, Victor. 1835-1868. Prague. BLA.

14312. CLELAND, John. 1835-1925. Br. BLA & Supp.; BLAF. Also Physiology*.

14313. EBERTH, Carl Joseph. 1835-1926. Ger. DSB; BMNH; BLA & Supp. Also Microbiology.

14314. ENGLISCH, Josef. 1835-1915. Aus. BLA.

14315. POLITZER, Adam. 1835-1920. Aus. BMNH; BLA.

14316. RANVIER, Louis Antoine. 1835-1922. Fr. DSB; ICB; BMNH; Mort.; BLA.
1. *Traité technique d'histologie.* Paris, 1875.
1a. ——— *Technisches Lehrbuch der Histologie.* Leipzig, 1888.
2. *Leçons sur l'histologie du système nerveux.* Paris, 1878.

14317. GILLETTE, Eugène Paulin. 1836-1886. Fr. BLA.

14318. HEITZMANN, Carl. 1836-1896. Hung./U.S.A. BMNH; Mort.; BLA; GC.

14319. LESSHAFT, Piotr Frantsovich. 1836-1909. Russ. BLAF.

14320. LOEWENBERG, Benjamin Benno. b. 1836. Aus./Fr. BLA.

14321. MÜNCH, Gregor N. 1836-1896. Russ. ICB; BLA & Supp.; BLAF.

14322. WALDEYER (-HARTZ), (Heinrich) Wilhelm (Gottfried) von. 1836-1921. Ger. DSB; ICB; BMNH; Mort.; BLAF. Also Embryology*.

14323. SKENE, Alexander Johnston Chalmers. 1837-1900. Br./U.S.A. Mort.; BLA; GE.

14324. ANTONELLI, Giovanni. 1838-1914. It. BLAF.

14325. FRITSCH, Gustav Theodor. 1838-1927. Ger. DSB; ICB; P III,IV;

BMNH; Mort.; BLA & Supp. Also Zoology and Physiology.

14326. BIESIADECKI, Alfred. 1839-1888. Cracow. BLA & Supp.

14327. COHNHEIM, Julius Friedrich. 1839-1884. Ger. ICB; Mort.; BLA & Supp.; GC. Also Microbiology.

14328. KIESSELBACH, Wilhelm. 1839-1902. Ger. BLA.

14329. PRUSSAK, Alexander. 1839-1897. Russ. BLA; BLAF.

14330. RUTHERFORD, William. 1839-1899. Br. BMNH; BLA & Supp.; GD1. Also Physiology*.

14331. TODARO, Francesco. 1839-1918. It. BMNH; BLA.

14332. VANLAIR, Constant François. 1839-1914. Belg. ICB; BLA; GF. Also Physiology.

14333. GIACOMINI, Carlo. 1840-1898. It. BLA & Supp.

14334. HELLER, Arnold Ludwig Gotthilf. 1840-1913. Ger. BLA.

14335. KOVALEVSKY, Nikolai Osipovich. 1840-1892. Russ. BMNH; BLA.

14336. LEBER, Theodor. 1840-1917. Ger. BLAF.

14337. TOLDT, Karl. 1840-1920. Prague/Aus. BLA.

14338. ALLEN, Harrison. 1841-1897. U.S.A. BMNH; BLA & Supp.; GE. Also Zoology.

14339. BAKER, Frank. 1841-1918. U.S.A. BLAF; GE.

14340. HAYEM, Georges. 1841-1933. Fr. Mort.; BLAF.

14341. JENSEN, Julius. 1841-1891. Ger. GC.

14342. KNOLL, Philipp. 1841-1900. Prague/Aus. BLAF. Also Physiology.

14343. RAUBER, August Antinous. 1841-1917. Dorpat. BLAF. Also Embryology.

14344. EBNER (VON ROSENSTEIN), Victor. 1842-1925. Aus. BMNH; BLA & Supp. Also Physiology.

14345. GAY, Alexander Heinrich. 1842-1907. Russ. BLAF.

14346. MAYER, Sigmund. 1842-1910. Ger. BMNH; BLAF. Also Physiology.

14347. PALADINO, Giovanni. 1842-1917. It. BLAF. Also Physiology.

14348. RETZIUS, (Magnus) Gustaf. 1842-1919. Swed. DSB; BMNH; Mort.; BLAF. Also Zoology and Embryology.

14349. ROSENBERG, Emil Woldemar. 1842-1925. Dorpat/Holl. BMNH; BLAF. Also Zoology and Embryology.

14350. SERTOLI, Enrico. 1842-1910. It. DSB; BLAF. Also Physiology.

14351. TROLARD, Paulin. 1842-1910. Fr./Algeria. BLAF.

14352. DWIGHT, Thomas. 1843-1911. U.S.A. ICB; BLA & Supp; GE.

14353. FLEMMING, Walther. 1843-1905. Ger. DSB; Mort.; BLAF. Also Cytology.

14354. GOLGI, Camillo. 1843-1926. It. DSB; ICB; Mort.; BLA & Supp. Also Cytology.

14355. POPOW, Mitrofan. 1843-1905. Russ. BLAF.

14356. THANHOFFER, Lajos (or Ludwig). 1843-1909. Hung. Mort.; BLAF. Also Physiology*.

14357. DUVAL, Mathias Marie. 1844-1907. Fr. DSB; BMNH; Mort.; BLA & Supp. Also Physiology and Embryology*.

14358. KLEIN, (Edward) Emanuel. 1844-1925. Br. BMNH; BLAF; GF. Also Physiology.
1. *Atlas of histology*. London, 1880. With E.N. Smith.
2. *Elements of histology*. London, 1883. X. Other eds: 1887, 1889, 1891, 1898.

14359. MACALISTER, Alexander. 1844-1919. Br. BMNH; BLAF; GF. Also Zoology*.

14360. MIHALKOVICS, Géza Victor von. 1844-1899. Hung. BMNH; BLAF. Also Embryology.

14361. MORRIS, Henry. 1844-1926. Br. BLAF.

14362. RENAUT, Joseph Louis. 1844-1917. Fr. DSB; BLAF.

14363. SATTLER, Hubert. 1844-1928. Ger. Mort.; BLAF.

14364. SCHWALBE, Gustav Albert. 1844-1916. Ger. BMNH; BLAF.

14365. WITKOWSKI, Gustave Jules (or Joseph). 1844-1923. Fr. BLAF.

14366. MERKEL, Friedrich Siegismund. 1845-1919. Ger. BMNH; Mort.; BLAF. Also Microscopy* and Embryology.

14367. TARENETZKY, Alexander Ivanovich. 1845-1905. Russ. BMNH; BLAF.

14368. WEIGERT, Carl. 1845-1904. Ger. DSB; ICB; BLAF.

14369. BIZZOZERO, Giulio Cesare. 1846-1901. It. DSB; ICB; Mort.; BLAF.

14370. BUDGE, Albrecht. 1846-1885. Ger. BLA.

14371. FÜRBRINGER, Max. 1846-1920. Ger. BMNH; BLAF. Also Zoology and Embryology.

14372. LUCAS, Richard Clement. 1846-1915. Br. BLAF.

14373. POZZI, Samuel Jean. 1846-1918. Fr. BMNH; BLAF.

14374. FLECHSIG, Paul Emil. 1847-1929. Ger. DSB; Mort.; BLA & Supp.

14375. GASSER, Emil. 1847-1919. Ger. BMNH; BLAF. Also Embryology.

14376. GIERKE, Hans Paul Bernard. 1847-1886. Ger./Japan. BLA.

14377. HELWEG, Hans Kristian Saxtroph. 1847-1901. Den. BLAF.

14378. HENSCHEN, Salomon Eberhard. 1847-1930. Swed. BMNH; Mort.; BLAF.

14379. LANGERHANS, Paul. 1847-1888. Ger. DSB; ICB; BMNH; Mort.; BLAF; GC.

14380. LEWIS, William Bevan. 1847-1929. Mort.

14381. THOMA, Richard. 1847-1923. Ger./Dorpat. BLAF.

14382. CHARPY, Adrien. 1848-1911. Fr. BLAF.

14383. FOREL, Auguste Henri. 1848-1931. Switz. DSB; ICB; BMNH; Mort.; BLAF. Also Zoology*#.

14384. LE DOUBLE, Anatole Félix. 1848-1913. Fr. DSB; BLAF.

14385. MACEWAN, William. 1848-1924. Br. ICB; Mort.; BLAF; GD4.

14386. WERNICKE, Carl. 1848-1905. Ger. DSB; Mort.; BLAF.

14387. BARDELEBEN, Karl (Heinrich) von. 1849-1918. Mort.; BMNH; BLAF.

14388. BARFURTH, Dietrich. 1849-1927. Ger. BLAF. Also Embryology.

14389. BOLL, Franz Christian. 1849-1878. Ger./It. P IV; BMNH; Mort.; BLA & Supp. Also Zoology and Physiology.

14390. BRUNN, Albert von. 1849-1895. Ger. BMNH; BLAF.

14391. DEJERINE, Joseph Jules. 1849-1917. Fr. Mort.; BLAF.

14392. DRASCH, Otto. 1849-1911. Aus. BLAF. Also Embryology.

14393. FRORIEP, August. 1849-1917. Ger. BLAF.

14394. GODLEE, Rickman John. 1849-1925. Br. BLA & Supp.; GD4.

14395. McCLELLAN, George. 1849-1913. U.S.A. BLAF; GE.

14396. OSLER, William. 1849-1919. Canada/Br. The celebrated physician. ICB; Mort.; BLAF; GD3.

14397. PRUDDEN, Theophil Mitchell. 1849-1924. U.S.A. DSB; BLAF; GE. Also Microbiology.

14398. SCHIEFFERDECKER, Paul. 1849-1931. Ger. BLAF.

14399. SEILER, Carl. 1849-1905. U.S.A. GE.
   1. *Micro-photographs in histology.* Philadelphia, 1876.

14400. SOLGER, Bernhard Friedrich. 1849-after 1902. Ger. BMNH; BLAF. Also Cytology.

14401. STÖHR, Philipp Adrian. 1849-1911. Ger. BMNH; BLAF. Also Embryology*.
   1. *Lehrbuch der Histologie und der mikroskopischen Anatomie des Menschen.* Jena, 1887. X. 3rd ed., 1889. 7th ed., 1896.

14402. TESTUT, (Jean) Léo. 1849-1925. Fr. DSB; BMNH; Mort.; BLAF.
   1. *Les anomalies musculaires chez l'homme expliquées par l'anatomie comparée.* Paris, 1884.

14403. ZUCKERKANDL, Emil. 1849-1910. Aus. BMNH; Mort.; BLAF; GF.

14404. FISCHER, Ernst. fl. 1876. Prague. Mort.

14405. CHIEVITZ, Johan Henrik. 1850-1901. Den. BLAF.

14406. CUNNINGHAM, Daniel John. 1850-1909. Br. BMNH; BLAF; GD2. Also Zoology.
   1. *The dissector's guide.* 3 parts. Edinburgh/London, 1879-87. X.
   1a. ——— 2nd ed. *A manual of practical anatomy.* 3 vols. Ib., 1889. Another ed., 2 vols, 1893-1903.

14407. FORTUNATOW, Alexius. 1850-1905. Russ. BLAF.

14408. MARCHANT, Gerard. 1850-1903. Fr. BLAF.

14409. NUSSBAUM, Moritz. 1850-1915. Ger. BLAF.

14410. SHARPEY-SCHÄFER, Edward Albert. 1850-1935. Br. DSB; ICB; BMNH (SCHAEFER); Mort.; BLAF; GD5. Also Physiology*.

14411. WOELFLER, Anton. 1850-1917. Aus./Prague. BLAF.

14412. ALBRECHT, (Carl Martin) Paul. 1851-1894. Ger./Belg. BMNH; BLA & Supp.; GC. Also Zoology.

14413. FUCHS, Ernst. 1851-1930. Aus. BLAF. Also Physiology.

14414. GERLACH, Leo. 1851-1918. Ger. BMNH; BLAF.

14415. MARCHI, Vittorio. 1851-1908. It. DSB; Mort.

14416. PICK, Arnold. 1851-1924. Prague. Mort.; BLAF.

14417. REID, Robert William. 1851-1939. Br. Mort.

14418. SERRANO, José Antonio. 1851-1904. Port. ICB; BLAF.

14419. SYMINGTON, Johnson. 1851-1924. Br. BLAF.

14420. ALTMANN, Richard. 1852-1900. Ger. P IV; Mort.; BLAF. Also Physiology and Cytology*.

14421. BRISSAUD, Edouard. 1852-1909. Fr. BLAF.

14422. DISSÉ, Joseph. 1852-1912. Ger. BLAF.

14423. DOGIEL, Alexander Stanislavovich. 1852-1922. Russ. BMNH (DOGHEL); Mort.; BLAF. Also Zoology.

14424. HOLL, Moritz. 1852-1920. Aus. BLAF.

14425. LANGLEY, John Newport. 1852-1925. Br. DSB; Mort.; BLAF; GD4. Also Physiology.

14426. MINOT, Charles Sedgwick. 1852-1914. U.S.A. DSB; BMNH; Mort.; BLAF; GE. Also Embryology*.

14427. QUENU, Edouard André Victor Alfred. 1852-1933. Fr. BLAF.

14428. RAMÓN Y CAJAL, Santiago. 1852-1934. Spain. DSB; ICB; BMNH; Mort.; BLAF. Also Autobiography*.
   1. *Textura del sistema nervioso del hombre y de los vertebrados.* 3 vols. Madrid, 1894-1904. X.
   1a. ——— *Les nouvelles idées sur la structure du système nerveux chez l'homme et chez les vertébrés. Edition française revue et augmentée par l'auteur.* Paris, 1895. Trans. by L. Azoulay.

14429. RUGE, Georg Hermann. 1852-1919. Ger./Holl./Switz. BMNH; BLAF. Also Zoology.

14430. SANDSTRÖM, Ivar Victor. 1852-1889. Swed. ICB; Mort.; BLAF.

14431. SIEBENMANN, Friedrich. 1852-1928. Switz. BMNH; BLAF.

14432. SPITZKA, Edward Charles. 1852-1914. U.S.A. BLAF; GE.

14433. STRASSER, Hans. 1852-1927. Switz. BMNH; BLAF. Also Embryology and Physiology.

14434. TENCHINI, Lorenzo. 1852-1906. It. BLAF.

14435. TOURNEUX, Frédéric. 1852-1922. Fr. BLAF.

14436. VIRCHOW, Hans. 1852-1940. Ger. BLAF.

14437. YOUNG, Alfred Harry. 1852-1912. Br. BLAF.

14438. BROESIKE, Gustav. 1853-after 1919. Ger. BLAF.

14439. FRASER, Alexander. 1853-1909. Br. BLAF. Also Embryology.

14440. MONAKOW, Constantin von. 1853-1930. Switz. BLAF.

14441. PFITZNER, Wilhelm. 1853-1903. Ger. BLAF. Also Cytology.

14442. POIRIER, Paul Julien. 1853-1907. Fr. BLAF.

14443. RABL, Carl. 1853-1917. Prague/Ger. DSB; BMNH; BLAF. Also Embryology and Cytology.

14444. TREVES, Frederick. 1853-1923. Br. BLAF; GD4.

14445. ZEISSL, Maximilian. 1853-1925. Aus. BLAF.

14446. BRUCE, Alexander. 1854-1911. Br. BLAF.

14447. EHRLICH, Paul. 1854-1915. Ger. DSB; ICB; Mort.; BLAF. Other entries: see 3361.

14448. ETERNOD, Auguste François Charles d'. b. 1854. Switz. BLAF.

14449. FÜRST, Carl Magnus. 1854-1935. Swed. BLAF.

14450. REES, Jacobus van. 1854-1928. Holl. BLAF.

14451. STARR, Moses Allen. 1854-1932. U.S.A. GE.
1. *Atlas of nerve cells*. New York, 1896.

14452. EDINGER, Ludwig. 1855-1918. Ger. ICB; BLAF.
1. *Zehn Vorlesungen über den Bau der nervösen Centralorgane des Menschen*. Leipzig, 1885. X.
1a. ——— *The anatomy of the central nervous system of man and of vertebrates in general*. Philadelphia, 1899. Trans. from the 5th German ed. by W.S. Hall.

14453. GUBAROFF, Alexander Petrovich. 1855-1931. Russ. BLAF (GUBAREW).

14454. ROBINSON, Frederick Byron. 1855-1910. U.S.A. BLAF; GE.

14455. SPEE, Ferdinand von. 1855-1937. Ger. BLAF. Also Embryology.

14456. JANOŠÍK, Jan. 1856-1927. Prague. BMNH; BLAF.

14457. KULTSCHITZKY, Nikolai. 1856-1925. Russ. BMNH; Mort. Also Zoology.

14458. LAUNOIS, Pierre Emile. 1856-1914. Fr. BLAF.

14459. LOCKWOOD, Charles Barrett. 1856-1914. Br. Mort.; BLAF.

14460. PIERSOL, George Arthur. 1856-1924. U.S.A. BLAF.
1. *A text-book of normal histology*. Philadelphia, 1893.

14461. SOMMER, Alfred Richard. b. 1856. Ger./Russ. BLAF. Also Embryology.

14462. AUERBACH, Leopold (second of the name). b. 1857. Ger. BLAF.

14463. BEKHTEREV, Vladimir Mikhailovich. 1857-1927. Russ. DSB; ICB; BLAF (BECHTEREW). Also Physiology.

14464. BENDA, Carl. 1857-1932. Ger. BLAF. Also Cytology.

14465. FUSARI, Romeo. 1857-1910. It. BLAF. Also Embryology.

14466. MARTINOTTI, Giovanni. 1857-1928. It. BLAF.

14467. PANETH, Joseph. 1857-1890. Aus. BLAF.

14468. KOGANEI, Yoshikiyo. b. 1858. Japan. BLAF.

14469. THOMSON, Arthur. 1858-1935. Br. BLAF; GD5.

14470. WINDLE, Bertram Coghill Alan. 1858-1929. Br. ICB; BLAF.

14471. CHIARUGI, Giulio. 1859-1944. It. DSB; ICB; BLAF. Also Embryology.

14472. MACKENRODT, Alwin Karl. 1859-1925. Ger. Mort.; BLAF.

14473. MAURER, Friedrich. 1859-after 1926. Ger. BMNH; BLAF.

14474. SCHULTZE, Oskar Maximilian Sigismund. 1859-1920. Ger. BMNH; BLAF. Also Embryology.

3.17 EMBRYOLOGY

Human and comparative.

Including "generation".

14475. MAÎTRE-JAN, Antoine. 1650-1730. Fr. ICB; BLA; GA.

14476. SCHURIG, Martin. 1656-1733. Ger. BLA. Also Physiology*.

14477. ANDRY (DE BOIS-REGARD), Nicolas. 1658-1742. Fr. ICB; BLA & Supp.; GA; GF.
1. *De la génération des vers dans le corps de l'homme.* Amsterdam, 1701.

14478. VALLISNIERI, Antonio. 1661-1730. It. DSB; ICB; P II; BMNH; Mort.; BLA; GA. Also Various Fields* and Geology*.

14479. RÉAUMUR, René Antoine Ferchault de. 1683-1757. Fr. DSB; ICB; P II,III; BMNH; Mort.; GA; GB. Also Physics, Chemistry*, and Zoology*.
1. *Art de faire éclore et d'élever en toute saison des oiseaux domestiques de toutes espèces.* Paris, 1749. X.
1a. ——— *The art of hatching and bringing up domestic fowls by artificial heat.* London, 1750. Trans. by A. Trembley.

14480. MAUPERTUIS, Pierre Louis Moreau de. 1698-1759. Fr./Ger. DSB; ICB; P II; GA; GB; GC. Other entries: see 3089.
1. *Dissertation physique à l'occasion du nègre blanc.* Leiden, 1744. RM.
2. *Vénus physique.* n.p., 1745.
2a. ——— *The earthly Venus.* New York, 1966. Trans. by S.B. Boas with introd. and notes by G. Boas.

14481. PAITONI, Giovanni Cristoforo Battista. 1703-1788. It. ICB; BLA.

14482. PARSONS, James. 1705-1770. Br. P II; BMNH; BLA; GA; GD.
1. *Philosophical observations on the analogy between the propagation of animals and that of vegetables, in which are answered some objections against the indivisibility of the soul, which have been inadvertantly drawn from the late curious and useful experiments upon the polypus and other animals.* London, 1752.

14482A THEMEL, Johann Christian. 1709-1767. Ger. BLA.
1. *Commentatio medica, qua nutritionem foetus in utero per vasa umbilicalia solum fieri....* Leipzig, 1751. X. Another ed., Amsterdam, 1764.

14483. NEEDHAM, John Turberville. 1713-1781. Br./Belg. DSB; ICB;

P II; BMNH; BLA; GA; GD. Also Microscopy*.
1. *Observations upon the generation, composition and decomposition of animal and vegetable substances*. London, 1749. RM.

14484. BONNET, Charles. 1720-1793. Switz. DSB; ICB; P I; BMNH; Mort.; BLA; GA; GB. Also Biology in General*#, Botany*, and Zoology.

14485. SPALLANZANI, Lazzaro. 1729-1799. It. DSB; ICB; P II; BMNH; Mort.; BLA; GA; GB. Other entries: see 3167.
1. *Dissertazioni di fisica animale e vegetabile*. 2 vols. Modena, 1780. X. (For English trans. see 14677/3a)
1a. ——— Trans. of the section on reproduction: *Expériences pour servir à l'histoire de la génération des animaux et des plantes*. Geneva, 1785. Trans. by J. Senebier. X. Another ed., Pavia/Paris, 1787.

14486. WOLFF, Caspar Friedrich. 1734-1794. Ger./Russ. DSB Supp.; ICB; BMNH; Mort.; BLA; GB; GC.
1. *Theoria generationis*. Halle, 1759. X. Reprinted, Hildesheim, 1966.
1a. ——— German trans. 2 parts. Leipzig, 1896. (OKEW)
2. *Theorie von den Generation in zwei Abhandlungen erklärt und bewiesen*. Berlin, 1764. X. Reprinted, Hildesheim, 1966.

14487. CRUIKSHANK, William Cumberland. 1745-1800. Br. DSB; P I (CRUICKSHANKS); Mort.; BLA & Supp.; GA; GD. Also Anatomy and Physiology.

14488. PROCHÁSKA, Georgius. 1749-1820. Prague/Aus. DSB; ICB; P II; Mort.; BLA; GC. Also Anatomy and Physiology*.

14489. BLUMENBACH, Johann Friedrich. 1752-1840. Ger. DSB; ICB; P I; BMNH; Mort.; BLA & Supp.; GA; GB; GC. Also Natural History*, Zoology*, and Physiology*.
1. *Über den Bildungsstrieb und das Zeugungsgeschäft*. Göttingen, 1781. X. Reprinted, Stuttgart, 1971.
1a. ——— *An essay on generation*. London, 1792. Trans. by A. Crichton.

14490. BARONIO, Giuseppe. 1759-ca. 1811. It. ICB; BLA & Supp.

14491. DÖLLINGER, (Johann) Ignaz (Josef). 1770-1841. Ger. DSB; ICB; P I; BLA & Supp.; GC. Also Physiology.

14492. ROSENMÜLLER. Johann Christian. 1771-1820. Ger. ICB; P II; BMNH; BLA; GC. Also Geology and Anatomy.

14493. RUSCONI, Mauro. 1776-1849. It. ICB; BMNH; BLA. Also Zoology.

14494. MECKEL,Johann Friedrich (second of the name). 1781-1833. Ger. DSB; ICB; BMNH; Mort.; BLA & Supp.; GA; GC. Also Zoology* and Anatomy*.

14495. TIEDEMANN, Friedrich. 1781-1861. Ger. DSB; ICB; P III; BMNH; Mort.; BLA; GA; GB; GC. Also Zoology and Physiology*.
1. *Anatomie und Bildungsgeschichte des Gehirns im Foetus des Menschen*. Nuremberg, 1816. X.

1a. ——— *Anatomie du cerveau, contenant l'histoire de son développement dans le foetus.* Paris, 1823. X. Trans. by A.J.L. Jourdan.
1b. ——— ——— *The anatomy of the foetal brain.* Edinburgh, 1826. Trans. from the French.

14496. SERRES, (Antonio) Etienne Reynaud Augustin. 1786-1868. Fr. DSB; ICB; BMNH; BLA; GA; Also Zoology.

14497. PURKYNĚ, Jan Evangelista. 1787-1869. Prague/Aus. DSB; ICB; BMNH. Under PURKINJE: P II,III; Mort.; BLA & Supp.; GC. Also Anatomy and Physiology*#.

14498. STEINHEIM, Salomon Levy. 1789-1866. Ger. BMNH; BLA; GF.

14499. HEROLD, Moriz Johann David. 1790-1862. Ger. BMNH; GA.

14500. PREVOST, Jean Louis. 1790-1850. Switz. DSB; ICB; P II; Mort.; BLA. Also Physiology.

14501. BAER, Karl Ernst von. 1792-1876. Ger./Russ. DSB; ICB; P I, III; Mort.; BMNH; BLA & Supp.; GA; GB; GC. Also Autobiography*.
1. *De ovi mammalium et hominis genesi.* Leipzig, 1827. RM. Reprinted, Brussels, 1967.
2. *Über die Entwickelungsgeschichte der Thiere.* 2 vols. Königsberg, 1828-37. RM/L. Reprinted, Brussels, 1967.
3. *Reden gehalten in wissenschaftlichen Versammlungen, und kleine Aufsätze vermischten Inhalts.* 3 vols. St. Petersburg, 1864-76.
4. [In Russian. *Von Baer's letters to members of the St. Petersburg Academy of Science.*] Leningrad, 1976. The letters are in German with a Russian trans. Ed. with commentary by T.A. Lukina.

14502. RATHKE, (Martin) Heinrich. 1793-1860. Ger. DSB; BMNH; Mort.; BLA; GC. Also Zoology*.
1. *Entwickelungsgeschichte der Wirbelthiere.* Leipzig, 1861.
2. *Untersuchungen über die Entwickelung und den Körperbau der Krokodile.* Brunswick, 1866.

14503. FLOURENS, (Marie Jean) Pierre. 1794-1867. Fr. DSB; ICB; BMNH; Mort.; BLA; GA; GB. Also Zoology*, Evolution*, Physiology*, and Biography*.
1. *Cours sur la génération, l'ovologie et l'embryologie.* Paris, 1836. Ed. by Deschamps.

14504. PANDER, Christian Heinrich von. 1794-1865. Russ. DSB; ICB; P II,III; BMNH; Mort.; BLA; GC. Also Geology* and Zoology.

14505. HUSCHKE, Emil. 1797-1858. Ger. DSB; BMNH; BLA; GC. Also Zoology*, Anatomy, and Physiology.

14506. AMMON, Friedrich August von. 1799-1861. Ger. BMNH; BLA & Supp. Also Anatomy.

14507. POUCHET, Félix Archimède. 1800-1872. Fr. DSB; ICB; BMNH; BLA; GA. Also Natural History*, Zoology*, Microbiology*, and History of Science*.
1. *Théorie positive de l'ovulation spontanée et de la fécondation des mammifères et de l'espèce humaine.* 2 vols. Paris, 1847. RM/L.

14508. MÜLLER, Johannes (Peter). 1801-1858. Ger. DSB; ICB; P II; BMNH; Mort.; BLA & Supp.; GB; GC. Also Zoology, Anatomy, and Physiology*#.

14509. VROLIK, Willem. 1801-1863. Holl. BMNH; BLA; GF. Also Zoology.
1. *Tabulae ad illustrandam embryogenesin hominis et mammalium.* Amsterdam, 1844-49.

14510. BARRY, Martin. 1802-1855. Br. DSB; BLA; GD. Also Histology.

14511. NEWPORT, George. 1803-1854. Br. DSB; BMNH; GD. Also Zoology*.

14512. LEREBOULLET, Dominique Auguste. 1804-1865. Fr. DSB; BMNH; BLA. Also Zoology.

14513. SIEBOLD, Carl Theodor Ernst von. 1804-1885. Ger. DSB; BMNH; BLA; GB; GC. Also Zoology*.
1. *Wahre Parthenogenesis bei Schmetterlingen und Bienen.* Leipzig, 1856. X.
1a. ——— *On a true parthenogenesis in moths and bees.* London, 1857. Trans. by W.S. Dallas.

14514. WAGNER, Rudolph. 1805-1864. Ger. DSB; BMNH; Mort.; BLA; GB; GC. Also Zoology*, Physiology*, and Biography*.
1. *Prodromus historiae generationis hominis atque animalium.* Leipzig, 1836.

14515. BISCHOFF, Theodor Ludwig Wilhelm. 1807-1882. Ger. DSB; BMNH; Mort.; BLA & Supp.; GA; GC. Also Zoology and Physiology.
1. *Entwickelungsgeschichte der Säugethiere und des Menschen.* Leipzig, 1842.
2. *Beweis der von der Begattung unabhängigen periodischen Reifung und Loslösung der Eier der Säugethiere und des Menschen.* Giessen, 1844.

14516. COSTE, (Jean Jacques Marie Cyprien) Victor. 1807-1873. Fr. BMNH; BLA; GA.

14517. THOMSON, Allen. 1809-1884. Br. BMNH; BLA; GD.

14518. VALENTIN, Gabriel Gustav. 1810-1883. Ger./Switz. DSB; ICB; P II,III; BMNH; BLA; GC. Also Anatomy, Physiology*, and Cytology.

14519. REICHERT, Karl Bogislaus. 1811-1883. Ger./Dorpat. DSB; ICB; BMNH; Mort.; BLA; GC. Also Zoology and Anatomy.

14520. ERDL, Michael Pius. 1815-1848. Ger. BMNH; BLA; GA. Also Zoology.

14521. KOCH, Heinrich. 1815-1881. Switz. ICB; BMNH.

14522. REMAK, Robert. 1815-1865. Ger. DSB; ICB; Mort.; BLA & Supp.; GC. Also Histology, Physiology, and Cytology.

14523. KOELLIKER, (Rudolf) Albert von. 1817-1905. Ger. DSB; ICB; BMNH; Mort.; BLA & Supp.; GA; GB. Also Zoology, Anatomy*, Physiology, and Cytology.
1. *Entwicklungsgeschichte des Menschen und der höheren Thiere.* Leipzig, 1861. RM. 2nd ed., 1879.
2. *Grundriss der Entwicklungsgeschichte des Menschen und der höheren Thiere.* Leipzig, 1880. A textbook. X. 2nd ed., 1884.

14524. ERCOLANI, Giovanni Battista. 1819-1883. It. BMNH; BLA.
1. [Trans. of various writings] *The reproductive process.* London, 1881.

14525. VARNEK, Nikolai Aleksandrovich. 1821-1876. Russ. ICB.

14526. THIERSCH, Karl. 1822-1895. Ger. BLA; GC.

14527. BALBIANI, Edouard Gérard. 1823-1899. Fr. DSB; BMNH. Also Cytology.
1. *Cours d'embryogénie comparée ... Leçons sur la génération des vertébrés.* Paris, 1879. Ed. by F. Hanneguy.

14528. RANSOM, William Henry. 1824-1907. Br. BLAF; GD2.

14529. SCHULTZE, Max Johann Sigismund. 1825-1874. Ger. DSB; ICB; P III; BMNH; Mort.; BLA; GB; GC. Also Microscopy, Zoology, Histology, and Cytology.

14530. GEGENBAUER, Karl. 1826-1903. Ger. DSB Supp; ICB; BMNH; Mort.; BLA & Supp.; GB. Also Zoology* and Anatomy.

14531. PFLÜGER, Eduard Friedrich Wilhelm. 1828-1910. Ger. DSB; P II, III; BMNH; Mort.; BLA. Also Physiology*.
1. *Über die Eierstöcke der Säugethiere und des Menschen.* Leipzig, 1863.

14532. KUPFFER, Carl Wilhelm von. 1829-1902. Ger. BMNH; Mort.; BLA. Also Histology.

14533. HIS, Wilhelm (first of the name). 1831-1904. Switz./Ger. DSB; ICB; BMNH; Mort.; BLA & Supp. Also Anatomy*.
1. *Untersuchungen über die erste Anlage des Wirbelthierlebens ... Die erste Entwickelung des Hünchens im Ei.* 2 parts. Leipzig, 1868-73.
2. *Über die Bedeutung Entwickelungsgeschichte für die Auffassung der organischen Natur. Rectoratsrede ... 1869.* Leipzig, 1870.
3. *Unsere Körperform und das physiologische Problem ihrer Entstehung.* Leipzig, 1875. Another ed., 1879.
4. *Anatomie menschlicher Embryonen.* 3 vols & atlas of 2 vols. Leipzig, 1880-85.

14534. LA VALETTE ST. GEORGE, Adolph von. 1831-1910. Ger. BMNH; BLA (under V); BLAF (under L).

14535. HAECKEL, Ernst Heinrich Philipp August. 1834-1919. Ger. DSB; ICB; BMNH; Mort.; BLAF; GB. Also Science in General*, Zoology, Evolution*#, and Personal Writings*.
1. *Zur Entwickelungsgeschichte der Siphonophoren.* Utrecht, 1869.
2. *Anthropogenie; oder, Entwickelungsgeschichte des Menschen. Gemeinverständliche wissenschaftliche Vorträge über die Grundzüge der menschlichen Keimes- und Stammes-Geschichte.* Leipzig, 1874. RM/L.

14536. FOSTER, Michael. 1836-1907. Br. DSB; ICB; BMNH; Mort.; BLAF; GB; GD2. Also Physiology* and Biography*. See also Index.
1. *The elements of embryology.* London, 1874. With F.M. Balfour. 2nd ed., by A. Sedgwick and W. Heape, 1883.

14537. WALDEYER (-HARTZ), (Heinrich) Wilhelm (Gottfried) von. 1836-

1921. Ger. DSB; ICB; BMNH; Mort.; BLAF. Also Anatomy.
1. *Eierstock und Ei. Ein Beitrag zur Anatomie und Entwicklungsgeschichte der Sexualorgane.* Leipzig, 1870.

14538. LANGHANS, Theodor. 1839-1915. Switz. Mort.; BLA.

14539. PACKARD, Alpheus Spring. 1839-1905. U.S.A. DSB; P III,IV; BMNH; GE. Also Zoology*.
1. *Life histories of animals, including man; or, Outlines of comparative embryology.* New York, 1876.

14540. GOETTE, Alexander Wilhelm. 1840-1922. Ger. DSB; BMNH; BLAF. Also Zoology.
1. *Die Entwickelungsgeschichte der Unke (Bombinator igneus) als Grundlage einer vergleichenden Morphologie der Wirbelthiere.* Text & atlas. Leipzig, 1875.
2. *Abhandlungen zur Entwickelungsgeschichte der Thiere.* 5 parts. Hamburg/Leipzig, 1882-90.

14541. KOVALEVSKY, Aleksander Onufrievich. 1840-1901. Russ. DSB; ICB; BMNH.

14542. SCHENK, (Samuel) Leopold. 1840-1902. Aus. BMNH; BLAF. Also Microbiology*.
1. *Lehrbuch der vergleichenden Embryologie der Wirbelthiere.* Vienna, 1874.
2. *Einfluss auf das Geschlechtsverhältniss.* 2nd ed. Magdeburg, 1898. X.
2a. ——— *Schenk's theory. The determination of sex.* London, 1898.

14543. HOFFMANN, Christian Karl. 1841-1903. Holl. BMNH; BLA. Also Zoology.

14544. PLATEAU, Félix Auguste Joseph. 1841-1911. Belg. ICB; P III; BMNH. Also Zoology and Physiology.
1. *Etudes sur la parthénogenèse.* Ghent, 1868.

14545. RAUBER, August Antinous. 1841-1917. Dorpat. BLAF. Also Anatomy.

14546. KLEINENBERG, Nicolaus. 1842-1897. It. DSB; BMNH. Also Zoology.

14547. RETZIUS, (Magnus) Gustaf. 1842-1919. Swed. DSB; BMNH; Mort.; BLAF. Also Zoology and Anatomy.

14548. ROSENBERG, Emil Woldemar. 1842-1925. Dorpat/Holl. BMNH; BLAF. Also Zoology and Anatomy.

14549. BRANDT, Aleksandr Fedorovich. 1844-after 1911. Russ. BMNH.
1. *Über das Ei und seiner Bildungsstätte. Ein vergleichend-morphologischer Versuch mit Zugrundelegung des Insecteneies.* Leipzig, 1878.

14550. DUVAL, Mathias Marie. 1844-1907. Fr. DSB; BMNH; Mort.; BLA & Supp. Also Histology and Physiology.
1. *Atlas d'embryologie.* Paris, 1889.

14551. MIHALKOVICS, Géza Victor von. 1844-1899. Hung. BMNH; BLAF. Also Anatomy.

14552. FOL, Hermann. 1845-1892. Switz. DSB; ICB; BMNH. Also Zoology* and Cytology.

14553. MERKEL, Friedrich Siegismund. 1845-1919. Ger. BMNH; Mort.; BLAF. Also Microscopy* and Anatomy.

14554. METCHNIKOFF, Elie. 1845-1916. Russ./Fr. DSB; ICB (MECHNIKOV); BMNH (MECHNIKOV); BLAF (METSCHNIKOW). Also Zoology and Microbiology.

14555. BENEDEN, Edouard van. 1846-1910. Belg. DSB; ICB; BMNH; Mort.; BLA & Supp. Also Zoology.
1. *Recherches sur la composition et la signification de l'oeuf, basées sur l'étude de son mode de formation et des premiers phénomènes embryonnaires.* Brussels, 1868.
2. *Recherches sur la maturation de l'oeuf, la fécondation et la division cellulaire.* Ghent/Liège, 1883.

14556. FÜRBRINGER, Max. 1846-1920. Ger. BMNH; BLAF. Also Zoology and Anatomy.

14557. GIARD, Alfred Mathieu. 1846-1908. Fr. DSB; ICB; BMNH; BLAF. Also Zoology.

14558. GASSER, Emil. 1847-1919. Ger. BMNH; BLAF. Also Anatomy.

14559. ZALENSKII, Vladimir Vladimirovich. 1847-1918. Russ. BMNH. Also Zoology.

14560. BROOKS, William Keith. 1848-1908. U.S.A. DSB; ICB; BMNH; Mort.; BLAF; GE. Also Zoology*.

14561. BARFURTH, Dietrich. 1849-1927. Ger. BLAF. Also Anatomy.

14562. DRASCH, Otto. 1849-1911. Aus. BLAF. Also Anatomy.

14563. HERTWIG, (Wilhelm August) Oscar. 1849-1922. Ger. DSB; ICB; BMNH; Mort.; BLAF. Also Zoology* and Cytology*.
1. *Der Organismus der Medusen und seine Stellung zur Keimblättertheorie.* Jena, 1878. With C.W.T.R. Hertwig.
2. *Studien zur Blättertheorie.* 5 parts. Jena, 1879-83. With C.W.T.R. Hertwig.
3. *Lehrbuch der Entwicklungsgeschichte des Menschen und der Wirbelthiere.* Jena, 1888. 3rd ed., 1890.
3a. ——— *Traité d'embryologie; ou, Histoire du développement de l'homme et des vertébrés.* Paris, 1891. Trans. from the 3rd German ed. by C. Julien.
3b. ——— *Textbook of the embryology of man and mammals.* London, 1892. Trans. from the 3rd German ed. by E.L. Mark.
4. *Zeit- und Streitfragen der Biologie.* 2 vols. Jena, 1894-99.
4a. ——— *The biological problem of to-day: preformation or epigenesis? The basis of a theory of organic development.* London, 1896.

14564. STÖHR, Philipp Adrian. 1849-1911. Ger. BMNH; BLAF. Also Anatomy*.
1. *Zur Entwicklungsgeschichte der Kopfskeletes der Teleostier.* Leipzig, 1882.

14565. USKOW, Nikolai Vasilevich. b. 1849. Russ. BMNH.

14566. VEJDOVSKÝ, František. 1849-1939. Prague. DSB Supp.; ICB; BMNH. Also Zoology and Cytology.

1. *Entwicklungsgeschichtliche Untersuchungen*. 4 parts. Prague, 1888-92.

14567. HERTWIG, (Karl Wilhelm Theodor) Richard von. 1850-1937. Ger. DSB; ICB; BMNH; Mort.; BLAF. Also Zoology* and Cytology. See also Index.

14568. ROUX, Wilhelm. 1850-1924. Ger. DSB; ICB; Mort.; BLAF.
1. *Der Kampf der Theile im Organismus. Ein Beitrag zur Vervollstandigung der mechanischen Zweckmässigkeitslehre*. Leipzig, 1881.
2. *Gesammelte Abhandlungen über Entwickelungsmechanik der Organismen*. 2 vols. Leipzig, 1895.

14569. SEESSEL, Albert. 1850-1910. U.S.A./Ger. BLAF.

14570. BALFOUR, Francis Maitland. 1851-1882. Br. DSB; BMNH; Mort.; BLA & Supp.; GB; GD.
1. *A treatise of comparative embryology*. 2 vols. London, 1880-81. RM/L.
2. *Studies from the Morphological Laboratory in the University of Cambridge*. 2 parts. London, 1880-82.
3. *Works*. 4 vols. London, 1885. Ed. by M. Foster and A. Sedgwick.
See also 14536/1.

14571. BONNETT, Robert. 1851-1921. Ger. BMNH (BONNET); BLAF.

14572. BORN, Gustav Jacob. 1851-1900. Ger. BLAF.

14573. MARSHALL, Arthur Milnes. 1852-1893. Br. BMNH; GD1. Also Zoology* and Evolution*.
1. *Vertebrate embryology*. 1893.

14574. MINOT, Charles Sedgwick. 1852-1914. U.S.A. DSB; BMNH; Mort.; BLAF; GE. Also Anatomy.
1. *Human embryology*. New York, 1892. Another ed., 1897.

14575. REPYAKHOV, Vasilii Mikhailovich. b. 1852. Russ. BMNH.

14576. STRASSER, Hans. 1852-1927. Switz. BMNH; BLAF. Also Anatomy and Physiology.

14577. FRASER, Alexander. 1853-1909. Br. BLAF. Also Anatomy.

14578. HUBRECHT, (Ambrosius Arnold) Willem. 1853-1915. Holl. DSB; BMNH; BLAF. Also Zoology and Evolution*.

14579. RABL, Carl. 1853-1917. Prague/Ger. DSB; BMNH; BLAF. Also Anatomy and Cytology.

14580. GROBBEN, Karl. 1854-after 1910. Aus. BMNH.

14580A CHABRY, Laurent. 1855-1893. Fr. DSB; ICB. Also Physiology.

14581. HADDON, Alfred Cort. 1855-1940. Br. ICB; BMNH; GD5. Also Zoology.
1. *An introduction to the study of embryology*. London, 1887.

14582. SPEE, Ferdinand von. 1855-1937. Ger. BLAF. Also Anatomy.

14583. HEIDER, Karl. 1856-1935. Aus. BMNH; Mort. See also 14590/1.

14584. SOMMER, Alfred Richard. b. 1856. Ger./Russ. BLAF. Also Anatomy.

14585. WILSON, Edmund Beecher. 1856-1939. U.S.A. DSB; ICB; BMNH; Mort. Also Cytology*.
1. *Atlas of the fertilization ... of the ovum.* New York, 1895. With E. Leaming.

14586. FUSARI, Romeo. 1857-1919. It. BLAF. Also Anatomy.

14587. MITROFANOV, Pavel Ilich. 1857-1920. Warsaw. BMNH; BLAF (MITROPHANOW). Also Zoology.

14588. STRAHL, Hans. 1857-1920. Ger. BMNH; BLAF.

14589. BLOCHMANN, Friedrich. 1858-after 1924. Ger. BMNH. Also Zoology and Microbiology.

14590. KORSCHELT, Eugen. 1858-1946. Ger. BMNH; Mort.; BLAF.
1. *Lehrbuch der vergleichenden Entwicklungsgeschichte der wirbellosen Thiere. Specieller Theil.* 3 vols in 2. Jena, 1890-93. With C. Heider.
1a. ——— *Allgemeiner Theil.* Several vols. Jena, 1902-.
1b. ——— *Text-book of the embryology of invertebrates.* 4 vols. London, 1895-1900. Trans. by E.L. Mark. Rev. and ed. by M.F. Woodward.

14591. CHIARUGI, Giulio. 1859-1944. It. DSB; ICB; BLAF. Also Anatomy.

14592. KERSCHNER, Ludwig. 1859-1911. Aus. BLAF.

14593. LOEB, Jacques. 1859-1924. Ger./U.S.A. DSB; ICB; Mort.; BLAF; GE. Also Physiology*.

14594. MacMURRICH, James Playfair. 1859-1939. Canada. BMNH; GD5.

14595. PETERS, Hubert. 1859-1934. Aus. Mort.; BLAF.

14596. SCHULTZE, Oskar Maximilian Sigismund. 1859-1920. Ger. BMNH; BLAF. Also Anatomy.

3.18 CELL THEORY AND CYTOLOGY

14597. SCHLEIDEN, Matthias Jacob. 1804-1881. Ger. DSB; ICB; BMNH; Mort.; BLA & Supp.; GB; GC. Also Science in General* and Botany*. See also Index.

14598. SCHWANN, Theodor Ambrose Hubert. 1810-1882. Ger./Belg. DSB; ICB; P II,III; BMNH; Mort.; BLA & Supp.; GB; GC. Also Physiology and Microbiology.

1. *Mikroskopische Untersuchungen über die Übereinstimmung in der Struktur und dem Wachsthume der Thiere und Pflanzen.* Berlin, 1839. Reprinted, Leipzig, 1910. (OKEW. 176)

1a. ——— *Microscopical researches into the accordance in the structure and growth of animals and plants.* [To which is added] *Contributions to phytogenesis.* London, Sydenham Society, 1847. Trans. by H. Smith. RM/L.

2. *Lettres.* Liège, 1960. Ed. with notes by M. Florkin.

14599. VALENTIN, Gabriel Gustav. 1810-1883. Ger./Switz. DSB; ICB; P II,III; BMNH; BLA; GC. Also Anatomy, Embryology, and Physiology*.

14600. REMAK, Robert. 1815-1865. Ger. DSB; ICB; Mort.; BLA & Supp.; GC. Also Histology, Embryology, and Physiology.

14601. KOELLIKER, (Rudolf) Albert von. 1817-1905. Ger. DSB; ICB; BMNH; Mort.; BLA & Supp.; GA; GB. Also Zoology, Histology*, Embryology*, and Physiology.

14602. VIRCHOW, Rudolf Ludwig Karl. 1821-1902. Ger. DSB; ICB; BMNH; Mort.; BLA & Supp.; GB. Also Science in General* and Biography*.

1. *Gesammelte Abhandlungen zur wissenschaftlichen Medicin.* Frankfurt, 1856.

2. *Die Cellularpathologie in ihrer Begründung auf physiologische und pathologische Gewebelehre.* Berlin, 1858. X. 3rd ed., 1862.

2a. ——— *Cellular pathology as based upon physiological and pathological histology.* London, 1860. Trans. from the 2nd German ed., with notes, by F. Chance. Reprinted, New York, 1971.

3. [Trans.] *Disease, life, and man. Selected essays.* Stanford, 1958. Trans. with an introd. by L.J. Rather.

14603. BALBIANI, Edouard Gérard. 1823-1899. Fr. DSB; BMNH. Also Embryology*.

14604. SCHULTZE, Max Johann Sigismund. 1825-1874. Ger. DSB; ICB; P III; BMNH; Mort.; BLA; GB; GC. Also Microscopy, Zoology, Histology, and Embryology.

14605. BURNETT, Waldo Irving. 1827-1854. U.S.A. ICB; BLA. Also Histology.

14606. FROMMANN, Carl Friedrich Wilhelm. 1831-1892. Ger. BMNH; BLA & Supp.; GC. Also Anatomy.

14607. SCHNEIDER, Anton Friedrich. 1831-1890. Ger. DSB; BMNH. Also Zoology*.

14608. SCHWEIGGER-SEIDEL, Franz. 1834-1871. Ger. Mort.; BLA. Also Histology.

14609. MONTGOMERY, Edmund Duncan. 1835-1911. U.S.A. DSB; ICB; GE.

14610. CARNOY, Jean Baptiste. 1836-1899. Belg. BMNH; BLA & Supp.

14611. FLEMMING, Walther. 1843-1905. Ger. DSB; Mort.; BLAF. Also Anatomy.

14612. GOLGI, Camillo. 1843-1926. It. DSB; ICB; Mort.; BLA & Supp. Also Histology.

14613. STRASBURGER, Eduard Adolf. 1844-1912. Ger. DSB; ICB; BMNH; Mort. Also Botany*.
1. *Studien über Protoplasma*. Jena, 1876.
2. *Zellbildung und Zelltheilung*. Jena, 1876. X. 3rd ed., 1880.
2a. ——— *Etudes sur la formation et la division des cellules*. 1876.
3. *Über Befruchtung und Zelltheilung*. Jena, 1878.
4. *Über den Theilungsvorgang der Zellkerne und das Verhältniss der Kerntheilung zur Zelltheilung*. Bonn, 1882.
5. *Die Controversen der indirecten Kerntheilung*. Bonn, 1884.
6. *Histologische Beiträge*. 7 vols. Jena, 1888-1909.

14614. FOL, Hermann. 1845-1892. Switz. DSB; ICB; BMNH. Also Zoology* and Embryology.

14615. BÜTSCHLI, (Johann Adam) Otto. 1848-1920. Ger. DSB; ICB; P IV; BMNH; BLAF. Also Zoology.
1. *Untersuchungen über microskopische Schäume und die Struktur des Protoplasmas*. Leipzig, 1892. X.
1a. ——— *Investigations on microscopic foams and on protoplasm*. London, 1894. Trans. by E.A. Minchin.
2. *Vorläufiger Bericht über fortgesetzte Untersuchungen an Gerinnungsschäumen, Sphärokrystallen und die Struktur von Cellulose- und Chitenmembranen*. Heidelberg, 1894.

14616. HERTWIG, (Wilhelm August) Oscar. 1849-1922. Ger. DSB; ICB; BMNH; Mort.; BLAF. Also Zoology* and Embryology*.
1. *Untersuchungen zur Morphologie und Physiologie der Zelle*. 6 parts. Jena, 1884-90. With C.W.T.R. Hertwig.
2. *Die Zelle und die Gewebe. Grundzüge der allgemeinen Anatomie und Physiologie*. 2 vols. Jena, 1893-98.
2a. ——— *The cell. Outlines of general anatomy and physiology*. London, 1895. Trans. by M. Campbell.

14617. SOLGER, Bernhard Friedrich. 1849-after 1902. Ger. BMNH; BLAF. Also Anatomy.

14618. VEJDOVSKÝ, František. 1849-1939. Prague. DSB Supp.; ICB; BMNH. Also Zoology and Embryology*.

14619. HENNEGUY, Louis Félix. 1850-1928. Fr. ICB; BMNH.
1. *Leçons sur la cellule. Morphologie et reproduction*. Paris, 1896.

14620. HERTWIG, (Karl Wilhelm Theodor) Richard von. 1850-1937. Ger. DSB; ICB; BMNH; Mort.; BLAF. Also Zoology* and Embryology. See also Index.

14621. MAGINI, Giuseppe. 1851-1916. It. BLAF.

14622. ALTMANN, Richard. 1852-1900. Ger. P IV; Mort.; BLAF. Also Anatomy and Physiology.
1. *Die Elementarorganismen und ihre Beziehungen zu den Zellen.* Leipzig, 1890. RM/L.

14623. PFITZNER, Wilhelm. 1853-1903. Ger. BLAF. Also Anatomy.

14624. RABL, Carl. 1853-1917. Prague/Ger. DSB; BMNH; BLAF. Also Anatomy and Embryology.

14625. GEDDES, Patrick. 1854-1932. Br. ICB; BMNH; GD5. Also Botany* and Evolution*.
1. *A re-statement of the cell theory.* Edinburgh, 1884.

14626. WILSON, Edmund Beecher. 1856-1939. U.S.A. DSB; ICB; BMNH; Mort. Also Embryology*.
1. *The cell in development and inheritance.* New York, 1897. RM/L. Reprinted, ib., 1966.

14627. BENDA, Carl. 1857-1932. Ger. BLAF. Also Histology.

14628. FISCHER, Alfred. 1858-1913. BMNH. Also Botany and Microbiology.
1. *Fixirung, Färbung und Bau des Protoplasmas. Kritische Untersuchungen über Technik und Theorie in der neueren Zellforschung.* Jena, 1899.

14629. HENKING, Hermann. 1858-1942. Ger. DSB; BMNH. Also Zoology.

3.19 PHYSIOLOGY

Human and comparative.

Plant physiology is included in Botany.

14630. SCHURIG, Martin. 1656-1733. Ger. BLA. Also Embryology.
1. *Sialogia historico-medica; hoc est, Salivae humanae consideratio physico-medico-forensis.* Dresden, 1723.

14631. HOFFMANN, Friedrich. 1660-1742. Ger. DSB; ICB; P I; BMNH; Mort.; BLA; GA; GB; GC. Also Chemistry*# and Part 2.

14632. STAHL, Georg Ernst. 1660-1734. Ger. DSB; ICB; P II; Mort.; BLA & Supp.; GA; GB; GC. Also Chemistry* and Part 2. See also 15505/1.
1. *Theoria medica vera, physiologiam et pathologiam ... sistens.* Halle, 1708. X. Another ed., with a biography, by L. Choulant, 3 vols, Leipzig, 1831-33.
2. *Über den mannigfaltigen Einfluss von Gemütsbewegungen auf den menschlischen Körper. (Halle, 1695) Über die Bedeutung des synergischen Prinzips für die Heilkunde. (Halle, 1695) Über den Unterschied zwischen Organismus und Mechanismus. (1714) Überlegungen zum ärztlichen Hausbesuch. (Halle, 1703).* Leipzig, 1961. Trans. from the Latin with introd. and notes by B.J. Gottlied.

14633. POURFOUR DU PETIT, François. 1664-1741. Fr. DSB; ICB; BMNH (DU PETIT); Mort. Under PETIT: P II; BLA; GA. Also Anatomy.

14634. ARBUTHNOT, John. 1667-1735. Br. DSB; ICB; P I; BMNH; BLA & Supp.; GA; GB; GD. Also Various Fields, Mathematics, and Part 2*.
1. *An essay concerning the effects of air on human bodies.* London, 1733. RM/L.

14635. TABOR, John. b. 1667. Br. DSB.

14636. BOERHAAVE, Hermann. 1668-1738. Holl. DSB; ICB; P I; BMNH; Mort.; BLA & Supp.; GA; GB. Also Chemistry*# and Botany*. See also Index.

14637. BROWNE, Joseph. fl. 1706-1721. Br. BLA & Supp.; GA; GD.

14638. KEILL, James. 1673-1719. Br. DSB; P I; BLA; GA; GD. Also Anatomy*. See also 2573/1c,1d.

14639. MEAD, Richard. 1673-1754. Br. ICB; P II; BMNH; BLA; GA; GD.
1. *A mechanical account of poisons.* London, 1702. X. 2nd ed., 1708.

14640. MICHELOTTI, Pietro Antonio. 1673-1740. It. BLA.
1. *De separatione fluidorum in corpore animali. Dissertatio physico-mechanico-medica.* Venice, 1721. RM.

14641. STUART, Alexander. 1673-1742. Br. DSB; BLA.

14642. HALES, Stephen. 1677-1761. Br. DSB; ICB; P I; BMNH; Mort.; BLA & Supp.; GA; GB; GD. Also Chemistry, Geology*, and Botany*.
1. *Statical essays*. 2 vols. London, 1733. Another ed., 1769.
1a. ——— Vol. 1. See 11349/1b.
1b. ——— Vol. 2. *Statical essays, containing Haemastaticks; or, An account of some hydraulick and hydrostatical experiments made on the blood and blood-vessels of animals*. London, 1733. RM/L. Reprinted, New York, 1964.

14643. JUNCKER, Johann. 1679-1759. Ger. DSB; P I; BLA (JUNKER); GA. Also Chemistry*.
1. *Conspectus physiologiae medicae et hygieines, in forma tabularum, repraesentatus et ad dogmata Stahliana potissimum adornatus*. Halle, 1735.

14644. ROBINSON, Bryan. 1680-1754. Br. GD. Also Mechanics*.
1. *A treatise of the animal oeconomy*. Dublin, 1732. X. 2nd ed., with additions, 1734. RM/L.
1a. ——— 3rd ed. *A continuation of a treatise of....* 1737.

14645. JURIN, James. 1684-1750. Br. ICB; P I; BLA; GA; GD. Also Physics.

14646. THEBESIUS, Adam Christian. 1686-1732. Holl. ICB; Mort.; BLA.

14647. GORTER, Johannes de. 1689-1762. Holl. ICB; BLA & Supp.; GA.
1. *De perspiratione insensibile*. Leiden, 1725. X. 2nd ed., rev., 1736.

14648. MORGAN, Thomas. d. 1743. Br. GB; GD.
1. *Mechanical practice of physick in which the specifick method is examin'd and exploded, and the Bellinian hypothesis of animal secretion and muscular motion consider'd and refuted*. London, 1735.

14649. PORTERFIELD, William. d. 1771. Br. P II; BLA & Supp.

14650. PEMBERTON, Henry. 1694-1771. Br. DSB; ICB; P II; BLA; GA; GD. Also Mechanics* and Chemistry*.

14651. MALOUIN, Charles. 1695-1718. Fr. P II; BLA.

14652. HAMBERGER, Georg Erhard. 1697-1755. Ger. ICB; P I; BMNH; Mort.; BLA; GA; GC. Also Physics*.

14653. FLEMYNG, Malcolm. d. 1764. Br. BLA & Supp.; GD.

14654. BERNOULLI, Daniel. 1700-1782. Switz./Russ. DSB; ICB; P I; BLA; GA; GB; GC. Also Mathematics* and Mechanics*.

14655. LE CAT, Claude Nicolas. 1700-1768. Fr. DSB; ICB; P I; BMNH; BLA & Supp.; GA. Also Anatomy.
1. *Traité des sens en particulier*. Rouen, 1740.
1a. ——— *A physical essay on the senses*. London, 1750. RM.
2. *Traité de la couleur de la peau humaine en général, de celle des nègres en particulier, et de la métamorphose de l'une de ces couleurs en l'autre, soit de la naissance, soit accidentellement*. Amsterdam, 1765. RM.
3. *Oeuvres physiologiques*. 3 vols. Paris, 1767-68. RM.

14656. MARTINE, George. 1702-1741. Br. ICB; BMNH; BLA; GA; GD.
Also Physics*.
1. *Essays, medical and philosophical.* London, 1740.

14657. LIEUTAUD, Joseph. 1703-1780. Fr. DSB; ICB; Mort.; BLA.
Also Anatomy.
1. *Elementa physiologiae, juxta solertiora novissimaque physicorum experimenta et accuratiores anatomicorum observationes concinnata.* Amsterdam, 1749.

14658. BELCHIER, John. 1706-1785. Br. Mort.; BLA; GD.

14659. HALLER, Albrecht von. 1708-1777. Switz./Ger. DSB; ICB; P I; BMNH; Mort.; BLA & Supp.; GA; GB; GC. Also Botany*, Anatomy*, and Personal Writings*.
1. *Primae lineae physiologiae.* Göttingen, 1747. X. Reprinted, Hildesheim, 1974. Another ed., 1751.
1a. ——— *Elémens de physiologie.* Paris, 1752.
1b. ——— *First lines of physiology.* Edinburgh, 1779. Ed. by W. Cullen. Another ed., with notes and illustrations, 1786; reprinted, New York, 1966.
2. [Trans.] *A dissertation on the sensible and irritable parts of animals.* London, 1755. Trans. from a French version of the Latin original--a periodical article of 1753. X. Reprinted, Baltimore, 1936.
3. [Trans.] *Deux mémoires sur le mouvement du sang et sur les effets de la saignée, fondés sur les expériences faites sur les animaux.* Lausanne, 1756.
4. *Elementa physiologiae corporis humani.* 8 vols. Lausanne, 1757-66.
5. *Opera minora.* 3 vols. Lausanne, 1763-68.

Correspondence

6. *A. von Haller, G. Morgagni: Briefwechsel, 1745-1768.* Bern, 1964. Ed. by E. Hintzsche.
7. *A. von Haller, I. Somis: Briefwechsel, 1754-1777.* Bern, 1965. Ed. by E. Hintzsche.
8. *A. von Haller, M.A. Caldani: Briefwechsel, 1756-1776.* Bern, 1966. Ed. by E. Hintzsche.
9. *Zwanzig Briefe A. von Hallers an J. Gessner.* Bern, 1972. Ed. by U. Boschung.

14660. LA METTRIE, Julien Offray de. 1709-1751. Fr. DSB; ICB; Mort.; BLA; GA; GB; GC. Also Science in General*#.

14661. CULLEN, William. 1710-1790. Br. DSB; ICB; P I; Mort.; BLA & Supp.; GA; GB; GD. Also Chemistry. See also Index.
1. *Institutions of medicine. Part I. Physiology.* Edinburgh, 1772. X. 2nd ed., London, 1777. 3rd ed., Edinburgh, 1785.
2. *Works.* 2 vols. Edinburgh, 1827. Ed. by J. Thomson.

14662. HOFMANN, Franz. 1711-1773. Prague. P I.

14663. WHYTT, Robert. 1714-1766. Br. DSB; ICB; Mort.; BLA; GD.
1. *Physiological essays, containing: I. An inquiry into the causes which promote the circulation of the fluids in the very small vessels of animals. II. Observations on the sensibility and irritability of the parts of men and other animals.* Edinburgh, 1755.

14664. KESSLER, Carl Gottlob. 1715-1753. Ger. BLA (KESLER).

14665. TANDON, Antoine. 1717-1806. Fr. BLA. Also Anatomy.

14666. PENROSE, Francis. 1718-1798. Br. GD. Also Physics*.
1. *A physical essay on the animal economy, wherein the circulation of the blood and its causes are particularly considered.* London, 1754.

14667. PETIT, Antoine. 1718-1794. Fr. ICB; BLA; GA. Also Anatomy.

14668. BORDEU, Théophile de. 1722-1776. Fr. DSB; ICB; Mort.; BLA & Supp.; GA. Also Anatomy.

14669. CAMPER, Peter. 1722-1789. Holl. DSB; ICB; P I; BMNH; Mort.; BLA & Supp.; GA; GB. Also Zoology*# and Anatomy.
1. *Dissertatio optica de visu.* Leiden, 1746. X.
1a. ——— *Optical dissertation on vision.* Nieuwkoop, 1962. Facsimile reprint with trans. and introd. by G. ten Doesschate. (Dutch classics on history of science. 3)

14670. PROTASOV, Aleksei P. 1724-1796. Russ. ICB; BLA.

14671. CALDANI, (Leopoldo) Marco Antonio. 1725-1813. It. DSB; ICB; P I; BMNH; BLA & Supp.; GA; GB. Also Anatomy. See also 14659/8.

14672. WALSH, John. 1725?-1795. Br./India. P II; Mort.; GD.

14673. LORRY, Anne Charles de. 1726-1783. Fr. DSB; ICB; BLA & Supp.; GA.

14674. UNZER, Johann August. 1727-1799. Ger. DSB; P II; BMNH; Mort.; BLA; GA; GC.
1. *Erste Gründe einer Physiologie der eigentlichen thierischen Natur thierischer Körper.* Leipzig, 1771. X.
1a. ——— *Principles of physiology.* London, Sydenham Society, 1851. Trans. and ed. by T. Laycock.

14675. BORDENAVE, Toussaint. 1728-1782. Fr. BLA; GA.

14676. ZIMMERMANN, Johann Georg. 1728-1795. Switz./Ger. ICB; BLA; GA; GB; GC.

14677. SPALLANZANI, Lazzaro. 1729-1799. It. DSB; ICB; P II; BMNH; Mort.; BLA; GA; GB. Other entries: see 3167.
1. *De' fenomeni della circolazione.* Modena, 1773. X.
1a. ——— *Expériences sur la circulation.* Paris, 1800. Trans. by J. Tourdes.
2. *Opuscoli di fisica animale e vegetabile.* 2 vols. Modena, 1776. RM.
2a. ——— *Opuscules de physique, animale et végétale.* 2 vols. Geneva, 1777. Trans. with an introd. by J. Senebier. X. Another ed., Pavia/Paris, 1787.
2b. ——— *Tracts on the nature of animals and vegetables.* 2 vols. Edinburgh, 1799. Trans. by J.G. Dalyell. RM.
2c. ——— ——— 2nd ed. *Tracts on the natural history of ....* 2 vols. Ib., 1803.
3. *Dissertazioni di fisica animale e vegetabile.* 2 vols. Modena, 1780. X.

3a. ——— *Dissertations relative to the natural history of animals and vegetables.* 2 vols. London, 1784-89. Trans. by T. Beddoes. RM.

Posthumous Works

4. [Trans.] *Mémoires sur la respiration.* Geneva, 1803. Trans. from the unpublished MS by J. Senebier.
5. [Trans.] *Rapports de l'air atmosphérique avec les êtres organisés.* 3 vols. Geneva, 1807. Extracted from his notebooks and trans. by J. Senebier.
6. *Opere.* 6 vols. Milan, 1825-26.

6a. ——— Reprint of Vol. 1. *Le opere. Volume primo. Circolazione, digestione, respirazione animale.* Milan, 1932.

7. *Epistolario.* Florence, 1958-64. Ed. by B. Biagi.

14678. FONTANA, Felice Gaspar Ferdinand. 1730-1805. It. DSB; ICB; P I; BMNH; Mort.; BLA & Supp.; GA. Also Chemistry* and Anatomy.

1. [Trans.] *Traité sur le vénin de la vipère ... et sur quelques autres poisons.* 2 vols. Florence, 1781. Trans. from the Italian MS by J. Gibelin. RM.

14679. JOHNSTONE, James. 1730-1802. Br. Mort.; BLA; GD.

14680. MENURET DE CHAMBAUD, Jean Jacques. 1733-1815. Fr. DSB; BLA; GA.

14681. BARTHEZ, Paul Joseph. 1734-1806. Fr. DSB; ICB; BLA & Supp.; GA; GB.

1. *Nouvelle mécanique des mouvements de l'homme et des animaux.* Carcassonne, 1798. RM.

14682. VANDELLI, Domenico. 1735-1816. It./Port. ICB; BMNH; BLA. Also Natural History.

14683. COTUGNO, Domenico Felice Antonio. 1736-1822. It. DSB; ICB; P I; Mort.; BLA & Supp.; GA. Also Anatomy*.

14684. FORDYCE, George. 1736-1802. Br. ICB; P I; BMNH; BLA & Supp.; GA; GD. Also Various Fields.

1. *A treatise on the digestion of food.* London, 1791.

14685. GALVANI, Luigi. 1737-1798. It. DSB; ICB; P I; Mort.; BLA & Supp.; GA; GB. Also Anatomy.

1. *De viribus electricitatis in motu musculari commentarius.* Bologna, 1791. RM.

1a. ——— Another ed. Cum J. Aldini dissertatione et notis. Modena, 1792. X. Reprinted, Berlin, 1925.

1b. ——— German trans. Leipzig, 1894. (OKEW. 52)

1c. ——— *Commentary on the effect of electricity on muscular motion.* Cambridge, Mass., 1953. Trans. by R.M. Green.

1d. ——— *Commentary on the effects of electricity on muscular motion.* Norwalk, Conn., 1953. Trans. by M.G. Foley, with introd. and notes by I.B. Cohen.

2. *Memorie sulla elettricita animale.* Bologna, 1797. RM.
3. *Opere edite ed inedite.* Bologna, 1841. RM. *Aggiunta*, ib., 1842. RM.

4. *Memorie ed experimenti inediti.* Bologna, 1937.
5. *Opere scelte.* Turin, 1967. Ed. by G. Barbensi.

14686. MARHERR, Philipp Ambrosius. 1738-1771. Prague. P II; BLA.

14687. PLENK, Joseph Jacob von. 1738-1807. Aus./Hung. P II. Under PLENCK: BMNH; BLA; GA. Also Chemistry and Botany.
1. *Hydrologia corporis humani; sive, Doctrina chemico-physiologica de humoribus.* Vienna, 1794. X.
1a. ——— *The hygrology or chemico-physiological doctrine of the fluids of the human body.* London, 1797. Trans. by R. Hooper.

14688. ARALDI, Michele. 1740-1813. It. P I; BLA. Also Physics.

14689. SENFT, Adam Andreas. 1740-1795. Ger. P II; BLA.

14690. LAVOISIER, Antoine Laurent. 1743-1794. Fr. DSB; ICB; P I; Mort.; GA; GB. Also Chemistry*#. See also Index.

14691. SAUNDERS, William. 1743-1817. Br. ICB; P II; BLA; GD.
1. *A treatise on the structure, economy and diseases of the liver, together with an inquiry into the properties and component parts of the bile and biliary concretions.* London, 1793. X. 3rd ed., 1803.

14692. CRUIKSHANK, William Cumberland. 1745-1800. Br. DSB; P I (CRUICKSHANKS); Mort.; BLA & Supp.; GA; GD. Also Anatomy and Embryology.

14693. RIGBY, Edward. 1747-1821. Br. P II; BLA; GA; GD.
1. *An essay on the theory of the production of animal heat.* London, 1785.

14694. ANDRIA, Nicola. 1748-1814. It. P I; BLA; GA. Also Chemistry.
1. *Osservazioni generali sulla theoria della vita.* Naples? 1805. X.
1a. ——— *Observations générales sur la théorie de la vie; ou, Appendix des leçons de physiologie.* Paris, 1805. RM.

14695. CRAWFORD, Adair. 1748-1795. Br. DSB Supp.; ICB; P I; Mort.; BLA & Supp.; GA; GD. Also Chemistry*.

14696. MIHELIČ, Anton. 1748-1818. ICB.

14697. PROCHÁSKA, Georgius. 1749-1820. Prague/Aus. DSB; ICB; P II; Mort.; BLA; GC. Also Anatomy and Embryology.
1. *Lehrsätze aus der Physiologie des Menschen.* Vienna, 1797. X.
1a. ——— *Principles of human physiology.* Prague, 1971. Trans. with notes by K. Resler.

14698. YPEY, Adolph. 1749-1820. Holl. BLA; GF.

14699. ADAMS, George (second of the name). 1750-1795. Br. P I; BMNH; BLA; GA; GD. Other entries: see 3217.
1. *An essay on vision, briefly explaining the fabric of the eye and the nature of vision.* London, 1789. RM.

14700. BLUMENBACH, Johann Friedrich. 1752-1840. Ger. DSB; ICB; P I; BMNH; Mort.; BLA & Supp.; GA; GB; GC. Also Natural History*, Zoology*, and Embryology*.

1. *Institutiones physiologicae*. Göttingen, 1787. X.
1a. ——— *Institutions of physiology*. 2nd ed. London, 1817. Trans. by J. Elliotson from the 3rd Latin ed.
1b. ——— ——— 4th ed. *Elements of physiology*. 1828.

14701. HAIGHTON, John. 1755-1823. Br. BLA; GD.

14702. STEVENS, Edward. ca. 1755-1834. Br./U.S.A. DSB; ICB; Mort.

14703. VALLI, Eusebio. 1755-1816. It. ICB; BLA.

14704. GALLINI, Stefano. 1756-1828 (or 1836). It. P I; BLA.

14705. GOODWYN, Edmund. 1756-1829. Br. ICB; BLA; GD.

14706. WELLS, William Charles. 1757-1817. Br./U.S.A. DSB; ICB; P II; BMNH; BLA; GD; GE. Other entries: see 3243.
1. *An essay upon single vision with two eyes*. London, 1792. RM. Another ed., 1818: see 3243/1.

14707. BARCLAY, John. 1758-1826. Br. DSB; BMNH; BLA & Supp.; GD. Also Anatomy*.
1. *Dissertatio inauguralis de anima seu principio vitali*. Edinburgh, 1796. RM.
2. *An enquiry into the opinions, ancient and modern, concerning life and organization*. Edinburgh, 1822. RM/L.

14708. GOLDWITZ, Sebastian. b. 1758. Ger. ICB; BLA & Supp.

14709. REIL, (Johann) Christian. 1759-1813. Ger. DSB; ICB; P II; BMNH; Mort.; BLA; GA; GC. Also Anatomy.

14710. PERROLLE, Etienne. ca. 1760-1838. Fr. P II; BLA (PÉROLLE). Also Physics.

14711. HELM, Jacob. 1761-1831. ICB.

14712. BRANDIS, Joachim Dietrich. 1762-1845. Ger./Den. P I; BMNH; BLA & Supp.; GA.

14713. DEMOURS, Antoine Pierre. 1762-1836. Fr. P I; BLA & Supp.; GA.

14714. HUFELAND, Christoph Wilhelm. 1762-1836. Ger. ICB; BLA & Supp.; GA; GB; GC.
1. *Die Kunst, das menschliche Leben zu verlängern*. Jena, 1797. (Expounds his theory of the life-force) X. 2nd ed., 1798.

14715. ARNEMANN, Justus. 1763-1806. Ger. BMNH; BLA & Supp.; GA; GC.

14716. ABERNETHY, John. 1764-1831. Br. P I; BMNH; BLA & Supp.; GA; GB; GD. Also Anatomy*.
1. *Physiological lectures, exhibiting a general view of Mr Hunter's physiology, and of his researches in comparative anatomy*. London, 1817. RM.

14717. DUMAS, Charles Louis. 1765-1813. Fr. BLA & Supp.; GA.

14718. KIELMEYER, Karl Friedrich von. 1765-1844. Ger. DSB; ICB; P I; BMNH; BLA; GA; GC. Other entries: see 3260.
1. *Über die Verhältnisse der organischen Kräfte unter einander in der Reihe der verschiedenen Organisationen, die Gesetze und Folgen dieser Verhältnisse*. Stuttgart, 1793. RM.

14719. GARNETT, Thomas. 1766-1802. Br. DSB; P I; BMNH; BLA; GA; GD.

Also Chemistry.
1. *Popular lectures on zoonomia; or, The laws of animal life in health and disease.* London, 1804. RM.

14720. WOLLASTON, William Hyde. 1766-1828. Br. DSB; ICB; P II; BLA; GA; GB; GD. Also Physics, Chemistry, and Crystallography.

14721. SÉGUIN, Armand. 1767-1835. Fr. DSB; ICB; P II; Mort.; BLA. Also Chemistry.

14722. CRÈVE, Johann Caspar Ignaz Anton. 1769-1853. Ger. P I; BLA.

14723. HUMBOLDT, Alexander von. 1769-1859. Ger./Fr. DSB; ICB; P I, III; BMNH; Mort.; G. Other entries: see 3268.
1. *Versuche über die gereizte Muskel- und Nerven-Faser, nebst Vermuthungen über den chemischen Process des Lebens in der Thier- und Pflanzen-Welt.* 2 vols. Posen/Berlin, 1797.
1a. ——— *Expériences sur le galvanisme, et en général sur l'irritation des fibres musculaires et nerveuses.* Paris, 1799. Trans. with additions by J.F.N. Jadelot.

14724. PROST, P.A. d. 1832. Fr. BLA.
1. *Essai physiologique sur la sensibilité.* Paris, 1805.

14725. DÖLLINGER, (Johann) Ignaz (Josef). 1770-1841. Ger. DSB; ICB; P I; BLA & Supp.; GC. Also Embryology.

14726. HALDAT DU LYS, Charles Nicolas Alexandre de. 1770-1852. Fr. P I; BLA; GA. Also Physics.

14727. LEGALLOIS, Julien Jean César. 1770-1814. Fr. DSB; P I; Mort.; BLA; GA.

14728. PHILIP, Alexander Philips Wilson. 1770-1851? Br. DSB; ICB; BLA; GD.
1. *An experimental inquiry into the laws of the vital functions.* London, 1817. X. 2nd ed., 1818. 3rd ed., 1826 and 1832.

14729. BICHAT, Xavier. 1771-1802. Fr. DSB; ICB; Mort.; BLA & Supp.; GA; GB. Also Anatomy*.
1. *Recherches physiologiques sur la vie et la mort.* Paris, 1800. X. Reprinted: Paris, 1856; Geneva, 1962; Verviers, 1973. 2nd ed., 1802. Another ed., 1805. 5th ed., rev. by F. Magendie, 1829.
1a. ——— *Physiological researches upon life and death.* Philadelphia, 1809. Trans. by T. Watkins from the 2nd French ed. RM/L.

14730. NYSTEN, Pierre Hubert. 1771-1818 (or 1774-1817). Fr. P II; BLA; GA.

14731. RUDOLPHI, Karl Asmund. 1771-1832. Ger. DSB; BMNH; BLA & Supp.; GA; GC. Also Botany and Zoology.

14732. AUTENRIETH, Johann Heinrich Ferdinand von. 1772-1835. Ger. P I & Supp.; BMNH; BLA & Supp.; GA; GC. Also Zoology.

14733. CARSON, James. 1772-1843. Br. BLA; GD.

14734. BOSTOCK, John. 1773-1846. Br. DSB; ICB; P I; BMNH; BLA & Supp.; GD.

1. *An account of the history and present state of galvanism.* London, 1818. RM.
2. *An elementary system of physiology.* 3 vols. London, 1824-27. Another ed., 1836.

14735. GIROU DE BUZAREINGUES, Louis François Charles. 1773-1856. Fr. P I; BLA; GA.

14736. ROLANDO, Luigi. 1773-1831. It. DSB; Mort.; BLA. Also Zoology and Anatomy.

14737. SORG, Franz Lothar August. 1773-1827. Ger. P II; BLA.

14738. YOUNG, Thomas. 1773-1829. Br. DSB; ICB; P II; Mort.; BLA; GA; GB; GD. Also Physics*#. See also Index.

14739. HEIDMANN, Johann Anton. 1775-1855. Aus. P III. Also Physics*.

14740. VIREY, Jules (or Julien) Joseph. 1775-1846. Fr. DSB; P III; BMNH; BLA. Also Zoology*.

14741. AUGUSTIN, Friedrich Ludwig. 1776-1854. Ger. P I; BLA & Supp. Also Physics.

14742. BUNTZEN, Thomas. 1776-1807. Den./Russ. P I; BLA & Supp.

14743. BURDACH, Karl Friedrich. 1776-1847. Ger. DSB; ICB; BMNH; Mort.; BLA & Supp.; GA; GC. Also Anatomy.

14744. DUTROCHET, René Joaquim Henri. 1776-1847. Fr. DSB; ICB; P I; BMNH; Mort.; BLA; GA; GB. Also Botany*.
Works on plant and animal physiology: see 11781/1-4.

14745. EDWARDS, (William) Frédéric. 1776-1842. Fr. DSB; P I; BMNH; BLA.
1. *De l'influence des agents physiques sur la vie.* Paris, 1824. RM.

14746. RITTER, Johann Wilhelm. 1776-1810. Ger. DSB; ICB; P II; BLA; GA; GC. Also Physics*# and Chemistry.

14747. TREVIRANUS, Gottfried Reinhold. 1776-1837. Ger. DSB; ICB; P II; BMNH; BLA; GB; GC. Also Biology in General* and Zoology.

14748. EMMERT, August Gottfried Ferdinand. 1777-1819. Ger. BLA; GC.

14749. BARTELS, Ernst Daniel August. 1778-1838. Ger. P I; BMNH; BLA; GA; GC.

14750. RICHERAND, Anthelme Balthasar. 1779-1840. Fr. BLA; GA.
1. *Nouveaux éléments de physiologie.* Paris, 1801. X. (There were 13 French eds and many in other languages.)
1a. ——— *Elements of physiology.* London, 1819. Trans. by G.J.M. de Lys. 4th ed., 1824. 5th ed., 1829.

14751. ROGET, Peter Mark. 1779-1869. Br. ICB; P II,III; BMNH; BLA; GD.
1. *Animal and vegetable physiology considered with reference to natural theology.* 2 vols. London, 1834. (Bridgewater treatises. 5 and 6) 2nd ed., 1836 and 1838. 3rd ed., 1840. 4th ed., 1867.
2. *Treatises on physiology and phrenology.* 2 vols. London, 1838. (Appeared originally as encyclopaedia articles.)

2a. ——— 1st American ed. *Outlines of physiology. With an appendix on phrenology.* Philadelphia, 1839.

14752. WALKER, Alexander. 1779-1852. Br. DSB. Also History of Neurology*.

14753. WILBRAND, Johann Bernhard. 1779-1846. Ger. DSB; BMNH; BLA; GC. Also Zoology.

14754. BARRY, David. 1780-1835. Br. BLA & Supp.; GD.
1. *Experimental researches on the influence exercised by atmospheric pressure upon the progression of the blood in the veins, upon that function called absorption....* London, 1826.

14755. TROXLER, Ignaz Paul Vitalis. 1780-1866. Switz. ICB; BLA; GC.

14756. WINKELMANN, August Stefan. 1780-1806. Ger. BLA; GC.

14757. TIEDEMANN, Friedrich. 1781-1861. Ger. DSB; ICB; P III; BMNH; Mort.; BLA; GA; GB; GC. Also Zoology and Embryology*.
1. *Physiologie des Menschen. I. Allgemeine Betrachtungen der organischen Körper.* Darmstadt, 1830. X.
1a. ——— *A systematic treatise on comparative physiology, introductory to the physiology of man.* Vol. 1. [No more publ.] London, 1834. Trans. with notes by J.M. Gully and J.H. Lane.

14758. WALTHER, Philipp Franz von. 1782-1849. Ger. BLA.

14759. YOUNG, John Richardson. 1782?-1804. U.S.A. DSB; ICB; Mort.; GE.
1. *An experimental inquiry into the principles of nutrition and the digestive process.* Philadelphia, 1803. X. Reprinted, Urbana, Ill., 1959.

14760. BRODIE, Benjamin Collins (first of the name). 1783-1862. Br. DSB; ICB; Mort.; BLA & Supp.; GA; GB; GD.
1. *Physiological researches.* London, 1851.

14761. LAWRENCE, William. 1783-1867. Br. DSB; ICB; BMNH; BLA; GA; GD. Also Zoology* and Anatomy.

14762. MAGENDIE, François. 1783-1855. Fr. DSB; ICB; P II; Mort.; BLA; GA.
1. *Précis élémentaire de physiologie.* 2 vols. Paris, 1816-17. 4th ed., 1836.
1a. ——— *An elementary compendium of physiology.* Edinburgh, 1823. Trans. with notes by E. Milligan. 3rd ed., 1829. 4th ed., 1831.
1b. ——— *An elementary summary of physiology.* London, 1825. Trans. by F.S. Forsyth.
2. *Recherches physiologiques et cliniques sur le liquide céphalo-rachidien ou cérébro-spinal.* Paris, 1842.
3. *Phénomènes physiques de la vie.* 4 vols. Paris, 1842.
See also 14729/1 and 15169.

14763. BEAUMONT, William. 1785-1853. U.S.A. DSB; ICB; Mort.; BLA & Supp.; GE.
1. *Experiments and observations on the gastric juice and the physiology of digestion.* Plattsburgh, Mo., 1833. RM. Other eds: Boston, 1834; Edinburgh, 1838.

14764. EWELL, Thomas. 1785-1826. U.S.A. GE. Also Chemistry*.

14765. PANIZZA, Bartolomeo. 1785-1867. It. BMNH; Mort.; BLA. Also Zoology.

14766. DEBREYNE, Pierre Jean Corneille. 1786-1867. Fr. ICB; BLA; GA.

14767. PRICHARD, James Cowles. 1786-1848. Br. The anthropologist. DSB; BMNH; Mort.; BLA; GA; GB; GD.
1. *A review of the doctrine of a vital principle ... with observations on the causes of physical and animal life.* London, 1829. RM/L.

14768. PURKYNĚ, Jan Evangelista. 1787-1869. Prague/Aus. DSB; ICB; BMNH. Under PURKINJE: P II,III; Mort.; BLA & Supp.; GC. Also Anatomy and Embryology.
1. *Opera omnia. Sebrané spisy.* 13 vols. Prague, 1919-197-?

14769. DUNGLISON, Robley. 1788-1869. Br./U.S.A. DSB; ICB; BLA & Supp.; GD; GE.

14770. GASPARD, Marie Humbert Bernard. 1788-1871. Fr. BLA.

14771. REICHENBACH, Karl Ludwig. 1788-1869. Aus. DSB; ICB; P II, III; GA; GC. Also Chemistry.
1. *Physikalisch-physiologische Untersuchungen über die Dynamide des Magnetismus, der Elektricität, der Wärme, des Lichtes, der Kristallisation, des Chemismus in ihren Beziehungen zur Lebenskraft.* 2 vols. Brunswick, 1850. X.
1a. ——— *Researches on magnetism ... in their relations to the vital force.* London, 1850. Trans. with introd. and notes by W. Gregory. 2nd ed., 1850.
1b. ——— *Physico-physiological researches on the dynamics of magnetism ... in their relations to the vital force.* London, 1851. Trans. by J. Ashburner.
1c. ——— ——— Another ed. *The dynamics of magnetism....* New York, 1853.

14772. BELLINGERI, Carlo Francesco. 1789-1848. It. P I; BLA & Supp.

14773. CARUS, Carl Gustav. 1789-1869. Ger. ICB; BMNH; BLA & Supp.; GA; GB; GC. Also Zoology*.
1. *Physis. Zur Geschichte des leiblichen Lebens.* Stuttgart, 1851.

14774. ALISON, William Pulteney. 1790-1859. Br. BLA & Supp.; GD.
1. *Outlines of physiology and pathology.* Edinburgh, 1831. X. 2nd ed., 1832 and 1836.
1a. ——— 3rd ed. *Outlines of human physiology.* Ib., 1839.

14775. BLUNDELL, James. 1790-1877. Br. ICB; BLA; GD.

14776. DAVY, John. 1790-1868. Br. DSB; ICB; P I,III; BMNH; BLA & Supp.; GD. Also Chemistry, Natural History, and Biography*. See also Index.
1. *Researches, physiological and anatomical.* 2 vols. London, 1839.
2. *Physiological researches.* London, 1863.

14777. HALL, Marshall. 1790-1857. Br. DSB; ICB; Mort.; BLA & Supp.; GB; GD.
1. *A critical and experimental essay on the circulation of*

the blood. London, 1831.
2. *Theory of the inverse ratio which subsists between the respiration and irritability in the animal kingdom. On hybernation. On the reflex function of the medulla oblongata and medulla spinalis*. Ib., 1832-33. Reprint of periodical articles.
3. *Lectures on the nervous system and its diseases*. Ib., 1836.
4. *Memoirs on the nervous system*. Ib., 1837.

14778. PREVOST, Jean Louis. 1790-1850. Switz. DSB; ICB; P II; Mort.; BLA. Also Embryology.

14779. WEDEMEYER, Georg Ludwig Heinrich Karl. ca. 1790-1829. Ger. BLA; GC.

14780. ELLIOTSON, John. 1791-1868. Br. ICB; BLA & Supp.; GB; GD.
1. *Human physiology*. London, 1840.
See also 14700/1a.

14781. MACKENZIE, William. 1791-1868. Br. Mort.; BLA; GD.
1. *Physiology of vision*. London, 1841.

14782. FLETCHER, John. 1792-1836. Br. BLA; GD.

14783. HEUSINGER (VON WALDEGG), Karl Friedrich. 1792-1883. Ger. BMNH; BLA; GA; GC. Also Zoology.

14784. LARDNER, Dionysius. 1793-1859. Br. ICB; P I & Supp.; BMNH; GA; GB; GD. Other entries: see 3303.
1. *Animal physics; or, The body and its functions familiarly explained*. London, 1857.
1a. ——— 2nd ed. *Handbook of animal physics*. Ib., 1873.
2. *Animal physiology for schools*. Ib., 1858.

14785. FLOURENS, (Marie Jean) Pierre. 1794-1867. Fr. DSB; ICB; BMNH; Mort.; BLA; GA; GB. Also Zoology*, Evolution*, Embryology* and Biography*.
1. *Recherches expérimentales sur les propriétés et les fonctions du système nerveux dans les animaux vertébrés*. Paris, 1824.
2. *Mémoires d'anatomie et de physiologie comparées*. Ib., 1844.
3. *Théorie expérimentale de la formation des os*. Ib., 1847.

14786. GRAHAM, Sylvester. 1794-1851. U.S.A. ICB; GB; GE.

14787. SCHULTZE, Carl August Sigmund. 1795-1877. Ger. ICB; BMNH; BLA.

14788. THACKRAH, Charles Turner. 1795-1833. Br. ICB; BLA.
1. *An inquiry into the nature and properties of the blood as existent in health and disease*. London, 1819.

14789. WEBER, Ernst Heinrich. 1795-1878. Ger. DSB; ICB; P II,III; BMNH; Mort.; BLA; GC. Also Anatomy.
1. *Wellenlehre auf Experimente gegründet; oder, Über die Wellen tropfbarer Flüssigkeiten, mit Anwendung auf die Schall- und Lichtwellen*. Leipzig, 1825. With his brother, Wilhelm Eduard Weber. (The work is concerned largely with the application of hydrodynamics to the circulation of the blood.) RM.
2. *Über die Anwendung der Wellenlehre auf die Lehre vom Kreis-*

laufe des Blutes und insbesondere auf die Pulslehre. [Periodical article, 1850] Leipzig, 1889. (OKEW. 6)
3. *Tastsinn und Gemeinfühlung.* Brunswick, 1851. X. Reprinted, Leipzig, 1905. (OKEW. 149)

14790. BOUILLAUD, Jean Baptiste. 1796-1881. Fr. ICB; Mort.; BLA & Supp.; GA.

14791. BOURDON, (Jean Baptiste) Isidore. 1796-1861. Fr. BLA; GA.

14792. LE BLANC, Urbain. 1796-1871. Fr. P I,III; GA.

14793. MAYO, Herbert. 1796-1852. Br. DSB; Mort.; BLA; GA; GD. Also Anatomy.
1. *Outlines of human physiology.* London, 1827. 2nd ed., 1829. 3rd ed., 1833. 4th ed., 1837.

14794. BÉRARD, Pierre Honoré. 1797-1858. Fr. BLA & Supp.; GA.

14795. BISHOP, John. 1797-1873. Br. BLA; GD.

14796. COMBE, Andrew. 1797-1847. Br. BLA & Supp.; GA; GB; GD.
1. *Principles of physiology.* Edinburgh, 1834. 10th ed., 1841. 16th ed., 1885.

14797. DOWLER, Bennett. 1797-1879. U.S.A. ICB; BLA & Supp.

14798. DUGÈS, Antoine Louis. 1797-1838. Fr. BMNH; BLA & Supp.; GA. Also Zoology.

14799. GERDY, Pierre Nicolas. 1797-1856. Fr. BLA & Supp.; GA. Also Anatomy.

14800. HUSCHKE, Emil. 1797-1858. Ger. DSB; BMNH; BLA; GC. Also Zoology*, Anatomy, and Embryology.

14801. POISEUILLE, Jean Léonard Marie. 1797-1869. Fr. DSB; ICB; P II,III; Mort.; BLA. Also Physics.

14802. SCHROEDER VAN DER KOLK, Jacobus Ludovicus Conradus. 1797-1862. Holl. DSB. Under KOLK: BMNH; BLA & Supp.; GF. Also Zoology and Anatomy.

14803. BARKOW, Hans Carl Leopold. 1798-1873. Ger. BMNH; BLA. Also Zoology.

14804. BRIGHAM, Amariah. 1798-1849. U.S.A. BLA; GE.
1. *An inquiry concerning the disease and functions of the brain, the spinal cord, and the nerves.* New York, 1840. X. Reprinted, ib., 1973.

14805. BUCHANAN, Andrew. 1798-1882. Br. Mort.; BLA & Supp.
1. *Forces which carry on the circulation of the blood.* 2nd ed. London, 1874.

14806. ESCHRICHT, Daniel Frederik. 1798-1863. Den. BMNH; BLA.
1. *Das physische Leben.* Berlin, 1852. X. 2nd ed., 1856.

14807. SCHULTZ (-SCHULTZENSTEIN), Karl Heinrich. 1798-1871. Ger. P II,III; BMNH; BLA; GC. Also Botany.
1. *Über die Organization des Blutes.* Stuttgart, 1836.
2. *Das System der Circulation in seiner Entwicklung durch die Thierreihe und in Menschen.* Stuttgart, 1836.

14808. WALKER, John (second of the name). d. 1847. Br. BLA.

1. *Philosophy of the eye, being a familiar exposition of its mechanism and of the phenomena of vision, with a view to the evidence of design.* London, 1837.

14809. HEINE, Bernhard. 1800-1846. Ger. ICB; BLA; GC.

14810. KILIAN, Hermann Friedrich. 1800-1863. Ger./Russ. BLA & Supp.; GC.
1. *Über den Kreislauf des Blutes im Kinde welches noch nicht geathmet hat.* Karlsruhe, 1826.

14811. VOLKMANN, Alfred Wilhelm. 1800-1877. Ger. P II,III; BMNH; Mort.; BLA; GC.

14812. DONNÉ, Alfred. 1801-1878. Fr. P III; Mort.; BLA. Also Microbiology.

14813. MÜLLER, Johannes (Peter). 1801-1858. Ger. DSB; ICB; P II; BMNH; Mort.; BLA & Supp.; GB; GC. Also Zoology, Anatomy, and Embryology.
1. *Über die phantastischen Gesichtserscheinungen. Eine physiologische Untersuchung.* Coblenz, 1826. X. Reprinted, Munich, 1967.
2. *Handbuch der Physiologie des Menschen.* Vol. I. Coblenz, 1833. 3rd ed., 1837-38. X. Vol. II, 1837-40. X.
2a. ——— *Elements of physiology.* 2 vols. London, 1838-42. Trans. with notes by W. Baly. 2nd ed., 1840-42. Another ed., Philadelphia, 1843.
2b. ——— *Manuel de physiologie.* Paris, 1845. Trans. by A.J.L. Jourdan from the 4th German ed.
3. *Briefe von Johannes Müller an Anders Retzius, 1830-1857.* Stockholm, 1900.

14814. SHARPEY, William. 1802-1880. Br. DSB; ICB; Mort.; BLA; GD. Also Anatomy.

14815. TOURTUAL, Kaspar Theobald. 1802-1865. Ger. P II; BLA; GC. Also Anatomy.

14816. BERTHOLD, Arnold Adolph. 1803-1861. Ger. DSB; ICB; P I & Supp.; BMNH; Mort.; BLA & Supp.; GA; GC. See also 12800/3a.

14817. GULLIVER, George. 1804-1882. Br. BMNH; BLA; GD. Also Anatomy.

14818. DEEN, Izaak van. 1805-1869. Holl./Ger. ICB; BMNH; BLA & Supp.

14819. MAISSIAT, Jacques Henri. 1805-1878. Fr. P II,III; BMNH; BLA.

14820. WAGNER, Rudolph. 1805-1864. Ger. DSB; BMNH; Mort.; BLA; GB; GC. Also Zoology*, Embryology*, and Biography*.
1. *Lehrbuch der Physiologie.* Erlangen, 1838. 7th ed., by A. Gruenhagen, 3 vols, Hamburg, 1885-87.
1a. ——— *Elements of physiology.* London, 1841. Trans. with additions by R. Willis. Another ed., 1844.
2. *Handwörterbuch der Physiologie.* 4 vols in 5. Brunswick, 1842-53.

14821. BAUDRIMONT, Alexandre Edouard. 1806-1880. Fr. DSB; P III,IV; BMNH. Also Chemistry*.

14822. DUCHENNE (DE BOULOGNE), Guillaume Benjamin Amand. 1806-1875. Fr. ICB; Mort.; BLA & Supp; GB.

1. *Physiologie des mouvements, démontrée à l'aide de l'expérimentation électrique.* Paris, 1867.
1a. ——— *Physiology of motion, demonstrated by means of electrical stimulation.* Philadelphia, 1959. Trans. and ed. by E.B. Kaplan.

14823. WEBER, Eduard Friedrich. 1806-1871. Ger. ICB; P II,III; Mort.; BLA; GC. See also 6326/1.

14824. BISCHOFF, Theodor Ludwig Wilhelm. 1807-1882. Ger. DSB; BMNH; Mort.; BLA & Supp.; GA; GC. Also Zoology and Embryology*.

14825. NASSE, Hermann. 1807-1892. Ger. BLA.

14826. ZAGORSKII, Aleksandr Petrovich. 1807-1888. Russ. BMNH.

14827. LISTING, Johann Benedict. 1808-1882. Ger. P I,III; Mort. Also Physics.
1. *Beitrag zur physiologischen Optik.* Göttingen, 1845. X. Reprinted, Leipzig, 1905. (OKEW. 147)

14828. STANNIUS, Hermann (Friedrich). 1808-1883. Ger. DSB; BMNH; Mort.; BLA; GC. Also Zoology*.

14829. GAVARRET, (Louis Dominique) Jules. 1809-1890. Fr. P III; BLA & Supp.

14830. KING, Thomas Wilkinson. 1809-1847. Br. Mort.

14831. REID, John. 1809-1849. Br. BLA; GD.
1. *Physiological, anatomical and pathological researches.* London, 1848.

14832. BIDDER, Friedrich Heinrich. 1810-1894. Dorpat. DSB; BMNH; Mort.; BLA & Supp.; GC. Also Anatomy.
1. *Die Verdauungssäfte und der Stoffwechsel. Eine physiologisch-chemische Untersuchung.* Mitau, 1852. With C. Schmidt.

14833. DUTTENHOFER, Friedrich Martin. 1810-1859. Ger. P I; BMNH.

14834. NOBLE, Daniel. 1810-1885. Br. BLA.
1. *The brain and its physiology.* London, 1846.
2. *The human mind in its relations with the brain and nervous system.* London, 1858.

14835. RUETE, Christian Georg Theodor. 1810-1867. Ger. P II,III; BMNH; BLA.

14836. SCHWANN, Theodor Ambrose Hubert. 1810-1882. Ger./Belg. DSB; ICB; P II,III; BMNH; Mort.; BLA & Supp.; GB; GC. Also Microbiology and Cell Theory*#.

14837. TÜRCK, Ludwig. 1810-1868. Aus. DSB; Mort.; BLA.

14838. VALENTIN, Gabriel Gustav. 1810-1883. Ger./Switz. DSB; ICB; P II,III; BMNH; BLA; GC. Also Anatomy, Embryology, and Cytology.
1. *Grundriss der Physiologie des Menschen.* Brunswick, 1846.
1a. ——— *A text-book of physiology.* London, 1853. Trans. by W. Brunton from the 3rd German ed.

14839. BUDGE, Julius Ludwig. 1811-1888. Ger. P III; Mort.; BLA & Supp.; GC.

14840. DRAPER, John William. 1811-1882. U.S.A. DSB; ICB; P I,III; BMNH; BLA & Supp.; GB; GD; GE. Also Physics*, Chemistry, and History of Science*.
1. *Human physiology, statical and dynamical*. New York, 1856. 7th ed., 1875 and 1885.

14841. HUTCHINSON, John. 1811-1861. Br. DSB; Mort.; BLA.

14842. ISAACS, Charles Edward. 1811-1860. U.S.A. DSB; ICB; Mort.; BLA & Supp.

14843. LONGET, François Achille. 1811-1871. Fr. BMNH; BLA; GA. Also Zoology.
1. *Traité de physiologie*. Paris, 1850-52. X. 2nd ed., 1857.

14844. MATTEUCCI, Carlo. 1811-1868. It. DSB; ICB; P II,III; BMNH; Mort.; BLA; GB. Also Physics.
1. *Essai sur les phénomènes électriques des animaux*. Paris, 1840. RM.
2. *Traité des phénomènes électro-physiologiques des animaux*. Paris, 1844.
3. *Leçons sur les phénomènes physiques des corps vivants*. Paris, 1847.
3a. ——— *Lectures on the physical phenomena of living beings*. London, 1847.

14845. MIESCHER, (Johann) Friedrich (first of the name). 1811-1887. Ger./Switz. BMNH; BLA. Also Anatomy.

14846. MIRAM, Eduard. 1811-1886. Russ. BLA.

14847. BENNETT, John Hughes. 1812-1875. Br. BLA & Supp.; GB; GD.
1. *Outlines of physiology*. Edinburgh, 1849. Another ed., 1858.
2. *A text-book of physiology, general, special and practical*. Edinburgh, 1872.

14848. GLUGE, Gottlieb. 1812-1898. Belg. BLA & Supp.

14849. LAYCOCK, Thomas. 1812-1876. Br. ICB; GD. See also 14674/1a.

14850. MERKEL, Carl Ludwig. 1812-1876. Ger. Mort.; BLA. Also Anatomy.

14851. BERNARD, Claude. 1813-1878. Fr. DSB; ICB; P III; Mort.; BLA & Supp.; GA; GB. See also 15200/5 and 15526/1.
1. *Leçons de physiologie expérimentale appliquée à la médicine*. 2 vols. Paris, 1855-56.
2. *Leçons sur les effets des substances toxiques et médicamenteuses*. Paris, 1857. Another ed., 1883.
3. *Leçons sur la physiologie et la pathologie du système nerveux*. 2 vols. Paris, 1858.
4. *Leçons sur les propriétés physiologiques et les altérations pathologiques des liquides de l'organisme*. 2 vols. Paris, 1859.
5. *Introduction à l'étude de la médecine expérimentale*. Paris, 1865. RM/L. Reprinted, ib., 1966.
5a. ——— *An introduction to the study of experimental medicine*. New York, 1927; reprinted, 1949 and 1957. Trans. by H.C. Greene.
6. *Leçons de pathologie expérimentale*. Paris, 1872. X. 2nd ed., 1880.

7. *Leçons sur les anesthésiques et sur l'asphyxie.* Paris, 1875.
8. *Leçons sur la chaleur animale.* Paris, 1876.
9. *Leçons sur le diabète et la glycogenèse animale.* Paris, 1877. Reprinted, ib., 1966.

Posthumous

10. *Leçons sur les phénomènes de la vie communs aux animaux et aux végétaux.* 2 vols. Paris, 1878-79. Ed. by A. Dastre. Reprinted, ib., 1966.

10a. ——— *Lectures on the phenomena of life common to animals and plants.* Springfield, Ill., 1974. Trans. by H.H. Hoff et al.

11. *La science expérimentale.* Paris, 1878. Another ed., 1890.
12. *Leçons sur la physiologie opératoire.* Paris, 1879. Ed. by M. Duval.
13. *Principes de médecine expérimentale.* Paris, 1947. Introd. and notes by L. Delhoume. Reprinted, ib., 1963.
14. *Cahier de notes, 1850-1860: Edition intégrale du Cahier rouge.* Paris, 1965. Ed. by M.D. Grmek.

14a. ——— *The Cahier rouge.* Cambridge, Mass., 1967. Trans. by H.H. Hoff et al.

Catalogues and Selections

15. *L'oeuvre de Claude Bernard. Introduction. Table analytique et alphabétique des oeuvres complètes. Bibliographie des travaux scientifiques.* Paris, 1881. Reprinted, Amsterdam, 196-?
16. *Catalogue des manuscripts. Avec la bibliographie de ses travaux imprimés.* Paris, 1967. By M.D. Grmek.
17. *Pages choisies.* Paris, 1961. Ed. with introd. and notes by E. Kahane.
18. *Claude Bernard et la médecine expérimentale ... Choix de textes, bibliographie, etc.* Paris, 1961. Ed. by R. Clarke.
19. *Notes, mémoires et leçons sur la glycogenèse animale et le diabète.* Paris, 1965. Ed. by M.D. Grmek.
20. *Ausgewählte physiologische Schriften.* Bern/Stuttgart, 1966. Trans. and ed. by N. Mani.

14852. CARPENTER, William Benjamin. 1813-1885. Br. DSB; ICB; P III; BMNH; BLA & Supp.; GB; GD. Other entries: see 3326.

1. *Principles of general and comparative physiology.* London, 1839. X. 4th ed., 1854.
2. *Principles of human physiology.* London, 1842. Another ed., 1846. 5th ed., 1855. Another ed., 1864. 8th ed., 1876. 9th ed., 1881.
3. *A manual of physiology, including physiological anatomy.* London, 1846. 3rd ed., 1856. 4th ed., 1865.
4. *Mechanical physiology.* London, 1848.
5. *Animal physiology.* London, 1851. Other eds: 1859, 1876.

14853. FICK, (Franz) Ludwig. 1813-1858. Ger. P I & Supp.; BLA & Supp. Also Anatomy.

14854. HALLMANN, Eduard. 1813-1855. Ger. ICB; P I; BMNH; BLA. Also Zoology.

14855. BALY, William. 1814-1861. Br. BMNH; BLA; GD.

1. *Recent advances in the physiology of motion and the senses, generation, and development.* London, 1848. With W.S. Kirkes.

See also 14813/2a.

14856. BECK, Thomas Snow. 1814-1877. Br. Mort.; BLA.

14857. MAYER, Julius Robert von. 1814-1878. Ger. DSB; ICB; P II, III,IV; Mort.; BLA & Supp.; GB; GC. Also Physics*#.

14858. WYMAN, Jeffries. 1814-1874. U.S.A. DSB; ICB; BMNH; Mort.; GE. Also Zoology.

14859. MEYER, Georg Hermann von. 1815-1892. Switz. P II,III. Also Anatomy.

14860. REMAK, Robert. 1815-1865. Ger. DSB; ICB; Mort.; BLA & Supp.; GC. Also Histology, Embryology, and Cytology.

14861. TOYNBEE, Joseph. 1815-1866. Br. BLA; GD. Also Anatomy.

14862. BENNET, James Henry. 1816-1891. Br. ICB; BLA & Supp.
1. *Nutrition in health and disease.* London, 1858.

14863. BOWMAN, William. 1816-1892. Br. DSB; ICB; Mort.; BLA & Supp.; GD1. Also Anatomy.

14864. ECKER, Alexander. 1816-1887. Switz. BMNH; BLA & Supp.; GC. Also Zoology*, Anatomy, and Biography*.
1. *Icones physiologicae. Erläuterungstafeln zur Physiologie und Entwicklungsgeschichte.* Leipzig, 1851-59.

14865. GIRAUD-TEULON, (Marc Antoine Louis) Félix. 1816-1887. Fr. P III; BMNH; BLA & Supp.
1. *Physiologie ... de la vision binoculaire.* Paris, 1861.

14866. LETHEBY, Henry. 1816-1876. Br. P III; GD.

14867. LUDWIG, Carl Friedrich Wilhelm. 1816-1895. Switz./Aus./Ger. DSB; ICB; P I,III,IV; BMNH; Mort.; BLA & Supp.; GB; GC.
1. *Lehrbuch der physiologie des Menschen.* 2 vols. Heidelberg, 1852-56. X. Another ed., Leipzig, 1858-61.

14868. WALLER, Augustus Volney. 1816-1870. Br. DSB; ICB; P III; Mort.; BLA; GD. Also Anatomy.

14869. BÉCLARD, Jules Auguste. 1817-1887. Fr. BMNH; BLA & Supp.

14870. BROWN-SÉQUARD, Charles Edouard. 1817-1894. Fr./Br./U.S.A. DSB; ICB; Mort.; BLA & Supp.; GB; GD1.
1. *A course of lectures on the physiology and pathology of the central nervous system.* Philadelphia, 1860.

14871. KOELLIKER, (Rudolf) Albert von. 1817-1905. Ger. DSB; ICB; BMNH; Mort.; BLA & Supp.; GA; GB. Also Zoology, Anatomy*, Embryology*, and Cytology.

14872. LEWES, George Henry. 1817-1878. Br. ICB; BMNH; GA; GB; GD. Also Zoology* and Biography*.
1. *The physiology of common life.* 2 vols. Edinburgh, 1859-60.
2. *Problems of life and mind. First series.* 2 vols. London, 1874-75. X.
2a. ——— *Second series.* Ib., 1877. Another ed., 1893.
2b. ——— *Third series.* 2 vols. Ib., 1879. X.

14873. DONDERS, Franciscus Cornelis. 1818-1889. Holl. DSB; ICB; P III,IV; BMNH; Mort.; BLA & Supp.
1. *De harmonie van het dierlijke leven.* 1848. X. Reprinted, Utrecht, 1972.
2. [Trans.] *On the anomalies of accomodation and refraction of the eye. With a preliminary essay on physiological dioptrics.* London, 1864. Trans. from the author's MS by W.D. Moore. Reprinted, Boston, 1972.

14874. DU BOIS-REYMOND, Emil Heinrich. 1818-1896. Ger. DSB; ICB; P I,III,IV (under B); BMNH; Mort.; BLA & Supp. (under B); GB; GC. Also Science in General*. See also Index.
1. *Untersuchungen über thierische Elektrizität.* Berlin, 1848-84. (Vol. 1 appeared in 1848, Vol. 2, Part 1, in 1849, and Vol. 2, Part 2, during 1860-84) RM/L.
2. *Gesammelte Abhandlungen zur allgemeinen Muskel- und Nervenphysik.* 2 vols. Leipzig, 1875-77. RM/L.

14875. MARSHALL, John. 1818-1891. Br. BMNH; BLA; GB; GD. Also Anatomy*.
1. *Outlines of physiology, human and comparative.* 2 vols. London, 1867.

14876. PETTENKOFER, Max Josef von. 1818-1901. Ger. DSB; ICB; P II, III,IV; Mort.; BLA & Supp.; GB. Also Chemistry*.

14877. SMEE, Alfred. 1818-1877. Br. P II,III; BMNH; BLA; GD. Also Chemistry*.
1. *Sources of physical science, being an introduction to the study of physiology through physics.* London, 1843. RM/L.
2. *Elements of electro-biology; or, The voltaic mechanism of man.* London, 1849. RM. Another ed., 1869.
3. *Instinct and reason, deduced from electro-biology.* London, 1850. RM/L.

14878. SMITH, Edward. 1818-1874. Br. DSB; BLA; GD.
1. *Health and disease as influenced by the daily, seasonal and other cyclical changes in the human system.* London, 1861.

14879. TRAUBE, Ludwig. 1818-1876. Ger. ICB; Mort.; BLA; GC.

14880. VIERORDT. Karl von. 1818-1884. Ger. ICB; P II,III; Mort.; BLA; GC.
1. *Die quantitative Spectralanalyse in ihrer Anwendung auf Physiologie, Physik, Chemie und Technologie.* Tübingen, 1876.

14881. BRÜCKE, Ernst Wilhelm von. 1819-1892. Ger./Aus. DSB; P I, III,IV; Mort.; BLA & Supp.; GC. Also Botany* and Anatomy.
1. *Untersuchungen über den Farbenwechsel des afrikanischen Chamäleons.* [Periodical article, 1851] Leipzig, 1893. (OKEW. 43)
2. *Grundzüge der Physiologie und Systematik der Sprachlaute.* 1856. X. 2nd ed., Vienna, 1876.
3. *Vorlesungen über Physiologie.* 2 vols. 1873-75. X. 3rd ed., Vienna, 1881-84.
4. *Briefe an Emil Du Bois-Reymond.* Graz, 1978. Ed. by Hans Brücke et al.

14882. FRERICHS, Friedrich Theodor. 1819-1885. Ger. P I,III; Mort.; BLA & Supp.; GC.

14883. HARLESS, Emil. 1820-1862. Ger. BLA.

14884. MÜLLER, Heinrich. 1820-1864. Ger. BMNH; Mort.; BLA. Also Histology.

14885. PANUM, Peter Ludvig. 1820-1885. Den. ICB; P II,III; BMNH; BLA & Supp.
1. *Physiologische Untersuchungen über das Sehen mit zwei Augen*. Kiel, 1858.

14886. VALÉRIUS, Hubert. 1820-1897. Belg. P III; BLA; GF. Also Physics.

14887. HAUGHTON, Samuel. 1821-1897. Br. P III,IV; BMNH; Mort.; GB; GD1. Other entries: see 3336.
1. *Principles of animal mechanics*. London, 1873.

14888. HELMHOLTZ, Hermann Ludwig Ferdinand von. 1821-1894. Ger. DSB; ICB; P I,III,IV; Mort.; BLA & Supp.; GB; GC. Also Science in General*#, Mechanics*, and Physics*. See also Index.
1. *Die Lehre von den Tonempfindungen als physiologische Grundlage für die Theorie der Musik*. Brunswick, 1863. RM. Another ed., 1870. 4th ed., 1877.
1a. ——— *On the sensations of tone as a physiological basis for the theory of music*. London, 1875. Trans. from the 3rd German ed., with notes, by A.J. Ellis. RM/L.
1b. ——— ——— 2nd English ed. Ib., 1885. Trans. from the 4th German ed. by A.J. Ellis. Reprinted, New York, 1954.
2. *Handbuch der physiologischen Optik*. Leipzig, 1867. RM/L. 2nd ed., 2 vols, Hamburg, 1896.
2a. ——— *Treatise on physiological optics*. Rochester, N.Y., 1924-25. Trans. from the 3rd German ed. Ed. by J.P.C. Southall. Reprinted, New York, 1962.

14889. PREVOST, Alexandre Pierre. 1821-1873. Switz. P II,III.

14890. WITTICH, Wilhelm Heinrich von. 1821-1884. Ger. P II,III; BMNH; BLA; GC.

14891. ECKHARD, Conrad. 1822-1915. Ger. P III; BMNH; Mort.; BLA & Supp.
1. *Experimentalphysiologie des Nervensystems*. Giessen, 1867. See also 14982/3.

14892. HINTON, James. 1822-1875. Br. ICB; BLA; GB; GD. Also Biology in General*.
1. *Physiology for practical use*. 2 vols. London, 1874. By various writers; ed. by Hinton. X. 3rd ed., 1880.

14893. MOLESCHOTT, Jacob Albert Willebrord. 1822-1893. Holl./Ger./Switz./It. DSB; ICB; P II,III; BMNH; Mort.; BLA; GA; GC. Also Biography*.
1. *Physiologie des Stoffwechsels in Pflanzen und Thieren. Ein Handbuch*. Erlangen, 1851. RM.

14894. RADCLIFFE, Charles Bland. 1822-1889. Br. BLA; GD.

1. *Dynamics of nerve and muscle.* 1871.
2. *Vital motion as a mode of physical motion.* London, 1876.

14895. WELCKER, Hermann. 1822-1897. Ger. BMNH; Mort.; BLA; GC.

14896. BRINTON, William. 1823-1867. Br. BLA; GD.

14897. KIRKES, William Senhouse. 1823-1864. Br. BMNH; BLA; GD1.
1. *Handbook of physiology.* London, 1848. With J. Paget. X. 2nd ed., 1851. 4th ed., 1860. 7th ed., 1869. 8th ed., 1872. 10th ed., 1880. 12th ed., 1888. 13th ed., 1892. 16th ed., 1900.

See also 14855/1.

14898. LECONTE, Joseph. 1823-1901. U.S.A. DSB; P III,IV; BMNH; BLA & Supp.; GB; GE. Also Geology*, Evolution*, and Autobiography*.
1. *Sight. An exposition of the principles of monocular and binocular vision.* London, 1881. 2nd ed., 1883. 3rd ed., 1895.

14899. SCHIFF, Moritz. 1823-1896. Switz./It. DSB; ICB; P III,IV; BMNH; Mort.; BLA; GC.
1. *Leçons sur la physiologie de la digestion.* 2 vols. Paris, 1867.

14900. BÜCHNER, (Friedrich Karl Christian) Ludwig. 1824-1899. Ger. DSB; P III,IV; BLA (B., Louis); GC. Also Science in General*.
1. *Physiologische Bilder.* Vol. I. Leipzig, 1861. X. Another ed., 1872. Vol. II, 1875.

14901. CAMPBELL, Henry Fraser. 1824-1891. U.S.A. ICB; Mort.; BLA; GE.

14902. CORVISART, (François Rémy) Lucien. 1824-1882. Fr. Mort.; BLA.

14903. NOLL, Friedrich Wilhelm. 1824-1889. Ger. Mort.; GC.

14904. ROUGET, Charles Marie Benjamin. 1824-1904. Fr. DSB. Also Histology.

14905. STENBERG, Sten. 1824-1884. Swed. P III; BLA.

14906. CHARCOT, Jean Martin. 1825-1893. Fr. The famous physician. DSB; ICB; Mort.; BLA & Supp.; GB.

14907. COCCIUS, Ernst Adolf. 1825-1890. Ger. BLA & Supp.; GC.
1. *Der Mechanismus der Accomodation des menschlichen Auges.* Leipzig, 1868.

14908. COLIN, Gabriel Constant. 1825-1896. Fr. BMNH; BLA.
1. *Traité de physiologie comparée des animaux domestiques.* 2 vols. Paris, 1854-56.

14909. DALTON, John Call. 1825-1889. U.S.A. DSB Supp.; Mort.; BLA & Supp.; GE. Also History of Physiology.
1. *A treatise on human physiology.* Philadelphia, 1859. 6th ed., London, 1876. 7th ed., 1882.

14910. HUXLEY, Thomas Henry. 1825-1895. Br. DSB; ICB; BMNH; Mort.; BLA; GB; GD1. Other entries: see 3340.
1. *Lessons in elementary physiology.* London, 1866. 3rd ed., 1869. 4th ed., 1870. 7th ed., 1873. 11th ed., 1878. Other eds: 1879, 1881, 1885, 1898.

2. *The elements of physiology and hygiene.* New York, 1873. With W.J. Youmans.

14911. VELLA, Luigi. 1825-1886. It. BLA.

14912. AUBERT, Hermann Rudolf. 1826-1892. Ger. P III; BMNH; BLA & Supp.; GC. See also 14982/3.

14913. SAVORY, William Scovell. 1826-1895. Br. BMNH; BLA; GB; GD.
1. *On life and death.* 1863.

14914. VULPIAN, (Edme Félix) Alfred. 1826-1887. Fr. ICB; BMNH; Mort.; BLA.
1. *Leçons sur la physiologie générale et comparée du système nerveux.* Paris, 1866.
2. *Leçons sur l'appariel vaso-moteur.* 2 vols. Paris, 1875.

14915. CHAUVEAU, (Jean Baptiste) Auguste. 1827-1917. Fr. DSB; BMNH; Mort.; BLA & Supp. Also Zoology*.

14916. KEYT, Alonzo Thrasher. 1827-1885. U.S.A. ICB; GE.

14917. LIEBIG, Georg von. 1827-1903. Ger. P III,IV. Also Meteorology.

14918. AUERBACH, Leopold (first of the name). 1828-1897. Ger. ICB; Mort.; BLA & Supp.; GC. Also Anatomy.

14919. BURDON-SANDERSON, John Scott. 1828-1905. Br. DSB; ICB; Mort.; GB; GD2. Under SANDERSON: P III,IV; BMNH; BLA.
1. *Handbook for the physiological laboratory.* London, 1873. With E. Klein, M. Foster, and T.L. Brunton.
2. *Translations of foreign biological memoirs. I. Memoirs on the physiology of nerve, of muscle, and of the electric organ.* Oxford, 1887.

14920. CLOETTA, Arnold Leonhard. 1828-1890. Switz. P III; BLA & Supp.

14921. CZERMAK, Johann Nepomuk. 1828-1873. Prague/Ger. DSB; ICB; P III; BMNH; Mort.; BLA & Supp.; GC. Also Histology.
1. *Über das physiologische Privat-Laboratorium an der Universität Leipzig.* Leipzig, 1873.
2. *Gesammelte Schriften.* 3 vols. Leipzig, 1879.

14922. FUNKE, Otto. 1828-1879. Ger. P I,III; BMNH; Mort.; BLA & Supp. Also Chemistry*.
1. *Lehrbuch der Physiologie.* 2 vols. Leipzig, 1863-66.

14923. LUYS, (Jules) Bernard. 1828-1897. Fr. ICB; Mort.; BLA. Also Anatomy.
1. *Le cerveau et ses fonctions.* Paris, 1875. X.
1a. ——— *The brain and its functions.* London, 1881. 2nd ed., 1883. 3rd ed., 1889.

14924. PFLÜGER, Eduard Friedrich Wilhelm. 1828-1910. Ger. DSB; P II, III; BMNH; Mort.; BLA. Also Embryology*.
1. *Untersuchungen über die Physiologie des Electrotonus.* Berlin, 1859.
2. *Die teleologische Mechanik der lebendigen Natur.* Bonn, 1877. X. Reprinted, Quickborn, 1971.

14925. RICHARDSON, Benjamin Ward. 1828-1896. Br. DSB; ICB; BLA; GD1.
1. *The cause of the coagulating of the blood.* London, 1858.
See also 15174.

14926\. FICK, Adolf Eugen. 1829-1901. Ger. DSB; ICB; P I,III,IV; BMNH; Mort.; BLA & Supp.
1. *Die medicinische Physik.* Brunswick, 1856. X. 3rd., 1885.
2. *Compendium der Physiologie des Menschen, mit Einschluss der Entwicklungsgeschichte.* Vienna, 1860. X. 2nd ed., 1874.
3. *Über die Wärmeentwicklung bei der Zusammenziehung der Muskeln.* Leipzig, 1874.

14927\. MEISSNER, Georg. 1829-1905. Ger. DSB; ICB; P III; Mort.; BLA. Also Anatomy*.

14928\. MITCHELL, Silas Weir. 1829-1914. U.S.A. DSB; ICB; BMNH; Mort.; BLA & Supp.; GB; GE.

14929\. PAVY, Frederick William. 1829-1911. Br. P III; Mort.; BLA; GD2.
1. *A treatise on food and dietetics physiologically and therapeutically considered.* London, 1874. X. 2nd ed., 1875.
2. *Physiology of the carbohydrates.* London, 1894.

14930\. POWER, Henry. 1829-1911. Br. BMNH; BLA.
1. *Elements of human physiology.* London, 1884. 2nd ed., 1885.
See also 14307/1a.

14931\. SECHENOV, Ivan Mikhailovich. 1829-1905. Russ. DSB; ICB; Mort.; BLA (SSETSCHENOW).
1. *Selected works.* Moscow/Leningrad, 1935. Ed. by A.A. Subkov. Reprinted, Amsterdam, 1968.

14932\. DOGIEL, Johann (or Ivan Mikhailovich). 1830-1905. Russ. P III,IV; BMNH (DOGHEL); Mort.; BLA & Supp.

14933\. LABORDE, Jean Baptiste Vincent. 1830-1903. Fr. BLAF.

14934\. MAREY, Etienne Jules. 1830-1904. Fr. DSB; ICB; P III,IV; BMNH; Mort.; BLA.
1. *Physiologie médicale de la circulation du sang.* Paris, 1863.
2. *Du mouvement dans les fonctions de la vie. Leçons....* Paris, 1868. RM/L.
3. *La machine animale. Locomotion terrestre et aérienne.* Paris, 1873. X. 2nd ed., 1878.
3a. ——— *Animal mechanism. A treatise on terrestriel and aerial locomotion.* London, 1874. 2nd ed., 1874. 3rd ed., 1883.
4. *Physiologie expérimentale, 1875-79.* 4 vols. Paris, 1876-80.
5. *La méthode graphique dans les sciences expérimentales, et principalement en physiologie et médecine.* Paris, 1878. RM.
6. *Physiologie du mouvement. Le vol des oiseaux.* Paris, 1890. RM.
7. *Le mouvement.* Paris, 1874. X.
7a. ——— *Movement.* London, 1895. Trans. by E. Pritchard.

14935\. MATTHIESSEN, Heinrich Friedrich Ludwig. 1830-1906. Ger. P II,III,IV. Also Mathematics and Physics.

14936\. ROBERTS, William (second of the name). 1830-1899. Br. ICB;

BLA; GD1. Also Microbiology.
1. *On the digestive ferments.* London, 1880.
2. *Collected contributions on digestion and food.* Ib., 1891.

14937. BRAUNE, Christian Wilhelm. 1831-1892. Ger. BMNH; Mort.; BLA & Supp.; GC. Also Anatomy.

14938. HEYNSIUS, Adriaan. 1831-1885. Holl. BLA; GF.

14939. HOLMGREN, (Alarik) Frithiof. 1831-1897. Swed. DSB; ICB; P III,IV; Mort.; BLA.

14940. HORNER, Johann Friedrich. 1831-1886. Switz. Mort.; BLA & Supp.

14941. HORWICZ, Adolf. 1831-1894.
1. *Physiologische Analysen auf physiologischer Grundlage.* Halle, 1872-78.

14942. MANTEGAZZA, Paolo. 1831-1910. It. DSB; ICB; BMNH; BLA; GB.

14943. SCHMIDT, Alexander. 1831-1894. Dorpat. Mort.; BLA.

14944. VOIT, Karl von. 1831-1908. Ger. DSB; ICB; BMNH; Mort.; BLA.
1. *Physiologie des allgemeinen Stoffwechsels und der Ernährung.* Munich, 1881.

14945. BOTKIN, Sergei Petrovich. 1832-1889. Russ. BLA & Supp.

14946. DICKINSON, William Howship. 1832-1913. Br. Mort.; BLA & Supp.

14947. MARMÉ, Wilhelm. 1832-1897. Ger. P III,IV; BMNH; BLA.

14948. NORRIS, Richard. 1832-after 1881. Br. P III.
1. *Physiology and pathology of the blood.* 1882.

14949. VINTSCHAU, Maximilian Heinrich Christof Leopold Carl von. 1832-1902. Aus./It. P III,IV; BMNH; BLA.

14950. WUNDT, Wilhelm Max. 1832-1920. Ger. The psychologist. DSB; ICB; P II,III,IV; Mort.; BLA & Supp.; GB.
1. *Lehrbuch der Physiologie des Menschen.* Erlangen, 1865. X. 2nd ed., ib., 1868. 4th ed., Stuttgart, 1878.
2. *Handbuch der medicinischen Physik.* Erlangen, 1867.

14951. BERT, Paul. 1833-1886. Fr. DSB; ICB; BMNH; Mort.; BLA & Supp.
1. *Leçons sur la physiologie comparée de la respiration.* Paris, 1870.
2. *La pression barométrique. Recherches de physiologie expérimentale.* Paris, 1878.

14952. DANNER, Léon. 1833-1907. Fr. BLAF.

14953. PIOTROWSKI, Gustav von. 1833-1884. Cracow. ICB; P III,IV; BLA.

14954. GOLTZ, Friedrich Leopold. 1834-1902. Ger. DSB; Mort.; BLA & Supp.

14955. HEIDENHAIN, Rudolf (Peter Heinrich). 1834-1897. Ger. DSB; P III,IV; BMNH; Mort.; BLA; GC. Also Histology.
1. *Der sogenannte thierische Magnetismus. Physiologische Betrachtungen.* Leipzig, 1879. X.
1a. ——— *Animal magnetism. Physiological observations.* London, 1880. Trans. by L.C. Woolbridge from the 4th German ed.

14956. HERING, (Karl) Ewald (Konstantin). 1834-1918. Aus./Prague/Ger. DSB; ICB; P III,IV; BMNH; Mort.; BLA & Supp.
1. *Die Lehre vom binokularen Sehen.* Leipzig, 1868. X.
1a. ——— *The theory of binocular vision.* New York, 1977. Trans. and ed. by B. Bridgeman et al.
2. *Über das Gedächtniss als eine allgemeine Funktion der organisierten Materie.* [An address, 1870] Leipzig, 1905. (OKEW. 148)
3. *Über das Gedächtniss als eine allgemeine Function der organisierten Materie.* Vienna, 1876.
4. *Zur Lehre vom Lichtsinne.* Vienna, 1878.
5. *Fünf Reden.* Leipzig, 1921. Ed. by H.E. Hering. X. Reprinted, Amsterdam, 1969.
See also 13902/3.

14957. PETTIGREW, James Bell. 1834-1908. Br. BMNH; BLA & Supp.; GD2. Also Anatomy.
1. *Animal locomotion; or, Walking, swimming and flying, with a dissertation on aeronautics.* London, 1873. 2nd ed., 1874. 3rd ed., 1883. 4th ed., 1891.
2. *Physiology of the circulation in plants, in the lower animals, and in man.* London, 1874.

14958. ROLLETT, Alexander. 1834-1903. Aus. P III,IV; BMNH; BLA.

14959. SNELLEN, Hermann. 1834-1908. Holl. BLA; GF.

14960. STRICKER, Salomon. 1834-1898. Aus. BMNH; Mort.; BLA; GC. Also Histology*.
1. *Neuro-elektrische Studien.* Vienna, 1883.

14961. WORM-MÜLLER, Jacob. 1834-1889. Nor. P III,IV; BLA.

14962. CLELAND, John. 1835-1925. Br. BLA & Supp.; BLAF. Also Anatomy.
1. *Animal physiology.* New York, 1873. X. Another ed., London, 1877.

14963. HENSEN, (Christian Andreas) Victor. 1835-1924. Ger. DSB; P IV; BMNH; BLA & Supp. Also Oceanography (plankton studies).

14964. JACKSON, John Hughlings. 1835-1911. Br. DSB; ICB; BLA; GD2.
1. *Clinical and physiological researches on the nervous system.* London, 1873.

14965. LOVÉN, (Otto) Christian. 1835-1904. Swed. Mort.; BLA.

14966. MAPOTHER, Edward Dillon. 1835-1908. Br. BLA; GD2.
1. *Physiology and its aids to the study and treatment of disease.* Dublin, 1862.
2. *Manual of physiology.* Dublin, 1882.

14967. MAUDSLEY, Henry. 1835-1918. Br. BLA; GF.
1. *Physiology of mind.* 1876.

14968. RINGER, Sydney. 1835-1910. Br. DSB; Mort.; BLA; GD2.

14969. BEZOLD, Albert von. 1836-1868. Ger. DSB; P III; Mort.; BLA.
1. *Untersuchungen über die electrische Erregung der Nerven und Muskeln.* Leipzig, 1861. RM.

14970. FLINT, Austin. 1836-1915. U.S.A. Mort.; BLA & Supp.; GE.
1. *Physiology of man.* 5 vols. New York, 1866-75. X. Another ed., 1870-75.

2. *Text-book of human physiology.* New York, 1876. X. 4th ed., London, 1888.
3. *On the source of muscular power.* New York, 1878.

14971. FOSTER, Michael. 1836-1907. Br. DSB; ICB; BMNH; Mort.; BLAF; GB; GD2. Also Embryology* and Biography*. See also Index.
1. *A course of elementary practical physiology and histology.* London, 1876. X. 3rd ed., 1878.
2. *A text-book of physiology.* London, 1877. 2nd ed., 1878. 3rd ed., 1879. 4th ed., 1883. 5th ed., 4 vols, 1881-91. 6th ed., 4 parts in 1 vol., 1900. 3rd American ed., Philadelphia, 1885.

14972. ROSE, Edmund. 1836-1914. Ger./Switz. P II,III,IV; BLA.

14973. ROSENTHAL, Isidor. 1836-1915. Ger. P III,IV; BMNH; BLA.
1. *Die Athembewegungen und ihre Beziehungen zum Nervus Vagus.* Berlin, 1862.
2. *Allgemeine Physiologie der Muskeln und Nerven.* Leipzig, 1877. X.
2a. ——— *General physiology of muscles and nerves.* London, 1881. Another ed., 1883. 4th ed., 1895.

14974. WOLFF, Julius. 1836-1902. Ger. BMNH; Mort.; BLA.

14975. BASCH, Samuel Siegfried Karl von. 1837-1905. Aus. ICB; BLA & Supp.

14976. BASTIAN, Henry Charlton. 1837-1915. Br. DSB; BMNH; BLA & Supp. Also Microbiology*.
1. *The brain as an organ of mind.* London, 1880.

14977. KÜHNE, Wilhelm Friedrich. 1837-1900. Ger. DSB; P IV; BMNH; Mort.; BLA; GB. Also Chemistry*.

14978. LANDOIS, (Christian Clement August) Leonard. 1837-1902. Ger. BMNH; BLA.
1. *Lehrbuch der Physiologie des Menschen.* Vienna, 1880. Another ed., 1883.
1a. ——— *A text-book of human physiology.* London, 1885. Trans. from the 4th German ed., with additions, by W. Stirling. 2nd ed., 1886. 4th ed., trans. from the 7th German ed., 1891.

14979. BROWN, Alexander Crum. 1838-1922. Br. DSB; P III,IV. Also Chemistry.

14980. FRITSCH, Gustav Theodor. 1838-1927. Ger. DSB; ICB; P III,IV; BMNH; Mort.; BLA & Supp. Also Zoology and Anatomy.

14981. GRÉHANT, (Louis François) Nestor. 1838-1910. Fr. BMNH; Mort.

14982. HERMANN, Ludimar. 1838-1914. Switz./Ger. P III,IV; BMNH; Mort.; BLA.
1. *Grundriss der Physiologie des Menschen.* Berlin, 1863. X.
1a. ——— *Elements of human physiology.* London, 1875. Trans. by A. Gamgee from the 5th German ed. 2nd ed., 1878.
2. *Der Einfluss der Descendzlehre auf die Physiologie.* Leipzig, 1879.
3. *Handbuch der Physiologie.* 6 vols in 12. Leipzig, 1879-83. With H. Aubert and C. Eckhard.

14983. HITZIG, (Julius) Eduard. 1838-1907. Ger. DSB; ICB; Mort.; BLAF.

14984. MACH, Ernst. 1838-1916. Aus./Prague. DSB; ICB; P III,IV. Also Mechanics*, Physics*, and History of Physics*.

14985. NAWROCKI, Felix. 1838-1902. Warsaw. BLA.
1. *Über den Einfluss des Blutdruckes auf die Häufigkeit der Herzschläge.* Leipzig, 1874.

14986. BERGERON, Georges. b. 1839. Fr. BLA.

14987. BERNSTEIN, Julius. 1839-1917. Ger. DSB Supp.; P III,IV; Mort.; BLAF
1. *Untersuchungen über den Erregungsvorgang im Nerven- und Muskelsysteme.* Heidelberg, 1871.
2. *Die fünf Sinne des Menschen.* Leipzig, 1876. X.
2a. ——— *The five senses of man.* London, 1876. 2nd ed., 1876.
3. *Lehrbuch der Physiologie des thierischen Organismus, im speciellen des Menschen.* Stuttgart, 1894.

14988. HERZEN, Alexander. 1839-1906. Switz./It. BMNH; BLAF.

14989. KRONECKER, (Karl) Hugo. 1839-1914. Ger./Switz. DSB; Mort.; BLA.
1. *Das charakteristische Merkmal der Herzmuskelbewegung.* Leipzig, 1874.

14990. MUNK, Hermann. 1839-1912. Ger. BLAF.

14991. RUTHERFORD, William. 1839-1899. Br. BMNH; BLA & Supp.; GD1. Also Anatomy.
1. *Text book of physiology.* Edinburgh, 1880.
2. *Notes of lectures on physiology.* Edinburgh, 1884.

14992. VANLAIR, Constant François. 1839-1914. Belg. ICB; BLA; GF. Also Anatomy.

14993. BOWDITCH, Henry Pickering. 1840-1911. U.S.A. DSB; ICB; Mort.; BLA & Supp.; GE.

14994. FUCHS, Friedrich (or Fritz). 1840-1911. Ger. P III,IV; BLA & Supp. Also Physics.

14995. LUCIANI, Luigi. 1840-1919. It. DSB; Mort.; BLAF.

14996. ONIMUS, Ernest Nicolas Joseph. 1840-after 1898. Fr. P III,IV.

14997. FRASER, Thomas Richard. 1841-1920. Br. BLA & Supp.; GD3.

14998. GAMGEE, Arthur. 1841-1909. Br. BMNH; BLA & Supp.; BLAF; GD2. Also Biochemistry*. See also 14982/1a.

14999. GARIEL, Charles Marie. 1841-1924. Fr. P III,IV; BLAF. Also Physics.
1. *Eléments de physique médicale.* Paris, 1884. With V. Desplats.

15000. HOORWEG, Jan Leendert. 1841-1919. Holl. P III,IV. Also Physics.

15001. KNOLL, Philipp. 1841-1900. Prague/Aus. BLAF. Also Histology.

15002. LAMANSKY, Sergei Ivanovich. 1841-1901. Russ. P III; BLAF.

15003. LE BON, Gustave. 1841-1931. Fr. P IV. Also Physics.
1. *La vie. Physiologie humaine.* 1874.

15004. McKENDRICK, John Gray. 1841-1926. Br. P IV; BMNH; BLAF. Also Biography*.
1. *Outlines of physiology.* Glasgow, 1878.
2. *General physiology.* Glasgow, 1888.
3. *Text-book of physiology.* 2 vols. Glasgow, 1888-89.
4. *Physiology of the senses.* 1893. Another ed., 1898.

15005. OLIVER, George. 1841-1915. Br. DSB; Mort.; BLAF.

15006. PLATEAU, Félix Auguste Joseph. 1841-1911. Belg. ICB; P III; BMNH. Also Zoology and Embryology*.

15007. PREYER, (Thierry) William. 1841-1897. Br./Ger. DSB; P III, IV; BMNH; Mort.; BLAF; GC.
1. *Specielle Physiologie des Embryo.* Leipzig, 1885.

15008. QUINQUAUD, Charles Emile (or Eugene). 1841-1894. Fr. Mort.; BLAF.

15009. BREUER, Josef. 1842-1925. Aus. DSB; ICB; Mort.; BLAF; GF.

15010. CATON, Richard. 1842-1926. ICB; Mort.

15011. CYON, Elie de. 1842-1912. Fr./Russ. BMNH; Mort.; BLA & Supp.
1. *Zur Hemmungstheorie der reflectorischen Erregungen.* Leipzig, 1874.
2. *Methodik der physiologischen Experimente und Vivisectionen.* Text & atlas. Giessen, 1876.

15012. EBNER (VON ROSENSTEIN), Victor. 1842-1925. Aus. BMNH; BLA & Supp. Also Anatomy.

15013. GAD, Johannes. 1842-1926. Ger./Prague. BLA & Supp.
1. *Über einige Beziehungen zwischen Nerv, Muskel und Centrum.* Leipzig, 1882.

15014. GRUENHAGEN, Alfred Wilhelm. 1842-1912. Ger. BLAF. See also 14820/1.

15015. GESCHEIDLEN, Richard. 1842-1889. Ger. BLA.
1. *Physiologische Methodik. Ein Handbuch der praktischen Physiologie.* Brunswick, 1876-79.

15016. MALASSEZ, Louis Charles. 1842-1909. Fr. Mort.; BLAF.

15017. MAYER, Sigmund. 1842-1910. Ger. BMNH; BLAF. Also Histology.

15018. PALADINO, Giovanni. 1842-1917. It. BLAF. Also Anatomy.

15019. PLACE, Thomas. 1842-1910. Holl. BLAF.

15020. SERTOLI, Enrico. 1842-1910. It. DSB; BLAF. Also Histology.

15021. CHRISTIANI, Arthur. 1843-1887. Ger. P III; BLA & Supp.; GC. Also Physics.

15022. ENGELMANN, Theodor Wilhelm. 1843-1909. Ger./Holl. DSB; P III, IV; BMNH; Mort.; BLAF.

15023. FERRIER, David. 1843-1928. Br. DSB; BMNH; Mort.; BLA; GD4.
1. *The functions of the brain.* London, 1876. 2nd ed., 1886.
2. *The Croonian lectures on cerebral localisation.* London, 1890.

15024. KRATSCHMER, Florian. 1843-1922. Aus. BLAF.

15025. THANHOFFER, Lajos (or Ludwig). 1843-1909. Hung. Mort,; BLAF. Also Histology.
1. *Grundzüge der vergleichenden Physiologie und Histologie.* Stuttgart, 1885.

15026. ATWATER, Wilbur Olin. 1844-1907. U.S.A. DSB; Mort.; GE.

15027. BRUNTON, Thomas Lauder. 1844-1916. Br. DSB; BMNH; BLA & Supp.; GD3. See also 14919/1.

15028. BUNGE, Gustav von. 1844-1920. Dorpat/Switz. DSB; ICB; P III, IV; Mort.; BLAF. Also Biochemistry*.

15029. CURTIS, John Green. 1844-1913. U.S.A. GE.

15030. DASTRE, (Jules Frank) Albert. 1844-1917. Fr. BMNH; BLAF. See also 14851/10.

15031. DUVAL, Mathias Marie. 1844-1907. Fr. DSB; BMNH; Mort.; BLA & Supp. Also Histology and Embryology*. See also 14851/12.

15032. KLEIN, (Eduard) Emanuel. 1844-1925. Br. BMNH; BLAF; GF. Also Histology*. See also 14919/1.

15033. MIESCHER, (Johann) Friedrich (second of the name). 1844-1895. Switz. DSB; ICB; Mort.; BLAF. Also Biochemistry.
1. *Die histochemischen und physiologischen Arbeiten.* 2 vols. Leipzig, 1897.

15034. ASHBY, Henry. 1845-1908. Br. BLAF; GD2.
1. *Notes on physiology.* London, 1878. Another ed., 1884.

15035. CHAPMAN, Henry Cadwalader. 1845-1909. U.S.A. BMNH; BLAF; GE. Also Evolution*.
1. *A treatise on human physiology.* Philadelphia, 1887. X. 2nd ed., 1899.

15036. CHARLES, John James. 1845-1912. Br. BLAF.

15037. GOWERS, William Richard. 1845-1915. Br. ICB; Mort.; BLA & Supp.; GD3.

15038. KLUG, Nándor. 1845-1909. Hung. BMNH; BLAF.

15039. KÜLZ, (Rudolph) Eduard. 1845-1895. Ger. Mort.; BLAF.

15040. PASCHUTIN, Victor Vasilievich. 1845-1901. Russ. Mort.; BLAF.

15041. YEO, Gerald Francis. 1845-1909. Br. BLAF; GD2.
1. *Manual of physiology.* London, 1884. 2nd ed., 1887.

15042. COLASANTI, Giuseppe. 1846-1903. It. BLAF.

15043. EXNER (-EWARTEN), Sigmund. 1846-1926. Aus. P III,IV; BMNH; Mort.; BLA & Supp.; GF.

15044. FLEISCHL VON MARXOW, Ernst. 1846-1891. Aus. P IV; BLA & Supp.

15045. HÉGER, Paul. 1846-1925. Belg. ICB; BLAF.

15046. LANDOLT, Edmund. 1846-1926. Fr. BLAF.

15047. MORAT, Jean Pierre. 1846-1920. Fr. DSB; BLAF.

15048. MOSSO, Angelo. 1846-1910. It. DSB; ICB; BMNH; Mort.; BLAF.

15049. MÜLLER, Johann Jacob. 1846-1875. Ger./Switz. P III. Also Physics.

15050. STEFANI, Aristide. 1846-1925. It. BLAF.

15051. TARCHANOFF, Ivan Romanovich. 1846-1908. Russ. Mort.; BLA; BLAF (TARKHANOW).

15052. WATTEVILLE. Armand de. 1846-1925. Switz./Br. Mort.; BLAF.

15053. GASKELL, Walter Holbrook. 1847-1914. Br. DSB; Mort.; GD3. Also Zoology.

15054. GRÜTZNER, Paul Friedrich Ferdinand. 1847-1919. Ger. P IV; BMNH; BLAF.

15055. LADD (-FRANKLIN), Christine. 1847-1930. U.S.A. P III; Mort. Also Mathematics.

15056. LATSCHENBERGER, Johann. 1847-1905. Aus. BLAF.

15057. OTT, Isaac. 1847-1916. U.S.A. DSB; Mort.; BLAF; GE.

15058. PAGLIANI, Luigi. 1847-1932. It. BLAF.
 1. *Cognizioni della fisiologia intorno al sistema nervoso.* Turin, 1872.

15059. TALMA, Sape. 1847-1918. Holl. Mort.; BLAF.

15060. TAPPEINER, (Anton Josef Franz) Hermann. 1847-1927. Ger. P III,IV; BLAF.

15061. ZUNTZ, Nathan. 1847-1920. Ger. BMNH; BLAF.

15062. ANDERSON, Richard John. 1848-1914. Br. P IV.

15063. ECK, Nikolai Vladimirovich. 1848-1908. Russ. Mort.

15064. KLEMENSIEWICZ, Rudolf. 1848-1922. Aus. BLAF.

15065. KUNKEL, Adam Joseph. 1848-1905. Ger. BLAF.

15066. MARTIN, Henry Newell. 1848-1896. Br./U.S.A. DSB; BMNH; Mort.; GE.

15067. PAGE, Frederick James Montague. 1848-1907. Br. Mort.

15068. PEKELHARING, Cornelis Adrianus. 1848-1922. Holl. DSB; ICB; BMNH; BLAF.

15069. ROMANES, George John. 1848-1894. Br. DSB; ICB; BMNH; Mort.; GD. Also Science in General*, Evolution*, and Biography*. See also 15607/1.

15070. ALBERTONI, Pietro. 1849-1933. It. BLAF.

15071. BLIX, Magnus Gustav. 1849-1904. Swed. ICB; Mort.; BLAF.

15072. BOLL, Franz Christian. 1849-1878. Ger./It. P IV; BMNH; Mort.; BLA & Supp. Also Zoology and Anatomy.

15073. DUBOIS, Raphael. 1849-after 1918. Fr. BMNH.
 1. *Leçons de physiologie générale et comparée.* Paris, 1898.

15074. EWALD, August. 1849-1924. Ger. BLAF.

15075. FRANÇOIS-FRANCK, Charles Emile. 1849-1921. Fr. Mort.; BLAF.

15076. GAULE, Justus Georg. b. 1849. Switz. BLAF.

15077. HARTMANN, Arthur. 1849-1931. Ger. Mort.; BLAF.

15078. LUCHSINGER, (Johann) Balthasar. 1849-1886. Switz. ICB; BLA.

15079. MALERBA, Pasquale. 1849-1917. It. BLAF.

15080. PAVLOV, Ivan Petrovich. 1849-1936. Russ. DSB; ICB; Mort.; BLAF (PAWLOW).
1. *Selected works.* Moscow, 1955. Trans. from the Russian by S. Belsky.
2. *Essential works.* New York, 1966. Ed. with an introd. by M. Kaplan.

15081. STEINER, Isidor. 1849-1914. Ger. BMNH; BLAF.

15082. ADAMKIEWICZ, Albert. 1850-1921. Cracow. BMNH; BLA & Supp.

15083. AFONASSJEW, Michael. 1850-1910. Russ. BLA & Supp.

15084. HARTWELL, Edward Mussey. 1850-1922. Mort.

15085. LIVON, Charles Marie. 1850-1917. Fr. BLAF.

15086. REGNARD, Paul Marie Léon. 1850-1927. Fr. BLAF.

15087. RICHET, Charles Robert. 1850-1935. Fr. DSB; ICB; BMNH; Mort.; BLAF.
1. *Dictionnaire de physiologie.* 10 vols. Paris, 1895-1928. With P. Langlois, L. Lapicque, et al.
2. *Bibliographia physiologica, 1896.* 2 vols in 1. Paris, 1896-97.

15088. SHARPEY-SCHÄFER, Edward Albert. 1850-1935. Br. DSB; ICB; BMNH (SCHAEFER); Mort.; BLAF; GD5. Also Histology.
1. *Text-book of physiology.* 2 vols. london, 1898-1900.

15089. ARSONVAL, Arséne d'. 1851-1940. Fr. DSB; ICB; P IV; BLAF.

15090. FREDERICQ, Léon. 1851-1935. Belg. DSB; ICB; BMNH; BLAF.
1. *Genérátion et structure du tissu musculaire.* 1875.

15091. FUCHS, Ernst. 1851-1930. Aus. BLAF. Also Anatomy.

15092. MELTZER, Samuel James. 1851-1920. U.S.A. DSB; ICB; Mort.; BLAF; GE.

15093. ÖHRVALL, Hjalmar. 1851-1929. Swed. BLAF.

15094. STIRLING, William. 1851-1932. Br. BMNH; Mort.; BLA. See also 14978/1a.

15095. ALTMANN, Richard. 1852-1900. Ger. P IV; Mort.; BLAF. Also Anatomy and Cytology*.

15096. BIEDERMANN, Wilhelm. 1852-1929. Prague/Ger. Mort.; BLAF.
1. *Elektrophysiologie.* Jena, 1895. X.
1a. ——— *Electro-physiology.* 2 vols. London, 1896-98. Trans. by F.A. Welby.

15097. BRUBAKER, Albert Philson. 1852-1943. U.S.A. BLAF.

15098. CHARPENTIER, (Pierre Marie) Augustin. 1852-1916. Fr. P IV; BLAF.

15099. FREY, Maximilian (Ruppert Franz) von. 1852-1932. Ger. DSB; P III,IV; Mort.; BLAF.

15100. KRUKENBERG, Carl Friedrich Wilhelm. 1852-1889. Ger. BMNH.
1. *Vergleichend-physiologische Studien an des Küsten der Adrea. Erste Reihe.* 2 parts. Heidelberg, 1880.
1a. ——— *Zweite Reihe.* 5 parts. Ib., 1882-88.

15101. LANGLEY, John Newport. 1852-1925. Br. DSB; Mort.; BLAF; GD4. Also Histology.

15102. MUNK, Immanuel. 1852-1903. Ger. BLAF.
1. *Physiologie des Menschen und der Säugethiere.* Berlin, 1881. X. 2nd ed., 1888.

15103. STRASSER, Hans. 1852-1927. Switz. BMNH; BLAF. Also Anatomy and Embryology.

15104. VOIT, Erwin. 1852-1932. Ger. BLAF.

15105. VVEDENSKY, Nikolai Evgenievich. 1852-1922. Russ. DSB; ICB.

15106. GOTCH, Francis. 1853-1913. Br. Mort.; BLAF.

15107. KOSSEL, (Karl Martin Leonhard) Albrecht. 1853-1927. Ger. DSB; ICB; P IV; Mort.; BLAF. Also Biochemistry.

15108. KRIES, Johannes Adolf von. 1853-1928. Ger. Mort.; BLAF.

15109. LANGENDORFF, Oskar. 1853-1908. Ger. BLAF.

15110. LEA, Arthur Sheridan. 1853-1915. Br.
1. *The chemical basis of the animal body.* London, 1892.

15111. LUNIN, Nicolai Ivanovich. 1853-1937. Mort.

15112. MACKENZIE, James. 1853-1925. Br. ICB; Mort.; BLAF; GD4.

15113. MEYER, Hans Horst. 1853-after 1930. Ger. P III,IV; BLAF.

15114. SINGER, Jacob. 1853-1926. Prague. BLAF.

15115. STADELMANN, Ernst. b. 1853. Ger. Mort.; BLAF.

15116. TIGERSTEDT, Robert Adolf Armand. 1853-1923. Fin./Swed. Mort.; BLAF.

15117. VIERORDT, (Karl) Hermann. 1853-1943. Ger. BLAF.

15118. BEEVOR, Charles Edward. 1854-1908. Br. DSB; BLAF; GD2.

15119. CASH, John Theodore. 1854-1936. Br. ICB; GD5.

15120. MARCACCI, Arturo. 1854-1915. It. BLAF.

15121. NEUMEISTER, Richard. 1854-1905. Ger. BLAF.

15122. RAEVSKII, Igor Sviatoslavovich. 1854-1879. Russ. ICB; BMNH.

15123. ROY, Charles Smart. 1854-1897. BMNH; Mort.

15124. RUBNER, Max. 1854-1932. Ger. DSB; ICB; Mort.; BLAF.

15125. BOHR, Christian. 1855-1911. Den. P IV; Mort.; BLAF.

15126. CHABRY, Laurent. 1855-1893. Fr. DSB; ICB. Also Embryology.

15127. EWALD, (Ernst Julius) Richard. 1855-1921. Ger. Mort.; BLAF.

15128. HAY, Matthew. 1855-1932. Mort.

15129. SEWALL, Henry. 1855-1936. U.S.A. ICB.

15130. STEIN, Stanislav Aleksandr Fyodorovich. 1855-after 1910. Russ. Mort.

15131. VARIGNY, Henry Crosnier de. 1855-after 1927. Fr. BMNH.
Also Evolution*.

15132. WINKLER, Cornelis. 1855-1941. Holl. Mort.; BLAF.
1. *Opera omnia.* Haarlem, 1918-27.

15133. FANO, Giulio. 1856-1930. It. ICB; BMNH; BLAF.

15134. GEPPERT, Julius. 1856-after 1907. Ger. BLAF.

15135. HILL, Alexander. 1856-1929. Br.
1. *The physiologist's note-book. A summary of the present state of physiological science.* London, 1893.

15136. KÖNIG, Arthur. 1856-1901. Ger. DSB; P III,IV; BLAF. Also Physics.

15137. NICOLAÏDÈS, Rigas. 1856-1928. Greece. BLAF.

15138. WALLER, Augustus Désiré. 1856-1922. Br. Mort.; BLAF.
1. *An introduction to human physiology.* London, 1891. X. 2nd ed., 1893. 3rd ed., 1896.
2. *Lectures on physiology. 1st series. On animal electricity.* London, 1897.

15139. ABEL, John Jacob. 1857-1938. U.S.A. DSB; ICB; Mort.; BLAF.

15140. BABINSKI, Joseph Jules. 1857-1932. Fr. ICB; BLAF.

15141. BEKHTEREV, Vladimir Mikhailovich. 1857-1927. Russ. DSB; ICB; BLAF (BECHTEREW). Also Anatomy.

15142. DONALDSON, Henry Herbert. 1857-1938. U.S.A. DSB; Mort.
1. *The growth of the brain. A study of the nervous system in relation to education.* London, 1895.

15143. GLEY, Eugène. 1857-1930. Fr. Mort.; BLAF.

15144. HAYCRAFT, John Berry. 1857-1922. Br. BLAF.

15145. HORSLEY, Victor Alexander Haden. 1857-1916. Br. DSB; BMNH; Mort.; BLAF; GD3.
1. *The structure and functions of the brain and spinal cord.* London, 1892.

15146. MacWILLIAM, John Alexander. 1857-1937. Br. ICB; BLAF.

15147. MAREŠ, František. 1857-1942. Prague. BLAF.

15148. RACHFORD, Benjamin Knox. 1857-1929. U.S.A. GE.

15149. SHERRINGTON, Charles Scott. 1857-1952. Br. DSB; ICB; Mort.; BLAF; GD7.

15150. ZWAARDEMAKER, Hendrick. 1857-1930. Holl. Mort.; BLAF.

15151. GAGLIO, Gaetano. 1858-1925. It. ICB; BLAF.

15152. GOLDSCHEIDER, (Johannes Karl August Eugen) Alfred. 1858-1935. Ger. Mort.; BLAF.

15153. MACALLUM, Archibald Byron. 1858-1934. Canada. DSB; ICB; Mort.; BMNH.

15154. METZNER, Rudolf. 1858-1935. Ger./Switz. BLAF.

15155. NOORDEN, Carl Harko von. 1858-1944. Ger. Mort.; BLAF.

15156. SCHULZ, Oskar. b. 1858. Ger. BLAF.

15157. THIERFELDER, Hans. 1858-1930. Ger. Mort.; BLAF.

15158. FUCHS, Sigmund. 1859-1903. Aus. BLAF.

15159. GRIFFITHS, Arthur Bower. 1859-after 1903. Br. P IV. Also Microbiology*.
1. *Physiology of the invertebrata.* London, 1892.

15160. HAMBURGER, Hartog Jacob. 1859-1924. Holl. P IV; Mort.; BLAF; GF.

15161. HEDIN, Sven Gustaf. 1859-1933. Swed. P IV; Mort.; BLAF.

15162. HEYMANS, Jean François. 1859-1932. Belg. BMNH; BLAF.

15163. ISCOVESCO, Henri. b. 1859. Fr. BLAF.

15164. LEE, Frederic Schiller. 1859-1939. U.S.A. BLAF.

15165. LOEB, Jacques. 1859-1924. Ger./U.S.A. DSB; ICB; Mort.; BLAF; GE. Also Embryology.
1. *Einleitung in die vergleichende Gehirn-Physiologie und vergleichende Psychologie. Mit besonderer Berücksichtigung der wirbellosen Thiere.* Leipzig, 1899.

15166. PATON, Diarmid Noel. 1859-1928. Br. BMNH; BLAF; GD4.

15167. WEISS, Georges. 1859-1931. Fr. BLAF.

Appendix: The Vivisection Controversy

15168. *DIE VIVISECTION. Ihr wissenschaftlicher Werth und ihre ethische Berechtigung.* Leipzig, 1877.

15169. MACAULAY, James. 1817-1902. Br. GD2. A physician who was a strenuous opponent of vivisection as a result of witnessing Magendie's experiments on animals.
1. *Vivisection. Is it scientifically useful or morally justifiable?* London, 1881.

15170. OWEN, Richard (first of the name). 1804-1892. Br. DSB; ICB; BMNH; Mort.; BLA; GA; GB; GD. Also Geology*, Natural History*, and Zoology*. See also Index.
1. *Experimental physiology. Its benefits to mankind.* London, 1882.

15171. *THE VIVISECTION controversy. A selection of speeches and articles.* London, 1883.

15172. *VIVISECTION. "The Times" correspondence, Dec.-Feb., 1884-85.* London, 1885.

15173. WILKS, Samuel. 1824-1911. Br. BLA; GD.
1. *The value and necessity of experiments for the acquirement of knowledge.* London, 1889.

15174. RICHARDSON, Benjamin Ward. 1828-1896. Br. DSB; ICB; BLA; GD1. Also Physiology*.
1. *Biological experimentation. Its function and limits.* London, 1896.

3.20 MICROBIOLOGY

15175. SPALLANZANI, Lazzaro. 1729-1799. It. DSB; ICB; P II; BMNH; Mort.; BLA; GA; GB. Other entries: see 3167.

15176. MÜLLER, Otto Frederik. 1730-1784. Den. DSB; ICB; BMNH; BLA; GA (MULLER). Also Botany and Zoology*.
  1. *Animalcula infusoria, fluviatilia et marina.* Copenhagen, 1786.

15177. WRISBERG, Heinrich August. 1739-1808. Ger. ICB; BMNH; Mort.; BLA; GC. Also Anatomy.

15178. PRÉVOST, Isaac Bénédict. 1755-1819. Fr. DSB; ICB; P II; BMNH; GA. Also Various Fields.

15179. BASSI, Agostino Maria. 1773-1856. It. DSB; ICB; BLA & Supp.

15180. WEISSE, Johann Friedrich. 1792-1869. Russ. BLA; GC.

15181. EHRENBERG, Christian Gottfried. 1795-1876. Ger. DSB; ICB; P I,III; BMNH; Mort.; BLA & Supp.; GA; GB; GC. Also Geology*, and Zoology.
  1. *Die Infusionthierschen als vollkommene Organismen.* 2 vols. Leipzig, 1838. RM/L.
  2. *Mikroskopische Analyse des curlandischen Meteorpapiers von 1686.* Berlin, 1839. RM.
  3. *Passat-Staub und Blut-Regen. Ein grosses organisches unsichtbares Wirken und Leben in der Atmosphäre.* Berlin, 1849. RM.

15182. LAURENT, Paul. 1798-1857. BMNH.
  1. *Etudes physiologiques sur les animalcules des infusions végétales comparés aux organes élémentaires des végétaux.* Nancy/Paris, 1854-58.

15183. POLLENDER, (Franz) Aloys (Antoine). 1800-1879. Ger. DSB; ICB; BMNH; BLA. Also Botany.

15184. POUCHET, Félix Archimède. 1800-1872. Fr. DSB; ICB; BMNH; BLA; GA. Also Natural History*, Zoology*, Embryology*, and History of Science*.
  1. *Hétérogénie; ou, Traité de la génération spontanée, basé sur des nouvelles expériences.* Paris, 1859. RM.
  2. *Nouvelles expériences sur la génération spontanée et la résistance vitale.* Paris, 1864. RM.

15185. DONNÉ, Alfred. 1801-1878. Fr. P III; Mort.; BLA. Also Physiology.

15186. PRITCHARD, Andrew. 1804-1882. Br. P III; BMNH; GD. Also Microscopy*.
  1. *A history of infusoria, living and fossil.* London, 1841. X. 2nd ed., 1849. 4th ed., 1861.

15187. BRAUELL, Friedrich August. 1807-1882. Ger./Russ. ICB; BLA & Supp.

15188. GRUBY, David. 1810-1898. Aus./Fr. DSB; ICB; BLA & Supp.

15189. SCHWANN, Theodor Ambrose Hubert. 1810-1882. Ger./Belg. DSB; ICB; P II,III; BMNH; Mort.; BLA & Supp.; GB; GC. Also Physiology and Cell Theory*#.

15190. BAILEY, Jacob Whitman. 1811-1857. U.S.A. P III; BMNH; GE.

15191. BUDD, William. 1811-1880. Br. DSB; BLA & Supp.; GD.

15192. DAVAINE, Casimir Joseph. 1812-1882. Fr. DSB; ICB; BMNH; BLA & Supp. Also Zoology*.

15193. KUTORGA, Stepan Semenovich. 1812-1861. Russ. P III; BMNH. Also Geology*.
1. [Trans.] *Naturgeschichte der Infusionthiere, vorzüglich nach Ehrenberg's Beobachtungen.* Karlsruhe, 1841.

15194. SCHULZE, Franz Ferdinand. 1815-1873. Ger. DSB; P II,III; BMNH; GC. Also Chemistry.

15195. BÉCHAMP, (Pierre Jacques) Antoine. 1816-1908. Fr. DSB Supp.; ICB; P III,IV; BMNH; BLA & Supp. Also Chemistry.

15196. COZE, Léon. 1817-1896. Fr. BMNH; BLA & Supp.

15197. STEIN, (Samuel) Friedrich (Nathaniel) von. 1818-1885. BMNH. Also Zoology.
1. *Der Organismus der Infusionthiere.* 3 vols in 4. Leipzig, 1859-83.

15198. HOFFMANN, (Heinrich Karl) Hermann. 1819-1891. Ger. ICB; P III,IV; BMNH; GC. Also Botany.

15199. TYNDALL, John. 1820-1893. Br. DSB; ICB; P II,III,IV; GB; GD. Also Science in General*, Physics*, Geology*, and Biography*. See also Index.
1. *Essays on the floating matter of the air in relation to putrefaction and infection.* London, 1881. RM. 2nd ed., 1883. American ed., New York, 1882; reprinted, 1966.

15200. PASTEUR, Louis. 1822-1895. Fr. DSB; ICB; P II,III; BMNH; Mort.; BLA; G. Also Chemistry*. See also 15566/1.
1. *Sur les corpuscules organisés qui existent dans l'atmosphère. Examen de la doctrine des générations spontanées.* [Periodical article, 1862] German trans. Leipzig, 1892. (OKEW. 39)
2. *Etudes sur le vin. Ses maladies, causes qui les provoquent ....* Paris, 1866. RM.
3. *Etudes sur la maladie des vers à soie.* 2 vols. Paris, 1870. RM/L.
4. *Etudes sur la bière. Ses maladies, causes qui les provoquent....* Paris, 1876. RM.
4a. ——— *Studies on fermentation. The diseases of beer.* London, 1879. Trans. by F. Faulkner and D.C. Robb. RM/L. Reprinted, New York, 1969.
5. *Examen critique d'un écrit posthume de Claude Bernard sur la fermentation.* Paris, 1879. RM.

Correspondence and Collected Works

6. *Jacob C. Jacobsen, Louis Pasteur: Lettres échangées au cours des années 1878-1882.* Copenhagen 1964. (Jacobsen was the originator of the Carlsberg Foundation; see ICB)
7. *Correspondence of Pasteur and Thuillier concerning anthrax and swine fever vaccinations.* University, Alabama, 1968. Trans. and ed. by R.M. Frank and D. Wrotnowska.
8. *Oeuvres.* 7 vols in 8. Paris, 1922-39. Ed. by P. Vallery-Radot.

15201. WILLEMS, Louis. 1822-1907. Belg. ICB; BLA; GF.

15202. COHN, Ferdinand Julius. 1828-1898. Ger. DSB; ICB; BMNH; BLA & Supp.; GB; GC. Also Botany*.

15203. ARCHER, William. 1830-1897. Br. BMNH; GD1. Also Microscopy.

15204. ROBERTS, William (second of the name). 1830-1899. Br. ICB; BLA; GD1. Also Physiology*.

15205. DE BARY, (Heinrich) Anton. 1831-1888. Ger. DSB; GB. Under BARY: ICB; BMNH; BLA & Supp.; GC. Also Botany*.
1. *Vorlesungen über Bacterien.* Leipzig, 1885. X. 2nd ed., 1887. X.
1a. ——— *Lectures on bacteria.* Oxford, 1887. Trans. by H.E.F. Garnsey. Rev. by I.B. Balfour. Another ed., 1898.

15206. HALLIER, Ernst Hans. 1831-1904. Ger. DSB; P III,IV; BMNH; BLA. Also Botany* and History of Science*.

15207. KLEBS, (Theodor Albrecht) Edwin. 1834-1913. Ger./Switz. ICB; Mort.; BLA. Also Microscopy.

15208. EBERTH, Carl Joseph. 1835-1926. Ger. DSB; BMNH; BLA & Supp. Also Anatomy.

15209. EVANS, Griffith. 1835-1935. Br./India. ICB; BLAF.

15210. FELTZ, Victor Timothée. 1835-1893. Fr. BLA & Supp.

15211. BASTIAN, Henry Charlton. 1837-1915. Br. DSB; BMNH; BLA & Supp. Also Physiology*.
1. *The modes of origin of lowest organisms. Including a discussion of the experiments of M. Pasteur.* London, 1871.
2. *The beginnings of life. Being some account of the nature, modes of origin and transformations of lower organisms.* 2 vols. London, 1872.

15212. CORNIL, André Victor. 1837-1908. Fr. BMNH; BLA & Supp.

15213. SCHRÖTER, Josef. 1837 (or 1835)-1894. Ger. BMNH; BLA; GC. Also Botany.

15214. DÖNITZ, (Friedrich Karl) Wilhelm. 1838-1912. Ger./Japan. BLAF.

15215. STERNBERG, George Miller. 1838-1915. U.S.A. ICB; BMNH; BLA; BLAF; GE. Also Microscopy*.
1. *A textbook of bacteriology.* London, 1897.
See also 15259/1a.

15216. COHNHEIM, Julius Friedrich. 1839-1884. Ger. ICB; Mort.; BLA & Supp.; GC. Also Anatomy.

15217. JOHNE, (Heinrich) Albert. 1839-1910. Ger. BLAF.

15218. DUCLAUX, (Pierre) Emile. 1840-1904. Fr. DSB; P III,IV & Supp.; BMNH; BLAF. Also Biochemistry.
1. *Traité de microbiologie*. 4 vols. Paris, 1891-1901.

15219. MAGGI, Leopoldo. 1840-1905. It. BMNH; BLAF.

15220. SCHENK, (Samuel) Leopold. 1840-1902. Aus. BMNH; BLAF. Also Embryology*.
1. *Grundriss der Bakteriologie*. Vienna, 1893. X.
1a. ——— *Manual of bacteriology*. London, 1893. Trans. by W.R. Dawson.

15221. HANSEN, (Gerhard Henrik) Armauer. 1841-1912. Nor. DSB; ICB; BMNH; BLA.

15222. LEWIS, Timothy Richards. 1841-1886. Br. DSB; BMNH; BLA.

15223. BIRCH-HIRSCHFELD, Felix Victor. 1842-1899. Ger. GC.

15224. DALLINGER, William Henry. 1842-1909. Br. BMNH. GD2. Also Microscopy.

15225. HANSEN, Emil Christian. 1842-1909. Den. DSB; BMNH. Also Botany.

15226. ROSENBACH, (Anton) Julius (Friedrich). 1842-1923. Ger. BLAF.

15227. TROUESSART, Edouard Louis. 1842-1927. Fr. BMNH. Also Zoology.
1. *Les microbes, les ferments et les moisissures*. Paris, 1886. X.
1a. ——— *Microbes, ferments and moulds*. London, 1886.

15228. BOLLINGER, Otto von. 1843-1909. Ger. ICB; BLAF.

15229. CUNNINGHAM, David Douglas. 1843-1914. Br./India. BMNH; BLA & Supp.

15230. FODÓR, Josef. 1843-1901. Hung. BLAF.

15231. KOCH, (Heinrich Hermann) Robert. 1843-1910. Ger. DSB; ICB; Mort.; BLAF; GB. See also 15318/1.

15232. SATTERTHWAITE, Thomas Edward. 1843-1934. U.S.A. BLAF.
1. *An introduction to practical bacteriology*. Detroit, 1887.

15233. MANSON, Patrick. 1844-1922. Br./China. DSB; ICB; BMNH; BLAF; GD4.

15234. OGSTON, Alexander. 1844-1929. Br. BLAF.

15235. LAVERAN, (Charles Louis) Alphonse. 1845-1922. Fr. DSB; ICB; BLAF.

15236. METCHNIKOFF, Elie. 1845-1916. Russ./Fr. DSB; ICB (MECHNIKOV); BMNH (MECHNIKOV); BLAF (METSCHNIKOW). Also Zoology and Embryology.

15237. STRAUS, Isidore. 1845-1896. Fr. BLAF.

15238. WEICHSELBAUM, Anton. 1845-1920. Aus. DSB; BLAF.

15239. WOLFFHÜGEL, Gustav. 1845-1899. Ger. BLAF.

15240. ARLOING, Saturnin. 1846-1911. Fr. BMNH; BLAF. Also Zoology.

15241. ZOPF, Wilhelm Friedrich. 1846-1909. Ger. P IV; BMNH; BLAF. Also Botany.

15242. FLÜGGE, Karl Georg Friedrich Wilhelm. 1847-1923. Ger. BMNH; BLAF.
1. *Fermente und Microparasiten*. 1883. X.
1a. ——— 2nd ed. *Die Mikroorganismen*. Leipzig, 1886.
1b. ——— ——— *Micro-organisms*. London, 1890.

15243. FRIEDLÄNDER, Carl. 1847-1887. Ger. BMNH; BLA; BLAF; GC. Also Microscopy*.

15244. MAFFUCCI, Angiolo. 1847-1903. It. BLAF.

15245. MARCHIAFAVA, Ettore. 1847-1935. It. DSB; BLAF.

15246. ORTH, Johannes Joseph. 1847-1923. Ger. BMNH; BLAF; GF.

15247. PERRONCITO, Edoardo. 1847-1936. It. DSB; BMNH; BLAF.

15248. SALOMONSEN, Carl Julius. 1847-1924. Den. DSB; BLAF.

15249. BANG, Bernhard Laurits Frederick. 1848-1932. Den. ICB; BLA.

15250. BAUMGARTEN, Paul von. 1848-1928. Ger. BMNH; BLAF.

15251. FRAENKEL, Albert. 1848-1916. Ger. BLAF.

15252. GÄRTNER, August Anton Hieronymus. 1848-1934. Ger. BLAF.

15253. ISRAEL, James. 1848-1926. Ger. BLAF; GF.

15254. JÖRGENSEN, Alfred Peter Carslund. 1848-1925. BMNH.

15255. PFEIFFER, August. 1848-1919. Ger. BLAF.

15256. FERRÁN Y CLUA, Jaime. 1849-1929. Spain. BLAF.

15257. PRUDDEN, Theophil Mitchell. 1849-1924. U.S.A. DSB; BLAF; GE. Also Histology.

15258. SCHOTTELIUS, Max. 1849-1919. Ger. BMNH; BLAF.

15259. MAGNIN, Antoine. fl. 1878-1902. Fr. BMNH. Also Biography*.
1. *Les bactéries*. Paris, 1878. X.
1a. ——— *Bacteria*. Boston, 1880. Trans. by G.M. Sternberg.

15260. BEHLA, Robert. 1850-1921. Ger. BLAF.

15261. BOSTROEM, Eugen. 1850-1928. Ger. BLAF.

15262. BUCHNER, Hans Ernst August. 1850-1902. Ger. BMNH; BLA & Supp. See also 12184/3.

15263. GAFFKY, Georg Theodor August. 1850-1918. Ger. DSB; BLAF.

15264. MIQUEL, Pierre. 1850-1922. Fr. BMNH.

15265. NOCARD, Edmond Isidore Etienne. 1850-1903. Fr. BLAF.

15266. TALAMON, Charles. 1850-1929. Fr. BLAF.

15267. UNNA, Paul Gerson. 1850-1929. Ger. ICB; BLAF.

15268. WELCH, William Henry. 1850-1934. U.S.A. DSB; ICB; BLAF; GE.

15269. BEIJERINCK, Martinus Willem. 1851-1931. Holl. DSB Supp.; ICB; BMNH. Also Botany*.
 1. *Verzamelde geschriften*. 5 vols. The Hague. 1921-22.

15270. CHAMBERLAND, Charles Edouard. 1851-1908. Fr. DSB; BLAF.

15271. CHANTEMESSE, André. 1851-1919. Fr. BLAF.

15272. DIXON, Samuel Gibson. 1851-1918. U.S.A. BLAF.

15273. PROSKAUER, Bernhard. 1851-1915. Ger. BLAF.

15274. CHEYNE, William Watson. 1852-1932. Br. ICB; BMNH; BLAF; GD5.

15275. EMMERICH, Rudolf. 1852-1914. Ger. BLAF.

15276. FISCHER, Bernhard (first of the name). 1852-1915. Ger. BMNH; BLAF.

15277. HUEPPE, Ferdinand Adolph Theophil. 1852-1938. Ger./Prague. BMNH; BLAF.
 1. *Die Methoden der Bakterien-Forschung*. Wiesbaden, 1885. X. 4th ed., 1889.

15278. KITASATO, Shibasaburo. 1852-1931. Japan. DSB; ICB; BLAF.

15279. LOEFFLER, Friedrich August Johannes. 1852-1915. Ger. DSB; ICB; BMNH; BLAF.

15280. PERTIK (or PERTICK), Otto. 1852-1913. Hung. BLAF.

15281. PETRI, Richard Julius. 1852-1921. Ger. BLAF.

15282. FRAENKEL, Eugen. 1853-1925. Ger. BLAF.

15283. GRAM, (Hans) Christian (Joachim). 1853-1938. Den. DSB; BLAF.

15284. GRUBER, (Friedrich) August. b. 1853. Ger. BMNH.

15285. GRUBER, Max von. 1853-1927. Aus./Ger. DSB; BLAF; GF.

15286. ROUX, (Pierre Paul) Emile. 1853-1933. Fr. DSB; ICB; BLAF.

15287. SPITTA, Edmund Johnson. 1853-1921. Br. BMNH. Also Microscopy*.

15288. TIZZONI, Guido. 1853-1932. It. BLAF (in Nachträge).

15289. BABÈS, Victor. 1854-1926. Bucharest. ICB; BMNH; BLA & Supp.

15290. BEHRING, Emil von. 1854-1917. Ger. DSB; ICB; BLAF.

15291. EHRLICH, Paul. 1854-1915. Ger. DSB; ICB; Mort.; BLAF.
 Other entries: see 3361.

15292. FEHLEISEN, Friedrich. 1854-1924. Ger. BLA & Supp.

15293. GÜNTHER, Carl Oscar. 1854-1929. Ger. BLAF.

15294. NEELSEN, Friedrich (Carl Adolf). 1854-1894. Ger. BLAF.

15295. SMITH, Erwin Frink. 1854-1927. U.S.A. DSB; ICB; BMNH; GE.

15296. BRUCE, David. 1855-1931. Br. DSB; ICB; BMNH; BLAF; GD5.

15297. ESMARCH, Erwin von. 1855-1915. Ger. BLAF.

15298. NEISSER, Albert Ludwig Sigesmund. 1855-1916. Ger. DSB; ICB.

15299. NETTER, Arnold. 1855-1936. Fr. BLAF.

15300. RIBBERT, Hugo. 1855-1920. Ger. BMNH; BLAF.

15301. WOODHEAD, German Sims. 1855-1921. Br. BMNH; BLAF.
    1. *Bacteria and their products.* London, 1892.

15302. CHARRIN, Albert. 1856-1907. Fr. BLAF.

15303. DUFLOCQ, Paul. 1856-1903. Fr. BLAF.
    1. *Leçons sur les bactéries pathogènes.* Paris, 1897.

15304. ERNST, Harold Clarence. 1856-1922. U.S.A. BLAF; GE.

15305. HAUSER, Gustav. 1856-1935. Ger. BMNH; BLAF. Also Zoology.

15306. SEMPLE, David. 1856-1937. Br. GF.

15307. VINOGRADSKY, Sergey Nikolaevich. 1856-1953. Russ./Fr. DSB; ICB (under W); BMNH.

15308. BUJWID, Odo Feliks Kazimirs. 1857-1932. ICB.

15309. ESCHERICH, Theodor. 1857-1911. Ger./Aus. DSB; Mort.; BLAF.

15310. GARRÉ, Karl. 1857-1928. Switz./Ger. BLAF.

15311. HEIM, Ludwig. 1857-after 1931. Ger. BLAF.

15312. LUSTIG, Alessandro. 1857-1937. It. DSB; BLAF.

15313. RATTONE, Giorgio. 1857-1929. It. BLAF.

15314. ROSS, Ronald. 1857-1932. Br. DSB; ICB; BMNH; BLAF; GD5.

15315. BLOCHMANN, Friedrich. 1858-after 1924. Ger. BMNH. Also Zoology and Embryology.

15316. BUMM, Ernst von. 1858-1925. Ger. BLAF.

15317. CORNET, Georg. 1858-1915. Ger. BLAF.

15318. CROOKSHANK, Edgar March. 1858-1928. Br. BMNH.
    1. *An introduction to practical bacteriology based upon the methods of Koch.* London, 1886. X.
    1a. ——— 2nd ed. *Manual of bacteriology.* Ib., 1887. 3rd ed., 1890.

15319. FISCHER, Alfred. 1858-1913. BMNH. Also Botany and Cytology*.

15320. FRANKLAND, Percy Faraday. 1858-1946. Br. DSB; P IV; GD6. Also Chemistry and Biography*.

15321. KITT, Theodor. 1858-1941. Ger. BLAF.

15322. KOCH, Alfred. 1858-1922. Ger. BLAF.

15323. LEHMANN, Karl Bernhard. 1858-1961. Ger. ICB; BMNH; BLAF.

15324. PALTAUF, Richard. 1858-1924. Aus. BLAF.

15325. PFEIFFER, Richard Friedrich Johannes. 1858-after 1912. Ger. BMNH; BLAF.

15326. PLAUT, Hugo Carl. 1858-1928. Ger. BMNH; BLAF.

15327. TAVEL, Ernest. 1858-1912. Switz. BLAF.

15328. ANDREWES, Frederick William. 1859-1932. Br. ICB; GD5.

15329. BOTKIN, Sergei Sergeievich. 1859-1910. Russ. BLAF.

15330. GAMALEYA, Nikolay Fyodorovich. 1859-1949. Russ. DSB.

15331. GRIFFITHS, Arthur Bower. 1859-after 1903. Br. P IV. Also Physiology*.
1. *Researches on micro-organisms*. London, 1891.

15332. MOORE, Veranus Alva. 1859-1931. U.S.A. BMNH; GE.

15333. RUFFER, Marc Armand. 1859-1917. Br. DSB; BLAF.

15334. SMITH, Theobald. 1859-1934. U.S.A. DSB; ICB; BMNH; BLAF; GE1.

15335. WERNICKE, Erich Arthur Emmanuel. 1859-1928. Ger. BLAF.

3.21 AUTOBIOGRAPHY
AND OTHER PERSONAL WRITINGS

15336. SLOANE, Hans. 1660-1753. Br. DSB; ICB; P II; BMNH; BLA; GA; GB; GD. Also Natural History and Botany*.
1. *The will of Sir Hans Sloane.* London, 1753.

15337. WHISTON, William. 1667-1752. Br. DSB; P II; BMNH; GA; GB; GD. Also Science in General*, Astronomy*, Mechanics*, Physics*, and Part 2*. See also Index.
1. *Memoirs of the life and writings of Mr William Whiston, written by himself.* 3 parts. London, 1749-50. X. 2nd ed., 1753.

15338. FRANKLIN, Benjamin. 1706-1790. U.S.A. The famous statesman. DSB; ICB; P I; G. Also Physics* and Earth Sciences. See also Index.
1. *Autobiographical writings.* New York, 1945. Ed. by C. Van Doren.
2. *Memoirs.* Berkeley/Los Angeles, 1949. Ed. by M. Ferrand.
3. *The autobiography.* New Haven, 1964. Ed. by L.W. Labaree et al.

15339. LINNAEUS (later VON LINNÉ), Carl. 1707-1790. Swed. DSB; ICB; P I; BMNH. Under LINNÉ: BLA & Supp.; GA; GB. Also Natural History*#, Botany*, and Zoology*. See also Index.
1. *Orbis eruditi judicium de Caroli Linnaei, M.D., scriptis.* Stockholm, 1741. RM. A rare pamphlet. It was published by Linnaeus in reply to the criticisms of J.G. Wallerius, a competitor for the chair of medicine in the University of Uppsala. The *Collectio epistolarum* of 1792 (item 10958/13) has added to it the following appendix which reprints the above work together with Wallerius' criticisms: *Opuscula pro et contra virum immortalem scripta, extra Sveciam rarissima.*

15340. HALLER, Albrecht von. 1708-1777. Switz./Ger. DSB; ICB; P I; BMNH; Mort.; BLA & Supp.; GA; GB; GC. Also Botany*, Anatomy*, and Physiology*#.
1. *Tagebuch seiner Beobachtungen über Schriftsteller und über sich selbst.* 2 vols. Bern, 1787. Ed. by J.G. Heinzmann. X. Reprinted, Frankfurt, 1971.
2. *Tagebuch seiner Studienreise nach London, Paris, Strassburg und Basel, 1727-1728.* 2nd ed., rev. and enl. Bern, 1968, Ed. with notes by E. Hintzsche.
3. *Tagebücher seiner Reisen nach Deutschland, Holland und England, 1723-1727.* Bern, 1971. Ed. by E. Hintzsche.

15341. FERGUSON, James (first of the name). 1710-1776. Br. DSB; P I; GA; GB; GD. Also Astronomy* and Physics*.
1. *Life of James Ferguson, F.R.S., in a brief autobiographical*

account, and further extended memoir. Edinburgh, 1867. By E. Henderson.

15342. LE SAGE, George Louis (second of the name). 1724-1803. Switz. DSB; ICB; P I; GA. Also Mechanics.
1. *Notice de la vie et des écrits ... Redigée d'après ses notes.* Geneva, 1805. By P. Prévost.

15343. PENNANT, Thomas. 1726-1798. Br. DSB; ICB; BMNH; GA; GB; GD. Also Natural History* and Zoology*.
1. *The literary life ... by himself.* London, 1793.

15344. PRIESTLEY, Joseph. 1733-1804. Br. DSB; ICB; P II; GA; GB; GD; GE. Also Various Fields*#, Physics*, and Chemistry*.
1. *Memoirs to the year 1795, written by himself. With a continuation to the time of his decease by his son Joseph Priestley* [1768-1833]. 2 vols. London, 1806-07. Another ed., Northumberland (U.S.A.), 1806. RM. Other eds, London, 1809 and 1833.
1a. ——— *The memoirs.* Washington, 1964. An abridgement. Ed. by J.T. Bayer.
1b. ——— *A scientific autobiography.* [Together with] *Selected correspondence.* Cambridge, Mass., 1966. Ed. with commentary by R.E. Schofield.
1c. ——— *Autobiography* ... [Together with] *An account of further discoveries on air* [Periodical article, 1775]. Bath, 1970. Ed. by J. Lindsay.

15345. WATSON, Richard. 1737-1816. Br. DSB; ICB; P II; GA; GB; GD. Also Chemistry*.
1. *Anecdotes of the life of Richard Watson written by himself at different intervals and revised in 1814.* London, 1817.

15346. RUSH, Benjamin. 1746-1813. U.S.A. DSB; ICB; P II; Mort.; BLA & Supp.; GA; GB; GE. Also Chemistry*#.
1. *The autobiography.* Princeton, 1948. "Now first printed in full from the original manuscripts." Ed. with introd. and notes by G.W. Corner.

15347. CASSINI, Jean Dominique. 1748-1845. Fr. DSB; ICB; P I; GA; GB. Also Astronomy, Geodesy*, and History of Astronomy*.
1. ... *la vie de J.D. Cassini, écrite par lui-même....* 1810. See 15642/1.

15348. WELLS, William Charles. 1757-1817. Br./U.S.A. DSB; ICB; P II; BMNH; BLA; GD; GE. Other entries: see 3243.
1. ... *With a memoir of his life, written by himself.* 1818. See 3243/1.

15349. AMPÈRE, André Marie. 1775-1836. Fr. DSB; ICB; P I; GA; GB. Also Science in General*, Mathematics*, and Physics*#. See also Index.
1. *Journal et correspondance (de 1793 à 1805).* Paris, 1869. Ed. by Mme H. Cheuvreux. X.
1a. ——— *The story of his love. Being the journal and early correspondence of A.M. Ampère with his family ... 1793-1804.* London, 1873.

15350. CANDOLLE, Augustin Pyramus de. 1778-1841. Switz. DSB; ICB; P I (DECANDOLLE); BMNH; GA; GB. Also Botany*.

1. *Mémoires et souvenirs, écrits par lui-même et publiés par son fils.* Geneva, 1862. RM.

15351. DAVY, Humphry. 1778-1829. Br. DSB; ICB; P I; GA; GB; GD. Also Chemistry*#. See also Index.
1. *Consolations in travel; or, The last days of a philosopher.* London, 1830. Ed. by John Davy. RM/L. Reprinted, ib., 1889.

15352. BERZELIUS, Jöns Jacob. 1779-1848. Swed. DSB; ICB; P I; BLA & Supp.; GA; GB. Also Chemistry*#.
1. *Selbstbiographische Aufzeichnungen.* Leipzig, 1903. Trans. from the Swedish by Emma Wöhler (the wife of F. Wöhler) and rev. by H.G. Söderbaum. Ed. by G.W.A. Kahlbaum.
1a. ——— *Autobiographical notes.* Baltimore, 1934. Trans. by O. Larsell.

15353. BROUGHAM, Henry. 1779-1868. Br. The lawyer and politician. P I,III; GA; GB; GD. Also Science in General*#, Mechanics*, Physics*, and Biography*.
1. *The life and times of Henry, Lord Brougham, written by himself.* 3 vols. Edinburgh, 1871-72.

15354. SOMERVILLE, Mary Fairfax Greig. 1780-1872. Br. DSB; ICB; P II,III; GB; GD. Also Science in General*, Astronomy*, and Geology*.
1. *Personal recollections, from early life to old age. With selections from her correspondence.* London, 1873. Ed. by her daughter, Martha Somerville.

15355. WATERTON, Charles. 1782-1865. Br. DSB; ICB; BMNH; GB; GD. Also Zoology*. See also 15444/1.
1. ... *With an autobiography* ... 1838-44. See 12902/1.

15356. RAFINESQUE, Constantine Samuel. 1783-1840. It./U.S.A. DSB; ICB; BMNH; GE. Also Natural History*, Botany*, and Zoology*.
1. *A life of travels and researches in North America and South Europe.* Philadelphia, 1836. RM.

15357. AUDUBON, John James. 1785-1851. U.S.A. DSB; ICB; BMNH; GA; GB; GE. Also Zoology*.
1. *The life and adventures ... Edited from materials supplied by his widow.* London, 1868. Ed. by R. Buchanan.
2. *Audubon, by himself. A profile from* [his own] *writings.* Garden City, N.Y., 1969. Ed. by A. Ford.

15358. KLÖDEN, Karl Friedrich von. 1786-1856. Ger. P I; BMNH; GC. Also Geology*.
1. *Jugenderinnerungen.* Leipzig, 1874. Ed. by M. Jahn. X.
1a. ——— *The self-made man: Autobiography.* 2 vols. London, 1876. Trans. by A.M. Christie.

15359. BABBAGE, Charles. 1792-1871. Br. DSB; ICB; P I,III; GA; GB; GD. Also Science in General*, Mathematics*, and Geology*.
1. *Passages from the life of a philosopher.* London, 1864. Reprinted, Farnborough, 1969.

15360. BAER, Karl Ernst von. 1792-1876. Ger./Russ. DSB; ICB; P I, III; Mort.; BMNH; BLA & Supp.; GA; GB; GC. Also Embryology*.
1. *Nachrichten über Leben und Schriften ... mitgetheilt von*

ihm selbst. St. Petersburg, 1866. X. 2nd ed., Brunswick, 1886; reprinted, Hanover/Döhren, 1972.

15361. AIRY, George Biddell. 1801-1892. Br. DSB; ICB; P I,III,IV; GB; GD1. Also Mathematics*, Physics*, and Astronomy*.
1. *Autobiography*. Cambridge, 1896. Ed. by W. Airy. RM/L.

15362. MILLER, Hugh. 1802-1856. Br. DSB; ICB; P II; BMNH; GA; GB; GD. Also Geology*. For biographies see Index.
1. *First impressions of England and its people*. London, 1846. X. Another ed., 1857.
2. *My schools and schoolmasters; or, The story of my education*. Edinburgh, 1854. X. 2nd ed., 1854. 3rd ed., 1854; reprinted, Farnborough, 1971. Another ed., 1858. 15th ed., 1869. 24th ed., 1869. Another ed., 1885.

15363. DARWIN, Charles Robert. 1809-1882. Br. DSB; ICB; P III; BMNH; G. Also Geology*, Natural History*, Botany*, Zoology*, and Evolution*#. See also Index.
1. *Autobiography*. See 13873/8,9.
2. *Autobiographies: Charles Darwin, T.H. Huxley*. New York, 1974. Ed. with an introd. by G. de Beer.

15364. QUATREFAGES DE BRÉAU, (Jean Louis) Armand de. 1810-1892. Fr. DSB; ICB; BMNH; Mort.; BLA; GA; GB. Also Zoology* and Biography*.
1. *Souvenirs d'un naturaliste*. 2 vols. Paris, 1854.

15365. GALOIS, Evariste. 1811-1832. Fr. DSB; ICB; P I; GB. Also Mathematics*.
1. *Un souvenir d'enfance*. Paris, 1974. Ed. by P. Berloquin.

15366. HARVEY, William Henry. 1811-1866. Br. DSB; BMNH; GD. Also Botany*.
1. *Memoir ... With selections from his journal and correspondence*. London, 1869. (The memoir is almost entirely concerned with his private letters.)

15367. SIEMENS, (Ernst) Werner von. 1816-1892. Ger. DSB; ICB; P II, III,IV; GB; GC. Also Physics*#.
1. *Lebenserinnerungen*. Berlin, 1892. X.
1a. ——— *Personal recollections*. London, 1893. Trans. by W.C. Coupland.
1b. ——— ——— Reprint: *Inventor and entrepreneur: Recollections of Werner von Siemens*. Ib., 1966.

15368. SPENCER, Herbert. 1820-1903. Br. The philosopher. DSB; ICB; BMNH; GB; GD2. Also Science in General*, Biology in General*, and Evolution*.
1. *An autobiography*. 2 vols. London, 1904. RM/L. Reprinted, Farnborough, 1968.

15369. CROLL, James. 1821-1890. Br. DSB; P III,IV; BMNH; GB; GD1. Also Geology*, Climatology*, and Evolution*.
1. *Autobiographical sketch of James Croll, with memoir of his life and work*. London, 1896. By J.C. Irons.

15370. GALTON, Francis. 1822-1911. Br. DSB; ICB; P III,IV; BMNH; Mort.; BLAF; GB; GD2. Also Meteorology* and Heredity*.
1. *Memories of my life*. London, 1908. 3rd ed., 1909.

15371. LE CONTE, Joseph. 1823-1901. U.S.A. DSB; P III,IV; BMNH; BLA & Supp.; GB; GE. Also Geology*, Evolution*, and Physiology*.
1. *The autobiography*. New York, 1903. Ed. by W.D. Armes.

15372. WALLACE, Alfred Russel. 1823-1913. Br. DSB; ICB; P III; BMNH; GB; GD3. Also Natural History*, Zoology*, Evolution*, and History of Science*.
1. *My life. A record of events and opinions*. London, 1905. RM/L. Reprinted, Farnborough, 1969.
2. *Letters and reminiscences*. London, 1916. Ed. by J. Marchant.

15373. HUXLEY, Thomas Henry. 1825-1895. Br. DSB; ICB; BMNH; Mort.; BLA; GB; GD1. Other entries: see 3340.
1. *Diary of the voyage of H.M.S. "Rattlesnake."* London, 1935. First publication. Ed. by Julian Huxley.
2. *Autobiographies: Charles Darwin, T.H. Huxley*. New York, 1974. Ed. by G. de Beer. Huxley's autobiography was first published in his *Method and results* which forms Vol. 1 of his *Collected essays* (2963/8).

15374. HIS, Wilhelm (first of the name). 1831-1904. Switz./Ger. DSB; ICB; BMNH; Mort.; BLA & Supp. Also Anatomy* and Embryology*.
1. *Lebenserinnerungen*. See 14285/2.

15375. ROSCOE, Henry Enfield. 1833-1915. Br. DSB; ICB; P II,III,IV; GB; GD3. Also Chemistry* and Biography*. See also Index.
1. *The life and experiences ... written by himself*. London, 1906.

15376. HAECKEL, Ernst Heinrich Philipp August. 1834-1919. Ger. DSB; ICB; BMNH; Mort.; BLAF; GB. Also Science in General*, Zoology*, Evolution*# and Embryology*.
1. *Entwicklungsgeschichte einer Jugend: Briefe an die Eltern, 1852-1856*. Leipzig, 1921. Ed. by H. Schmidt. X.
1a. ——— *The story of the development of a youth: Letters to his parents, 1852-1856*. New York/London, 1923. Trans. by G.B. Gifford.
2. [Trans.] *The love letters of Ernst Haeckel* [and Franziska von Altenhausen], *written between 1898 and 1903*. New York/London, 1930. Arranged by J. Werner. Trans. by Ida Zeitlin.
3. *Ernst Haeckel in Selbstzeugnissen und Bilddokumenten*. Reinbek bei Hamburg, 1964. Ed. by J. Hemleben.

15377. AGASSIZ, Alexander. 1835-1910. U.S.A. DSB; ICB; BMNH; BLAF; GB; GE. Also Zoology* and Oceanography*.
1. *Letters and recollections of A. Agassiz, with a sketch of his life and work*. Boston, 1913. Ed. by G.R. Agassiz.

15378. GEIKIE, Archibald. 1835-1924. Br. DSB; ICB; P III,IV; BMNH; GB; GD4. Also Geology* and Biography*. See also Index.
1. *A long life's work. An autobiography*. London, 1924. RM/L.

15379. NEWCOMB, Simon. 1835-1909. U.S.A. DSB; ICB; P III,IV; GB; GE. Also Astronomy*.
1. *Reminiscences of an astronomer*. 1903.

15380. MUIR, John. 1838-1914. U.S.A. ICB; BMNH; GE. Also Natural History.
1. *The story of my boyhood and youth.* Boston/New York, 1913.

15381. BALL, Robert Stawell. 1840-1913. Br. P III,IV; GD3. Also Astronomy*, Mechanics*, and Biography*.
1. *Reminiscences and letters.* London, 1915. Ed. by W.V. Ball.

15382. SHALER, Nathaniel Southgate. 1841-1906. U.S.A. DSB; ICB; P III,IV; BMNH; GE. Also Geology*.
1. *The autobiography. With a supplementary memoir by his wife.* Boston, 1909.

15383. BUCHANAN, John Young. 1844-1925. Br. DSB; P III,IV; BMNH. Also Oceanography*.
See 10731/2,3.

15384. FLEMING, (John) Ambrose. 1849-1945. Br. DSB; ICB; P III,IV; GD6. Also Physics*.
1. *Fifty years of electricity. The memories of an electrical engineer.* London, 1921.

15385. KOVALEVSKY, Sonya. 1850-1891. Swed. DSB; ICB; P III,IV (KOWALEWSKY, Sophie); GB. Also Mathematics.
1. *Souvenirs d'enfance, écrits par elle-même et suivis de sa biographie.* Paris, 1895. By A.C. Leffler. X. Another ed., 1907.
2. [In Russian. *Recollections of childhood and autobiographical notes.*] Moscow, Academy of Sciences, 1945.
3. [In Russian. *Reminiscences and letters.*] Ib., 1961.

15386. STERNBERG, Charles Hazelius. 1850-1943. U.S.A. ICB; BMNH.
1. *The life of a fossil hunter.* New York, 1909.

15387. LODGE, Oliver Joseph. 1851-1940. Br. DSB; ICB; P III,IV & Supp.; GB; GD5. Also Physics* and Biography*.
1. *Past years. An autobiography.* London, 1931.
2. *Letters. Psychical, religious, scientific and personal.* London, 1932. Ed. by J.A. Hill.

15388. FISCHER, Emil. 1852-1919. Ger. DSB; ICB; P III,IV; Mort.; BLAF. Also Chemistry*#.
1. *Aus meinem Leben.* Berlin. 1922.

15389. RAMÓN Y CAJAL, Santiago. 1852-1934. Spain. DSB; ICB; BMNH; Mort.; BLAF. Also Anatomy*.
1. [Trans.] *Recollections of my life.* Philadelphia, 1937. Trans. by E.H. Craigie from the 3rd Spanish ed. of 1923. Reprinted, Cambridge, Mass., 1966.
2. [Trans.] *Precepts and counsels on scientific investigation. Stimulants of the spirit.* Mountain View, Calif., 1951. Trans. by J.M. Sanchez-Perez. Ed. by C.B. Courville.

15390. THOMSON, Joseph John. 1856-1940. Br. DSB; ICB; P III,IV; GD5. Also Physics*.
1. *Recollections and reflections.* London, 1936.

15391. HERTZ, Heinrich Rudolph. 1857-1894. Ger. DSB; ICB; P III,IV; GB; GC. Also Mechanics* and Physics*#.
1. *Erinnerungen, Briefe, Tagebücher.* Leipzig, 1927. Ed. by

Johanna Hertz. X.
1a. ——— *Memoirs, letters, diaries.* San Francisco, 1977. 2nd enlarged ed. by M. Hertz and C. Susskind. Trans. by L. Brenner. With a biographical introd. by M. von Laue. Text in German and English.

15392. PICTET, Amé. 1857-1937. Switz. ICB; P III,IV. Also Chemistry.
1. *Souvenirs et travaux d'un chimiste.* 1941.

15393. PLANCK, Max Karl Ernst Ludwig. 1858-1947. Ger. DSB; ICB; P III,IV. Also Physics*#.
1. *Wissenschaftliche Selbstbiographie.* Leipzig, 1948. X. 3rd ed., 1955.
1a. ——— *Scientific autobiography, and other papers.* New York, 1949. Trans. by F. Gaynor. Another ed., 1968.

Collection of Autobiographies

15394. *GEOLOGICAL stories. A series of autobiographies in chronological order.* London, 1873. Ed. by J.E. Taylor. 3rd ed., 1876.

3.22 BIOGRAPHY

Individual and collective.

The arrangement is by the authors of the biographies. The persons about whom the biographies are written are indicated in the Main Index at the end of the volume. Some of the authors are not scientists; in this section authors are included even if no information about them was found. The term 'biography' is taken to include studies of an individual's scientific work even if they contain no biographical material in the ordinary sense.

15395. FONTENELLE, Bernard Le Bovier de. 1657-1757. Fr. DSB; ICB; P I; GA; GB. Also Science in General*#, Mathematics*, Mechanics*, History of Science*, and Part 2*.
1. *Histoire du renouvellement de l'Académie Royale des Sciences en 1699, et les éloges de tous les académiciens morts depuis ce renouvellement.* 2 vols. Amsterdam, 1709-20. X. Reprinted, Brussels, 1969.
2. *Eloges des académiciens de l'Académie Royale des Sciences.* 2 vols. The Hague, 1731. X. Another ed., 1740; reprinted, Brussels, 1969.
3. *Eloges de Fontenelle.* Paris, n.d. Introd. and notes by F. Bouillier.
4. *Eloge de Monsieur le chevalier Neuton.* Paris, 1728. X.
4a. ——— *The life of Sir Isaac Newton.* London, 1728.

15396. NELLI, Giovanni Battista Clemente de. 1661-1725. It. P II.
1. *Vita e commercio letterario di Galileo Galilei.* 2 vols. Losanna, 1793. RM.

15397. MAIRAN, Jean Jacques d'Ortus de. 1678-1771. Fr. DSB; ICB; P II; GA. Also Mechanics*, Physics*, and Meteorology*.
1. *Eloges des académiciens de l'Académie Royale des Sciences, morts dans les années 1741, 1742 et 1743.* Paris, 1747. RM.

15398. WARD, John. 1679?-1758. Br. GD.
1. *The lives of the professors of Gresham College.* London, 1740. RM/L. Reprinted, New York, 1967.

15399. BOUGEREL, Joseph. 1680-1753. Fr. GA.
1. *Vie de Pierre Gassendi.* Paris, 1737. X. Reprinted, Geneva, 1970.

15400. STUKELEY, William. 1687-1765. Br. ICB; P II; BMNH; BLA; GA; GB; GD.
1. *Memoirs of Sir Isaac Newton's life ... being some account of his family, and chiefly of the junior part of his life.* London, 1936. Ed. from the MS of 1752 by A.H. White.

15401. MACLAURIN, Colin. 1698-1746. Br. DSB; ICB; P II; GA; GB; GD. Also Mathematics*.
1. *An account of Sir Isaac Newton's philosophical discoveries.* London, 1748. Ed. from the MS of 1728 by P. Murdoch. RM/L. Reprinted, New York, 1968, and Hildesheim, 1971. 2nd ed., 1750.

15402. MARTIN, Benjamin. 1704?-1782. Br. DSB; ICB; P II; BMNH; GA; GD. Other entries: see 3099.
1. *Biographia philosophica, being an account of the lives, writings and inventions of the most eminent philosophers and mathematicians who have flourished from the earliest ages of the world to the present time.* London, 1764. RM/L. Reprinted, New York, 1970.

15403. BIRCH, Thomas. 1705-1766. Br. ICB; P I; GA; GB; GD. Also History of Science*. See also Index.
1. *The life of the Hon. Robert Boyle.* London, 1744.

15404. FOUCHY, Jean Paul Grandjean de. 1707-1788. Fr. DSB. Under GRANDJEAN DE F.: ICB; P I; GA. Also Astronomy.
As secretary of the Académie des Sciences from 1743 to 1776 he wrote over sixty éloges of deceased academicians.

15405. BURTON, John. 1710-1771. Br. BLA & Supp.; GA; GD.
1. *An account of the life and writings of Herman Boerhaave.* London, 1743.

15406. TARGIONI-TOZZETTI, Giovanni. 1712-1783. It. DSB; ICB; P II; BMNH; BLA; GA. Also Geology, Natural History*, and History of Science*.
1. *Notizie della vita e delle opere di Pier' Antonio Micheli, botanico fiorentino.* Florence, 1858. RM.

15407. PULTENEY, Richard. 1730-1801. Br. BMNH; GA; GD. Also History of Botany*.
1. *A general view of the writings of Linnaeus.* London, 1781. RM/L. 2nd ed., rev., by W.G. Maton, 1805.

15408. SMALL, Robert. 1732-1808.
1. *An account of the astronomical discoveries of Kepler.* London, 1804. Reprinted, Madison, Wis., 1963.

15409. JAGEMANN, Christian Joseph. 1735-1804. Ger. P I.
Biography of Galileo, 1783.

15410. BAILLY, Jean Sylvain. 1736-1793. Fr. DSB; ICB; P I; GA; GB. Also Astronomy* and History of Astronomy*.
1. *Discours et mémoires.* 2 vols. Paris, 1790. Includes éloges of various scientists of the 17th and 18th centuries. RM.

15411. BALDINGER, Ernst Gottfried. 1738-1804. Ger. ICB; P I; BMNH; BLA & Supp. Also Botany.
1. *Biographien jetztlebender Aerzte und Naturforscher.* 1768. X.

15412. STEWART, David, *Earl of* BUCHAN. 1742-1829.
1. *An account of the life, writings and inventions of John Napier.* Perth, 1778. X. Another ed., 1787.

15413. CONDORCET, Marie Jean Antoine Nicolas Caritat, *Marquis* de.

1743-1794. Fr. DSB; ICB; P I,III; GA; GB. Also Science in General*# and Mathematics*. See also Index.
1. *Eloges des académiciens de l'Académie Royale des Sciences, morts depuis 1666 jusqu'en 1699*. Paris, 1773. X. Reprinted, ib., 1968.

15414. GENTY, Louis. 1743-1817.
1. *L'influence de Fermat sur son siècle, relativement aux progrès de la haute géométrie et du calcul....* Orléans, 1784.

15415. BERNOULLI, Johann (third of the name). 1744-1807. Ger. DSB; ICB; P I; GA; GB; GC. Also Astronomy.
Biographies of contemporary astronomers, 1776.

15416. DUNCAN, Andrew. 1744-1828. Br. BLA & Supp.; GD.
1. *A short account of the life of ... Sir Joseph Banks*. Edinburgh, 1821.

15417. FOOT, Jesse. 1744-1826. Br. BLA & Supp.; GD.
Biography of John Hunter, 1794.

15418. TRAIL, William. 1746-1831. Br. P II.
1. *An account of the life and writings of Robert Simson, late professor of mathematics*. Bath, 1812.

15419. SEWARD, Anna. 1747-1809. Br. GB; GD.
1. *Memoirs of the life of Dr* [Erasmus] *Darwin*. London, 1804.

15420. KERR, Robert. 1755-1813. Br. P I; BMNH; BLA; GA; GD. See also Index.
1. *Memoirs of the life, writings and correspondence of William Smellie*. Edinburgh, 1811.

15421. ADAMS, Joseph. 1756-1818. Br. P I & Supp.; BLA & Supp.; GA; GD.
1. *Memoirs of the life and doctrines of the late John Hunter*. London, 1817.

15422. THOMSON, John. 1765-1846. Br. BLA; GD.
1. *An account of the life, lectures and writings of William Cullen*. 2 vols. 1859.

15423. RIXNER, Thaddaeus Anselm. 1766-1838. Ger. P II; GC.
1. *Leben ... berühmter Physiker am Ende der 16. und Anfang der 17. Jahrhunderts*. 7 vols. 1819. X.

15424. STÖVER, Dietrich Johann Heinrich. 1767-1822. BMNH.
1. *Leben des Ritters Carl von Linné*. Hamburg, 1792. X.
1a. ——— *Life of Sir Charles Linnaeus*. London, 1794. Trans. by J. Trapp.
See also 10958/13.

15425. OTTER, William. 1768-1840. Br. BMNH; GD.
1. *The life and remains of Edward Daniel Clarke, professor of mineralogy*. London, 1825.

15426. CUVIER, Georges. 1769-1832. Fr. DSB; ICB; P I; BMNH; GA; GB. Also Science in General*, Geology*, Zoology*, and History of Science*. See also Index.
1. *Recueil des éloges historiques*. 3 vols. Strasbourg, 1819-27. Covers the period 1800-27. RM. Reprinted, Brussels, 1969. 3rd ed., 1874.

2. *Eloge historique de M. Banks.* n.p., 1821.

15427. GAMAUF, Gottlieb. 1772-1841. Hung. P I.
Reminiscences of G.C. Lichtenberg, 1808-19.

15428. BAILY, Francis. 1774-1844. Br. DSB; ICB; P I; GA; GB; GD.
Also Astronomy*.
1. *An account of the Revd John Flamsteed.* London, 1835. RM/L. Reprinted, ib., 1966.

15429. GURNEY, Hudson. 1775-1864. Br. GD.
1. *Memoir of the life of Thomas Young.* London, 1831.

15430. DUVERNOY, Georges Louis. 1777-1855. Fr. BMNH; BLA; GA.
Also Zoology.
1. *Notice historique sur les ouvrages et la vie de M. le Baron Cuvier.* Paris, 1833. RM/L.

15431. BROUGHAM, Henry. 1779-1868. Br. The lawyer and politician. P I,III; GA; GB; GD. Also Science in General*#, Mechanics*, Physics*, and Autobiography*.
1. *Lives of men of letters and science who flourished in the time of George III.* London, 1845.
2. *Lives of philosophers of the time of George III.* Ib., 1855.

15432. BREWSTER, David. 1781-1868. Br. DSB; P I,III; GA; GB; GD. Also Science in General* and Physics*. See also Index.
1. *The life of Sir Isaac Newton.* London, 1831.
2. *Martyrs of science; or, The lives of Galileo, Tycho Brahe and Kepler.* London, 1841. 3rd ed., 1856. 7th ed., 1870. Another ed., 1874.
3. *Memoirs of the life, writings and discoveries of Sir Isaac Newton.* 2 vols. Edinburgh, 1855. RM/L. Reprinted, New York, 1965. 2nd ed., 1860.

15433. DARLINGTON, William. 1782-1863. U.S.A. DSB; ICB; BMNH; BLA & Supp.; GE. Also Botany.
1. *Memorials of John Bartram and Humphrey Marshall.* 1849. X. Reprinted, New York, 1967.

15434. DUPIN, (Pierre) Charles (François). 1784-1873. Fr. DSB; P I, III; GA; GB. Also Mathematics.
1. *Essai historique sur les services et les travaux scientifiques de Gaspard Monge.* Paris, 1819. RM. Reprinted, ib., 1965; see item 15436/4.

15435. PARIS, John Ayrton. 1785-1856. Br. P II; BLA; GA; GD. Also Science in General* and Chemistry*.
1. *The life of Sir Humphrey Davy.* 2 vols. London, 1831.

15436. ARAGO, (Dominique) François (Jean). 1786-1853. Fr. DSB; ICB; P I; GA; GB. Also Astronomy*#, Physics, and Earth Sciences*.
1. *Eloge historique de James Watt.* Paris, 1834. X.
1a. ——— *Historical eloge of James Watt.* London, 1839. Trans. with additions by J.P. Muirhead.
2. *Biographie de Condorcet.* Paris, 1841.
3. *Analyse de la vie et des travaux de Sir William Herschel.* Paris, 1843.
4. *Biographie de Gaspard Monge.* [Periodical article, 1846] Paris, 1965. This reprint also includes item 15434/1.

5. *Notices biographiques*. 3 vols. Paris, 1853. (Part of the *Oeuvres complètes*) Reprinted, ib., 1964.

5a. ——— *Biographies of distinguished scientific men*. London, 1857. RM/L. Reprinted, 2 vols. Freeport, N.Y., 1972.

15437. CALLISEN, Adolph Carl Peter. 1787-1866. Den. BLA & Supp.; GA; GC.

1. *Medicinisches Schriftsteller-Lexicon der jetzt lebenden Aerzte ... und Naturforscher*. 33 vols. Copenhagen, 1830-45. X. Reprinted, (place?), 1962-64.

15438. DAVY, John. 1790-1868. Br. DSB; ICB; P I,III; BMNH; BLA & Supp.; GD. Also Chemistry, Natural History, and Physiology*. See also Index.

1. *Memoirs of the life of Sir Humphry Davy*. 2 vols. London, 1836. RM/L.

15439. HENSCHEL, August Wilhelm Eduard Theodor. 1790-1856. Ger. BMNH; BLA. Also Botany and History of Botany.

1. *Clavis Rumphiana botanica et zoologica. Accedunt vita G.E. Rumphii, Plinii Indici....* Breslau, 1833. Re G.E. Rumpf (1627-1702).

15440. BOWDICH (later LEE), Sarah, *Mrs*. 1791-1856. Br. BMNH; GD (LEE). Also Natural History*.

1. *Memoirs of Baron Cuvier*. London, 1833. By Mrs R. Lee.

15441. PEACOCK, George. 1791-1858. Br. DSB; ICB; P II; GB; GD. Also Mathematics*.

1. *Life of Thomas Young*. London, 1855.

15442. FLOURENS, (Marie Jean) Pierre. 1794-1867. Fr. DSB; ICB; BMNH; Mort.; BLA; GA; GB. Also Zoology*, Evolution*, Embryology*, and Physiology*.

1. *Eloge historique de Georges Cuvier*. Paris, [1833?]. X.

1a. ——— 3rd ed. *Histoire des travaux de Georges Cuvier*. Paris, 1858.

2. *Analyse raisonnée des travaux de Georges Cuvier, précedée de son éloge historique*. Paris, 1841.

3. *Buffon. Histoire de ses travaux et de ses idées*. Paris, 1844. RM/L.

3a. ——— 2nd ed. *Histoire des travaux et des idées de Buffon*. Paris, 1850. X. Reprinted, Geneva, 1971. 3rd ed., 1870.

4. *Fontenelle; ou, De la philosophie moderne relativement aux sciences physiques*. Paris, 1847. RM. Reprinted, Geneva, 1971.

5. *Recueil des éloges historiques*. 3 vols. Paris, 1856-62.

6. *Des manuscripts de Buffon*. Paris, 1860. RM/L. Reprinted, Geneva, 1971.

See also 10954/6.

15443. PORTLOCK, Joseph Ellison. 1794-1864. Br. P III; BMNH; GB; GD. Also Geology*.

1. *Memoir of the life of Major-General Colby, together with a sketch of the origin and progress of the Ordnance Survey*. London, 1869. Re Thomas F. Colby (1784-1852).

15444. HOBSON, Richard. 1795-1868. Br. GD.
1. *Charles Waterton*. London, 1866.

15445. MACGILLIVRAY, William. 1796-1852. Br. BMNH; GA; GB; GD. Also Geology*, Botany*, and Zoology*.
1. *Lives of eminent zoologists, from Aristotle to Linnaeus*. Edinburgh, 1834.

15446. PICHOT, Amédée. 1796-1877. Fr. GA.
1. *Sir Charles Bell. Histoire de sa vie et de ses travaux*. Paris, 1858. X.
1a. ——— *The life and labours of Sir Charles Bell*. London, 1860.

15447. POGGENDORFF, Johann Christian. 1796-1877. Ger. DSB; ICB; P II,III; GA; GC. Also Physics and History of Physics*.
1. *Biographisch-literarisches Handwörterbuch zur Geschichte der exacten Wissenschaften*. Vols I-IV. Leipzig, 1863-1904. Reprinted, Amsterdam, 1965.

15448. CHASLES, (Victor Euphémien) Philarète. 1798-1873. Fr. GA; GB.
1. *Galileo Galilei. Sa vie, son procès et ses contemporains, d'après les documents originaux*. Paris, 1862.

15449. NAPIER, Mark. 1798-1879. Br. GD.
1. *Memoirs of John Napier ... with a history of the invention of logarithms*. Edinburgh, 1834.
See also 6115/4.

15450. WILLIS, Robert. 1799-1878. Br. BMNH; BLA; GD. See also Index.
1. *William Harvey*. London, 1878.

15451. JARDINE, William. 1800-1874. Br. BMNH; GD. Also Zoology*.
1. *Memoirs of Hugh Edwin Strickland*. London, 1858. RM/L.

15452. JENYNS (after 1871 BLOMEFIELD), Leonard. 1800-1893. Br. BMNH; GD1 (BLOMEFIELD). Also Meteorology* and Zoology*.
1. *A memoir of the Rev. John Stevens Henslow*. London, 1862.

15453. LOEWENBERG, Julius. 1800-1893.
1. *Alexander von Humboldt. Bibliographische Übersicht*. [Extract from item 15514/1] Reprinted, Stuttgart, 1960.

15454. PHILLIPS, John. 1800-1874. Br. DSB; ICB; P II,III; BMNH; GA; GB; GD. Also Geology*.
1. *Memoirs of William Smith* [the geologist, 1769-1839]. London, 1844.

15455. SUNDEVALL, Carl Jacob. 1801-1875. Swed. BMNH. Also Zoology.
1. [Trans. from Swedish] *Die Thierarten des Aristoteles von den Klassen der Säugethiere, Vögel, Reptilien und Insekten*. Stockholm, 1863.

15456. HENRY, William Charles. 1804-1892.
1. *Memoirs of the life and scientific researches of John Dalton*. London, 1854.

15457. SHAW, Alexander. 1804-1890. Br. BLA; GD.
1. *Narrative of the discoveries of Sir Charles Bell in the nervous system*. London, 1839.

15458. GEOFFROY SAINT-HILAIRE, Isidore. 1805-1861. Fr. DSB; BMNH; BLA; GA; GB. Also Science in General* and Zoology*.

1. *Vie, travaux et doctrine scientifique d'Etienne Geoffroy Saint-Hilaire. Par son fils.* Paris, 1847. RM. Reprinted, Brussels, 1968.

15459. WAGNER, Rudolph. 1805-1864. Ger. DSB; BMNH; Mort.; BLA; GB; GC. Also Zoology*, Embryology*, and Physiology*.
1. *Gespräche mit C.F. Gauss in den letzten Monaten seines Lebens.* Göttingen, 1975. Ed. by H. Rubner.
See also 14051/1a.

15460. DE MORGAN, Augustus. 1806-1871. Br. DSB; ICB; P I (under D); P III (under M); GA; GB; GD. Also Mathematics*, Astronomy*, and History of Mathematics*. See also Index.
1. *Newton, his friend and his niece.* London, 1885. Ed. by his wife and A.C Raynard. Reprinted, ib., 1968.
2. *Essays on the life and work of Newton.* London, 1914. Ed. by P.E.P. Jourdain.

15461. KLEIN, Karl. 1806-1870. Ger. GC.
1. *Georg Forster in Mainz, 1788 bis 1793.* Gotha, 1863.

15462. STOUGHTON, John. 1807-1897. Br. GB; GD.
1. *Worthies of science.* London, [1879?].

15463. TOMLINSON, Charles. 1808-1897. Br. P III,IV; GD. Also Physics.
1. *Sir Joseph Banks and the Royal Society.* London, 1844.

15464. DE MORGAN, Sophia Elizabeth (née FRIEND). 1809-1892.
1. *A memoir of Augustus De Morgan.* London, 1882.

15465. GRAY, Asa. 1810-1888. U.S.A. DSB; ICB; BMNH; GA; GB; GE. Also Botany*# and Evolution*. See also Index.
1. *Sir J.D. Hooker. A short sketch, together with his notes on the botany of the Rocky Mountains.* London, 1877.

15466. MAILLY, (Nicolas) Edouard. 1810-1891. Belg. P III; BMNH. Also Science in General* and History of Science*.
1. *Essai sur la vie et les ouvrages de L.A.J. Quetelet.* Brussels, 1873. RM/L.

15467. QUATREFAGES DE BRÉAU, (Jean Louis) Armand de. 1810-1892. Fr. DSB; ICB; BMNH; Mort.; BLA; GA; GB. Also Zoology* and Autobiography*.
1. *Charles Darwin et ses précurseurs français. Etude sur le transformisme.* Paris, 1870. X. 2nd ed., 1892.

15468. BRIGHTWELL, Cecilia Lucy. 1811-1875. Br. BMNH; GD.
1. *Life of Linnaeus.* London, 1858.

15469. HOEFER, (Jean Chrêtien) Ferdinand. 1811-1878. Fr. ICB; P I, III; BMNH; BLA; GA. Also Chemistry* and History of Science*.
1. *La chimie enseignée par la biographie de ses fondateurs: R. Boyle, Lavoisier, Priestley, Scheele, Davy, etc.* Paris, 1865.

15470. HAURÉAU, (Jean) Barthélemy. 1812-1896. Fr. GA; GB.
1. *Les oeuvres de Hugues de Saint Victor. Essai critique.* Nouv. éd. Paris, 1886. X. Reprinted, New York, 1963.

15471. SMILES, Samuel. 1812-1904. Br. ICB; GB; GD2.
1. *Lives of the engineers.* London, 1862. RM/L.
1a. ——— Selections, ed. by T.P. Hughes. Cambridge, Mass., 1966.

2. *Lives of Boulton and Watt.* London, 1865. RM/L.
3. *Life of a Scotch naturalist, Thomas Edward.* London, 1876. 7th ed., 1879. Another ed., 1882.
4. *Robert Dick, baker of Thurso, geologist and botanist.* London, 1878.

15472. KLENCKE, (Philipp Friedrich) Hermann. 1813-1881. Ger. BLA.
1. [Trans.] *Lives of the brothers Humboldt, Alexander and William.* London, 1852. Trans. from the German of Klencke and Gustav Schlesier and arranged by J. Bauer.

15473. MARTIN, Thomas Henri. 1813-1884. Fr. P II,III. Also Science in General* and History of Science*.
1. *Galilée: Les droits de la science et la méthode des sciences physiques.* Paris, 1868. RM.

15474. MUIRHEAD, James Patrick. 1813-1898. Br. GD1.
1. *The origin and progress of the mechanical inventions of James Watt.* 3 vols. London, 1854. RM (under Watt)/L.
2. *The life of James Watt.* London, 1858. 2nd ed., 1859.
See also 7149/1 and 15436/1a.

15475. ELLIS, George Edward. 1814-1894. U.S.A. BMNH; GE.
1. *Memoir of Sir Benjamin Thompson, Count Rumford.* Philadelphia, 1871. Reprinted, Boston, 1972.
2. *Memoir of J. Bigelow.* Cambridge, 1880.

15476. HAMILTON, James. 1814-1867. Br. GD.
1. *Memoirs of the life of James Wilson.* London, 1859.

15477. JONES, Henry Bence. 1814-1873. Br. ICB; P I,III; BMNH; BLA; GB (BENCE-JONES); GD. Also Chemistry* and History of Science*.
1. *The life and letters of Faraday.* 2 vols. London, 1870. 2nd ed., 1870.

15478. LANKESTER, Edwin. 1814-1874. Br. BMNH; BLA; GD. Also Microscopy* and Zoology*. See also Index.
1. *Memorials of John Ray, consisting of his life by Dr Derham, biographical and critical notes by Sir J.E. Smith and Cuvier and Dupetit Thouars, with his itineraries, etc.* London, 1846. (Ray Society publn. 3) Ed. by Lankester. RM/L.

15479. ECKER, Alexander. 1816-1887. Switz. BMNH; BLA & Supp.; GC. Also Zoology*, Anatomy*, and Physiology*.
1. *Lorenz Oken. Eine biographische Skizze.* Stuttgart, 1880. X.
1a. ——— *Lorenz Oken. A biographical sketch.* London, 1883. Trans. by A. Turk.

15480. STEINSCHNEIDER, Moritz. 1816-1907. Ger. ICB; P III,IV; BLA; BLAF; GB. Also History of Science*.
1. *Al-Farabi.* 1869. X. Reprinted, Leipzig, 1966.

15481. LEWES, George Henry. 1817-1878. Br. ICB; BMNH; GA; GB; GD. Also Zoology* and Physiology*.
1. *Aristotle. A chapter from the history of science, including analyses of Aristotle's scientific writings.* London, 1864.

15482. MONTESSUS DE BALLORE, Ferdinand Bernard de. 1817-1899.
1. *Martyrologe et biographie de Commerson, médecin-botaniste et naturaliste du roi au XVIIIe siècle.* Châlon-sur-Saône, 1890.

15483. PROST, Auguste. 1817-1896.
1. *Les sciences et les arts occultes au XVIe siècle: Corneille Agrippa.* 2 vols. Paris, 1881-82. X. Reprinted, Lisse (Holland), 1965.

15484. SMITH, Robert Angus. 1817-1884. Br. DSB; P III,IV; BLA; GD. Also Chemistry* and History of Science*.
1. *Memoir of John Dalton, and history of the atomic theory up to his time.* London, 1856.

15485. HOFMANN, August Wilhelm von. 1818-1892. Ger./Br. DSB; ICB; P I,III,IV; GB; GC. Also Chemistry* and History of Chemistry*.
1. *The life-work of Liebig.* London, 1876.
2. *Zur Erinnerung an vorangegangene Freunde.* Brunswick, 1888.

15486. WILSON, George. 1818-1859. Br. P II; GD. Also Chemistry*.
1. *The life of the Hon. Henry Cavendish.* London, 1851. RM/L.
2. *Memoir of Edward Forbes, professor of natural history.* Cambridge, 1861. With A. Geikie. RM/L.
See also 15581/1.

15487. FIGUIER, Louis Guillaume. 1819-1894. Fr. P I,III; BMNH; GA. Other entries: see 3332.
1. *Vie des savants illustres.* 5 vols. Paris, 1866-70. (Ranges from antiquity to the 18th century.) X. 2nd ed., 2 vols, 1873.

15488. SHAIRP, John Campbell. 1819-1885. Br. BMNH; GB; GD.
1. *Life and letters of James David Forbes.* London, 1873. With P.G. Tait and A. Adams-Reilly.

15489. WADDINGTON, Charles Tzaunt. 1819-1914.
1. *Ramus (Pierre de La Ramée): Sa vie, ses écrits et ses opinions.* Paris, 1855. X. Reprinted, Dubuque, Iowa, 1962.

15490. BERTI, Domenico. 1820-1897. It. GF.
1. *Il processo originale di Galileo Galilei, pubblicato per la prima volta.* Rome, 1876. RM.

15491. DESMAZE, Charles Adrien. b. 1820.
1. *P. Ramus. Sa vie, ses écrits, sa mort (1515-1572).* Paris, 1864. X. Reprinted, Geneva, 1970.

15492. TODHUNTER, Isaac. 1820-1884. Br. DSB; P III; GB; GD. Also Mathematics*, Mechanics*, and History of Mathematics*.
1. *William Whewell ... an account of his writings.* 2 vols. London, 1876. Reprinted, Farnborough, 1970, and New York, 1970.

15493. TYNDALL, John. 1820-1893. Br. DSB; ICB; P II,III,IV; GB; GD. Also Science in General*, Physics*, Geology*, and Microbiology*. See also Index.
1. *Faraday as a discover.* London, 1868. RM/L. Another ed., 1870. 4th ed., 1884.

15494. DUNCAN, Peter Martin. 1821 (or 1824)-1891. Br. P III,IV; BMNH; GB; GD1. Also Palaeontology and Zoology*.
1. *Heroes of science. Botanists, zoologists and geologists.* London, 1882.

15495. DUNKIN, Edwin. 1821-1898. Br. P III. Also Astronomy.
1. *Obituary notices of astronomers.* London, 1879. RM/L.

15496. PROWE, Leopold Friedrich. 1821-1887. Thorn. P II,III.
Author of several works on Copernicus.

15497. VIRCHOW, Rudolf Ludwig Karl. 1821-1902. Ger. DSB; ICB; BMNH; Mort.; BLA & Supp. Also Science in General* and Cytology*.
1. *Goethe als Naturforscher*. Berlin, 1861. X. Reprinted, Wiesbaden, 1971.

15498. WALKER, William. b. 1821.
1. *Memoirs of the distinguished men of science of Great Britain living in the years 1807-8*. London, 1862. X. 2nd ed., 1864.

15499. AGASSIZ, Elizabeth Cabot (Cary), *Mrs*. 1822-1907. U.S.A. ICB; BMNH; GE. See also 11177/6.
1. *Louis Agassiz*. 2 vols. Boston, 1885. Another ed., 1887. 9th ed., 1890.

15500. BERTRAND, Joseph Louis François. 1822-1900. Fr. DSB; ICB; P I,III,IV; GA. Also Mathematics*, Physics*, and History of Science*.
1. *D'Alembert*. Paris, 1889.
2. *Eloges académiques*. Paris, 1889. X.
3. *Eloges académiques, nouvelle série*. Paris, 1902. With a biography of Bertrand by G. Darboux.

15501. MOLESCHOTT, Jacob Albert Willebrord. 1822-1893. Holl./Ger./Switz./It. DSB; ICB; P II,III; BMNH; Mort.; BLA; GA; GC. Also Physiology*.
1. *Georg Forster*. Frankfurt, 1854.

15502. MORLEY, Henry. 1822-1894. Br. ICB; GB; GD.
1. *Jerome Cardan*. 2 vols. London, 1854.

15503. WITTWER, Wilhelm Constantin. 1822-1908. Ger. P II,III,IV. Also Physics.
1. *Alexander von Humboldt*. Leipzig, 1861.

15504. DICKERSON, Edward Nicoll. 1824-1889. U.S.A. GE.
1. *Joseph Henry and the magnetic telegraph*. New York, 1885.

15505. LEMOINE, Albert. 1824-1874.
1. *Le vitalisme et l'animisme de Stahl*. Paris, 1864.

15506. MARCOU, Jules Belknap. 1824-1898. Fr./Switz./U.S.A. DSB; P III,IV; BMNH; GB; GE. Also Science in General* and Geology*.
1. *Life, letters and works of Louis Agassiz*. 2 vols. London, 1896. Reprinted, 2 vols in 1, Westmead, 1972.

15507. PROSSER, Francis Richard Wegg-. b. 1824.
1. *Galileo and his judges*. London, 1889.

15508. BJERKNES, Carl Anton. 1825-1903. Nor. DSB; ICB; P III,IV. Also Mechanics*.
1. *Niels Henrik Abel*. Paris, 1885. French original.

15509. VALSON, Claude Alphonse. 1826-after 1904. Fr. P III,IV.
1. *La vie et les travaux du Baron Cauchy*. Paris, 1868. X. Reprinted, ib., 1970.

15510. FISHER, George Park. 1827-1909. U.S.A. GB; GE.
1. *Life of Benjamin Silliman* [senior]. 2 vols. New York, 1866.

15511. GLADSTONE, John Hall. 1827-1902. Br. DSB; P III,IV; GB; GD2. Also Chemistry.
   1. *Michael Faraday*. London, 1872. 2nd ed., 1873. 3rd ed., 1874.

15512. VEITCH, John. 1829-1894. Br. GB; GD.
   1. *Lucretius and the atomic theory*. Glasgow, 1875.

15513. BAYNE, Peter. 1830-1896. Br. GD1.
   1. *The life and letters of Hugh Miller*. 2 vols. London, 1871.

15514. BRUHNS, Karl Christian. 1830-1881. Ger. DSB; P I,III; GC. Also Astronomy and History of Astronomy*.
   1. *Alexander von Humboldt. Eine wissenschaftliche Biographie*. 3 vols. Leipzig, 1872. With several collaborators. X. Reprinted, Osnabrück, 1969.
   1a. ——— *Life of Alexander von Humboldt*. 2 vols. London, 1873. Trans. by J. and C. Lassell.
   See also 3268/6.

15515. CAMPBELL, Lewis. 1830-1908. Br. GB; GD2.
   1. *The life of James Clerk Maxwell*. London, 1882. With W. Garnett. Reprinted, New York, 1969. Another ed., rev., 1884.

15516. GILMAN, Daniel Coit. 1831-1908. U.S.A. ICB; GB; GE.
   1. *The life of James Dwight Dana*. New York, 1899. RM/L.

15517. McCLURE, James Baird. 1832-1895.
   1. *Edison and his inventions*. Chicago, 1879. RM.

15518. BONNEY, Thomas George. 1833-1923. Br. DSB; P III,IV; BMNH; GB; GD4. Also Geology*.
   1. *Charles Lyell and modern geology*. London, 1895.

15519. CLARK, John Willis. 1833-1910. Br. BMNH; GD2. Also Zoology*.
   1. *The life and letters of the Rev. Adam Sidgwick* [the geologist, 1785-1873]. 2 vols. Cambridge, 1890. With T.M. Hughes. Reprinted, Farnborough, 1970.

15520. PRESTWICH, Grace Anne, *Lady*. 1833?-1899. BMNH.
   1. *Life and letters of Sir Joseph Prestwich*. Edinburgh, 1899.

15521. ROSCOE, Henry Enfield. 1833-1915. Br. DSB; ICB; P II,III,IV; GB; GD3. Also Chemistry* and Autobiography*. See also Index.
   1. *John Dalton and the rise of modern chemistry*. London, 1895.
   2. *A new view of the origins of Dalton's atomic theory*. London, 1896. With A. Harden. Reprinted, New York, 1970.

15522. ELLIS, Robinson. 1834-1913. Br. GB; GD3.
   1. *Noctes Manilianae; sive, Dissertationes in "Astronomia" Manilii. Accedunt conjecturae in Germanici "Aratea."*. Oxford, 1891. Re Marcus Manilius (entry 128) and Aratus of Soli (entry 83).

15523. LANGLEY, Samuel Pierpoint. 1834-1906. U.S.A. DSB; ICB; P III, IV; GB; GE. Also Astronomy* and Physics*.
   1. *Memoir of George Brown Goode, 1851-1896*. Washington, 1897.

15524. GEIKIE, Archibald. 1835-1924. Br. DSB; ICB; P III,IV; BMNH; GB; GD4. Also Geology* and Autobiography*. See also Index.
   1. *The life of Sir Roderick I. Murchison*. 2 vols. London,

1875. RM/L. Reprinted, Farnborough, 1970.
2. *Memoir of Sir Andrew Crombie Ramsay*. London, 1895. RM/L.
3. *The founders of geology*. London, 1897. RM/L. 2nd ed., 1905; reprinted, New York, 1962.

15525. GRIMAUX, (Louis) Edouard. 1835-1900. Fr. P III,IV. Also Chemistry*.
1. *Lavoisier, 1743-1794*. Paris, 1888. 2nd ed., 1896. 3rd ed., 1899.
2. *Charles Gerhardt. Sa vie, son oeuvre, sa correspondance, 1816-1856*. Paris, 1900.

15526. FOSTER, Michael. 1836-1907. Br. DSB; ICB; BMNH; Mort.; BLAF; GB; GD2. Also Embryology* and Physiology*. See also Index.
1. *Claude Bernard*. London, 1899.

15527. FRIIS, Frederik Reinholdt. 1836-after 1902. Den. P III,IV. Publications on Tycho Brahe and other astronomers.

15528. HIPLER, Franz. 1836-1898. Ger. P III; GC. Publications on Copernicus.

15529. KÖNIGSBERGER, Leo. 1837-1921. Aus./Ger. DSB; P III,IV. Also Mathematics*.
1. *Helmholtzs Untersuchungen über die Grundlagen der Mathematik und Mechanik*. Leipzig, 1896. X. Reprinted, Wiesbaden, 1971.

15530. KRAUSE, Ernst Ludwig. 1839-1903. Ger. DSB; BMNH. Also Evolution.
1. *Erasmus Darwin und seine Stellung in der Geschichte der Descendenz-Theorie. Mit seinem Lebens und Charakterbilde von Charles Darwin*. Leipzig, 1880.
1a. ——— *Erasmus Darwin*. London, 1879. Trans. by W.S. Dallas. With a preliminary notice by Charles Darwin. Reprinted, Farnborough, 1971.

15531. ROTH, Moritz. 1839-1914. Switz. BLA.
1. *Andreas Vesalius*. Berlin, 1892. Reprinted, Amsterdam, 1965.

15532. BALL, Robert Stawell. 1840-1913. Br. P III,IV; GD3. Also Astronomy*, Mechanics*, and Autobiography*.
1. *Great astronomers*. London, 1895. RM/L.

15533. PAYNE, Joseph Frank. 1840-1910. Br. BMNH; GD.
1. *Harvey and Galen*. London, 1897.

15534. McKENDRICK, John Gray. 1841-1926. Br. P IV; BMNH; BLAF. Also Physiology*.
1. *H.L.F. von Helmholtz*. London, 1899.

15535. DARBOUX, (Jean) Gaston. 1842-1917. Fr. DSB; ICB; P III,IV. Also Mathematics*. See 15500/3.

15536. FISKE, John. 1842-1901. U.S.A. GB; GE.
1. *E.L. Youmans, interpreter of science for the people*. New York, 1894. Reprinted, Freeport, N.Y., 1972.

15537. MIALL, Louis Compton. 1842-1921. Br. BMNH; GF. Also Zoology*.
1. *Life and work of Charles Darwin*. 1883.

15538. REBIÈRE, Alphonse. 1842-1901.

1. *Les femmes dans la science.* Paris, 1894. X. 2nd ed., 1897.

15539. REID, Thomas Wemyss. 1842-1905. Br. GD2.
1. *Memoirs and correspondence of Lyon Playfair.* London, 1899.

15540. NICHOLSON, Henry Alleyne. 1844-1899. Br. BMNH; GB; GD1. Also Geology*, Zoology*, and History of Science*.
1. *Lives and labours of leading naturalists.* London, [1893?].

15541. THORPE, (Thomas) Edward. 1845-1925. Br. DSB; P III,IV; GD4. Also Chemistry* and History of Chemistry*.
1. *Humphry Davy.* London, 1896.

15542. WATERHOUSE, Frederick Herschel. 1845-1919. Br. BMNH. Also Zoology.
1. *The dates of publication of some of the zoological works of ... John Gould.* London, 1885. Includes a biographical sketch.

15543. JACKSON, Benjamin Daydon. 1846-1927. Br. BMNH. Also Botany*.
1. *A catalogue of plants cultivated in the garden of John Gerard* [1545-1612] *and a life of the author* [i.e. of Gerard]. London, 1876.
See also 1451/1a.

15544. FAVARO, Antonio Nobile. 1847-1922. It. ICB; P III,IV. Also History of Science.
1. *Galileo Galilei e Suor Maria Celeste.* Florence, 1891.

15545. ALLEN, Grant. 1848-1899. Br. BMNH (A., Charles Grant); GB; GD1. Also Botany* and Evolution*.
1. *Charles Darwin.* London, 1885. Another ed., 1888.

15546. ROMANES, George John. 1848-1894. Br. DSB; ICB; BMNH; Mort.; GD. Also Science in General*, Evolution*, and Physiology. See also 15607/1.
1. *Charles Darwin, character and life, work in zoology, work in psychology.* London, 1882.

15547. GOSSE, Edmund William. 1849-1928. Br. The well known writer. GB; GD4.
1. *The life of Phillip Henry Gosse, by his son.* London, 1890.

15548. MAGNIN, Antoine. fl. 1878-1902. Fr. BMNH. Also Microbiology*.
1. *Claret de la Tourrette, ses recherches sur les lichens du Lyonnais.* Lyons, 1883.

15549. BETTANY, George Thomas. 1850-1891. BMNH. Also Natural History.
1. *Life of Charles Darwin.* London, 1887.

15550. GEBLER, Karl von. 1850-1878. Aus. P III.
1. *Galileo Galilei und die römische Curie.* 2 vols. Stuttgart, 1876-77. X. Reprinted, Wiesbaden, 1968.
1a. ——— *Galileo Galilei and the Roman Curia.* London, 1879. RM/L. Trans. by Mrs G. Sturge.

15551. SHENSTONE, William Ashwell. 1850-1908. Br. P IV; GD2. Also Chemistry.
1. *Justus von Liebig. His life and work.* London, 1895.

15552. FINK, Karl. 1851-1898. Ger. P IV Supp.
Biography of Lazare Carnot.

15553. FITZGERALD, George Francis. 1851-1901. Br. DSB; ICB; P III, IV; GD2. Also Physics*#.
1. *Lord Kelvin*. Glasgow, 1899.

15554. HOLDER, Charles Frederick. 1851-1915. U.S.A. BMNH; GE. Also Zoology.
1. *Charles Darwin*. New York, 1891.
2. *Louis Agassiz*. New York, 1893.

15555. LODGE, Oliver Joseph. 1851-1940. Br. DSB; ICB; P III,IV & Supp.; GB; GD5. Also Physics* and Autobiography*.
1. *Pioneers of science*. London, 1893. Reprinted, ib., 1922.

15556. THOMPSON, Silvanus Philipps. 1851-1916. Br. DSB; ICB; P III, IV; GD3. Also Physics*. See also Index.
1. *Michael Faraday*. London, 1898.

15557. DREYER, John Louis Emil. 1852-1926. Br. DSB; ICB; P III,IV; GD4. Also Astronomy.
1. *Tycho Brahe*. Edinburgh, 1890. Reprinted, New York, 1963.

15558. KAHLBAUM, Georg Wilhelm August. 1853-1905. Ger./Switz. P III,IV; BLAF. Also Chemistry and History of Chemistry. See also Index.
1. *Christian Friedrich Schönbein, 1799-1868*. 2 vols. Leipzig, 1899-1901. With E. Schaer. X. Reprinted, ib., 1970.

15559. WASSILJEW, Alexander Vassilievich. 1853-after 1922. Russ. P III,IV.
Biographies of Russian mathematicians.

15560. GLAZEBROOK, Richard Tetley. 1854-1935. Br. DSB; ICB; P III, IV; GD5. Also Physics*.
1. *James Clerk Maxwell and modern physics*. New York, 1896. RM/L.

15561. PAGET, Stephen. 1855-1926. Br. ICB; GD4.
1. *John Hunter. Man of science and surgeon, 1728-1793*. London, 1897.

15562. POWER, D'Arcy. 1855-1941. Br. ICB; GD6.
1. *William Harvey*. London, 1897.

15563. POULTON, Edward Bagnall. 1856-1943. Br. ICB; BMNH; GD6. Also Zoology* and Evolution*.
1. *Charles Darwin and the theory of natural selection*. London, 1896.

15564. LEASK, William Keith. 1857-after 1917. Br.
1. *Hugh Miller*. Edinburgh, 1896.

15565. PEARSON, Karl. 1857-1936. Br. DSB; ICB; P IV; GD5. Also Science in General*, Mathematics*, and Evolution*.
1. *The elastical researches of Barré de Saint-Venant*. Cambridge, 1889.

15566. FRANKLAND, Percy Faraday. 1858-1946. Br. DSB; P IV; GD6. Also Chemistry and Bacteriology.
1. *Pasteur*. London, 1898. With Mrs Frankland.

Imprint Sequence

15567. *MEMOIRS of the life and writings of Mr William Whiston.* London, 1749. RM.

15568. BARBIERI, Matteo.
1. *Notizie istoriche dei mattematici e filosofi del regno di Napoli.* Naples, 1778. RM.

15569. BAJAMONTI, Giulio.
1. *Elogio dell'abate Ruggiero Giuseppe Boscovich.* 2nd ed. Naples, 1790. RM.

15570. PALEY, Edmund.
1. *An account of the life and writings of William Paley.* London, 1825. X. Reprinted, Farnborough, 1970.

15571. MARCOZ, Jean Baptiste Philippe.
1. *Astronomie solaire d'Hipparque, soumise à une critique rigoureuse et ensuite rendue à sa vérité primordiale.* Paris, 1828. RM.

15572. *THE NATURALIST'S Library.*
1. *Lives of eminent naturalists.* Edinburgh, 1841. See 13852/2.

15573. HEWARD, Robert.
1. *A biographical sketch of the late Allan Cunningham.* London, 1842.

15574. ROYAL Society of London.
1. *Sir Joseph Banks and the Royal Society. A popular biography, with an historical introduction and sequel.* London, 1844.

15575. WILLIAMSON, George.
1. *Memorials of the lineage, early life, education and development of the genius of James Watt.* Edinburgh, 1856. RM.

15576. CROSSE, Cornelia A.H.
1. *Memorials, scientific and literary, of Andrew Crosse, the electrician.* London, 1857. Includes letters and numerous memoranda of his experiments.

15577. BROWN, Thomas N.
1. *Labour and triumph. The life and times of Hugh Miller.* New York, 1858.

15578. KING, Edmund Fillingham.
1. *A biographical sketch of Sir Isaac Newton.* 2nd ed. Grantham, 1858.

15579. KOENIG, Heinrich.
1. *Georg Forster's Leben in Haus und Welt.* 2nd ed. Leipzig, 1858.

15580. WHATTON, Arundell Blount.
1. *Memoirs of the life and labors of the Rev. Jeremiah Horrox. To which is appended a translation of his celebrated discourse upon the transit of Venus across the sun.* London, 1859.

15581. WILSON, Jessie Aitken.
1. *Memoir of George Wilson* [the chemist, 1818-1859]. Edinburgh, 1860.

15582. *THE HUMBOLDT Library: A catalogue of the library of Alexander von Humboldt. With a bibliographical and biographical memoir.* London, 1863. By Henry Stevens. X. Reprinted, Leipzig, 1967.

15583. *PHILOSOPHIE des deux Ampères.* Paris, 1866. Ed. by J. Barthélemy Saint-Hilaire. Contents: Introduction à la philosophie de mon père [i.e. André Marie Ampère, 1775-1836], par Jean Jacques Ampère [1800-1864].-- Philosophie de A.M. Ampère: Lettres à M. de Biran. Fragments.

15584. ANDREW, Alexander.
1. *Memoir of Dr John Rankine* [1801-1864]. Glasgow, 1866.

15585. McILRAITH, John. BMNH.
1. *Life of Sir John Richardson.* London, 1868.

15586. MOILLET, Amelia.
1. *Sketch of the life of James Keir, Esq., F.R.S., with a selection of his correspondence.* London, 1868.

15587. GORDON, Margaret Maria, *Mrs* (née BREWSTER).
1. *The home life of Sir David Brewster.* Edinburgh, 1869. 2nd ed., 1870.

15588. ALLAN-OLNEY, Mary.
1. *The private life of Galileo. Compiled principally from his correspondence and that of his eldest daughter, Sister Maria Celeste.* London, 1870. RM/L.

15589. SCHANZ, Paul.
1. *Der Cardinal Nicolaus von Cusa als Mathematiker.* 1872. X. Reprinted, Wiesbaden, 1967.

15590. HERSCHEL, Mary, *Mrs* (née CORNWALLIS). d. 1876.
1. *Memoir and correspondence of Caroline Herschel.* London, 1876. 2nd ed., 1879.

15591. COCHRANE, Robert.
1. *Heroes of invention and discovery. Lives of eminent inventors and pioneers in science.* Edinburgh, 1879.

15592. SMITHSONIAN Institution.
1. *A memorial of Joseph Henry.* Washington, 1880.

15593. DOUGLAS, Janet Mary.
1. *Life and selections from the correspondence of William Whewell.* London, 1881.

15594. LYELL, Katherine Murray, *Mrs.* BMNH.
1. *The life, letters and journals of Sir Charles Lyell.* 2 vols. London, 1881. By his sister-in-law. RM/L. Reprinted, Farnborough, 1970.

15595. *NATURE.* (The periodical)
1. *Charles Darwin. Memorial notices reprinted from "Nature."* London, 1882.

15596. GRAVES, Robert Perceval.
1. *The life of Sir William Rowan Hamilton, including selections from his poems, correspondence and miscellaneous writings.* 3 vols. Dublin, 1882-89.

15597. WOODALL, Edward.
1. *Charles Darwin.* London, 1884.

15598. BOMPAS, George C.
1. *Life of Frank Buckland* [the zoologist, 1826-1880]. London, 1885. By his brother-in-law.

15599. HARTMANN, Franz. d. 1912.
1. *The life of Philippus Theophrastus Bombast of Hohenheim, known by the name of Paracelsus, and the substance of his teaching.* London, 1887. X. 2nd ed, 1896.

15600. JEANS, William T.
1. *Lives of the electricians: Professors Tyndall, Wheatstone, and Morse. First series.* London, 1887.

15601. ALBERG, Albert.
1. *The floral king. A life of Linnaeus.* London, 1888.

15602. CUDWORTH, William.
1. *The life and correspondence of Abraham Sharp, the Yorkshire mathematician and astronomer, and assistant of Flamsteed.* London, 1889.

15603. GORDON, Elizabeth Oke, *Mrs* (née BUCKLAND).
1. *Life and correspondence of William Buckland* [the geologist, 1784-1856]. London, 1894.

15604. OWEN, Richard Startin. BMNH.
1. *The life of Richard Owen, by his grandson ... Also an essay on Owen's position in anatomical science by T.H. Huxley.* 2 vols. London, 1894-95. Reprinted, Farnborough, 1970.

15605. CARLI, Alarico.
1. *Bibliografia Galileiana, 1568-1895.* Rome, 1896. With A. Favaro. X. Reprinted, Bologna, 1972.

15606. CRAMER, Frank.
1. *The method of Darwin. A study in scientific method.* Chicago, 1896.

15607. ROMANES, Ethel, *Mrs* (née DUNCAN).
1. *The life and letters of G.J. Romanes.* London, 1896. By his wife.

15608. WATERS, William George.
1. *Jerome Cardan.* London, 1898.

3.23 HISTORY OF SCIENCE

In this section authors are included even if no information about them was found. Some of the authors are not scientists.

Authors in Part 2 (Early Modern Period) who wrote books which, from their titles, appear to be historical in the modern sense of the word are as follows: Baldi, B. (1738); Vossius, G.J. (1877); Nottnagel, C. (2145); Dati, C. (2242); Borel, P. (2267); Borrichius, O. (2341); Bernard, E. (2471); Zahn, J. (2532); Hartmann, P.J. (2600).

The following is an analysis of the aspects of the history of science that were written about by the authors listed in this section. The numbers are the entry numbers less the prefix 15. The classification is perforce only rough. One convention adopted is that the ancient and medieval periods are given precedence, e.g. a history of mathematics in antiquity is classified under antiquity, not under mathematics. Many authors are included in more than one category.

(a) Fields of Science

Science in General. 623, 628, 629, 638, 641, 651, 697, 702, 718, 734, 735, 766, 781, 785, 801, 818, 827, 888, 892, 893, 902, 915.

Mathematics. 611, 617, 619, 622, 625, 627, 664, 667, 668, 693, 706, 711, 719, 726, 732, 736, 743, 755, 758, 759, 764, 771, 772, 775, 777, 780, 798, 810, 829, 833, 839, 844, 846, 847, 851, 854, 856, 858, 860, 861, 865, 874, 875, 877, 881, 884, 887, 889, 906, 907.

Astronomy. 610, 615, 620, 624, 631, 635, 636, 639, 642, 644, 645, 646, 650, 664, 669, 681, 695, 698, 704, 737, 741, 747, 774, 779, 796, 804, 813, 814, 822, 824, 833, 841, 843, 844, 857, 884, 916, 918.

Mechanics. 678, 720, 758, 808.

Physics. 626, 635, 637, 649, 655, 658, 694, 701, 702, 706, 711, 755, 759, 774, 790, 793, 806, 808, 821, 828, 837, 840, 852, 873, 880, 885, 887.

Chemistry. 610, 612, 626, 632, 633, 634, 643, 659, 670, 684, 687, 723, 733, 749, 751, 752, 756, 773, 791, 805, 825, 826, 827, 830, 836, 838, 845, 849, 866, 879, 882, 917.

Mineralogy. 654, 713.

Crystallography. 696.

Geology. 682, 817.

Geodesy. 758, 876.

Microscopy. 738.

Botany. 630, 665, 690, 692, 716, 760, 770, 784, 786, 799, 807, 842, 895, 898.

Zoology (and Natural History). 647, 666, 675, 746, 757, 763, 834, 835, 850, 883, 905.

15609. FONTENELLE, Bernard Le Bouvier de. 1657-1757. Fr. DSB; ICB; P I; GA; GB. Also Science in General*#, Mathematics*, Mechanics*, Biography*, and Part 2*.
   1. *Histoire du renouvellement de l'Académie Royale des Sciences en 1699....* See 15395/1.

15610. FRISCH, Johann (or Jodocus) Leonhard. 1666-1743. Ger. P I; BMNH; GA; GC. Also Zoology*. History of astronomy and chemistry.

15611. DOPPELMAYR, Johann Gabriel. 1671?-1750. Ger. DSB; P I; GA; GC. Also Astronomy and Physics*. History of Mathematics.
   1. *Historische Nachricht von den Nürnbergischen Mathematicis und Künstlern.* Nuremberg, 1730. RM. Reprinted, Hildesheim, 1971.

15612. LENGLET DU FRESNOY, Nicholas. 1674-1755. Fr. P I; GA. History of chemistry and alchemy.
   1. *Histoire de la philosophie hermétique. Accompagnée d'un catalogue raisonné des écrivains de cette science.* Paris, 1842. RM.

15613. DOUGLAS, James. 1675-1742. Br. DSB; BMNH; Mort.; BLA & Supp.; GD. Also Natural History and Anatomy.
   1. *Bibliographiae anatomicae specimen; sive, Catalogus omnium pene auctorum qui ab Hippocrate ad Harveum rem anatomicam ex professo, vel obiter, scriptis illustrarunt.* London, 1715. X. Reprinted?

15614. LEBEUF, Jean. 1687-1760. Fr. GA; GB.
1. *De l'état des sciences dans l'étendue de la monarchie françoise sous Charlemagne*. Paris, 1734. RM.

15615. WEIDLER, Johann Friedrich. 1691-1755. Ger. P II; GA; GC. Also Astronomy*.
1. *Historia astronomiae; sive, De ortu et progressu astronomiae*. Wittenberg, 1741. RM. A supplement to this work is appended to item 2.
2. *Bibliographia astronomica temporis. Quo libri ... ordine servato, ad supplendam et illustrandam astronomiae historiam digesta*. Wittenberg, 1755. RM.

15616. STIERNMAN, Anders Anton von. 1695-1765. Swed. ICB. History of science in Sweden.

15617. CRAMER, Gabriel. 1704-1752. Switz. DSB; ICB; P I; GA. Also Mathematics*. See also Index. History of mathematics.

15618. BIRCH, Thomas. 1705-1766. Br. ICB; P I; GA; GB; GD. Also Biography*. See also Index.
1. *The history of the Royal Society of London*. 4 vols. London, 1756-58. RM/L. Reprinted, Brussels, 1967-68, and New York, 1968. Another ed., 1760.

15619. HEILBRONNER, Johann Christoph. 1706-1745. Ger. P I,III; GA. History of mathematics.
1. *Historia matheseos universae, a mundo condito ad seculum P.C.N. XVI, praecipuorum mathematicorum vitas, dogmata, scripta et manuscripta complexa*. Leipzig, 1742. RM.

15620. COSTARD, George. 1710-1782. Br. P I; GA; GD. History of Astronomy.
1. *A letter ... concerning the rise and progress of astronomy amongst the antients*. London, 1746.
2. *The history of astronomy, with its applications to geography, history and chronology*. London, 1767.

15621. TARGIONI-TOZZETTI, Giovanni. 1712-1783. It. DSB; ICB; P II; BMNH; BLA; GA. Also Geology, Natural History*, and Biography*.
1. *Notizie degli aggrandimenti delle scienze fisiche accaduti in Toscana nel corso di anni LX del secolo XVII*. Florence, 1780. RM. Reprinted, 3 vols in 4, Bologna, 1967.

15622. KAESTNER, Abraham Gotthelf. 1719-1800. Ger. DSB; ICB; P I; GA; GC. Also Mathematics, Mechanics, and Physics. History of mathematics.
1. *Geschichte der Mathematik seit der Wiederherstellung der Wissenschaften bis an das Ende des achtzehnten Jahrhunderts*. 4 vols. Göttingen, 1796-1800. RM. Reprinted, Hildesheim, 1970.

15623. SAVÉRIEN, Alexandre. 1720-1805. Fr. ICB; P II; GA.
1. *Dictionnaire universel de mathématique et de physique, où l'on traite de l'origine, du progrès de ces deux sciences et des arts qui en dépendent, et des divers révolutions qui leur sont arrivées jusqu'à notre temps*. 2 vols. Paris, 1753. RM.
2. *Histoire des philosophes modernes*. 4 vols. Paris, 1760-73. X. 3rd ed., 8 vols, 1773.

1. *Histoire des progrès de l'esprit humain dans les sciences exactes, et dans les arts qui en dépendent.* Paris, 1766. RM/L.

15624. HEATHCOTE, Ralph. 1721-1795. Br. P I; GA; GD. History of astronomy.

15625. MONTUCLA, Jean Etienne. 1725-1799. Fr. DSB; ICB; P II; GA.
1. *Histoire des mathématiques.* 2 vols. Paris, 1758. X.
1a. ——— 2nd ed., much enlarged, 4 vols, ib., 1799-1802. RM/L. Reprinted, ib., 1960. Deals with astronomy, mechanics, optics, and applied mathematics generally as well as pure mathematics.
See also 2516/2a.

15626. AMEILHON, Hubert Pascal. 1730-1811. Fr. P I; GA; GF. History of physics and chemistry.

15627. BOSSUT, Charles. 1730-1814. Fr. DSB; P I; GA. Also Mechanics*.
1. *Essai sur l'histoire générale des mathématiques.* 2 vols. Paris, 1802. RM/L.
1a. ——— *A general history of mathematics.* London, 1803. RM/L.
See also 2312/3.

15628. DUTENS, Louis. 1730-1812. Fr./Br. P I & Supp.; GA; GB; GD.
1. *Recherches sur l'origine des découvertes attribuées aux modernes, où l'on démontre que nos plus célèbres philosophes ont puisé la plupart de leurs connoissances dans les ouvrages des anciens.* Paris, 1766. X. Reprinted, ib., [1968?]. 4th ed., 2 vols, 1812.
1a. ——— *An inquiry into the origin of the discoveries attributed to the moderns.* London, 1769.

15629. LOYS, Charles de. ca. 1730-1789. Switz. P I.
1. *Abrégé chronologique pour servir à l'histoire de la physique.* 4 vols. Paris, 1786-89. X.

15630. PULTENEY, Richard. 1730-1801. Br. BMNH; GA; GB. Also Biography*.
1. *Historical and biographical sketches of the progress of botany in England, from its origin to the introduction of the Linnean system.* 2 vols. London, 1790. RM/L.

15631. LALANDE, Joseph Jérome Lefrançais de. 1732-1807. Fr. DSB; ICB; P I; GA; GB. Also Astronomy*.
1. *Bibliographie astronomique. Avec l'histoire de l'astronomie depuis 1781 jusqu'à 1802.* Paris, 1803. RM/L. Reprinted, Amsterdam, 1970.

15632. WIEGLEB, Johann Christian. 1732-1800. Ger. DSB; P II; GA; GC. Also Chemistry.
1. *Historisch-kritische Untersuchung der Alchemie.* Weimar, 1777. X. Reprinted, Leipzig, 1965.
2. *Geschichte des Wachsthums und der Erfindungen in der Chemie.* 3 vols. Berlin/Stettin, 1790-92. X. Reprinted, Leipzig, 1966.

15633. MURR, Christoph Gottlieb von. 1733-1811. Ger. P II; BMNH; GA; GC. History of alchemy.

15634. SCHRÖDER, Friedrich Joseph Wilhelm. 1733-1778. Ger. P II; BLA. History of alchemy and chemistry.

15635. OSTERTAG, Johann Philipp. 1734-1801. Ger. P II; GC. History of astronomy and physics.

15636. BAILLY, Jean Sylvain. 1736-1793. Fr. DSB; ICB; P I; GA; GB. Also Astronomy* and Biography*.
1. *Histoire de l'astronomie ancienne, depuis son origine jusqu'à l'établissement de l'école d'Alexandrie.* Paris, 1775. RM. 2nd ed., 1781.
2. *Lettres sur l'origine des sciences, et sur celle des peuples de l'Asie. Adressées à M. de Voltaire ... et précédées de quelques lettres de M. de Voltaire à l'auteur.* Paris/London, 1777. RM/L.
3. *Histoire de l'astronomie moderne, depuis la fondation de l'école d'Alexandrie.* 3 vols. Paris, 1779-82. RM/L. Another ed., 1785.
4. *Traité de l'astronomie indienne et orientale.* Paris, 1787. RM/L.

15637. ROMÉ DE L'ISLE, Jean Baptiste Louis. 1736-1790. Fr. DSB; P II; BMNH; GB. Also Crystallography* and Geology*.
1. *Métrologie; ou, Tables pour servir à l'intelligence des poids et mesures des anciens.* Paris, 1789. RM.

15638. BECKMANN, Johann. 1739-1811. Ger. DSB; ICB; BMNH; GA; GB; GC.
1. *Beyträge zur Geschichte der Erfindungen.* 5 vols. Leipzig, 1783-1805. RM. Reprinted, Hildesheim, 1965.
1a. ——— *A history of inventions and discoveries.* 3 vols in 4. London, 1797. Trans. by W. Johnston. RM/L. 4th ed., rev. and enlarged, 1846.

15639. DUPUIS, Charles François. 1742-1809. Fr. P I; GA; GB. History of astronomy.

15640. PORTAL, Antoine. 1742-1832. Fr. DSB; ICB; BLA; GA. Also Anatomy. History of anatomy.

15641. SABATIER, Antoine de Castres. 1742-1817. Fr. GA.
1. *Dictionnaire des origines, découvertes, inventions et établissemens; ou, Tableau historique de l'origine et des progrès de tout ce qui a rapport aux sciences et aux arts.* Paris, 1777. RM.

15642. CASSINI, Jean Dominique. 1748-1845. Fr. DSB; ICB; P I; GA; GB. Also Astronomy and Geodesy*.
1. *Mémoires pour servir à l'histoire des sciences et à celle de l'Observatoire de Paris. Suivi de la vie de J.D. Cassini, écrite par lui-même, et des éloges de plusieurs académiciens morts pendant la révolution.* Paris, 1810. RM.

15643. GMELIN, Johann Friedrich. 1748-1804. Ger. P I; BMNH; BLA & Supp.; GA; GC. Also Chemistry* and Botany.
1. *Geschichte der Chemie.* 3 vols. Göttingen, 1797-99. RM. Reprinted, Hildesheim, 1965.

15644. DELAMBRE, Jean Baptiste Joseph. 1749-1822. Fr. DSB; ICB; P I; GA; GB. Also Mathematics*, Astronomy*, and Geodesy*.

1. *Histoire de l'astronomie ancienne.* 2 vols. Paris, 1817. RM/L. Reprinted, New York, 1965.
2. *Histoire de l'astronomie du moyen âge.* Paris, 1819. RM/L. Reprinted, New York, 1965.
3. *Histoire de l'astronomie moderne.* 2 vols. Paris, 1821. RM/L. Reprinted, New York, 1969.
4. *Histoire de l'astronomie au dix-huitième siècle.* Paris, 1827. RM/L. Reprinted, New York, 1969.

See also 2897/1.

15645. LAPLACE, Pierre Simon de. 1749-1827. Fr. DSB Supp.; ICB; P I; GA. Also Mathematics*, Astronomy*#, and Physics. See also Index.
1. *Précis de l'histoire de l'astronomie.* Paris, 1821. RM.

15646. BRADY, John. d. 1814. Br. GD.
1. *Clavis calendaria; or, A compendious analysis of the calendar.* 2 vols. London, 1812. X. 2nd ed., 1812. 3rd ed., 1815.

15647. SCHNEIDER, Johann Gottlob. 1750-1822. Ger. P II; BMNH; GA; GB; GC. Also Zoology*. History of natural history, metallurgy, etc.

15648. ASSEMANI, Simone. 1752-1821. It. GA.
1. *Globus caelestis Cufico-Arabicus Veliterni Musei Borgiani illustratus. Praemissa ejusdem De Arabum astronomia dissertatione.* Padua, 1790.

15649. LIBES, Antoine. 1752-1832. Fr. P I; GA. Also Physics*.
1. *Histoire philosophique des progrès de la physique.* 4 vols. Paris, 1810-13. RM/L.

15650. BARRETT, John. 1753-1821. Br. P III; GA; GD.
1. *An enquiry into the origin of the constellations that compose the zodiac.* Dublin, 1800.

15651. DONNDORFF, Johann August. 1754-1837. Ger. P I; BMNH. Also Physics and Zoology.
1. *Geschichte der Erfindungen in allen Theilen der Wissenschaften und Künste ... In alphabetischer Ordnung.* Quedlinburg, 1817-21. RM.

15652. FIELD, Henry. 1755-1837. Br. BMNH; GD. Also Botany.
1. *Memoirs, historical and illustrative, of the Botanick Garden at Chelsea, belonging to the Society of Apothecaries of London.* London, 1820. Another ed., "corrected and continued to the present time", by R.H. Semple, 1878.

15653. HALMA, Nicolas B. 1755-1828. Fr. ICB; P I; GA. History of ancient mathematics and astronomy.

15654. SCHWARZE, Christian August. 1755-1809. Ger. P II; BMNH; History of mineralogy.

15655. FORTIA D'URBAN, A.J.F.X.P.E.S.P.A. *Marquis* de. 1756-1843. Fr. P I; GA. History of optics, etc.

15656. LAUTH, Thomas. 1758-1826. Ger./Fr. ICB; BLA; GC. Also Anatomy. History of anatomy.

15657. GUIGNES, Chrétien Louis Joseph de. 1759-1845. Fr./China. P I; GA. History of Chinese astronomy.

15658. FISCHER, Johann Karl. 1760-1833. Ger. P I.
1. *Geschichte der Physik*. 8 vols. Göttingen, 1801-08. RM.

15659. FUCHS, Georg Friedrich Christian. 1760-1813. Ger. P I; BLA. Also Chemistry.
1. *Repertorium der chemischen Litteratur von 494 vor Christi Geburt bis 1806, in chronologischer Ordnung*. 2 vols. Jena/Leipzig, 1806-08. X. Reprinted, Hildesheim, 1974.

15660. BUTTMANN, Philipp Karl. 1764-1829. Ger. P I; GA; GB; GC. History of science in antiquity.

15661. SCHAUBACH, Johann Konrad. 1764-1849. Ger. P II; GC. History of Greek and Indian astronomy.
1. *Geschichte der griechischen Astronomie bis auf Eratosthenes*. Göttingen, 1802. RM.

15662. YUAN YUAN. 1764-1849. China. ICB. History of Chinese mathematics and astronomy (in Chinese).

15663. COLEBROOKE, Henry Thomas. 1765-1837. Br./India. P I; GA; GB; GD. History of Indian mathematics and astronomy. See also 866.

15664. IDELER, (Christian) Ludwig. 1766-1846. Ger. P I; GA; GB; GC. History of astronomy and mathematics.
1. *Historische Untersuchungen über die astronomischen Beobachtungen der Alten*. Berlin, 1806. RM.
2. *Handbuch der mathematischen und technischen Chronologie*. Berlin, 1825. RM.

15665. SPRENGEL, Kurt Polycarp Joachim. 1766-1833. Ger. Well known as a historian of medicine. DSB; BMNH; BLA & Supp.; GA; GB; GC. Also Botany*.
1. *Historia rei herbariae*. 2 vols. Amsterdam, 1807-08.
1a. ——— Another ed. *Geschichte der Botanik*. 2 vols. Altenburg, 1817-18. RM.

15666. CUVIER, Georges. 1769-1832. Fr. DSB; ICB; P I; BMNH; GA; GB. Also Science in General*, Geology*, Zoology*, and Biography*. See also Index.
1. *Histoire des progrès des sciences naturelles depuis 1789 jusqu'à ce jour* [i.e. to 1831]. See 2897/3.
2. *Histoire des sciences naturelles depuis leur origine jusqu'à nos jours*. 3 vols in 5. Paris, 1841-45. Completed and ed. by Magdeleine de Saint-Agy. RM. Reprinted, Brussels, 1969, and Farnborough, 1970.

15667. CHRISTIANI, Johann Wilhelm. b. 1771. Ger. P III. History of mathematics.

15668. REIMER, Nicolaus Theodor. 1772-1832. Ger. P II. History of mathematics.

15669. RICHTER, Johann Andreas Leberecht. b. 1772. Ger. P II. History of the cultural influence of astronomy.

15670. THOMSON, Thomas. 1773-1852. Br. DSB; ICB; P II; BMNH; GB; GD. Also Physics*, Chemistry*, and Mineralogy*.
1. *History of the Royal Society*. London, 1812. RM/L.
2. *The history of chemistry*. 2 vols. London, 1830-31. RM/L.

15671. BIOT, Jean Baptiste. 1774-1862. Fr. DSB; ICB; P I,III; GA; GB. Also Mathematics*, Astronomy*, and Physics*.
1. *Recherches sur plusieurs points de l'astronomie égyptienne, appliquées aux monumens astronomiques trouvés en Egypte.* Paris, 1823. RM.
2. *Etudes sur l'astronomie indienne et sur l'astronomie chinoise.* Paris, 1862. RM. Reprinted, ib., 1969.

15672. RIGAUD, Stephen Peter. 1774-1839. Br. ICB; P II; GD.
1. *Historical essay on the first publication of Sir Isaac Newton's "Principia."* Oxford, 1838. Reprinted, New York, 1972.
See also 2832/2 and 4509/1,1a.

15673. HAUBER, Karl Friedrich. 1775-1851. Ger. P I. History of Greek mathematics.

15674. LÜDERS, Ludwig. 1776-1822. Ger. P I. History of Greek mathematics.

15675. BLAINVILLE, Henri Marie Ducrotay de. 1777-1850. Fr. DSB; P I; BMNH; BLA; GA; GB. Also Zoology*.
1. *Histoire des sciences de l'organisation et de leurs progrès, comme base de la philosophie.* 3 vols. Paris, 1847. Ed. from his lecture course by F.L.M. Maupied. RM/L.

15676. SEDILLOT, Jean Jacques Emmanuel. 1777-1832. Fr. P II. History of Arabic astronomy.

15677. WALKER, Alexander. 1779-1852. Br. DSB. Also Physiology.
1. *Documents and dates of modern discoveries in the nervous system.* London, 1839.

15678. PLANA, Giovanni (or Jean). 1781-1864. It. DSB; ICB; P II,III. Also Mathematics, Astronomy, and Physics*.
1. *Mémoire sur la découverte de la loi du choc direct des corps durs.* Turin, 1843.

15679. SCORZA, Giuseppe. 1781-1844. It. P II. History of ancient mathematics.

15680. MOORE, Nathaniel Fish. 1782-1872. U.S.A. BMNH; GE.
1. *Ancient mineralogy; or, An inquiry respecting mineral substances mentioned by the ancients.* 1834. X. Reprinted, New York, 1968.

15681. NARRIEN, John. 1782-1860. Br. P III; GA; GD.
1. *Historical account of the origin and progress of astronomy.* London, 1833.

15682. KEFERSTEIN, Christian. 1784-1866. Ger. P I; BMNH; GA; GC. Also Geology*.
1. *Geschichte und Litteratur des Geognosie. Ein Versuch.* Halle, 1840. RM.

15683. BOECKH, (Philipp) August. 1785-1867. Ger. P I,III; GA; GB. History of Greek science.

15684. CHEVREUL, Michel Eugène. 1786-1889. Fr. DSB; ICB; P I,III, IV; GA; GB. Also Physics* and Chemistry*.
1. *Introduction à l'histoire des connaissances chimiques.*

Connexion des sciences au domaine de la philosophie naturelle. Paris, 1866. RM/L. Sometimes listed as • *Histoire des connaissances chimiques.*
2. *Resumé d'une histoire de la matière depuis les philosophes grecs jusqu'à Lavoisier.* Paris, 1878.

15685. PARAVEY, Charles Hippolyte de. 1787-1871. Fr. P II,III; GA. History of ancient astronomy and of Chinese astronomy.

15686. STUHR, Peter Feddersen. 1787-1851. Ger. GC.
1. *Untersuchungen über die Ursprünglichkeit und Alterthümlichkeit der Sternkunde unter den Chinesen und Indiern, und über den Einfluss der Griechen auf den Gang ihrer Ausbildung.* Berlin, 1831. RM.

15687. BRANDE, William Thomas. 1788-1866. Br. DSB; ICB; P I,III; BMNH; GB; GD. Also Science in General*, Chemistry*, and Geology*.
1. *Dissertation third. Exhibiting a general view of the progress of chemical philosophy from the earliest ages to the end of the eighteenth century.* London, 1817. (Separate issue of an encyclopaedia article) X. Another ed., Boston, 1818. RM.

15688. DIERBACH, Johann Heinrich. 1788-1845. Ger. BMNH; BLA & Supp. Also Botany.
1. *Flora mythologica; oder, Pflanzenkunde in Bezug auf Mythologie und Symbolik der Greichen und Römer. Ein Beitrag zur ältesten Geschichte der Botanik, Agricultur und Medizin.* Frankfurt, 1833. X. Reprinted, Wiesbaden, 1970.

15689. JOURDAIN, Amable Louis Marie Michel Bréchillet. 1788-1818. Fr. GA.
1. *Recherches critiques sur l'âge et l'origine des traductions latines d'Aristote et sur les commentaires grecs ou arabes employés par les docteurs scolastiques.* Paris, 1819. X. Another ed., 1843; reprinted, New York, 1960.

15690. HENSCHEL, August Wilhelm Eduard Theodor. 1790-1856. Ger. BMNH; BLA. Also Botany and Biography*. History of Botany.

15691. CHOULANT, (Johann) Ludwig. 1791-1861. Ger. ICB; P I & Supp.; BLA & Supp.; GA.
1. *Handbuch der Bücherkunde für die ältere Medizin.* Leipzig, 1841. X. Reprinted, Graz, 1956.
2. *Bibliotheca medico-historica; sive, Catalogus librorum historicorum de re medica et scientia naturali systematicus.* Leipzig, 1842. X. Reprinted, Hildesheim,1960.
3. *Geschichte und Bibliographie der anatomischen Abbildung nach ihrer Beziehung auf anatomische Wissenschaft und bildene Kunst.* Leipzig, 1852. X.
3a. ——— *History and bibliography of anatomical illustration in its relation to anatomic science, etc.* Chicago, 1920. Trans. by M. Frank. Another ed., with various additions, New York, 1945; reprinted, ib., 1962.
4. *Graphische Incunabeln für Naturgeschichte und Medicin. Enthaltend Geschichte und Bibliographie der ersten naturhistorischen und medicinischen Drucke des XV. und*

XVI. Jahrhunderts, welche mit illustrirenden Abbildungen versehen sind. Leipzig, 1858. X. Reprinted, Hildesheim, 1963.

See also 14632/1.

15692. MEYER, Ernst Heinrich Friedrich. 1791-1858. Ger. P II; BMNH; GC. Also Botany.
1. *Geschichte der Botanik. Studien.* 4 vols in 2. Königsberg, 1854-57. RM/L. Reprinted, Amsterdam, 1965.

15693. CHASLES, Michel. 1793-1880. Fr. DSB; ICB; P I,III; GA. Also Mathematics*.
1. *Aperçu historique sur l'origine et le développement des méthodes en géometrie.* Brussels, 1837. 2nd ed., Paris, 1875. 3rd ed., Paris, 1889.

15694. WILDE, (Heinrich) Emil. 1793-1859. Ger. P II. Also Physics.
1. *Geschichte der Optik.* 2 vols. Berlin, 1838-43.

15695. MÄDLER, Johann Heinrich von. 1794-1874. Ger./Dorpat. DSB; P II,III; GA; GC. Also Astronomy.
1. *Geschichte der Himmelskunde.* 2 vols. Brunswick, 1873. RM. Reprinted, Wiesbaden, 1973.

15696. MARX, Karl Michael. 1794-1864. Ger. P II & Supp.; BMNH. Also Crystallography.
1. *Geschichte der Krystallkunde.* Karlsruhe, 1825. X. Reprinted, Wiesbaden, 1970.

15697. WHEWELL, William. 1794-1866. Br. DSB; ICB; P II,III; BMNH; GB; GD. Other entries: see 3304.
1. *History of the inductive sciences.* 3 vols. London, 1837. RM/L. 2nd ed., 1847. 3rd ed., 1857; reprinted, ib., 1967.

15698. ROTHMAN, Richard Wellesley. d. 1856. Br. P II. History of astronomy.

15699. DAUBENY, Charles Giles Bridle. 1795-1867. Br. DSB; ICB; P I, III; BMNH; GB; GD. Other entries: see 3305.
1. *Essay on the trees and shrubs of the ancients.* Oxford, 1865.

15700. JUNGE, Ernst Friedrich. ca. 1796-1840. Ger. P I & Supp. History of Greek astronomy.

15701. POGGENDORFF, Johann Christian. 1796-1877. Ger. DSB; ICB; P II,III; GA; GB; GC. Also Physics and Biography*.
1. *Geschichte der Physik.* Leipzig, 1879. RM/L. Reprinted, ib., 1964.

15702. POWELL, Baden. 1796-1860. Br. DSB; P II,III; GD. Also Science in General* and Physics*.
1. *An historical view of the progress of the physical and mathematical sciences from the earliest ages to the present time.* London, 1834. RM/L.
2. *History of natural philosophy from the earliest periods to the present time.* London, 1842. RM.

15703. QUETELET, (Lambert) Adolphe (Jacques). 1796-1874. Belg. DSB; ICB; P II,III; GA; GB. Other entries: see 3306.

1. *Histoire des sciences mathématiques et physiques chez les belges*. Brussels, 1864. RM/L.
2. *Les sciences mathématiques et physiques chez les belges au commencement du XIXe siècle*. Brussels, 1866. RM/L.

See also 15466/1.

15704. GRESWELL, Edward. 1797-1869. Br. GD.
1. *Origines kalendariae Italicae. Nundinal calendars of ancient Italy*. Oxford, 1854.
2. *Origines kalendariae Hellenicae; or, The history of the primitive calendar among the Greeks*. Oxford, 1862.

15705. WILLIAMS, John. 1797-1874. Br. P III.
1. *Observations of comets from B.C. 611 to A.D. 1640, extracted from the Chinese annals*. London, 1871.

15706. FUSS, Paul Heinrich von (or Pavel Nicolaievich). 1798-1855. Russ. ICB; P I. History of mathematics and physics. See also 3447/8,15,16.

15707. HITCHCOCK, Ethan Allen. 1798-1870. U.S.A. GE.
1. *Remarks upon alchemy and the alchemists, indicating a method of discovering the true nature of hermetic philosophy*. Boston, 1857. RM.

15708. JULLIEN, (Marcel) Bernard. 1798-1881. Fr. GA.
1. *De quelques points des sciences dans l'antiquité (physique, métrique, musique)*. Paris, 1854. RM.

15709. LENZ, Harald Othmar. 1799-1870. Ger. BMNH. Also Zoology*.
1. *Zoologie der alten Griechen und Römer*. Gotha, 1856. RM/L. Reprinted, Wiesbaden, 1966.
2. *Botanik der alten Griechen und Römer*. 1859.

15710. POUCHET, Félix Archimede. 1800-1872. Fr. DSB; ICB; BMNH; BLA; GA. Also Natural History*, Zoology*, Embryology*, and Microbiology*.
1. *Histoire des sciences naturelles au moyen âge; ou, Albert le Grand et son époque, considérés comme point de départ de l'école expérimentale*. Paris, 1853. RM/L.

15711. GHERARDI, Silvestro. 1802-1879. It. ICB; P III. Also Physics. History of mathematics and physics.

15712. BIOT, Edouard Constant. 1803-1850. Fr. P I. History of Chinese astronomy, mathematics, etc.
1. *Catalogue général des étoiles filantes et des autres météores observés en Chine depuis le VIIe siècle avant J.C. jusqu'au milieu du XVIIe de notre ère*. Paris, 1841.

15713. KOBELL, (Wolfgang Xaver) Franz von. 1803-1882. Ger. P I,III; BMNH; GB; GC. Also Mineralogy*.
1. *Geschichte der Mineralogie, 1650-1860*. Munich, 1864. RM/L. Reprinted, New York, 1965.

15714. LIBRI(-CARRUCCI DELLA SOMMAIA), Guglielmo. 1803-1869. It./Fr. ICB; P I,III; GA. Also Mathematics.
1. *Histoire des sciences mathématiques en Italie depuis la renaissance des lettres jusqu'à la fin du dix-septième siècle*. 4 vols. Paris, 1838-41. RM/L. Reprinted, New York, 1967. 2nd ed, 1865; reprinted, Hildesheim, 1965.

15715. LASAULX, Ernst von. 1805-1861. Ger. P III; GA. History of geology in antiquity.

15716. PICKERING, Charles. 1805-1878. U.S.A. BMNH; GB; GE. Also Botany.
1. *Chronological history of plants.* Boston, 1879.

15717. BURGGRAVE, Adolphe Pierre. 1806-1902. Belg. BLA & Supp.
1. *Précis de l'histoire de l'anatomie.* Ghent, 1840.

15718. CANDOLLE, Alphonse (Louis Pierre Pyramus) de. 1806-1893. Switz. DSB; ICB; P I,III (DECANDOLLE); BMNH. Also Botany*.
1. *Histoire des sciences et des savants depuis deux siècles.* Geneva, 1873. X. 2nd ed., 1885.

15719. DE MORGAN, Augustus. 1806-1871. Br. DSB; ICB; P I (under D); P III (under M); GA; GB; GD. Also Mathematics*, Astronomy*, and Biography*. See also Index.
1. *Arithmetical books from the invention of printing to the present time.* London, 1847. Reprinted, ib., 1967.

15720. DIRCKS, Henry. 1806-1873. Br. P III; GD.
1. *Perpetuum mobile; or, The search for self-motive power during the 17th, 18th, and 19th centuries.* London, 1861. RM/L.
1a. ——— *Series I and II.* 2 vols. Ib., 1870. Series I is a new ed. of the original work; Series II was first published in 1870.
2. *Contributions to a history of electro-metallurgy.* London, 1863.

15721. LACROIX, Paul. 1806-1884. Fr. ICB; GA; GB.
1. *Sciences et lettres au moyen âge et à l'époque de la renaissance.* Paris, 1877. X.
1a. ——— *Science and literature in the Middle Ages and at the period of the Renaissance.* London, 1878. RM/L.
2. *XVIIIme siècle. Lettres, sciences et arts. France 1700-1789.* Paris, 1878. RM.

15722. LEWIS, George Cornewall. 1806-1863. Br. GA; GB; GD.
1. *An historical survey of the astronomy of the ancients.* London, 1862. RM/L.

15723. TROUESSART, Joseph Louis. 1806-1870. Fr. P III.
1. *Essai historique sur la théorie des corps simples ou élémentaires.* Paris, 1854.

15724. COCKAYNE, Thomas Oswald. 1807-1873. Br. BMNH; GD.
1. *Leechdoms, wort-cunning and starcraft of early England. Being a collection of documents ... illustrating the history of science before the Norman conquest.* London, 1858.

15725. BRETSCHNEIDER, Carl Anton. 1808-1878. Ger. P I,III. Also Mathematics.
1. *Geometrie und die Geometer vor Euklides. Ein historischer Versuch.* Leipzig, 1870. Reprinted, Wiesbaden, 1968.

15726. SÉDILLOT, Louis Pierre Eugène Amélie. 1808-1875. Fr. P II,III.
1. *Matériaux pour servir à l'histoire comparée des sciences*

mathématiques chez les grecs et les orientaux. Paris, 1845-49. RM.
2. *Les professeurs de mathématiques et de physique générale au Collège de France.* Rome, 1869.

15727. WÜSTENFELD, (Heinrich) Ferdinand. 1808-1899. Ger. GC.
1. *Geschichte der arabischen Ärzte und Naturforscher.* Göttingen, 1840. X. Reprinted, Hildesheim, 1963.

15728. IDELER, Julius Ludwig. 1809-1842. Ger. P I; BLA; GA; GC.
History of meteorology in antiquity.

15729. FRAAS, Karl Nikolas. 1810-1875. Ger. BMNH; GB; GC.
1. *Synopsis plantarum florae classicae; oder, Übersichtliche Darstellung der in den klassischen Schriften der Griechen und Römer vorkommenden Pflanzen.* Munich, 1845. X. Another ed., Berlin, 1870.

15730. HYRTL, (Carl) Joseph. 1810-1894. Prague/Aus. DSB; ICB; BMNH; Mort.; BLA. Also Zoology* and Anatomy.
1. *Onomatologia anatomica. Geschichte und Kritik der anatomischen Sprache der Gegenwart.* Vienna, 1880. Reprinted, Hildesheim, 1970.
2. *Die alten deutschen Kunstworte der Anatomie.* Vienna, 1884.

15731. MAILLY, (Nicolas) Edouard. 1810-1891. Belg. P III; BMNH. Also Science in General* and Biography*.
1. *Précis de l'histoire de l'astronomie aux Etats-Unis d'Amérique.* Brussels, 1860. RM/L.
2. *Histoire de l'Académie Impériale et Royale des Sciences et Belles-Lettres de Bruxelles.* 2 vols. Brussels, 1883.

15732. OFTERDINGER, Ludwig Felix. 1810-1896. Ger. P II,III,IV; GC.
History of mathematics.

15733. WITTSTEIN, Georg Christian. 1810-1887. Ger. P II,III; BMNH. Also Chemistry*.
1. *Vollständiges etymologisch-chemisches Handwörterbuch mit Berücksichtigung der Geschichte und Literatur der Chemie.* 3 vols. Munich, 1849-58. X. Reprinted, Hildesheim, ca. 1970.

15734. DRAPER, John William. 1811-1882. U.S.A. DSB; ICB; P I,III; BMNH; BLA & Supp.; GB; GD; GE. Also Physics*, Chemistry, and Physiology*.
1. *History of the conflict between religion and science.* London, 1875. X. Reprinted, Farnborough, 1970. 3rd ed., 1875. 7th ed., 1876. 8th ed., 1877. 13th ed., 1879. 17th ed., 1883.

15735. HOEFER, (Jean Chrétien) Ferdinand. 1811-1878. Fr. ICB; P I, III; BMNH; BLA; GA. Also Chemistry* and Biography*.
1. *Histoire de la chimie.* 2 vols. Paris, 1842-43. X. 2nd ed., 1866-69. RM/L.
2. *Histoire de la botanique, de la minéralogie et de la géologie.* Paris, 1872. RM.
3. *Histoire de la physique et de la chimie.* Paris, 1872. RM.
4. *Histoire de l'astronomie.* Paris, 1873. RM.
5. *Histoire de la zoologie.* Paris, 1873. X. Another ed., 1890. RM.
6. *Histoire des mathématiques.* Paris, 1874. RM.

15736. NESSELMANN, Georg Heinrich Ferdinand. 1811-1881. Ger. P II, III.
1. *Versuch einer kritischen Geschichte der Algebra. T.1. Die Algebra der Griechen*. Berlin, 1842. X. Reprinted, Frankfurt, 1969.

15737. APELT, Ernst Friedrich. 1812-1859. Ger. DSB; P I & Supp.; GC. Also Science in General. History of astronomy and cosmology.

15738. HARTING, Pieter. 1812-1885. Holl. DSB; P I,III; BMNH; Mort.; BLA. Also Zoology and Microscopy.
1. *Het mikroskoop*. 3 vols. Utrecht, 1848-50. X.
1a. ——— *Das Mikroscop*. Brunswick, 1859. Trans. by F.W. Thiele. 2nd ed., 1866; reprinted, Amsterdam, 1970.

15739. MARTIN, Thomas Henri. 1813-1884. Fr. P II,III. Also Science in General* and Biography*. History of mathematics, astronomy, and physics in antiquity.
1. *Histoire de l'arithmétique. Recherches*. Paris, 1857. RM.
2. *La foudre, l'électricité et le magnétisme chez les anciens*. Paris, 1866. RM.

15740. WELD, Charles Richard. 1813-1869. Br. GD.
1. *A history of the Royal Society, with memoirs of the presidents*. 2 vols. London, 1848. RM/L.

15741. GRANT, Robert. 1814-1892. Br. P III; GB; GD1. Also Astronomy.
1. *History of physical astronomy*. London, 1852. RM/L. Reprinted, New York, 1966.

15742. JONES, Henry Bence. 1814-1873. Br. ICB; P I,III; BMNH; BLA; GB (BENCE-JONES); GD. Also Chemistry* and Biography*.
1. *The Royal Institution, its founders and its first professors*. London, 1871.

15743. GERHARDT, Karl Immanuel. 1816-1899. Ger. P I,III,IV; GC.
1. *Geschichte der Mathematik in Deutschland*. Munich, 1877. RM/L. Reprinted, New York, 1965.
See also 242/1 and 2583/14,17.

15744. LECLERC, Lucien. 1816-1893. Fr. ICB; BLA.
1. *Histoire de la médecine arabe. Exposé complet des traductions du grec. Les sciences en Orient, leur transmission à l'Occident par les traductions latines*. Paris, 1876. X. Reprinted, 2 vols, New York, 1961.

15745. STEINSCHNEIDER, Moritz. 1816-1907. Ger. ICB; P III,IV; BLA; BLAF; GB. Also Biography*.
1. *Mathematik bei den Juden*. 1893-99. X. Reprinted, Hildesheim, 1964.

15746. STRICKER, Wilhelm Friedrich Carl. 1816-1891. Ger. ICB; P III; BMNH; BLA; GC. History of zoology.

15747. WOLF, (Johann) Rudolf. 1816-1893. Switz. DSB; ICB; P II,III, IV; GC. Also Astronomy.
1. *Geschichte der Astronomie*. Munich, 1877. RM/L. Reprinted, New York, 1965.
2. *Handbuch der Astronomie. Ihrer Geschichte und Litteratur*. 2 vols. Zurich, 1890-93. RM/L. Reprinted, Hildesheim, 197-?

15748. DAREMBERG, Charles Victor. 1817-1872. Fr. ICB; BLA & Supp.; GA.
1. *Histoire des sciences médicales, comprenant l'anatomie, la physiologie....* Paris, 1870.
See also 129/1 and 159/2.

15749. KOPP, Hermann. 1817-1892. Ger. DSB; ICB; P I,III,IV; GB; GC. Also Chemistry.
1. *Geschichte der Chemie.* 4 vols. Brunswick, 1843-47. RM/L. Reprinted, Hildesheim, 1966.
2. *Beiträge zur Geschichte der Chemie.* 2 vols. Brunswick, 1869-75. RM/L.
3. *Die Entwickelung der Chemie in der neueren Zeit.* Munich, 1873. RM/L. Reprinted, Hildesheim, 1966.
4. *Die Alchemie in alterer und neuerer Zeit. Ein Beitrag zur Culturgeschichte.* 2 vols. Heidelberg, 1886. RM/L. Reprinted, 2 vols in 1. Hildesheim, 1962.

15750. SMITH, Robert Angus. 1817-1884. Br. DSB; P III,IV; BLA; GD. Also Chemistry* and Biography*.
1. *A centenary of science in Manchester.* London, 1883.

15751. WURTZ, Charles Adolphe. 1817-1884. Fr. DSB; ICB; P II,III, IV (WÜRTZ, Karl A.); BLA; GB. Also Chemistry*.
1. *Histoire des doctrines chimiques depuis Lavoisier jusqu'à nos jours.* Paris, 1868. RM/L.
1a. ——— *A history of chemical theory from the age of Lavoisier to the present time.* London, 1869. Trans. and ed. by H. Watts. RM/L.

15752. HOFMANN, August Wilhelm von. 1818-1892. Ger./Br. DSB; ICB; P I,III,IV; GB; GC. Also Chemistry* and Biography*.
1. *Chemische Erinnerungen aus der Berliner Vergangenheit.* Berlin, 1882.
2. *Berliner Alchemisten und Chemiker. Rückblick auf die Entwickelung der chemischen Wissenschaft in der Mark.* Berlin, 1882. X. Reprinted, Wiesbaden, 1965.

15753. ZUCKERMANN, Benedict. 1818-1891. Ger. GF. History of Jewish mathematics and astronomy.

15754. FIGUIER, Louis Guillaume. 1819-1894. Fr. P I,III; BMNH; GA. Other entries: see 3332.
1. *L'alchimie et les alchimistes.* Paris, 1854. X. 2nd ed., 1856. 3rd ed., 1860.

15755. MARIE, (Charles François) Maximilien. 1819-1891. Fr. P III, IV. Also Mathematics*.
1. *Histoire des sciences mathématiques et physiques.* 12 vols. Paris, 1883-88.

15756. GERDING, Theodor. 1820-after 1875. Ger. P I,III. Also Chemistry.
1. *Geschichte der Chemie.* Leipzig, 1867. RM. Reprinted, Wiesbaden, 1973.

15757. PIZZETTA, Jules. 1820-1900. Fr. BMNH.
1. *Galerie des naturalistes. Histoire des sciences naturelles.* Paris, 1891. RM.

15758. TODHUNTER, Isaac. 1820-1884. Br. DSB; P III; GB; GD. Also Mathematics*, Mechanics*, and Biography*.

1. *A history of the progress of the calculus of variations during the nineteenth century.* Cambridge, 1861. Reprinted, New York, 1961.
2. *A history of the mathematical theory of probability, from the time of Pascal to that of Laplace.* Cambridge, 1865. RM/L. Reprinted, New York, 1949.
3. *A history of the mathematical theories of attraction and the figure of the earth, from the time of Newton to that of Laplace.* 2 vols. London, 1873. RM/L. Reprinted, 2 vols in 1, New York, 1962.
4. *A history of the theory of elasticity and of the strength of materials, from Galilei to the present time.* 2 vols in 3. Cambridge, 1886-93. Completed and ed. by Karl Pearson.

15759. BONCOMPAGNI, Baldassarre. 1821-1894. It. DSB; ICB; P III,IV; GA. History of mathematics and physics. See also 573/1.

15760. JESSEN, Karl Friedrich Wilhelm. 1821-1889. Ger. ICB; BMNH; GC. Also Botany. History of Botany.

15761. BERTRAND, Joseph Louis François. 1822-1900. Fr. DSB; ICB; P I,III,IV; GA. Also Mathematics*, Physics*, and Biography*.
1. *L'Académie des Sciences et les académiciens de 1666 à 1793.* Paris, 1869. X. Reprinted, Amsterdam, 1969.

15762. BIERENS DE HAAN, David. 1822-1895. Holl. ICB; P III,IV; GF (HAAN). Also Mathematics*.
1. *Bibliographie néerlandaise historique-scientifique des ouvrages importants dont les auteurs sont nés aux 16e, 17e et 18e siècles, sur les sciences mathématiques et physiques, avec leurs applications.* Rome, 1883. X. Reprinted, Nieukoop, 1960.

15763. CARUS, Julius Victor. 1823-1903. Ger. DSB Supp.; BMNH; BLA & Supp. Also Zoology*.
1. *Geschichte der Zoologie bis auf Joh. Müller und Charl. Darwin.* Munich, 1872. RM/L. Reprinted, New York, 1965.

15764. MARRE, Eugène Aristide. 1823-after 1897. Fr. P III,IV. History of mathematics.

15765. STRUTHERS, John. 1823-1899. Br. ICB; BLA; GD1. Also Anatomy.
1. *Historical sketch of the Edinburgh Anatomical School.* Edinburgh, 1867.

15766. WALLACE, Alfred Russel. 1823-1913. Br. DSB; ICB; P III; BMNH; GB; GD3. Also Natural History*, Zoology*, Evolution*, and Autobiography*.
1. *The wonderful century. Its successes and its failures.* London, 1898. Reprinted, Farnborough, 1970.
1a. ——— Another ed., rev. *The wonderful century. The age of new ideas in science and invention.* Ib., 1903.

15767. ALLMAN, George Johnston. 1824-1904. Br. P III,IV; GD2. Also Mathematics.
1. *Greek geometry from Thales to Euclid.* Dublin, 1889. Reprinted, New York, 1976.

15768. DALTON, John Call. 1825-1889. U.S.A. DSB Supp.; Mort.; BLA & Supp.; GE. Also Physiology*. History of physiology.

15769. LANGKAVEL, Bernhard August. 1825-1902. BMNH.
1. *Botanik der späteren Griechen vom 3. bis zum 13. Jahrhundert*. Berlin, 1866. X. Reprinted?
See also 236/1.

15770. SAINT-LAGER, Jean Baptiste. 1825-1912. Fr. BMNH. Also Botany. History of Botany.

15771. GIESEL, Carl Franz. 1826-1892. Ger. P I,III,IV; GC. History of mathematics.

15772. WOEPCKE, Franz. 1826-1864. Ger. DSB; P II,III; GA; GC. Also Mathematics. History of mathematics. See also 456/1.

15773. BERTHELOT, (Pierre Eugène) Marcellin. 1827-1907. Fr. DSB; ICB; P I,III,IV; GB. Also Science in General* and Chemistry*.
1. *Les origines de l'alchimie*. Paris, 1885. RM/L. Reprinted, ib., 1938, and Osnabrück, 1966.
2. *Introduction à l'étude de la chimie des anciens et du moyen âge*. Paris, 1889. RM/L. Reprinted, ib., 1938.
3. *Notice historique sur Lavoisier*. Paris, 1889. RM/L.
4. *La révolution chimique. Lavoisier*. Paris, 1890. 2nd ed., 1902. RM/L.
5. *La chimie au moyen âge*. 3 vols. Paris, 1893. RM/L. Reprinted, Osnabrück, 1967.
6. *Archéologie et histoire des sciences. Avec publication nouvelle du papyrus grec chimique de Leyde, et impression originale du "Liber de Septuaginta" de Geber*. Paris, 1906. RM. Reprinted, Amsterdam, 1968.
See also 218/1,3.

15774. WOLF, Charles Joseph Etienne. 1827-1918. Fr. DSB; P III,IV. Also Astronomy*. History of physics and astronomy.

15775. FRIEDLEIN, Johann Gottfried. 1828-1875. Ger. P III; GC. History of mathematics. See also 192/1.

15776. RICCARDI, Pietro. 1828-1898. It. P III,IV.
1. *Biblioteca matematica italiana dalla origine della stampa ai primi anni del secolo XIX*. 3 vols & supp. Modena, 1870-80.

15777. CANTOR, Moritz Benedikt. 1829-1920. Ger. DSB; ICB; P III,IV.
1. *Mathematische Beiträge zum Kulturleben der Völker*. Halle, 1863.
2. *Die römischen Agrimensoren und ihre Stellung in der Geschichte der Feldmesskunst*. Stuttgart, 1875. X. Reprinted, Wiesbaden, 1968.
3. *Vorlesungen über Geschichte der Mathematik*. 4 vols. Leipzig, 1880-1908. 2nd ed., 1894-1908. RM/L. Another ed., 1900-09; reprinted, Stuttgart, 1965.

15778. JORET, Charles. 1829-1914. Fr. BMNH.
1. *Les plantes dans l'antiquité et au moyen âge. Histoire, usages et symbolisme*. 2 vols. Paris, 1897-1904.

15779. BRUHNS, Karl Christian. 1830-1881. Ger. DSB; P I,III; GC. Also Astronomy and Biography*.
1. *Die astronomische Strahlenbrechnung in ihrer historischen Entwicklung*. Leipzig, 1861. RM.

15780. WEISSENBORN, (Friedrich Wilhelm Heinrich Christian) Hermann. 1830-1896. Ger. P III.
1. *Die Principien der höheren Analysis in ihrer Entwicklung von Leibniz bis auf Lagrange.* Halle, 1856. X. Reprinted, Leipzig, 1972.

15781. HALLIER, Ernst Hans. 1831-1904. Ger. DSB; P III,IV; BMNH; BLA. Also Botany* and Microbiology.
1. *Kulturgeschichte des neunzehnten Jahrhunderts in ihren Beziehungen zu der Entwickelung der Naturwissenschaften.* Stuttgart, 1889. RM.

15782. NARDUCCI, Enrico. 1832-1893. It. P III,IV. History of medieval science.

15783. NORDENSKIÖLD, (Nils) Adolf Erik. 1832-1901. Swed. DSB; ICB; P II,III,IV; BMNH; GB. Also Geology.
1. *Facsimile-atlas to the early history of cartography.* 1889. X. Reprinted, U.S.A., 1963.

15784. SACHS, (Ferdinand Gustav) Julius von. 1832-1897. Ger. DSB; ICB; BMNH; GB; GC. Also Botany*.
1. *Geschichte der Botanik.* Munich, 1875. RM/L. Reprinted, Hildesheim, 1966.
1a. ——— *History of botany, 1530-1860.* Oxford, 1890. Trans. by E.F. Garnsey. Rev. by I.B. Balfour. Reprinted, ib., 1906. RM/L. Another reprint, New York, 1967.

15785. WHITE, Andrew Dickson. 1832-1918. U.S.A. GB; GE.
1. *The warfare of science.* London, 1876. Preface by J. Tyndall. RM/L. 2nd ed., 1877.
2. *A history of the warfare of science with theology in Christendom.* 2 vols. London, 1896. RM/L. Reprinted, New York, 1955 and 1960.

15786. BRETSCHNEIDER, Emil. 1833-1901. Russ./China. BMNH; BLAF. Also Botany*.
1. *History of European botanical discoveries in China.* London, 1898. Reprinted, Leipzig, 1962.

15787. HULTSCH, Friedrich Otto. 1833-1906. Ger. ICB; P III,IV. History of ancient mathematics and astronomy. See also 172/10.

15788. POUCHET, (Henri Charles) Georges. 1833-1894. Fr. ICB; BMNH; BLA; BLAF. Also Zoology.
1. *Les sciences pendant la Terreur.* Paris, Société de l'Histoire de la Révolution Française, 1896. RM.

15789. SCOTT, Robert Henry. 1833-1916. Br. P III,IV; BMNH. Also Geology and Meteorology*.
1. *The history of the Kew Observatory.* London, 1885.

15790. BERTHOLD, Gerhard. 1834-1918. Ger. ICB; P IV; BLAF. History of physics.

15791. SCHORLEMMER, Carl. 1834-1892. Br. DSB; ICB; P III,IV; GD. Also Chemistry*.
1. *The rise and development of organic chemistry.* London, 1879. X. Another ed., 1894.

15792. SCHIAPARELLI, Giovanni Virginio. 1835-1910. It. DSB; ICB;

P III,IV; GB. Also Astronomy*#.
1. *Origine del sistema planetario eliocentrico presso i greci*. 1898.

15793. WOHLWILL, Emil. 1835-1912. Ger. P III,IV. History of physics.

15794. BERGER, Ernst Hugo. 1836-1904.
1. *Geschichte der wissenschaftlichen Erdkunde der Griechen*. Leipzig, 1887-93. X. 2nd ed., 1903. RM.

15795. DEL MAR, Alexander. 1836-1926. U.S.A. GE.
1. *A history of precious metals from the earliest times to the present*. London, 1880. RM.

15796. LOCKYER, Joseph Norman. 1836-1920. Br. DSB; ICB; P III,IV; GB; GD3. Also Astronomy* and Physics*.
1. *Stargazing, past and present*. London, 1873. Another ed., 1878.
2. *The dawn of astronomy. A study of the temple-worship and mythology of the ancient Egyptians*. London, 1894. RM/L. Reprinted, Cambridge, Mass., 1964.

15797. CAVERNI, Raffaello. 1837-1900. ICB.
1. *Storia del metodo sperimentale in Italia*. Florence, 1891-1900. RM. Reprinted, 6 vols, New York, 1972.

15798. CURTZE, (Ernst Ludwig Wilhelm) Maximilian. 1837-1903. Thorn. DSB; P III,IV & Supp. History of mathematics.
See also 708/1 and 4016/1.

15799. JACKSON, John Reader. 1837-1920. Br. BMNH. Also Botany.
1. *Commercial botany in the nineteenth century*. London, 1890.

15800. KEEN, William Williams. 1837-1932. U.S.A. Mort. BLA; GE1. History of anatomy.

15801. SCUDDER, Samuel Hubbard. 1837-1911. U.S.A. DSB; BMNH; GE. Also Zoology.
1. *Catalogue of scientific serials ... in the natural, physical, and mathematical sciences. 1633-1876*. Cambridge, Mass., 1879. Reprinted, New York, 1965.

15802. STIEDA, (Christian Hermann) Ludwig. 1837-1918. Dorpat/Ger. BMNH; BLA. Also Zoology.
1. *Geschichte der Entwickelung der Lehre von den Nervenzellen und Nervenfasern während des XIX. Jahrhunderts*. Jena, 1899.

15803. WATSON, Robert Spence. 1837-1911. Br. GD2.
1. *The history of the Literary and Philosophical Society of Newcastle-upon-Tyne, 1793-1896*. London, 1897. X. Reprinted, Farnborough, 1970.

15804. ALLEN, Richard Hinckley. 1838-1908.
1. *Star-names and their meanings*. New York, 1899. Reprinted, ib., 1963.

15805. BEILSTEIN, Friedrich Konrad. 1838-1906. Russ. DSB; P III,IV. Also Chemistry*.
1. *Beilstein-Erlenmeyer: Briefe zur Geschichte der chemischen Dokumentation und des chemischen Zeitschriftenwesens*. Munich, 1972. Ed. by O. Kraetz.

15806. GERLAND, (Anton Werner) Ernst. 1838-1910. Ger. P III,IV.
1. *Geschichte der physikalischen Experimentierkunst.* Leipzig, 1899. With F. Traumüller. Reprinted, Hildesheim, 1965.
See also 2375/9.

15807. KIRCHHOFF, (Karl Reinhold) Alfred. 1838-1907. Ger. P III,IV; BMNH.
1. *Die Idee der Pflanzen-Metamorphose bei Wolff und bei Göthe.* Berlin, 1867.

15808. MACH, Ernst. 1838-1916. Aus./Prague. DSB; ICB; P III,IV.
Also Mechanics*, Physics*, and Physiology.
See 5703/1 and 6632/4,5.

15809. MAINDRON, Ernest. 1838-1908.
1. *L'Académie des Sciences, histoire....* Paris, 1888. RM/L.

15810. REYE, (Carl) Theodor. 1838-1919. Switz./Ger. DSB; ICB; P III, IV. Also Mathematics*.
1. *Die synthetische Geometrie im Alterthum und in der Neuzeit.* Strasbourg, 1886.

15811. SCHWARZ, Julius. 1838-1900. Hung. P III; BMNH (SCHVARCZ, G.).
History of Greek science.

15812. YOUMANS, William Jay. 1838-1901. U.S.A. GE. See also 14910/2.
1. *Pioneers of science in America.* 1896. X.

15813. BLAKE, John Frederick. 1839-1906. BMNH.
1. *Astronomical myths. Based on Flammarion's "History of the heavens."* London, 1877.

15814. FRITSCHE, Hermann Peter Heinrich. 1839-1913. Russ./China.
P III,IV. Also Geomagnetism.
1. *On chronology and construction of the calendar, with special regard to the Chinese computation of time compared with the European.* St. Petersburg, 1886.

15815. HANKEL, Hermann. 1839-1873. Ger. DSB; P III; GC. Also Mathematics*.
1. *Zur Geschichte der Mathematik in Alterthum und Mittelalter.* Leipzig, 1874.

15816. ZEUTHEN, Hieronymus Georg. 1839-1920. Den. DSB; ICB; P III, IV & Supp. Also Mathematics.
1. *Die Lehre von den Kegelschnitten in Altertum.* Copenhagen, 1886. X. Reprinted, Hildesheim, 1966.
2. *Geschichte der Mathematik in Alterthum und Mittelalter.* Copenhagen, 1896. RM. Reprinted, New York/Stuttgart, 1966.

15817. ZITTEL, Karl Alfred von. 1839-1904. Ger. DSB; P III,IV; BMNH; GB. Also Geology*.
1. *Geschichte der Geologie und Paläontologie.* Munich, 1899. RM/L.
1a. ——— *History of geology and palaeontology.* London, 1901. Trans. by M.M. Ogilvie-Gordon. RM/L. Reprinted, Weinheim, 1962.

15818. BUCKLEY, Arabella (*Mrs* A.B. FISHER). 1840-1929. Br. BMNH.
1. *Short history of natural science, and of the progress of discovery from the time of the Greeks to the present day.* London, 1876. See also 2907/1.

15819. HOUDAS, Octave Victor. 1840-1916.
    1. *L'alchimie arabe au moyen âge.* 1893. X. Reprinted?

15820. SCHLEGEL, Gustave. 1840-1903. Holl. P III,IV & Supp.; BMNH. History of Chinese science.
    1. *Uranographie chinoise; ou, Preuves directes que l'astronomie primitive est originaire de la Chine et qu'elle a été empruntée par les anciens peuples occidentaux à la sphère chinoise.* The Hague, 1875. Reprinted, Taipei, 1967.

15821. MENDENHALL, Thomas Corwin. 1841-1924. U.S.A. ICB; P III,IV; GE. Also Physics.
    1. *A century of electricity.* Boston, 1887.

15822. ANDRÉ, Charles (Louis François). 1842-1912. Fr. DSB; P III, IV. Also Astronomy and Meteorology.
    1. *L'astronomie pratique et les observatoires en Europe et en Amérique depuis le milieu du 17e siècle.* 5 vols. Paris, 1874-78.

15823. BOUCHÉ-LECLERCQ, Auguste. 1842-1923.
    1. *L'astrologie grecque.* Paris, 1899. X. Reprinted, Brussels, 1963.

15824. CLERKE, Agnes Mary. 1842-1907. Br. GB; GD2. Also Astronomy*.
    1. *A popular history of astronomy during the nineteenth century.* Edinburgh, 1885. 2nd ed., 1887. 3rd ed., 1893.
    2. *The Herschels and modern astronomy.* London, 1895.

15825. LADENBURG, Albert. 1842-1911. Ger. DSB; ICB; P III,IV. Also Chemistry.
    1. *Vorträge über die Entwicklungsgeschichte der Chemie in den letzten hundert Jahren.* Brunswick, 1869. RM/L. 2nd ed., 1887.
    1a. ——— *Lectures on the history of the development of chemistry.* Edinburgh, 1900. RM/L.

15826. TILDEN, William Augustus. 1842-1926. Br. DSB; P III,IV. Also Chemistry*.
    1. *A short history of the progress of scientific chemistry in our own times.* London, 1899. RM/L.

15827. BOLTON, Henry Carrington. 1843-1903. U.S.A. P III,IV; GE. Also Chemistry.
    1. *A catalogue of scientific and technical periodicals, 1665-1882.* Washington, 1885. 2nd ed., 1897; reprinted, New York, 1965.
    2. *A catalogue of chemical periodicals.* New York, 1885.
    3. *A select bibliography of chemistry, 1492-1892.* With supplements. Washington, 1893-1902. RM/L. Reprinted, New York, 1966.
    See also 3174/3.

15828. HELLER, Agost. 1843-1902. Hung. P III,IV.
    1. *Geschichte der Physik.* 2 vols. Stuttgart, 1882-84. Reprinted, Wiesbaden, 1965.

15829. MÜLLER, Hermann Felix. 1843-1928. Ger. P III,IV. History of mathematics.

15830. RODWELL, George Farrer. 1843-after 1883. Br. P III. Also

Science in General*.

1. *The birth of chemistry*. London, 1874.

15831. TANNERY, Paul. 1843-1904. Fr. DSB; ICB; P III,IV.

1. *La géometrie grecque*. Paris, 1887.
2. *Pour l'histoire de la science hellène. De Thales à Empedocle*. Paris, 1887. RM. Reprinted, ib., 1930.
3. *Recherches sur l'histoire de l'astronomie ancienne*. Paris, 1893.
4. *Mémoires scientifiques*. 17 vols. Paris, 1912-50. Ed. by Mme Tannery et al.

See also 2032/10 and 2087/2.

15832. BROWN, Robert (third of the name). b. 1844.

1. *Researches into the origin of the primitive constellations of the Greeks, Phoenicians and Babylonians*. 2 vols. London, 1899-1900.

15833. MANSION, Paul. 1844-1919. Belg. DSB; ICB; P III,IV. Also Mathematics*. History of mathematics and astronomy.

15834. NICHOLSON, Henry Alleyne. 1844-1899. Br. BMNH; GB; GD1. Also Geology*, Biology in General*, Zoology*, and Biography*.

1. *Natural history, its rise and progress in Britain*. 1886.

15835. PERRIER, (Jean Octave) Edmond. 1844-1921. Fr. DSB; BMNH; BLAF. Also Zoology.

1. *La philosophie zoologique avant Darwin*. Paris, 1884. X. 3rd ed., 1896.

15836. JAGNAUX, Raoul. b. 1845. Fr. BMNH.

1. *Histoire de la chimie*. 2 vols. Paris, 1891.

15837. ROSENBERGER, (Johann Karl) Ferdinand. 1845-1899. Ger. DSB; P III,IV; GC.

1. *Die Geschichte der Physik*. 3 vols. Brunswick, 1882-90. RM. Reprinted, 3 vols in 2, Hildesheim, 1965.

15838. THORPE, (Thomas) Edward. 1845-1925. Br. DSB; P III,IV; GD4. Also Chemistry* and Biography*.

1. *Essays in historical chemistry*. London, 1894. RM/L.

15839. TREUTLEIN, Josef Peter. 1845-1912. Ger. P III,IV. History of mathematics.

15840. FAHIE, John Joseph. 1846-1934. ICB.

1. *A history of wireless telegraphy, 1838-1899*. New York, 1899. RM.

15841. LEBON, (Desiré) Ernest. 1846-1922. Fr. P IV. Also Mathematics.

1. *Histoire abrégée de l'astronomie*. Paris, 1899. RM.

15842. MOWAT, John Lancaster Gough. 1846-1894. Br. History of botany. See also 764.

15843. WITTSTEIN, Armin Arthur Eginhard. 1846-after 1903. Ger. P IV. History of astronomy.

15844. FAVARO, Antonio Nobile. 1847-1922. It. ICB; P III,IV. Also Biography*. History of mathematics and astronomy.

15845. MEYER, Ernst (Sigismund Christian) von. 1847-1916. Ger.

P III,IV. Also Chemistry.
1. *Geschichte der Chemie*. Leipzig, 1889. RM.
1a. ——— *A history of chemistry*. London, 1891. Trans. by G. McGowan. RM/L. 2nd ed., 1898.

15846. STRINGHAM, (Washington) Irving. 1847-1909. U.S.A. P III,IV; GE. Also Mathematics.
1. *The past and present of elementary mathematics*. Berkeley, 1893.

15847. GUENTHER, (Adam Wilhelm) Siegmund. 1848-1923. Ger. DSB; P III,IV & Supp. Also Mathematics and Earth Sciences*.
1. *Geschichte des mathematischen Unterrichts im deutschen Mittelalter bis zum Jahre 1525*. Berlin, 1887. X. Reprinted, Wiesbaden, 1969.

15848. LASSWITZ, (Carl Theodor Victor) Kurd. 1848-1910. Ger. ICB; P III,IV. Also Science in General*.
1. *Geschichte der Atomistik vom Mittelalter bis Newton*. 2 vols. Hamburg, 1890. RM. Reprinted, Hildesheim, 1964.

15849. MUIR, Matthew Moncrieff Pattison. 1848-1931. Br. DSB; P III, IV. Also Chemistry.
1. *The alchemical essence and the chemical element*. London, 1894.
2. *The story of the chemical elements*. London, 1897.

15850. SEAGER, Herbert West. b. 1848.
1. *Natural history in Shakespeare's time*. London, 1896.

15851. SUTER, Heinrich. 1848-1922. Switz. DSB; ICB; P III,IV.
1. *Geschichte der mathematischen Wissenschaften*. 2nd ed. Zurich, 1873-75. (The 1st ed. was his doctoral thesis, 1871.) RM.
2. *Die Mathematiker und Astronomen der Araber und ihre Werke*. Leipzig, 1900. X. Reprinted, New York, 1972.

15852. BENJAMIN, Park. 1849-1922. U.S.A. GE.
1. *The intellectual rise in electricity*. London, 1895.
1a. ——— American ed. *A history of electricity ... from antiquity to the days of Benjamin Franklin*. New York, 1898. RM.

15853. BOBYNIN, Victor Victorovich. 1849-1919. Russ. ICB. History of mathematics in Russia.

15854. BALL, Walter William Rouse. 1850-1925. Br. ICB (ROUSE BALL); P IV. Also Mathematics.
1. *A short account of the history of mathematics*. London, 1888. 2nd ed., 1893. RM. 4th ed., 1908; reprinted, New York, 1960.
2. *A history of the study of mathematics at Cambridge*. Cambridge, 1889. RM/L.
3. *Mathematical recreations and problems of past and present times*. London, 1892. X. 3rd ed., 1896.
4. *An essay on Newton's "Principia."* London, 1893. Reprinted, New York, 1972.

15855. FAIRCHILD, Herman Leroy. 1850-1943. U.S.A. ICB; P IV. Also Geology.
1. *A history of the New York Academy of Sciences*. New York, 1887. RM.

15856. RODET, Léon. ca. 1850-after 1884. Fr. P III,IV. History of mathematics in India, Islam, etc.

15857. BIGOURDAN, (Camille) Guillaume. 1851-1932. Fr. DSB; ICB; P III,IV. Also Astronomy. History of astronomy, etc.

15858. DICKSTEIN, Samuel. 1851-1939. Warsaw. DSB; P IV. Also Mathematics. History of mathematics.

15859. GOODE, George Brown. 1851-1896. U.S.A. ICB; BMNH; GE. Also Zoology. See also 15523/1.
1. *An account of the Smithsonian Institution. Its origin, history, objects and achievements.* Washington, 1895.
2. *The Smithsonian Institution, 1846-1896. The history of its first half century.* Washington, 1897.

15860. BOSMANS, Henri. 1852-1928. ICB. History of mathematics in Europe and China.

15861. ENESTRÖM, Gustaf. 1852-1923. Swed. ICB; P IV. History of mathematics.

15862. KÜHNERT, Franz. 1852-1918. Aus. P III,IV. History of Chinese astronomy.

15863. URBANITZKY, Alfred von. 1852-1906.
1. *Elektricität und Magnetismus im Alterthume.* Vienna, 1887. X. Reprinted, Wiesbaden, 1967.

15864. WIEDEMANN, Eilhard Ernst Gustav. 1852-1928. Ger. ICB; P III, IV. Also Physics.
1. *Aufsätze zur arabischen Wissenschaftsgeschichte.* Several vols. Hildesheim, 1970-.

15865. BRAUNMÜHL, Anton von. 1853-1908. Ger. DSB; P IV. History of mathematics.

15866. KAHLBAUM, Georg Wilhelm August. 1853-1905. Ger./Switz. P III,IV; BLAF. Also Chemistry and Biography*. See also Index. History of chemistry.

15867. MABILLEAU, Léopold. b. 1853.
1. *Histoire de la philosophie atomistique.* Paris, 1895. RM.

15868. PETRIE, (William Matthew) Flinders. 1853-1942. Br. The archaeologist. DSB; GB; GD6. See also entry 9.
1. *Inductive metrology; or, The recovery of ancient measures from the monuments.* London, 1877.

15869. SUDHOFF, Karl Friedrich Jakob. 1853-1938. Ger. DSB; ICB; Mort.; BLAF. Well known historian of medicine who also contributed much to the history of science.

15870. GAIZO, Modestino del. 1854-1921. It. BLAF. History of anatomy and physiology.

15871. GOW, James. 1854-1923.
1. *A short history of Greek mathematics.* Cambridge, 1884. Reprinted, New York, 1968.

15872. HEIBERG, Johan Ludvig. 1854-1928. Den. ICB; P III,IV. See also Index. History of Greek mathematics.

15873. HOPPE, Edmund. 1854-1928. Ger. ICB; P III,IV. Also Physics.
1. *Geschichte der Elektricität.* Leipzig, 1884. RM/L.

15874. REIFF, Richard August. 1855-1908. Ger. P IV. Also Physics.
1. *Geschichte der unendlichen Reihen*. Tübingen, 1889.

15875. VASILIEV, Aleksandr Vasilievich. 1855-1929. Russ. ICB. History of mathematics.

15876. GORE, James Howard. 1856-1939. U.S.A. P IV. Also Geodesy*. History of geodesy.

15877. RUDIO, Ferdinand. 1856-1929. Switz. DSB; P IV. Also Mathematics. History of mathematics.

15878. TÖPLY, Robert von. 1856-1947. Aus. Mort.; BLAF. History of anatomy.

15879. VENABLE, Francis Preston. 1856-1934. U.S.A. P IV; GE. Also Chemistry.
1. *The development of the periodic law*. Easton, Pa., 1896.

15880. VOLKMANN, Paul Oskar Eduard. 1856-1938. Ger. DSB; P III,IV. Also Physics and Philosophy of Science. History of physics.

15881. CONANT, Levi Leonard. 1857-1916.
1. *The number concept, its origin and development*. New York, 1896.

15882. LIPPMANN, Edmund Oskar von. 1857-1940. Ger. ICB; P III; IV. Also Chemistry. History of chemistry.

15883. OSBORN, Henry Fairfield. 1857-1935. U.S.A. DSB; ICB; BMNH; GE1. Also Zoology.
1. *From the Greeks to Darwin. An outline of the development of the evolution idea*. New York, 1894. RM/L.

15884. STAIGMÜLLER, Hermann Christian Otto. 1857-1908. Ger. P IV. History of mathematics and astronomy.

15885. ALBRECHT, Gustav. b. 1858.
1. *Geschichte der Elektricität*. Vienna, 1885. RM.

15886. MACKINTOSH, Robert. 1858-1933.
1. *From Comte to Benjamin Kidd. The appeal to biology or evolution for human guidance*. London, 1899.

15887. CAJORI, Florian. 1859-1930. U.S.A. ICB; P IV; GE1.
1. *The teaching and history of mathematics in the United States*. Washington, 1890.
2. *A history of mathematics*. New York, 1894. RM. Another ed., 1895.
3. *History of elementary mathematics*. New York, 1896.
4. *A history of physics in its elementary branches, including the evolution of physical laboratories*. New York, 1899. RM/L.

15888. DANNEMANN, Johann Friedrich. 1859-1936. Ger. P IV. History of science in general.

15889. HENRY, Charles. 1859-1926. Fr. ICB; P III,IV. See also Index. History of mathematics.

Imprint Sequence

15890. ISTITUTO delle Scienze. Bologna.
    1. *Notizie dell'origine dell'Istituto delle Scienze di Bologna.* Bologna, 1780.

15891. BENTLEY, John.
    1. *A historical view of the Hindu astronomy, from the earliest ... to the present time.* London, 1825. RM/L. Reprinted, Osnabrück, 1970.

15892. MORELL, Thomas. fl. 1817-1836.
    1. *Elements of the history of philosophy and science from the earliest authentic records to the commencement of the eighteenth century.* London, 1827.

15893. WHITE, Francis Sellon.
    1. *A history of inventions and discoveries, alphabetically arranged.* London, 1827. RM/L.

15894. FOURCY, A. d. 1842.
    1. *Histoire de l'Ecole Polytechnique.* Paris, 1828. RM.

15895. CARR, Daniel C.
    1. *Linnaeus and Jussieu; or, The rise and progress of systematic botany.* London, 1844.

15896. HOOPER, Francis John Bodfield.
    1. *Palmoni. An essay on the chronographical and numerical systems in use among the ancient Jews.* London, 1851.

15897. GUMPACH, Johannes von.
    1. *Die Zeitrechnung der Babylonier und Assyrier.* Heidelberg, 1852. X. Reprinted, Wiesbaden, 1972.

15898. WINCKLER, Emil. BMNH.
    1. *Geschichte der Botanik.* Frankfurt, 1854. RM.

15899. BIERNATZKI, Karl L.
    1. *Die Arithmetik der Chinesen.* Wiesbaden, 1973. Reprint of a periodical article of 1855.

15900. UHLEMANN, Maximilian Adolph.
    1. *Grundzüge der Astronomie und Astrologie der Alten, besonders der Aegypter.* Leipzig, 1857. RM.

15901. SAIGEY, Emile. d. 1875.
    1. *Les sciences au XVIIIe siècle. La physique de Voltaire.* Paris, 1873.

15902. ROUTLEDGE, Robert.
    1. *Discoveries and inventions of the nineteenth century.* London, 1876. X. 9th ed., 1891.

15903. SOCIÉTÉ Hollandaise des Sciences. Harlem.
    1. *Notice historique. Liste des membres ... et ... des publications de la Société depuis sa fondation en 1752.* Harlem, 1876.

15904. FELLNER, Stefan Karl.
    1. *Compendium der Naturwissenschaften an der Schule zu Fulda im IX. Jahrhundert.* Berlin, 1879. RM.

15905. PHIPSON, Emma, *Miss.*
1. *The animal-lore of Shakespeare's time.* London, 1883. Reprinted, New York, 1973.

15906. JOHN, Vincenz.
1. *Geschichte der Statistik.* Stuttgart, 1884. X. Reprinted?

15907. HEPPEL, G.
1. *Curiosities in early English mathematical teaching.* 1885.

15908. ROBERTS, William W.
1. *The pontifical decrees against the doctrine of the earth's movement, and the ultramontane defence of them.* Oxford, 1885.

15909. WATKINS, Morgan George.
1. *Gleanings from the natural history of the ancients.* London, 1885. X. Another ed., 1896.

15910. *L'ANCIENNE France. L'Ecole et la science jusqu'à la renaissance.* Paris, 1887.

15911. INSTITUT de France. Paris.
1. *Lois, statuts et règlements concernant les anciennes académies et l'Instut de 1635 à 1889. Tableau des fondations.* Paris, 1889.

15912. BRENNAND, W.
1. *Hindu astronomy.* London, 1890. RM. Another ed., 1896.

15913. JOHNSON, Valentine Edward.
1. *Our debt to the past; or, Chaldean science.* London, 1890? RM.

15914. BERENDES, Julius.
1. *Die Pharmazie bei den alten Kulturvölkern.* 2 vols. Halle, 1891. X. Reprinted, Hildesheim, 1965.

15915. MARMERY, J. Villin.
1. *The progress of science. Its origin, course, promoters and results.* London, 1895.

15916. ORCHARD, Thomas Nathaniel.
1. *The astronomy of Milton's "Paradise Lost."* London, 1896. X. Reprinted, New York, 1966.

15917. CHEMICAL Society. London.
1. *The jubilee of the Chemical Society of London. A record of the proceedings together with an account of the history and development of the Society, 1841-1891.* London, 1896.

15918. MACDONALD, James Cecil.
1. *Chronologies and calendars.* London, 1897.

MAIN INDEX

Including persons, institutions, and ships.

References are to entry numbers (page numbers are not used). Main references are in roman type and subsidiary references, if any, are in italics. The alphabetical arrangement is on the letter-by-letter principle, with a few small modifications.

Except in special cases, editors and translators were not indexed unless they already appeared as authors. Thus twentieth-century editors and translators were not indexed.

Only limited reliance can be put on initials of forenames because some have various spellings or forms, e.g. Emmanuel and Immanuel, Caspar and Kaspar, Girolamo and Hieronymus, etc. Because the spellings 'Carl' and 'Karl' can be a frequent source of confusion they are given in full where it seemed desirable. Confusion can also arise from the fact that many authors who had several forenames often did not use them all, e.g. F.A.P. Barbier generally gave his name as P. Barbier. When these rarely used forenames have been identified they are put in brackets; thus the entry: Barbier, (F.A.) P. Likewise, parts of surnames that were infrequently used are put in brackets.

INDEX OF SELECTED TITLES

Chiefly of anonymous works

INDEX OF DICTIONARIES AND ENCYCLOPAEDIAS

INDEX OF BIBLIOGRAPHIES AND BOOK CATALOGUES